iF design award 2004

Impressum Imprint

Herausgeber, Editor:
iF International Forum Design GmbH, Messegelände, D-30521 Hannover, Germany
fon +49 511 8932400, fax +49 511 8932401
www.ifdesign.de, info@ifdesign.de

iF Team:
Geschäftsführung Managing Director: Ralph Wiegmann

iF design award 2004:
Anne-Kathrin Aumann, Dirk Bartelsmeier, Gabriele Bertemann
Rainer Schwarz, Frank Zierenberg, Joachim Möllmann
Anja-Martina Kirschning, Heike Meier

iF press office:
Silke Becker, Claudia Neumann
Neumann + Luz, Roonstraße 108, D-50674 Köln, Germany
fon +49 221 2584773, fax +49 221 2584774
team@neumann-luz.de

Produktion Production:
bangert verlag, Verlagsbüro Schopfheim
Corinna Haneke (Production Manager), Margot Bangert (Art Director)
www.bangertverlag.com, studio@bangertverlag.com
Textredaktion der englischen Texte Editing of English texts: Dr. Jeremy Gaines, Frankfurt
Textredaktion der deutschen Texte Editing of German texts: Petra Gaines-Soukup, Frankfurt
Fotografie Jury Jury Photography: Rüdiger Buhl, Kirchzarten
Gedruckt auf säurefreiem Papier Printed on acid-free paper produced from chlorine-free pulp
Magno Matt Classic 150 gsm, produced by Sappi, www.sappi.com

Verlag, Publisher:
Birkhäuser – Publishers for Architecture, P.O.Box 133, CH-4010 Basel, Switzerland
Member of Springer Science+Business Media, www.birkhauser.ch

A CIP catalogue record for this book is available
from the Library of Congress, Washington D.C., USA.

Bibliographical information published by Die Deutsche Bibliothek
Die Deutsche Bibliothek lists this publication in the Deutsche Nationalbibliografie;
detailed bibliographical data is available in the internet at http://dnb.ddb.de.

Printed in Italy by Printer Trento S.r.l. - www.printertrento.it
ISBN 3-7643-6855-1

iF design award 2004

Birkhäuser – Verlag für Architektur
Birkhäuser – Publishers for Architecture
Basel · Boston · Berlin

Inhalt contents

Design denken
Ralph Wiegmann

Runde Geburtstage sind wie Meilensteine: Man blickt zurück und voraus, zieht Bilanz und sucht Perspektiven – sowohl für sich selbst als auch für den Kontext, in dem man sich bewegt. Design ist angetreten, um auf der Basis wirtschaftlichen und gestalterischen Handelns komplexe Zusammenhänge sichtbar und verständlich zu machen und hat sich mittlerweile zu einer Disziplin entwickelt, die sich selbst kritisch beleuchten muss. Dazu gehören Produkte, grafische und mediale Arbeiten ebenso wie Dienstleistungen und nicht zuletzt die im Design agierenden Personen selber. Es ist an der Zeit, nicht nur innezuhalten und zu reflektieren, sondern Ideen zu entwickeln, Maßnahmen zu ergreifen und interdisziplinäre Kommunikation zu betreiben, wollen wir in Deutschland nicht in der Design-Pubertät verhaften bleiben. Design muss Fragen stellen, einen Standpunkt bekommen, Bezug nehmen – und fordert alle daran Beteiligten auf, sich dem Diskurs zu stellen.

50 JAHRE DESIGNFÖRDERUNG – EINE KURZE BILANZ UND ERMUTIGENDE PERSPEKTIVEN

2003 war für iF in seinem 50. Jahr gefüllt mit Wettbewerben, Konferenzen, einer guten Zahl an Projekten als Dienstleister für die Wirtschaft und, nicht zu vergessen, dem ersten ICSID Design Congress in Deutschland. „Designed in Germany" ist – auch durch iF – zu einer internationalen Marke mit hohem Qualitätsstandard geworden. Die Anforderungen an ein Designzentrum haben sich geändert, sind vielfältiger geworden. Heute ist iF Kommunikator, Netzwerker und sogar Agentur, mit allen Chancen und Herausforderungen, die sich daraus ergeben. Design ist viel mehr und eindeutiger mit der Wirtschaft verknüpft als früher, ist Triebfeder und Ideengeber. iF hat sich zur Aufgabe gemacht, über den Tellerrand hinauszublicken und kann neben den neu ausgerichteten zwei großen Designwettbewerben, dem „iF PRODUCT design award" und dem „iF communication design award", nun auch auf den „iF design award.CHINA" blicken, der mitten im asiatischen Zentrum der Entwicklung und des wirtschaftlichen Hot-Spots steht.

DESIGN IM ELFENBEIN-TURM?

Dennoch ist dieser Teilerfolg noch lange kein Grund, sich auszuruhen. Deutsche Designer haben Weltruf, sind international gefragt und haben viel zu viel Nachwuchs. Aber wo geht es hin, wie geht es dem Gros der Designer, denen es nicht an Talent, aber vielleicht an Mangel an Gelegenheiten fehlt, dieses unter Beweis zu stellen? Zudem: Auch nach fünf Jahrzehnten deutschen Designs haben wir nicht erreicht – oder nicht zugelassen? –, dass Design aus der elitären Ecke herauskommt. Auf der anderen Seite wird mehr und mehr alles zum Design, verkommt bei unvorsichtigem Gebrauch und im fremden Kontext die Bezeichnung gar zum Schimpfwort für schlechte Produkte und Dienstleistungen. Wie schaffen wir diesen Balance-Akt, wie kann Design seine Fähigkeiten unter Beweis stellen, was sind die relevanten und zukünftigen Themen der eigenen Zunft? Einiges liegt auf der Hand und ist zu tun:

– Die zunehmende Alterung der Gesellschaft bedingt, dass wir uns in den kommenden Jahren verstärkt mit Universal Design auseinandersetzen müssen: Design, das für alle die gleichen Chancen bietet, Dinge wie Vorgänge zu er- und begreifen.
– Die Entdeckung von Material als wirkliche Innovationsquelle wird zunehmend wichtiger. Die Forschung ist wichtiger Bestandteil der Produktherstellung und sollte im Hinblick auf Design gefördert werden.
– Die Medien und deren differente Peripheriegeräte erleichtern bisher nicht zwingend die Kommunikation – sei es von Mensch zu Mensch, von Mensch zu Maschine oder von Maschine zu Maschine. Ein Zuviel an Möglichkeiten, ein Zuwenig an Inhalten bedeuten nur eine weitere Liane im Informationsdschungel. Leitsysteme, eindeutige Standards und die Rückbesinnung auf Wesentliches können Abhilfe schaffen – und benötigen Designer, die sich der Problematiken annehmen.

– Design und Wissenschaft kommen bisher, von wenigen Ausnahmen abgesehen, kaum zusammen. Die Theorie, die Auseinandersetzung mit der eigenen Disziplin ist wichtig, um neue Schubkraft zu gewinnen. Es lohnt sich, kritische Fragen zu stellen und auf die Suche nach den Antworten zu gehen.
– Nach der Globalisierung und deren lokaler Gegenbewegung stellt sich die Frage, wo nunmehr die Ressourcen liegen: vor der eigenen Haustür oder eher in fernen Landen? Design sollte sich vor Gleichmacherei hüten und kulturelle Divergenzen analysieren, akzeptieren und mit ihnen arbeiten.

DESIGNZENTREN – RELIKT ODER WEGWEISER?

Die deutschen Designzentren verdanken ihre Entstehung der föderalistischen Struktur der Bundesrepublik Deutschland. In der Vergangenheit waren dies sicher zu viele, die zu unterschiedlich aufgestellt und ausgerichtet waren. Welche Modelle sind denkbar, welche Kooperationen können eingangen werden, welche Vergleiche haben wir?
In Großbritannien gibt es mit dem British Design Council eine gut ausgestattete Einzelinstitution, PR– und kommunikationsstark, federführend – für Deutschland eine nicht diskutable oder wünschenswerte Art der Designförderung. Aber sind viele, schlecht geförderte Zentren mit geringer Effizienz und Reichweite ein Zustand, den es zu tragen gilt? Könnten vier Zentren in Nord, Süd, Ost und West mit einem gemeinsamen, wirtschaftlichen und kulturellen Konzept die Lösung sein? Oder liegt die Zukunft der Designvermittlung ganz woanders, beispielsweise in Bremen, im Universum Science Center. 1,4 Millionen Besucher in drei Jahren sprechen für sich, die experimentelle Ausrichtung und Vermittlung dichter Wissensgebiete in einfacher und animierender Form scheint die Zukunft – ist das Design also dort bereits integriert, sind designspezifische und polarisierende Themen nicht mehr von Nöten? Oder machen wir die Vermittlung von Design zum Stolperstein, ist es zu schwer, sich mit der Disziplin vertraut zu machen? Sollte Design in den Ausbildungsprozess eingebunden werden, können bereits Kinder und Jugendliche sich das Wissen um Gestaltung zu eigen machen, sollten wir an dieser Stelle einhaken und Design nicht immer wieder und allzu gern unter uns diskutieren?

WAS KOMMT, WAS BLEIBT?

2004 wird ein spannendes Jahr: Der sehnlichst erwartete Ansprung des wirtschaftlichen Motors ist spürbar, erste Vibrationen sind zu bemerken. Manche knurren noch, harren der Dinge. Aber aus dem Summen wird ein Brummen werden. Aufträge werden wieder erteilt, Staus in Köpfen und bei der Entwicklung lösen sich auf. Der Export ist noch immer die treibende Kraft, Design wird sich noch stärker als wesentliches Differenzierungsmerkmal durchsetzen – selbst wenn es nicht immer als solches benannt oder erkannt wird.

iF hat eine große Anzahl Projekte für 2004 und 2005 initiiert. Wir werden weiter dafür sorgen, dass Gestalter bei der Kommunikation ihrer Dienstleistung wahrgenommen werden, wir werden weiterhin ausgezeichnete Unternehmen bei ihrer Kommunikationsarbeit unterstützen, wir werden im Design unerfahrene Unternehmen stärker zu ersten Gehversuchen animieren, wir werden als e.V. unseren Mitgliedern zur Verfügung stehen und mit ihnen und in ihrem Namen den Nachwuchs fördern.
Wir bei iF waren und bleiben optimistisch – für uns selbst, aber vor allem für die Anbieter und Nutzer von gestalterischen Spitzenleistungen.

Unser Dank gilt allen Kunden, Partnern, Freunden, Mitgliedern und auch unseren Wettbewerbern für Kooperation, Kommunikation und Inspiration.

Thinking Design
Ralph Wiegmann

On a birthday, round figures are like milestones – you look back and you look ahead, take stock of your situation and search for new horizons, both for yourself and for the context in which you find yourself. Design has set itself the goal of using business and creative activities to render complex contexts both visible and comprehensible and has long since emerged as a discipline that needs to examine itself critically. In the process, it must take into account products, graphic issues and media tasks as well as the services and, not least, the people who work in design themselves. The time has come not only to stop and reflect, but also to devise ideas, initiate measures and engage in interdisciplinary communication if we in Germany want to grow out of our design puberty. Design has to ask questions, to take a stance, to be relevant – and challenges everyone involved to enter the discursive fray.

50 YEARS OF PROMOTING DESIGN – A BRIEF SURVEY AND AN ENCOURAGING OUTLOOK

For iF in its 50th year, 2003 was filled with competitions, conferences, a good number of projects as a service provider for trade and industry and, not to be forgotten, the first ICSID Design Congress in Germany. Among others, thanks to iF "Designed in Germany" has now become an international brand signaling outstanding quality. And over the years, customers' requirements of design centers have changed, become more complex. Today, iF is a communicator, a networker and even an agency, with all the opportunities and challenges that this involves. Design is now much more closely and more clearly linked to trade and industry, acts as a driving force and as a source of ideas. It is iF's mission to look beyond the immediate present and, as well as organizing two major new design competitions, the iF PRODUCT design award and the iF communication design award, there is now the iF design award.CHINA, right at the heart of the Asian development center and the region's economic hotspot.

DESIGN IN AN IVORY TOWER?

However, these successes to date by no means allow us to sit back on our laurels. German designers enjoy a global reputation, are in demand throughout the world, and there are far too many of them in the upcoming generation. But where is this all leading? How is the main body of designers doing, those people who are not lacking in talent but who are perhaps not being given the opportunity to prove themselves? What's more, even after five decades of German design we have still not succeeded in shedding our elitist image – or perhaps we simply have not wanted to. At the same time, increasingly everything is becoming a matter of design. Indeed, used carelessly and in the wrong context, the word is even deteriorating into a term of abuse for bad products and services. How can we keep up the balancing act, how can design prove its worth, what topics are and will become relevant to the designer's trade? A number of issues are obvious and need to be addressed:

- As the average age of Germans increases, we will need to more strongly focus on universal design, the kind of design that offers everyone the same opportunities to comprehend and use both things and processes.
- The discovery of new materials is becoming ever more important as a real source of innovation. Research is an important element in manufacturing and should be promoted with regard to design.
- To date, media technology and the various peripherals involved have not necessarily facilitated interperson, man/machine or machine/machine communications. Too many features and too little content simply spell another tangle in the information jungle. Guidance systems, clear standards and a return to basics could be of help here and require the kind of designers who are willing to deal with the problems involved.

- To date, with few exceptions, design and science have lived in separate worlds. Theory, the process of getting to grips with your own discipline, is important if new stimuli are to arise. It is worthwhile asking critical questions and going in search of the answers.
- After globalization and the various local countermovements the question arises: Where are the resources? On our own doorstep, or in more distant countries? Design should beware of standardization, analyze, accept and exploit cultural diversity.

DESIGN CENTERS – RELICS OR SIGNPOSTS?

The German design centers owe their genesis to Germany's federal structure. In the past, there were doubtless too many of them, with great discrepancies in terms of approach and focus. What models are feasible, what kind of cooperation agreements can be entered upon, what can we use as comparisons?
Great Britain has as its leading light the British Design Council, a single well-equipped institution, strong on PR and communication – a form of design promotion that would not come into question or be desirable in Germany. But are a large number of centers with little sponsorship, low efficiency and a limited scope the kind of situation we need? Could four centers in the north, south, east and west with a joint business and cultural concept be the solution? Or does the future of communicating design lie somewhere completely different, for example in Bremen, in the Universum Science Center? 1.4 million visitors in three years would seem to speak a clear language, and the experimental approach which communicates complex fields of knowledge in a simple and animated form appears to be the future. So are we making a stumbling block out of communicating design? Is it too difficult to get to grips with the discipline? Should design be included in the educational process? Can children and teenagers master the concept of design? Should we chip in at this point and not carry on discussing design among ourselves, as we are so fond of doing?

WHAT IS TO COME, WHAT WILL REMAIN?

2004 is going to be an exciting year. After waiting with bated breath we can feel the wheels of industry starting to turn again, the economy is picking up. Some people are still grumbling, waiting to see what will happen. But the creaking is becoming a whirling. Orders are being placed again. Hurdles are being cleared away, both in our heads and in the development process. The export market is still the driving force, design will become an even more important factor for product differentiation – even when it is not always mentioned or recognized as such.

iF has initiated a large number of projects for 2004 and 2005. We will continue to make sure that designers are heard when they communicate their services, we will keep offering our support to excellent companies in their communication activities, we will do more to encourage fledgling companies' first steps on the way, we will be there for our members as a registered association and promote the upcoming generation with them and on their behalf.

We at iF have remained and continue to remain optimistic – for ourselves, but especially for those people who provide and use superb design achievements.

Our thanks go to all our customers, partners, friends, members – not to mention to our competitors for their cooperation, communication and inspiration.

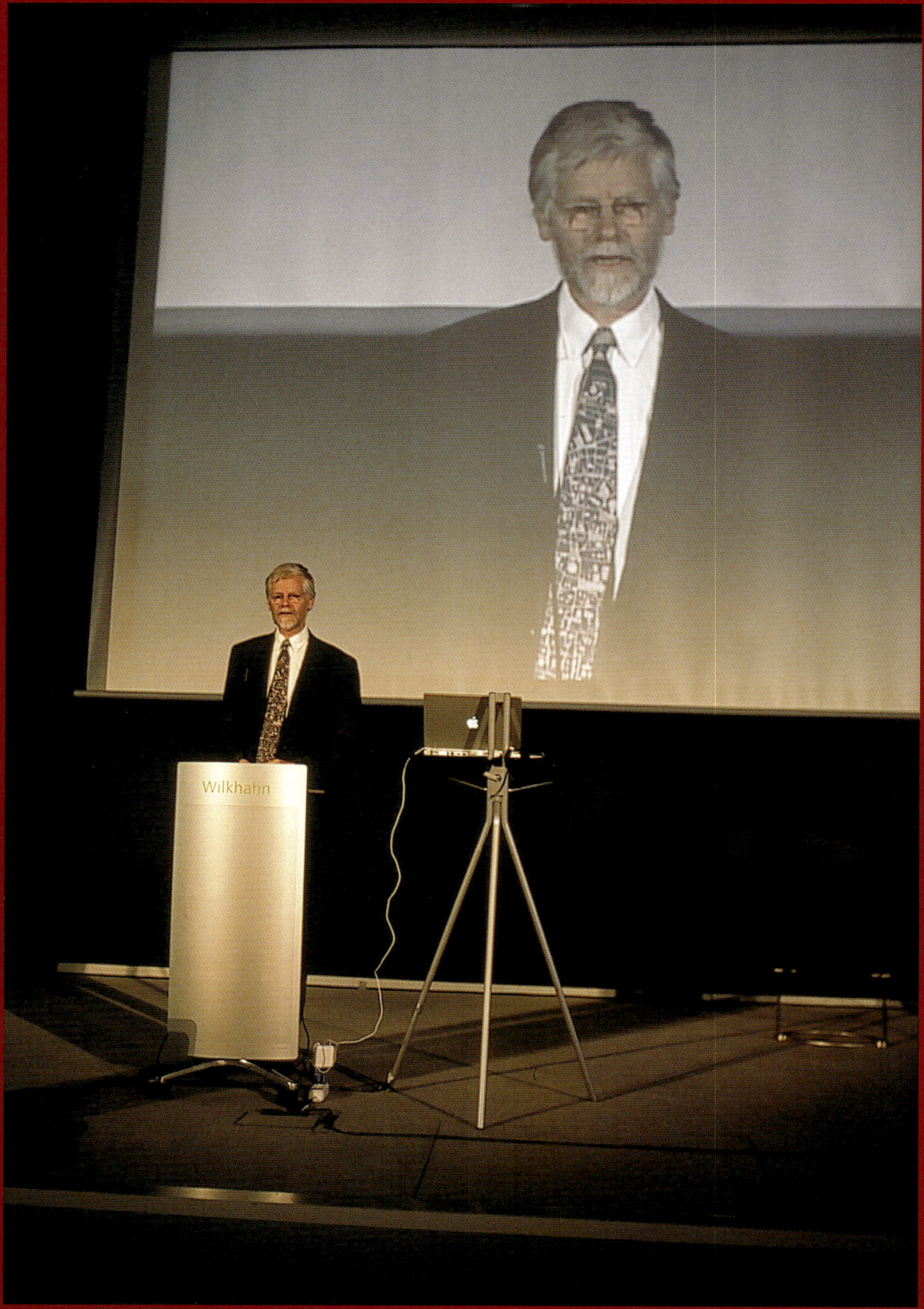
Wilkhahn

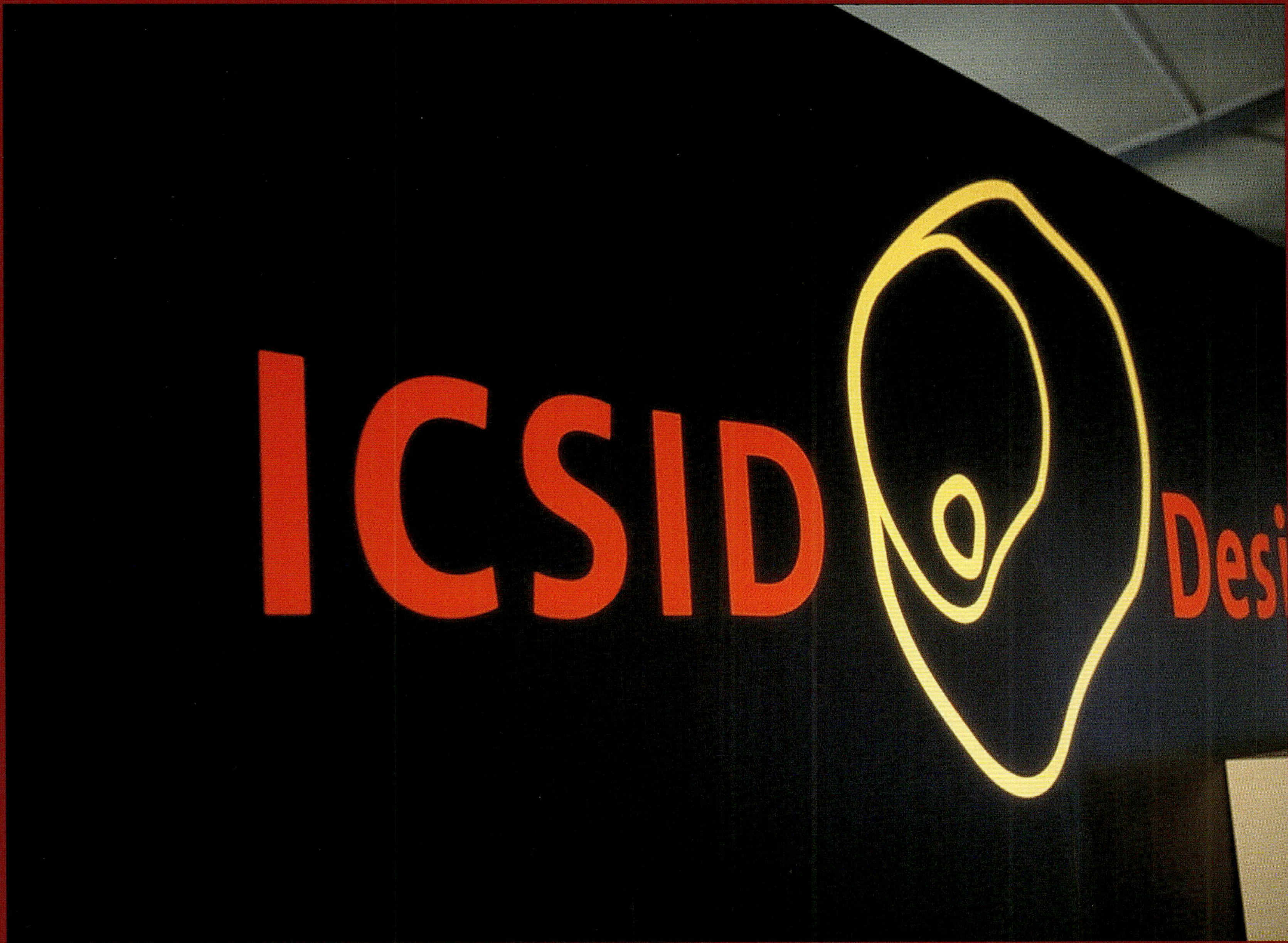

Begrüßungsrede zum ersten ICSID Congress in Deutschland durch Peter Butenschøn, ICSID President (linke Seite).
Peter Butenschøn, ICSID President gives the welcoming address at the very first ICSID Congress to be held in Germany (left).

Weich schwingende Linien kennzeichnen das Corporate Design des ICSID Design Congress 2003, Design Helke Brandt und Partner Hannover.
Soft flowing lines characterize the corporate design of the ICSID Design Congress 2003, Design Helke Brandt and Partner Hanover.

Reflecting Experience
Der ICSID Design Congress 2003 (8.-10.9.2003)

DIGITALISIERUNG + GLOBALISIERUNG = DESIGNISIERUNG?

Digitalisierung und Globalisierung haben unsere Wirklichkeit in den letzten Jahren stark verändert: Labels, Pixel und Icons beherrschen die Welt von Hannover bis Hanoi, der Bildschirm ist längst zum sprichwörtlichen Fenster zur Welt geworden. Gleichzeitig erlauben moderne, Computer gesteuerte Präzisionsinstrumente operative Eingriffe in codierte Genstrukturen wie in fremde Volkskörper und stehen am Anfang (und am Ende?) einer universalen Neugestaltung – nicht nur des guten alten Menschen, sondern der gesamten Welt.

DAS DESIGN IM INTERDISZIPLINÄREN DISKURS

Das betrifft vor allem auch das Design als Wirklichkeit prägende Macht, die an der Schnittstelle von materieller, medialer oder auch imaginärer Welt operiert und damit völlig neuen Aufgaben und Herausforderungen entgegensieht. Zwischen Produkten, Medien und Services: Wo ist der Designer von heute zu verorten – als Mediator oder auch – im besseren Fall – als Katalysator zwischen den genau zu analysierenden Bedürfnissen der User-Massen und den auf Alleinstellung bedachten Marken generierenden Weltkonzernen? Ist er Erfinder oder sogar Erzähler, Gebrauchsforscher oder genialer Ingenieur? Und: An welchen Werten orientiert sich das Design in einer überkomplexen, zunehmend diffusen Welt? In Anlehnung an die sechs Thesen, die der italienische Schriftsteller Italo Calvino für seine Disziplin auf der Schwelle in ein neues Jahrtausend formuliert hat, versuchte der ICSID Design Congress 2003, Antworten auf diese und andere Fragen zu finden; das Terrain wurde markiert durch die Begriffe Leichtigkeit, Schnelligkeit, Genauigkeit, Sichtbarkeit, Vielfältigkeit und Beständigkeit.

WILDES DENKEN UND DIGITALES ARBEITEN – DIE KEYNOTE SPEAKER DES KONGRESSES

Wie weit reicht der Gestaltungsauftrag und die Gestaltungskompetenz des Designs heute – was kann das Design aus seinen Erfahrungen aus 100 Jahren technologischer Innovation, aber auch aus der Erfahrung anderer Gestaltungsdisziplinen lernen? Der Kongress suchte den Dialog mit diesen Disziplinen, um die Anforderungen an seine eigenen zu präzisieren. Der slowenische Philosoph Slavoj Zizek beispielsweise, bekannt für sein „wildes Denken", lotete die Kategorie des Designs zwischen den Kategorien des Sichtbaren und Unsichtbaren, des Immanenten und des Anderen, des Subjektiven und des Objektiven – ganz im Sinne einer Positiv und Negativ-Form – aus. Der japanische Wirtschaftsautor und Unternehmensberater Kenichi Ohmae sah unsere Welt auf der Schwelle zu einer zunehmenden Unsichtbarkeit (invisibility), wobei gerade dem Designer die wichtige Aufgabe zukommt, die Brücke zwischen sichtbarer und unsichtbarer Welt zu schlagen und somit die Menschen mit dieser Entwicklung zu versöhnen. Li Edelkoort untersuchte die aktuellen Trends und Tendenzen anhand ihrer „Oberflächenphänomene" wie Mode und Lifestyle und zieht davon ausgehend ein Zukunftsfenster auf. John Maeda brachte seine Erfahrungen am MIT Medialab in Boston ein und gab einen Überblick über aktuelle Entwicklungen an der Schnittstelle zu einer zunehmend digitalisierten Welt. Und die international renommierte Sozialwissenschaftlerin Saskia Sassen erweiterte die Perspektive und lenkte den Blick auf die politischen, wirtschaftlichen und sozialen Umbrüche durch die Globalisierung, speziell im Kontext der Stadt und ihrer veränderten Topographie. Am Beispiel der Globalisierungsgegner und New Media und Art Aktivisten untersuchte sie die Möglichkeiten des Individuums jenseits des klassisch-politischen Feldes, sich in dieser veränderten Welt zu behaupten und seine Interessen zu artikulieren. Über die Potenziale des menschlichen Gehirns zur permanenten Verbesserung reflektierte der deutsche Neurologe und Gehirnforscher Detlef. B. Linke.

REFLECTING DESIGN – DIE SICHT AUF DIE EIGENE DISZIPLIN

Sechs Keynote Speaker, sechs „disziplinfremde" Sichten auf das Design. Der ICSID Design Congress hat versucht, Positionen zu verändern, Standpunkte zur Diskussion zu stellen. Ob dieses Vorhaben gelungen ist, werden die Besucher selbst entschieden haben. Fakt ist, dass Design nicht in seinen althergebrachten Strukturen verhaften darf. Es ist Zeit, die Dinge von außen zu betrachten und selbstkritisch zu hinterfragen. Wo geht die Reise hin, wird Design zukünftig in anderen Disziplinen aufgehen, noch kooperativer mit anderen Disziplinen zusammenarbeiten, verstärkt andere Disziplinen integrieren und in sich versammeln? Was muss Design heute leisten, was müssen Designer leisten? Gibt es eine Design-Ethik? Wo besteht die Gefahr, dass Design instrumentalisiert wird? Brauchen wir bald die designfreie Zone, um uns vor allzu „lebenseingreifendem" Design zu schützen?

Der Diskurs geht weiter, er muss weitergehen. Design ist nicht allein Gestaltung außerhalb von uns – Design beginnt im Kopf, wo es nicht verbleiben sollte. Kommunikation mag eine abgedroschene Worthülse für den Austausch von Inhalten, manches Mal auch inhaltsleer sein. Mit vielen Ideen und energetisch eingesetzt, kann Kommunikation allerdings viel bewegen. Selbstreflexiv auch das Design.

Reflecting Experience The ICSID Design Congress 2003 (Sept. 8-10, 2003)

DIGITALIZATION + GLOBALIZATION = DESIGNIZATION?

In recent years, digitalization and globalization have strongly changed the world in which we live: Labels, pixels and icons predominate from Hanover to Hanoi, the monitor has long since become the proverbial window on the world. At the same time, modern computer-controlled precision instruments enable operative intervention in coded genetic structures and in foreign peoples alike, and we are just at the beginning (or the end?) of a universal re-design – not only of the good old human, but also the entire world.

DESIGN IN THE INTERDISCIPLINARY DISCOURSE

This applies above all to design as the power shaping reality, functioning at the interface of the material, the media-based or the imaginary world and thus accessing completely new tasks and challenges. Between products, media and services – where should we locate the designer today – as an intermediary or (in the best scenario) as the catalyst between the needs of the user masses (still to be carefully analyzed) and the brands of world corporations bent on becoming the sole players. Is the designer an inventor or even a storyteller or ingenious engineer? And: from what values does design take its cue in an overly complex, increasingly diffuse world? Borrowing from the six hypotheses which Italian author Italo Calvino put forward for his discipline at the threshold of a new millennium, the ICSID Design Congress 2003 endeavored to find answers to these and other questions; the terrain was staked out by the concepts of lightness, swiftness, exactness, visibility, multiplicity and durability.

WILD THOUGHT, DIGITAL WORK – THE CONGRESS' KEYNOTE SPEAKERS

What is the scope of the mission and competence of design today? How can design learn from its experiences of 100 years of technological innovation, not to mention the experiences of other creative disciplines? The congress sought to engage these fields in dialogue, to hone the requirements its own discipline must face. Slovenian philosopher Slavoi Zizek, for example, renowned for his "wild thought", explored the category of design between the categories of the visible and the invisible, the immanent and the other, the subjective and the objective – quite in line with the notion of a positive and negative shape. Japanese economic writer and management consultant Kenichi Ohmae viewed our world on the threshold of increasing invisibility, in which context the designer has the crucial task of building a bridge between the visible and invisible worlds and thus reconciling people to this development. Li Edelkoort investigated current trends and tendencies by studying their "surface phenomena", such as fashion and lifestyle, and from this vantage point opened a window on the future. John Maeda brought his experience at MIT's Medialab in Boston to bear and offered an overview of current developments along the interface to an increasingly digitized world. And internationally renowned social scientist Saskia Sassen broadened the perspective and directed attention to the political, economic and social upheaval as a product of globalization, specifically in the context of the city and its changed topography. Taking the example of opponents of globalization and the new media and art activists she investigated personal opportunities for asserting oneself in this changed world and articulating one's interests outside the classical realm of politics. German neurologist and brain specialist Detlef B. Linke reflected on the potential of the human brain for permanent improvement.

REFLECTING DESIGN – THE INSIDE VIEW

Six keynote speakers, six views of design "from the outside". The ICSID Design Congress tried to change received positions and open viewpoints up for discussion. Whether it succeeded is something the visitors must decide for themselves. The fact is that design must not get lodged in a rut of tradition. It is time to view things from the outside and subject them to self-critical examination. So in what direction will things go: Will design in future merge with other disciplines, become more cooperative in its collaboration with other fields, increasingly integrate other areas and pool their resources? What must design achieve today, and with it designers? Is there a design ethics? Is there a danger of design being instrumentalized? Will we soon need a design-free zone in order to protect ourselves against "live-embracing" design?

The debate continues and must continue. Design is not just a creative approach to shaping things outside ourselves, but starts in our heads, and should not get stuck there. Communication may be a commonplace term for exchanging concepts, and at times be empty. Energetically applied with a diversity of ideas communication can move much. And self-reflectively, so can design.

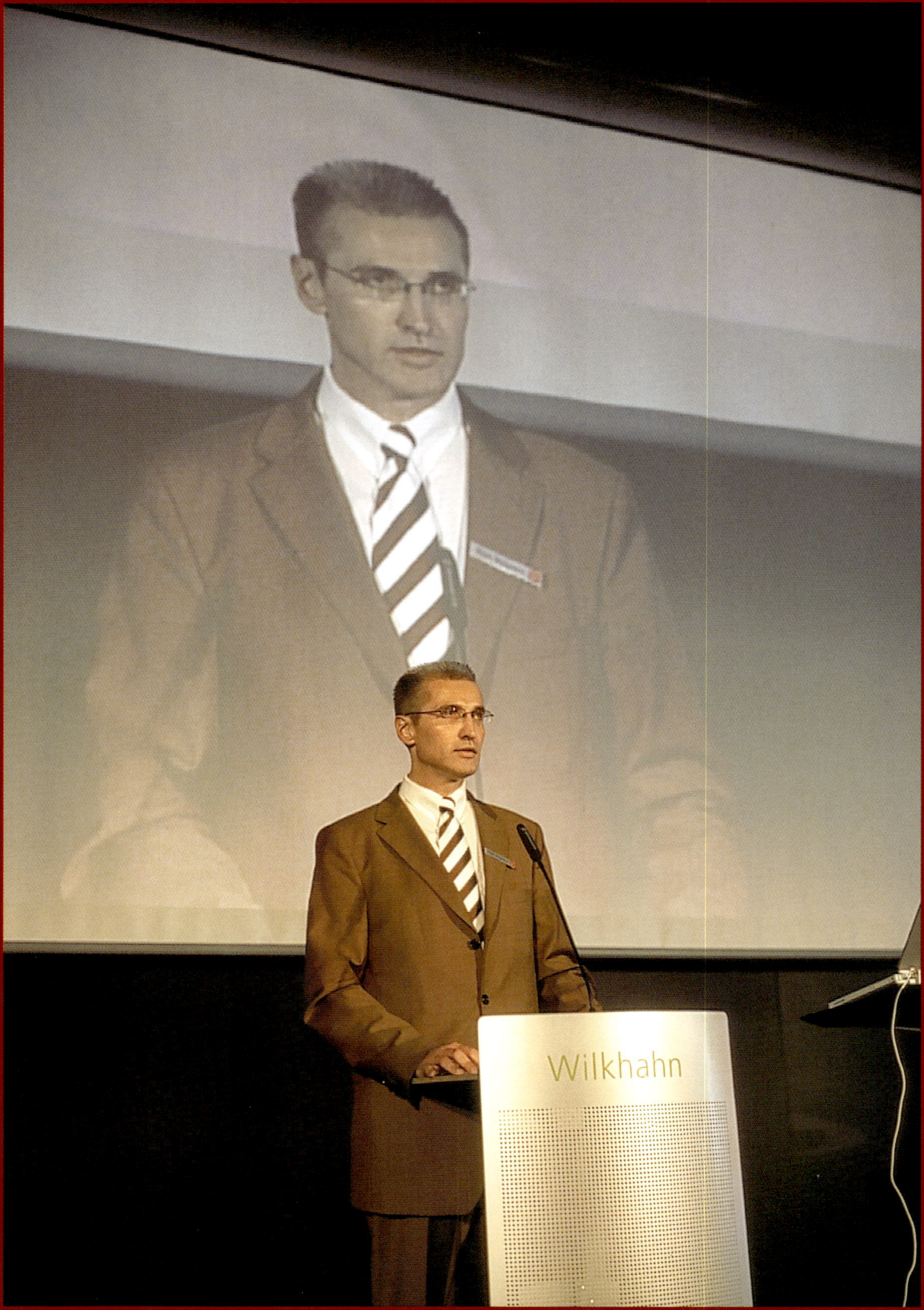
Wilkhahn

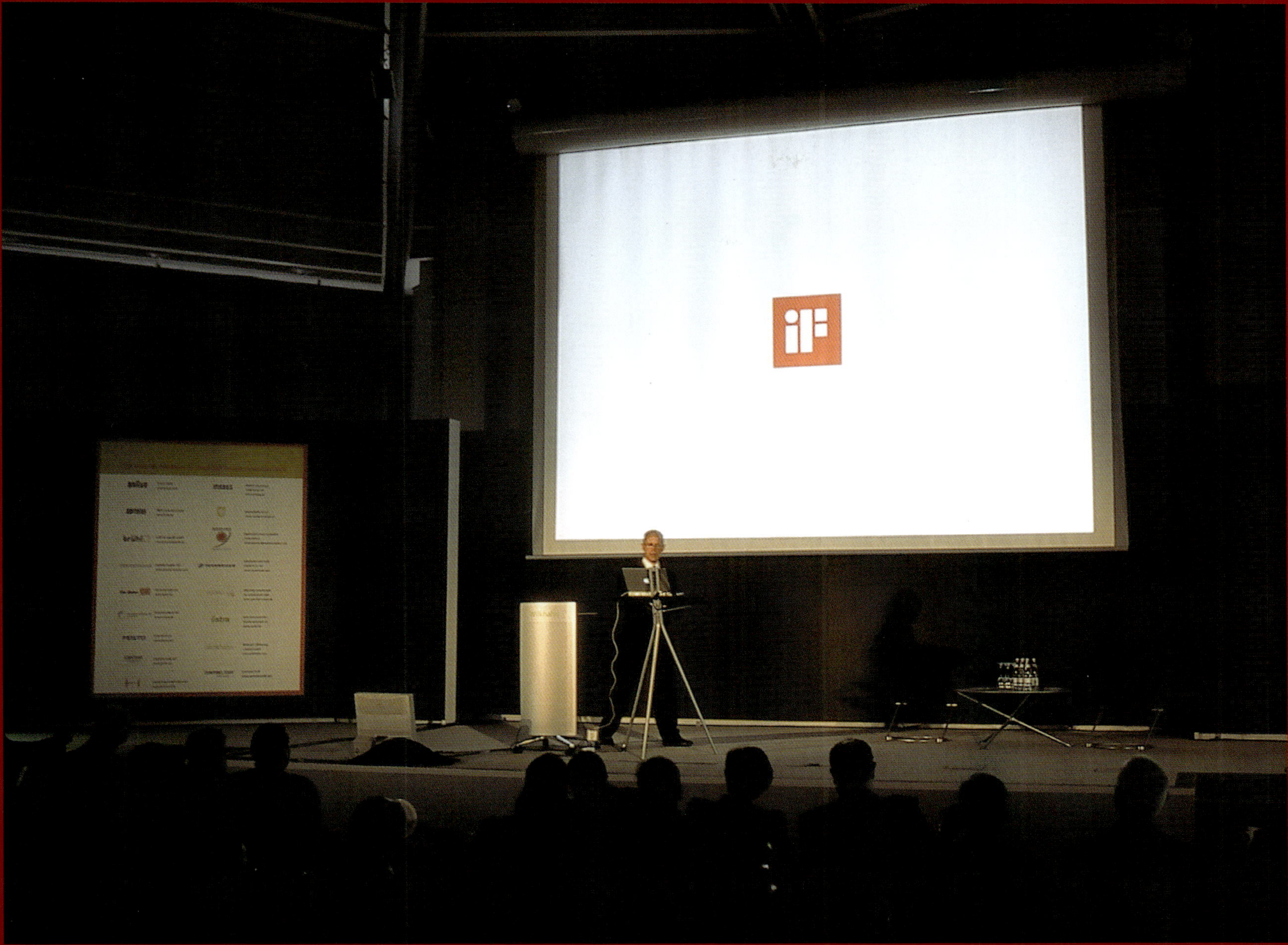

Ralph Wiegmann, iF Geschäftsführer und Gastgeber des Kongresses, bei seiner Begrüßungsansprache (linke Seite).
Ralph Wiegmann, iF Managing Director of iF and Congress host giving his welcoming address (left).

iF hat in Hannover und Berlin die internationale Veranstaltung des ICSID Design Congress vom 8. bis10. September 2003 ausgerichtet.
iF hosted the ICSID international design congress in Hanover and Berlin from Sept. 8-10, 2003.

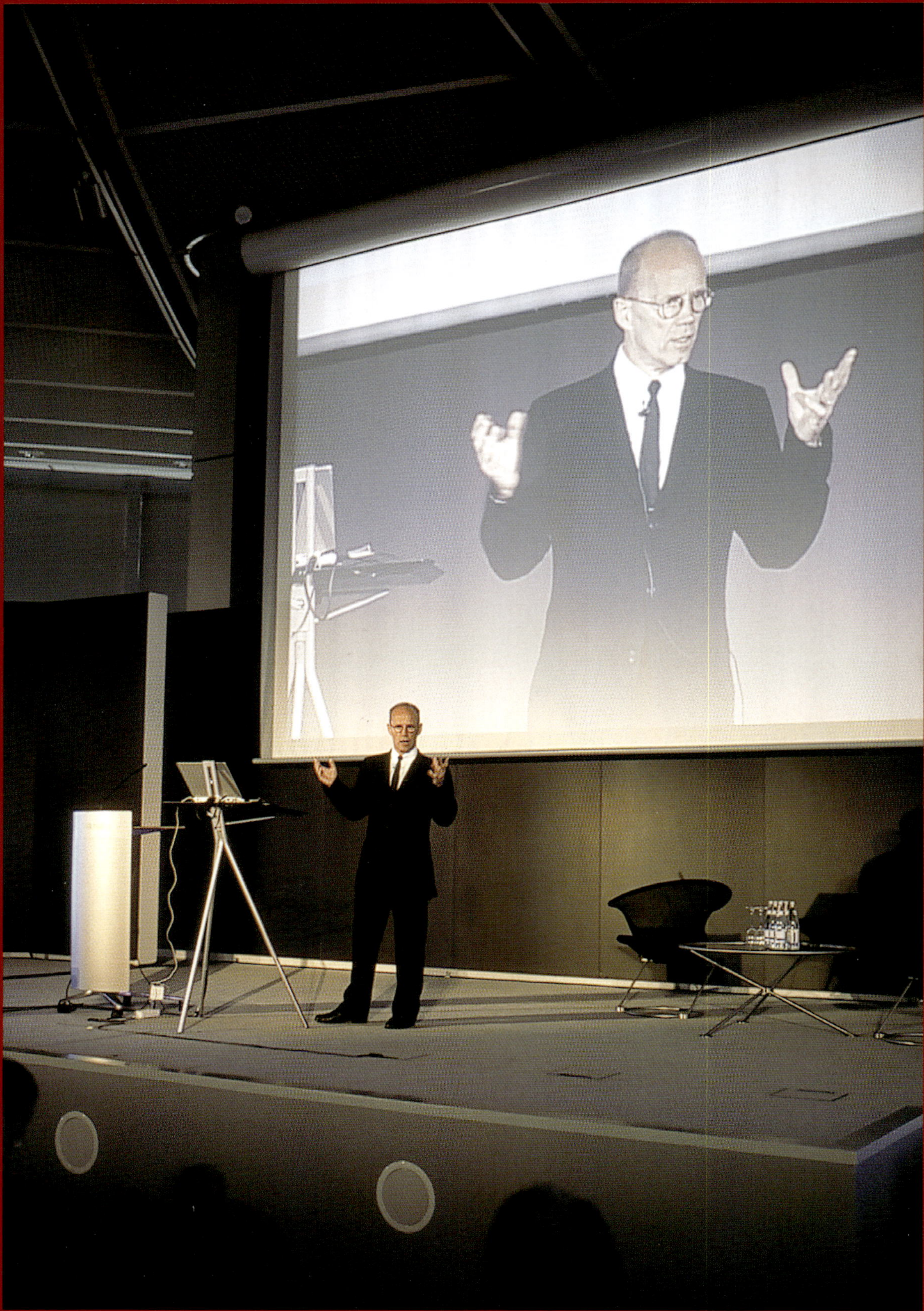

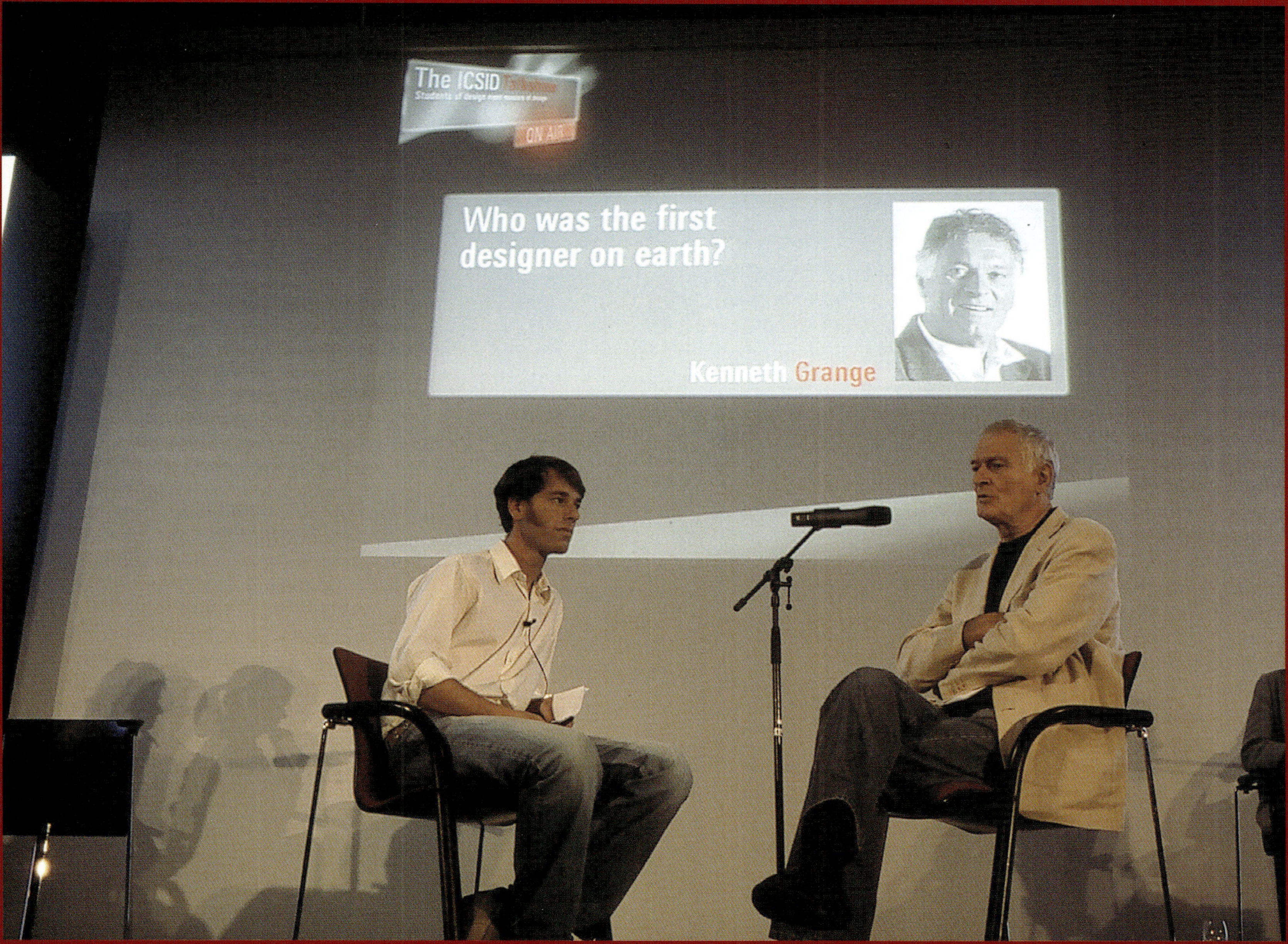

Erik Spiekermann, ICSID Beirat, moderierte die Keynote Speaker (linke Seite).
Erik Spiekermann, ICSID Advisory Council, chairing one of the keynote speech sessions (left).

Generationen im Dialog, Studenten diskutieren mit den Grand Masters, hier der Student Carsten Schelling im Gespräch mit dem britischen Designer Kenneth Grange.
Generations in dialogue - students discuss with the Grand Masters: the photo shows student Carsten Schelling in conversation with British designer Kenneth Grange.

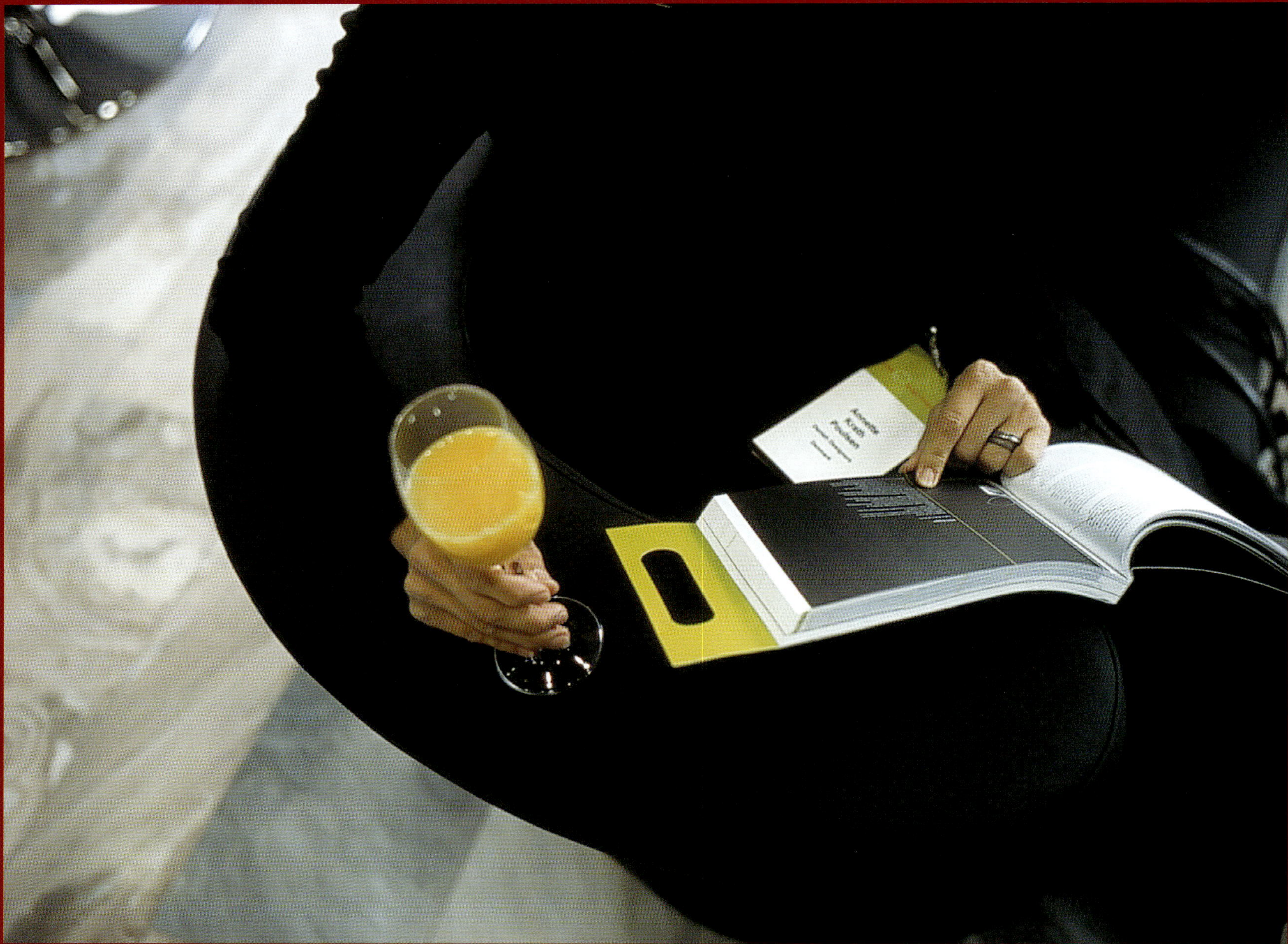

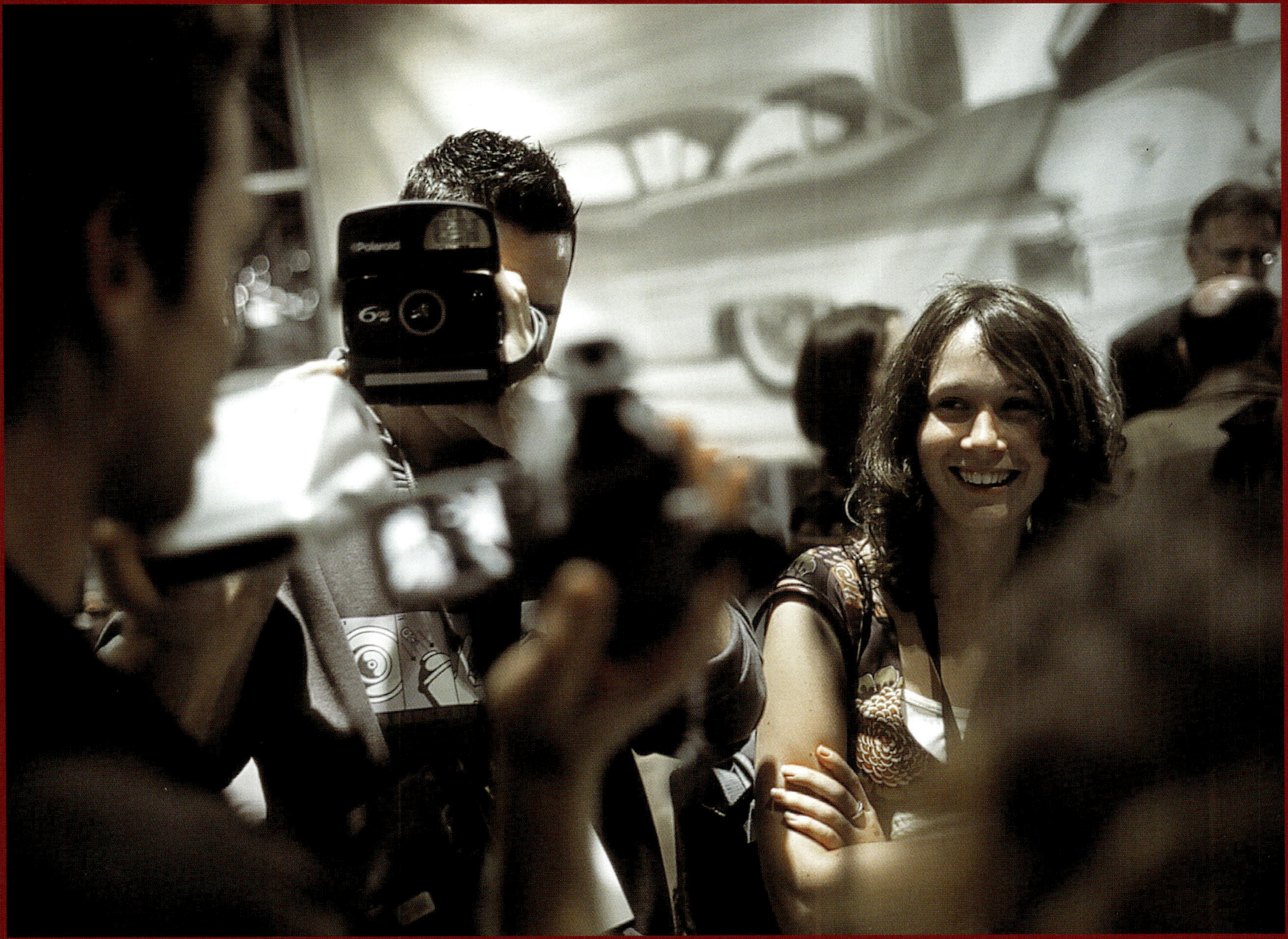

Studenten und Professionals aus der ganzen Welt nahmen an dem viertägigen Gipfeltreffen in Hannover und Berlin teil.
Students and professionals from the world over participated in the 4-day summit in Hanover and Berlin.

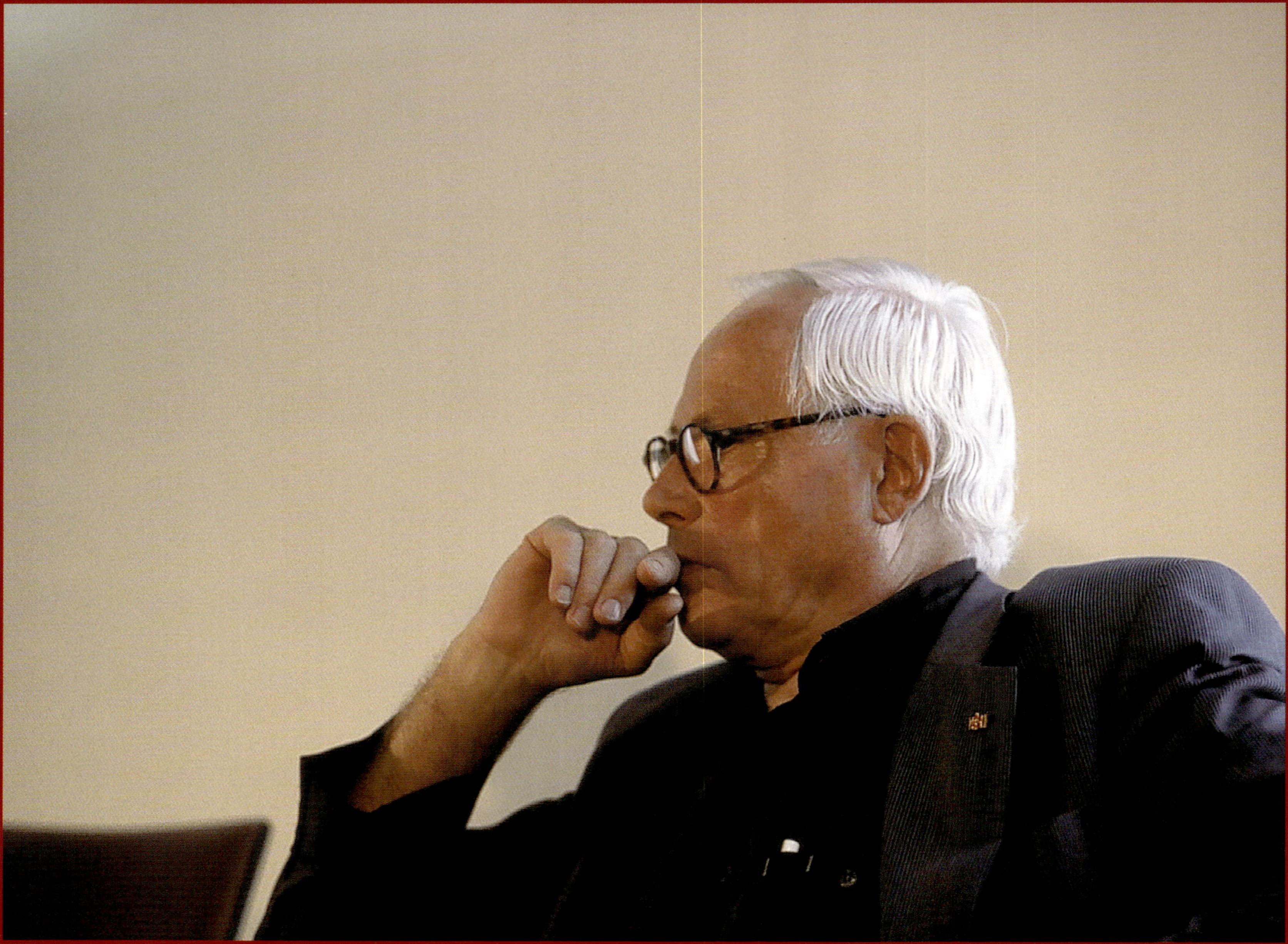

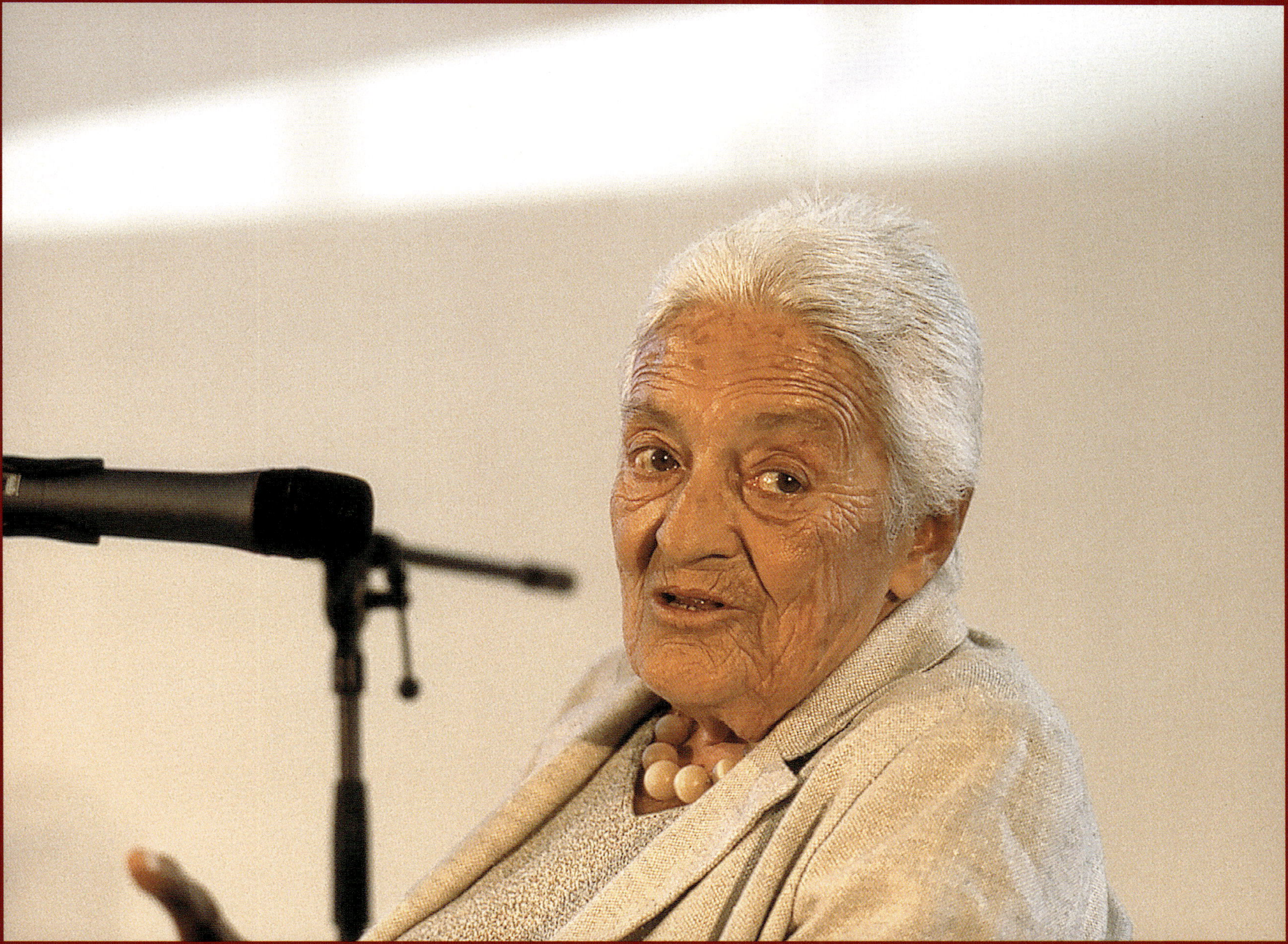

Der deutsche Designer Dieter Rams (linke Seite) in der Gesprächsrunde der Grand Masters of Design zusammen mit der italienischen Architektin und Designerin Anna Castelli Ferrieri. Sie ist eine der wenigen Frauen im Design, deren Entwürfe von Italien aus um die Welt gegangen sind.

German designer Dieter Rams (left) during the panel discussion for the Grand Masters of Design together with Italian architect and designer Anna Castelli Ferrieri. She is one of the few women in design whose creations have, starting in Italy, made their way the whole way round the world.

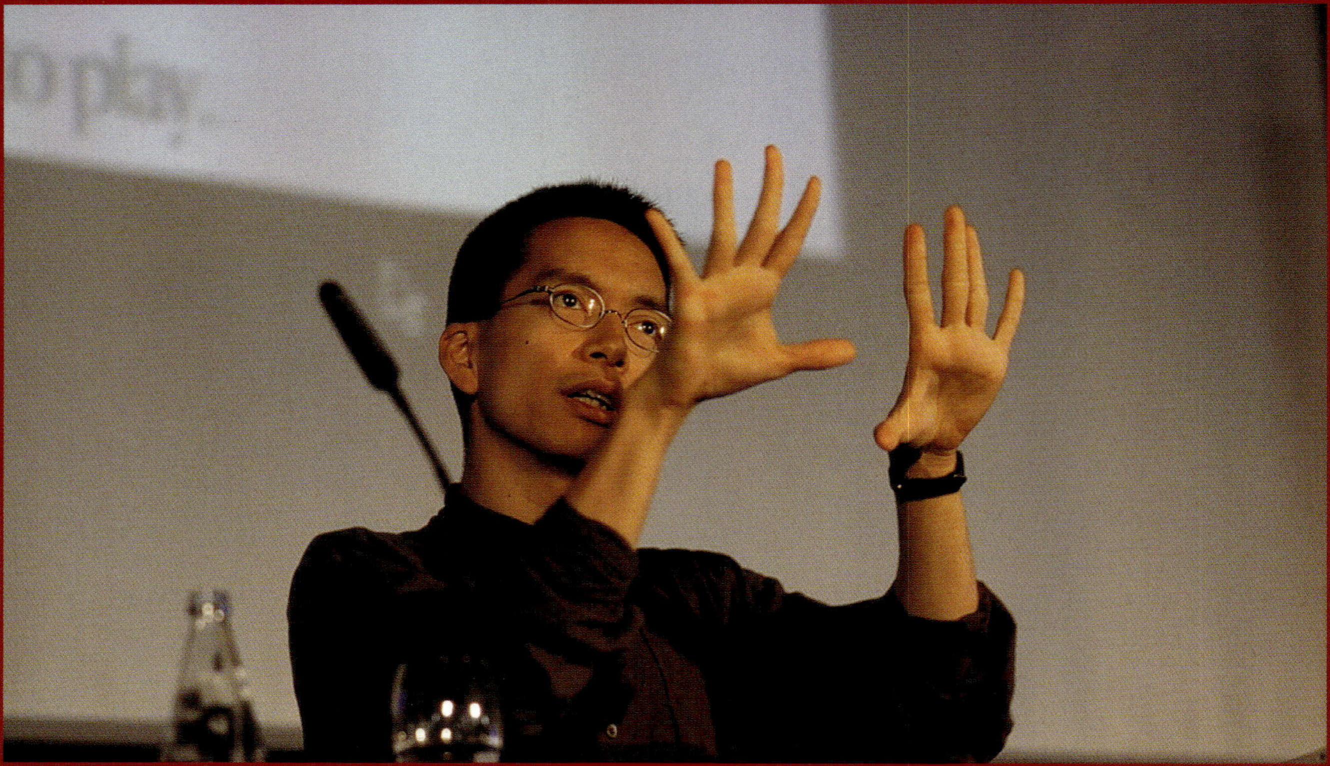

Der MIT Professor John Maeda referierte als Keynote Speaker zum Thema „Undigital".
MIT professor John Maeda gave one of the keynote addresses on the theme of "Undigital".

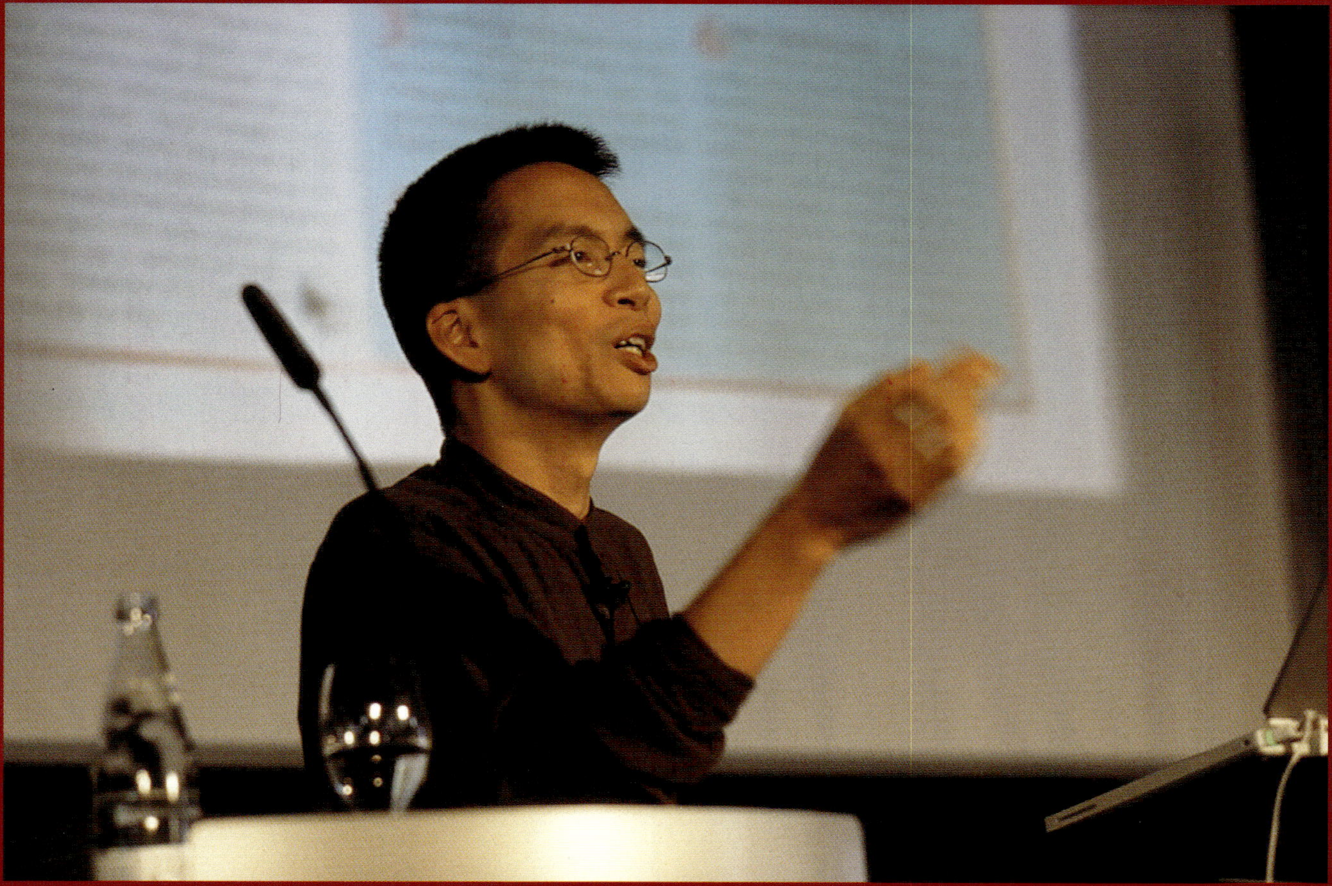

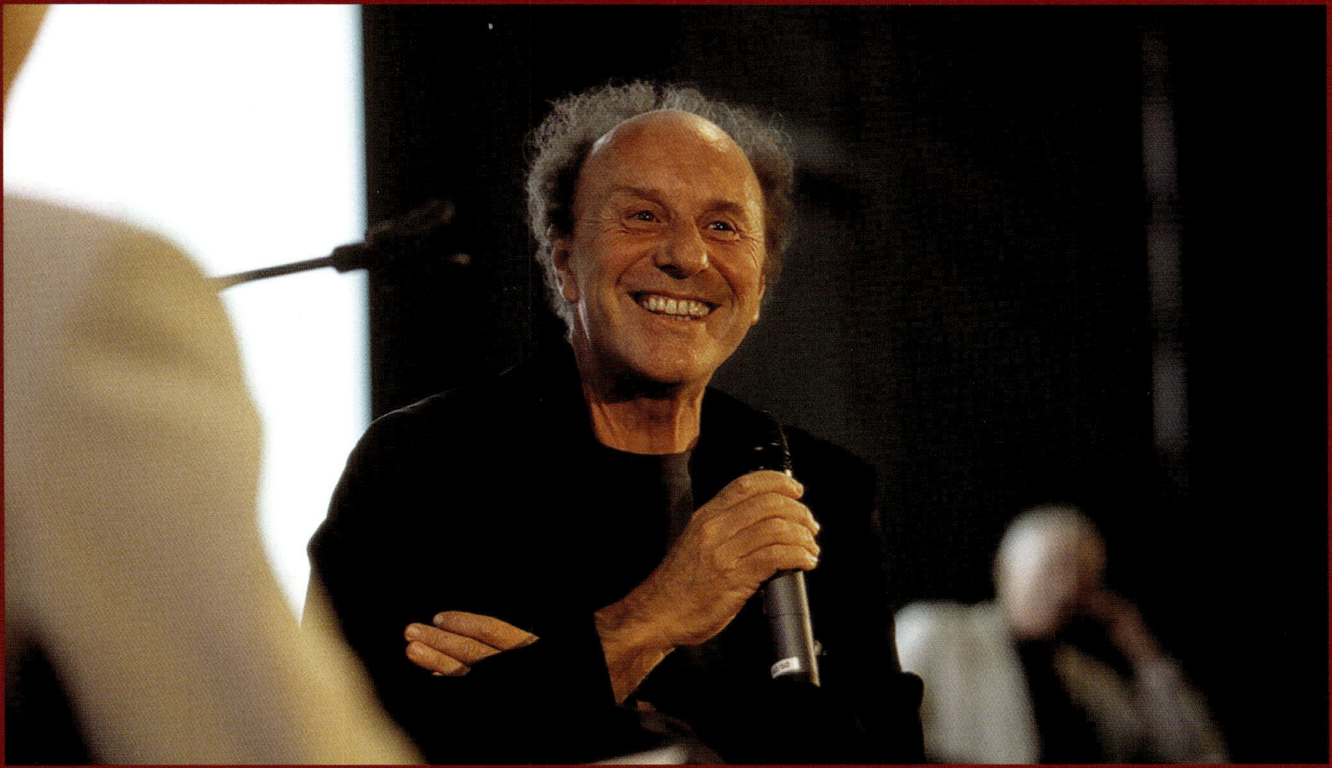

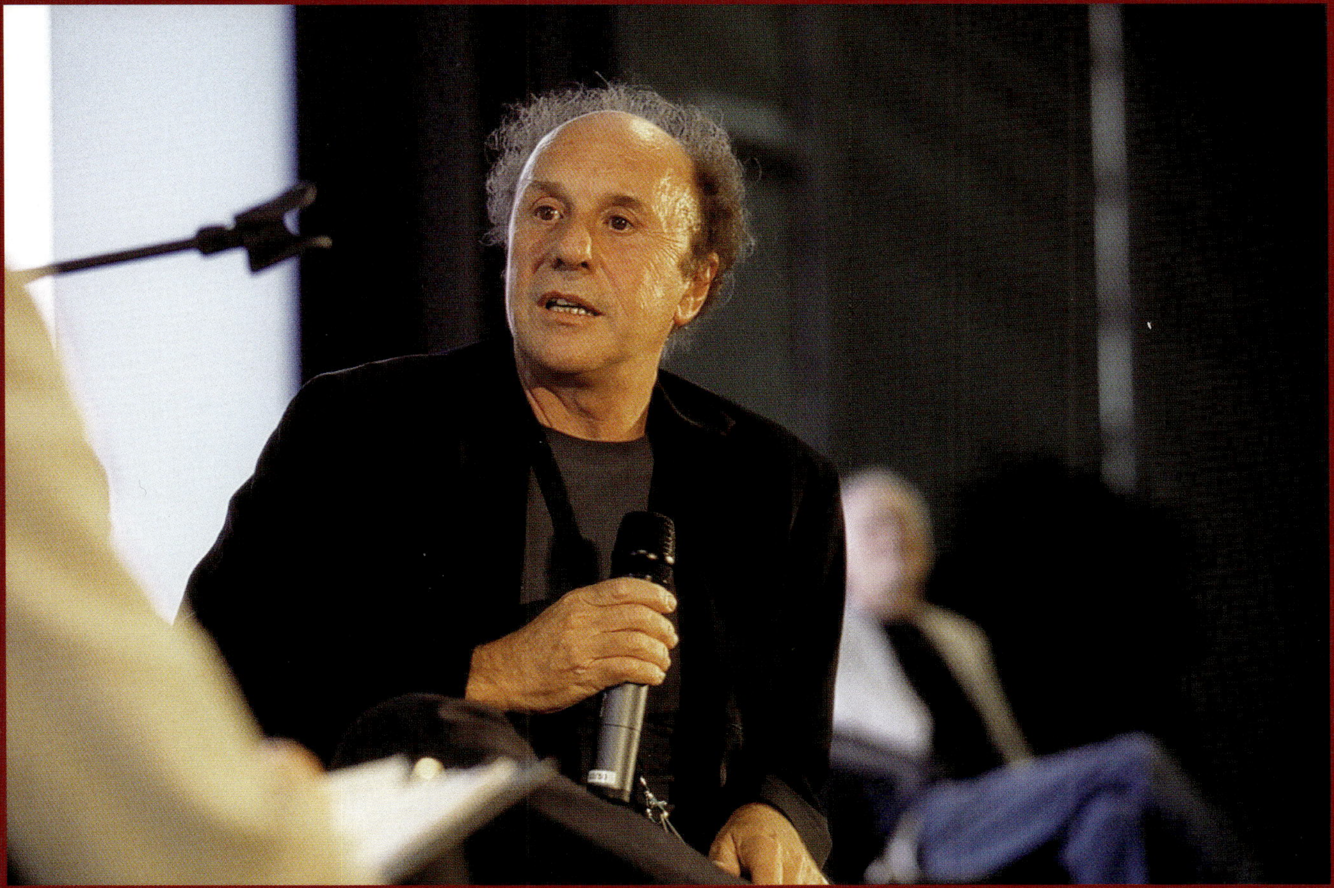

Der spanische Designer Jorge Pensi traf in der Runde der Grand Masters auf kritisch fragende Studenten.

Spanish designer Jorge Pensi during the Grand Masters discussions, fielding critical questions from students.

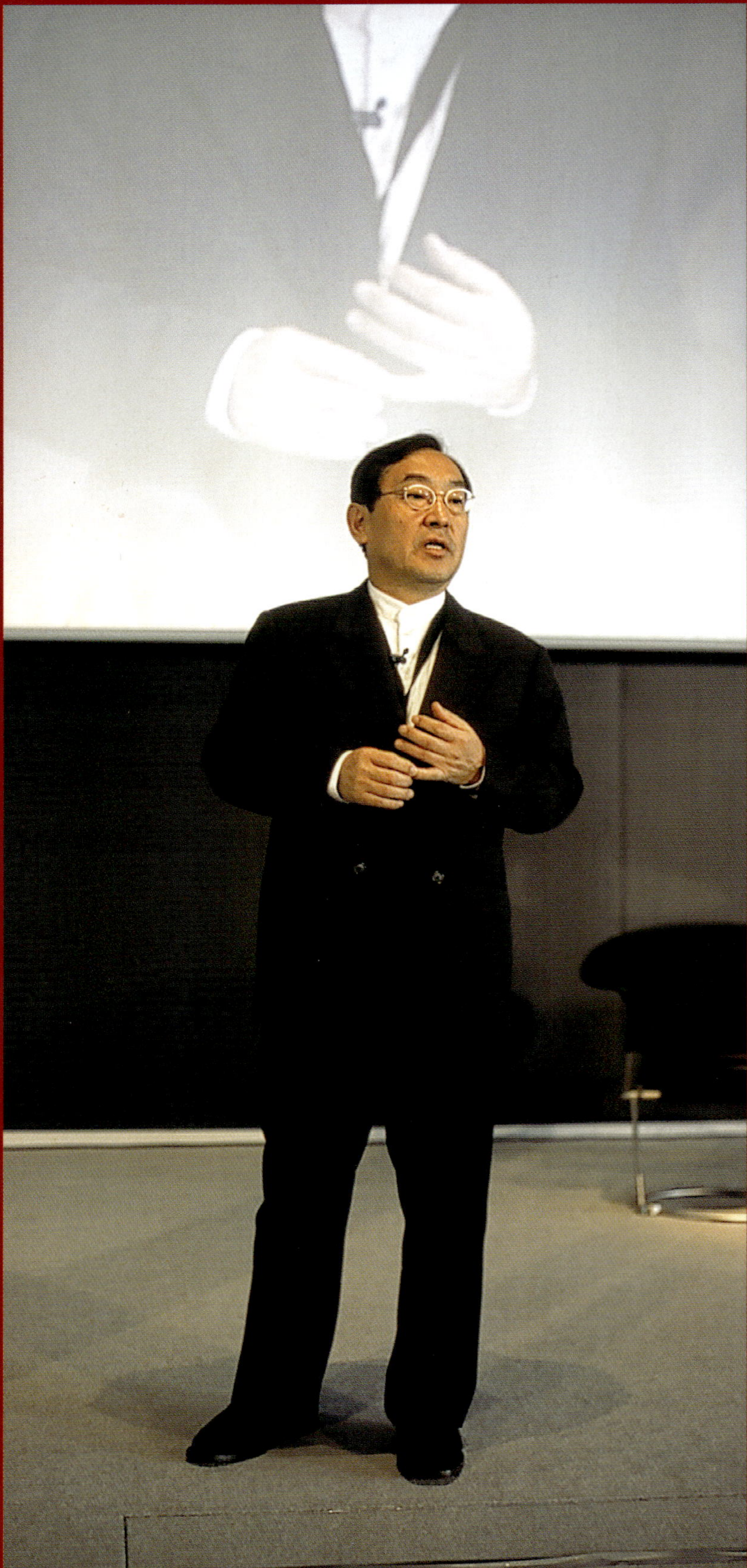
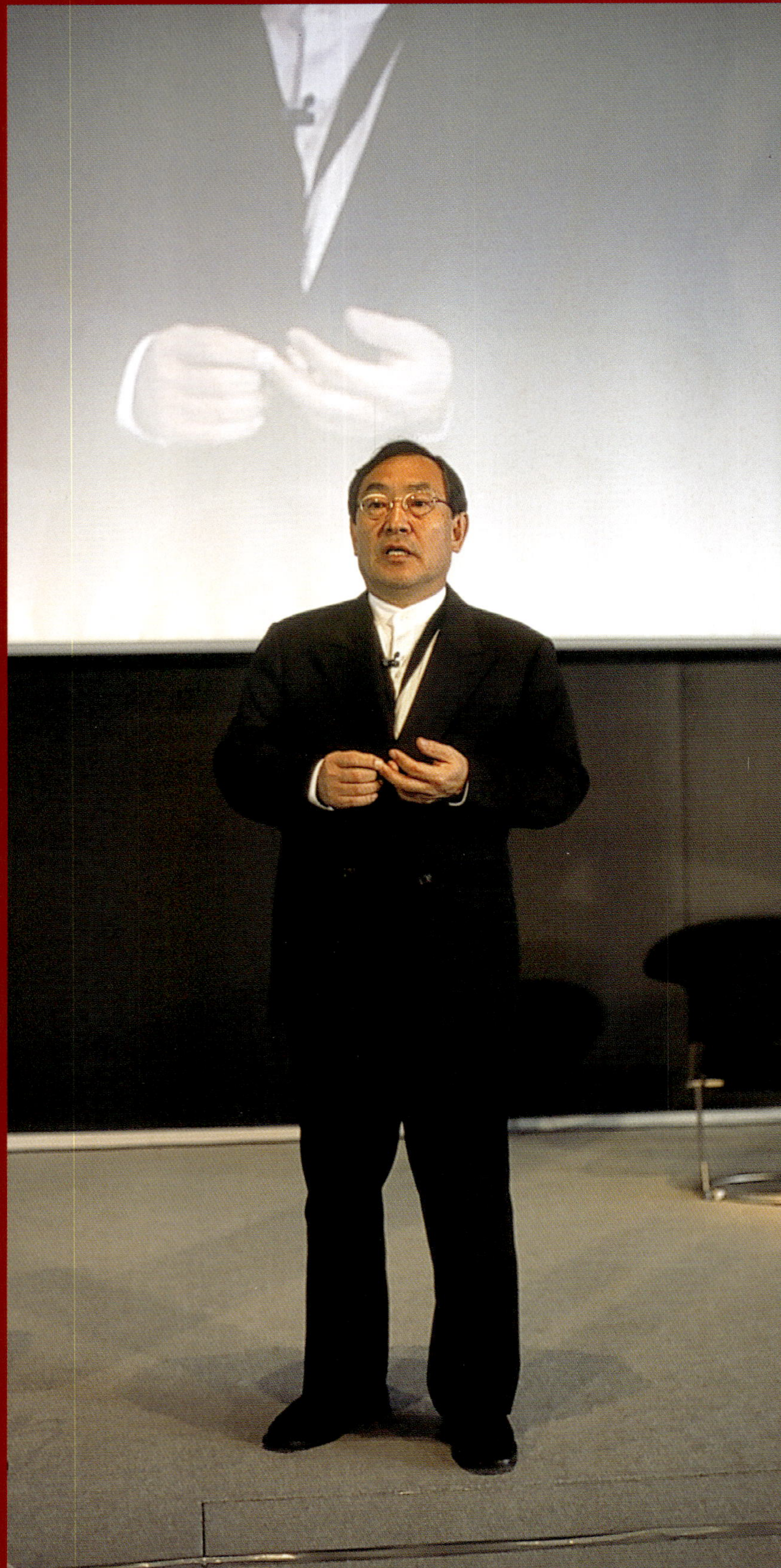

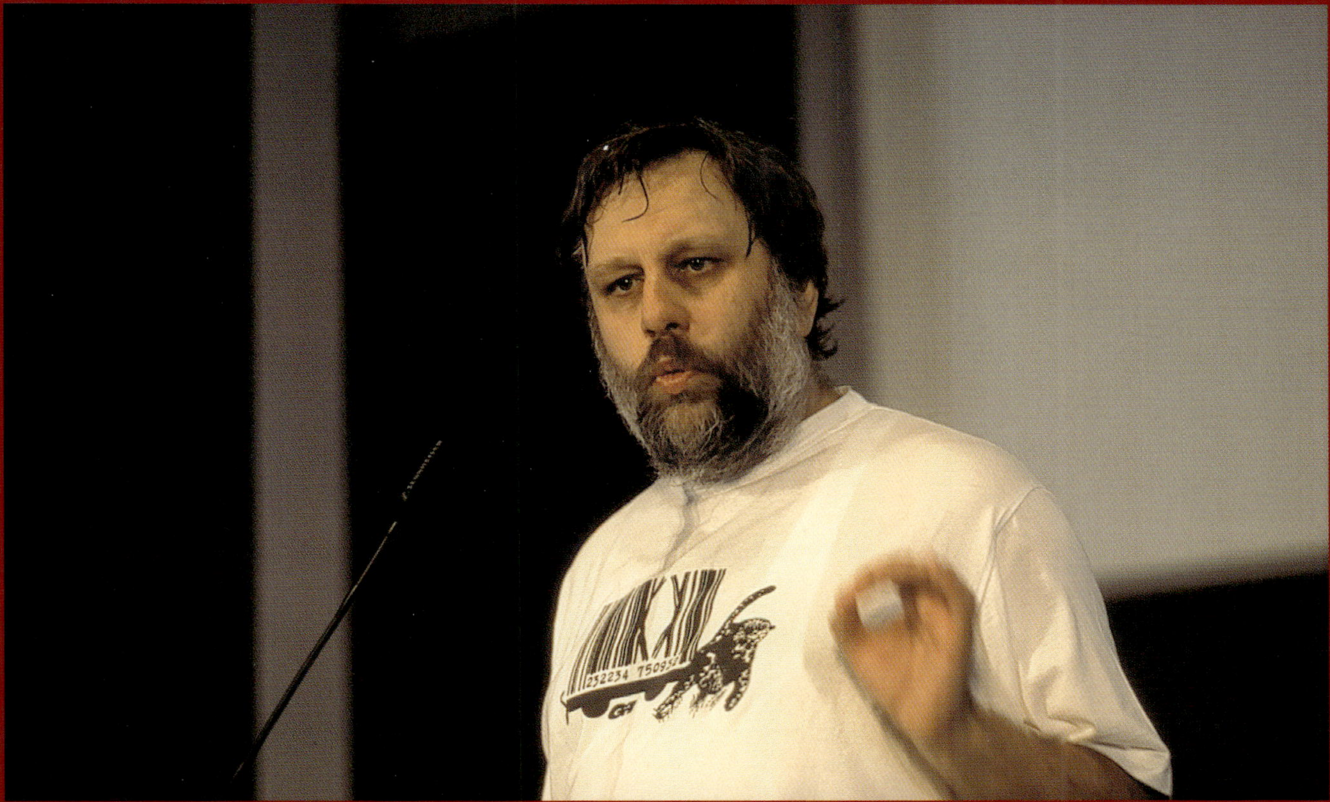

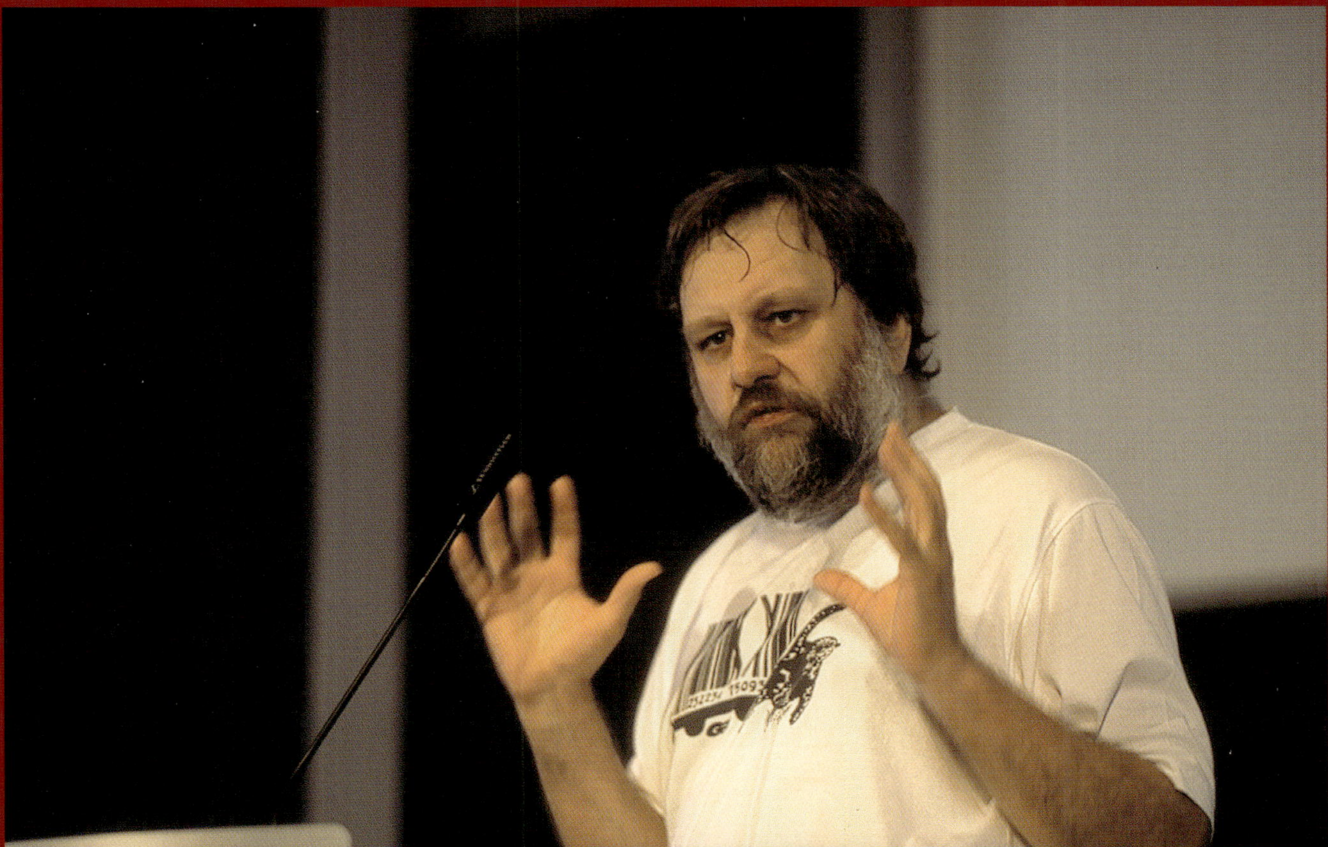

Der japanische Wirtschaftsautor und Unternehmensberater Kenichi Ohmae referierte über „Strategien global agierender Unternehmen im 21. Jahrhundert" (linke Seite).
Japanese economics writer and management consultant Kenichi Ohmae lectured on strategies for global corporate players in the 21st century (left).

Der slowenische Philosoph und Psychoanalytiker Slavoj Zizek referierte zu dem Thema „Unerträgliche Leichtigkeit des Virtuellen".
Slovenian philosopher and psychoanalyst Slavoi Zizek lectures on the "Intolerable Lightness of Virtuality".

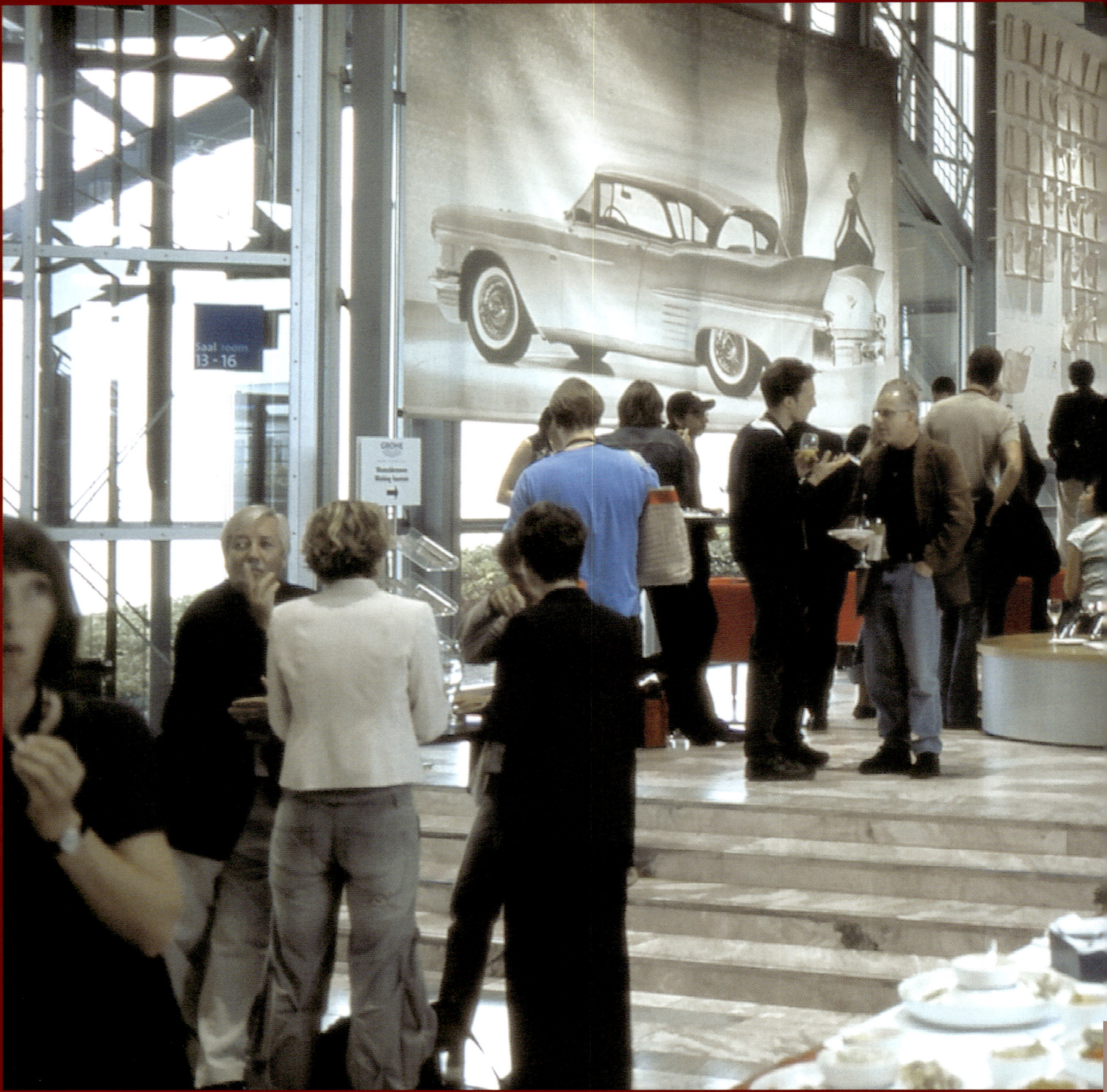
Saal room
13 - 16

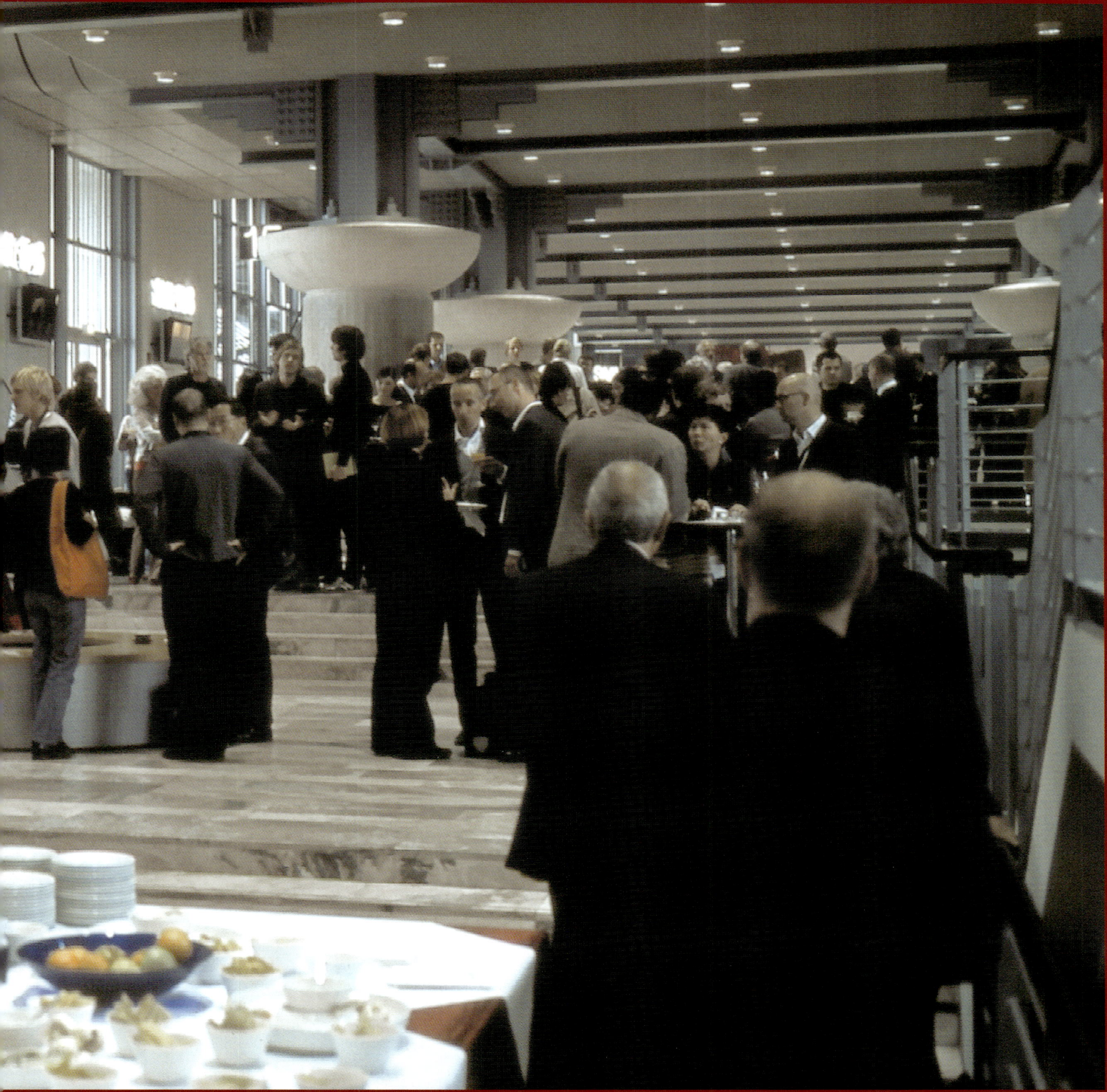

iF design award 2004

COMMUNICATION DESIGN

Disziplin 1 Discipline 1

COMMUNICATION DESIGN

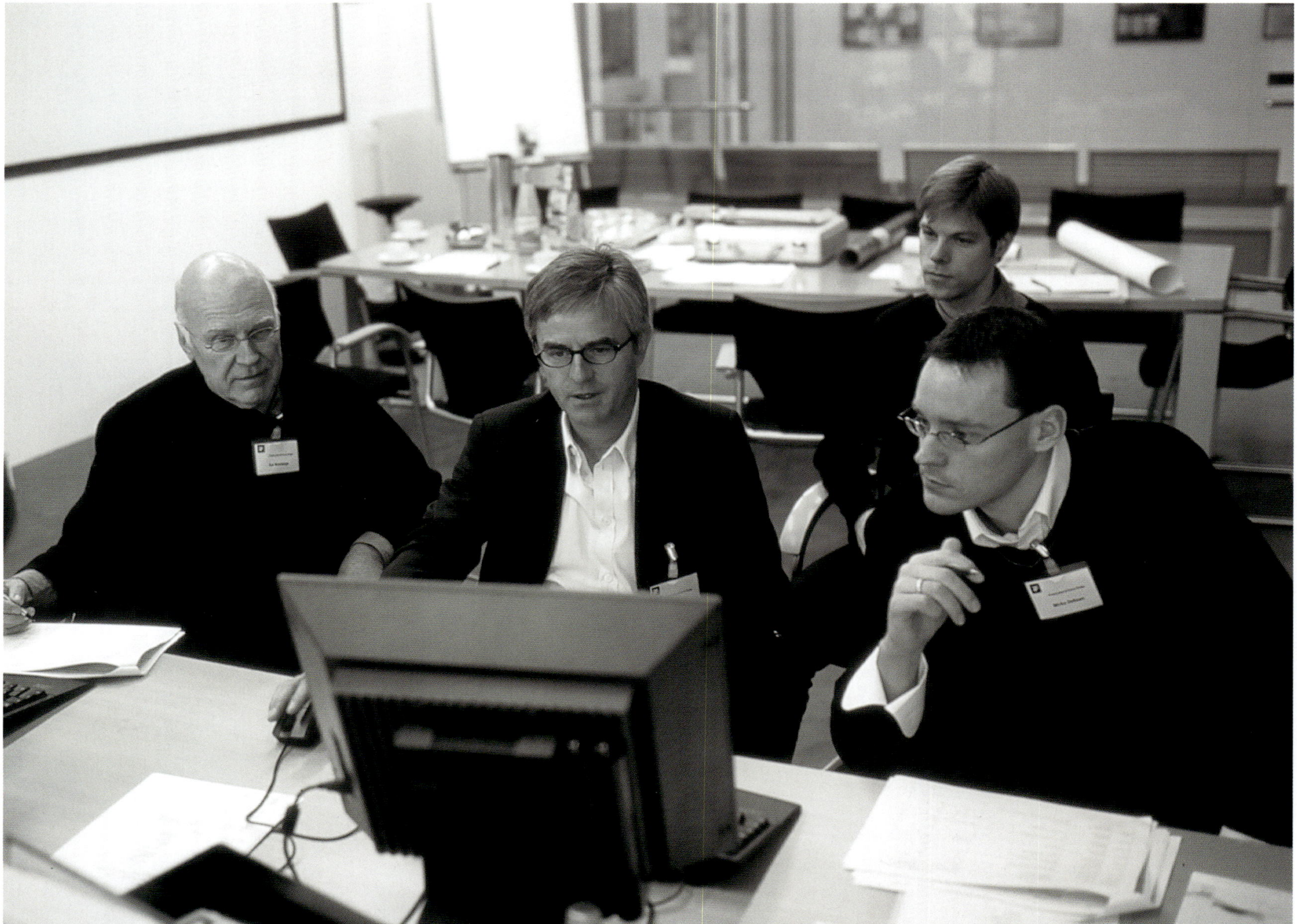

Digital Media
Online Media
TV and Film
Offline Media

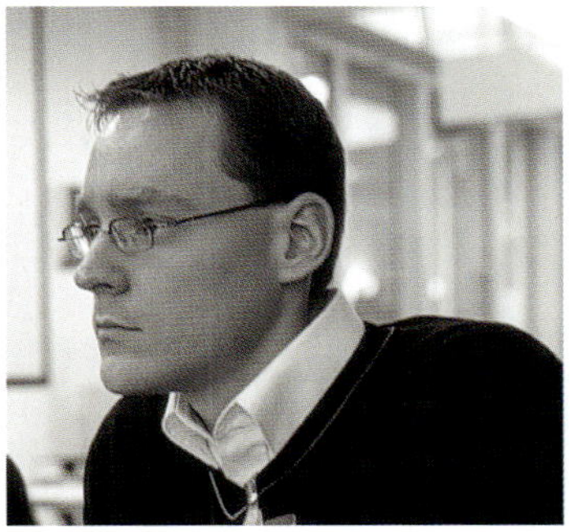

MIRKO DEBUAN studierte Technische Dokumentation an der Fachhochschule Hannover. 1995 brachte er mit vier Mitstreitern die weltgrößte Computermesse CeBIT „ans Netz". Seit 1996 ist er Geschäftsführer und Mitgesellschafter der Agentur DEVICE/N, die Informationssysteme auf der Basis von Internettechnologien entwickelt. Als Ergebnis eines Joint Ventures entstand 2001 die Agentur ABIAN. Bei beiden Firmen ist Mirko Debuan in beratender und konzeptioneller Funktion tätig.

Mirko Debuan studied Technical Documentation at the Hanover Polytechnic. In 1995, with four companions he hooked CeBIT, the world's largest computer trade fair, up to the Web. Since 1996 he has been Managing Director and partner of the DEVICE/N agency, which develops information systems that run on Internet technologies. As part of a joint venture, in 2001 the ABIAN agency first saw the light of day. Mirko Debuan acts in an advisory and conceptual capacity for both corporations.

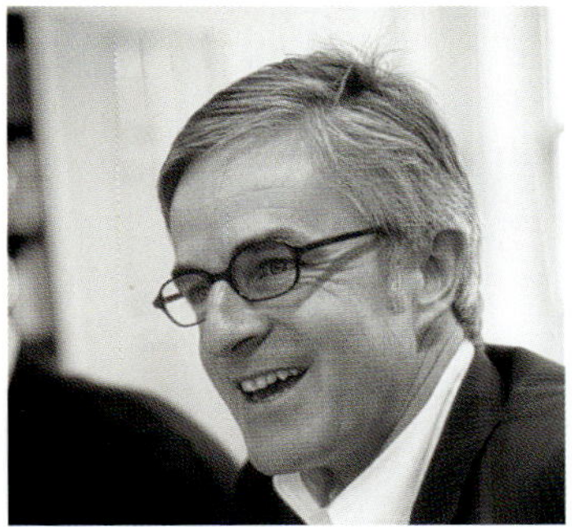

RUEDI ALEXANDER MÜLLER-BEYELER ist Gründungsmitglied, Partner und CEO von NOSE Applied Intelligence AG in Zürich. Der ETH-Ingenieur ist seit mehr als zwanzig Jahren als Markenstratege und Designmanager tätig. Zudem ist er Präsident des Schulrates der Hochschule für Gestaltung und Kunst in Zürich.

Ruedi Alexander Müller-Beyeler is founding member, partner and CEO of NOSE Applied Intelligence AG in Zurich. A graduate from Zurich's renowned ETH Polytechnic, for more than 20 years he has worked as a branding strategist and design manager. He is also President of the College Council of the Zurich Academy of Art and Design.

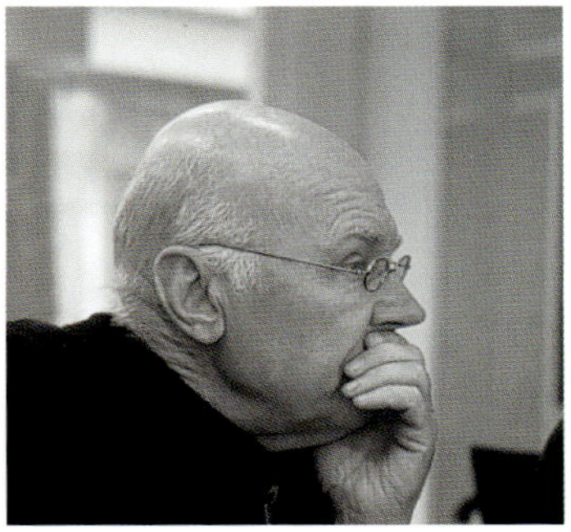

PROF. GUI BONSIEPE ging nach seinem Studium an der Hochschule für Gestaltung in Ulm und dortiger Lehr- und Entwicklungstätigkeit nach Lateinamerika. Von 1987-1990 war er Chef-Designer in einem Software House in Emeryville (Kalifornien). Seit 1993 ist er Professor für Interface-Design und Hypermedien an der FH Köln.

Prof. Gui Bonsiepe graduated from the hfg in Ulm, thereafter taking a teaching and development post there before moving to Latin America. From 1987 to 1990 he was chief designer at a software house in Emeryville (CA). Since 1993 he has been professor of interface design and hypermedia at the Cologne Polytechnic.

Eins zu Null für die Werbung.
Mirko Debuan

0110100101000110 – die Zeichenfolge „iF“, dargestellt als Binärwert. Nicht unbedingt ein Design, das vom Hocker haut. Und auch der Informationscharakter ist vielleicht ein wenig zu subtil.

Und dennoch, auf einer beliebigen Anzahl dieser beiden Ziffern bauen alle Ergebnisse auf, die im Bereich Digital Media angetreten sind. Und was man erreichen kann, wenn man mit den Einsen und Nullen entsprechend umgeht, das haben die diesjährigen Beiträge eindrucksvoll demonstriert.

Zur Bewertung der eingereichten Beiträge reichte die binäre Ziffernvielfalt längst nicht aus – die digitalen Medien sind den Kinderschuhen entwachsen. Als Konsequenz daraus liegt allerdings auch die Messlatte für eine überdurchschnittliche Leistung wieder ein Stück höher.

Die technische Qualität der Beiträge war durchweg hoch, was dafür spricht, dass die beteiligten Personen ihr Handwerk und ihre Werkzeuge beherrschen. Um jedoch herausragend zu sein, bedarf es inzwischen mehr als nur einer ansprechenden Gestaltung oder einer interessanten Animation.

Als Innovationsträger spielt dabei die Produktwerbung eine erkennbare Rolle. Die Auftraggeber haben offensichtlich erkannt, dass Werbung in den digitalen Medien mehr ist als ein bewegtes Plakat. Und die Gestalter und Entwickler verstehen es immer besser, den Begriff Multimedia entsprechend den Anforderungen der Aufgabenstellung zu definieren – immer wieder anders, immer wieder neu: Animationen, die den Inhalt umfangen und nicht davon ablenken, musikalische Untermalungen, die die Atmosphäre des Dargestellten verstärken, Navigationsstrukturen, die sehr schnell erlernbar sind und nicht einfach nur anders. Um in der binären Welt zu bleiben: Diese Entwicklung ist eine „Eins“ wert.

Leider musste die Jury aber auch die Null akzeptieren: Die klassischen Informationssysteme, ob Unternehmensportale oder multimediale Bedienungshilfen, waren – leider – diesmal unterrepräsentiert. Deshalb sei an dieser Stelle auch der Wunsch der Jury geäußert, für den kommenden neuen iF communication award auch wieder mutig Beiträge für diesen Bereich einzureichen.

Sehen wir es doch mal sportlich und nicht technisch – ein 1:0 ist noch zu packen, schließlich befinden wir uns gerade in der ersten Halbzeit.

One-Zero for Advertising.

0110100101000110 – the "iF" logo as a binary number. Not exactly a design that bowls you over. And perhaps it is even a touch too subtle in terms of the information content.

Yet all the products that made an appearance in the Digital Media category are based on a random number of these two digits. And the projects submitted this year are an impressive example of just what can be achieved with all those ones and zeros if they are dealt with in the appropriate manner.

The multitude of binary digits was by no means sufficient to evaluate all the projects submitted – long gone are the days when digital media was a domain in its infancy. As a result, the benchmark for something to be judged above average has been raised once again.

Throughout, the technical quality of the products was high, a sure sign that their creators are skilled in their trade and know how to handle their tools. However, nowadays, to be outstanding calls for more than an appealing design or interesting animations.

With regard to innovation, product advertising plays a very clear role. Those commissioning digital projects have obviously come to realize that when it comes to digital media, advertising is more than an emotionally-charged poster. And designers and developers are becoming ever more expert at defining the term multimedia to suit the requirements of the task at hand. They are constantly finding ever new and ever different solutions: animations, which highlight rather than detract from the subject matter; background music to enhance the atmosphere of what is being displayed; navigation structures that can be learnt quickly and are not simply different. Keeping up the binary theme: this trend definitely deserves a "one".

Unfortunately however, the jury also had to accept a zero: classical information systems, be they company portals or multimedia instructions were – unfortunately – insufficiently represented this year. This is a good opportunity to make known the jury's wish that people should not be afraid to submit entries in this category for the forthcoming new communications award.

And putting things in sporting terms, as opposed to technical – a 1:0 can still be turned around; after all, it's not even half-time yet.

Digital Media

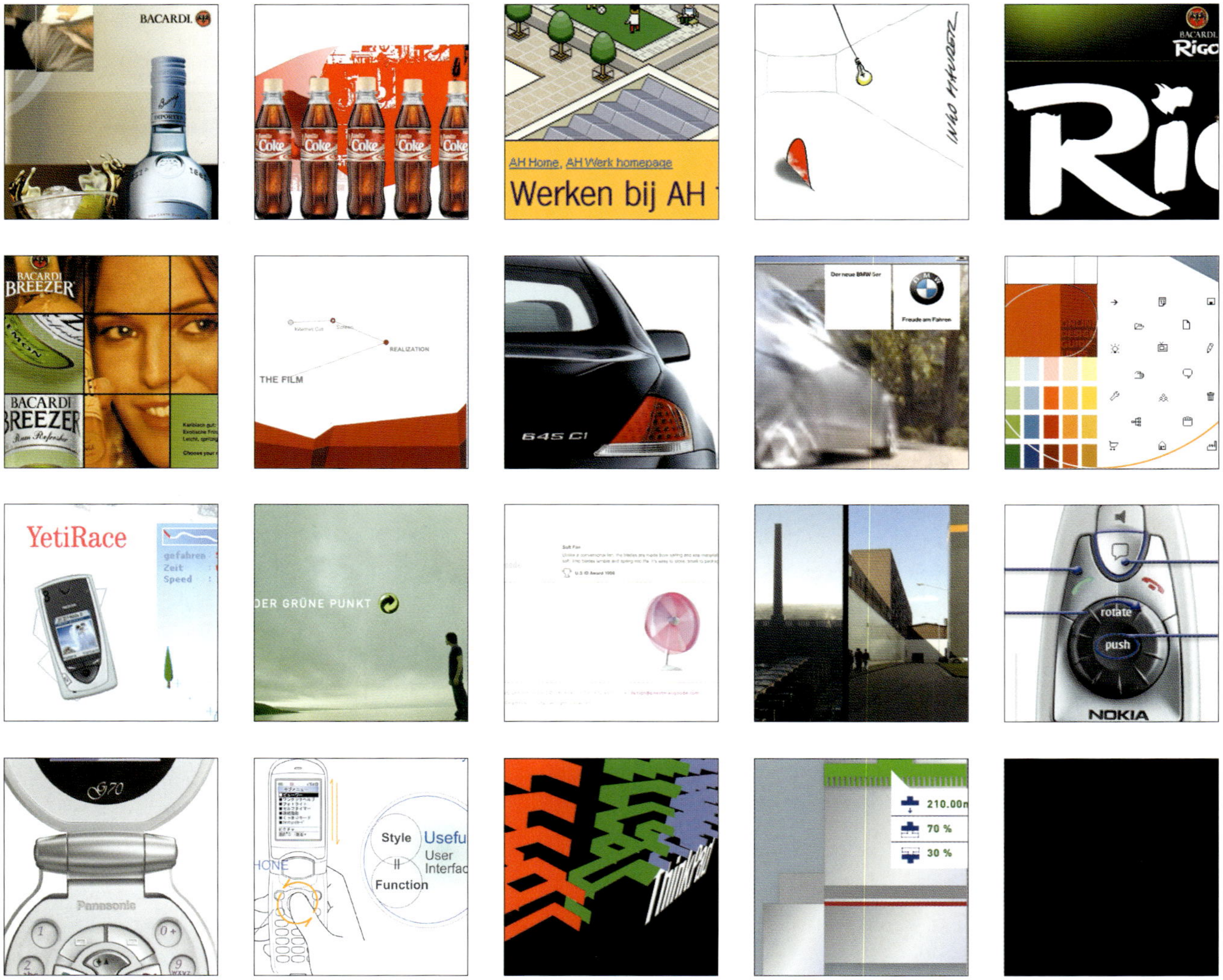

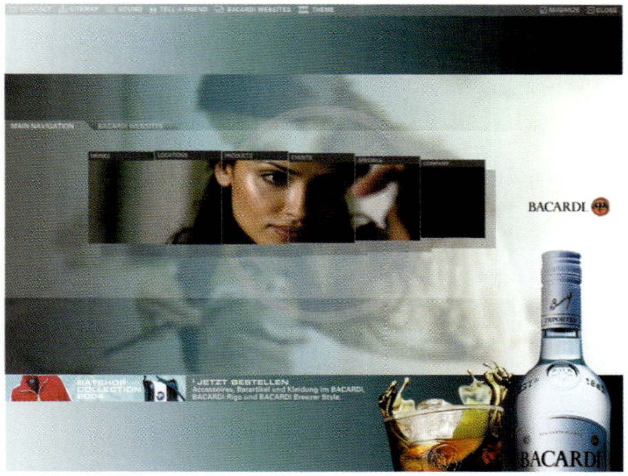

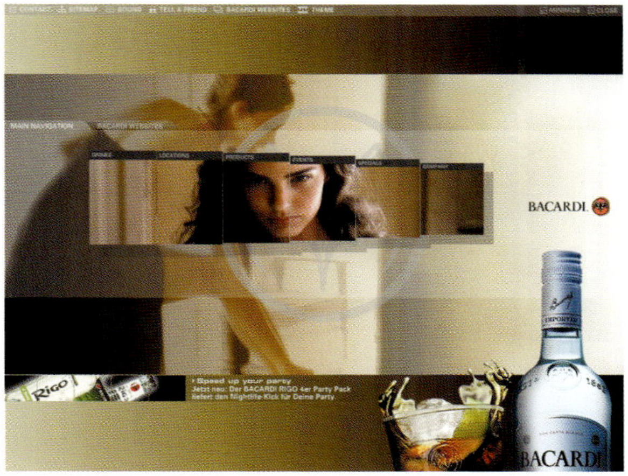

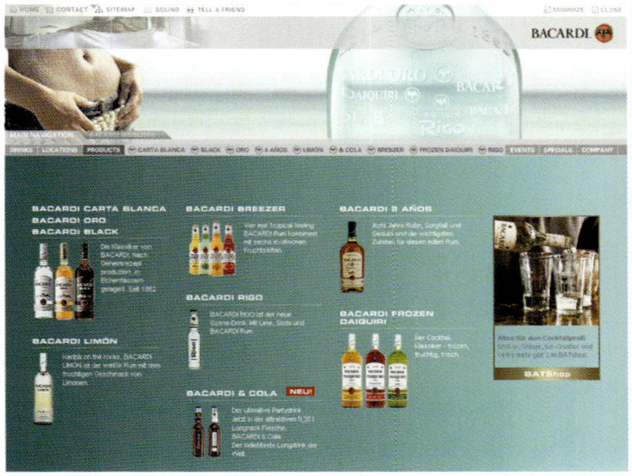

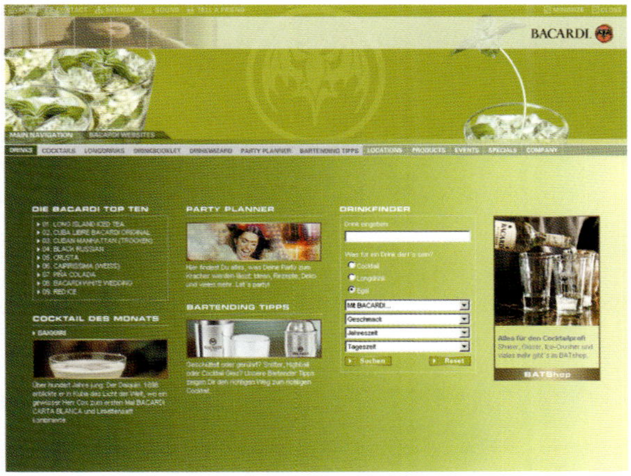

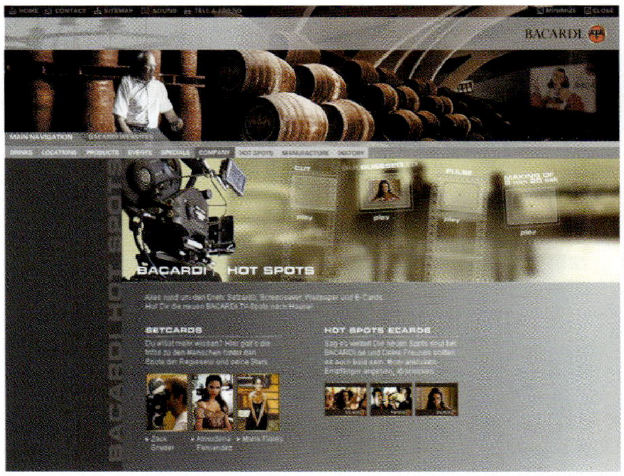

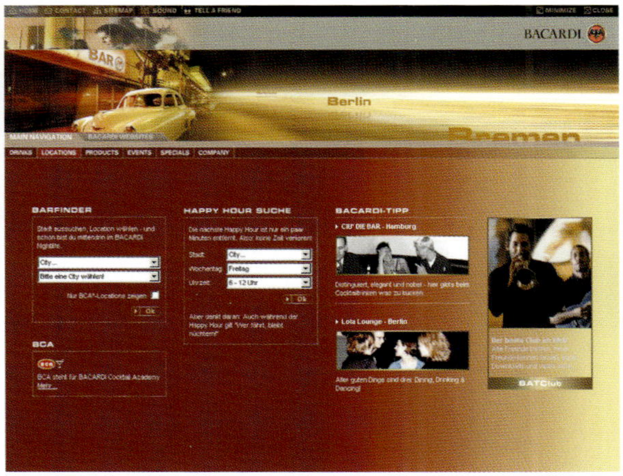

Auftraggeber Client
BACARDI Deutschland GmbH
Hamburg, Germany

Produkt Product
www.bacardi.de
Relaunch BACARDI
BACARDI site relaunch

Design
Ruth Heinzelmann (SAD)
Sven Küster (CD)
Alexandra Plaschnik (JAD)
argonauten360 GmbH
Berlin, Germany

Die neue Marken-Website von BACARDI Deutschland steht unter dem Motto: Discover the BACARDI World. Die karibische Lebensfreude, für die die Marke steht, wurde als bestimmendes Motiv in Wort, Bild und Musik auf die gesamte Site übertragen. Die Website ist in sechs Bereiche gegliedert: „Drinks", „Locations", „Products", „Events", „Specials" und „Company". Der User findet hier Informationen und Interaktion rund um den BACARDI: Von der Cocktail-Datenbank bis zum Event-Kalender, von der Produktdarstellung bis zum neuesten TV-Spot. Karibische Rhythmen – als Soundteaser von der Doppel-CD „Ritmo de BACARDI" – sowie eine umfangreiche Event-Berichterstattung runden das Angebot ab.

The motto of the new BACARDI Deutschland Web site is: Discover the BARCADI World. Caribbean vivacity, which the brand stands for, is the characteristic motif of the site, and is communicated in words, images and music. The Web site is divided into six areas: "Drinks", "Locations", "Products, "Events", "Specials" and "Company". Here the user has access to information and interaction on the topic of BACARDI – from the cocktail database to event calendars, from product depictions to the newest TV ad. Caribbean rhythms – a taste of the double CD "Ritmo de BACARDI" – and comprehensive events coverage round out the offerings.

iF Gold Award

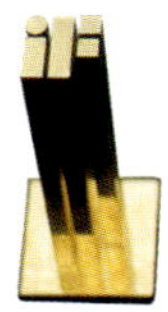

Auftraggeber Client
Coca-Cola GmbH
Berlin, Germany

Produkt Product
Vanilla-Coke
Launch www.vanilla-coke.de
Web site launch

Design
Ruth Heinzelmann (SAD)
Sven Küster (CD)
argonauten360° GmbH
Berlin, Germany

Die Website von Vanilla Coke richtet sich an Jugendliche und junge Erwachsene im Alter von 14 bis 22 Jahren. Anstelle einer klickbaren, klassisch aufgebauten Navigation wurde für die Website ein räumliches Nutzungskonzept umgesetzt: Der User steuert einen Avatar über die (Homepage-) Ebene. Linien zeigen ihm den Weg zum nächsten Contentmodul. Hat der Avatar den Hotspot des jeweiligen Moduls erreicht, öffnet sich ein Layer mit den entsprechenden Inhalten. Dieses neue Konzept sorgt nicht nur für viel Freiheit bei der Inszenierung der Inhalte, sondern auch bei der redaktionellen Betreuung und Erweiterung der Website: Da es außer der Homepage nur eine Gliederungsebene (Contentmodule) gibt, ist die Ergänzung um weitere Elemente sehr einfach und führt dennoch nicht zu Unübersichtlichkeit für den User.

The Vanilla Coke Web site targets youth and young adults in the 14-22 year age bracket. Instead of clickable, classically composed navigation, the Web site employs a spatial concept: The user maneuvers an avatar over the top level of the homepage. Lines direct him to the next content module. When the avatar reaches the hotspot of the particular module, a layer opens divulging its content. Not only does this concept allow plenty of freedom to arrange the contents, but also to edit and expand the Web site: Apart from the homepage level there is only one organizational level, the content module, making it very easy to add other elements while not rendering the site confusing.

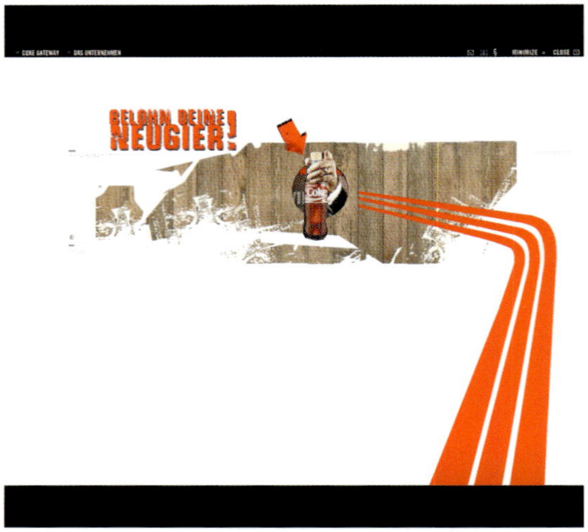

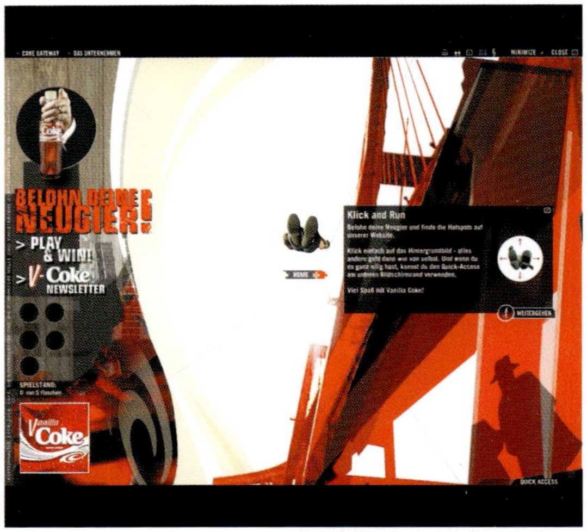

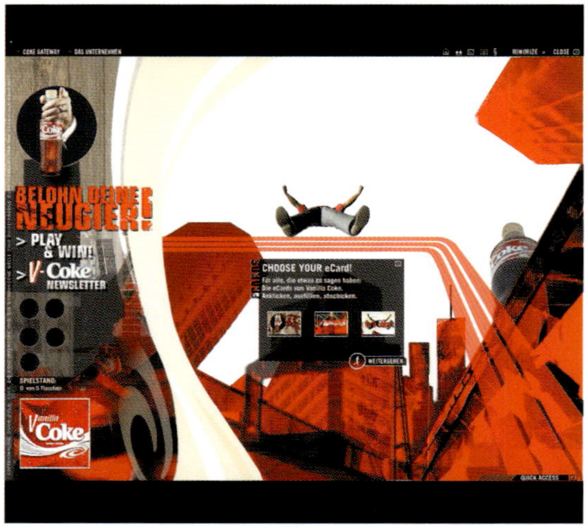

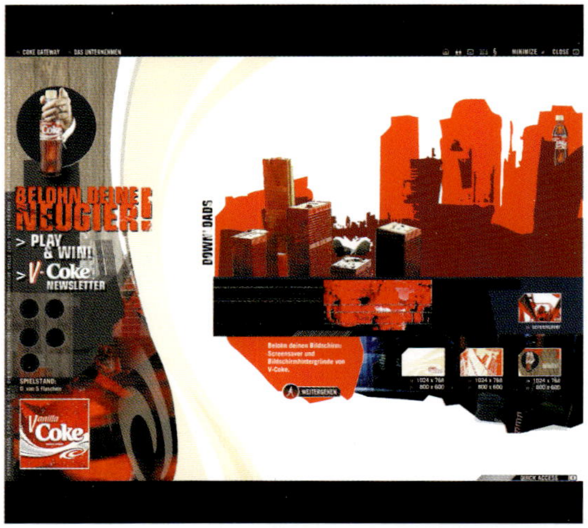

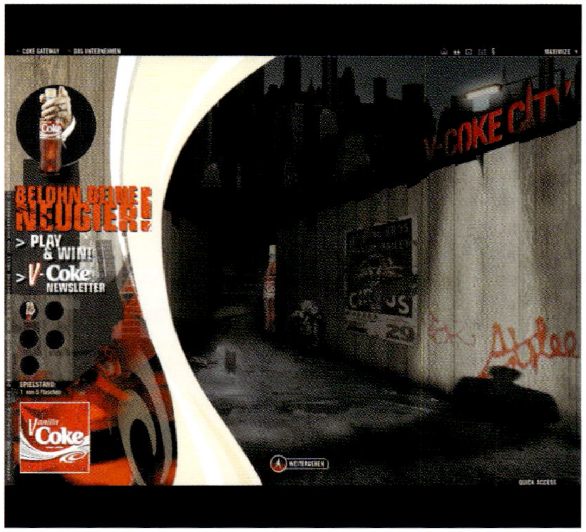

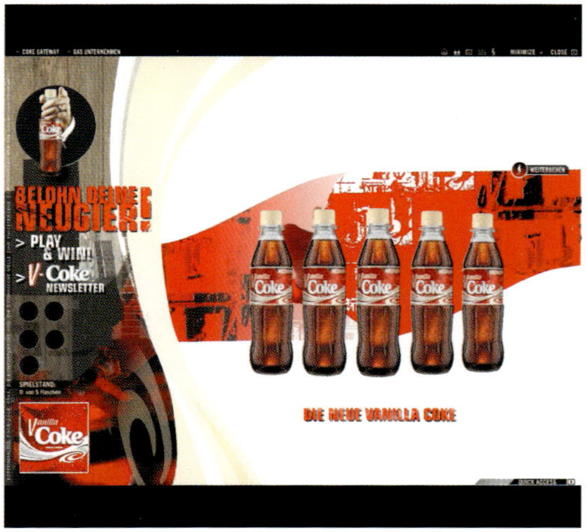

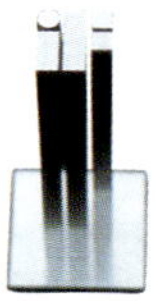

iF Silver Award

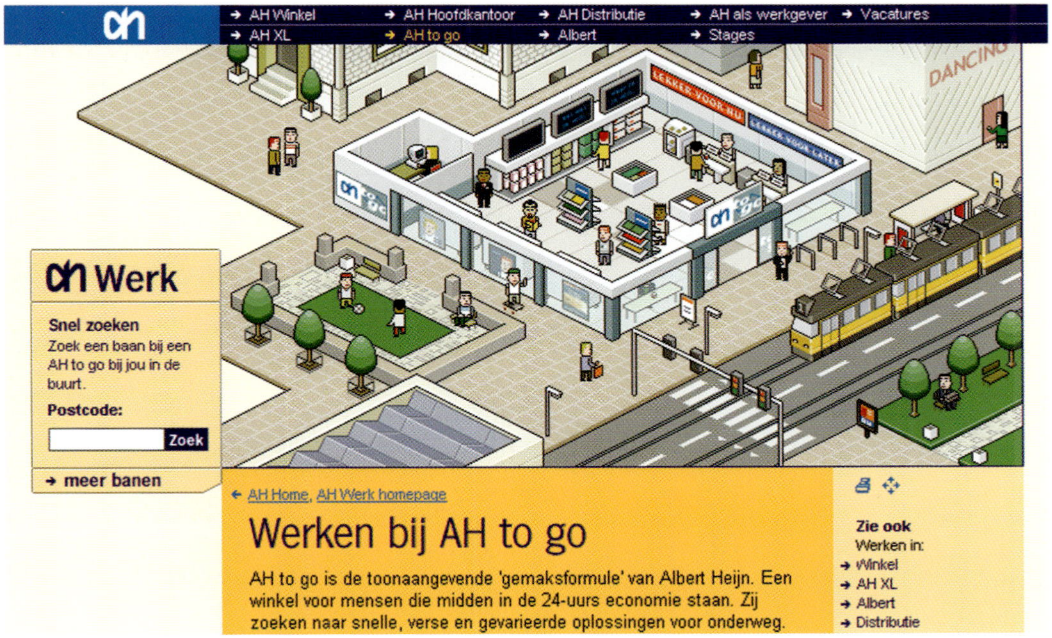

Auftraggeber Client
Albert Heijn
Zaandam, Netherlands

Produkt Product
www.ah.nl/werk
Website
Web site

Design
Renate Frotscher
Matthijs Kamstra
Panc
Wim Wepster
Fabrique (design, communication & new media)
Delft, Netherlands

www.ah.nl/werk ist die Website, mit der sich die niederländische Supermarktkette Albert Heijn auf dem Arbeitsmarkt profiliert. Was an den Illustrationen sofort auffällt, ist der verwendete Pixelstil. Dieser Retrostil, der sich vor allem an die junge Zielgruppe wendet, steht stellvertretend für das digitale Zeitalter, für das Internet. Hiermit wird hervorgehoben, dass Albert Heijn in seiner Arbeitsmarkt-Strategie dem Internet künftig Vorrang einräumt. Die Website stellt schwerpunktmäßig vor allem die Attraktivität der Tätigkeiten und Arbeitsumgebungen des Supermarkts dar. Website-Besucher können ihren Traumjob nach Tätigkeit oder Standort geordnet suchen.

www.ah.nl/werk is the Web site with which the Dutch Albert Heijin supermarket chain presents itself on the job market. What immediately catches the eye about the illustrations is the pixel style. This retro-style, directed at a young target group, is supposed to represent the digital age of the Internet. It underlines the fact that Albert Heijin considers the Internet the way to go in its job market strategy. The Web site focuses on the attractive nature of the supermarket as a work environment. Visitors clicking onto it can search for their ideal job by activity or location.

iF Silver Award

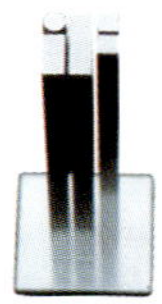

Auftraggeber Client
Ingo Maurer GmbH
München, Germany

Produkt Product
www.ingo-maurer.com
Website
Web site

Design
Christa Heinold (AD)
Dennis Oswald (Screendesign)
Anette Scholz (CD)
Marion Stolz (Screendesign)
Jörg Waldschütz (AD)
SCHOLZ & VOLKMER GmbH
Wiesbaden, Germany

Ingo Maurers Lampen und Installationen stehen im Mittelpunkt des vollständig überarbeiteten Internetauftritts. Die in großflächigen Bildern dargestellten Produkte und Projekte bilden einen Kontrast zum skizzenhaften Charakter der restlichen Website. Die großflächigen Bilder sind an den Rändern verschwommen, um die Datenmengen gering zu halten und trotzdem Fullscreen-Bilder zu suggerieren. Alle Produkte sind in mindestens drei Ansichten (Gesamt- und Detailaufnahmen) verfügbar. Das illustrative Intro basiert auf einer Originalskizze von Ingo Maurer.

Luminaires and installations are the focus of Ingo Maurer's completely revamped Internet site. The products and projects are shown in large pictures contrasting with the sketchy character of the rest of the Web site. The large pictures have muted edges to keep the data quantity low while suggesting full-screen pictures. There are at least three different views of every one of the products, showing them in their entirety as well as in detail. The illustrated introduction is based on an original drawing by Ingo Maurer.

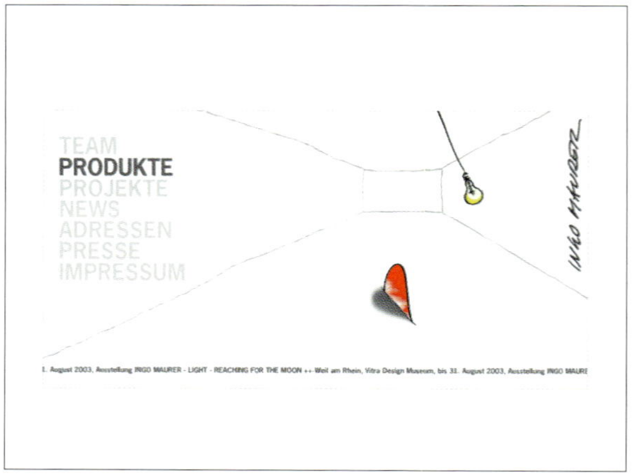

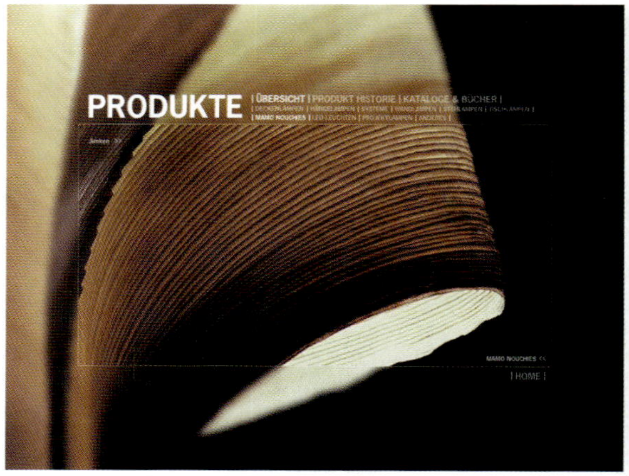

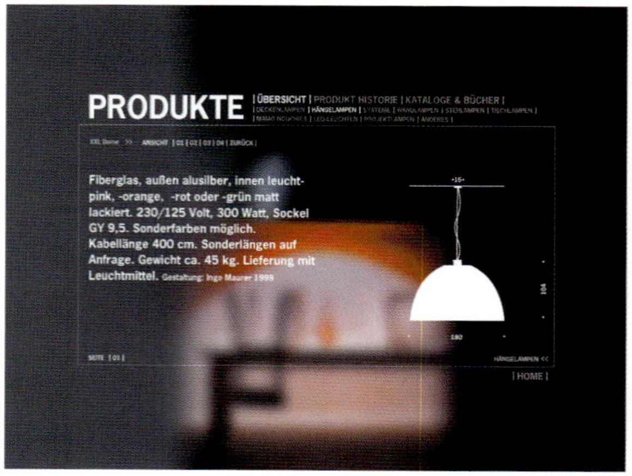

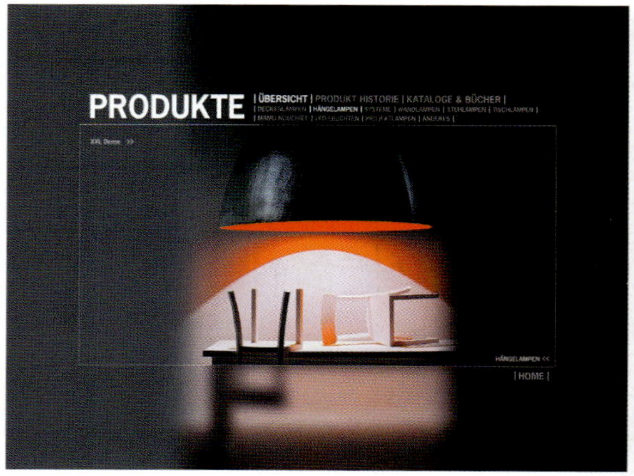

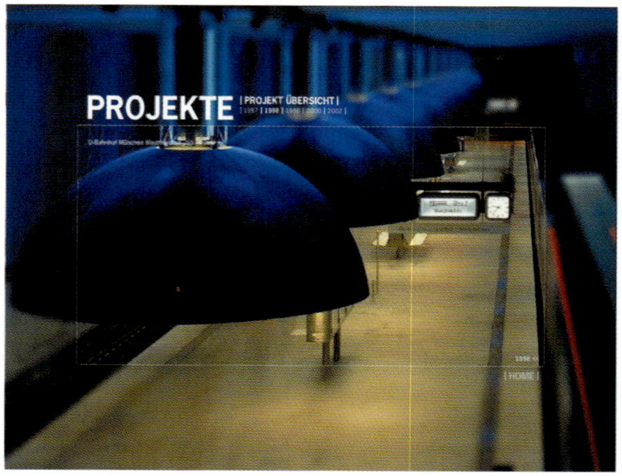

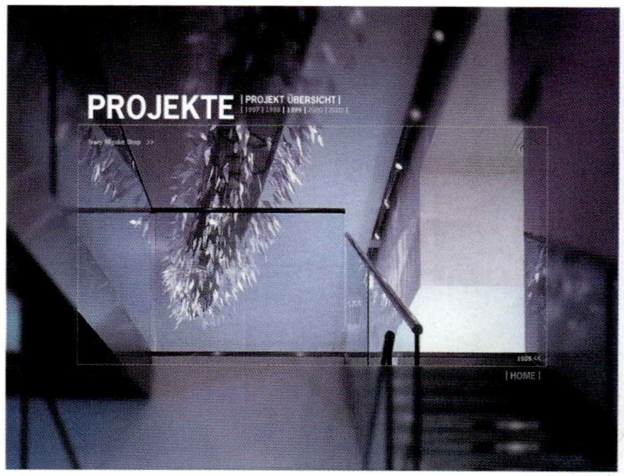

iF Silver Award

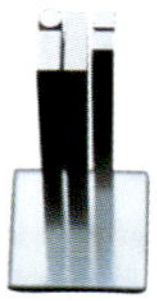

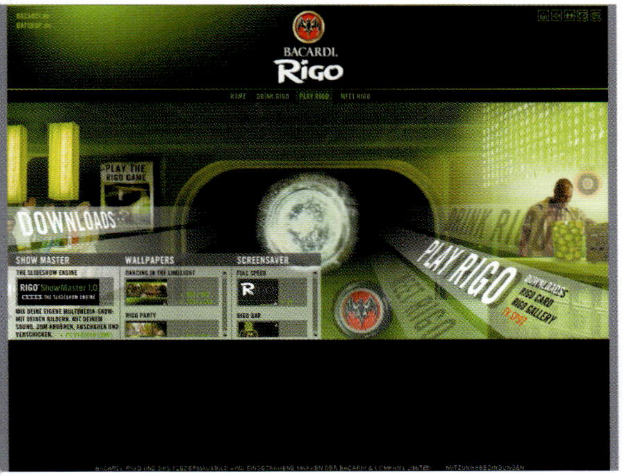

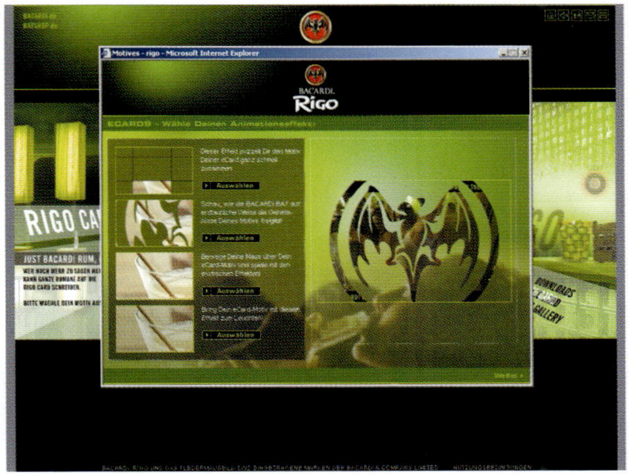

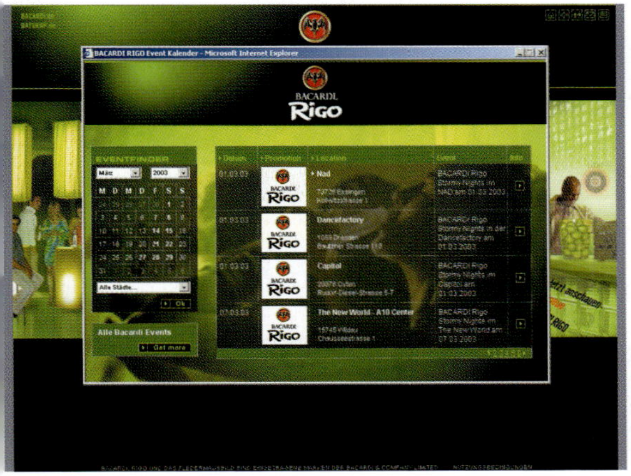

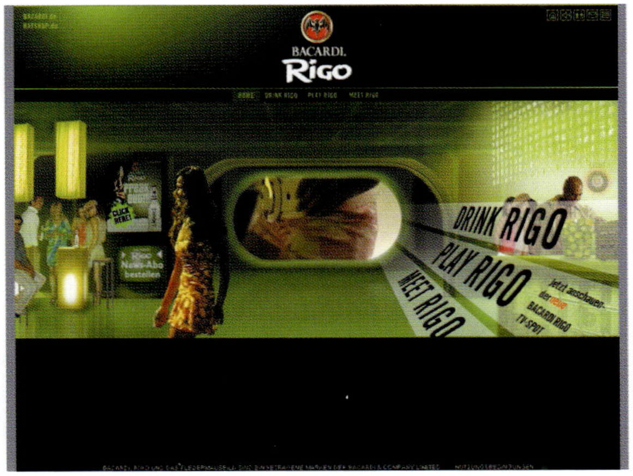

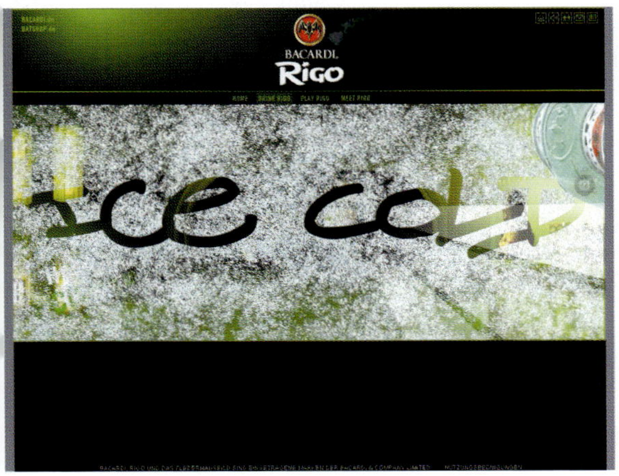

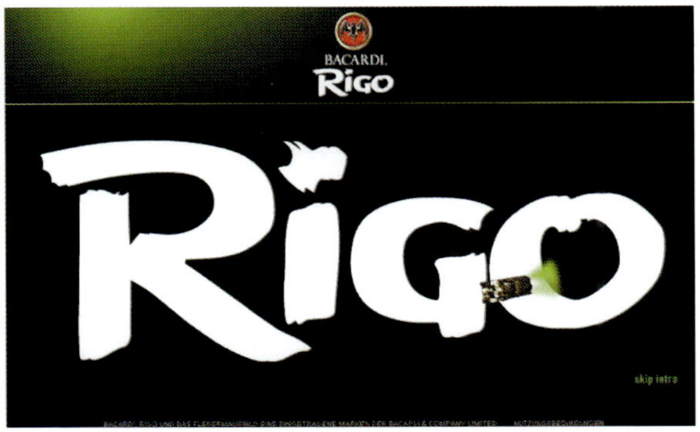

Auftraggeber Client
BACARDI Deutschland GmbH
Hamburg, Germany

Produkt Product
www.bacardi.rigo.de
Relaunch BACARDI RIGO
BACARDI RIGO site relaunch

Design
Ruth Heinzelmann (SAD)
Sven Küster (CD)
Alexandra Plaschnik (JAD)
argonauten360 GmbH
Berlin, Germany

Die Microsite für BACARDI RIGO wurde in Anlehnung an den BACARDI RIGO TV-Spot entwickelt. Durch die Anpassung des Screendesigns an die Szenerie des TV-Spots, die Verwendung der TV-Charaktere auf der Website und die Übernahme der Latino-House-Musik wurde die Atmosphäre des TV-Spots erhalten. Die Microsite ist in drei Bereiche – „Drink RIGO", „Play RIGO" und „Meet RIGO" – unterteilt. „Drink RIGO" bietet die Möglichkeit zur Beschäftigung mit dem Produkt, im Bereich „Play RIGO" kann mit dem Showmaster eine eigene Diashow zusammengestellt werden und „Meet RIGO" ist ein datenbankgesteuerter Event-Kalender.

The microsite for BACARDI RIGO was developed on the basis of the BACARDI RIGO TV commercial. The atmosphere of the commercial is retained by matching the screen design to the scenery in the ad, endowing the Web site with a TV feel and by integrating Latino house music. The microsite is divided into three areas: "Drink RIGO", "Play RIGO" and "Meet RIGO". At Drink RIGO you can get to know the product, at "Play RIGO" you can put together your own slideshow, and in the latter category you are greeted with a data-based events calendar.

Auftraggeber Client
BACARDI Deutschland GmbH
Hamburg, Germany

Produkt Product
www.bacardi-breezer.de
Relaunch BACARDI BREEZER
Web site relaunch

Design
Ruth Heinzelmann (SAD)
Sven Küster (CD)
Alexandra Plaschnik (JAD)
argonauten360 GmbH
Berlin, Germany

Die Website von BACARDI BREEZER ist ein wichtiger Baustein der gesamten BACARDI-Markenwelt im Internet. Die Microsite ist in vier Bereiche gegliedert: „Taste" bietet Produktinformationen, „Specials" stellt Screensaver, Downloads und den aktuellen TV-Spot zur Verfügung. „Live" bietet neben dem Zugriff auf die Event-Datenbank mehr Informationen zu den drei wichtigsten, von BACARDI BREEZER unterstützten Veranstaltungen und Vote schließlich sorgt für Preise und spezielle Spielmechanik. Bei der grafischen Gestaltung stand die Adaption des TV-Spots im Vordergrund, bei der inhaltlichen und funktionalen Konzeption galt das besondere Augenmerk der Verknüpfung von On- und Offline-Promotions.

The BACARDI BREEZER Web site is an important component in the overall Internet presence of the BACARDI brand world. The micro-site is divided into four areas: "Taste" provides infos on the product, "Specials" offers screensaver, downloads and the current TV commercial, with "Live" you can access the events database and more information on the three main events backed by BACARDI BREEZER, and "Vote" supplies the prices and special game functions. The graphic design is based on the TV commercial, while content and functional concepts focus on linking online and offline promotions.

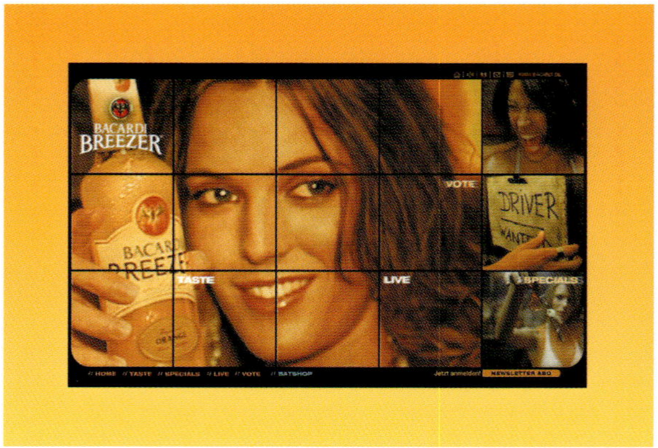

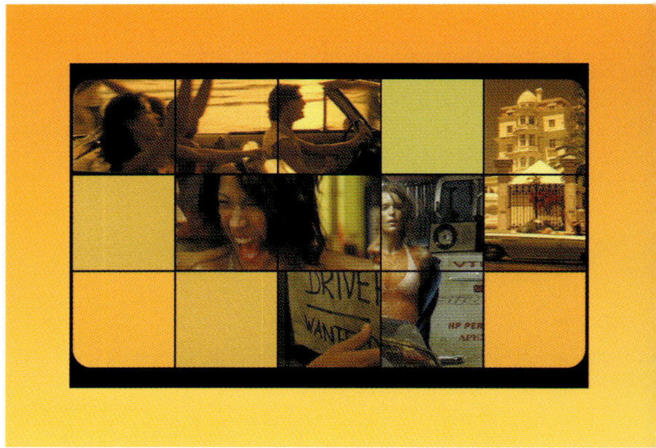

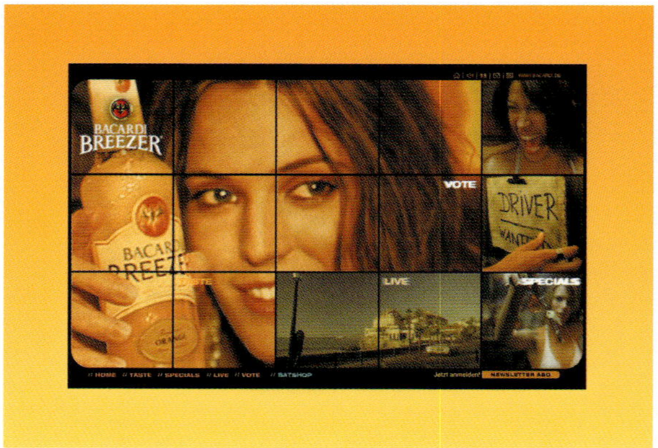

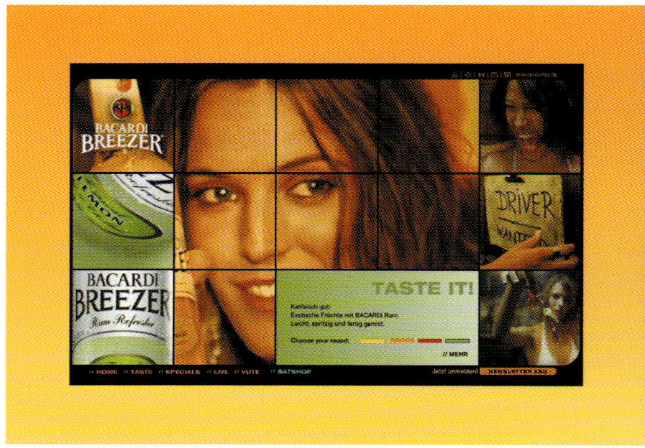

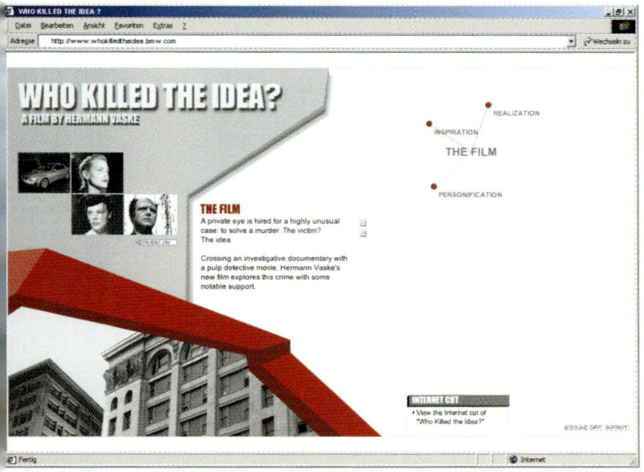

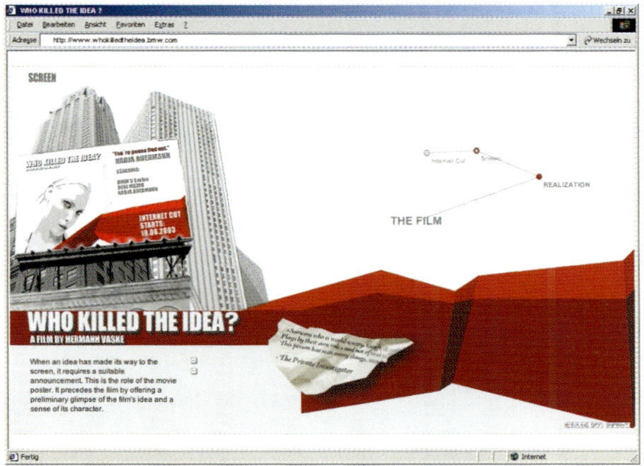

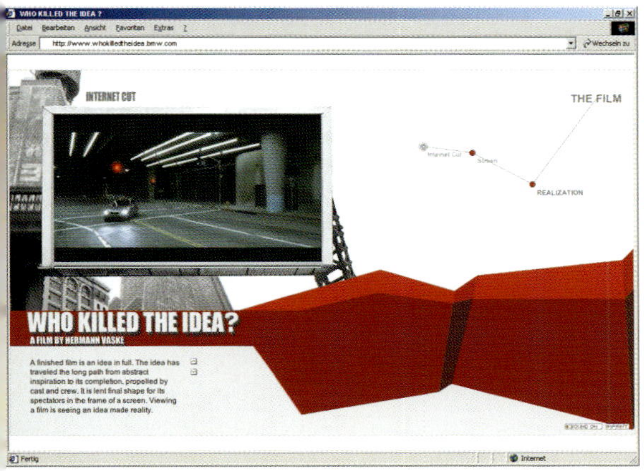

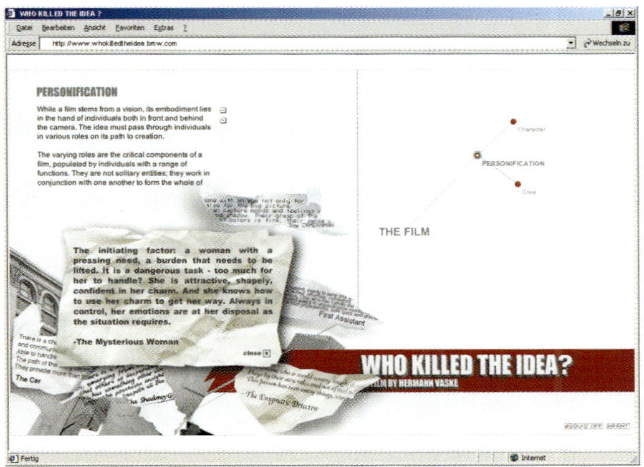

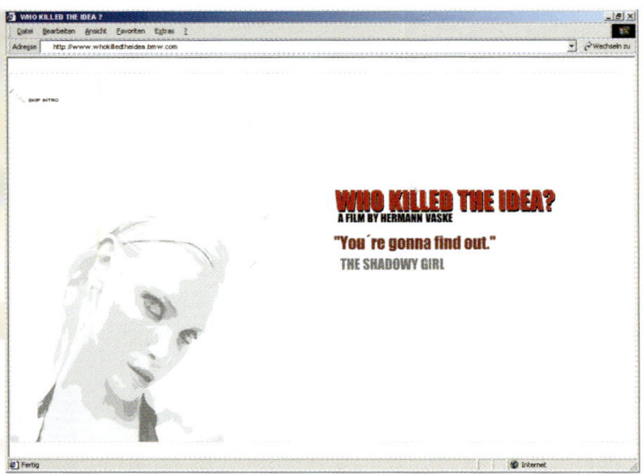

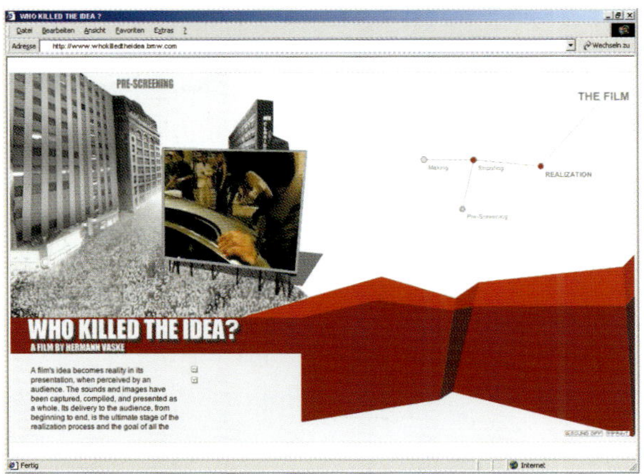

Auftraggeber Client
BMW AG
München, Germany

Produkt Product
Who killed the idea?
www.whokilledtheidea.bmw.com
Website
Web site

Design
BBDO InterOne GmbH
Hamburg, Germany

Der Webauftritt begleitet die Entwicklung des Films „Who killed the idea?" von Hermann Vaske, der während des Filmfestivals von Cannes 2003 vorgestellt wurde. Neben Debi Mazar und Nadja Auermann spielt auch der neue BMW 5er eine Hauptrolle in dieser klassischen Detektivgeschichte, die gleichzeitig auch das investigative Vorgehen dokumentarisch begleitet. Bei den Nachforschungen geht es um die Schöpfung neuer Ideen und deren Schicksal. Wie wird eine Idee realisiert und in einem Film umgesetzt? Die Antworten auf diese Frage bilden in den Zweigen der Mindmap das Rückgrat der Navigation und ermöglichen dem Nutzer seine individuelle Forschungsreise durch die Inhalte des Webauftritts. Neben Bild und Text liefern Videos, Sound-Dateien und Animationen dem User ein multimediales Erlebnis.

The Web site was developed to accompany the production of Hermann Vaske's film "Who killed the idea?", which débuted during the 2003 Cannes Film Festival. Along with Debi Mazar and Nadja Auermann, the new BMW 5 Series also plays a major role in this classic detective story fused with a documentary-style investigation. The investigation focuses on the creation of new ideas and their fate. How does an idea take form and then become realized in a film? The answers to these questions form the backbone of the site's navigation in the branches of the mindmap and allow users to embark on their own investigative journey through the Web site's content. Image and text as well as videos, audio and animation all offer the user a true multimedia experience.

Auftraggeber Client
BMW AG
München, Germany

Produkt Product
BMW 6er Coupé Einführungskampagne
BMW 6 Series Coupé Launch campaign
www.bmw.com/6ercoupe
Website
Web site

Design
BBDO InterOne GmbH
Hamburg, Germany

Das BMW 6er Coupé wird im Webauftritt in den drei Themenwelten Sportlichkeit, Präzision und Stil inszeniert. Der Hintergrund der Präsentation mit der mediterranen und sportlichen Anmutung ist an die Verkaufsliteratur angelehnt. In Flash-Filmsequenzen wird zu jeder Themenwelt eine Story erzählt. Den drei Sequenzen werden jeweils passende Themenblöcke zugeordnet, die Sportlichkeit, Präzision und Stil in Hinblick auf das Fahrzeug interpretieren und die zentralen Produkt-Highlights thematisieren. Parallel dazu werden in einer Übersicht die gesamten Fahrzeug-Details erläutert.

The Web site created for the launch of the BMW 6 Series Coupé is divided into sections based on three themes: athletic character, precision and aesthetics. The underlying character of the entire Web site is Mediterranean and emphasises the car's sporting style - this approach parallels the description of the BMW 6 Series in the product catalogues. Each section of the Web site starts with a Flash film which stars the 6 Series, and each section is subdivided into topics which illustrate the corresponding theme (athletic character, precision and aesthetics) as embodied by the car. The core product highlights are also showcased. Parallel to the three main themes, the Web site also contains a section which supplies an overview of all the car's features and technology.

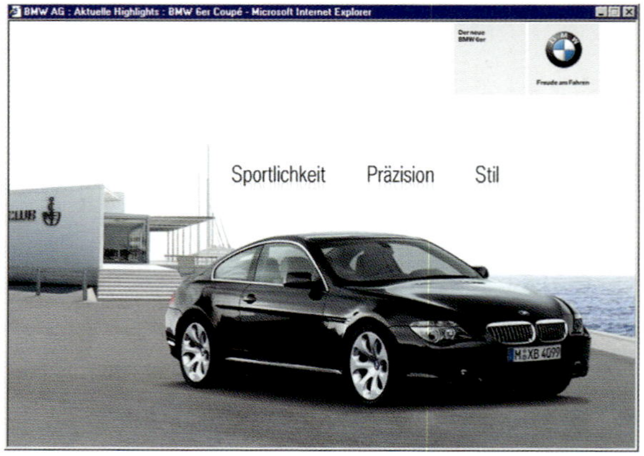

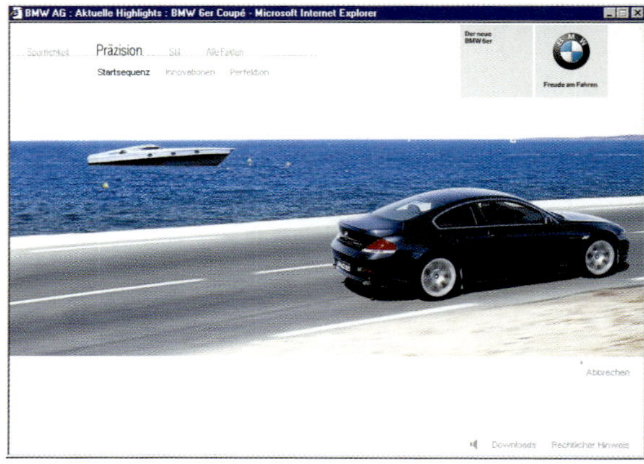

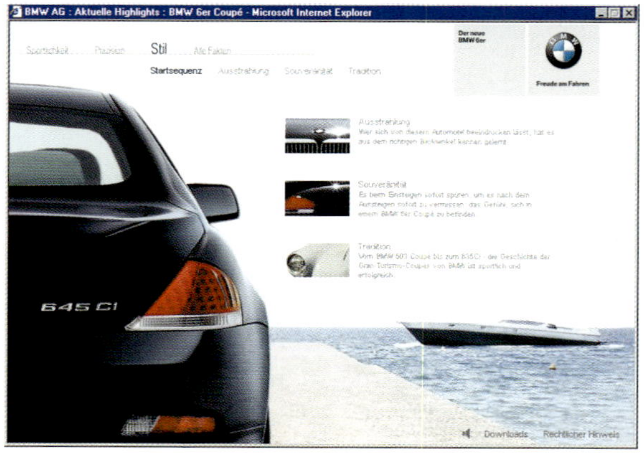

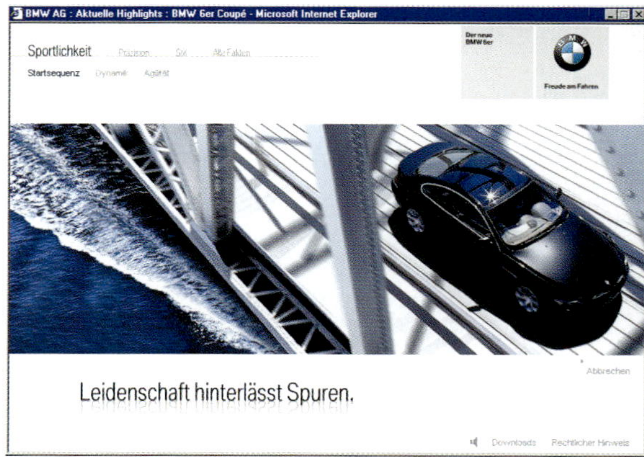

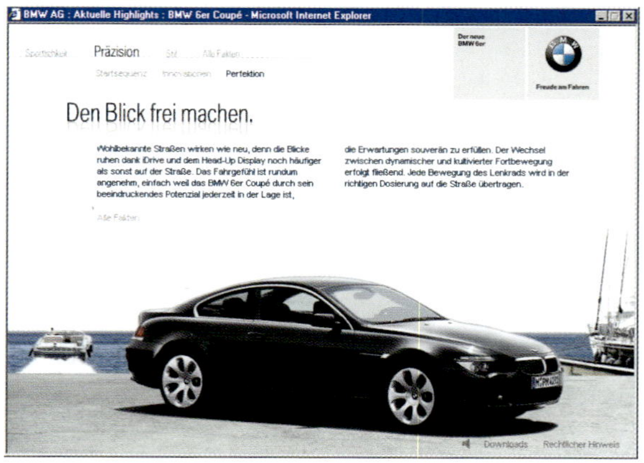

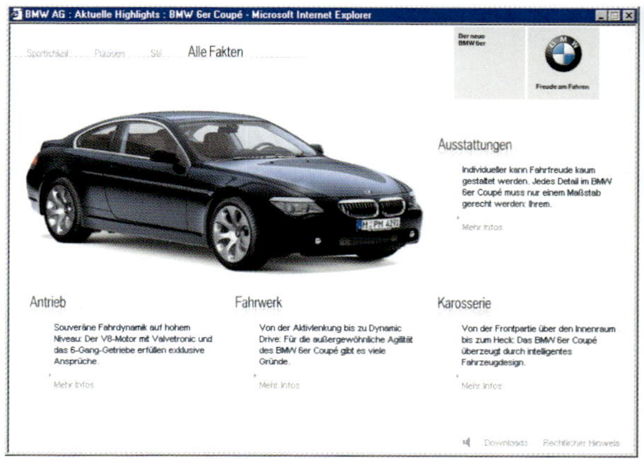

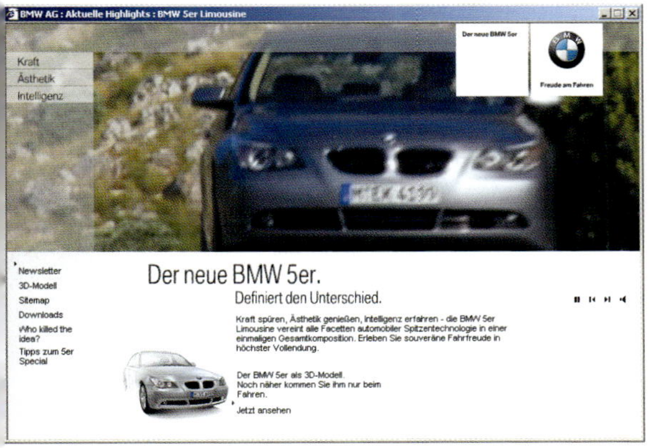

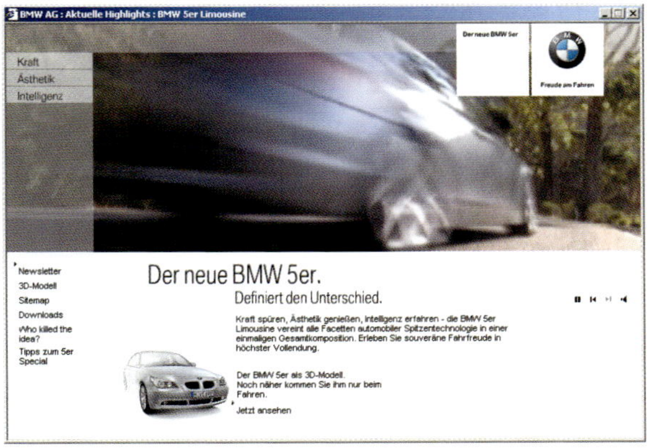

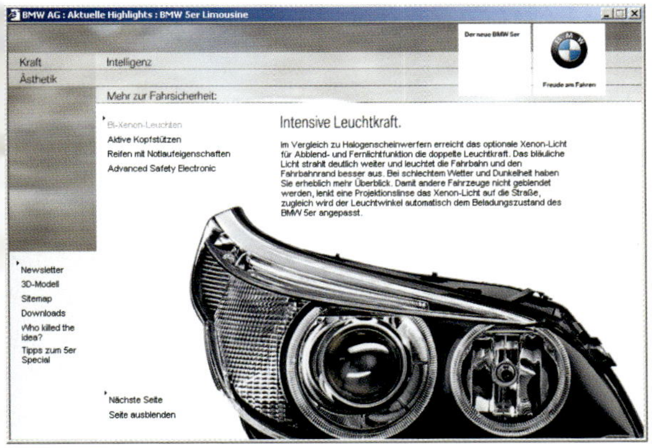

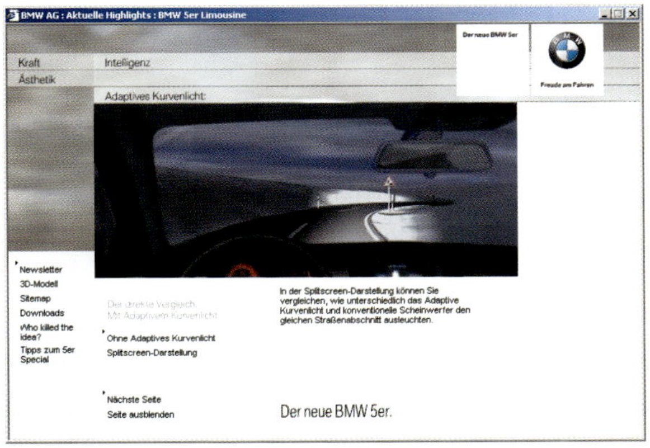

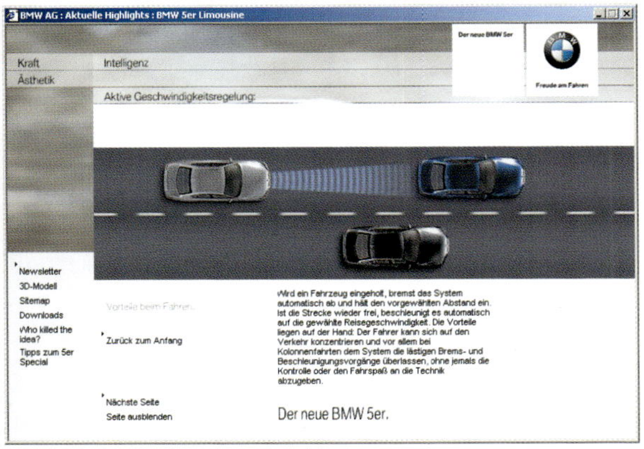

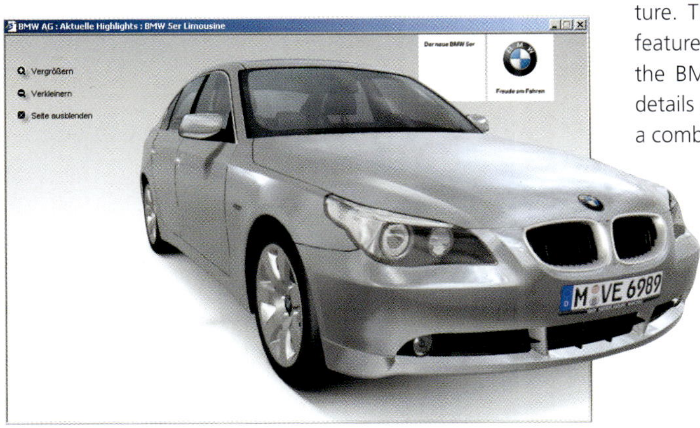

Auftraggeber Client
BMW AG
München, Germany

Produkt Product
BMW 5er Limousine Einführungskampagne
BMW 5 Series Sedan Launch campaign
www.bmw.com/5er
Website
Web site

Design
BBDO InterOne GmbH
Hamburg, Germany

Im Vordergrund des Webauftritts für den BMW 5er steht der offensive Einsatz innovativer Webtechniken wie Video-Streaming und 3D-Technologien, die Kommunikation der Fahrzeugstärken und ein ästhetisches Design, das nicht nur dem Fahrzeug gerecht wird, sondern auch die Usability konsequent umsetzt. Das Special wird von einem selbstablaufenden Film mit dem fahrenden BMW 5er dominiert. Alle zentralen Fahrzeugdetails werden durch aufwändig produzierte 3D-Darstellungen und informative Bild-Text-Kombinationen veranschaulicht.

The Web site of the BMW 5 Series is based on the eye-catching implementation of innovative internet technology, such as video streaming and 3D media. The use of this technology not only communicates the car's strengths, it complements the vehicle's aesthetic design and it is the key to the site's user-friendly structure. The internet highlight's single most important feature is the film which runs continuously, displaying the BMW 5 Series on the move. All significant car details are illustrated using top-quality 3D images and a combination of highly informative texts and images.

Auftraggeber Client
Michael Rossa
Siemens AG
München, Germany

Produkt Product
Siemens Online Design

Design
Nhan Nguyen
Oliver Habboub
Bernhard Schwab
Virtual Identity
Freiburg, Germany

Als globales Informationssystem setzt sich das Siemens Web aus einer Vielzahl von Websites für unterschiedliche Zielgruppen zusammen. Das Siemens Online Corporate Design reflektiert diese Vielfalt und stellt einen flexiblen und charakteristischen Rahmen für die Gestaltung von Siemens Websites zur Verfügung. Als Teil der übergreifenden Markenstrategie definiert das Online Corporate Design so einen wichtigen Teil der Benutzer-Erfahrung im Siemens Web.

The Siemens Web is a global information system, consisting of a multitude of Web sites for a wide range of target audiences. The Siemens Online Corporate Design takes this variety into account offering a flexible, colourful and characteristic framework for the design of Siemens branded Web sites. As an integral part of the overall brand strategy, the Online Corporate Design defines an important part of the Siemens Web user experience.

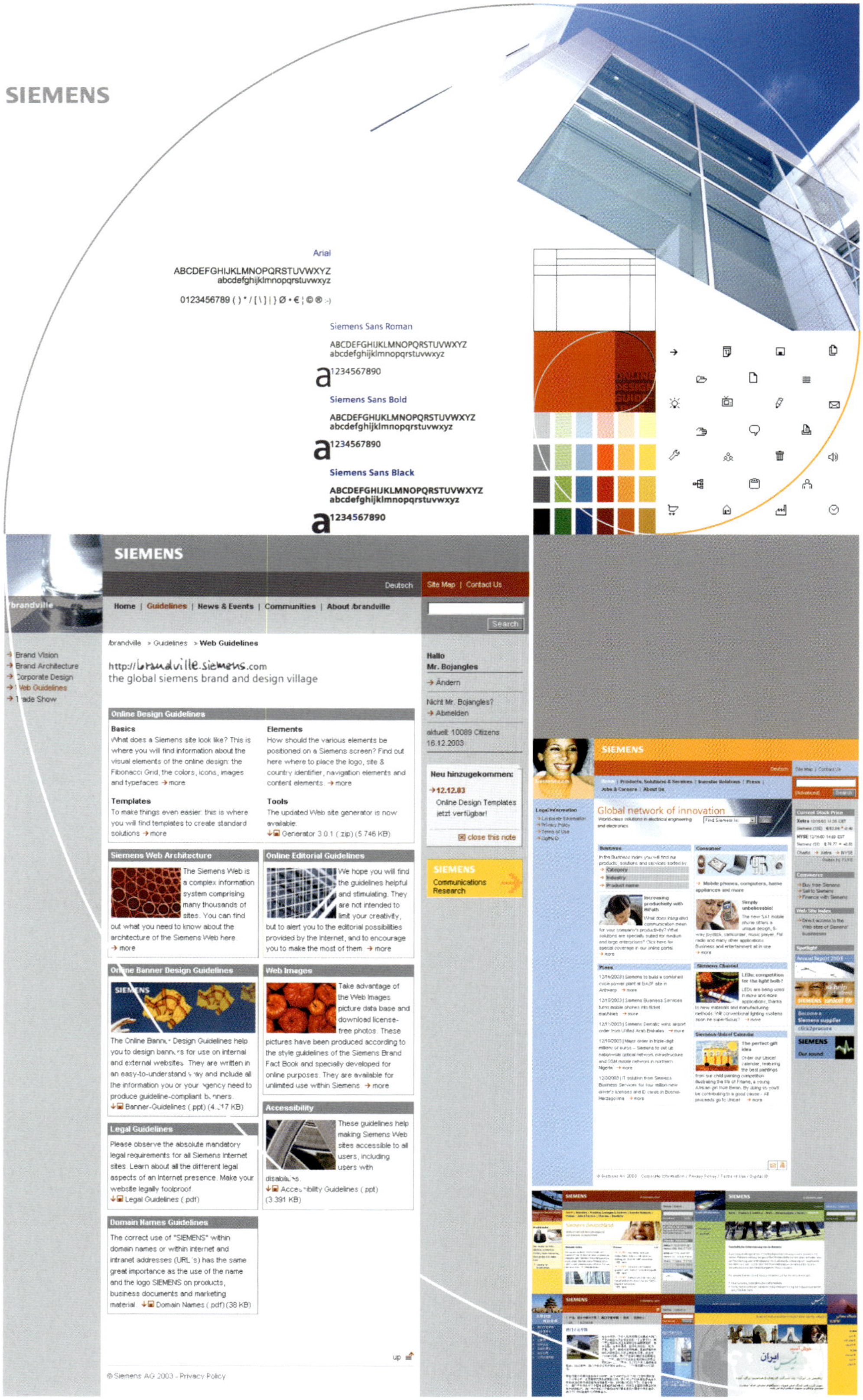

Auftraggeber Client
T-Mobile Deutschland GmbH
Bonn, Germany

Produkt Product
YetiRace
T-Mobile Wintergame
www.i-dmedia.com/t-mobile/wintergame/

Design
Regine Hachka (Creative Director)
Carsten Druck (Sound)
Patrick de Jong (Animation)
Tony Ferreira (Screendesign)
Yannick Hervy (Graphics)
I-D Media AG
Berlin, Germany

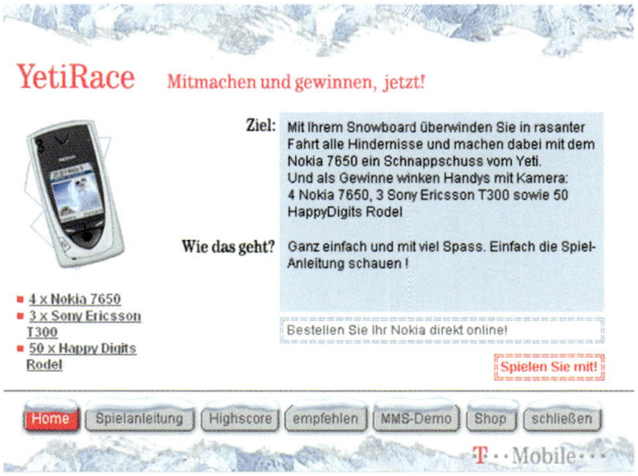

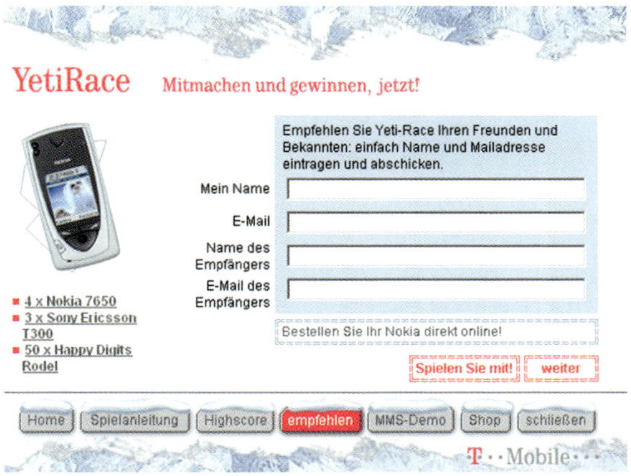

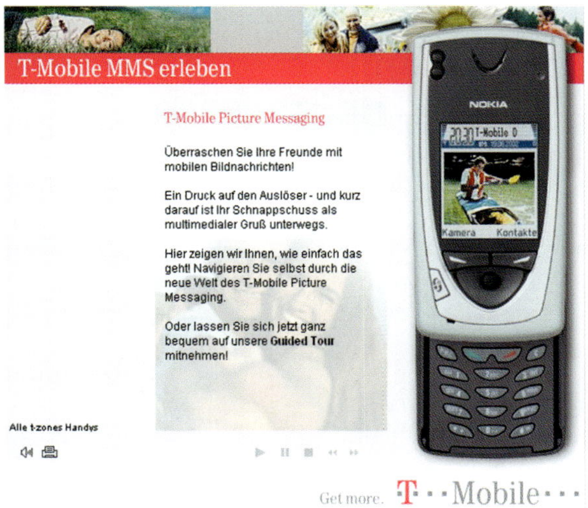

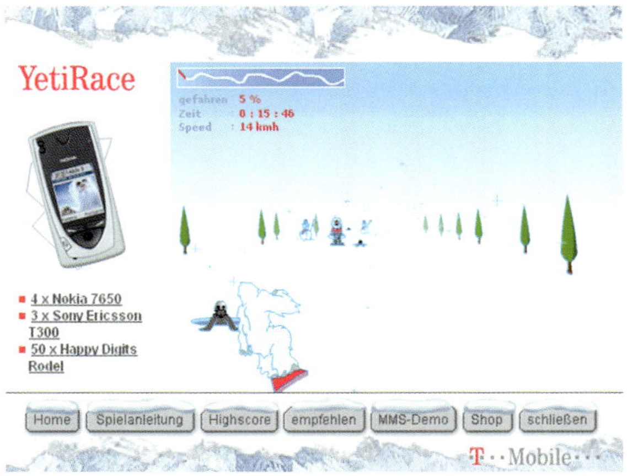

YetiRace war teil der Winterkampagne 2002 von T-Mobile Deutschland. Es handelt sich um ein innovatives Spiel, das mit State-of-the-Art Flash und Action Script-Technologie entwickelt wurde. Das Action Script ist dabei sehr flexibel. Anzahl, Art und Position der Hindernisse, Landschaftstypen und Akteure sowie Abmessung und Form der Rennbahn können durch Setzen von Parametern und Anpassung der Grafik innerhalb des Scripts gesteuert werden. Um die eigentliche Spielseite gruppieren sich Informations- und Serviceseiten, die dem User die Möglichkeiten der MMS-Technologie nahe bringen.

YetiRace was part of the 2002 T-Mobile Deutschland winter campaign. It is an innovative game which was developed with state-of-the-art Flash and Action script technology. The Action script is very flexible. Number, type and position of the obstacles, types of countryside and protagonists as well as size and shape of the racing track can be controlled by setting parameters and adjusting graphics within the script. The actual game site is connected to service sites and information pages which offer the user a chance to familiarize himself with MMS technology.

Auftraggeber Client
Der Grüne Punkt –
Duales System Deutschland AG
Köln, Germany

Produkt Product
Relaunch 2003 www.gruener-punkt.de
Website
Web site

Design
Mashiba Badura
Thomas Baumhoff
Christian Boros
BOROS.INTERAKTIV GmbH
Wuppertal, Germany

Die neue Website der Duales System Deutschland AG informiert in einem zeitgemäßen Design verbrauchernah und dialogorientiert über den Grünen Punkt. Das Layout zeigt im oberen Drittel die Standard- und die Hauptnavigation sowie emotionale Themenbilder in Verbindung mit der Wort/Bildmarke des Grünen Punkts. Der Zugang zu den Themen rund um den Grünen Punkt ist dank strenger Navigationsführung wesentlich erleichtert, Sachverhalte sind anschaulich und einfach erklärt. Idee und Funktionsweise des Grünen Punkts, Geschichte und Struktur des Dualen Systems, Abfallvermeidung und Ressourcenschonung: Die Sprache auch komplexer Zusammenhänge ist transparent und allgemein verständlich.

The new Web site of Duales System Deutschland AG is consciously consumer-friendly and dialogue-oriented, and presents its information in a up-to-the-minute design. The layout places the standard and main navigation as well as thematic emotional images connected with the word and image logo of 'Grüne Punkt' in the upper third. Thanks to rigorous navigational system it is considerably easier than before to access the various 'Grüne Punkt'-related themes, while content is descriptive and simply explained. Be it the idea behind Grüne Punkt and the way it functions, the history and structure of the dual system, garbage avoidance and resource preservation - the dialogue is transparent and generally comprehensible even when the associations are more complex.

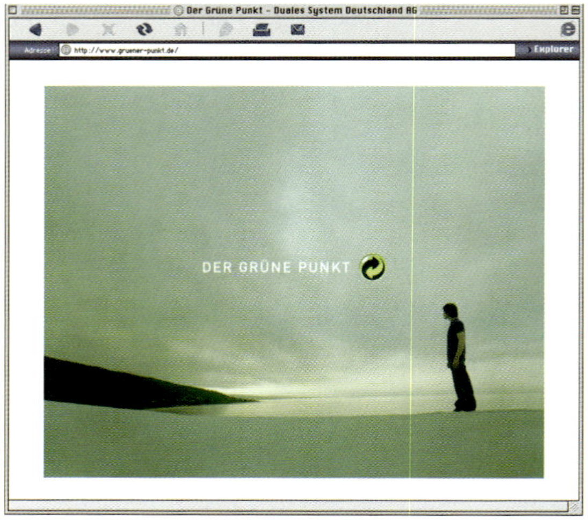

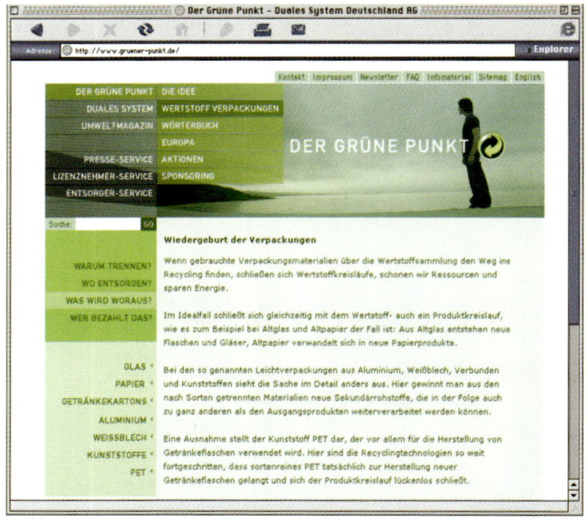

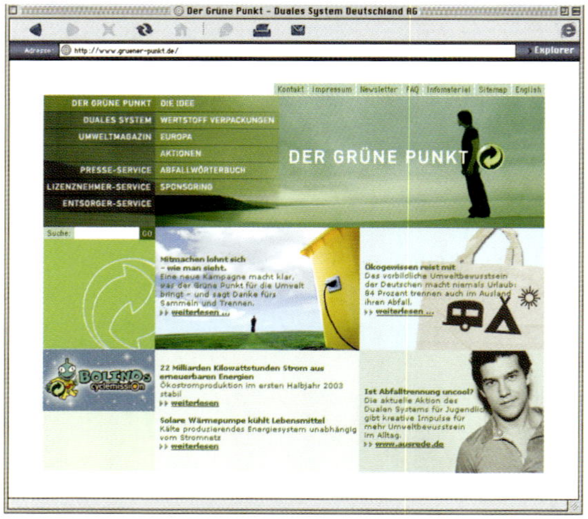

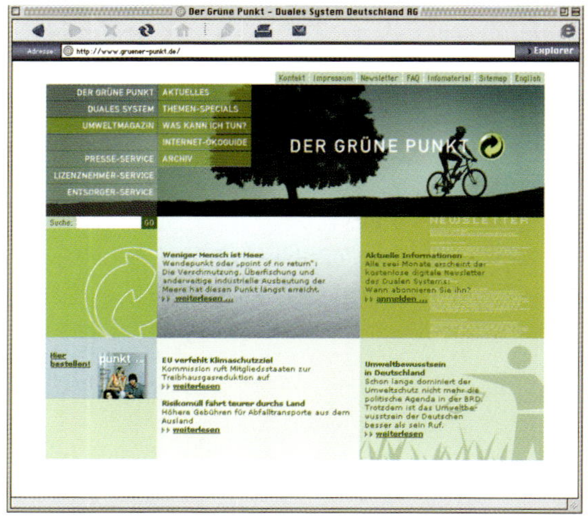

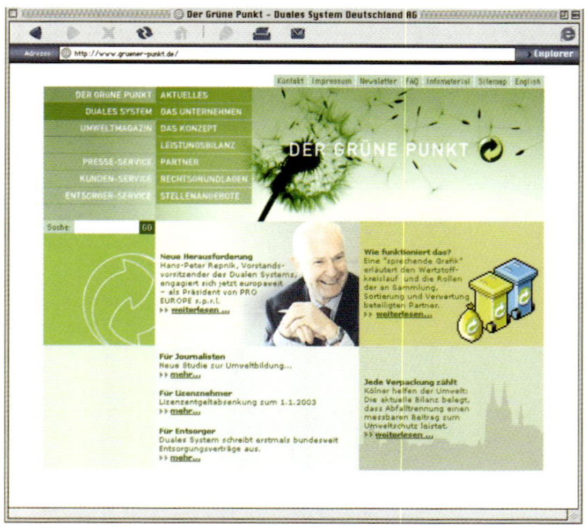

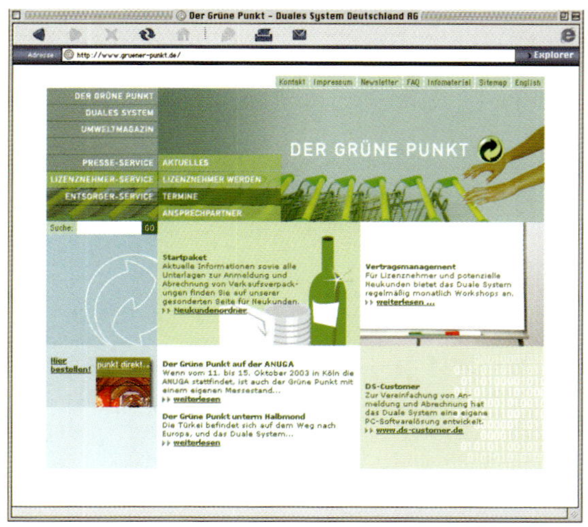

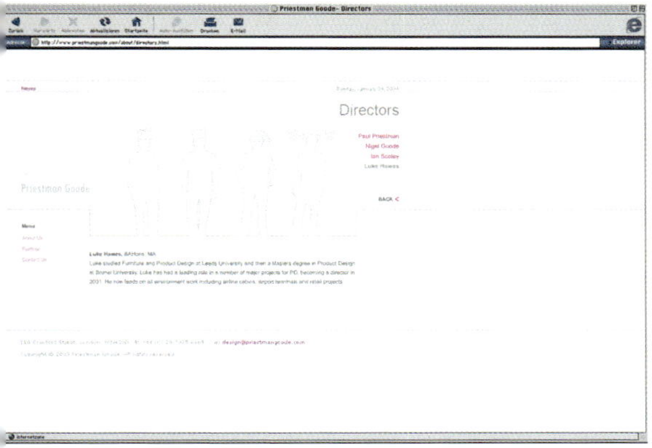

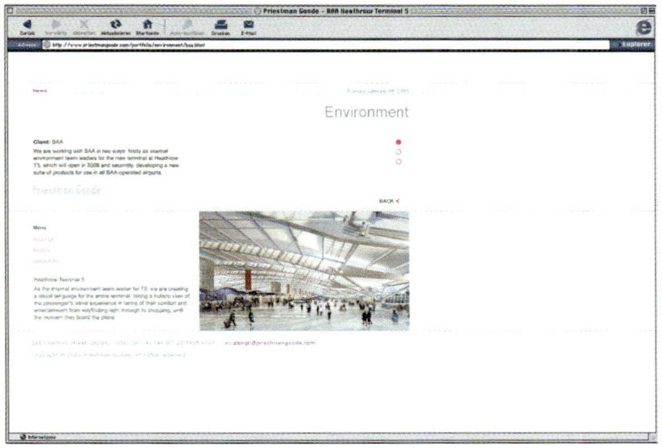

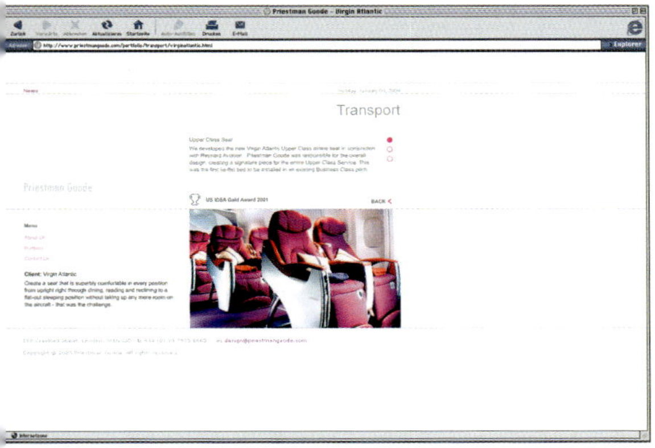

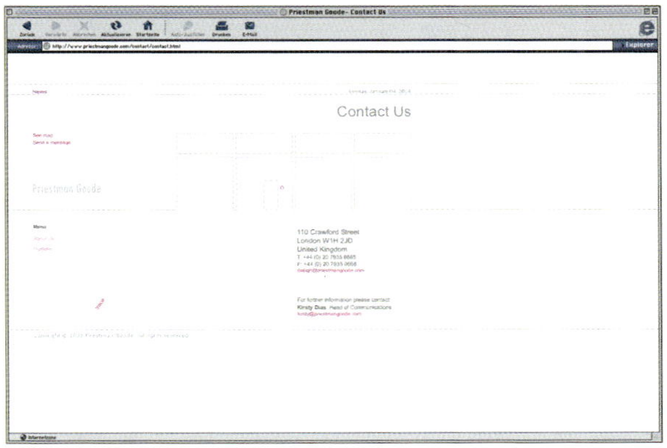

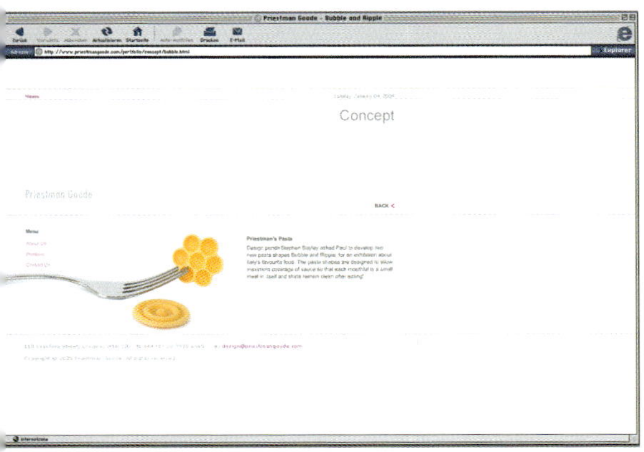

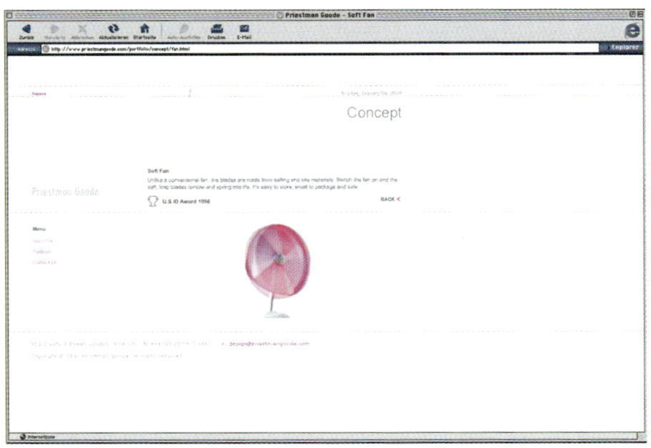

Auftraggeber Client
Priestman Goode
London, UK

Produkt Product
Priestman Goode Website
www.priestmangoode.com
Website
Web site

Design
Emma Lewis
Lisa Tse
Paul Priestman
Priestman Goode
London, UK

Die Website der Designfirma Priestman Goode ist in ihrer grafischen Gestaltung äußerst schlicht und einfach zu navigieren. Gezeigt wird nicht das gesamte Portfolio, sondern lediglich eine Auswahl aus dem Spektrum der Projekte und Produkte des Unternehmens. Aufgrund der geistreichen witzigen Animationen und unerwarteten Soundeffekte ist die Site sehr unterhaltsam.

The Web site of design company Priestman Goode is extremely clear and easy to navigate. Instead of presenting the firm's entire product portfolio, it features just a few examples from the overall spectrum of projects and products. The site is very entertaining with witty animations and unexpected sound effects.

Auftraggeber Client
Autostadt Wolfsburg GmbH
Wolfsburg, Germany

Produkt Product
ZUSAMMEN
Film
Movie

Produzent Producer
Cine plus Media Service GmbH & Co. KG
produktion 2 / schöne filme
Berlin, Germany

Design
Frank Schulte
Frank Schulte Corporate Design
Berlin, Germany

Der Film ‚Zusammen' ist eine Auftragsproduktion für eines der vier Themenkinos in der VW Konzernwelt innerhalb der Autostadt. Es handelt sich um einen dokumentarischen Imagefilm über soziale Kompetenz bei Volkswagen. Dargestellt werden verschiedene Mitarbeiter aus allen Ebenen des Konzerns. Wichtige Designaspekte sind die Typografie-Animation sowie das durchgehende Split-Screen-Verfahren (Bild im Bild) bei der Bildgestaltung, das sich variabel vom Vollbild in bis zu 9 verschiedene Bildteile unterteilt und mit dem die Gleichzeitigkeit und Parallelität von Personen und Ereignissen dargestellt werden konnten. Dies ermöglichte eine komplexere Dramaturgie und einen höheren Informations- und Unterhaltungswert des Films.

The film "Zusammen" (Together) was commissioned for one of the four theme cinemas in the VW Group World within the Autostadt complex. It is a documentary image film about Volkswagen's social commitment and expertise. Various staff from all levels within the Group appear in the film, whose key design features include animated typography and split-screen effects, whereby the latter runs throughout the film. The screen can be used for full-screen images or divided in up to nine sections, enabling different people and experiences to be shown at one and the same time or parallel to one another. This enhances the dramatic potential and makes the film more informative and entertaining.

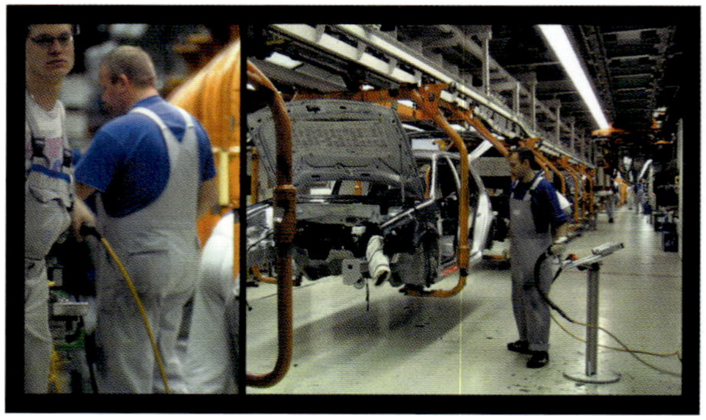

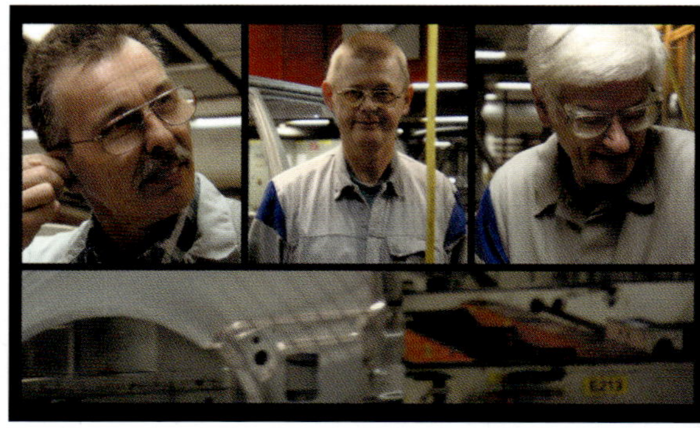

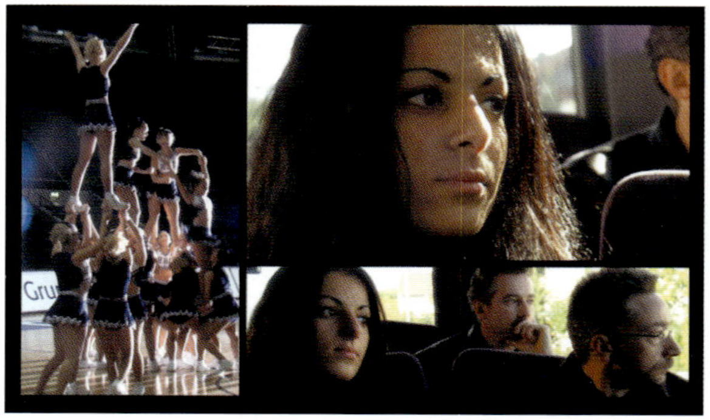

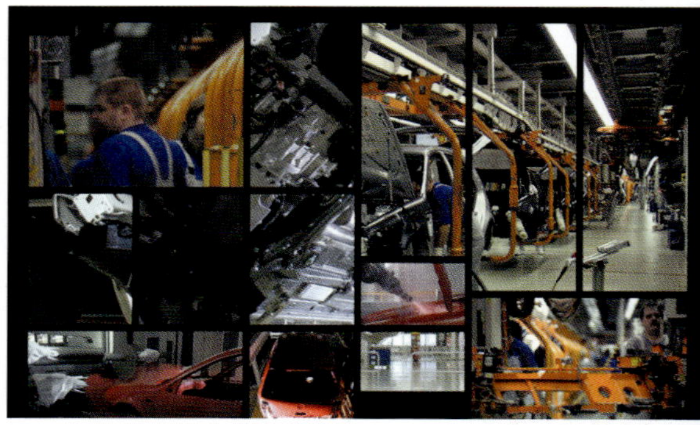

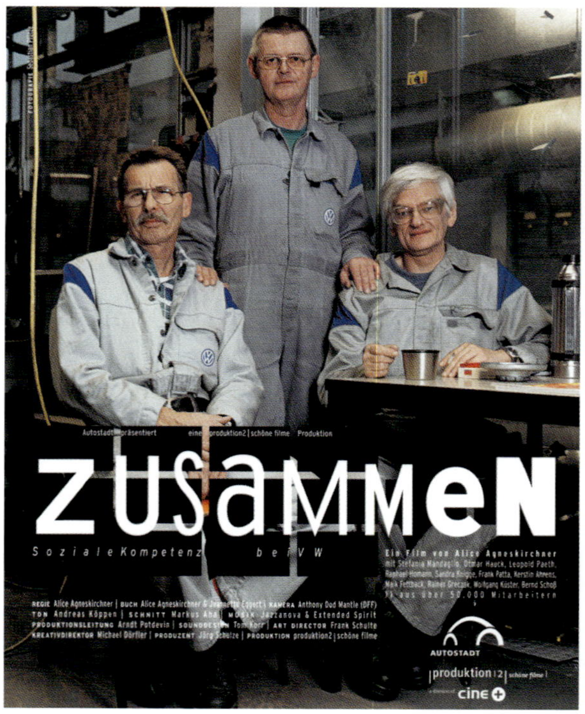

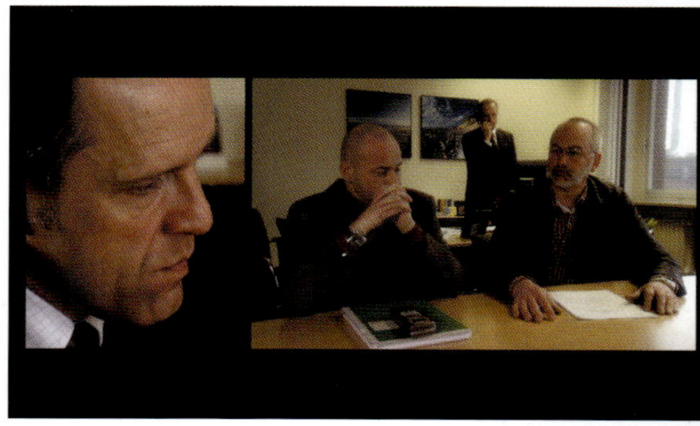

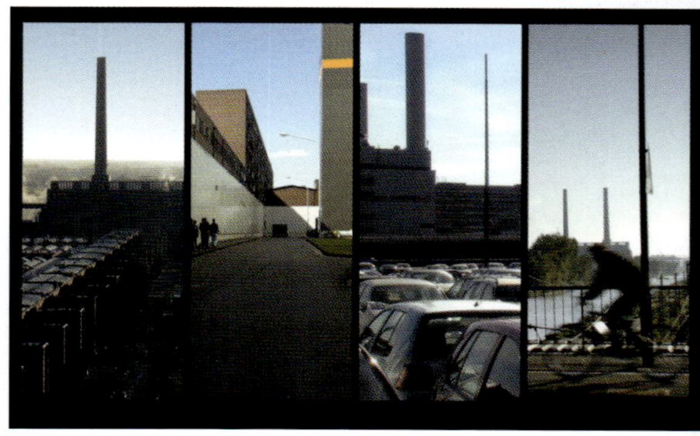

Die wichtigsten Funktionen ausgehend vom Start
Key functionality from Idle

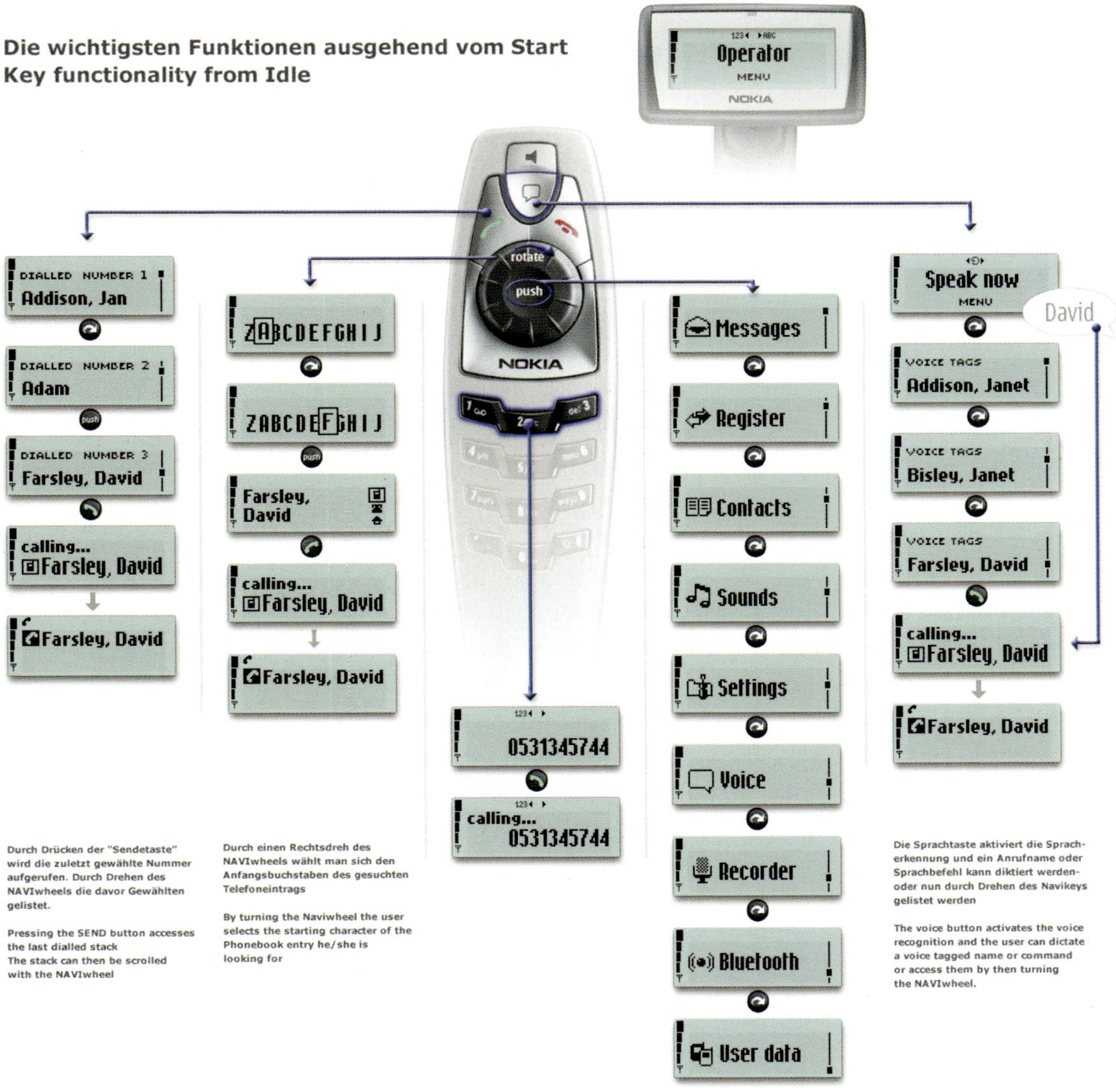

Durch Drücken der "Sendetaste" wird die zuletzt gewählte Nummer aufgerufen. Durch Drehen des NAVIwheels die davor Gewählten gelistet.

Pressing the SEND button accesses the last dialled stack
The stack can then be scrolled with the NAVIwheel

Durch einen Rechtsdreh des NAVIwheels wählt man sich den Anfangsbuchstaben des gesuchten Telefoneintrags

By turning the Naviwheel the user selects the starting character of the Phonebook entry he/she is looking for

Die Sprachtaste aktiviert die Spracherkennung und ein Anrufname oder Sprachbefehl kann diktiert werden- oder nun durch Drehen des Navikeys gelistet werden

The voice button activates the voice recognition and the user can dictate a voice tagged name or command or access them by then turning the NAVIwheel.

Auftraggeber Client
Nokia GmbH
Bochum, Germany

Produkt Product
NOKIA 810
Autotelefon
Car phone

Design
Marcus Anlauff
Tobias Pfauter
MANGO DESIGN
Braunschweig, Germany

Das Nokia Autotelefon 810 wurde speziell für den Einsatz in Fahrzeugen konzipiert. Die eigens für diesen Zweck entwickelte Benutzeroberfläche nur mit den wichtigsten Funktionen und der Einsatz moderner Technik ermöglichen hohen Bedienkomfort. Ein separates Display mit großer Text- und Grafikanzeige kann im Blickfeld des Fahrers angebracht werden. Weitere Features: Hörer mit neuem Navi™-Wheel, Bluetooth-Unterstützung und Unterstützung für individuelle Profile für zwei verschiedene Benutzer, einschließlich Kontaktdaten, 12 Aufnahmen und persönlichen Einstellungen. Der Passagier im Fond kann dank Bluetooth schnurlos surfen.

The Nokia 810 car phone is a dedicated device boasting a user interface designed specifically for this purpose. It includes only the most important functions – but based on ultra-modern technology and thus ensures prime user comfort. A separate screen with large text and graphic displays can be shifted into the driver's view. Other features: a receiver with new Navi™-Wheel, Bluetooth as well as support for individual profiles for two different users, including contact data, 12 recording and personalized settings. Back-seat passengers can enjoy cordless Web surfing thanks to Bluetooth.

Hersteller Manufacturer
Panasonic Mobile Communications Co. Ltd.
Yokohama-City, Kanagawa, Japan

Produkt Product
Panasonic EB-G70
Display design
User Interface Design

Design
Natsuko Nishigaki
Corporate Multimedia Design Group
Yokohama-City, Kanagawa, Japan

Das 4096 Farbdisplay des Mobiltelefons EB-G70 von Panasonic verwandelt sich in ausgeschaltetem Zustand in einen Kosmetikspiegel. Die Gestaltung der neun Icons harmoniert mit der runden Form des Displays, wobei die Farbgebung die Anmutung einer Lidschatten-Palette hat. Das Benutzermenü ist trotz des kleinen Displays intuitiv und leicht zu bedienen. Den jeweiligen Anrufern können 10 verschiedenen Panasonic-Bilder zugeordnet werden.

Panasonic's EB-G70 mobile phone boasts a 4,096-color display that turns into a make-up mirror when the phone is switched off. The design of the nine icons harmonizes with the display's round shape, with the color tones resembling those of a range of eye-shadows. The user interface is intuitive and easy to use, despite the display's small dimensions. The respective callers can be assigned to 10 different Panasonic images.

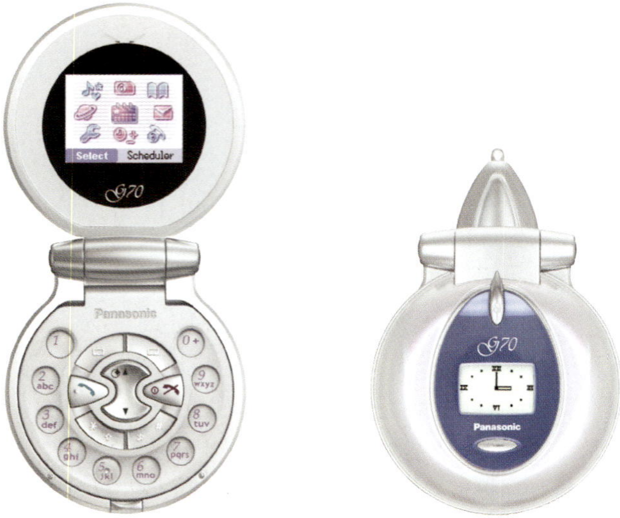

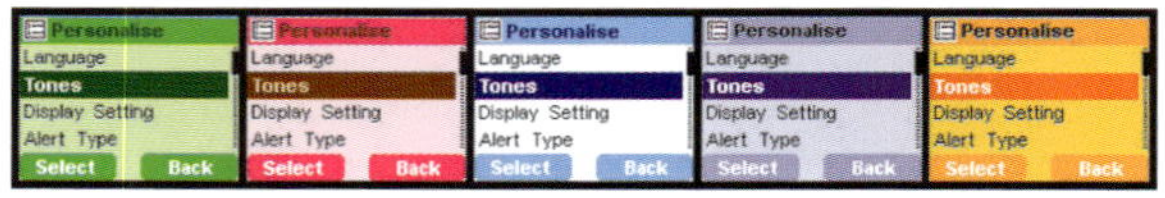

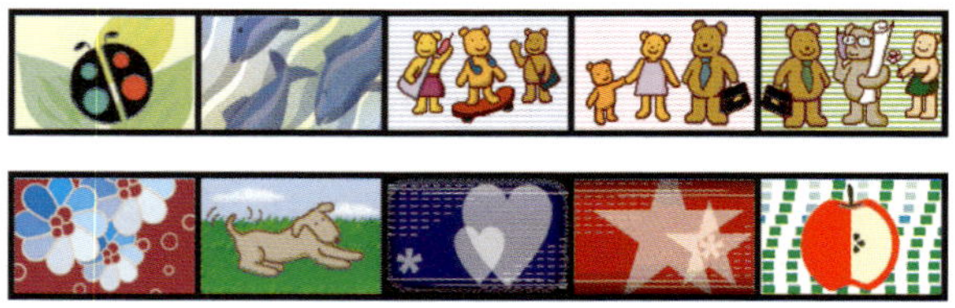

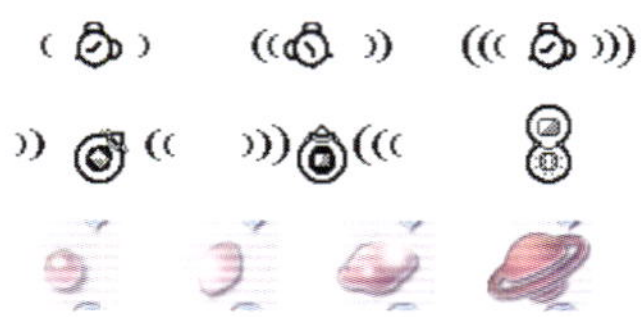

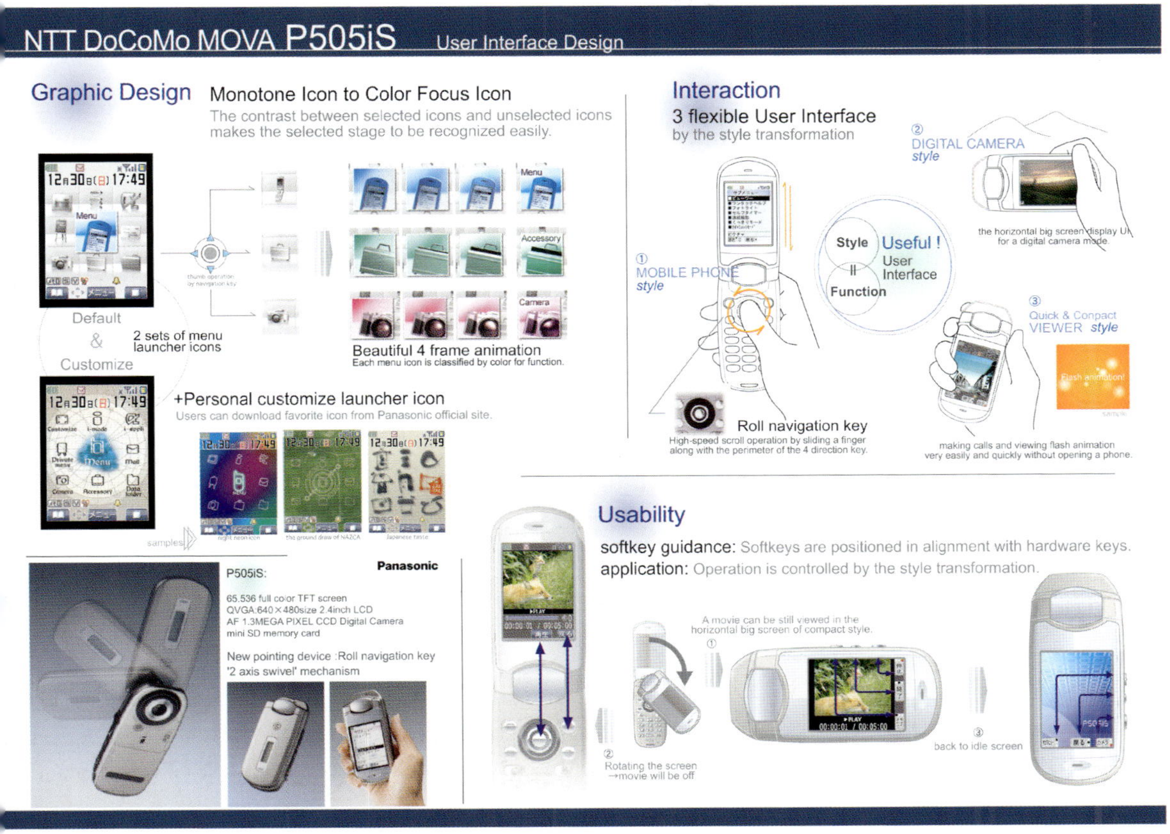

Auftraggeber Client
Panasonic Mobile Communications Co., Ltd.
Yokohama-City, Kanagawa, Japan

Produkt Product
MOVA P505iS
Benutzeroberfläche für Handy
User Interface for mobile phone

Design
Yuichi Ishihara
Yoshito Ishikawa
Junichi Oshiro
Corporate Multimedia Design Group
Yokohama-City, Kanagawa, Japan

Das User Interface Design des Mobiltelefons MOVA P505iS bietet eine klare und leicht verständliche Grafik, wobei die Icons, wenn sie aktiviert sind, farblich hervorgehoben werden. Zudem können individuelle Icons von der Panasonic Site heruntergeladen werden. Aufgrund des drehbaren Displays kann das Interface im Mobile Phone Modus ausgeklappt und als Digital-Kamera horizontal genutzt werden, oder es können Flash-Animationen angeschaut bzw. Anrufe getätigt werden, während das Gerät geschlossen ist. Mittels Roll Navigation Key und Softkey ist eine einfache und schnelle Bedienung möglich.

The MOVA P505iS is a mobile phone with a user interface structured with clear and easily comprehensible graphics. The icons when activated are emphasized by a specific color, and further individual icons can be downloaded from the Panasonic Web site. Thanks to the rotatable display, the interface folds out in mobile phone mode or used horizontally as a digital camera or as a display for Flash animations – when closed calls can be made. Thanks to the roll navigation key and soft-key, the handset is very easy to use swiftly.

Auftraggeber Client
IBM Corporation
Research Triangle Park, NC, USA

Produkt Product
IBM ThinkPad Features Screensaver
Bildschirmschoner
Screensaver

Design
IBM Design
Research Triangle Park, NC, USA

Die IBM-Notebooks der Think-Pad-Reihe sind mit einem Screensaver ausgestattet, der auf unterhaltsame Weise Informationen zu den Produkteigenschaften des Computers vermittelt. Anhand von einfachen animierten Zeichnungen, die der minimalistischen Geometrie des Designs des Notebooks entsprechen, und speziellen Effekten erfährt der User wichtige Einzelheiten und erhält Anwendungsbeispiele, die er normalerweise nur im Manual nachlesen kann. Es werden insgesamt zehn verschiedene Funktionen auf diese Art und Weise dargestellt, u.a. zum Thema Track Point, zum IBM Embedded Security Subsystem oder IBM Rapid Restore.

The IBM Think Pad series notebooks all come with a screensaver that offers an entertaining run-through of information on the notebooks' features and specifications. By means of simple animated drawings that reflect the minimalist geometry of the notebooks' design, not to mention special effects, users are familiarized with key details and shown sample applications that they would normally only be able to peruse in a user manual. In this way, a total of ten different functions are presented, among others on issues such as Track Point, the IBM Embedded Security Subsystem or IBM Rapid Restore.

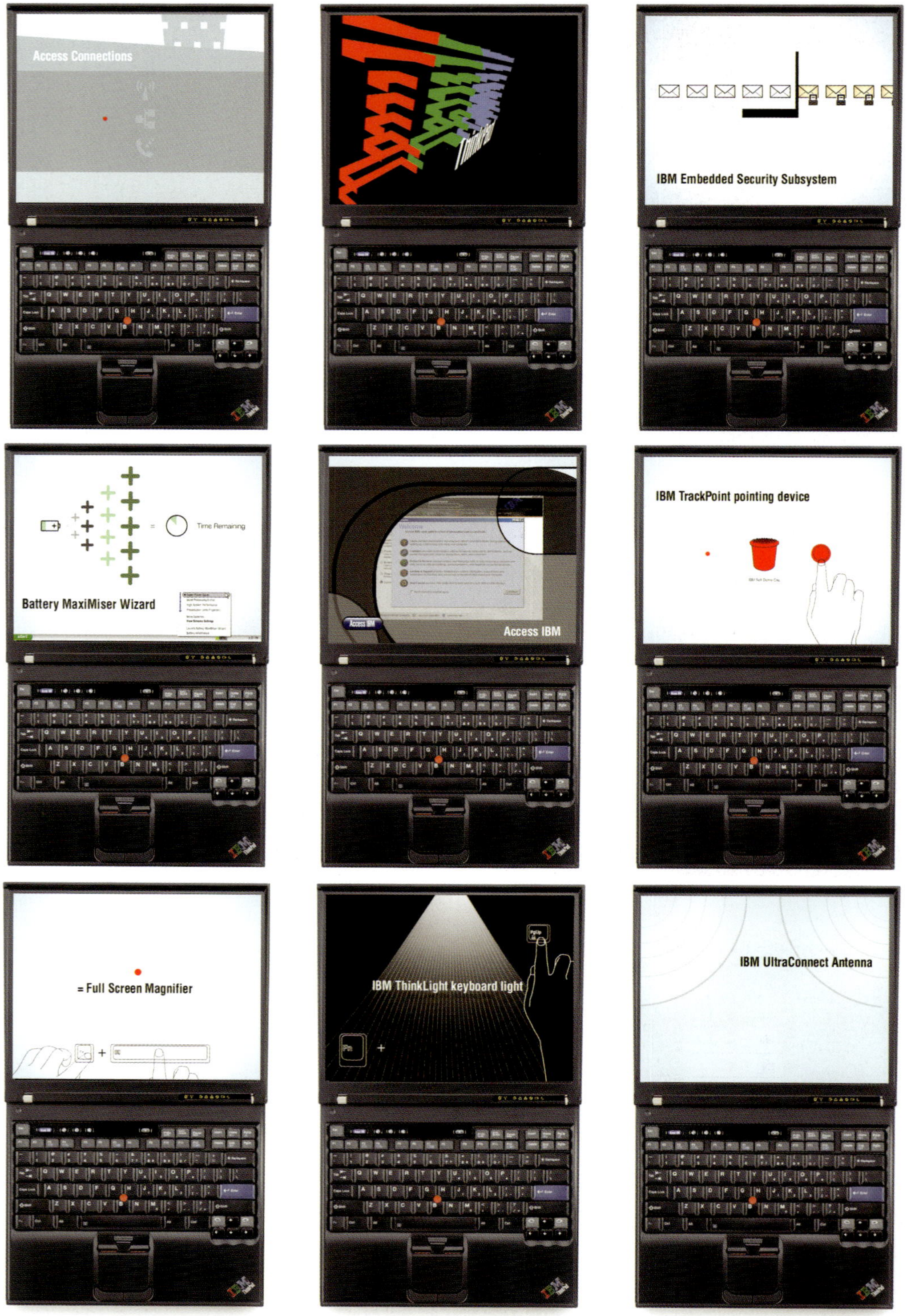

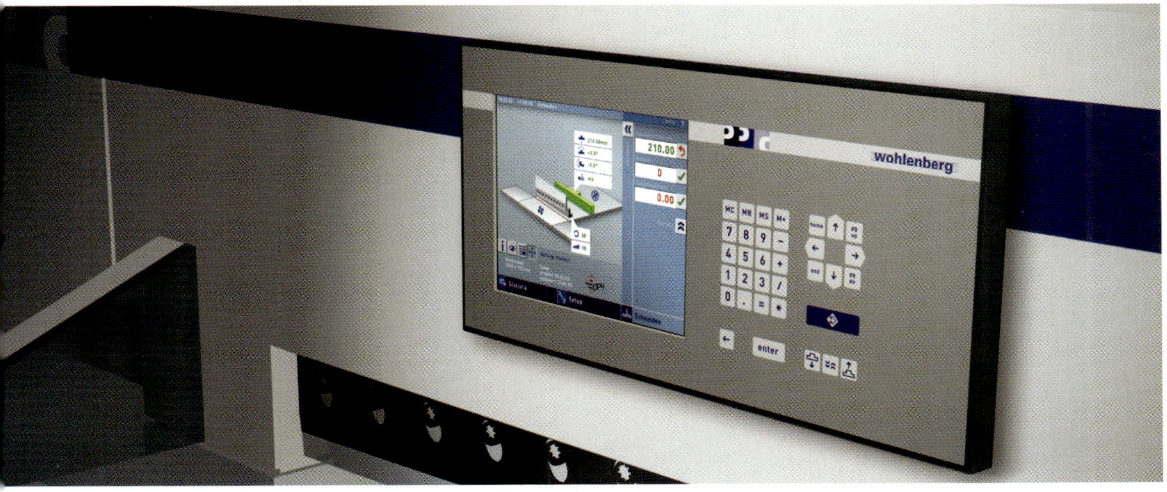

Auftraggeber Client
Wohlenberg Schneidesysteme GmbH
Hannover, Germany

Produkt Product
Interaktive Bedienoberfläche
für eine Schneidemaschine
Interactive user interface
for a guillotine

Design
Prof. Detlef Rahe
Riad Hamadmad
Michael Wilde
Frank Lieser
:i/i/d Institut für Integriertes Design
Bremen, Germany

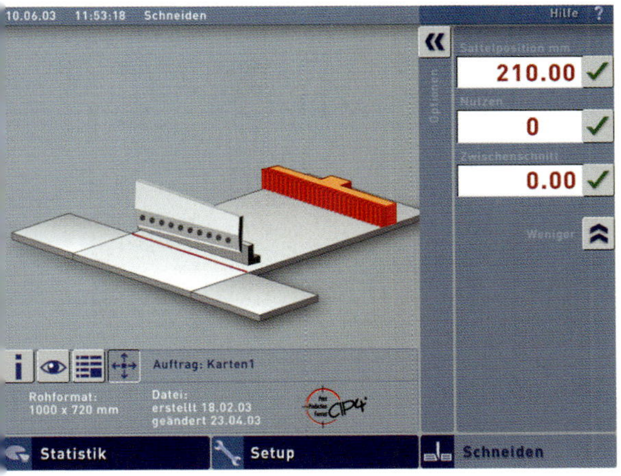

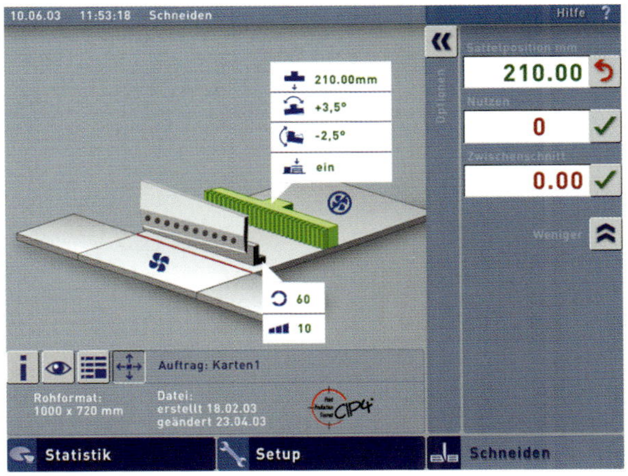

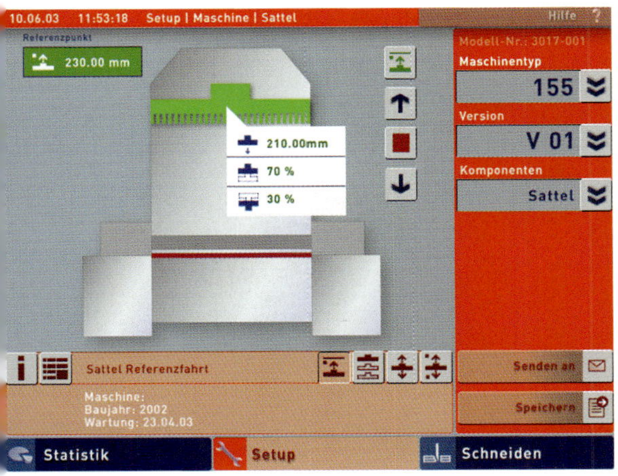

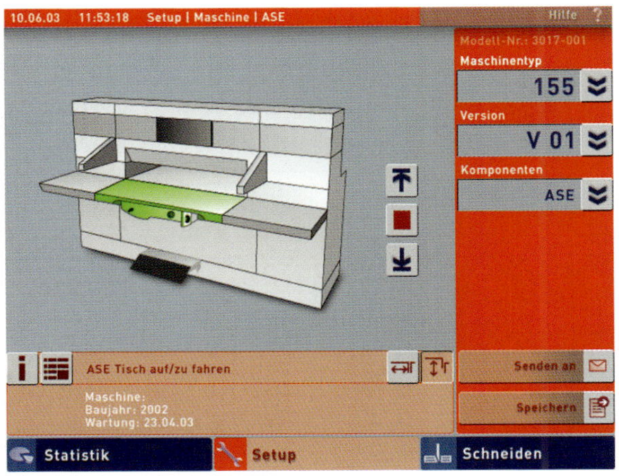

Das User-Interface der neu entwickelten Papierschneidemaschine von Wohlenberg wurde im Hinblick auf optimierte Benutzerfreundlichkeit und Usability gestaltet. Grundprinzip der Steuerung ist eine manuelle Eingabe über einen Touch-Screen und ein Tastenfeld. Die Wege für den Benutzer sind kurz und logisch nachvollziehbar, es gibt wenige Bildwechsel. Nur aktuell relevante und ausführbare Aktionen werden angezeigt. Farben bilden einen wesentlichen Bestandteil des Steuerungskonzeptes und helfen Redundanzen in der Darstellung zu vermeiden. Jedem Hauptmenüpunkt ist eine Farbwelt zugeordnet, um dem Benutzer Orientierung im System zu geben. Die Oberfläche löst sich von der Windows-Erscheinung und bietet eine eigene Ästhetik und Funktionalität, die dem Corporate Design des Unternehmens entspricht.

The user interface was devised for the new guillotine developed by Wohlenberg – with a view to optimizing user-friendliness and usability. Manual inputs via the keypad and Touchscreen constitute the basic principle underlying the controls. The paths involved are short and logically intuitive, and images only change when absolutely necessary. Only actions that are both relevant at that moment and possible are displayed. Colors form a fundamental part of the control concept, helping avoid any redundant images. Each main menu item is allocated to a specific color coding to assist the user in finding his way in the system. The GUI severs any links with Windows icons and instead has an aesthetics and functionality of its own that concurs with company's corporate design.

COMMUNICATION DESIGN

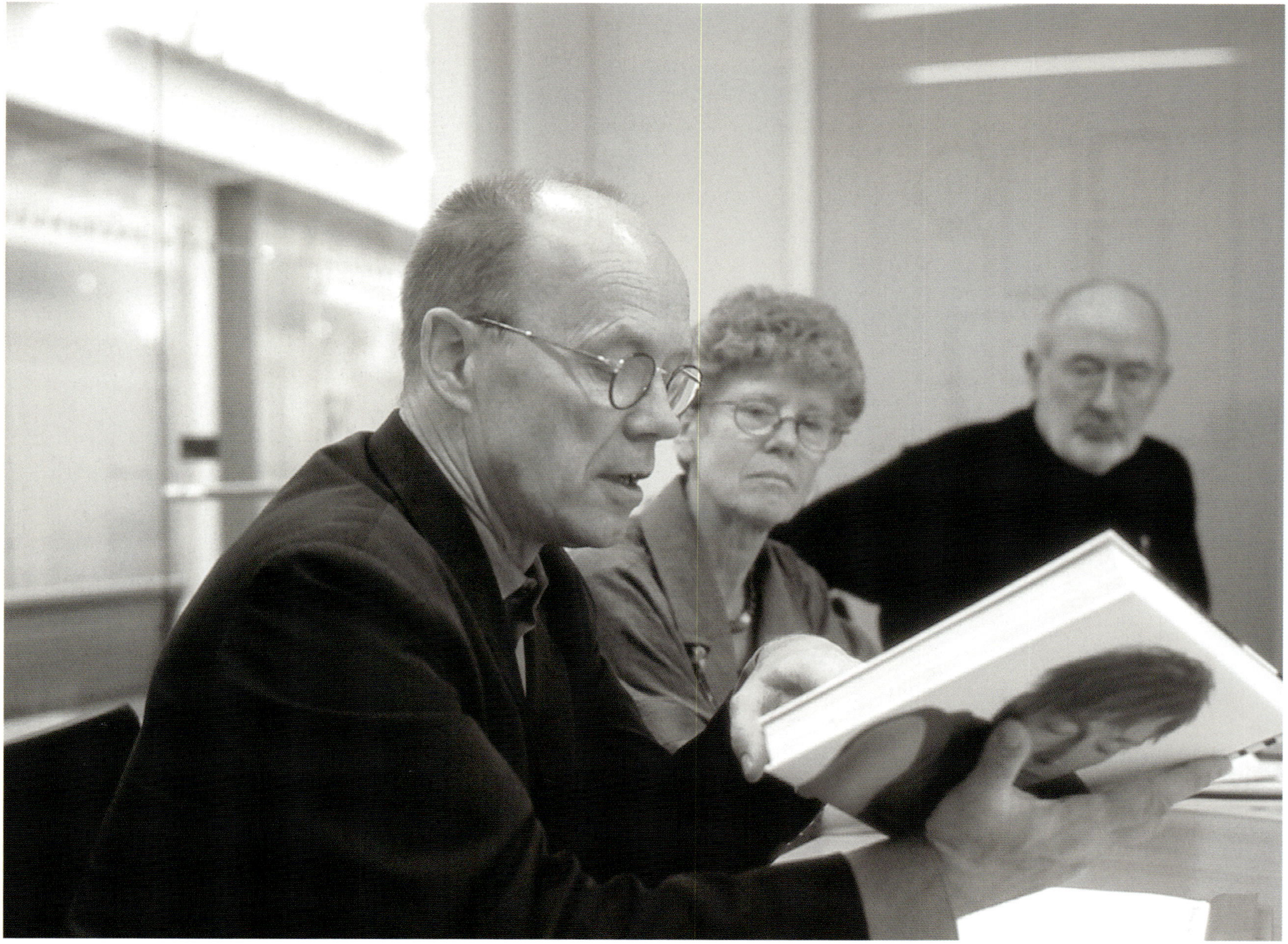

Grafic Design
Publishing Media
Corporate Communication
Marketing Communication

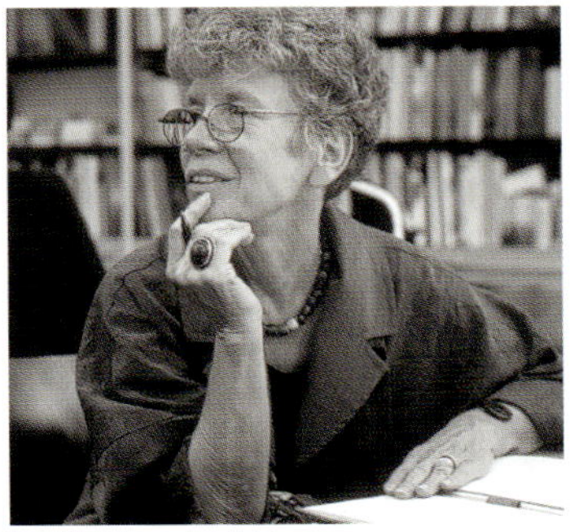

HEIDE HACKENBERG erwarb an der Werbefachlichen Akademie in Hamburg einen Abschluss als Kommunikationsfachwirtin. Sie arbeitete viele Jahre für Werbeagenturen im Bereich Text und Konzeption. Seit 1989 ist sie als Texterin/Journalistin selbstständig. Sie ist Autorin des Buches „Was kostet Grafikdesign?" und Pressesprecher in der AGD, Alliance of German Designers.

Heide Hackenberg graduated from the Advertising College in Hamburg in communications. She worked for many years in ad agencies in copy-writing and conceptualization. Freelance copywriter and journalist since 1989 and author of "Was kostet Grafikdesign?"; press spokesperson of AGD, Alliance of German Designers.

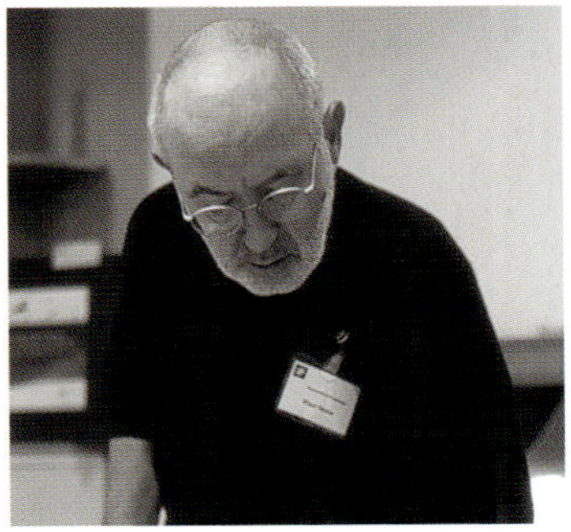

KLAUS OBERER gründete nach seiner Ausbildung zum Grafikdesigner an der Schule für Gestaltung in Basel zusammen mit Pierre Mendell die Graphic Design Studios Mendell & Oberer in München. Er erhielt zahlreiche nationale und internationale Auszeichnungen. Zugleich ist er als Fotograf tätig und stellt seine Arbeiten in Museen und Galerien aus.

Klaus Oberer trained as a graphic designer at the Basel College of Design and then, together with Pierre Mendell, founded the Mendell & Oberer Graphic Design Studios in Munich. He has won numerous national and international awards. He also works as a photographer and regularly exhibits in museums and galleries.

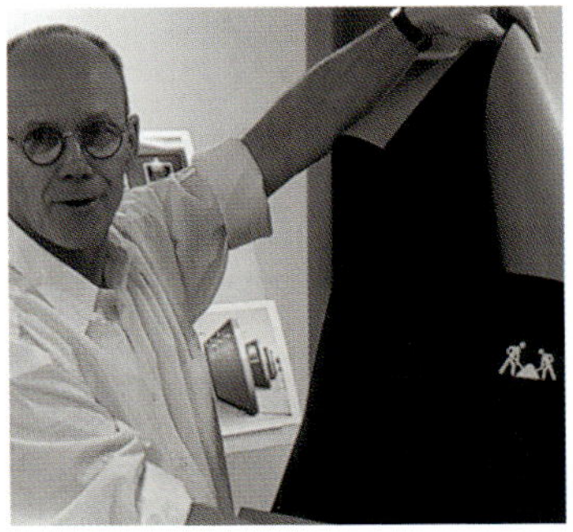

ERIK SPIEKERMANN ist Fachautor, typografischer Gestalter, Schriftenentwerfer. Er gründete 1979 MetaDesign und 1988 Font-Shop. An der HfK Bremen hält er eine Honorarprofessur, ist Präsidiumsmitglied im Rat für Formgebung und Präsident des Internationalen Instituts für Informationsdesign und der International Society for Typographic Designers. Seit 2000 arbeitet er als freier Gestalter unter dem Label United.

Erik Spiekermann is a specialist author, typographical designer and font inventor. In 1979 he founded MetaDesign and in 1988 he established FontShop. He is Honorary Professor at the Bremen Academy of Arts, a member of the Presiding Council of the German Design Council, and President of the International Institute for Information Design and the International Society for Typographic Designers. Since 2000 he has freelanced for his United label.

The End of Print? Keineswegs.
Heide Hackenberg

Um mit Gold oder Silber ausgezeichnet zu werden, müssen Arbeiten ein außergewöhnliches Profil besitzen. Was die Jury im Wettbewerb um die höchste grafische und kreative Qualität bewog, vier dieser Auszeichnungen zu vergeben, kann mit Sicherheit etwas über den Status quo des Metiers aussagen.

Die einzige Gold-Auszeichnung erhielt ein Buch. Die Arbeit ist im akademischen Umfeld entstanden und dokumentiert vier Jahrzehnte Fotografielehre an der Hochschule für Künste in Bremen. Das Fotobuch wurde von einem Studententeam unter Mitarbeit ihres Professors konzipiert und umgesetzt.

Das Besondere am iF Wettbewerb 2004 ist vielleicht, dass ein autonomes Werk im Wettbewerb siegte und zudem ein Buch.

Auch der Blick auf die zweite Auszeichnung – Silber – offenbart einen interessanten Aspekt. Sie stammt nicht aus einem europäischen Land, sondern aus Brasilien, wie noch einige ebenfalls prämierte Arbeiten. Aus Brasilien kommt eine frische, ungezwungene, leicht folkloristische Grafik, die ideen- und farbenreich ist.

Auch der dritte Preis ging an ein Buch. Es handelt sich um eine handwerklich akkurat gestaltete Arbeit zur Architektur des Bundeslandes Baden-Württemberg. Dass der vierte Gewinner ebenfalls mit einer Buchproduktion richtig lag, ist kein Zufall. Das frei gestaltete Buch für ein Unterhaltungselektronikunternehmen dokumentiert eine Reise zu den entlegendsten Orten der Welt und der jeweiligen Musikszene vor Ort. Es folgt in Form und Anmutung dem klassischen Typ des Fotoalbums.

Auf diesen Siegeszug des Buches folgen die anderen visuellen Medien wie Geschäftsberichte, Firmendarstellungen auf Messen, Plakate und Begleitpublikationen. Sie wirken gegenüber den Büchern eher wie eine Form der angewandten und angepassten Kunst.

Das Buch als grafische Kunstform hat 2004 vierfach gewonnen und taucht in interessanten Varianten noch mehrmals auf: einmal als Umweltedition mit thermoaktivem Umschlag, der die Körperwärme der Hand temporär nachzeichnet und einmal in einem mit Wasser gefüllten transparenten Kunststoffbeutel, der das Buch mit der Abbildung eines Kanus des italienischen Künstlers Fabrizio Plessi auf dem Umschlag einschließt.

Haben nicht kluge Leute schon früher gesagt, dass ein Medium, wenn es seine unmittelbare Bedeutung verliert, zur Kunstform werden kann? Wenn man dem Ergebnis des iF design awards Glauben schenken mag, kommt das totgesagte Buch als fantasievolle Kunstform wieder.

The End of Print? By No Means.

In order to receive a Gold or a Silver award, entries must have a quite unique and outstanding profile. And the reasons why, when judging the competition on graphic and creative quality, the jury decided to bestow four of these awards most certainly says something about the status quo in the profession.

The only Gold award went to a book. The publication in question arose in an academic environment and documents four decades of teaching photography at the Bremen Academy of Arts. The book of photos was designed and realized by a team of students, acting under the guidance of their professor.

Perhaps this illustrates why the iF competition is so special: a non-commercial project won the day, and it was an example of a traditional medium, namely a book.

A glance at the "second line" prizes, namely the Silver awards, also offers an interesting insight. Silver went not to a European country, but to Brazil, which also picked up a number of iF awards this year. Brazilian graphics has a fresh, lively and slightly folkloristic feel to it, brimming over with ideas and colors.

The Silver award was again garnered by a book. And surprisingly, another Silver award was also won by a book: a beautifully crafted piece on architecture in the Federal State of Baden Württemberg. It can thus evidently not be coincidence that the third of the Silvers likewise went to a book, one with a very free design created on behalf of an entertainment electronics corporation and documenting a trip to very remote parts of the world, where the local music scene was observed. In terms of form and feel, it follows the classical medium of a photo album.

In the wake of the victorious onward march of books come the other visual media, such as annual reports, corporate image presentations for trade fairs, posters and accompanying publications. Compared with the books, they seem to be examples of applied art, conforming to what is expected of them.

The book as a graphic form of art thus won four distinctions in 2004 – and it puts in an appearance in numerous other interesting variants. For example, as an environmental edition complete with a thermo-active cover that temporarily traces the body-heat of the hand that holds it. Or as a transparent plastic pouch filled with water, which contains the book with a cover portrait of a canoe created by Italian artist Fabrizio Plessi.

Did not smart minds once conclude that a medium, once it loses its immediate importance, can emerge as art? If you take a glance at the results of the iF design award, then you could be forgiven thinking that the book, once declared dead, has resurfaced as a marvelously imaginative art form.

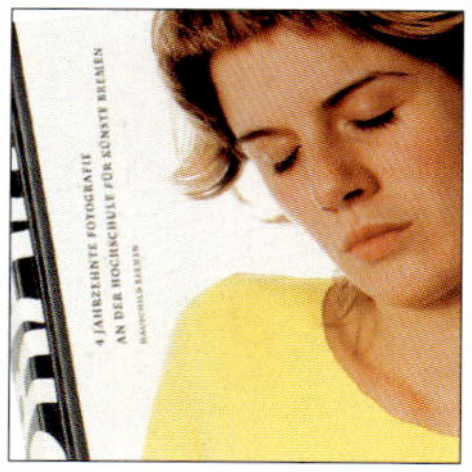
4 JAHRZEHNTE FOTOGRAFIE
AN DER HOCHSCHULE FÜR KÜNSTE BREMEN
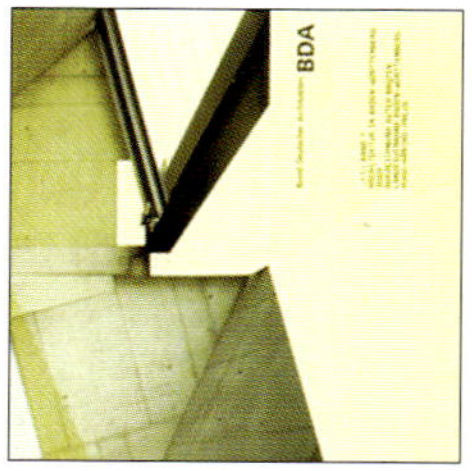
BDA
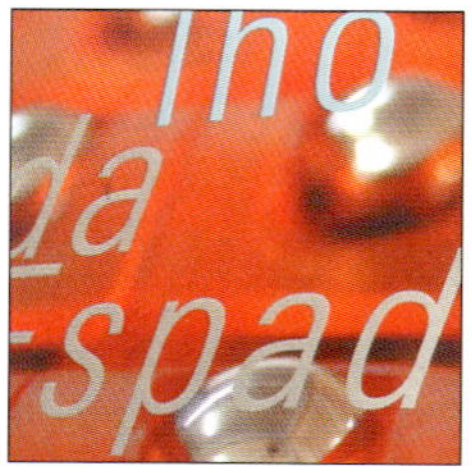
lno
da
spad
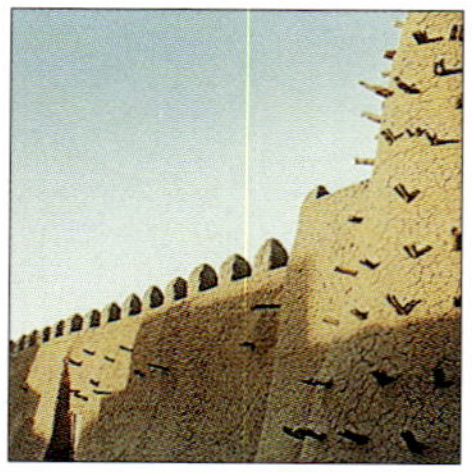
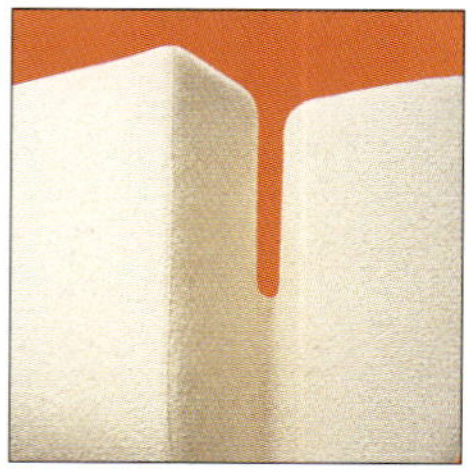
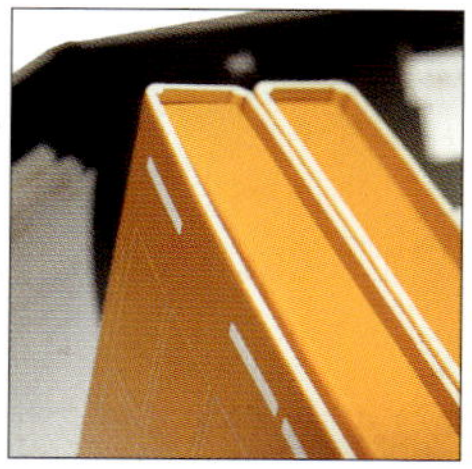
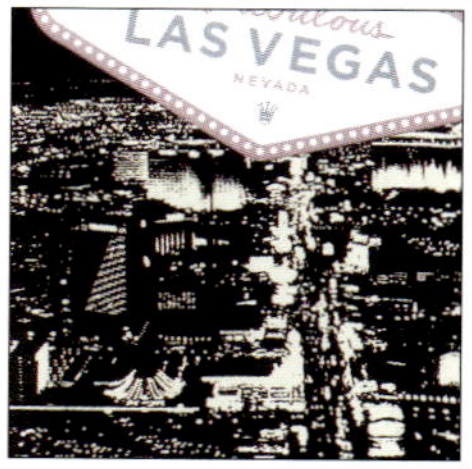
LAS VEGAS
NEVADA
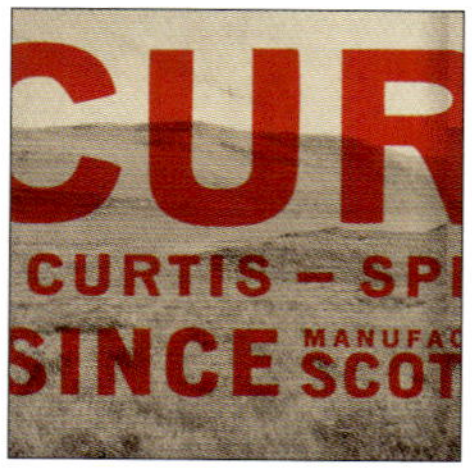
CUR
CURTIS – SP
SINCE
SCOT
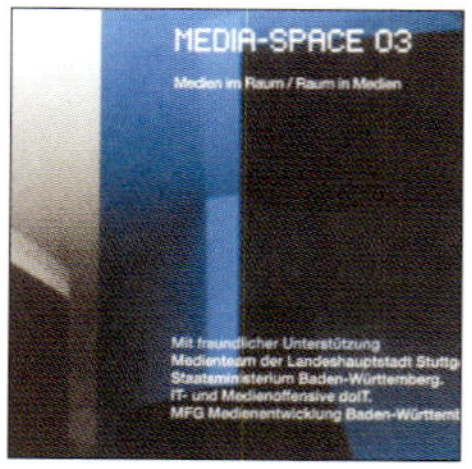
MEDIA-SPACE 03
Medien im Raum / Raum in Medien
Mit freundlicher Unterstützung
Medienteam der Landeshauptstadt Stuttg
Staatsministerium Baden-Württemberg,
IT- und Medienoffensive doIT,
MFG Medienentwicklung Baden-Württem
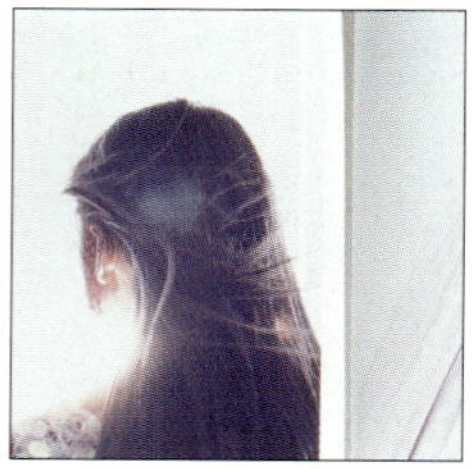
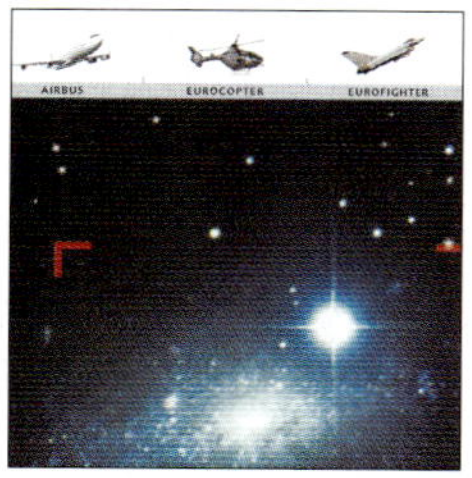
AIRBUS
EUROCOPTER
EUROFIGHTER
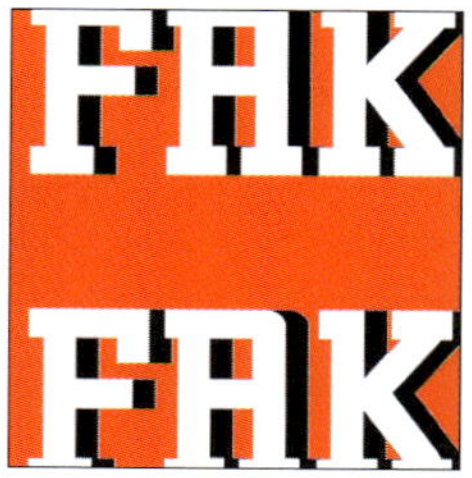
FAK
FAK
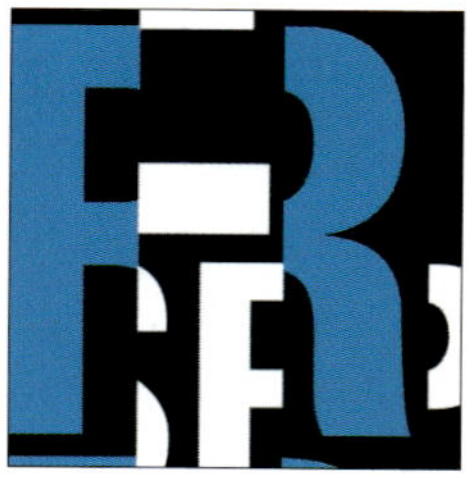
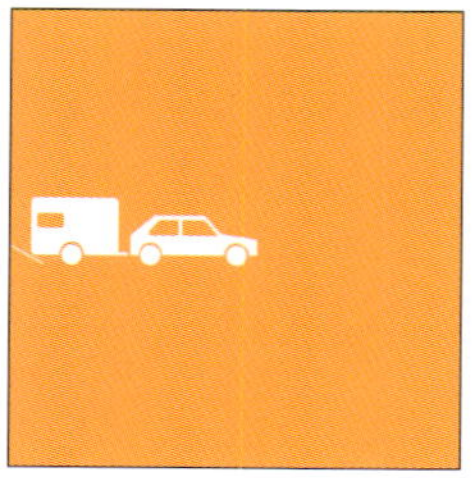
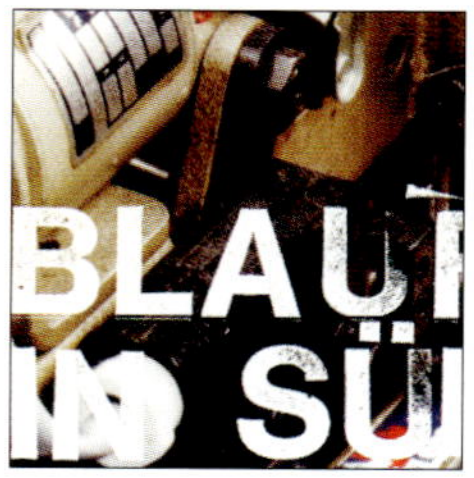
BLAU
SÜ
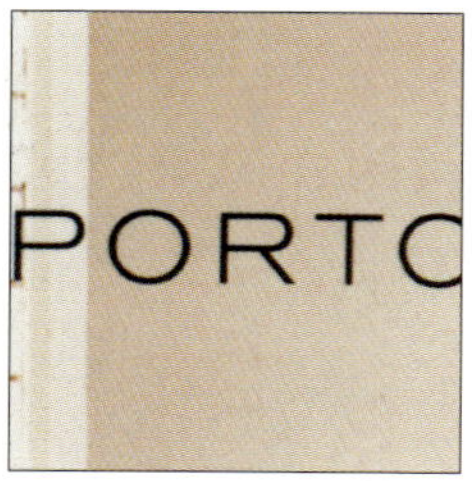
PORT
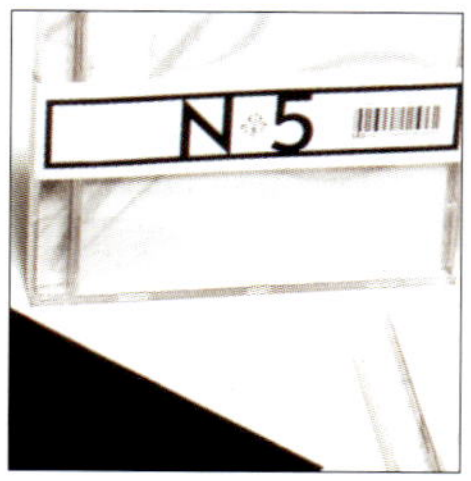
N 5
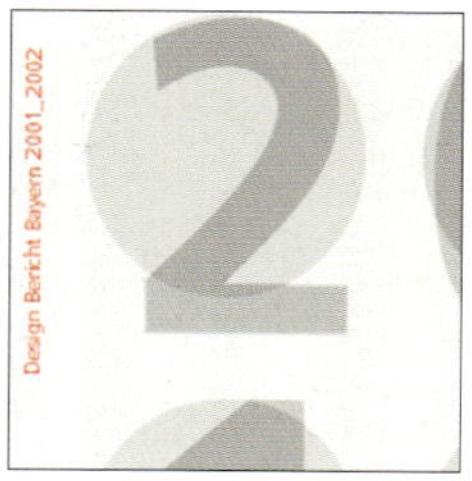
Design Bericht Bayern 2001_2002
2
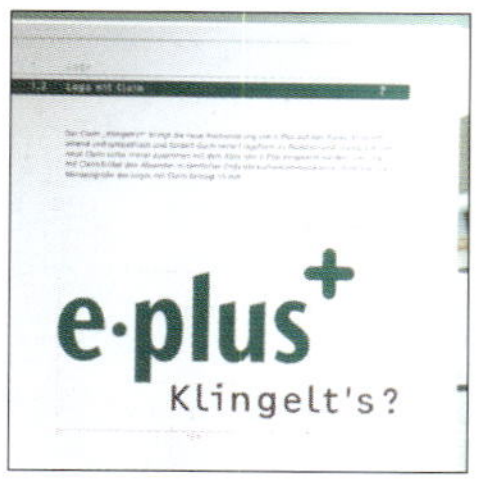
e·plus
Klingelt's?
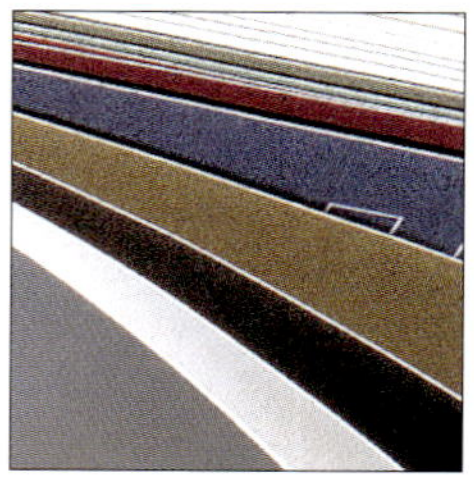

Grafic Design

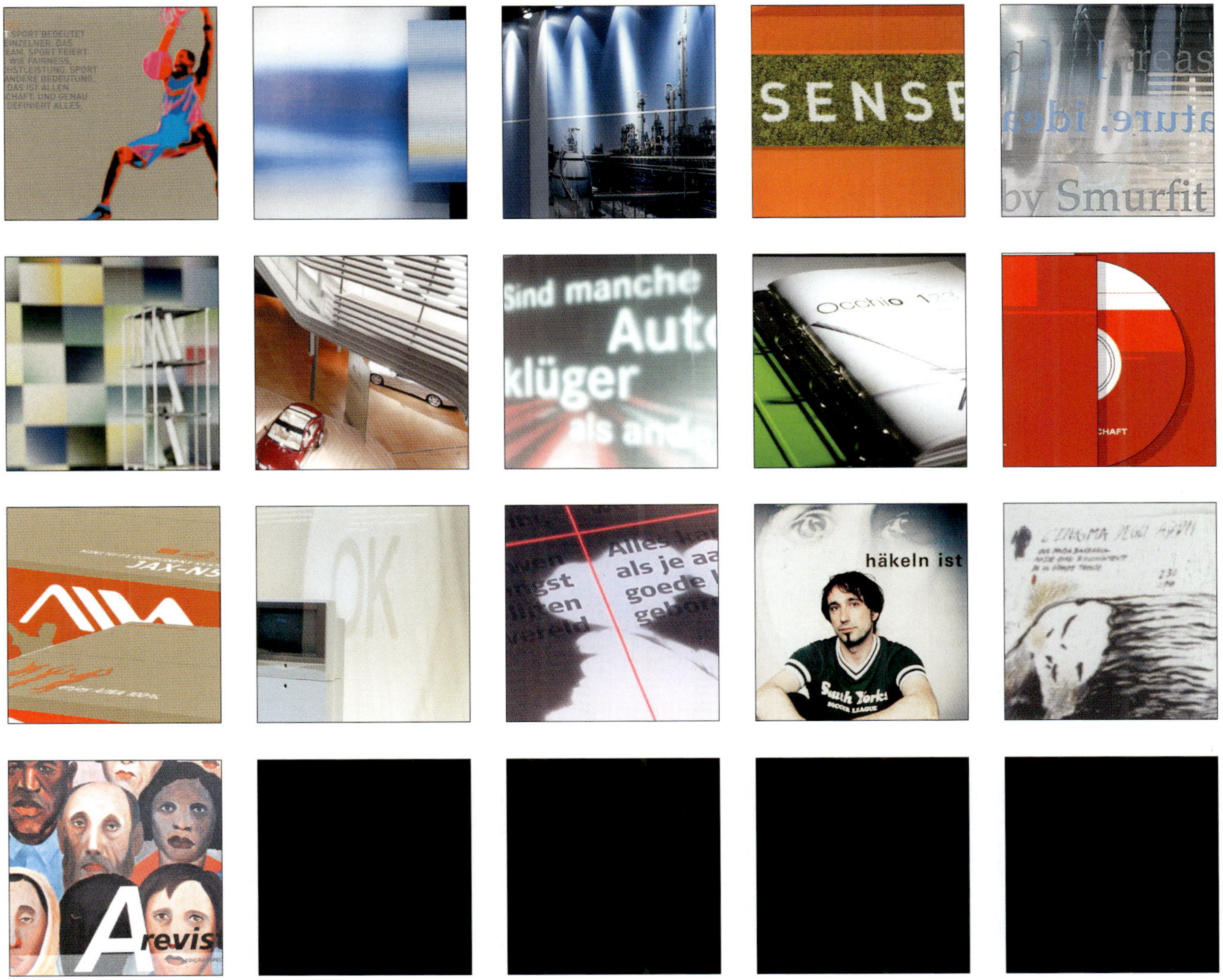

Auftraggeber Client
Hochschule für Künste Bremen
Bremen, Germany

Produkt Product
Wandrahm – 4 Jahrzehnte Fotografie
an der Hochschule für Künste Bremen
Buch
Book

Design
Prof. Fritz Haase
Siena Jakobi
Jack Kraska
Hochschule für Künste Bremen
Bremen, Germany

Das Buch „Wandrahm" ist eine sehr persönliche Betrachtung fotografischer Entwicklung an der Hochschule für Künste Bremen von Professor Fritz Haase. Vier Jahrzehnte Fotografie dokumentieren im Rückblick, Trends, Tendenzen, Zeitgeist und insgesamt ein avanciertes Medium in der Werbung für Reportagen und die Erfassung von kulturellen und politischen Zeitläufen.

The Wandrahm book is a very personal view by Professor Fritz Haase of how the field of photography has developed at the Bremen Academy of Arts down through the years. Four decades of photography provide a retrospective document on trends, tendencies, and the zeitgeist while also paying tribute to what overall is a cutting-edge medium for advertising, photo journalism and the recording of cultural and political conjunctures.

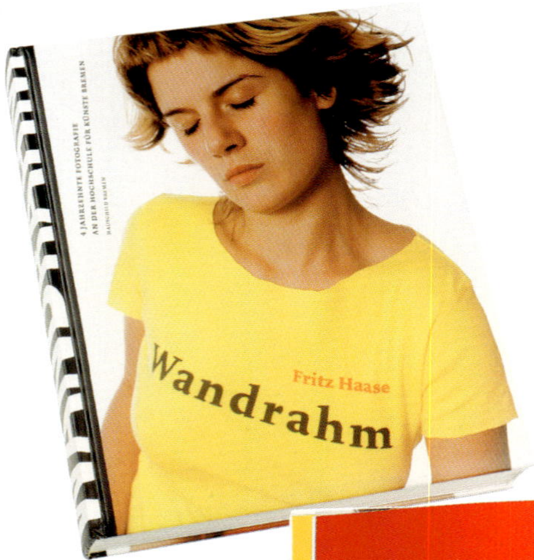

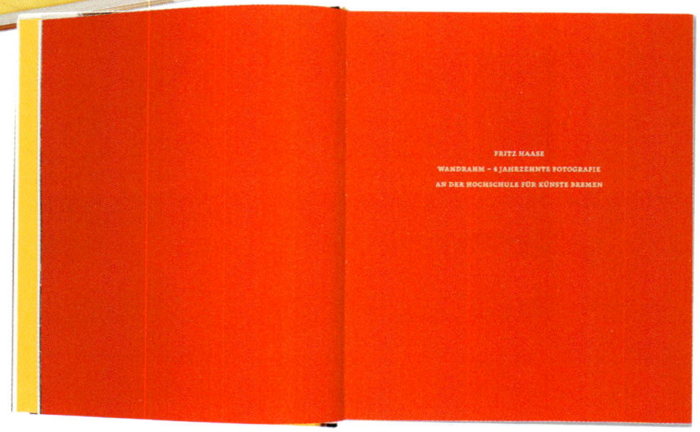

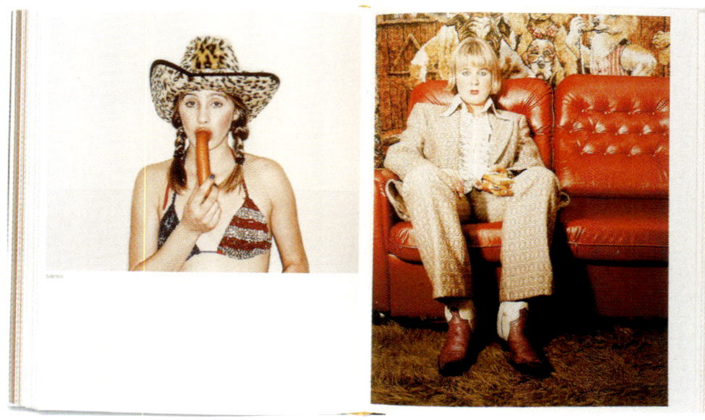

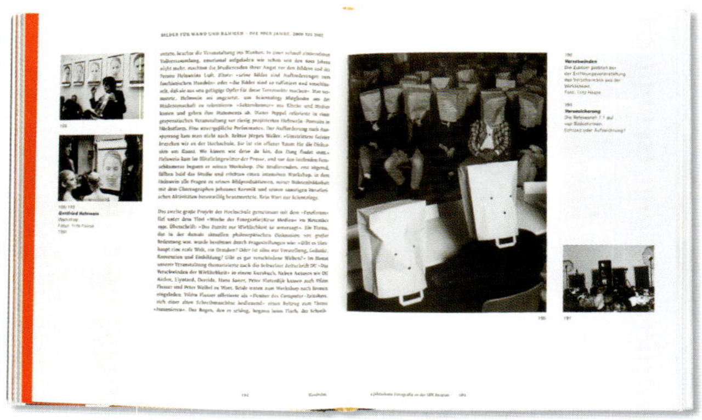

iF Gold Award

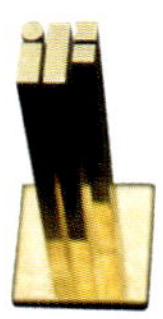

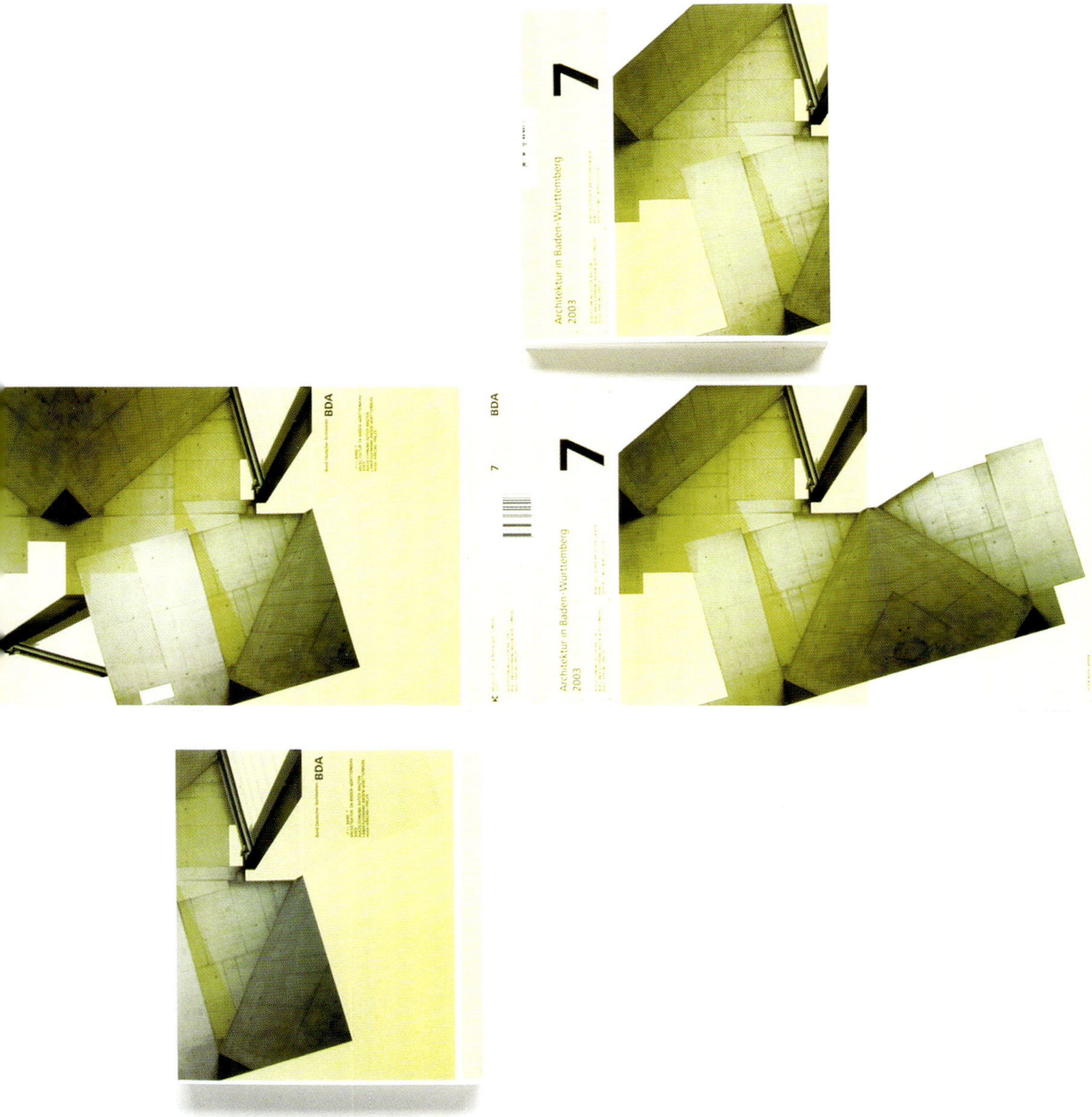

Auftraggeber Client
Bund Deutscher Architekten BDA
Stuttgart, Germany

Produkt Product
Architektur in Baden-Württemberg 2003
Buch
Book

Design
büro diffus GmbH
Stuttgart, Germany

Eine klare, sachliche und informative Dokumentation von 142 Bauten in Baden-Württemberg. Es geht in diesem Buch sowohl um eine prägnante Darstellung der sehr unterschiedlichen Gebäude als auch um eine Neupositionierung des BDA in Baden-Württemberg. Ein strenges Gestaltungsraster, einheitlicher Bildaufbau sowie die klare Typografie machen das Buch zu einem interessanten und höchst informativen Nachschlagewerk. Die Kapitel-Introseiten funktionieren als Leitsystem (Typografie auf Sichtbeton). Ein Farbcode zeigt das Spektrum an Baumaterialien von Beton bis Holz auf. Die Landkarten der BDA-Kreisgruppen Baden-Württemberg bilden eine ständige Legende für den Inhalt des Buches. Das Format ist durch die 6 vorausgegangenen Ausgaben vorgegeben, der Farbcode für den Titel wird den Seriencharakter in den darauf folgenden Ausgaben fortführen.

The book offers clear, objective and informative documentation on 142 buildings in Baden Württemberg. It features a succinct representation of very different buildings and re-positions the BDA German Architects Association in the State of Baden Württemberg. A strict layout grid, uniform image structures and clear typography make the book an interesting and highly informative reference material. The chapter intros function as a guidance system (typography on fair-faced concrete). The spectrum of construction materials, from concrete to wood, is indicated by a color code. BDA local district maps of Baden Württemberg form a constant key to the book's content. The format was defined by the six predecessor books, while the color code for the title will be continued as a series backbone in forthcoming issues.

iF Silver Award

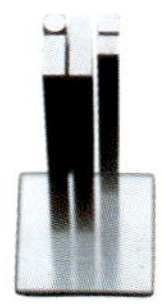

Auftraggeber Client
Metal Design Ltda.
Rio de Janeiro, Brazil

Produkt Product
Tátil Book

Design
Frederico Gelli
Ana Laet
Fernanda Saboia
Tátil Design Ltda.
Rio de Janeiro, Brazil

Das Tátil Book ist eine Präsentation unterschiedlicher Verpackungsmaterialien in höchst ansprechender Form. Die Umverpackung, ein Flechtwerk aus Kautschuk, bildet einen spannungsreichen Gegensatz zum orangefarbenen Katalog-Umschlag. Im Innenteil die Präsentation aller Materialien von Kautschuk bis Wellpappe, die Tátil in 13 Jahren entwickelt hat: optisch reizvoll, sympathisch und produktgerecht, vor allem typisch für die Zielgruppe kreativer Gestalter in Südamerika.

The Tátil Book presenting various different packaging materials in a very impressive form. The surrounding package, woven from rubber, forms an exciting contrast to the orange-colored catalogue jacket. Inside are all the materials which Tátil has developed over a full 13 years - from rubber to corrugated cardboard. It is visually very attractive, congenial and does justice to the products - and above all it is typical for the target group in question, namely creative designers in South America.

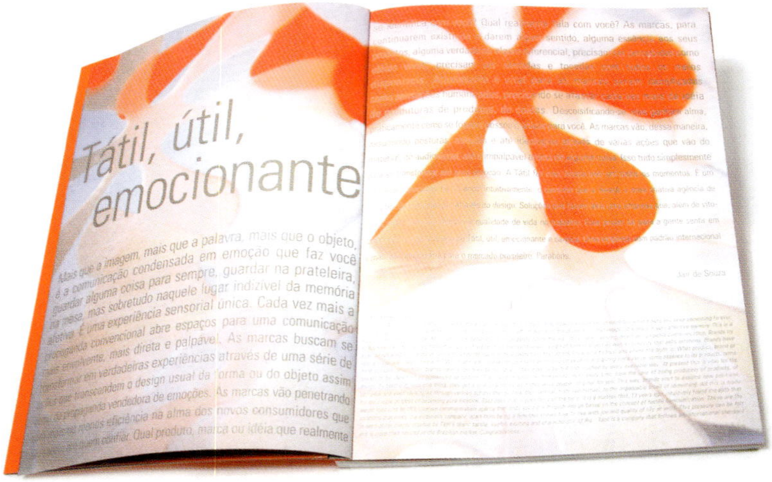

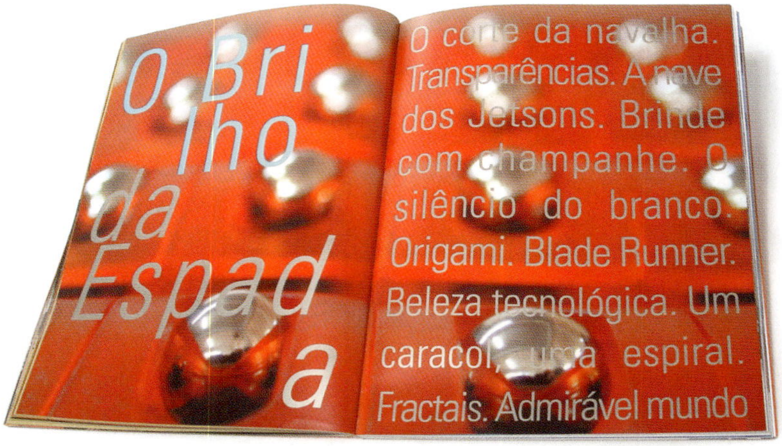

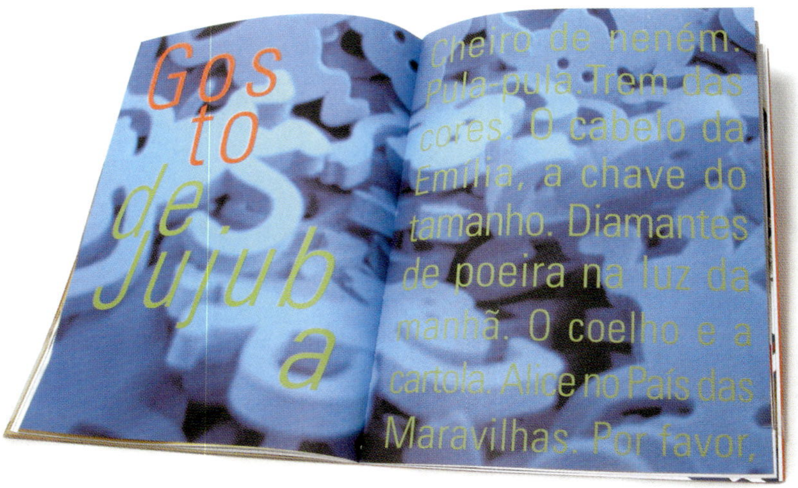

iF Silver Award

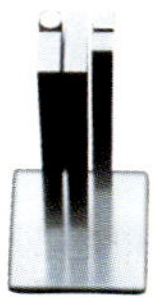

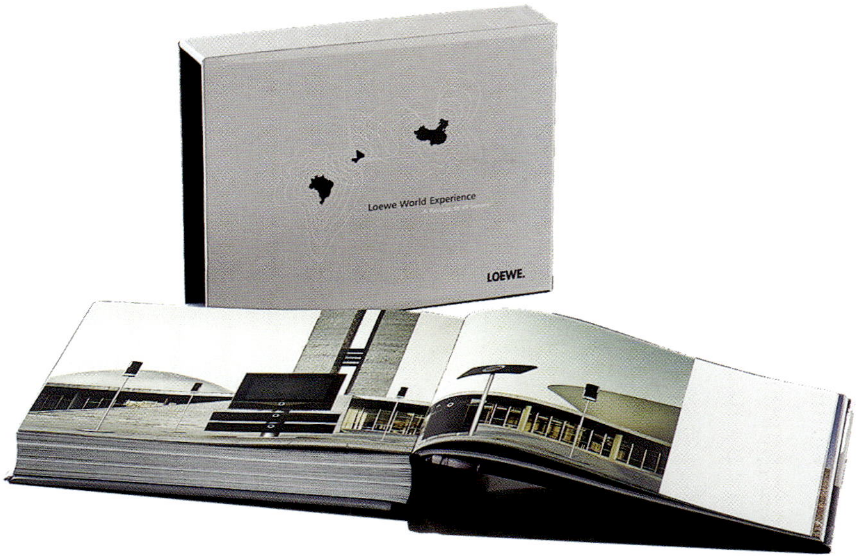

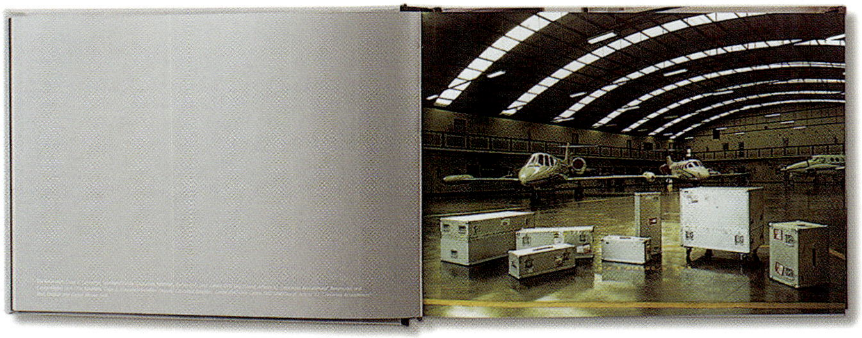

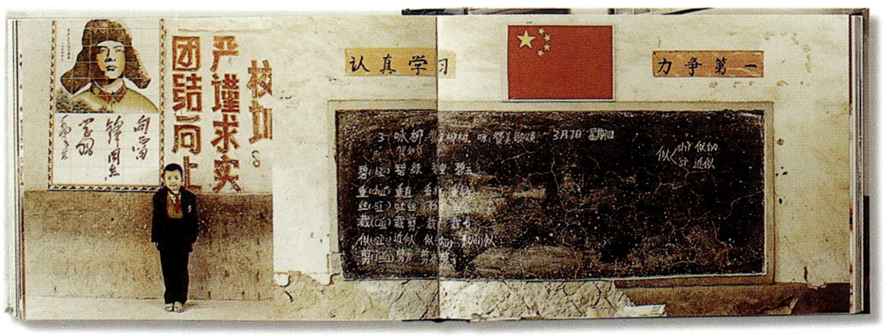

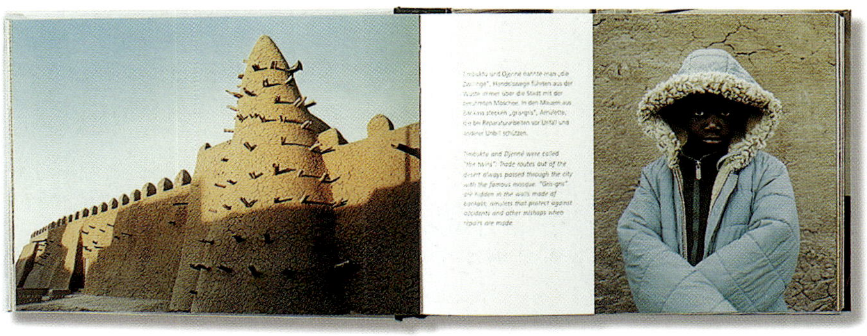

Auftraggeber Client
Loewe AG
Kronach, Germany

Produkt Product
Loewe World Experience
Buch und DVD
Book and DVD

Design
KW43, brandbuilding and design
Düsseldorf

Loewe World Experience ist der Name der Einführungskampagne für Systems, einer Premium-Home-Cinema-Anlage. Das Reisetagebuch ist die Zusammenfassung einer audio-visuellen Dokumentation eines kulturellen Staffellaufs durch drei Kontinente. Verfolgen konnte man die Reise durch Mali, China und Brasilien und die entsprechenden musikalischen Eindrücke erst im Internet, dann bei der Demonstration in Form einer einstündigen DVD im Dolby Surround Sound und schließlich in Form des Reisetagebuchs.

Loewe World Experience is the name of the launch campaign for Systems, the company's premium home-cinema complex. The travelogue is a summarized audio-visual documentation of a cultural relay race across three different continents. Initially, the trip was on the Web, where it could be followed across Mali, China and Brazil along with the corresponding musical impressions – after which, it came out as a demo in the form of a one-hour DVD complete with Dolby Surround Sound and then, finally, in the form of a written travelogue.

iF Silver Award

Auftraggeber Client
COR Sitzmöbel
Helmut Lübke GmbH & Co. KG
Rheda-Wiedenbrück, Germany

Produkt Product
COR Corporate Design

Design
Markus Blankenburg
Jan Kruse
Olaf Stein
Factor Design AG
Hamburg, Germany

Möbel von COR bauen auf ein einfaches Prinzip: Nur das Beste ist gut genug. Dem folgt das Corporate Design und positioniert COR zwischen Moderne und Zukunft. Die Massnahmen umfassten eine Marken- und Positionierungsanalyse, die Überarbeitung des Logos, Grundlagen des Erscheinungsbildes, Hausschriften, Farbsystem, Illustrationsstil, Unternehmensausstattung, die gesamte Literatur, Internetauftritt, Messe- und Shopausstattungen und die internationale Anzeigenkampagne.

COR furniture relies on a simple principle: only the best is good enough. The company's corporate design obeys the same credo and positions COR between Modernism and the future. The measures involved analyzing the brand and positioning, revising the logo, laying the foundations for a corporate identity, for corporate fonts, color codes, illustrations styles, company interior design, all the printed matter it produces, the Internet presence, trade-fair and shop interiors, and the international ad campaign.

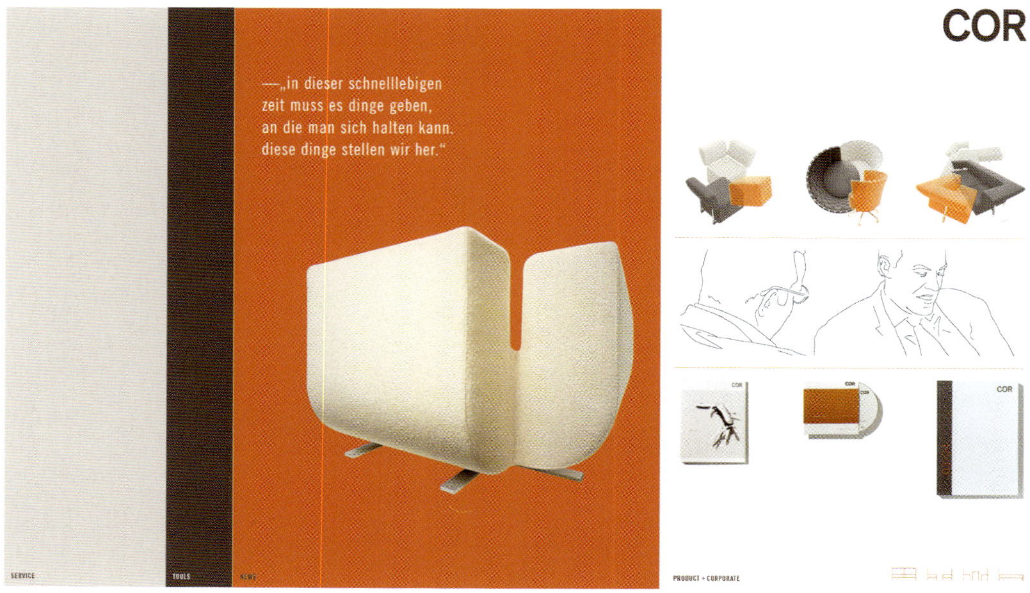

iF Silver Award

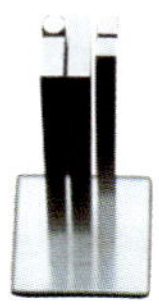

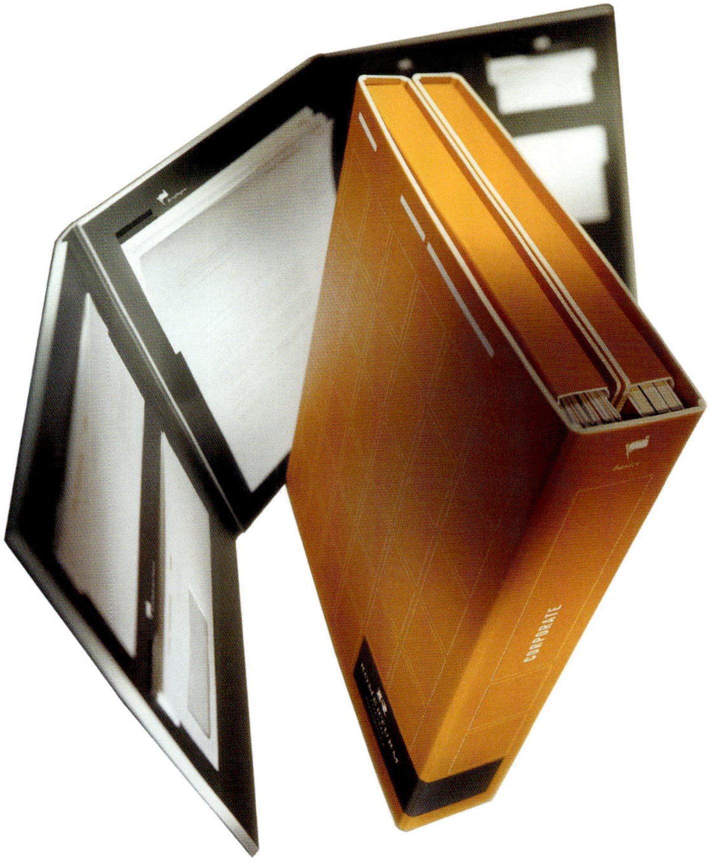

Auftraggeber Client
RÖMERTURM Feinstpapier
Frechen, Germany

Produkt Product
RÖMERTURM Corporate Basics

Design
Verena Baumhögger
Daniel Sorge
Olaf Stein
Factor Design AG
Hamburg, Germany

Corporate Basics ist eine fiktive Firma, deren Name stellvertretend für alle Firmen steht, die ihre Corporate Communications überzeugend gestalten wollen. Das Basics-Paket ist ein neuartiges Planungstool, das Gestaltern und Kommunikationsprofis einen vollständigen Überblick über die gesamte Bandbreite typischer Print-Kommunikation bietet. Die Mappe enthält zahlreiche Dummies von Briefbögen, -hüllen, Visitenkarten, Grußkarten, Adressaufklebern und Broschüren, anhand derer sich die Anmutung und Wirkung unterschiedlicher Papiere am Objekt selbst veranschaulichen lässt.

Corporate Basics is a fictitious company whose name substitutes for those of all other firms who wish to find a compelling face for their corporate communications. The Basics package is a new type of planning instrument, offering designers and communications professionals a complete overview of the entire spectrum of typical print communications. The portfolio includes countless dummies of letterheads and envelopes, business cards, greetings cards, address labels and brochures. They aptly offer a tangible visualization of the possible feel and effect of different types of paper.

Auftraggeber Client
RÖMERTURM Feinstpapier
Frechen, Germany

Produkt Product
RÖMERTURM Portfolio Las Vegas

Design
Verena Baumhögger
Olaf Stein
Factor Design AG
Hamburg, Germany

Basis für das achte Römerturm Portfolio ist ein Aufenthalt des Fotografen Philip Koschel in Las Vegas und seine Einblicke hinter die Kulissen. Der Text stammt von Anna Zuber, das Konzept und die Gestaltung von Factor Design.

The eight in the series of Römerturm portfolios is based on a trip photographer Philip Koschel made to Las Vegas and his views behind the scenes there. The text stems from the pen of Anna Zuber, and the concept and design were handled by Factor Design.

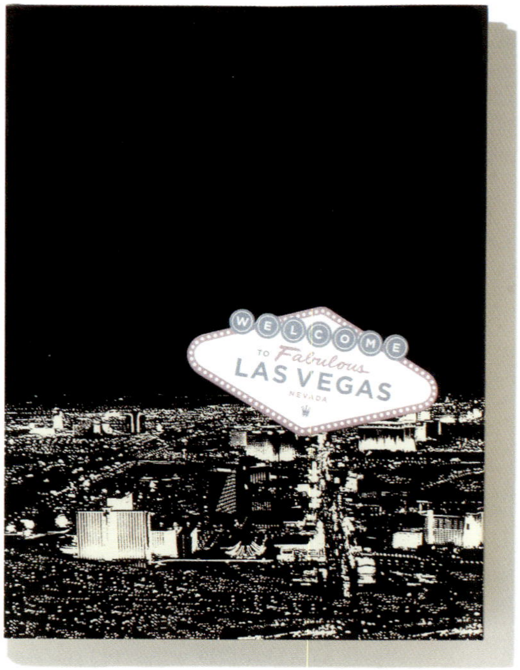

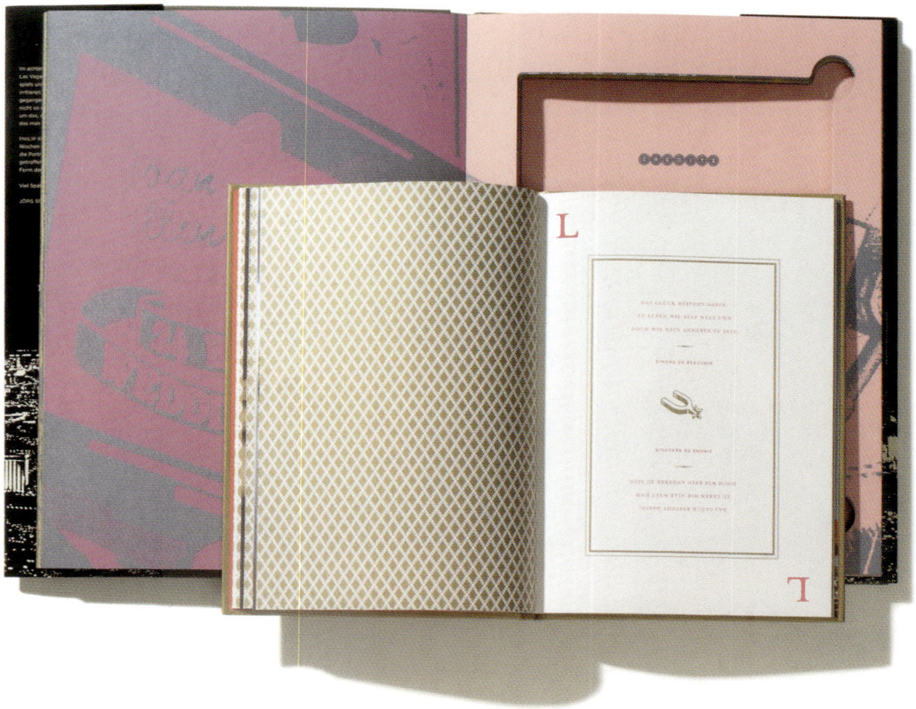

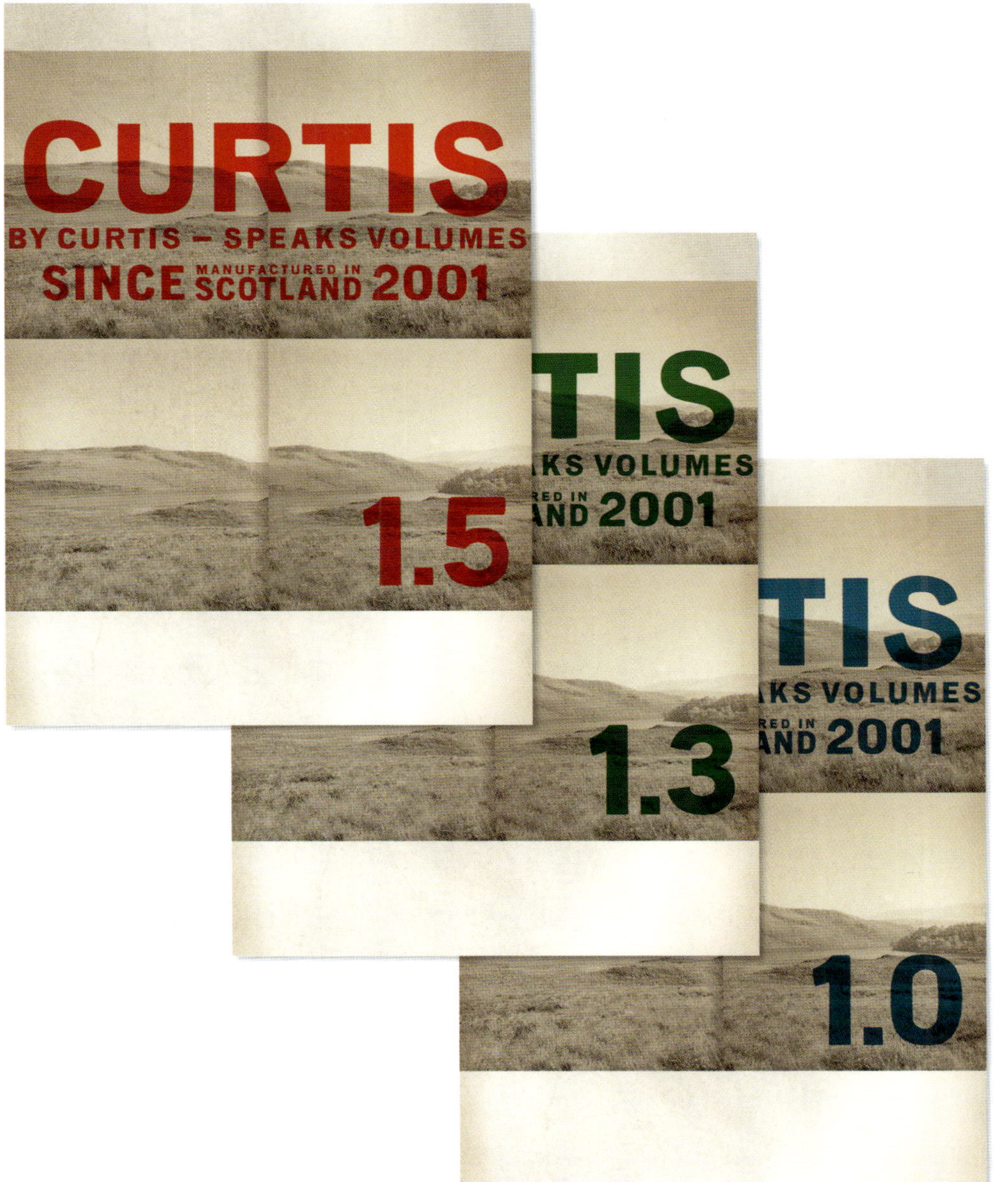

Auftraggeber Client
Curtis Fine Paper Ltd.
Guardbridge, Scotland

Produkt Product
Curtis by Curtis
Demonstrationsposter
Sample poster

Design
Tanja Maus
Daniel Sorge
Olaf Stein
Factor Design AG
Hamburg, Germany

Das von der schottischen Papiermühle Curtis Fine Papers in Guardbridge, St. Andrews, stammende Papier Curtis by Curtis hat ein besonders griffiges Volumen, das Bücher, Broschüren oder Kataloge auch mit wenigen Seiten bereits sehr wertig wirken lässt. Trotz seines starken Volumens wiegt das Papier nicht einmal ein Gramm mehr als normales Papier. Mit drei verschiedenen Oberflächen, fünf Grammaturen (von 115 bis 360 g) und seiner mattweißen, typofreundlichen Farbigkeit lässt es sich vielseitig als Text-and-Cover-Papier einsetzen. Die drei Poster demonstrieren die unterschiedlichen Wirkungen der Volumina des Materials (1.0, 1.3 und 1.5). Auf einer Microsite www.curtisbycurtis.com wird das Produkt ausführlich präsentiert.

The "Curtis by Curtis" paper manufactured by Scottish paper mill Curtis Fine Papers in Guardbridge, St. Andrews, has an especially textured volume, imbuing books, brochures or catalogs that even have only a few pages with an especially high-quality feel. Despite its strong volume, the paper weighs less than a gram more than normal paper. Available with three different surfaces, five different weights (ranging from 115 g to 360 g) and a satin white, typo-friendly color tone, it can be used for many applications as text-and-cover paper. The three posters demonstrate the different effects of the paper's volume (1.0, 1.3 and 1.5). The special www.curtisbycurtis.com micro-site offers a detailed portrait of the product.

Auftraggeber Client
Wand 5 e.V.
Stuttgart, Germany

Produkt Product
Media-Space 03
Kommunikationsmedien zur Veranstaltung
Media-Space 03
Communication Media promoting
the event Media-Space 03

Design
büro diffus GmbH
Stuttgart, Germany

Das Leitmotiv für die Entwicklung des Corporate Designs und die Gestaltung der Werbemittel für die Veranstaltung Media-Space 03 ist eine Collage aus mehreren einzelnen Bildelementen, die einen Eindruck aus verschiedenen Perspektiven synthetisieren. Das Thema Perforation zieht sich im doppelten Sinne durch die Gestaltung. Es werden immer zwei Plakate mit Vorder- und Rückseite kombiniert nebeneinander gehängt. Durch Abreißen entlang einer Perforation kann das Plakat in jeweils 14 Flyer gesplittet werden, die zwar die gleichen Informationen enthalten, sich aber visuell unterscheiden. Die Idee spielt mit den jeweiligen Raumverhältnissen und wird durch die Mitnahme von Flyern stetig verändert. Diese Gestaltungsidee wurde auch auf den Katalog übertragen. Die Perforationen erlauben es, den Katalog in drei nach den Veranstaltungstagen getrennte Einzelkataloge zu teilen. Außerdem lässt sich jede Veranstaltung als Flyer einzeln entnehmen.

The leitmotif when developing the corporate design and advertising design for the Media-Space 03 event was a collage composed of a number of picture elements which together synthesize to give the impression of various different perspectives. The theme perforation permeates the design in several ways. Two double-sided posters always hang together. By tearing it along a dotted and perforated line each of the posters sub-divides into 14 flyers which contain the same information yet differ visually. The idea plays on the respective spatial relationships and is in constant flux as flyers are taken away. This design idea has been transposed onto the catalogue, where perforations enable the catalogue to be divided into three individual catalogues each corresponding to a day of the event. Each event can also be extracted from the catalogue in the form of a flyer.

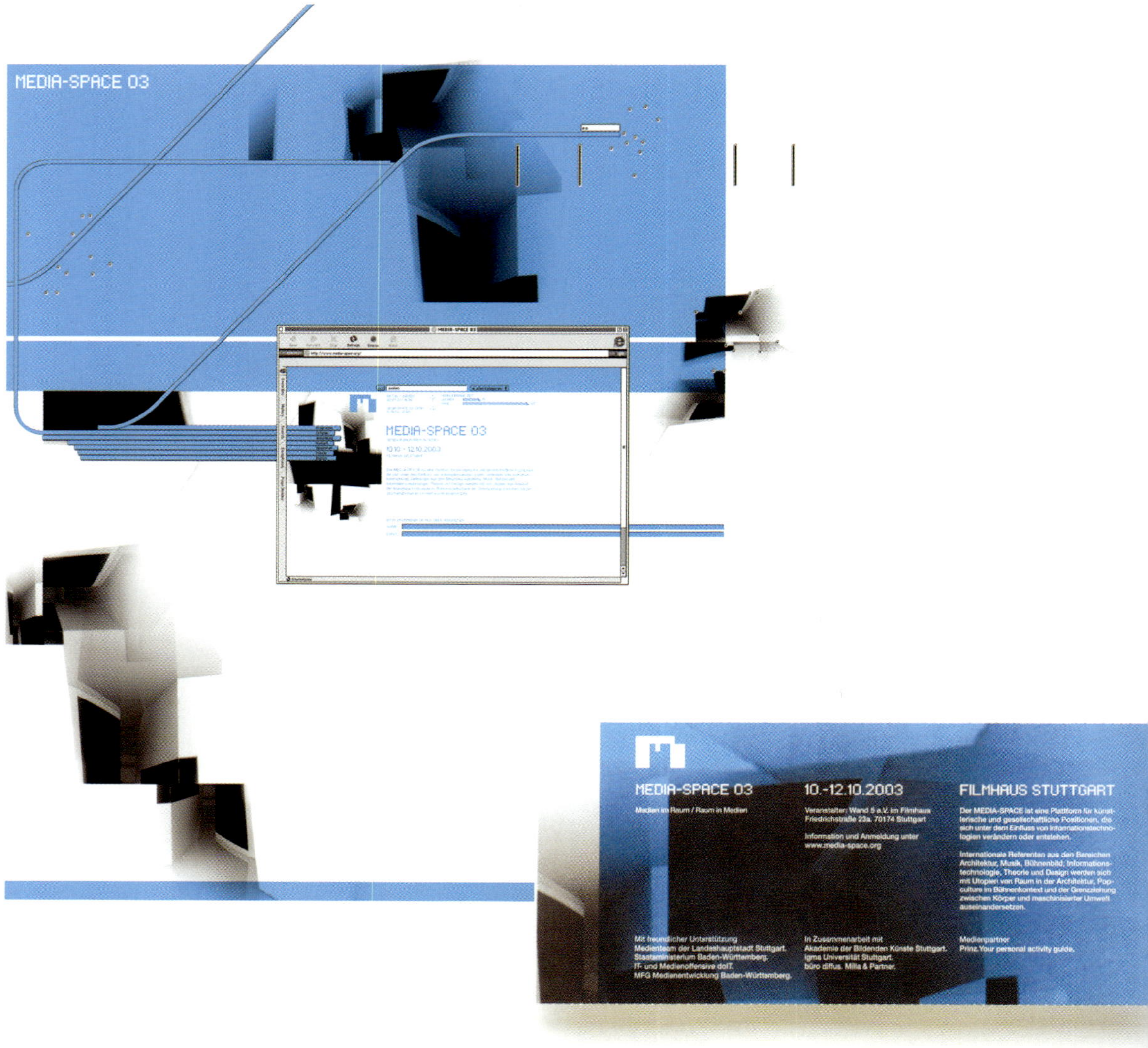

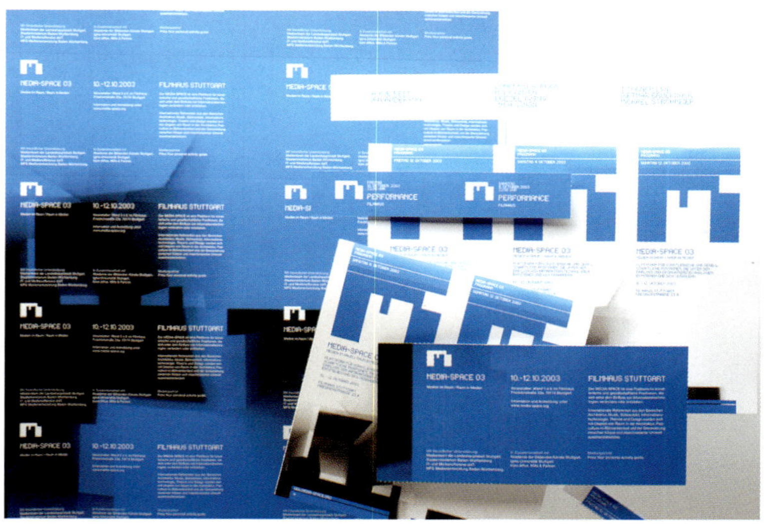

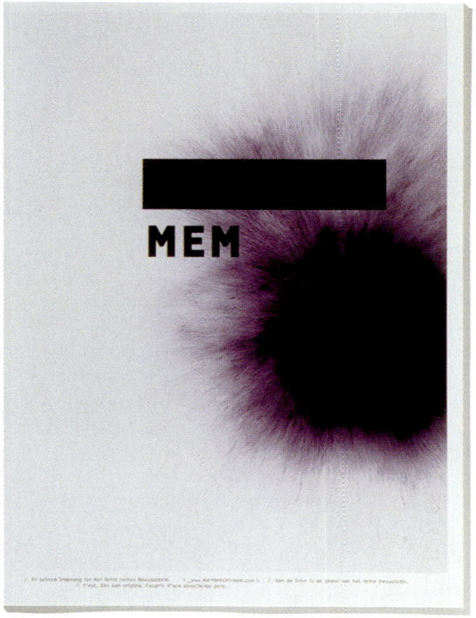

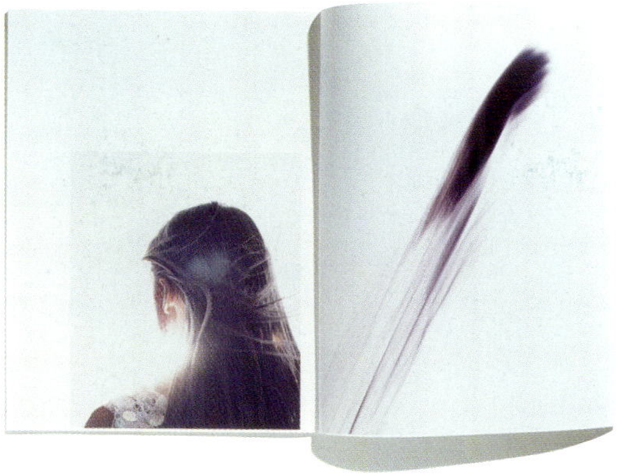

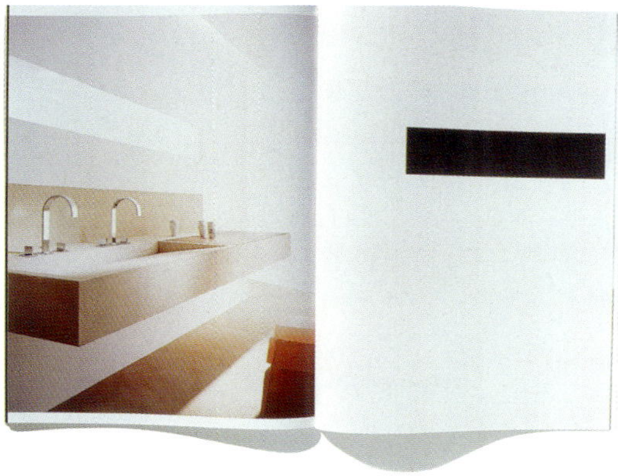

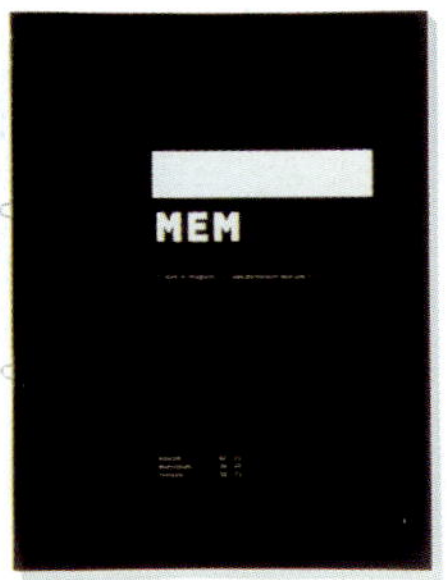

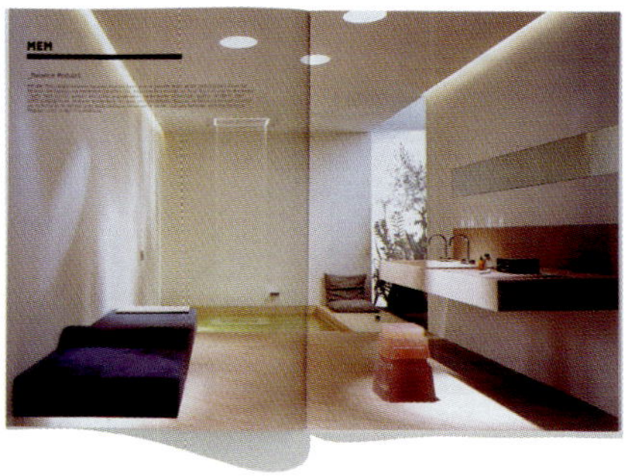

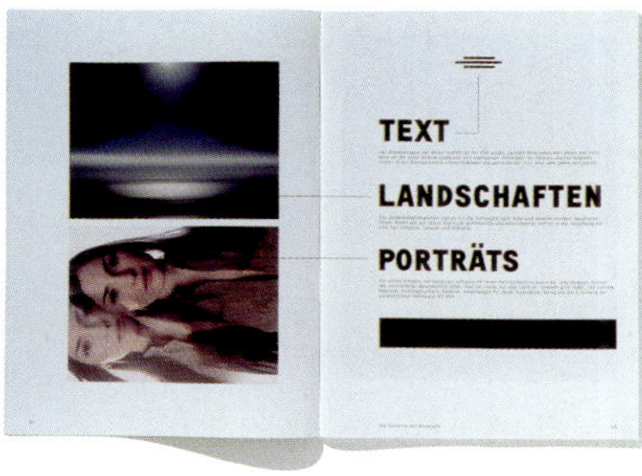

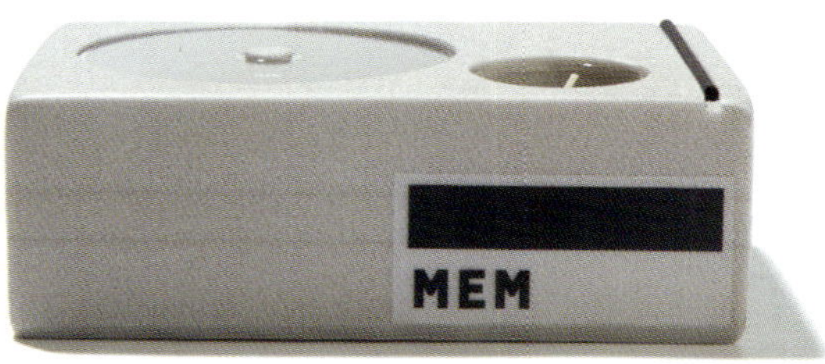

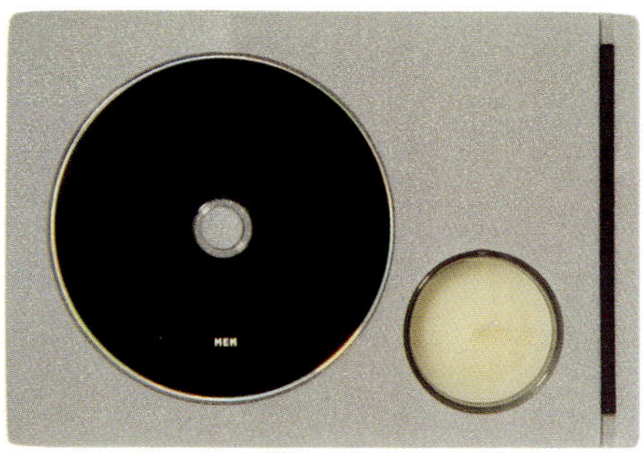

Auftraggeber Client
Aloys F. Dornbracht GmbH & Co. KG
Armaturenfabrik, Iserlohn, Germany

Produkt Product
MEM
Imagebroschüre, Sales manual, Press Kit
Image Folder, Sales manual, Press Kit

Design
Mike Meiré CD
Katja Fössel CD
Kerstin-Anna Hielscher Photo Direction
Alexandra Grossmann AD Press Kit
Tanja Wintels RZ
Meiré und Meiré AG
Köln, Germany

Die Imagebroschüre, das Sales Manual und Press Kit für die Armaturenfabrik Aloys F. Dornbracht, reflektieren formal wie inhaltlich die Welt des neuen Bad-Architektur-Konzepts MEM. Bezugnehmend auf die memetische Philosophie, in deren Mittelpunkt das Erspüren unserer Selbst steht, setzt die Imagebroschüre auf das atmosphärische, wortlose Zusammenspiel assoziativer Tagtraum-Bilder und zarter Porträts in Mehrfachbelichtung. Das Sales Manual erläutert in präziser Wort- und Bildsprache den konzeptionellen Gedanken, die Gestaltungsprinzipien der Broschüre sowie die Produktwelt von MEM. Das Press Kit nimmt die Idee mit einem Materialmix aus Räucherstäbchen, CD mit memetischem Soundtrack und einem 3D-Film, Duftkerze und einem Block aus schwarzer Steckmasse auf.

The image brochure, the sales manual and the press kit for bathroom fixtures manufacturer Aloys F. Dornbracht reflects the world of the new MEM bathroom architecture concept in terms of both its form and contents. Taking its cue from the memetic philosophy, which hinges on gaining a feel for our selves, the image brochure concentrates on offering an atmospheric, no-copy associative series of day-dream-like images and soft portrait shots based on multiple exposures. The sales manual provides precise texts and images to explain the concepts and design principles underlying the brochure as well as the MEM product world. The press kit takes up this idea – with a materials mix including joss sticks, a CD with a memetic soundtrack and a 3D film, scented candle and a block of black putty to hold the sticks.

Auftraggeber Client
European
Aeronautic Defence & Space Company
Ottobrunn, Germany

Produkt Product
EADS
Anzeigenkampagne 2003
Ad Campaign 2003

Design
Lowe Lintas & Partners
Hamburg, Germany

Ziel der Printkampagne 2003 von EADS ist die Etablierung des Unternehmens mit einer klaren und unverwechselbaren Identität. EADS stellt die Dachmarke für bekannte Marken wie Airbus, Eurocopter, Eurofighter, MBDA dar und versteht es, Synergien zu nutzen und somit dieses Gefüge sinnvoll miteinander zu vereinen. Die Kampagne positioniert EADS als ein starkes, führendes, selbstbewusstes Unternehmen mit weltweiter Präsenz. Mit dem Einsatz warmer Farben und ansprechender Texte soll die Marke EADS menschlicher und weniger distanziert dargestellt werden. Trotzdem ist die Herausstellung des High-Tech-Charakters wichtig, für den die EADS steht.

The goal behind the EADS 2003 print campaign was to establish the corporation with a clear, unmistakable identity of its own. EADS is the umbrella label for well-known brands like Airbus, Eurocopter, Eurofighter, MBDA – and is a prime platform for exploiting synergies and thus meaningfully linking these units to form a whole. The campaign positions EADS as a strong, leading and self-confident corporation with a worldwide presence. Thanks to the use of warm colors and appealing copy, the idea is to present the EADS brands with a human face and less as an abstract entity. That said, the campaign still emphasizes the important high-tech side to EADS.

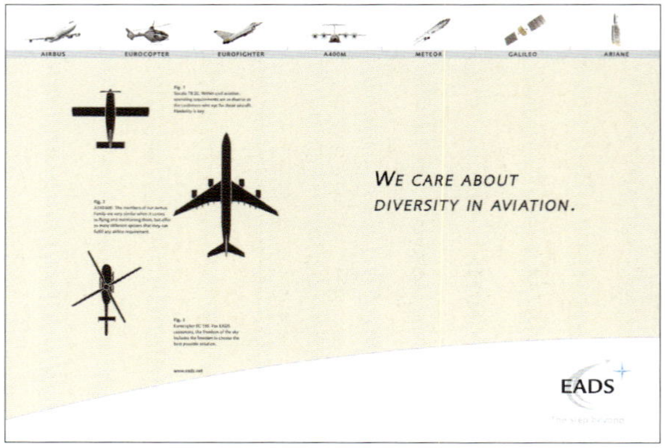

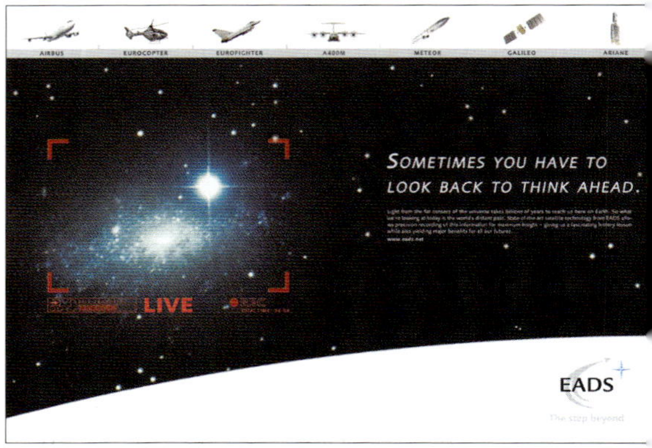

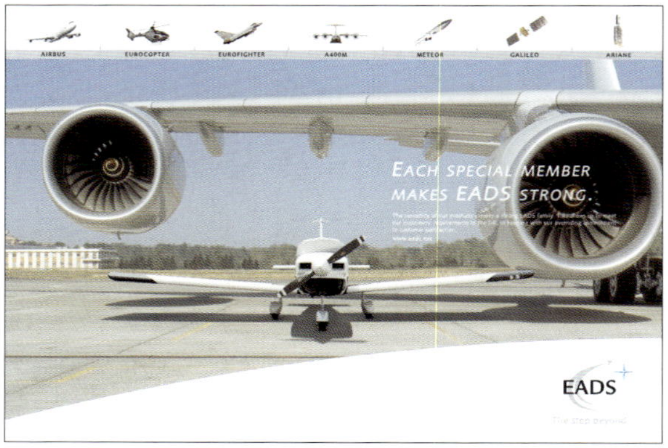

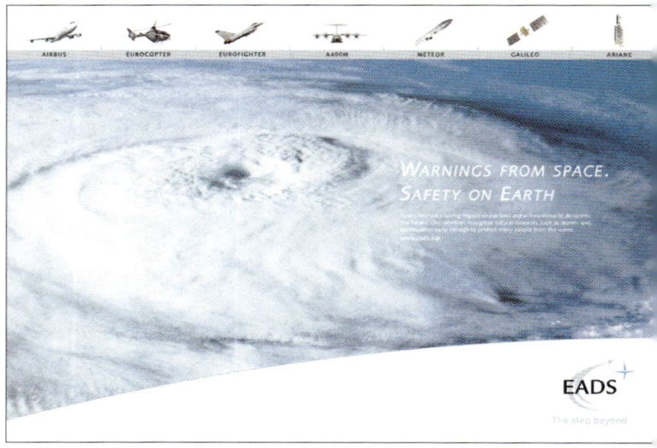

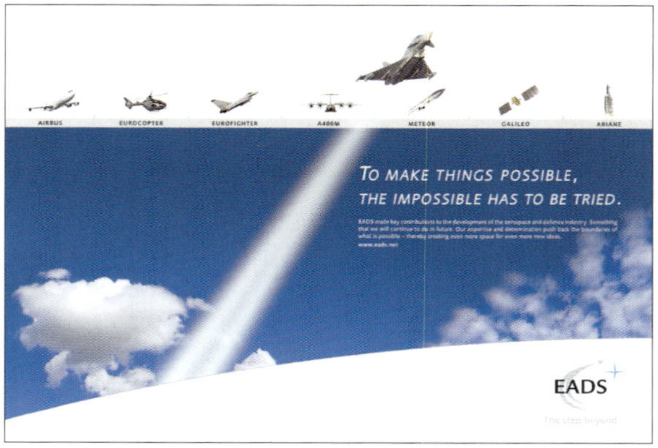

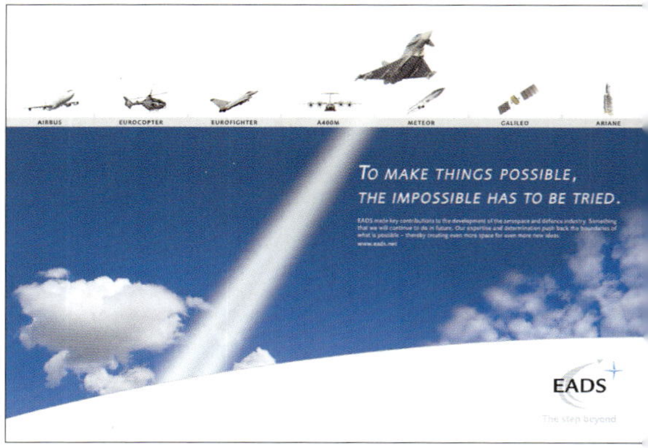

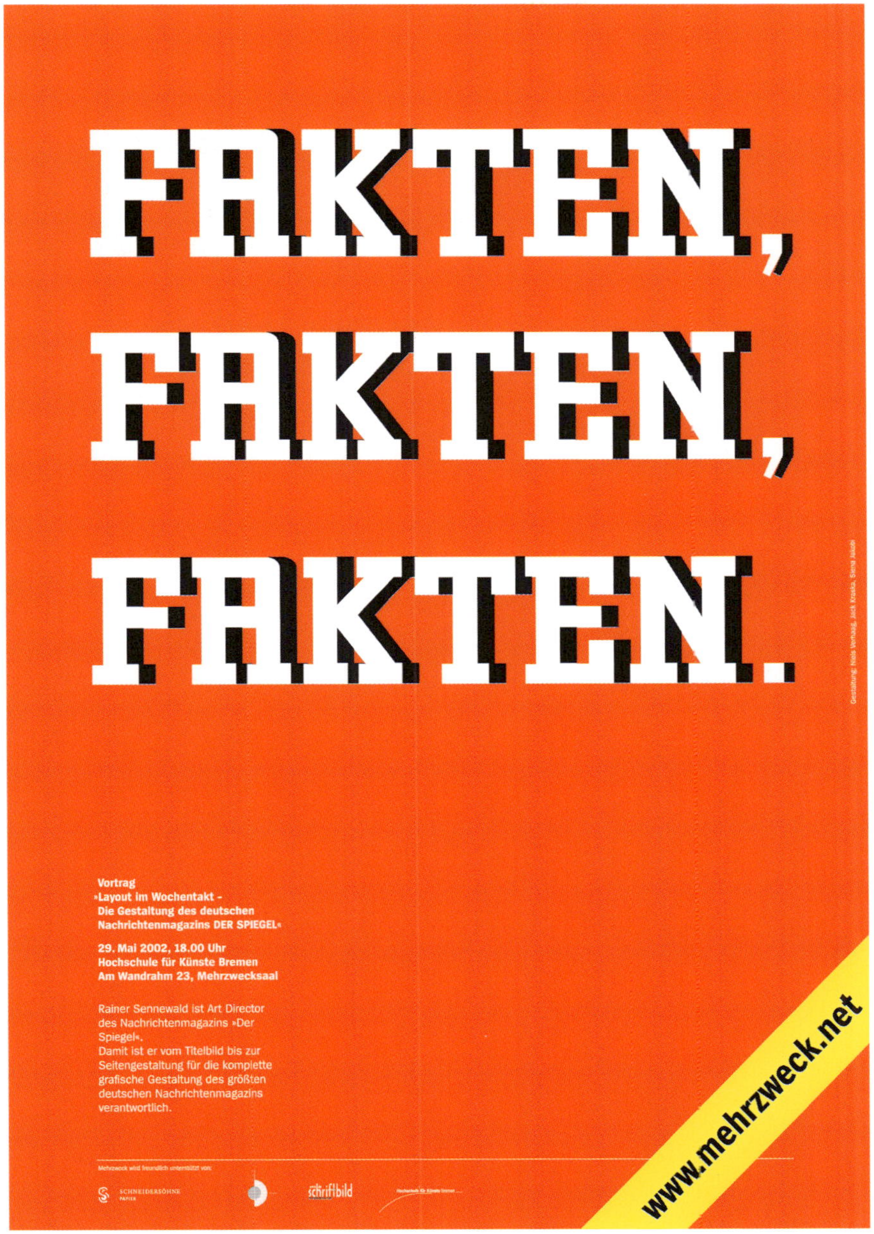

Auftraggeber Client
Hochschule für Künste Bremen
Bremen, Germany

Produkt Product
Fakten, Fakten, Fakten
Plakat
Poster

Design
Siena Jakobi
Jack Kraska
Niels Verhaag
Hochschule für Künste Bremen
Bremen, Germany

Mehrzweck ist eine studentisch organisierte Vortragsreihe an der Hochschule für Künste Bremen, die sich zum Ziel gesetzt hat, hochkarätige Gestalter einzuladen, um einen Einblick in deren Arbeiten zu bekommen. Das Plakat „Fakten, Fakten, Fakten" kündigt den Vortrag des Art-Directors des Nachrichtenmagazins „Der Spiegel" an.

Mehrzweck (Multipurpose) is a series of lectures organized by students at the Bremen Academy of Art. The object of the series is to invite expert designers as speakers destined to provide an insight into their respective oeuvre. The poster entitled "Fakten, Fakten, Fakten" (Facts, facts, facts) announces the lecture held by the Art Director of the "Der Spiegel" news magazine.

Auftraggeber Client
Hochschule für Künste Bremen
Bremen, Germany

Produkt Product
Spiekermann/Rambow
Faltplakat
Folded poster

Design
Siena Jakobi
Jack Kraska
Niels Verhaag
Hochschule für Künste Bremen
Bremen, Germany

Mehrzweck ist eine studentisch organisierte Vortragsreihe an der Hochschule für Künste Bremen, die sich zum Ziel gesetzt hat, hochkarätige Gestalter einzuladen, um einen Einblick in deren Arbeiten zu bekommen. Das Faltplakat „Erik Spiekermann / Gunter Rambow" kündigt die Vorträge der beiden Gestalter an.

Mehrzweck (Multipurpose) is a series of lectures organized by students at the Bremen Academy of Art. The object of the series is to invite expert designers as speakers destined to provide an insight into their respective oeuvre. The folded poster "Erik Spiekermann / Gunter Rambow" announces the lectures held by the lecturers.

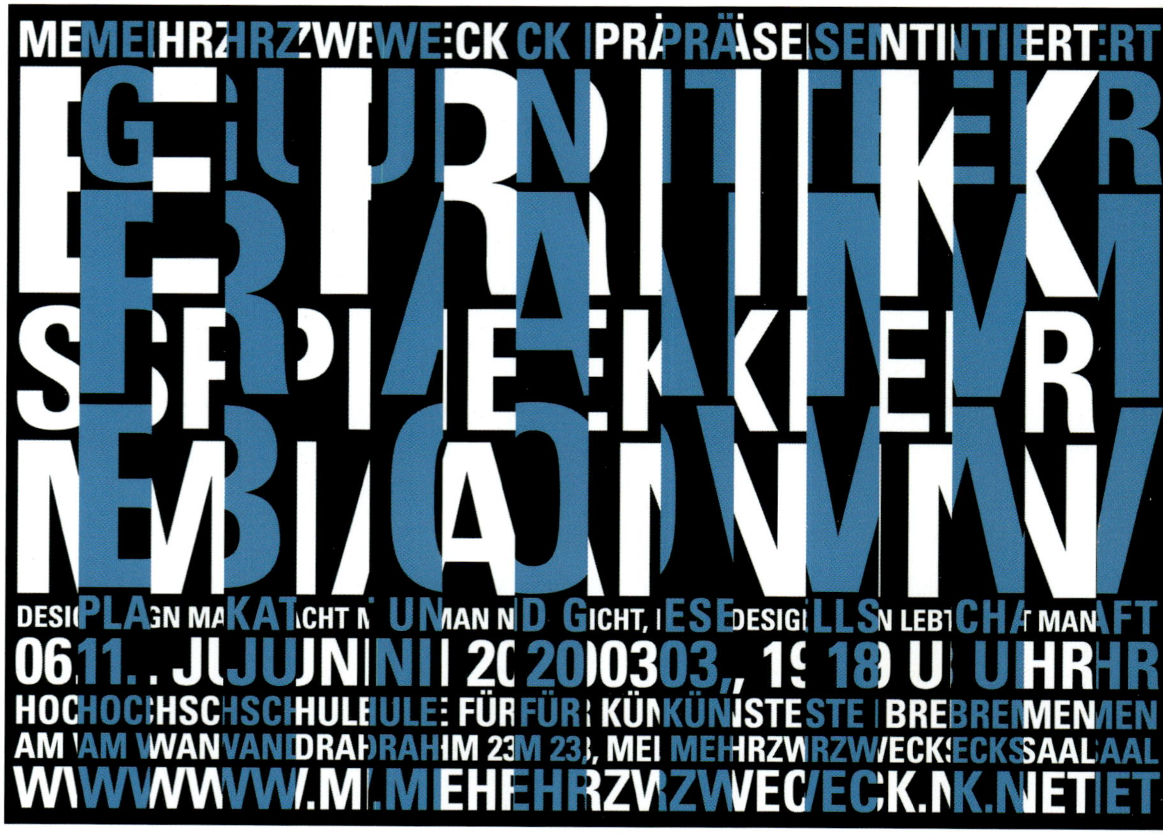

Auftraggeber Client
Hochschule für Künste Bremen
Bremen, Germany

Produkt Product
Baumann & Baumann / Total Identity
Wendeplakat
Reversible Poster

Design
Siena Jakobi
Jack Kraska
Niels Verhaag
Hochschule für Künste Bremen
Bremen, Germany

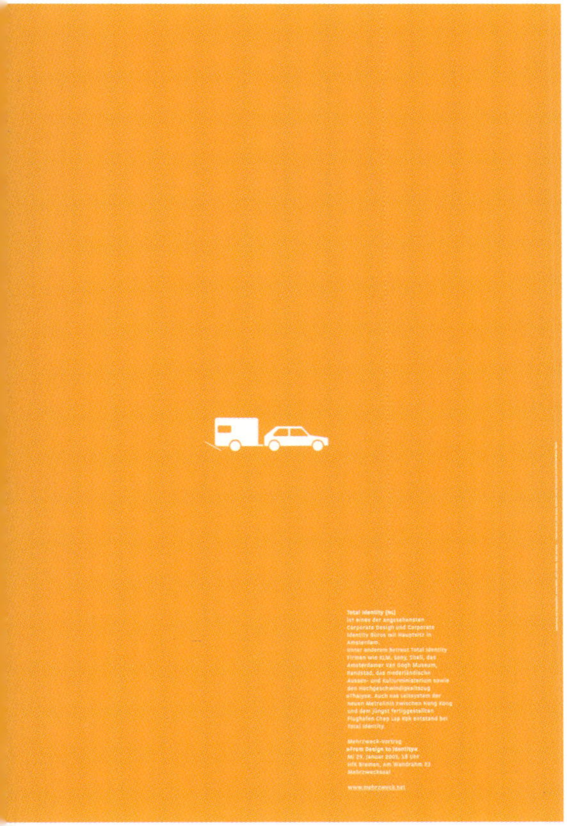

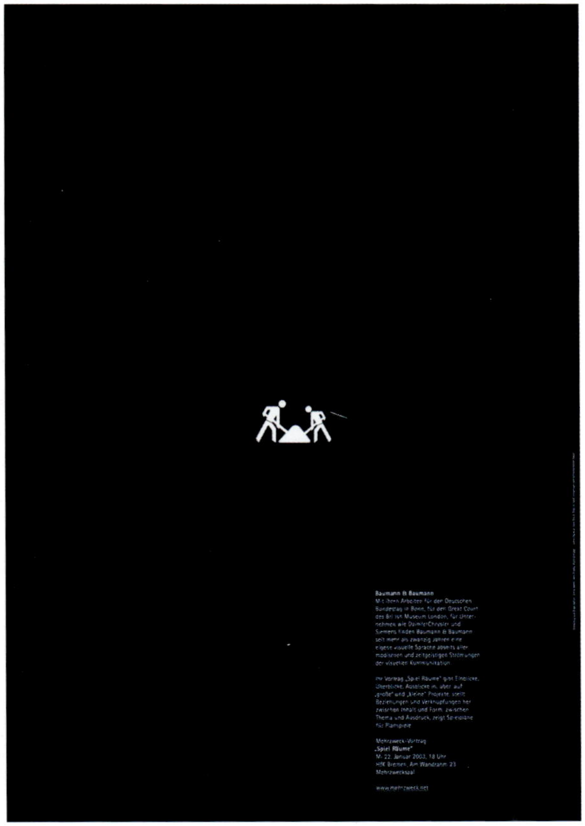

Mehrzweck ist eine studentisch organisierte Vortragsreihe an der Hochschule für Künste Bremen, die sich zum Ziel gesetzt hat, hochkarätige Gestalter einzuladen, um einen Einblick in deren Arbeiten zu bekommen. Das Wendeplakat „Baumann & Baumann / Total Identity" kündigt die Vorträge der beiden Gestaltungsbüros an.

Mehrzweck (Multipurpose) is a series of lectures organized by students at the Bremen Academy of Art. The object of the series is to invite expert designers as speakers destined to provide an insight into their respective oeuvre. The double-sided poster "Baumann & Baumann / Total Identity" announces the lectures held by the two architectural/design agencies.

Auftraggeber Client
Mustang Jeans GmhH
Künzelsau, Germany

Produkt Product
Mustang Imagebuch
Imagebuch
Image book

Design
Alexander Rötterink
Springer & Jakoby
Design GmbH & Co. KG
Hamburg, Germany

Anlässlich des fünfzigsten Jubiläums der Damen-Jeans (eine Mustang Erfindung) wurde dieses Markenbuch zur Jeansmesse 2003 produziert. Gezeigt wird neben der Familienhistorie Sefranek die Geschichte der deutschen Jeans und somit die Geschichte der Marke Mustang. Auf 164 Seiten wird im Collagen-Stil das Lebenswerk der Familie Sefranek dargestellt, die in den letzten 80 Jahren die Jeans nach Deutschland gebracht hat und in diesem Markt stets Vorreiter geblieben ist. Der aufwändig produzierte Einband ist mit einer Prägung versehen, der Buchtitel wurde eingestickt. Gedruckt wurde auf hochwertiges, ungestrichenes Papier, die Typografie kombiniert eine Antiqua-Schrift mit handgestempelten Headlines. Trotz der ausschließlichen Verwendung von Archivmaterial ist das Bildlayout sehr modern.

This brand book was produced for the Jeans Fair 2003 on the occasion of the 50th anniversary of the Ladies Jeans (a Mustang invention). Alongside the story of the Sefranek family, the history of jeans in Germany is told and with it that of the Mustang brand. The book sports 164 pages designed with a collage in mind, presenting the life's work of the Sefranek family, who over the last 80 years not only introduced jeans to Germany but were always the trailblazers in this market. The elaborately produced hardcover has an embossed cover, and the book title is actually stitched into it. It is printed on high-quality, coarse-fiber paper, and the typography combines an Antiqua typeface with hand-stamped headlines. Despite the exclusive reliance on archive material, the layout chosen for the images is decidedly modern.

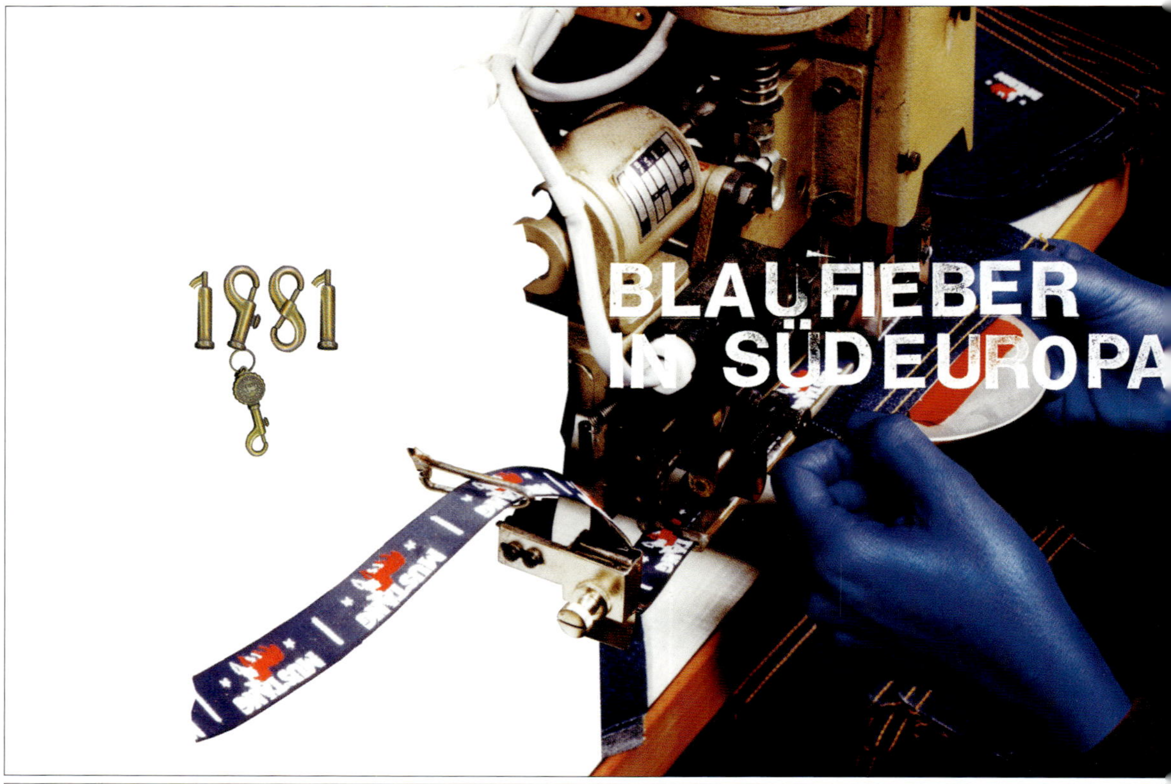

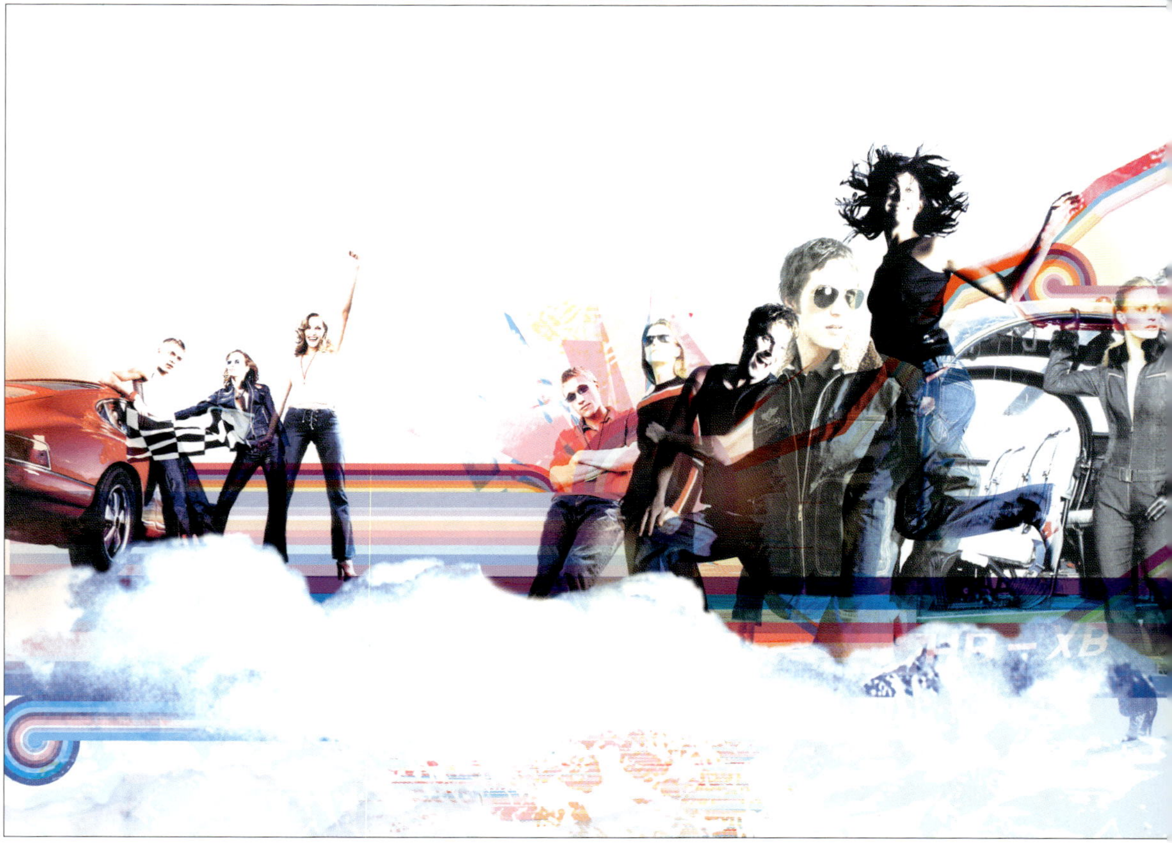

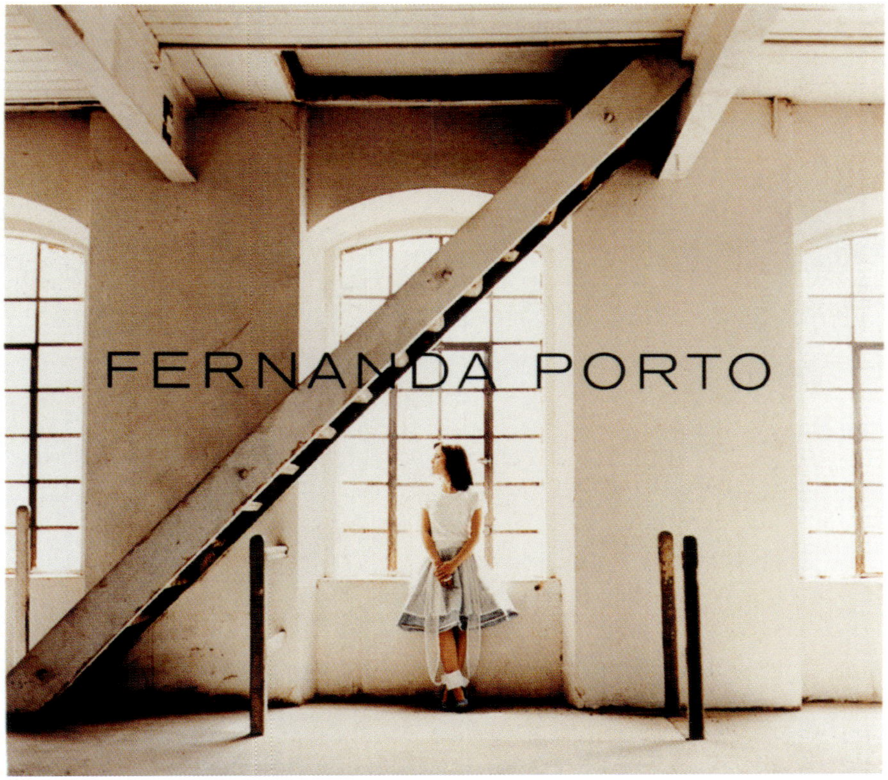

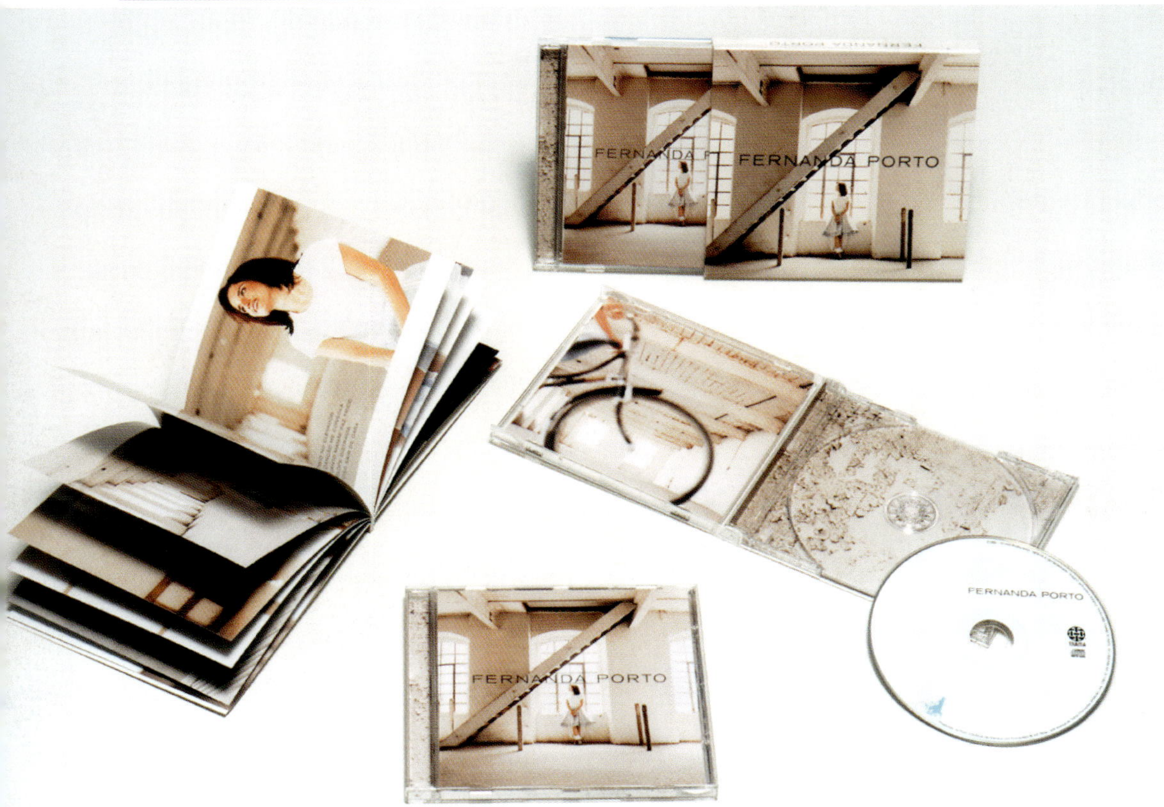

Auftraggeber Client
Trama Promoções Artísticas Ltda.
São Paulo, Brazil

Produkt Product
Fernanda Porto Spezialedition mit Fotobuch
Fernanda Porto special edition with photo book
CD

Design
Simone Drago Kail
Trama Promoções Artísticas Ltda.
São Paulo, Brazil

Die umfassende Spezialedition der brasilianischen Musikerin Fernanda Porto enthält eine CD und ein dazugehöriges Fotobuch. Mit der Edition möchte die Künstlerin ihre neuesten Produktionen bekannt machen, mit dem Fotobuch wird ihre Musik atmosphärisch ins Bild umgesetzt.

The special edition set for Brazilian musician Fernanda Porto is made up of a CD and the relevant photo-book. She wishes by means of the edition to increase awareness of her latest recordings and the photos aptly convey the atmosphere of the artist's music.

Auftraggeber Client
Trama Promoções Artísticas Ltda.
São Paulo, Brazil

Produkt Product
São Paulo Fashion Week CD Vol. 05
CD und Promotions Material

Design
Alexandre Suannes
Trama Promoções Artísticas Ltda.
São Paulo, Brazil

Zur São Paulo Fashion Week wurden eine begleitende CD und Promotion-Material entwickelt. Die CD enthält eine Vertonung des Ereignisses und dient als Subprodukt der Marke SPFW. Die Gestaltung des Titels bezieht sich mit N 5 auf das Veranstaltungskonzept und das nunmehr fünfte von Trama für dieses Ereignis gestaltete Album.

In conjunction with the São Paulo Fashion Week a CD and promotional material were also developed. The CD contains a sound recording of the event and serves as a sub-product for the brand SPFW. The N 5 in the design of the title refers to the event concept and the fact that it is the fifth SPFW album by Trama.

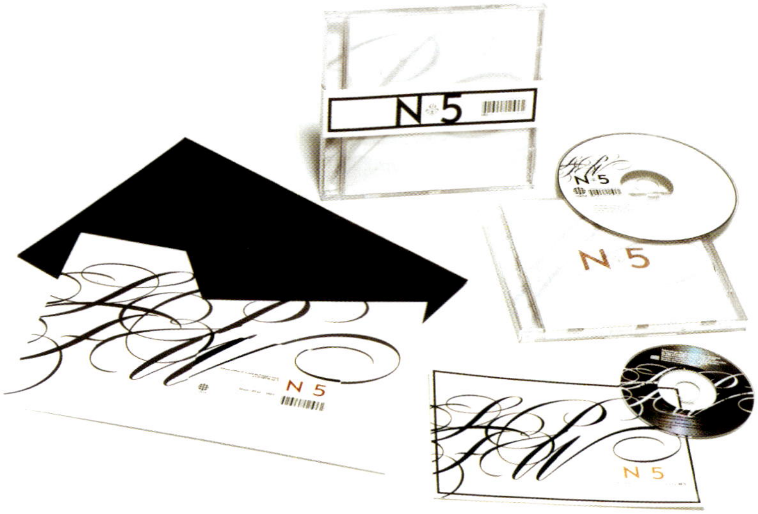

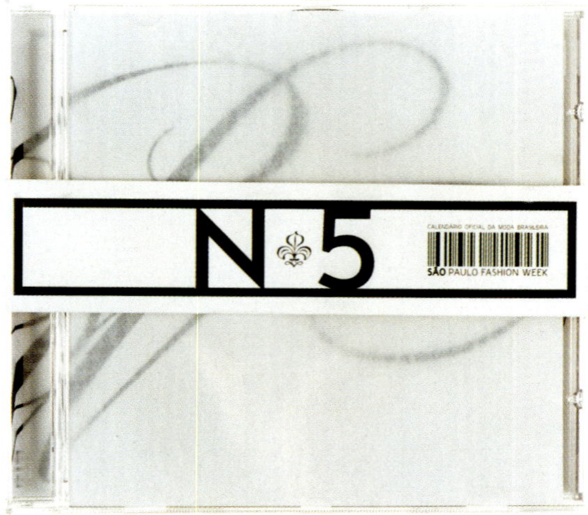

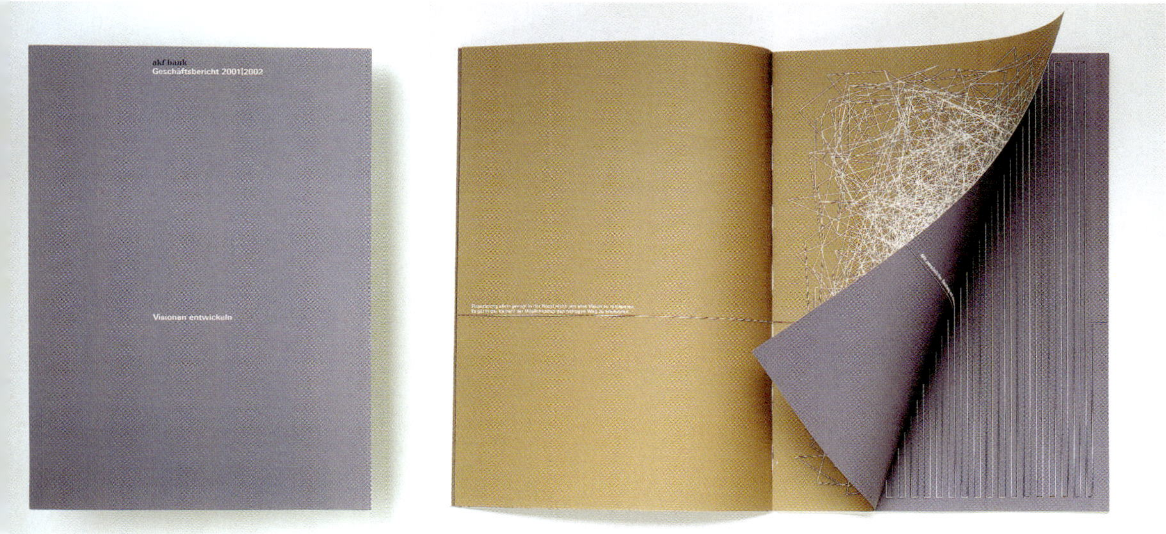

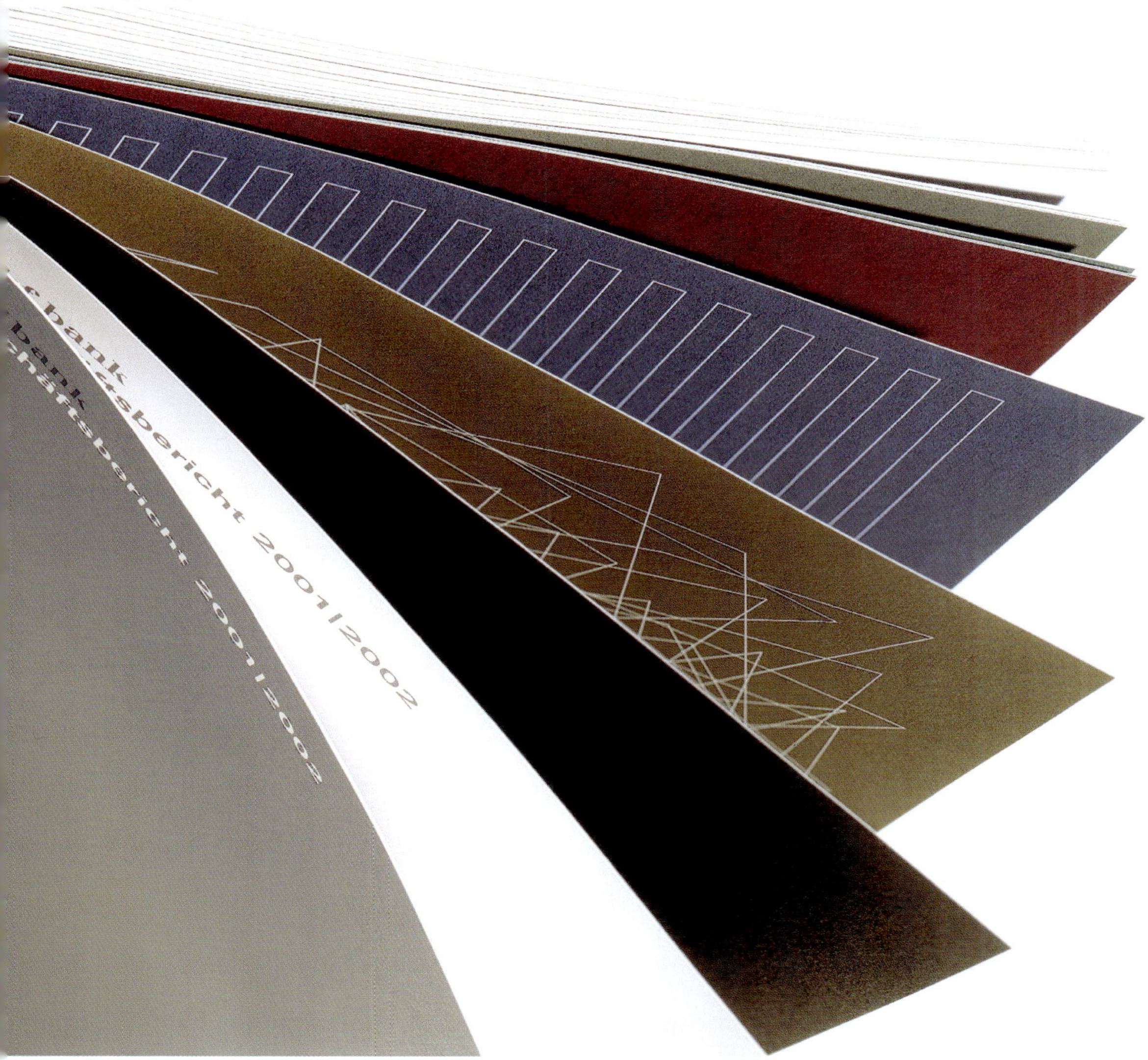

Auftraggeber Client
akf Bank
Wuppertal, Germany
Büro für Design Petra Hattab
Düsseldorf, Germany

Produkt Product
akf Geschäftsbericht 2002
Geschäftsbericht
Annual report

Design
Herzogenrathsaxler Kommunikationsdesign
Düsseldorf, Germany

Der Geschäftsbericht 2002 der akf Bank zeichnet sich durch eine grafische Gestaltung aus, bei der insbesondere die rhythmisierende Typografie ins Auge fällt. Ein System aus Farbe, Schriftschnitten und Einzügen kennzeichnet den Aufbau der Seiten. Betonungen erfolgen durch signifikante Brüche im Rhythmus. Durch eine fortlaufende Linie wird die Gesamtheit der ineinander greifenden Dienstleistungen illustriert.

The akf Bank's Annual Report 2002 catches the eye for a graphic design that spotlights a special rhythmic use of typography. The layout of the pages is defined by a system of colors, cropped fonts and indents. Emphasis is achieved by significantly breaking up the rhythm at particular points. Thanks to the continuous line, the totality of the interwoven services is lent visual form.

Auftraggeber Client
E-Plus Mobilfunk GmbH & Co. KG
Düsseldorf, Germany

Produkt Product
Erfrischend anders! Der neue Auftritt von E-Plus
Refreshingly different! The new appearance of E-Plus

Design
Claus Koch, Michael Mehler, Jörg Hemker,
Susanne Büker, Jürgen Adolph, Birgit Geisler
Claus Koch Corporate Communications
Düsseldorf, Germany

Die neue Markenkampagne „Ein + verbindet" verwendet als zentrales verbindendes grafisches Element ein Pluszeichen. Es steht für die Verbindung zwischen Menschen, die E-Plus täglich viele Millionen Mal schafft. Das E-Plus-Logo wird ebenfalls durch das hochgestellte Pluszeichen ergänzt. Es soll den Mehrwert an Ideen, Innovationen, Kundennähe und Glaubwürdigkeit kommunizieren. Der Claim „Klingelt's?" fordert durch seine Frageform zu Reaktion und Dialog auf. Die Farbe Grün ist als Identitätsmerkmal konzentriert eingesetzt. Bildtypik und Motive greifen die Idee des Verbindens auf und zeigen Personen in entsprechenden Lebensmomenten.

The new "Ein + verbindet" (a + connects) brand campaign features a plus sign as the central, connecting graphic element. It stands for the connection between people which E-Plus makes several million times a day. The E-Plus logo will also now feature an upright plus sign. It is designed to communicate added value in terms of ideas, innovations, customer-friendliness and credibility. The enquiry "Klingelt's?" (Ringing yet?) prompts a response, kindling a dialogue with its question form. The color green is used as a highlight and identifying feature. Imagery and motifs take up the "connecting" idea and show people at various moments in their lives.

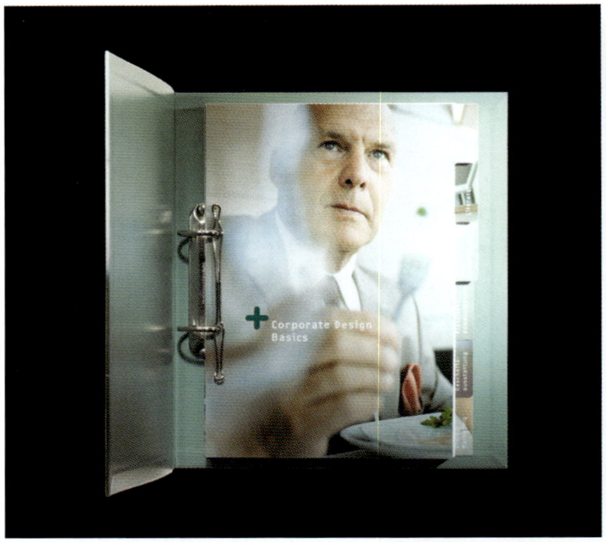

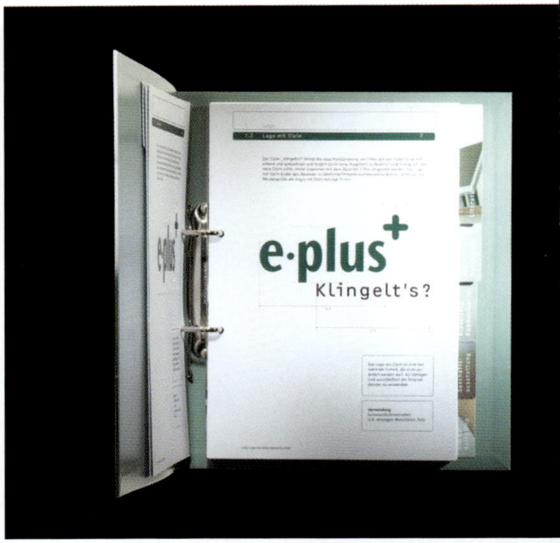

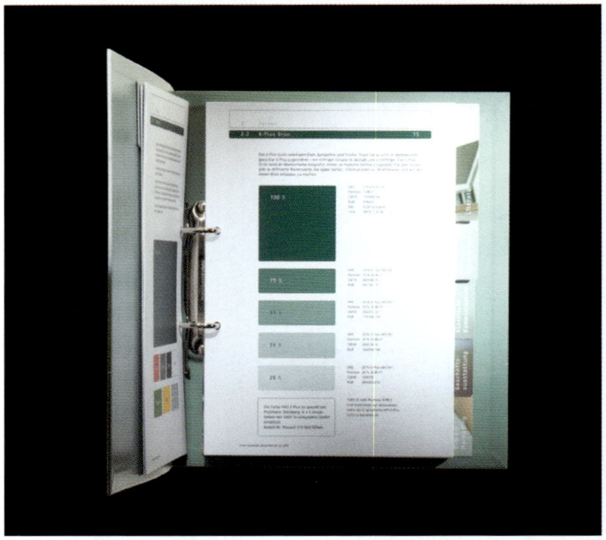

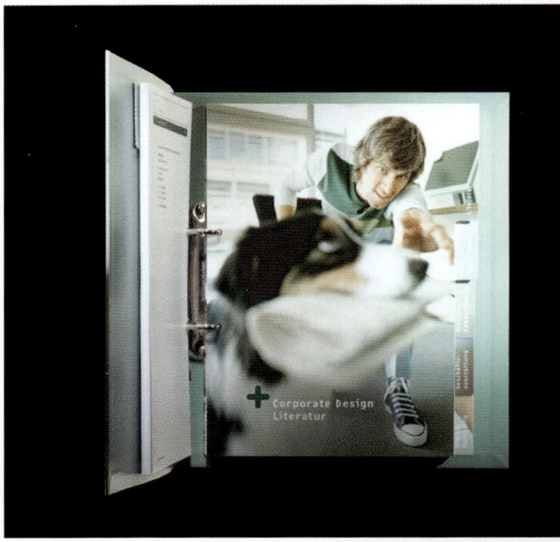

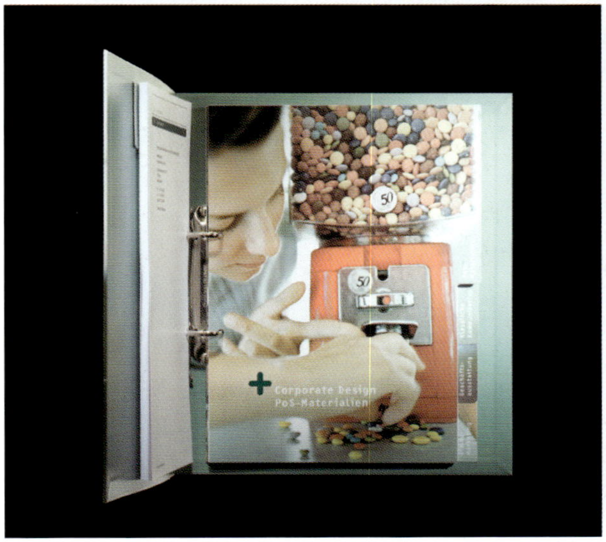

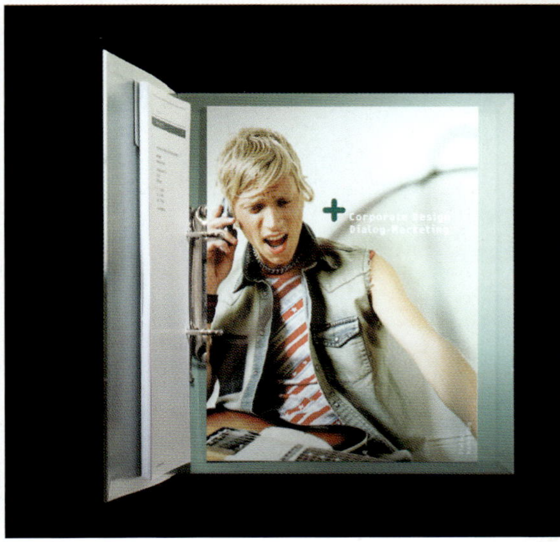

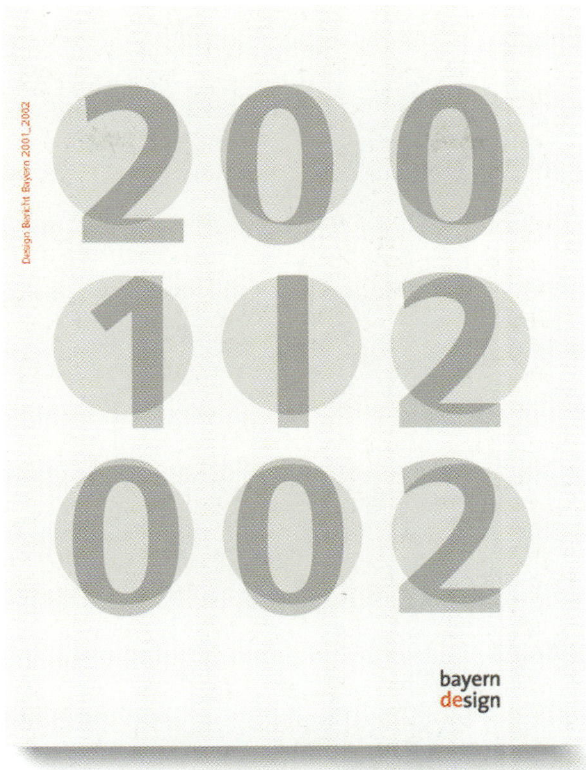

Auftraggeber Client
Bayern Design
München, Germany

Produkt Product
Design Bericht Bayern 2001_2002
Tätigkeitsbericht
Annual Report

Design
häfelinger+wagner design
München, Germany

Als zentrale Einrichtung zur Designförderung in Bayern besteht eine der wesentlichen Leistungen von Bayern Design GmbH im Aufbau und in der Koordination einer vernetzten Infrastruktur, die es erlaubt, bei der Durchsetzung von Projekten und Beratungsleistungen all jene Ressourcen, Fachleute und möglichen Partner anzusprechen, die für einen optimalen Forschungs-, Wissens-, und Informationstransfer wichtig erscheinen. Vorliegende Publikation ist der Tätigkeitsbericht für das Jahr 2001/2002.

One of the important services of Bayern Design GmbH as a central agency for the advancement of design in Bavaria is to establish and co-ordinate a networked infrastructure that enables people to contact all those resources, experts and possible partners who are potentially important to guarantee optimum research, knowledge and information transfer when implementing projects and consulting services.

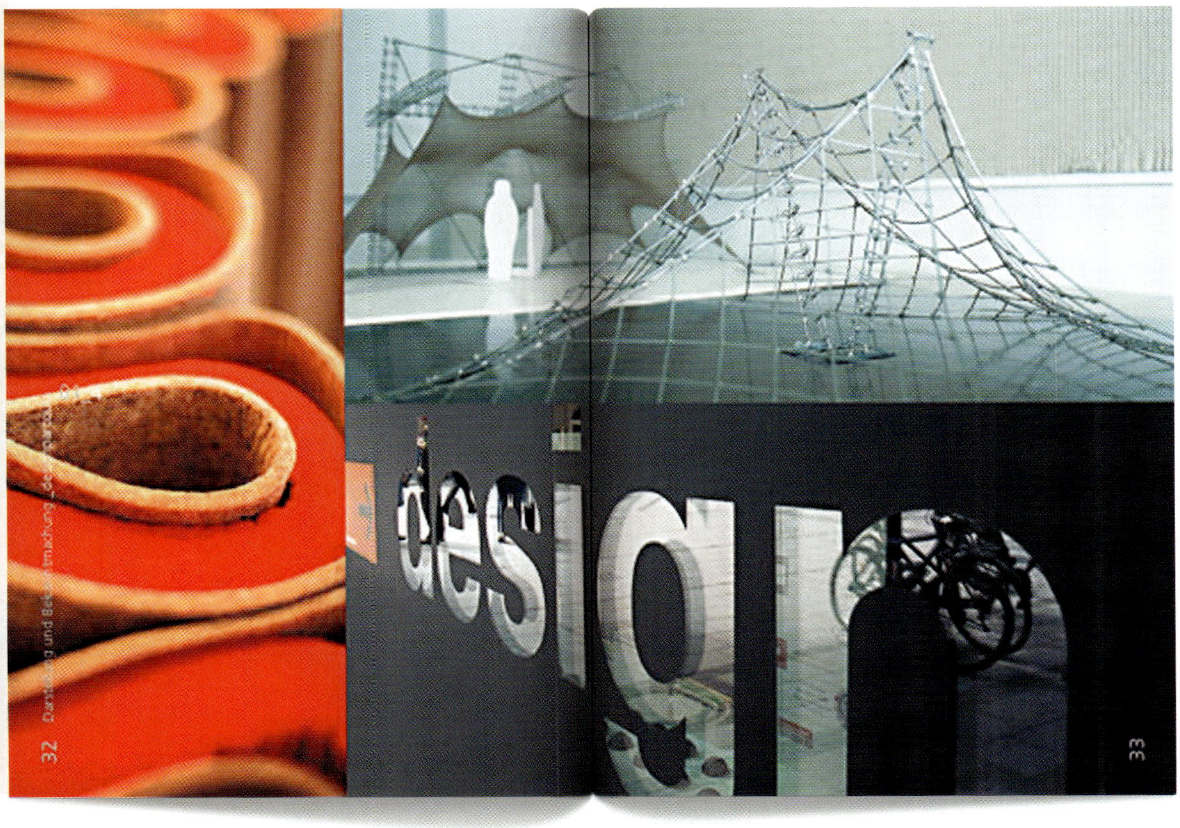

Auftraggeber Client
adidas-Salomon AG
Herzogenaurach, Germany

Produkt Product
Passion of Sport
Geschäftsbericht 2002
Annual report 2002

Design
häfelinger+wagner design
München, Germany

Mit der Neukonzeption zum Geschäftsbericht 2002 bekennt sich adidas-Salomon zu seinen traditionellen Werten Leidenschaft, Technologie und Leistung. Der Geschäftsbericht visualisiert diese Kernwerte illustrativ in einer Imagestrecke, den Kapiteleinstiegsseiten sowie auf den Produkt-Highlight-Seiten. Der hochwertige moderne Auftritt setzt die inhaltlich strategischen Ansätze mit trendiger Modernität überzeugend um. Es gelingt, die traditionellen Werte Leidenschaft für Sport und Leistung visuell attraktiv darzustellen. Mit der Erkenntnis, dass sich die Stärke von adidas-Salomon auf authentische Werte bezieht, aus denen adidas seit Firmengründung Sinn und Vision gezogen hat.

With the new concept for the annual business report 2002, adidas-Salomon is professing its traditional values passion, technology and performance. The business report visualises these core values illustratively by means of an image sequence, the chapter openers as well as the product highlight pages. The high-quality modern appearance implements the contextually strategic approaches with trendy modernity. It is visually and attractively able to depict the traditional values passion for sport and performance. With the knowledge that the strengths from adidas-Salomon are related to the authentic values from which adidas has derived a sense of purpose and vision since the establishment of the corporation.

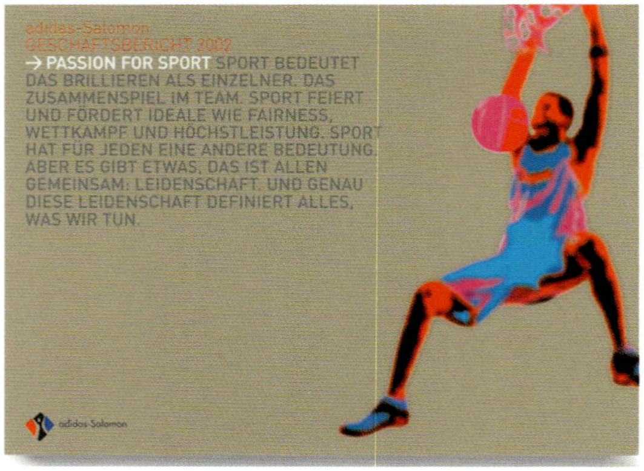

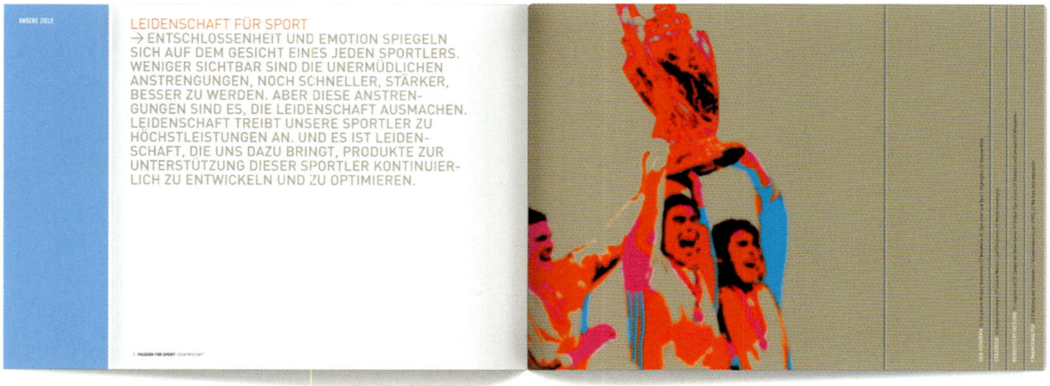

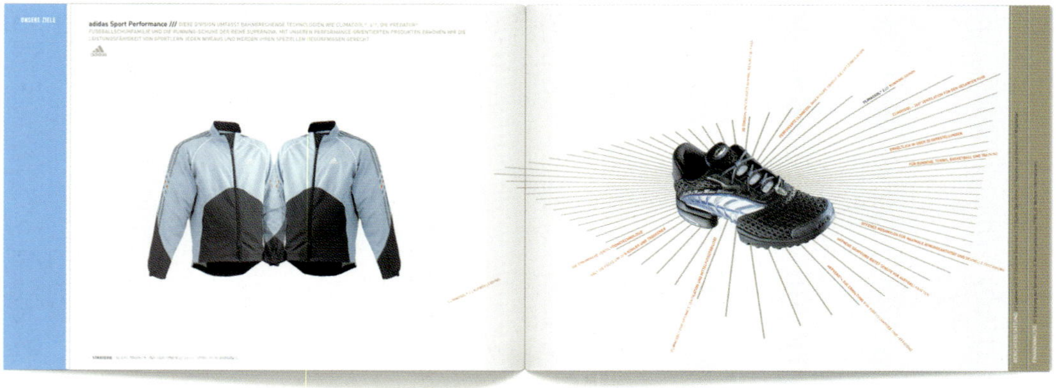

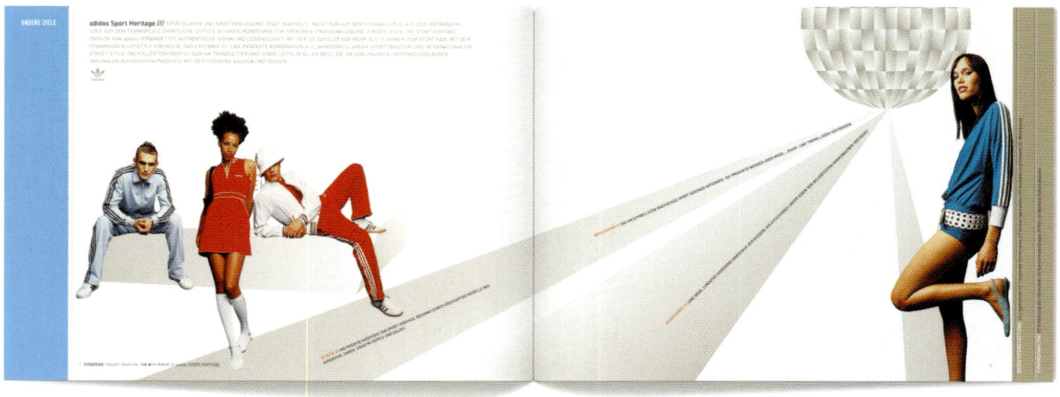

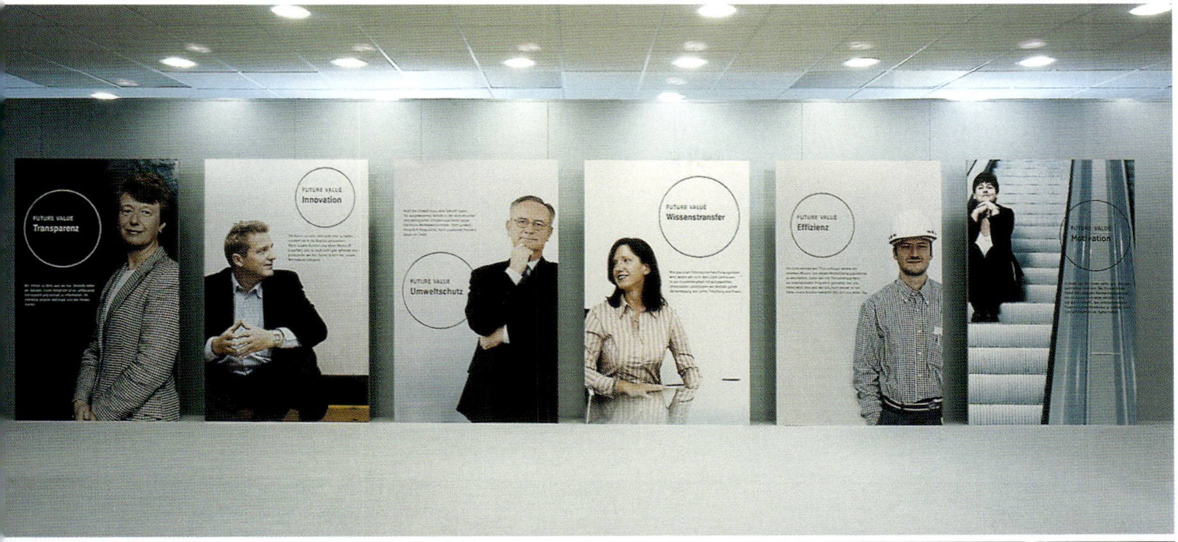

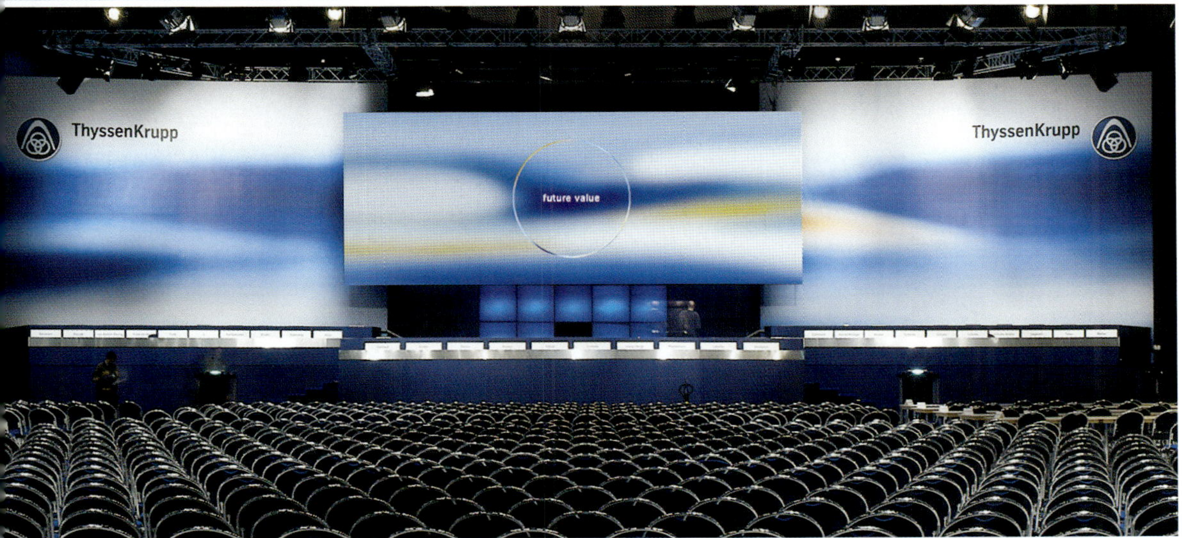

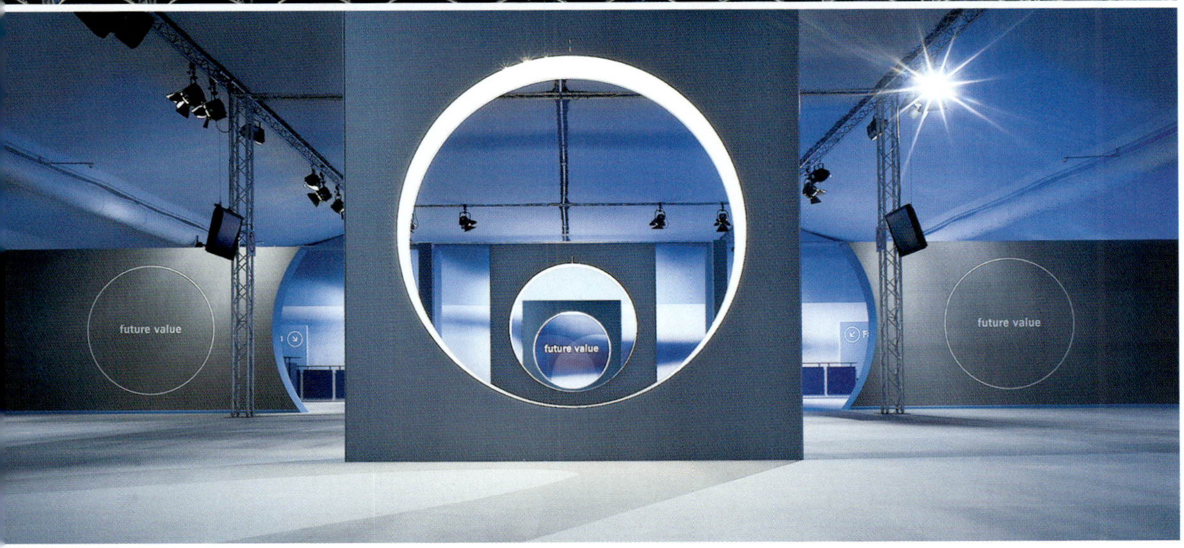

Auftraggeber Client
ThyssenKrupp AG
Düsseldorf, Germany

Produkt Product
Grugahalle, Essen
Hauptversammlung 2003
Annual Stockholders' Meeting

Design
Kommunikationskonzept- und Design
häfelinger+wagner design
München, Germany

Bei der ThyssenKrupp Hauptversammlung in der Grugahalle in Essen wurde das visuelle Erscheinungsbild des Geschäftsberichtes mit dem Thema „Future Value" in eine räumliche Dimension übertragen. Dabei wurden sämtliche Aufenthaltsbereiche, wie Foyer, Catering, Presse-, VIP-Bereiche sowie die Halle mit Bühnenkonstruktion gestaltet. Der Auftritt betraf das gesamte Erscheinungsbild der Halle sowie temporär aufgestellte Gebäude auf einer Fläche von insgesamt rund 9000 qm.

For the ThyssenKrupp Annual Stockholders' Meeting in Essen, the visual appearance of the annual business report was lent a spatial dimension in the form of the topic of "future value". To this end, all non-conference areas in the convention complex, such as the lobby, the catering zones, as well as the press and VIP rooms, and even the hall itself featured a stage-like structure. The presentation extended to the overall appearance of the hall as well as the temporary facilities that together embraced a surface area of some 9,000 sq. m.

Auftraggeber Client
Uhde GmbH
Dortmund, Germany

Produkt Product
Achema 2003, Frankfurt
Messedesign
Trade fair design

Design
Kommunikationskonzept und -design
häfelinger+wagner design
München, Germany

Der Messeauftritt von Uhde – einer Tochter der ThyssenKrupp AG – wurde auf der Achema 2003 in Frankfurt durch vier überdimensionale U-Displays bestimmt. Auf jedem der Displays wurden zwei der Unternehmensbereiche durch Texte, Bilder und jeweils einen Image-Flashfilm präsentiert. Durch Umkehrung der Displayform entstanden spezifische Sitzgelegenheiten.

Uhde is a subsidiary of ThyssenKrupp AG and for its trade-fair presence at the Achema 2003 in Frankfurt opted in favor of four out-sized U-shaped displays. Each of the displays presented two of the company's divisions in the form of copy, images and an image film in Flash. The display stands themselves swiveled to provide special seating arrangements.

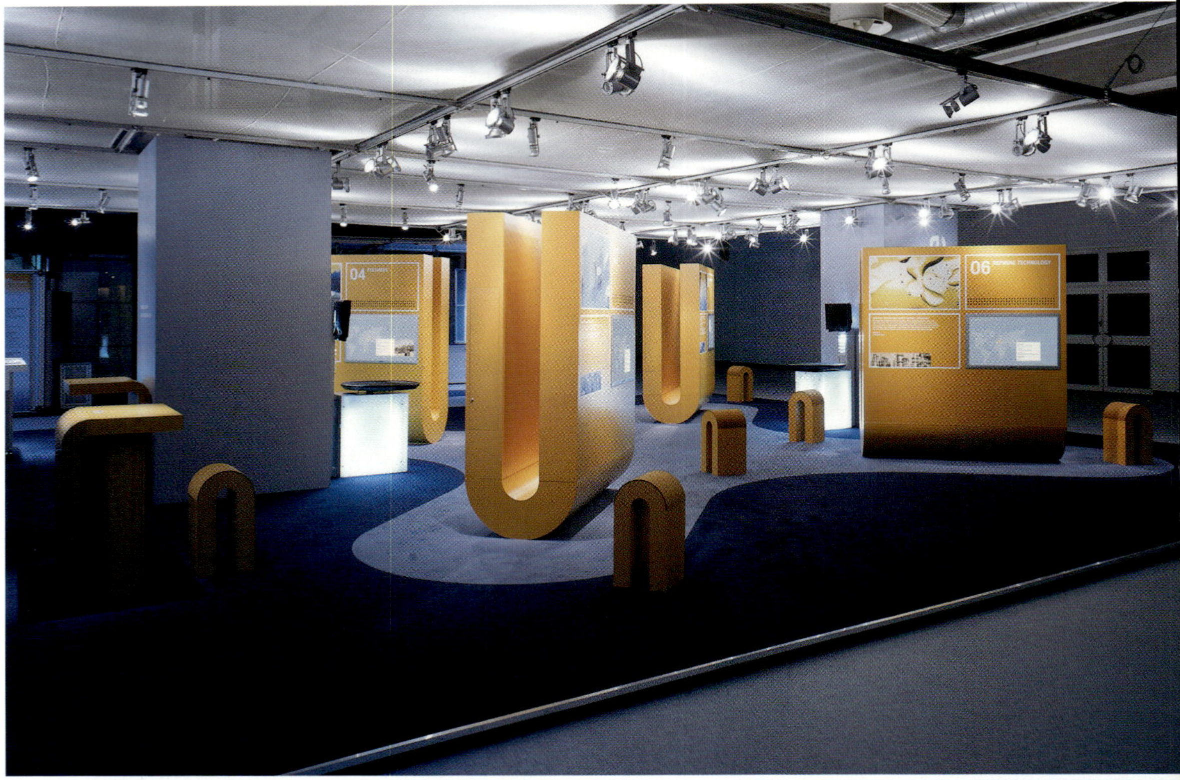

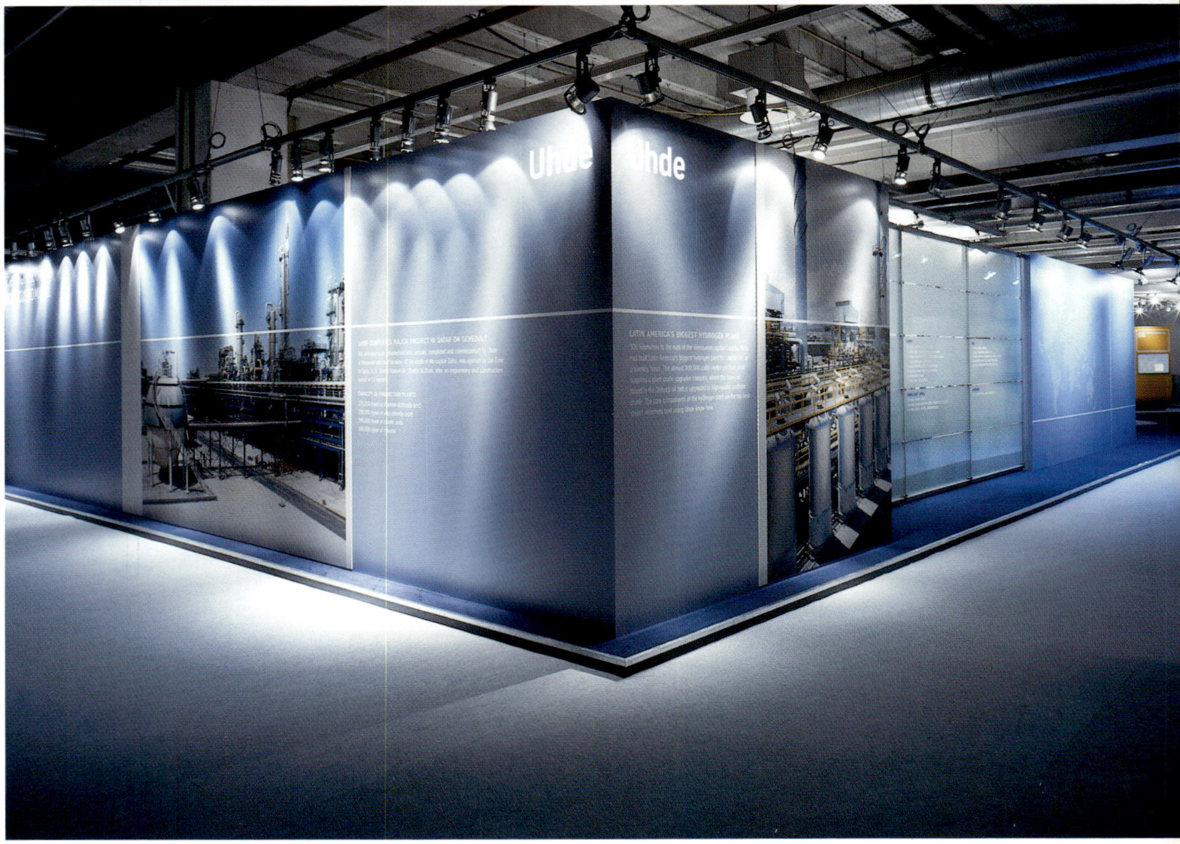

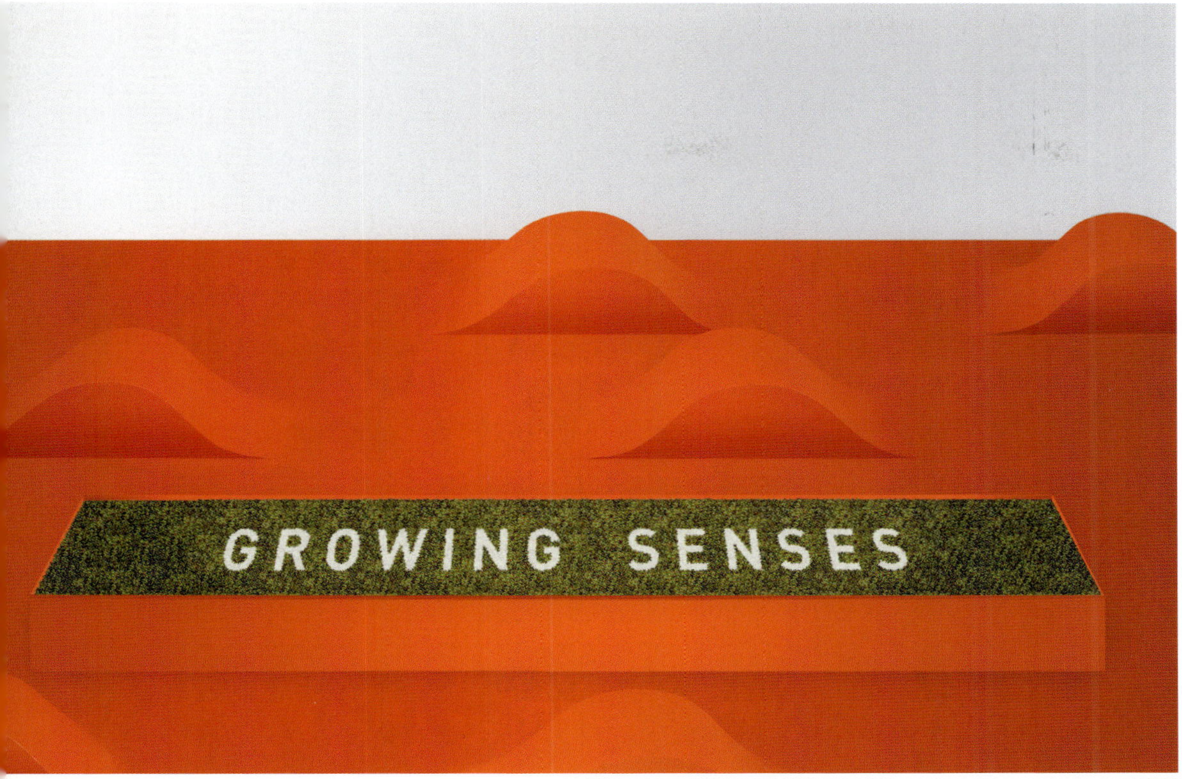

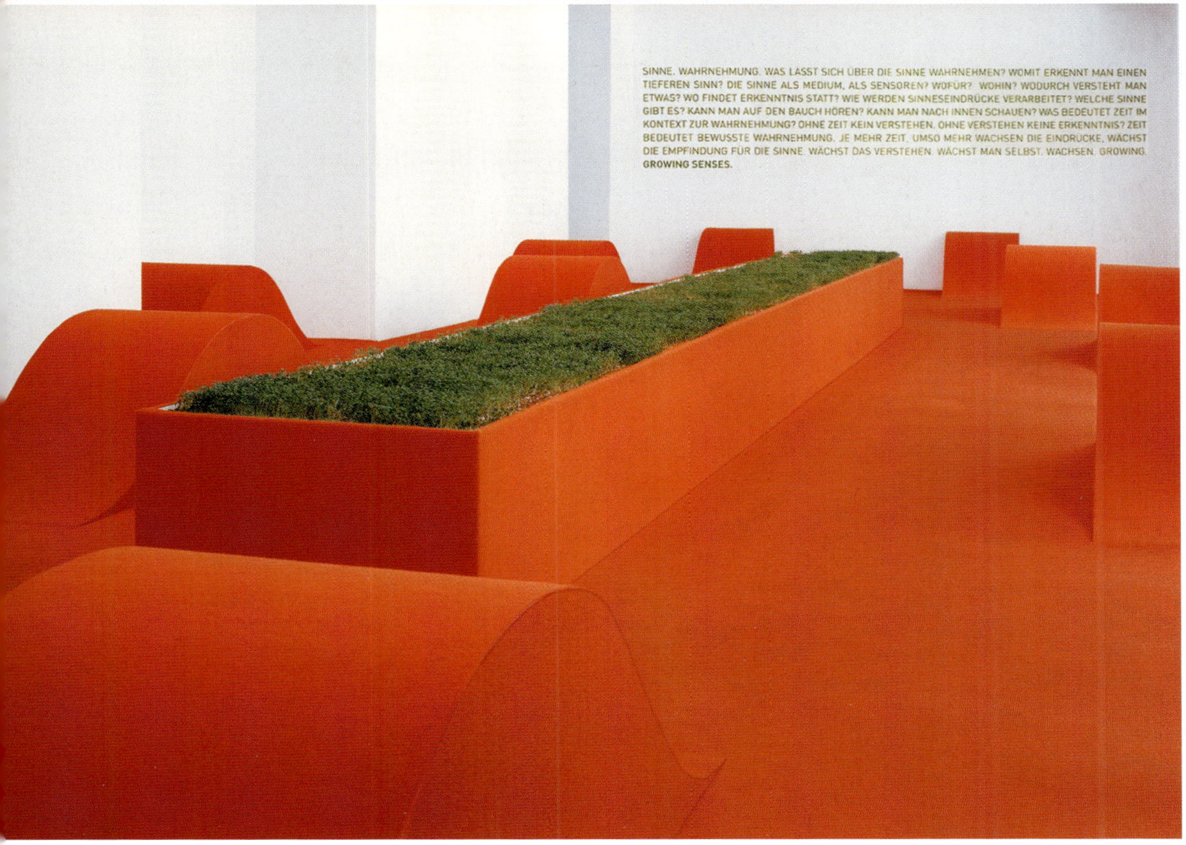

Auftraggeber Client
häfelinger+wagner design
München, Germany

Produkt Product
Installation GROWING SENSES
Projekt zum designparcours 2003, München

Design
häfelinger+wagner design
München, Germany

häfelinger+wagner design beteiligte sich am designparcours 2003 in München mit einer Raum-Installation, die das vorgegebene Thema Design & Sinne in den Kontext von Zeit und Reflexion stellt. Die Bodenwellen sind Sitz- und Liegemöglichkeiten. Sphärische Musik lädt zum Verweilen ein, zur Entspannung und Wahrnehmung. Ein während des designparcours stetig wachsendes Kressebeet zeigt allmählich den zentralen Begriff der Installation GROWING SENSES.

häfelinger+wagner design played a strong role in the Munich designparcours 2003, contributing a spatial installation that placed the fair's theme of Design & the Senses in the context of time and reflection. The floor was structured in waves to provide seats and recliners. Atmospheric music encouraged visitors to tarry a while, to relax and open their senses. In the course of the designparcours, a bed of cress gradually grew greener and lusher, showcasing the central concept underlying the installation: GROWING SENSES.

Auftraggeber Client
Smurfit Munksjö Paper AB
Aalen, Germany

Produkt Product
Interzum 2003, Köln, Germany
Messedesign
Trade fair design

Design
Kommunikationskonzept- und Design
häfelinger+wagner design
München, Germany

Architektur
Schwebius Gestaltung
München, Germany

Der Messeauftritt von Smurfit Munksjö Paper auf der Interzum 2003 stand unter dem Imagethema Reflexionen, wobei zwischen verbalen und visuellen Reflexionen unterschieden wurde, die beide Inspirationsquelle für neue Dekore und Oberflächen darstellen. Am Messestand wurden visuelle Reflexionen durch Schattenwürfe natürlicher Materialien an den Messewänden und verbale Reflexionen durch Assoziationsketten – Textbänder auf einem spezifischen Polypropylen-Wandsegel – umgesetzt. Die Dualität der verbalen und visuellen Reflexion setzte sich bei der Wahl der Materialien und der Formensprache am Messestand fort: weiche organische versus konstruierte geometrische Formen.

The Smurfit Munksjö Paper trade fair stand for Interzum 2003 focused on an image message: reflections. In this context, a distinction was drawn between verbal and visual reflections, the two sources of inspiration for new decors and surface textures. The stand presented visual reflections in the form of shadows cast by natural materials on the walls of the stand. The verbal reflections were shown in the forms of series of associations – lines of copy transposed onto specific polypropylene wall sails. The duality of verbal and visual reflection also informed the choice of materials and formal idiom for the stand: soft organic shapes versus structured geometric forms.

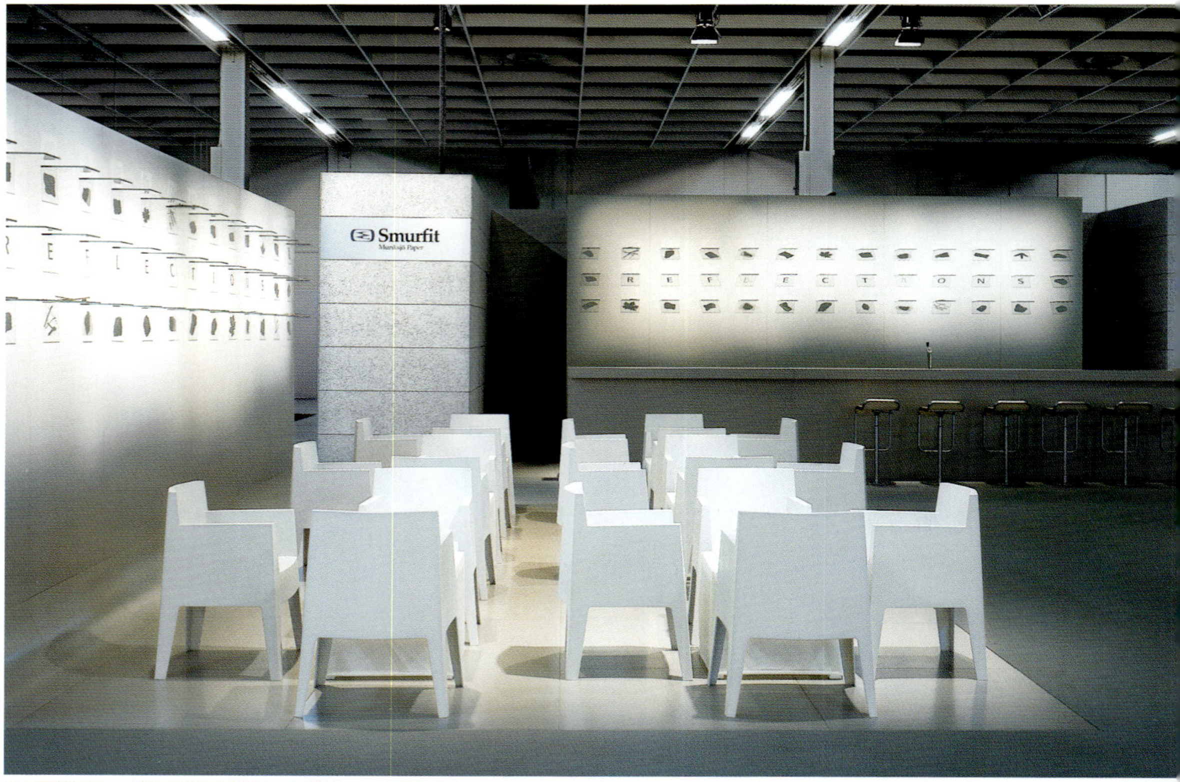

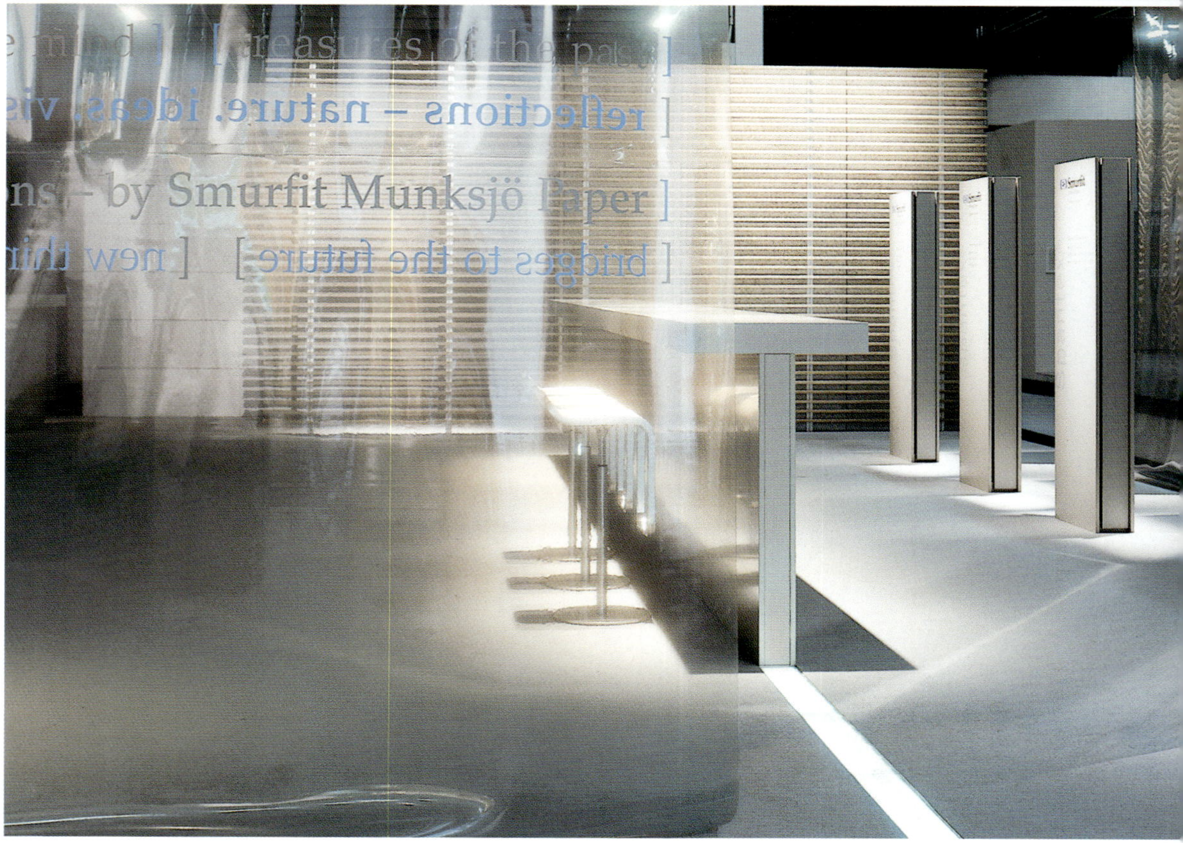

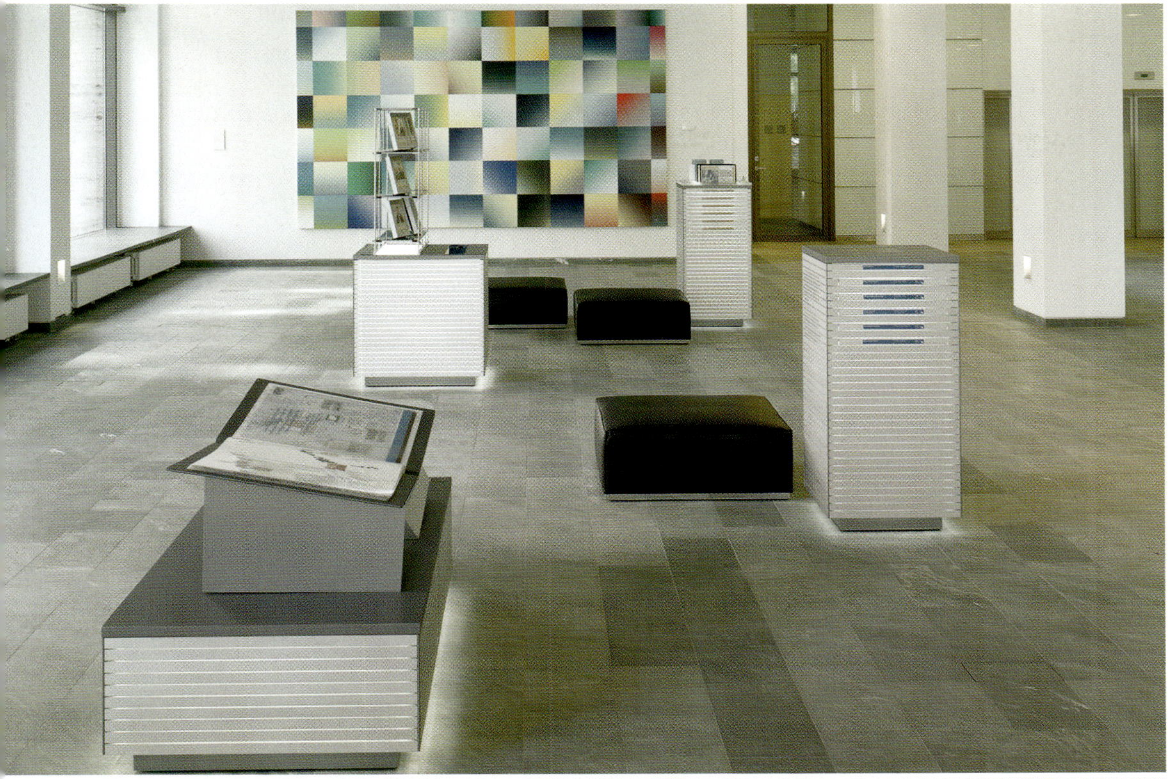

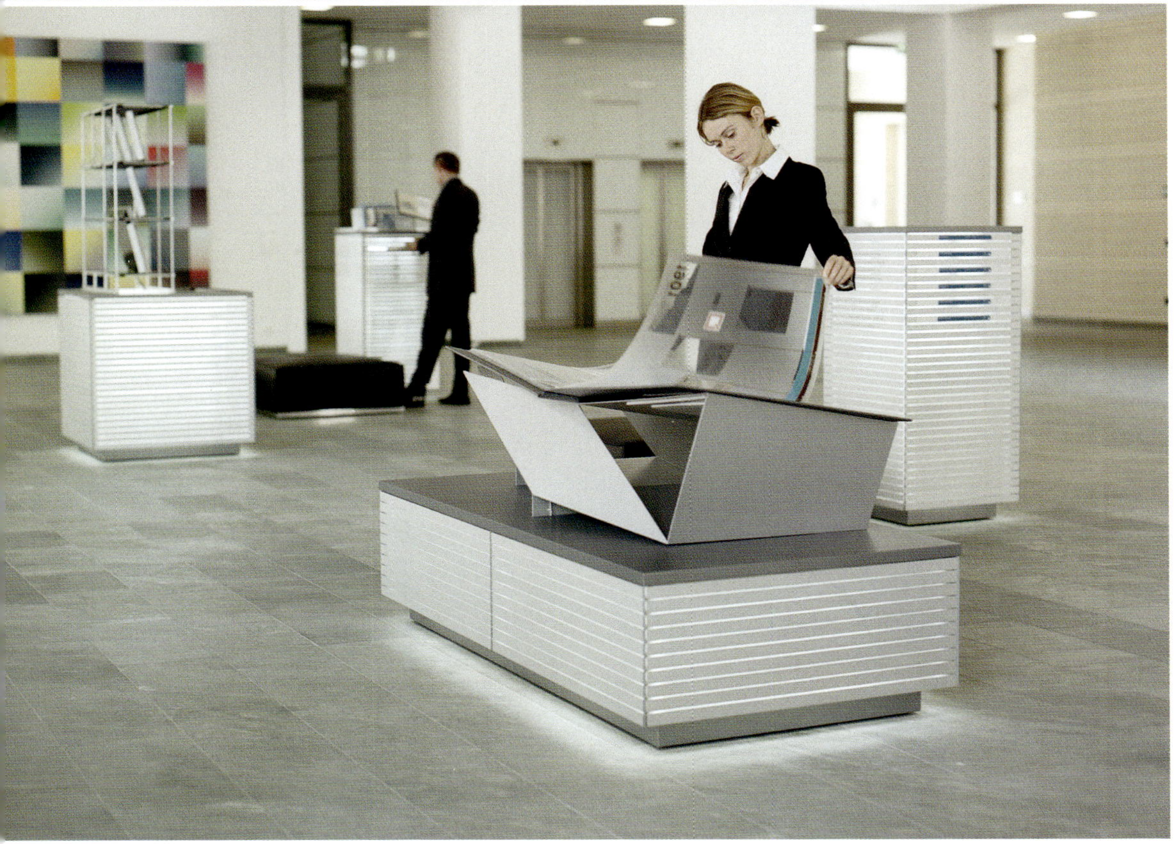

Auftraggeber Client
Robert Bosch GmbH
Stuttgart, Germany

Produkt Product
Ausstellung Robert Bosch Repräsentanz, Berlin
Foyergestaltung
Lobby design

Design
Milla und Partner GmbH
Jean-Louis Vidière (Konzept und Gestaltung)
Marion Kerckhoff (Projektleitung)
Björn Peters, Design Klinik (Grafik Design)
Stuttgart, Germany

Die Ausstellung im Foyer der Robert-Bosch-Repräsentanz in Berlin anlässlich einer Bilanz-Pressekonferenz im April 2003 basiert auf dem Konzept des Bosch-Pavillons auf der Expo 2000 in Hannover. Gegenstand der Ausstellung sind nicht die Bosch-Produkte selbst, sondern Würfel und Podeste aus mehreren Schichten, die gleich Supercomputern, Archivschränken, Energiekuben oder geheimnisvollen Speichern Wissen bündeln und anschaulich machen, was es zu entdecken gilt. Die Vielfalt und Vielschichtigkeit von Bosch wird auf mehreren Ebenen veranschaulicht. Zu den ausgestellten Themen zählen: Bosch Internationalität – durch Drehen des Rings können die weltweiten Standorte von Bosch erkundet werden; Bosch-Werte – ein sprechendes Buch; Bosch-Innovationen – ein visuelles Labyrinth; Bosch Stiftung – Porträts von Menschen der Stiftung.

The show in the lobby of the Robert Bosch representative office in Berlin on the occasion of the press conference presenting the annual financial statements in April 2003 is based on the concept for the Bosch Pavilion at Expo 2000 in Hanover. The object of the exhibition: not Bosch products themselves, but rather multi-layered cubes and plinths: these combine supercomputers, archive cabinets, energy cubes or mysterious storage media for knowledge, rendering it visible and setting viewers on the track of new insights. The diversity and many-layered nature of Bosch is thus visualized across several different layers. The topics included in the exhibition include: Bosch's international reach – by swiveling the ring, Bosch's numerous hubs worldwide are shown; Bosch's values – a talking book; Bosch's innovations – a visual labyrinth; the Bosch Foundation – portraits of renowned members of the foundation.

Auftraggeber Client
DaimlerChrysler AG, Stuttgart, Germany

Produkt Product
Mercedes-Benz auf der IAA 2003
Mercedes-Benz at IAA 2003

Architektur Architecture:
Kauffmann Theilig & Partner
Freie Architekten BDA, Ostfildern, Germany
Kommunikationsdesign Communication Design:
Atelier Markgraph GmbH, Frankfurt a.M., Germany
Grafikdesign Graphic Design:
Design Hoch Drei, Stuttgart, Germany
Lichtplanung Light Planning:
TLD Planungsgruppe GmbH, Wendlingen, Germany

Mercedes-Benz präsentiert sich auf der IAA 2003 traditionsgemäß in der Festhalle Frankfurt. Imposanter Blickfang des Auftritts ist eine scheinbar schwebende Ringfigur. Ihre Lamellenfassade ist mit LED-Streifen bestückt und trägt im Zusammenspiel mit der großen LED-Fläche im Zentrum die zentralen Technologiethemen in den Raum. Gezielte Öffnungen schaffen Einblicke in die Ausstellungsräume hinter der Lamellenfassade: Auf 12.000 qm Ausstellungsfläche und über drei Ebenen verteilt erwartet die Besucher dort eine dramaturgische Abfolge von Inhalten und Atmosphären. Architektur, Material, Licht- und Mediendesign tauchen die Ausstellungsräume in eine für die darin ausgestellten Baureihen und Themen charakteristische Stimmung. Der Messestand wird zur Bühne, auf der sich Kompetenz, Technologie und Design in der für Mercedes-Benz typischen Art und Weise miteinander verbinden.

At the IAA 2003 auto show, the Mercedes-Benz presentation was, in keeping with tradition, in the Festival Hall at the Frankfurt trade fair complex. The impressive eye-catcher was the ring that seemingly floated in the center. Its lamella façade was coated with LED strips and, together with the large LED screen in the center, created a presence for the major technology themes. Carefully positioned openings offer an insight into the exhibition spaces behind the lamella façade. Across a total of 12,000 sq. m. of exhibition space and three different levels, the visitor is greeted by a dramatic sequence of content and atmospheres. The architecture, materials, light and media design all imbue the exhibition area with a mood characteristic of the marques on show and the related themes. The trade-fair stand becomes a stage on which expertise, technology and design are linked in the manner typical for Mercedes-Benz.

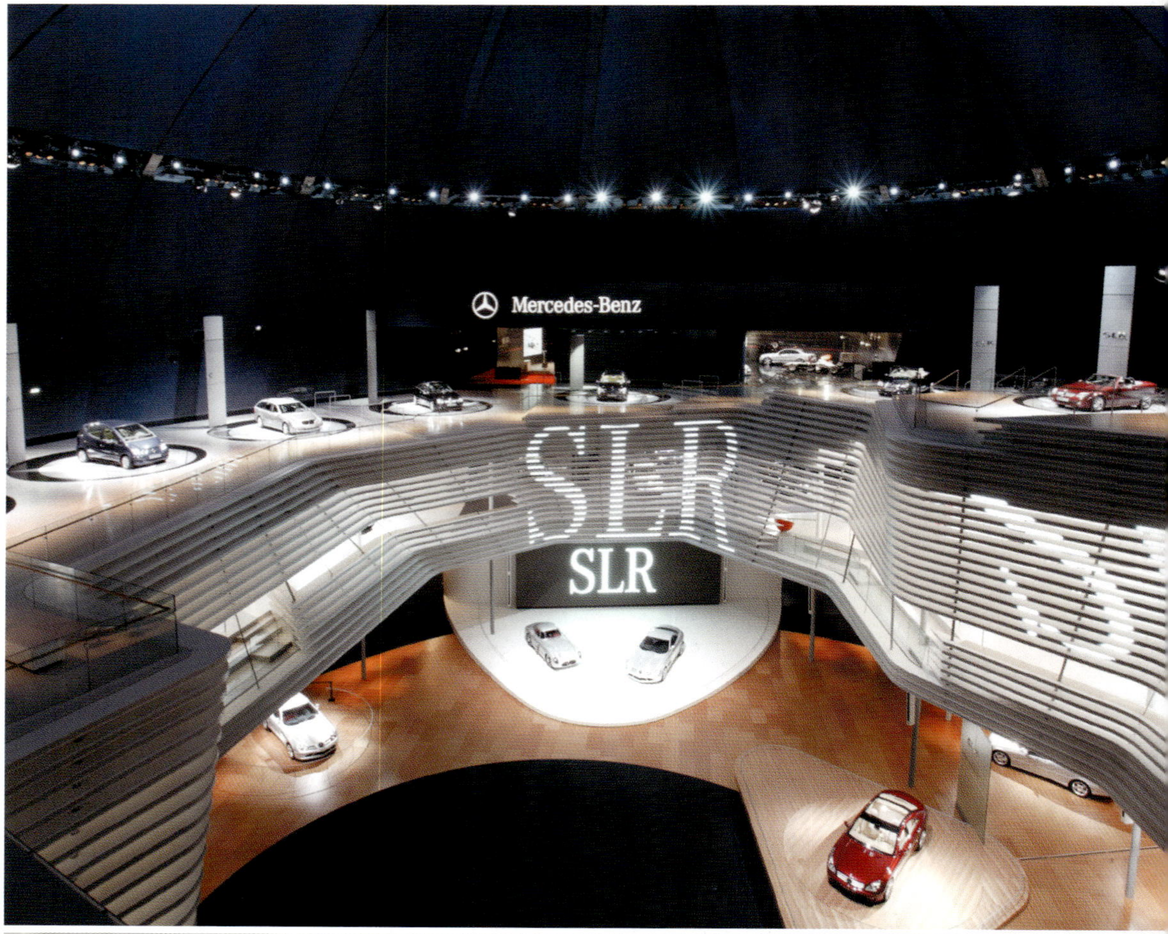

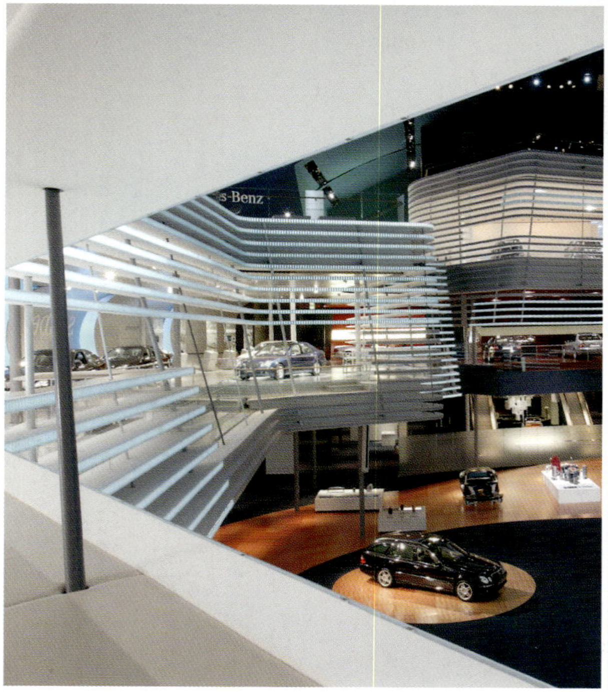

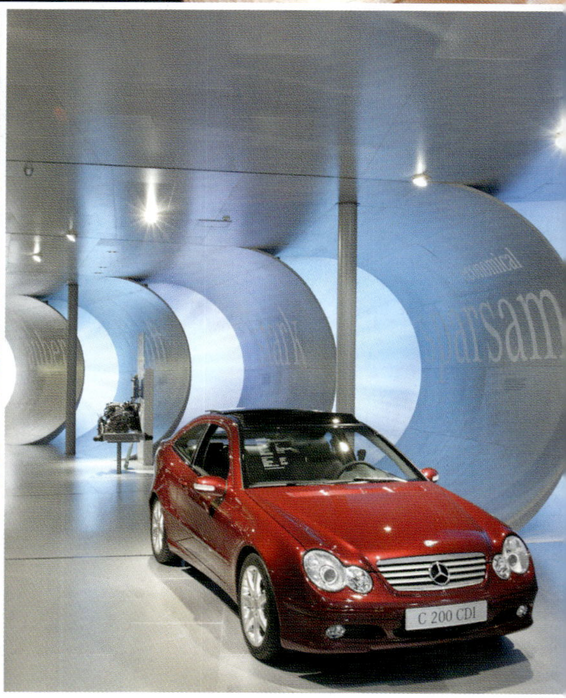

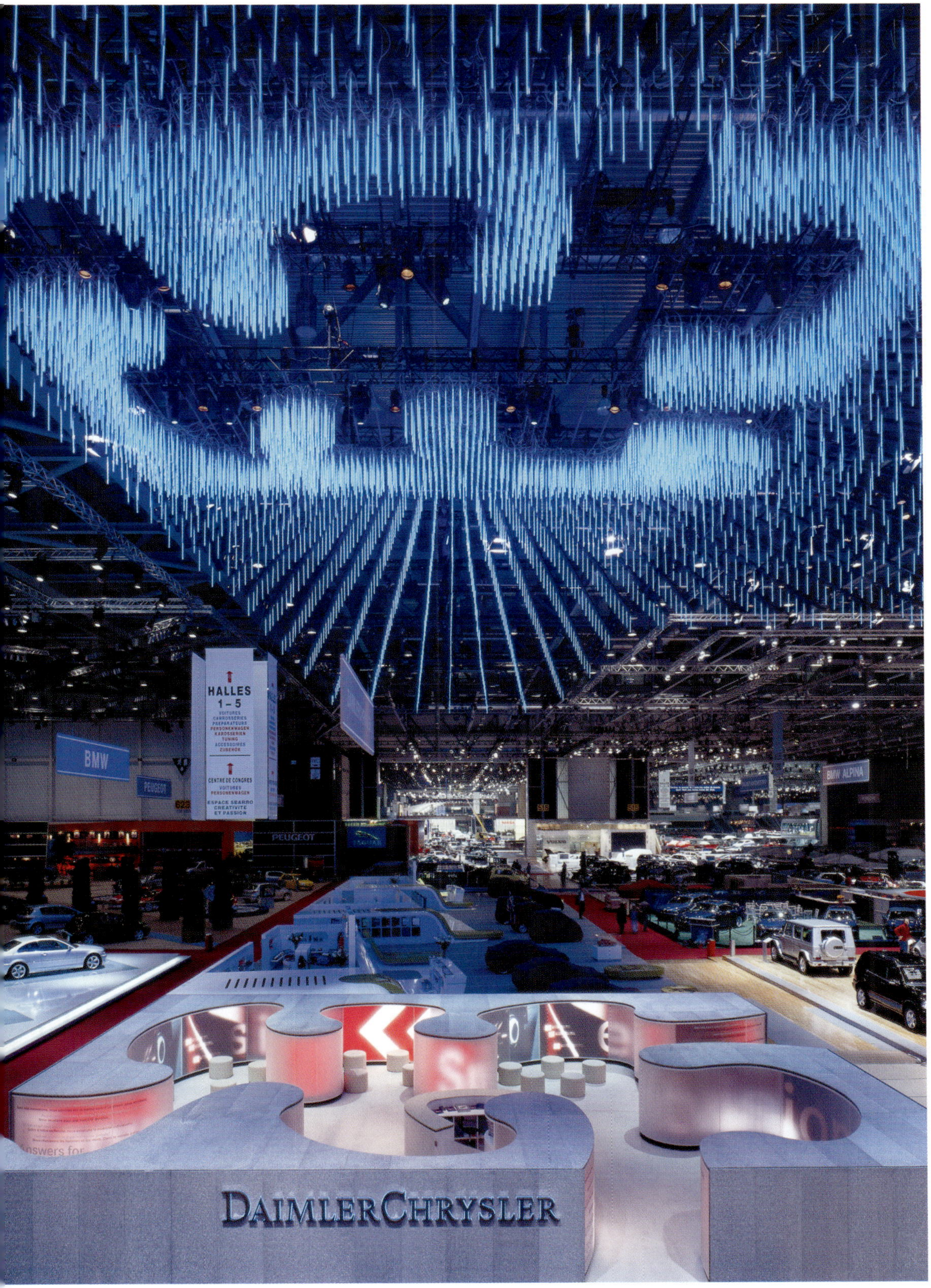

Auftraggeber Client
DaimlerChrysler AG Stuttgart, Germany

Produkt Product
DaimlerChrysler Corporate Exhibition
73. Automobilsalon Genf
73rd Geneva Motor Show

Kommunikationsdesign Communication Design:
Atelier Markgraph GmbH, Frankfurt a. M., Germany
Architektur Architecture:
Kauffmann Theilig & Partner
Freie Architekten BDA, Ostfildern, Germany
Lichtplanung Light Planning:
TLD Planungsgruppe GmbH, Wendlingen, Germany

2003 präsentiert sich DaimlerChrysler erstmals mit einem eigenen Unternehmensauftritt inmitten seiner Marken Mercedes-Benz, Maybach, smart®, Chrysler und Jeep®. Ziel ist es, das hohe Technologiepotenzial zu kommunizieren und ein Passepartout für die einzelnen Marken zu schaffen. Eine Deckengestaltung aus 5.000 strahlenförmig angeordneten Neonleuchten verbindet die aneinander angrenzenden Stände zu einem „Corporate Space". Das Zentrum der Lichtinstallation spiegelt den Grundriss der Standfläche wider und markiert die Position des DaimlerChrysler-Forums. Das Forum entsteht durch amöbenhafte Einschnitte in ein 1.65 m hohes Volumen. Die Innenfassade setzt mit einer rundum laufenden Medieninstallation den Konzernclaim „Answers for questions to come" in ein Raumerlebnis um. In diesem „Software-Driven Space" lädt der Konzern die Besucher ein, sich auf die Spur aktueller Forschungs- und Entwicklungsfragen zu begeben.

At the 73rd International Auto Salon in Geneva, DaimlerChrysler presented itself for the first time in 2003 as a corporate entity alongside its marques Mercedes-Benz, Maybach, Smart®, Chrysler and Jeep®. The goal of the presentation: to communicate the company's great technological expertise and create a frame for the individual brand presentations. A ceiling design featuring 5,000 ray-like neon lamps and links the adjacent stands to form a single corporate space. The heart of the light installation reflects the layout of the stand and pinpoints the location of the DaimlerChrysler Forum. The latter arises through amoeba-like inserts into a 1.65 m high body. The façade of the interior boasts a wraparound media installation that serves to lend physical form to the Group's claim: "Answers for questions to come". In this software-driven space, the Group invites the visitor to trace the path taken to address current research and development topics.

Auftraggeber Client
AML Licht + Design GmbH
München, Germany

Produkt Product
Leuchtensystem Occhio von axelmeiselicht
Neupositionierung der Produktmarke Occhio
Brand repositoning of Occhio
Brand Concept, Corporate Identity, Corporate Design

Design
Peter Martin (Design)
Birgit Keuerleber (Design)
Daniel Karczinski (Konzept)
Martin et Karczinski GmbH
München, Germany

Für das Leuchtensystem Occhio von axelmeiselicht hat die Designfirma Martin et Karczinski ein neues Brand Concept und Corporate Design sowie eine neue Corporate Identity entwickelt. Die feine Linienführung ist ein zentrales Element im Baukasten des Occhio-Erscheinungsbildes. Der Rahmen fungiert als visuelle Klammer für die Marke Occhio, in dem er Produkte, Fotos, Texte und Zeichnungen elegant verbindet und als identitätsstiftendes Wiedererkennungsmerkmal funktioniert. Die Natur als Ideengeber für das Leuchtensystem bildet die Grundlage für den Claim, „Inspiration durch Form und Funktion" und die Bildkonzepte mit Fotos von Blättern. Die Eigenschaften, Systematik und Kombinierbarkeit werden in den Medien über Imageflächen in der vertrauten Occhio-Farbe Grün kommuniziert.

For the Occhio luminaire line produced by axelmeiselicht, design agency Martin et Karczinski came up with a new brand concept and a corporate design as well as a new CI. The fine lines form a central element in the set of principles underlying the Occhio image. The frame functions as a visual bracket for the Occhio brand, elegantly marrying products, images, copy and drawings to generate an unmistakable image that forges an identity for the company. Nature as the source of inspiration for the family of luminaires likewise acts as the basis for the claim "Inspiration through form and function" and the image concept – photos of leaves. The properties of systematic method and combinable elements are communicated in the different media by means of image sections in the familiar Occhio green.

axelmeiselicht

Occhio 123

OCCHIO
OCCHIO PRO
OCCHIO SISTEMA

Occhio 123

Occhio 123

Occhio system

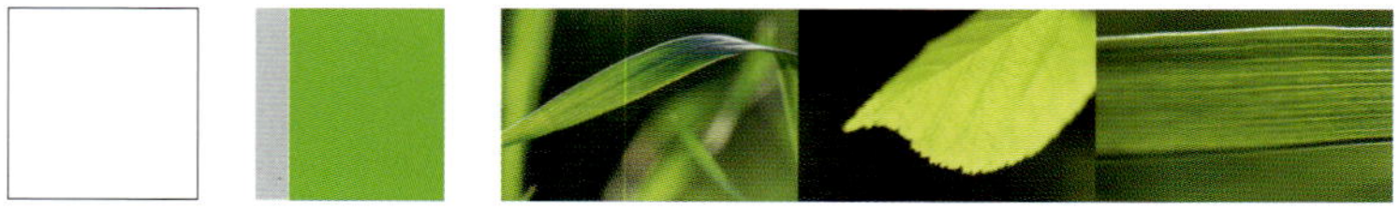

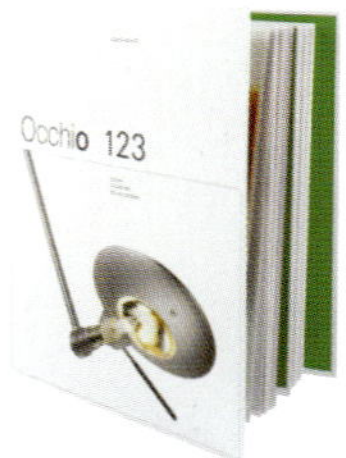

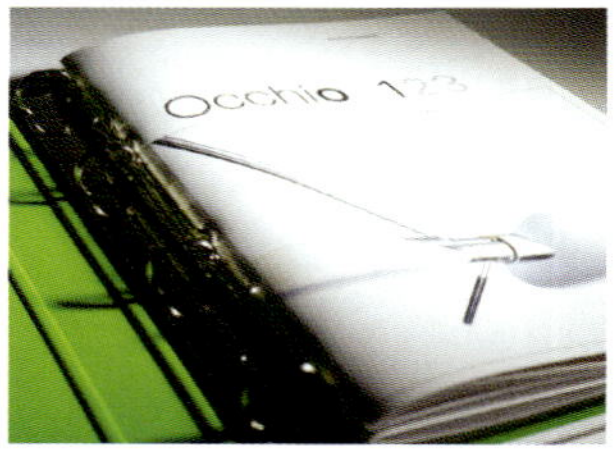

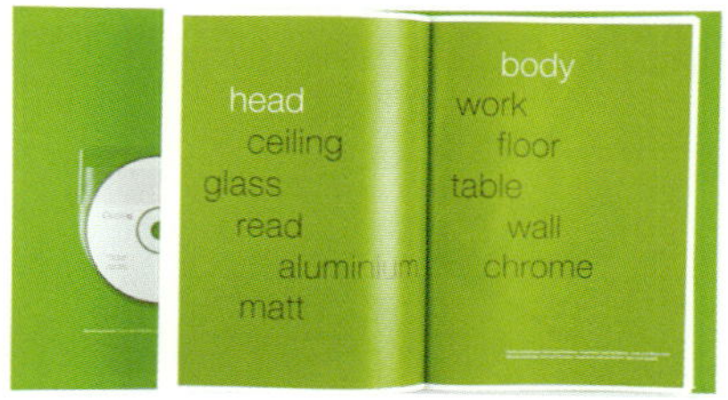

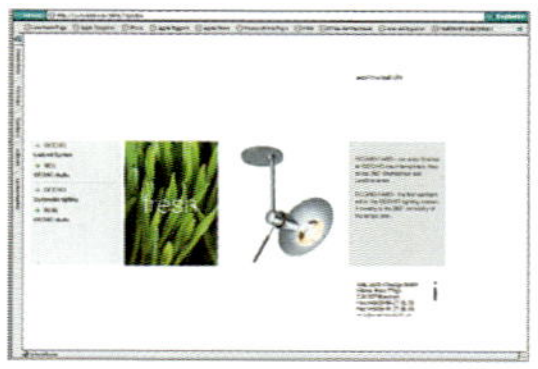

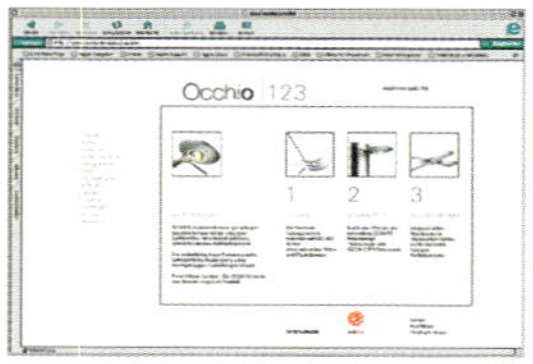

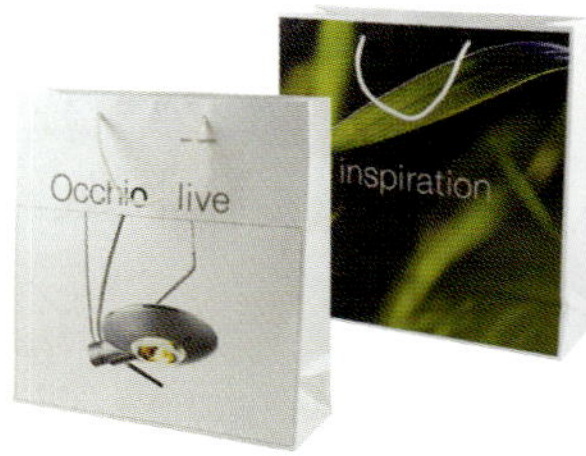

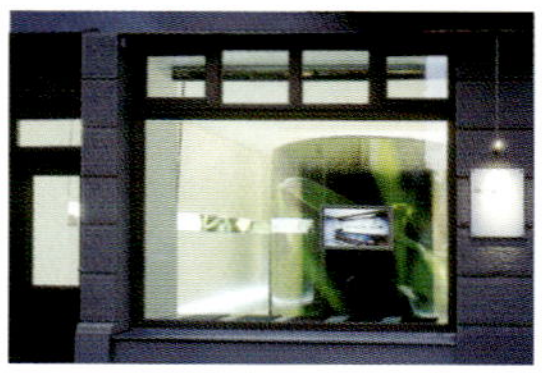

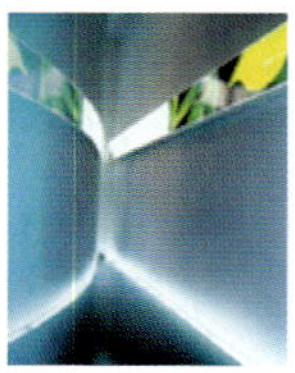

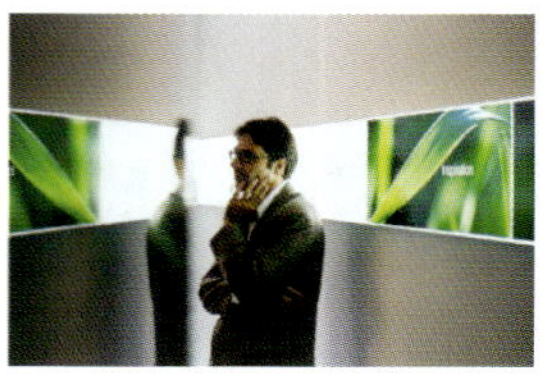

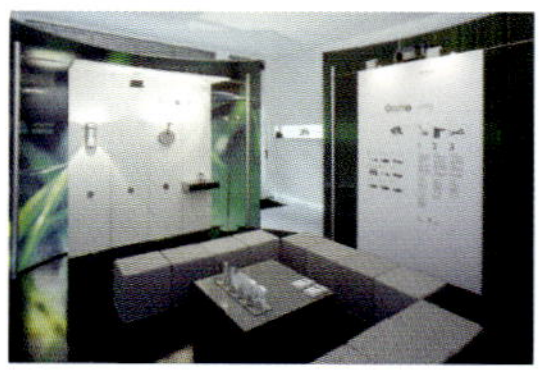

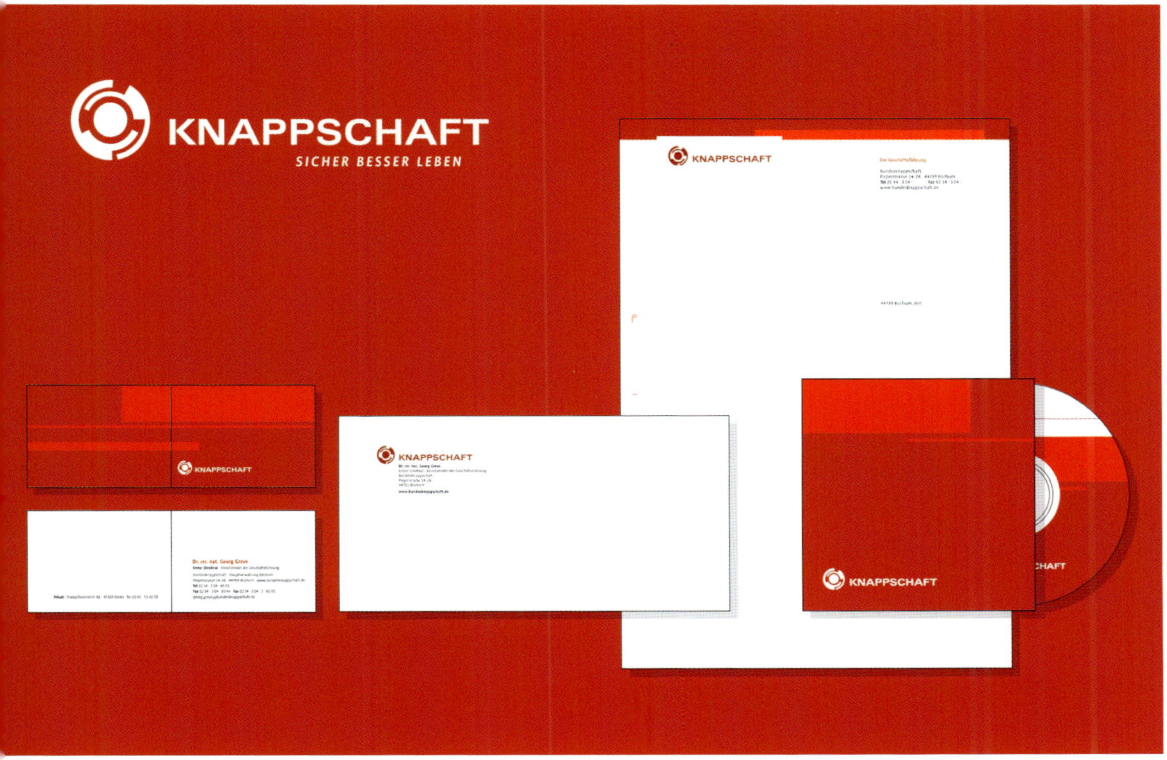

Auftraggeber Client
Die Bundesknappschaft
Bochum, Germany

Produkt Product
Die Bundesknappschaft
Corporate Design

Design
Ligalux GmbH
Hamburg, Germany

Die Bundesknappschaft ist der zweitgrößte Sozialversicherungsträger in Deutschland. Ursprünglich als Sozialversicherungssystem der Bergleute gegründet, bietet die Knappschaft heute ein modernes Verbundsystem aus Kranken- und Pflegeversicherung, Rentenversicherung und einem eigenen medizinischen Netz. Ziel war es, die Bundesknappschaft, ein ebenso traditionsreiches wie innovationsfreudiges Unternehmen, im Gesundheits- und Rentenmarkt neu zu positionieren. Auf Basis des neuen Unternehmensprofils gestaltete Ligalux für die Bundesknappschaft eine visuelle Identität und realisierte die optische Umsetzung des Positionierungsprozesses.

Bundesknappschaft is Germany's second largest social-security agency. Originally set up as a social-security system for miners, today the Bundesknappschaft is a modern cooperative organization handling health and healthcare insurance, pension insurance and with a medical network of its own. The goal here was to reposition the Bundesknappschaft, which is a both long-standing and innovative entity, in the healthcare and pension market. On the basis of the new company profile, Ligalux created a visual identity for the Bundesknappschaft and handled visual realization of the concept during the positioning process.

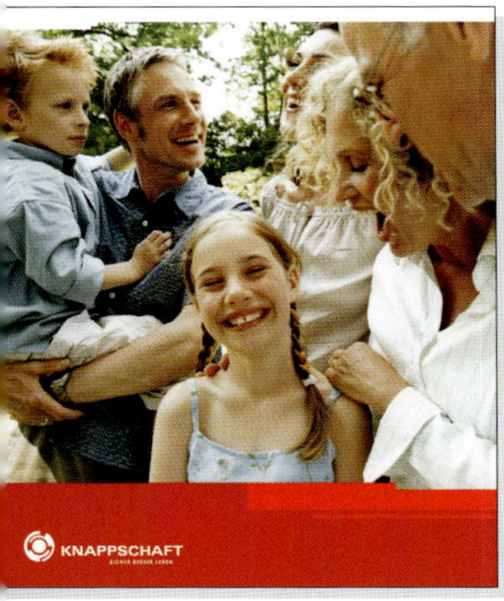

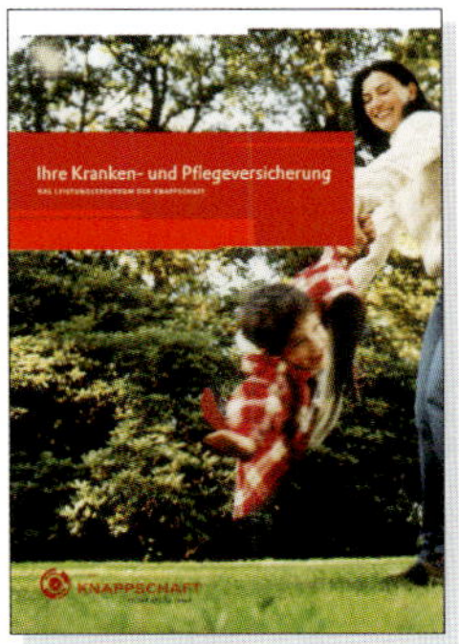

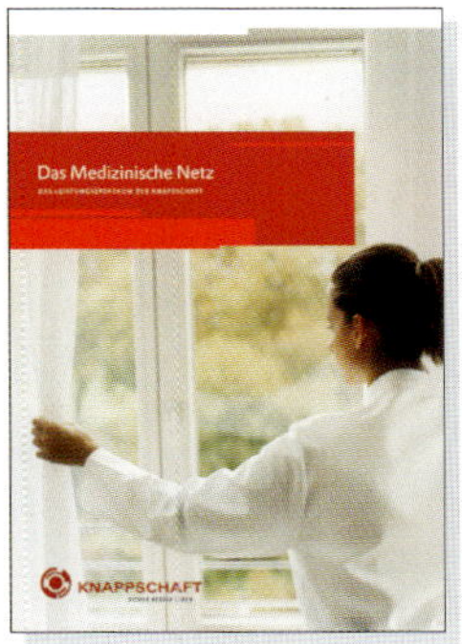

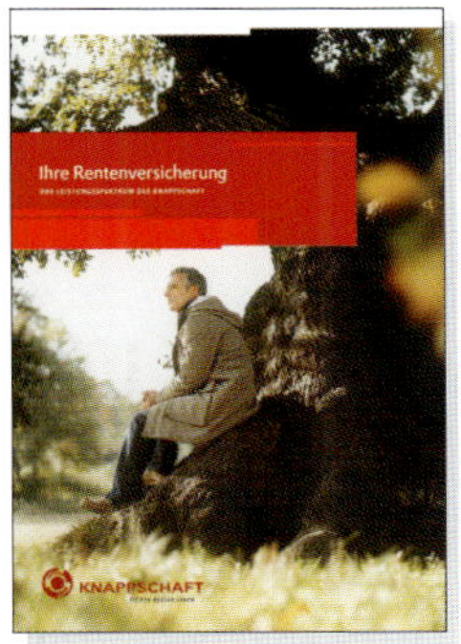

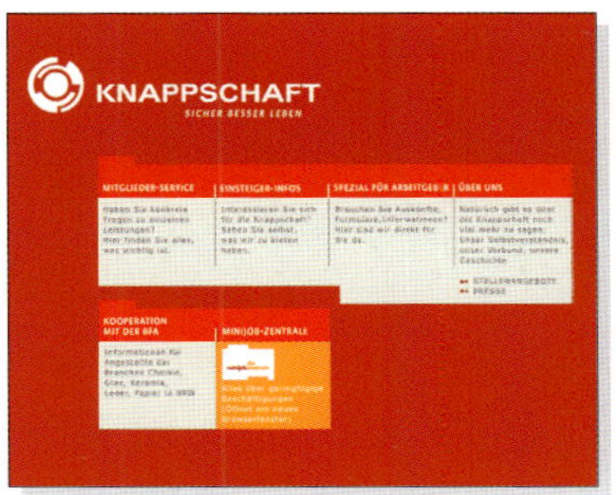

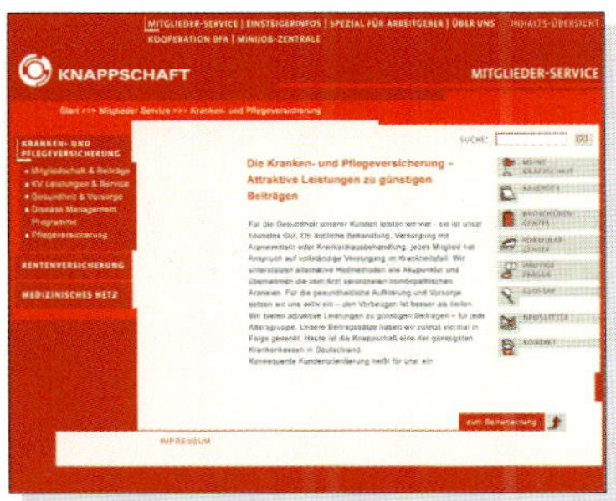

Auftraggeber Client
Sony Corporation
Tokyo, Japan

Produkt Product
AIWA-Markenidentität
AIWA Corporation CI

Design
Minako Gemma
Natsuki Kimura
Tokyo, Japan

Die Wellenform des neuen AIWA-Logos steht für das Reiten auf der Welle der Zeit und die Kraft, Dinge entschlossen in Bewegung zu setzen sowie die Flexibilität, neue Chancen zu suchen und sich anzupassen. Die von links nach rechts kräftiger werdende Linienführung soll den Willen kommunizieren, die Stärke der Marke mit der Zeit auszubauen, die Spiralform des Schriftzugs soll auf eine vertrauensvolle Beziehung zwischen AIWA und seinen Kunden anspielen. Die Farbe Rot schließlich steht für die Energie des Unternehmens und die Freude an seinen Produkten.

The wave shape of the new AIWA logo symbolizes the wish to ride the waves of time and the power to resolutely set things in motion, not to mention the flexibility to search for new opportunities and adapt along the way. The lines increase in emphasis from left to right, communicating the will to communicate, to expand on the brand's strengths over time, and the spiral shape of the logo is intended to allude to the spirit of trust which bonds AIWA and its clients. Finally, the color red stands for the company's energy and the joy brought by its products.

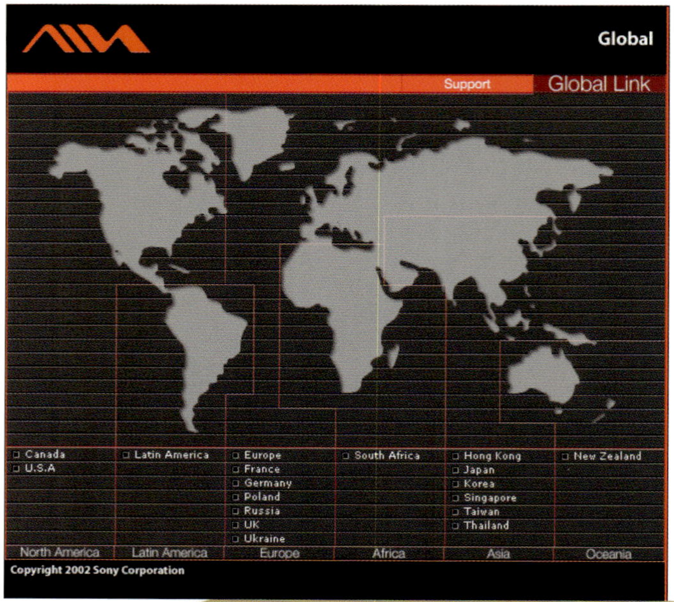

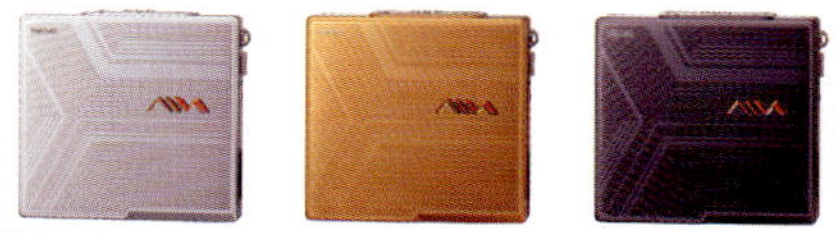

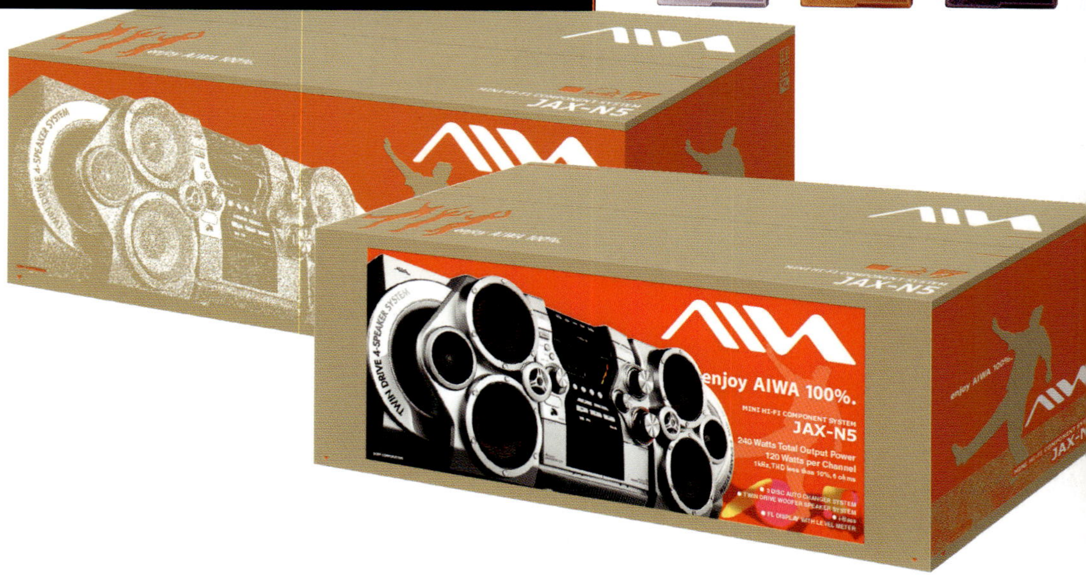

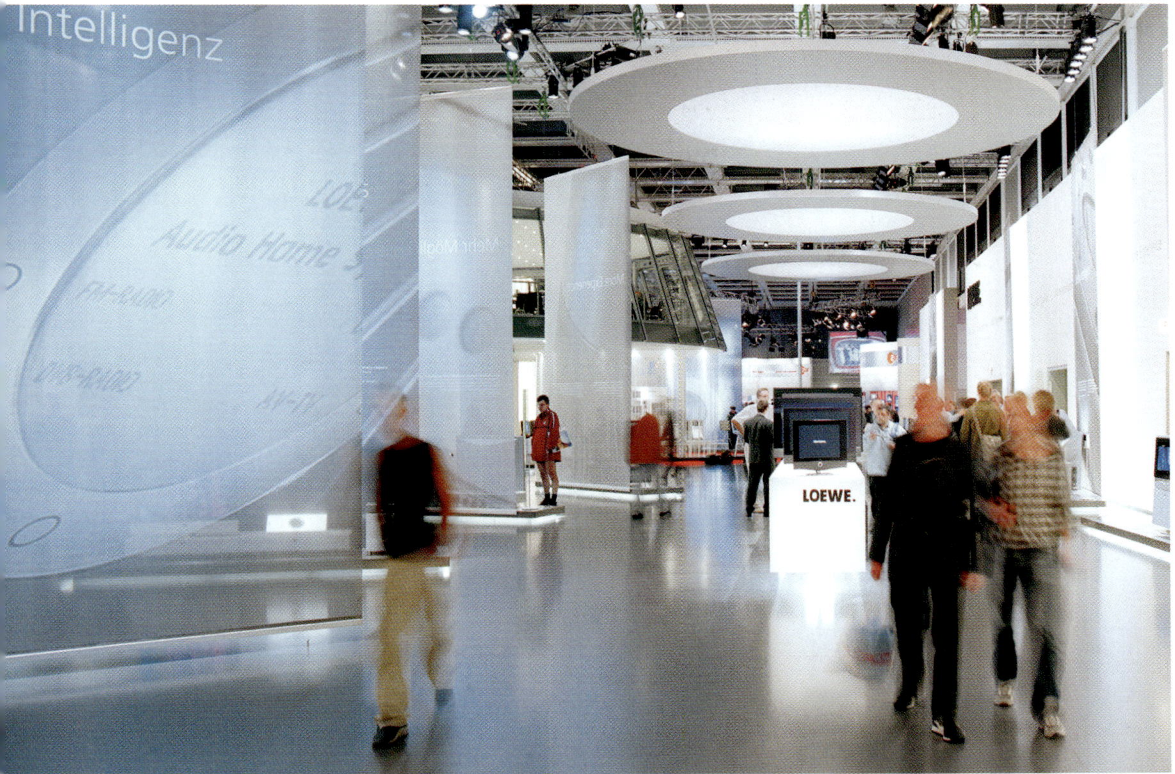

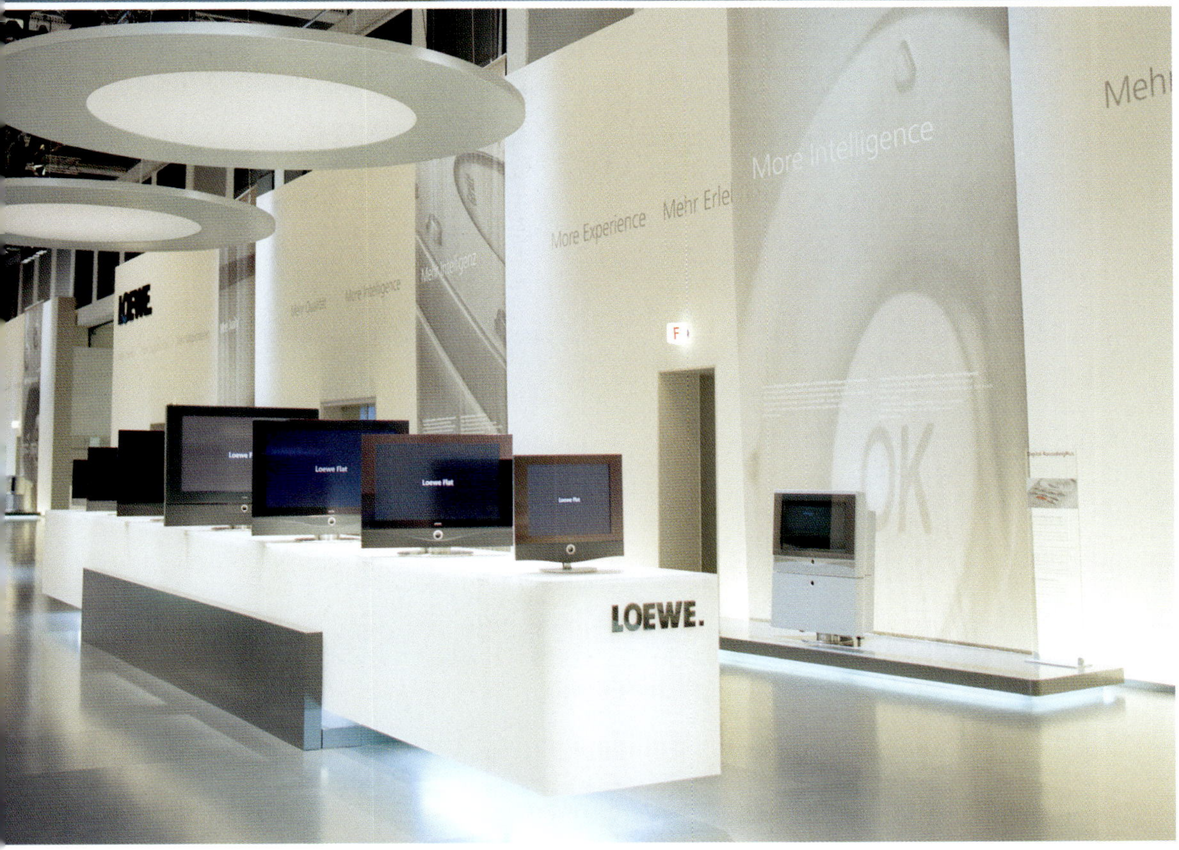

Auftraggeber Client
Loewe AG
Kronach, Germany

Produkt Product
Loewe Messestand IFA 2003
Messestand
Trade fair design

Design
KMS (Kommunikation)
München, Germany
Braun Wagner (Architektur)
Aachen, Germany
MAC (Messebau)
Langenlonsheim, Germany

Für den Auftritt des Unterhaltungselektronik-Herstellers Loewe wurde auf der diesjährigen IFA ein 3000-qm-Messestand entworfen. Sein Fokus lag auf den Bereichen Flat-TV und der High-End-Produktlinie „Systems". Das Motto „Mehr Loewe" thematisierte den Mehrwert von Loewe-Produkten und wurde über Motivfahnen und vielfältige Inszenierungen eingelöst. So befand sich im Zentrum die Highlight-Inszenierung mit den Flat-TVs, die sich um die eigene Achse drehten und über den Bildschirm Loewe-Produktfeatures kommunizierten. Zwei angrenzende Lounges zu den Themen Home Cinema sowie zu Home Multimedia und zum 'intelligenten Zuhause' boten Möglichkeiten zur vertiefenden Information. Abgeleitet aus der Produktsprache wurde der Kreis zur Grundform der Architektur; insgesamt dominierten Offenheit und Transparenz. Das Design kontrastierte zwischen organisch weicher Linienführung und klaren Architekturkomponenten.

A trade-fair stand was created for Loewe the home-entertainment company for the 2003 and IFA offering a full 3,000 sq. m. of space. The thematic focus: Flat TVs and the company's high-end "Systems" product line. A crucial part of the trade-fair presence was therefore given over to product displays. The architectural and communications idiom chosen highlight a sense of openness and transparency. The form of a circle plays a key role in Loewe's product language and as a consequence is also the basic shape taken for the architecture. The design juxtaposes soft organic lines and clear architectural elements.

Auftraggeber Client
Essent
Arnhem, Netherlands

Produkt Product
Beleef 2030
Buch
Book

Design
Fabrique (design, communication & new media)
Delft, Netherlands

Um einen besseren Einblick in die Zukunft der Energiebranche zu erhalten, organisierte der niederländische Energieversorger Essent im September 2002 die „Pieterpad-Konferenz". Drei Tage kamen die Teilnehmer wandernd miteinander ins Gespräch. Am Ende der Konferenz wurden vier Zukunftsvisionen als Grobszenarien in Form des Buchs „Beleef 2030" (2030 erleben) zu Papier gebracht. Um den Kontrast zwischen den Visionen zu betonen, wurde jede Vision im Buch auf einer anderen Papiersorte gedruckt. Der Umschlag ist mit schwarzer Thermotusche bedruckt, die sich durch Körperwärme weiß verfärbt. Hierdurch erscheinen Kerntexte der vier Szenarien, die zunächst bedeckt waren.

In order to gain a better insight into the future of the energy industry, in September 2003 Dutch power utility Essent organized the "Pieterpad Conference". For three days, the participants conversed with one another while hiking. At the end of the conference, four visions of the future were put down on paper as rough future scenarios: in the form of the book "Beleef 2030" (Experiencing 2030). In order to emphasize the contrasts between the different visions, each is printed on a different type of paper. The cover is printed with black thermo-ink with turns white when it comes into contact with body heat. In this way, the core texts of the four scenarios appear before the eye, whereas they had previously remained hidden from view.

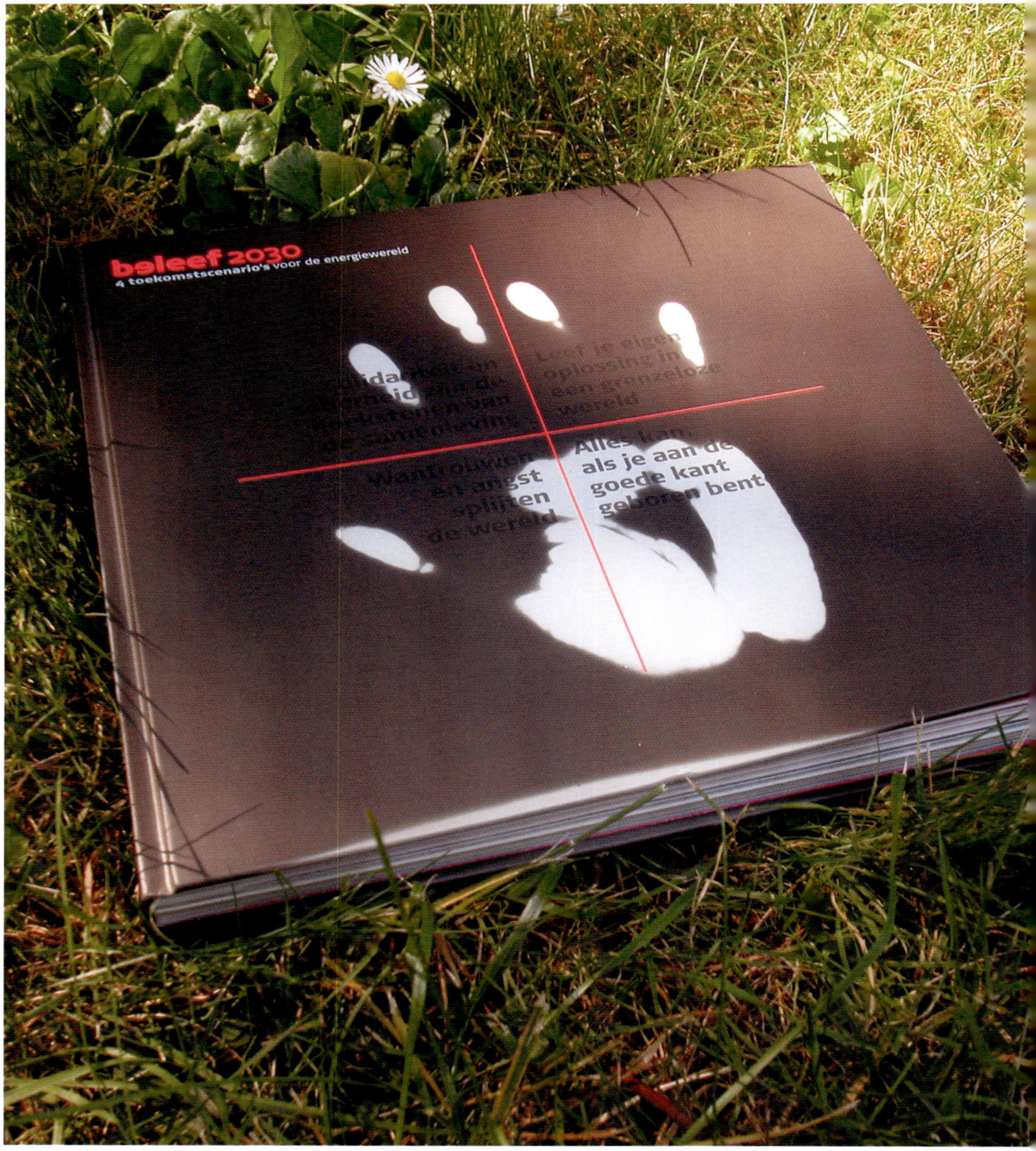

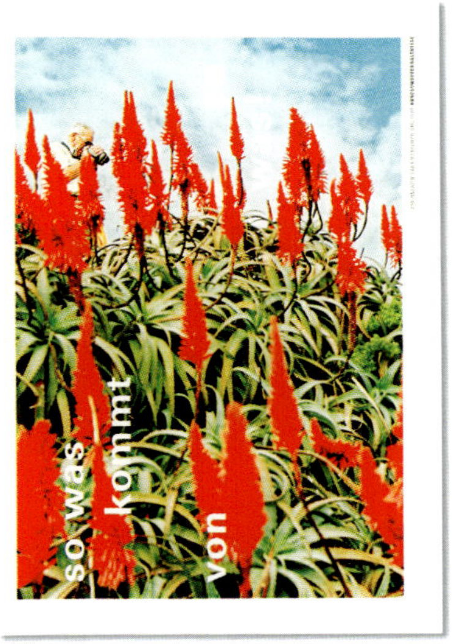

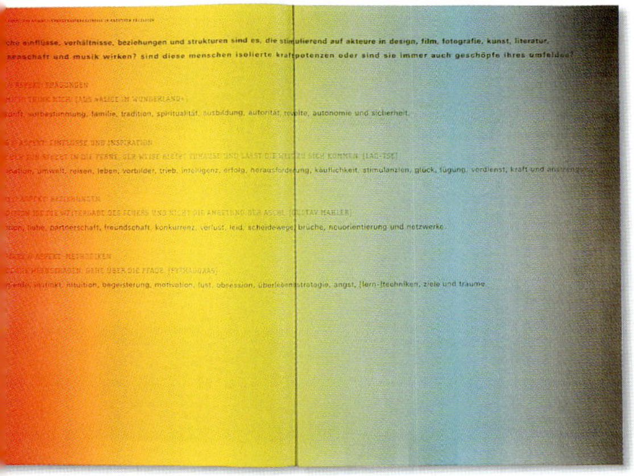

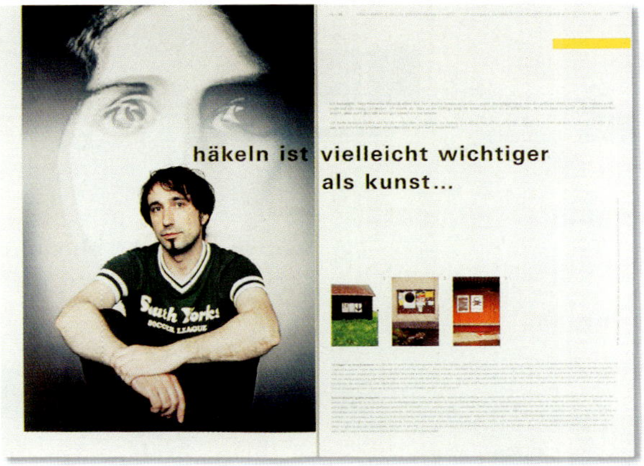

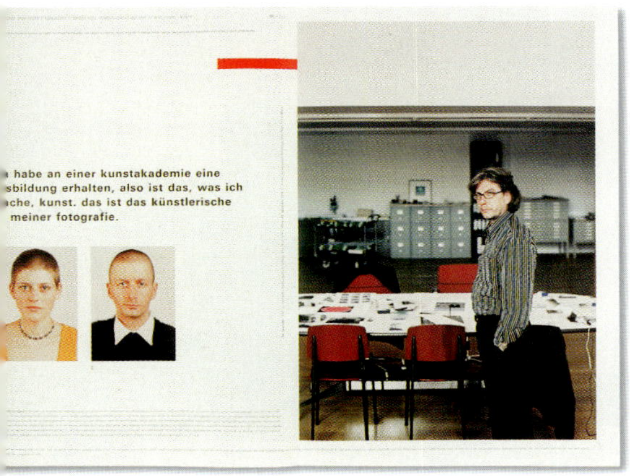

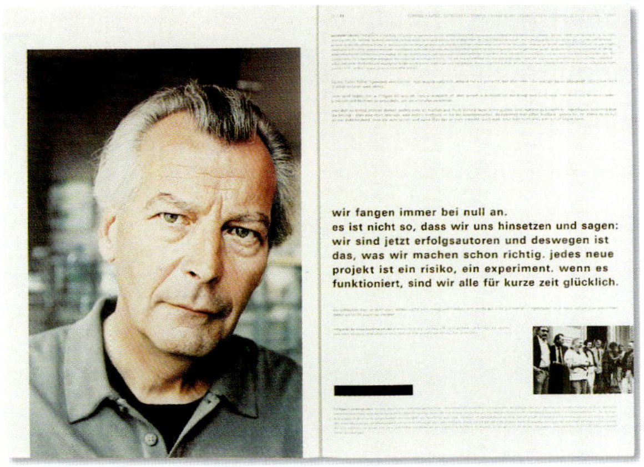

Auftraggeber Client
Hochschule für Künste Bremen
Bremen, Germany

Produkt Product
„so was kommt von so was"
anregungsverhältnisse in kreativen prozessen
Magazin
Magazine

Design
Siena Jakobi
Diplomarbeit bei Prof. Eckard Jung
Foto: Thomas Kleiner
Hochschule für Künste Bremen
Bremen, Germany

„so was kommt von so was" ist eine Diplomarbeit an der Hochschule für Künste Bremen, die versucht, dem Geheimnis von Anregungsverhältnissen in kreativen Prozessen auf die Spur zu kommen. Welche Einflüsse, Verhältnisse, Beziehungen und Strukturen sind es, die stimulierend auf Akteure in Design, Fotografie, Kunst, Literatur, Musik und Wissenschaft wirken? Sind diese Menschen isolierte Kraftpotenzen oder sind sie immer auch Geschöpfe ihres Umfeldes? Exemplarische Geschichten von Menschen aus unterschiedlichen Disziplinen geben in dieser Arbeit Anregungen aus ihren Tätigkeitsfeldern, erzählen ihren ganz persönlichen Lebensweg und verraten die Quellen, wie, wo und warum Wissen und Können entsteht. Unter anderem mit Beiträgen von Fettes Brot, Thomas Ruff, Mike Meiré, Till Brönner, Gert Dumbar und Thomas Bouchard.

"so was kommt von so was" is a graduate project submitted at the Bremen Academy of Art. It attempts to trace the secret of what prompts and stimulates creative processes. What influences, conditions, relationships and structures serve to fire the imagination of people at work in design, photography, art, literature, music and science? Are these people isolated powerful minds or are they the product of their environment? By way of example, here stories are offered of people from a variety of disciplines, who provide encouragement and ideas from their particular field, narrate their own personal approach to life, and reveal the sources on how, where and why knowledge and ability arise. Among others, with contributions by Fettes Brot, Thomas Ruff, Mike Meiré, Till Brönner, Gert Dumbar and Thomas Bouchard.

Auftraggeber Client
RWE AG
Essen, Germany

Produkt Product
agenda edition no.1
Image- und Unternehmensbuch
Image and corporate book

Design
Dirk Linke
für HOFFMANN UND CAMPE VERLAG GMBH
Corporate Publishing
Hamburg, Germany

Die „agenda edition no.1", ein Image- und Unternehmensbuch der RWE AG, ist als erste Ausgabe in einer Reihe dem Thema Wasser gewidmet. Sie enthält Originalbeiträge renommierter Politiker, Künstler, Philosophen und Wissenschaftler. Der italienische Künstler Fabrizio Plessi hat die Zeichnung eines Einbaums eigens für dieses Buch geschaffen. Das Motiv findet sich auf dem Umschlag und als Altarfalz im Buch. Eine Fotoserie der 16 größten Flüsse der Welt ist als doppelter vierseitiger Zickzackfalz eingebunden. Die Bilder wurden alle zur gleichen Zeit von 16 Fotografen aufgenommen. Bereits die Buchverpackung signalisiert das Thema Wasser: Eine „flüssige" Tasche, in einen Kunstschwamm eingebettet, schützt das Buch.

"agenda edition no.1" is an image book presenting the RWE AG company and, as the first issue in a planned series, is dedicated to the topic of water. It showcases original essays and statements by renowned politicians, artists, philosophers and scientists. Italian artist Fabrizio Plessi created the drawing of a dug-out specially for this book. The motif stands proud on the dust jacket and in the gatefold of the book. A photo series of the world's 16 largest rivers is inserted as a double four-page spread using an accordion fold. The pictures were all taken at the same time – by 16 different photographers. The book's packaging itself highlights the theme of water: the book is protected by a "fluid" pouch, embedded in artificial sponge.

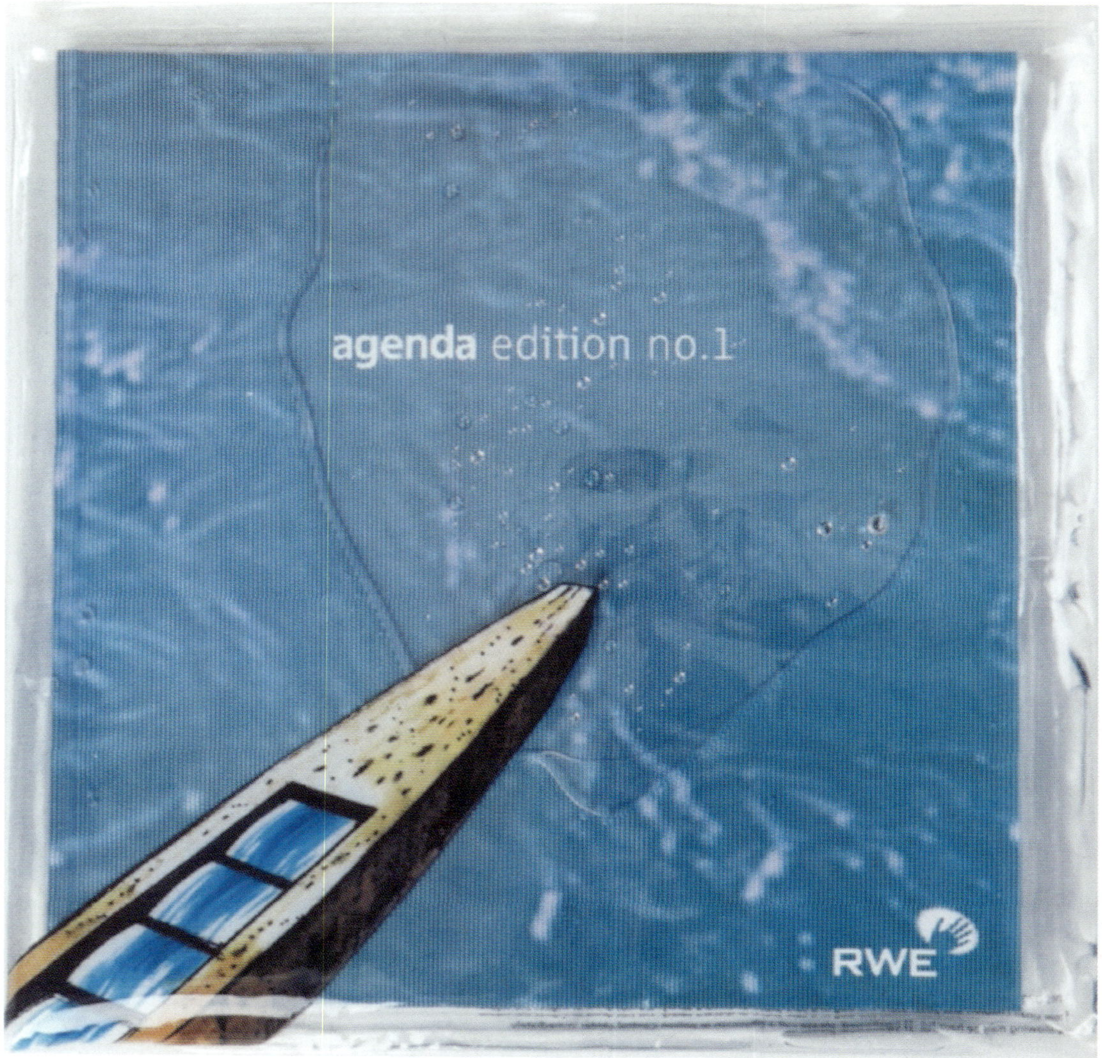

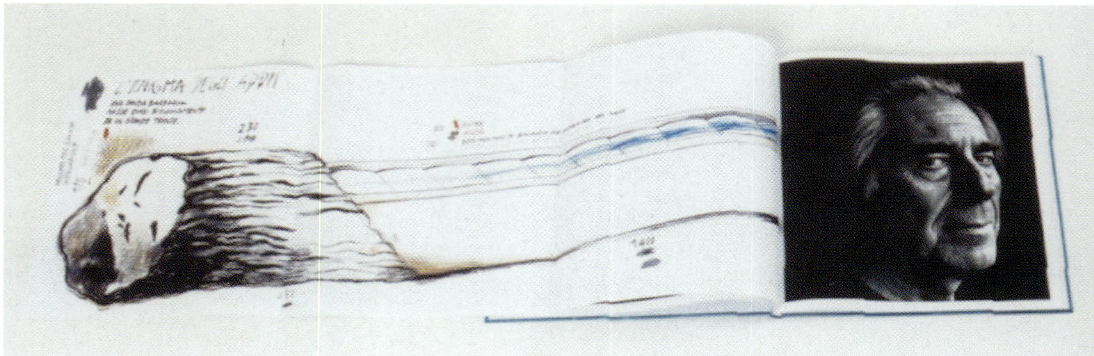

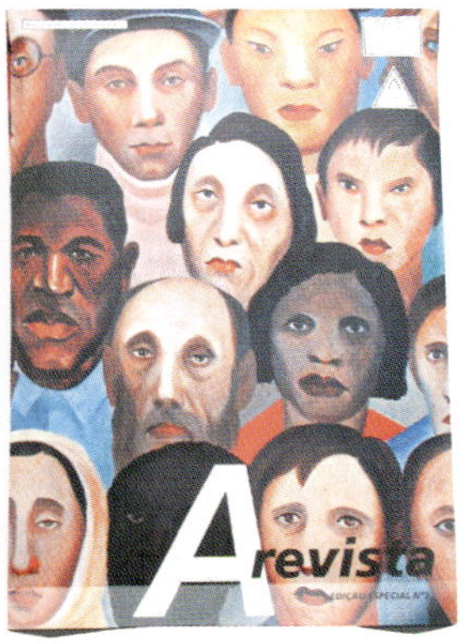

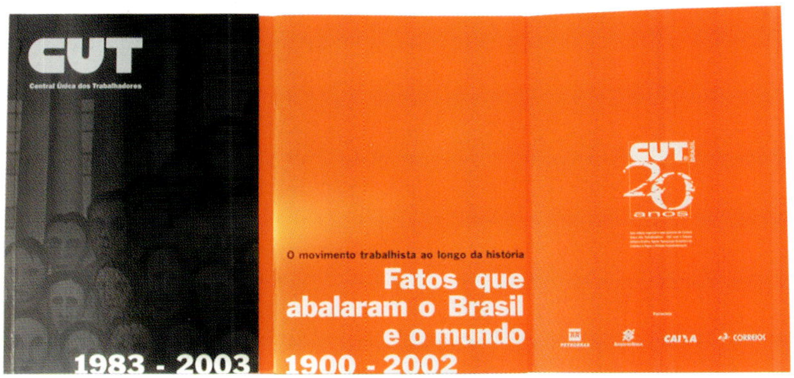

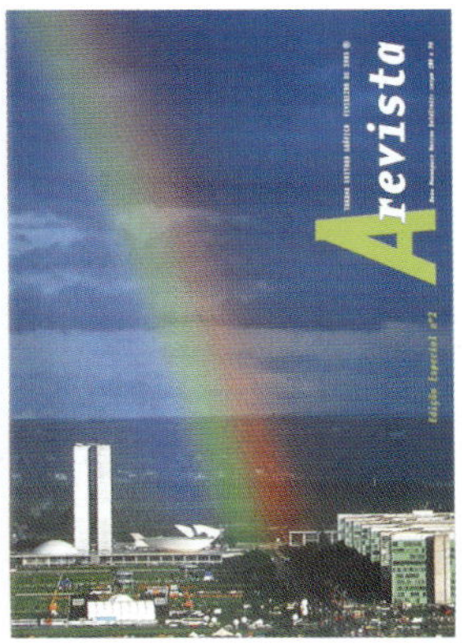

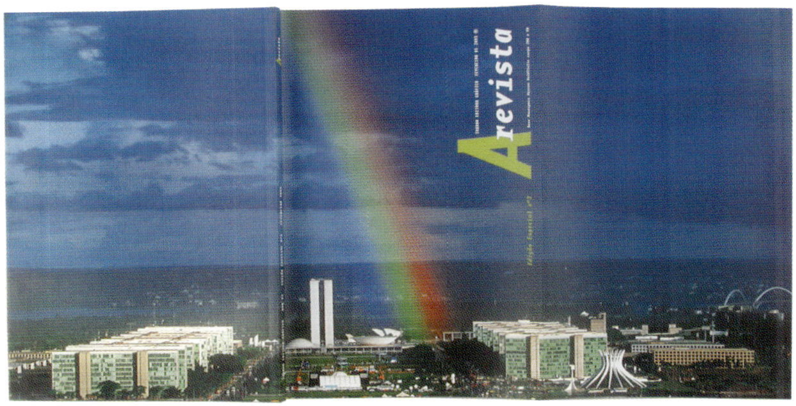

Auftraggeber Client
Takano Editora Gráfica
São Paulo, Brazil

Produkt Product
A Revista Special edition # 2
A Revista Special edition # 3
Zeitschrift
Magazine

Design
Fernanda Sarmento (Graphic Design, Design Direction)
Gisele Dias (Art Direction)
Melissa Roviriego, Carla Arbex, Rogério Alves
Takano Editora Gráfica
São Paulo, Brazil

„A Revista Special edition #2" und „A Revista Special edition #3" sind Ausgaben einer Zeitschrift der brasilianischen Druckerei Takano, die für Kunden, Unternehmer, Künstler, Grafiker und Meinungsbildner gedacht ist. Die Zeitschrift hat experimentellen Charakter und stellt sich durch ein klar definiertes gestalterisches Konzept mit wiederkehrenden Elementen wie dem Aufbau des Logos und der Vignetten der einzelnen Abteilungen dar, das sich aber von Ausgabe zu Ausgabe z.B. in Form des Einbands und der zur Anwendung kommenden technischen Mittel verändert. Der experimentelle Charakter entsteht durch die Darstellung unterschiedlicher grafischer Techniken, Abänderungen des Drucksatzes und der Behandlung der Bebilderung. Jede Ausgabe verwendet eine andere Schriftart, wobei das Logo trotz seiner festen Struktur diese Variationen begleitet.

"A Revista Special edition #2" and "A Revista Special edition #3" are editions of the magazine brought out by Braizilian printing company Takano and intended for a readership of clients, entrepreneurs, artists, graphic artists and opinion makers. The magazine is experimental in character and stands out for a clearly-defined design concept that hinges on recurrent elements, such as the structure of the logos and vignettes of the individual company departments. The concept otherwise changes from one issue to the next in terms of elements such as the form of cover chosen or the technical means applied. The experimental character arises from the presentation of different graphic techniques, alterations in type face and the treatment of illustrations. Each issue makes use of a different font, and for all its fixed structure the logo also undergoes these variations.

iF design award 2004

PRODUCT DESIGN

Disziplin 2 Discipline 2

Wo stehen wir heute?
Fritz Frenkler

Lucius Burckhardt formulierte 1980 den Satz „Design ist unsichtbar". Damit wollte er auf die sozialen und gesellschaftspolitischen Verflechtungen des Designs aufmerksam machen, die unter der Oberfläche wirken.

Gleichzeitig hat mit dem Beginn der Postmoderne das Design als sichtbare Oberfläche eine nie zuvor gekannte breite gesellschaftliche Akzeptanz gefunden. Es wurde das Potenzial als strategisches Instrument für Unternehmenspositionierung und als Produktdifferenzierung erkannt und durch die „Neuen Medien" und Globalisierung die Parameter unserer Wahrnehmung verändert. Aber ist das „Unsichtbare" dadurch „sichtbarer" geworden?

Fragen, um die es heute im Pluralismus des Designs geht: Über welche spezifischen Kompetenzen verfügen Designer? Erfüllt das Design jenseits der Gestaltung von Oberflächen heute tatsächlich soziale und gesellschaftspolitische Aufgaben? Welche Bedeutung haben Designer in den komplexen unternehmerischen Abläufen? Wo liegen künftig die zentralen Aufgaben in der Designausbildung? Ist das Design in der Lage, zwischen lokalen Bedürfnissen und globalen Anforderungen zu moderieren?

Die wirtschaftlichen Unterschiede und somit der Lebensstandard der Menschen zwischen den Industrieländern (USA, Japan, Europa) und sogenannten Schwellenländern oder Entwicklungsländern werden trotz Wirtschaftskrisen immer größer. So sind politische und soziale Spannungen vorprogrammiert. Zumal die Naturrohstoffe, die die Grundlage für das „gute Leben" in allen „überentwickelten Überflussländern" bilden, stammen zum größten Teil aus den industriell unterentwickelten Ländern. Im Wettbewerb der global agierenden Unternehmen haben sich die Entwicklungszeiten von technisch hochkomplexen Geräten ständig halbiert. Man spricht heute nicht nur bei den mobilen Telefonen von Saisonwaren.

Jedes Vierteljahr werden neue Geräte auf den Markt geworfen und mit riesigem Werbeaufwand an die Frau oder den Mann gebracht. Die „alten Geräte", auch die aus den eigenen Fabriken, verlieren so nach jeder Saison dramatisch an Wert.

Darüberhinaus erscheint es als realitätsfremd, dass versucht wird, eine immer älter werdende Weltbevölkerung mit immer „jünger" werdenden Produkten und mit immer „jünger" werdender Werbung zum Kauf dieser Produkte zu bringen.

Wir sind an dem Punkt angekommen, wo wir uns fragen: Kann dieses alles so weiter gehen? Es gibt also wieder Ziele im Leben! Wir umgeben uns wieder mit sinnvollen Produkten. Design kann dieses leisten und iF zeichnet solche Produkte aus.

Where do we stand today?

In 1980, Lucius Burckhardt stated that: "Design is invisible". He wished with this adage to draw attention to how design's social context and how its socio-political impact both function beneath the surface of what we see.

At the same time, with the advent of Post-Modernism, design as a visible outer surface has come to enjoy an unprecedented degree of acceptance in society. People became aware of its potential as a strategic instrument with which companies can position themselves and as a tool for product differentiation. Moreover, the "new media" and globalization have changed the parameters of how we perceive things. Does this, however, mean that the "invisible" has thus become "more visible"?

Questions that are important in today's pluralistic design world are: What specific expertise do designers have? Beyond the design of outer surfaces, does design nowadays actually fulfill any social function or play a socio-political role? Of what importance are designers in complex corporate processes? In future, what should the main focus in designers' training be? Is design in a position to interface between local requirements and global challenges?

Despite economic crises, the economic gap and as such the differences in standards of living between people in the industrialized countries (US, Japan, and Europe) and those in the threshold countries, as they have become known, and the developing countries is becoming ever greater. As such, political and social tensions are preprogrammed into the course of things, especially since for the most part those raw materials on which the "good life" in "over-developed lands of plenty" is based are in large part sourced in industrially under-developed countries. In the battle between the global corporate players, the lead times for developing highly sophisticated technical equipment have continually been halved.

Nowadays it is not just with regard to cell phones that we talk about seasonal goods. New devices are launched on the market every three months, and gigantic marketing campaigns persuade consumers to buy them. The "old devices", including those made by the same producer, therefore lose value dramatically after just one season.

Furthermore, the attempt by means of advertising that is becoming "ever younger" to persuade a world population that is becoming ever older to buy products that are becoming "ever younger" would appear to have little to do with reality.

We have reached a point at which we need to ask ourselves: Can things just carry on like this? In other words, there are indeed still goals in life! We are surrounding ourselves once again with products that have some sort of meaning. Design can achieve this and iF honors these products.

PRODUCT DESIGN

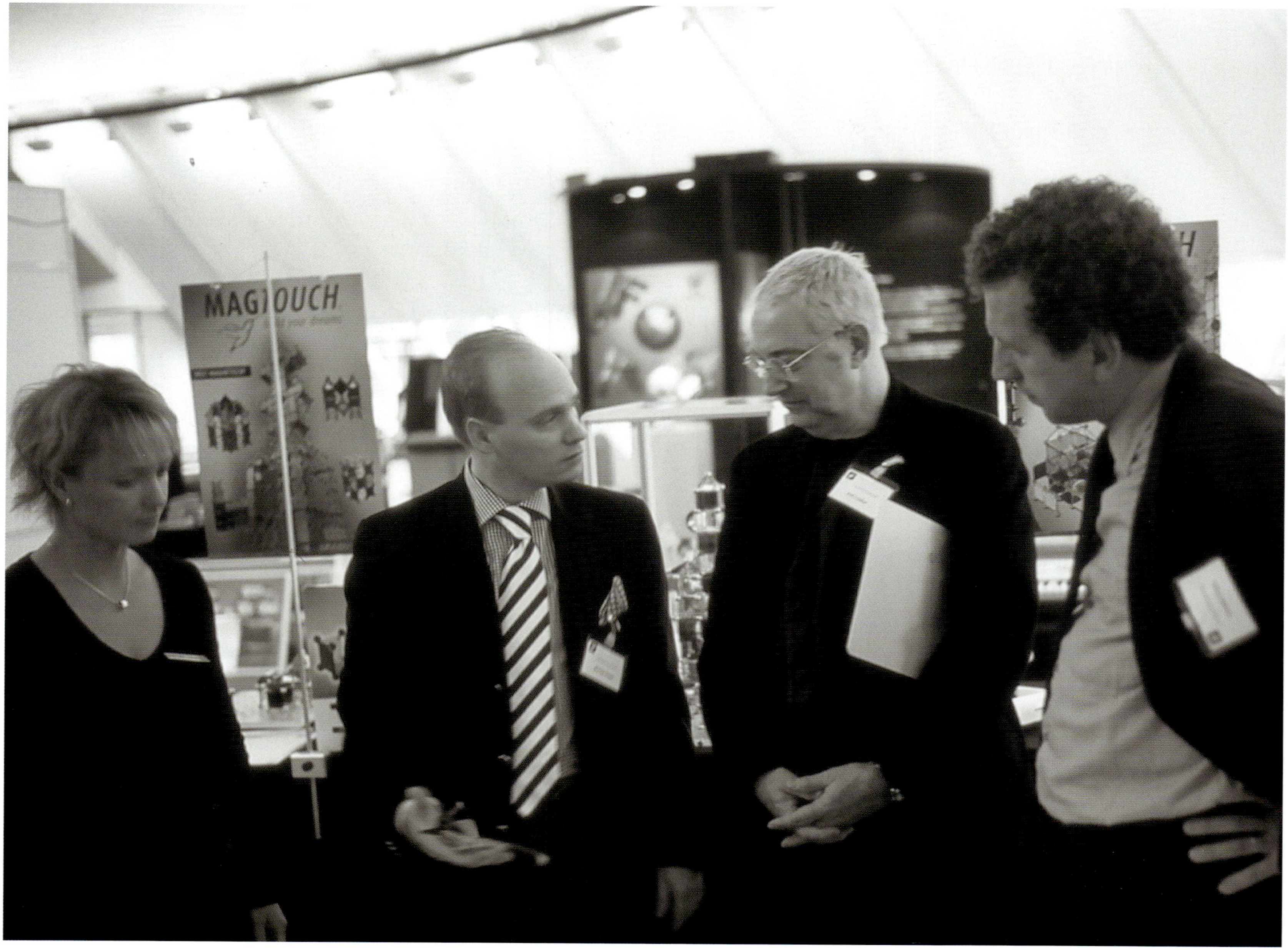

Freizeit Leisure
Lifestyle
Medizin Medicine
Office Business

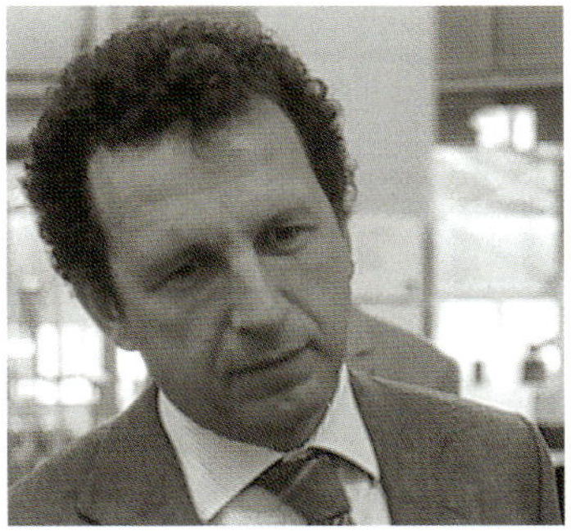

TORSTEN FRITZE studierte Industriedesign und Marketing am St. Martins College in London und in Mailand an der SDA Bocconi. Er ist neben Michele De Lucchi und Nicolas Bewick Mitbegründer von Studio & Partners, seit 2000 ist er mit Nicolas Bewick Alleininhaber. Von 1994 bis 1999 war er zudem für das Esprit-Projekt der Europäischen Kommission tätig sowie von 1999 bis 2001 als Dozent an der Hochschule für bildende Künste Hamburg.

Torsten Fritze studied industrial design and marketing at St. Martins, London and in Milan at the SDA Bocconi. Together with Michele de Lucchi and Nicolas Bewick he founded Studio & Partners, solely owned since 2000 by himself and Nicolas Bewick. From 1994-9, he handled the European Commission's Esprit project, and from 1999 to 2001 lectured at the Hamburg Academy of the Applied Arts.

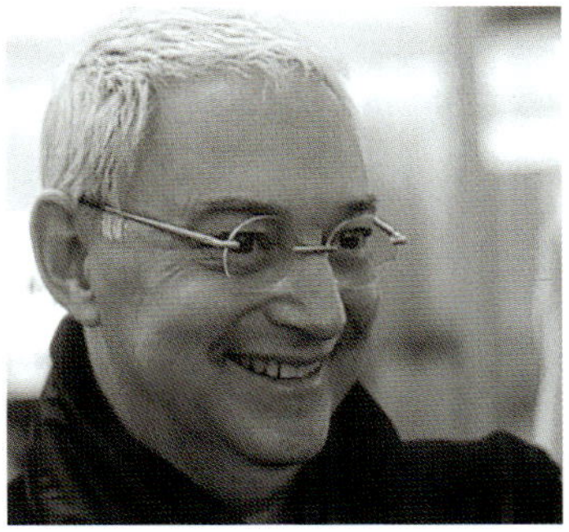

NEIL FRANKEL ist Mitinhaber der Designfirma Frankel & Coleman und Gastprofessor an der University of Wisconsin, Milwaukee, zum Thema Designkritik. Außerdem ist er Mitglied des American Institute of Architects, der International Interior Design Association wie auch des Design Futures Council.

Neil Frankel is co-partner of the Frankel & Coleman design company and as a design critic Visiting Professor at the University of Wisconsin, Milwaukee. He is a member of the American Institute of Architects, the International Interior Design Association and the Design Futures Council.

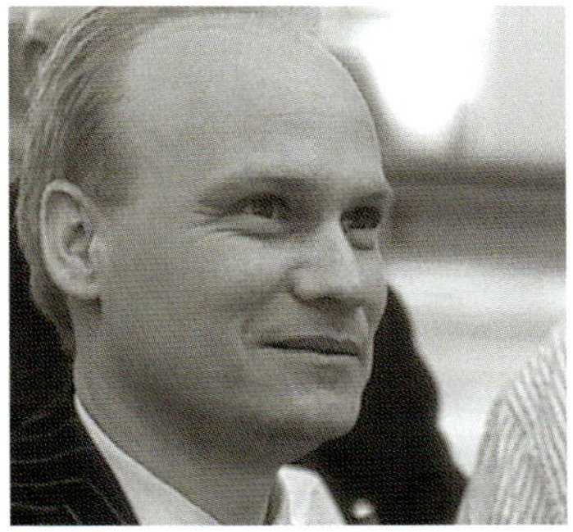

MICHAEL SIEGER ist Design-Unternehmer auf Schloss Harkotten im Münsterland. Nach ersten Projekten schon zur Schulzeit und dem Eintritt in das Familienunternehmen erhielt er Designpreise und Auszeichnungen für Grafik, Fotografie und Produktdesign. Heute ist er verantwortlich für Markenkonzepte wie Ritzenhoff, Arzberg und Octopus.

Michael Sieger is a design entrepreneur based at Harkotten Castle in Central Germany. Initial projects while still at school, after which he entered his family's company, winning design prizes and awards for graphic design, photography and product design. Today he handles brand concepts for companies such as Ritzenhoff, Arzberg and Octopus.

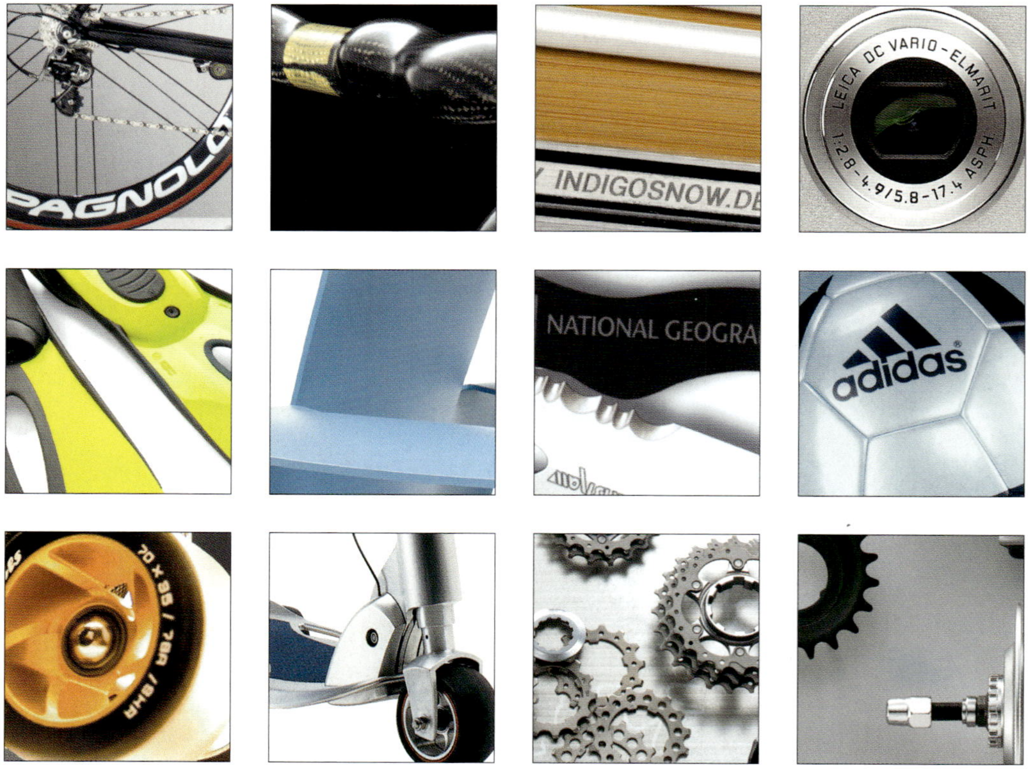
PAGNOL
INDIGOSNOW.DE
LEICA DC VARIO-ELMARIT
1:2.8-4.9/5.8-17.4 ASPH
NATIONAL GEOGRA
adidas

Freizeit Leisure

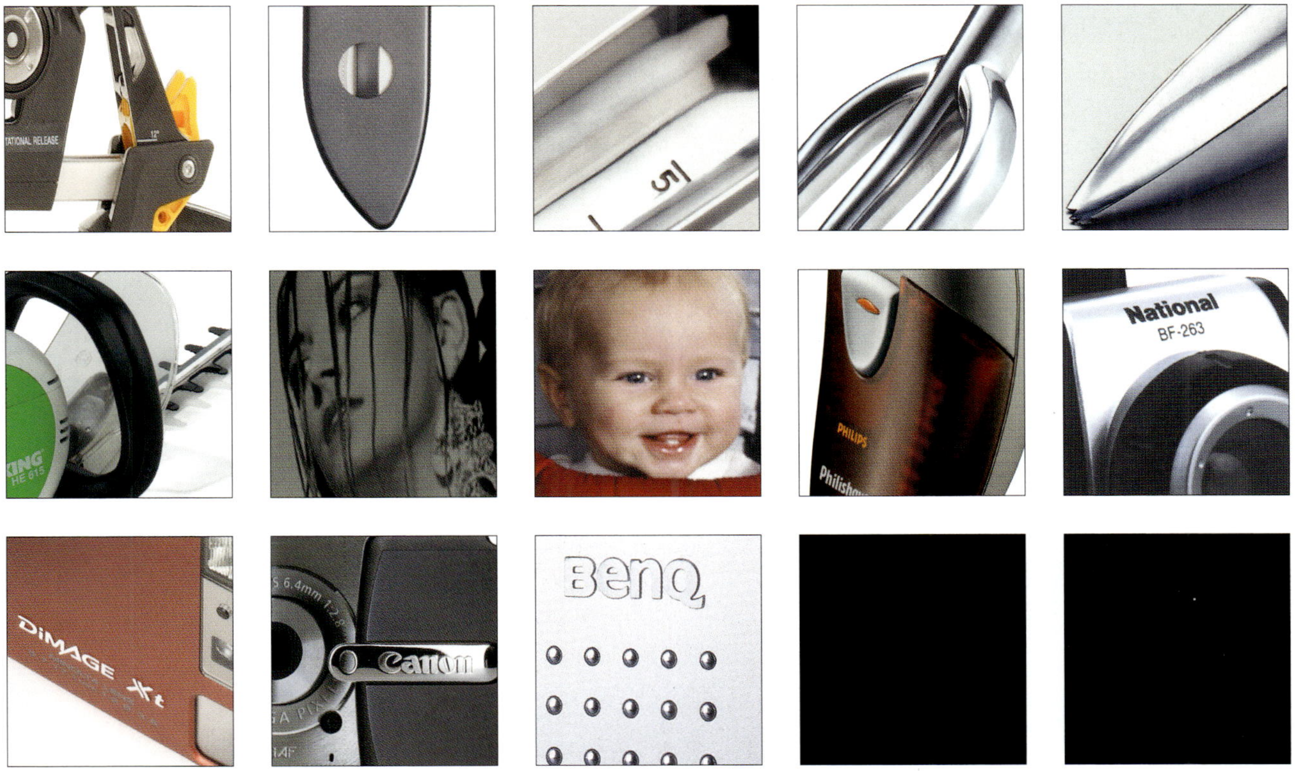

Hersteller Manufacturer
Storck Bicycle
Bad Camberg, Germany

Produkt Product
Scenario Carbon
Fahrrad
Race bike

Design
Storck Bicycle
Bad Camberg, Germany

Das Scenario Carbon hat trotz seines außerordentlich geringen Gewichts von unter 1000 g sehr hohe Steifigkeitswerte. Aufgrund seiner Kompaktgeometrie ist es in vier verschiedenen Rahmenhöhen erhältlich, die alle Fahrergrößen abdecken. Das klassische Rennradrahmen-Konzept wurde insbesondere für ambitionierte Rennradfahrer und den Renneinsatz konzipiert.

Despite its exceptional low weight of less than 1000 g, the Scenario Carbon frame boasts superlative STW ratios. With its compact geometry, it is available in four different frame sizes and different downtube diameters to accommodate all riders. The classic race bike concept is destined specifically for ambitious amateurs and dedicated racing.

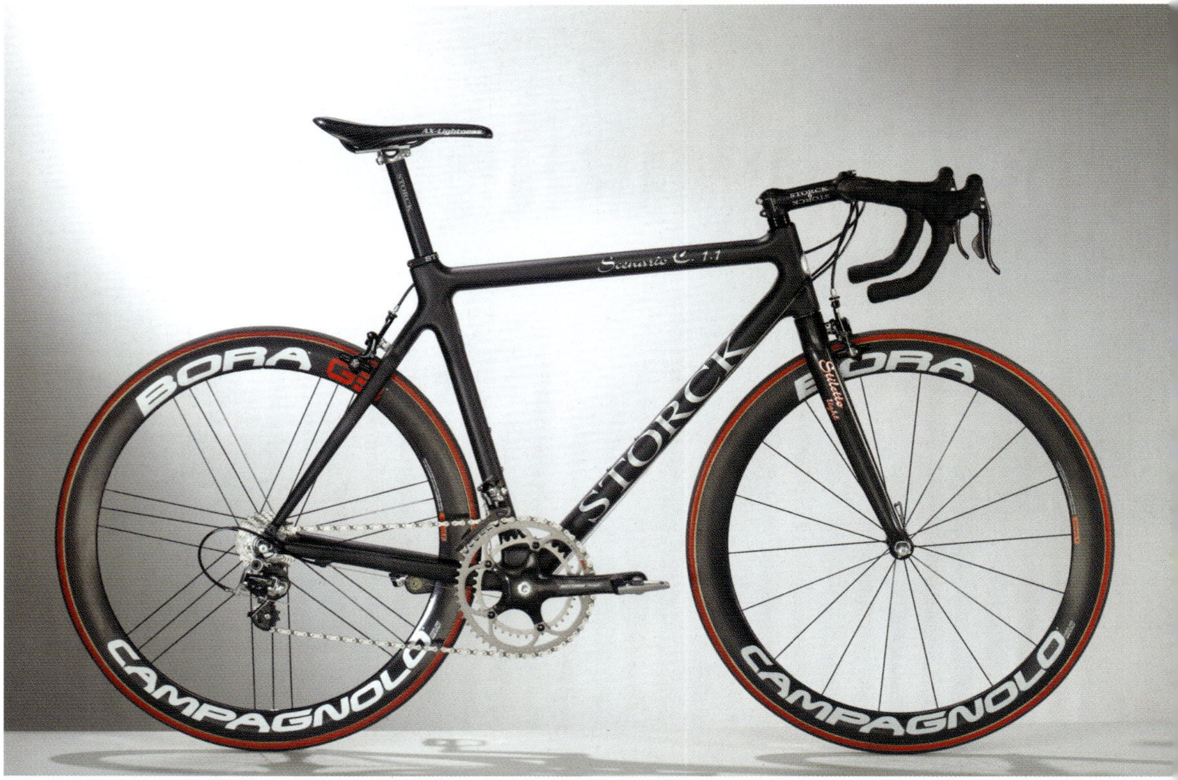

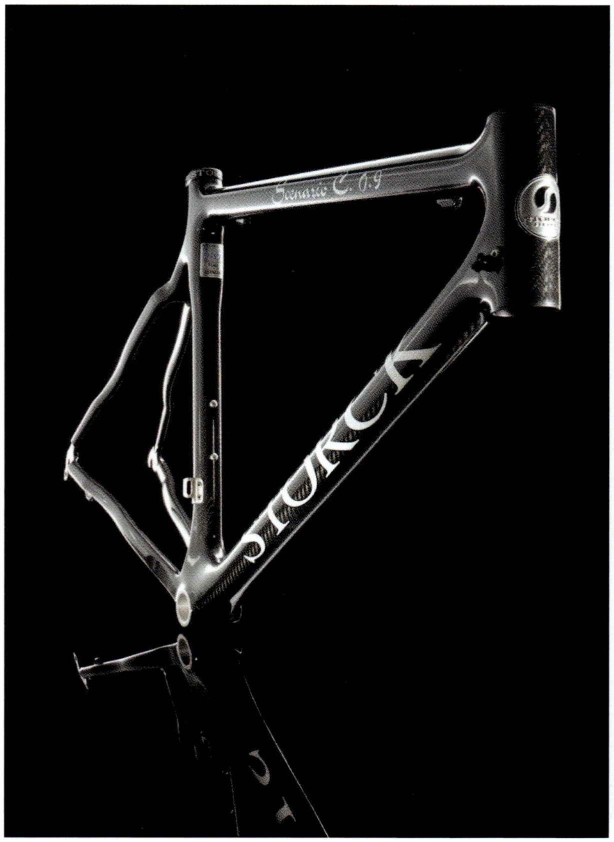

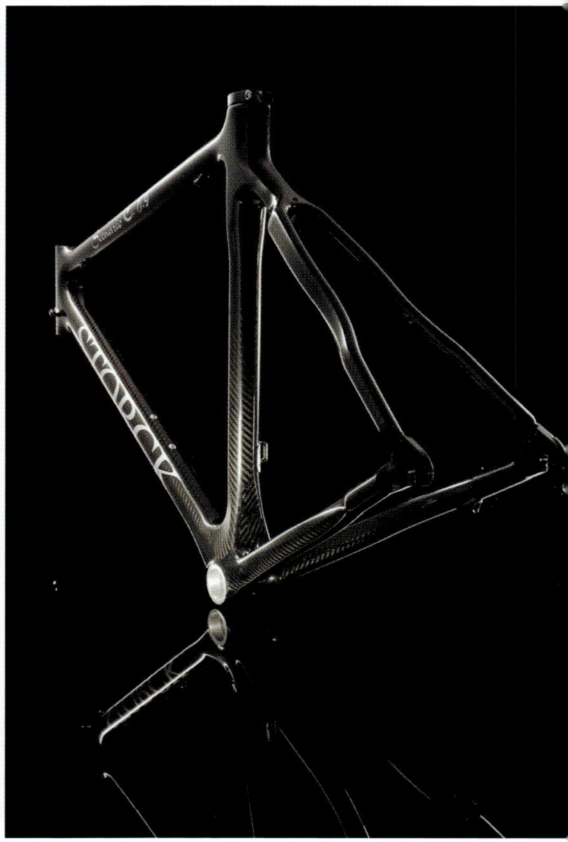

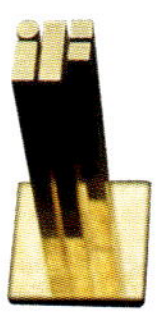

iF Gold Award

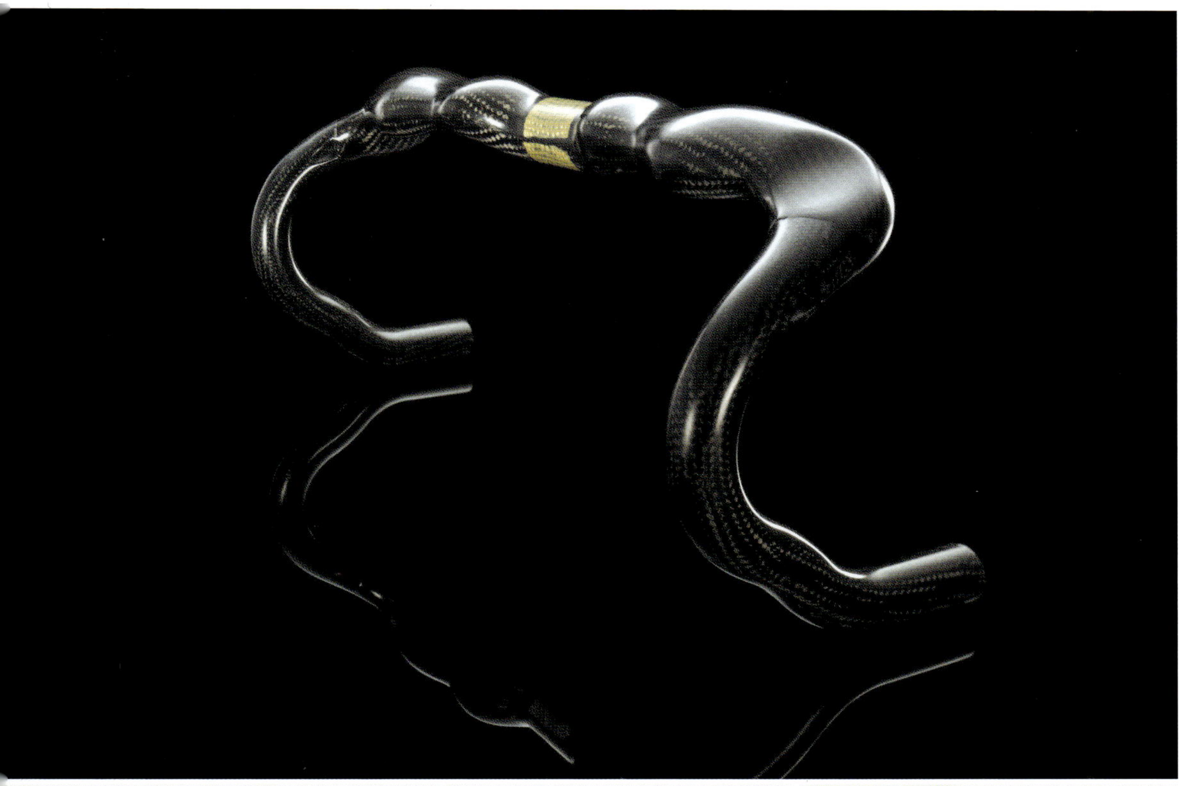

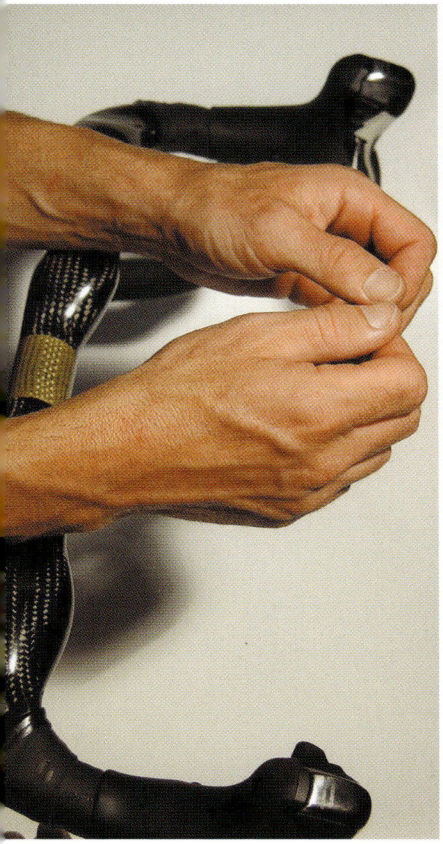

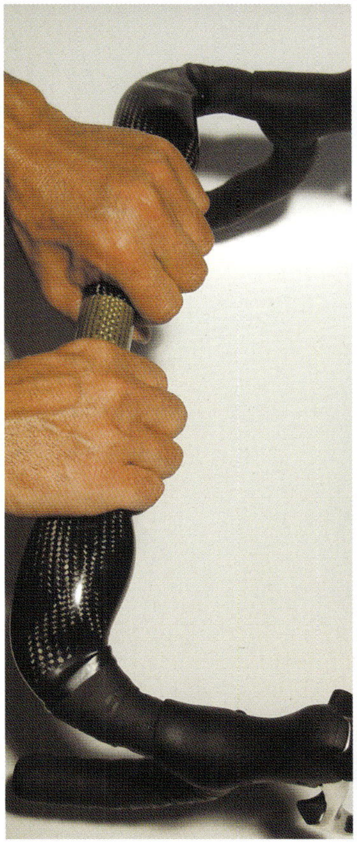

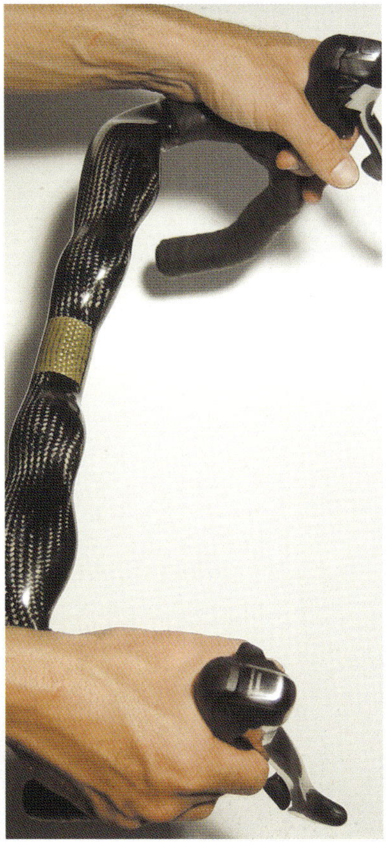

Hersteller Manufacturer
Storck Bicycle
Bad Camberg, Germany
THM- Faserverbundtechnologie GmbH
Alt Duvenstedt, Germany

Produkt Product
Ergo Lenker Carbon
Fahrradlenker
Cycle handlebars

Design
Storck Bicycle
Bad Camberg, Germany

Der Ergo Lenker zählt zu den neuesten Produktentwicklungen von Storck. Durch den Werkstoff Carbon konnte ein Rennlenker entwickelt werden, der eine ergonomische Form hat und sechs verschiedene Griffpositionen erlaubt. Trotz des niedrigen Gewichts von nur 200 g bietet er eine sehr hohe Festigkeit.

The Ergo handlebars are among Storck's latest innovative products. By using carbon as the material of choice, handlebars have been developed that have an ergonomic shape and offer riders six different grip positions. Despite their modest weight of only 200 g, the bars provide optimal stiffness.

iF Gold Award

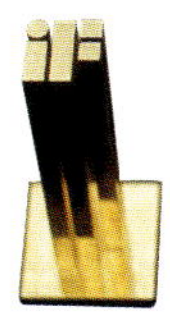

Hersteller Manufacturer
BOGNER SKI
Schwabe & Baer Entwicklungs GmbH
München, Germany

Produkt Product
BOGNER SKI
Bambus-Ski
Bamboo ski

Design
Willy Bogner
und Schwabe & Baer Entwicklungs GmbH
München, Germany

Natural-high-tech. Bei der Entwicklung des BOGNER SKI suchte man gezielt nach einem natürlichen, schnell nachwachsenden Rohstoff, welcher als Alternative zu branchenüblichen Werkstoffen eingesetzt werden kann. Durch die besonderen, materialspezifischen Eigenschaften von Bambusgras, wie hohe Belastbarkeit auf Zug, Druck und Torsion bei verhältnismäßig geringem Gewicht, ist es vergleichbaren Materialien weit überlegen. Es entstand somit eine neuartige technische Konstruktion eines traditionellen Produktes. Über die Verwendung von Bambus in der Kernkonstruktion hinaus wird der Werkstoff auch als Oberfläche eingesetzt. Kratzer und oberflächliche Gebrauchsspuren können leicht behoben werden. Durch die jeweils unterschiedliche Maserung des Bambus wird jedes Produkt zu einem einmaligen Zeichen seiner Zeit.

Natural high-tech. When developing the BOGNER SKI the emphasis was on using a natural, fast-growing regenerative material as an alternative to the materials customary in the industry. Thanks to the special properties of bamboo grass, such as high tensile strength, torsion and pressure load, not to mention its light weight, it is decidedly superior to comparable materials. In this way, an innovative technical construction has been adopted for a traditional product. Not only can bamboo be used in the ski's core but also as the surface finish. Scratches and traces of usage on the surface can easily be removed and the grain of the wood means each ski is unique.

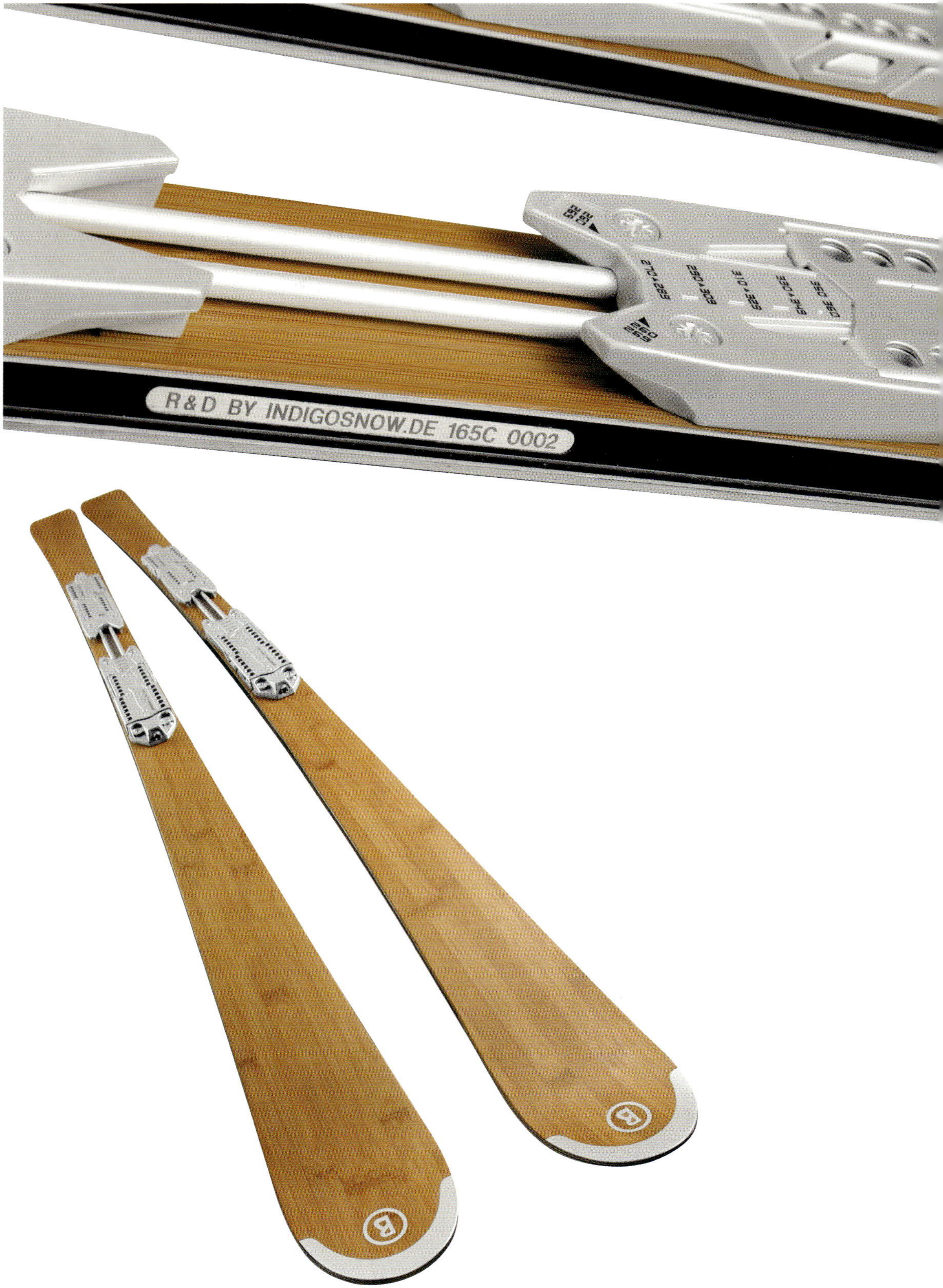

iF Gold Award

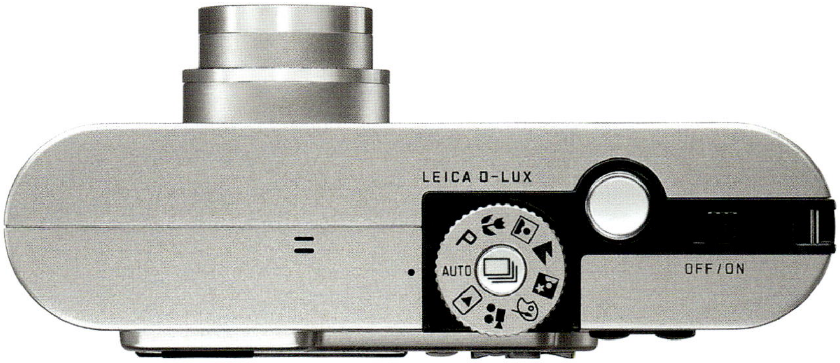

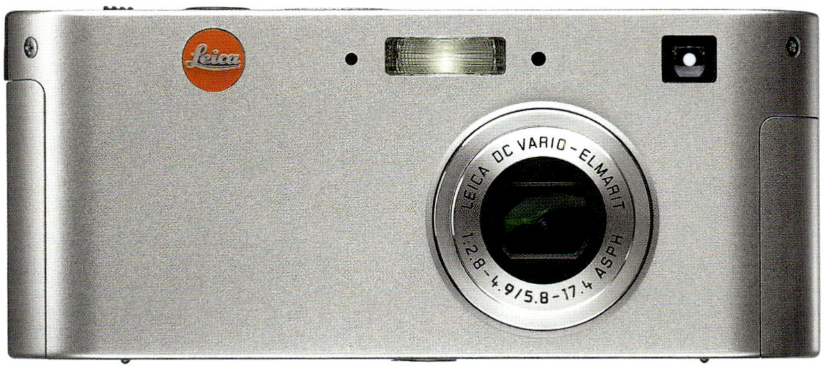

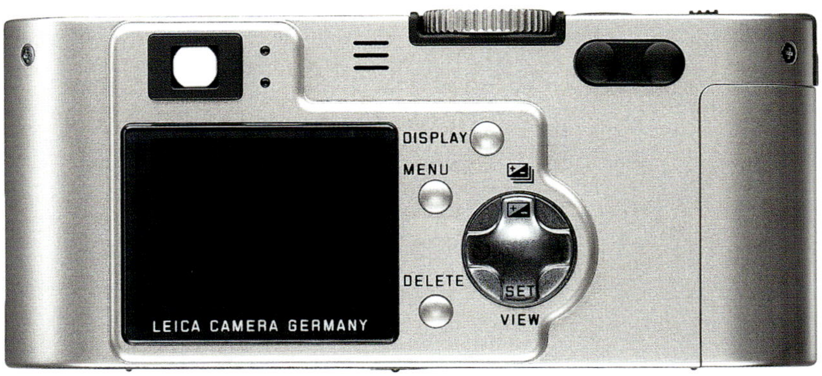

Hersteller Manufacturer
Leica Camera AG
Solms, Germany

Produkt Product
Leica D-Lux
Digitalkamera
Digital camera

Design
Achim Heine
Berlin, Germany

In ihrem Design lehnt sich die neue Leica D-Lux mit 3fach-Zoomobjektiv bewusst an die zeitlose Formensprache der klassischen Leica Reportagekameras an. Das Metallgehäuse aus Aluminium unterstützt diesen Look. Details wie das Einstellrad sowie die beiden Abdeckklappen für Batterie, Speicherkarte und Anschlüsse sind ebenfalls aus massivem Metall gearbeitet. Die Symbole des Einstellrades sind – dauerhaft gegen Abrieb geschützt – in das Aluminium eingraviert und schwarz ausgelegt. Die Kamera lässt sich einfach bedienen: Alle Betriebsarten sind mit dem übersichtlichen Einstellrad auszuwählen und die wichtigsten Funktionen mit einem Knopfdruck zu steuern.

The design of the new Leica D-Lux with its 3x zoom lens deliberately borrows from the timeless visual vocabulary of the classic Leica photojournalist cameras. The aluminum body reinforces this image. Details such as the adjusting dial and the two flap covers for the battery, memory card and sockets are likewise made of solid metal. The symbols are engraved into the aluminum of the dial and then dyed black to ensure they remain legible for the camera's entire service life. The camera has well laid out functions and is simple to use. All operational modes can be selected using the dial and the key functions are push-button controlled.

iF Gold Award

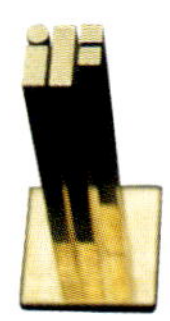

Hersteller Manufacturer
Decathlon Sportartikel GmbH & Co.KG
Dortmund, Germany

Produkt Product
Flossen / Tribord FLP 500
Flossen
Flipper

Design
Decathlon Design
Villeneuve d'Ascq, France

Das Design der FLP 500 Flossen ist von der Ästhetik der Unterwasserwelt inspiriert. Der Fußteil wurde zusammen mit Profis aus dem Sportschuhbereich entwickelt. Er vereint Tragekomfort und ergonomische Qualitäten mit einer optimierten Übertragung der Muskelkraft auf das Flossenblatt. Das Flossenblatt ist vor dem Fuß durchbrochen, da dieser Bereich einen unnötigen Widerstand beim Vortrieb erzeugt. Aufgrund der hydrodynamischen Gestaltung muss der Schwimmer für die gleiche Geschwindigkeit entsprechend weniger Kraft aufwenden.

The design of the FLP 500 flippers was inspired by the aesthetics of the underwater world. The foot section was developed with the assistance of professionals from the sports shoe segment. It combines wearer comfort and ergonomic qualities with optimized transmission of muscle power to the flipper blades. The latter is split in front of the foot as the area would otherwise cause unnecessary resistance to forward thrust. Thanks to the hydrodynamic design, the swimmer requires less power than with conventional blades to achieve the same speed.

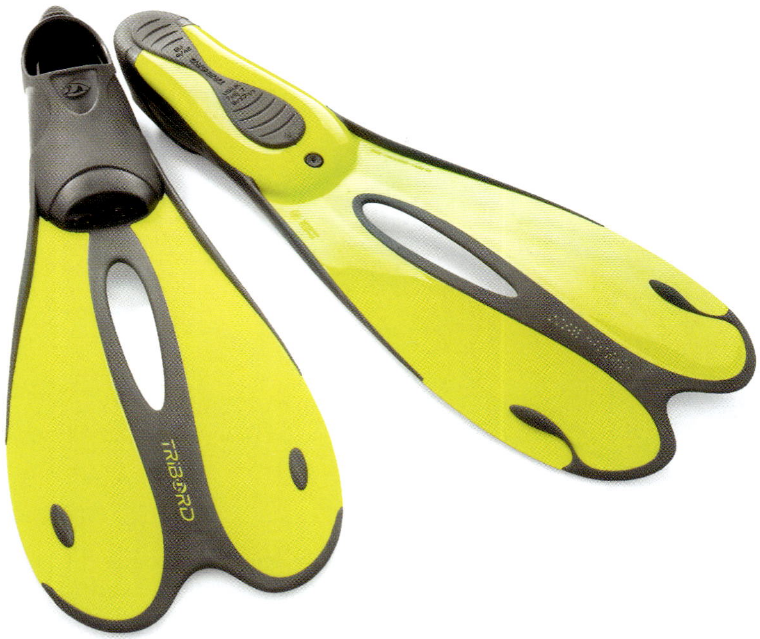

iF Silver Award

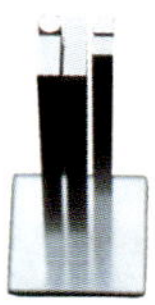

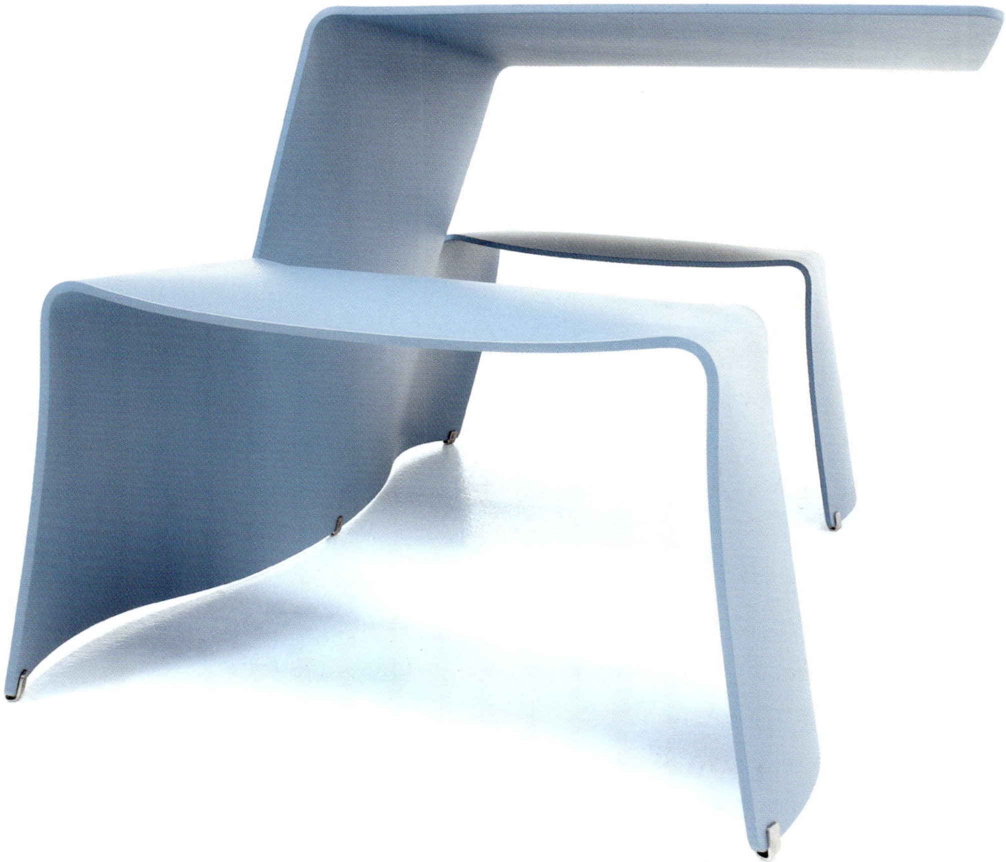

Hersteller Manufacturer
Extremis
Gijverinkhove, Belgium

Produkt Product
PicNik
Sitzmöbel und Tisch für 2
Seating furniture with integrated table for 2

Design
Xavier Lust
Dirk Wynants

PicNik ist ein Möbelstück, das von Dirk Wynants und Xavier Lust insbesondere für kleine Terrassen und Outdoor-Flächen, wie man sie oft im städtischen Umfeld findet, entworfen wurde. PicNik ist kompakt (er passt durch alle Türen und Fahrstühle) und stapelbar und kann das ganze Jahr über draußen stehen bleiben. Der Tisch und die zwei Sitze sind aus einem Stück einer Standardaluminiumplatte (pro Platte können zwei PicNiks hergestellt werden). Es gibt so gut wie keinen Abfall und Aluminium ist zudem korrosionsbeständig und recyclingfähig. Außer den fünf Standardfarben sind auch andere Farben auf Wunsch erhältlich.

PicNik is a piece of furniture created by Dirk Wynants and Xavier Lust with those small patios and outdoor areas in mind, such as are often encountered in urban settings. PicNik is compact (it fits through all doors and into elevators) is easy to stack, and can be left outside all year round. The table and the two seats are made of a single piece of standard sheet aluminum (two PicNiks can be made from one sheet). There is as good as no waste produced and, into the bargain the aluminum is corrosion-resistant and can be recycled. In addition to five standard colors, other customized colors can be provided on request.

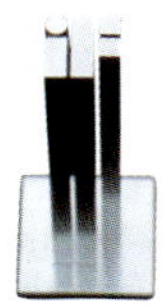

iF Silver Award

Hersteller Manufacturer
Kai Corporation
Tokio, Japan

Produkt Product
Kershaw 1004
National Geographic Tool
Taschenmesser
Carabiner knife

Design
National Geographic Society

Das Kershaw 1004 National Geographic Tool ist durch die Verbindung von Karabinerhaken und Taschenmesser ein sehr handliches und nützliches Outdoortool. Es kann durch die ergonomische Gestaltung in den unterschiedlichsten Positionen eingesetzt werden, wobei der Benutzer durch den Karabinerhaken hindurch greifen oder auch den ganzen Griff umfassen kann. Der Karabinerhaken macht die üblichen Befestigungsmechanismen überflüssig und erlaubt eine schnelle und leichte Handhabung des Messers. Zusätzlich wurden ein Flaschenöffner, ein Schraubenzieher und eine Notfallklinge integriert. Der Griff des Messers wurde aus Aluminium gefertigt, die Klinge besteht aus Edelstahl.

The Kershaw 1004 National Geographic Tool is a striking combination of carabiner fastener and knife as well as other useful outdoor tools. Thanks to its ergonomic shape it can be used in all manner of positions, and the user can place their fingers through the aperture of the body for increased one-hand dexterity or hold it the traditional way. The carabiner hook renders a side clip unnecessary, thanks to the quick-fastener at the heart of the tool, ensuring swift and easy availability. The tool features an integrated bottle opener, Phillips and flathead screwdrivers, and a Rescue-style knife blade. The body comes in aluminum, while the blade is stainless steel.

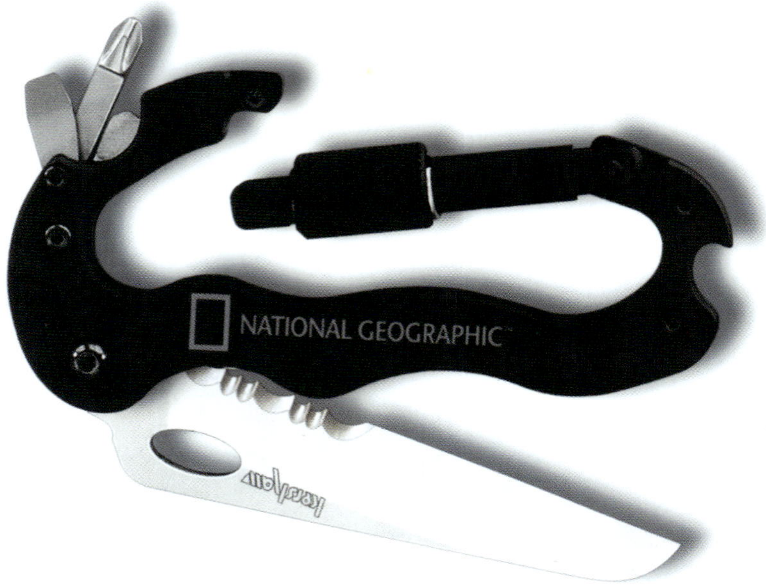

iF Silver Award

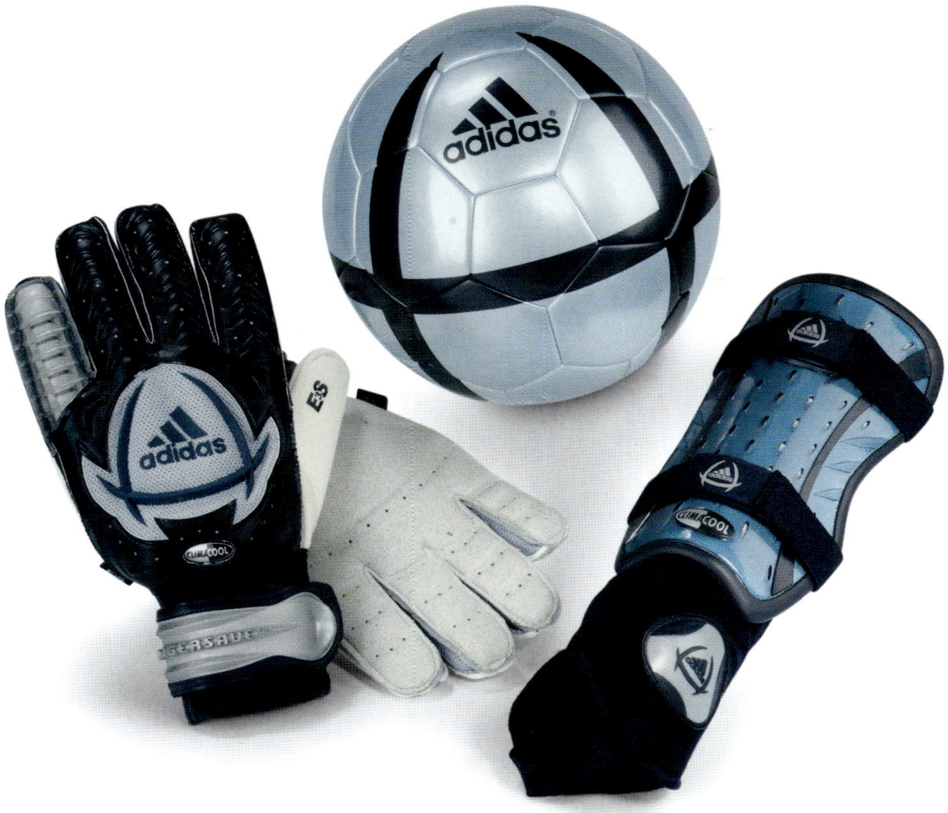

Hersteller Manufacturer
adidas-Salomon AG
Herzogenaurach, Germany

Produkt Product
The Roteiro Family
Fußball, Schienbeinschützer,
Torhüterhandschuhe
Football, goalkeeper's gloves,
Shin guards

Design
adidas-Salomon AG
Herzogenaurach, Germany

Die Roteiro Family von adidas besteht aus Fußball, Schienbeinschützern und Torhüterhandschuhen. Der Fußball hat aufgrund thermisch verstärkter Außenhautränder eine nahtlose Oberfläche, die extreme Strapazierfähigkeit und Wasser abweisende Eigenschaften gewährleistet. Die natürliche Latexblase sorgt für eine optimierte Energiewiedergabe. Der äußerst leichte, kompakte Schienbeinschutz verfügt über flexible integrierte Schichten, um dem Spieler eine individuelle gute Passform zu bieten. Statt Festschaum wurde ein luftdurchlässiges Gewebe verwendet, wodurch das Gewicht verringert werden konnte. Die Torhüterhandschuhe für Profis schützen durch eine Spritzgussschiene vor dem Verbiegen der Finger und eine spezielle ClimaCool Technology hält die Hände kühl und trocken.

adidas's Roteiro Family consists of a soccer ball, shin guards and goalie's gloves. Thanks to the new thermal bonding technique instead of hand-stitching the ball has a smooth seamless surface, substantially minimizing water uptake and maximizing abrasion resistance. The natural latex bladder delivers optimal rebound characteristics. The ultra-light compact shin guards come with flexible integrated layers for a customized wraparound fit. The use of an aerated 3D mesh instead of foam reduces weight significantly. The professional goalie's gloves feature a resilient molded spine in each finger that stiffens and resists pressure if pushed backwards, not to mention palm foam for cushioning as well as special ClimaCool technology to keep the hands cool and dry.

Hersteller Manufacturer
K2 Ski Sport Mode GmbH
Penzberg, Germany
Playmaker
Taiwan, ROC

Produkt Product
K2 Rollerskates
K2 roller skates

Design
Zeug Design GmbH
Detlev Magerer
Erwin Weitgasser
Salzburg, Austria

Die K2 Rollerskates verfügen über einen patentierten StepIn®-Mechanismus, der ein leichtes Anlegen der Schiene erlaubt. Der Rollschuh wird durch das Abnehmen der Schiene zu einem vollwertigen, flexiblen und dem Trend entsprechenden Straßenschuh. Die groß dimensionierten Leichträder garantieren ein neuartiges Fahrgefühl. Überflüssige Konstruktionselemente wurden eliminiert, durch den Einsatz von High-Tech-Materialien konnte das Gesamtgewicht stark reduziert werden.

The K2 roller skates feature a patent-protected StepIn® mechanism, whereby the shoe clicks onto the quad wheel track. After taking off the track, the skate shoe thus becomes a high-grade, flexible and trendy street shoe. The large lightweight roller wheels deliver an entirely new ride feeling. All superfluous elements have been eliminated and thanks to the use of high-tech materials, the shoe's overall weight has been substantially reduced.

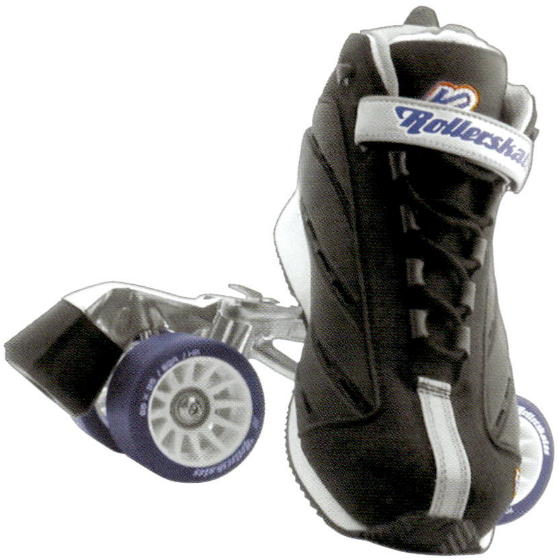

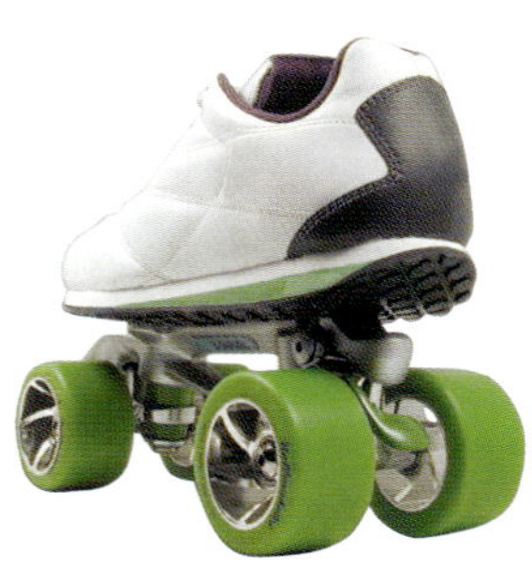

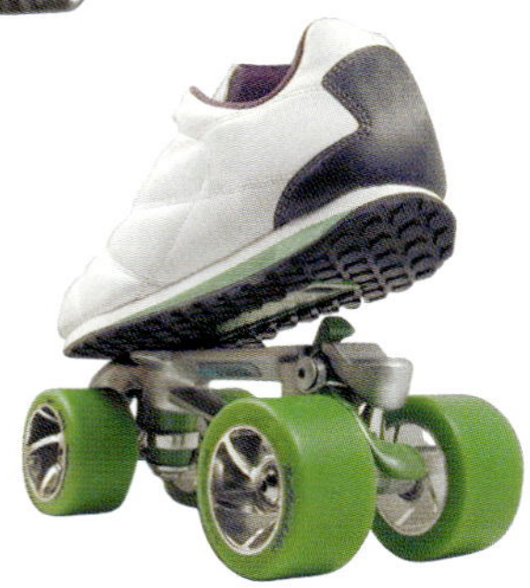

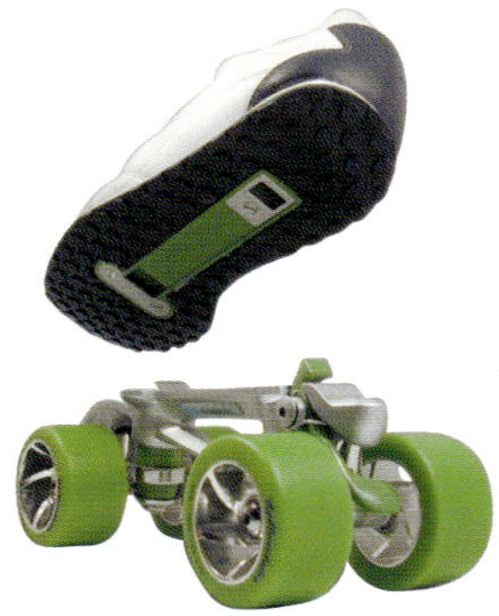

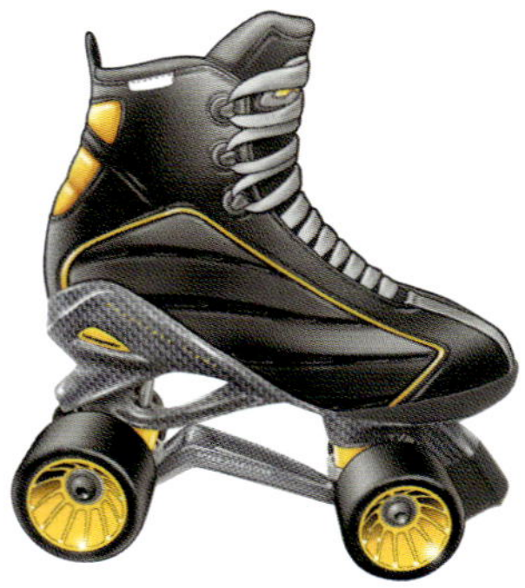

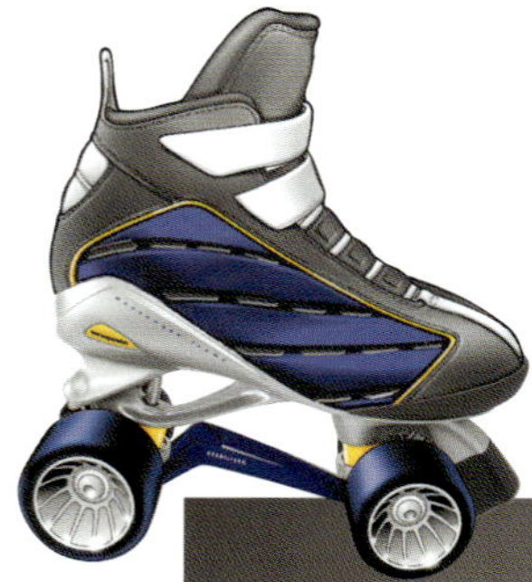

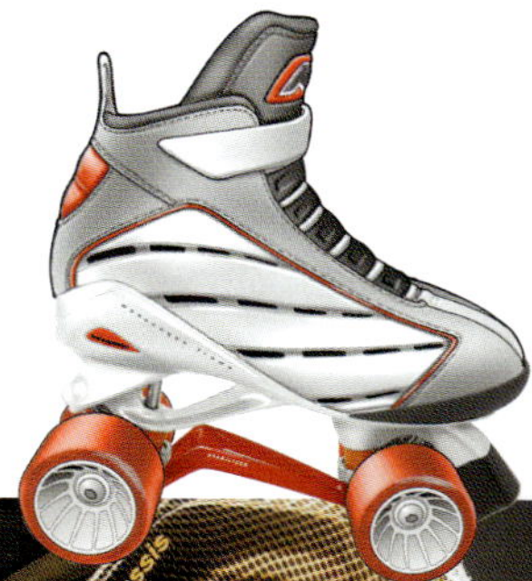

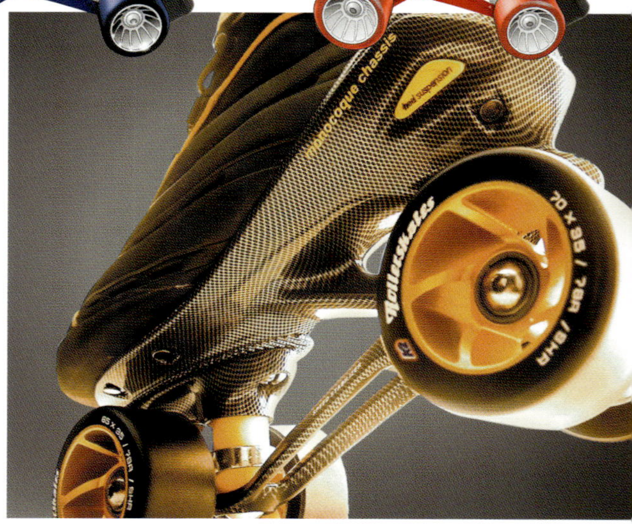

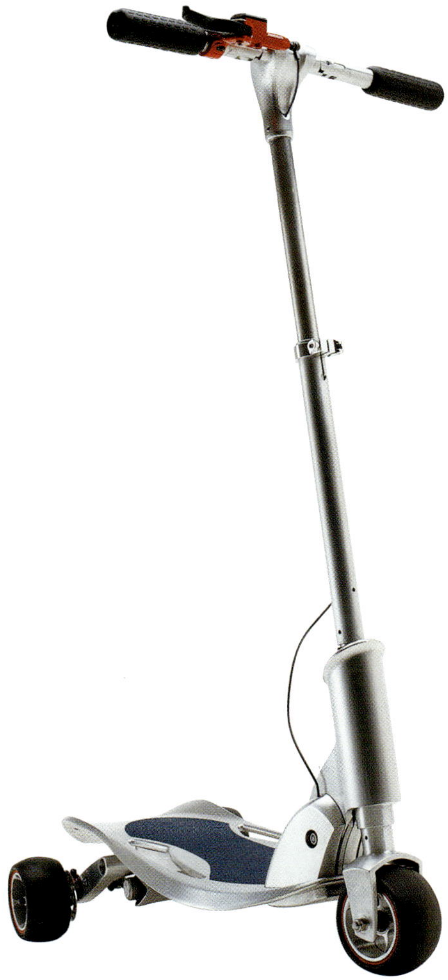

Hersteller Manufacturer
BMW AG
München, Germany

Produkt Product
BMW Slidecarver

Design
Designteam der BMW Group
München, Germany
DesignworksUSA
Newbury Park, CA, USA

Der für die Innenstadt gedachte High-Tech-Tretroller BMW Slidecarver verfügt über ein speziell entwickeltes Fahrwerk sowie aus der Automobiltechnik stammende Pendelstützen. Gebremst wird der Slidecarver durch Doppelscheibenbremsen an den Hinterrädern. Die neu entwickelte Einhandbedienung ermöglicht ein einfaches Umklappen von Lenkergriffen und Lenkstock, sodass der Roller zum Beispiel im Kofferraum des Autos leicht zu transportieren ist.

Created for urbanites, the high-tech BMW Slidecarver scooter boasts a specially developed wheel system and rocker bearings taken on from the auto industry. The Slidecarver has double disc brakes for the back wheels to deliver precision stops. The scooter is developed for single-hand use, with the handle-bars and steering column folding up easily making it simple to tuck the Slidecarver away in a car trunk, for example.

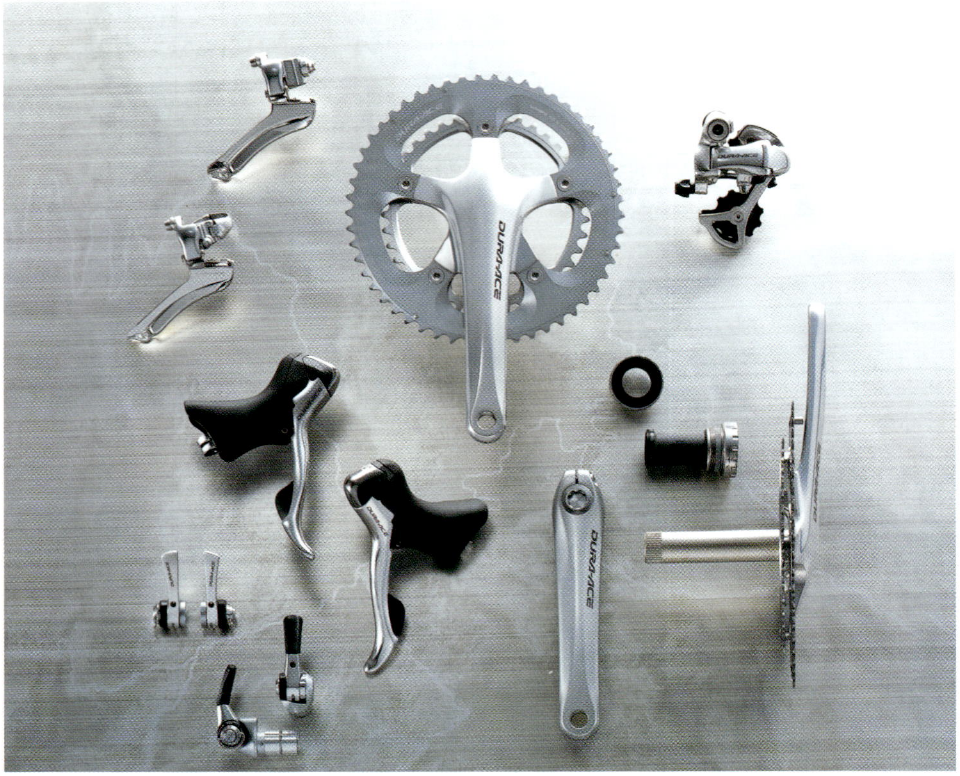

Hersteller Manufacturer
Shimano
Osaka, Japan

Produkt Product
DURA-ACE
Kettenblatt
Gear chain ring

Design
Paul Lange & Co.
Stuttgart, Germany

Das DURA-ACE-System für Rennräder wurde vor allem unter den Gesichtspunkten Geschwindigkeit, Komfort und Stabilität entwickelt, um ein müheloses Schalten, verbessertes Bremsen und eine gute Ergonomie zu gewährleisten. Trotz des Leichtbaus mussten keine Kompromisse hinsichtlich Steifigkeit und Haltbarkeit eingegangen werden. Die Hollowtech II-Technik (innen hohl, breite Lagerabstützung) ermöglicht eine optimierte Kraftübertragung bei nur sehr geringem Gewicht (740 g mit Innenlager und 175 mm Kurbel).

The DURA-ACE system for race bikes was developed with a view to speed, comfort and stability: to provide easy gear changes, enhanced braking and good ergonomics. Despite the lightweight structure, no compromises have been made as regards rigidity and durability. The Hollowtech II technology (hollow inside, broad bearing seal) delivers optimum power transmission despite the set's very light weight of 740 g with bearing race and 175 mm crankshafts).

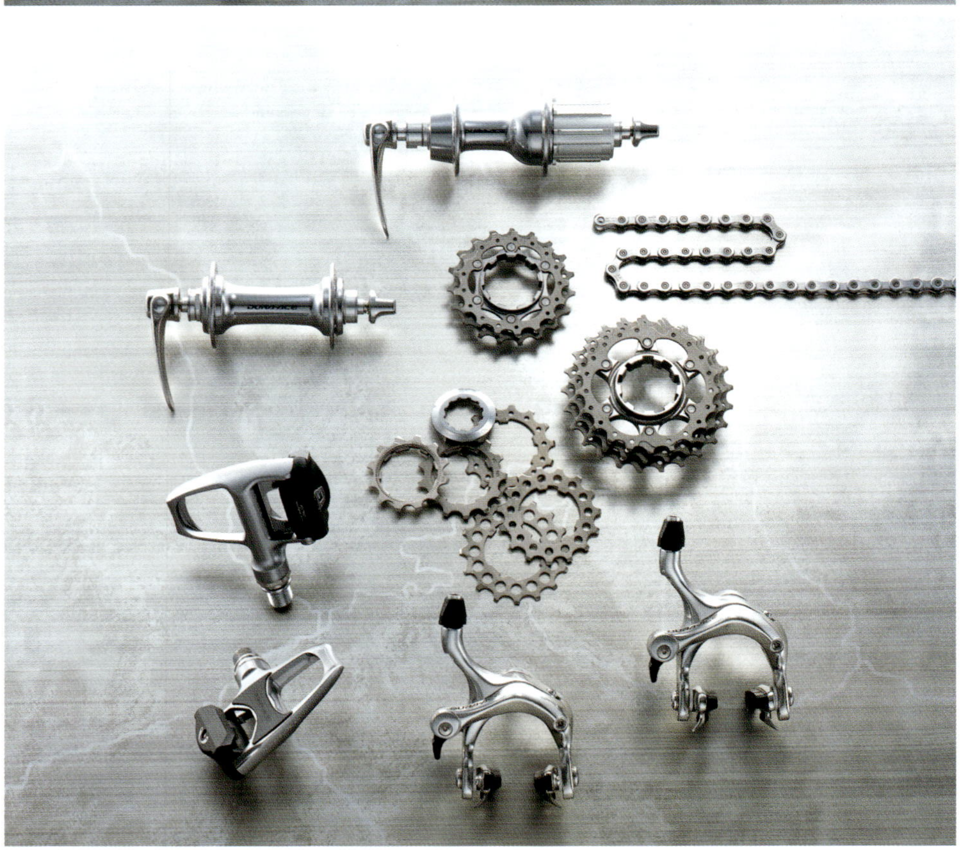

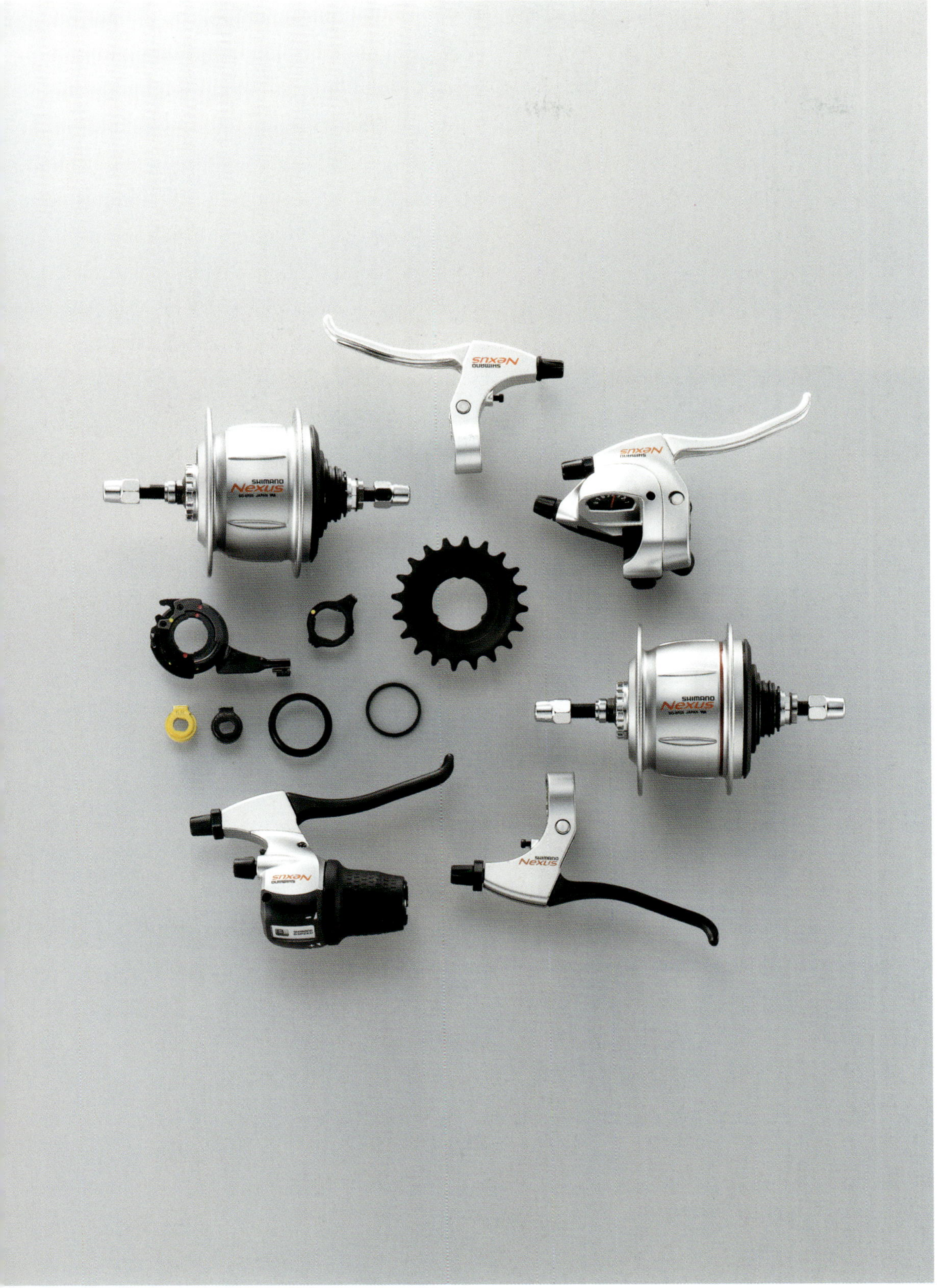

Hersteller Manufacturer
Shimano
Osaka, Japan

Produkt Product
Nexus Inter 8
Nabenschaltung
In-hub gear set

Design
Shimano
Osaka, Japan

Nexus Inter 8 ist eine besonders leicht gängige, servounterstützte (mit Pedalkraft) Nabenschaltung mit großem Übersetzungsbereich (307 %) und verschiedenen Schalter-Optionen (3 Finger Drehgriff und 2 Finger Rapid Fire). Sie wurde für komfortorientierte Trekkingfahrer konzipiert, die die einfache Handhabung einer Nabenschaltung zu schätzen wissen und auf einen großen Übersetzungsbereich nicht verzichten wollen. Als Normal- bzw. gewichtsoptimierte Premiumversion lieferbar kann sie mit V-Bremsen oder Rollenbremsen kombiniert werden.

Nexus Inter 8 is an especially light-running servo-supported (by pedal power) in-hub gear set that offers a wide range of gearings (307%) and different gear-switch options (3-finger gripshift and 2-finger Rapid Fire shift). It has been devised with comfort-oriented trekking bikers in mind who expect easy gear-shifts and smooth running gears across a broad range. Available as a normal-run and weight-optimized premium version, it can be combined with V-brakes or drum brakes.

Hersteller Manufacturer
Naxo AG
Gwatt (Thun), Switzerland

Produkt Product
Naxo nx 01
Skitourenbindung
Touring ski binding

Design
Christian Benz
Christian Harbeke
Nose Applied Intelligence
Zürich, Switzerland

Naxo nx 01 ist eine neuartige Skitourenbindung mit verbesserter Ergonomie. Durch das Virtual Rotation System entsteht ein ergonomisch runder Bewegungsablauf, der dem normalen Gehen sehr nahe kommt. Man spart bis zu 35 Prozent Kraft und der Fersen- und Schienbeinbereich wird spürbar entlastet. In der Gestaltung der Bindung wurde Wert auf Gewichtsreduktion gelegt, zudem wurde auf den Einsatz farbiger Pigmente verzichtet, um die Festigkeitswerte der Hightech-Kunststoffe nicht zu beeinträchtigen.

Naxo nx 01 is an innovative type of ski touring binding with enhanced ergonomic specifications. Thanks to the virtual rotation system, the skier's motion is ergonomic with little energy loss and approximates the movement of walking. Up to 35 percent of the power applied is saved and the strain taken off the heels as well as shins. In design terms, the focus has been on reducing the weight of the binding, and no colored pigments used in order not to impair the rigidity of the high-tech plastics used.

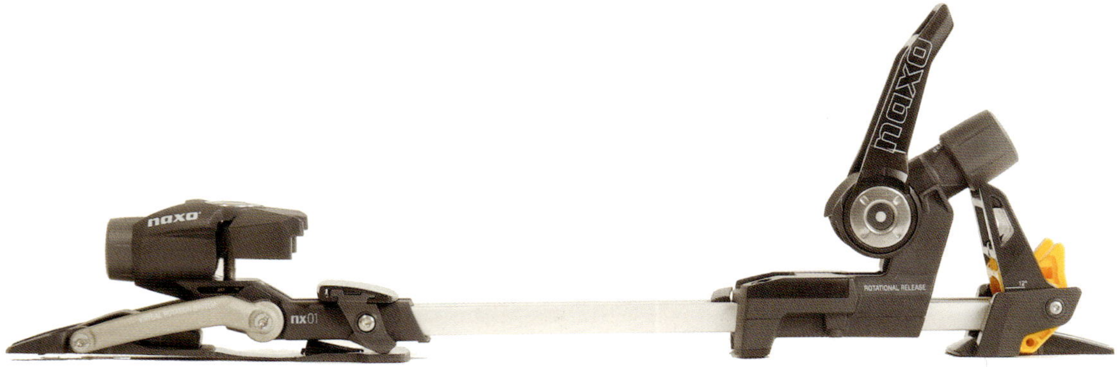

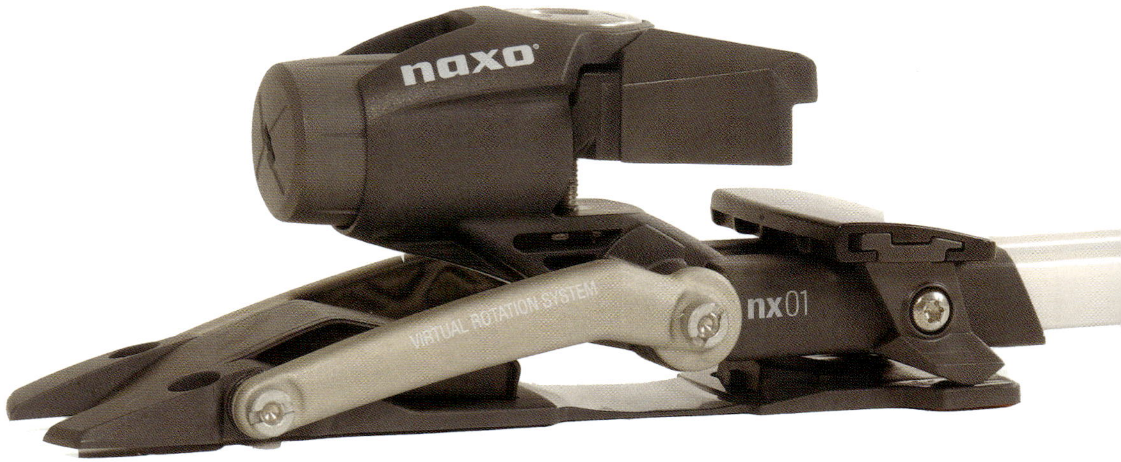

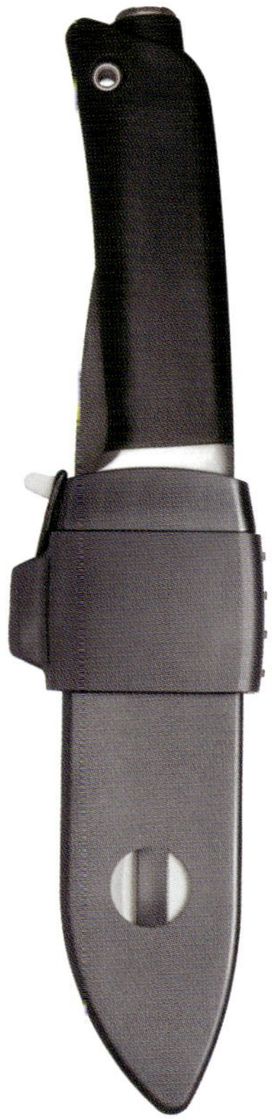

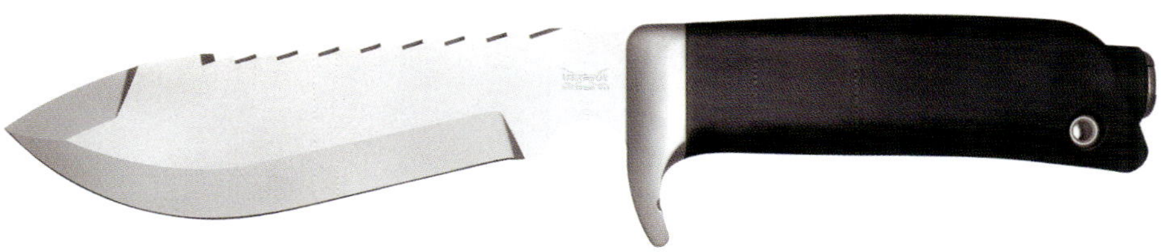

Hersteller Manufacturer
Wilkinson Sword Limited
London, UK

Produkt Product
Wilkinson Sword CSK 185
Dartmoor survival knife

Design
Tangerine direction and design
London, UK

Das Wilkinson Sword CSK 185 Dartmoor survival knife (Überlebensmesser) besitzt eine 185 mm „Drop Point"-Klinge aus 440C rostfreiem Stahl. Das Sägeprofil wurde entlang des Dorns der Klinge angelegt, um eine perfekte Schneidegeometrie zu gewährleisten. Die ergonomische Gestaltung des Griffs erlaubt verschiedene Greifvarianten zum Hacken, Spalten, Sägen, feinem Schneiden etc. Der steife Kern des Griffs ist von einem weichen, rutschfreien Gummiüberzug umgeben. Im Griff ist eine kleine Kammer untergebracht, in der sich ein Überlebens-Kit mit Kompass, Zündhölzern, Fischausrüstung sowie wesentlichen anderen Utensilien befindet. Die Gummihalterung für die Schneide kann sowohl von Rechts- als auch Linkshändern getragen werden.

The Wilkinson Sword CSK 185 Dartmoor survival knife comes with a 185 mm drop-point blade made of 440C stainless steel. The saw profile along the spine of the blade guarantees perfect cutting geometry. The ergonomic detailing of the handle allows a range of grip configurations for tasks such as chopping, splitting, sawing, fine cutting, etc. The rigid core of the handle has a softer non-slip rubber grip overmolded onto it and contains a small chamber to store the survival kit, including a compass metal match, fishing equipment and other basic survival essentials. The rubber scabbard holder can be configured for both left- and right-handed users.

Hersteller Manufacturer
Pao Shin Garden Tools Co. Ltd.
Taipei, Taiwan, ROC

Produkt Product
XPERT 2004–2
Gartenwerkzeuge
Garden digging tools

Design
Pao Shin Garden Tools Co. Ltd.
Taipei, Taiwan, ROC

Die Gartengeräteserie XPERT 2004–2 besteht aus drei Handwerkzeugen mit unterschiedlich geformten Klingen zum Umgraben und Bearbeiten von Erde. Die aus Aluminium bestehenden Geräte sind leicht, bruchsicher und besonders haltbar. Der ergonomisch geformte Griff mit Softgrip wurde der ihn umschließenden Hand nachempfunden.

The XPERT 2004–2 garden digging tools collection consists of three hand-polished aluminum tools with different functionally shaped blades for scooping and loosening soil. Thanks to the aluminum-alloy casing, each if the digging tools is lightweight, durable and will not break. The ergonomically rounded-over soft grip is designed to lie perfectly in the hand and is optimally balanced.

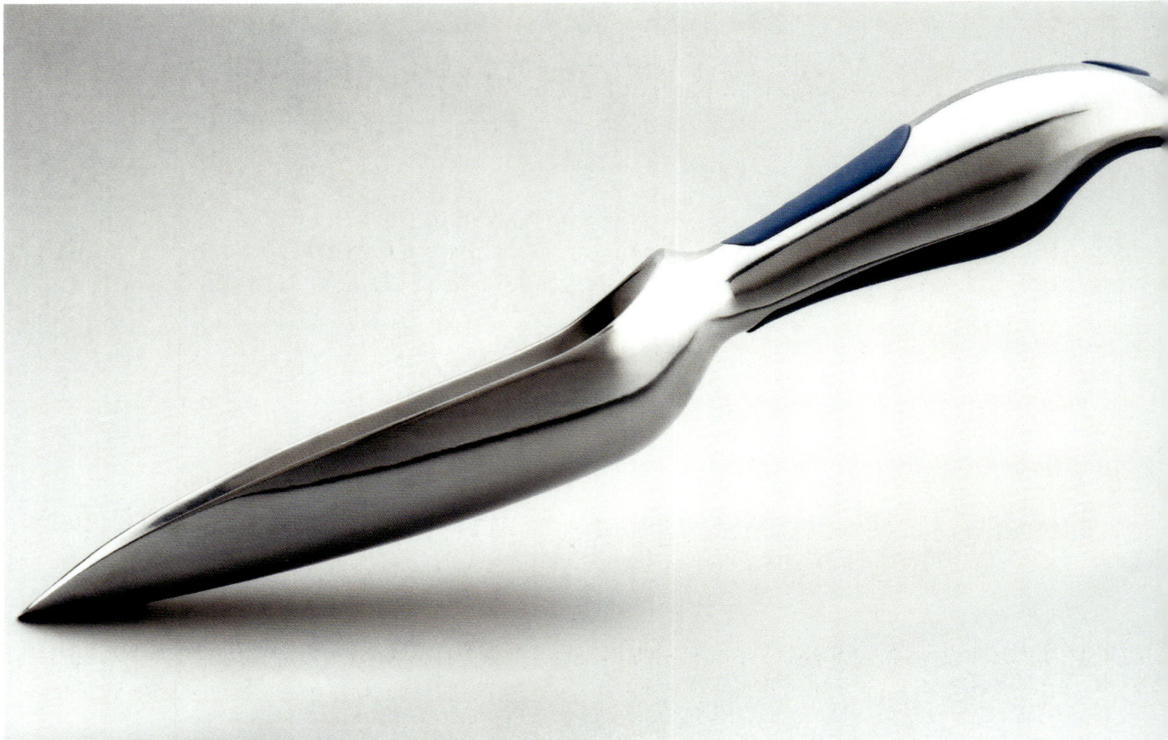

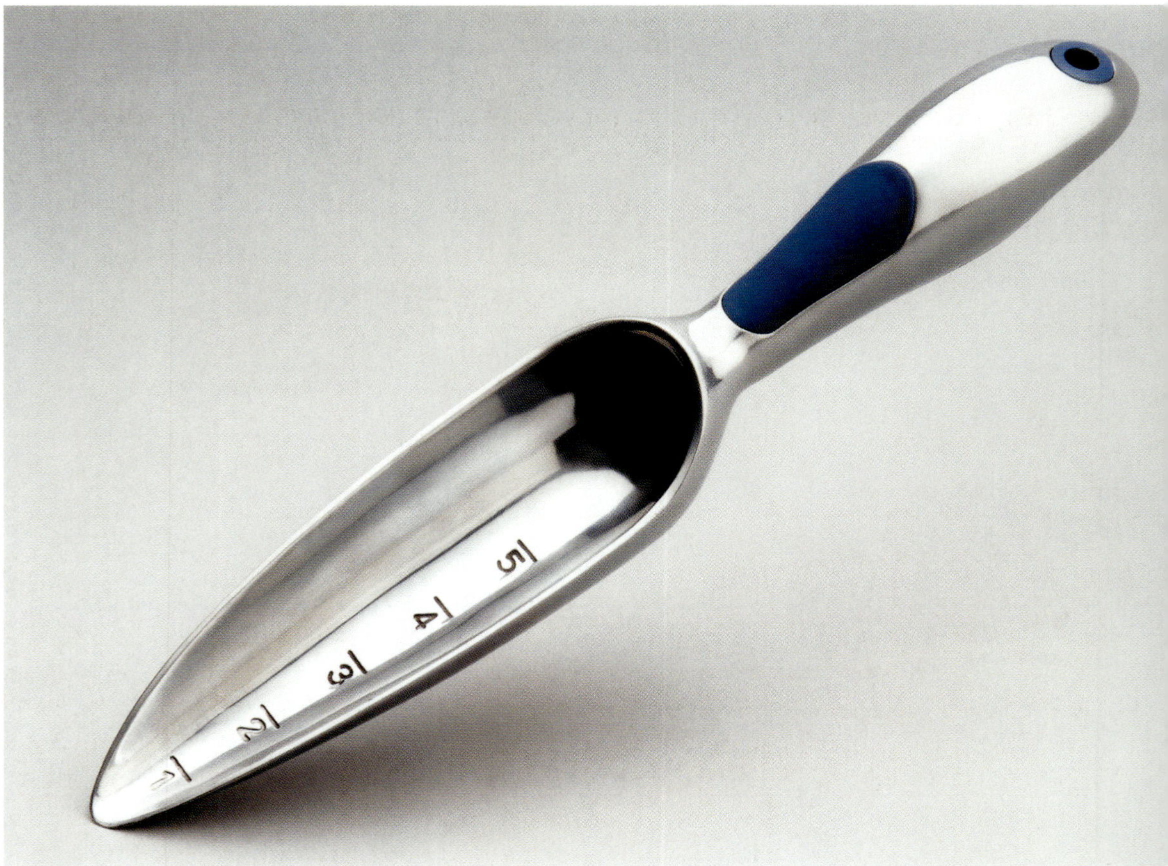

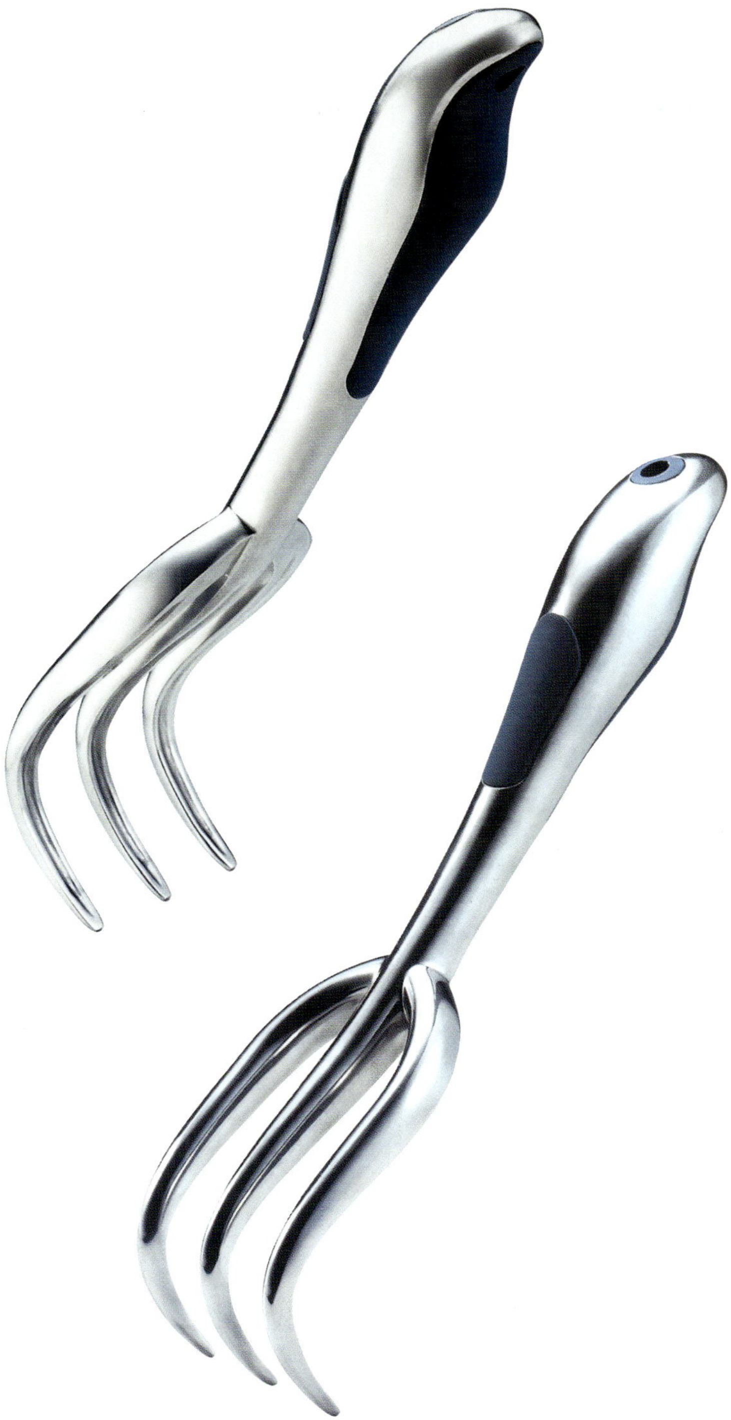

Hersteller Manufacturer
Pao Shin Garden Tools Co. Ltd.
Taipei, Taiwan, ROC

Produkt Product
XPERT 2004–1
Gartenwerkzeuge
Garden digging tools

Design
Pao Shin Garden Tools Co. Ltd.
Taipei, Taiwan, ROC

Die Gartengeräteserie XPERT 2004–1 besteht aus drei Handwerkzeugen mit unterschiedlich geformten Klingen zum Umgraben und Bearbeiten von Erde. Die aus Aluminium bestehenden Geräte sind leicht, bruchsicher und besonders haltbar. Der ergonomisch geformte Griff mit Softgrip wurde der ihn umschließenden Hand nachempfunden.

The XPERT 2004–1 garden digging tools collection consists of three hand-polished aluminum tools with different functionally shaped blades for scooping and loosening soil. Thanks to the aluminum-alloy casing, each if the digging tools is lightweight, durable and will not break. The ergonomically rounded-over soft grip is designed to lie perfectly in the hand and is optimally balanced.

Hersteller Manufacturer
Pao Shin Garden Tools Co. Ltd.
Taipei, Taiwan, ROC

Produkt Product
XPERT 2004–3
Gartenwerkzeuge
Garden digging tools

Design
Pao Shin Garden Tools Co. Ltd.
Taipei, Taiwan, ROC

Die Gartengeräteserie XPERT 2004–3 besteht aus drei Handwerkzeugen mit unterschiedlich geformten Klingen zum Umgraben und Bearbeiten von Erde. Die aus Aluminium bestehenden Geräte sind leicht, bruchsicher und besonders haltbar. Der ergonomisch geformte Griff mit Softgrip wurde der ihn umschließenden Hand nachempfunden.

The XPERT 2004–3 garden digging tools collection consists of three hand-polished aluminum tools with different functionally shaped blades for scooping and loosening soil. Thanks to the aluminum-alloy casing, each if the digging tools is lightweight, durable and will not break. The ergonomically rounded-over soft grip is designed to lie perfectly in the hand and is optimally balanced.

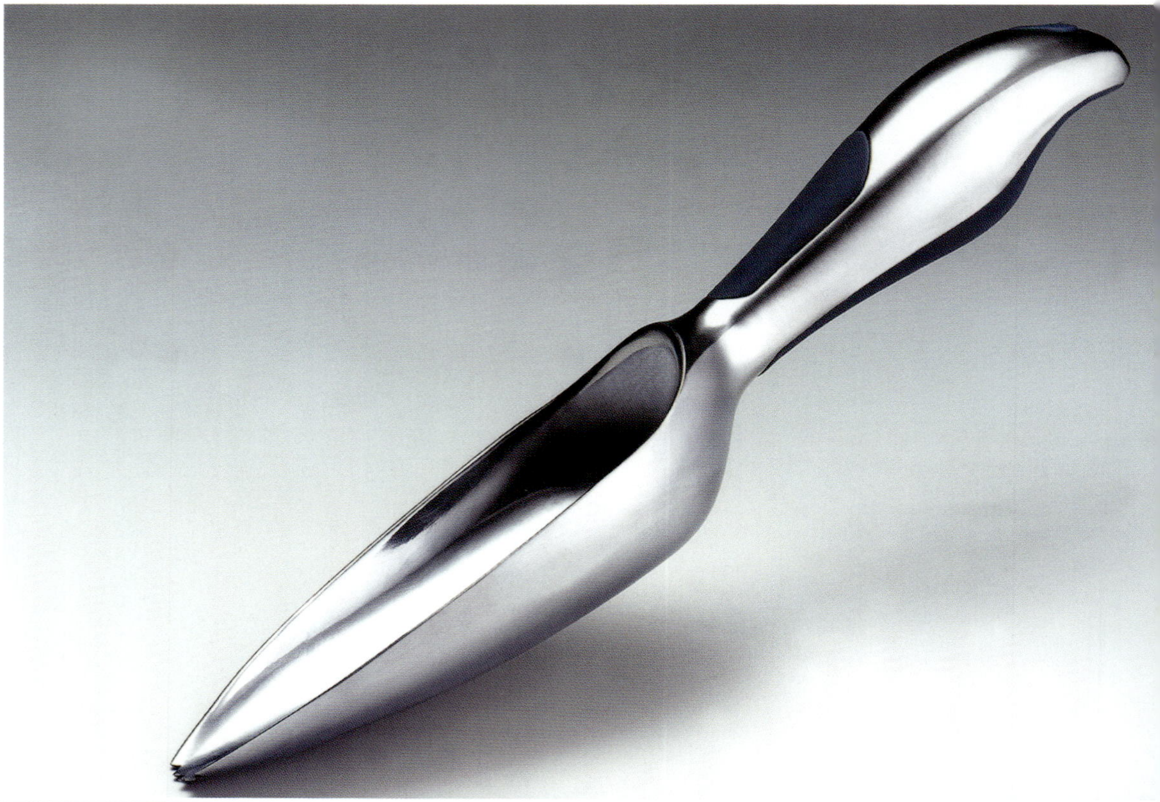

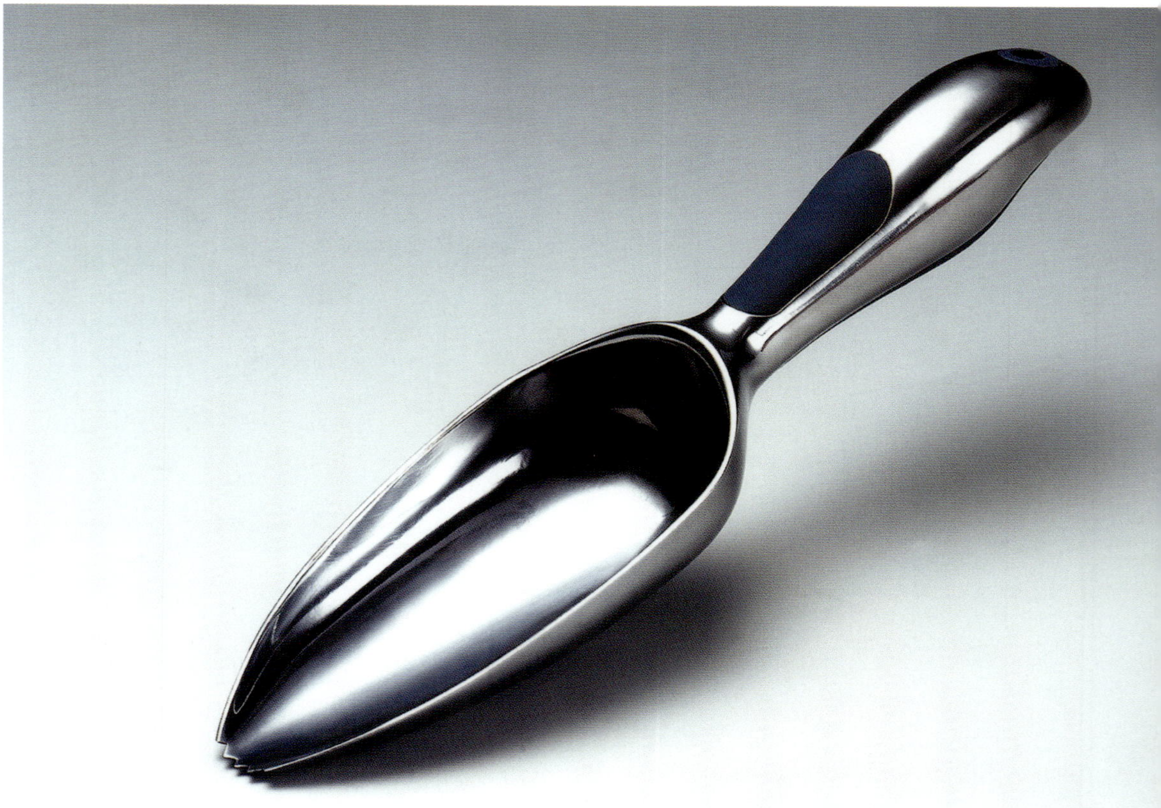

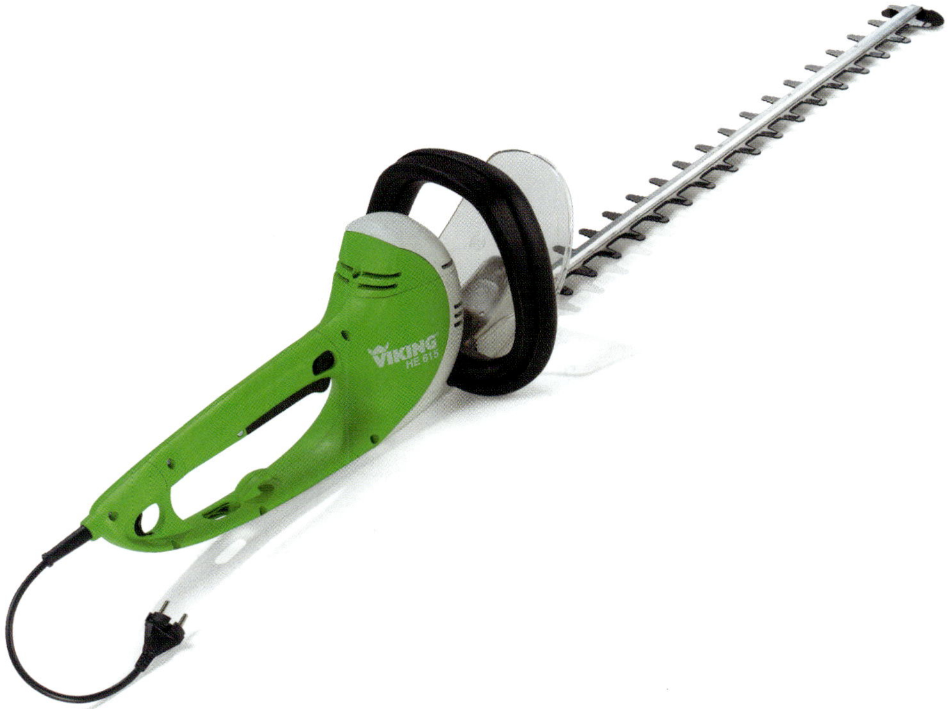

Hersteller Manufacturer
VIKING GmbH
Langkampfen/Kufstein, Austria

Produkt Product
VIKING HE 615
Heckenschere
Hedge trimmer

Design
Busse Design Ulm GmbH
Elchingen, Germany

Die Heckenschere VIKING HE 615 zeichnet sich durch eine ausgewogene Gewichtsverteilung und eine optimierte Griffergonomie für ermüdungsarmes Arbeiten in jeder Position aus. Das Antriebskonzept und die lasergeschnittenen Messer sorgen für ruhigen, vibrationsarmen Schnitt. Ein synchroner Sicherheitsmesser-schnellstopp und der Messeranstoßschutz ermöglichen eine sichere Handhabung.

The VIKING HE 615 hedge trimmers stand out for their balanced weight distribution and optimized handling ergonomics, and non-fatiguing operation in any position. The drive system and laser-cut blades deliver quiet and low-vibration cutting. A synchronized safety-blade quick-stop and a blade impact protector ensure secure cutting performance.

Hersteller Manufacturer
R & L Thomas
Essen, Germany

Produkt Product
Waschsessel 580
Haarwaschsessel
Wash basin seat

Design
Martin Hänel
modern.need gmbh
München, Germany

Mit dem Waschsessel 580 wurden Waschbeckenaufnahme und Sitzmöbel zu einer formalen und funktionalen Einheit miteinander verschmolzen. Das Design berücksichtigt sowohl die einfache Montage des Waschbeckens als auch eine komfortable Sitzposition des Kunden. Das angewendete Verfahren ermöglichte die Herstellung eines preisgünstigen Kunststoffkörpers, die selbst tragende Form macht eine aufwendige Unterkonstruktion überflüssig. Zudem ist das Material resistent gegen Bleich- und Färbemittel und lässt sich durch Pigmentzugabe beliebig durchfärben. Das einfache Einlegen der Sitz- und Rückenpolster gewährleistet ein leichtes Reinigen des Waschsessels.

The hairdresser's chair 580 blends the wash basin section and seat to create one formal and functional unit. The design ensures easy assembly of the wash basin and guarantees the client a comfortable seating position. By means of molding, a cost-effective plastic shell is made and the standalone carcass renders any elaborate lower frame unnecessary. Moreover, the material is bleach- and dye-resistant, and can be given any color throughout by adding pigments during the molding process. The seat and backrest upholstery simply slots into place, spelling easy cleaning of the chair after use.

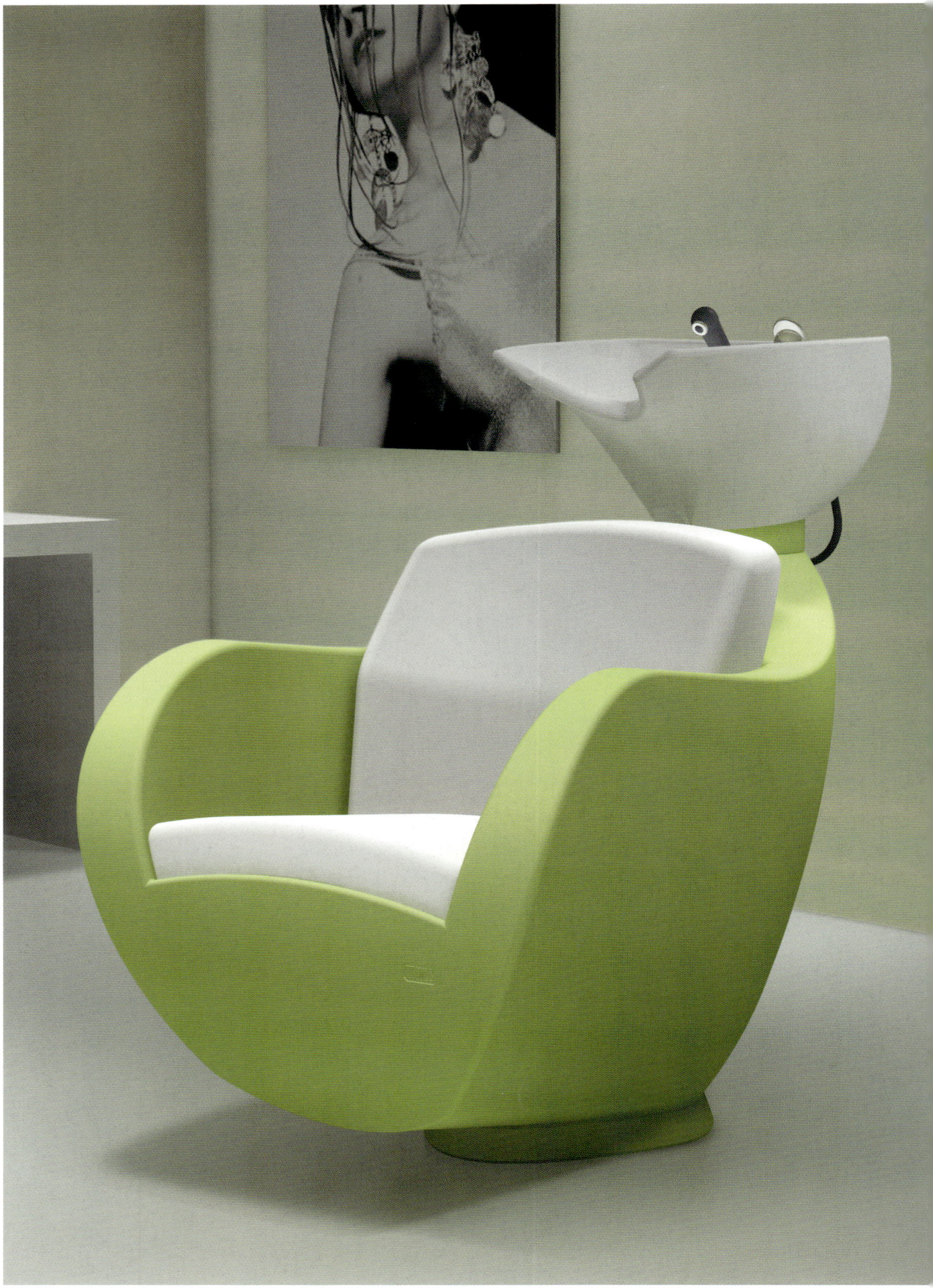

Hersteller Manufacturer
BabyBjörn AB
Danderyd, Sweden

Produkt Product
BabyBjörn Baby Carrier Active
Tragesitz
Baby carrier

Design
Håkan Bergqvist
Ergonomi Design Gruppen
Bromma, Sweden
Lisen Elmberg
Anders Nyberg, Lillemor Jakobson
BabyBjörn AB
Danderyd, Sweden

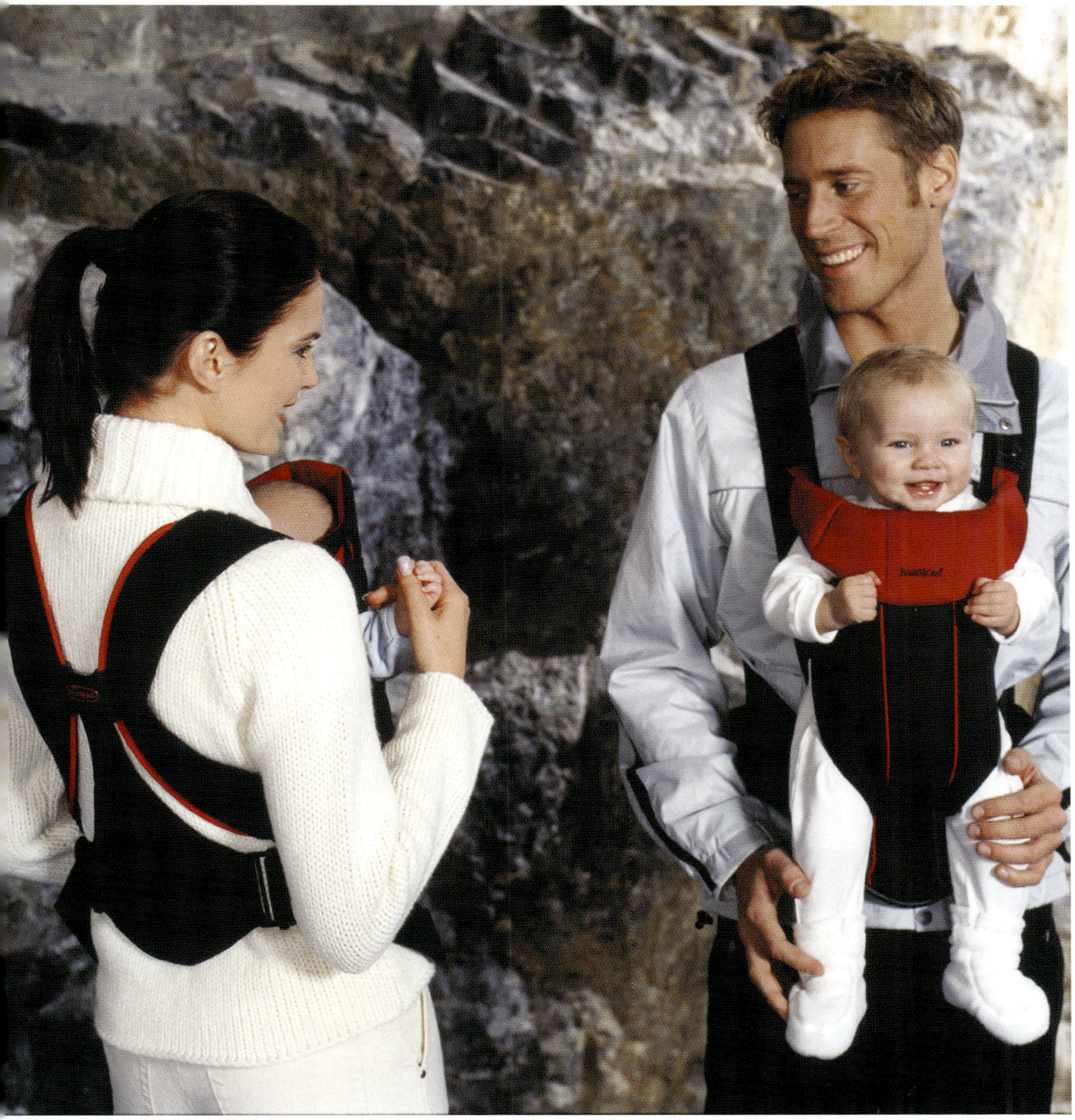

Der BabyBjörn Baby Carrier Active wächst durch seine unterschiedlichen Funktionskomponenten mit dem Kind mit und ist als Alternative zum Kinderwagen gedacht. Die zusätzliche Lendenrückenstütze, die Kopfstütze und die seitlichen Tragegurte entlasten die Schultern beim Tragen des Kindes. Die Kopfstütze kann erweitert werden und an die Wachstumsstadien des Kindes sowie eine stärker werdende Nackenmuskulatur angepasst werden. Dies ermöglicht entweder Augenkontakt mit dem Kind oder ein Herunterklappen der Kopfstütze, so dass es nach vorne sehen kann.

Thanks to the great variety of functional components, the BabyBjörn Baby Carrier Active grows along with the child and is intended as an alternative to a pram. The additional lumbar support, the head support and the side straps take the strain off the shoulders when carrying child. The head support can be broadened as the child gets older and the neck stronger. This enables eye contact between child and parent, or can be folded down so the child can face forward and look forward.

Hersteller Manufacturer
Philips Domestic Appliances & Personal Care
Amersfoort, Netherlands

Produkt Product
Philishave Micro+ 40 Series
Rasierer
Shaver

Design
Esther Beckers
Martyn Beedham
Philips Design

Die Philishave Micro+ 40 Serie umfasst eine Reihe von leichten, kompakten und bedienungsfreundlichen Rasierern. Merkmale wie die Lift & Cut Technologie und eine spezielle Micro-Groove-Funktion sorgen für eine hautnahe und komfortable Rasur. Das Design der Rasierer ist modern, ergonomisch und funktionsbezogen. Die Abdeckung des Akkufachs ist mit unterschiedlichen Oberflächen, von transparent bis metallic und in verschiedenen Farben erhältlich.

The Philishave Micro+ 40 series is a line of lightweight, compact and user-friendly shavers. Special features such as the Lift-&-Cut technology and the micro-groove function deliver a very close and comfortable shave. The shavers boast a modern, ergonomic and functional design. The chargeable battery lid is available with different finishes, ranging from transparent to metallic, as well as in a variety of colors.

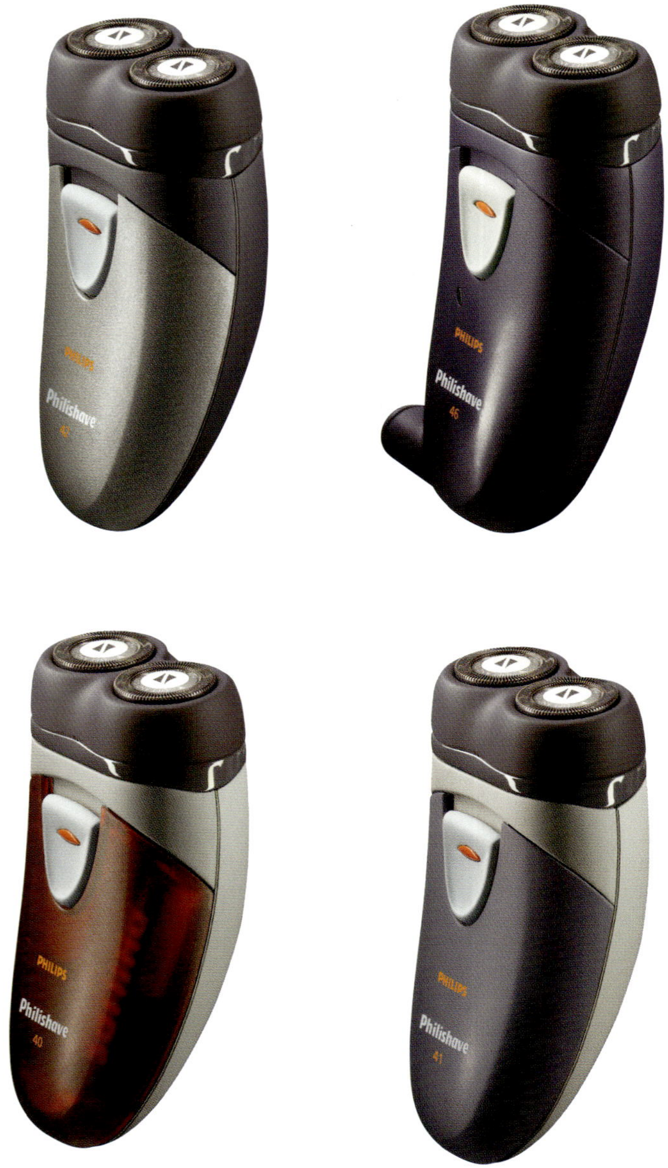

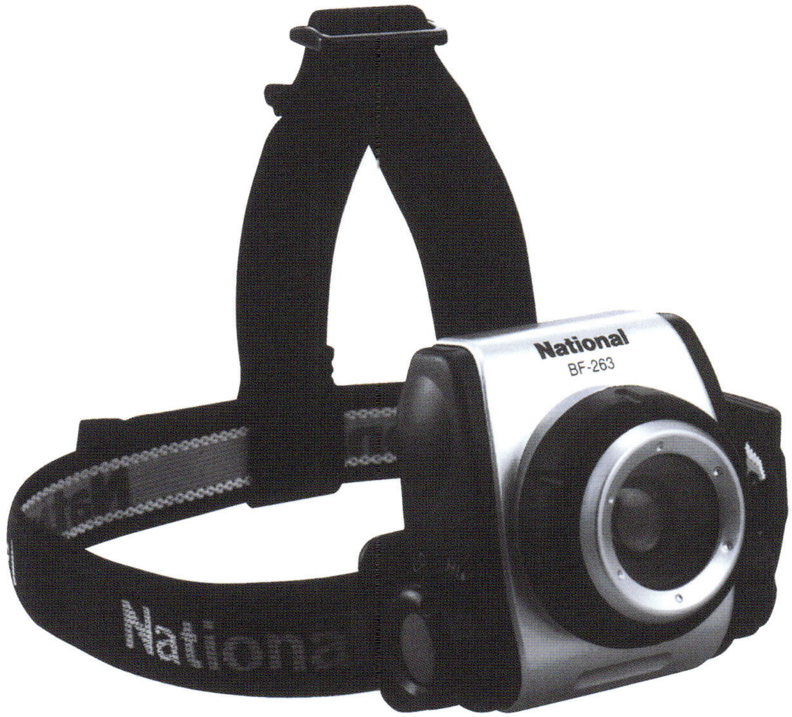

Hersteller Manufacturer
Matsushita Battery Industrial Co. Ltd.
Osaka, Japan

Produkt Product
BF-263
Stirnlampe
Headlamp

Design
Osamu Fukuda
Panasonic Design Company
Osaka, Japan

Die Stirnlampe BF-263 wurde insbesondere für den professionellen Einsatz für Bergsteiger oder -führer entwickelt. Das wasserdichte Gehäuse enthält eine weiße Hochleistungs-LED mit der zehnfachen Helligkeit einer herkömmlichen LED. Durch den Blinkmodus kann die Leuchte auch als Signal- und Positionsgeber genutzt werden. Dank des eingebauten Mikrocomputers ist die Helligkeit über ein Rad an der Stirnleuchteninnenseite in sechs Stufen einstellbar. Durch Drehen der Gummifassung der Linse wird das Licht entweder gebündelt oder gestreut. Das Kopfband besteht aus einem Schweiß absorbierenden und schnell trocknenden Material.

The BF-263 headlamp was designed specially with advanced and professional climbers and mountain guides in mind. The waterproof body contains a high-power white LED that is ten times brighter than a conventional LED. The flashing mode enables the lamp to be used as a signal and as a marker. By dint of the integrated microcomputer, six levels of brightness can be selected by the turning the dial on the side of the body. By rotating the rubberized lens ring, it can be adjusted to emit condensed or diffused light. The headband belt is made of sweat-absorbent quick-dry material.

Hersteller Manufacturer
Konica Minolta Camera, Inc.
Tokio, Japan

Produkt Product
Minolta DiMAGE Xt
Digitalkamera
Digital camera

Design
Kenji Baba

Die äußerst flache Minolta DiMAGE Xt Kompakt-Digitalkamera zeichnet sich durch ein minimalistisches und klares Design aus, bei dem bewusst auf dekorative Akzente verzichtet wurde. Durch das ungewöhnliche Vertikaldesign liegt die Kamera sehr gut in der Hand, zudem ermöglicht die Anordnung der Bedienelemente eine intuitive und bequeme Bedienung. Die Kamera verfügt über 3,2 Megapixel, einen 12fach Zoom (3fach optisch, 4fach digital), eine Makrofunktion, einen eingebauten Blitz, USB-Schnittstelle, Webcam-Funktion und Mehrzonenbelichtungsmessung. Im Farbmodus kann auch zwischen Schwarz-Weiß und einem speziellen Sepiamodus für nostalgische Fotos gewählt werden.

The ultra-slim Minolta DiMAGE Xt compact digital still camera catches the eye with its minimalist and uncluttered design – deliberately bereft of decorative frills. By virtue of its unusual vertical design, the camera lies perfectly in the hand, and the layout of the controls ensures intuitive ease of use. The camera offers 3.2 Megapixels, a 12x (3x optical, 4x digital) zoom, a macrofunction, an integrated flash, USB interface, Webcam functions and multi-zone light meter. Moreover, in color mode users can choose black-and-white or special sepia pictures for a nostalgic touch.

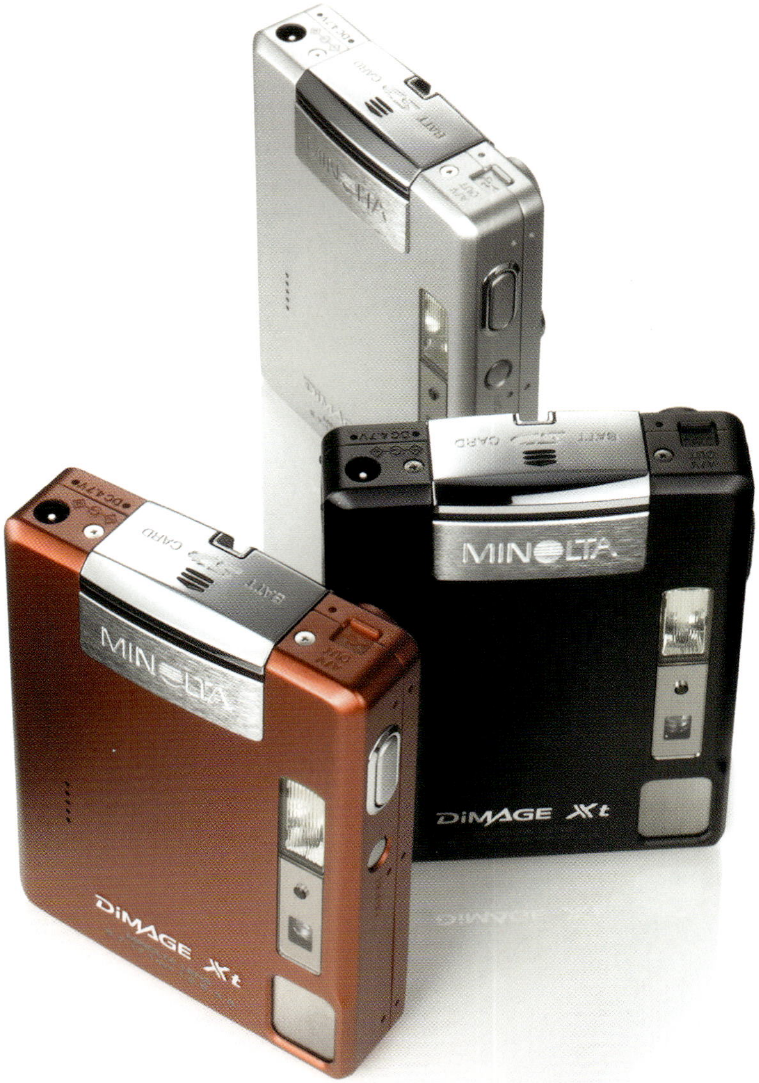

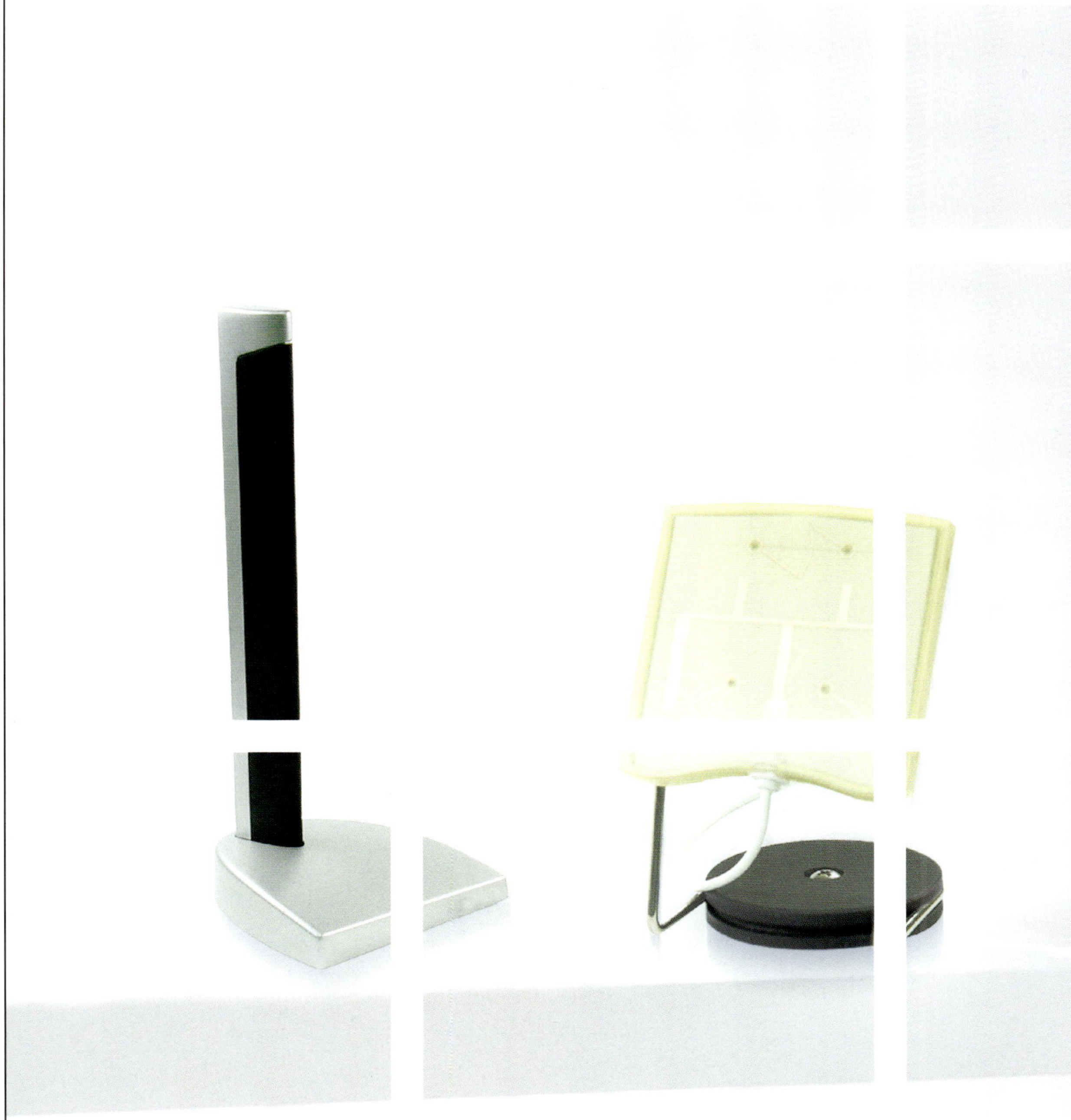

Having incorporated capabilities and aesthetics in its design, the introduction of SmartAnt's Multi-Band Directional Antenna for Wi-Fi applications in 2003 has transformed antenna from a component traditionally with boring appearance however essential for wireless communication into an adorable stuff that is even decorative to daily life. The newly released antenna is revolutionarily designed with attention-getting, neutrally harmonious colors to attract communication product users mainly of younger generations. Its multi-band design applicable for both 2.4GHz and 5GHz enables users to adapt the antenna to various circuit modules. Further, the lightweight antenna equipped with an angle-adjustable mounting feature could easily be fixed anywhere on wall, ceiling, or desk.

With gorgeous look and fancy colors, SmartAnt's Multi-Band Directional Antenna is more than just a medium for signal transmission. By incorporating aesthetics into antenna design, SmartAnt has successfully amazed antenna users and overturned their typical impression on antennas.

Hersteller Manufacturer
Canon Inc.
Tokio, Japan

Produkt Product
Digital IXUS i
Digitalkamera
Digital camera

Design
Masataka Isomoto

Die Digitalkamera Digital IXUS i von Canon mit 4 Megapixel Auflösung ist für die Hemdtasche konzipiert. Ihre kompakten Gehäuseabmessungen betragen lediglich 9,03 x 4,7 x 1,85 cm und sie wiegt nur ca. 122 g inkl. Akku und Speicherkarte. Beim Oberflächenfinish hat der Käufer die Wahl zwischen den Farben Platinum Silber, Piano Schwarz, Classic Bronze und Perlmutt Weiß. Zum Lieferumfang gehören eine 32 MB SD Speicherkarte und ein Lithium-Ionen-Akku. Die Kamera unterstützt den neuen, herstellerübergreifenden PictBridge-Standard für den Direktdruck, sodass der Fotodruck auf entsprechenden Druckern auch ohne Umweg über den PC möglich ist.

Canon's digital Digital IXUS i camera delivers 4 Megapixels resolution and is designed to fit in a shirt pocket. The compact body is sized a mere 9.03 x 4.7 x 1.85 cm and weighs only approx. 122 g incl. battery and memory stick. In terms of surface finish, users can choose from Platinum Silver, Piano Black, Classic Bronze and Mother-of-Pearl White. The standard version includes a 32 MB SD memory card and a lithium-ion battery. The camera supports the new manufacturer-independent PictBridge standard for direct printouts, meaning photos can be output directly via a printer rather than through a PC.

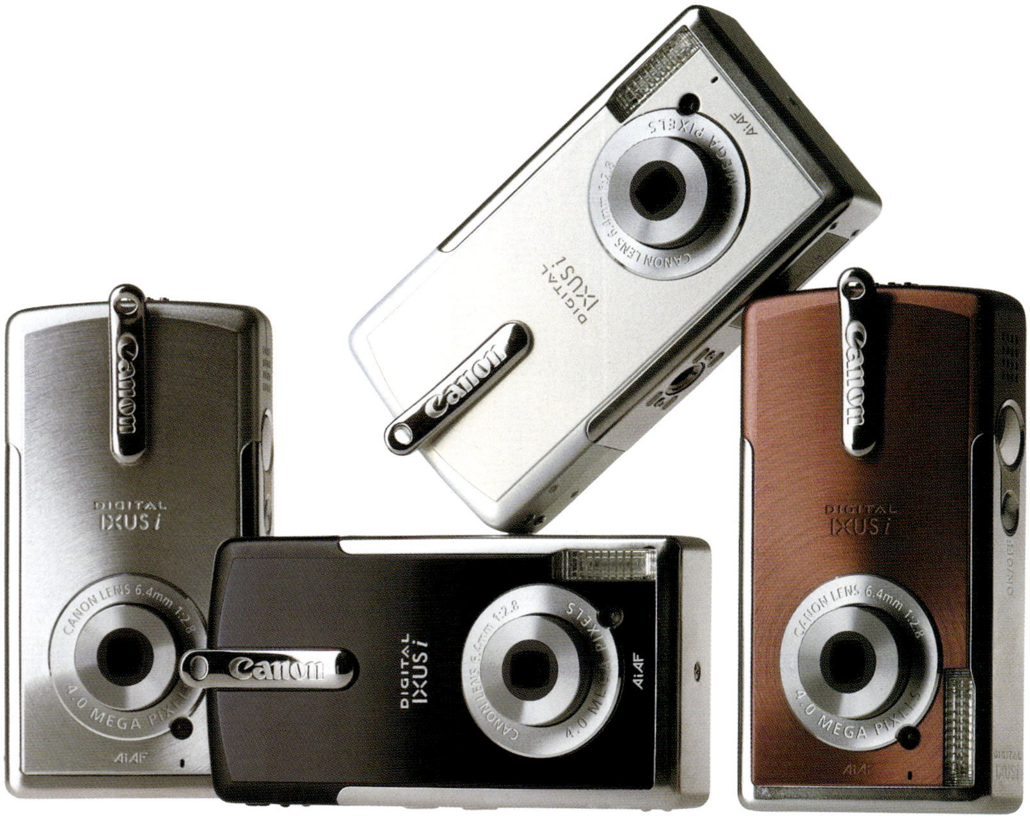

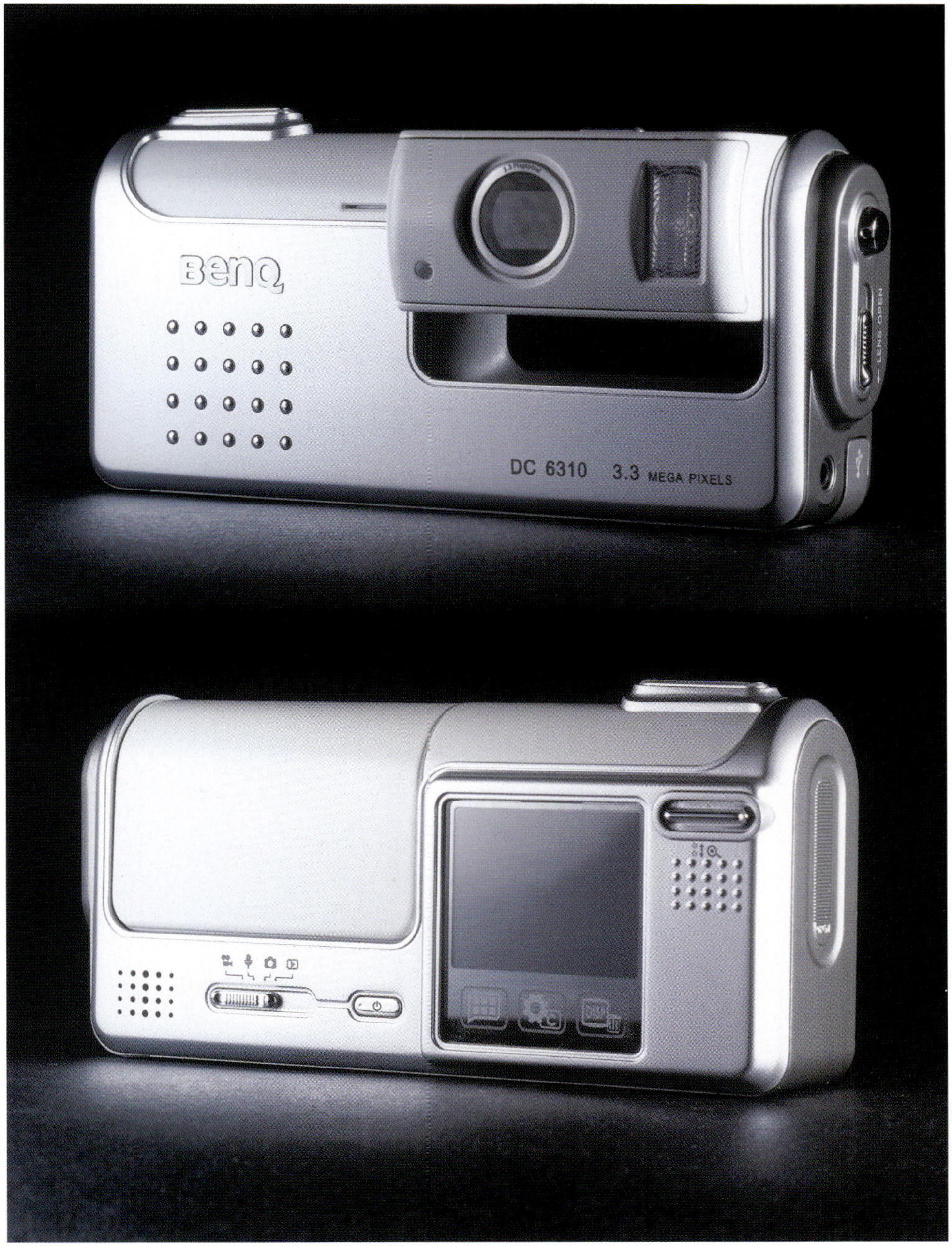

Hersteller Manufacturer
BENQ Corporation
Taipei, Taiwan, ROC

Produkt Product
BenQ Digital Camera – DC6310
Digital-Kamera
Digital Camera

Design
BENQ Corporation
Lifestyle Design Center
Taipei, Taiwan, ROC

DC6310 von BenQ ist Digitalkamera und MP3-Player in einem. Sie wurde für eine jüngere Zielgruppe konzipiert und dementsprechend mit einem trendigen Aluminiumlegierungsgehäuse versehen. Wenn die Kamerafunktion durch Knopfdruck aktiviert wird, bewegt sich das drehbare Objektiv automatisch in die richtige Position. Wird das Objektiv geschlossen, wird die Digitalkamera im Nu wieder zum tragbaren MP3-Player. Durch den Drehmechanismus kann der Benutzer zudem den Aufnahmewinkel flexibel bestimmen. Über einen Touch-Control-Monitor werden die einzelnen Funktionen gesteuert.

DC6310 von BenQ is a 3 million-pixel digital camera and MP3 player rolled into one. It has been developed specially for younger users and therefore also boasts a bright-face aluminum alloy outer cover. If the camera function is actuated by the press of a button, the rotary lens automatically springs up into the general photographing position. If the lens is shut again, the digital camera morphs into a portable MP3 player. Thanks to the rotating mechanism, the lens angle can be set as the user requires. And the touch-control monitor integrates all the complex functions offered.

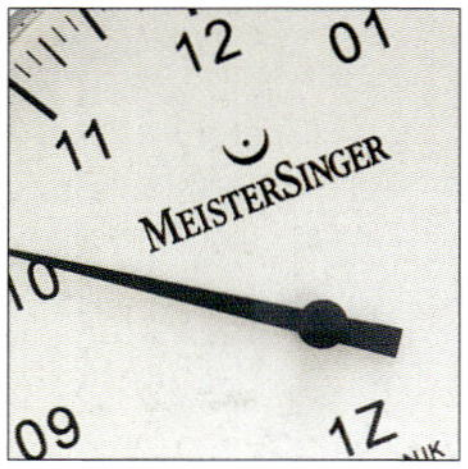
12
01
11
MeisterSinger
09
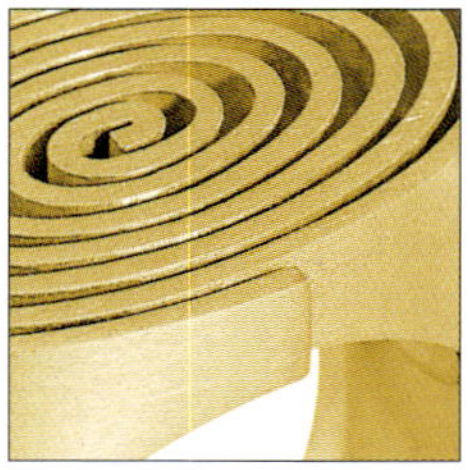
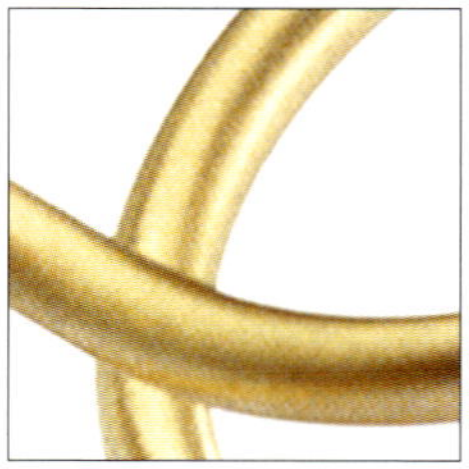
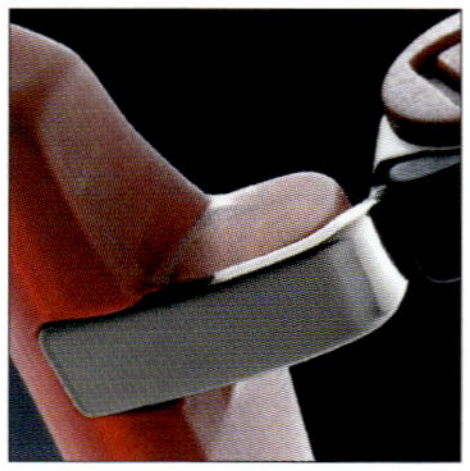
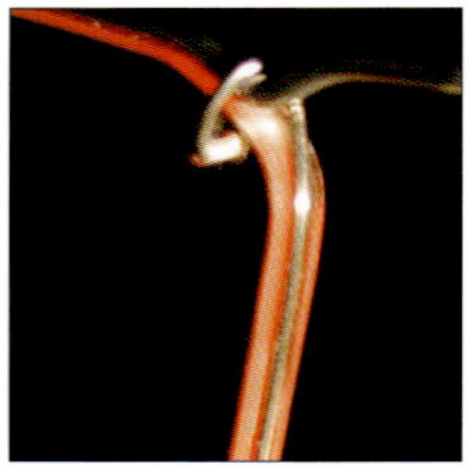

Lifestyle

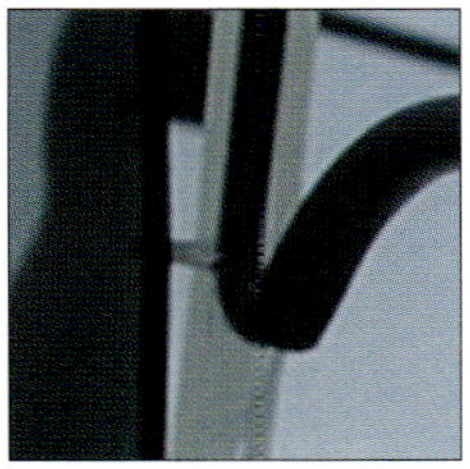
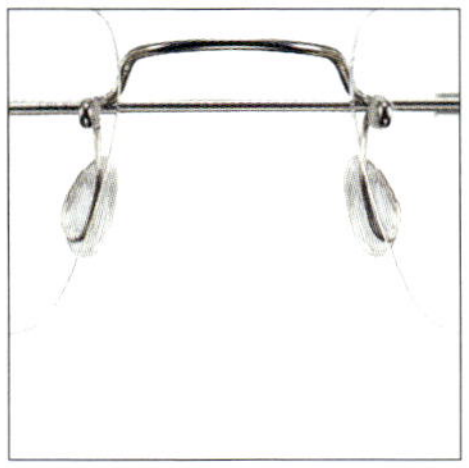
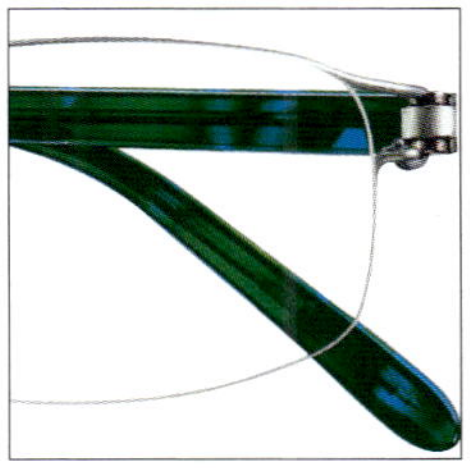
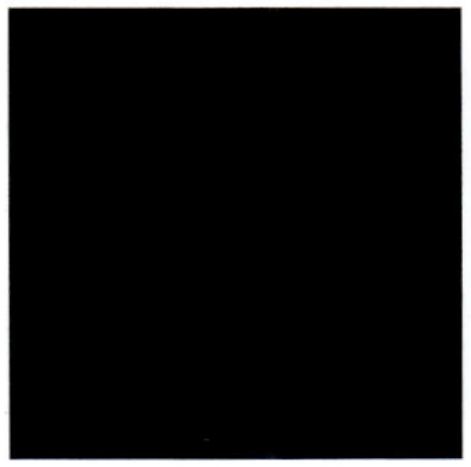

Hersteller Manufacturer
Meistersinger, Manfred Brassler
Münster, Westf., Germany

Produkt Product
Scrypto 1Z
Mechanische Armbanduhr
Mechanical wristwatch

Design
Meistersinger, Manfred Brassler
Münster, Westf., Germany

Die Scrypto 1Z ist eine mechanische Armbanduhr, die trotz der ungewöhnlichen Ein-Zeiger-Technik präzise abgelesen werden kann. Die nüchterne Bezifferung mit vorangestellter Null bei den Einerzahlen und die kühle Typografie verleihen der Uhr eine sehr sachliche, klare Anmutung. Sie ist bis 50 m wasserdicht und besitzt ein Armband aus Kalbsleder mit Dornschließe.

The Scrypto 1Z is a mechanical wristwatch which despite its single-hand system offers accurate time readings. The sober numerals with the prefixed zero in the case of single digits and the cool typography lend the watch a very matter-of-fact and crystalline appearance. It is water resistant up to 50 meters and comes with a calfskin strap with a buckle.

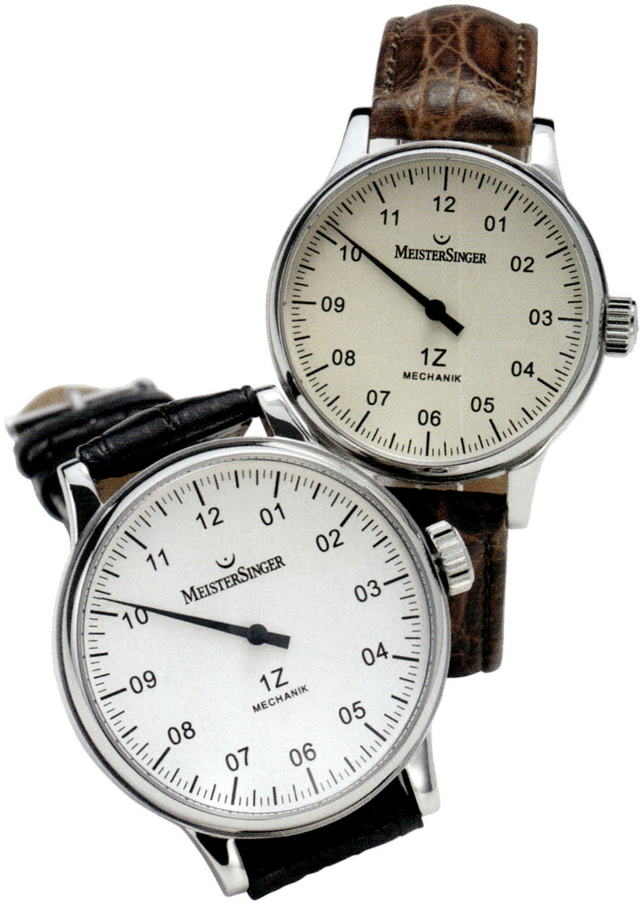

iF Silver Award

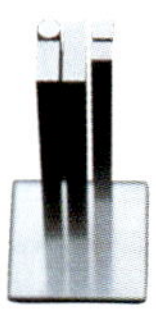

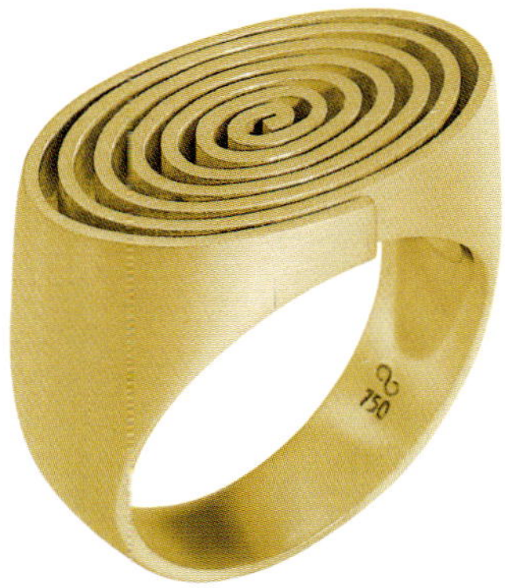

Hersteller Manufacturer
Antonio Bernardo
Rio de Janeiro, Brazil

Produkt Product
Expand Ring
Ring

Design
Antonio Bernardo
Rio de Janeiro, Brazil

Der Expand Ring von Antonio Bernardo vermittelt den Eindruck einer laufenden Bewegung, indem seine Masse bei der Umdrehung um sich selbst stetig zunimmt. Dieses Prinzip ist typisch für Himmelskörper, die sich im Laufe ihrer Bahn durch das Weltall ständig ausdehnen – daher die Bezeichnung Expand. Der Werkstoff ist mittels herkömmlicher Juwelierverfahren bearbeitetes Gold.

The Expand Ring by Antonio Bernando suggests continuous motion in that its mass increases as it swirls around itself. This principle is typical of celestial bodies which in the course of their orbit through space constantly expand – hence the product name. The ring is worked with gold using conventional jewelry techniques and offers a striking balance of matt and polished gleam.

iF Silver Award

Hersteller Manufacturer
Antonio Bernardo
Rio de Janeiro, Brazil

Produkt Product
Cycles Ring
Ring

Design
Antonio Bernardo
Rio de Janeiro, Brazil

Der Cycles Ring ist aus einem röhrenförmigen Metallrohling gefertigt, der sich dreimal in entgegengesetzter Richtung windet und dabei immer wieder an den gleichen Punkt zurückkehrt, wodurch drei Ringe entstehen, die beliebig auf den Finger gesteckt werden können. Durch den Positionswechsel bricht dieser Ring mit dem Konzept „Ring und Tisch". Er vereint Form und Anwendung, da er beliebig aufgesteckt werden kann.

The Cycles Ring is made from a tubular-shaped metal blank which takes three turns, each time in an opposing direction, and always passes through the same point. As a result, three rings are created that can be worn on the finger in the order desired, uniting form and function and enabling a departure from the 'ring and table' concept. It melds form and function, as the finger can be placed in any one of the three curves offered.

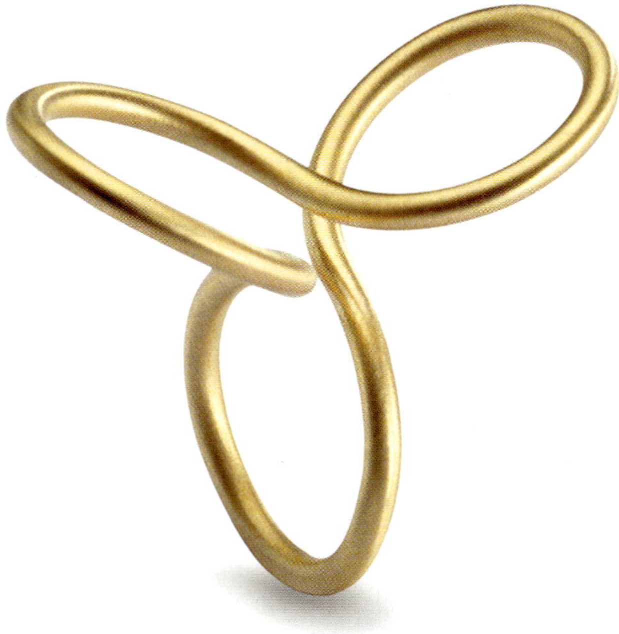

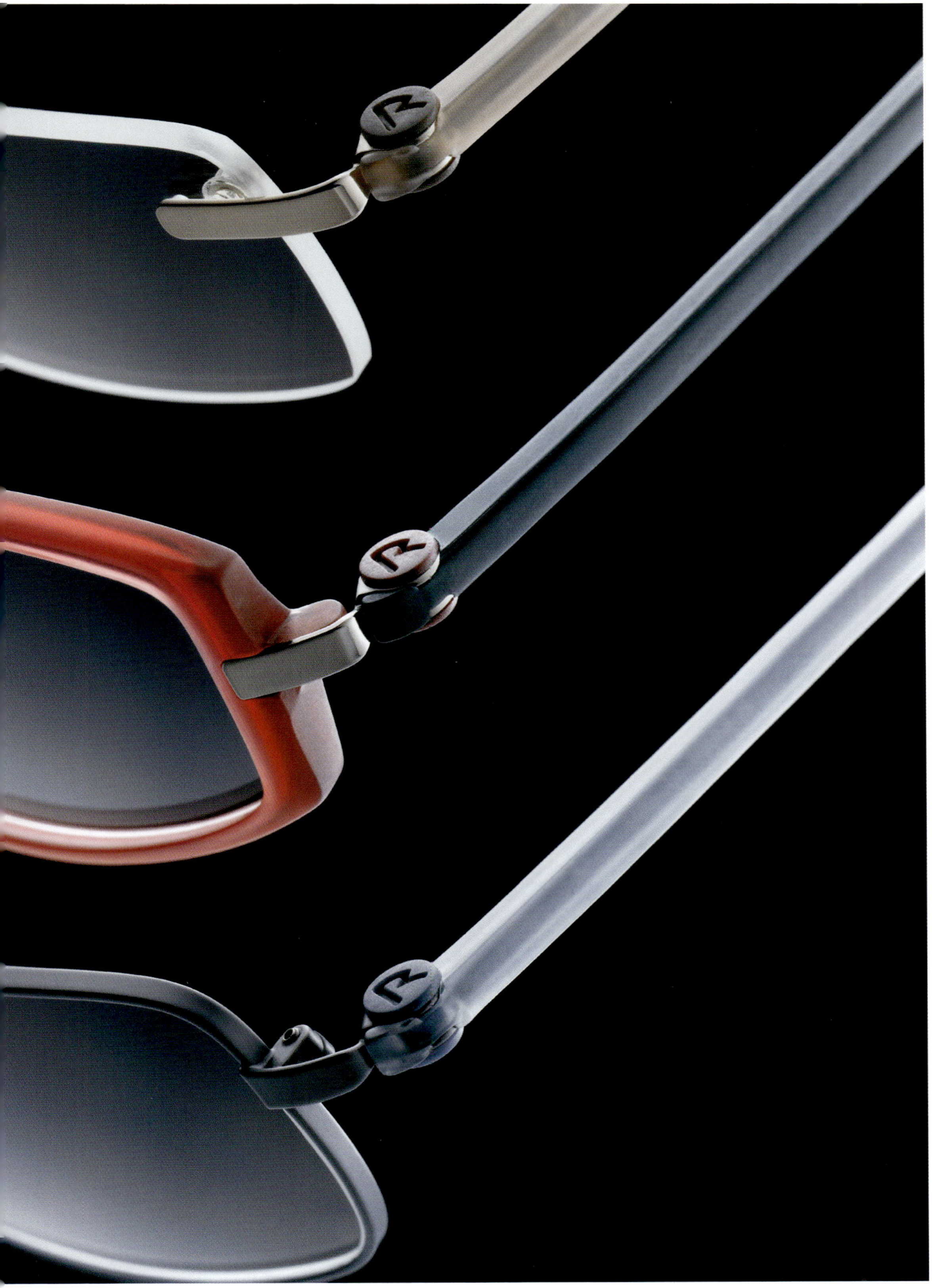

Hersteller Manufacturer
Rodenstock GmbH
München, Germany

Produkt Product
Trendkollektion mit extra flexibler Baugruppe
Brillenfassungen
Spectacles frames featuring ultra-flexible materials

Design
Rodenstock GmbH
München, Germany

Die Brillenfassungen aus der Trendkollektion mit extra flexibler Baugruppe von Rodenstock sind aus einem neuartigen Polyamid-Werkstoff gefertigt, der bisher bei Brillenfassungen nicht verwendet wurde. Die besonderen Eigenschaften des Materials erlauben die Herstellung eines Bauteils ohne die üblichen Metallarmierungen zur Unterstützung der mechanischen Stabilität. Die Gestaltung der Teile verknüpft gezielt die Rahmenbedingungen der verwendeten Werkstoffe mit Formen, Details und Proportionen für eine jüngere Zielgruppe.

The spectacles frames featured in Rodensstock's Trend Collection boast ultra-flexible components made using an innovative polyamide material utilized for frames for the very first time. The material's special properties mean that the temple sections can be made without any of the metal reinforcements generally used to enhance mechanical stability and adjustability. The design of the components deliberately combines the basic features of the materials used with shapes, details and proportions that appeal to a younger target group.

Hersteller Manufacturer
deSIGN Markus T. GmbH
Gütersloh, Germany

Produkt Product
best of Markus T.
Titanbrille
Titanium glass frames

Design
deSIGN Markus T. GmbH
Gütersloh, Germany

Die Brillenkollektion best of Markus T. beruht auf einem patentierten Stecksystem, das Bügel und Mittelteil ohne Schrauben verbindet. Damit ist eine lange Haltbarkeit und Gängigkeit der Bügel gewährleistet. Die ebenfalls patentierte Glashalterung verzichtet auch auf Verschraubungen, hier wird das Glas lediglich eingehängt und der Faden über die Fassung gelegt. Die Brillen sind aus reinem Titan gefertigt und von Hand poliert.

The best of Markus T glass frames collection is based on a patented system with which the temple sections and central pieces of the frame can be plugged together without screws. This makes the side pieces durable and continuous flexibility. The mounting of lenses is realized with a patented technique which requires no screws. The lenses are simply positioned and the nylon thread placed over the frame. best of Markus T. is made of pure titanium and polished by hand.

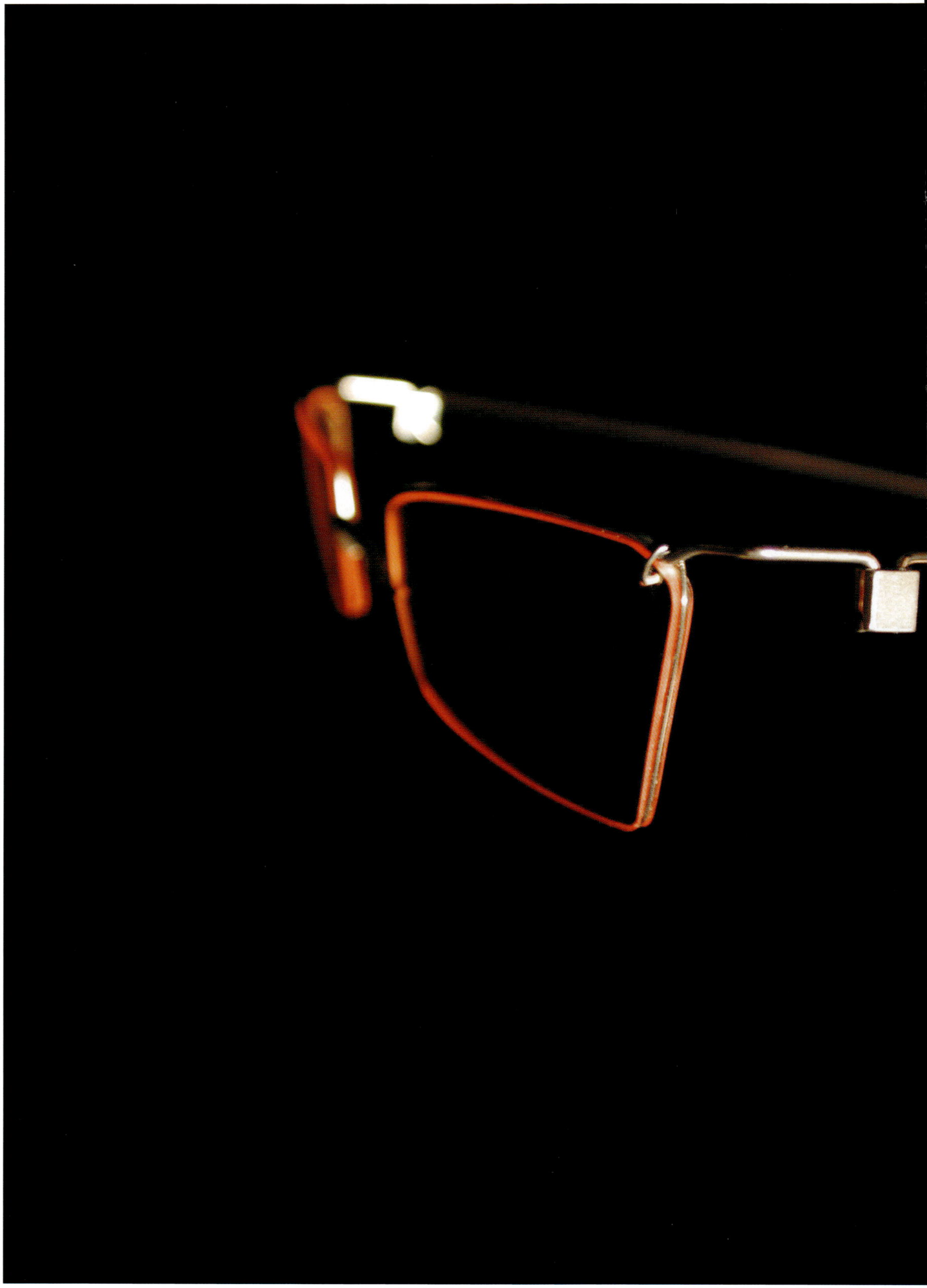

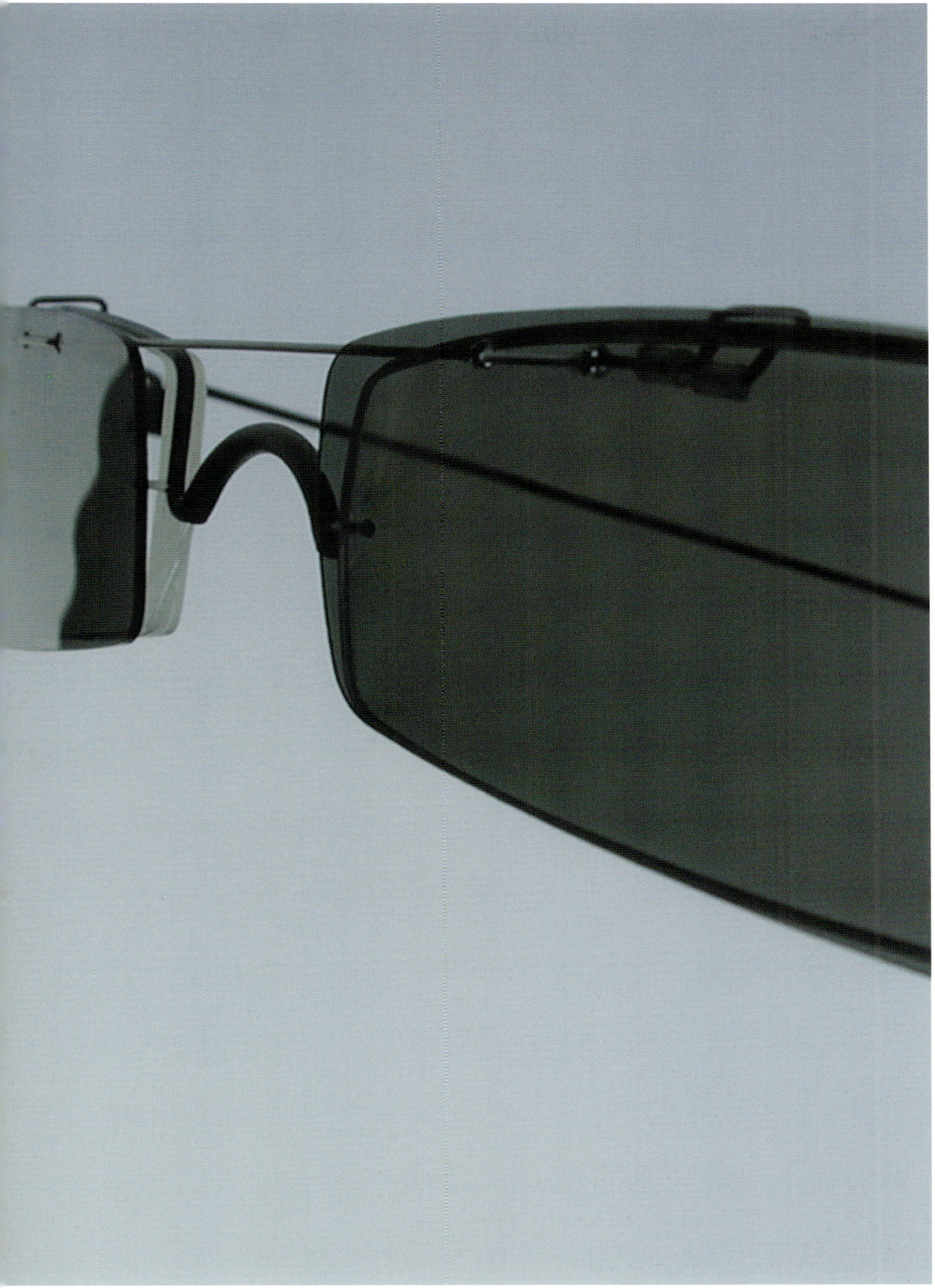

Hersteller Manufacturer
deSIGN Markus T. GmbH
Gütersloh, Germany

Produkt Product
clip-on Markus T.
Sonnenclip
clip-on sunglasses

Design
deSIGN Markus T. GmbH
Gütersloh, Germany

Die clip-on von Markus T. beruht auf einer minimalistischen Befestigungskonstruktion, die zum Patent angemeldet wurde. Ein einziger dünner Titandraht dient zur Befestigung an der Brillenfassung und gleichzeitig als Verbindung der Gläser. Der Draht wird durch das Glas gefädelt, weitere Bauteile sind überflüssig. Bei den Gläsern handelt es sich um äußerst dünne Polarisationsfolien mit einem sehr geringen Gewicht. Je nach Modell wiegt die clip-on zwischen 1,8 und 2,5 g.

The Markus T. clip-on is based on a patented, reduce-to-the-minimum attachment system. A single thin titanium wire serves to fix the frame and also to connect the lenses, eliminating the need for additional components. The ultra-thin plastic lenses have an extremely thin polarized film and are exceptionally lightweight. Depending on the model, clip-ons weigh between 1.8 and 2.5 g.

Hersteller Manufacturer
Lindberg A/S
Aabyhoj, Denmark

Produkt Product
Lindberg Spirit
Titan Brillengarnitur
Rimless titanium spectacles

Design
Lindberg A/S
Aabyhoj, Denmark

Die Brillenkollektion Lindberg Spirit besteht aus ultraleichten Titanbrillen mit einem Gewicht von nur 1,9 g. Die randlosen Brillen sind äußerst minimalistisch, sie besitzen keine Verschraubungen und im Bereich des Sehfeldes befinden sich keine sichtbaren Teile. Die einzelnen Komponenten sind für eine individuelle Anpassung an die Physiognomie des Brillenträgers in unterschiedlichen Längen erhältlich. 30 verschiedene Brillenfarben stehen zur Auswahl.

The Lindberg Spirit spectacles collection is composed of ultra-light titanium-wire glass frames weighing just 1.9 g. The frameless spectacles are truly minimalistic, include no screws let alone obstructive components in the field of vision. Individual components are available in a variety of lengths to match the particular physiognomy of the wearer. The frames, with their strong hinges, also come in a choice of 30 colors.

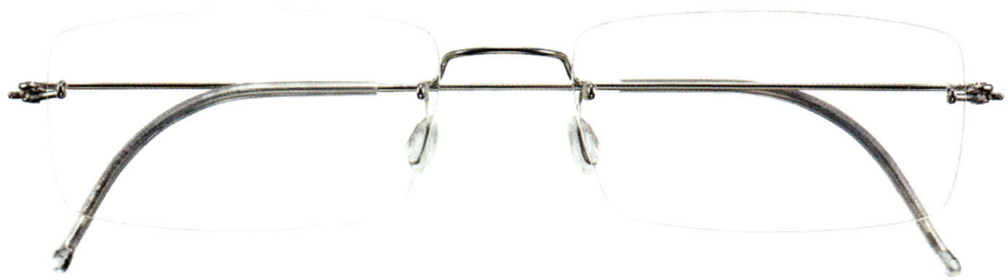

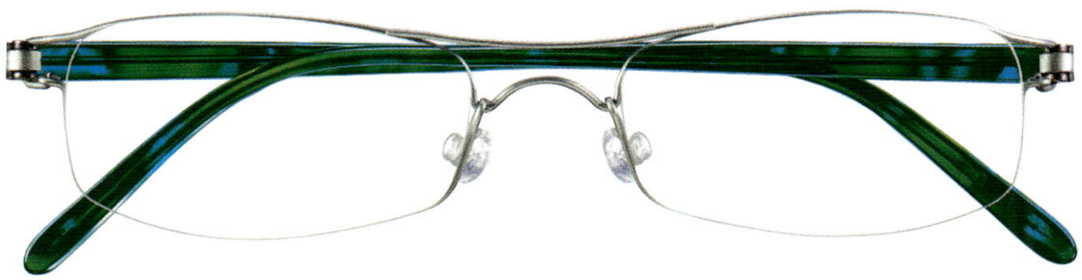

Hersteller Manufacturer
Lindberg A/S
Aabyhoj, Denmark

Produkt Product
Lindberg 9000 StripWire
Titan/Azetat Brillenfassung
Titanium/acetate frames

Design
Lindberg A/S
Aabyhoj, Denmark

Bei der Brillenkollektion Lindberg 9000 StripWire handelt es sich um leichte und elegante Titan/Azetat Brillenfassungen mit einem Gewicht von lediglich 3,9 g. Das Scharnier besteht aus einem stabilen, mit Titanium überzogenen Metallkern.

The Lindberg 9000 series features light and elegant featherweight titanium/acetate glass frames weighing only 3.9 g. The hinge has a stable, titanium-coated metal core.

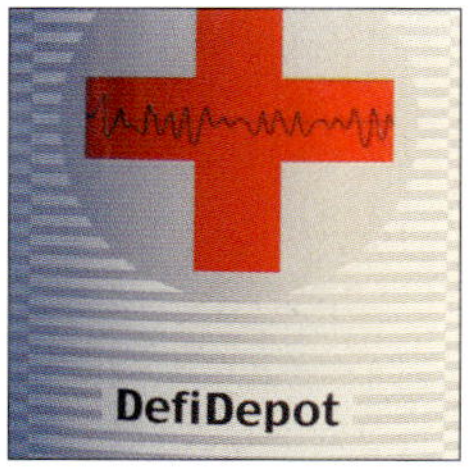
DefiDepot
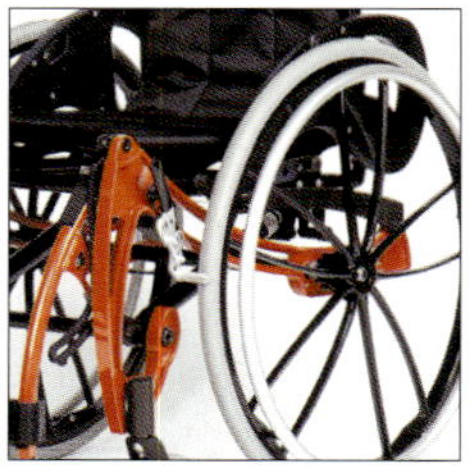
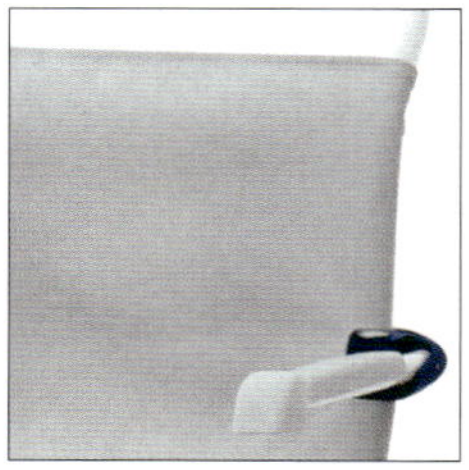
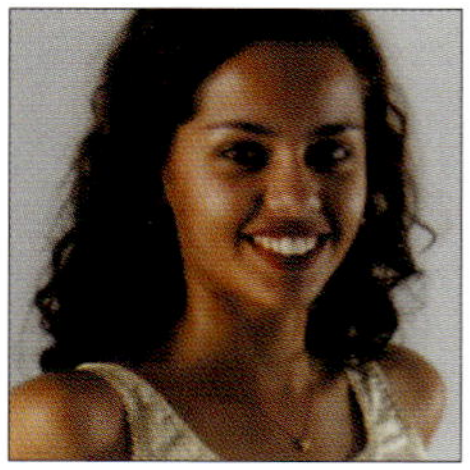
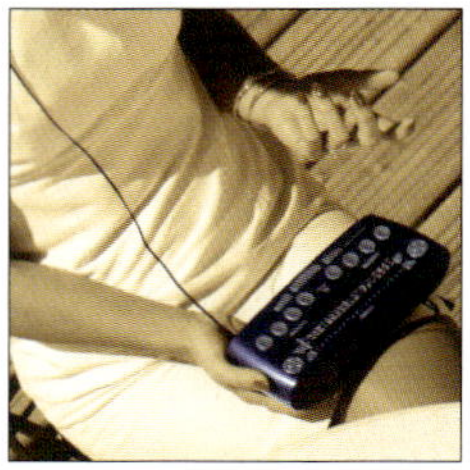
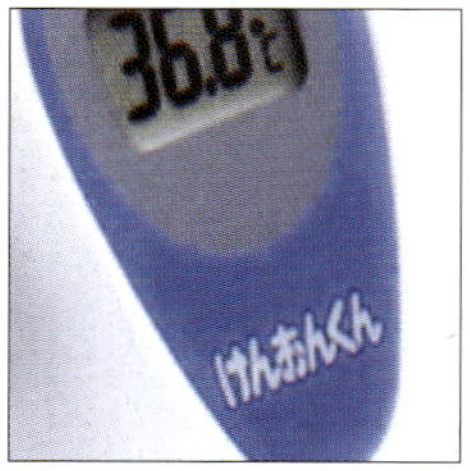
けんおんくん
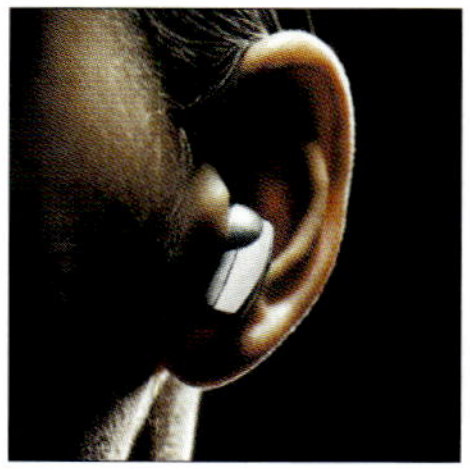
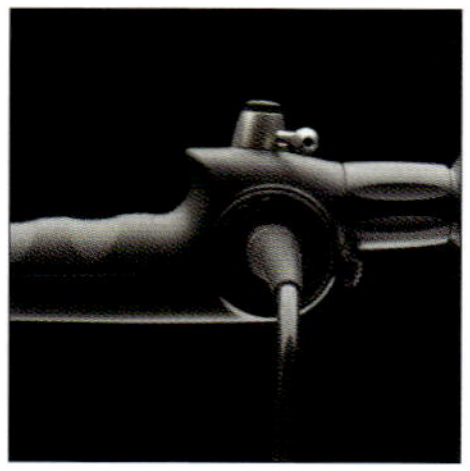
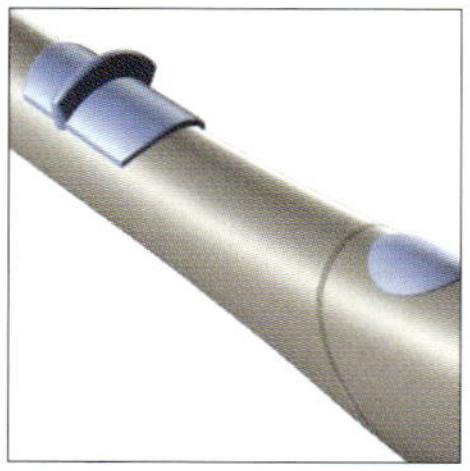
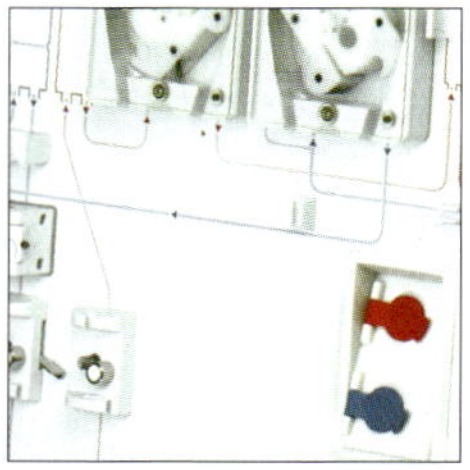

Medizin Medicine

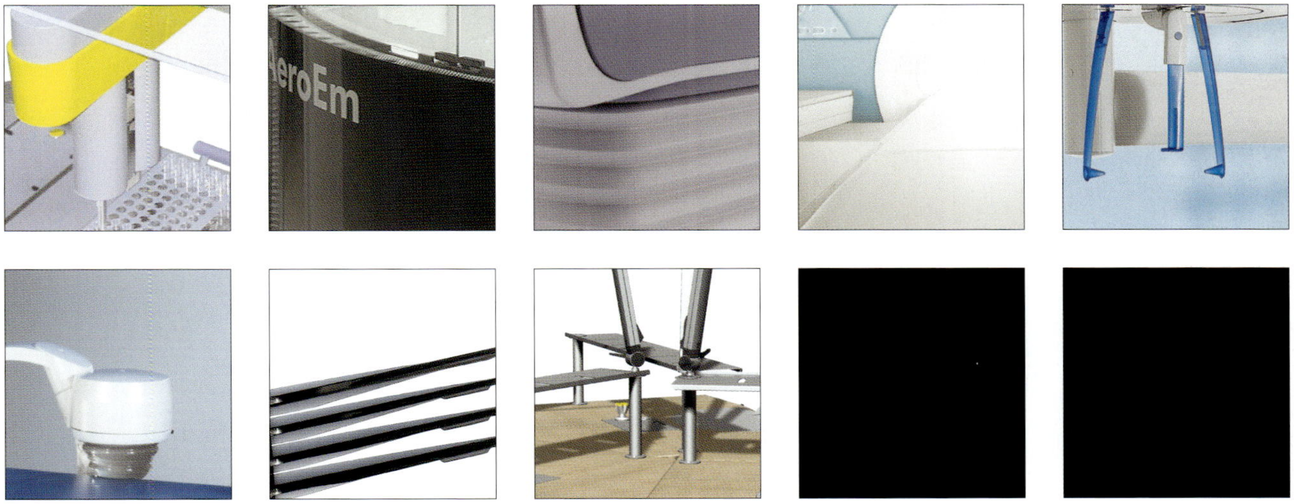

Hersteller Manufacturer
ENGELUNDLÖWENTWICKLUNGEN
Stuttgart, Germany

Produkt Product
DefiDepot
Outdoor-Gehäuse für Defibrillatoren
Outdoor casing for defibrillation

Design
Leo Christmann
Angelika Hagen
Stuttgart, Germany

Das DefiDepot ist eine Outdoorbox für Defibrillatoren, das es ermöglicht, die sensible Gerätschaft im Außenbereich einsetzen zu können und gleichzeitig die Benutzung mit Hilfe der integrierten Transpondertechnik zu steuern. Im Falle eines Herzstillstandes sind die Überlebenschancen am höchsten, wenn innerhalb der ersten paar Minuten, noch vor dem Eintreffen des Notarztes, eine „Automatisierte Externe Defibrillation" (AED) durch ausgebildete Ersthelfer angewendet wird. Voraussetzung ist, dass Defibrillatoren jederzeit und überall zur Verfügung stehen. Da das DefiDepot für den medizinischen Notfall geschaffen wurde, hat es als Hauptmerkmal ein rotes Kreuz überlagert von einer symbolischen EKG-Kennlinie, die dem plötzlichen Herztod vorausgeht. Das kombinierte Zeichen steht vor einem hellen Hintergrund, der großteils aus reflektierender Folie besteht, um auch bei Nacht die entsprechende Signalwirkung zu erzielen.

The DefiDepot is an outdoor box for defibrillators enabling the use of such sensitive instruments outdoors while also providing integrated transponder-based control of their use. The survival chances of patients who have suffered heart attacks is greatest if within the very first few minutes (i.e., before the paramedics arrive) "automated outdoor defibrillators" (AOD) are immediately activated by trained first-aid staff. This means, of course, that defibrillators must always be at hand, everywhere. Since the DefiDepot was created as a first-aid station, its main feature is a red cross, and the typical ECG waveline stands symbolically for the ventricular flutter prior to sudden cardiac death. The combined sign is positioned on a bright background, mostly composed of a reflecting foil, so that it radiates the right signal at night, too.

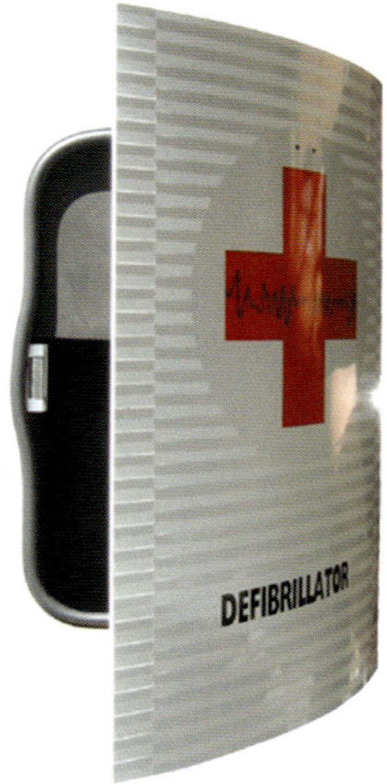

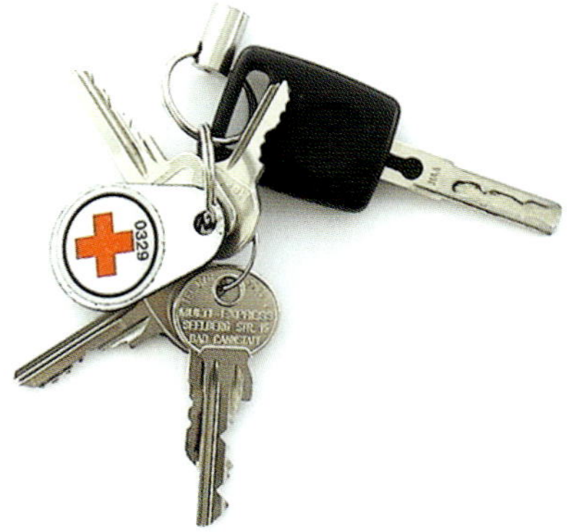

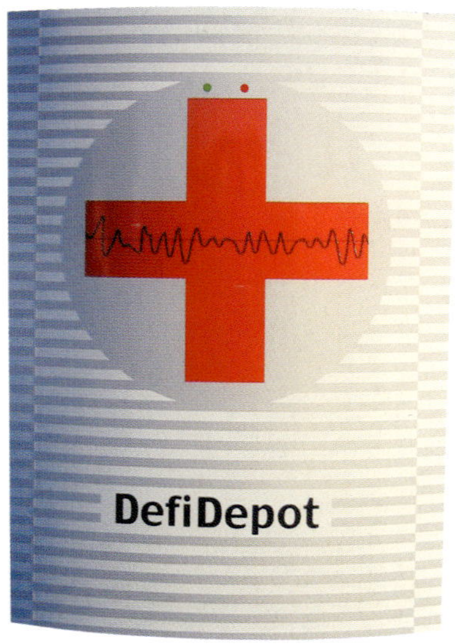

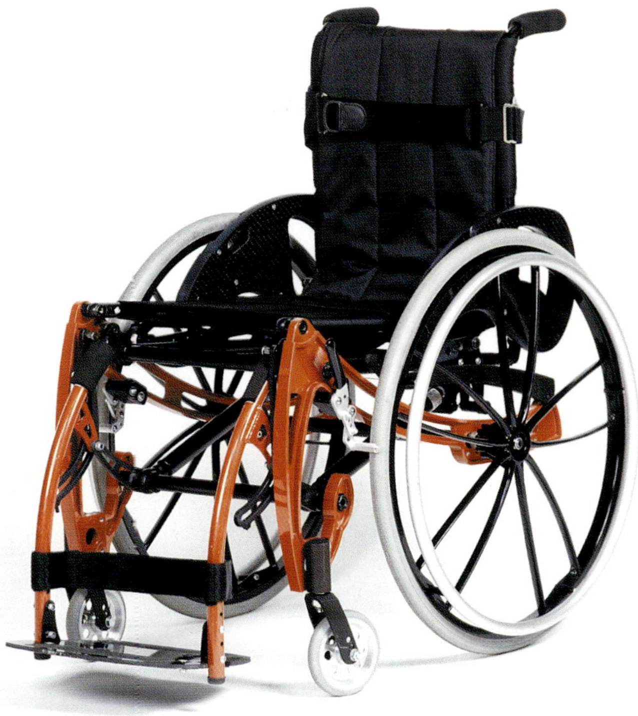

Hersteller Manufacturer
LifeStand
Les Echêts, France

Produkt Product
LifeStand Active
Stehrollstuhl
Standing wheel chair

Design
Gilles Damon
Damon Design
Chaponost, France

Der LifeStand Stehrollstuhl verbindet die Funktionen des selbstbestimmten Stehens mit den Fahreigenschaften eines kompakten Aktivrollstuhls. Eine in den Stuhl integrierte Funktion ermöglicht es dem Nutzer, seinen Bedürfnissen entsprechend aus der sitzenden in die stehende Position zu wechseln. Die patentierte Stehfunktion ermöglicht es, zumeist aus eigener Kraft aufzustehen. Das Körpergewicht wird durch Gasdruckzylinder ausgeglichen. Die zum Aufrichten erforderliche Kraft entspricht in Intensität und Bewegungsablauf dem Fahren eines leichtgängigen Rollstuhls. Das Fixierungssystem des LifeStand stabilisiert den Ober- und Unterkörper ohne Druck auf Verdauungsorgane und Brustkorb auszuüben.

The LifeStand standing wheel chair combines the functions of autonomous standing with the driving capacity of a compact active wheelchair. For example, the chair features a function that enables the user to shift from standing to sitting and vice versa as required. The patent-protected standing function enables most users to stand up on their own. The user's body weight is compensated by a pneumatic cylinder. In terms of intensity, the power needed to stand up corresponds to that involved in terms of motion to propelling a conventional light wheelchair. The LifeStand's fixing bands help stabilize the user's upper and lower body without exerting pressure on the stomach or chest.

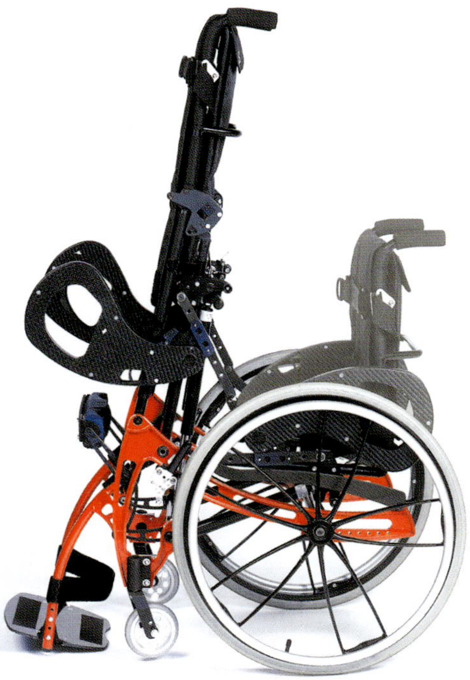

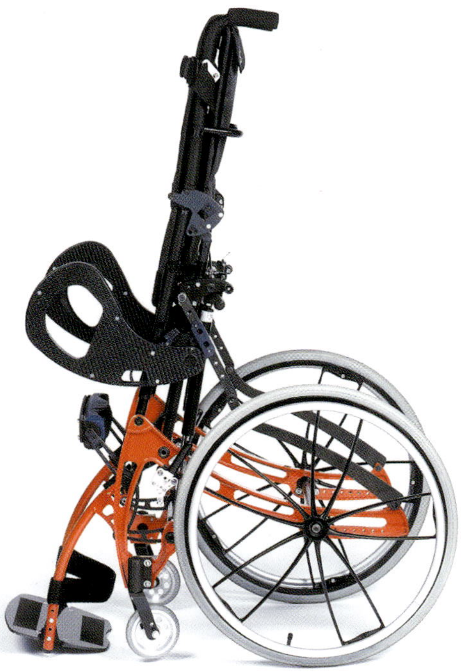

Hersteller Manufacturer
Aquatec GmbH + Co. KG
Isny, Germany

Produkt Product
Ocean
Dusch- und Toilettenstuhl
Shower/Toilet chair

Design
Rupert Bläsi
Marc Weidt
Roland Wendt
Einmaleins – Büro für Gestaltung
Burgrieden, Germany

Der Dusch- und Toilettenstuhl Ocean hilft Menschen, die eine Dusche ohne Hilfsmittel nicht nutzen können. Für den Einsatz im Nassbereich ist der robuste Rahmen aus rostfreiem Edelstahl aufgebaut. Der Rahmen dient gleichzeitig als Schiebehilfe und erleichtert dem Pflegepersonal das einfache Rangieren des Stuhls. Die spiegelsymmetrischen Seitenteile aus Kunststoff reduzieren die Herstellkosten und ermöglichen die patentierte Verstellung der Sitzhöhe in 5 Schritten. Alle Verstellelemente sind farbig codiert und leicht zu finden. Das „Knock-Down"-Prinzip gestattet den Einsatz des Ocean-Stuhls auf Reisen.

The Ocean shower and toilet chair assists people unable to use a shower without bathroom aids. The sturdy, non-corrosive stainless steel frame is destined for use under wet conditions. The frame doubles up as a walking frame and likewise enables care personnel to push the chair with a minimum of effort. The mirror-image plastic side components reduce manufacturing costs and mean the seat height can be adjusted in five steps using the patented adjustment system. All the controls for adjustments are color-coded and easy to spot. Since the Ocean can be easily "knocked down", it is ideal for travel use, too.

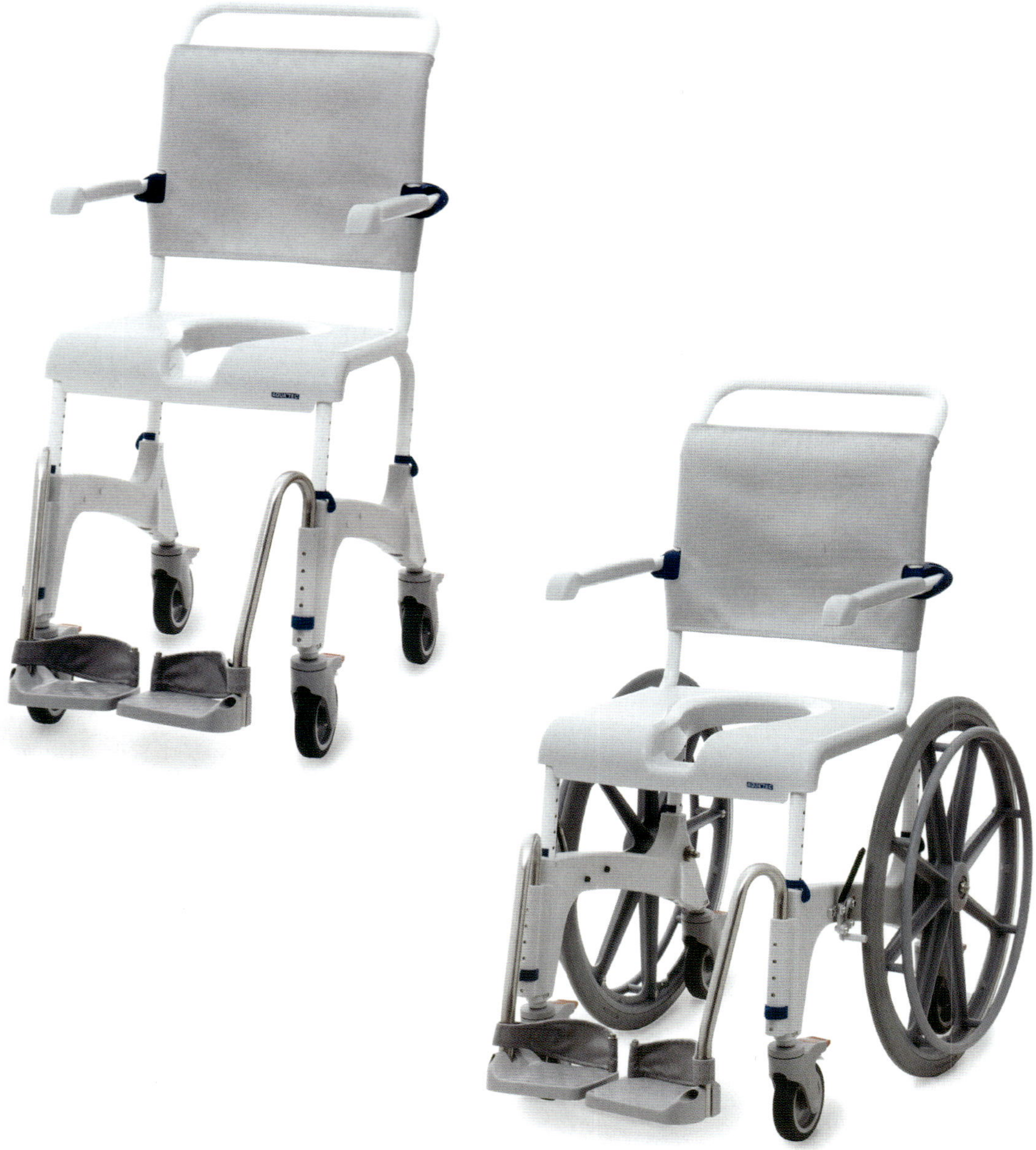

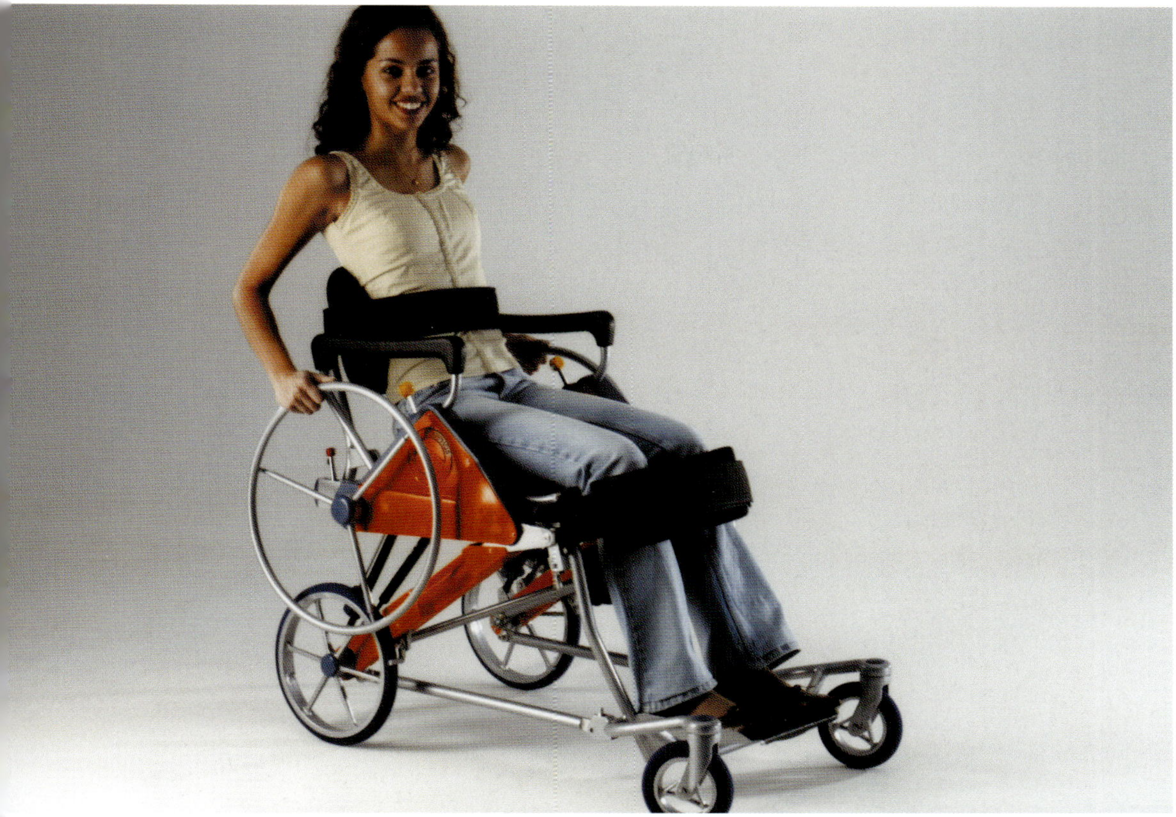

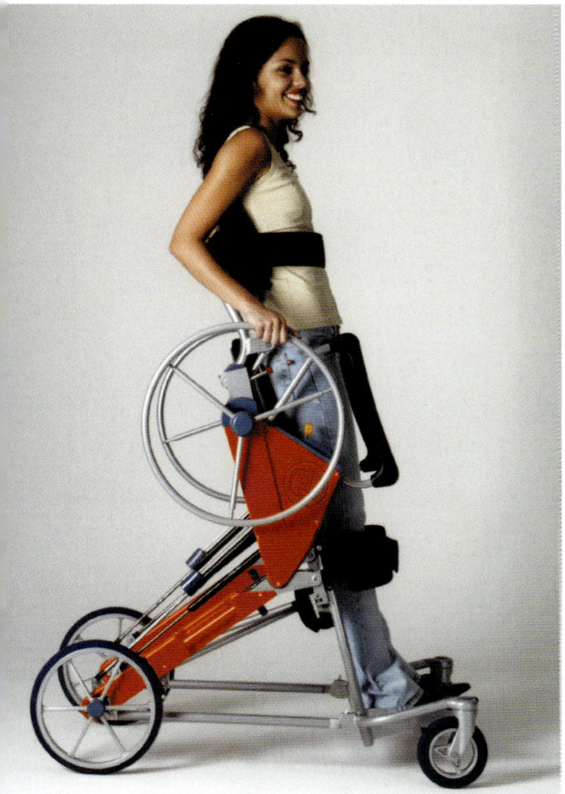

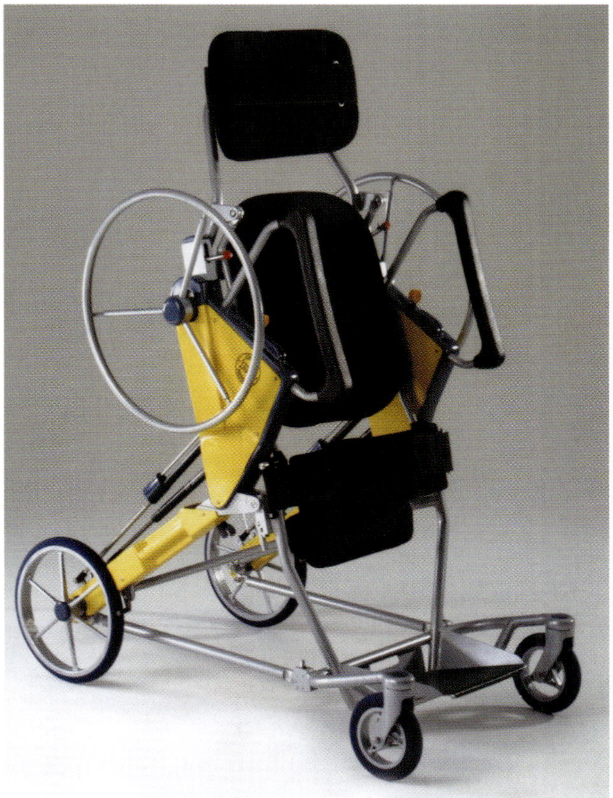

Hersteller Manufacturer
The Sarah Network of Hospitals for Rehabilitation
Brasilia, Brazil

Produkt Product
Ortomovel
Orthomobil
Orthomobile

Design
Antonio Cutolo
Bruno Matos
Claudio Duarte
Henry Macario
Jose Carlos Ferraz
Manoela Palmeiro
Paulo Fernandes
William Dias
EquipHos – Hospital Technology Research Group
Brasilia, Brazil

Das Orthomobil ermöglicht dem Gehbehinderten, sich stehend fortzubewegen. Mit einer einfachen Bewegung kann der Benutzer mit dem Orthomobil aufstehen und in dieser Position ohne fremde Hilfe umhergehen. Der Patient wird mit Gurten an Knien und Rumpf angeschnallt, was ihm Standsicherheit bietet. Zum Transport wird das Orthomobil in drei kompakte Teile zerlegt. Durch die Anwendung von Flugzeugaluminium, -stahl und Titan konnten höchste Verwindungssteifigkeit und Minimalgewicht erreicht werden. Die Kissen von Sitz, Rückenlehne und allen anderen gepolsterten Teilen sind aus eingespritztem Polyurethanschaum hergestellt.

The Orthomobil enables individuals with functional disability of the lower limbs the freedom to move in a standing position. By means of a simple movement, the Orthomobil user can change from a sitting to a standing position and then move around without outside assistance. The patient is held safely in place by belts that fasten across the knees and chest, and parking brake levers are in easy reach. The Orthomobil knocks down into three compact parts for transport. Thanks to the use of airplane aluminum, steel and titanium, it delivers maximum strength combined with minimum weight. The seat and backrest cushions and all other upholstered parts are made of injected polyurethane foam.

Hersteller Manufacturer
Alva B.V.
Arnheim, Netherlands

Produkt Product
MPO 5500
Mobile Phone Organizer für sehbehinderte Menschen
Mobile phone organizer for visually impaired

Design
PEMSTAR B.V.
Christian Suurmeijer
Almelo, Netherlands

Der Alva Mobile Phone Organizer kombiniert GSM-Handy, mobile Internetdienste und wichtige Sprachfunktionen. Es ist das erste schnurlose Endgerät seiner Art für sehbehinderte Menschen und ist besonders kompakt und leicht zu transportieren. Sämtliche Funktionen des Geräts wie die verschiedenen Menüs, das Adressbuch, den SMS-Versand, die Mobilfunkdienste und die Systemeinstellungen bedient der Anwender entweder mit der integrierten Braille-Tastatur oder per Sprache über eine neuartige menügesteuerte Benutzerführung. Die taktilen Qualitäten des MPO erlauben eine schnelle und leichte Navigation in verschiedenen Lebenslagen.

The Alva MPO 5500 Mobile Phone Organizer is an especially compact all-in-one GSM mobile phone, mobile Internet service link and essential notetaker for use on the move. It is the first wireless device of its kind for the visually impaired. All the MPO's features, such as the menus, contact information, SMS message system, mobile phone services and system information can be accessed either through the integrated Braille keypad or via speech input/output using the innovative menu-based user interface. The MPO's tactile qualities ensure swift and simple navigation in all those everyday situations where it is needed.

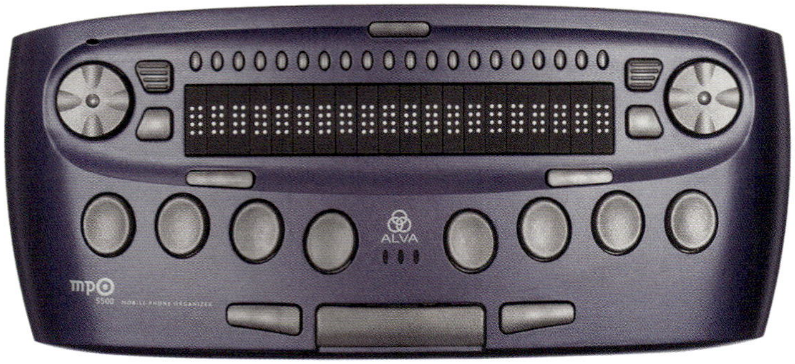

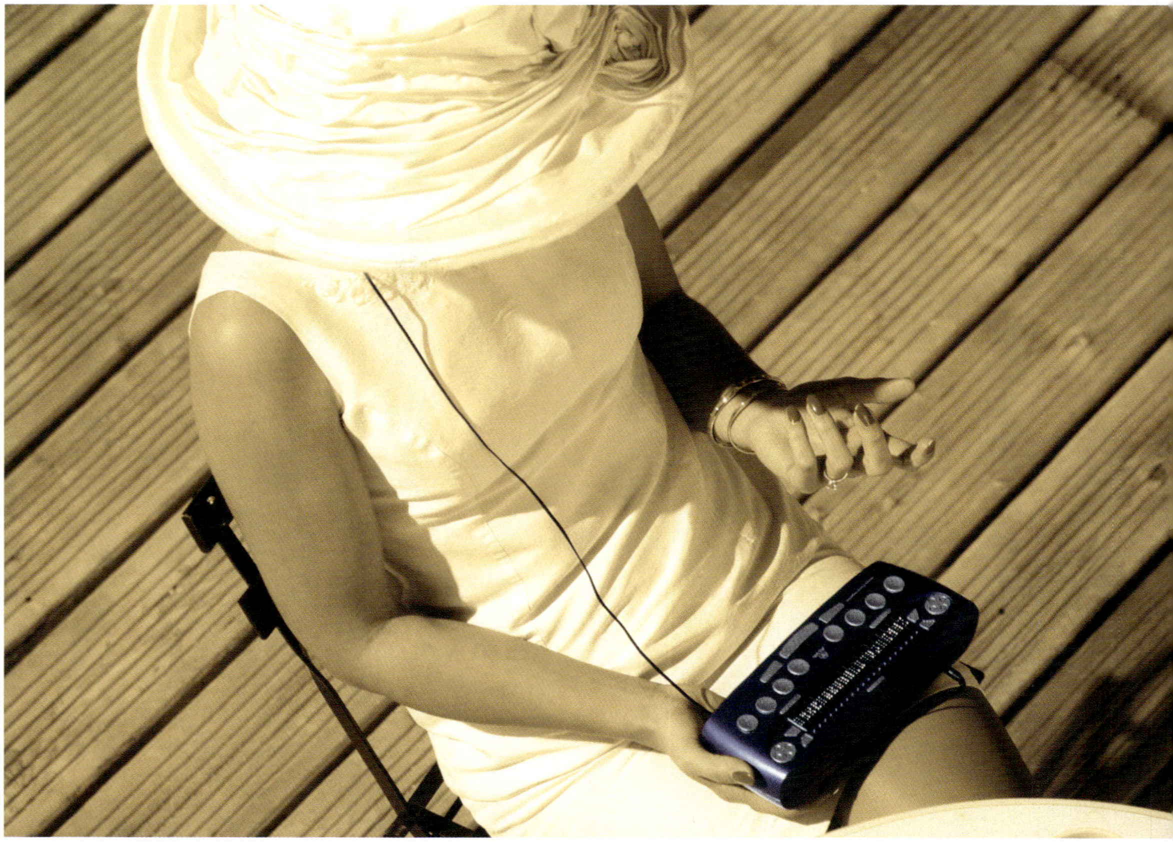

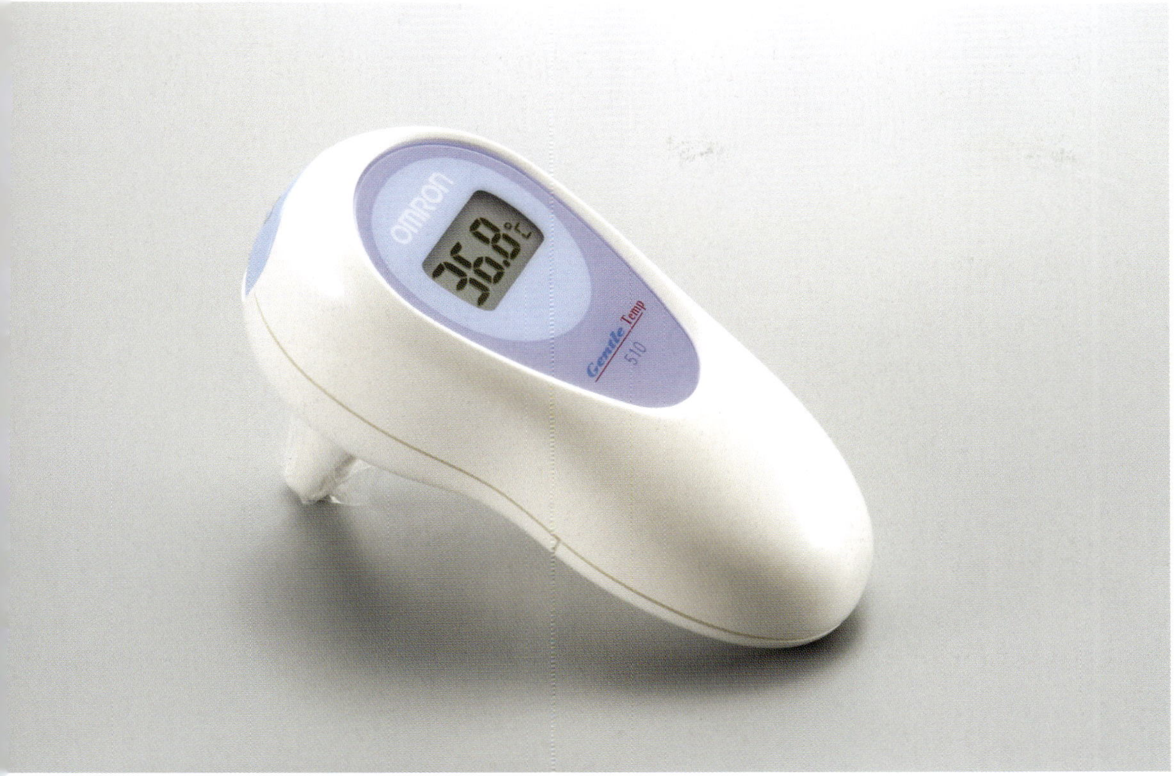

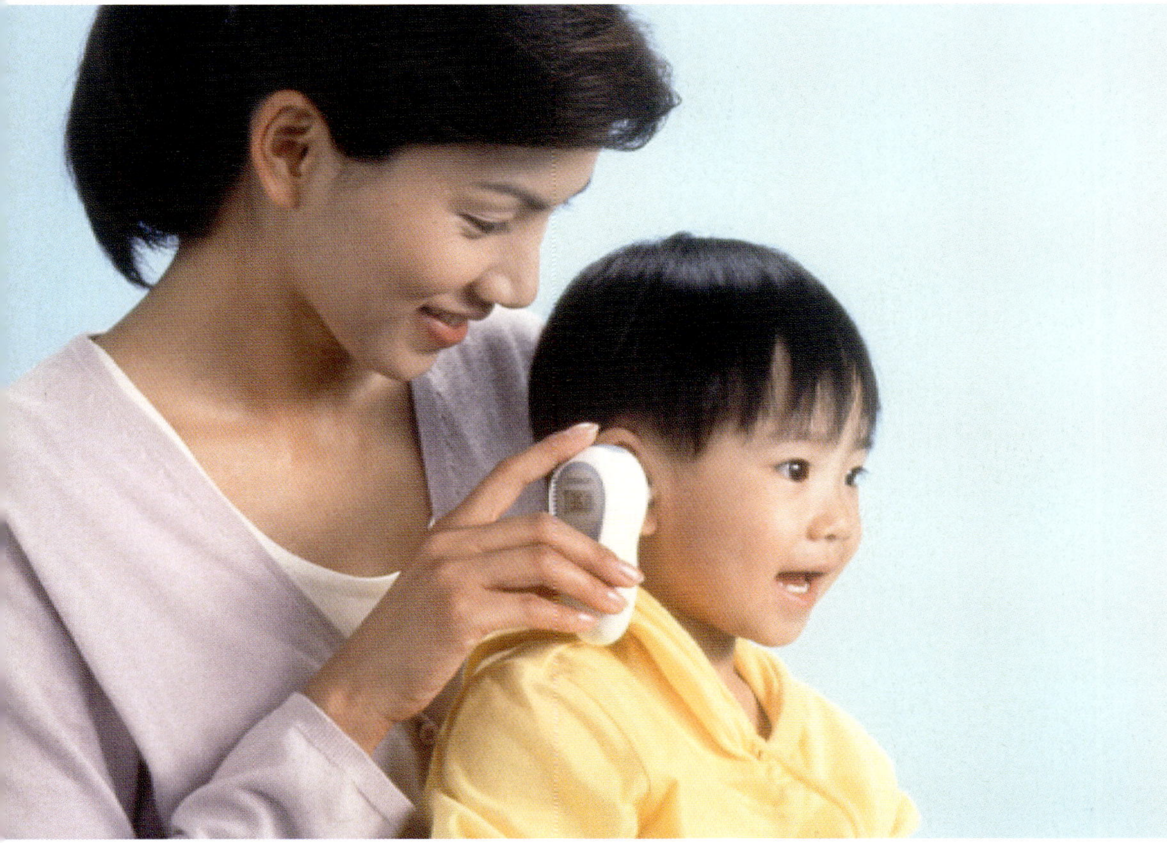

Hersteller Manufacturer
Omron Corporation
Kyoto, Japan

Produkt Product
GT-510
Ohrthermometer
Ear thermometer

Design
Noguchi Sou
YS Design inc.
Osaka, Japan

Das „Gentle Temp" Ohrthermometer GT-510 wurde insbesondere für Familien mit Kindern oder Säuglingen entwickelt. Das Thermometer hat eine kinderfreundliche, runde Form, der abgerundete untere Anzeigeteil passt genau in die Hand. Die Seitenfläche weist eine Einbuchtung für leichteren Griff und korrekte Halteposition auf. Durch bloßen Tastendruck, Displayanzeige und Kontrollton ist das Gerät besonders einfach zu bedienen. Die Messung erfolgt in nur einer Sekunde, es wird nur der höchste Messwert angezeigt, wodurch variierende Ablesewerte während Messwiederholungen vermieden werden.

The "Gentle Temp" ear thermometer GT-510 has been developed specifically with families with children or infants in mind. The thermometer has a child-friendly round shape, and the rounded lower display section sits perfectly in the hand. The side is indented for easier grip and correct holding position. The device is extremely easy to use: one-touch one-button operation, easy-to-read display and beep signal. It requires only one second to measure the temperature and indicates only the highest value, eliminating the need to repeat measurements owing to varying figures.

Hersteller Manufacturer
Microsound A/S
Silkeborg, Denmark

Produkt Product
Microsound Pilot
Hörgerät
Hearing aid

Design
Designit A/S
Aarhus, Denmark

Bei der Entwicklung des Microsound Pilot Hörgeräts wurde die Philosophie verfolgt, dass es sichtbar getragen werden sollte – so wie Brillen bei Menschen mit Sehschwächen. Ziele des Konzepts: Die Hörgeräte sollen nicht nur exklusiv über Hörgeräte-Akustiker vertrieben werden, sondern auch in Geschäften in städtischen Einkaufsbereichen erhältlich sein. Bei Microsound Pilot handelt es sich um eines der ersten aufladbaren digitalen Hörgeräte, bei dem die Möglichkeit besteht, aktuelle Software herunterzuladen. Dank der verschiedenen Größen der Ohrbügel lässt sich das Gerät schnell an seine/n Träger/in anpassen. Um ganz nach eigenem Geschmack in Erscheinung zu treten, lässt sich die farbige Muschel problemlos austauschen.

The Microsound Pilot hearing aid is meant to be visible, just as glasses are on people with impaired vision – this is the philosophy behind the design. The concept's goal: to distribute the hearing aids not exclusively via hearing-aid centres but also through outlets in the main city shopping areas. Microsound Pilot is one of the world's first rechargeable digital hearing aids with the possibility of downloading new software. Thanks to the different sizes of the eartips, the unit can swiftly be adjusted to fit the user. The colored shell can easily be changed to give the desired look.

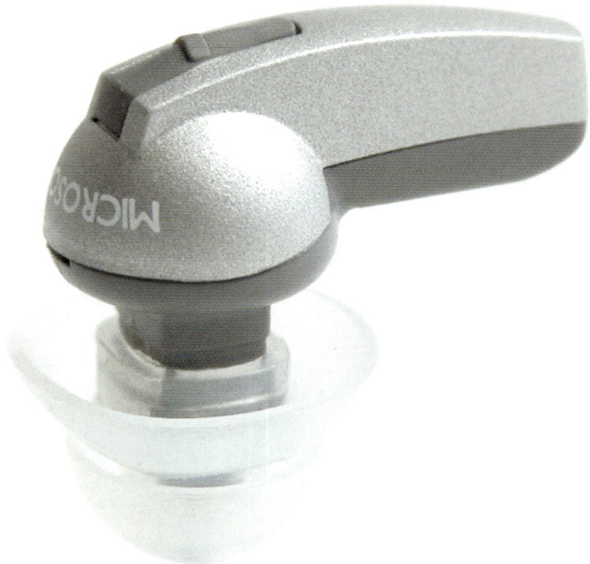

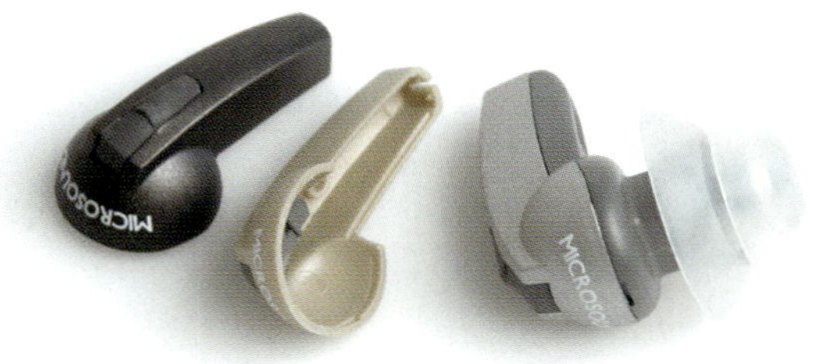

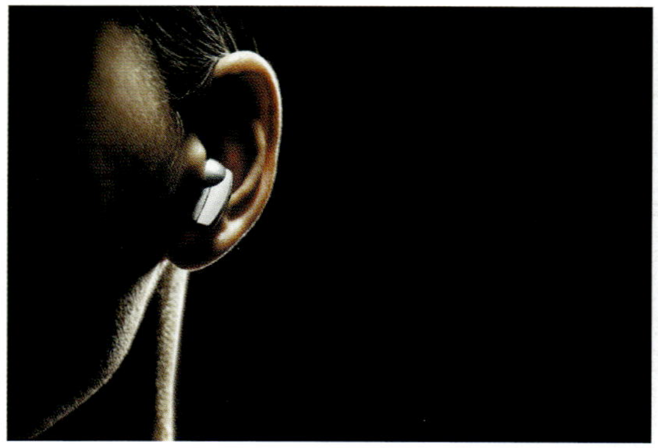

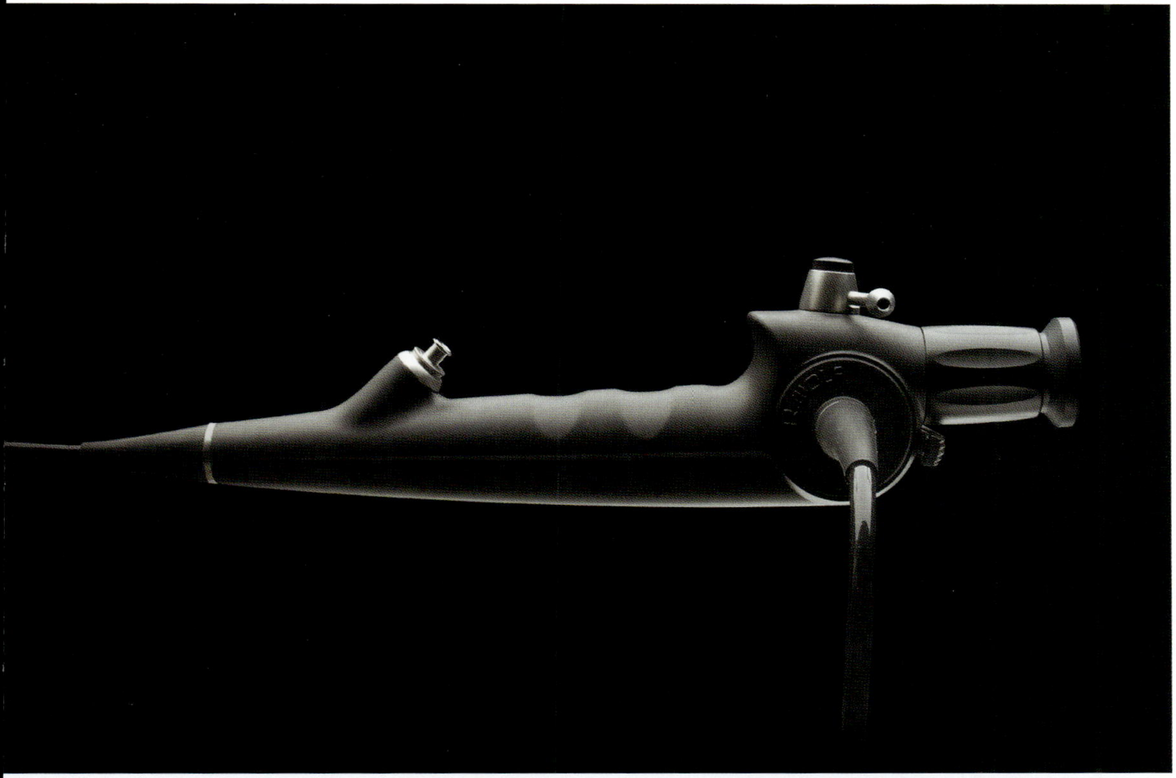

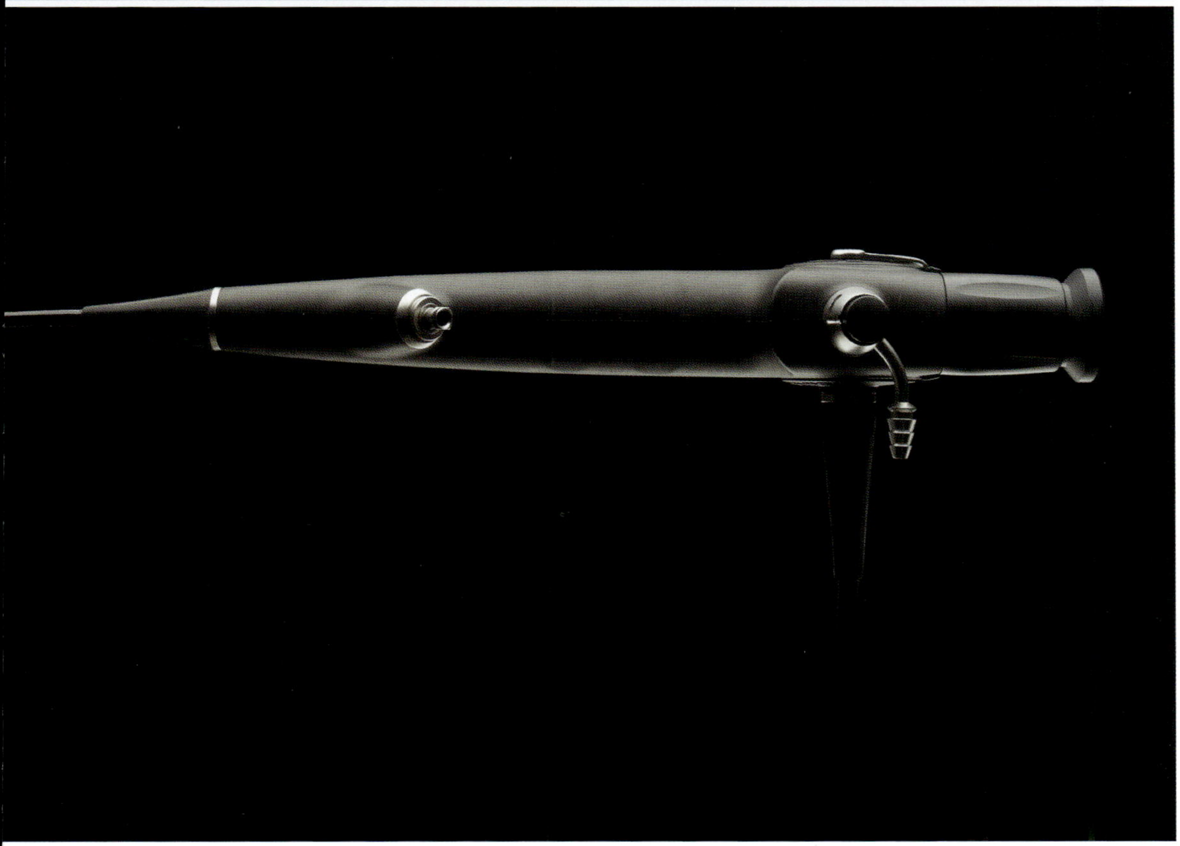

Hersteller Manufacturer
Richard Wolf GmbH
Knittlingen, Germany

Produkt Product
Flexibles Endoskop
Steuergehäuse
Steeringcase

Design
designafairs GmbH
Erlangen, Germany

Die Anwender des flexiblen Endoskops sind diagnostizierende und therapierende Ärzte in diversen Fachgebieten wie z.B. HNO, Gynäkologie, Urologie etc. Die ergonomische Gestaltung des Handhabungsteiles erlaubt ein ermüdungsfreies Arbeiten, d.h. das Instrument ist ohne Kraftschluss durch die Hand in einer Lage fixiert. Zum anderen bestimmen die Längsrillen taktil die Position des Gehäuses zur Hand. Durch die Halbschalenbauweise mit fixierenden Schraubverschlüssen an den Durchführungen können spritztechnisch alle notwendigen Komponenten und Aufnahmen für Funktionsteile montageoptimiert und preiswert hergestellt werden.

Physicians active in diagnostics and therapies in various fields (such as ENT, gynecology, urology etc.) are the main users of flexible endoscopes. The ergonomic design of this handset case ensures non-fatiguing use, i.e., the instrument can be held firmly in position without the hand having to exert power to steer it. Moreover, the lengthwise grooves provide tactile definition for how the hand holds the case. Thanks to the case's composition – it is made of half-shells held in place by screws at the outlets – all the requisite components and insets for functional elements can be injection molded and assembled in a cost-effective manner.

Hersteller Manufacturer
Sirona Dental Systems
Bensheim, Germany

Produkt Product
Sirocam C
Intraoralkamera
Intraoral camera

Design
Alexander Müller
designafairs GmbH
München, Germany

Die Intraoralkamera Sirocam C für den Zahnarzt besteht aus einem einteiligen und daher leicht zu reinigenden Titangehäuse. Die ergonomische Gestaltung erlaubt eine ermüdungsfreie Handhabung. Im Kopf der Kamera befindet sich eine Leuchtdiode.

The Sirocam C intraoral camera for dental applications consists of a single-piece titanium casing and is therefore easy to clean. Thanks to its ergonomic design, the strain is taken off dental practitioners' hands. An LED is positioned in the camera's head to facilitate use.

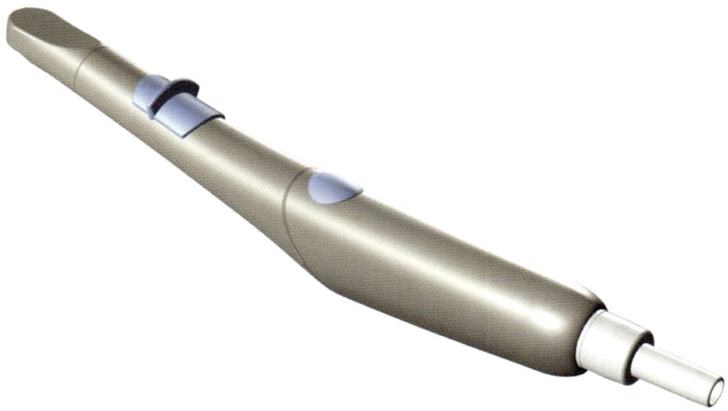

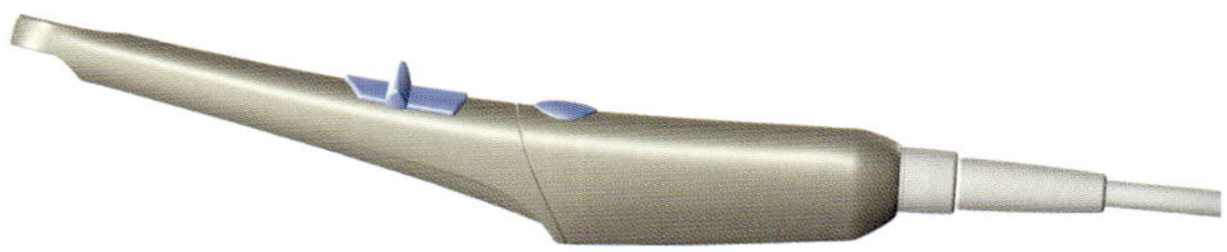

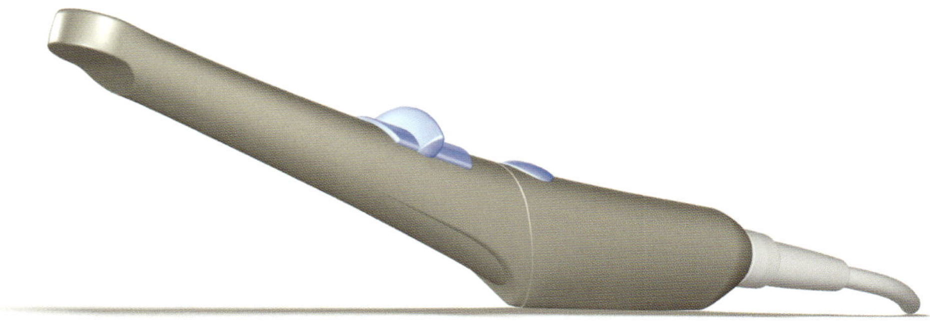

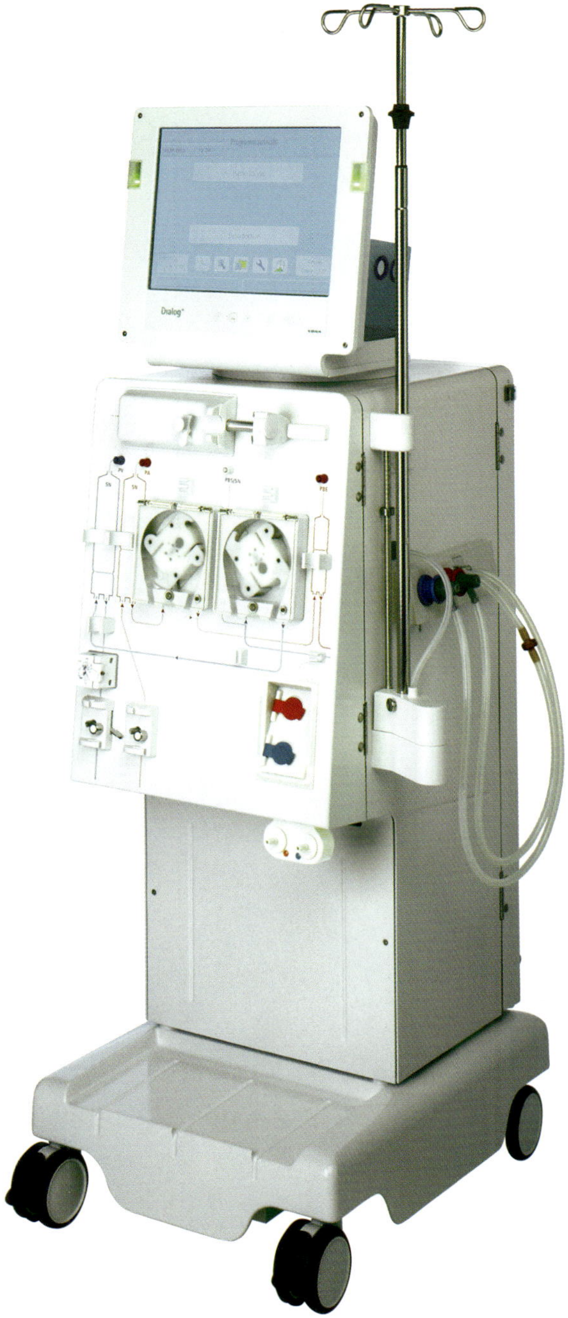

Hersteller Manufacturer
B.Braun Medizintechnologie GmbH
Melsungen, Germany

Produkt Product
Dialog+
Dialysegerät
Dialysis unit

Design
Meyer-Hayoz Design Engineering Group
Konstanz, Germany
Winterthur, Switzerland

Bei der Entwicklung des Dialysegeräts Dialog+ wurden insbesondere bei der ergonomischen Gestaltung die Anforderungen des Praxisalltags zugrunde gelegt. So ermöglicht der übersichtliche Touchscreen eine schnelle Handhabung im hektischen Klinikbetrieb, der drehbare Monitor und die geneigte Frontplatte sorgen zusätzlich für ein vereinfachtes Handling. In einer multifunktionalen Box können Patientenunterlagen, Verbrauchsmaterialien und Zubehör abgelegt werden. Die farbige Betriebszustandsanzeige ist gut sichtbar im Monitorgehäuse positioniert. Durch die integrierte Softwarelösung wird die Effizienz im Therapiealltag deutlich gesteigert.

The development work for the Dialog+ dialysis unit focused specially on an ergonomic design in light of everyday treatment needs. Thus, the clear touchscreen ensures swift-and-sure use in the bustle of doctor's offices, the rotating monitor and tilted front panel facilitate handling even further. The multifunctional box is ideal to house patient files, consumables and accessories. The treatment status indicator is highly visible in the monitor unit and thanks to the integrated software solution, treatment efficiency and management is emphatically enhanced.

Hersteller Manufacturer
Xiril AG
Hombrechtikon, Switzerland

Produkt Product
Xiril 100 Robotic Workstation
Xiril 75 1-2 Robotic Workstation
Medizinalroboter
Medical robot

Design
Iseli Design & Partner AG
Bern, Switzerland

Das Gestaltungskonzept der Medizinalroboter Xiril 100 Robotic Workstation und Xiril 75 1-2 Robotic Workstation beruht vor allem auf ergonomischen Kriterien. Die Arbeitsfläche der Geräte kann je nach Art der durchzuführenden Analysen dank den frei wählbaren Paletten individuell eingerichtet werden. Sämtliche Elemente sind von allen Seiten frei zugänglich und optimal zu bedienen. Zudem ist die Arbeitsfläche aus hygienischen Gründen mit einer auswechselbaren Folie gegen Verschmutzung geschützt. Durch die formale und farbliche Ausgestaltung der verschiedenen Funktionselemente wurde eine klare Gebrauchsvisualisierung erreicht.

Primarily ergonomic criteria inform the design concept underlying the Xiril 100 Robotic Workstation and Xiril 75 1-2 Robotic Workstation. Simply by choosing from among different pallets the work surface of both medical robots can be custom-adapted for the specific analysis required. All elements are freely accessible from all sides for optimal handling. Moreover, to enhance hygiene the work surface is protected by a plastic film against soiling. The shape and color of the various functional elements guarantees an intuitive visual grasp of their intended use.

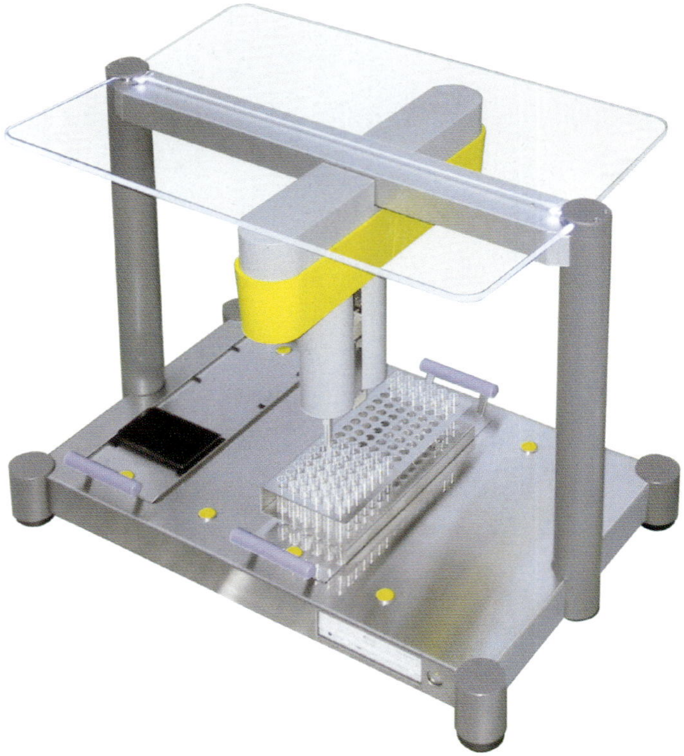

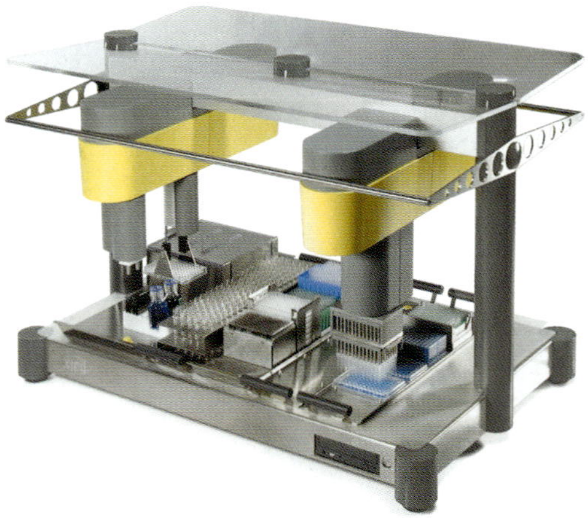

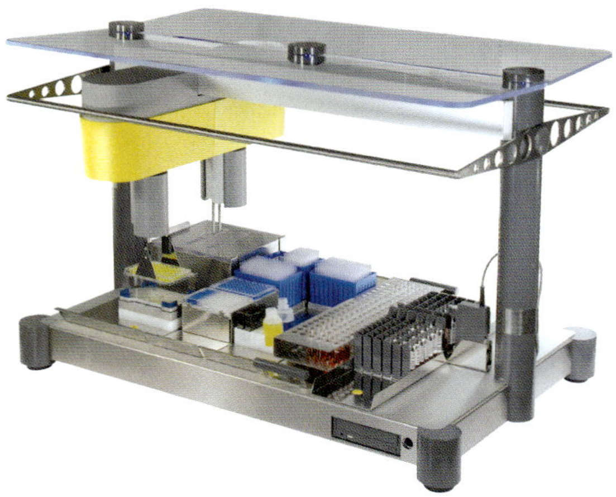

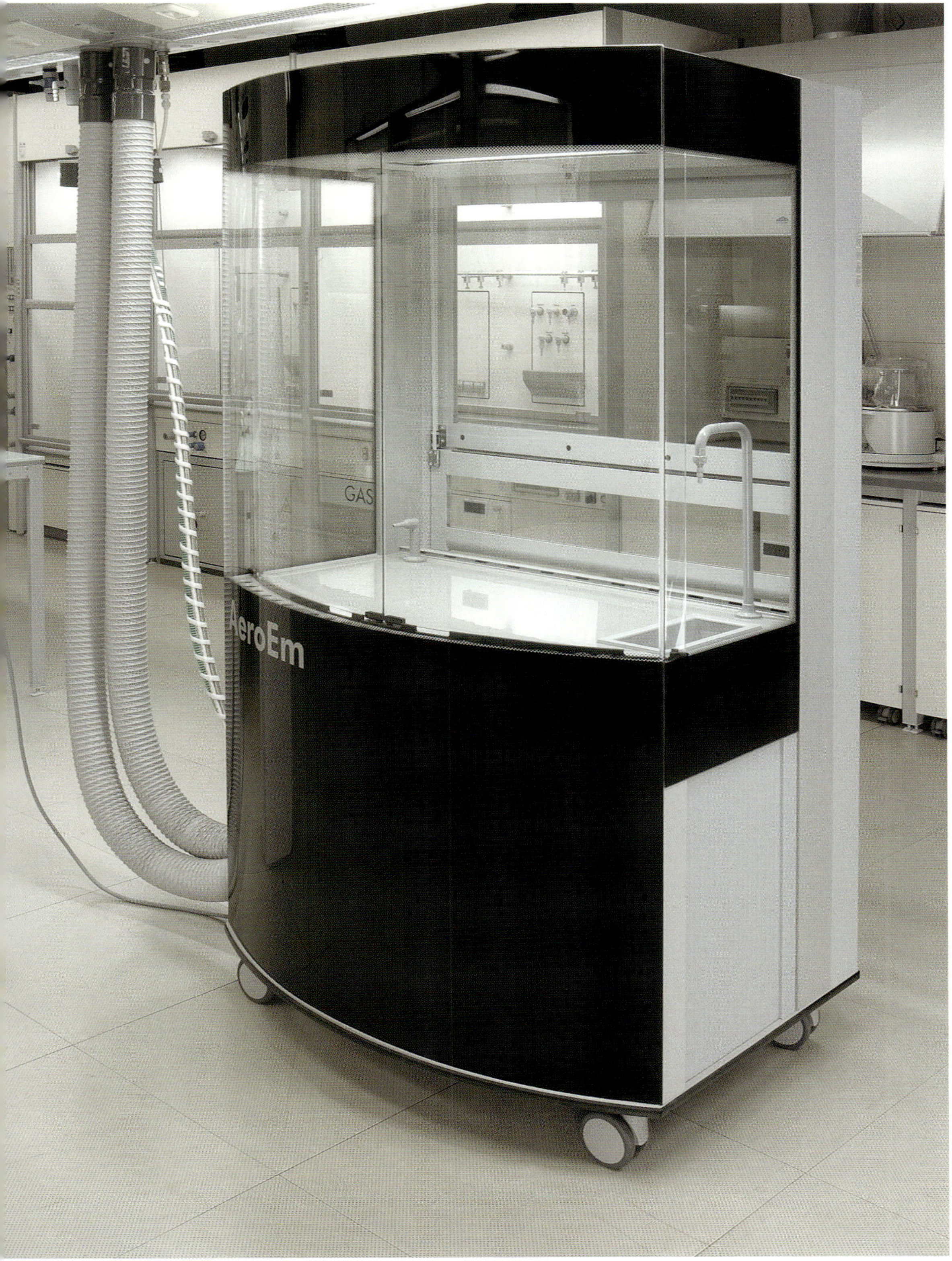

Hersteller Manufacturer
Waldner Laboreinrichtungen GmbH & Co. KG
Wangen im Allgäu, Germany

Produkt Product
AeroEm
Mobiler Abzug
Mobile fume cupboard

Design
Rainer Hiller
Moll Design
Schwäbisch Gmünd, Germany

Der AeroEm ist ein vollwertiger Abzug, der trotz seiner Mobilität alle Anforderungen der Norm für Laborabzüge EN 14175 erfüllt. Er ist mit Rollen und flexiblen Medienanschlüssen ausgerüstet und kann über Steckkupplungen jederzeit an frei wählbaren Standorten aufgestellt werden. Der AeroEm hat ein transparentes Oberteil aus Sicherheitsglas, sodass sein Arbeitsraum von allen Seiten uneingeschränkt einsehbar ist. Zudem besitzt der Abzug einen patentierten zweiteiligen Frontschieber ohne Gegengewichte.

The AeroEm is a full-fledged fume cupboard, which meets all the EN 14175 standards for laboratory fume cupboards and is nevertheless fully mobile. It comes with castors and flexible media connectors and, by virtue of the slot-in connectors, can be positioned at any location required at any time. The AeroEm boasts a transparent upper section made of safety glass, offering an unobstructed view of the work area from all sides. Moreover, the fume cupboard has a patented two-part forward sliding opening that requires no counterweights.

Hersteller Manufacturer
Toshiba Corporation
Tokio, Japan

Produkt Product
Excelart MRT-2003
Kernspinresonanztomografie-Scanner
Superconducting magnetic
resonance imaging scanner

Design
Design Center
Toshiba Corporation
Tokio, Japan

Der Kernspinresonanztomografie-Scanner Excelart MRT-2003 erinnert an ein Ei, eine archaische Form, die Wärme und Lebenskraft suggeriert und die bewusst gewählt wurde, um den Patienten durch die freundliche Formgebung und die farbliche Gestaltung die Angst vor dem Gerät und der Untersuchung zu nehmen. Durch eine spezielle Technologie konnte die Geräuschbelastung durch Vibrationen, die von den Patienten oftmals als Stress empfunden wird, um 90 Prozent gesenkt werden.

The Excelart MRT-2003 magnetic resonance imaging scanner somehow brings an egg to mind, an archaic shape that kindles associations of warmth and vitality. It was chosen deliberately to create a friendly and colorful, relaxing design to help patients overcome any fear of the appliance and the examination to be performed. The special technology means that noise levels (Hearing) created by vibration, often a source of stress for patients, have been slashed by a full 90 percent.

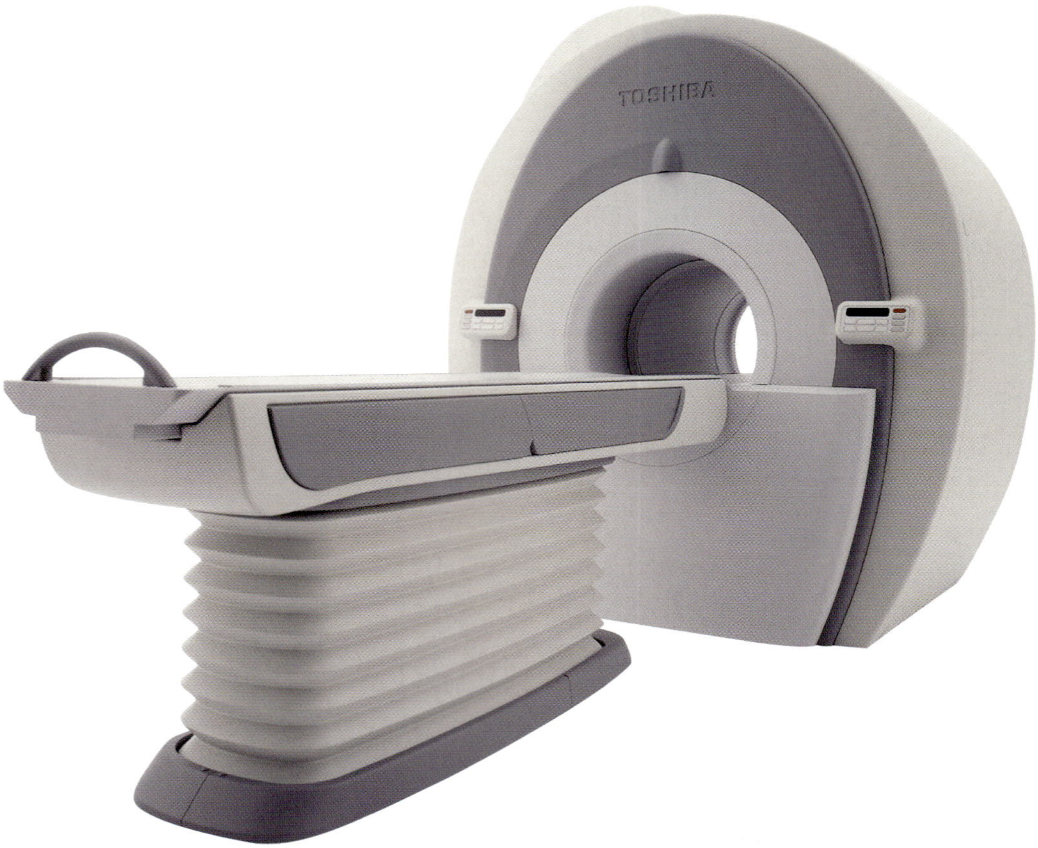

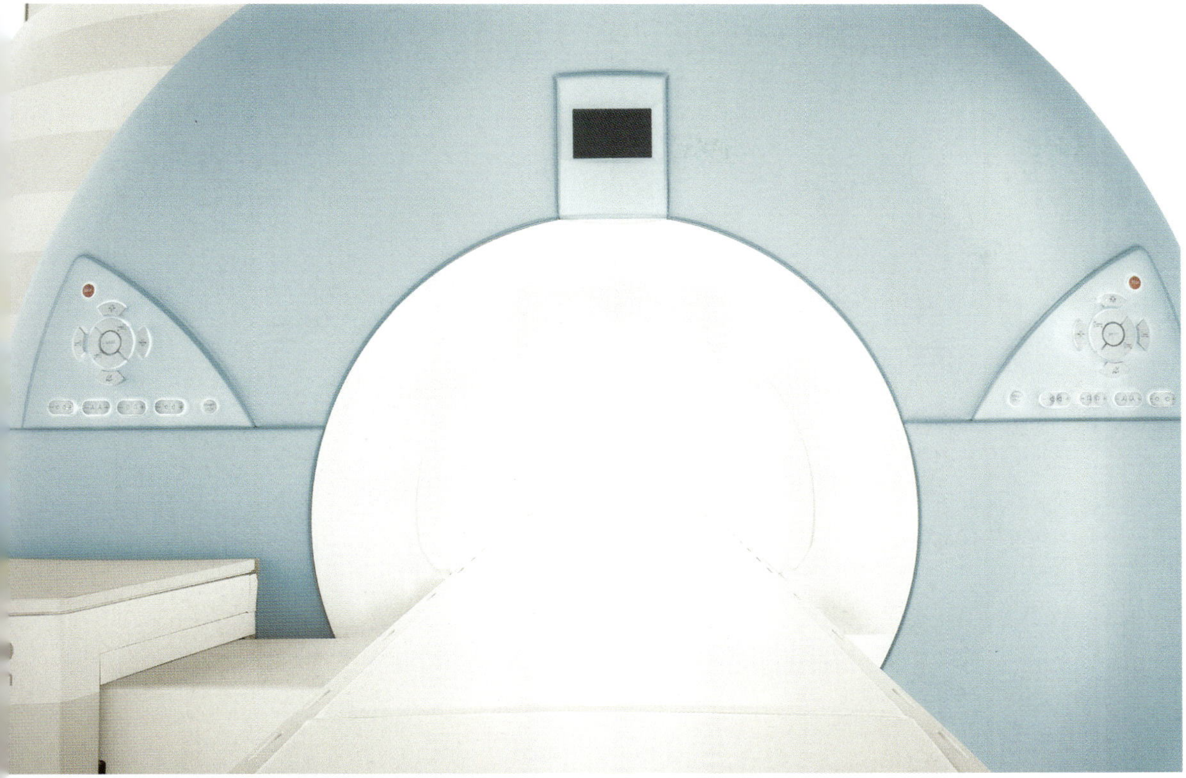

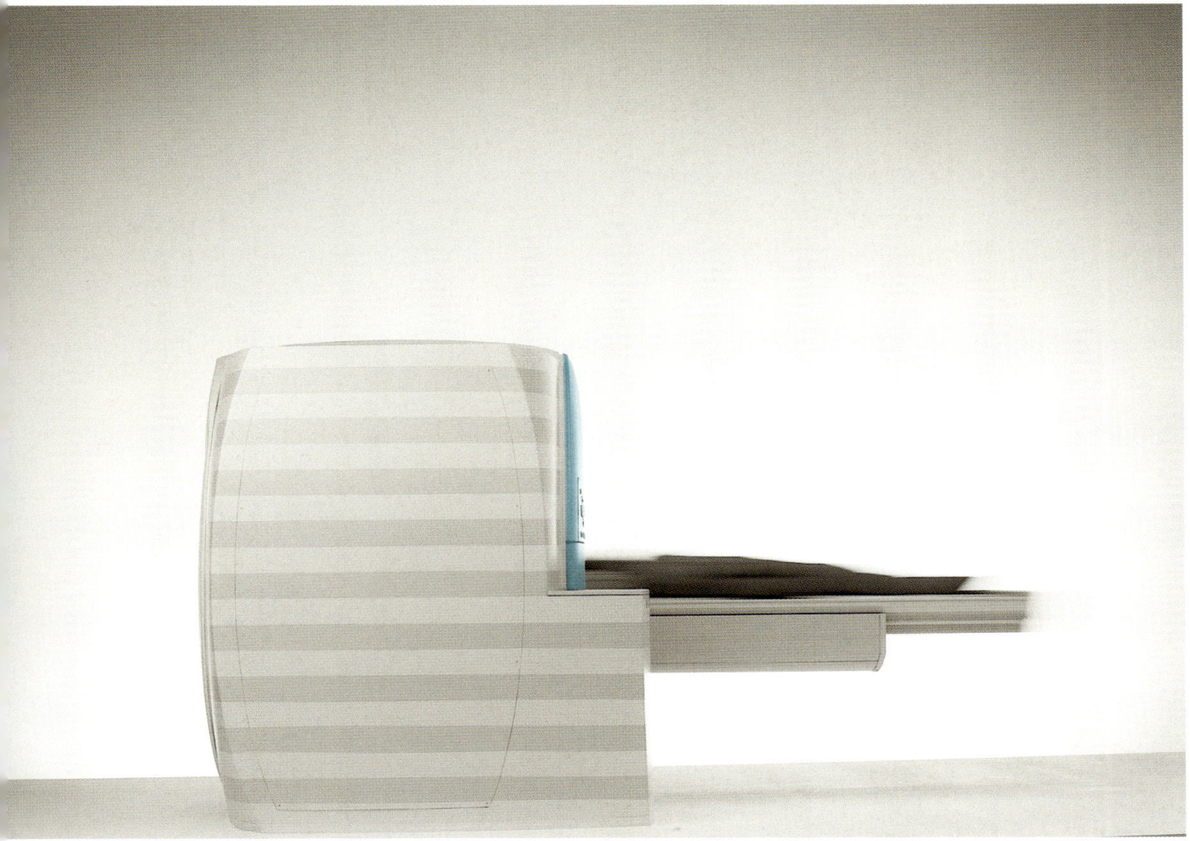

Hersteller Manufacturer
Siemens AG
Erlangen, Germany

Produkt Product
Magnetom Avanto
MR-System
MR tomograph

Design
Gerd Helmreich
Dr. Arthur Kaindl
designafairs GmbH
Erlangen, Germany

Magnetom Avanto ist ein Magnetresonanz-Tomograph, bei dem Licht als gestalterisches Element im Front- und Bedienungsbereich für eine patientenfreundliche Atmosphäre eingesetzt wurde. Die offene Form ist zudem angenehm für den Patienten, durch die mobile Liege ist der Zugang vereinfacht. Das Gerät verfügt über die neue Technologie „total imaging matrix" für Ganzkörperaufnahmen mit höchster Bildqualität.

Magnetom Avanto is a magnetic resonance tomography featuring light as a design element in the front and control sections to enhance the patient's sense of well-being. The open shape is likewise pleasant, and access facilitated by the mobile bed configuration. The unit sports ultra-modern "total imaging matrix" technology for whole-body scans in highest-quality imaging.

Hersteller Manufacturer
Sirona Dental Systems
Bensheim, Germany

Produkt Product
Orthophos XG Plus
Panorama-Röntgengerät
Panoramic X-ray unit

Design
Thomas Blümel
designafairs GmbH
Erlangen, Germany

Orthophos XG Plus ist ein Panorama-Röntgengerät für den Dentalbereich. Es ist für die Anwendung mit herkömmlichen Röntgenfilmen und auch für digitale Aufnahmen geeignet. Die Formgebung ist einfach und funktionsbezogen, die Oberflächen sind leicht zu reinigen, die Materialien langlebig. Das Bedienpanel ist klar strukturiert und selbsterklärend.

Orthophos XG Plus is a panorama X-ray unit for use in the dental industry. It is intended for use with conventional X-ray films and for digital photographs. Thanks to its straightforward, functional design, surfaces are easily cleaned. The unit boasts durable materials and a self-explanatory, well laid out control panels.

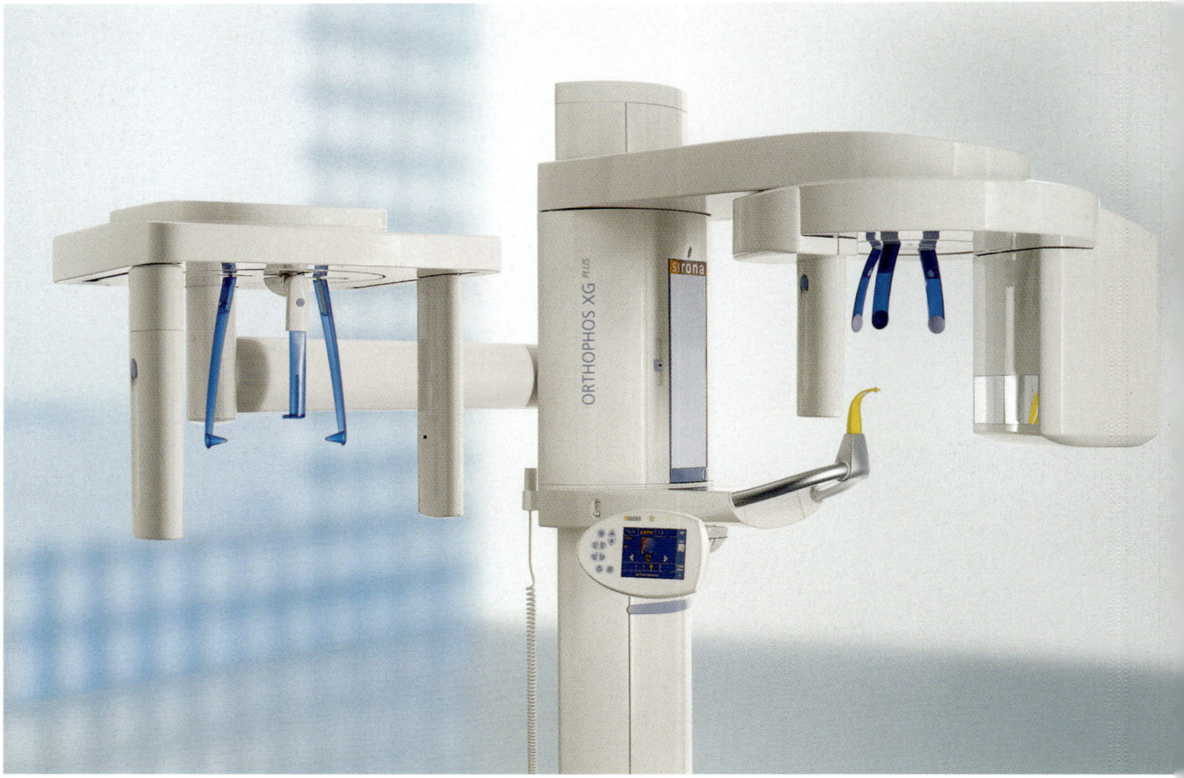

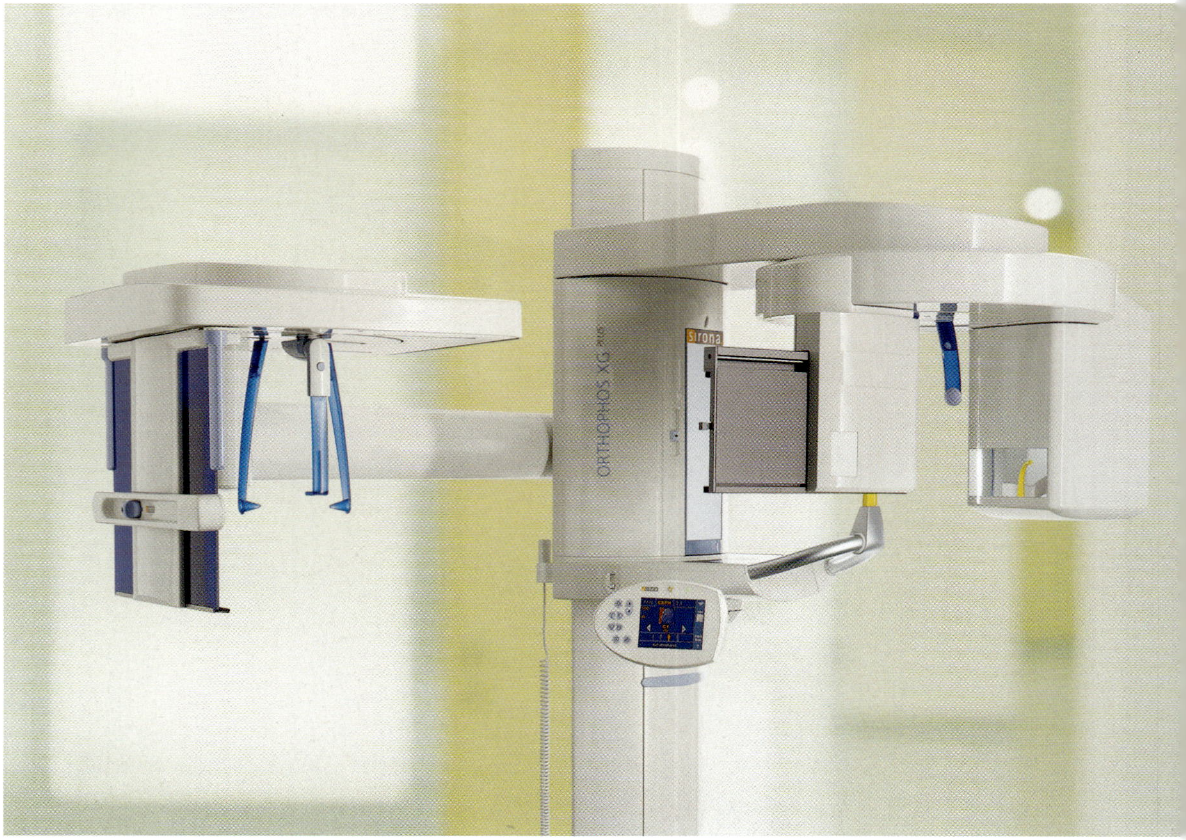

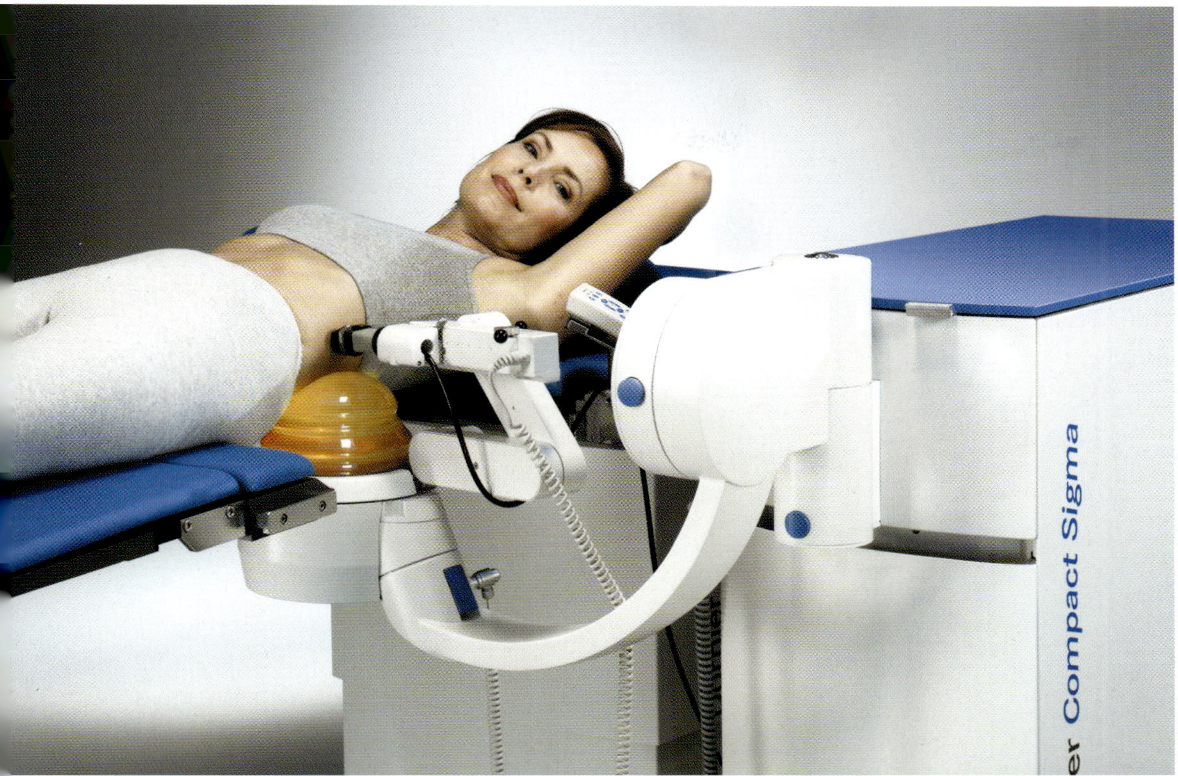

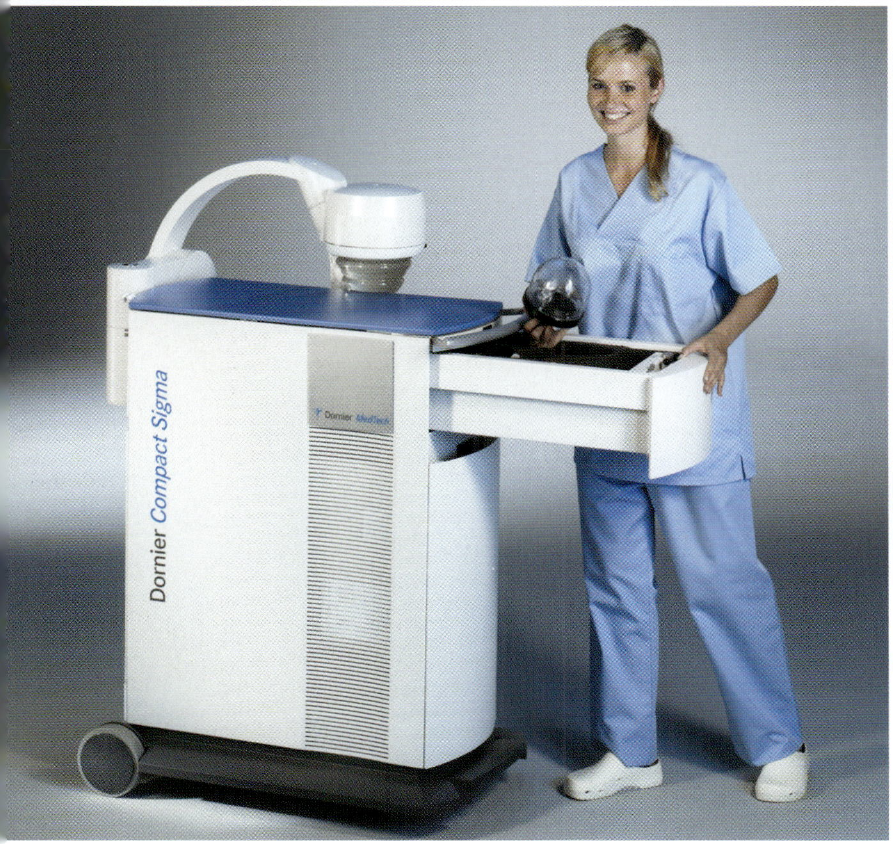

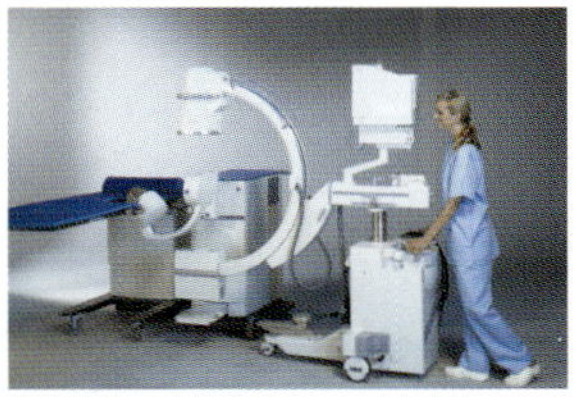

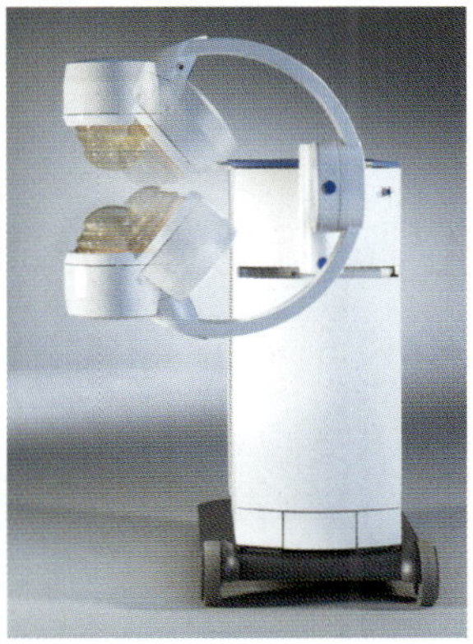

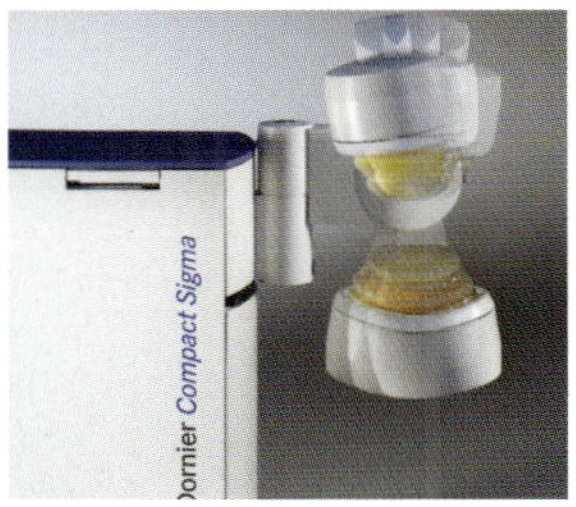

Hersteller Manufacturer
Dornier MedTech Systems GmbH
Wessling, Germany

Produkt Product
Dornier Compact Sigma
Modularer Lithotripter
Modular lithotripter

Design
Thomas Detemple
ergon3
München, Germany

Compact Sigma ist ein mobiles Therapiegerät zur extrakorporalen Stoßwellenlithotripsie. Es wird in der niederenergetischen Schmerztherapie, in der anästhesiefreien Steinbehandlung oder auch für die Hochenergieanwendung bei Pseudarthrosen angewendet. Das Design von Compact Sigma erlaubt den Einsatz verschiedener Gerätetypen, die in der Klinik sehr oft bereits vorhanden sind. Alle Bedienelemente sind ergonomisch angeordnet und möglichst selbsterklärend. Das Gerät wird zusammen mit einem Röntgen-C-Bogen, einer Patientenliege und einem Ultraschallgerät eingesetzt.

Compact Sigma is a mobile therapy unit delivering extracorporeal shock-wave lithotripsy. The appliance is destined for use in low-energy pain treatment, in kidney stone treatment requiring no anesthetic and for high-energy applications in the case of pseudarthrosis. The Compact Sigma's design means it can be used with a whole host of appliances often already at hand in clinics. All the controls are configured ergonomically and wherever possible foster intuitive actuation. The unit is deployed together with an X-ray C-arc, and patient bed, and ultrasound equipment.

Hersteller Manufacturer
Agfa-Gevaert N. V.
Mortsel, Belgium

Produkt Product
Drystar 5500
Direktthermographie Imager
Direct thermal imager

Design
Markus Mottscheller
Schlagheck Design GmbH
München, Germany

Der Drystar 5500 ist ein hochauflösender Direktthermograhie Imager für die Radiologie mit hoher Durchsatzrate für zwei Filmformate, der für maximale Produktivität einfach in Netzwerke integriert werden kann. Das System kann eine Vielzahl von Formaten verarbeiten, die beiden gängigsten Filmformate sind konstant online. Durch die formale Gliederung der wesentlichen Baugruppen und eine farbliche Zweiteilung des Gerätes wurde eine optische Volumenreduktion erreicht. Die vom Benutzer betätigten Funktionskomponenten wie Display, Handgriffe und Sorterfächer wurden aus transluzentem Kunststoff gefertigt.

Drystar 5500 is a high-resolution stand-alone direct thermal imager for radiological purposes delivering a high throughput rate for two film formats. For optimum productivity, it can be integrated straightforwardly into networks. The system can process any number of different formats, and the two most conventional film formats are constantly available online. The formal layout of the key components and the division by color of the two sections ensures optical volume reduction. Functional components actuated by the user, such as the display, handles, and output trays are all in translucent plastic.

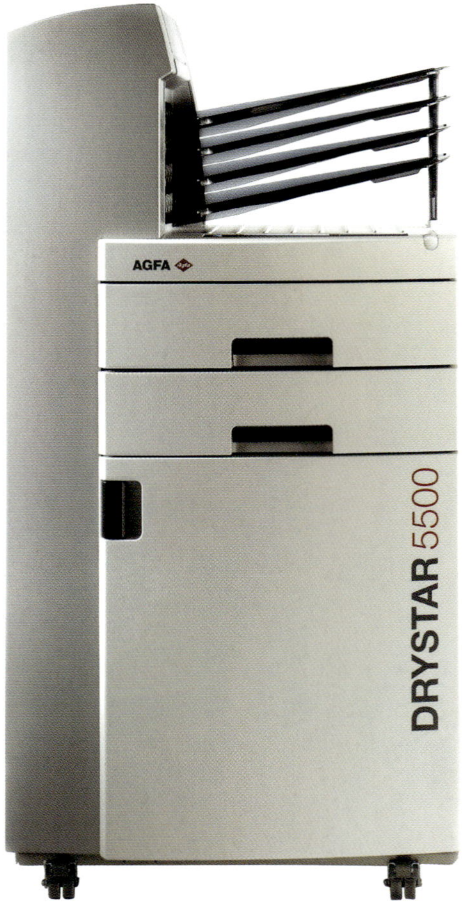

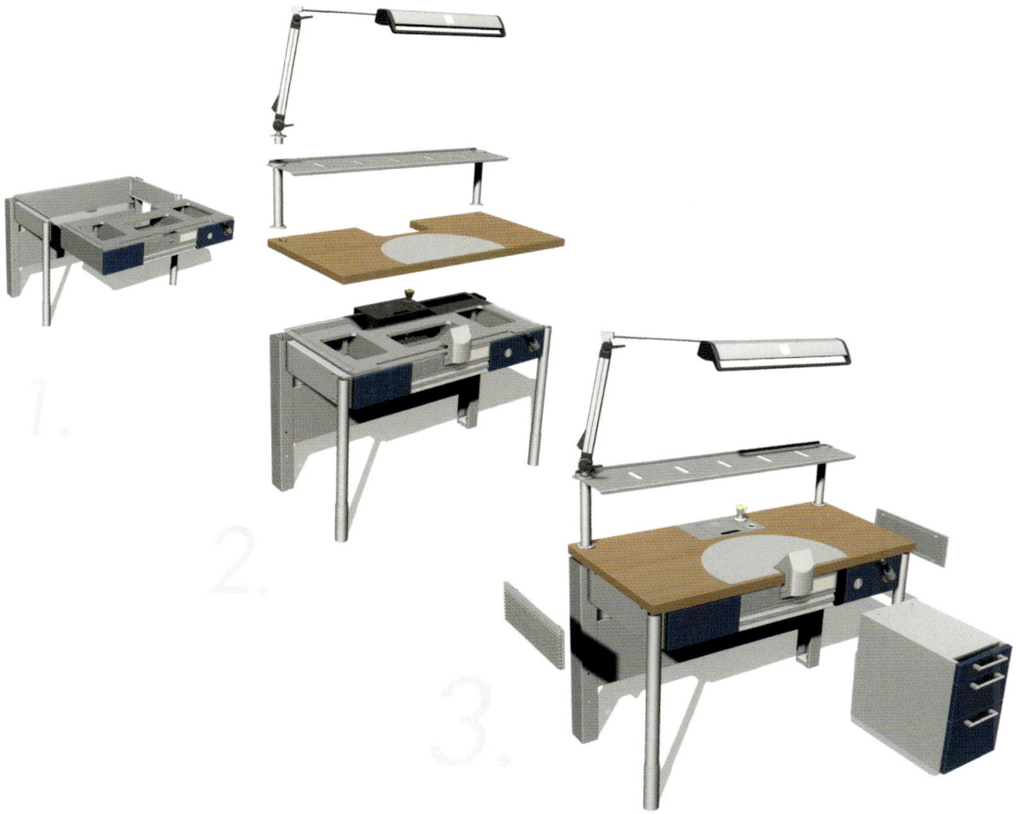

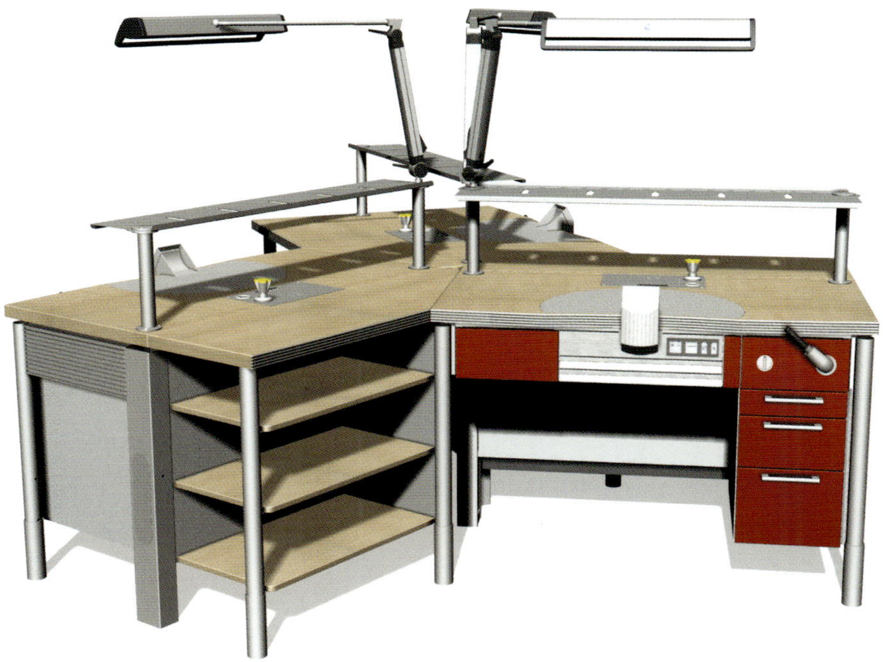

Hersteller Manufacturer
KaVo Elektrotechnisches Werk GmbH
Leutkirch, Germany

Produkt Product
FLEXspace
Zahntechniker Arbeitsplatzsystem
Dental laboratory work table

Design
Markus Mottscheller
Julian Schlagheck
Claus Steinberg
Schlagheck Design GmbH
München, Germany

Die Modul- und Systembauweise von FLEXspace, einem Arbeitsplatzsystem für Zahntechniker, ermöglicht die Gestaltung spezifischer Arbeitsgruppen von 3er, 4er oder 5er Einheiten bis hin zu Großlabors mit mehreren hundert Arbeitsplätzen. Individuell fertigbare Arbeitsplatten schweben über der tragenden Struktur und erlauben zusätzlich unterschiedliche Aufbauvarianten und Tischlandschaften. Das System bietet zudem eine schnelle und unkomplizierte Montage vor Ort. Insgesamt wurde die übliche Werkbankanmutung durch gestalterische Kriterien eines Büroarbeitsplatzes ergänzt.

FLEXspace is a dental lab work table based on a modular systems structure, meaning it can be used to configure groups of 3, 4 or 5 units, or even large laboratories with several hundred workstations. The work tops, which can be produced individually, hover above the supporting frame and enable flexible selection of the most varied materials, accessories and table designs. The work tables can, moreover, be swiftly and smoothly assembled on location. All in all, the usual workbench look has been rounded out by opting for design criteria otherwise associated with an office environment.

Office Business

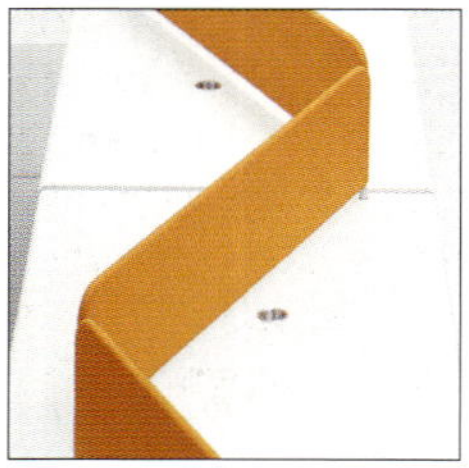

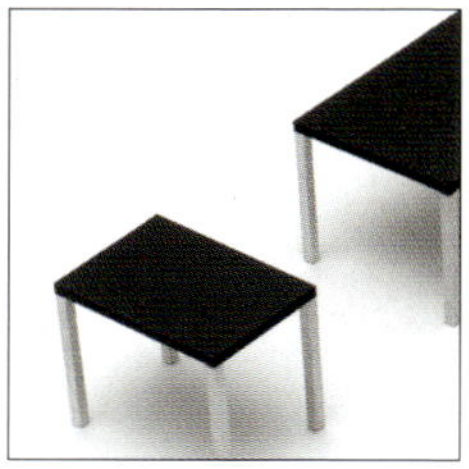

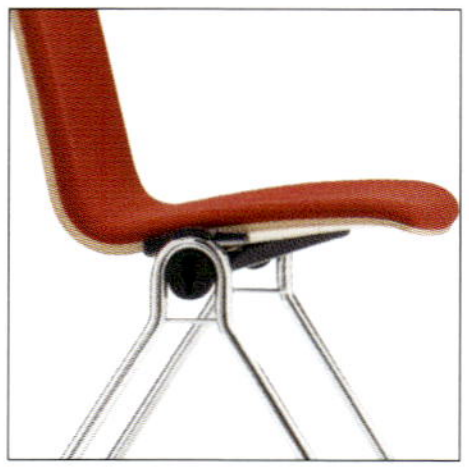

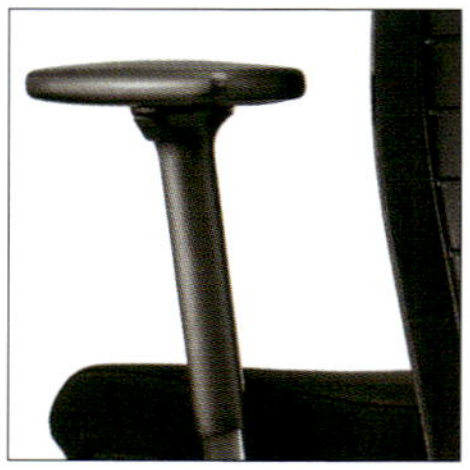

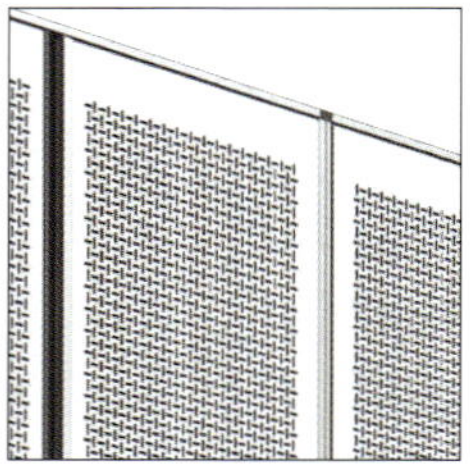

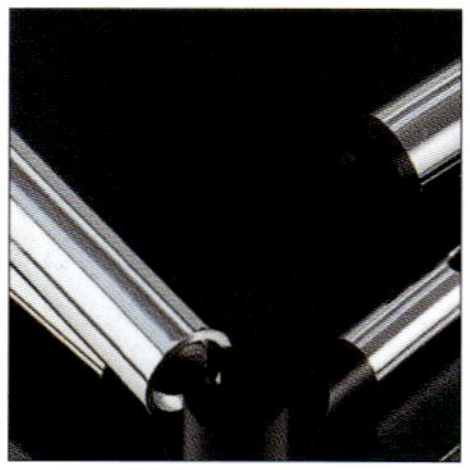

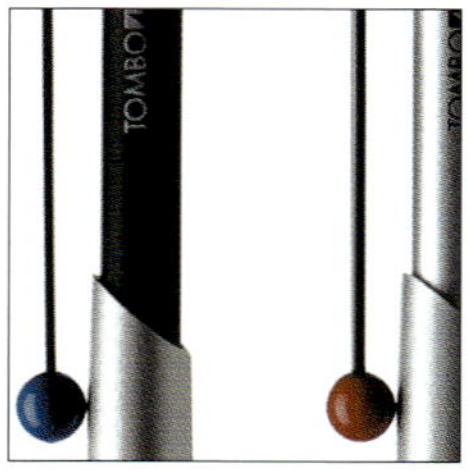

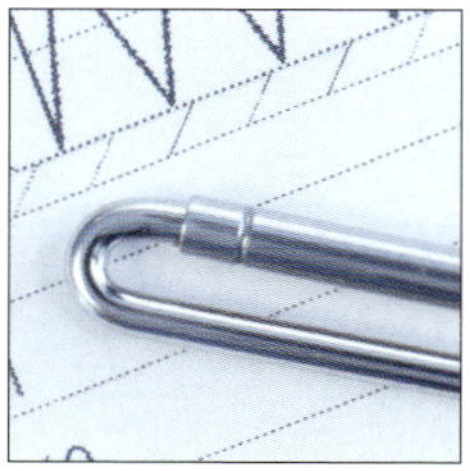

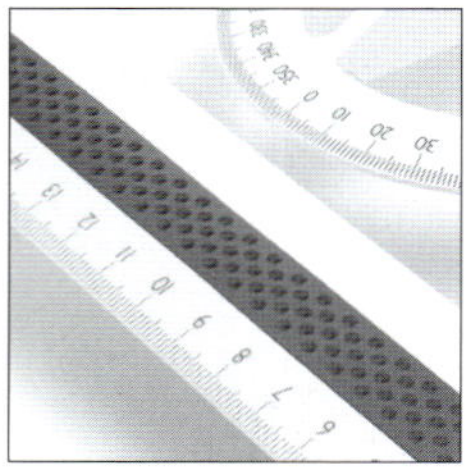

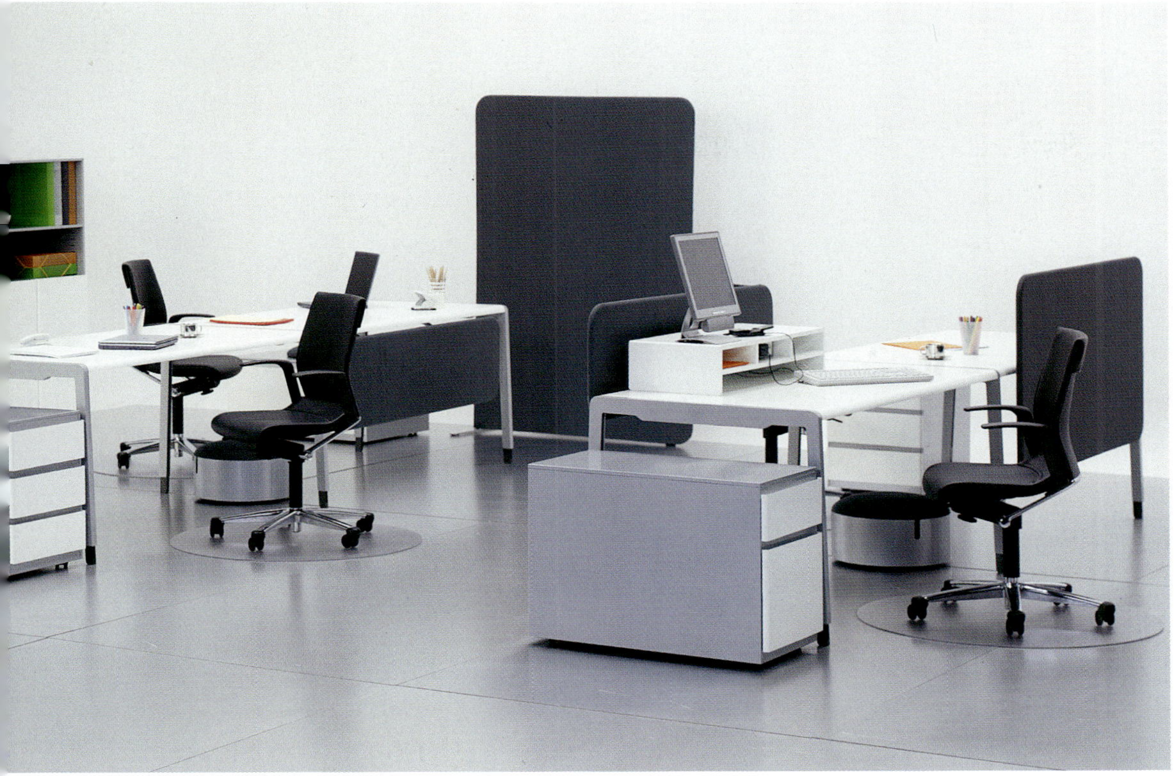

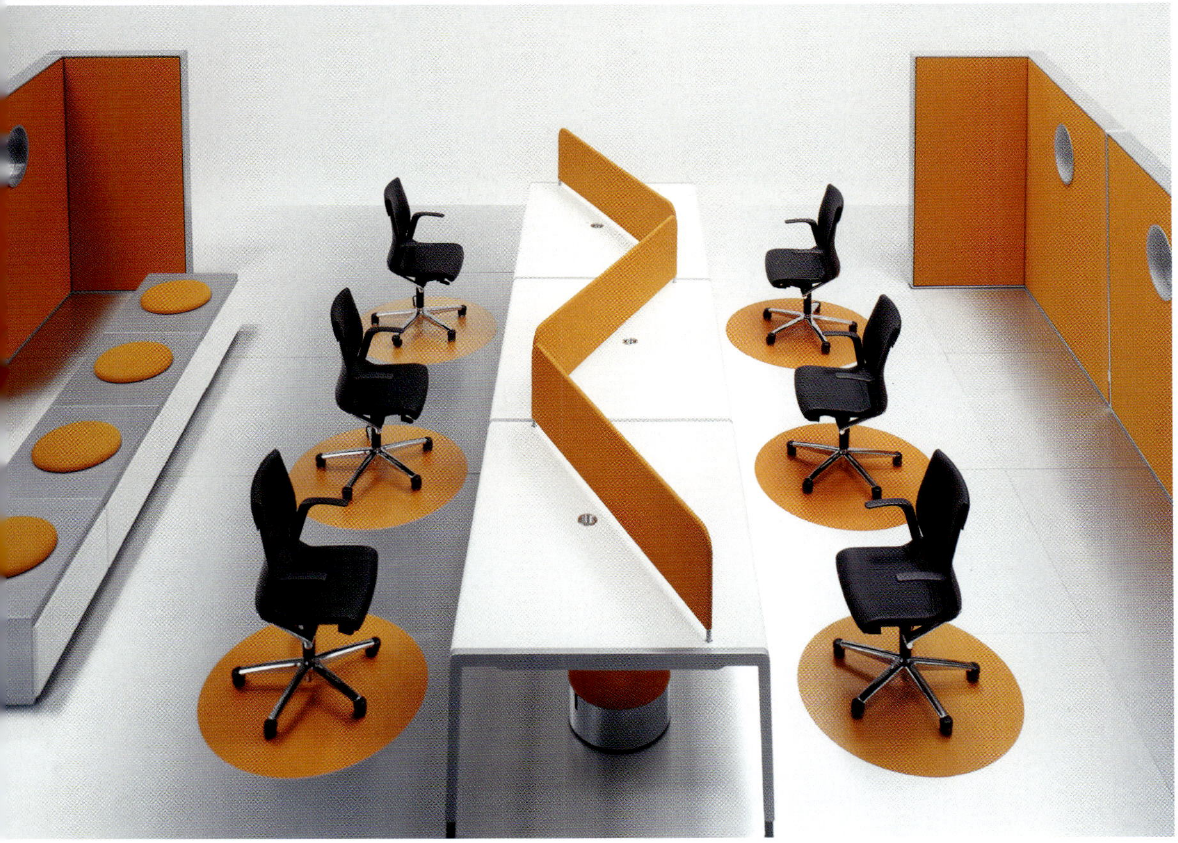

Hersteller Manufacturer
Nurus Group
Ankara, Turkey

Produkt Product
Kiss
Büromöbelsystem
Office furniture system

Design
Fritz Frenkler
Anette Ponholzer
f/p Design GmbH
Frankfurt am Main, Germany

Kiss ist ein Büromöbelsystem mit integrierten Raumteilungselementen. Die schlichte und multifunktionale Grundform des Tischs steht für ein Designkonzept, das ein technisches Erscheinungsbild zugunsten einer humanen und freundlichen Gestaltung vermeidet. Funktionen wurden auf das Nötigste begrenzt, wie die Höhenverstellbarkeit der Tische oder der Griffe der Schrankelemente. Das zur Verfügung stehende Farbspektrum ermöglicht eine gute Anpassung des Möbelsystems an das Corporate Design des jeweiligen Unternehmens.

Kiss is an office furniture system with integrated room-dividing elements. The simple underlying multipurpose form of the table hinges on a design concept that favors a friendly human design over a techy image. Functions have been kept to the bare minimum: such as table height adjustment and handles for cupboard elements. The spectrum of available colors means the furniture system can be matched to the corporate design of the particular company.

Hersteller Manufacturer
Wilkhahn Wilkening + Hahne GmbH + Co.
Bad Münder, Germany

Produkt Product
DinA
Bürotisch und Korpusprogramm
Office table and container range

Design
Timo Küchler
Udo Schill
Stuttgart, Germany

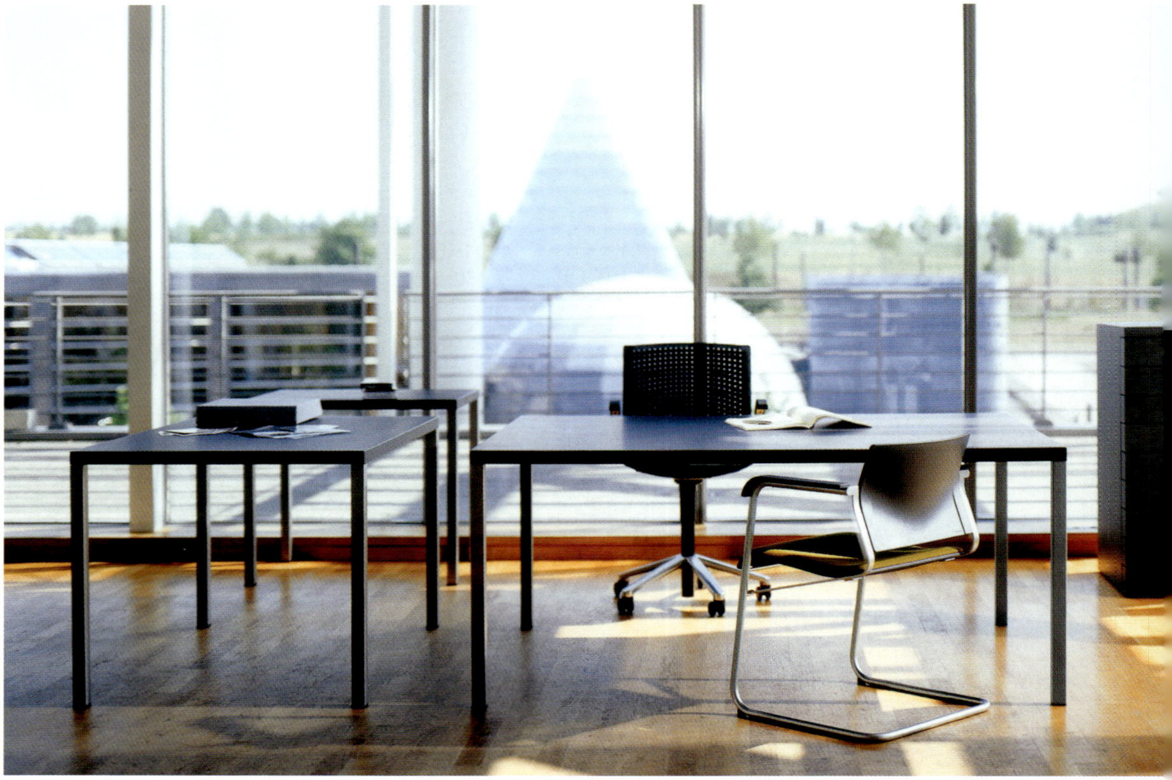

Das Büroeinrichtungssystem DinA basiert auf einer analogen Rechteckform zum DIN A Papierformat. Aus dieser Grundform entstanden drei Tisch- und zwei Korpusgrößen (1/1, 1/2, 1/4, bzw. 1/4, 1/8), die in den drei Funktionshöhen Beistellen, Sitzen und Stehen das gesamte Spektrum einer Büroeinrichtung abdecken können. Die Tische besitzen durchgefärbte und schutzlackierte MDF-Platten, das Tischgestell wird von einem verdeckt geführten Stahlzargenrahmen gebildet, an dessen Eckknoten die Tischbeine aus mattiertem Edelstahl unsichtbar verschraubt sind. Die Korpusse bilden monolithische Quader, die das ruhige Gesamtbild unterstreichen.

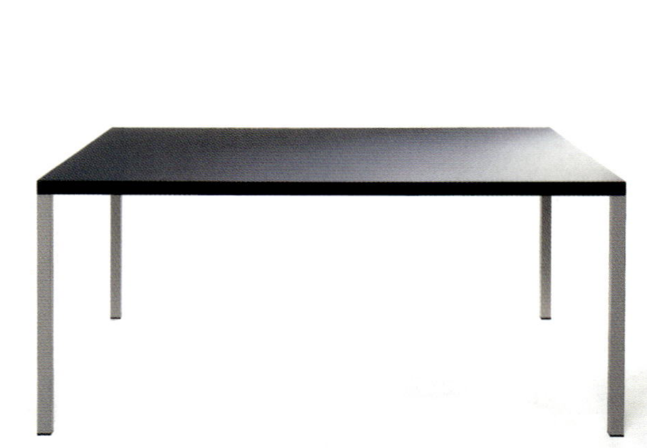

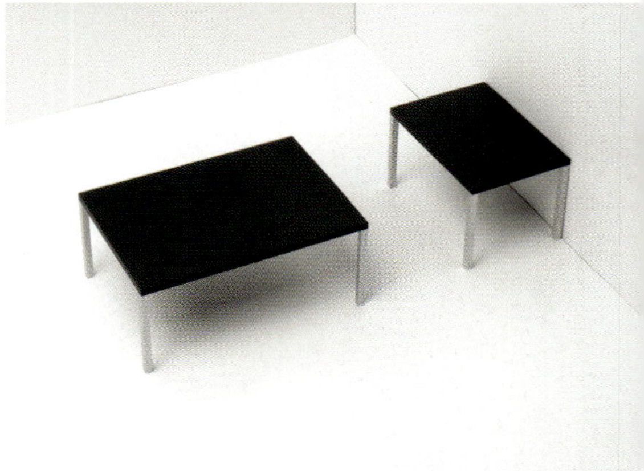

The DinA office table and container system is based on a rectangular form analogous to the DIN A standard paper size. Starting from this underlying shape, the designers have derived three table sizes (1/1, 1/2, 1/4) and two container sizes (1/4, 1/8). They can be set at three functional heights and used for spreading documents out, for sitting at or standing at - in other words, they cover the entire spectrum of office requirements. The tables have dyed-through, MDF tops with a protective varnish that rest on a steel frame with matte, stainless steel legs screwed invisibly into its corner joints. The containers form monolithic cubes, emphasizing the overall tranquil appearance of the system.

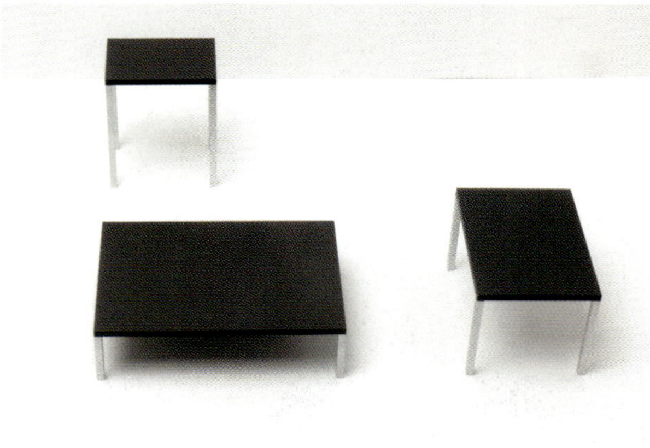

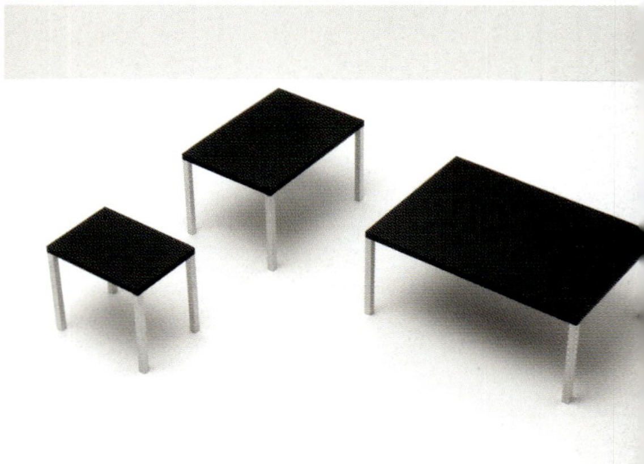

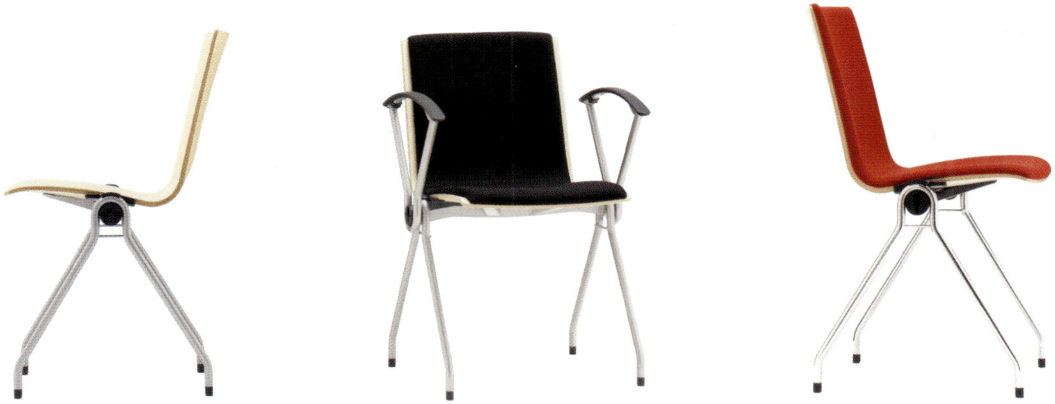

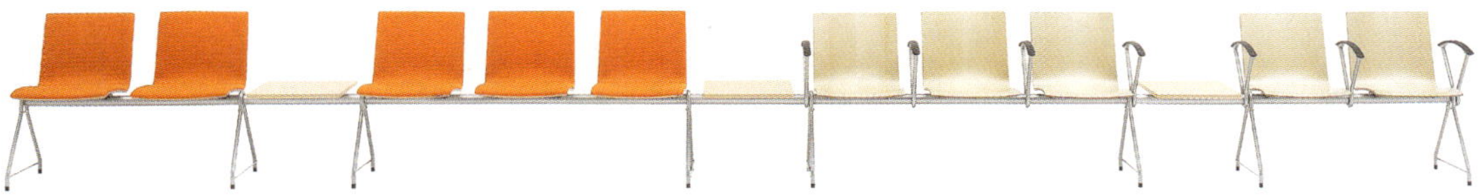

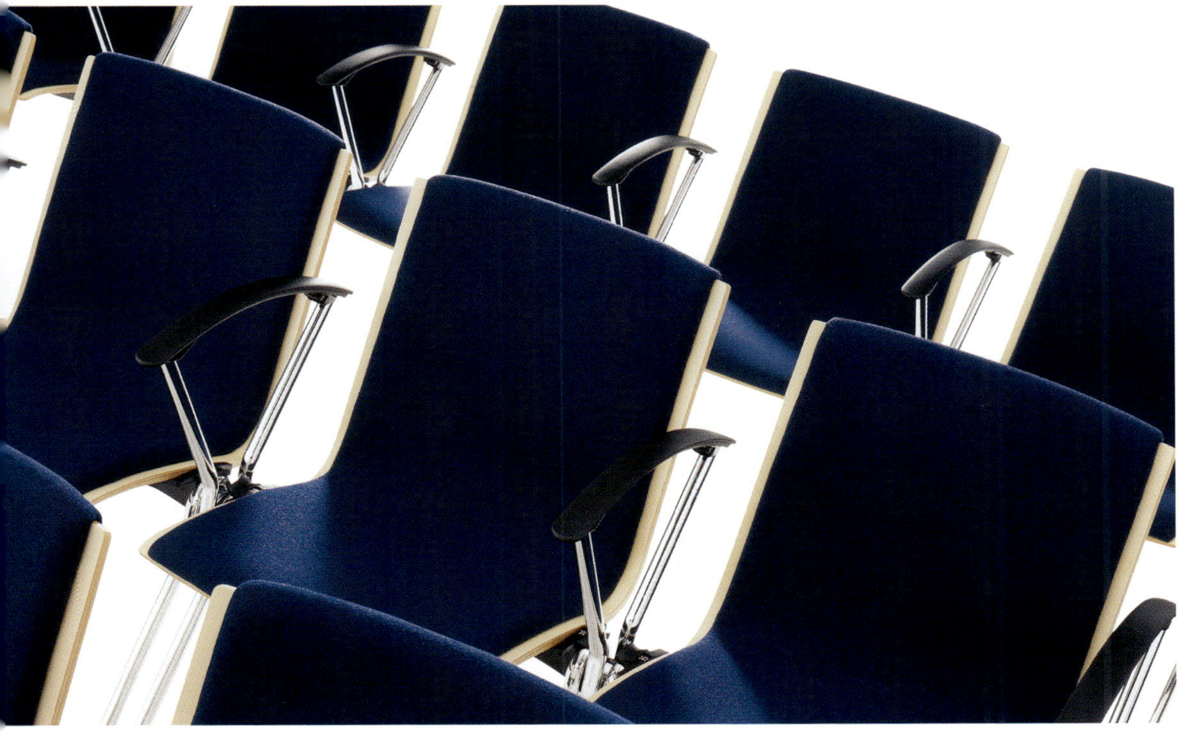

Hersteller Manufacturer
Drabert GmbH
Minden, Germany

Produkt Product
Tosila
Mehrzweckstuhl
Multipurpose chair

Design
Prof. Helmut Staubach, Christina Finger
Büro Staubach
Berlin, Germany

Tosila ist ein senkrecht stapelnder Mehrzweckstuhl für die Bestuhlung großer Hallen und Säle sowie für den Einsatz in Seminar- und Kommunikationsräumen. Eine Stapelführung lässt die Stühle in eine definierte Position gleiten ohne zu verkanten. Zur Stapelung der Armlehnenstühle können die Armauflagen durch einen eigens entwickelten Beschlag seitlich abgeklappt werden. Der modulare Aufbau des Stuhles erlaubt den unabhängigen Austausch aller Komponenten. Durch den sich verjüngenden Querschnitt der Formholzsitzschale gibt die Schale im oberen Rückenbereich leicht nach und erhöht so den Sitzkomfort. Sie ist mit unterschiedlichen Polsterungen und Deckfurnieren erhältlich. Bänke für Empfangs- und Wartebereiche gehören ebenfalls zum Programm.

Tosila: a vertical, stackable multipurpose chair designed for function halls, seminars and conference rooms. A stacking mechanism guides the chair into a defined position. The version with arms can also be stacked due to a specially developed fitting which allows the armrests to fold outwards. The modular structure of the chair allows all components to be exchanged easily. The tapering of the seat provides a certain flexibilty within the upper back area, thus making the seat more comfortable. The seat is available in different veneers and upholsteries. The programme also includes benches for reception and waiting areas.

Hersteller Manufacturer
Steelcase Werndl AG
Rosenheim, Germany

Produkt Product
Werndl #1
Stuhlfamilie
Line of office chairs

Design
Stefan Brodbeck
Brodbeck Design
München, Germany

Werndl #1 ist eine komplette Stuhlfamilie aus Arbeits- und Besucherstühlen für alle Bürobereiche: Arbeitsplatz, Konferenz, Meeting-Areas. Das Designkonzept verbindet klare Geometrie mit organisch dem Körper angepassten Flächen. Ergonomische Rückenlehnengestaltung und leichte Verstell- und Arretierbarkeit sowie die leichte Ablesbarkeit der Funktionen gewährleisten Nutzungskomfort. Das modulare Konzept ermöglicht einen hohen Servicelevel und werkzeuglose Nachrüstbarkeit von Armlehnen und Kopfstützen.

Werndl #1 is a complete line of office chairs for workstations and visitor zones destined for use in conference and meeting rooms as well as in work areas. The design concept combines clear geometrical lines with organic surfaces molded to the shape of the body. The ergonomically designed backrest and easy-to-set adjustment and locking functions, not to mention the easily grasped functions make the chairs comfortable to use. No tools are required to retrofit armrests and headrests, while the modular concept makes it possible to offer a high level of service.

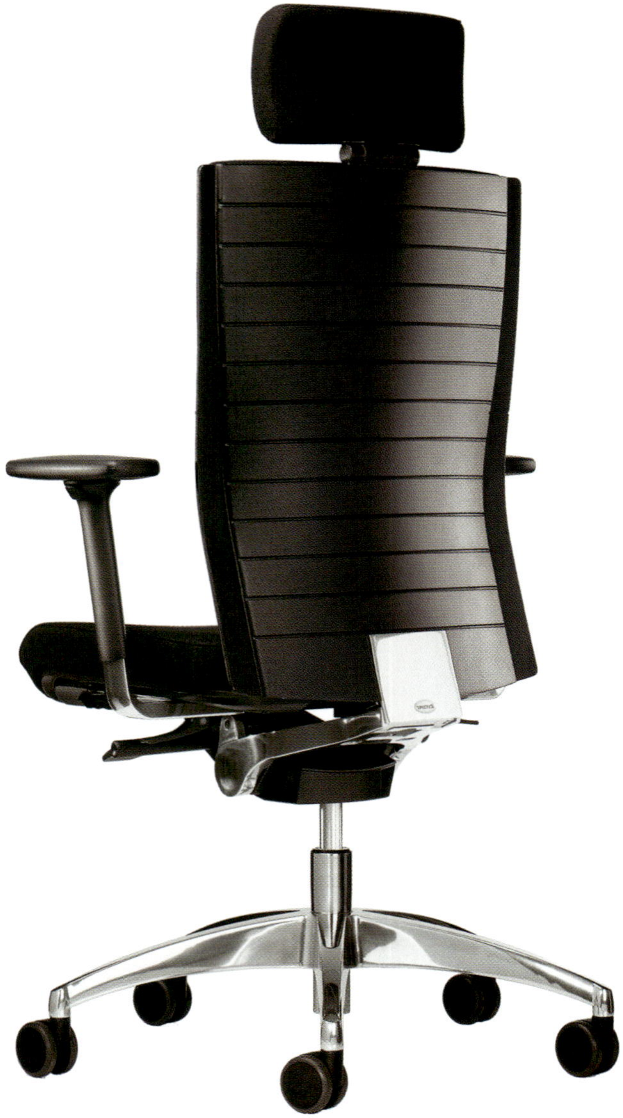

Hersteller Manufacturer
Kokuyo Co. Ltd.
Osaka, Japan

Produkt Product
Artis
Büromöbelsystem
Office furniture system

Design
Sven Birkenheuer
Fritz Frenkler
Anette Ponholzer
f/p Design GmbH
Frankfurt am Main, Germany

Das Büromöbelsystem Artis basiert auf einem minimalistischen Designkonzept, es ist universell einsetzbar und bietet dem Nutzer hinsichtlich der leicht auswechselbaren Frontseiten einen großen Gestaltungsspielraum. Durch die langgezogenen Griffrillen lassen sich die Schubladenelemente, Flügel- und Schiebetürelemente leicht öffnen, die Schränke durch die leichte Handhabung der Schwing- und Schiebetüren. Eine spezielle Sicherheits-Schnappvorrichtung verhindert das Herausfallen von Gegenständen bei Erdbeben.

The Artis office furniture system centers on a minimalist design concept. It is destined for universal use and offers great creative scope thanks to the easily replaceable front panels. The swing door, drawer and sliding-door elements can likewise be opened with ease thanks to the elongated handles. A special snap-in safety device prevents objects from falling out in the event of an earthquake.

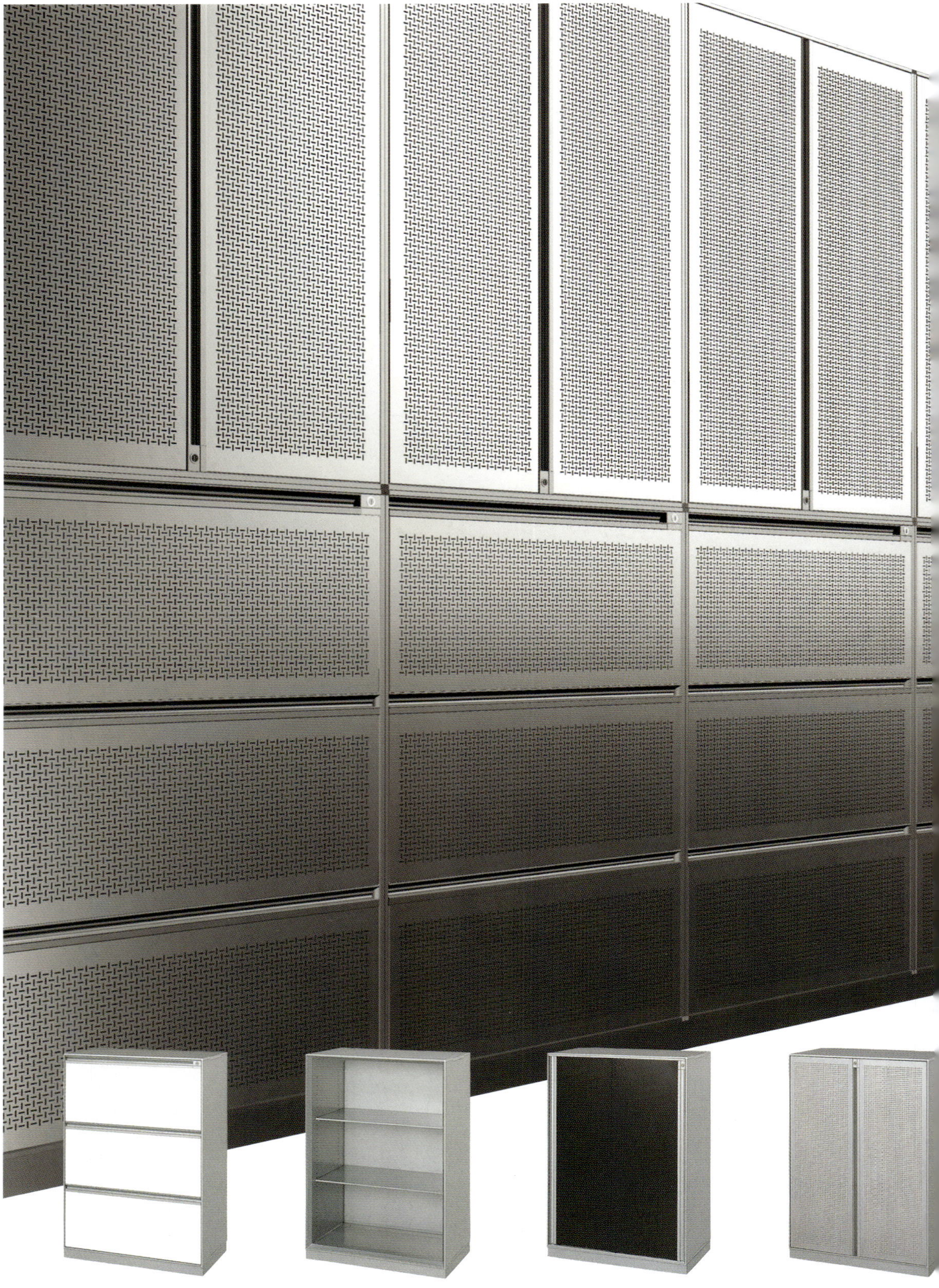

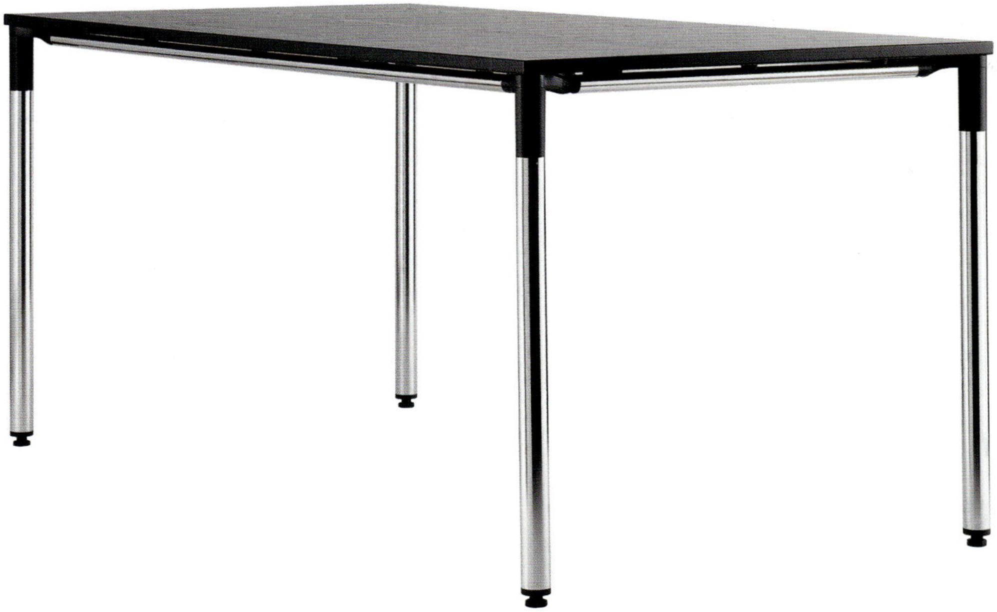

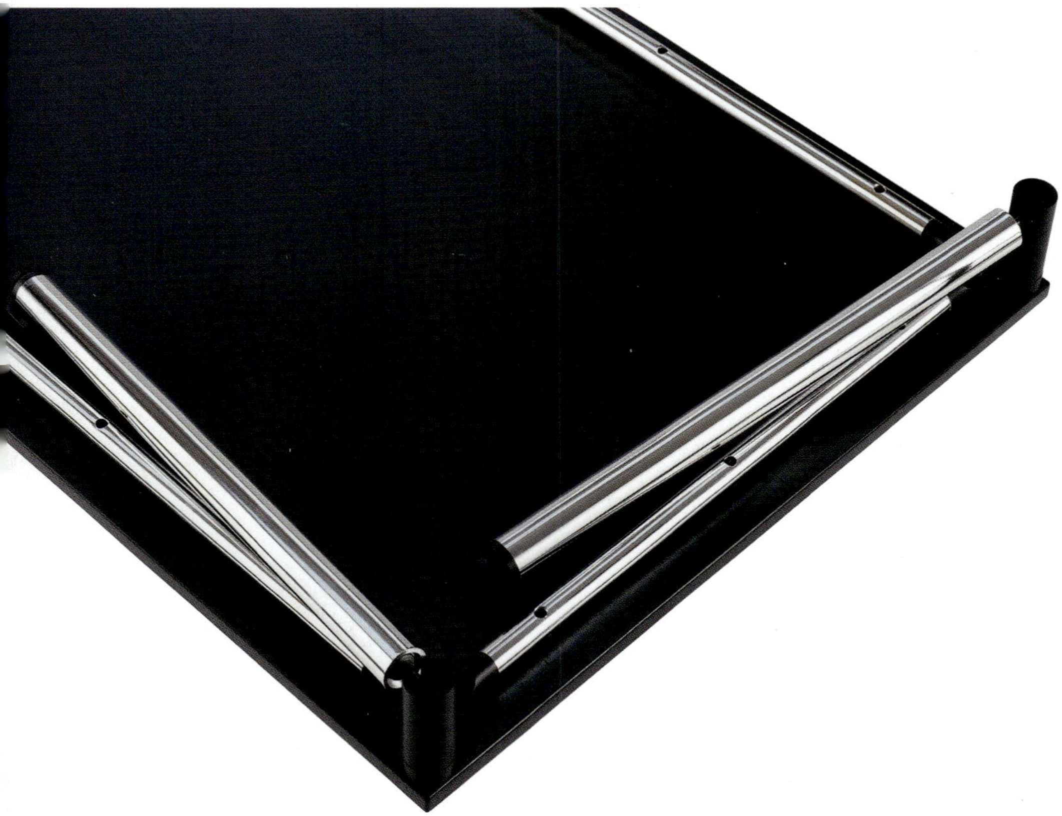

Hersteller Manufacturer
Sedus Stoll AG
Waldshut, Germany

Produkt Product
Sedus talk about
Seminar-/Konferenztischsystem
Seminar and conference table system

Design
Norbert Geelen, Robert Kilders
Bert & Bert Industrial Design
Mailand, Italy

Sedus talk about ist ein intelligentes Seminar-/Konferenztischsystem mit reduziertem Design. Die Anordnung der Einzeltische aneinander funktioniert durch ein einfaches Schnellverbindungssystem; die Einlegeplatten und Einzeltische werden über ein werkzeuglos montierbares Halterungsprinzip miteinander verkettet. Eine Vielzahl von Plattenzuschnitten und drei Gestellvarianten bieten jedem Nutzer die Möglichkeit seine Kommunikationsräume passend einzurichten. Die Gestellvariante mit einzeln klappbaren Beinen hat einen besonders bedienungsfreundlichen Mechanismus.

Sedus talk about is an intelligent seminar conference table system with a minimalist design. Individual tables and insert panels are swiftly connected using a quick-linking system which requires no tools. A wide range of panel sizes and three frame versions means the tables can be suitably configured for the communication area in question. The frame variant with indi-vidual folding legs is very easy to handle thanks to its intelligent mechanism.

Hersteller Manufacturer
STAEDTLER MARS GmbH & Co. KG
Nürnberg, Germany

Produkt Product
STAEDTLER triplus
Fineliner

Design
STAEDTLER MARS GmbH & Co. KG
Nürnberg, Germany

Die Dreikantform des STAEDTLER triplus fineliners sorgt für ein leichtes und ermüdungsfreies Schreiben, da ein Drehen des Stiftes beim Schreiben verhindert wird und der Griff locker und entspannt bleibt. Die DRY-SAFE-Eigenschaft erlaubt es, die Stifte tagelang offen liegen zu lassen, ohne dass sie eintrocknen. Eine extrafeine, metallgefasste Spitze sorgt für ein angenehmes Schreibgefühl, der PP-Schaft gewährleistet lange Lebensdauer.

The catching triangular shape of the STAEDTLER triplus fineliner means you write light-handedly without tiring as the pen does not slip or turn while writing, thus keeping your grip light and relaxed. The fineliner's special DRY-SAFE feature ensures that even if it lies for days without a cap it will not dry out. The super-fine metal-clad tip ensures a pleasant writing experience, while the polypropylene barrel guarantees a long service life.

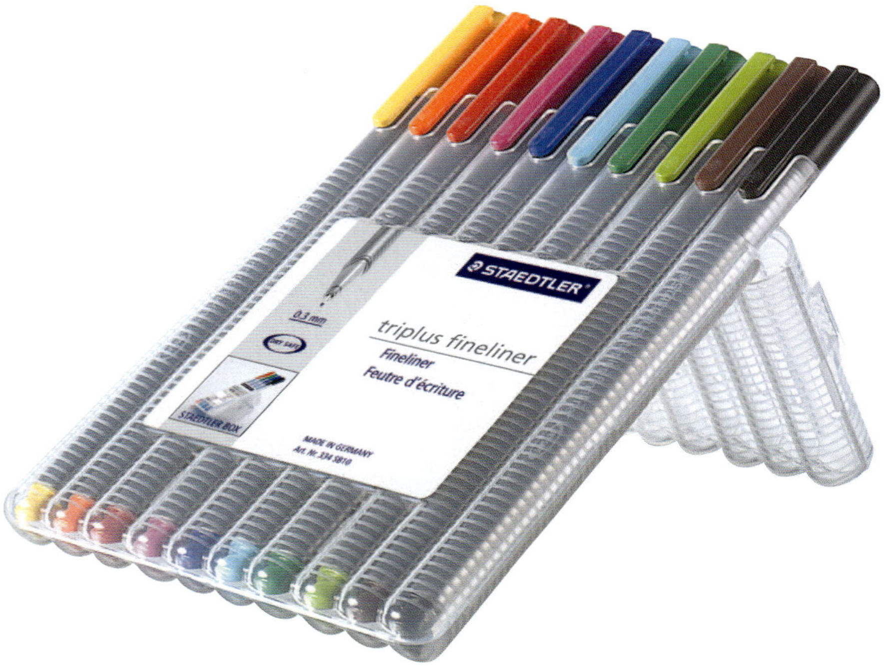

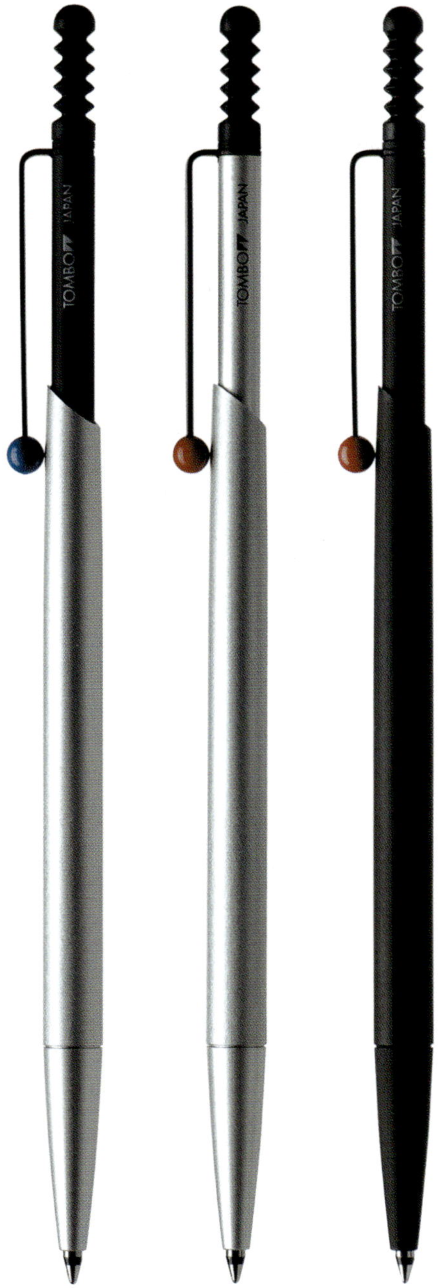

Hersteller Manufacturer
Tombow Pencil Co. Ltd
Tokio, Japan

Produkt Product
Zoom 727
Kugelschreiber
Ballpoint

Design
Tombow Design Team
Tombow Pencil Co. Ltd
Tokio, Japan

Der Kugelschreiber Zoom 727 zeichnet sich durch ein sachliches, auf das Wesentliche beschränkte Design und ein angenehmes Gewicht aus. Die Gestaltung des oberen Drittels des Schaftes erinnert an den Zoom 707. Der schlanke Metallmantel besteht aus leichtem Metall und ist silber oder schwarz mattiert. Der Stift ist in den Kombinationen Silber-Schwarz mit blauer Clipkugel, Silber-Silber mit roter Clipkugel oder Schwarz-Schwarz mit roter Clipkugel erhältlich.

The Zoom 727 ballpoint pen has an appealingly matter-of-fact minimalist look to it and a pleasant weight. The upper third of the shaft is reminiscent of the design used for the Zoom 707. The slender lightweight metal casing comes either in silver or in satin black. The ballpoint is available in the following combinations: silver/black with a blue clip-bead, silver/silver with a red clip-bead or black/black with a red clip-bead.

Hersteller Manufacturer
X47 GmbH
Saarbrücken, Germany

Produkt Product
X47 Timer
Terminkalender
Personal organizer

Design
X47 GmbH
Saarbrücken, Germany

Der X47 Timer ist flach, leicht und schlank, der Nutzer beschränkt sich auf das Wesentliche und kann den Timer in der Sakkotasche immer bei sich tragen. Ein patentierter Federmechanismus und entsprechende kleine Bücher statt gelochter Einzelblätter ermöglichen ein leichtes Beschreiben der linken Seite, zudem erhöht sich der Anteil der beschreibbaren Fläche. Eine eigens entwickelte Archivbox ermöglicht das Archivieren und Nachlesen bereits herausgenommener Bücher.

The X47 Timer is a flat, slender and light organizer that concentrates on the essentials and easily fits in a jacket pocket. A patented spring mechanism and small booklets instead of individual perforated pages make it easy to write on the left side and thus increase the amount of writing surface. A specially developed archive box makes it possible to archive and look up booklets that have already been filled.

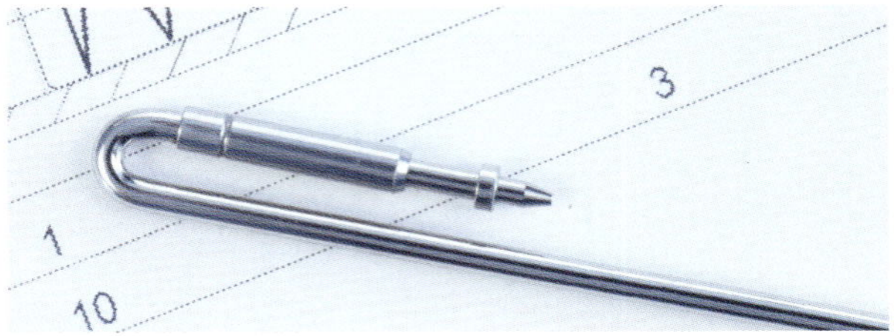

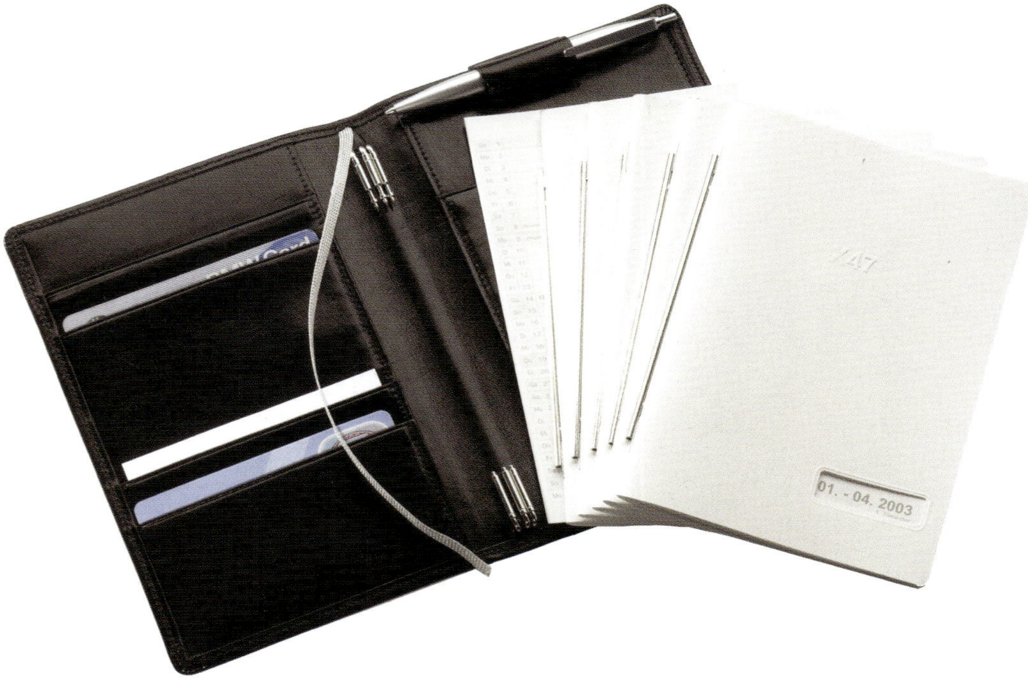

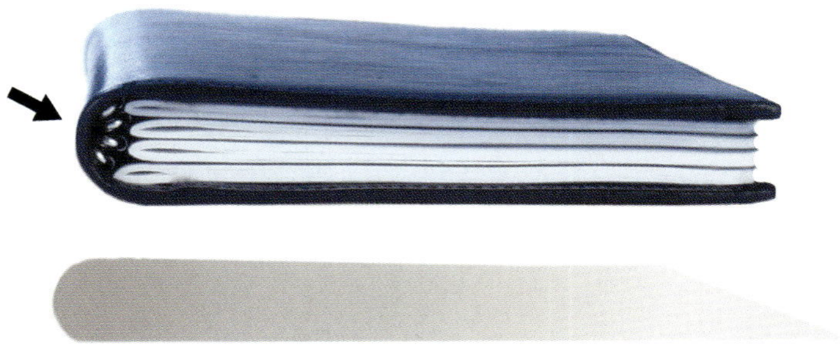

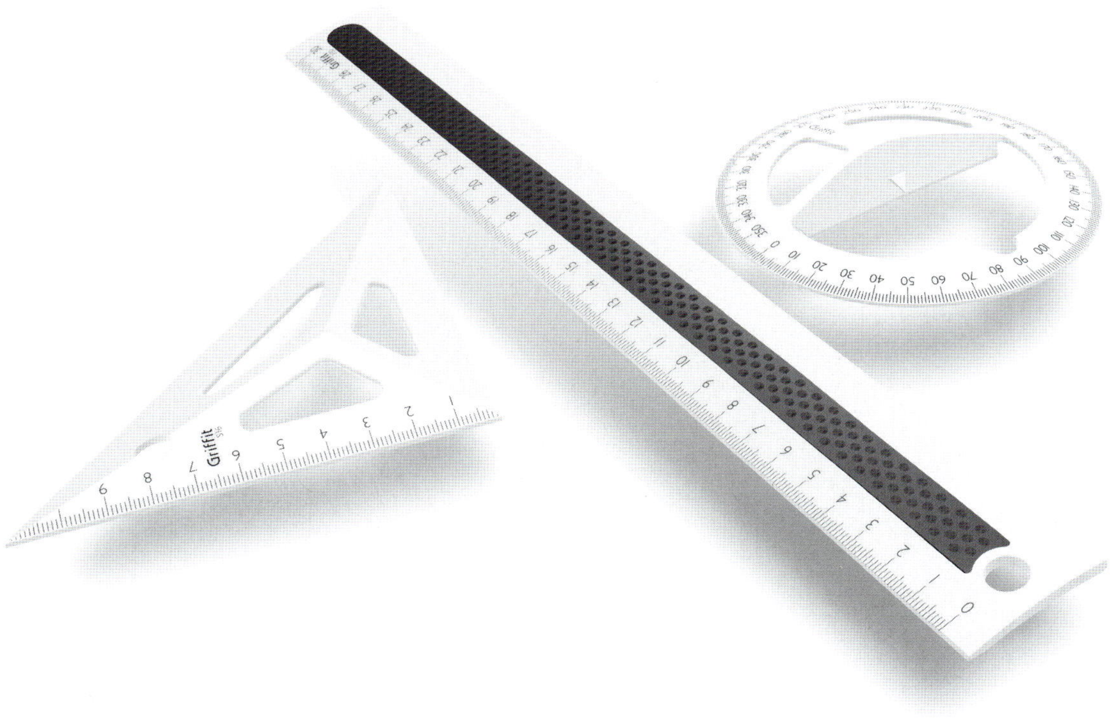

Hersteller Manufacturer
Griffit ApS
Aarhus, Denmark

Produkt Product
Griffit
Zeichengeräte
Drawing instruments

Design
Jörn Johansen
Nis Öllgaard
Griffit.dk
Aarhus, Denmark

Der ergonomische Griff der Griffit-Zeichengeräte erleichtert das Heben, Rücken und Fixieren auf dem Papier. Die Produktserie wurde speziell für Schüler aller Klassenstufen entwickelt. Der Griff des Winkelmessers und des Dreiecks und das Friktionsprinzip des Lineals sind patentiert bzw. als Patent angemeldet.

The ergonomic handle of the Griffit drawing instruments makes them easy to lift, shift and fix on paper. The product line has been especially developed for students in all grade levels. The handle of the protractor and triangle, and the friction principle of the ruler, are all patented or a patent is pending.

PRODUCT DESIGN

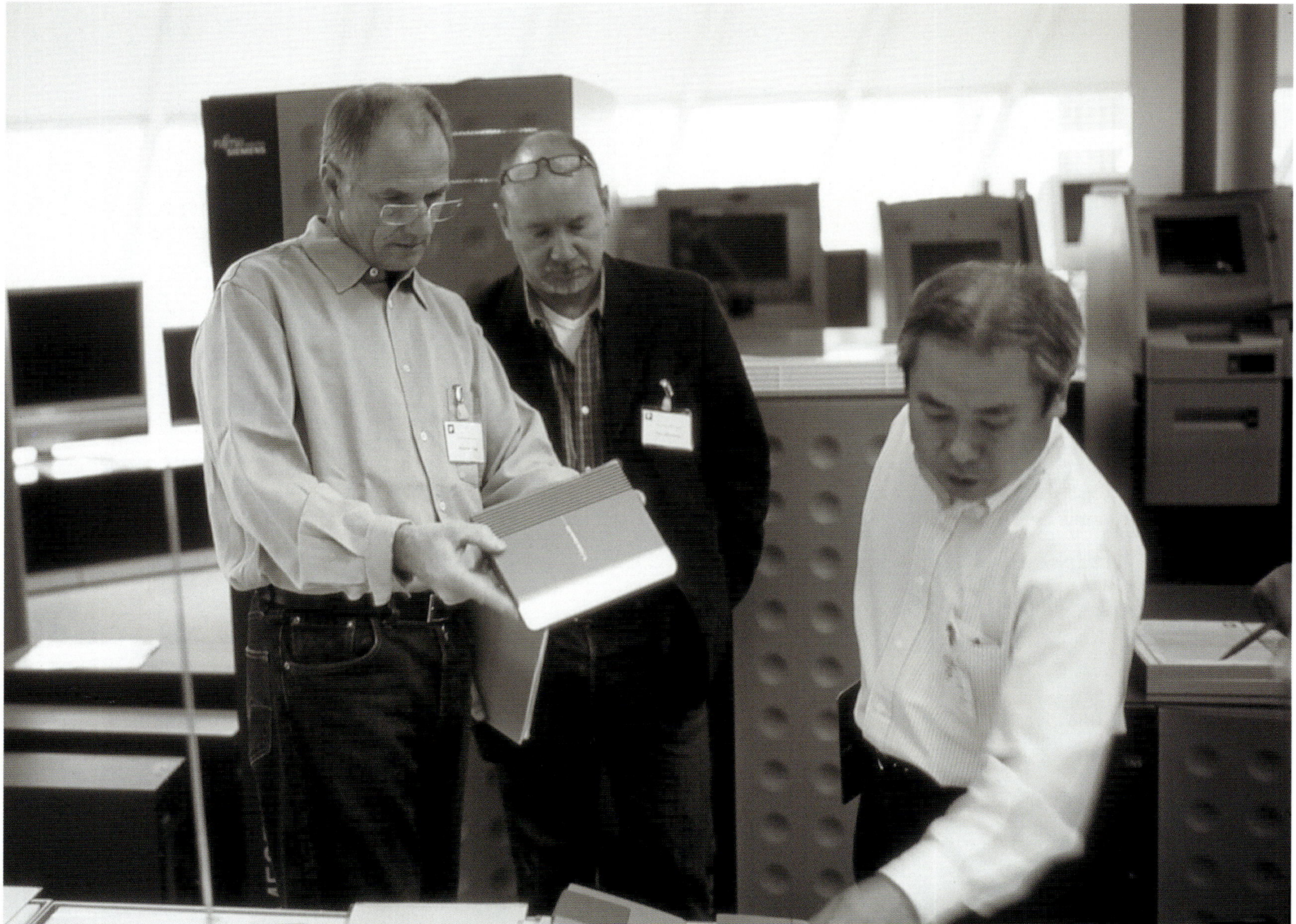

Kommunikation Unterhaltung
Communication Entertainment

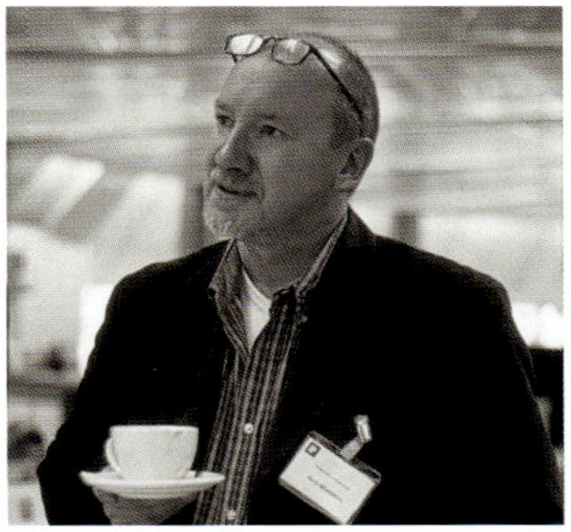

EERO MIETTIENEN hat nach seinem Industriedesign-Studium in Helsinki für verschiedene Unternehmen als Chef-Designer gearbeitet und ist nunmehr seit 1998 Design Director der Nokia Corporation, zudem hat er seit 1994 einen Lehrstuhl für Design an der Helsinki University of Technologies.

Eero Miettinen studied in Helsinki industrial design and went on to work for various corporations as head designer. Since 1998 he has been Design Director at the Nokia Corporation, and since 1994 has lectured on design at the Helsinki University of Technologies.

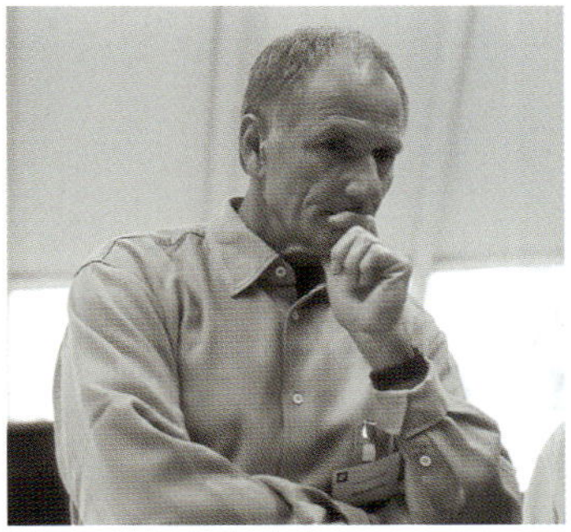

ANDREAS HAUGG war nach seinem Designstudium an der Staatlichen Akademie der Bildenden Künste, Stuttgart, für Esslinger Design und Frogdesign tätig. 1987 gründete er zusammen mit Tom Schönherr Phoenix Product Design.

Andreas Haug studied design at the State Academy of Applied Arts in Stuttgart, and then worked for Esslinger Design and Frogdesign. In 1987, he founded Tom Schönherr Phoenix Product Design.

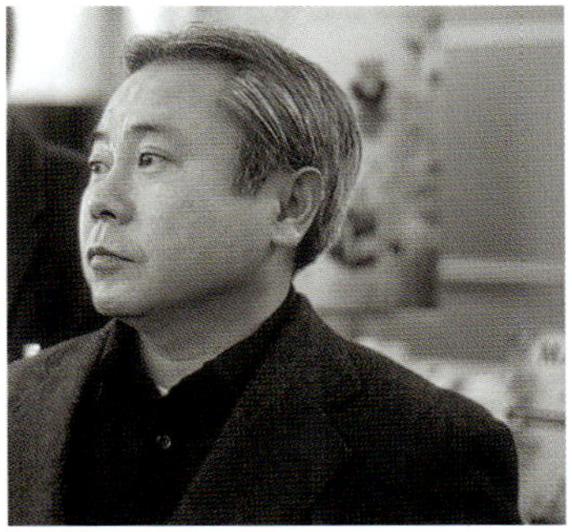

YOSHINORI KATAGAMI studierte in Tokio und Mailand Industriedesign und ist seit 1975 für die Toshiba Corporation tätig, mittlerweile ist er General Manager des Toshiba Design Centers.

Yoshinori Katagami studied industrial design in Tokyo and Milan. Since 1975, he has been a member of the Toshiba Corporation and now he is General Manager of the Toshiba Design Center.

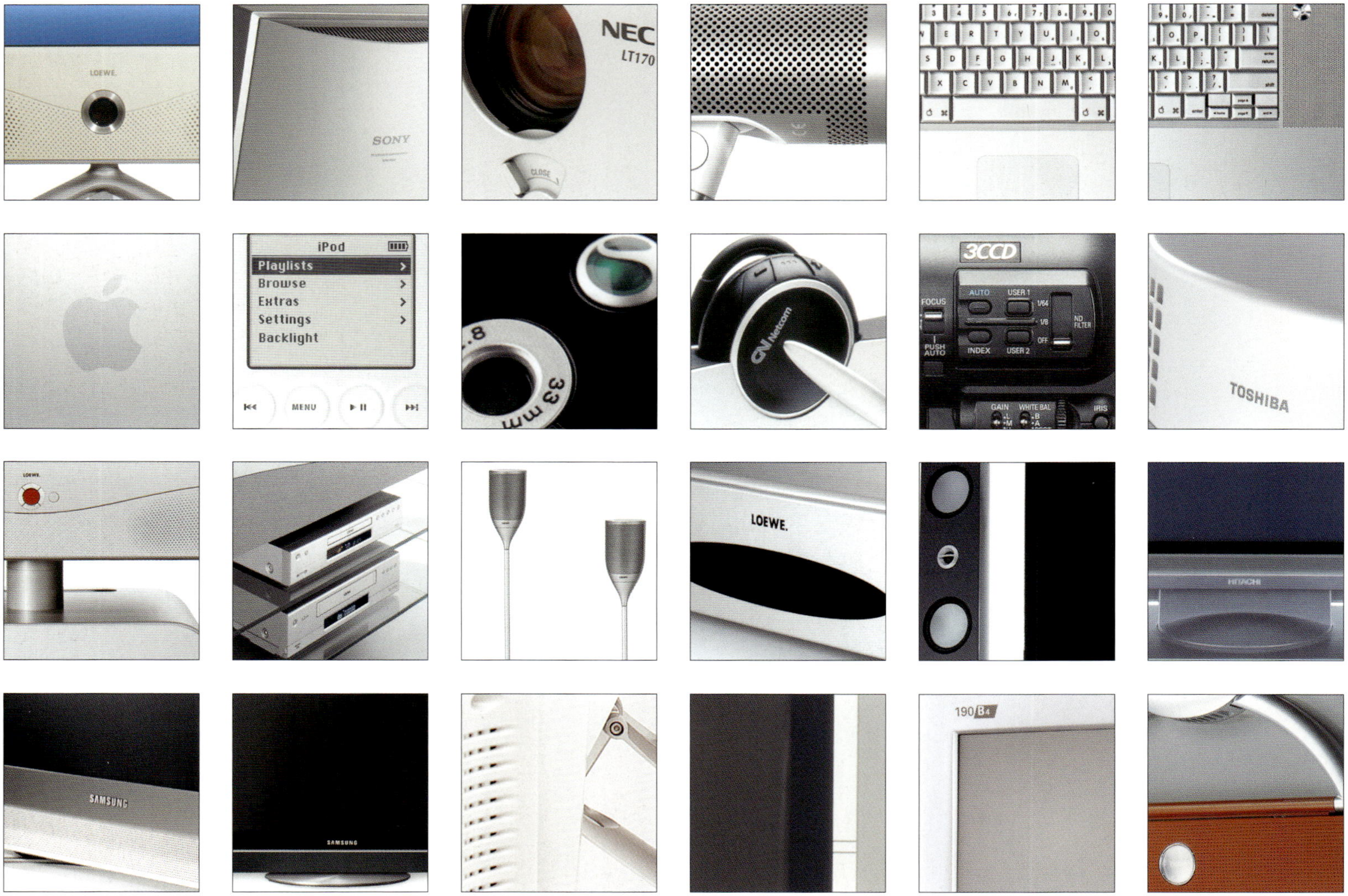
LOEWE.
SONY
NEC
LT170
iPod
Playlists
Browse
Extras
Settings
Backlight
MENU
33 mm
GN Netcom
3CCD
FOCUS
AUTO
USER 1
1/64
1/8
ND FILTER
OFF
PUSH AUTO
INDEX
USER 2
GAIN
WHITE BAL
IRIS
TOSHIBA
LOEWE.
HITACHI
SAMSUNG
SAMSUNG
190B4

Kommunikation Unterhaltung
Communication Entertainment

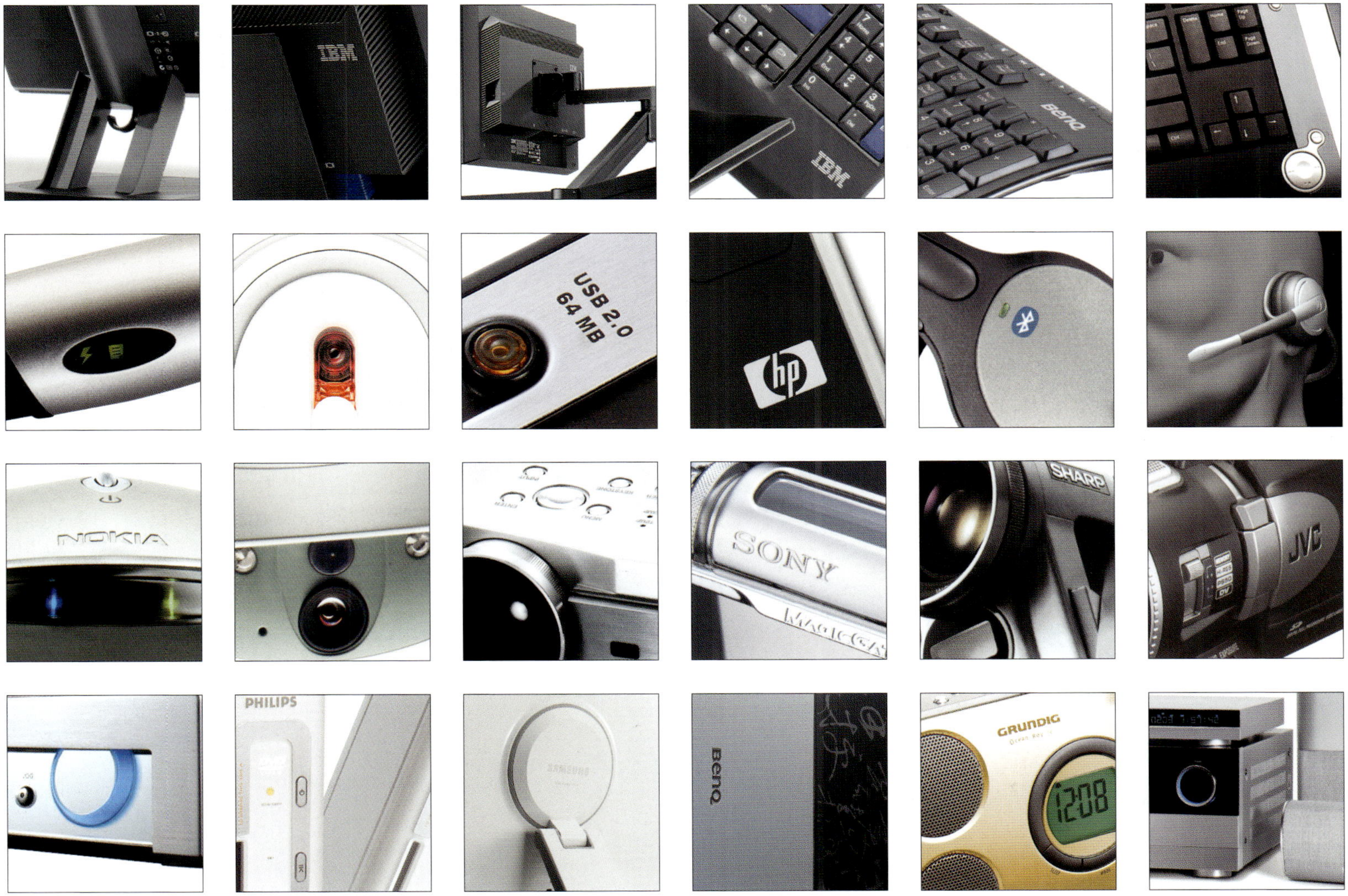

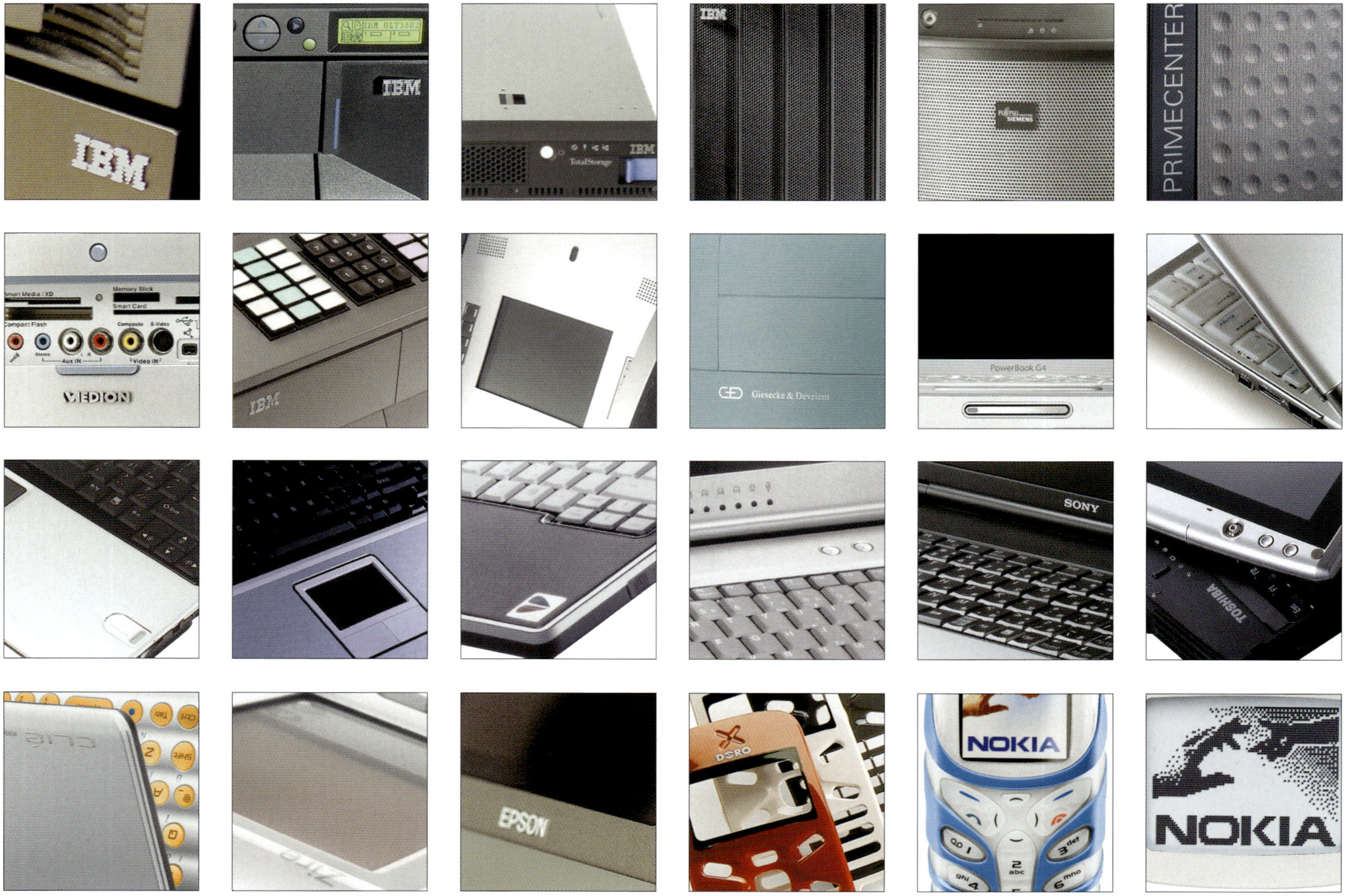
IBM
IBM
IBM
TotalStorage
IBM
SIEMENS
PRIMECENTER
MEDION
IBM
Giesecke & Devrient
PowerBook G4
SONY
TOSHIBA
EPSON
NOKIA
NOKIA

Kommunikation Unterhaltung
Communication Entertainment

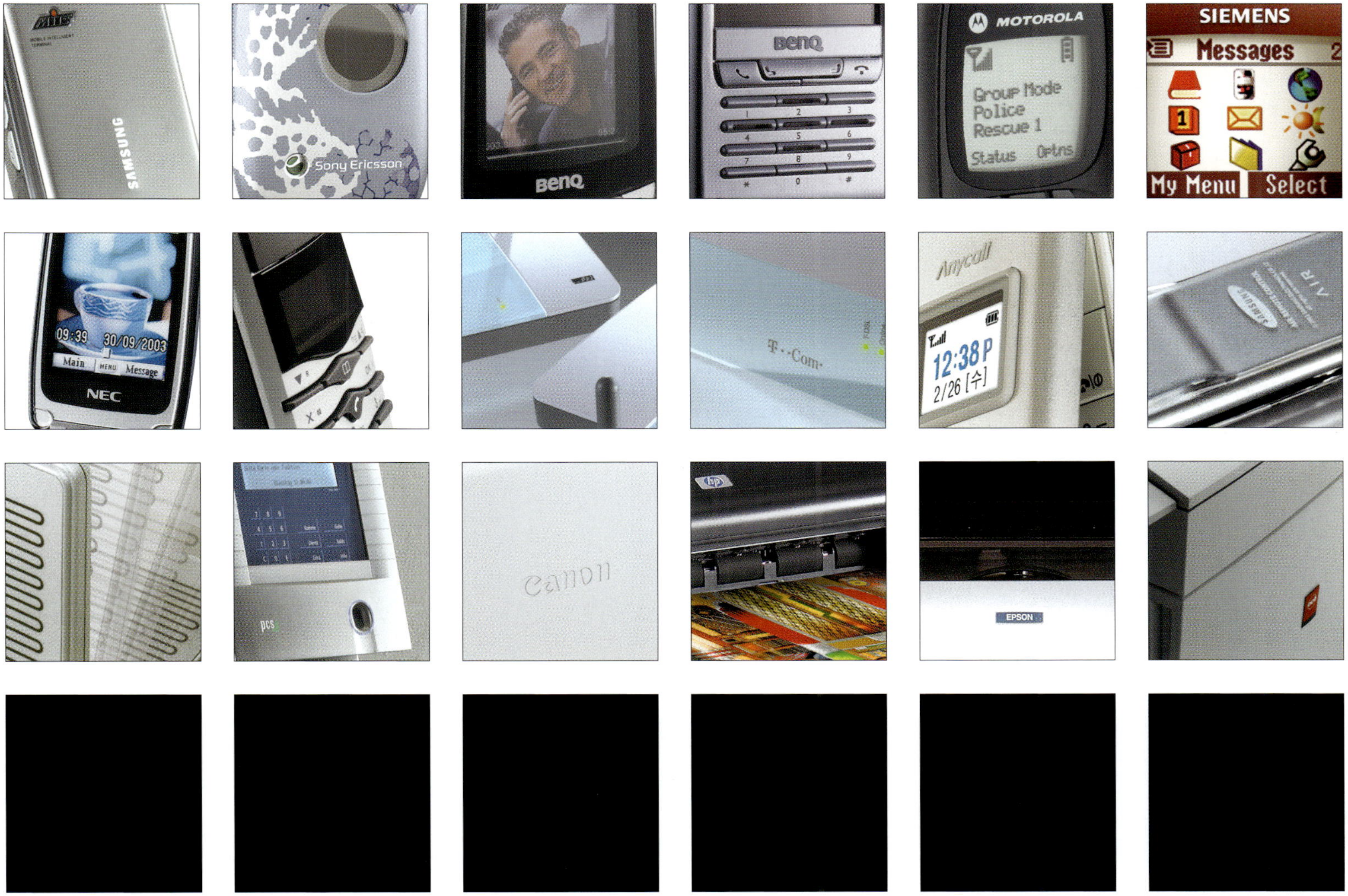

Hersteller Manufacturer
Loewe AG
Kronach, Germany

Produkt Product
Mimo 15
Fernsehgerät
TV set

Design
Phoenix Design
Stuttgart, Germany

Der Fernseher Mimo 15 von Loewe ist äußerst flach und besitzt mit den großen Radien und geschwungenen Linien eine weiche Formensprache. Die Gestaltung wird über die Rückwand fortgeführt, wo alle Stecker und Kabelanschlüsse durch eine Kabelabdeckung verborgen sind. Die LCD-Flatscreen-Technik gewährleistet hohe Bild- und Farbqualität, das eingebaute Radio und die integrierte Weckfunktion vervollständigen das Leistungsprofil. Der Bodenstand in Stativform ist durch verschiedene Zwischenstücke höhenverstellbar und ermöglicht eine von vorne nicht sichtbare Kabelführung im hinteren Aluminiumprofil. Zudem können unterschiedlich farbige Gehäuse-Wechselblenden gewählt werden.

Loewe's Mimo 15 TV is extremely flat and with its large radiuses and rolling lines presents a soft formal vocabulary. The design continues around to the back cover, which conceals all the plugs and cable connections. The LCD flatscreen technology delivers distortion-free pictures and outstanding color quality, while the built-in radio and integrated alarm function round out the performance profile. A tripod-style stand solution can adjusted in height by means of various adaptors, and cables can be channeled invisibly through the rear aluminum profile.

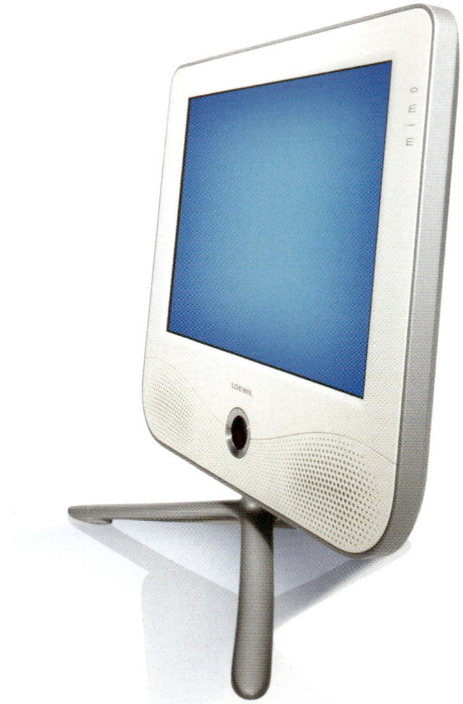

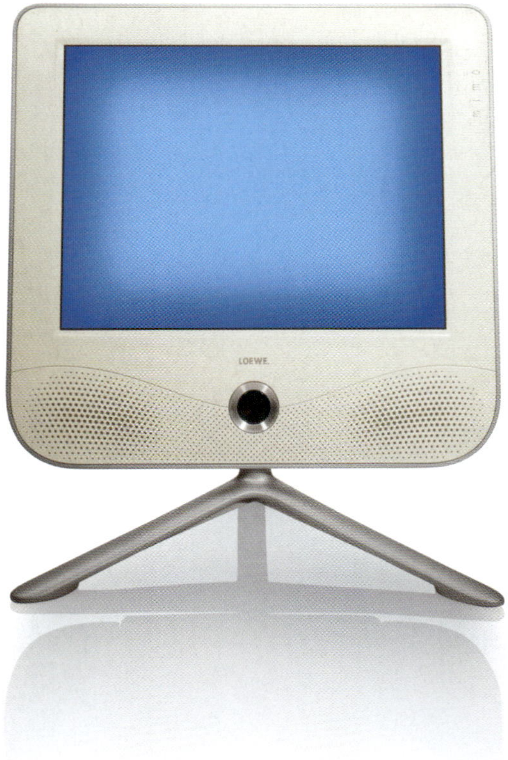

iF Gold Award

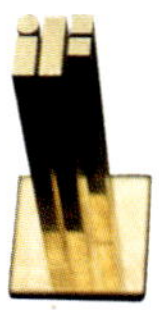

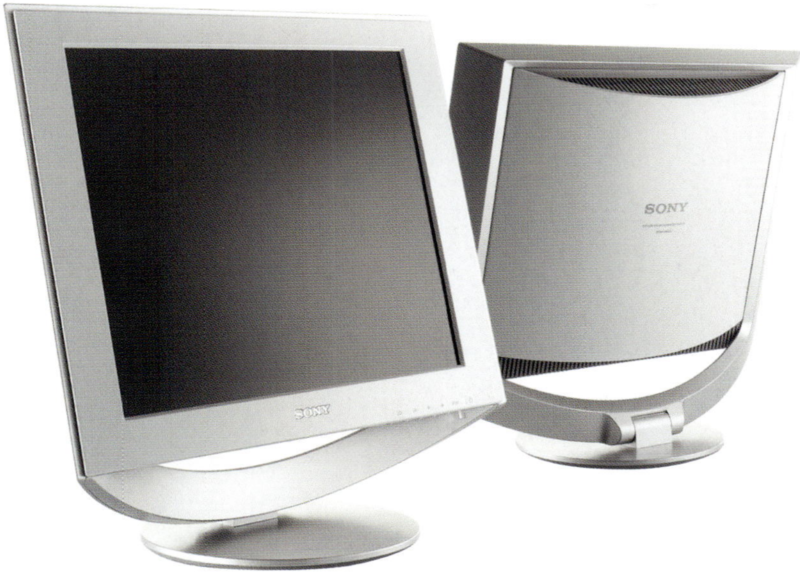

Hersteller Manufacturer
Sony Corporation
Tokio, Japan

Produkt Product
SDM-HS53 / HS73 / HS93
LCD-Monitor
LCD monitor

Design
Yuuki Kubota
Sony Corporation
Tokio, Japan

Der SDM-HS53 / HS73 / HS93 LCD-Monitor ist für das private Umfeld konzipiert und dementsprechend im Design auf ein Wohnambiente abgestimmt. Ein bogenförmiger Arm stützt den Bildschirm von unten und geht als Einheit in den Rahmen über. Der Bildschirm wird durch den spiegelbearbeiteten Rahmen betont, insgesamt entsteht der Eindruck von Schwerelosigkeit und Helligkeit. Schrauben und Netzkabel sind mit der Rückplatte abgedeckt und die Entlüftungsöffnungen sind so angeordnet, dass sie nicht sichtbar sind. Der Monitor ist in den Farben Weiß, Dunkelblau und Grau erhältlich.

The SDM-HS53 / HS73 / HS93 LCD monitor is intended for home use and therefore designed to blend in with residential interiors. A bow-shaped arm supports the screen from the bottom and merges with the frame to create a single unit. The screen is emphasized by the mirror-finished frame, and the overall appearance is one of weightlessness and brightness. Screws and the power cord are tucked away behind a rear panel and the heat dissipation holes placed such as to be invisible. The monitor is available in white, dark blue and grey.

iF Gold Award

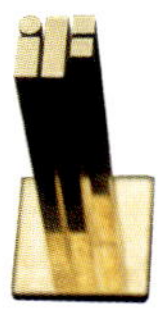

Hersteller Manufacturer
NEC Viewtechnology Ltd.
Kanagawa, Japan

Produkt Product
LT170
Multimedia-Projektor
Multi-media projector

Design
Ricca Tezuchi
Naoto Fukasawa Design
Tokio, Japan

Das quaderartige Design des Multimedia-Projektors LT170 wird zusätzlich betont, wenn die Linse hinter dem Projektionsschutz verschwindet. Das Gerät ist für die Ein-Hand-Bedienung konzipiert und bietet einen entsprechenden Bedienkomfort. Mit einem Knopfdruck lassen sich die Füße ausfahren, um den optimalen Neigungswinkels des Projektors einzustellen. Die automatische Trapezkorrektur zieht ein verzogenes Bild wieder gerade und über einen Drehknopf können Einstellungen mit einer Hand gewählt und per Knopfdruck bestätigt werden. Der LT170 misst lediglich 12 auf 28 cm bei einer Höhe von 8,5 cm und ist leicht mit einer Hand zu umfassen.

The cuboid design of NEC's LT170 mobile mini projector is especially emphasized when the lens is concealed behind the projector cover. The unit is designed for single-hand operation with exceptional control convenience. The feet extend at the press of a button to set the optimum angle of inclination for the projector. The automatic keystone correction straightens distorted images swiftly. A control dial and menu scroll function means settings can be selected by one hand and then confirmed by the press of button. The LT170 measures only 12 x 28 cm and is a mere 8.5 cm high – and light enough to carry easily with one hand.

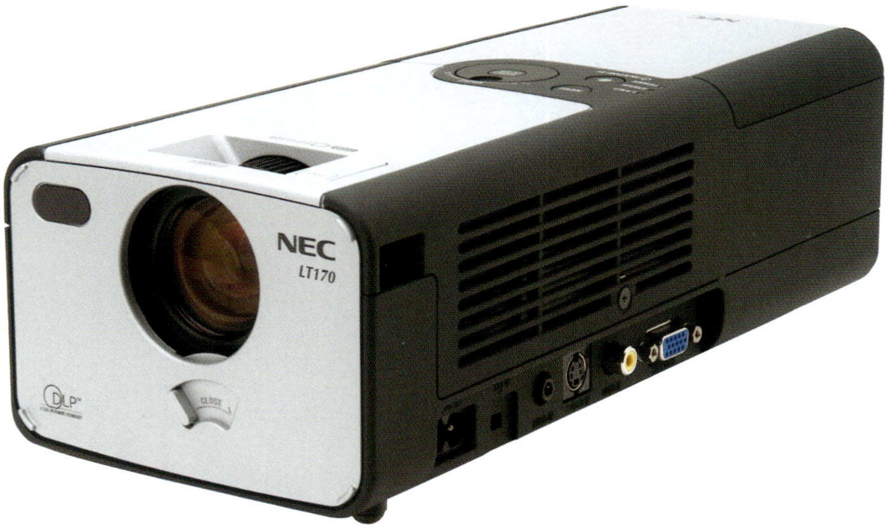

iF Gold Award

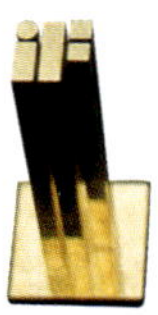

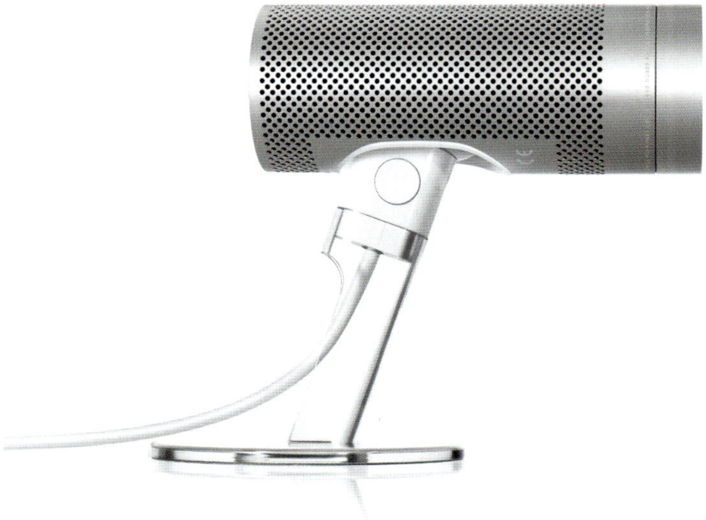

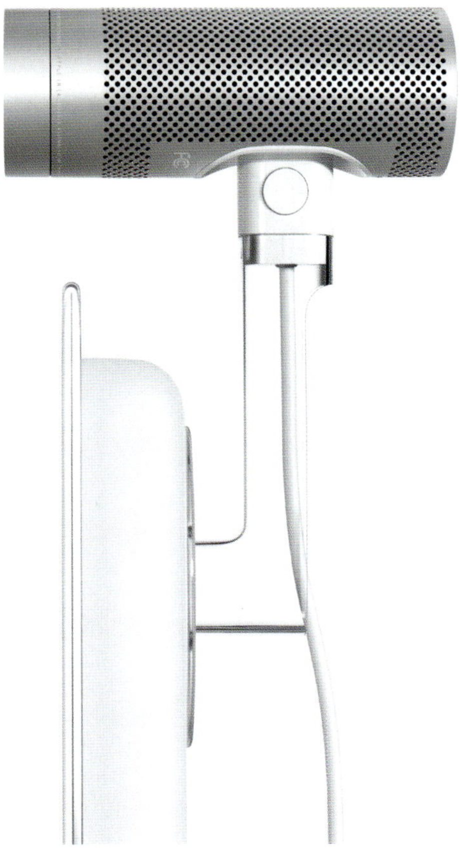

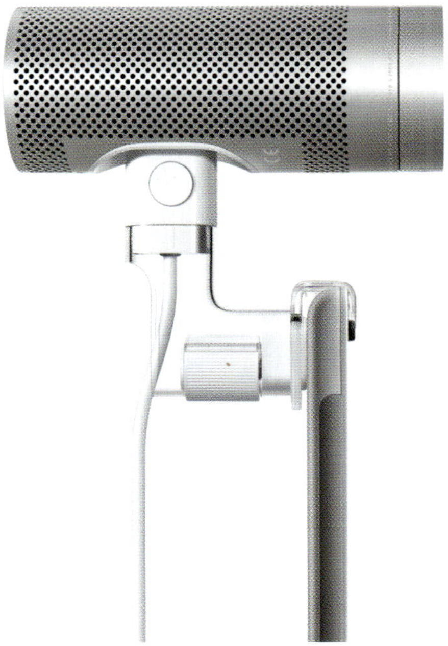

Hersteller Manufacturer
Apple Computer Inc.
Cupertino, CA, USA

Produkt Product
iSight
Videokamera
Video camera

Design
Bart Andre, Danny Coster
Daniele De Iuliis, Richard Howarth
Jonathan Ive, Duncan Kerr
Matthew Rohrbach, Doug Satzger
Cal Seid, Christopher Stringer, Eugene Whang
CAD Team: Carlos Ragudo, Carter Multz
Christopher Hood, Fred Simon
Irene Chan-Jones, Ken Provost, Mas Watanabe
Apple Computer, Inc.
Cupertino, CA, USA

Die iSight Videokamera von Apple besitzt ein F/2,8-Objektiv mit Autofokus und automatischer Belichtung, das selbst bei ungünstigen Lichtverhältnissen Bilder in hoher Qualität aufnimmt. Sie ist vielseitig, widerstandsfähig, leicht und handlich. Zudem ist ein Dual-Element-Mikrofon in das kompakte und stabile Aluminium-Gehäuse der Kamera integriert. Ein einziges FireWire Kabel überträgt Video- und Audiosignale sowie Strom.

Apple's iSight audio and video camera has a F/2.8 lens with an autofocus and automatic exposure times that delivers high-quality images even when ambient light conditions are not favorable. The versatile robust camera is light and handy. Integrated into the compact and stable aluminum body is a dual-element microphone. A single FireWire cable streams video and audio signals as well as powering the camera.

iF Gold Award

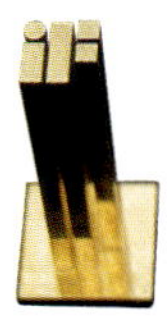

Hersteller Manufacturer
Apple Computer Inc.
Cupertino, CA, USA

Produkt Product
15" PowerBook G4
Notebook

Design
Bart Andre, Danny Coster
Daniele De Iuliis, Richard Howarth
Jonathan Ive, Duncan Kerr
Matthew Rohrbach, Doug Satzger
Cal Seid, Christopher Stringer, Eugene Whang
CAD Team: Carlos Ragudo, Carter Multz
Christopher Hood, Fred Simon
Irene Chan-Jones, Ken Provost, Mas Watanabe
Apple Computer Inc.
Cupertino, CA, USA

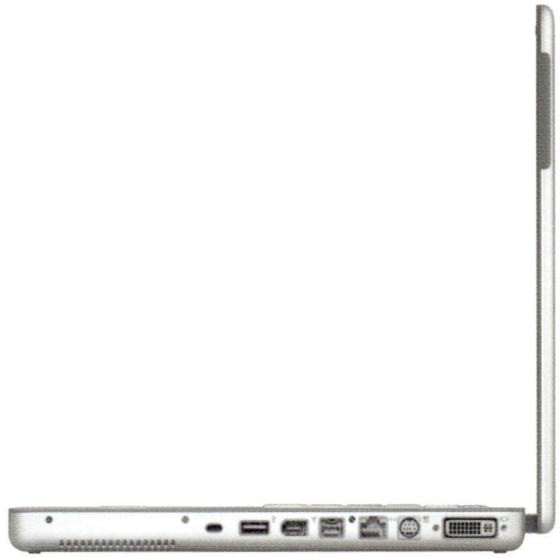

Das neue 15" PowerBook G4 von Apple wiegt lediglich 2,5 kg und zeichnet sich mit 2,8 cm durch sein äußerst flaches Design aus. Das Gehäuse besteht aus einer kratzfesten und widerstandsfähigen Aluminiumlegierung und hat keinerlei vorstehende Verriegelungen und Hebel, die abbrechen, keine scharfen Kanten und keine außen angebrachten Tasten, die versehentlich gedrückt werden könnten. Das Display-Scharnier wurde für das neue Modell nach innen verlegt und ist somit ebenfalls durch das Gehäuse geschützt. Die Tastatur ist hintergrundbeleuchtet, zudem passt ein Sensor die Helligkeit des Bildschirms automatisch an das Umgebungslicht an. Das Leistungsspektrum umfasst u.a.:1,25 GHz PowerPC G4, 512 KB L2 Cache, AirPort Extreme Karte, Radeon Grafikprozessor, Super-Drive Laufwerk.

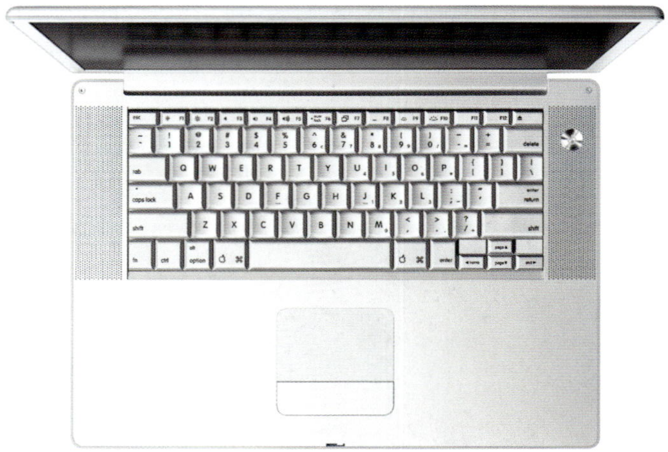

Apple's new 15" PowerBook G4 weighs only 2.5 kg and is a mere 1.1" thin. The flat body is made of scratch-resistant, extremely tough aluminum alloy and has no protruding locks or levers, sharp edges, or outer buttons that might be depressed accidentally. The display hinge is in-board and thus likewise protected within the product's envelope. The keyboard is backlit, and a sensor ensures that screen brightness adjusts automatically to ambient light. Features include a 1.25 GHz PowerPC G4, 512 KB L2 Cache, AirPort Extreme card, Radeon graphic processor and a SuperDrive.

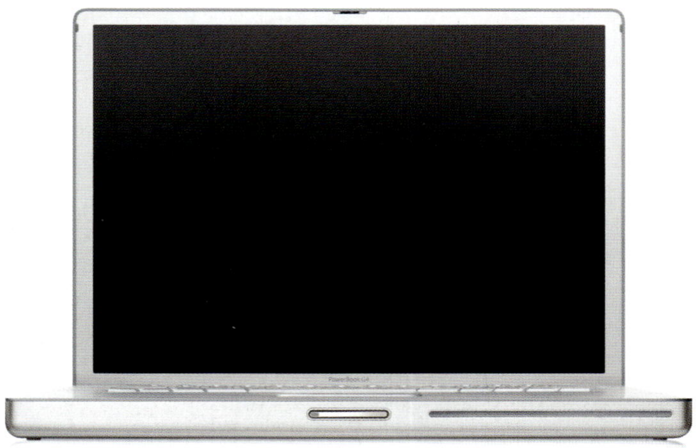

iF Gold Award

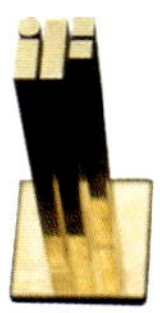

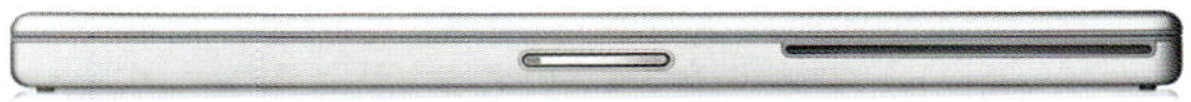

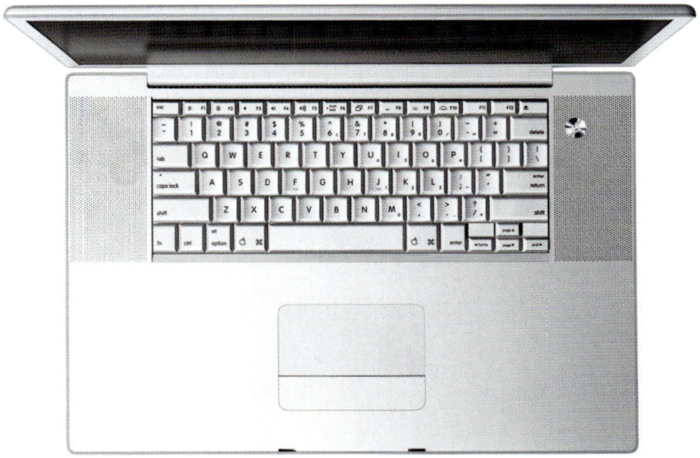

Hersteller Manufacturer
Apple Computer Inc.
Cupertino, CA, USA

Produkt Product
17″ PowerBook G4
Notebook

Design
Bart Andre, Danny Coster
Daniele De Iuliis, Richard Howarth
Jonathan Ive, Duncan Kerr
Matthew Rohrbach, Doug Satzger
Cal Seid, Christopher Stringer, Eugene Whang
CAD Team: Carlos Ragudo, Carter Multz
Christopher Hood, Fred Simon
Irene Chan-Jones, Ken Provost, Mas Watanabe
Apple Computer Inc.
Cupertino, CA, USA

Das neue 17″ PowerBook G4 von Apple besitzt einen auffallend breiten Bildschirm mit einer Diagonale von 43,18 cm. Das Gehäuse ist trotzdem mit 2,6 cm äußerst flach, das Gewicht beträgt lediglich 3,1 kg. Es besteht aus einer kratzfesten und widerstandsfähigen Aluminiumlegierung und hat keinerlei vorstehende Verriegelungen und Hebel, die abbrechen, keine scharfen Kanten und keine außen angebrachten Tasten, die versehentlich gedrückt werden könnten. Die Tastatur verfügt über eine Hintergrundbeleuchtung, zudem passt ein Sensor die Helligkeit des Bildschirms automatisch an das Umgebungslicht an. Das Leistungsspektrum umfasst u.a.:1,33 GHz PowerPC G4, bis zu 2 GB DDR-Arbeitsspeicher, 1440 x 900 Pixel Auflösung, AirPort Extreme Karte, Radeon Grafikprozessor, SuperDrive Laufwerk.

Apple's new 17″ PowerBook G4 has a strikingly broad 17″ screen and yet a 2.6 cm ultra-thin body and weighs in at only 3.1 kg. The outer skin is made of a scratch-resistant and sturdy aluminum alloy and has no protruding locks or levers that could break off, no sharp edges or outer buttons that might be activated by accident. The keyboard is backlit and a sensor adjusts screen brightness to ambient light levels. Features include a 1.33 GHz PowerPC G4, up to 2 GB DDR RAM, 1440 x 900 pixel resolution, high-speed FireWire, an AirPort Extreme card, a Radeon graphic processor and a SuperDrive.

iF Gold Award

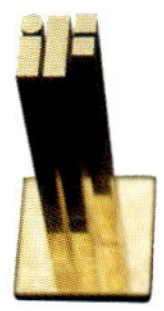

Hersteller Manufacturer
Apple Computer Inc.
Cupertino, CA, USA

Produkt Product
Power Mac G5
Personal computer

Design
Bart Andre, Danny Coster
Daniele De Iuliis, Richard Howarth
Jonathan Ive, Duncan Kerr
Matthew Rohrbach, Doug Satzger
Cal Seid, Christopher Stringer, Eugene Whang
CAD Team: Carlos Ragudo, Carter Multz
Christopher Hood, Fred Simon
Irene Chan-Jones, Ken Provost, Mas Watanabe
Apple Computer Inc.
Cupertino, CA, USA

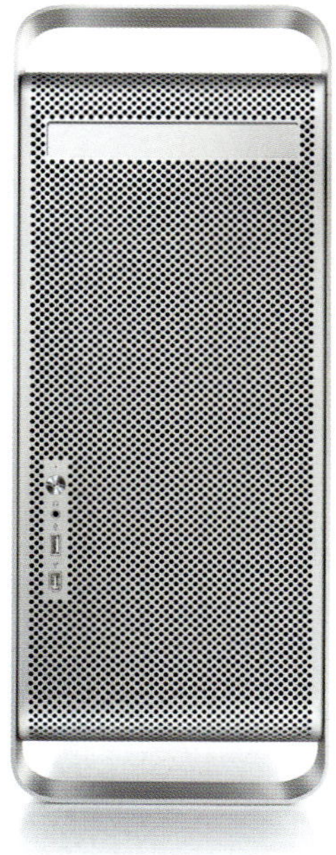

Das Gehäuse des Power Mac G5 von Apple wurde um vier interne Wärmezonen herum konstruiert, wobei das spezielle Belüftungssystem einen besonders geräuscharmen Betrieb ermöglicht. Die einfach zu öffnende Gehäuseabdeckung ermöglicht einen schnellen Zugriff auf die internen Komponenten, die integrierten Handgriffe erleichtern den Transport. Der Power Mac G5 verfügt über einen 64 Bit G5 Prozessor, der mit einer Systemarchitektur mit hoher Bandbreite sowie AGP 8X und PCI-X Unterstützung ausgestattet ist, dabei ist ein Hauptspeicher von bis zu 8 Gigabyte möglich. Das SuperDrive Laufwerk liest DVDs mit 8facher Geschwindigkeit und beschreibt 4,7 GB DVD-Rs mit 4facher Geschwindigkeit.

The casing of Apple's Power Mac G5 is constructed around four internal thermal zones and the special air cooling system enables quiet operation. The easy-to-open access panel offers quick access to internal components and the integrated handles make it easier to move the Power Mac G5. The computer boasts a 64-bit G5 processor, and the systems architecture support a broad range of applications such as AGP 8X and PCI-X; main memory is as much as 8 GB; the SuperDrive reads DVDs at 8x and writes 4.7 GB DVD-Rs at 4x.

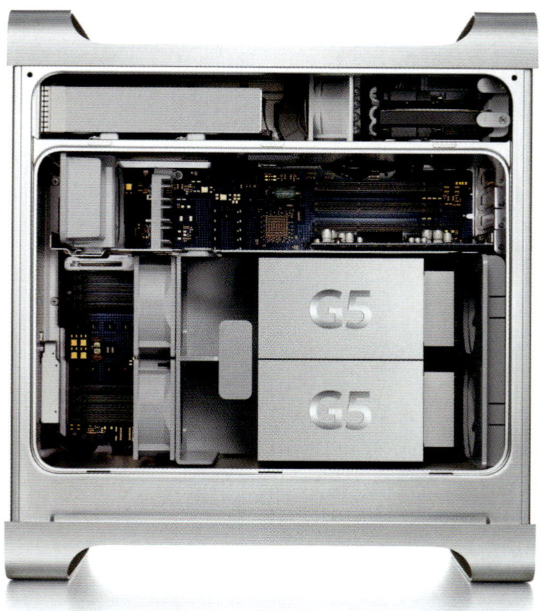

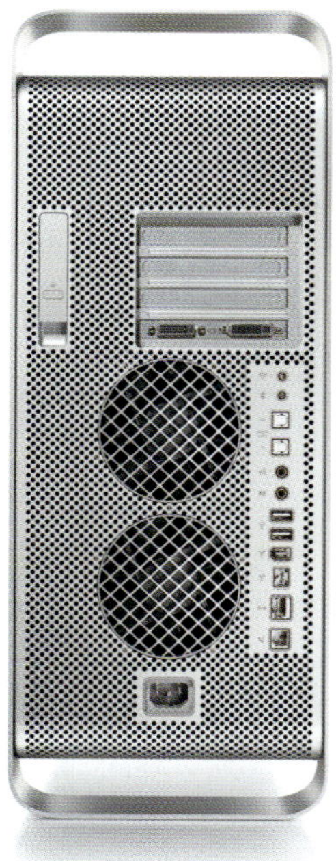

iF Silver Award

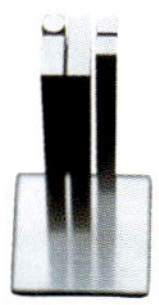

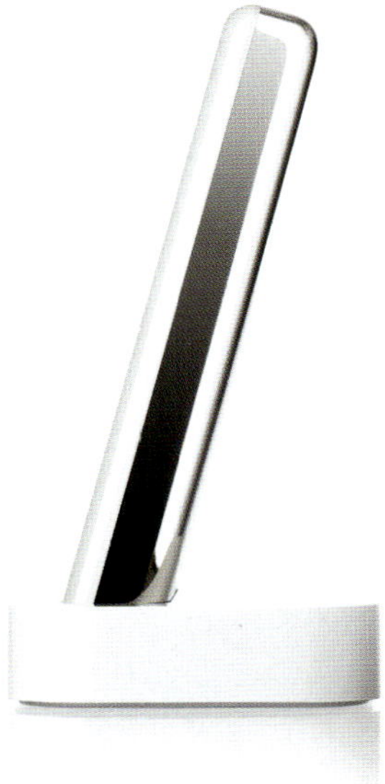

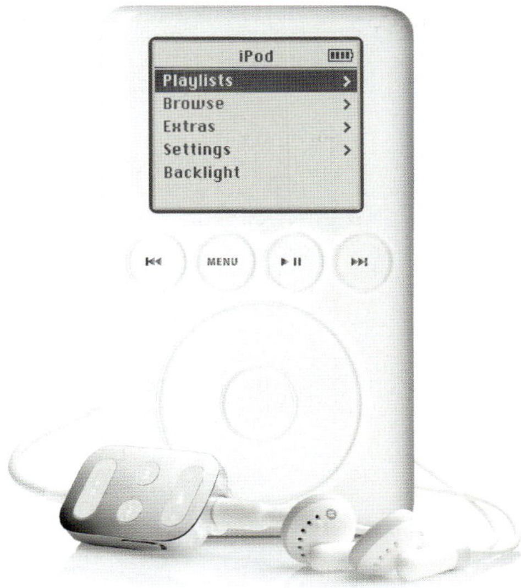

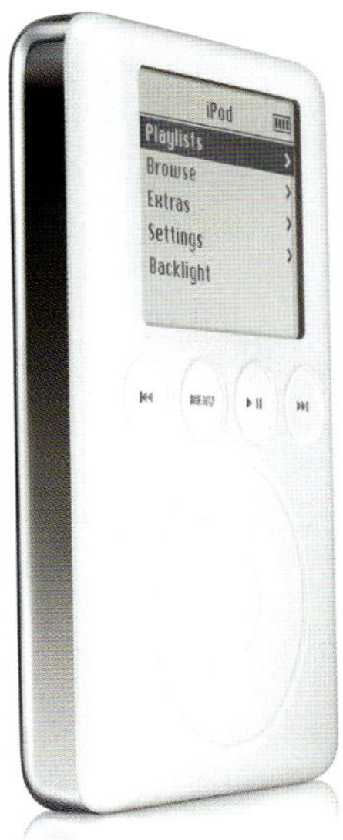

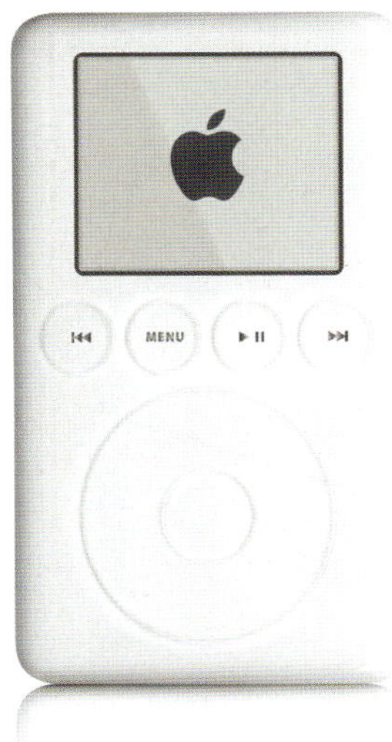

Hersteller Manufacturer
Apple Computer Inc.
Cupertino, CA, USA

Produkt Product
Next Generation iPod
iPod und MP3 Player
iPod and MP3 player

Design
Bart Andre, Danny Coster
Daniele De Iuliis, Richard Howarth
Jonathan Ive, Duncan Kerr
Matthew Rohrbach, Doug Satzger
Cal Seid, Christopher Stringer, Eugene Whang
CAD Team: Carlos Ragudo, Carter Multz
Christopher Hood, Fred Simon
Irene Chan-Jones, Ken Provost, Mas Watanabe
Apple Computer Inc.
Cupertino, CA, USA

Der neue iPod ist leichter als 2 CDs, kann bis zu 10.000 Songs speichern und ist mit 1,57 cm extrem flach. Er liegt angenehm in der Hand und passt in jede Tasche. Aufbauend auf dem berührungsempfindlichen Scrollrad bietet der iPod eine übersichtliche Oberfläche, LCD-Bildschirm und Tasten sind durch die Hintergrundbeleuchtung auch bei geringem Licht gut sichtbar, zudem kann das Gerät leicht mit einer Hand bedient werden. Der iPod unterstützt die gängigsten Audioformate und das AAC-Format, mit dem Audio-Material in CD-Qualität codiert werden kann. Er ist mit 15 GB, 20 GB oder 40 GB Festplatte erhältlich.

The new iPod weighs less than 2 CDs, can hold up to 10,000 songs and is only 1.57 cm thick. It rests comfortably in the hand and fits in any pocket. Based around the touch-sensitive scroll dial, the iPod boasts a well laid-out surface, an LCD and backlit buttons, visible even when light is poor. Moreover, the player is designed for single-hand use. The iPod supports the customary audio formats and the AAC format with which audio material can be coded in CD quality. Available with a 15 GB, 20 GB or 40 GB hard drive.

iF Silver Award

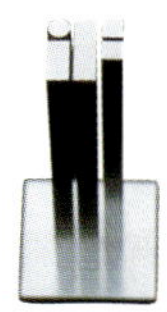

Hersteller Manufacturer
Sony Ericsson Mobile Communications AB
Lund, Sweden

Produkt Product
T610
Mobiltelefon
Mobile phone

Design
Erik Ahlgren
Sony Ericsson
Creative Design Center
Lund, Sweden

Das Design des T610 Mobiltelefons von Sony Ericsson ist in seiner Farbgebung vom Kontrast von Dunkelheit und Licht inspiriert. Es hat eine integrierte Digitalkamera und ein großes Farbdisplay, zudem bietet es Multimedia Messaging, polyphone Klingeltöne und eine Bluetooth-Schnittstelle.

The design of the Sony Ericcson T610 mobile phone is inspired by the contrast of light and dark in its color scheme. It features an integrated digital camera and a large color display. In addition it offers multimedia, messaging, polyphonic ring tones and Bluetooth.

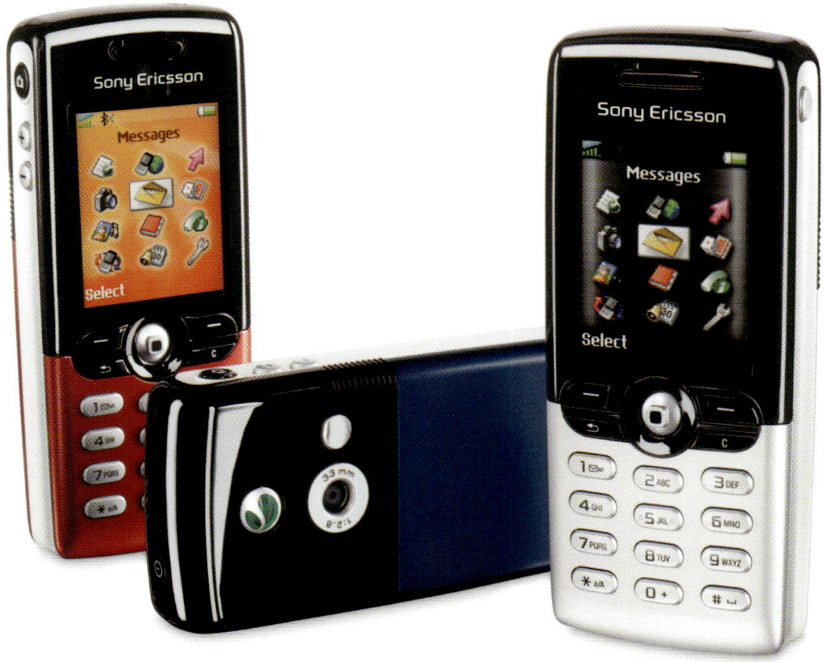

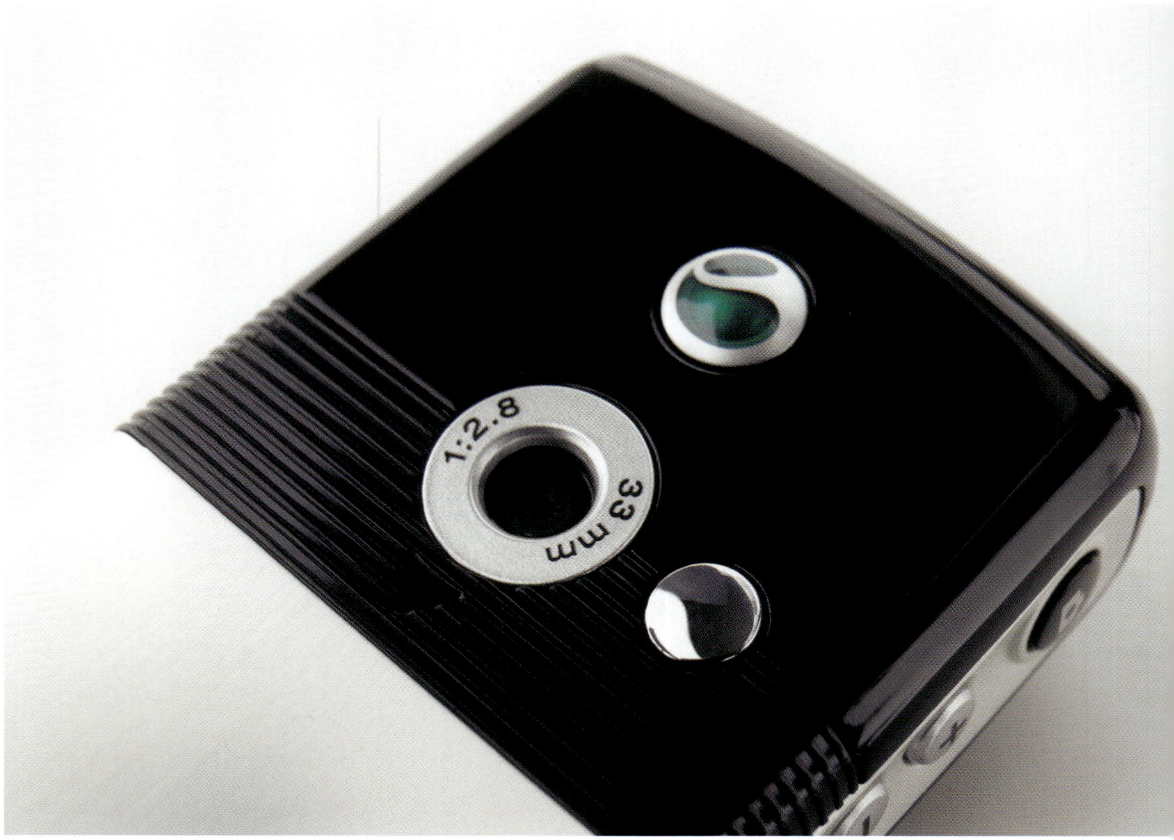

iF Silver Award

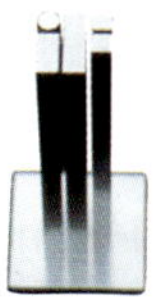

Hersteller Manufacturer
GN Netcom
Ballerup, Denmark

Produkt Product
GN 9120
Kopfhörer + Mikrofon
Headset

Design
Klaus Rath
Gentofte, Denmark

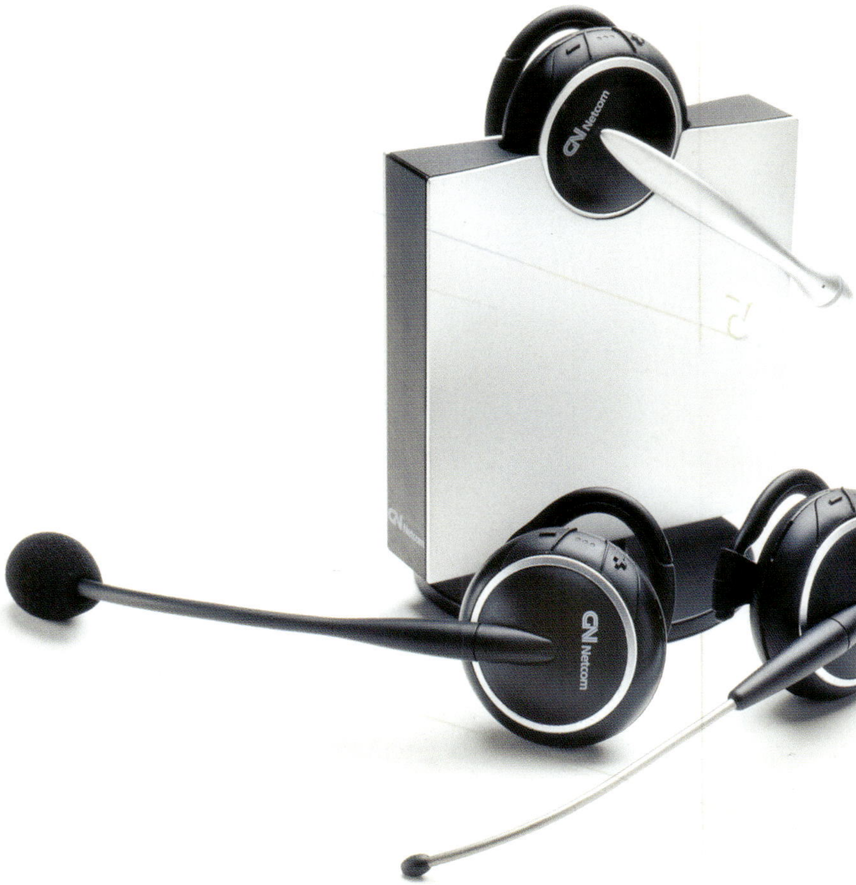

Das GN 9120 ist eine völlig schnurlose Headset-Lösung mit einem aufladbaren Lithium-Akku. Das Headset ermöglicht einen Bewegungsradius bis zu 150 m um den Schreibtisch und eine Sprechzeit von bis zu 12 Stunden. Es stehen drei verschiedene Mikrofonarme zur Auswahl: Flex-Boom und Midi-Boom sind mit einem geräuschreduzierenden (NC) Mikrofon ausgestattet. Die mit einem Standard-Mikrofon ausgestattete Mikro-Boom-Version eignet sich vor allem für ruhige Bereiche. Die Trageform des Headsets kann frei gewählt werden zwischen Überkopfbügel oder Ohrhaken.

The GN 9120 is a truly cordless headset solution with a chargeable lithium battery. The headset permits you to move in a 150 m radius of the desk with up to 12 hours of talk time. Three different microphone arm options are available: the flex-boom and midi-boom are both fitted with a noise-canceling microphone while the micro-boom version is fitted with a standard microphone which makes it particularly suitable for quieter environments. The headset can be worn over the head using the headband or with an ear-hook for on-ear use.

iF Silver Award

Hersteller Manufacturer
Matsushita Electric Industrial Co.Ltd
Osaka, Japan

Produkt Product
AG-DVX100
Digitale Profi-Videokamera
Digital professional video camera

Design
Ryuji Miki
Hirofumi Takemasa
Panasonic Design Company
Osaka, Japan

Die AG-DVX100 bietet die im Film übliche Aufnahmetechnik mit 25 Vollbildern pro Sekunde. Er verfügt über ein Weitwinkelobjektiv F32,5-325 mm mit optischem 10fach-Zoom und 3IT-CCD 1/3-Zoll-Sensoren mit 470.000 Bildpunkten (interlaced und progressive Scan). Die Kamera wurde nach ergonomischen Gesichtspunkten entwickelt, so wurde z.B. der Schwerpunkt der Kamera so gewählt, dass sie auch bei längeren Aufnahmezeiten gut im Gleichgewicht zu halten ist, zudem können verschiedene Funktionskomponenten blind bedient werden.

The AG-DVX100 is destined for professional users, shooting 25 progressive images a section for cinema quality shots. The wide angle F32.5-325 mm lens with its optical 10x zoom has 3IT-CCD 1/3" sensors with 470,000 pixels (interlaced and progressive scan). The camera was developed with a view to ergonomic criteria, and its point of gravity is thus such that it remains balanced even when held for longer periods of time. Various functional elements can be actuated blind.

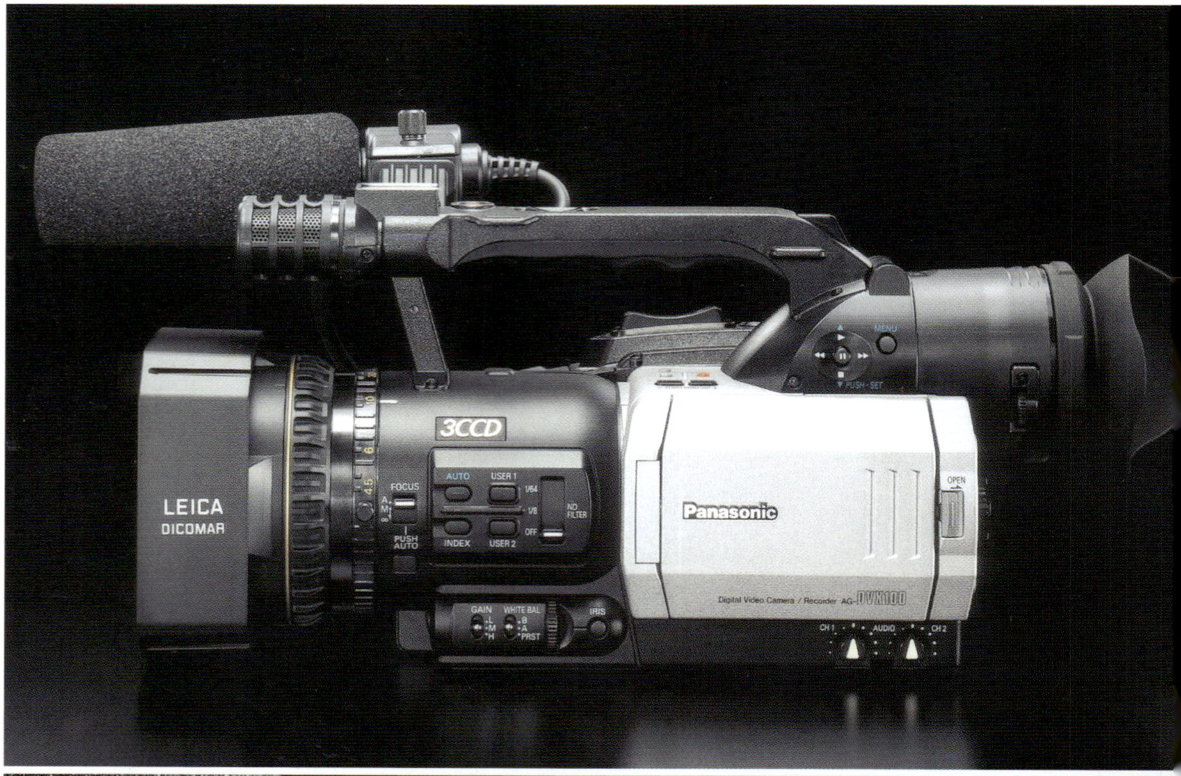

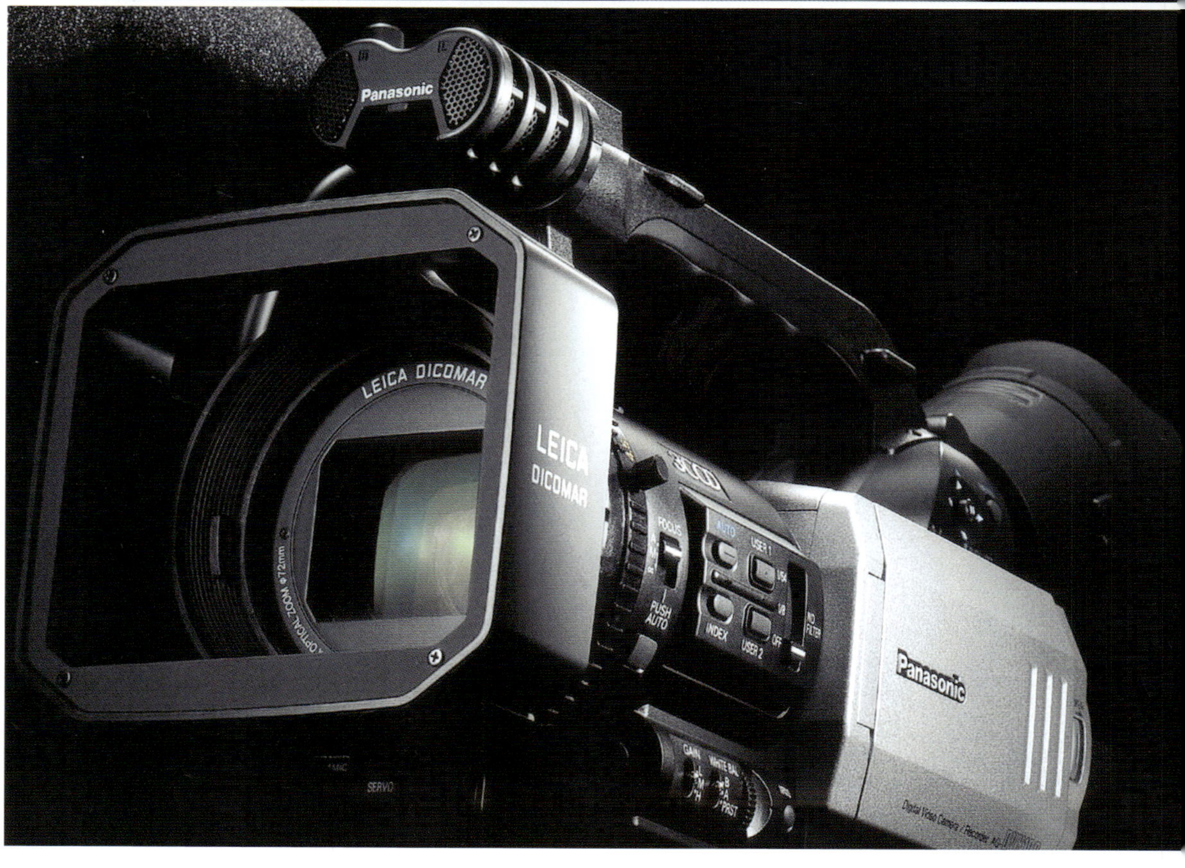

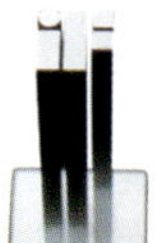

iF Silver Award

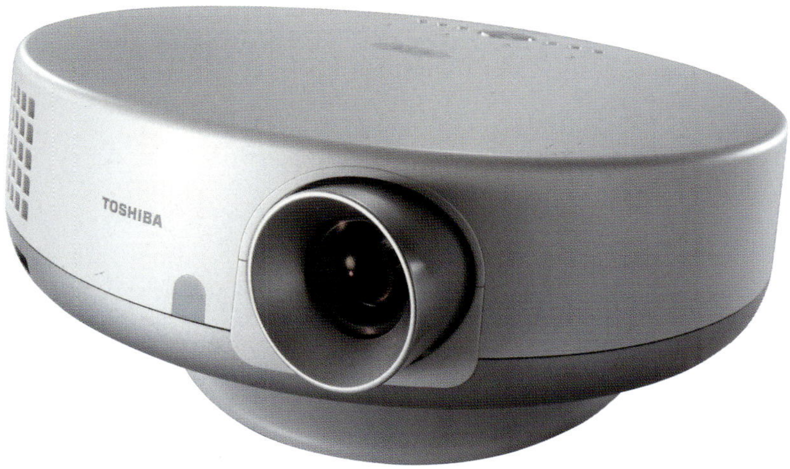

Hersteller Manufacturer
Toshiba Corporation
Tokio, Japan

Produkt Product
TLP-ET1
Heimkino-Projektor
Home projector

Design
Ryusuke Kurimoto
Tadashi Kurokawa
Design Center
Toshiba Corporation
Tokio, Japan

Der Heimkino-Projektor TLP-ET1 von Toshiba verfügt über eine spezielle Linse, die auch in kleineren Räumen scharfe Bilder gewährleistet. Die integrierten Stereo-Lautsprecher liefern ohne die Notwendigkeit zusätzlicher externer Lautsprecher guten Sound, zudem zeichnet sich der Projektor durch ein sehr leises Betriebsgeräusch aus. Die runde, ungewöhnliche Formgebung soll eine Alternative zum herkömmlichen Kastendesign dieser Produktgruppe bieten.

Toshiba's TLP-ET1 home projector features a special lens which achieves sharper pictures thanks to the close-focus function, even in smaller rooms. The built-in stereo loudspeakers deliver good sound without the need for additional external speakers and the projector also significantly reduces fan noise. The unusual, round shape offers an alternative to the conventional box-like design otherwise typical of such products.

iF Silver Award

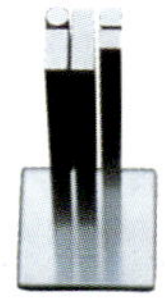

Hersteller Manufacturer
Loewe AG
Kronach, Germany

Produkt Product
Mimo 28
Fernsehgerät
TV set

Design
Phoenix Design
Stuttgart, Germany

Der Loewe Fernseher Mimo 28 hat zwar die Anmutung eines konventionellen Fernsehgerätes, jedoch kennzeichnet ihn eine für diese Produktgruppe neue Formensprache. Auffallend sind die weiche Behandlung der Kanten, die geschwungene Linie und die Lochstruktur des Lautsprechergitters. Um die Front reduziert zu halten, sind die Bedienknöpfe in einem chromfarbenen Ring zusammengefasst. Neuartig ist auch die Gestaltung der Rückwand, die Front und Kabelabdeckung in einer fließenden Form umfasst und dadurch von vorne sichtbar bleibt. Der Geräteaufbau ermöglicht interessante Farbkombinationen: Passend zur Rückwand in Platin kann zwischen Fronten in Cream, Platin oder Anthrazit gewählt werden.

Loewe's Mimo 28 TV set looks like a conventional television, but is nevertheless characterized by a formal idiom which is new for this product group. Prominent features are the soft edges, curved lines and perforated structure of the loudspeaker grate. To keep the front minimalist in tone, user buttons are reduced to one chromed ring round a red dot. The design of the back cover is also innovative – it encloses the front and a cable cover as part of a flowing form and thus is also visible from the front. The color combinations are also novel: the front is available in a choice of cream, platinum or anthracite front to match a platinum back cover.

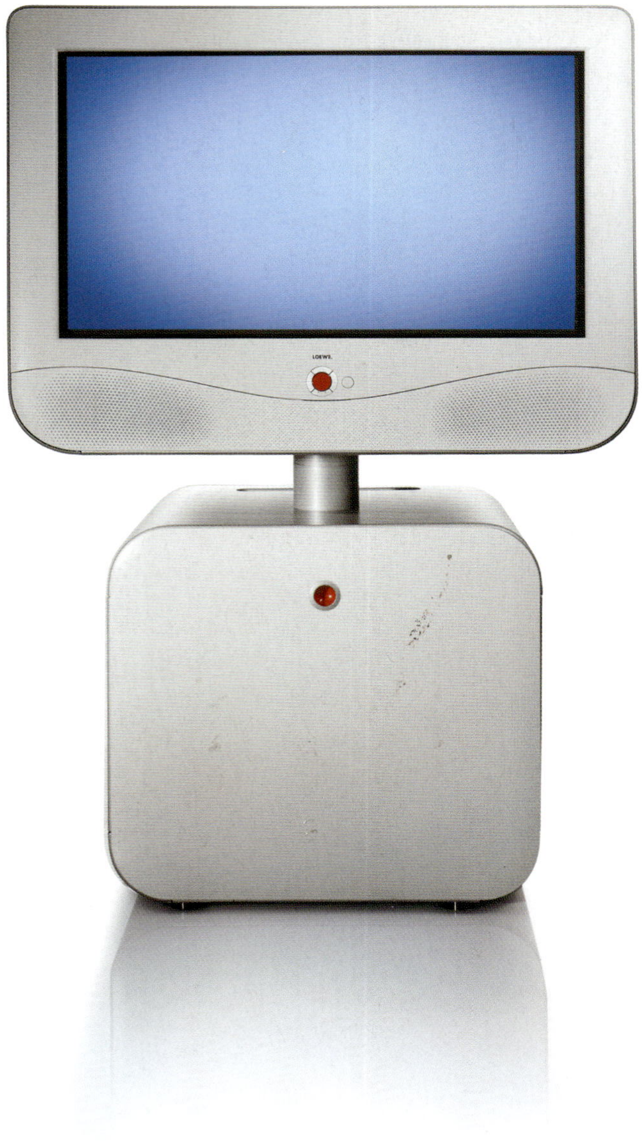

iF Silver Award

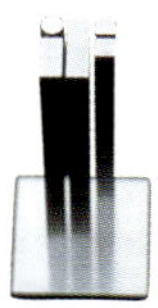

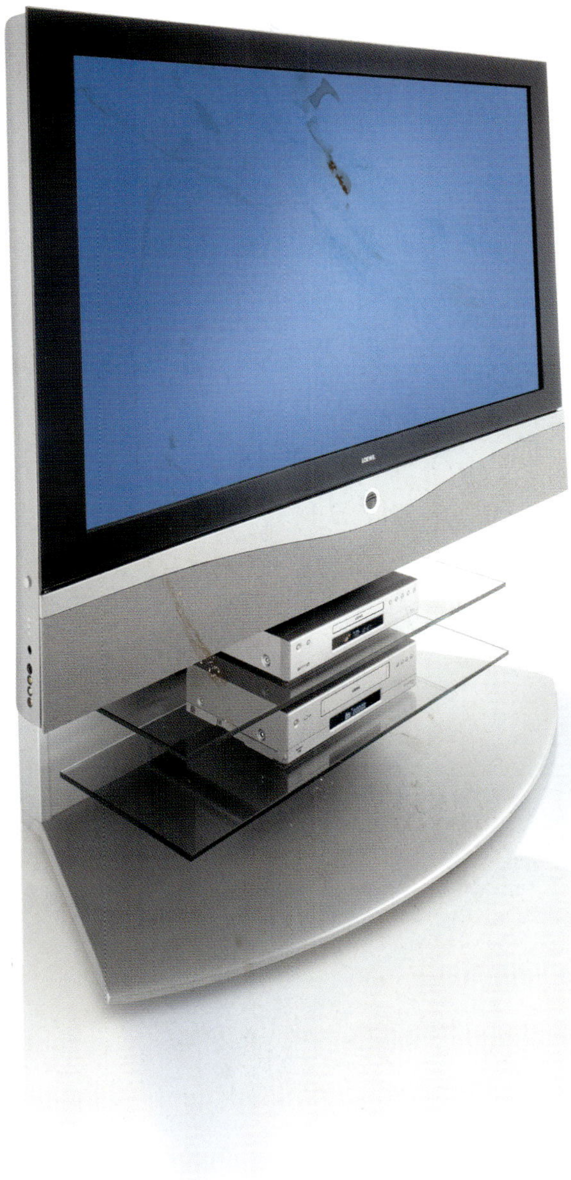

Hersteller Manufacturer
Loewe AG
Kronach, Germany

Produkt Product
Articos 55 HD
Fernsehgerät
TV set

Design
Phoenix Design
Stuttgart, Germany

Articos 55 HD von Loewe ist ein hochauflösender Projektions-TV mit High-Definition-Displaytechnologie und 140 cm Bilddiagonale, der in Kooperation mit Carl Zeiss Jena und Texas Instruments entwickelt wurde. Er besitzt eine minimale Bautiefe, wobei das aus Projektions- und Filterscheibe vorgesetzte Glas-Laminat mechanischen Schutz und verzerrungsfreie Projektionsbilder bietet. Mit einem geschlossenen und einem offenen Rack sowie dem Tischstand sind drei Aufstellungen möglich, die individuell mit anderen Loewe Komponenten aus der Systems-Linie kombiniert werden können.

The Loewe Articos 55 HD projection TV features high-definition display technology and a 140 cm screen (diagonal), and was developed together with Carl Zeiss Jena and Texas Instruments for home cinema. The screen has a minimal depth and features a pre-positioned glass laminate consisting of a projection and a filter screen, affording mechanical protection and distortion-free projection images. With a closed and an open rack as well as a table stand there are offer three storage/display possibilities and the set blends well with other components from the Loewe Systems line.

Hersteller Manufacturer
Loewe AG
Kronach, Germany

Produkt Product
Tremo
Subwoofer-Satelliten-System
Subwoofer satellite system

Design
Phoenix Design
Stuttgart, Germany

Das Subwoofer-Satelliten-System Tremo von Loewe zeichnet sich durch ein innovatives Materialkonzept aus. Der natürliche Werkstoff Keramik als Grundlage für den Subwoofer bietet neue Gestaltungsmöglichkeiten bei kostengünstiger Fertigung. Als in sich abgeschlossener Tonkörper besitzt das Subwoofer-Gehäuse sehr gute akustische Eigenschaften, wobei der Werkstoff zusätzlich zur Kühlung der Elektronik beiträgt. Um eine optimale Klangcharakteristik zu ermöglichen haben auch die Satelliten-Lautsprecher eine zylindrische Grundform mit weichen Formverläufen. Das System ist in drei Farbvarianten erhältlich.

Loewe's Tremo subwoofer satellite system stands out for its revolutionary materials concept. Ceramic as a natural material forms the basis of the subwoofer and generates new design possibilities and allows for inexpensive manufacturing. The subwoofer cabinet, which forms a self-contained sound body, possesses prime acoustic properties while the material also contributes to cooling the electronics. In order to deliver optimal tonality, the satellite loud speakers also have a basic cylindrical form with smooth shapes. The system is available in three different colors.

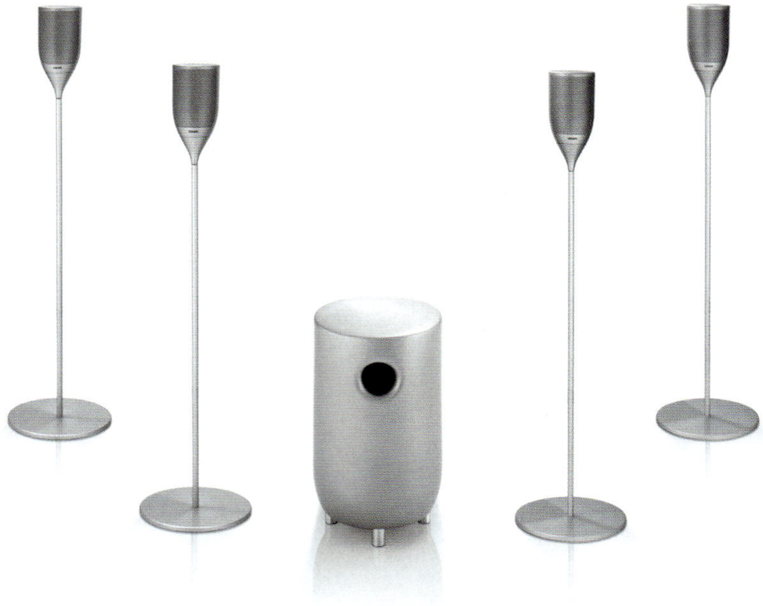

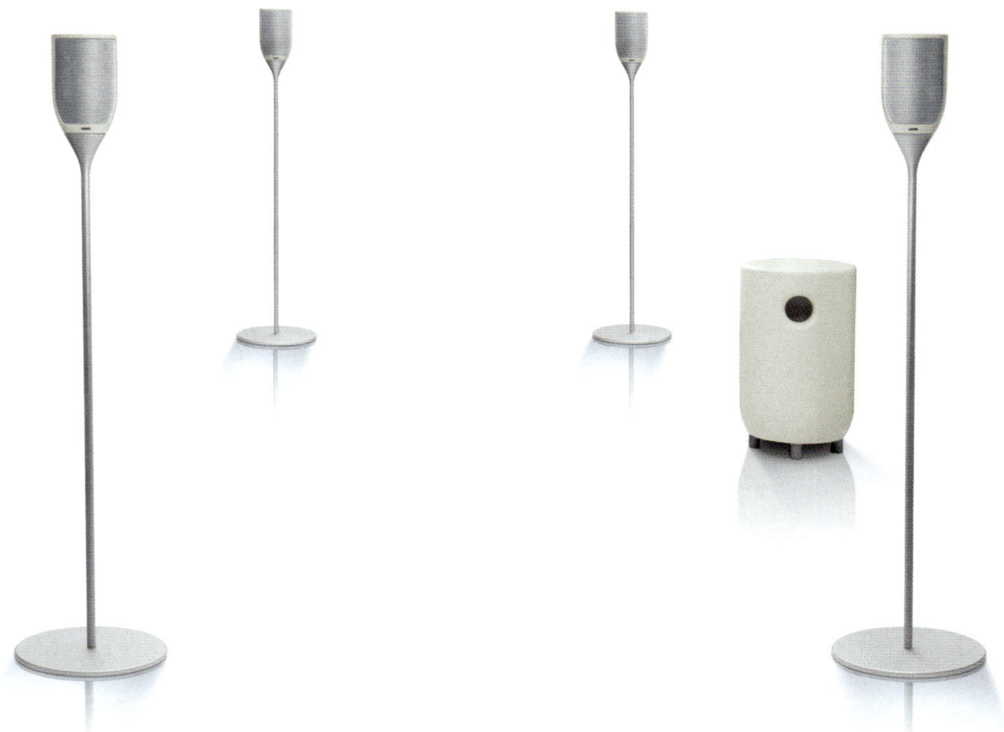

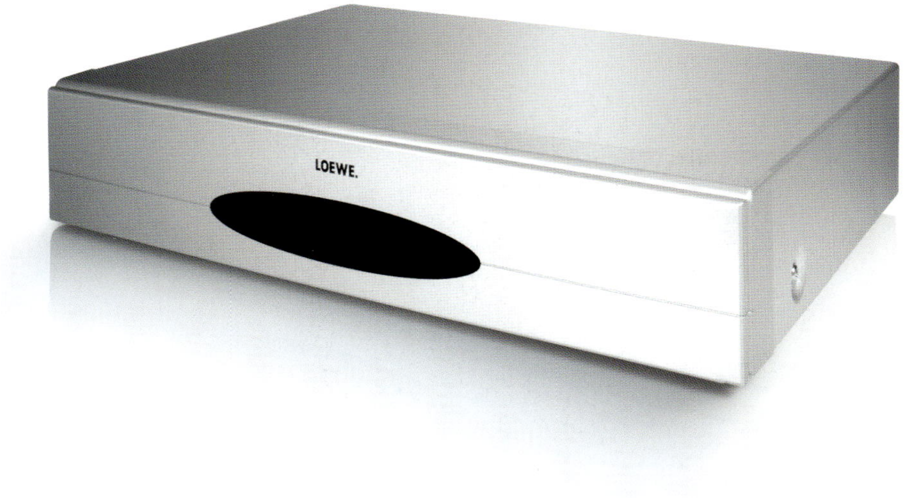

Hersteller Manufacturer
Loewe AG
Kronach, Germany

Produkt Product
Xemix 8122 DA
DVD-Player
DVD player

Design
Phoenix Design
Stuttgart, Germany

Das Design des Xemix 8122 DA DVD-Players von Loewe sowie der nachrüstbaren Frontblenden ist auf andere Loewe Geräte abgestimmt. Das voll integrierte Loewe PCS (Personal Control System) ermöglicht die Steuerung sämtlicher Funktionen mit der Fernbedienung über ein einheitliches Bildschirmmenü am TV-Gerät. Die kompakten Abmessungen dienen bewusst als Lösung für enge Aufstellsituationen.

The design of Loewe's Xemix 8122 DA DVD player and the retrofit front covers match the appearance of all the other Loewe appliances. The fully integrated Loewe PCS personal control system enables all functions to be controlled by the remote control via a standardized on-screen menu on the TV. The compact dimensioning is conceived as a solution to situations when space for positioning becomes tight.

Hersteller Manufacturer
LG Electronics Inc.
Gumi-City, Gyung Buk, Korea

Produkt Product
RZ-48SZ40RB
48″ LCD-Projektionsfernseher
48″ LCD Projection TV

Design
Joe Sung Gu
Ahn Yoo Jung
Corporate Design Center
LG Electronics
Seoul, Korea

Der Projektionsfernseher RZ-48SZ40RB von LG Electronics für Heimkino-Anwendungen besitzt zwei vom Gerät getrennte Lautsprecher, die Stereo-Sound ermöglichen. Das Design ist einfach und klar ohne zusätzlichen Rahmen, der Bildschirm äußerst flach. PIP und POP ermöglichen das gleichzeitige Schauen von 2 bzw. 4 Programmen, zudem bietet das Gerät PC-Direktempfang und einen digitalen Anschluss.

The LG Electronics' RZ-48SZ40RB projection TV for home cinema applications comes with two separate loudspeakers for stereo sound. The design is simple and clear without an additional frame and a very flat screen. PIP and POP enable 2 or 4 channels to be watched at once, and also direct PC reception and a digital connector.

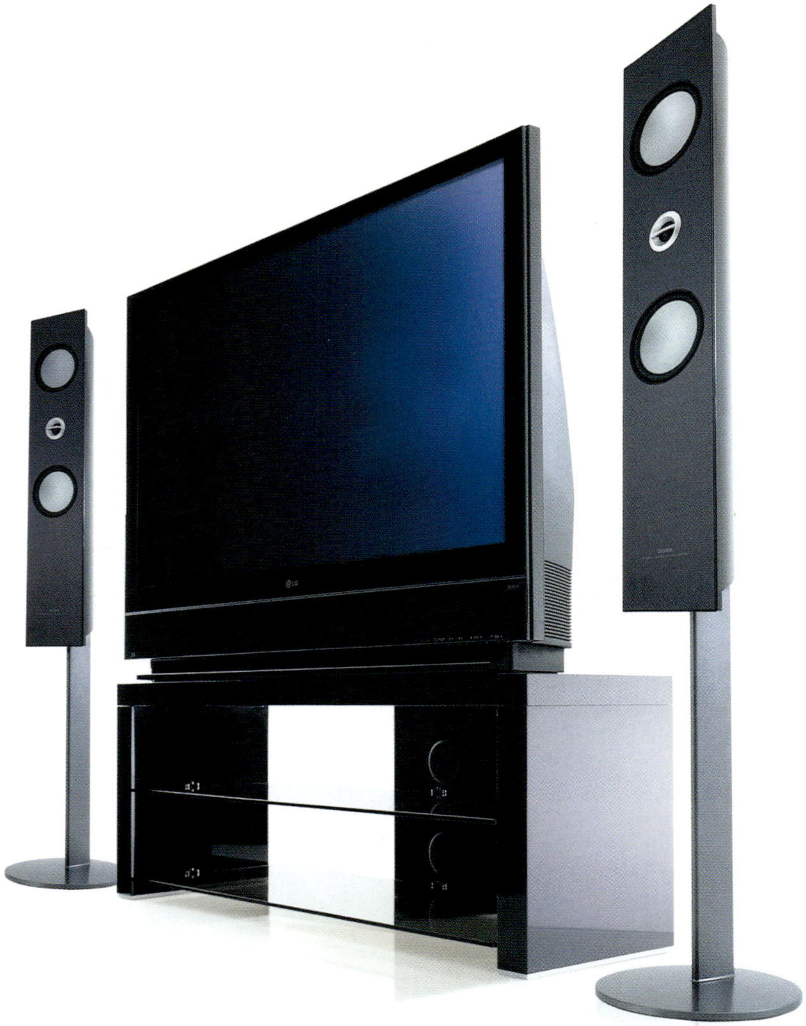

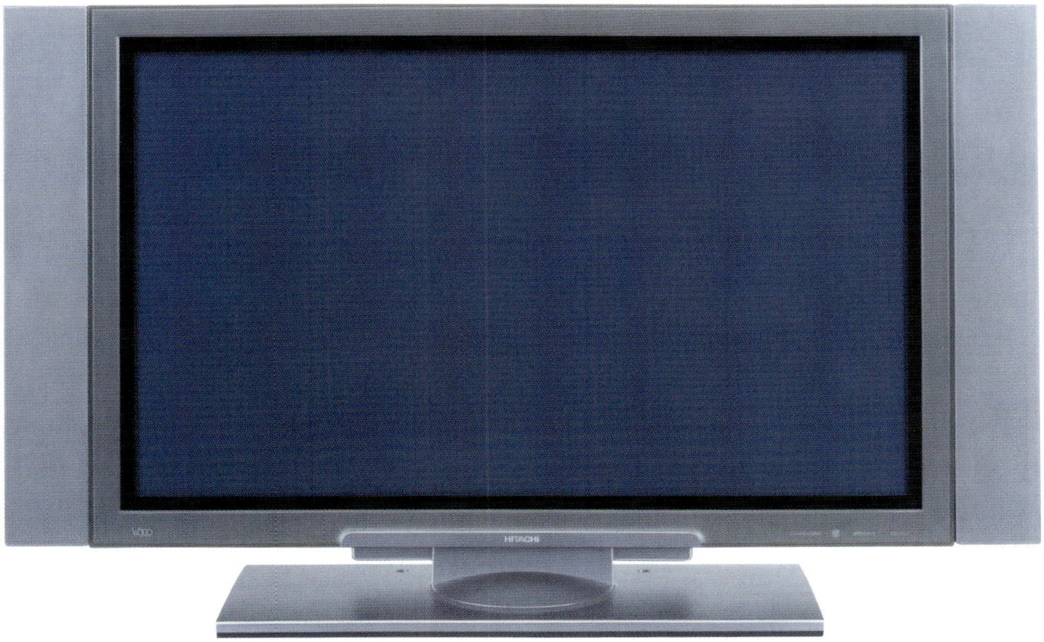

Hersteller Manufacturer
Hitachi Ltd.
Tokio, Japan

Produkt Product
Plasma TV 5000 series
Fernsehgerät mit Plasmabildschirm
TV with plasma display

Design
Akira Omachi
Takashi Yamamoto
Tokio, Japan

Plasma TV 5000 series ist eine weltweit vertriebene Serie von Plasmadisplay-Fernsehgeräten, die verschiedene digitale Medien unterstützen. Das Bildschirmdesign ist funktional, sodass es bei Aufstellung im Wohnbereich oder anderen Innenräumen nicht zu sehr hervorsticht. Der Bildschirmrahmen ist glänzend transparent, um die Glasstruktur des Plasmadisplays optisch zu erweitern. Der untere Abschluss des Bildschirmrahmens wird durch ein Bedienfeld aus Aluminium akzentuiert. Die Serie ist in den Bildschirmgrößen 32 Zoll, 37 Zoll und 42 Zoll verfügbar.

Plasma TV 5000 is a series of televisions with plasma displays marketed worldwide which support a variety of digital media. The screen design is functional so as not to dominate the living area or other interior space in which it is displayed. The frame of the screen is shiny and transparent with the optical effect of expanding the glass structure of the plasma display. The lower edge of the frame is accented with an aluminum panel. The series is available with 32", 37" and 42" screens.

Hersteller Manufacturer
Samsung Electronics Co., Ltd.
Seoul, Korea

Produkt Product
LT-17N2, LT-22N2
17", 22" Wide LCD TV
Flachbildfernseher
LCD TV

Design
Yang Jun-Ho
Won Dong-Hee
VD Design Group
Samsung Electronics Co., Ltd.
Seoul, Korea

Das schmale Design der LCD Fernseher LT-17N2 bzw. LT-22N2 von Samsung betont die Charakteristik des Flachbildfernsehers. Der halbrunde Fuß nimmt die abgerundete Linie des Profils auf, die Anschlüsse auf der Rückseite sind in einem Gehäuse versteckt. Die schraubenlose Konstruktion des Außengehäuses macht den Zusammenbau und die Reparatur leicht.

The slender outlines of the Samsung LT-17N2 and LT-22N2 LCD TV sets highlight the characteristic features of these flatscreen appliances. The semi-circular stand takes up the rounded lines of the profile; all connectors on the rear are tucked away out of sight behind the back panel. The outer skin has no screw connections, making assembly and repair access easy.

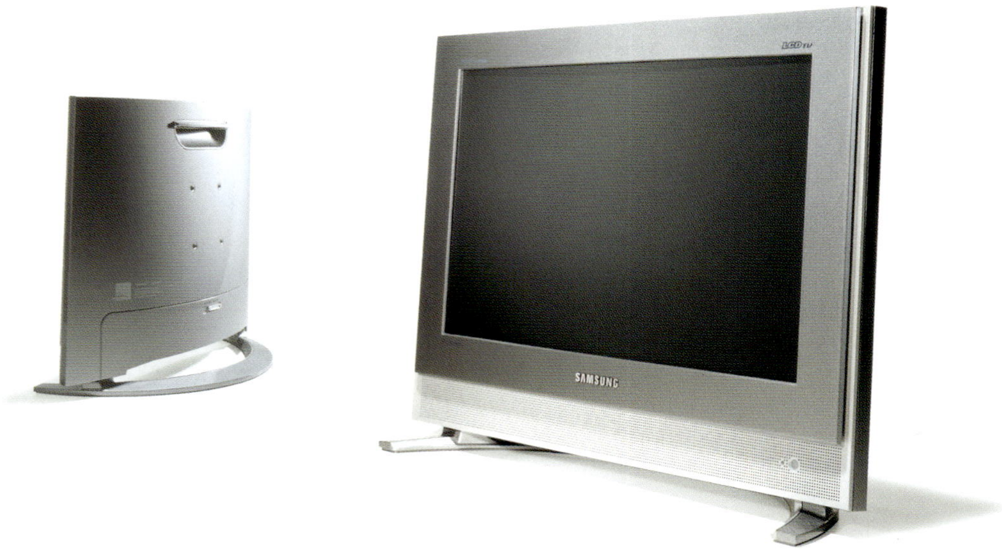

Hersteller Manufacturer
Samsung Electronics Co., Ltd.
Seoul, Korea

Produkt Product
50 inch Flat Panel PDP TV System (SPD 50P4H)
Plasma-TV-System
PDP TV System

Design
Yang Jun-Ho
Kim Young-Jun
VD Design Group
Samsung Electronics Co., Ltd.
Seoul, Korea

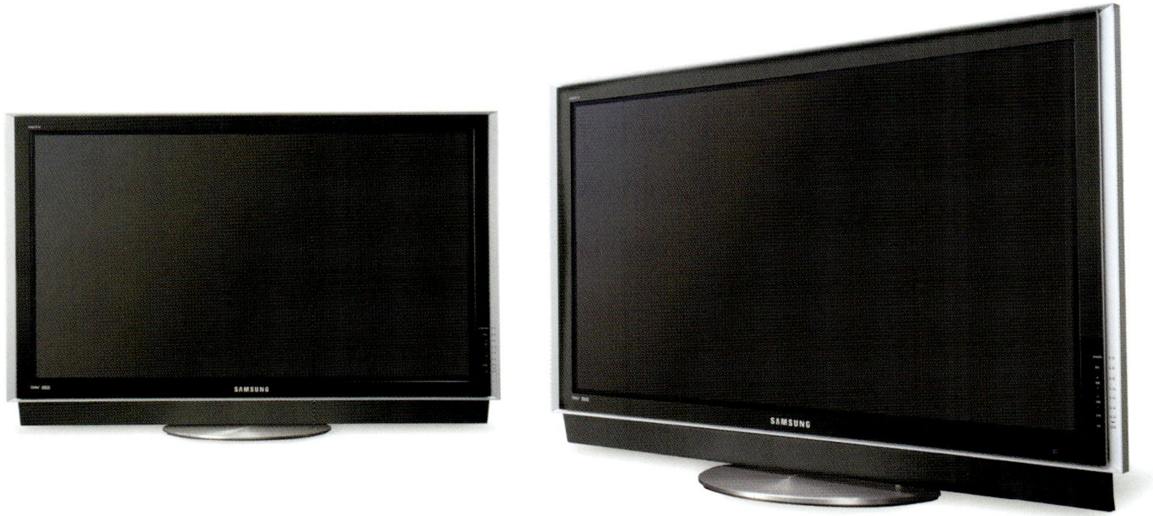

Der 50 inch Flat Panel PDP TV System (SPD 50P4H) von Samsung ist ein hochauflösender Plasma-Fernseher im Frame-Design mit Aluminium und Glas, der auch an der Wand aufgehängt werden kann. Er steht auf einem schlichten runden und drehbaren Fuß, im unteren Bereich wurden zwei äußerst schmale Lautsprecher integriert. Zudem bietet er eine Bildauflösung von 1366 x 768 und dreidimensionalen Dolby-Klang mit dem 5.1-Surround-Lautsprechersystem mit integriertem Verstärker (Dolby Theater System).

Samsung's 50 inch flat panel PDP TV System (SPD 50P4H) is a high-resolution plasma TV set with an aluminum-and-glass frame design that can also be hung from the wall. It has a rotatable round stand and two very slender loudspeakers are incorporated into the lower section. The set offers an HD pixel resolution of 1366 x 768 and 3D Dolby virtual sound with the 5.1 Surround system and integrated amplifier (Dolby Theater System).

Hersteller Manufacturer
Nakamichi Research Pte Ltd.
Singapore, SGP

Produkt Product
VU29 Digital LCD Multimedia Display
LCD-Bildschirm
LCD monitor

Design
Edwin Beck, Ton Borsboom
Nick Field, Ben Rowan
GE/Fitch Singapore P/L
Mechanical Design
Hoe Sei Sing
Vincent Wee
Gao Peng Fei
Richard Soh
Nakamichi Research Pte Ltd.
Singapore, SGP

Der VU29 Digital LCD Multimedia Display verfügt über ein Gasfedergelenk, aufgrund dessen der Bildschirm nach rechts und links, oben und unten in alle erdenklichen Positionen gedreht werden kann, unabhängig davon, ob er an der Wand montiert ist oder auf dem dazugehörigen Fuß. Eine spezielle LCD-Technologie bietet ungewöhnlich weite Blickwinkel und hohe Farbbrillanz.

The VU29 Digital LCD Multimedia Display has a gas spring enabling the screen to be swiveled left or right, up and down in all conceivable angles, irrespective of whether it is hung on a wall or on its slender base. Special LCD technology delivers an unusually wide viewing angle and high-definition images in superb color.

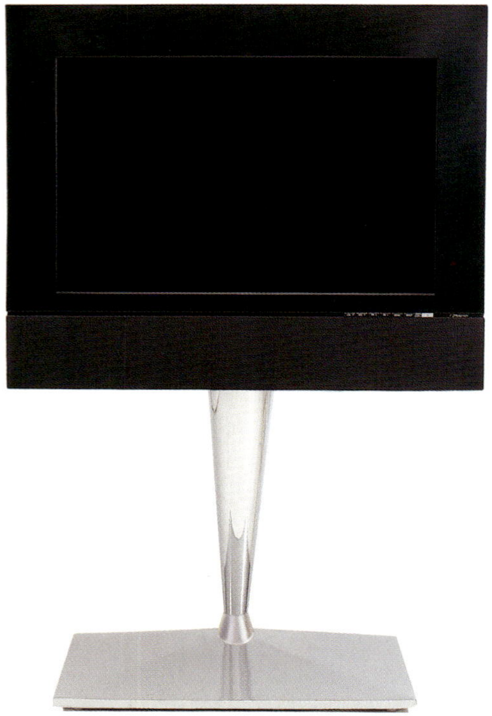

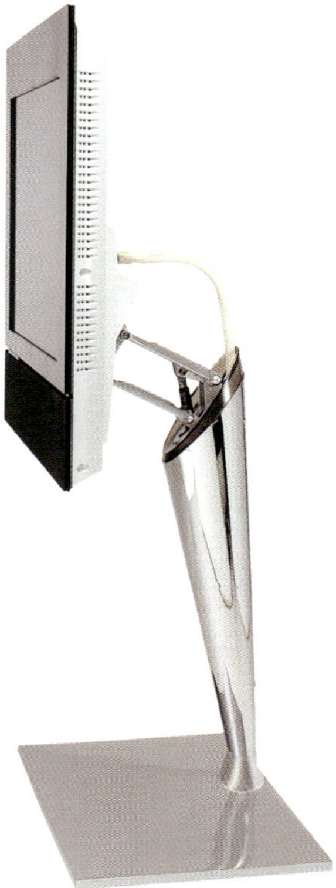

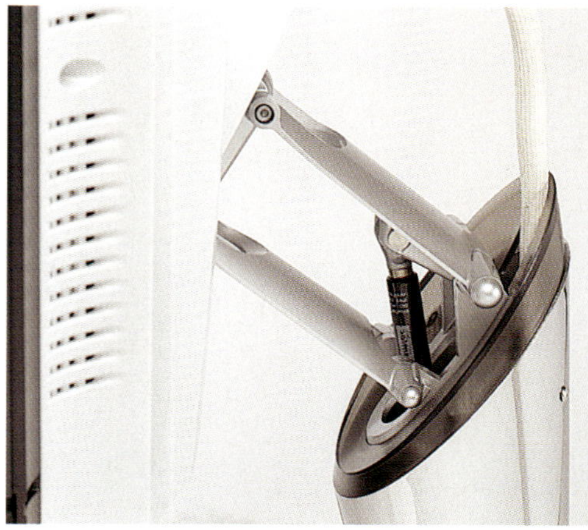

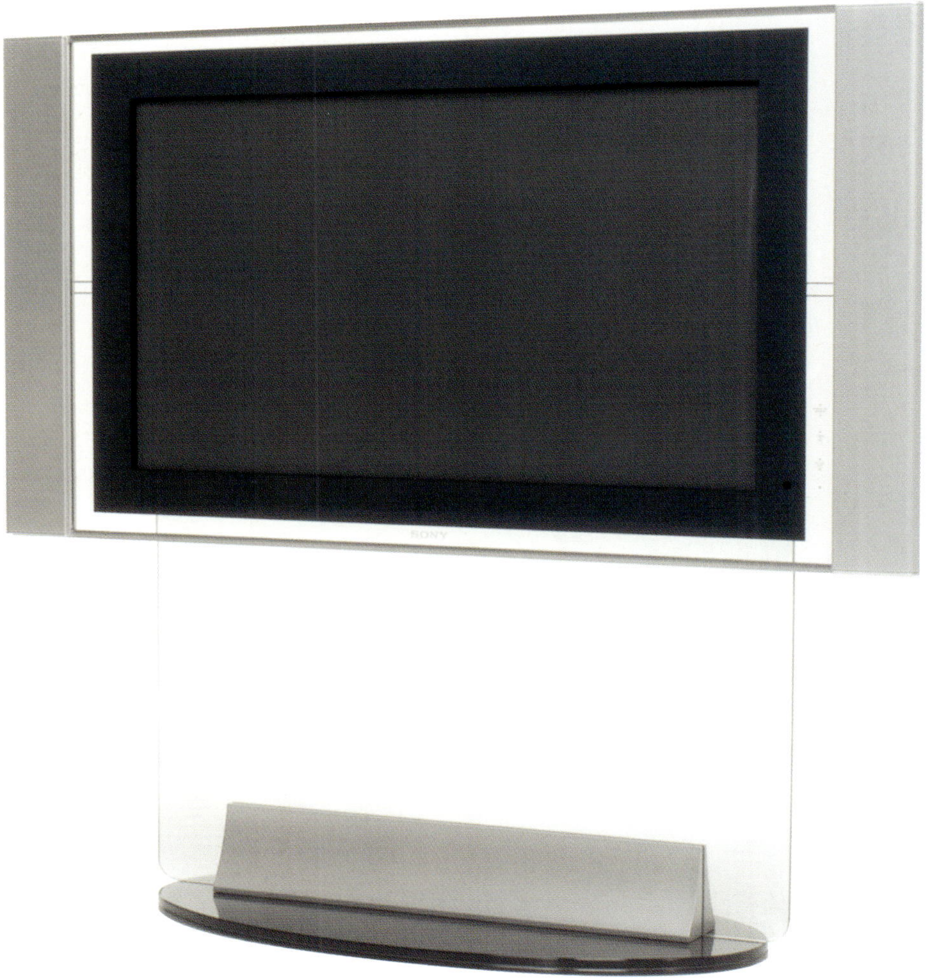

Hersteller Manufacturer
Sony Corporation
Tokio, Japan

Produkt Product
KE-42MR1
Plasma-Fernseher
Plasma TV

Design
Takuya Niitsu
Sony Corporation
Tokio, Japan

Aufgrund des schlanken Designs bietet der Plasma-Fernseher KE-42MR1 flexible Installationsoptionen von der Wandmontage bis zum Glasständer. Die aus strukturellen Gründen notwendige Glasfläche wurde so vergrößert, dass es den Anschein hat, als würde ein an der Wand hängendes Bild in der Luft schweben. Die Verwendung von Aluminium im Gehäuse ermöglicht die flache Gestaltung des Displays. Als Teil des schwebenden Designkonzepts wurden die LED-Anzeige und das Sony-Logo im transparenten Teil der Frontplatte positioniert, wobei das Logo, wenn das Gerät ein- oder ausgeschaltet wird, noch 8 Sekunden lang leuchtet und den Effekt des Schwebens unterstreicht.

By virtue of its slim design, the KE-42MR1 plasma TV can be installed in a variety of ways, ranging from wall mounting to glass stand. The glass panel required for structural reasons is extended such that it looks like a picture floating in the air on the wall. The use of aluminum in the housing makes such a flat design possible. As part of the floating design concept, the LED and Sony logo are positioned in the transparent part of the front plate – when the TV is switched off or on, the logo is illuminated for a period of eight seconds underlining the floating effect.

Hersteller Manufacturer
Philips Electronics Industries (Taiwan) Ltd.
Taoyuan, Taiwan, ROC

Produkt Product
Philips 190B4 LCD Monitor
LCD-Monitor
LCD monitor

Design
Angus Dick
Philips Design

Der Philips LCD-Monitor 190B4 gehört zu einer Reihe, die in 15, 17, 19, 20 und 23 Zoll Ausführung erhältlich ist. Der Monitor wurde für die geschäftliche Nutzung in einer Back- oder Front-Office-Umgebung konzipiert. Die Sandwich-Farbkonstruktion in zwei Tönen sowie die fließenden Linien auf der Rückseite sorgen für ein schlankes Erscheinungsbild. Die runde Basis ist besonders kompakt und harmoniert mit den Gehäusedetails. Das Kabelmanagement wurde in das Basisdesign integriert, um eine aufgeräumte Schreibfläche zu gewährleisten.

The Phillips 190B4 LCD monitor belongs to a series of screens, features 15", 17", 19", 20" and 23" versions. The display was conceived for commercial use in back- or front-office environments. The 2-tone sandwich color construction and the flowing lines on the back give it a slim and sleek appearance. The round base is particularly compact and harmonizes with the casing details. Cable management is integrated into the base design leaving the desktop clear and uncluttered.

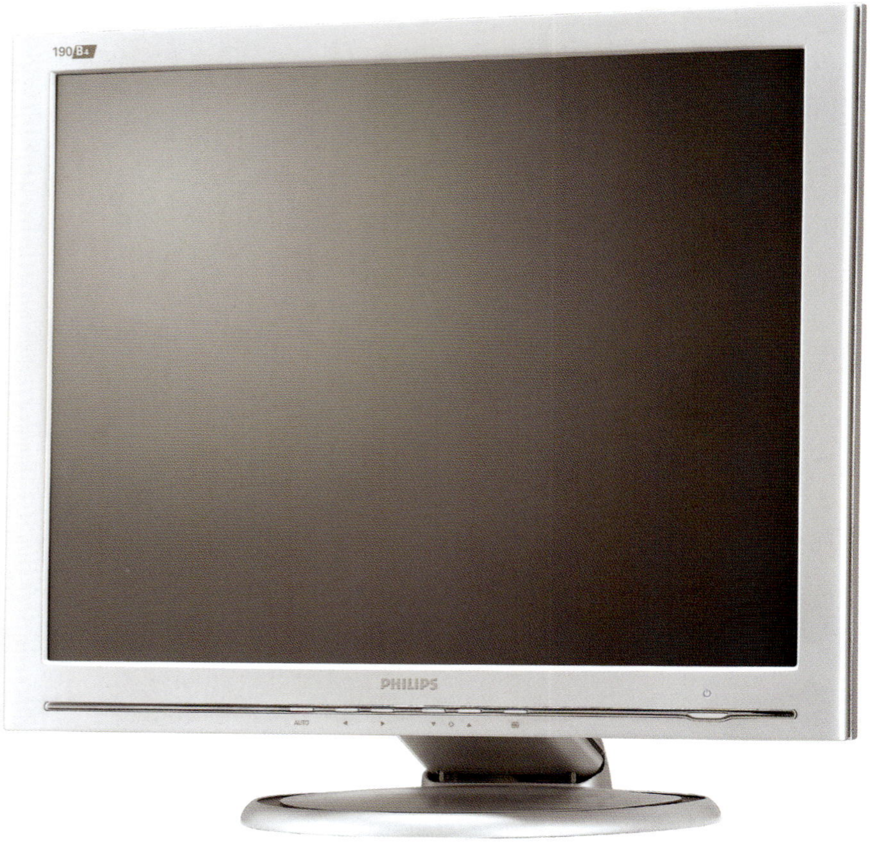

Hersteller Manufacturer
BENQ Corporation
Taipei, Taiwan, ROC

Produkt Product
FP785 LCD Monitor
LCD-Monitor
LCD monitor

Design
BENQ Corporation
Lifestyle Design Center
Taipei, Taiwan, ROC

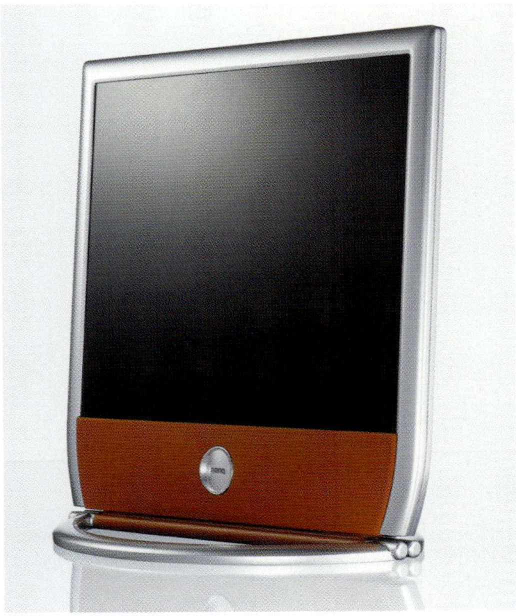

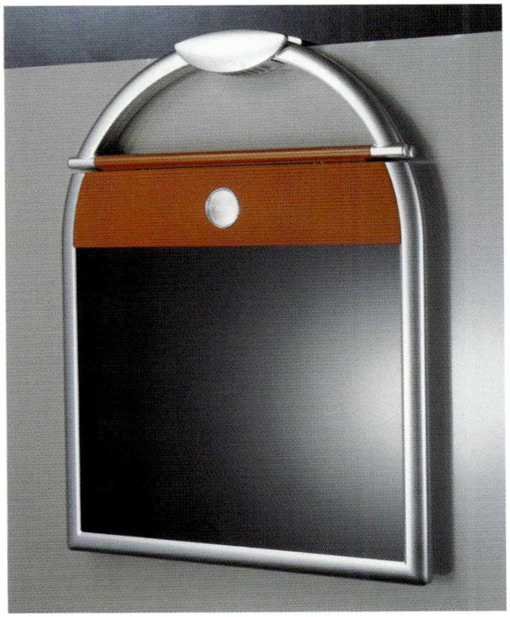

Das ungewöhnliche Design des 17-Zoll LCD-Monitors FP785 ist von der Form einer Damenhandtasche inspiriert und dementsprechend auch für eine weibliche Zielgruppe konzipiert. Der nur 3,5 cm dicke Monitor kann am halbrunden Bügel getragen werden und bei Gebrauch formt er, nach vorne geklappt, den Standfuß des Monitors oder er kann dank einer speziellen Vorrichtung auch gehängt werden.

The unusual design of the 17″ FP785 LCD monitor draws on ladies handbags for its inspiration and is thus also intended for female target groups. A mere 3.5 cm thick, the monitor can also be carried on its semi-circular double rotary shaft that, for utilization, folds forward to create a monitor footprint. Thanks to a special hanging hook accessory and its 180° rotation display, the monitor can also be hung on the wall.

Hersteller Manufacturer
IBM Corp.
Armonk, NY, USA

Produkt Product
IBM ThinkVision L170p
Flachbildschirm
Flat panel monitor

Design
IBM Design
Research Triangle Park, NC, USA

Der Flachbildschirm ThinkVision L170p (17-Zoll LCD) zählt hinsichtlich seiner ergonomischen Eigenschaften und seiner grafischen Performance zu den Spitzenmodellen der ThinkVision Reihe von IBM. Durch die Abschrägungen an den Seiten wirkt der Bildschirm sehr flach, was mit der Formgebung des Standfußes korrespondiert. Er kann in einem Winkel von bis zu 135 Grad gedreht werden und lässt sich auch auf der vertikalen Achse mit einem Spielraum von ca. 10 cm nach den individuellen Bedürfnissen des Nutzers ausrichten. Die Bedienleiste in Form einer Schlangenlinie am unteren Rand setzt einen diskreten ironischen Kontrast zur ansonsten sachlich funktionalen Gestaltung. Dank seiner Hybrid-Konnektivität eignet sich der Flachbildschirm sowohl für den Anschluss an analoge als auch digitale Systeme.

IBM's ThinkVision L170p (17" LCD) flatpanel monitor is one of the premier flat panel monitors in the Think Vision series thanks to its ergonomic features and graphic performance. The deep chamfers on the sides create a very flat appearance, in keeping with the shape of the stand. The monitor can swivel up to 135° and also offers a 100 mm vertical adjustment range to adapt to user needs. The controls along the bottom of the frame mimic a snake, creating a discretely ironic contrast to the otherwise functionally purposeful design. Thanks to its hybrid connectivity, the L170p can be connected to both analogue and digital systems.

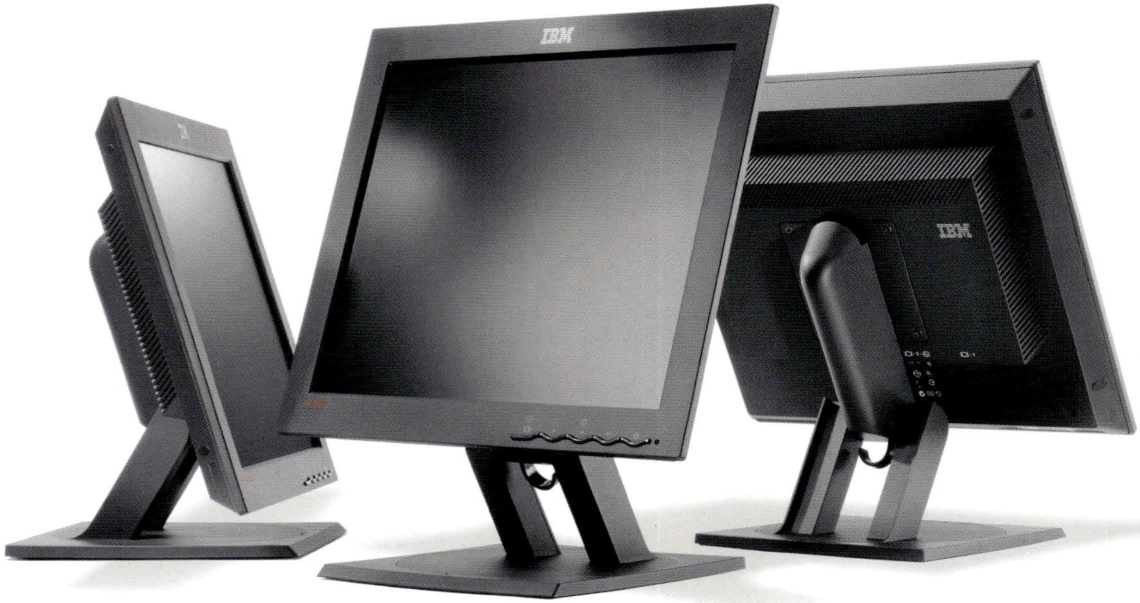

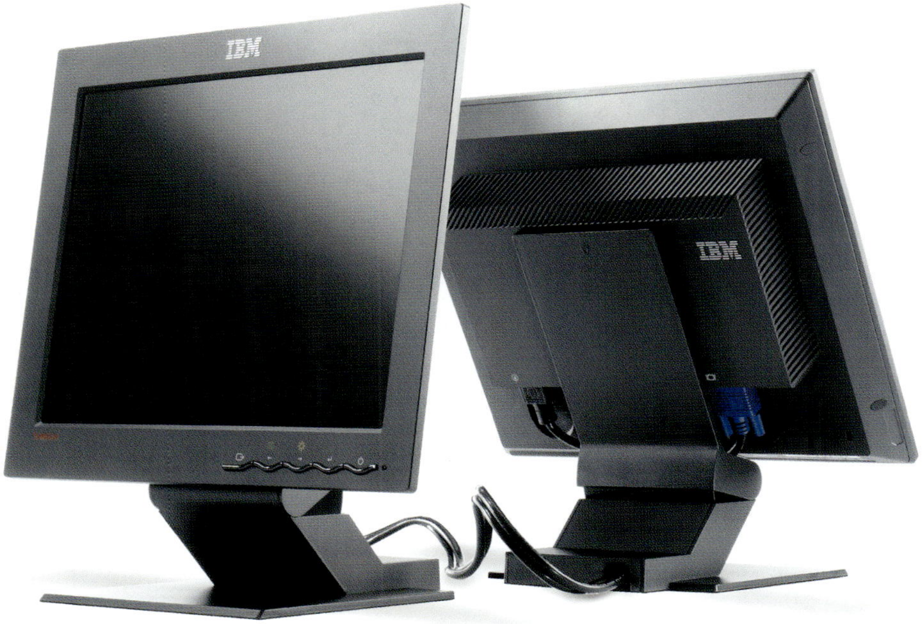

Hersteller Manufacturer
IBM Corp.
Armonk, NY, USA

Produkt Product
IBM ThinkVision L150
Flachbildschirm
Flat panel monitor

Design
IBM Design
Research Triangle Park, NC, USA

Der IBM ThinkVision L150 Flachbildschirm vermittelt durch seinen schrägen Neigungswinkel die Anmutung eines frei im Raum schwebenden Blattes Papier. Der integrierte Zylinder-Neigungsmechanismus gewährleistet eine feine ergonomische Anpassung, wobei die Formgebung visuell auf den Neigungswinkel verweist. Die Kabel werden vom Bildschirm durch den schlichten und eleganten Fuß nach unten geführt. Die Abschrägungen an den Seiten lassen den Bildschirm wie eine dünne Scheibe erscheinen.

The IBM ThinkVision L150 flat panel display angles away from the user making the entire display present itself to the user like a floating sheet of glowing information. An integrated cylindrical tilt mechanism provides critical ergonomic adjustment and gives a visual cue to the range of tilt provided by the device. Cables from the head unit are passed neatly through this simple yet elegant stand. The deep chamfers around the sides of the display help to remove visual mass and give the panel the appearance of a thin-edged plate.

Hersteller Manufacturer
IBM Corp.
Armonk, NY, USA

Produkt Product
IBM ThinkVision L200p
Flachbildschirm
Flatpanel monitor

Design
IBM Design
Research Triangle Park, NC, USA

Der ThinkVision L200p Flachbildschirm ist das High-End-Gerät der Spitzenmodelle der ThinkVision Reihe von IBM. Der Monitor hat eine Bildschirmdiagonale von 20,1 Zoll und eignet sich besonders für den Einsatz im grafischen Bereich. Die Handhabung des Monitors ist leicht, der Benutzer kann schnell auf individuell eingestellte Funktionen zugreifen oder von den flexiblen Aufstellmöglichkeiten Gebrauch machen. Optional kann der Bildschirm auf dem IBM Radial Arm angebracht werden, was eine fast unbegrenzte Beweglichkeit des Bildschirms in alle Richtungen, auch hochkant, ermöglicht. Die Bedienleiste in Form einer Schlangenlinie am unteren Rand setzt einen diskreten ironischen Kontrast zur ansonsten sachlich funktionalen Gestaltung.

IBM's ThinkVision L200p flatpanel monitor is the high-end flat panel monitor in the ThinkVision series. The monitor boasts a 20.1" diagonal screen and is ideal for graphic applications. The monitor is easy to operate, and users can enjoy using individually set functions or the flexible positioning. The monitor can optionally be placed on an IBM Radial Arm, delivering almost unlimited flexibility in all directions, even in portrait format. The controls along the bottom of the frame mimic a snake, creating a discretely ironic contrast to the otherwise functionally purposeful design.

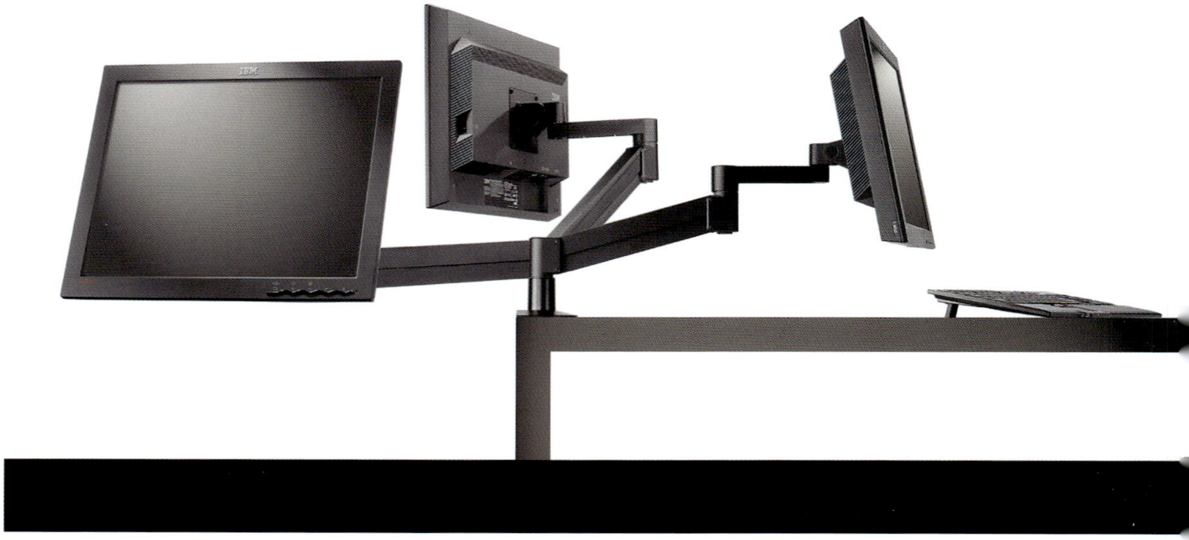

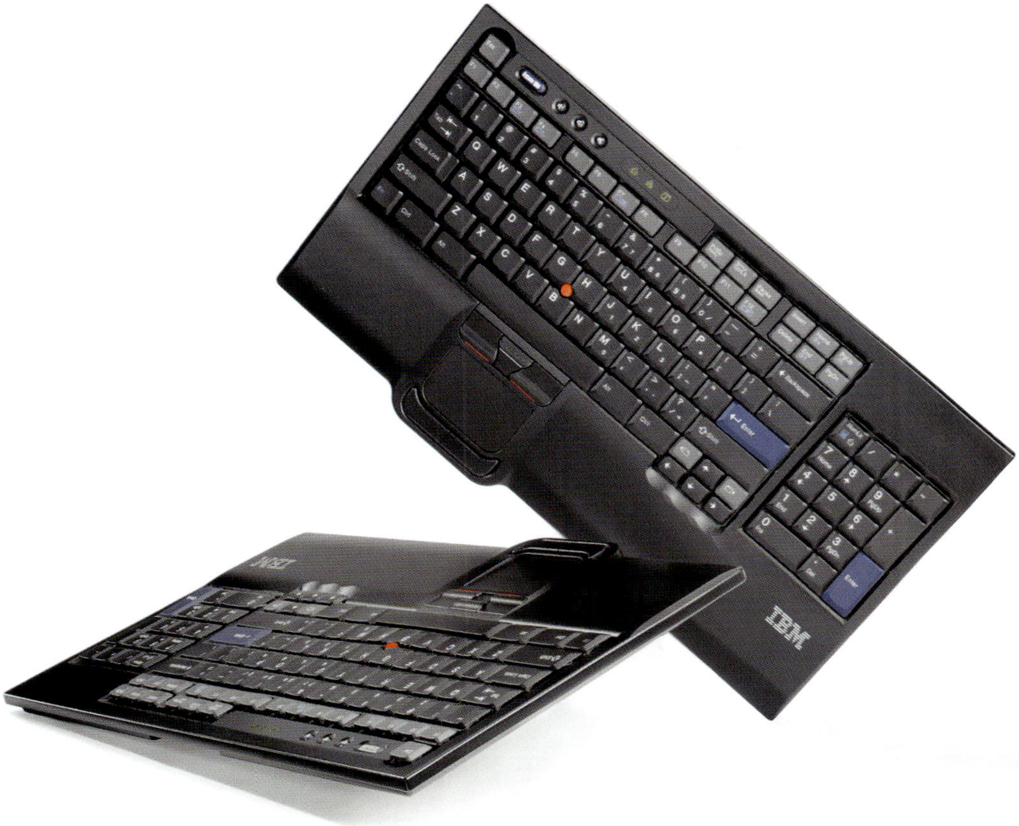

Hersteller Manufacturer
IBM Corp.
Armonk, NY, USA

Produkt Product
IBM USB Keyboard with UltraNav
Tastatur
Keyboard

Design
IBM Design
Research Triangle Park, NC, USA

Das IBM USB Keyboard with UltraNav wurde für Notebook User konzipiert, die ihr Notebook regelmäßig in einer Docking Station nutzen und es deshalb vorziehen, in einer komfortablen Position zu arbeiten. Bei der Designentwicklung wurde besonderen Wert darauf gelegt, dass sich die Desktop-Tastatur in ihrer Anordnung und Gestaltung möglichst nicht von der Notebook-Tastatur unterscheidet (bis auf das zusätzliche Zahlenfeld), da der beim Nutzer automatisch auftretende Gewöhnungseffekt nicht gestört werden sollte. Die integrierte Navigationsmöglichkeit macht zudem die Verwendung einer Maus überflüssig. Der Kursor wird über die Fingerspitzen mittels eines UltraNav-Touch-Pads und mit einem Track Point Stick bewegt.

The IBM USB Keyboard with UltraNav was designed for Notebook users who regularly use their Notebook in a docking station and thus prefer to work in a comfortable position. The design emphasizes having a desktop keyboard which is as close as possible to the layout and design of the Notebook keyboard (apart from the additional numerical zone), meaning that the user does not need to alter working reflexes and habits. The integrated navigation features also render mouse use redundant. The cursor can be moved with fingertips on an UltraNav Touch Pad or with a Track Point Stick.

Hersteller Manufacturer
Darfon Electronics Corp.
Taoyuan, Taiwan, ROC

Produkt Product
BenQ AM805
Drahtlose Tastatur und Maus
Wireless multimedia keyboard & mouse kit

Design
Chun-Chung Huang
Jihuang Chen
Shu-Wen Tzeng
Darfon Electronics Corp.
Taoyuan, Taiwan, ROC

Das drahtlose BenQ AM805 Keyboard & Mouse Kit kann hinsichtlich des Eingabewinkels der Hand zur Tastatur mit vier unterschiedlichen Positionen individuell angepasst werden. Die geschwungene Formgebung des Keyboards gewährleistet zudem eine ermüdungsfreie Eingabe. Eine spezielle Scherenkonstruktion der Tasten ermöglicht ein vibrationsfreies leises Schreiben sowie einen sicheren Stand der Tastatur. Die hochempfindliche dünnwandige Maus registriert jede Bewegung der Hand.

The wireless BenQ AM805 keyboard and mouse kit can be set in four different positions by adjusting the feet to support the angle of the user's hands. The curved form of the keyboard helps prevent fatigued wrists. Special scissors plunger key-cap technology permits noiseless, vibration-free writing and stabilizes the keyboard. The highly sensitive thin-walled mouse registers all hand movements.

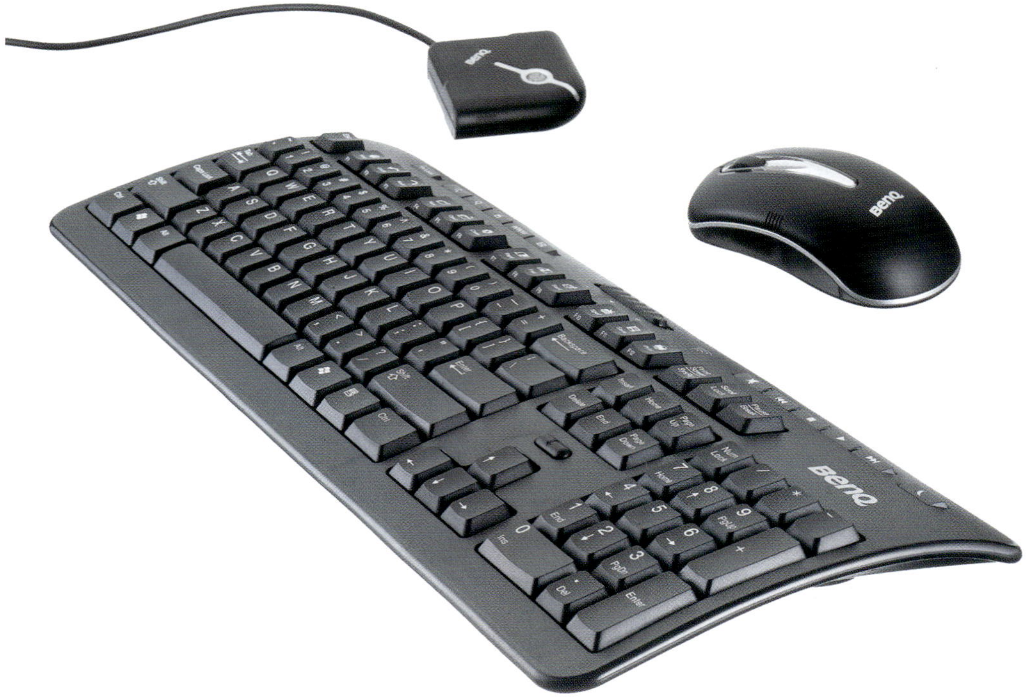

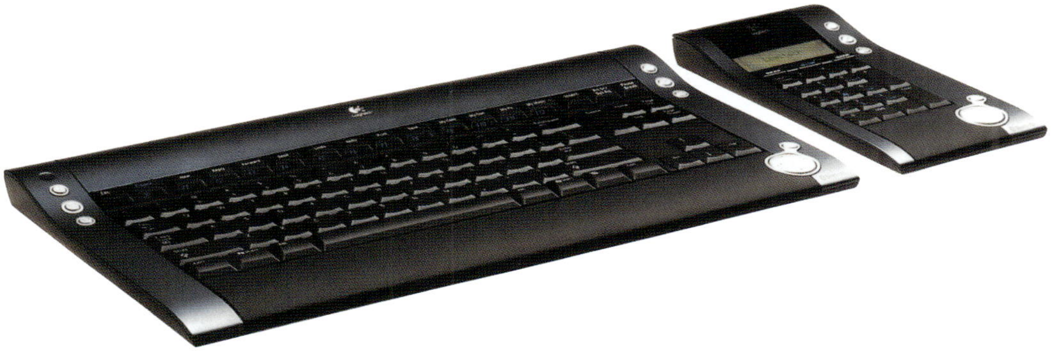

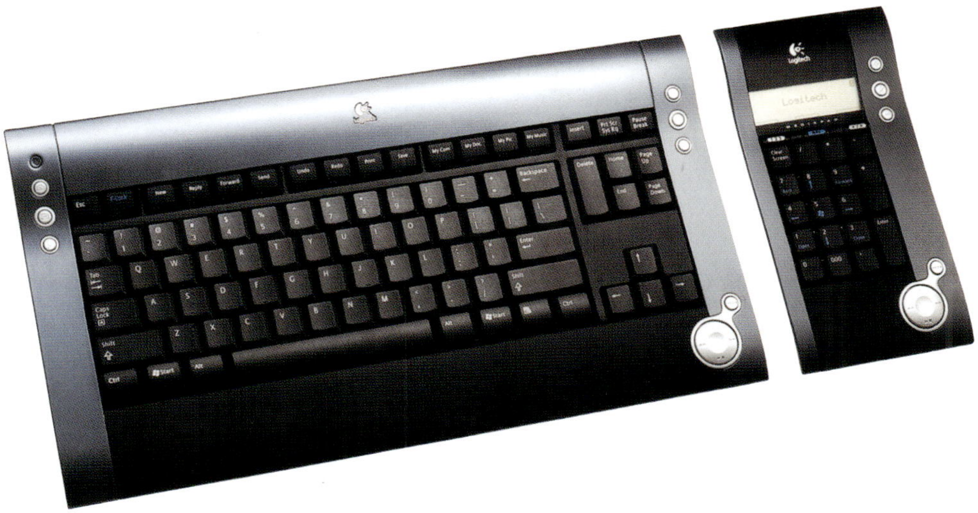

Hersteller Manufacturer
Logitech USA
Fremont, CA, USA

Produkt Product
Logitech diNovo
Media-Tastatur
Media keyboard

Design
Cathal Loughnane
Diarmuid McMahon
Peter Sheehan
Keith Tritschler,
Design Partners
Bray, Co. Wicklow, Ireland

Die Logitech diNovo Media-Tastatur zeichnet sich durch ihr absolut flaches und kompaktes Design aus und ist als Ergänzung zu den immer häufiger verwendeten modernen Flachbildschirmen gedacht. Die Gestaltung des Tastaturfeldes ist klar und funktionsbezogen, die ergonomische, leicht geschwungene Formgebung ermöglicht eine komfortable Betätigung. Das Zahlenfeld ist eine eigenständige Einheit, sodass die Maus näher an der Tastatur bedient werden kann. Zum Logitech diNovo Media Desktop gehören vier Komponenten: die Tastatur, eine aufladbare optische Maus, ein Multifunktions-MediaPad und die in das Aufladegerät der Maus integrierte Bluetooth-Technologie.

The Logitech diNovo Media keyboard stands out for the fact it is as absolutely flat as possible, boasts a compact design and is conceived for use with increasingly popular, modern flat panel monitors. The design of the keyboard is clear and functional with an ergonomic, slightly curved form which makes it comfortable to use. The numeric keypad zone can be slid aside so that the mouse can be used in closer proximity to the keyboard. The Logitech di Novo Media Desktop consists of four components: the keyboard, a rechargeable optical mouse, a multifunctional MediaPad, and a Bluetooth wireless hub integrated in the mouse's recharging module.

Hersteller Manufacturer
Logitech USA
Fremont, CA, USA

Produkt Product
Logitech iO Digital Pen
Digitale papierbasierte Schreiblösung
Cordless digital pen

Design
Mathew Bates
Eugene Canavan
Peter Sheehan
Design Partners
Bray, Co. Wicklow, Ireland

Der iO Digital Pen von Logitech erfasst handschriftliche Informationen während des Schreibens auf Papier und speichert diese mit Hilfe seines 1 MB Flash Memorys. Die gespeicherten Aufzeichnungen können dann auf einen PC übertragen werden, in dem der Digital Pen in den USB-Anschluss gesteckt wird. Bei der Designentwicklung wurde besonderen Wert darauf gelegt, dass der Digital Pen den eigentlichen Charakter eines Stiftes behält, zudem wurden ergonomische Kriterien zugrunde gelegt.

Logitech's iO Digital Pen remembers handwritten information while you write on paper and stores this with the help of a 1 MB flash memory. The stored recordings can then be transferred to a PC by simply slotting the digital pen in the USB cradle. In designing the digital pen the focus was on retaining the character of a regular pen while also taking ergonomic factors into account.

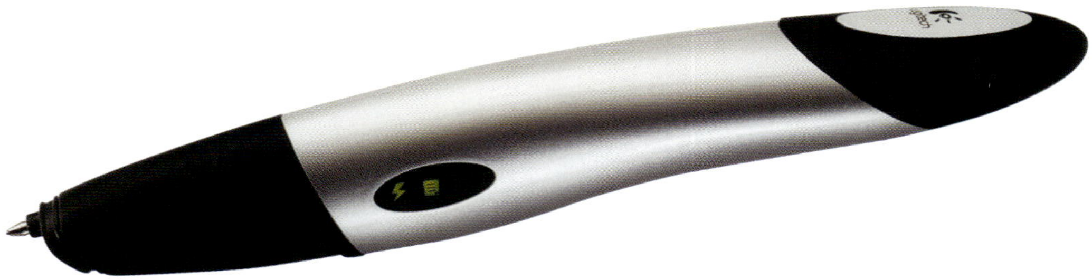

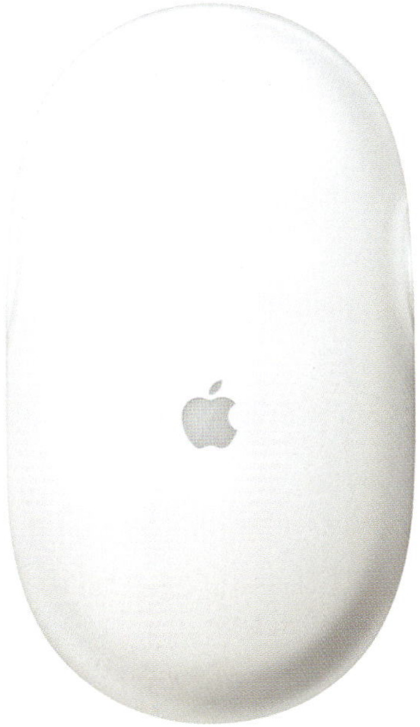

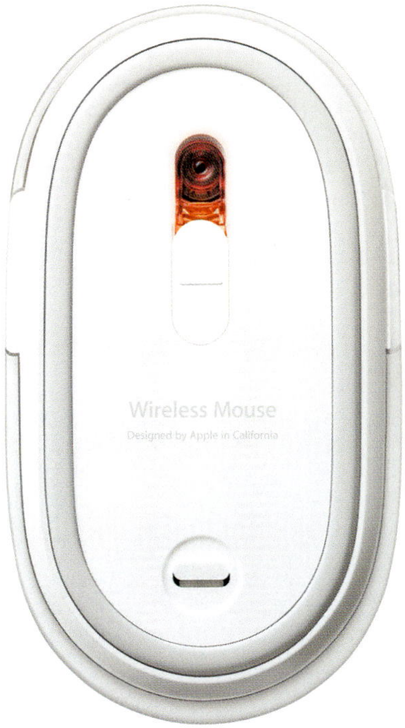

Hersteller Manufacturer
Apple Computer Inc.
Cupertino, CA, USA

Produkt Product
Apple Wireless Mouse
Drahtlose Maus
Wireless Mouse

Design
Bart Andre, Danny Coster
Daniele De Iuliis, Richard Howarth
Jonathan Ive, Duncan Kerr
Matthew Rohrbach, Doug Satzger
Cal Seid, Christopher Stringer, Eugene Whang
CAD Team: Carlos Ragudo, Carter Multz
Christopher Hood, Fred Simon
Irene Chan-Jones, Ken Provost, Mas Watanabe
Apple Computer Inc.
Cupertino, CA, USA

Mit der Apple Wireless Mouse können die Vorteile der Bluetooth-Technologie optimal genutzt werden. In einem Umkreis von 10 m kann mit dem Computer ohne Kabel kommuniziert werden. Eine spezielle Apple Software vermeidet Interferenzen mit anderen kabellosen Geräten und drahtlosen AirPort Netzwerken. Die Maus ist mit einem optischen Mechanismus ausgestattet, der für eine präzise Zeigersteuerung auf den meisten Oberflächen sorgt. Das komfortable und benutzerfreundliche Design der Maus passt sich der Hand optimal an, sie ist sowohl für Rechts- als auch Linkshänder geeignet.

The Apple Wireless Mouse enables you to optimally use the advantages of Bluetooth technology. The mouse can communicate with the computer cordlessly in a radius of 10 m from it. Special Apple software avoids interferences with other cordless devices and cordless AirPort networks. The mouse possesses an optical mechanism that ensures precise tracking on most surfaces. The comfortable and user-friendly Mouse design fits the hand sweetly and the top shell is shaped for ambidextrous use.

Hersteller Manufacturer
BENQ Corporation
Taipei, Taiwan, ROC

Produkt Product
BenQ U-Disk
High-Speed-Flash Laufwerk
High-speed Flash disk drive

Design
BENQ Corporation
Lifestyle Design Center
Taipei, Taiwan, ROC

BenQ U-Disk ist ein sehr kompaktes USB-kompatibles High-Speed-Flash Laufwerk, das speziell für Geschäftsleute entwickelt wurde, die auf den Austausch größerer Datenmengen zwischen Computern angewiesen sind. Das Gerät ist stoßfest und robust, der herausschiebbare USB-Anschluss macht eine zusätzliche Schutzkappe überflüssig – ein wichtiger Aspekt, da diese üblicherweise eingesetzten Schutzkappen während des Gebrauchs des Geräts gerne verlegt werden. Das Gehäuse hat eine schwarze Gummibeschichtung, die zusätzlich den Bedienkomfort verbessert.

BenQ U-Disk is a very compact USB-compatible high-speed Flash disk drive designed specially for business-people and professionals who need data portability and to shift large volumes of data back and forth between computers. The drive is shockproof and robust, while the retractable USB connector mechanism renders the conventional protective caps obsolescent – an important feature, as the latter tend to get lost. The aluminum body is encased in black rubber, making the unit easier to hold and handle.

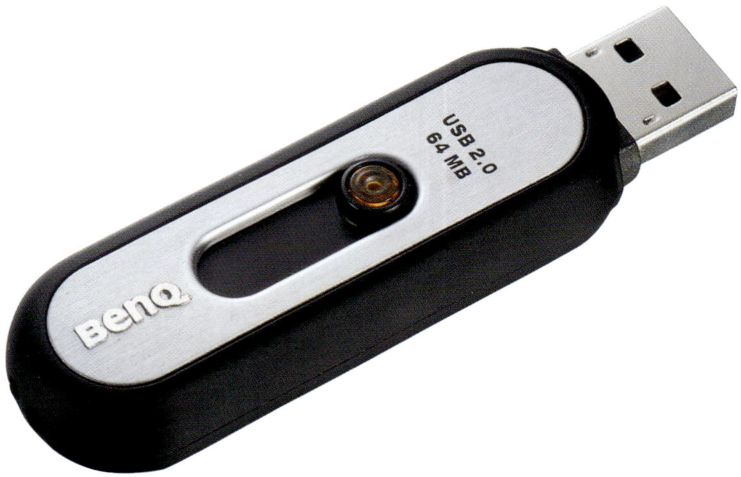

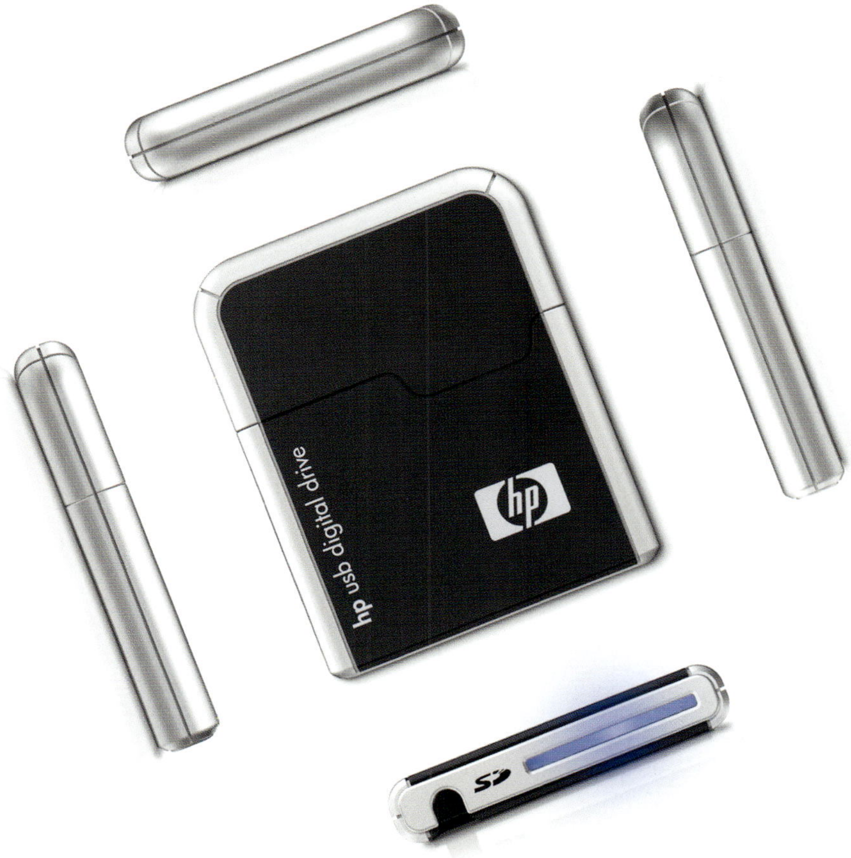

Hersteller Manufacturer
Hewlett Packard
Houston, Texas, USA

Produkt Product
HP USB Digital Drive
USB-Laufwerk
Digital USB Drive

Design
Paul Doczy, Jon Harris
Kevin Massaro, Mark Solomon
Stacy Wolff, Memphis Yin
HP Mobile Industrial Design Group
Houston, Texas, USA

Das digitale USB-Laufwerk in Taschenformat für den schnellen Datentransfer zwischen verschiedenen Systemen ist so kompakt, dass es in bestimmte HP Notebooks hineinpasst. Es hat eine Speicherkapazität von 128 MB, 256 MB oder 512 MB für Fotos, Musik und Daten und eine ausklappbare Kabellösung, die den Anschluss an unterschiedlichste Computer vereinfacht.

This pocket-sized digital USB drive ensures swift data transfer between different computer systems. It is so compact that it even fits into some HP notebooks. The drive can provide storage capacity of 128 MB, 256 MB or 512 MB – ideal for photos, music and data. It boasts a flip-out cable integrated into the unit which allows access to difficult connection areas in a variety of computers.

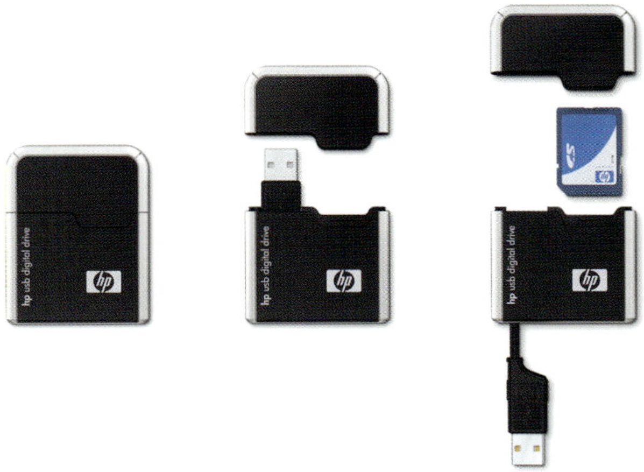

Hersteller Manufacturer
Logitech
Vancouver, WA, Canada

Produkt Product
Logitech Mobile Cordless Cell Phone Headset
Kopfhörer
Headset

Design
Ziba Design
Portland, OR, USA

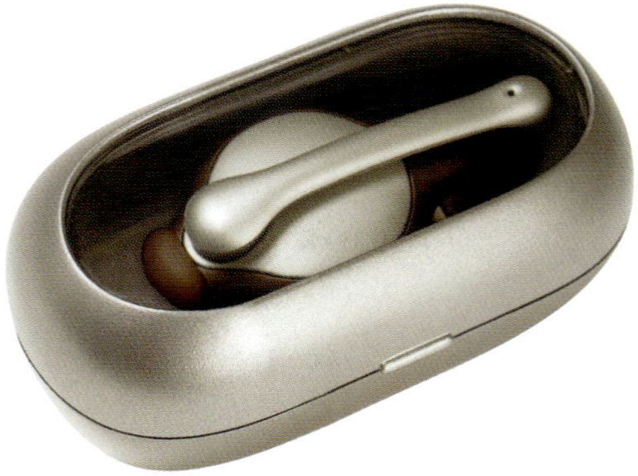

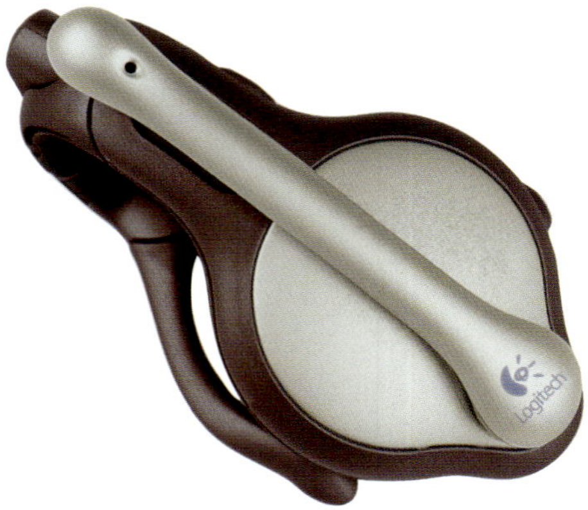

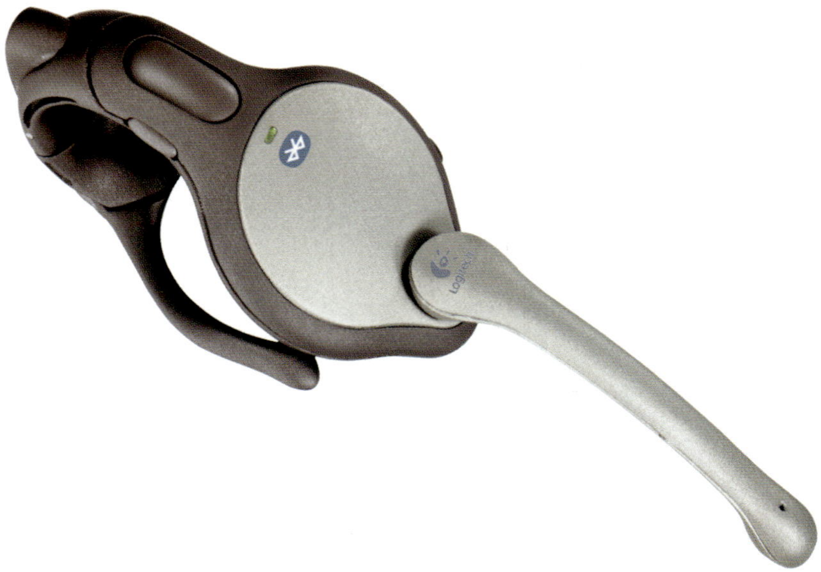

Das Logitech Mobile Cordless Cell Phone Headset erleichtert aufgrund des Nichtvorhandenseins jeglicher Kabelverbindungen insbesondere das synchrone Arbeiten mit anderen digitalen Eingabemedien wie z.B. Laptops etc., zudem ist es mit den Spracherkennungssystemen der gängigsten Mobiltelefone kompatibel. Alle Komponenten des Produkts, die die Haut berühren, sind ergonomisch optimiert. Die gummierte Halterung, die das Ohr umschließt, ist durch eine Mechanik flexibel auf unterschiedlichste Ohrtypen einstellbar und erleichtert das Auf- und Absetzen des Gerätes. Durch die Verwendung des Materials Gummi, die organische Gesamtform und eine metallische Lackierung werden die Attribute menschlich und professionelle Funktion miteinander verbunden.

The absence of connection cables means the Logitech Mobile Cordless Cell Phone can be used simultaneously with other digital recording media such as a laptop and it is, moreover, compatible with the voice recognition systems of the most popular mobile phones. All the product components which come into contact with the skin are ergonomically optimized. The rubberized clip which fits snugly behind the ear and clicks open at the press of a button features a mechanism which allows it to be adjusted for different ear shapes and makes the headset light to wear. Use of rubber as well as the overall organic shape and metallic paint give the design a human as well as a professionally functional feel.

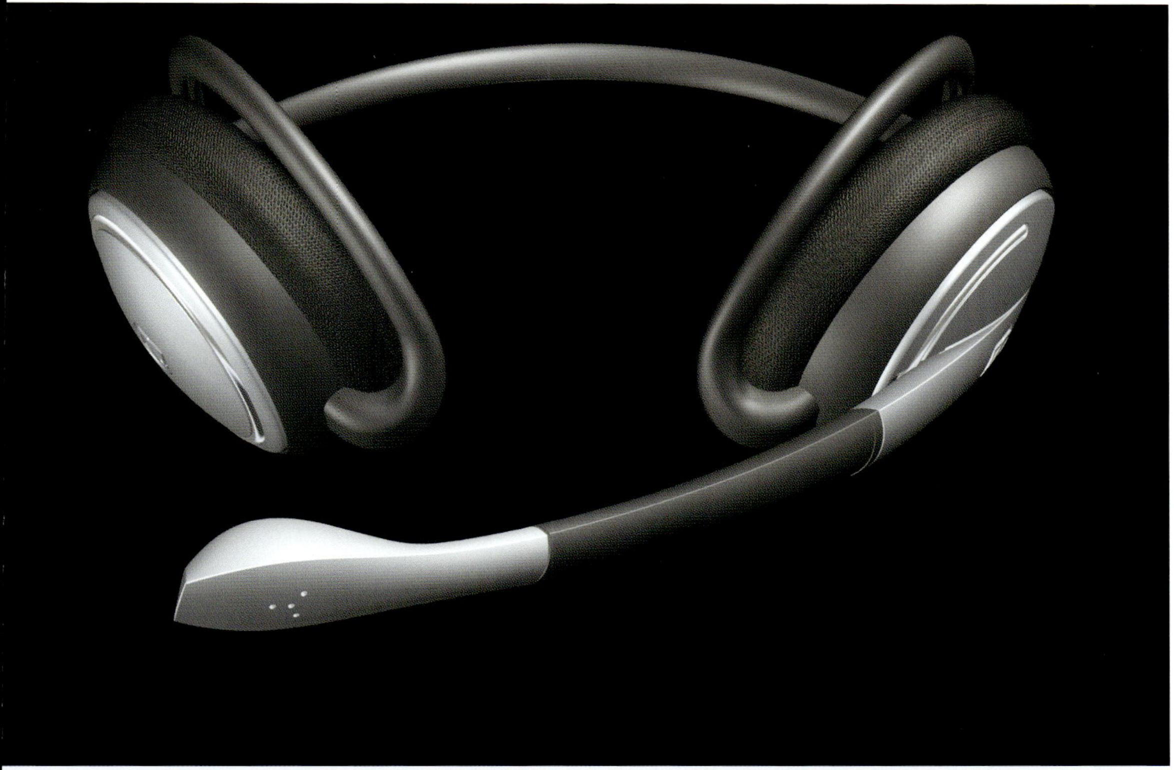

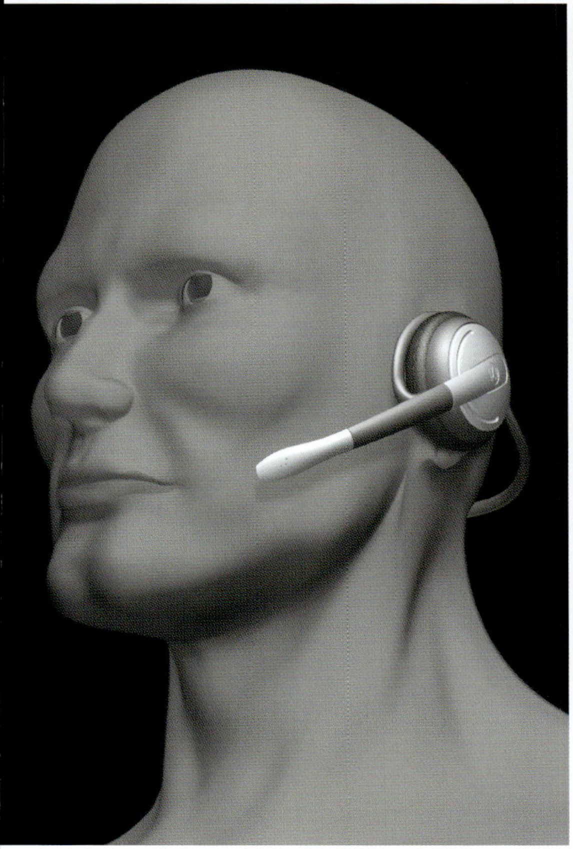

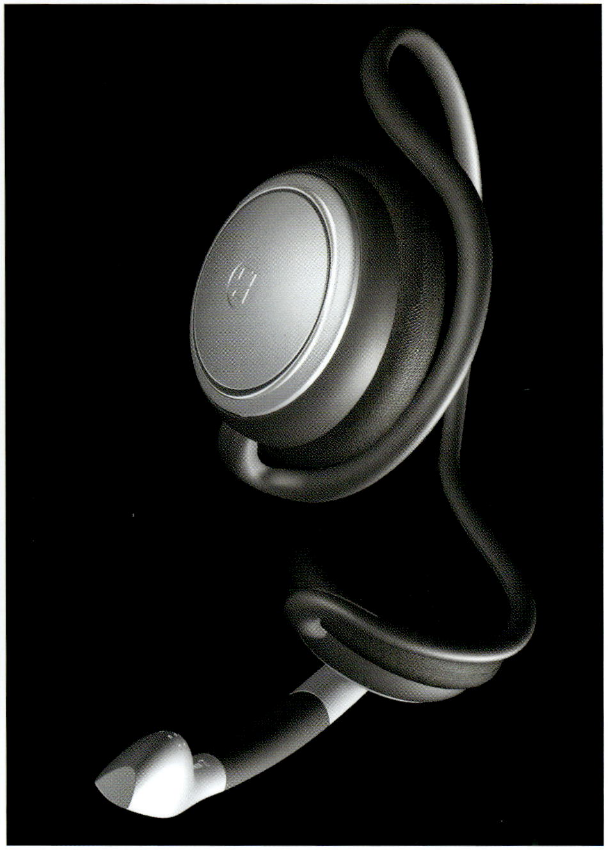

Hersteller Manufacturer
Sennheiser Communications A/S
Solrød Strand, Denmark

Produkt Product
PC 140
Stereo Headset

Design
Jörg Brennwald
Christian Geiger
Brennwald Design
Kiel, Germany

Das PC 140 ist ein Stereo-Headset für die PC-Nutzung, welches sich durch einen komfortablen Nackenbügel, ein schwenkbares Mikrofon, austauschbare Ohrpolster und guten Sitz auszeichnet. Durch sein geringes Gewicht und die Stereo-Anschlussmöglichkeit kann es aber auch gut an portablen Abspielgeräten genutzt werden. Das Mikrofon ist plastisch verformbar für die individuelle Anpassung. Zum Headset gehört eine ins Kabel integrierte Lautstärkeregelung, die kombiniert ist mit einer Mikrofonstummschaltung.

The PC 140 is a stereo headset for PC use, characterized by a comfortable neckband, a tiltable microphone, exchangeable ear pads and a perfect fit. Thanks to its minor weight and stereo connections, it can also be utilized for portable playing devices. The mike is made of ductile plastic to enhance personal adjustment. The headset comes with a volume control integrated into the cable along with a mike mute.

Hersteller Manufacturer
Nokia Group
Espoo, Finland

Produkt Product
Nokia Image Viewer SU-2
Fotobetrachter
Image viewer

Design
Nokia Design Team
Calabasas, CA, USA

Mit dem Nokia Image Viewer SU-2 kann der Nutzer Fotos, die er mit seinem Mobiltelefon aufgenommen hat, auf einem Fernsehbildschirm oder Videobeamer betrachten. Die Fotos werden über Bluetooth kabellos von einem kompatiblen Mobiltelefon an den Fotobetrachter übertragen. Die Verbindung zwischen dem Fernseher oder Videobeamer und dem Image Viewer erfolgt über ein Cinch-Kabel, das im Ruhezustand bequem aufgewickelt werden kann. Die Bilder lassen sich im Einzelbild-Modus oder als Diashow anschauen.

With a Nokia SU-2 Image Viewer you can view photos taken with a mobile phone on a television screen or video beamer. The photos can be transmitted to the viewer wirelessly via Bluetooth from any compatible mobile phone. The television or video beamer is simply hooked up to the image viewer with a cinch cable which can be effortlessly rolled up when not in use. The photos can be shown separately or as a slide show.

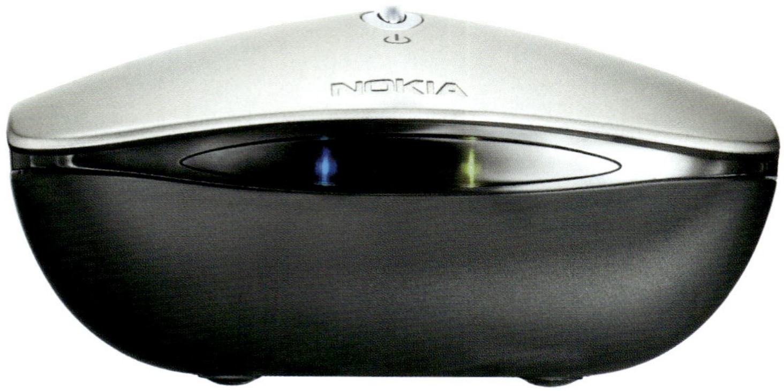

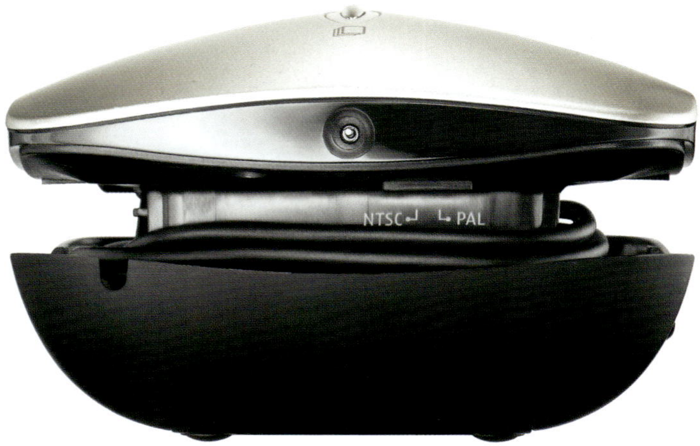

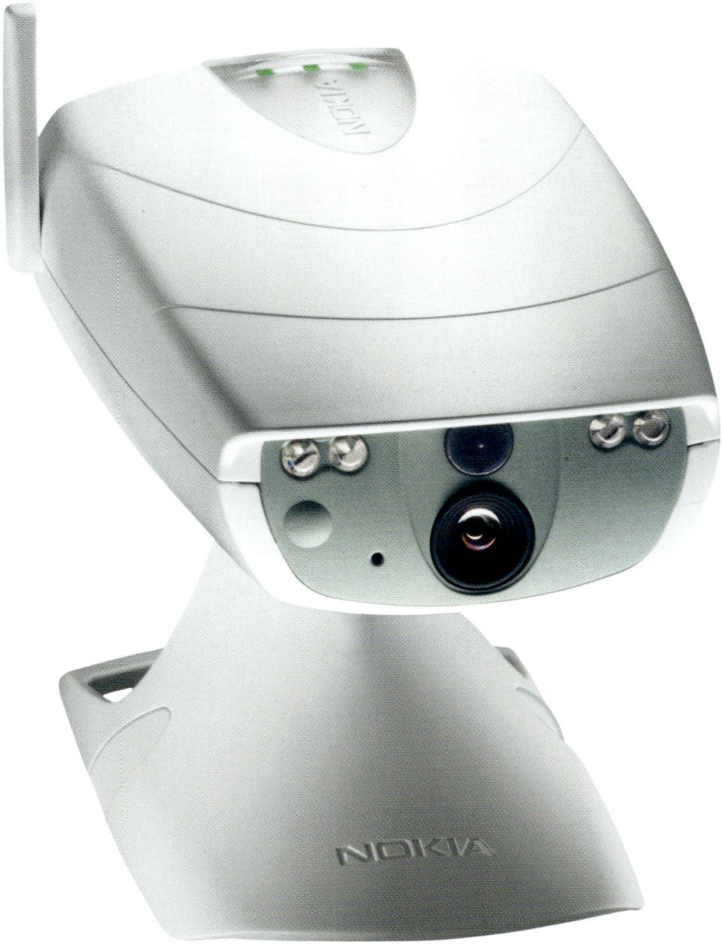

Hersteller Manufacturer
Nokia Group
Espoo, Finland

Produkt Product
Nokia Observation Camera
Überwachungskamera
Surveillance camera

Design
Nokia Design Team
Calabasas, CA, USA

Die Nokia Observation Camera ist ein eigenständig arbeitendes Imaging-Gerät, mit dem Bildinformationen angefordert und empfangen werden können. Sie unterstützt Dualband-Betrieb in (E) GSM 900-/1800 Netzen und eignet sich sowohl für den privaten als auch den beruflichen Einsatz. Die kompakte und portable Kamera vereint eine VGA-Digitalkamera, ein GSM-Sende-/Empfangsmodul und unterstützt MMS.

The Nokia Observation Camera is an independently functioning image device with which image data can be requested and received. It operates with Dualband in (E) GSM 900/1800 networks and is suitable for both private and commercial use. The compact and portable camera combines a VGA digital camera with a GSM sender/receiver module and supports MMS.

Hersteller Manufacturer
Matsushita Electric Industrial Co.Ltd
Osaka, Japan

Produkt Product
PT-AE500E
LCD-Videoprojektor
LCD video projector

Design
Yoshihito Morishita
Azuma Sadayoshi
Panasonic Design Company
Osaka, Japan

Der PT-AE500E LCD-Videoprojektor ist mit einer Auflösung von 1280 x 720 Pixeln HDTV-tauglich und sorgt durch die New Smooth Screen Technologie, einem Kontrastverhältnis von 1300:1 und einer Lichtstärke von 850 ANSI-Lumen, für ein äußerst scharfes und klares Bild. Ungewollten Verzerrungen wirkt die horizontale und vertikale Trapezkorrektur entgegen. So kann das Gerät beliebig im Raum positioniert werden. Mehrere Anschlussmöglichkeiten wie DVI-D-Eingang, Scart-Anschluss, YUV-Komponenteneingang u.a. stehen zur Verfügung.

The PT-AE500E LCD video projector delivers 1280 x 720 pixels and is suitable for HDTV reproduction. Thanks to the New Smooth Screen technology, a contrast ratio of 1300:1 and brightness of 850 ANSI lumens, the projector delivers a superbly sharp and clear image. Distortion is eliminated by the horizontal and vertical keystone correction mode, meaning the unit can be situated anywhere in the room. The projector has numerous terminals such as DVI-D input, Scart I/O, video input, etc.

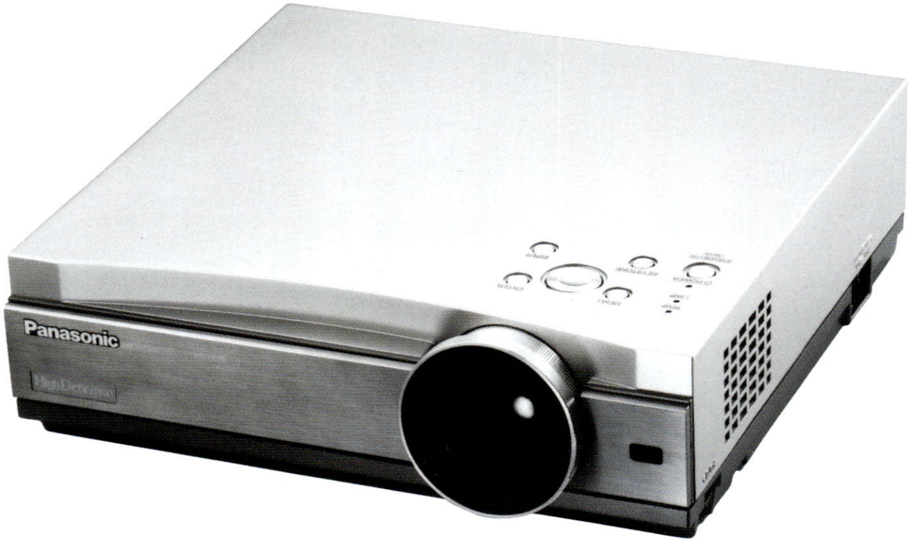

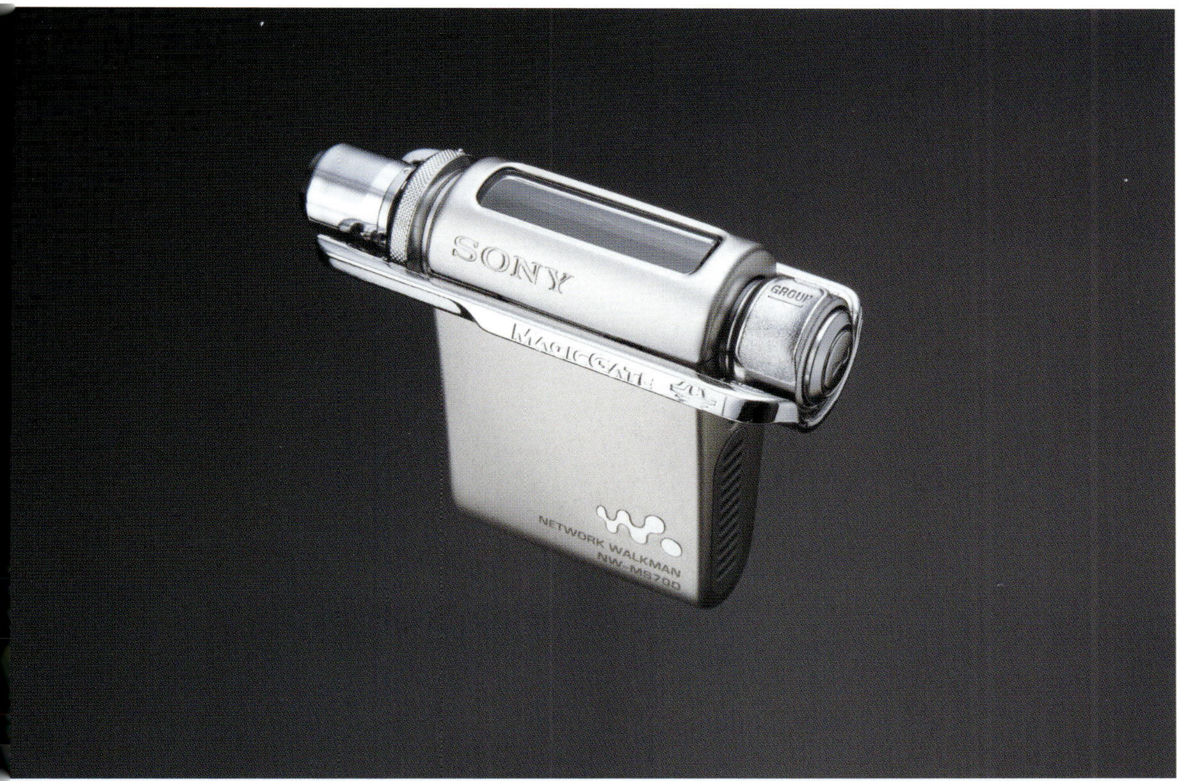

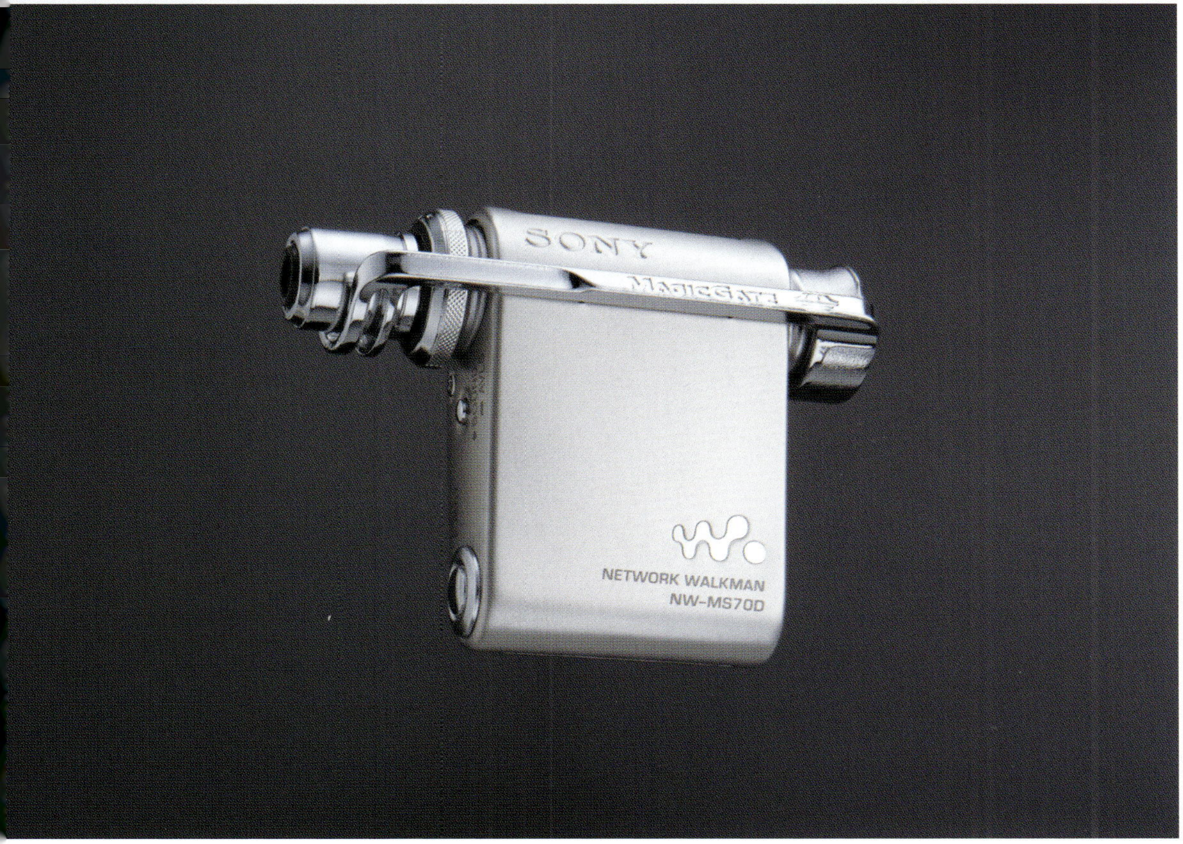

Hersteller Manufacturer
Sony Corporation
Tokio, Japan

Produkt Product
NW-MS70D
Network Walkman

Design
Noriaki Takagi
Sony Corporation
Tokio, Japan

Der NW-MS70D Network Walkman wiegt lediglich 54 g und ist nur 36,4 x 48,5 x 18 mm groß und eignet sich deshalb perfekt zum Hören von digitaler Musik unterwegs. 256 MB of Embedded Memory erlauben das Speichern von bis zu 11 Musik-CDs, bei Anwendung des Magic Gate Memory Stick noch mehr und der eingebaute Akku ermöglicht bis zu 33 Stunden kontinuierlicher Wiedergabe. Bedienungsfreundlichkeit wird durch die intuitiv bedienbare Jog-Scheibe, über die alle Hauptfunktionen gesteuert und die auch per Einhandbedienung betätigt werden kann, gewährleistet. Das Gehäuse des Walkmans ist aus Titan gefertigt, das Kopfhörerband zweigt für eine verbesserte Tragbarkeit vom Halsband ab, zudem ist die Kopfhörerbuchse so entworfen, dass der Stecker nicht versehentlich abgezogen werden kann.

Sony's NW-MS70D Network Walkman weighs in at only 54 g and is sized a mere 36.4 x 48.5 x 18 mm – and thus ideally suited for listening to digital music recordings while on the road. 256 MB of embedded memory means up to 11 music CDs can be stored to memory, and even more if the Magic Gate Memory Stick is used. The built-in battery provides up to 33 hours of continuous playback. User friendliness is enhanced by the intuitive dial that covers all the main functions, and can be used by a single hand. The Walkman boasts a titanium casing, and the headphone strap branches off from the neckband for improved portability; moreover, the headphone jack is designed to ensure that it locks in when plugged in to prevent its accidental removal.

Hersteller Manufacturer
Sharp Electronics
Hamburg, Germany

Produkt Product
VL-Z 7S
Digitaler Camcorder
Digital camcorder

Design
Sharp Electronics

Der neue kompakte digitale Camcorder VL-Z 7S im Miniformat verfügt über eine View-Turn-Einhandbedienung, einen High-Transmissive-Advanced-System-LCD-Monitor (CG Silicon), eine Turbo-Starttaste und weitere Funktionen und Features, die das Filmen in jeder Situation an jedem Ort vereinfachen.

The new compact digital Camcorder VL-Z 7S features a one-hand view-turn function, a high-transmissive advanced system LCD monitor (CG silicon), a turbo-start button and other functions and features, all designed to make filming easier whatever the situation or location.

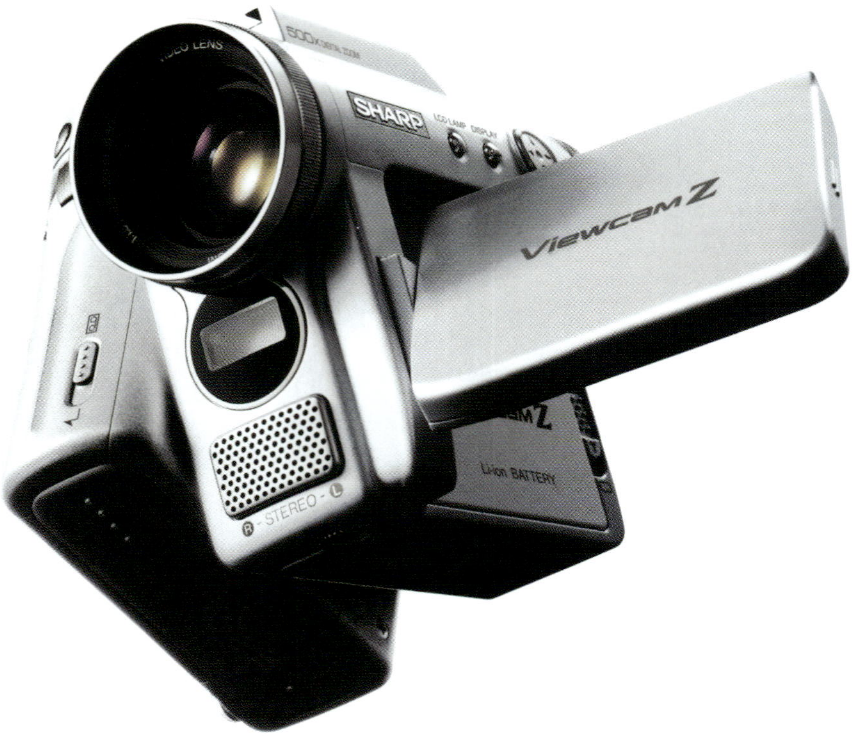

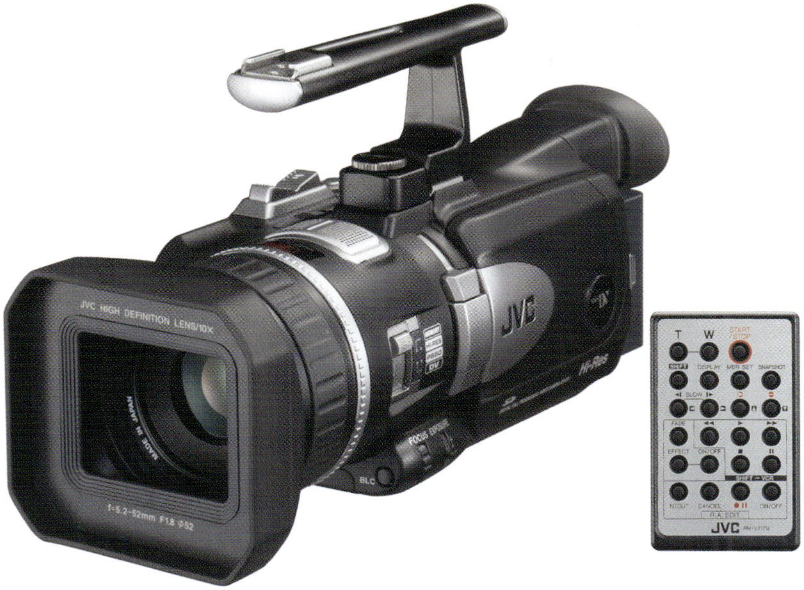

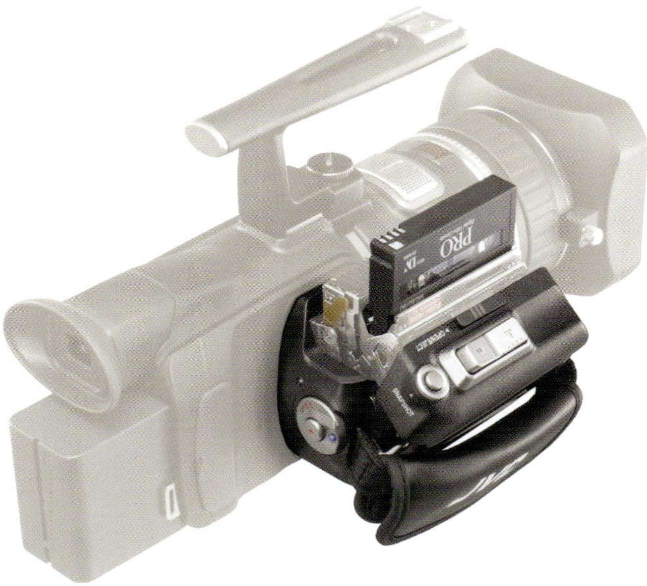

Hersteller Manufacturer
Victor Company of Japan Ltd.
Yokohama, Japan

Produkt Product
GR-PD1
PAL Progressive DV-Camcorder
Digital Full Progressive Video Camera

Design
Yutaka Nakanishi
Tomohide Inomata
Atsushi Kurita
Akihiko Miyamori
Hajime Toyoguti
JVC Design Center
Yokohama, Japan

Durch sein ergonomisches Design sowie der wegweisenden PAL Progressive Technologie ist der GR-PD1 sowohl für den professionellen Kameramann also auch für den High-End-Amateurfilmer konzipiert. Neben dem DV-Format kann der Filmer hochwertige 16:9-Aufnahmen im MPEG2-Format in Vollbild-Technologie (625/25p, 625/50p) auf MiniDV-Kassette aufzeichnen. Dank 16:9-CCD mit 1,18 Megapixel, besonderem Farbfilter und Pixelmanagement beweist der GR-PD1 professionelle Bildqualität bei jeder Aufnahme. Der um 90° drehbare Multifunktionsgriff mit integriertem Kassetten-Laufwerk ermöglicht bei ungewöhnlichen Aufnahmepositionen eine optimale Kameraführung.

With its ergonomic design as well as the revolutionary full progressive technology the GR-PD1 is conceived for professional cameramen as well as for high-end-videofilmers. Beside the normal recording in DV-format, the GR-PD1 is able to record premium 16:9 progressive films in MPEG2 format (625/25p, 625/50p) on MiniDV. Due to the 16:9-CCD with 1.18 Megapixel, JVC's special Digital Colour Filter and pixel management the GR-PD1 offers high picture quality for every shooting. The grip, which rotates up to 90° and integrates a cassette drive, allows perfect camera work even from unusual angles.

Hersteller Manufacturer
Samsung Electronics Co., Ltd.
Seoul, Korea

Produkt Product
DVD-HD931
DVD-Player
DVD player

Design
Lim Hyun Taek
Byun Won Gab
DVS Design Group
Samsung Electronics Co., Ltd.
Seoul, Korea

Der DVD-Player HD931 ist äußerst dünn und lässt sich aufgrund seines eleganten Objektcharakters gut in das Wohnumfeld integrieren. Die intuitiv zu bedienende Benutzeroberfläche ermöglicht eine einfache und bequeme Steuerung. Mit Hilfe des DVI-Anschlusses lässt sich der Player direkt an den hochauflösenden Fernseher anschließen, ohne dass eine Analog-Digital-Wandlung und umgekehrt erforderlich ist. Es können alle Datenträgerformate wie DVD, DVD Audio, CD-RW, DVD-RW, MP3 etc. wiedergegeben werden.

The HD931 DVD player is extremely slim and thanks to its elegant appearance can be smoothly integrated into home environments. The intuitive GUI makes for simple and effortless operation. The player can be directly connected to a high-definition TV through the DVI without the need for DAC/ADC conversions. All data formats can be played on the DVD player including DVD, DVD Audio, CD-RW, DVD-RW and MP3s.

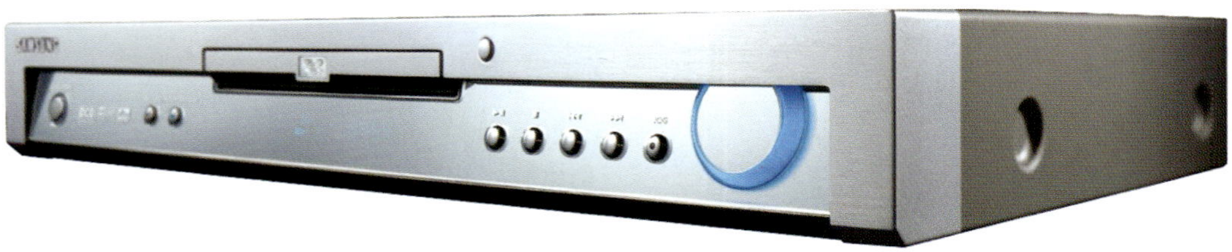

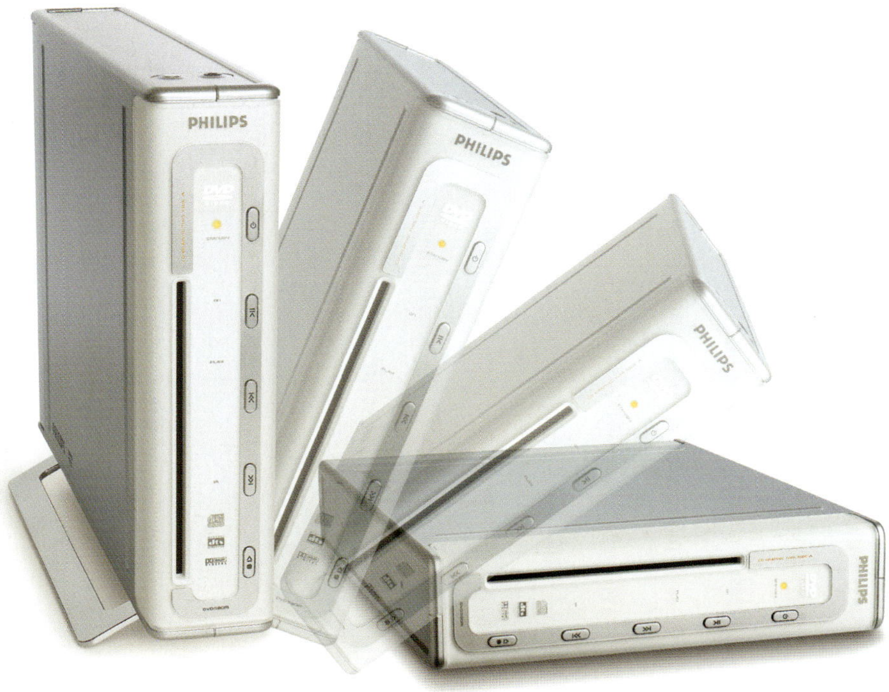

Hersteller Manufacturer
Philips Electronics Asia Pacific Pte Ltd
Singapore, SGP

Produkt Product
Philips DVD 590
DVD Player
DVD player

Design
Pernilla Johansson, Charis Ng
David Ng, Leon Yoong
Philips Design

Der kompakte Philips DVD 590 kann auf drei Weisen positioniert werden: vertikal oder horizontal in einem Rack oder horizontal gestapelt. Die weiche Formensprache und die Weiß/Silber-Töne sorgen dafür, dass sich das Produkt harmonisch in die Raumausstattung einfügt, das Aluminiumgehäuse ist mit einer fein mattierten Textur versehen. Der voll ausgestattete DVD-Player verfügt über eine benutzerfreundliche Oberfläche, die seine Bedienung einfach macht.

The compact Phillips DVD 590 player can be positioned in three ways: vertically or horizontally on a stand, or horizontally stacked without the stand. The soft formal language and the white/silver tones spell harmonious integration into residential interiors. The aluminum casing has a soft satin matte front. The feature-packed DVD player has an easy-to-use interface to keep operation simple.

Hersteller Manufacturer
Samsung Electronics Co., Ltd.
Seoul, Korea

Produkt Product
SyncMaster 173P
LCD-Monitor
LCD monitor

Design
Kim Eui-Seok
Kim Young-Jun
VD Design Group
Samsung Electronics Co., Ltd.
Seoul, Korea

Der SyncMaster 173P ist ein schmaler und kompakter LCD-Monitor mit lediglich 26 mm Dicke für die 17-Zoll-Version und 32 mm für die 19-Zoll-Version. Eine neuartige Steuerung mit der SOFT-OSD-Methode ermöglicht eine Gestaltung des Monitors ohne zusätzliche Knöpfe. Das Gehäuse und der Fuß bestehen aus Aluminium, eine frei wählbare Höhen- und Winkeleinstellung sowie die Portrait-Stellung (mit Kippfunktion) erlauben eine individuelle Anpassung des Bildschirmwinkels an den Nutzer.

The SyncMaster 173P is a slender and compact LCD monitor only 26 mm thick for the 17" screen and 32 mm for the 19" variant. An innovative control system featuring SOFT-OSD enables the monitor to get by without additional buttons. The front cabinet and footprint are made of aluminum, while height and angle can be adjusted as required, and together with the portrait (pivot) function mean the screen angle can be customized to the particular user's needs.

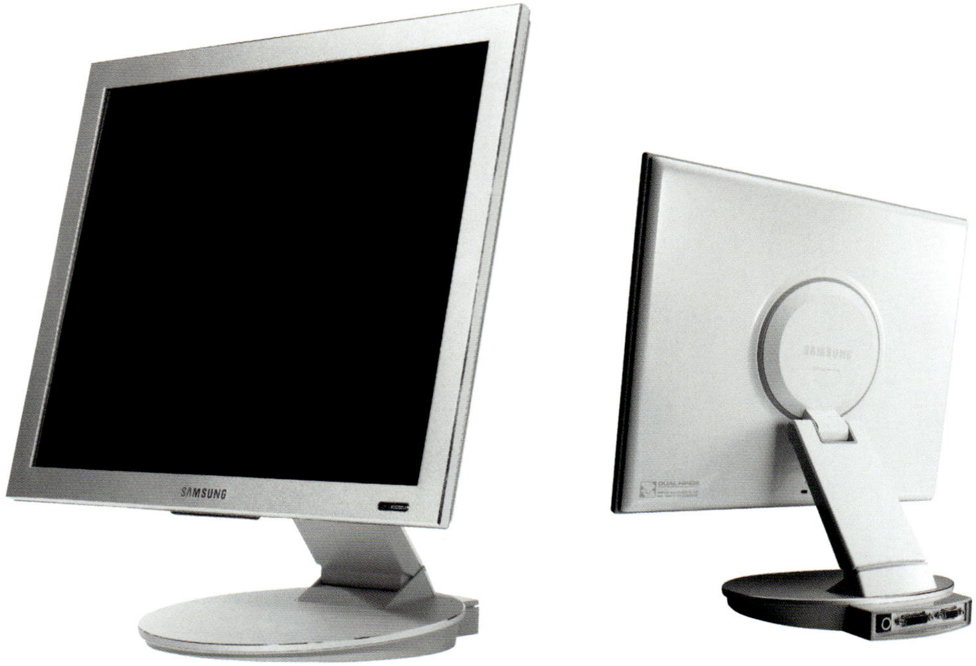

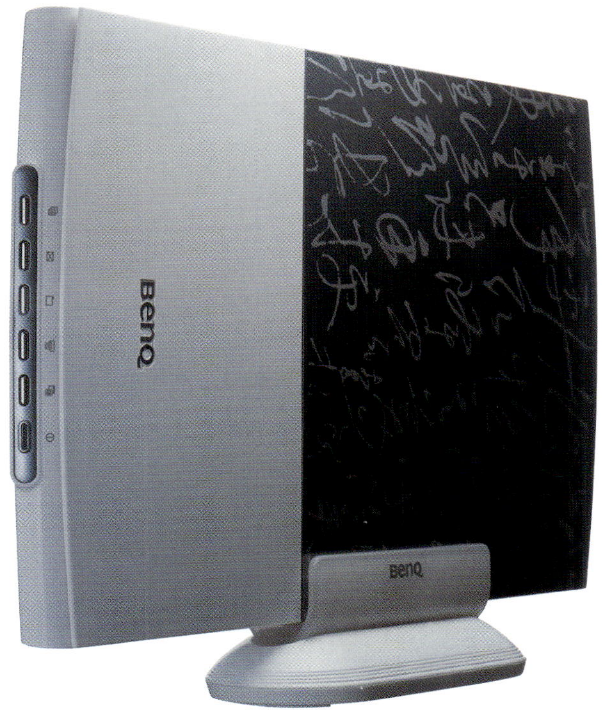

Hersteller Manufacturer
BENQ Corporation
Taipei, Taiwan, ROC

Produkt Product
BenQ Scanner 5250C
Scanner
Scanner

Design
BENQ Corporation
Lifestyle Design Center
Taipei, Taiwan, ROC

Der BenQ Scanner 5250C ist mit seinen 38 mm außerordentlich schlank, zudem ermöglicht der Ständer eine Platz sparende vertikale Aufstellung. Das Design vermittelt die Anmutung eines aufgeschlagenen Buches, wobei die dunkle Seite des Gehäuses mit kursiv geschriebenen chinesischen Schriftzeichen versehen ist. Für optimierte Farb- und Bildqualität ist der Scanner mit einer speziellen von BenQ entwickelten Technologie und versilberten Sensoren ausgestattet. Mit den 5 Scan-Tasten für die wichtigsten Funktionen, dem Auto-Positioning für schief eingelegte Vorlagen und der Multi-Scan-Funktion, ist er einfach zu bedienen und benutzerfreundlich.

The BenQ Scanner 5250C is astonishingly slender, only 38 mm thick, and the upright stand enables a small footprint. The design has the feel of an open book about it, with a pure clean white outer cover and a dark cover plate with its cursive style Chinese characters. The scanner comes with a special BenQ technology to optimize color and image quality, and includes silverized sensitization components for fast film speeds. With 5 scan buttons for the core functions, an auto-positioning feature for off-cue documents and a multi-scan function, the 5250C is easy to operate and user-friendly.

Hersteller Manufacturer
Grundig AG
Nürnberg, Germany

Produkt Product
Ocean Boy 70
Tragbares Radio
Portable radio

Design
Alexis Pepall
Pepall Design
Hong Kong, China

Der Ocean Boy 70 von Grundig ist ein PLL-Synthesizer-Radio für LW-, MW-, UKW- und KW-Empfang. Das tragbare Radio verfügt über 20 Senderspeicher, ein beleuchtbares Multifunktions-LC-Display sowie Weck- und Timerfunktionen. Das Design wird von dem runden Display, dem unterteilten Lautsprecher und den abgerundeten Ecken bestimmt.

The Ocean Boy 70 by Grundig is a PLL synthesizer radio for short, middle and longwave reception. The portable radio can store over 20 stations to memory, and sports an illuminated multifunctional LCD as well as alarm and timer functions. The design is defined by the round display, the sub-divided loudspeaker and the rounded corners.

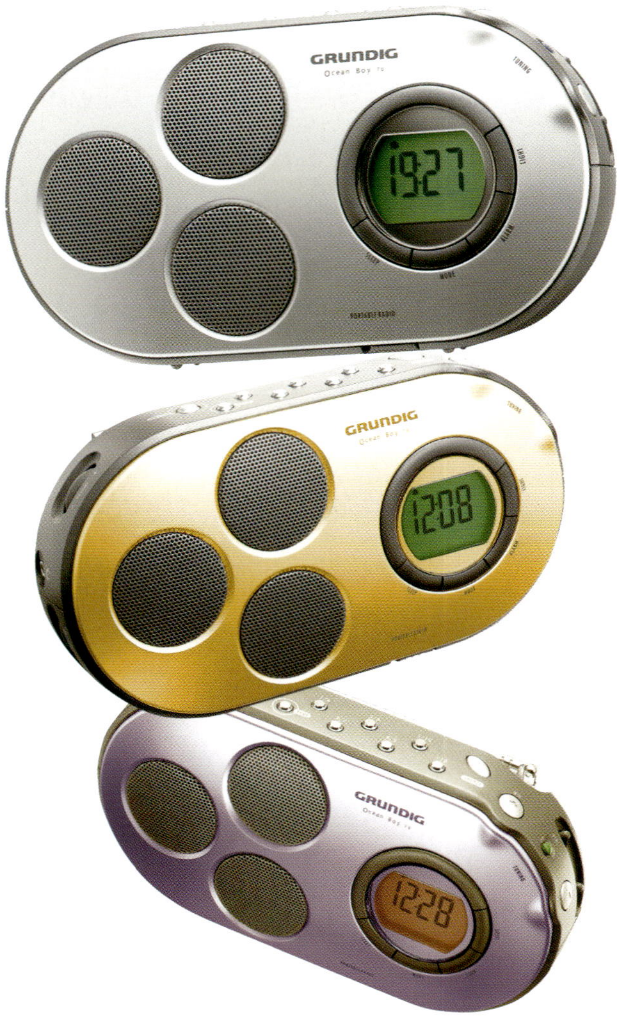

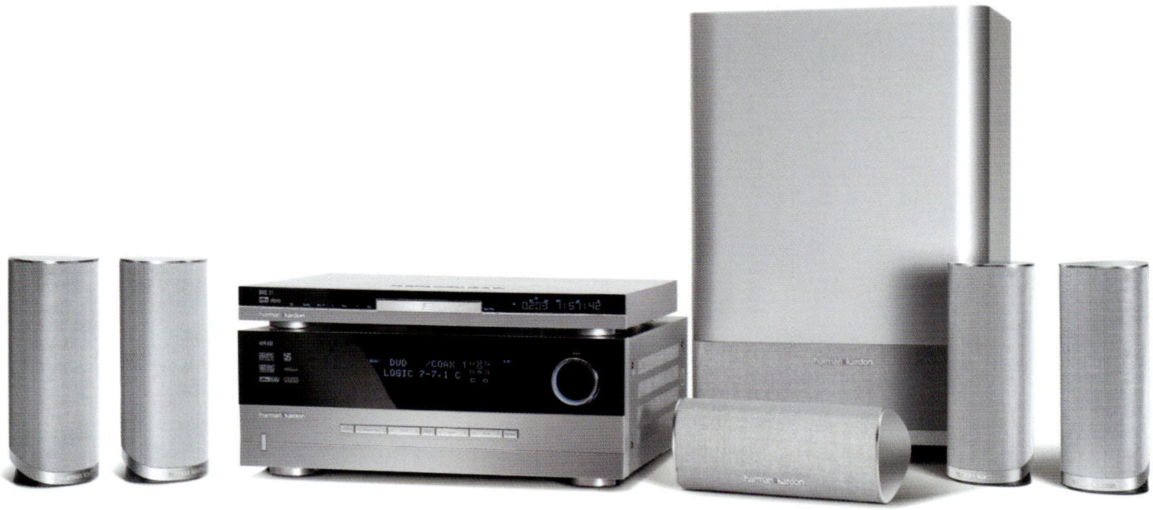

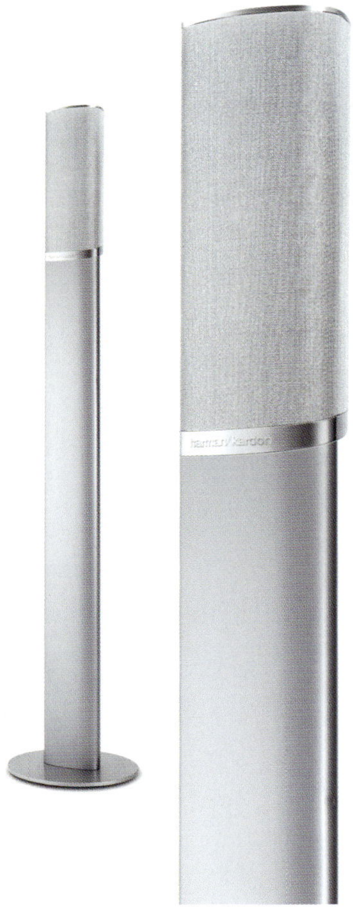

Hersteller Manufacturer
Harman Kardon
Woodbury, New York, USA

Produkt Product
Harman Kardon TS 14
Heimkino-System
Home theater system

Design
Daniel Ashcraft
Erin Napolitano
David Wathen
Ashcraft Design
Los Angeles, CA, USA

Das Harman Kardon TS 14 ist ein Heimkino-System, das Receiver, Lautsprecher und DVD-Player in einer gemeinsamen digitalen Lösung verbindet. Bei der Designentwicklung standen insbesondere die Kompaktheit des Systems und ein hoher ästhetischer Anspruch für eine harmonische Integration in das innenarchitektonische Umfeld im Vordergrund. Die Anlage bietet Voraussetzungen zur Befestigung an der Wand, sie kann aber auch auf den Boden oder ins Bücherregal gestellt werden.

The Harman Kardon TS 14 is a home theater system that combines receiver, loudspeaker and DVD player in a single digital solution. The design focuses on the creating a compact system of a high aesthetic standard so that it can be harmoniously integrated into residential interiors. The system is such that it can be mounted on the wall or alternatively stood on the floor or placed on a book shelf.

Hersteller Manufacturer
IBM Corp.
Rochester, MN, USA

Produkt Product
IBM eServer xSeries 365
Serverfamilie
Server line

Design
Roland Alo
ISG Industrial Design
IBM Corporation
Rochester, MN, USA

Der IBM eServer xSeries 365 ist ein kompakter, Platz sparender, 4-Wege-fähiger 3U Rack Server für On-Demand Enterprise Anwendungen. Markante Designmerkmale des Servers sind die funktionale Farbcodierung und die hexagonale Rasterlochung für gute Luftzirkulation. Auf die internen Komponenten kann ohne Zuhilfenahme von Werkzeugen zugegriffen werden, die sechs Festplatteneinheiten können während des Betriebs ausgewechselt werden.

The IBM eServer xSeries 365 is a compact high-performance, 3U 4-way rack server for on-demand enterprise applications. Striking design features include the functional color coding and the hexagonal honeycomb perforation for inherently enhanced airflow. Tool-free access to the interior is guaranteed, and six removable hotplug hard-disk trays are available for storage upgrade during operation.

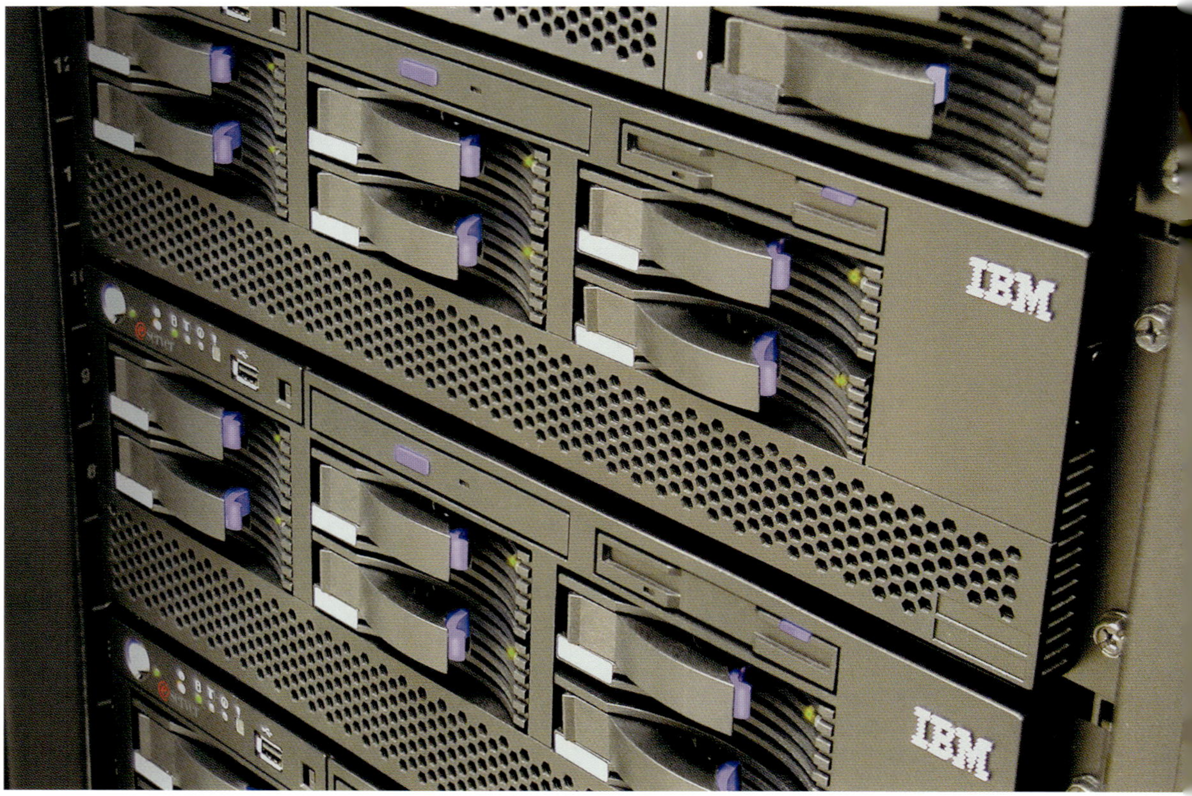

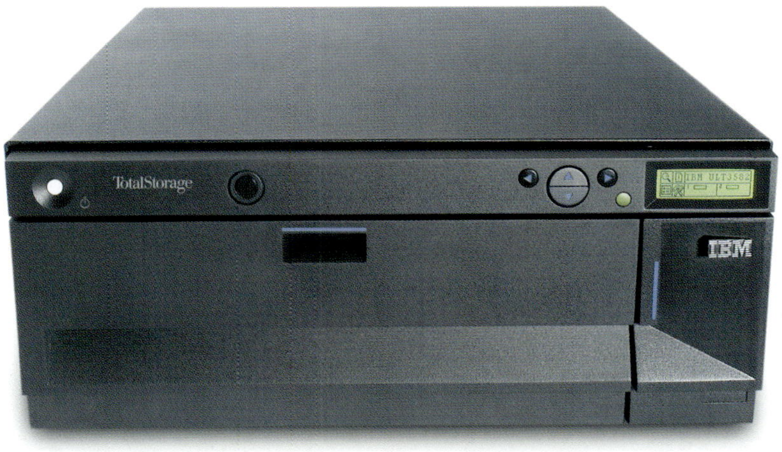

Hersteller Manufacturer
IBM Corp.
Tucson, AZ, USA

Produkt Product
IBM TotalStorage Ultrium Tape Library 3582
Bandkassettenarchiv
Tape library

Design
Brian McClain
IBM Industrial Design
Tucson, AZ, USA

IBM TotalStorage Ultrium Tape Library 3582 ist ein Midrange Linear Tape Open (LTO) Bandkassettenarchiv, das kleinen und mittelgroßen Firmen eine hohe Leistung und Zuverlässigkeit bei der Bandspeicherautomatisierung bietet. Die kompakte 4U Einschub-Version bietet eine maximale Ausnutzung des Platzes im Rack. Das Modell 3582 spiegelt die IBM TotalStorage Designsprache wider und ist mit farbcodierten Berührungspunkten, intuitiven Kontrollelementen und einer Statusanzeige ausgestattet.

IBM TotalStorage Ultrium Tape Library 3582 is a midrange linear tape open (LTO) tape library that offers small and midsized businesses excellent tape-storage automation performance and reliability. The compact 4U rack-mounted version maximizes rack usage. The 3582 model reflects the IBM TotalStorage design vocabulary and features recognizable color-coded touch points, intuitive controls and local status displays.

Hersteller Manufacturer
IBM Corp.
Tucson, Z, USA

Produkt Product
IBM Total Storage NAS 100
Speichereinheit
Storage unit

Design
Aaron Cox
Hiroki Hirano
IBM Industrial Design
Tucson, AZ, USA

Die einfach zu installierende und zu konfigurierende IBM Total Storage Network Attached Storage NAS 100 Lösung bietet eine kostengünstige Möglichkeit für Nutzer, die ein NAS zur Speicherkonsolidierung, zum Datenaustausch oder zur schrittweisen Speichererweiterung benötigen. Die maximale Installationsdauer für ein IBM NAS 100 beträgt 30 Minuten und man kann zwischen einer 1U Tisch- oder Rackinstallation wählen. Die einfach zu bedienenden, farbkodierten Hot-Swap Laufwerke ermöglichen den Austausch von Festplatten, während der ATA Bus aktiv ist, was in einer niedrigeren Ausfalldauer und -rate resultiert.

Easy to install and configure, the IBM Total Storage Network Attached Storage NAS 100 provides a cost-effective solution for users who require NAS for storage consolidation, file sharing or simply as added incremental storage for a network. The maximum deployment uptime for an IBM NAS 100 is merely 30 minutes – and users can choose from a 1U tabletop or a rack-mounted option. The easy-to-use color-coded hot-swappable drives support plug-in or plug-out of hard drives while the ATA bus is active, resulting in less system downtime.

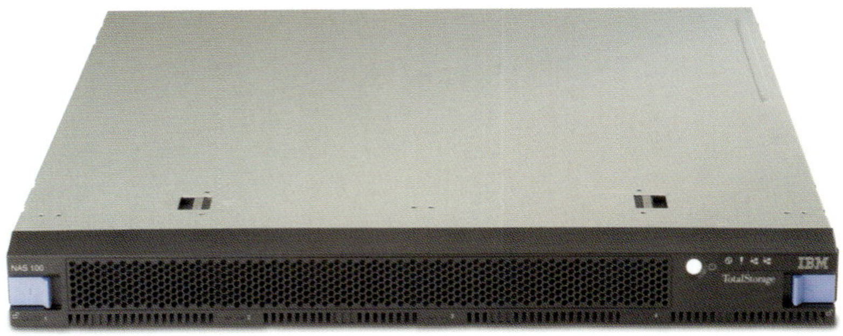

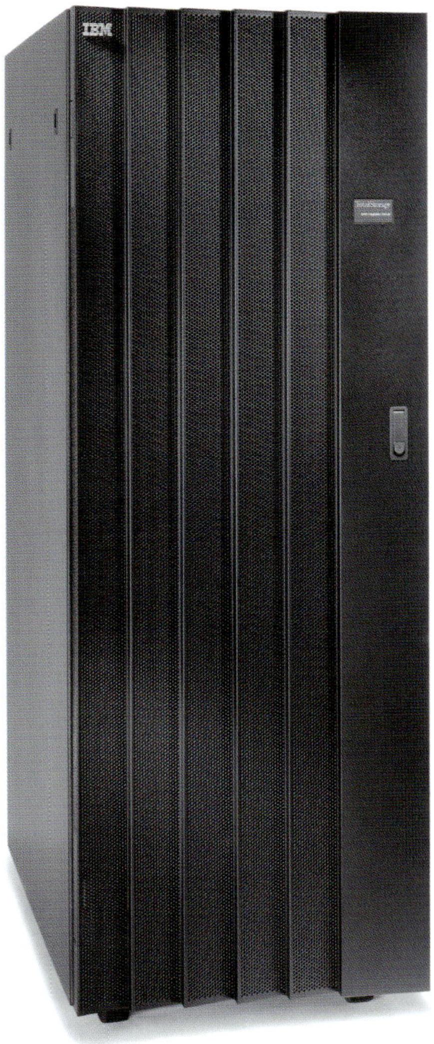

Hersteller Manufacturer
IBM Corp.
Tucson, AZ, USA

Produkt Product
IBM Total Storage SAN
Integration Server
Integration server

Design
Aaron Cox
Hiroki Hirano
IBM Industrial Design
Tucson, AZ, USA

IBM Total Storage SAN Integration Server wird als eine Einheit geliefert und installiert und bietet Erweiterungsmöglichkeiten für Konnektivität, Speicherkapazität und Leistungsstufen. Es ist eine sofort einsetzbare Storage Area Network (SAN) Komplettlösung, die die Vorteile eines SANs mit der einfachen Handhabung eines Einzelsystems verknüpft. Der SAN Integration Server kann bis zu einer Speicherkapazität von über 100 Terabytes (TB) erweitert werden und mit bis zu 42 Hosts verbunden werden. Das schlanke Gehäuse und die einfache, pulverbeschichtete Metallkonstruktion verdecken die dicht angeordneten internen Komponenten.

IBM's Total Storage SAN integration server is supplied and installed as a single unit. The server offers upgrade options for connectivity, storage capacity and performance levels. It is an out-of-the-box complete SAN solution that combines the benefits of an SAN with the ease of single-system manageability. The SAN integration server is capable of scaling to over 100 Terabytes (TB) storage capacity and connecting up to 42 hosts. The sleek enclosure and simple, powder-coated metal construction shields users from the complexity of the densely-packed internal components.

Hersteller Manufacturer
Fujitsu Siemens Computers
München, Germany

Produkt Product
TX 150/200/600
Primergy Server Family
Serverreihe
Server range

Design
Bernd Eigenstetter
Georg Trost
designafairs GmbH
München, Germany

Primergy TX ist eine Serverreihe vom Mono-Prozessor über Dual-Prozessor Entry und Business Server bis zum 4-way Midrange Server für unterschiedlichste Anforderungen und nahezu alle Einsatzgebiete. Die Floorstand-Systeme sind auch als Rack-Variante verfügbar bzw. nachträglich umrüstbar.

Primergy TX is a server range including mono-processors via dual processor entry and business server right through to 4-way midrange servers – for almost all fields of application. The floorstand systems are also available as rack versions or can be subsequently retrofitted to this end.

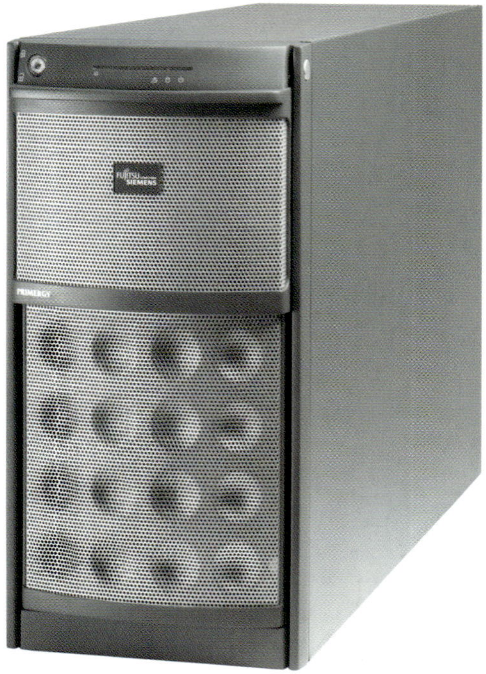

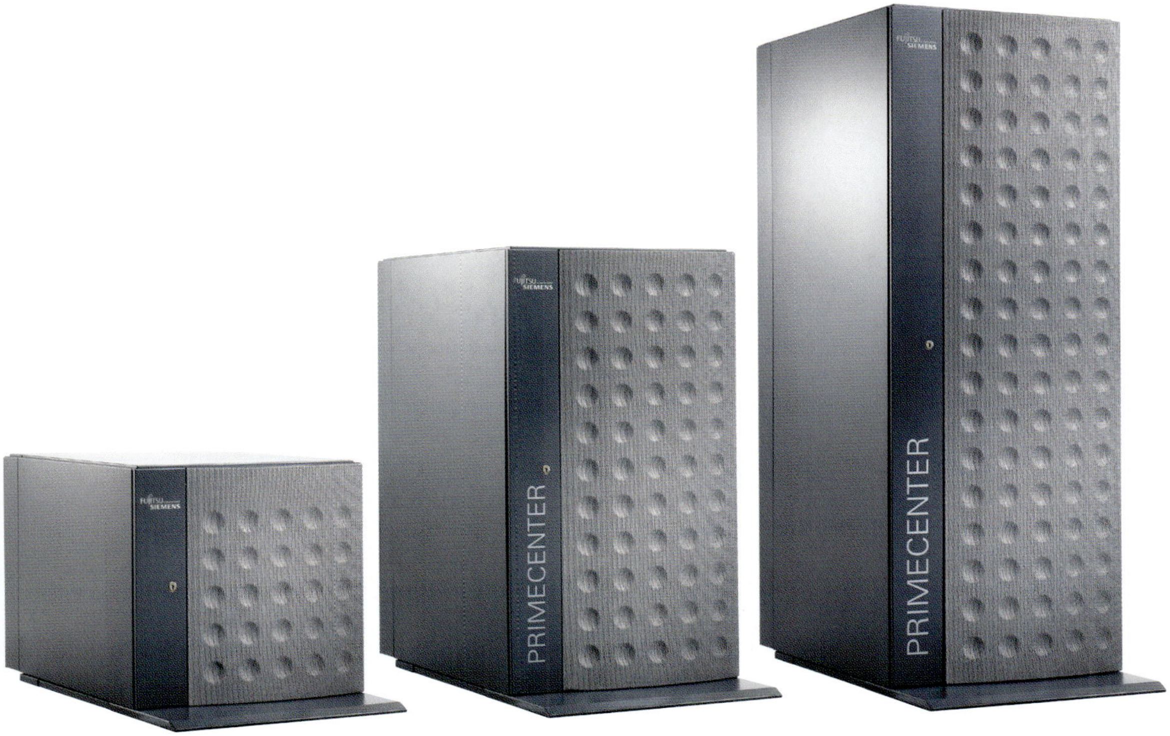

Hersteller Manufacturer
Fujitsu Siemens Computers
München, Germany

Produkt Product
Primecenter Rack Family
Serverschrank
Server rack system

Design
Bernd Eigenstetter
Georg Trost
designafairs GmbH
München, Germany

Primecenter ist ein Schranksystem für Server mit 12 U, 24 U und 38 U. Ein spezielles Kabelmanagement erleichtert das Handling der Einschübe. Die Lochblech-Struktur signalisiert die Zugehörigkeit zur Fujitsu-Siemens-Server-Familie.

Primecenter is a rack system for 12U, 24U and 38U servers. Special cable management facilitates handling of the stacks. The perforated plate design highlight the fact that the racks are members of the Fujitsu Siemens Server family.

Hersteller Manufacturer
Medion AG
Essen, Germany

Produkt Product
MD Sion
Multimedia Internet PC

Design
Bernhard Geisen, Lidan Liu,
Yorgo Liebsch, André Hanisch
Uwe Spannagel, Andreas Reineke
Essen, Germany

Das Design des MD Sion Multimedia Internet PCs zeichnet sich durch seine weichen Kanten und die symmetrische Anordnung der Funktionskomponenten aus. Die Blende ist als wichtiges Gestaltungselement in den Farben Silber und Blau gehalten, die silbernen Tasten weisen den Betrachter auf das Connect X-Modul hin, welches das schnelle und bedienungsfreundliche Verbinden mit externen Geräten ermöglicht. Zu diesem Zweck gleitet die Tür nach unten und gibt das Modul frei. Der zentrale Power-Schalter sitzt prominent in der Fläche und trägt entscheidend zum Gesamtcharakter des PCs bei.

The MD Sion Multimedia Internet PC has an eye-catching design, with soft edges and a symmetrical layout for the functional components. The screen surround is an important design element and maintains the silver and blue color scheme – silver buttons draw attention to the Connect X module which enables fast and user-friendly connections to external devices when the door of the module is slid down, exposing it. The central on/off switch is prominently placed in the front and plays a decisive part in setting the overall tone of the PC.

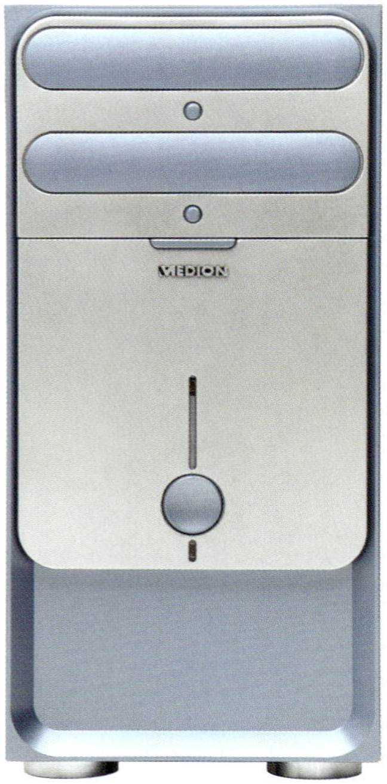

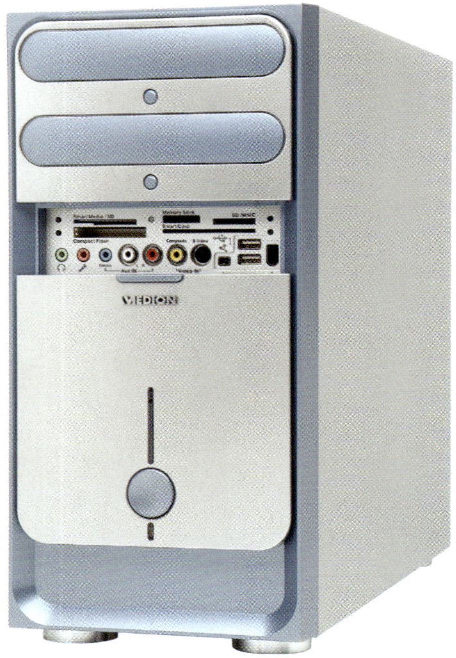

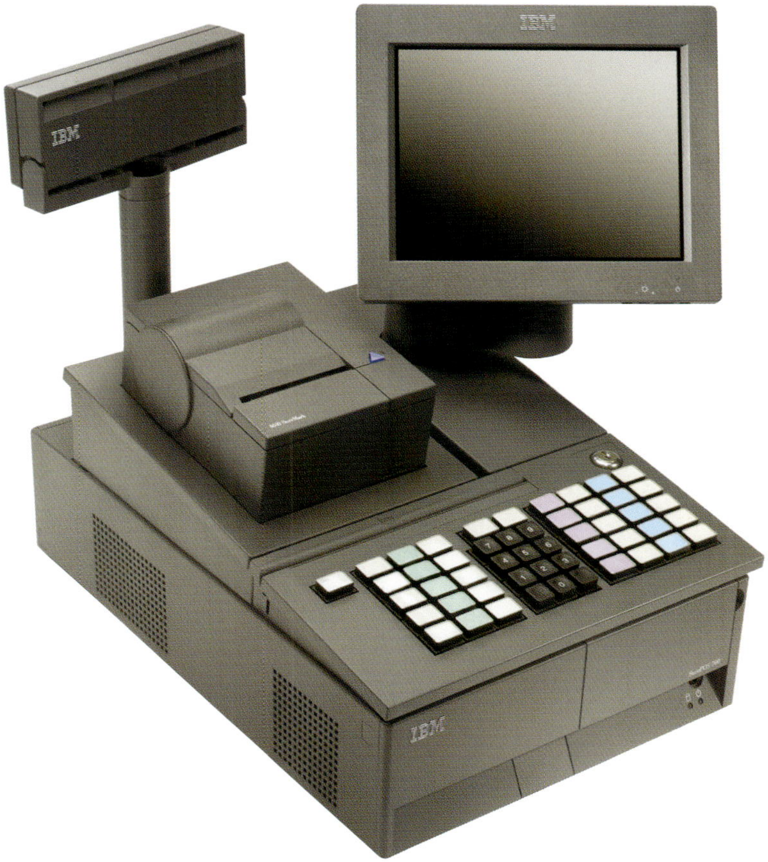

Hersteller Manufacturer
IBM Corp.
Armonk, NY, USA

Produkt Product
IBM SurePos 700
Kassensystem
Point-of-Sales System

Design
IBM Design
Research Triangle Park, NC, USA

SurePos 700 von IBM ist ein Kassensystem für den Einzelhandel, das mit bereits existierenden Geräten wie Bildschirmen, Tastaturen, Druckern etc. kombiniert werden kann. Die Einheit ist so robust, dass andere Geräte auf ihr abgestellt werden können und Staub, Temperaturschwankungen etc. sie nicht beeinträchtigen. Sie ist in einer schmalen und einer breiteren Version in den Farben Hellgrau oder Dunkelgrau erhältlich und kann den Anforderungen des Kunden entsprechend individuell zusammengestellt werden.

IBM's SurePos 700 is a point-of-sales system which can be integrated into existing systems, enabling customers to continue to use all the displays, keyboards, printers etc. they already own. The unit is retail-hardened so that not only can other devices be placed on top of it, but it is also impervious to such things as dust and temperature changes. It is available in a small and a broader version and in either light grey or dark grey, and can be configured to suit individual requirements.

Hersteller Manufacturer
Itautec Philco S.A.
São Paulo, Brazil

Produkt Product
Mini Cash
Geldautomat
Banking self-service product

Design
Edson Danta Dias

Der leichte und Platz sparende Geldautomat Mini Cash ist eine preisgünstige und praktische Alternative für geschützte Umgebungen wie z. B. Einkaufszentren, Flughäfen, Supermärkte, Hotels. Er verzichtet auf das in anderen Umgebungen notwendige Sicherheitssystem und ist somit kleiner, leichter und einfacher zu transportieren. Die Tastaturkonsole und die Seitenflächen bestehen aus Polyurethanschaumstoff, das Gehäuse aus 1,5 mm dickem Stahlblech. Das Gerät trägt den unterschiedlichen Körpergrößen der Anwender Rechnung, sodass ergonomische Nutzung, Komfort, Lesbarkeit und Diskretion gewährleistet sind.

Banking self-service machine developed to provide the following financial services: cash dispensing up to two denominations, bill payment, inquiries, EFT, investments, print documents and slips. Light and compact, its reduced footprint results in a cost-effective and practical alternative for high-traffic secure environments such as shopping malls, airports, supermarkets, schools, hotels among others. Mini Cash was dimensioned within ergonomic parameters taking into account the differences in height of the users so that everyone could comfortably access and view it.

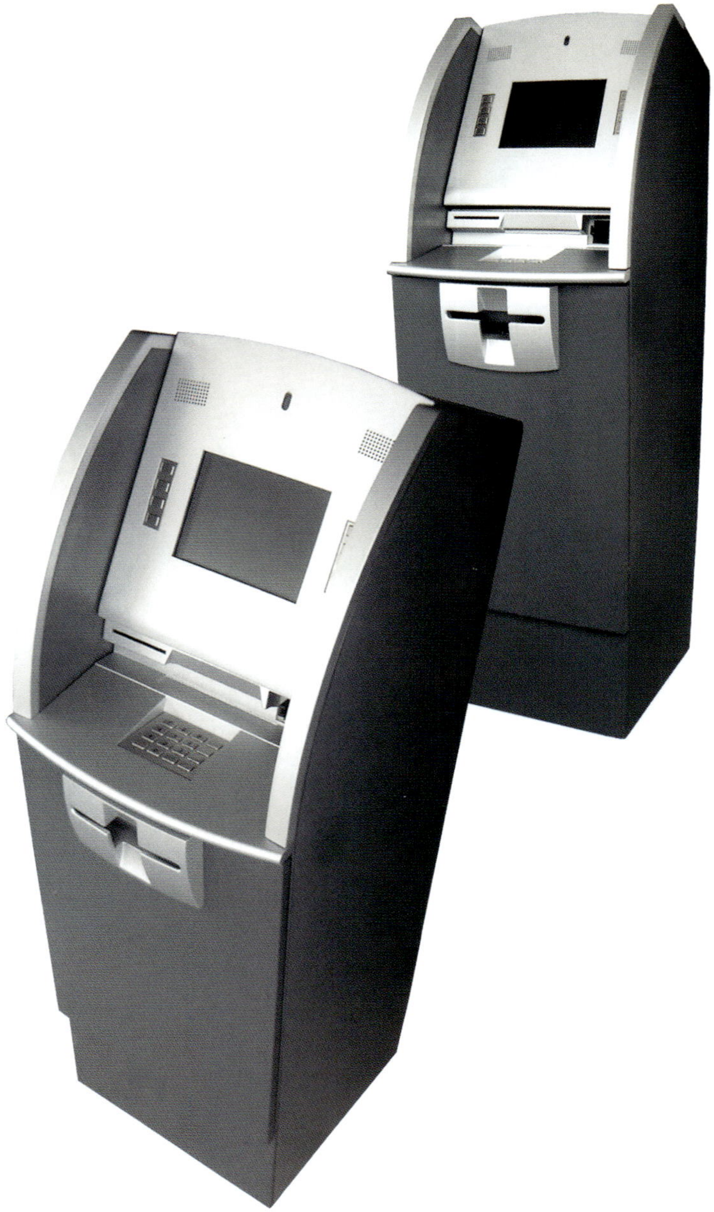

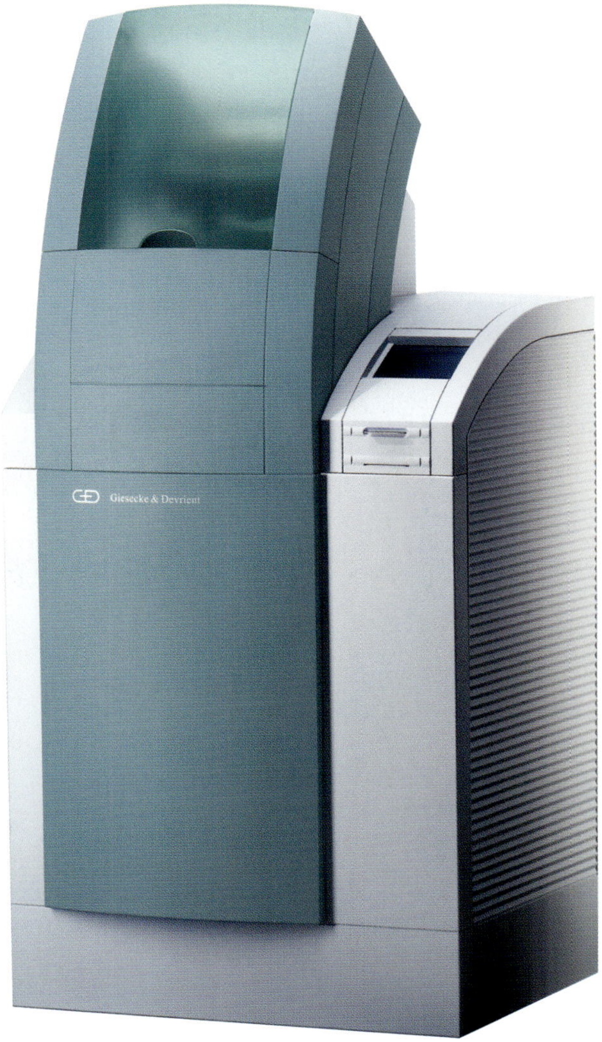

Hersteller Manufacturer
Giesecke & Devrient GmbH
München, Germany

Produkt Product
Castella®
Einzahlungsterminal
Cash deposit terminal

Design
Gerhard Nüssler
ergon3
München, Germany

Castella® optimiert den bislang kostenintensiven Geldbearbeitungsprozess zwischen Kassenbüro, Cash-Center und Werttransportunternehmen und macht ihn effizienter. Direkt am Point of Sale können die Banknoten gezählt, geprüft und anschließend im Tresor aufbewahrt werden. Safebags mit Gutscheinen oder Münzen können in einem separaten Fach eingeworfen und abgerechnet werden. Die Abrechnungsdaten werden direkt an übergeordnete Software-Systeme weitergeleitet, das Unternehmen kennt zu jedem Zeitpunkt seine Umsätze. Das optisch geschlossene Erscheinungsbild vermittelt Sicherheit und schafft Vertrauen in die zugriffsichere Verwahrung der dem System anvertrauten Werte. Die displaygestützte Bedienungsführung mit LCD-Touchscreen erlaubt eine einfache und intuitive Benutzung des Gerätes.

Castella® optimizes cash handling between back offices, cash centers and CIT companies, making this previously costly phase of the cash cycle more efficient. Banknotes can be counted, verified and then securely stored in a tamper-proof safe directly at the point of sale or in the back office. Safebags containing vouchers or coins can be inserted through a different slot and the deposited amount recorded. The accounting data is transferred directly to superordinate software systems, the enterprise can keep track of its receipts at all times. The closed, self-contained appearance of the unit projects security and generates confidence in reliably tamper-proof storage of the deposited valuables. A display-supported user interface with LCD touchscreen permits simple, intuitive operation.

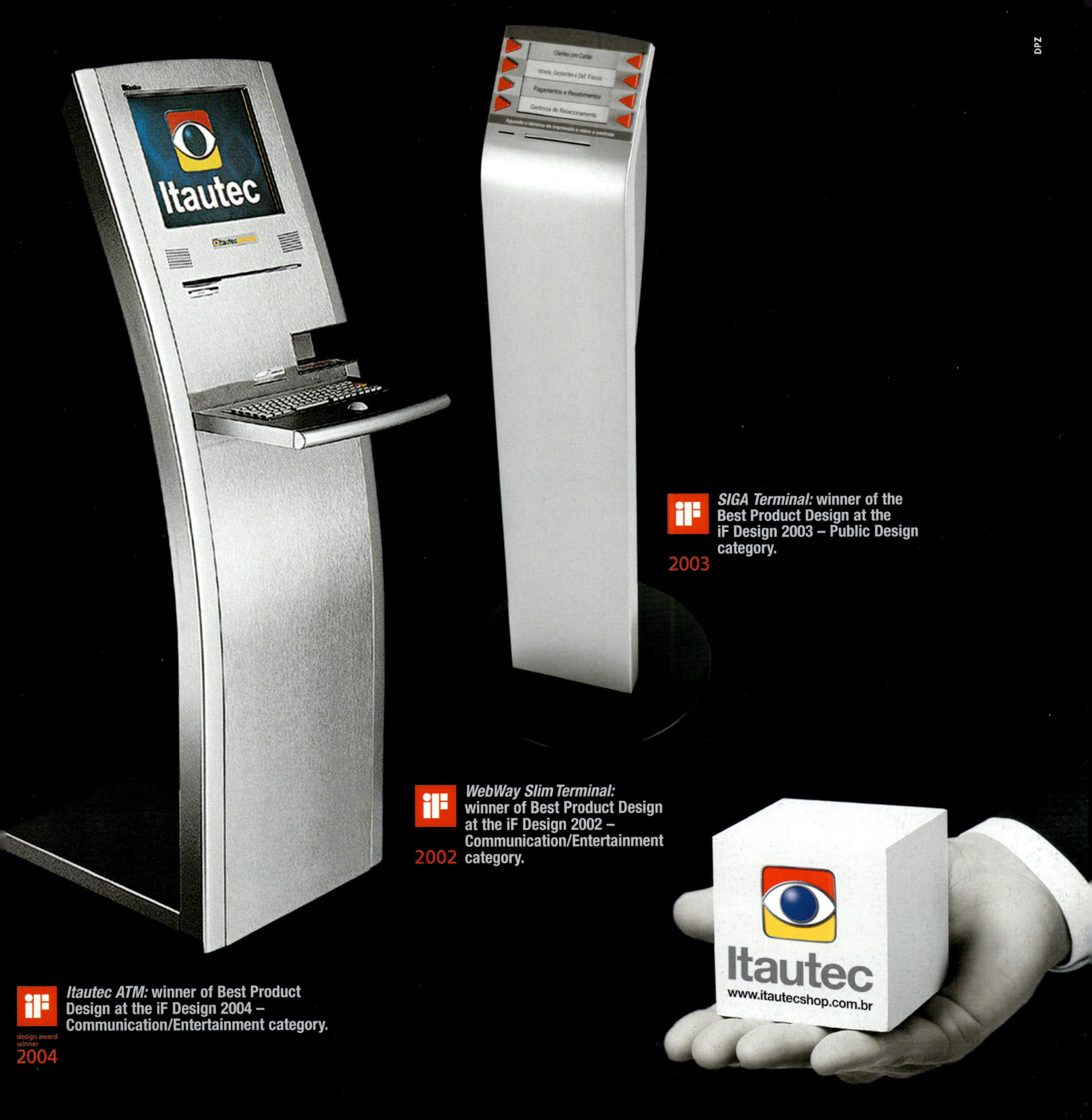
Itautec
SIGA Terminal: winner of the Best Product Design at the iF Design 2003 – Public Design category.
iF
2003
WebWay Slim Terminal: winner of Best Product Design at the iF Design 2002 – Communication/Entertainment category.
iF
2002
Itautec ATM: winner of Best Product Design at the iF Design 2004 – Communication/Entertainment category.
iF
design award winner
2004
Itautec
www.itautecshop.com.br
DPZ

Hersteller Manufacturer
Apple Computer Inc.
Cupertino, CA, USA

Produkt Product
12" PowerBook G4
Notebook
Notebook

Design
Bart Andre, Danny Coster
Daniele De Iuliis, Richard Howarth
Jonathan Ive, Duncan Kerr
Matthew Rohrbach, Doug Satzger
Cal Seid, Christopher Stringer, Eugene Whang
CAD Team: Carlos Ragudo, Carter Multz
Christopher Hood, Fred Simon
Irene Chan-Jones, Ken Provost, Mas Watanabe
Apple Computer Inc.
Cupertino, CA, USA

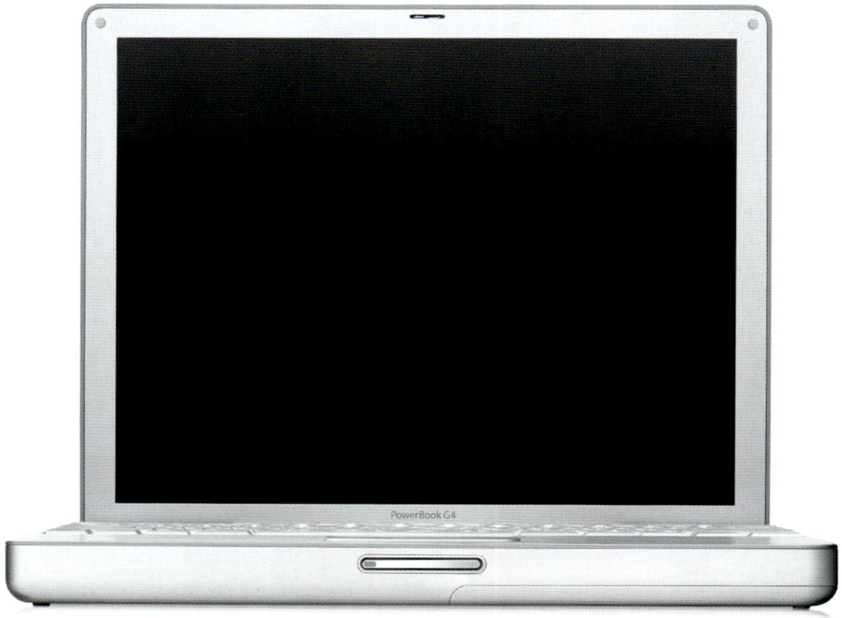

Das neue 12" PowerBook G4 wiegt lediglich 2,1 kg und hat einen 12-Zoll-Aktiv-Matrix-Bildschirm in einem kratzfesten und widerstandsfähigen Gehäuse aus einer Aluminiumlegierung. Das Gehäuse hat keinerlei vorstehende Verriegelungen und Hebel, die abbrechen, keine scharfen Kanten und keine außen angebrachten Tasten, die versehentlich gedrückt werden könnten. Das PowerBook ist mit einem SuperDrive Laufwerk, Bluetooth, AirPort Extreme, FireWire, dem Velocity Engine Prozessor und einem speziellen Grafikprozessor ausgestattet. Die Lithium-Ionen-Batterie liefert bis zu 5 Stunden Batteriestrom.

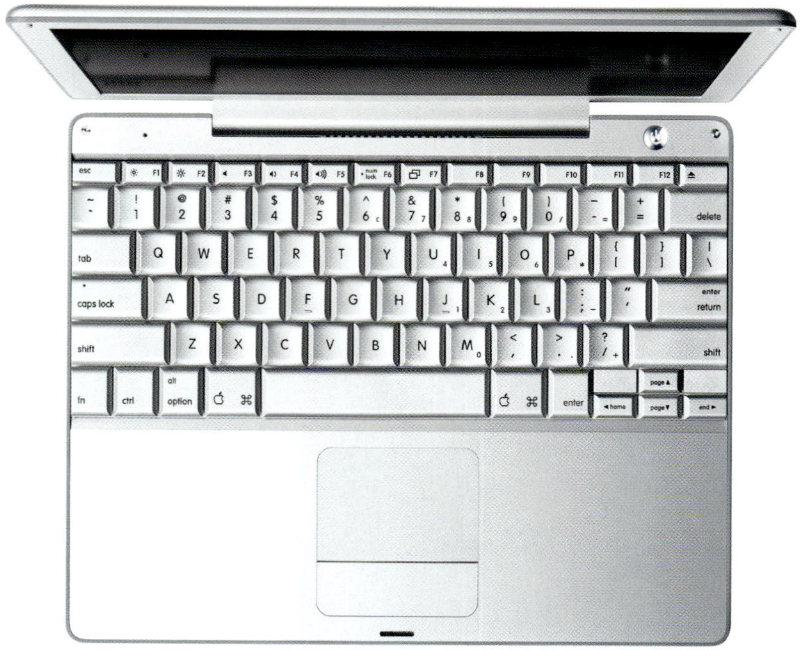

The new 12" PowerBook G4 weighs as little as 2.1kg and comes with a 12" active-matrix display in a scratch-resistant, sturdy aluminum-alloy outer skin. The frame has no outer locks or levers that might get broken, no sharp edges and no outside buttons that might be depressed by accident. The PowerBook comes with a SuperDrive, Bluetooth, AirPort Extreme slot, FireWire, a Velocity Engine processor and a special graphic processor. The lithium ion battery provides power for as long as 5 hours.

Hersteller Manufacturer
AsusTek Computer Inc.
Taipei, Taiwan, ROC

Produkt Product
S5200, M5200
PC-Notebook
PC notebook

Design
Devin Lin, Ken Wang, HW Wei
Asus In-House Design Team
Taipei, Taiwan, ROC

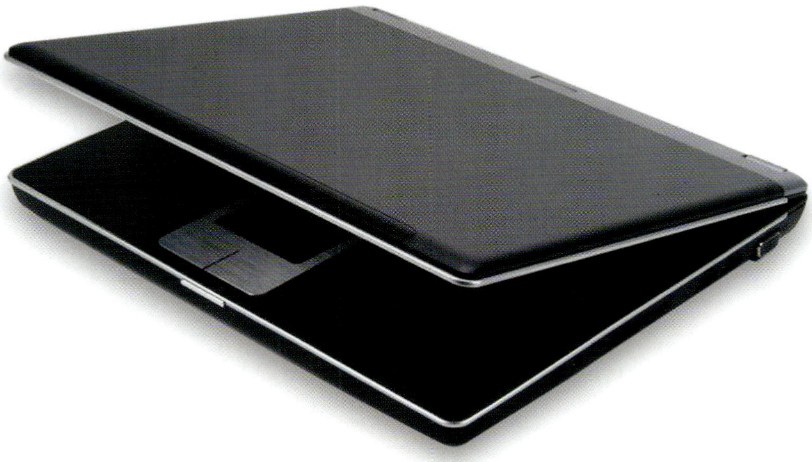

Sowohl das Asus S5200 als auch das M5200 Notebook zeichnen sich durch ihr geringes Gewicht und ihre äußerst dünne Bauweise aus. Beide Geräte verfügen über die neueste Centrino-Technologie von Intel und kabelloses LAN. Während es sich beim M5 um ein voll ausgestattetes Notebook mit 24-fach CD-ROM Laufwerk handelt, ist das S5 ein zugunsten von Tragbarkeit konzipiertes „Sub-Notebook", das für ein geringes Gewicht auf ein optisches Laufwerk verzichtet. Die Notebooks haben die Anmutung eines Lederetuis, das von Silberklammern gehalten wird. Das Design ist klar und modern und von einfachen geometrischen Proportionen geprägt.

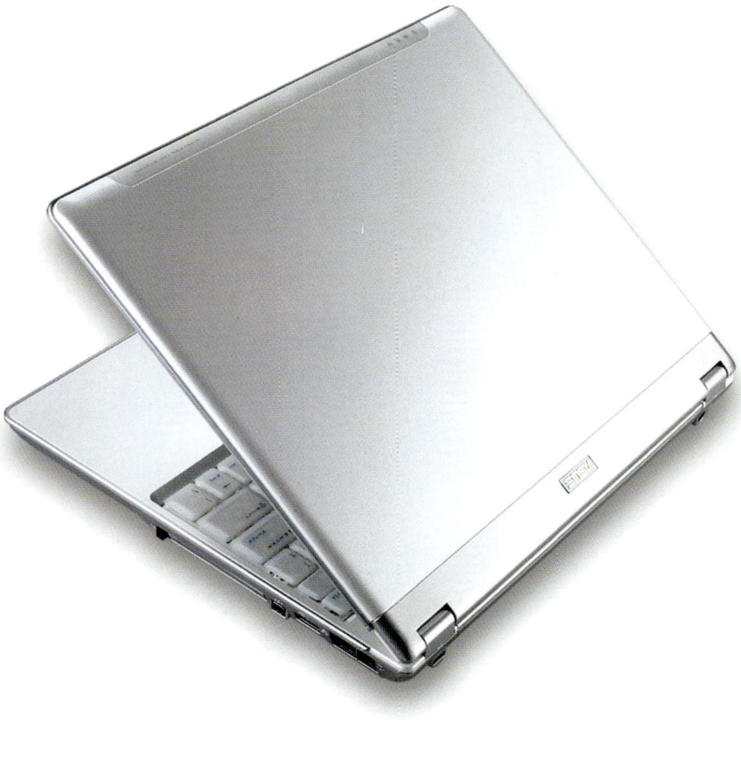

Both the Asus S5200 and the M5200 Notebook are extremely thin and true lightweight portables. Both come powered by the latest Intel Centrino technology and boast state-of-the-art wireless LAN. While the M5 is a full-fledged Notebook with a 24 x CD ROM drive, the S5 is designed as an ultra-portable sub-notebook – it has no built-in optical drives in order to reduce its weight. The notebooks have the appeal of a leather briefcase with a special silver frame. The appearance sports clear lines and is modern, based on simple geometric proportions.

Hersteller Manufacturer
LG Electronics Inc.
Pyungtaik-City, Kyunggi-Do, Korea

Produkt Product
Mckinley 15" XNOTE 15
Notebook PC

Design
Huh Byung Mu
Lee Ji Yeon
Kim Kyung Min
Corporate Design Center
LG Electronics
Seoul, Korea

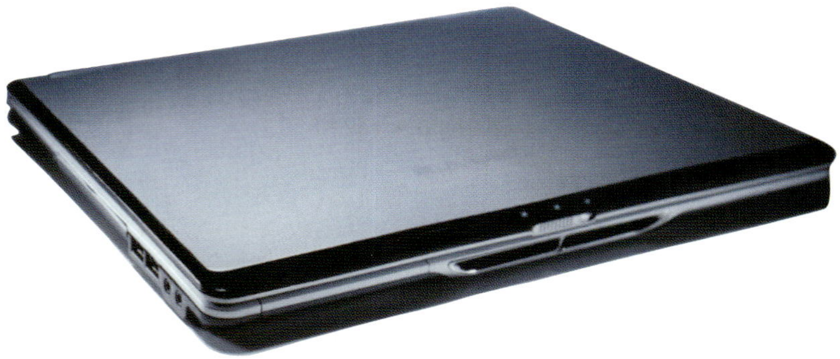

Das Mckinley 15" XNOTE 15 Notebook von LG Electronics ist mit seinem betont sachlichen und klaren Design vor allem für die Zielgruppe junger erfolgreicher Geschäftsleute mit hohen Anforderungen an Mobilität entwickelt worden, dabei werden verschiedene Funktionsaspekte wie Wireless, Multimedia durch einzelne Gestaltungselemente (Beleuchtung, Lautsprecher) besonders hervorgehoben. Der Batteriezustand kann für verbesserten Benutzungskomfort auch abgelesen werden, wenn das Notebook zugeklappt ist.

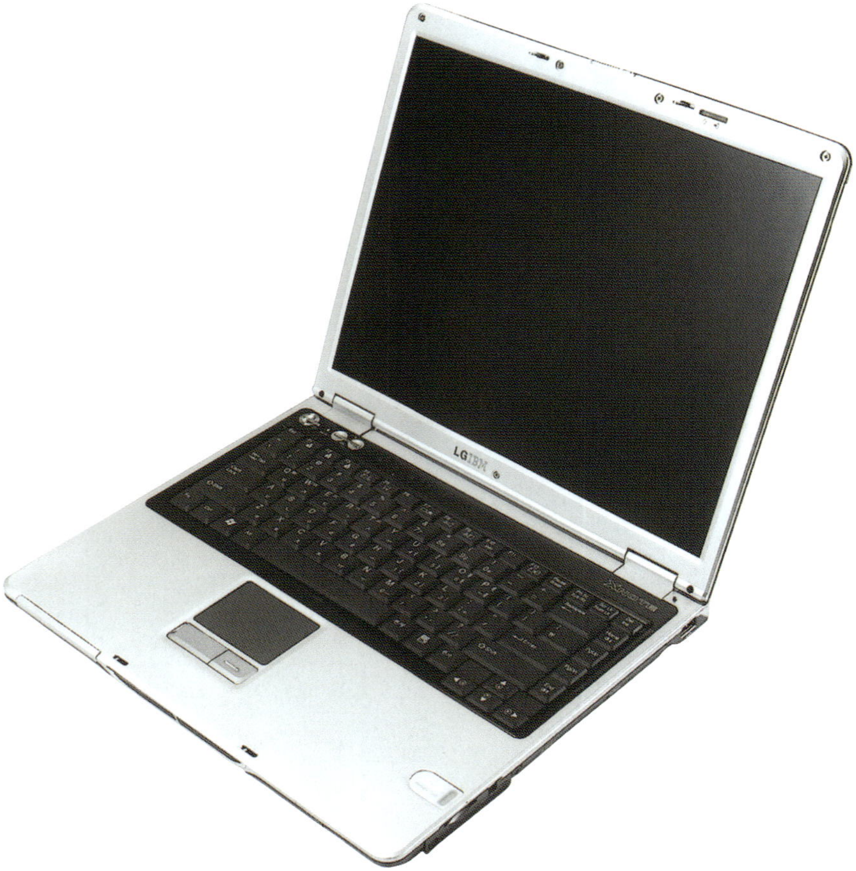

The LG Electronics' Mckinley 15" XNOTE 15 Notebook has a deliberately matter-of-fact and clear look to it, appealing above all to younger successful professionals who expect high portability and thus individual design elements (lighting, loudspeakers) emphasize specific functionalities, such as wireless and multimedia. Battery charge level is immediately apparent for enhanced user comfort even when the Notebook is shut.

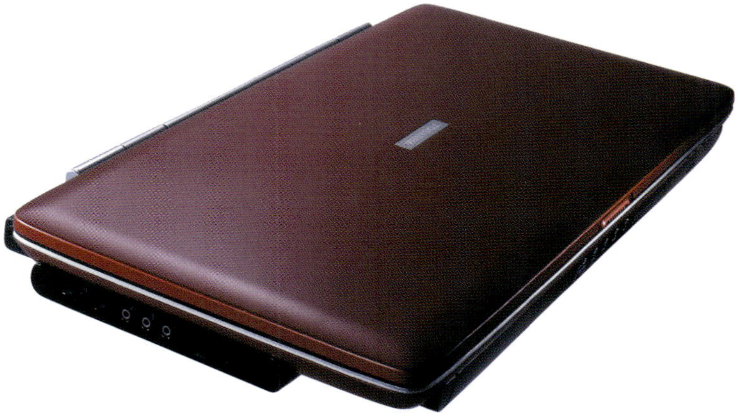

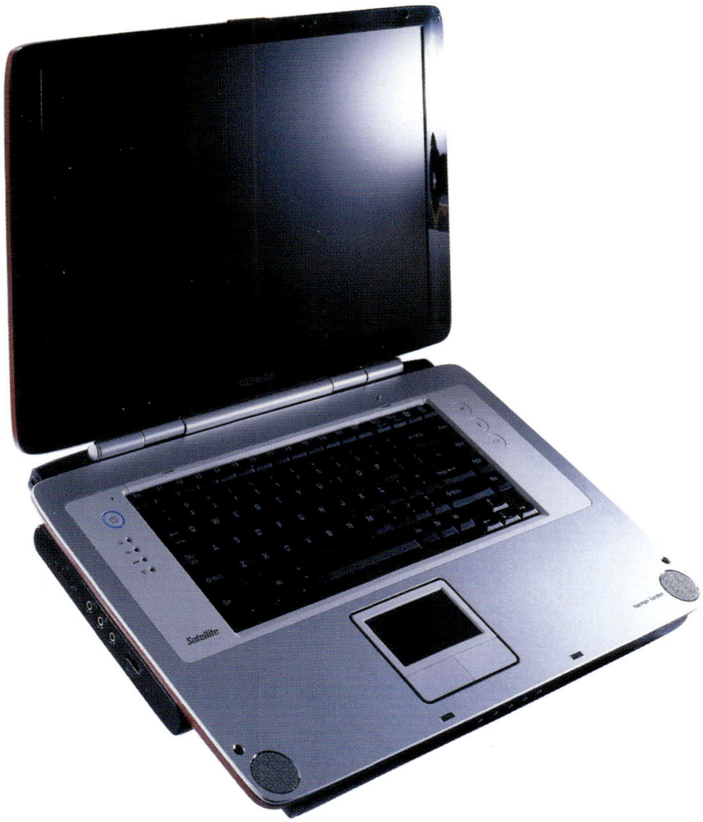

Hersteller Manufacturer
Toshiba Information Systems (UK) Ltd.
Weybridge, UK

Produkt Product
Satellite P 20
PC-Notebook
PC notebook

Design
Osam Kondo
Design Center
Toshiba Corporation
Tokio, Japan

Das Satellite P 20 Notebook von Toshiba ist mit seinem rot akzentuierten Design an das CD von Toshiba angepasst. Es besitzt ein 17-Zoll Display mit 1440 x 900 Pixel Auflösung und eignet sich ebenso für Computerspiele wie für aufwendige Grafikapplikationen und Office-Anwendungen. Weitere Features sind ein DVD-Player, ein Realtek ALC202A Soundchip mit 3D-Soundunterstützung sowie Stereo-Lautsprecher und Subwoofer des Harman/Kardon Soundsystems. Die Mediensteuerung erfolgt über fünf Steuerungstasten an der Vorderseite des Notebooks direkt neben dem optischen Laufwerk. Diese erlauben zudem die Wiedergabe von MP3-Files und Audio-CDs, auch während der Rechner abgeschaltet ist.

The Toshiba Satellite P 20 notebook with its red-and-silver color contrasts is a unique design statement that conforms with Toshiba's corporate design. The Notebook features a 17" screen offering 1440 x 900 pixel resolution and is suitable for computer games as well as graphic and office applications. Other features include a DVD player, a Realtek ALC202A soundchip with 3D sound support as well as stereo loudspeakers and a subwoofer from the Harman/Kardon sound system. Five buttons on the front side of the notebook adjacent to the optical disk drive control the media. They also allow MP3 files and audio CDs to be played even when the computer is turned off.

Hersteller Manufacturer
First International Computer Inc.
Portable Computing Group
Taipei, Taiwan, ROC

Produkt Product
MB 02
Notebook PC
notebook PC

Design
First International Computer Inc.
Portable Computing Group
Industrial Design Dept.
Taipei, Taiwan, ROC
IDEO Design
Palo Alto, CA, USA

Bei der Designentwicklung des Notebooks MB 02 wurde versucht, alle Spezifikationen eines Business-Notebooks mit geringst möglichem Platzaufwand sowie größtmöglicher Komprimierung und Batteriekapazität anzuordnen. Die Formgebung ist somit äußerst kompakt und das Design reduziert, funktionsbezogen und schlicht. Der Kontrast zwischen glatten und angerauten Oberflächen unterstreicht das Gestaltungskonzept.

In designing the Notebook MB 02 the aim was to fit all the specifications of a notebook into as small and compact a portable unit as possible. The resulting notebook is thus extremely compact and the minimalist design is functional and simple, with extremely low power depletion features. The contrast between flat and rough surfaces emphasizes the design concept.

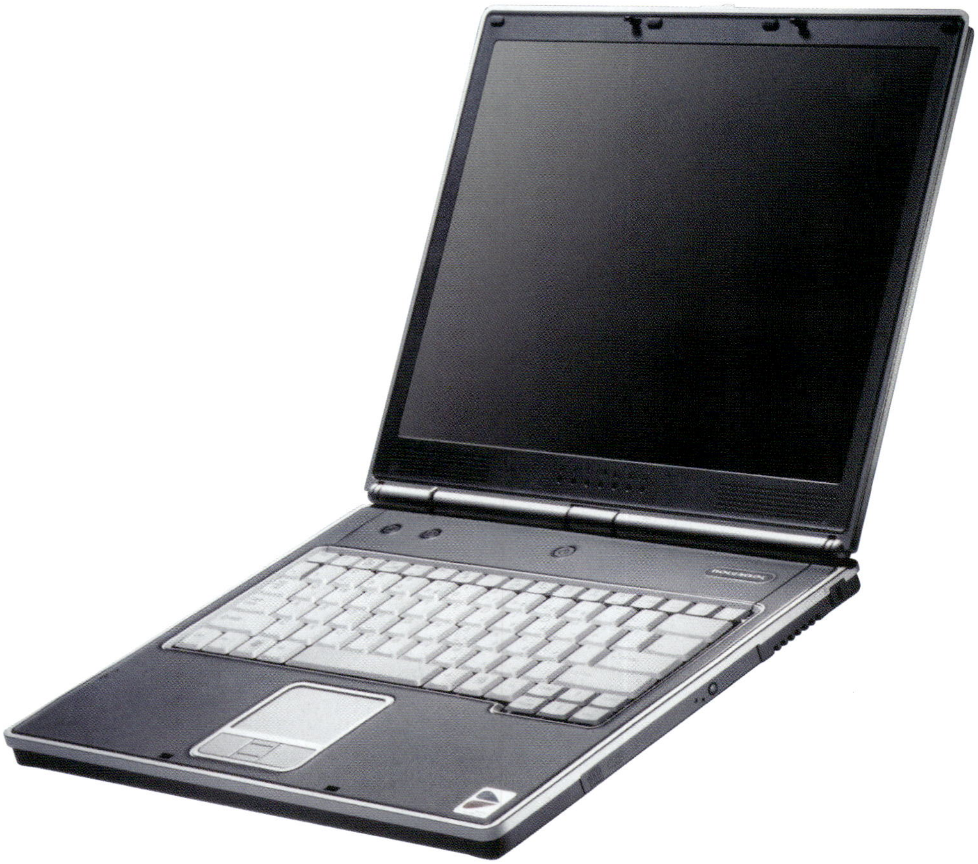

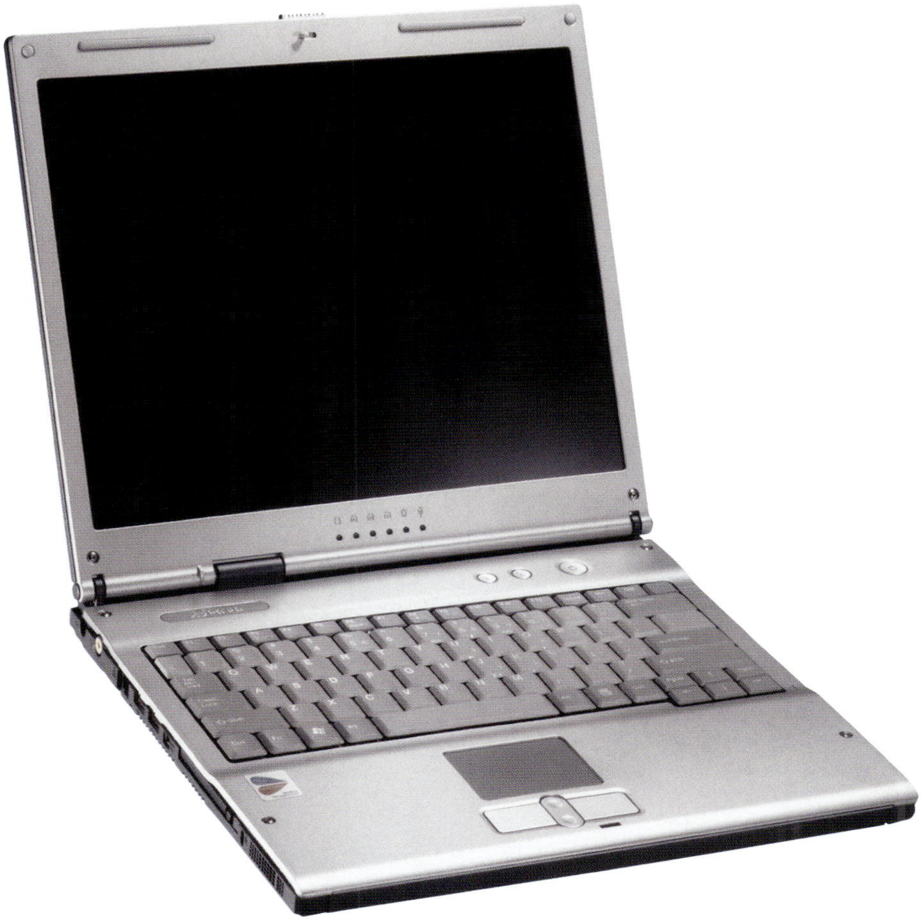

Hersteller Manufacturer
First International Computer Inc.
Portable Computing Group
Taipei, Taiwan, ROC
Legend Group LTD.
Shanghai, China

Produkt Product
E360/Y300 Series
Notebook PC
notebook PC

Design
First International Computer Inc.
Portable Computing Group
Industrial Design Dept.
Taipei, Taiwan, ROC

Das Design des E360/Y300 Series Notebooks ist minimalistisch und schlicht mit einem besonderen Augenmerk für gute Tragbarkeit – es wiegt lediglich 2,2 kg – und Langlebigkeit. Markantes Gestaltungsmerkmal ist der Hell-Dunkel-Kontrast zwischen dem farbig akzentuierten Außengehäuse des High End-Notebooks und dem hellen Silber des Aluminiums, wenn der Bildschirm aufgeklappt wird. Eine spezielle Energiespartechnologie gewährleistet eine besonders lange Lebensdauer der Batterie. Das Notebook beruht auf der Intel Centrino®-Technologie und beinhaltet u.a. eine drahtlose LAN Funktion.

The design of the E360/Y300 Series notebook is minimalist and simple with particular attention paid to portability (it weighs mere 2.2 kg) and durability. The prominent design feature is the color contrast between the darker external casing of this high-end notebook and the light silver of the screen when it is opened. A special energy-saving technology offers batteries a longer life. The notebook uses Intel's Centrino® technology and boasts such features as a wireless LAN utility.

Hersteller Manufacturer
Sony Corporation
Tokio, Japan

Produkt Product
PCG-Z1
PC-Laptop
Laptop computer

Design
Tetsu Kataoka
Sony Corporation
Tokio, Japan

Für den PCG-Z1 Laptop von Sony wurde ein schlichtes Gehäuse mit einem konsistenten, alle Bauteile umfassenden Design auf Basis einer einzigen gerundeten Linie entwickelt. Ein Zweiton-Farbschema unterstreicht diese Formgebung, da der Silberteil des Außenpaneels und die Ballenauflage den schwarzen Teil des LCDs und der Tastatur umfassen. Da die Buchsen und Tasten für Außenverbindungen keine Abdeckung haben, sind sie leicht zu erreichen, zudem sind die entsprechenden Symbole so angebracht, dass man nicht von der Seite schauen muss, um sie zu identifizieren.

The PCG-ZI laptop from Sony is designed with a simple body featuring a consistent design which encompasses all components and centers around a single curved line. A two-tone color scheme further enhances the form, with the silver section of the outer panel and the palm rest section gently surrounding the black segment of the LCD and keyboard. The connectors and buttons for outer connection terminals have no covers, so they can be easily accessed. Moreover, the respective icons are arranged so that they can be identified without having to look from the side.

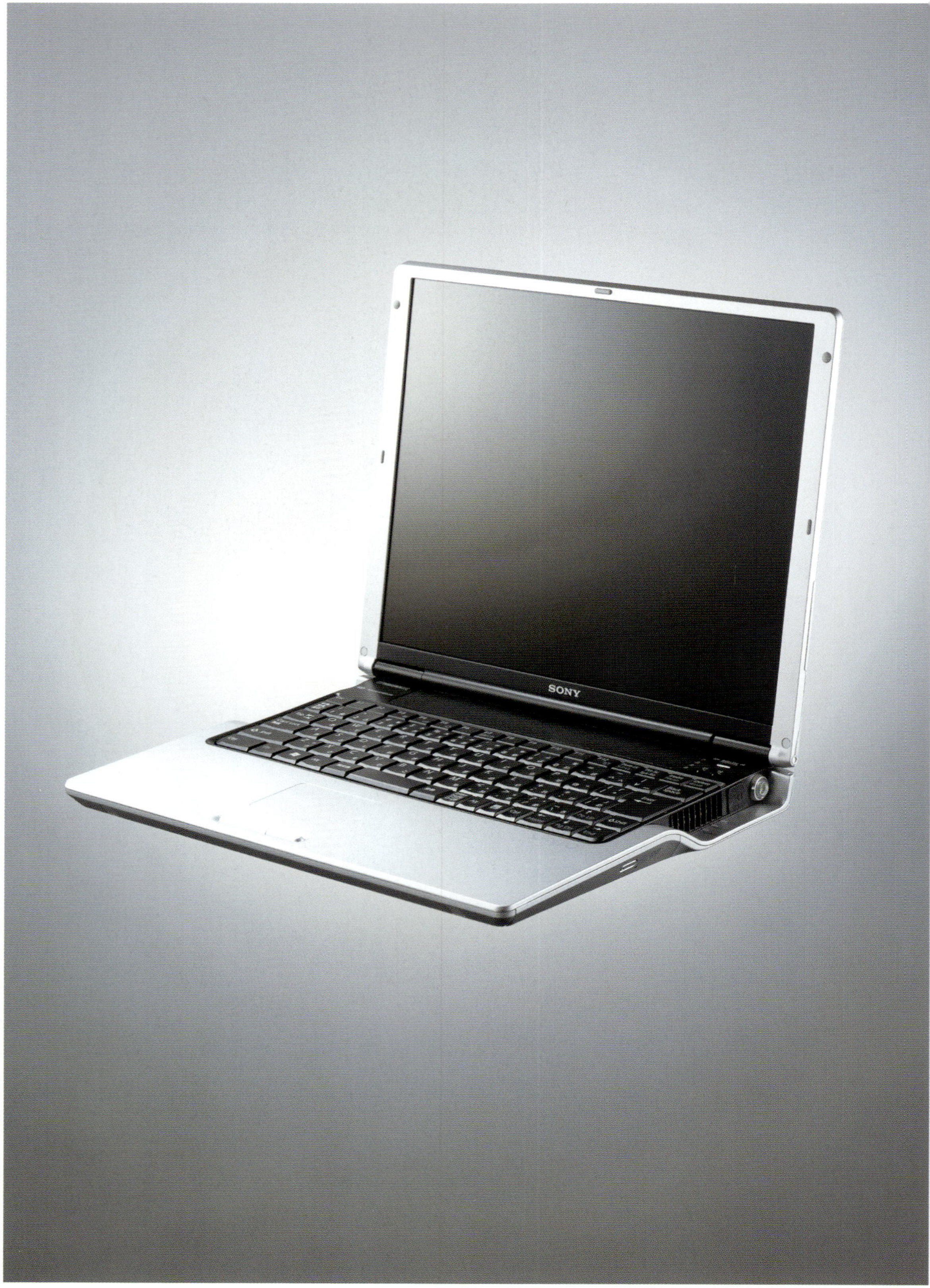

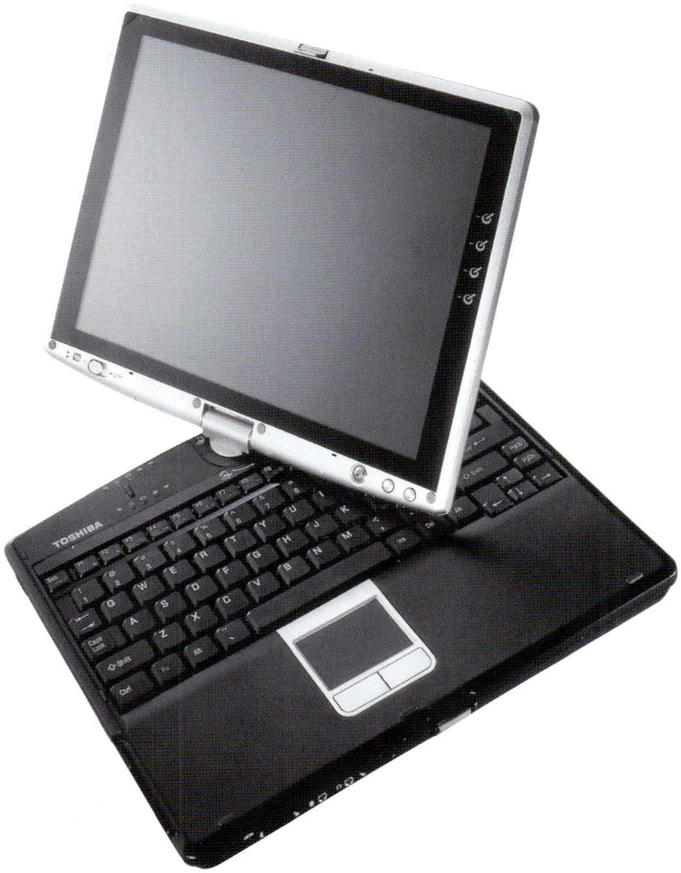

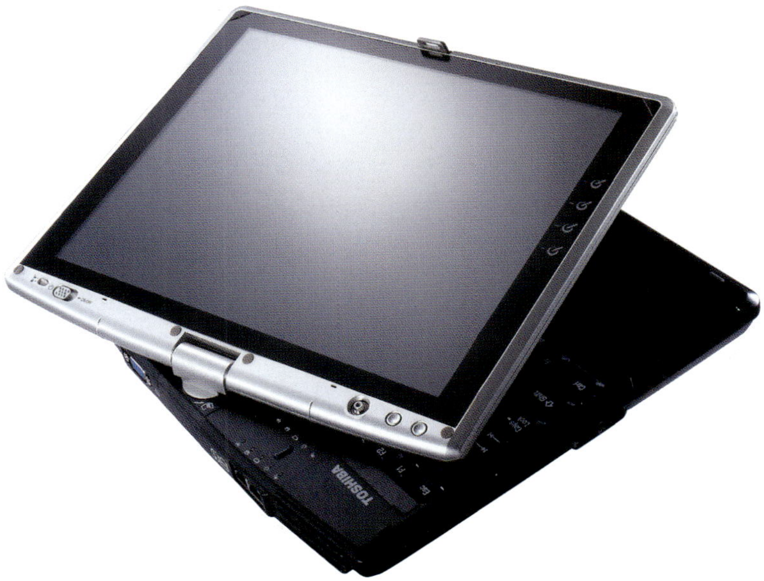

Hersteller Manufacturer
Toshiba Corporation
Tokio, Japan

Produkt Product
Portege M 200
PC-Notebook
PC notebook

Design
Satoshi Araki
Design Center
Toshiba Corporation
Tokio, Japan

Das Toshiba Notebook Portege M 200 besitzt eine Tablet-Funktion mit drehbarem LCD-Bildschirm. Es ist ein leistungsstarker und vielseitiger PC mit Intel® Centrino™ Mobiltechnologie, Microsoft® Office OneNote™ 2003 sowie erweiterten Funktionalitäten und verbesserter Benutzerfreundlichkeit. Die Eingabe kann über die Tastatur oder mit dem Stift erfolgen. Das optionale Toshiba Tablet Multi Dock ermöglicht verschiedene Anwendungsszenarien mit umfangreichen Dockingmöglichkeiten.

Toshiba's Portege M 200 notebook PC features a tablet function with a rotating LCD panel that can be turned to face out. It is a high-powered and multifunctional PC boasting Intel® Centrino™mobile technology, Microsoft® Office OneNote™ 2003, expanded functionality and improved user friendliness. Input is via keyboard or pen. The revolutionary Toshiba Tablet Multi Dock as an extra enables a variety of applications with extensive docking options.

Hersteller Manufacturer
Sony Corporation
Tokio, Japan

Produkt Product
CLIÉ PEG-UX50
Mobiler Netzwerk-Kommunikator
Mobile network communicator

Design
Shinichi Ogasawara
Sony Corporation
Tokio, Japan

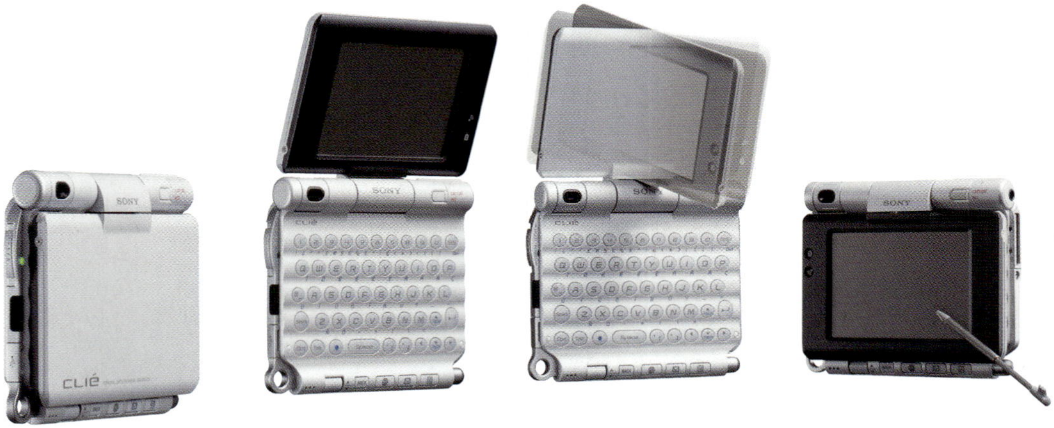

Das neue CLIÉ PEG-UX50 Handheld-Gerät wurde als mobiler Netzwerk-Kommunikator entworfen, der über die Rolle des typischen Tool-Aspekts hinausgeht. Mit sowohl Wi-Fi® und Bluetooth™ Wireless-Technologien erlaubt das Gerät den Anwendern, auf web-basierende Inhalte zuzugreifen und leicht über das Internet zu kommunizieren. Das Gerät wiegt lediglich 175 g und hat die äußerst reduzierten Abmessungen von 103 x 86,6 x 17,9 mm. Aus ergonomischen Gründen wurden die Tasten wellenförmig angelegt, um eine fehlerfreie Betätigung trotz der geringen Abmessungen zu gewährleisten. Das widerstandsfähige Magnesiumgehäuse schützt das Display vor Kratzern und Staub.

The new CLIÉ PEG-UX50 handheld device was designed as a mobile network communicator that goes beyond the typical tool role such appliances are conventionally assigned. Boasting both Wi-Fi® and Bluetooth™ wireless technology, the device enables users to access Web-based content and communicate effortlessly via the Internet. The device weighs a mere 175 g and is extremely small, measuring a mere 103 x 86.6 x 17.9 mm. The keys are integrated into the waveform style for ergonomic reasons, namely so they can be used precisely despite the size of the device. The durable magnesium casing protects the LCD from scratches and dust.

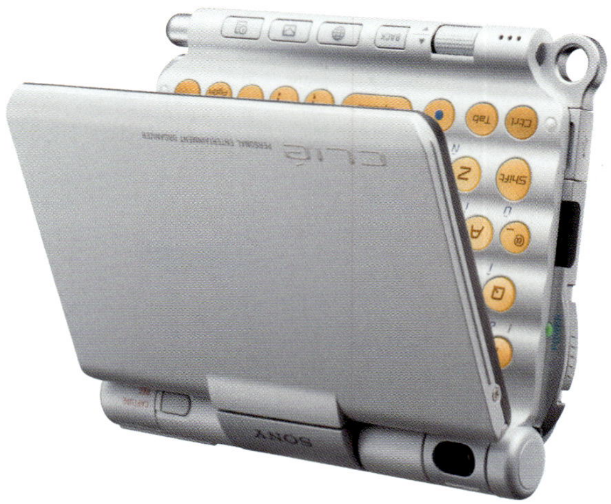

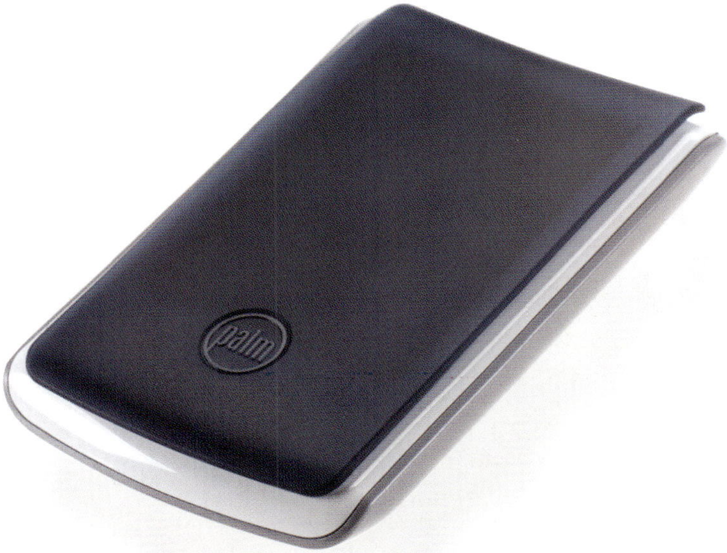

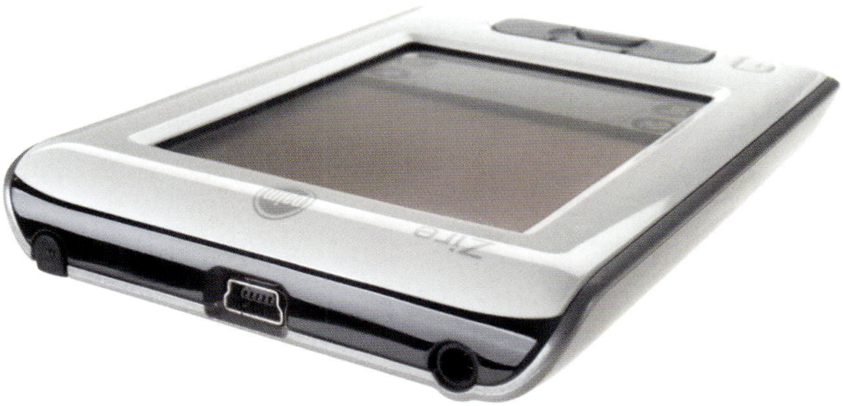

Hersteller Manufacturer
palmOne Inc.
Milpitas, CA, USA

Produkt Product
Palm™ Zire™
Handheld Computer

Design
Gadi Amit, Josh Morenstein
Mike Massucco, Chris Lenart
Yoshi Hoshino
Newdealdesign
San Francisco, CA, USA
Rich Gioscia
Lawrence Lam, Zita Netzel
Steven Shiozaki, Mark Babella
palmOne, Inc.
Milpitas, CA, USA

Der auf dem PalmOS® Betriebssystem basierende Palm™ Zire™ Handheld Computer bietet ein Personal Information Management mit Funktionen wie Adressbuch, Aufgabenliste und Kalender. Er ist klein und leicht genug, um in einer Hemdentasche Platz zu finden. Er besitzt eine polierte, weiße Front und einen Kunststoffklappdeckel aus einem lichtdurchlässigen Elastomer. Über vier Bedienknöpfe und ein vereinfachtes Interface werden die wichtigsten Informationen zur Verfügung gestellt.

The Palm™ Zire™ handheld computer, based on the PalmOS® operating system, offers personal information management with functions such as address book, task list and calendar. It is small and light enough to fit in a shirt pocket. The handheld device features a polished, white front and a plastic spring cover made of transparent elastomer. The most important information can be accessed via four user buttons and a simplified interface.

Hersteller Manufacturer
Seiko Epson Corporation
Suwa-shi
Nagano-ken, Japan

Produkt Product
EPSON Photo PC Player P-1000
Bildbetrachter- und Speichergerät
Photo storage viewer

Design
Kenji Yanagisawa
Matsumoto-shi
Nagano-ken, Japan

Mit dem EPSON Photo PC Player P-1000 können Bilder gespeichert und angesehen werden. Er verfügt über eine 10 GB Festplatte, auf der etwa 7500 Bilder gespeichert und in der Tasche mitgenommen werden können. Das Gerät besitzt eine blaue LED im Zentrum der Kreuztaste, die bei Anschluss an einen Drucker oder andere Peripheriegeräte blinkt. Bei der schlichten, rechteckigen Formgebung des Photo PC Players wurden insbesondere eine einfache Bedienung (Tastengestaltung) und Tragbarkeit (Rundung der Rückseite) berücksichtigt. Er besitzt ein Aluminiumgehäuse, verchromte Tasten und für die Front wurde ein dunkler Farbton gewählt, um die Fotos besser zur Wirkung kommen zu lassen.

The EPSON Photo PC Player P-1000 is designed for storing and viewing photos. It features a 10GB hard disk with storage space for about 7,500 pictures and is small enough to be slipped into a pocket. In the center of the X-button is a small blue LED which blinks when the device is attached to a printer or other peripheral device. The simple rectangular form of the Photo PC Player emphasizes user friendliness (key design) and portability (rounded rear side). It comes in an aluminum casing with chrome keys and in order that photos can be seen better, the front is a dark color.

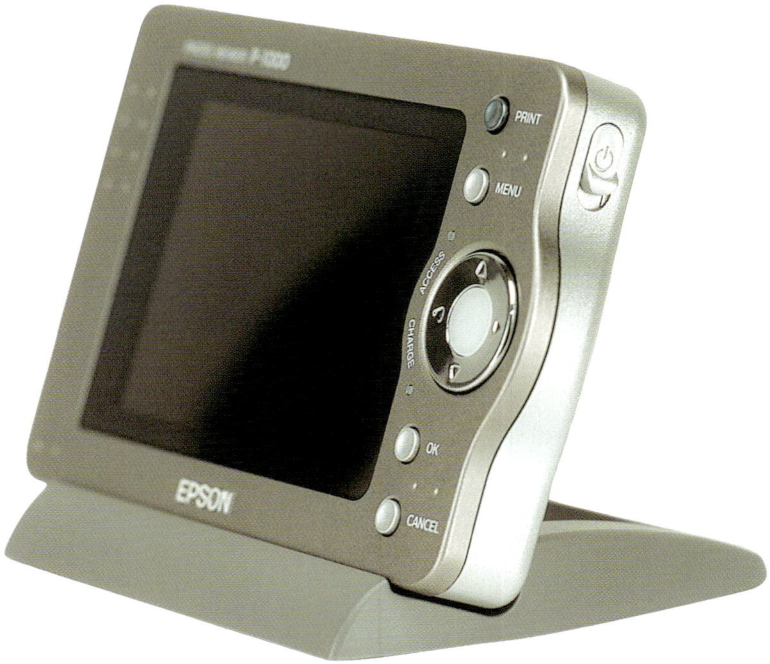

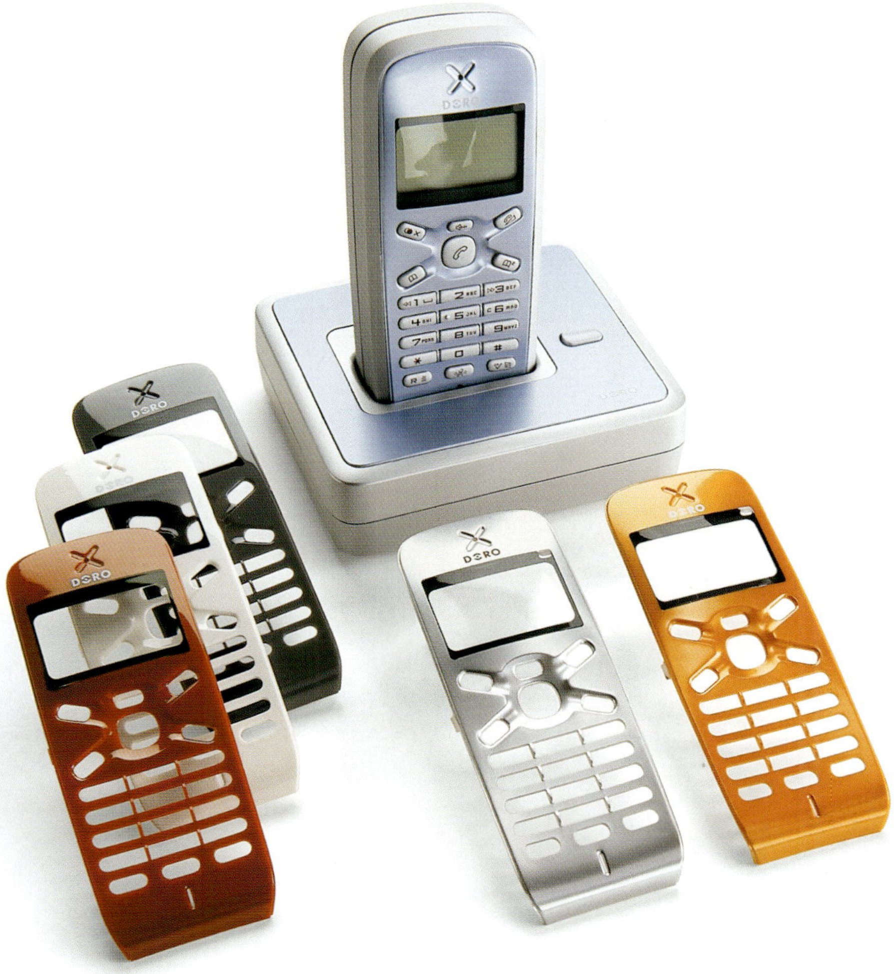

Hersteller Manufacturer
Doro AB
Lund, Sweden

Produkt Product
Doro 800 Series – 820/830R
Digitale schnurlose Haustelefone
Digital cordless home phones

Design
Fredrik Magnusson, Filip Sauer
Anna Stern, Fredrik Hallander
Propeller
Stockholm, Sweden

Die digitalen Schnurlostelefone 820 und 830R mit Anrufbeantworter aus der Doro 800-Serie gehören zu den kleinsten auf dem Markt. Das Design ist modern und trendorientiert, wobei neun unterschiedliche austauschbare Gehäusevarianten gewählt werden können. Über ein intuitiv zu bedienendes überschaubares Menü kann eine Vielzahl von Funktionen aktiviert werden. Die Benutzeroberfläche wurde so gestaltet, dass der Anwender sich mühelos darauf zurecht findet.

The digital cordless 820 and 830R home phones with integrated answering machines in Doro's 800 series are among the smallest of their kind on the market. Their design is modern and trendy, offering a choice of nine different interchangeable shells. A large variety of functions can be accessed through an intuitive, easy-to-understand menu. The user interface is designed in such a way that it is simple and effortless to operate.

Hersteller Manufacturer
Nokia Group
Espoo, Finland

Produkt Product
Nokia 5100
Mobiltelefon
Mobile phone

Design
Nokia Design Team
Calabasas, CA, USA

Das Nokia 5100 Mobiltelefon ist ein Triband-Handy (EGSM 900 sowie GSM 1800 und GSM 1900 Netze) und garantiert dem Nutzer weltweite Erreichbarkeit. Es verfügt über ein Farbdisplay, ist MMS-fähig und unterstützt Java™ Technik. Das Design wurde speziell für Menschen mit einem sportlich-aktiven Lebensstil entwickelt. So besitzt es eine integrierte Taschenlampe, ein Thermometer, eine Count-Down-Funktion, einen eingebauten Kalorienzähler, ein Stereo-UKW-Radio und eine Stoppuhr.

The Nokia 5100 mobile phone is a Triband phone (EGSM 900 as well as GSM 1800 and GSM 1900 networks) and guarantees that the user is available the world over. It features a color display, supports MMS and Java™ technology. The design was especially developed for people with an active lifestyle and includes a flashlight, a thermometer, a count-down function, a built-in calorie counter, a stereo short-wave radio and a stopwatch.

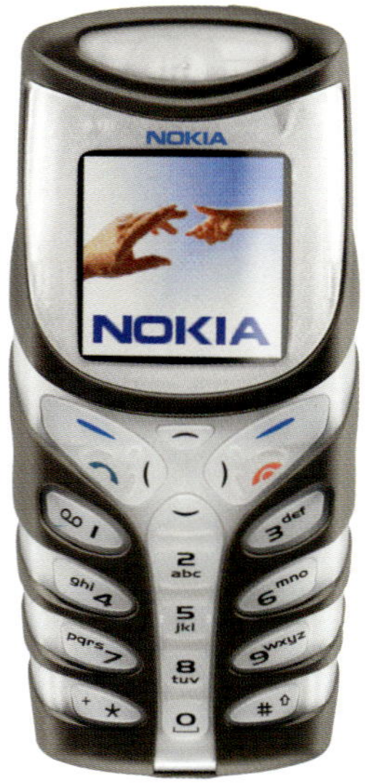

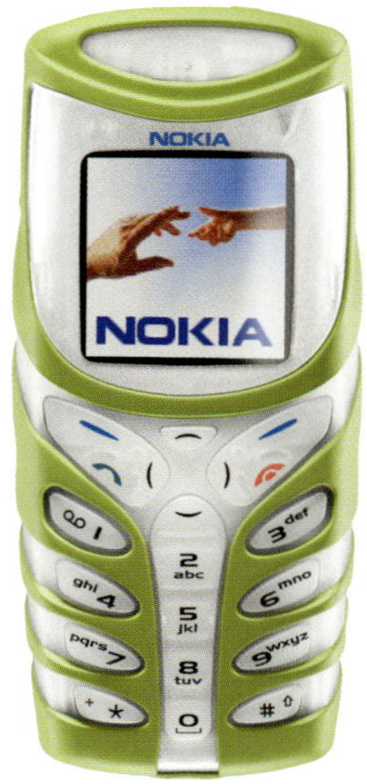

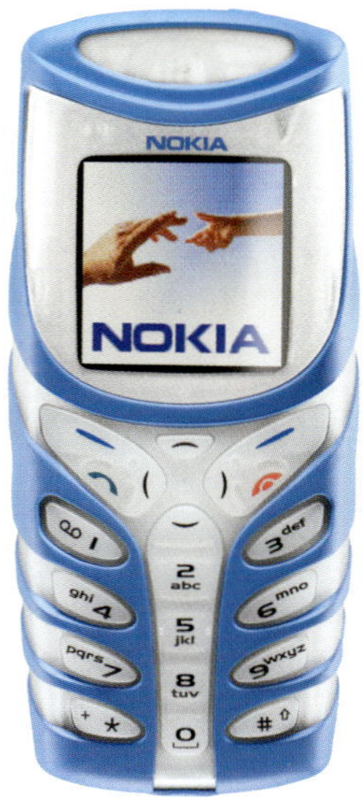

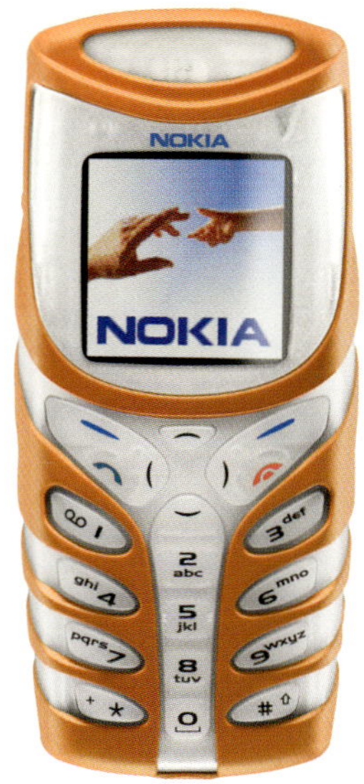

Hersteller Manufacturer
Nokia Group
Espoo, Finland

Produkt Product
Nokia 2100
Mobiltelefon
Mobile phone

Design
Nokia Design Team
Calabasas, CA, USA

Zum modernen Design des Nokia 2100 Mobiltelefons gehören transparente Xpress-on™ Cover in 6 verschiedenen frischen Farben wie z.B. Lightblue, Orange, Fuchsia. Das Handy verfügt über eine übersichtliche Tastatur mit griffigen Tasten, ein integriertes Sichtfenster im rückseitigen Cover und über spezielle Funktionen wie einen animierten Bildschirmschoner, Stoppuhr, SMS-Verteilerlisten, Wecker etc.

The modern design of the Nokia 2100 mobile phone includes a transparent Xpress-on™ cover in six different fresh colors, such as light blue, orange, and fuchsia. This cell phone has a well thought-out keypad with easy-to-touch buttons and an integrated window in the back cover. It also boasts special functions such as an animated screen saver, stop watch, SMS mailing lists, alarm etc.

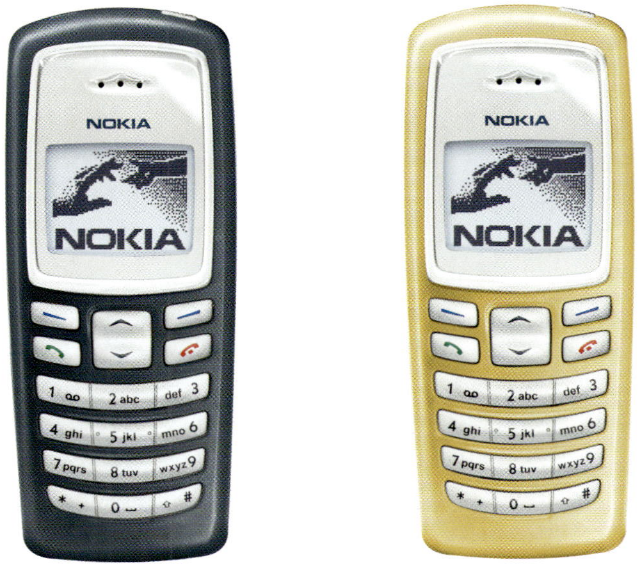

Hersteller Manufacturer
Samsung Electronics Co., Ltd.
Seoul, Korea

Produkt Product
Bluechip PDA Phone (SPH-i500)
PDA Multimedia-Telefon
PDA multimedia phone

Design
Park Bum Soo
Lee Jong Seong
Hyun Sang Min
Mobile Design Group
Samsung Electronics Co., Ltd.
Seoul, Korea

Bluechip PDA Phone (SPH-i500) ist ein Mobiltelefon und Palm-OS in einem äußerst schmalen und klappbaren PDA-Multimedia-Endgerät. Das Design ist schlicht und funktionsbezogen, ohne überflüssige Linien, unterstützt wird dieser Ansatz auch durch die versteckte mechanische Funktion (Klappbereich). Das Gerät ist mit einer Touch-Screen Schnittstelle (2.3-Zoll-TFT-LCD Touch-Screen) und einem Farb-LCD (176 x 140) mit 65.000 Farben ausgestattet. Es ist 54 mm x 87 mm x 21,5 mm groß.

Bluechip PDA Phone (SPH-i500) is a mobile phone and Palm-OS in one, providing a very slender and fold-out PDA multimedia terminal. It has a straightforward functional design, without any superfluous lines and is supported by the concealed mechanical function (in the folder). The device has a touchscreen interface (2.3" TFT LCD touchscreen) and a color LCD (176 x 140 dots) delivering 65,000, although it is sized 54 x 87 x 21.5 mm.

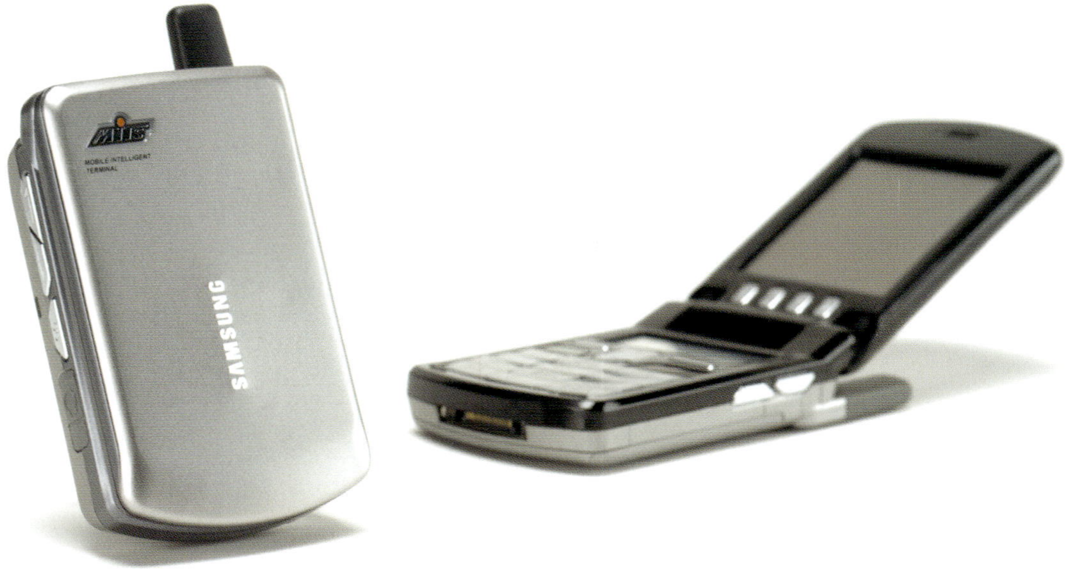

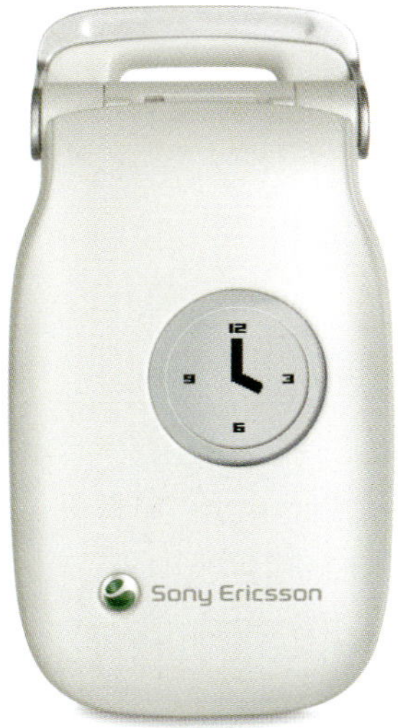

Hersteller Manufacturer
Sony Ericsson Mobile Communications AB
Lund, Sweden

Produkt Product
Z200
Mobiltelefon
Mobile phone

Design
John Leong
Eiji Shintani
Sony Ericsson
Creative Design Center
Lund, Sweden

Das Z200 von Sony Ericsson ist als Trend-Accessoire konzipiert. Es kombiniert clevere, benutzerfreundliche Details mit unkonventionellem und witzigem modischen Design und vielen Möglichkeiten zur individuellen Gestaltung wie auswechselbare Oberschalen, Hintergründe und polyphone Klingeltöne.

The Z200 from Sony Ericcson is conceived as a trend accessory; it combines clever user-friendly details with unconventional and comical, modern design. There are countless possibilities for personalizing the design such as interchangeable outer covers, backgrounds and polyphonic ring tones.

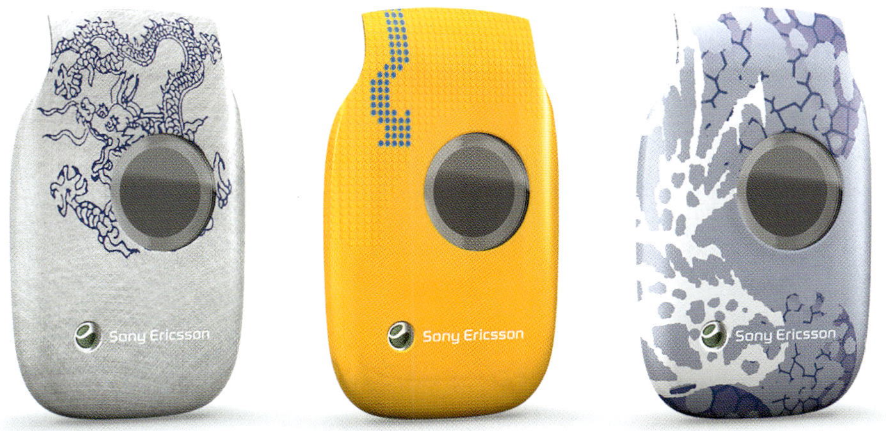

Hersteller Manufacturer
BENQ Corporation
Taipei, Taiwan, ROC

Produkt Product
BenQ Bar phone – M780
Mobiltelefon
Mobile phone

Design
BENQ Corporation
Lifestyle Design Center
Taipei, Taiwan, ROC

Das BenQ Bar phone – M780 ist aus einem High-Tech-Keramik-Material gefertigt, das nicht nur extreme Widerstandsfähigkeit, sondern auch eine besondere ästhetische Oberflächenqualität bietet. Das bislang vorwiegend für industrielle Zwecke eingesetzte Material bringt einen positiven ökologischen Aspekt mit sich, denn es macht eine Lackierung des Gehäuses überflüssig. Zwischen den silbernen Digitaltasten und der schwarzen Keramikoberfläche tritt als zusätzlicher Farbkontrast das blaue LED-Licht hervor.

The BenQ Bar phone – M780 is made of high-tech ceramics which ensures it is extremely heat- and chemical-resistant, while also being very aesthetically appealing. To date, such ceramic tends to be used for industrial purposes – and here it also means favorable ecological properties, as there is no need for a coating of paint. Between the silver digit buttons and the black ceramics, the blue LED light emerges and above it the highly scratch-resistant sapphire crystal monitor gleams for extra color contrasts.

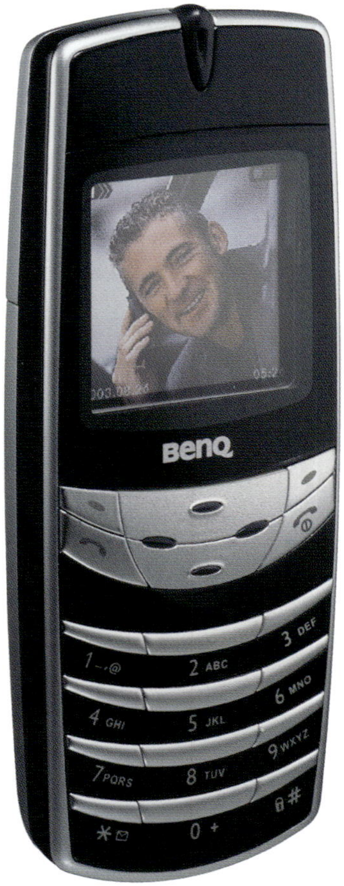

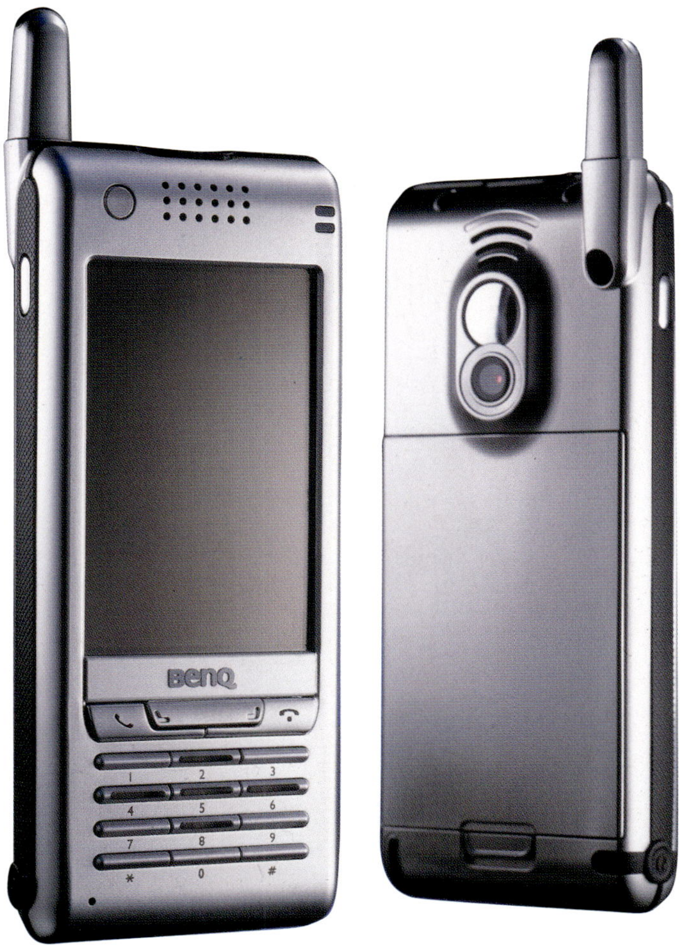

Hersteller Manufacturer
BENQ Corporation
Taipei, Taiwan, ROC

Produkt Product
BenQ Smart Phone p30
Mobiltelefon
Mobile phone

Design
BENQ Corporation
Lifestyle Design Center
Taipei, Taiwan, ROC

Das Smart Phone p30 von BenQ bietet Kommunikation, PDA und Digitalkamera in einem digitalen Endgerät. Schwerpunkte der Designentwicklung waren einfache Bedienung und entsprechend klare und geometrische Linien. Das Gehäuse des Gerätes besteht aus Aluminium, die schwarzen strukturierten Streifen an den Seiten sorgen für ergonomischen Griffkomfort. Der Nutzer kann für die Eingabe zwischen Touch-Control mit Pen und Tastenbetätigung wählen, das Display ist sowohl horizontal als auch vertikal zu benutzen.

BenQ's Smart Phone p30 offers communications, PDA and digital camera all rolled into one digital handset. The design focuses on simplifying use and creating correspondingly clear and geometrical lines. The phone sports an aluminum metal shell, with black stripes on the sides for comfortable and ergonomic grip. Users can toggle between inputting data by touch-control using a Pen and by digital buttons; the display can be used laterally and horizontally.

Hersteller Manufacturer
Motorola Inc.
Fort Lauderdale, FL, USA

Produkt Product
Motorola MTH650
Handsprechfunkgerät
Handheld walkie-talkie

Design
Scott Richards
Bruce Claxton
Avihu Cohen
Aviv Shoval
Mark Palmer
Noam Shani
Motorola Inc.
Fort Lauderdale, FL, USA

Das Motorola MTH650 Tetra Handsprechfunkgerät ist für den öffentlichen und privaten Sicherheitsbereich, den Rettungsdienst oder das Transportwesen entwickelt worden. Es bietet eine komfortable Anwendung, zuverlässigen Betrieb, Sicherheit und flexible und schnelle Kommunikation überall da, wo es auf Sekunden ankommt. Das ergonomische Design gewährleistet eine optimale Trageweise am Gürtel oder in der Hand. Die Bedienung ist einfach und sicher. Das Funkgerät verfügt zudem über ein großes Anzeigenfeld, einen fest zugewiesenen, oben liegenden Notrufknopf, einen Einstellknopf für die Kanal-Suche oder Lautstärken-Einstellung, außerdem individuell programmierbare Knöpfe.

The Motorola MTH650 handheld walkie-talkie was developed for use by police, fire and public safety professionals, for rescue services and in transport. It is comfortable to use, operates reliably and offers fast and flexible communication in situations where every second counts. The ergonomic design means it can be carried by hand or worn on a belt. The radiophone is easy and safe to use. In addition it features a large display, a specific emergency button, a dual-mode rotary knob for channel select or volume, as well as individually programmable buttons.

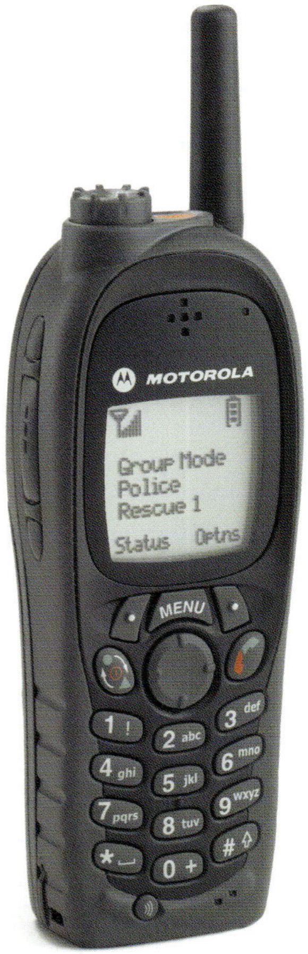

Hersteller Manufacturer
Siemens AG
ICM MP UC SL
München, Germany

Produkt Product
SL 55
Mobiltelefon
Mobile phone

Design
Graham Keevy
designafairs GmbH
München, Germany

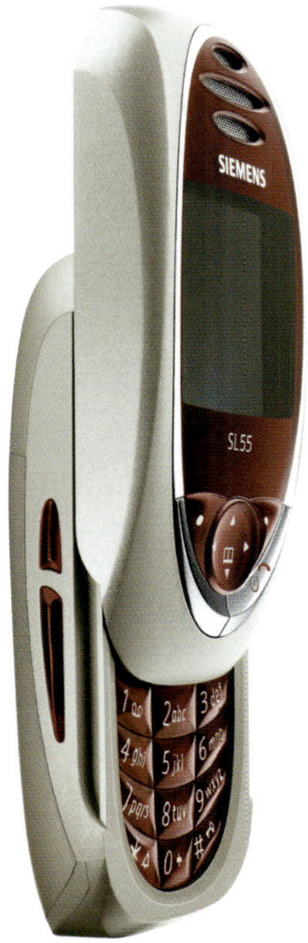

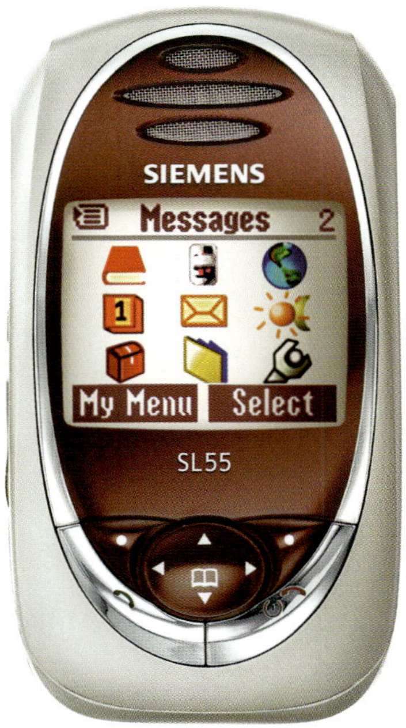

Das SL 55 Mobiltelefon ist für eine modebewusste Zielgruppe im Alter von 25 bis 35 Jahren konzipiert. Durch die Schiebefunktion ist es besonders klein. Es besitzt ein gut beleuchtetes Tastaturfeld, ein gutes Farbdisplay und ist in den Farben Rubinrot oder Titan erhältlich.

The SL 55 mobile phone was developed with a fashion-conscious target group in the 25-35-year-old age bracket in mind. The slide-out function makes it particularly small, and it features a well-lit key pad, a strong color display and is available in two colors: ruby red or titanium.

Hersteller Manufacturer
NEC Corporation
Tokio, Japan

Produkt Product
e525
Mobiltelefon
Mobile phone

Design
Masakatsu Takizawa
NEC Design
Tokio, Japan

Das e525 Mobiltelefon von NEC ist von der Ästhetik japanischen Industriedesigns geprägt. Das Dualband-Gerät (GSM 900/1800) hat eine interne Antenne, eine digitale Kamera und eine Auflösung von 162 x 216 Pixel. Das Design wurde für eine junge, trend- und stilbewusste Zielgruppe entwickelt.

NEC's e525 mobile phone has an aesthetic appearance representative of sophisticated Japanese industrial design. The Dualband phone (GSM 900/1800) boasts an internal antenna, a digital camera and display resolution of 162 x 216 pixels. The design was developed with a young, trend- and style-conscious target group in mind.

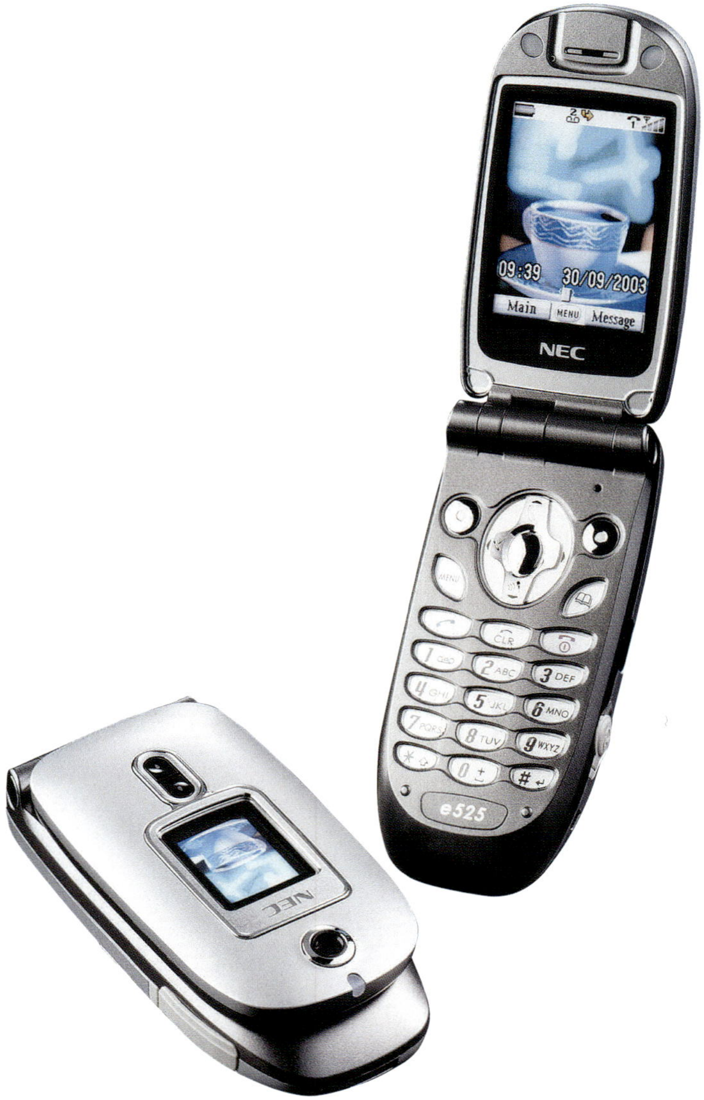

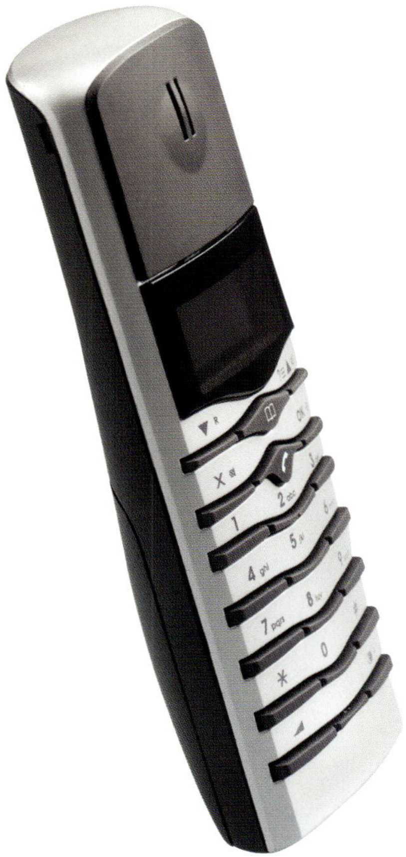

Hersteller Manufacturer
DRS Electronics Ltd.
Hong Kong, China

Produkt Product
CT 7020
DECT Schnurloses Telefon
DECT cordless phone

Design
Henrik Andersen
Frederick Rickmann
Design/Nord IS
Aarhus, Denmark

Das Design des CT 7020 Cordless telephone trägt der Tatsache Rechnung, dass ein schnurloses Telefon ein Funktionsgegenstand der Einrichtung und kein Handy ist. Insofern wurde der ergonomischen Gestaltung besondere Aufmerksamkeit gewidmet. Die Bedienfunktionen haben eine puristische Linienführung und sind intuitiv zu verstehen. Die alphanumerischen Display-Informationen werden in klarem Blau im Display sichtbar. Die Tasten sind aus grifffreundlichem Silikongummi gefertigt mit einem präzisen Drukkpunkt.

The design of the CT 7020 cordless phone takes into account that such telephones are functional objects for household or office use and not mobile phones. To this extent, especial attention was paid to delivering prime ergonomic qualities. The interface functions have a purity of line and are intuitively understandable. The alphanumeric display appears through the screen in precise clear blue. The keyboard is made of comfortable silicone rubber delivering optimal tactile response.

Hersteller Manufacturer
Deutsche Telekom AG
Darmstadt, Germany

Produkt Product
T-Sinus 111-154 Produktfamilie
WLAN-Produktserie
WLAN product range

Design
Ma Design
Kiel, Germany

Die WLAN-Modems T-Sinus 111 DSL und T-Sinus 154 DSL integrieren DSL-Modem, Router und WLAN-Station. Das Design ist auf unterschiedliche Gehäusegrößen übertragbar und erlaubt eine flexible Nutzung der LED-Positionen. Die hinterdruckte Plexiglas-Blende deckt in den Produktvarianten die jeweils nicht benötigten LEDs ab und weist in der Farbe Eisblau auf das integrierte DSL-Modem hin, in Grau auf ein ISDN-Modem. Die Buchsenbelegung an der Rückseite ist über einen Wechseleinsatz variierbar. Die Formensprache ist insgesamt klar und vermittelt die Einfachheit der Handhabung sowie die Qualität der Verbindung.

The WLAN T-Sinus 111 DSL and T-Sinus 154 DSL modems integrate a DSL modem, router and WLAN terminal. The design is transferable to different housing sizes and ensures the LED positions can be used flexibly. The back-printed Plexiglas blind covers unused LED holes and with its ice blue highlights the integrated DSL modem, whereas the grey version is used for the ISDN modem. The flexible inset at the back makes for simple reconfiguration of plugs. Overall, the formal vocabulary is clear and conveys a sense of how easy the unit is to use for high-quality connections.

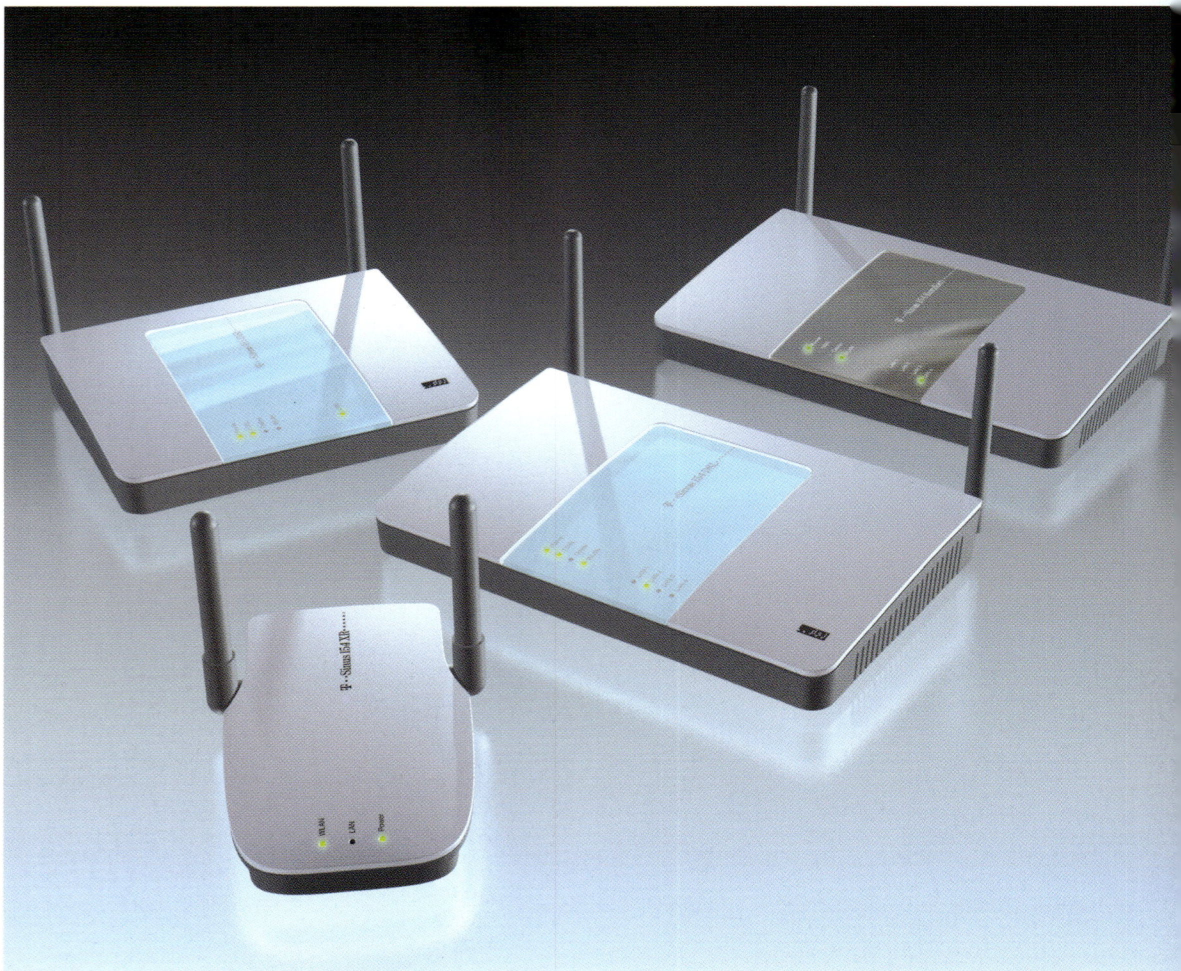

Hersteller Manufacturer
Samsung Electronics Co., Ltd.
Seoul, Korea

Produkt Product
SCH-X800, SPH-X9100
Mobiltelefon
Mobile phone

Design
Mark Delaney
Park In-Gon
Lee Chang-Soo
Mobile Design Group
Samsung Electronics Co., Ltd.
Seoul, Korea

Das Mobiltelefon SCH-X800, SPH-X9100 zeichnet sich durch ein modisches Accessoire-Design aus, das speziell für den trendbewussten Verbraucher konzipiert wurde. Es ist aus recyclebarem unlackiertem Plastikmaterial (PC) und Aluminium hergestellt. Benutzeroberfläche und Tastatur auf der Frontseite sind einfach zu bedienen.

The SCH-X800 and SPH-X9100 mobile phones stand out for the elegant fashionable accessories, designed specifically with the fashion-conscious in mind. It is made from recyclable plastic (PC) with no paint used, and aluminum. The simple interface and keypad, which takes up the entire front, are easy to operate.

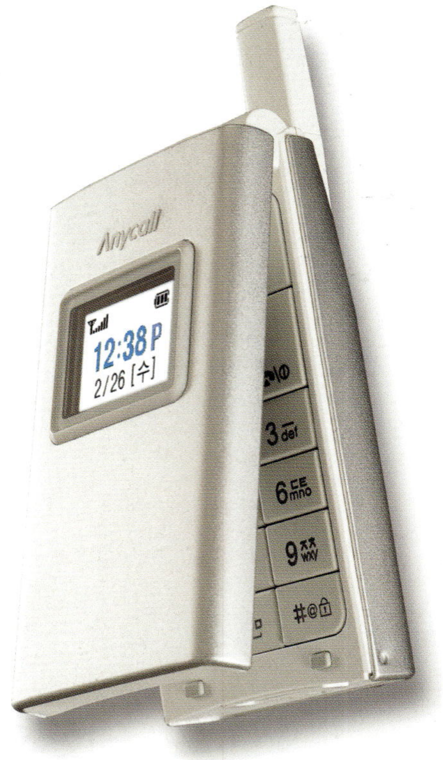

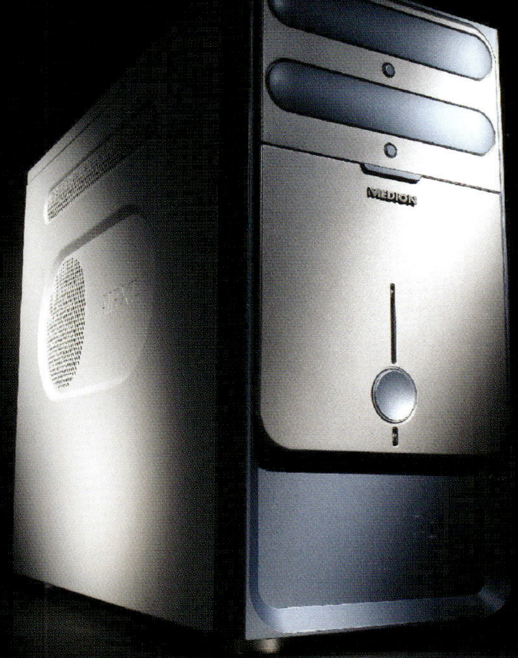
MEDION

Hersteller Manufacturer
Samsung Electronics Co. Ltd.
Seoul, Korea

Produkt Product
AIR Remote
Fernbedienung
Remote control

Design
Ok Joon-Ho
Kim Mi-Jeong
Samsung Electronics Co. Ltd.
Seoul, Korea

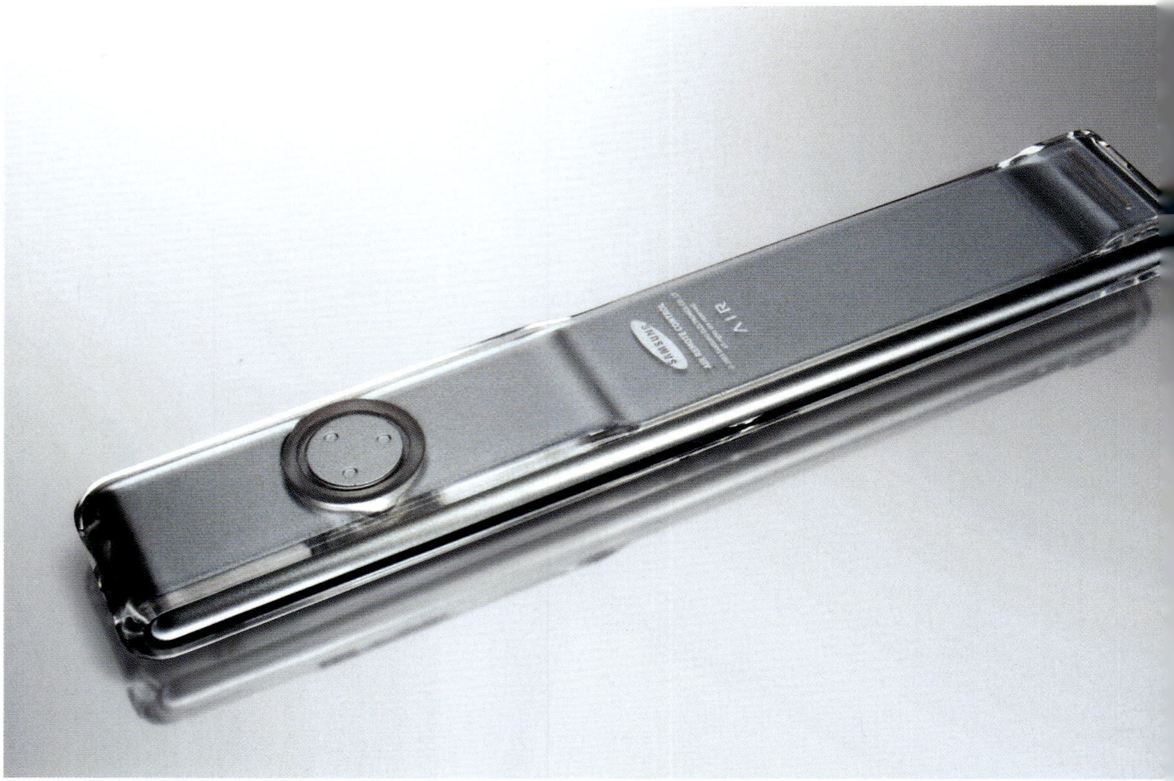

Die Fernbedienung AIR Remote funktioniert aufgrund natürlicher Handbewegungen und erfordert keine Knopfbetätigung. Sie ist mit drei Schwingungssensoren und drei Kreiselsensoren sowie Steuerungsfunktionen ausgestattet. Die von Experten ermittelten einzelnen Bewegungssequenzen sind auf der Fernbedienung dargestellt, wobei durch ein Touch-Sensor die Hintergrundbeleuchtung aktiviert werden kann, zudem werden auch alphanumerische Zeichen erkannt. Die Handzeichen müssen vom Nutzer nicht eingeübt, sondern können spontan ausgeführt werden. Das transparente Gehäuse ist kompakt und liegt gut in der Hand.

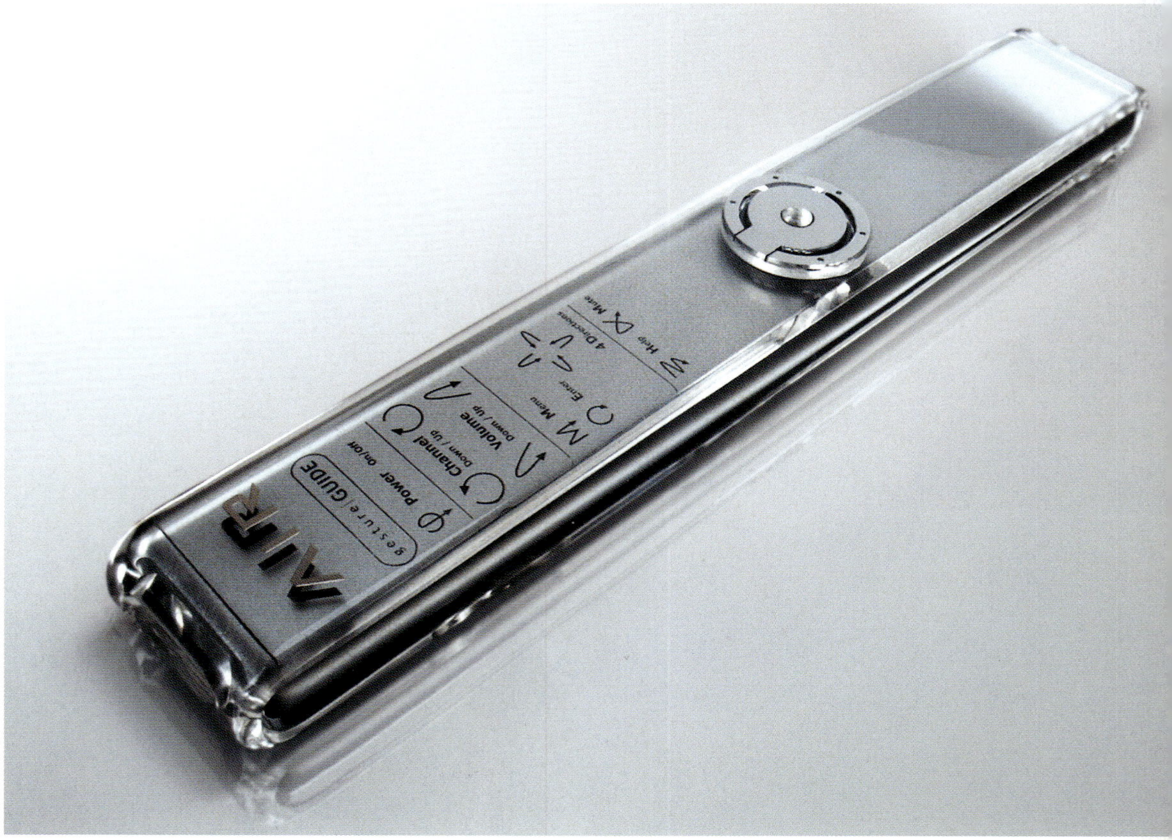

The AIR Remote control functions on the basis of natural hand gestures and requires no button actuation. It has three gyro sensors and three accelerometers as well as control appliances. Identified by gesture experts, the individual gestural sequences are represented on the remote control, and the back lighting is activated by a touch sensor. The remote also recognizes alphanumeric characters. Users need not practice the hand gestures, but can act spontaneously. The transparent outer body is compact and lies well in the hand.

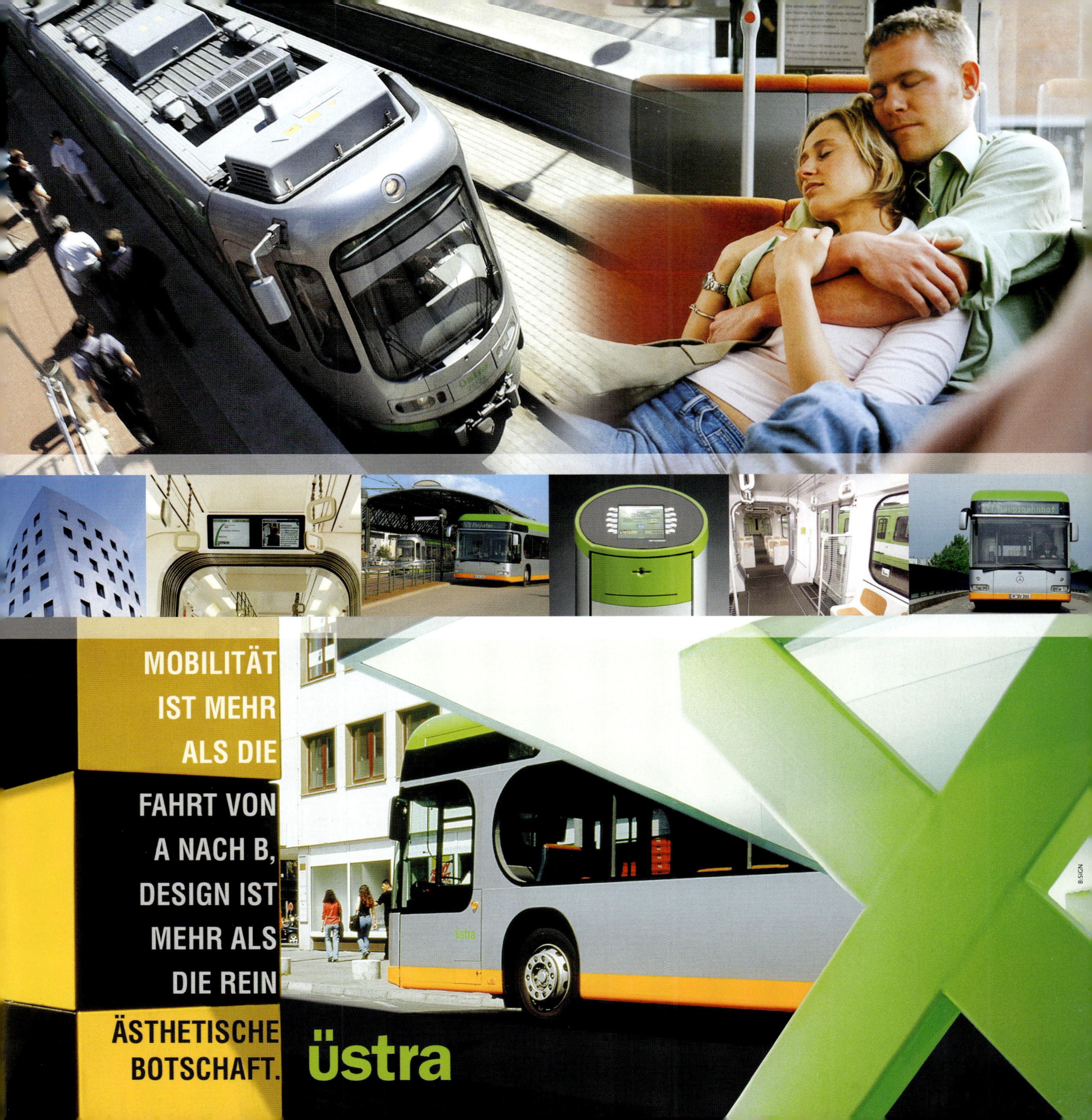
MOBILITÄT
IST MEHR
ALS DIE
FAHRT VON
A NACH B,
DESIGN IST
MEHR ALS
DIE REIN
ÄSTHETISCHE
BOTSCHAFT.
üstra

Hersteller Manufacturer
Cisco Systems
San Jose, CA, USA

Produkt Product
Cisco Aironet 1200 Series
Wireless Access Point

Design
Sandrine Lebas
Kenneth Wood
Lunar Design
Palo Alto, CA, USA
Mike Chmura, Ron Seide
Gary McCavitt, Luis Arreola
Tim Cox, Jim Mass, Tim Frank
Cisco Systems
San Jose, CA, USA

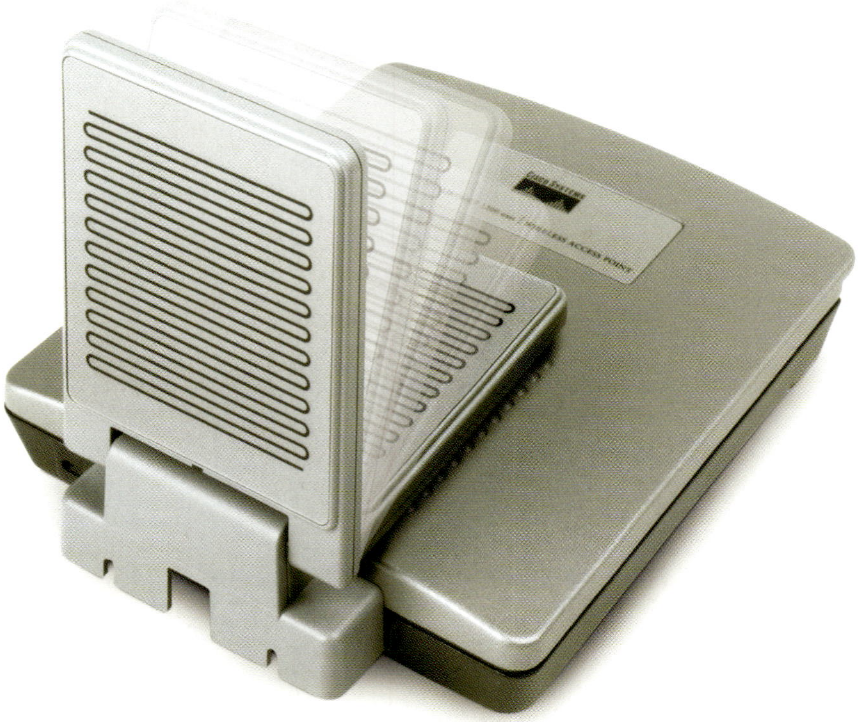

Die Cisco Aironet 1200 Series erlaubt es den Anwendern, ein zuverlässiges, drahtloses, lokal begrenztes Netzwerk (WLAN) zu installieren. Als drahtlose Auflage integriert es sich nahtlos in ein bereits existierendes Netzwerk oder schafft frei stehende, komplett drahtlose Netzwerke und ermöglicht somit rasche Mobilität und Flexibilität. Überall, wo ein drahtloser Zugang installiert ist, gewährleistet es eine konstante Verbindung zu allen Netzwerkquellen. Das modulare Design der Serie unterstützt IEEE 802.11a und 802.11b Technologien im Einzel- und Dual-Mode-Verfahren, gekoppelt mit externen 2.4 GHz und integrierten 5 GHz-Antennen. Die Gestaltung ist schlicht und neutral, um die Serie in Einklang mit verschiedenen Büroeinrichtungen zu bringen.

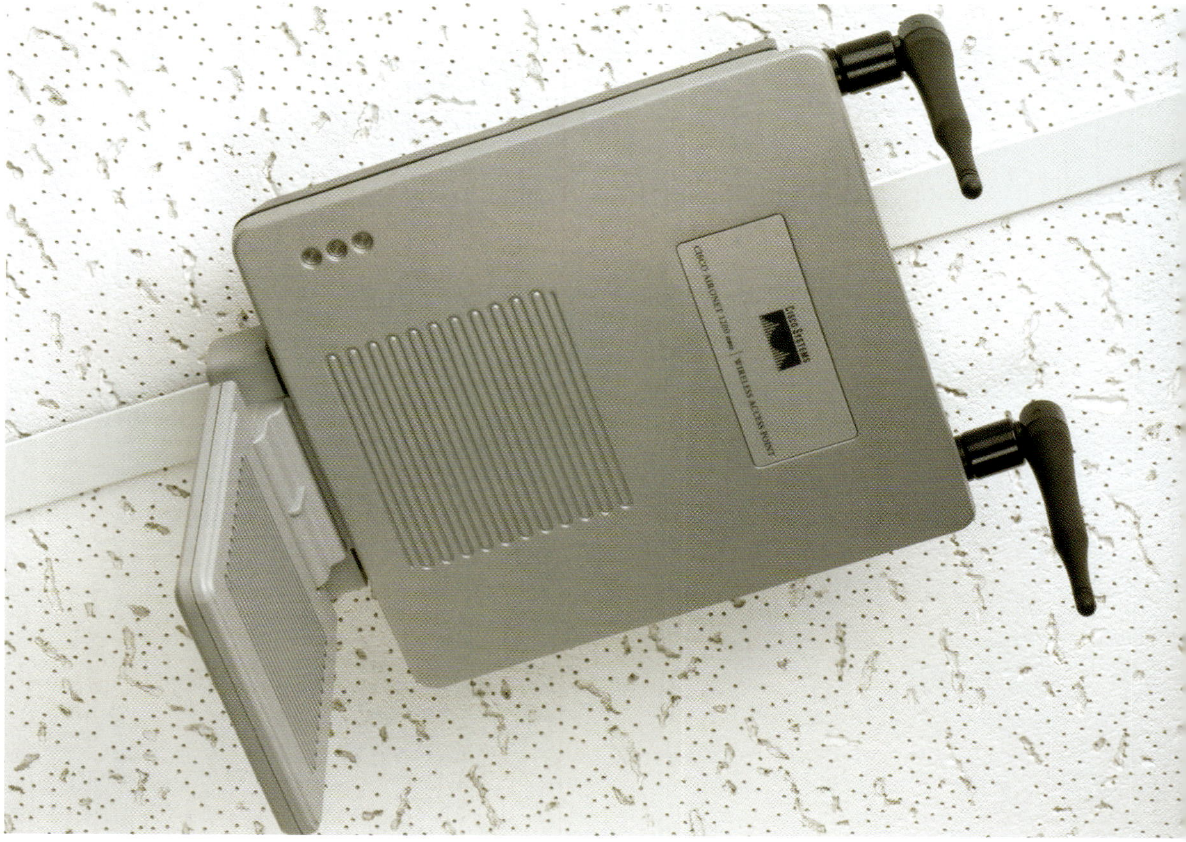

The Cisco Aironet 1200 Series enables users to install a reliable wireless local area network (WLAN). As a wireless overlay it integrates smoothly into an existing network or creates standalone all-wireless LANS for swift and cost-effective mobility. It guarantees constant connection to all network sources wherever a wireless access point is available. The series' modular upgradeable design supports IEEE 802.11a and 802.11b technologies in both single- and dual-mode operation, coupled with external 2.4 GHz and integrated 5 GHz antennas. The appearance has been kept straightforward and neutral in order to ensure it blends with a variety of different office settings.

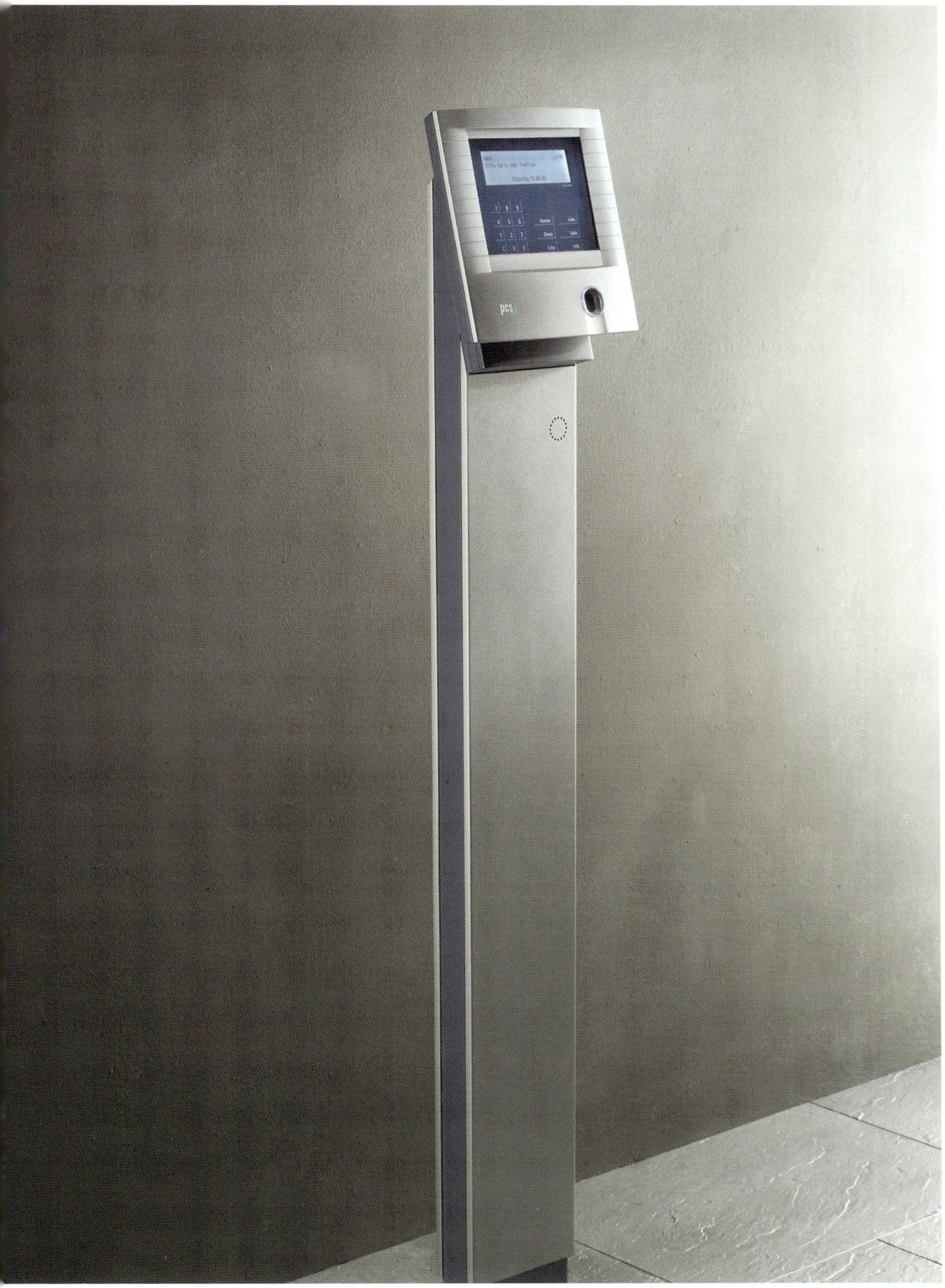

Hersteller Manufacturer
PCS Systemtechnik GmbH
München, Germany

Produkt Product
INTUS 3400
Datenterminal
Data terminal

Design
Thomas Detemple
ergon3
München, Germany

INTUS 3400 ist ein modernes Zeitwirtschafts- und Human Resources-Terminal für den Einsatz in der Industrie, bei Banken, Versicherungen, im Gesundheits- und Dienstleistungsbereich und im Handel. Das Terminal nimmt Kommen- und Gehen-Meldungen auf, gibt Auskünfte über Arbeitszeit und Urlaubskonten und informiert Mitarbeiter über Neues in der Firma oder Abteilung. Es besitzt ein großes Grafik-Display in einer ergonomisch geneigten Front, eine Touch-Tastatur, einen berührungslosen Proximity-Identleser und eine optische Signalisierung in Form des Magic Eye. Durch Lackieren oder Bearbeiten der Frontblende kann das Terminal an das Corporate Design des jeweiligen Unternehmens angepasst werden.

INTUS 3400 is a modern time/access and human resources terminal for use in industry, banks, insurance companies, and in the healthcare, service and retail sectors. The terminal handles clocking in and out, provides information on working hours and holiday accounts, and informs staff on news on the company or section. The stele features a large graphic display in an ergonomically titled front, a touchkey pad, a touch-free Proximity-Ident reader and optical signaling in the form of a Magic Eye. The front panel can be painted or processed to match the terminal with the corporate design of the respective company.

Hersteller Manufacturer
Canon Inc.
Tokio, Japan

Produkt Product
Canon i70
Mobiler Farbtintenstrahldrucker
Portable color bubble-jet printer

Design
Hiroyuki Fukushima
Canon Design Center
Tokio, Japan

Der Canon i70 ist ein mobiler Farbdrucker mit einem äußerst kompakten Design. Er druckt bis zu 13 Seiten pro Minute, in Farbe bis zu 9 Seiten, 4.800 x 1.200 dpi Auflösung und 5 Picoliter Tröpfchengröße gewährleisten hochwertige Fotodrucke. Alle Komponenten des Druckers wurden für den transportablen Einsatz optimiert und aufeinander abgestimmt. Dank der Direct Print Funktion ist Fotodruck auch ohne PC möglich. Die Infrarotschnittstelle ermöglicht zudem den kabellosen Druck.

The Canon i70 is a mobile color bubble-jet printer with an extraordinarily compact design that can create up to 13 pages per minute, and up to 9 pages in color. It is capable of printing extremely high resolutions camera art at 4,800 x 1,200 dpi and 5 Picoliter droplets. All the printer's components have been optimized for portability and are fine-tuned to interact accurately. Thanks to the Direct Print function, photos can be printed without a PC being required, and the IR interface also enables cordless printing.

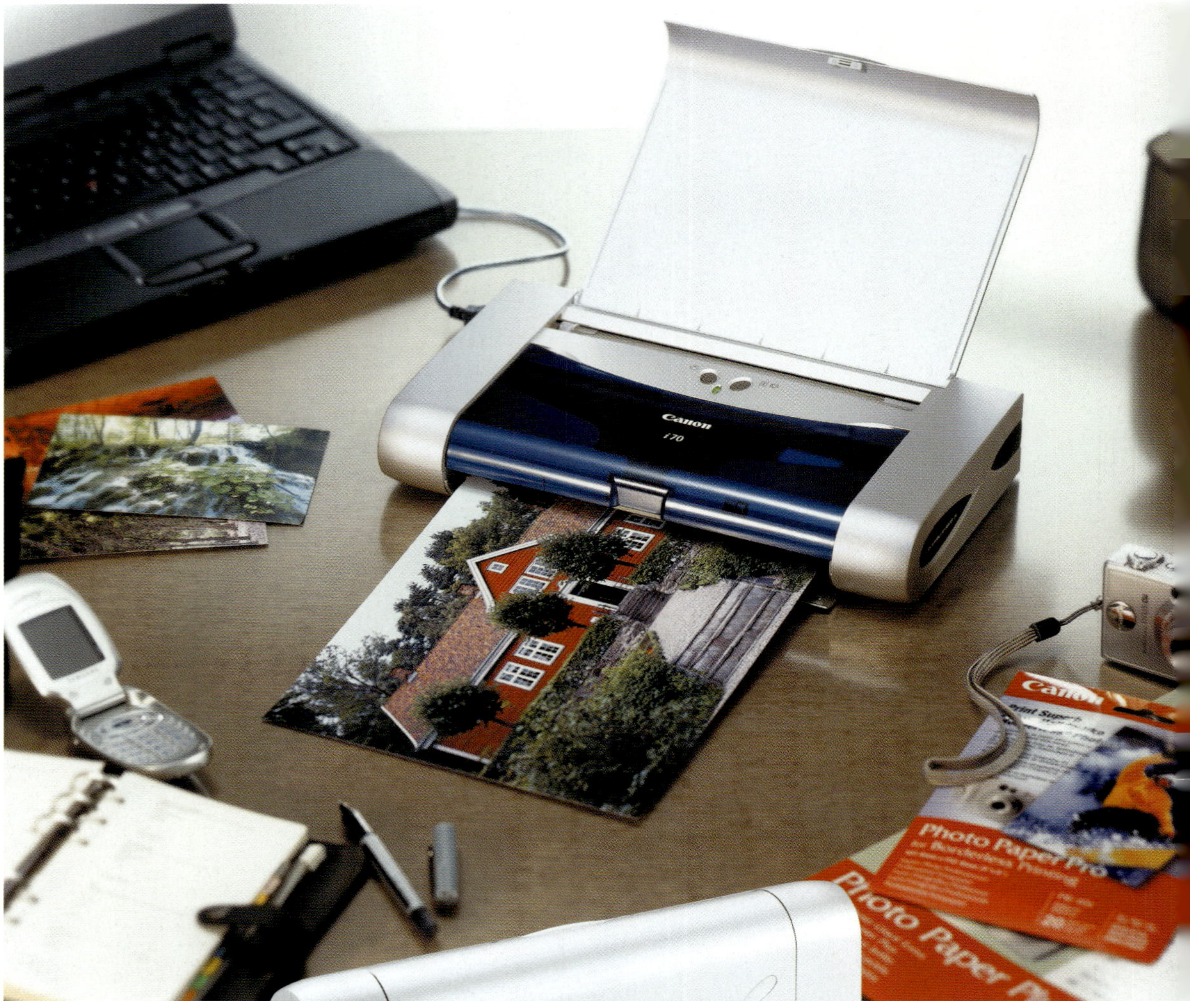

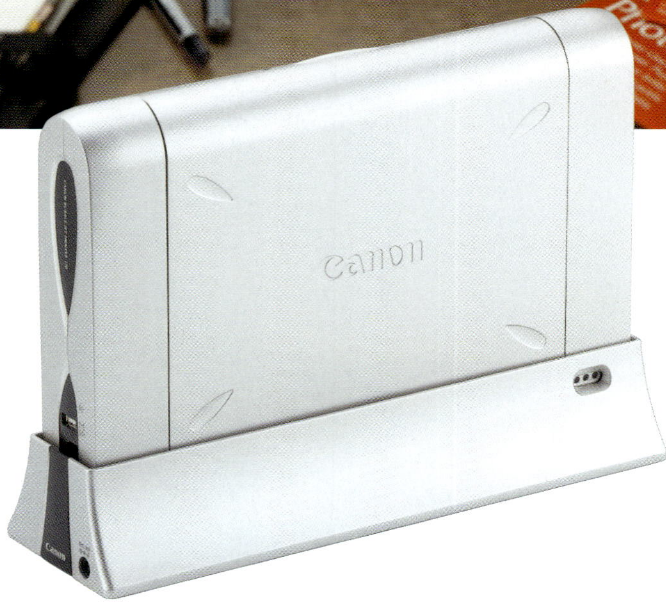

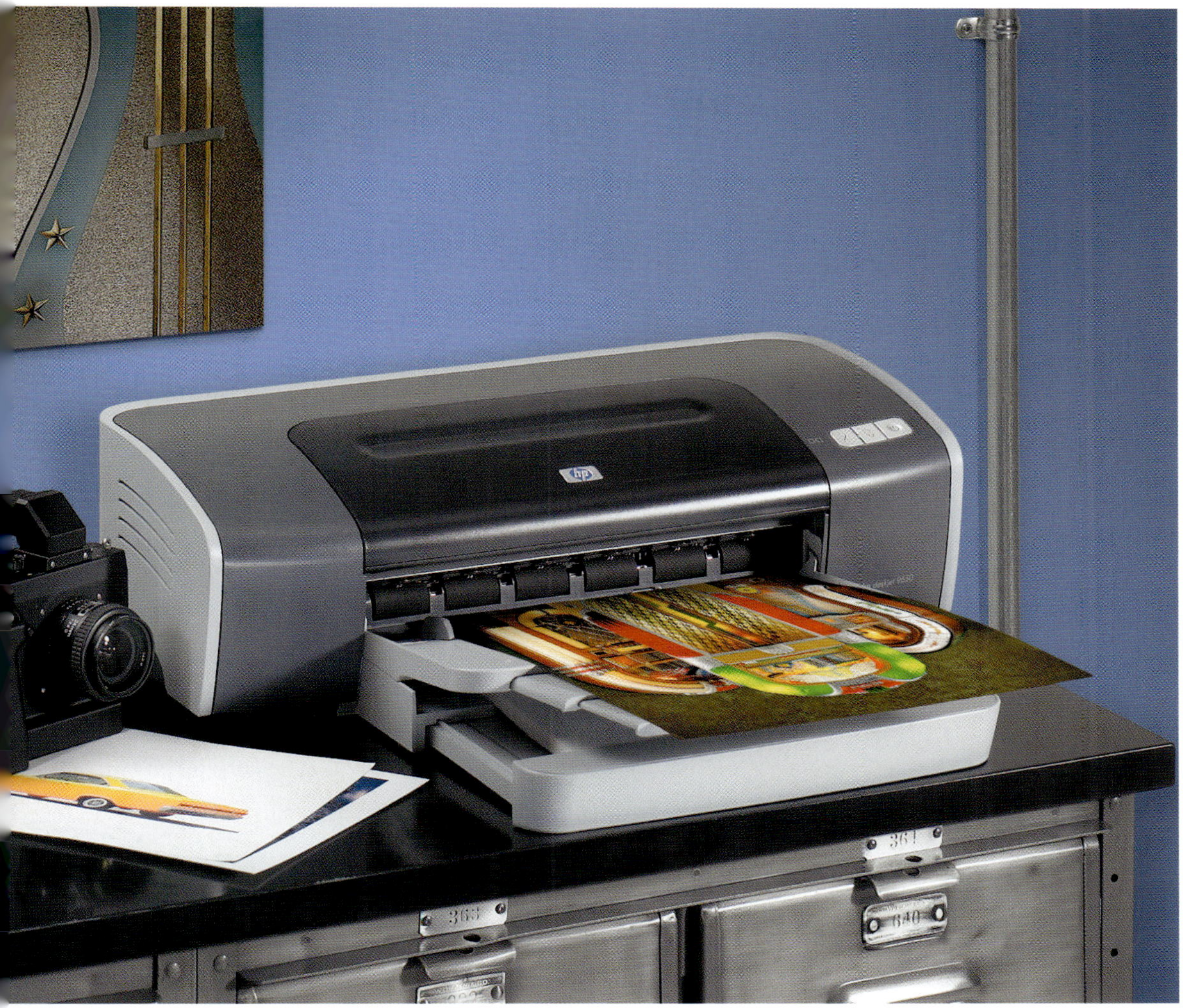

Hersteller Manufacturer
Hewlett Packard Far East Pte Ltd
Singapore, SGP

Produkt Product
HP Deskjet 9600 Serie
Farbdruckerserie
Color printer series

Design
Timothy Hoo
Kee Hong Song
Mylene Tjin
Design Exchange Pte Ltd.
Singapore, SGP

Die Farbdruckerserie Deskjet 9600 (9650, 9670 und 9680) von Hewlett Packard ist für kleine Firmen oder Büros und kreative Anwender konzipiert, die hochwertige professionelle Dokumente und Fotos auf einer Vielzahl von Medienarten und -formaten bis DIN A3-Übergröße erstellen wollen. Die Serie bietet hohe Druckgeschwindigkeiten von bis zu 20 Seiten/Minute in Schwarz-Weiß und bis zu 15 Seiten/Minute in Farbe DIN A4 sowie 6-Farb-Tintendrucktechnologie von HP und eine Auflösung von bis zu 4800 dpi auf Fotopapier. Zudem ist randloses Drucken von 10 x 15 cm bis DIN A3-Format möglich.

Hewlett Packard's Deskjet 9600 (9650, 9670 and 9680) color printer series has been specially conceived with smaller companies and offices as well as creative users in mind, who require high-grade professional documents and photos on a variety of different media and in differing formats through to outsize A3 sheets. The series combines high printer speeds of up to 20 pages/min. black-and-white through to a max. of 15 pages/min. in color A4 or HP 6-color-inkjet technology and a quality of up to 4800 dpi on photo paper. Furthermore, are not only wide-formats but borderless printing of 10 x 15 cm through to A3-sizes possible.

Hersteller Manufacturer
Seiko Epson Corporation
Suwa-shi
Nagano-ken, Japan

Produkt Product
EPSON STYLUS PRO 4000
A2-Drucker
A2 format printer

Design
Hirokazu Yamano
Matsumoto-shi
Nagano-ken, Japan

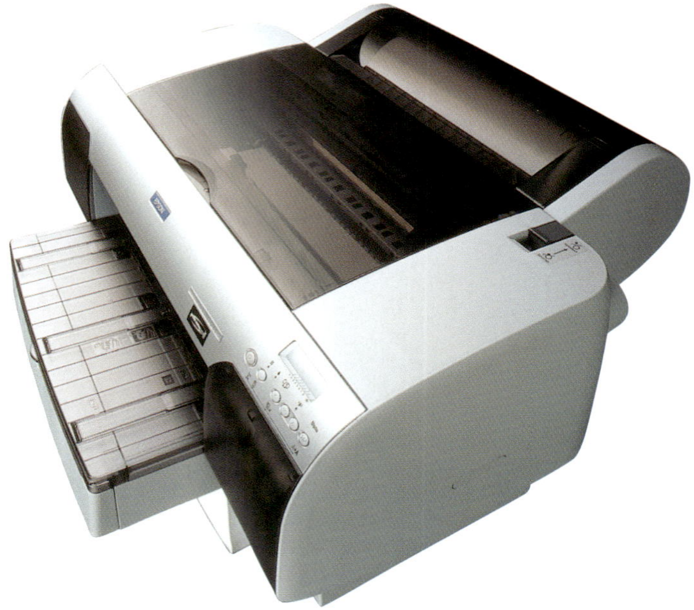

Der EPSON STYLUS PRO 4000 ist ein Drucker für A2-Formate, der insbesondere für den Bereich professionelle Grafik entwickelt wurde. Bei der Gestaltung standen vor allem Bedienfreundlichkeit und Minimierung des Platzbedarfs im Vordergrund. Das Bedienfeld und die Funktionselemente für den Wechsel der Farbpatrone sind auf der Vorderseite angeordnet, um dem Benutzer den Zugang zu erleichtern. Der Papiervorschub und Druckabschnitt sind mit einem transparenten Gehäuseteil versehen, sodass der Benutzer den Druckvorgang visuell überprüfen kann.

The EPSON STYLUS PRO 4000 is an A2 format printer which was developed especially with professional graphics applications in mind. The design concentrates on user friendliness and minimizing the footprint. The user control panel and elements for changing ink cartridges are located on the front to enhance access. The paper feed section and print area are housed in a transparent cover, meaning the user can keep an eye on the print status.

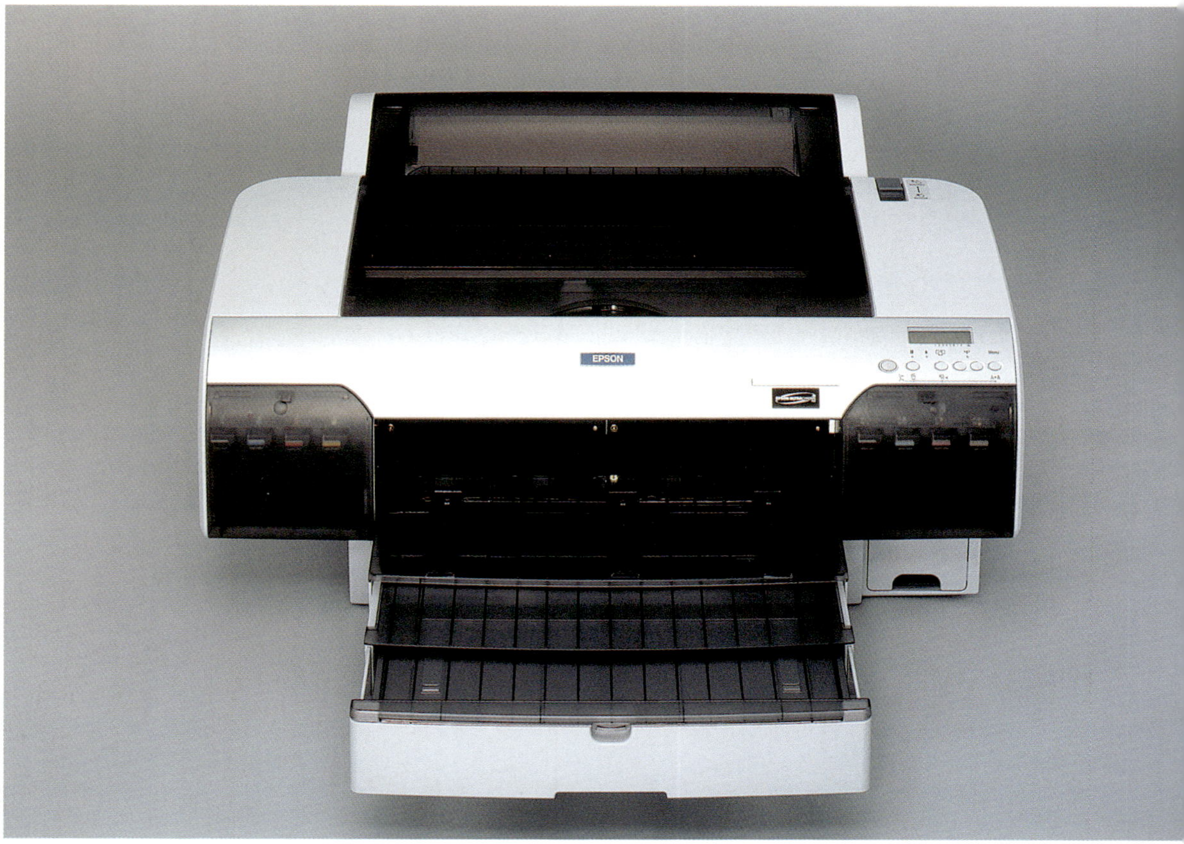

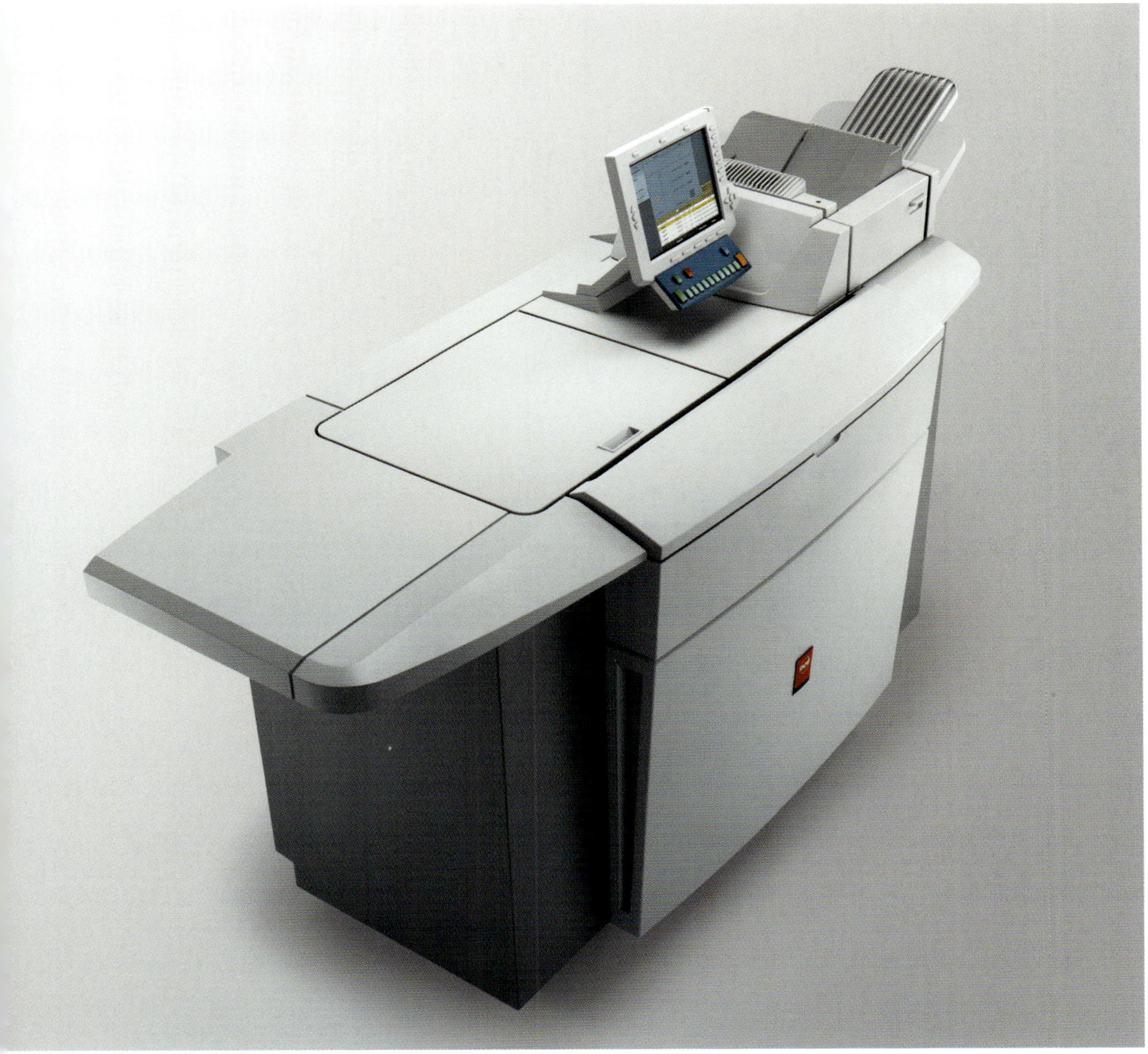

Hersteller Manufacturer
Océ Technologies bv.
Venlo, Netherlands

Produkt Product
Océ CPS900
Hochleistungs-Farbdrucker
Full-color production printer

Design
Océ Design Team
Venlo, Netherlands

Der Océ 900 ist ein vielseitiger Produktionsfarbdrucker der speziell für den grafischen Einsatz ausgelegt ist. Es werden Papierformate bis zu 12 Zoll x 18 Zoll unterstützt. Dies zusammen mit den integrierten Endverarbeitungskomponenten ermöglicht eine Vielzahl von grafischen Anwendungen in Kleinauflage wie Bürodokumente, Wurfsendungen, Broschüren, Umschläge, Visitenkarten, Bücher und sogar Offset-Andrucke. Das Gerät wurde so entworfen, dass es dem natürlichen Verhalten seiner Anwender entspricht, so wurde zum Beispiel für Ablagefläche auf dem Drucker gesorgt. Hier können Monitor und Tastatur eines von Dritten gelieferten Controllers abgestellt werden. Das mit Softkeys ausgestattete Interface hat ein farbiges Display und ist schwenkbar ausgelegt, um einen optimalen ergonomischen Blickwinkel zu gewährleisten.

The Océ 900 is a versatile full-color production printer dedicated for graphic applications. It can take paper sizes up to 12" x 18". Together with the integrated standard finishing components, this allows for a whole host of short-run graphic applications such as office documents, flyers, brochures, covers, business cards, books and even offset proofing. The device is designed for the natural way the users work, an example being the shelf where next to work material the monitor and keyboard can be placed, belonging to the controller of a third party supplier. The soft-key full-color interface can be angled to offer an optimal, ergonomic viewing position.

PRODUCT DESIGN

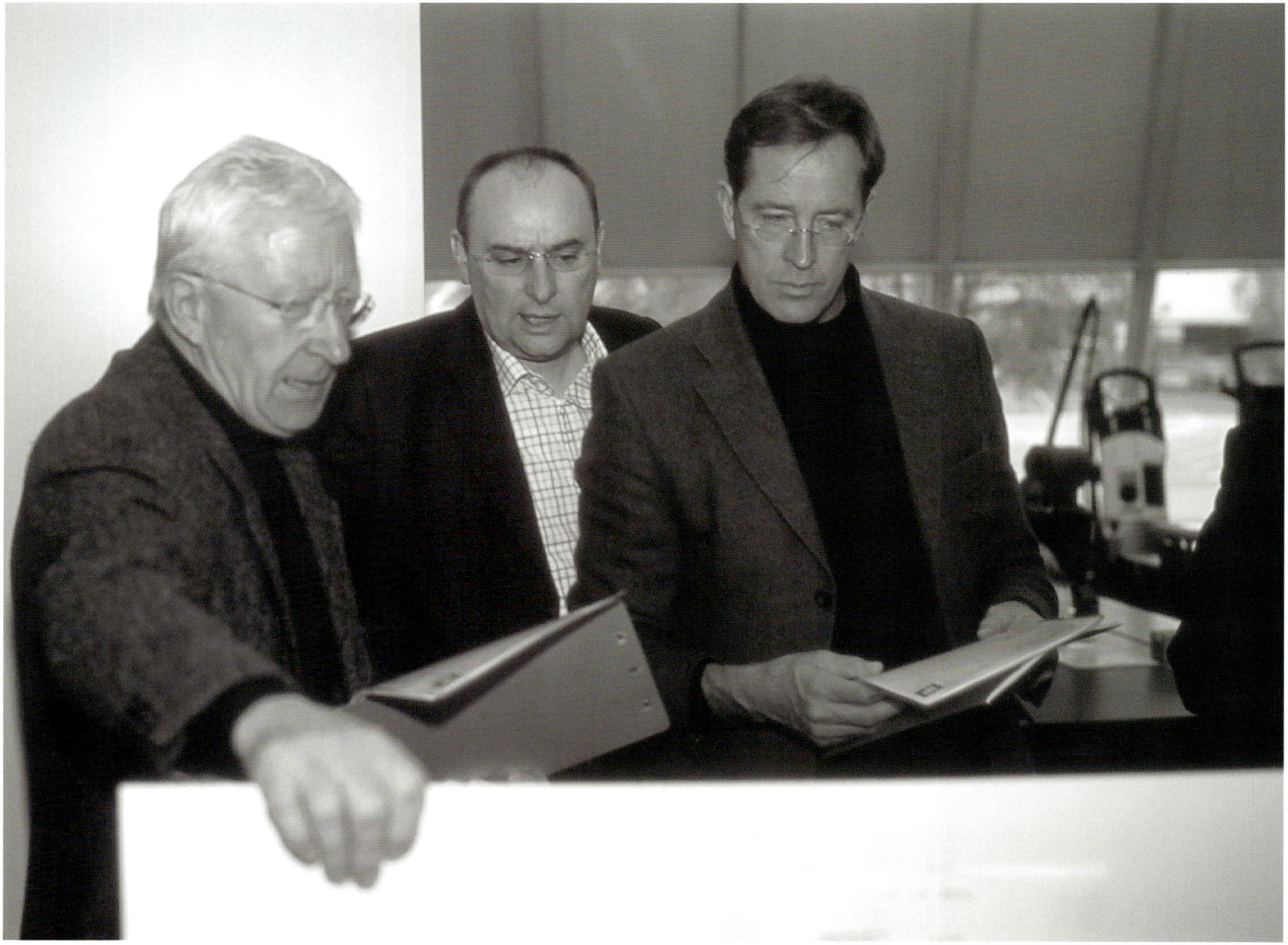

Gebäude Buildings
Industrie Industry
Licht Lighting

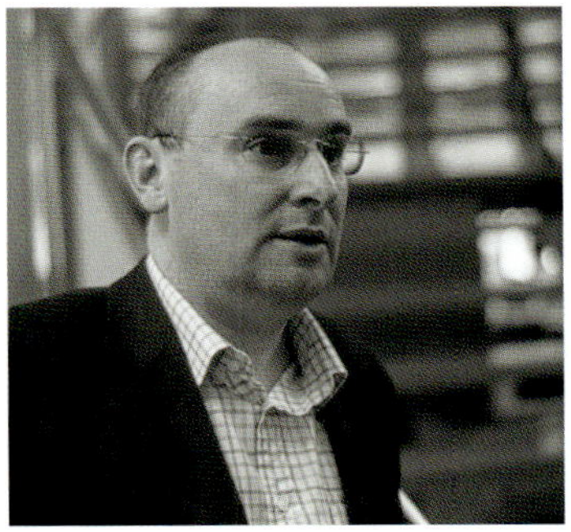

MARTIN TOPEL hat nach seinem Industriedesign-Studium im Jahre 1990 Topel und Pauser Industrial Design gegründet. Seit 1999 lehrt er an der Universität Wuppertal, wo er zusammen mit Prof. Gert Trauernicht Visionlabs, ein Labor für visionäre Produktgestaltung ins Leben gerufen hat. 2002 gründete er zudem in Düsseldorf die Agentur Squareone.

Martin Topel graduated in industrial design and in 1990 set up Topel und Pauser Industrial Design. Since 1999 he has lectured at Wuppertal University, where, together with Prof. Gert Trauernicht, he has created Visionlabs, a laboratory for visionary product design. In 2002 he founded Düsseldorf-based agency Squareone.

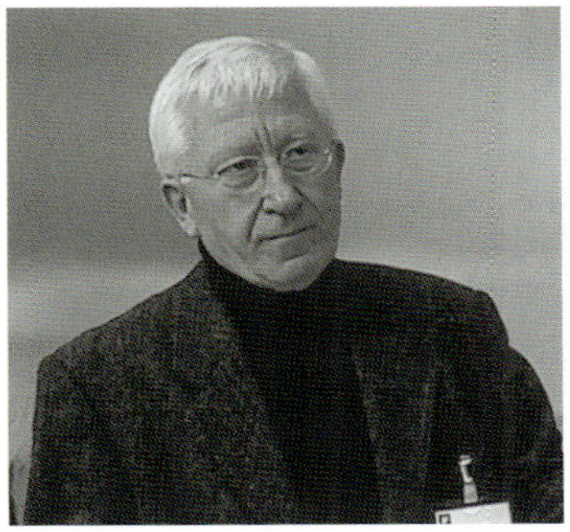

JÜRGEN W. BRAUN hat in Bonn und Paris Rechtswissenschaften studiert und, nachdem er bei Daimler-Benz, Stuttgart, und NTN, Düsseldorf, in leitenden Positionen tätig war, ist er seit 1981 Geschäftsführer des Türklinkenherstellers FSB, Brakel.

Jürgen W. Braun studied law in Bonn and Paris; since 1981, having held executive positions at Daimler-Benz, Stuttgart, and then NTN, Düsseldorf, he has been Managing Director of door-handle manufacturer FSB in Brakel.

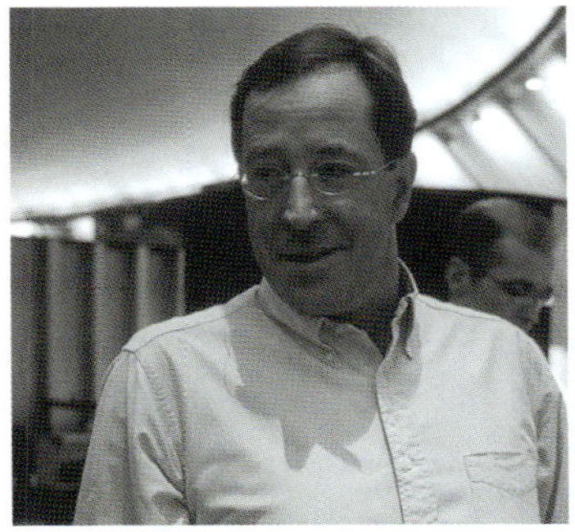

ACHIM NAGEL studierte Architektur an der TU Hannover und gründete im Jahr 2000 TWO Product Development and Integrated Design zusammen mit Claudia de Bruyn. Zwischenzeitlich arbeitete er als Architekt unter anderem für die Bertelsmann AG und zudem als Partner im Architekturbüro Ingenhoven Overdiek und Partner, Düsseldorf.

Achim Nagel studied architecture at the Hanover Polytechnic and in 2000 together with Claudia de Bruyn founded TWO Product Development and Integrated Design. He has since also worked as an architect, among others for Bertelsmann AG and as a partner in the architects office of Ingenhoven Overdiek und Partner, Düsseldorf.

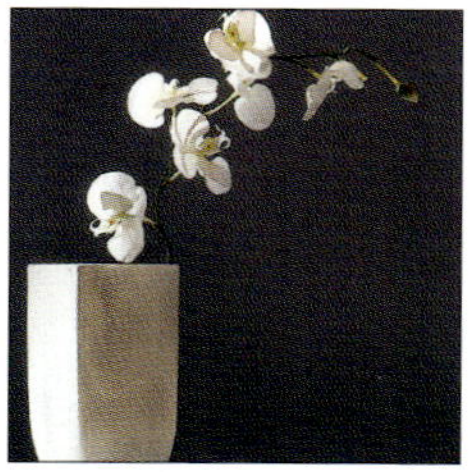

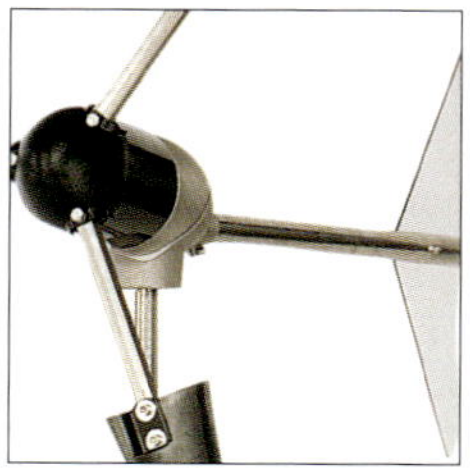

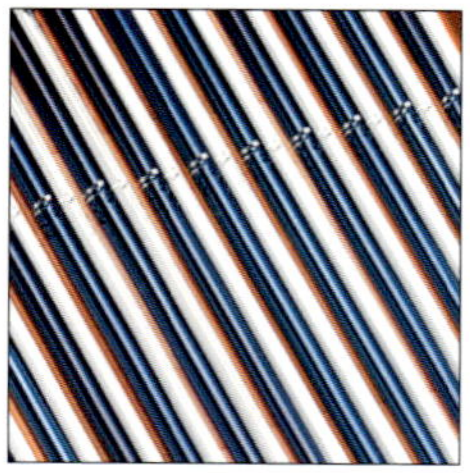

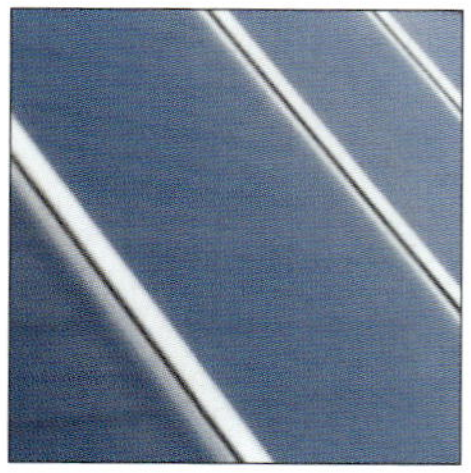

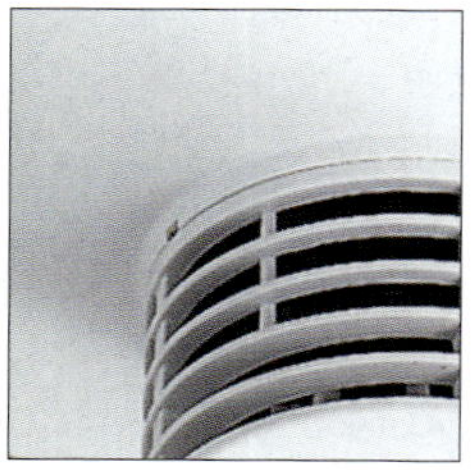

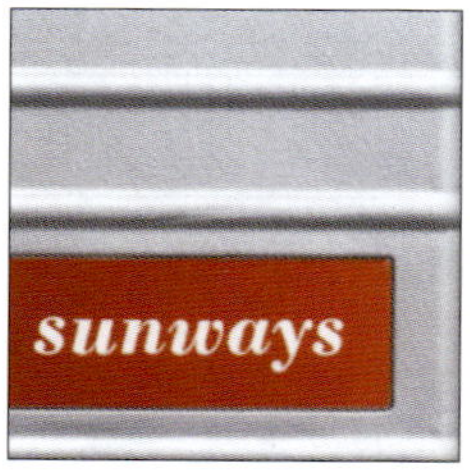
sunways

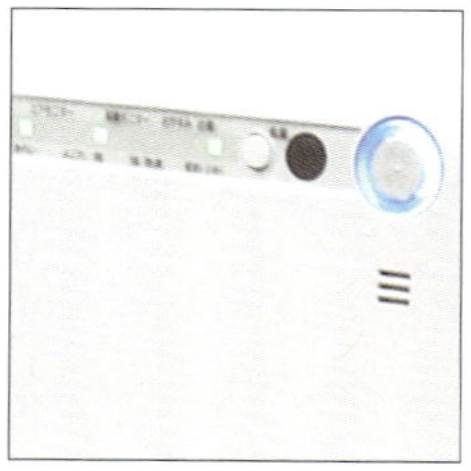

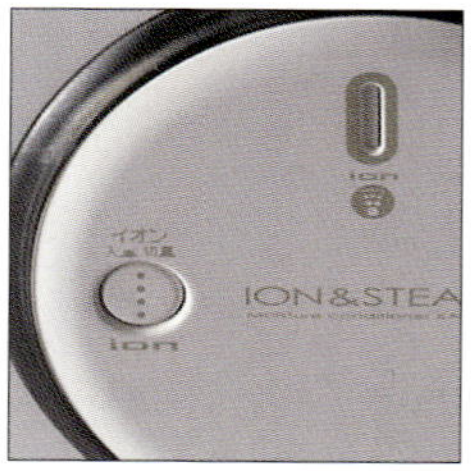
ion

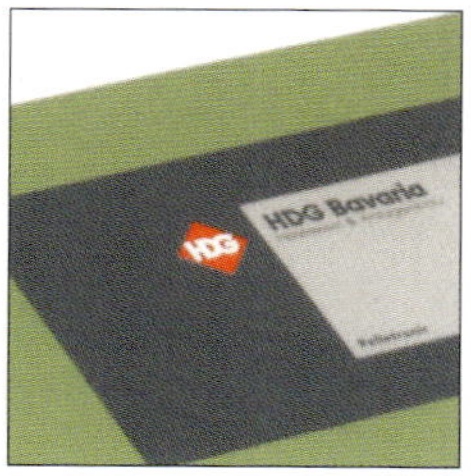
HDG
HDG Bavaria

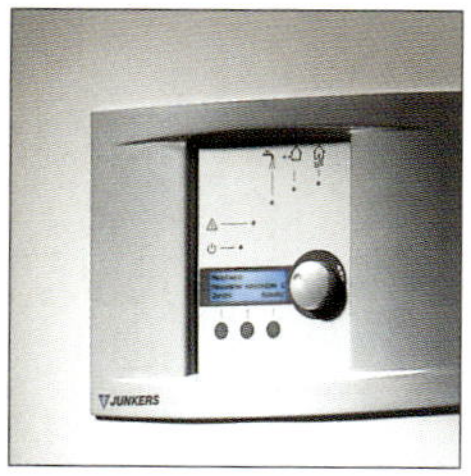
JUNKERS

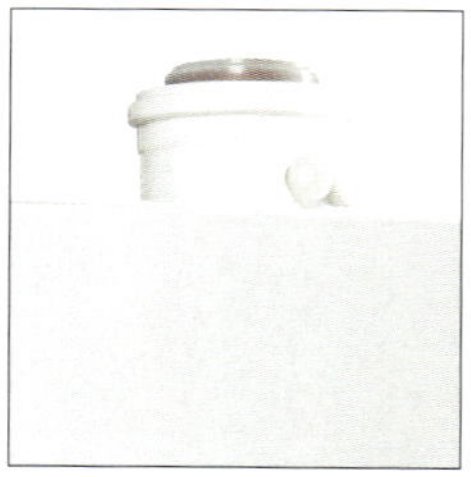

Gebäude Buildings

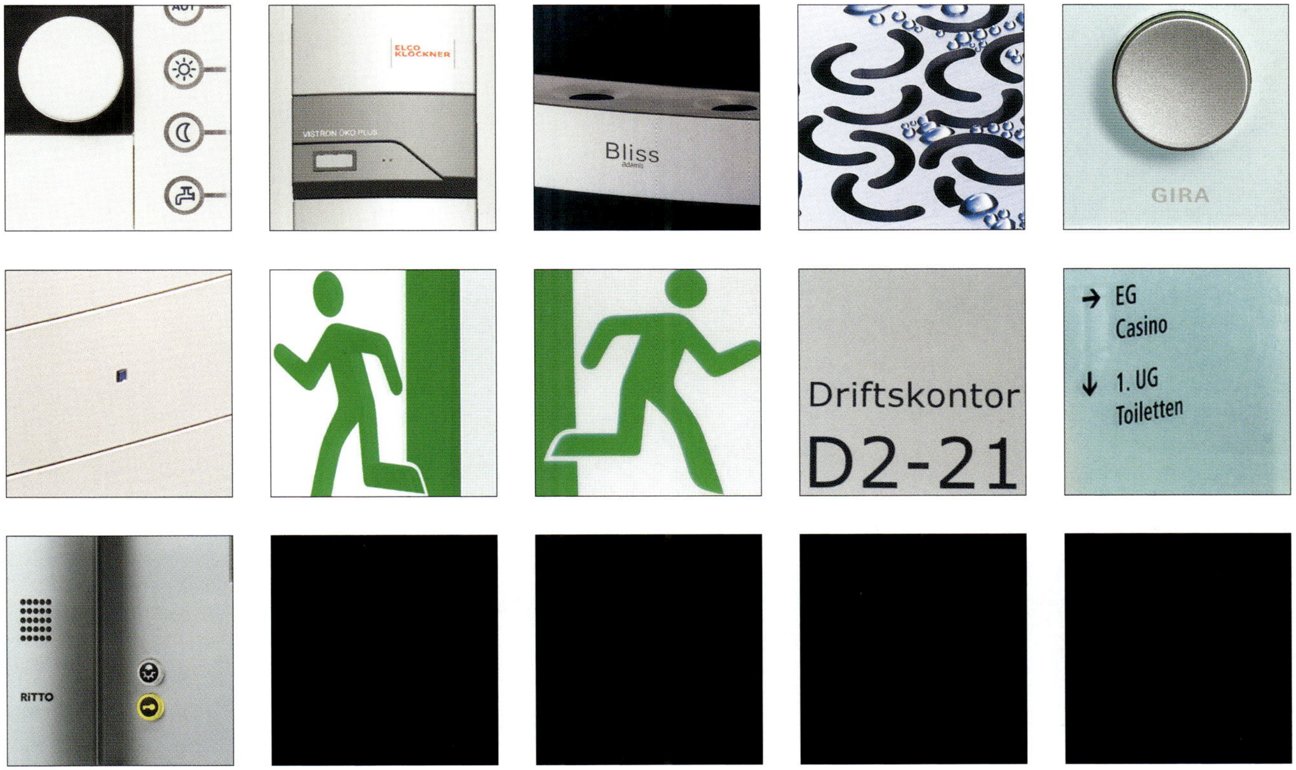

Hersteller Manufacturer
Kinon Aachen GmbH
Aachen, Germany

Produkt Product
SGG Thermovit Elegance
Transparenter Glasheizkörper
Transparent glass radiator

Design
Kinon Aachen GmbH
Aachen, Germany

Der SGG Thermovit Elegance ist ein Glasheizkörper, der aufgrund seiner Transparenz für das Interior Design von Wohnräumen ideale Voraussetzungen und großen Gestaltungsspielraum eröffnet. Ein Funktemperaturregler regelt in Verbindung mit dem im Scheibenanschluss untergebrachten Regelsystem der Glasheizung über Funk die Temperatur, um so die optimale Raumtemperatur zu erreichen. Durch eine achtstellige Flüssigkristallanzeige sind alle Werte alphanumerisch ablesbar und mittels vier Tasten einstellbar.

The SGG Thermovit Elegance is a glass radiator, which thanks to its transparency creates ideal conditions and great scope for interior designers when configuring residences. A radio temperature controller is linked to the temperature regulator located in the edge of the glass panel for remote temperature control, enabling an optimal ambient temperature. The eight-character LCD presents all the operating data in alphanumeric form and new settings are inputed through the four press buttons.

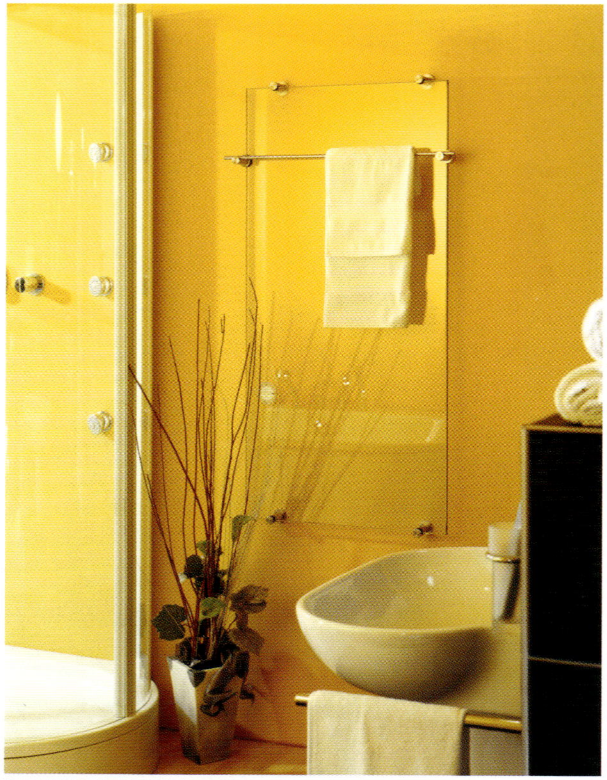

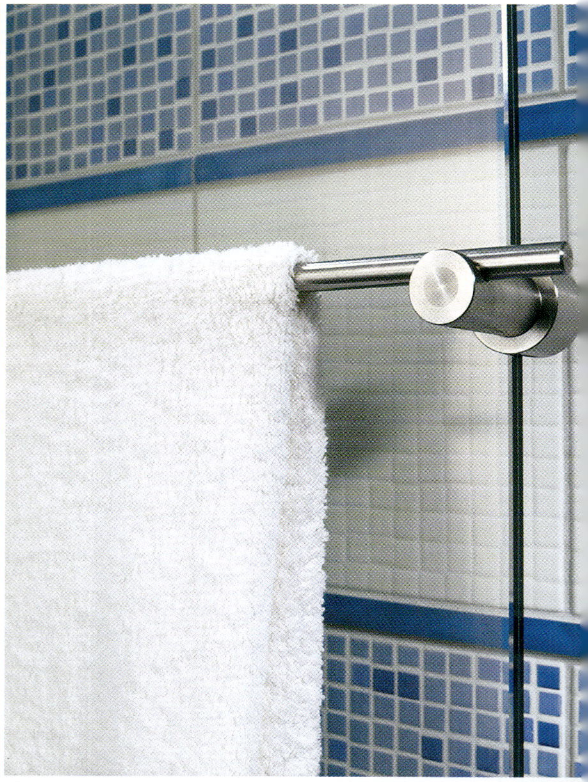

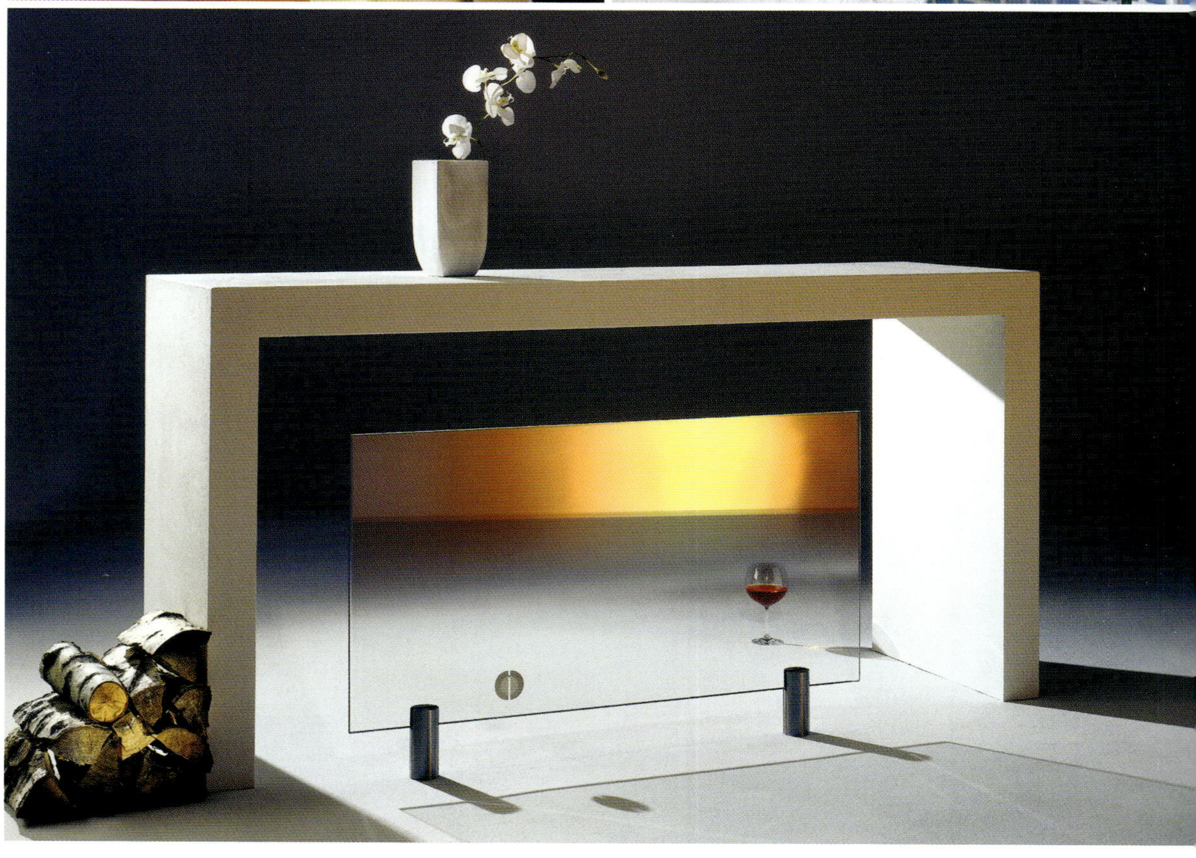

iF Gold Award

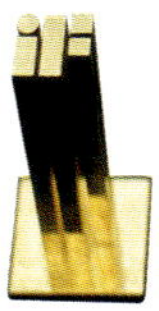

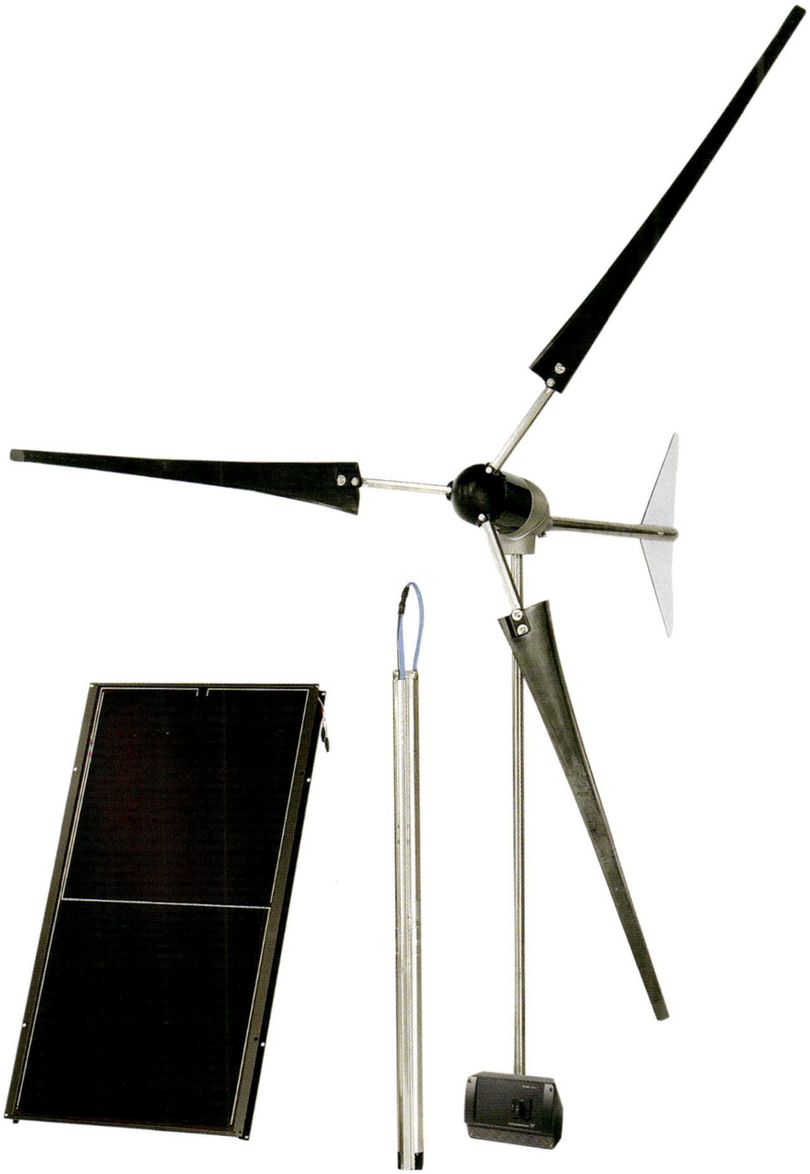

Hersteller Manufacturer
Grundfos A/S
Bjerringbro, Denmark

Produkt Product
SQFlex
Wasserversorgung
Water supply system

Design
Niels Due Jensen

Das SQFlex System von Grundfos ist ein umfassendes Wasserversorgungskonzept für alle Einsatzbereiche, bei denen die eigene, völlig unabhängige Wasserversorgung sichergestellt werden muss. Die Solarmodule von SQFlex Solar sichern die komplette Energieversorgung des Pumpensystems in abgelegenen Gebieten. Wo mehr Wind als Sonnenschein zu erwarten ist, kann die Energieversorgung über den kompakten Windgenerator SQFlex Wind erfolgen. SQFlex Kombi verbindet beide Optionen, während SQFlex Back-up einen Notstromgenerator vorsieht, falls Sonne oder Wind nicht zur Verfügung stehen.

Grundfos' SQFlex System is a total water supply concept for use wherever a completely independent energy-efficient water supply system is called for. The SQFlex Solar modules guarantee complete powering for the pumping system in remote areas. Where more wind than sunshine is to be expected, the pump can be driven using the compact SQFlex Wind generator. SQFlex Combi combines both options, while SQFlex Back-up includes a fall-back generator, should the sun and wind fail to be sufficient on occasion.

iF Silver Award

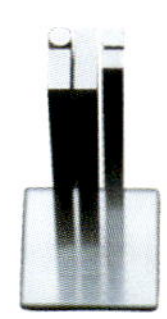

Hersteller Manufacturer
Viessmann Werke
Allendorf (Eder), Germany

Produkt Product
Vitosol 250
Röhrenkollektor
Tube collector

Design
Phoenix Design
Stuttgart, Germany

Die Viessmann Röhrenkollektoren wurden vor allem unter Berücksichtigung architektonischer Gesichtspunkte entworfen. Dabei stand die Übertragung architektonischer Strukturen auf die Gestaltung der Kollektoren im Vordergrund. Reduziert auf die wesentlichen Bestandteile des Produktes werden die Röhren als vertikales Element betont und ohne untere sichtbare Querverbindung belassen. So scheinen die Röhren über dem Dach zu schweben. Die quer liegende Versorgungseinheit unterstreicht mit ihrer feinen Profilierung die Leichtigkeit des Designs.

The Viessmann tube collectors were developed specifically with a view to architectural criteria. The focus was on transposing architectural structures onto the design of the collectors. Reduced to their core components, the vertical nature of the tubes is now emphasized and there are no visible cross connections. As a result, the tubes seem to hover above the roof. The fine profiles of the cross-wise supply unit underscore the light feel of the design.

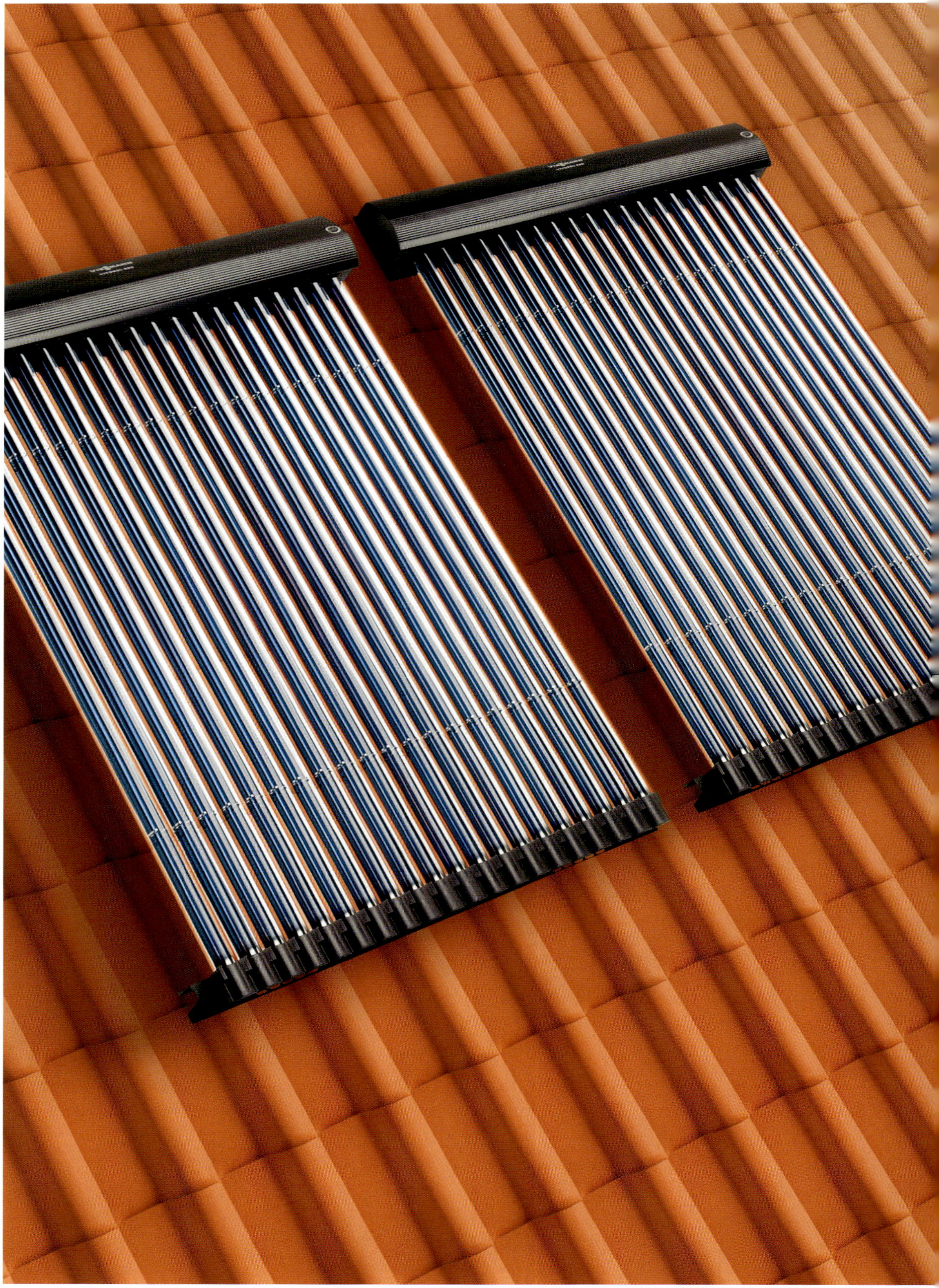

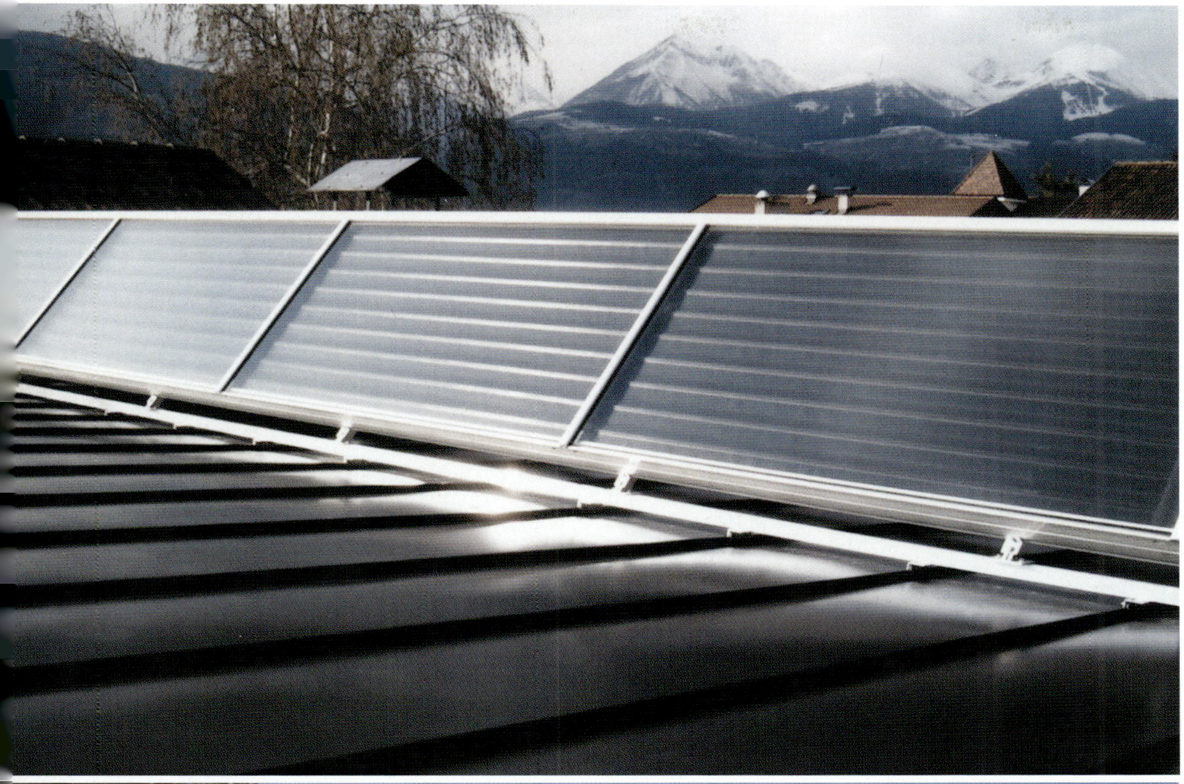

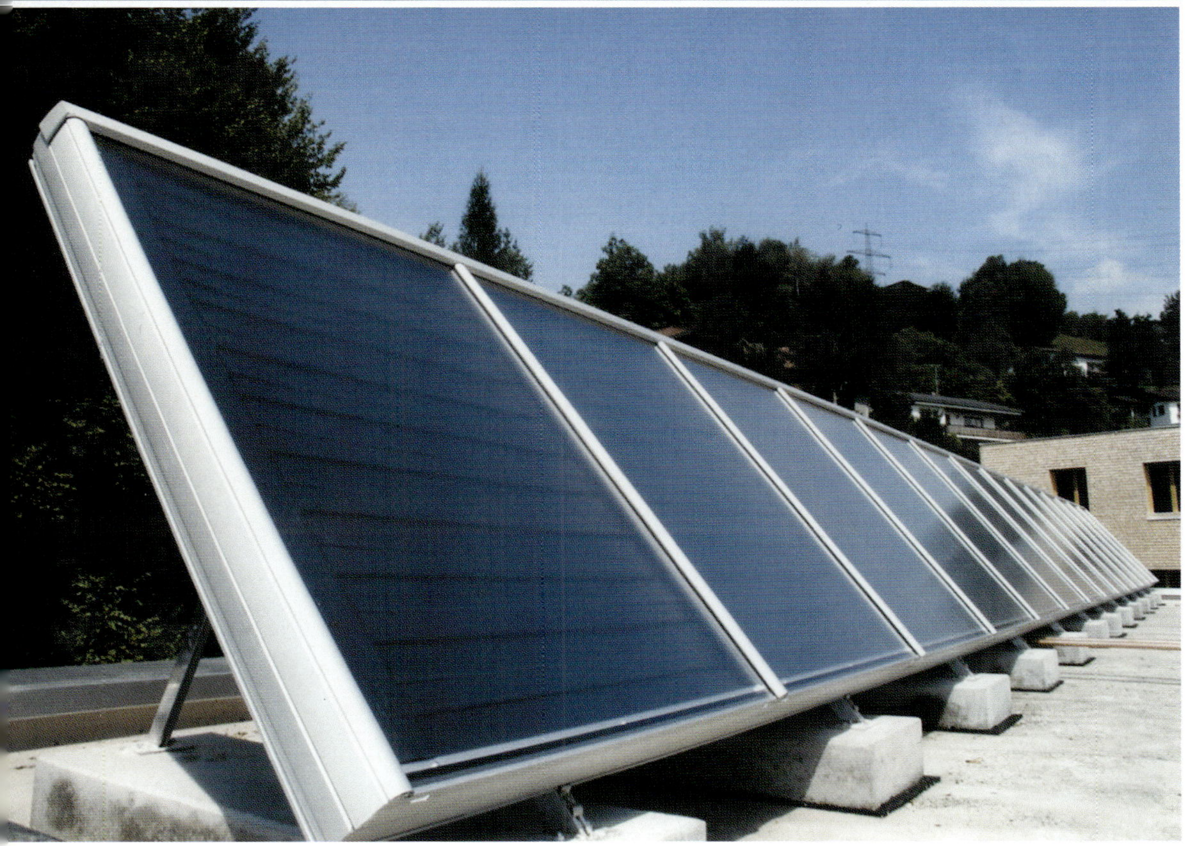

Hersteller Manufacturer
Winkler Solarsysteme Spenglerei GmbH
Feldkirch, Austria

Produkt Product
VarioSol A
Großflächen-Sonnenkollektor
Wide-area solar collector

Design
Winkler Solarsysteme Spenglerei GmbH
Feldkirch, Austria

VarioSol A ist ein thermischer Sonnenkollektor in Großflächen-Bauweise, d. h. Elemente bis zu 24 m^2 werden an einem Stück hergestellt und montiert. Die umlaufenden, eloxierten Alu-Rahmenprofile sind komplett durchgehend ohne Schrauben oder andere Befestigungspunkte angebracht. Die Profile sind abgerundet und bilden damit einen harmonischen Übergang der Seiten zur vorderen Glasfläche. Trotz dieser schraubenlosen Oberflächen ist der Kollektor ohne Spezialwerkzeug leicht zu öffnen und wieder dicht zu schließen. Die Größe und Form des Kollektors kann den baulichen Gegebenheiten angepasst werden.

VarioSol A is a wide-area thermal solar collector, meaning that elements of up to 24 sq. m. are produced and mounted on a single frame. The frame's anodized aluminum profiles clip into place free of any screws and are rounded in order to create a harmonious transition from the sides to the front glass. Despite these bolt-free surfaces, the collector can easily be opened and sealed perfectly without the use of special tools. Size and shape of the collectors can be customized to the demands of the particular building or setting.

Hersteller Manufacturer
Matsushita Electric Works Ltd.
Osaka, Japan

Produkt Product
Smoke Senser
Feuermeldesensor
Fire alarm sensor

Design
Tsuyoshi Kadono
Information Equipments & Wiring Products
Design Office

Der Smoke Senser ist ein kleiner Feuermeldesensor für Einrichtungen, die außerhalb von Wohngebieten liegen. Indem er die Veränderungen der Rauchdichte analysiert, erkennt er Brände in ihren Anfangsstadien und reduziert die Zahl der Falschalarme. Das Design des Sensors ist schlicht und auf seine Funktion beschränkt, zudem ist er klein und lässt sich somit in jede Umgebung integrieren.

The Smoke Senser is a small fire alarm sensor destined to detect fires in non-residential facilities. In analyzes changes in smoke density and can thus identify fires in their early stages while likewise reducing the number of non-fire alarms. The sensor boasts a sleek design and is restricted to functionality; moreover, thanks to its compact size it blends smartly with any environment.

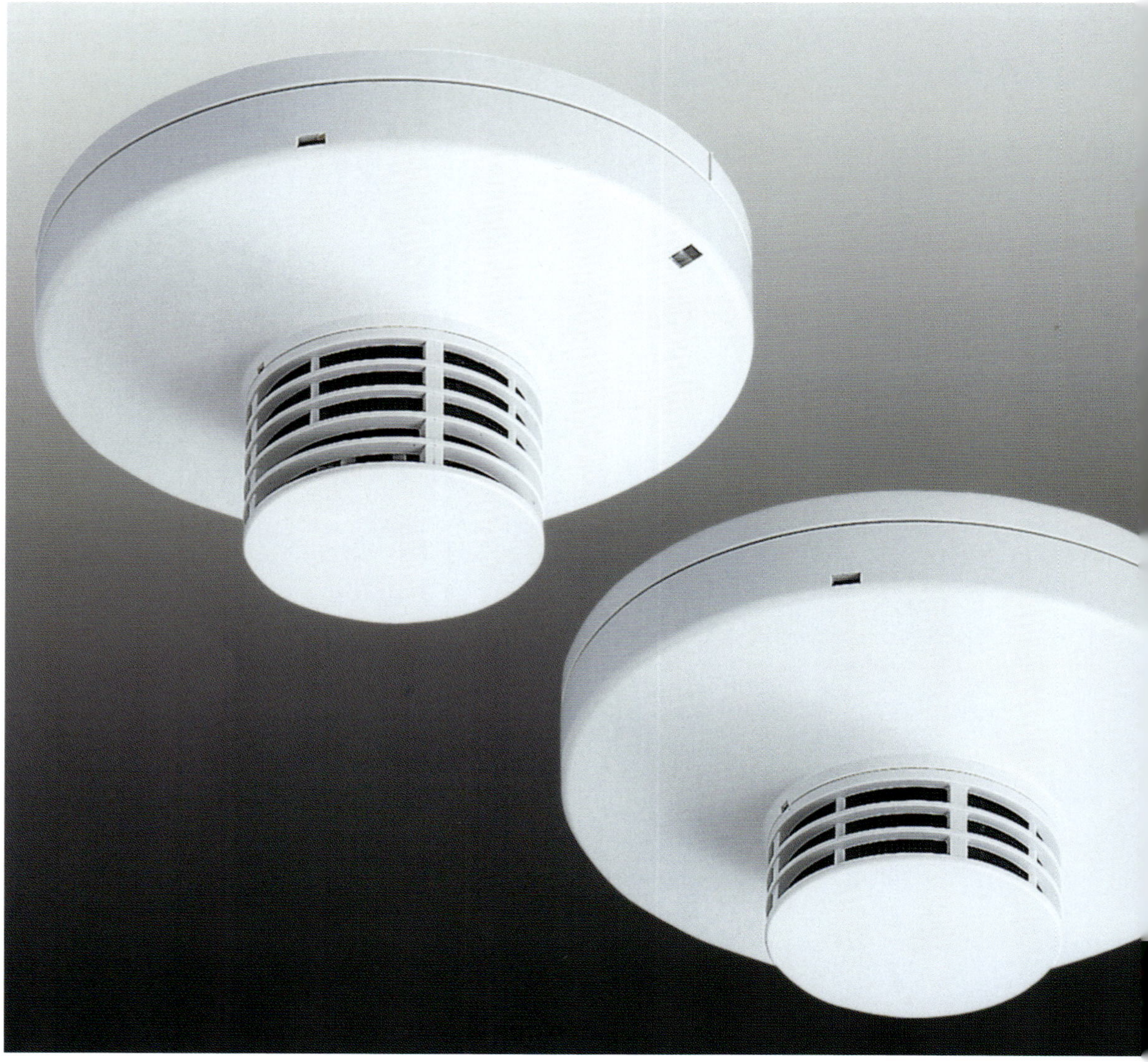

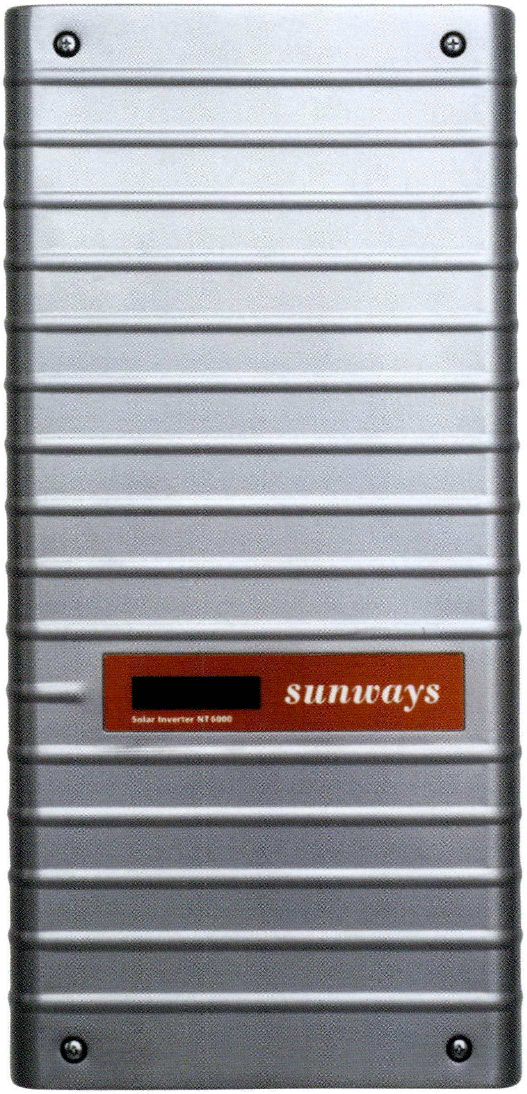

Hersteller Manufacturer
Sunways AG, Photovoltaic Technology
Konstanz, Germany

Produkt Product
Sunways Solar Inverter NT
Wechselrichter
Solar Inverter

Design
Melzer-Müller id
Kreuzlingen, Switzerland

Der Sunways Solar Inverter NT beinhaltet ein zum Patent angemeldetes neues Technologiekonzept (HERIC-Topologie = Highly Efficient and Reliable Inverter Concept), das zusammen mit dem Fraunhofer Institut für Solare Energiesysteme entwickelt wurde. Das Design des Inverters ist klar, überschaubar und funktional. Die Bedienung wurde mit einem einzigen übersichtlichen Display ebenfalls auf das Wesentliche reduziert. Die Kühlrippen auf der Rückseite bewirken einen effizienten Wärmeaustausch und halten die Betriebstemperaturen im Inneren niedrig. Die spezifische Auslegung der Bauteile und die stabile Konstruktion machen das Gerät leise und ermöglichen auch die Montage in öffentlichen Zonen oder im privaten Wohnbereich.

The Sunways Solar Inverter NT relies on a new patent-pending technological concept, namely HERIC topology, which stands for the Highly Efficient and Reliable Inverter Concept developed together with the Fraunhofer Institute for Solar Energy Systems. The inverter sports a clear, functional and well structured design, whereby operations are streamlined around a single display that presents all operational data and a data logger. The cooling fins at the back ensure efficient heat exchange and keep the operating temperature inside low. The special design of the components and its sturdy structure ensure that it is whisper-quiet, encouraging use in public areas and private residences.

Hersteller Manufacturer
Toshiba Carrier Corporation
Tokio, Japan

Produkt Product
CAF-C4
Luftreiniger
Air purifier

Design
Design Center
Toshiba Corporation
Tokio, Japan

Der Air Purifier CAF-C4 reinigt die Raumluft, indem er Mikropartikel wie Pollen und Hausstaub sammelt und Gerüche, wie sie durch Haustiere, Zigaretten, Abfall etc. entstehen, beseitigt. Er produziert eine hohe Dichte negativer Ionen, durch die freie Radikale aus der Luft gefiltert werden können. Form, Farbe und grafische Elemente des Luftreinigers wurden so gewählt, dass er sich gut in alle Wohninterieurs integrieren lässt. Die Betriebsanzeige in Form eines blauen Rings soll die mit dem Gerät assoziierten Aspekte von Gesundheit und Sauberkeit symbolisch ausdrücken.

The CAF-C4 is an air purifier that functions by collecting extremely small particles such as airborne pollens, house dust, and removes the unpleasant odors associated with pets, cigarette smoke, waste etc. It features a healthy ion charger that produces high-density negative ions to remove free radicals from the air for healthier rooms. The unit's shape, color and visual elements were chosen to blend well with interiors. The blue ring, which lights when the unit is operational, symbolizes positive aspects such as health and cleanliness.

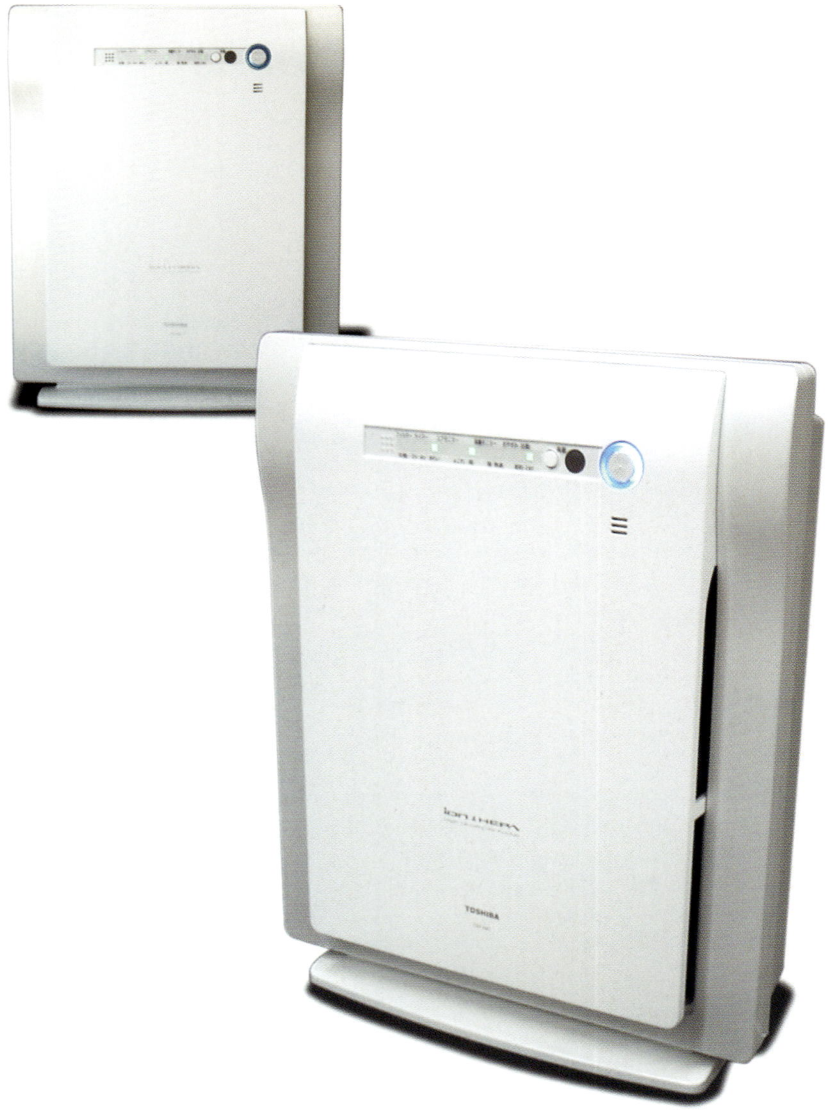

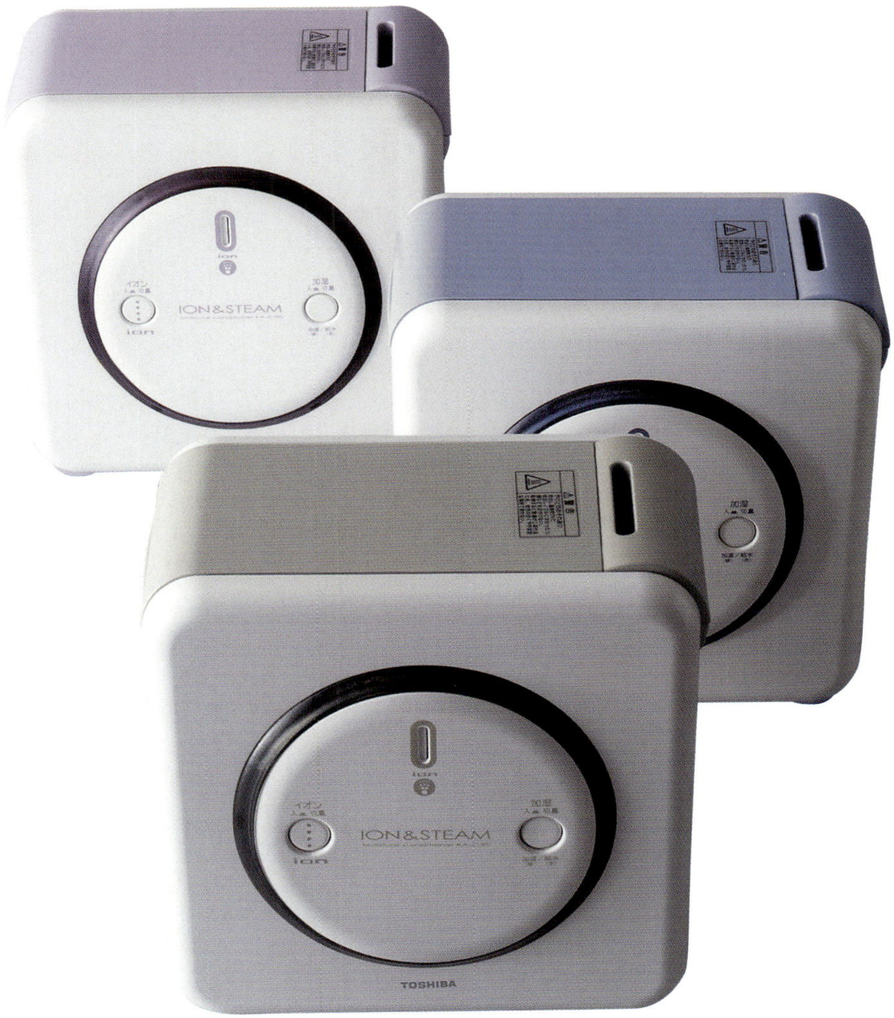

Hersteller Manufacturer
Toshiba Home Technology Corporation
Tokio, Japan

Produkt Product
KA-C3S
Dampf-Luftbefeuchter
Steam-type humidifier

Design
Design Center
Toshiba Corporation
Tokio, Japan

Der Luftbefeuchter KA-C3S produziert negative Ionen, um eine gesunde Raumluft zu schaffen. Er besitzt einen antibakteriellen Filter sowie einen antibakteriellen Wassertank. Die Ionenfunktion kann unabhängig von der Luftbefeuchtungsfunktion aktiviert werden, so dass das Gerät von gesundheitsbewussten Nutzern das ganze Jahr über betrieben werden kann. Die Gestaltung basiert auf den geometrischen Grundformen von Quadrat und Kreis. Sie wurde insgesamt schlicht gehalten, um eine gute Anpassung an das Wohnumfeld zu gewährleisten. Die Betriebsanzeige in Form eines Rings kann auch aus der Distanz oder in einem wenig beleuchteten Raum gut erkannt werden.

The KA-C3S air humidifier produces negative ions to create "people-friendly" ambient air. It relies on an antibacterial filter and an antibacterial water tank. The ion function can be actuated independently of the air humidifier, meaning that the appliance can be operated all year round by health-conscious users. The design is based on two basic geometrical shapes: a square and a circle. The overall appearance has been kept plain and simple to ensure it melds well with living interiors. The ring shows operating status and can be easily discerned from a distance or in a poorly lit room.

Hersteller Manufacturer
HDG Bavaria GmbH
Massing, Germany

Produkt Product
Pelletmaster
Pelletheizkessel
Pellet heating system

Design
Plato Designgroup
München, Germany

Der Pelletmaster ist ein ökologisch verträgliches Heizsystem, das mit Holz in Form von Pellets (kleinen gepressten zylindrischen Röllchen aus Hobel- und Sägespänen) betrieben wird. Eine sinnfällige Ergonomie des Gehäuses und eine einfache Bedienung (Folientastatur und Steuerung) waren wichtige Kriterien der Designkonzeption. Zudem wurde versucht die aufwendige und robuste Technik sehr kompakt zu halten. Die glatten und geschlossenen Formen sollen dem zeitgemäßen Anspruch an Ökologie, Sauberkeit und Ästhetik gerecht werden, die klare und funktionale Gliederung reduziert optisch die Größe der Anlage.

The Pelletmaster is an eco-compatible heating system that runs on wood in the form of pellets (small pressed cylindrical tubes of sawdust or wood shavings). The convincing ergonomic shape of the body and its simple controls (foil-covered keypad and regulator) were key criteria underlying the design concept. Moreover, another emphasis was on keeping the elaborate and robust technology as compact as possible. The smooth and closed structure of the casing is intended to point up the contemporary thrust of ecology while combining it with cleanliness and aesthetics; the clear and functional subdivision of the unit means the system looks smaller than it actually is.

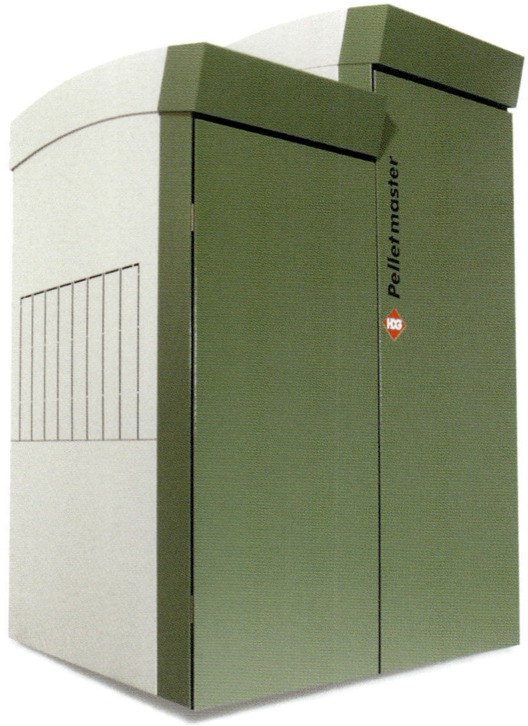

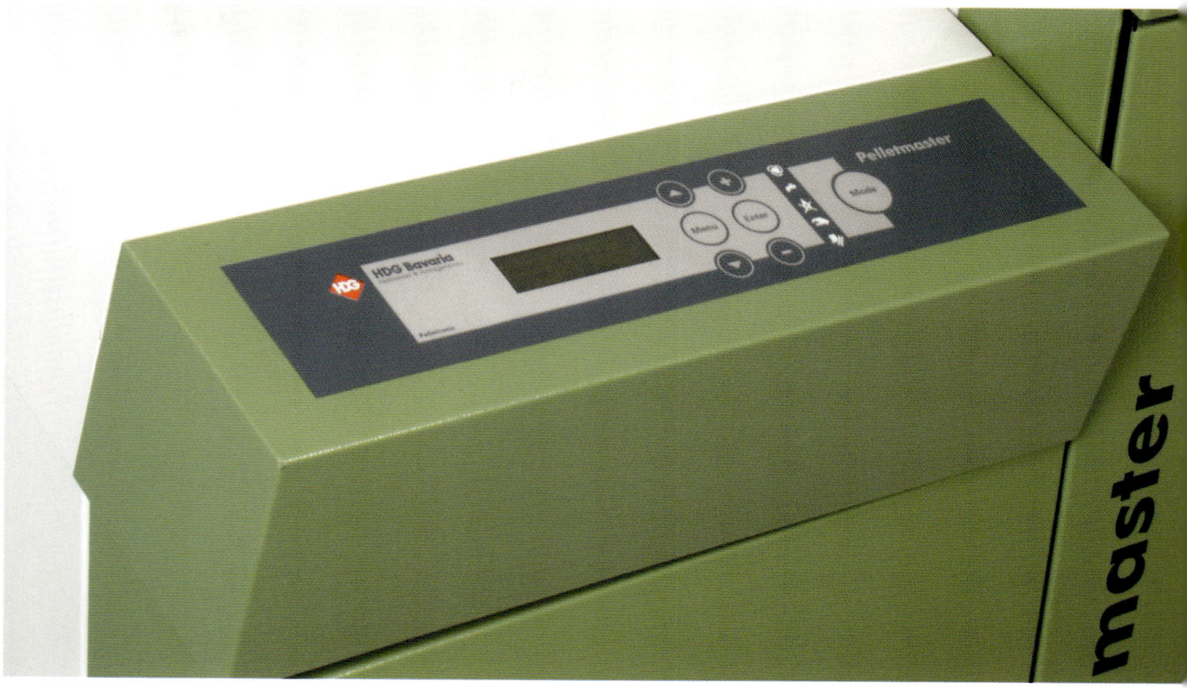

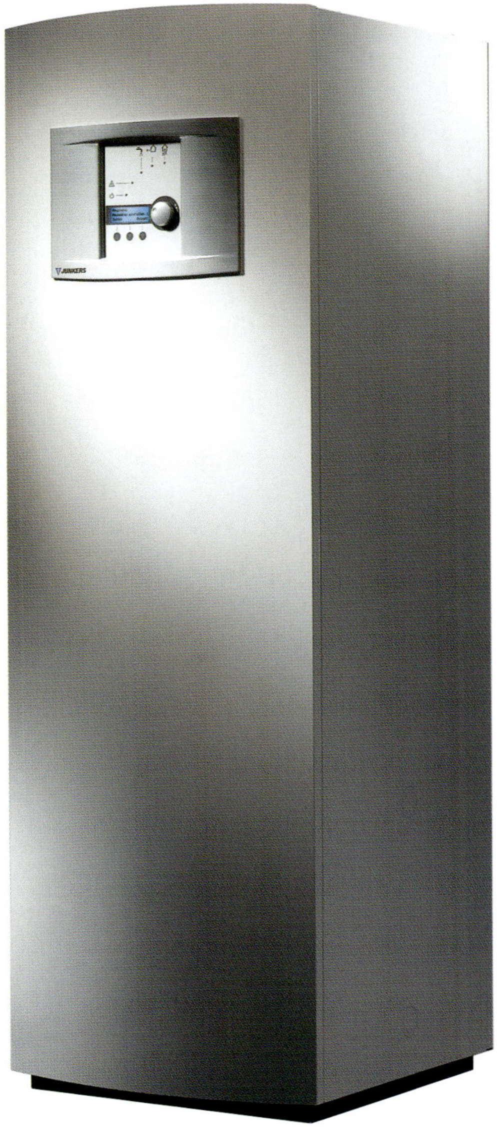

Hersteller Manufacturer
Robert Bosch GmbH
Wernau, Germany

Produkt Product
Junkers Erdwärmepumpe
Heizungswärmepumpe
Heat pump

Design
TEAMS Design GmbH
Esslingen, Germany

Die Junkers Erdwärmepumpe ist für Ein- bis Zwei-Familienhäuser konzipiert, deren Bauherren umweltfreundliche, regenerative Energien nutzen wollen. Das schlichte, zeitlose Design erlaubt eine Integration ins Wohnumfeld. Heizung und Brauchwasserspeicher zur Warmwasserbereitung, die bei herkömmlichen Geräten getrennt installiert werden, sind hier in einem einzigen Gehäuse untergebracht. Das zentrale Bedienrad und ein Klartextdisplay erlauben eine einfache Bedienung durch Menüführung. Die Wärmepumpe ist durch den Einsatz hochwertiger Materialien und Komponenten, wie z.B. dem Warmwasserspeicher aus Edelstahl, auf Langlebigkeit ausgerichtet. Im Sommer kann das System zur Raumkühlung eingesetzt werden.

The Junkers heat pump is intended for detached or semi-detached houses, whose developers wish to tap eco-friendly regenerative energy. The straightforward, timeless design means the appliance can be smoothly integrated into domestic settings. Heating and domestic hot water tank, which are customarily separate units, are housed here in a single casing. The central operating dial and an alphanumeric display ensure ease of use by menu guidance. The heat pump itself is highly durable, thanks to the use of high-grade materials and components, examples being the stainless steel hot-water storage tank. In summer, the system can be set to deliver air conditioning, too.

Hersteller Manufacturer
Buderus Heiztechnik GmbH
Lollar, Germany

Produkt Product
Logamax plus GB 142
Gas-Brennwert-Wandheizkessel
Wall-mounted condensing furnace

Design
Ottenwälder und Ottenwälder
Schwäbisch Gmünd, Germany

Der Gas-Brennwert-Wandheizkessel Logamax plus GB 142 wurde mit seiner wohnraumadäquaten Gestaltung für das private Umfeld konzipiert. Integriert ist das Logamatic EMS Regelsystem mit Bedieneinheit RC 30. Der Wandheizkessel bietet eine optimale Energieausnutzung, gute Umwelteigenschaften und er ist wartungsfreundlich.

The Logamax plus GB 142 gas-fired high-efficiency wall-mounted heating furnace has been given an attractive design to meld with private living areas. The Logamatic EMS regulation system and the RC 30 control panel both form integral parts of the unit. The wall-mounted heating furnace delivers optimal fuel use, is eco-friendly and very easy to service.

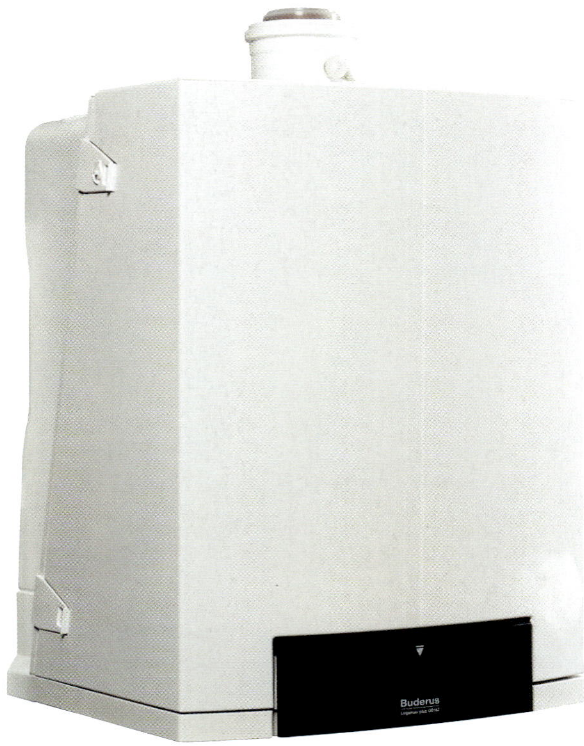

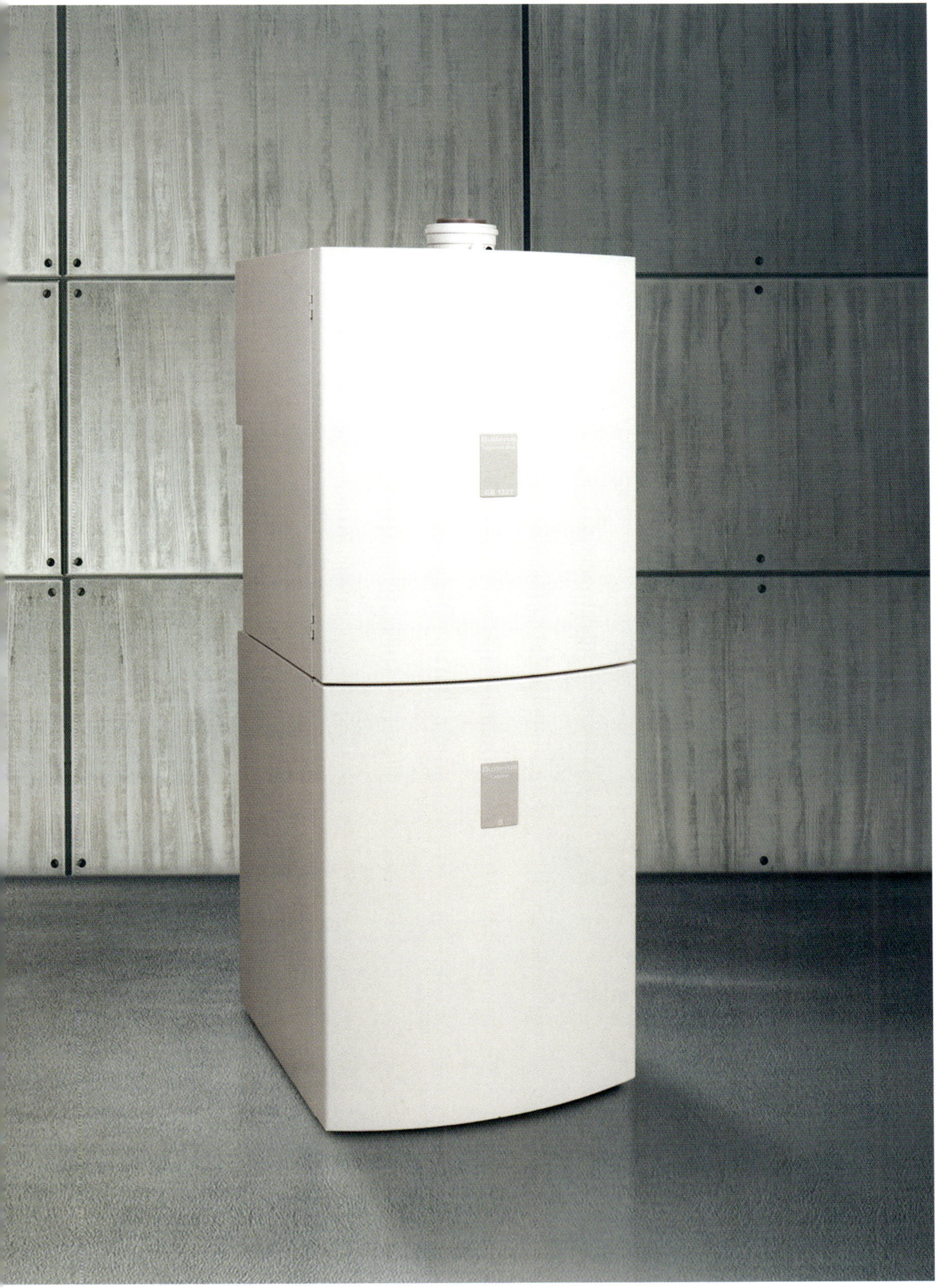

Hersteller Manufacturer
Buderus Heiztechnik GmbH
Lollar, Germany

Produkt Product
Logamax GB 132 T
Brennwert-Kompaktheizkessel für Gas
Compact gas-fired condensing furnace

Design
Ottenwälder und Ottenwälder
Schwäbisch Gmünd, Germany

Der Logamax GB 132 T Brennwert-Kompaktheizkessel für Gas hat einen äußerst geringen Platzbedarf und ist mit seiner wohnraumadäquaten Gestaltung für die Aufstellung außerhalb von Kellerräumen geeignet. Er ist Teil eines modularen Gesamtkonzepts und lässt sich durch seinen entsprechend modularen Aufbau an unterschiedliche Anforderungen anpassen.

The Logamax GB 132 T high-efficiency gas-fired condensing furnace takes up a minimum amount of space and thanks to its appealing look by no means needs to be hidden away in a cellar. It is part of a modular overall concept and thanks to this structure can be expanded to match a variety of specifications.

Hersteller Manufacturer
Buderus Heiztechnik GmbH
Lollar, Germany

Produkt Product
Logamatic RC 10, RC 20, RC 30
Raumbedieneinheit
Room thermostats

Design
Ottenwälder und Ottenwälder
Schwäbisch Gmünd, Germany

Die Raumbedieneinheiten Logamatic RC 10, RC 20, RC 30 wurden für Privathaushalte konzipiert. Sie sind im Wohnbereich montierbar und an Buderus Heizungsanlagen adaptierbar. Die Bauweise ist kompakt und funktionsbezogen, die Bedienung wurde durch eine Klartextanzeige und einfache Programmierung verbessert.

The Logamatic RC 10, RC 20, RC 30 room thermostats were devised with private households in mind. They are ready for installation in living areas and for connection to Buderus heating systems. The thermostats are compact and highly functional, with usage facilitated by the alphanumeric display and straightforward convenient programming options.

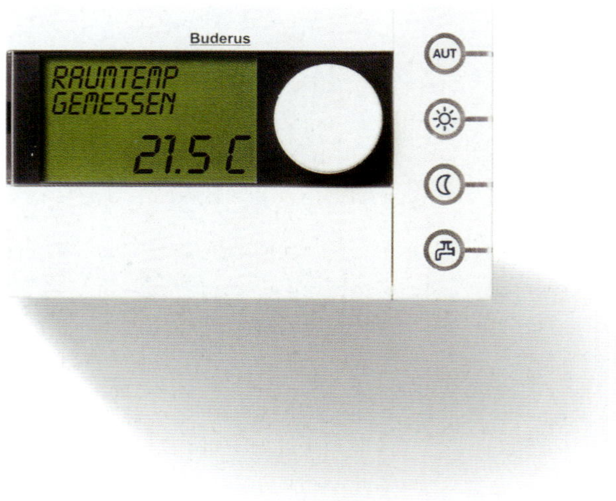

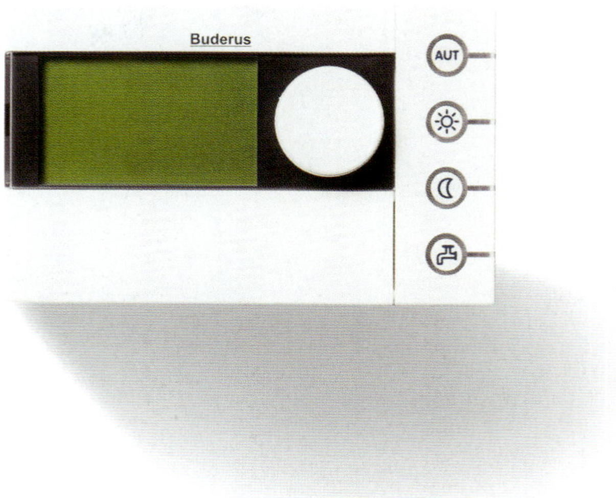

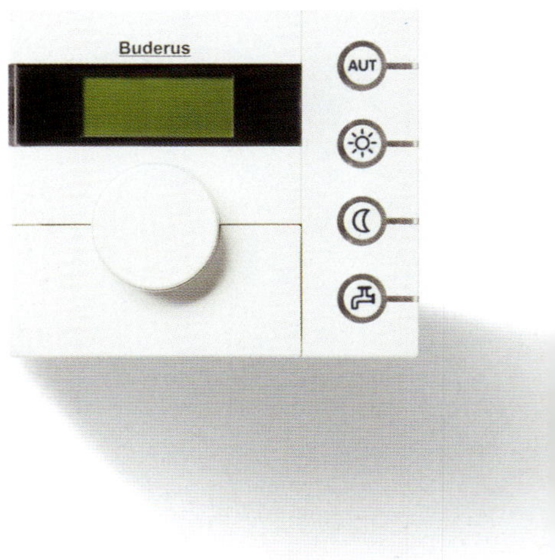

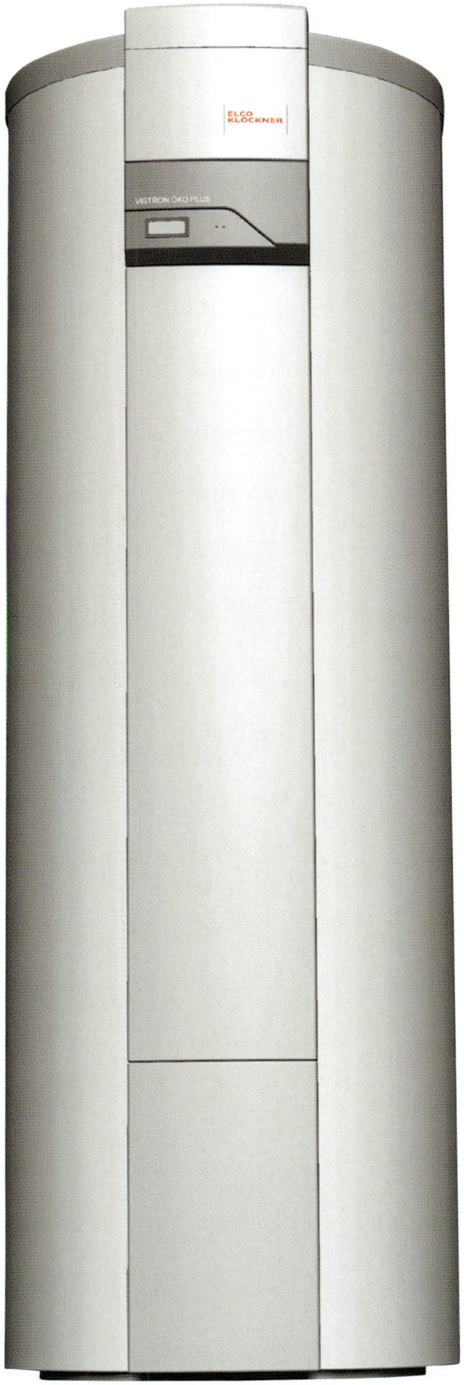

Hersteller Manufacturer
Elco Klöckner Heiztechnik GmbH
Hechingen, Germany

Produkt Product
Vistron-Öko-Plus
Warmwasserspeicher
Hot water storage unit

Design
Weinberg & Ruf
Produktgestaltung
Filderstadt, Germany

Der Vistron-Öko-Plus Warmwasserspeicher besitzt einzelne modulare Abdeckungssegmente, die eine einfache Anpassung an die jeweilige Ausstattung erlauben. Die umweltfreundliche Gehäuseisolierung vermeidet Wärmeverluste und lässt sich getrennt transportieren. Die Vistron Öko Plus Speicher stehen in fünf bedarfsgerechten Größen von 150 bis 500 Liter zur Verfügung.

The Vistron-Öko-Plus warm-water storage unit features individual modular casing sections. As a consequence, the unit can be configured simply to individual requirements. The environmentally friendly insulation to the body of the unit averts any heat dissipation and can be transported separately. The Vistron Öko Plus storage unit is available in five handy sizes ranging from 150l through 500l.

Hersteller Manufacturer
Adamis Co. Ltd.
Tokio, Japan

Produkt Product
Bliss
Wasseraufbereitungsgerät
Water purifying device

Design
Mikio Shiba
Ippei Noji
Kenichi Abe
Geo Design Inc.
Tokio, Japan

Bliss ist ein hochwertiges Wasseraufbereitungsgerät für Boutiquen, Schauräume, Vorstandsetagen etc. Das Design des Metallkorpus ist schlicht und symbolisch, die polierten Wasserhähne bestehen aus Edelstahl, die darüber positionierte LED informiert über den Zustand des Wassers.

Bliss is a high-quality water purifying device for use in boutiques, showrooms, executive offices, etc. The metal body has a decidedly sleek and symbolic look, heightened by the polished stainless steel faucets. The LED positioned above the outlet informs users of the state of the water.

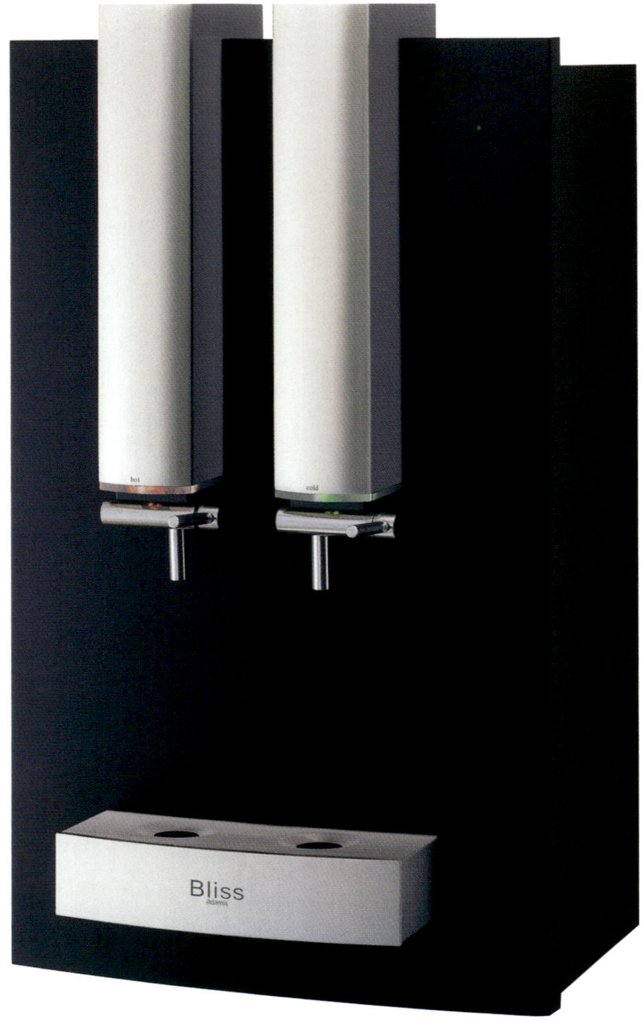

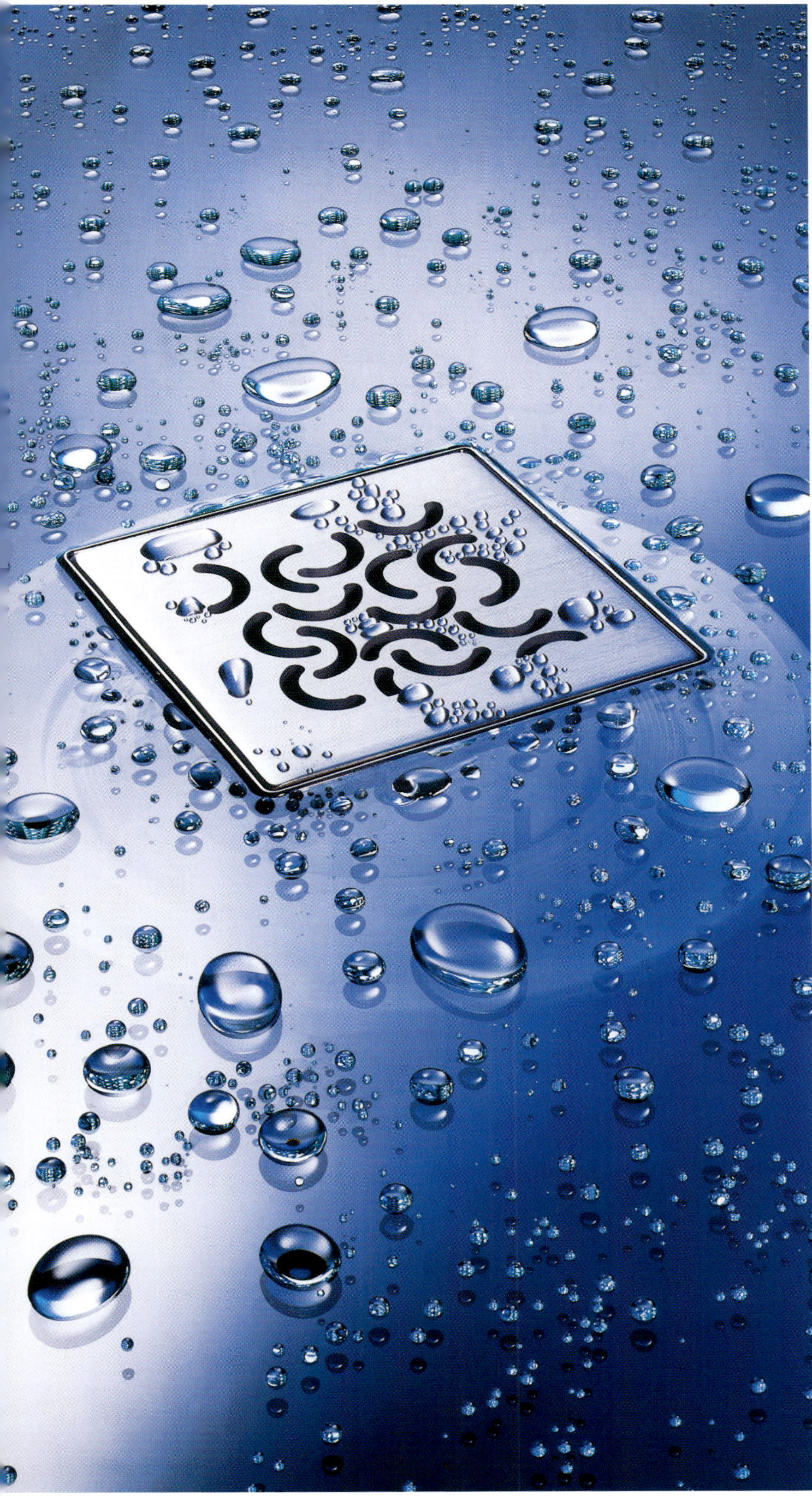

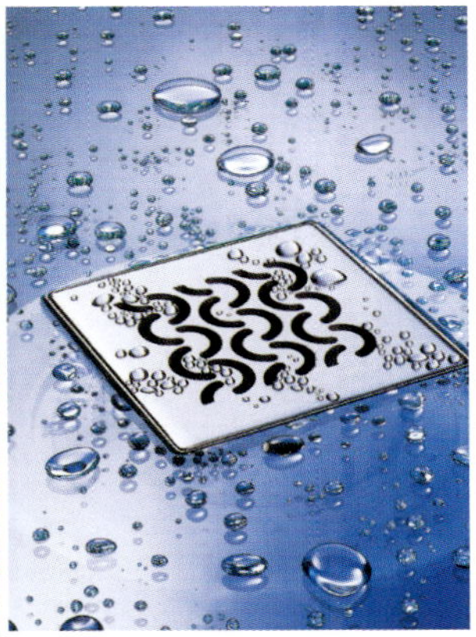

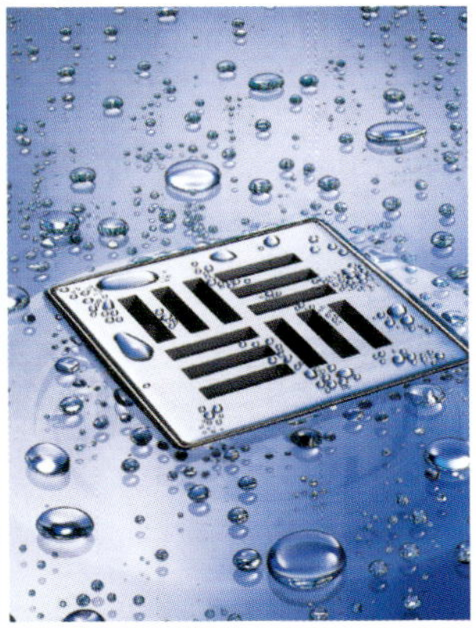

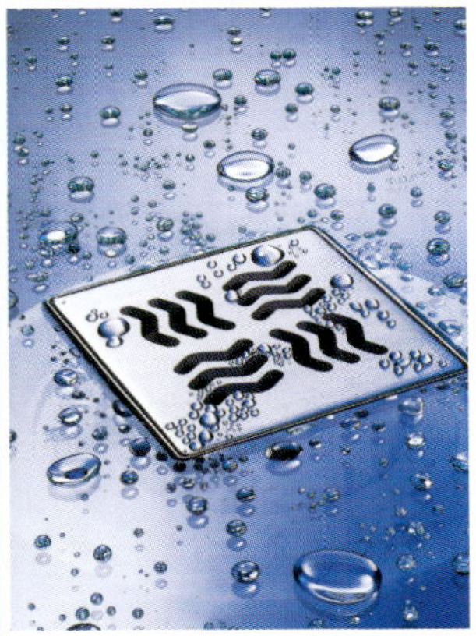

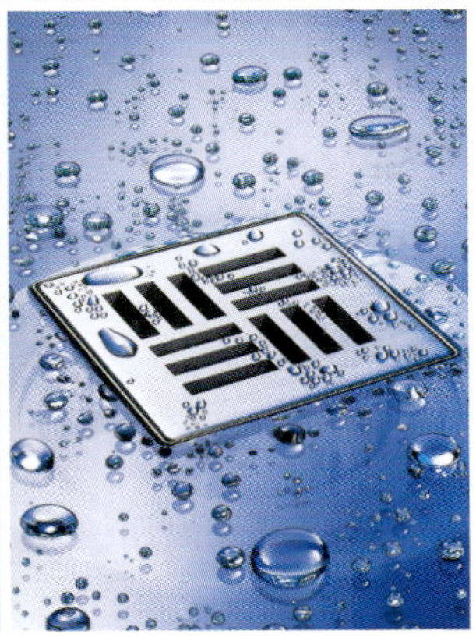

Hersteller Manufacturer
Dallmer GmbH + Co. KG
Arnsberg, Germany

Produkt Product
Flow
Ablaufroste für Bodenabläufe
Grates for floor drains

Design
Dallmer GmbH + Co. KG
Arnsberg, Germany

Bei der Serie Flow handelt es sich um Ablaufroste mit prägnanten Mustern aus Halbkreisen, geschwungenen Linien und geometrischen Anordnungen für die Einlauföffnungen. Die Ablaufroste sind aus massivem, hochwertigem Edelstahl V4A (5 mm stark) gefertigt, die Einlauföffnungen sind mit Laserstrahl geschnitten, die Oberfläche ist matt satiniert. Alle Roste sind Teil eines Baukastensystems und mit verschiedenen Ablaufgehäusen und Aufsatzrahmen kombinierbar. Zur leichteren Montage und zur Anpassung an das Fliesenbild sind Aufsatzrahmen und Rost stufenlos höhen- und seitenverstellbar.

The Flow series is a line of grates for floor drains with striking designs, such as semi-circles, curved lines and geometric patterns for the drain slits. The grates are all made of solid, high-grade stainless steel V4A (5 mm thick), and the grating slits are cut by laser. The grates boast a satin finish and together constitute a modular system, ready for combination with different Dallmer drain sections and frames. To ensure easy installation, the frames in which they lie can be adjusted in height and laterally to fit specific tiling configurations.

Hersteller Manufacturer
Gira Giersiepen GmbH & Co.KG
Radevormwald, Germany

Produkt Product
Gira SmartTerminal
Terminal für Instabus und Internet
Terminal for Instabus and Internet

Design
Phoenix Design
Stuttgart, Germany

Das Gira SmartTerminal ist die zentrale Steuer-, Melde- und Kontrolleinheit für die gesamte Instabus-Installation in einem Gebäude: Mit ihm sind z.B. Leuchten, Jalousien, Heizungen und die Alarmanlage steuer- und programmierbar. Zugleich dient das Gira SmartTerminal als zentrales Terminal für den Empfang von E-Mails und den Zugriff auf nützliche Online-Dienste, z.B. aktuelle Nachrichten, Wettervorhersagen und Staumeldungen. Die Bedienung ist besonders einfach und intuitiv. Sie erfolgt wie bei den heutigen Navigationssystemen im Auto menügestützt durch Drehen und Drücken eines zentralen Bedienknopfes. Die Frontplatte des Gira SmartTerminals besteht aus hochwertigem Echtglas.

The Gira SmartTerminal is the central control, signalling and regulating unit for the entire Instabus installation in a building: You can control and program lights, blinds, heaters, the alarm system etc. with it. The Gira SmartTerminal also functions as a central terminal for the reception of e-mails and access to useful online services, e.g. current news, weather forecasts and traffic reports. Operation is especially simple and intuitive. You simply turn and push a main operating button to make selections in menus, similar to modern navigation systems in automobiles. The front panel of the Gira SmartTerminal is made of high-quality glass.

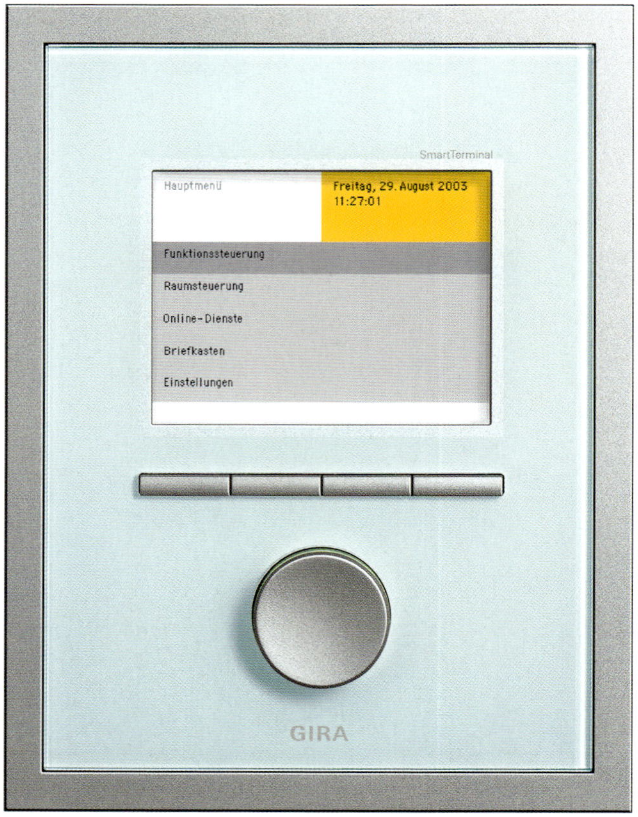

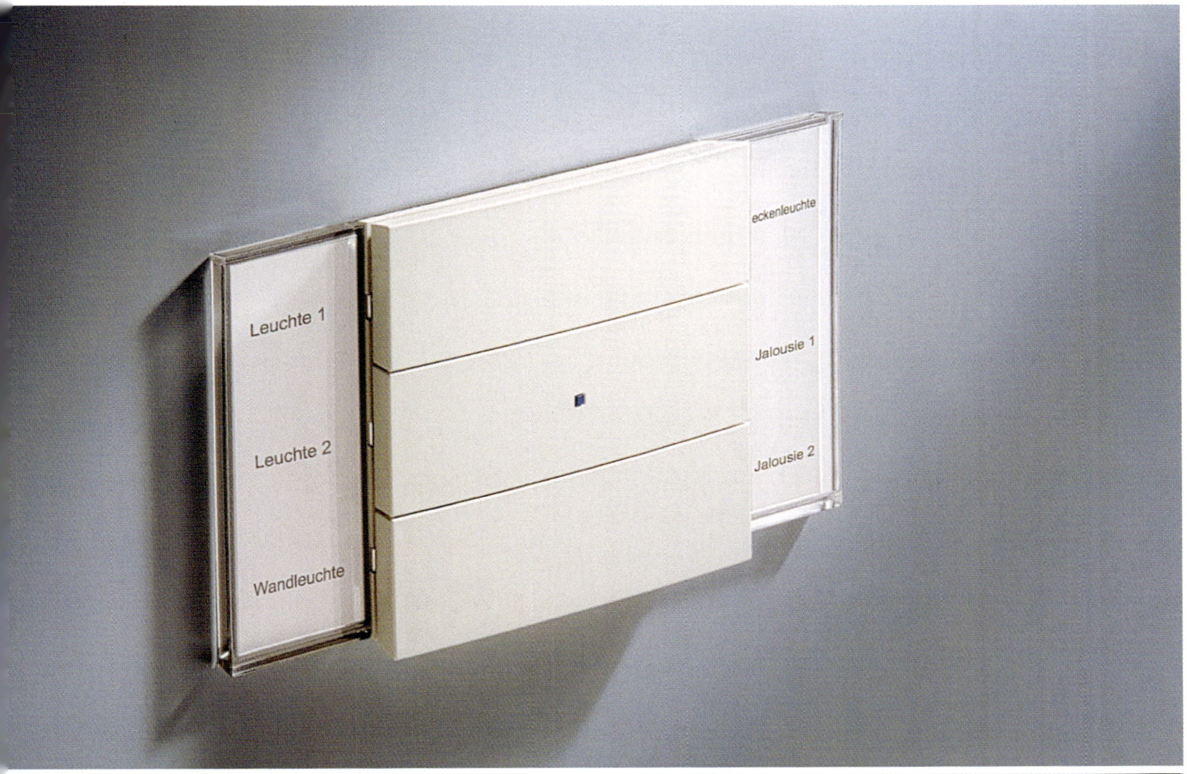

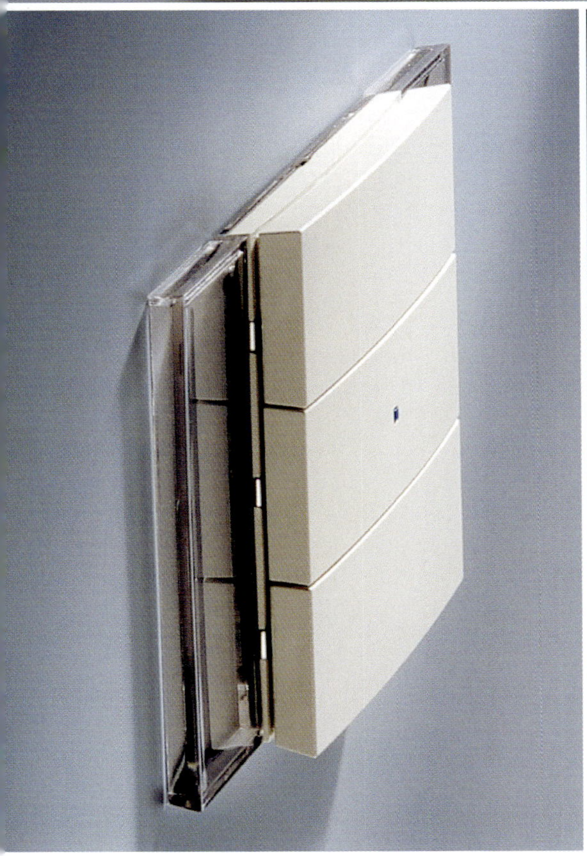

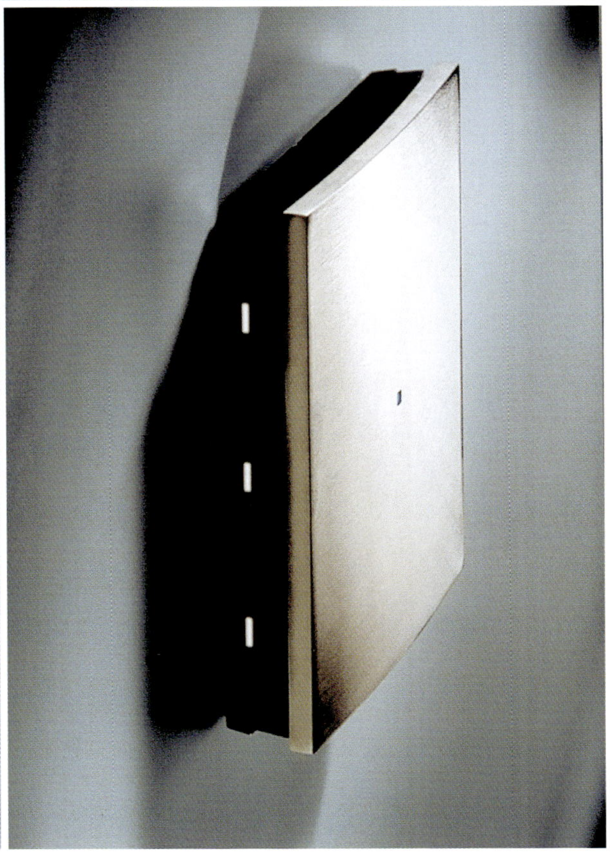

Hersteller Manufacturer
Berker GmbH & Co. KG
Schalksmahle, Germany

Produkt Product
Berker B.IQ
Tastsensor
Pushbutton

Design
OCO-Design
Münster, Germany

Das B.IQ Tastsensor-System orientiert sich an den Bedürfnissen der Benutzer: einfache Bedienung, flexibel in der Anwendung, vom Anwender individuell anpassbar in der Bedienung und systemoffen (Instabus EIB, LON, IR...). Grundlagen der Gestaltung sind entsprechend Klarheit, deutliche Linienführung, ganzflächige, rahmenlose Tastsensor-Elemente und einfache Integrationsmöglichkeiten in moderne Architektur. Neben der Standardausführung in hochwertigem Kunststoff gibt es eine Linie mit gleicher Funktionalität in den Ausführungen Edelstahl und Glas in massivem Material.

The B.IQ tactile sensor system takes its cue from user requirements. It is very easy to use, flexible in terms of applications and the functionality can be individually adapted; it is an open system (Instabus EIB, LON, IR...). The design correspondingly focuses on clarity, distinct contours, frameless all-over tactile sensor elements and convenient integration into interiors. Alongside the standard high-quality plastic version, the sensors are also available with the same functions but in solid stainless steel or glass.

Hersteller Manufacturer
Zumtobel Staff GmbH
Dornbirn, Austria

Produkt Product
Onlite Artsign central EB-HO RZ-2S
Rettungszeichenleuchte
Escape sign luminaire

Design
Studio Matteo Thun
Mailand, Italy

Die Rettungszeichenleuchte Onlite Artsign wurde für den Einsatz in Büros, Konferenzräumen, Boutiquen und Bereichen mit eher geringen Blickdistanzen entwickelt. Mit einer Piktogramm-Abmessung von 75 x 150 mm ist sie kleiner als eine Postkarte und doch noch auf eine Entfernung von 15 Metern deutlich erkennbar. Die Transparentscheibe ist aus Acryl, die Abdeckung aus Alu-Druckguss, das Gehäuse aus Polycarbonat. Das Design der Leuchte ist formal reduziert und funktionsbezogen, um hinter der entsprechenden architektonischen Umgebung zurückzutreten.

Die Onlite Artsign escape sign luminaire has been designed with applications in offices, conference rooms, boutiques and zones with limited visibility in mind. The pictogram measures 75 x 150 mm, making it smaller than a postcard, yet it is nevertheless clearly identifiable from a distance of 15 m. The transparent pane is made of acrylic, the flaps of die-cast aluminum, and the casing of polycarbonate. The luminaire's design has been formally reduced and is decidedly functional, ensuring that it recedes behind the corresponding architectural setting.

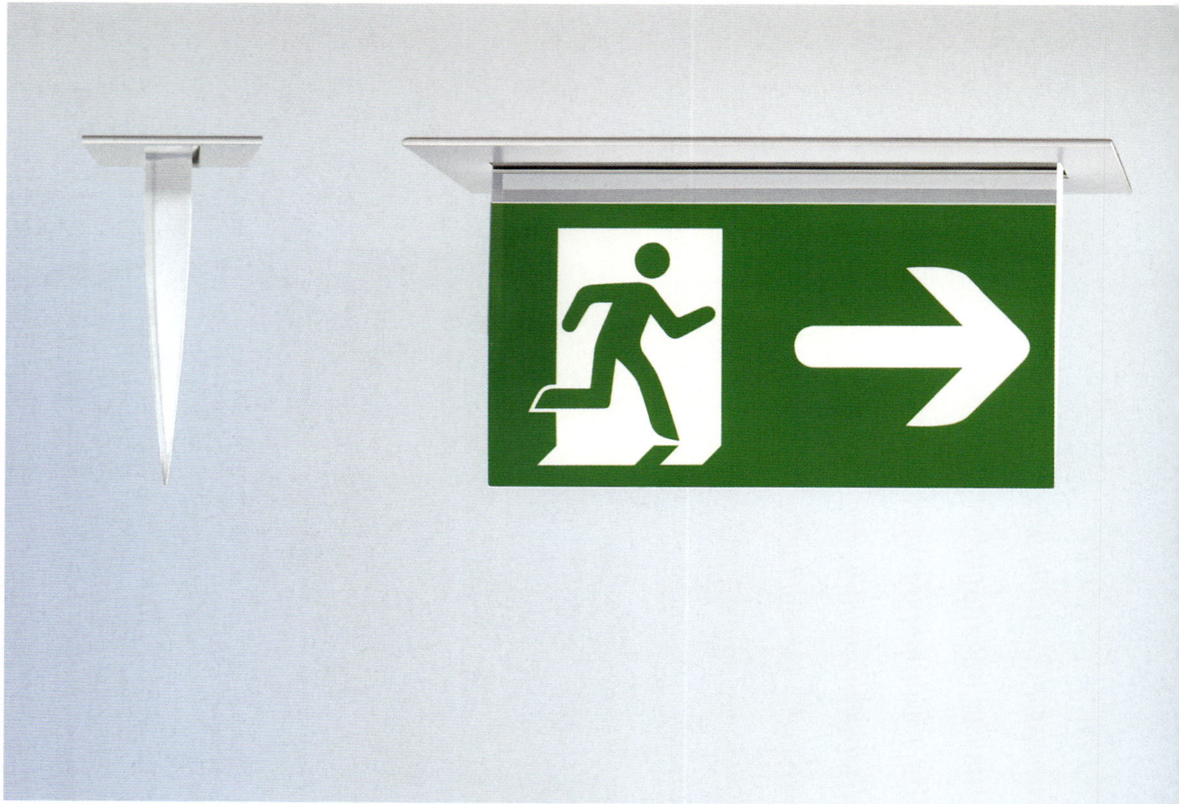

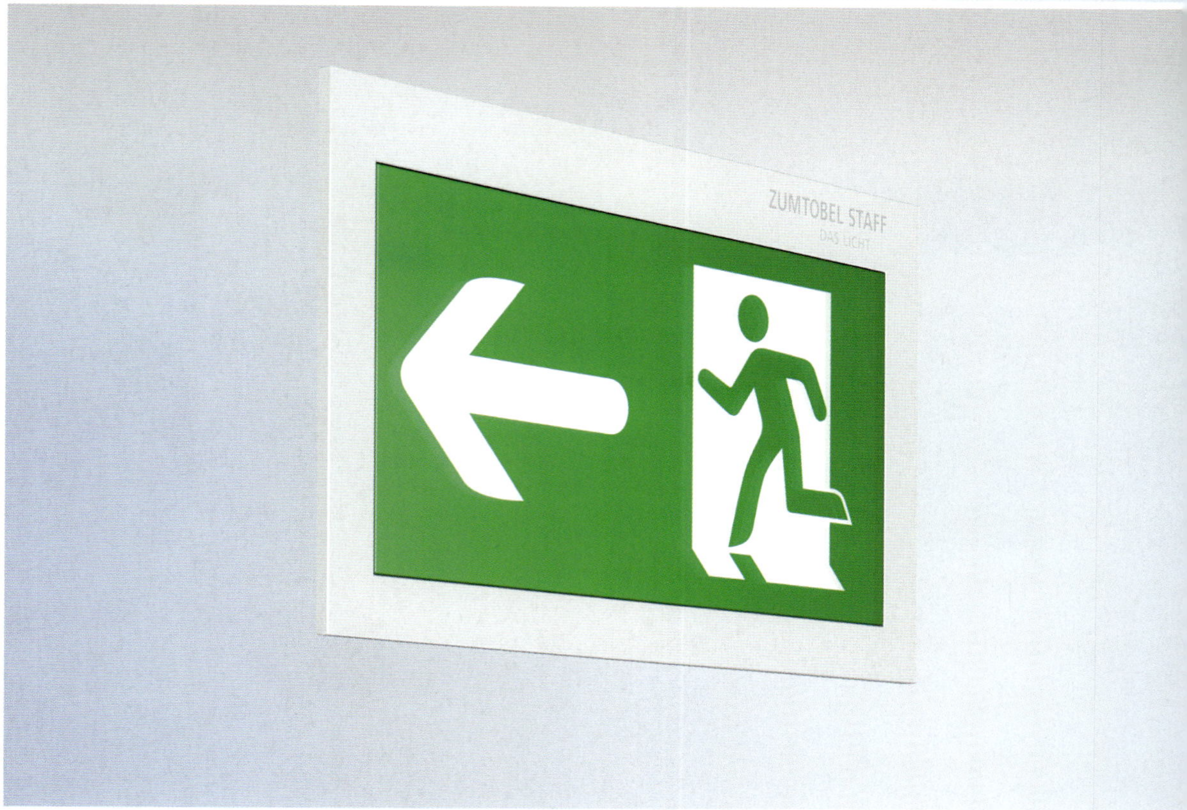

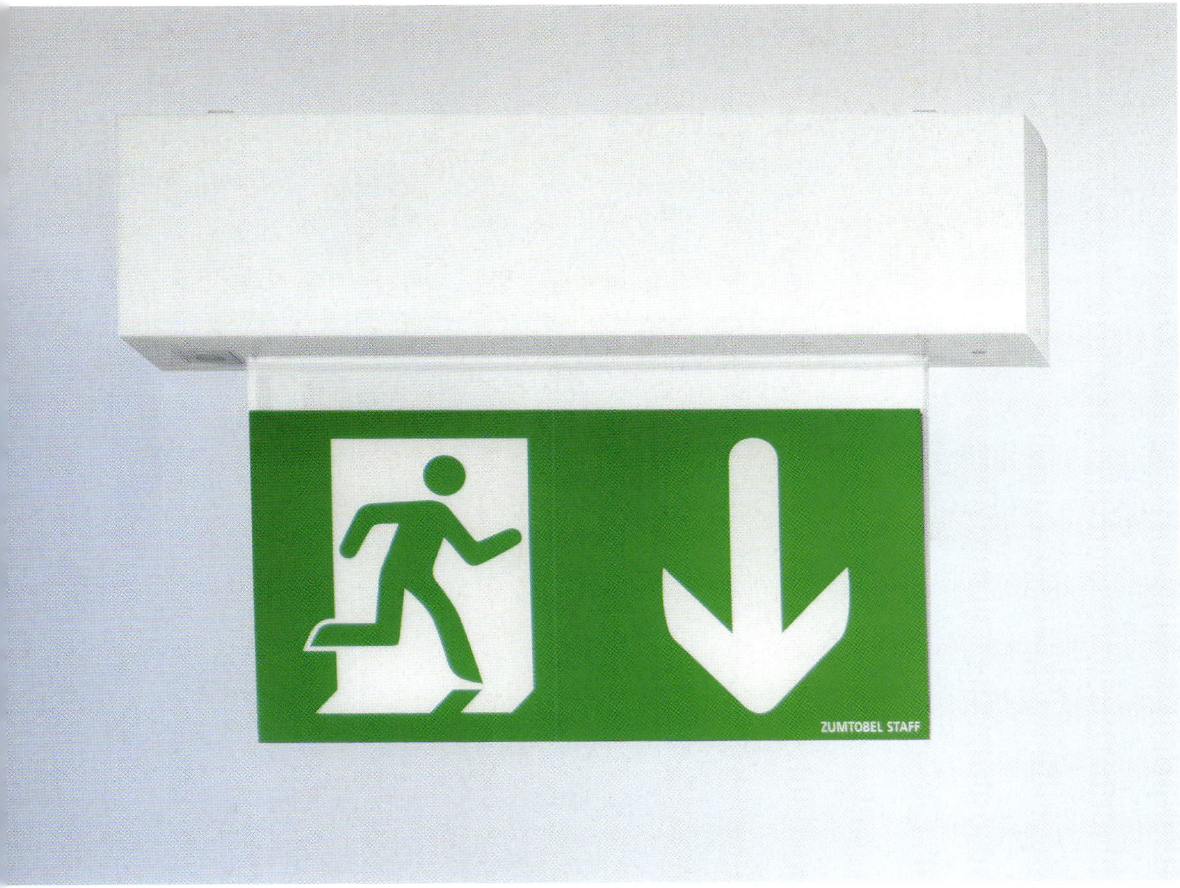

Hersteller Manufacturer
Zumtobel Staff GmbH
Dornbirn, Austria

Produkt Product
Onlite Comsign EB central 1/8W
Rettungszeichenleuchte
Escape sign luminaire

Design
Studio Matteo Thun
Mailand, Italy

Die Rettungszeichenleuchte Onlite Comsign EB central 1/8W hat eine leichte und transparente Ausstrahlung. Die Transparentscheibe besteht aus Acryl, das Abdeckcover aus Alu-Druckguss, das Gehäuse aus Polycarbonat. Die Leuchte ist in jeder Umgebung leicht zu montieren und mit dem TECTON-Lichtbandsystem kompatibel. Comsign gibt es in der An- und Einbauvariante.

The Onlite Comsign EB central 1/8W escape sign luminaire has a light and transparent appearance. The transparent pane is made of acrylic, and the cover flaps of die-cast aluminum, while the housing consists of polycarbonate. The luminaire is easy to assemble in any environment and is compatible with the TECTON continous-row lighting system. Comsign is available in a mounted or recessed version.

Hersteller Manufacturer
DanSign A/S
Roedovre, Denmark

Produkt Product
PlanSign
Beschilderungssystem
Signage system

Design
PLH Design A/S
Kopenhagen, Denmark

Zum Beschilderungssystem PlanSign gehören Tischschilder, Wandschilder, Fahnen-, Hänge- und Paperflexschilder sowie freistehende Infoständer und Pylonen für den Außenbereich. Die Schilder werden aus 5 mm starkem, massivem Aluminium hergestellt. Das versteckte Schloss, das in das Schild integriert ist, wird mit Hilfe eines Magneten geöffnet (patentiertes Verschlusssystem). PlanSign kann sowohl im Innen- als auch im Außenbereich eingesetzt werden.

The PlanSign signage system covers a wide range of applications such as door or wall signs, banners, suspended and Paperflex signs, and includes free-standing electronic information stands and pylons for outdoor use. The signs are made of 5mm solid aluminum. The concealed lock, which is integrated into the sign, is opened by means of a magnet (patented locking system). PlanSign is designed for use indoors as well as out.

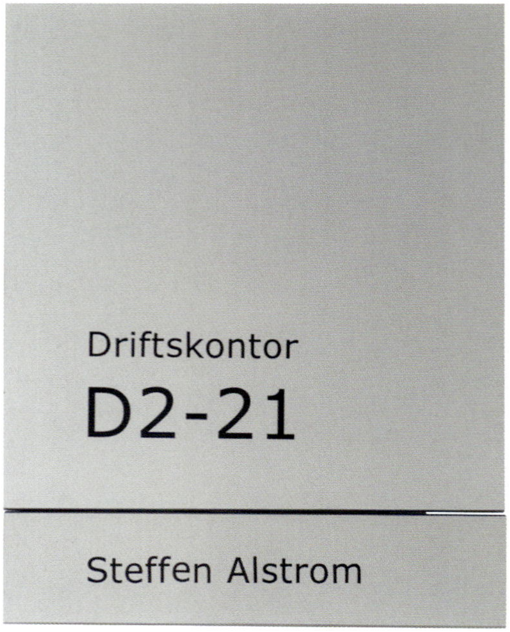

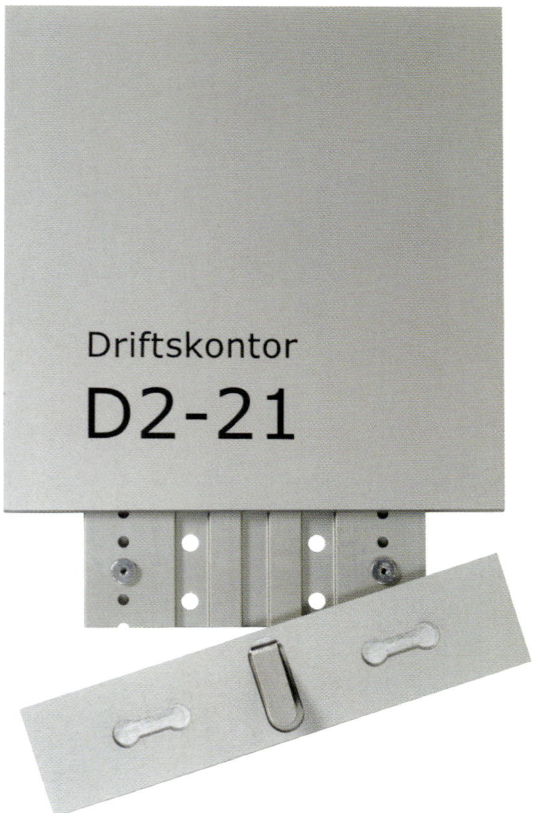

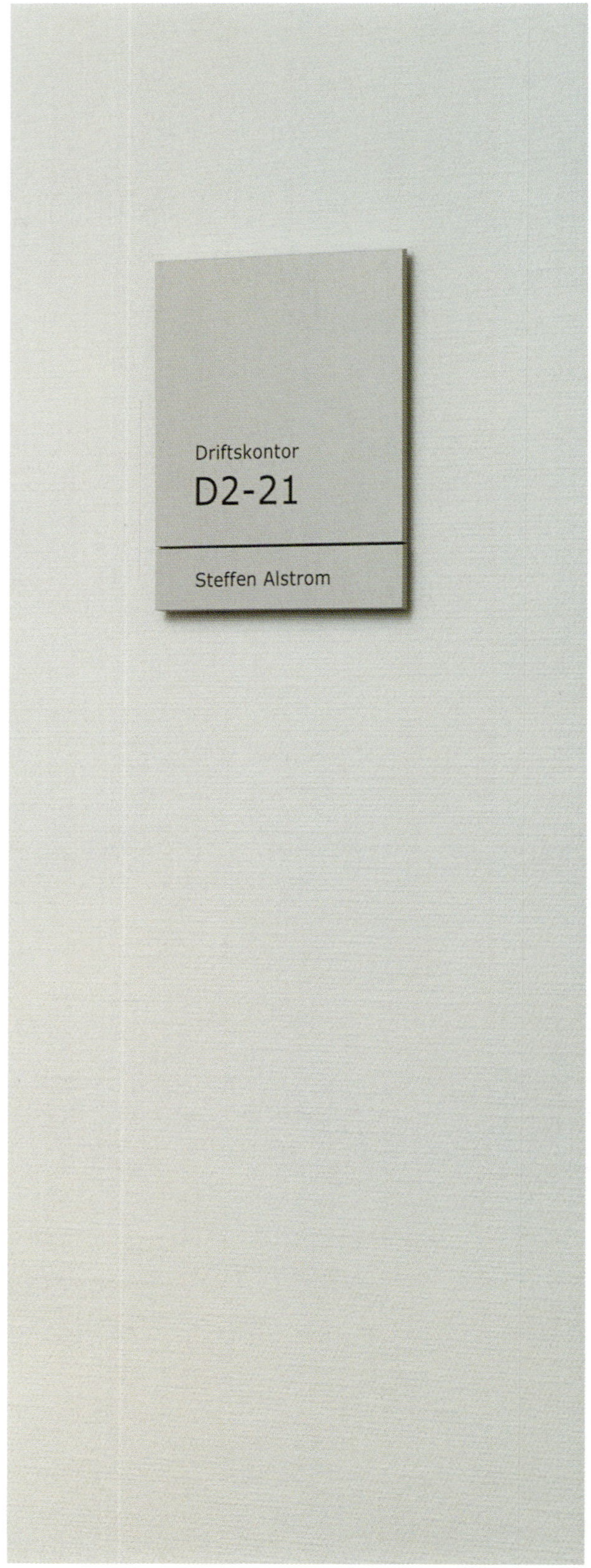

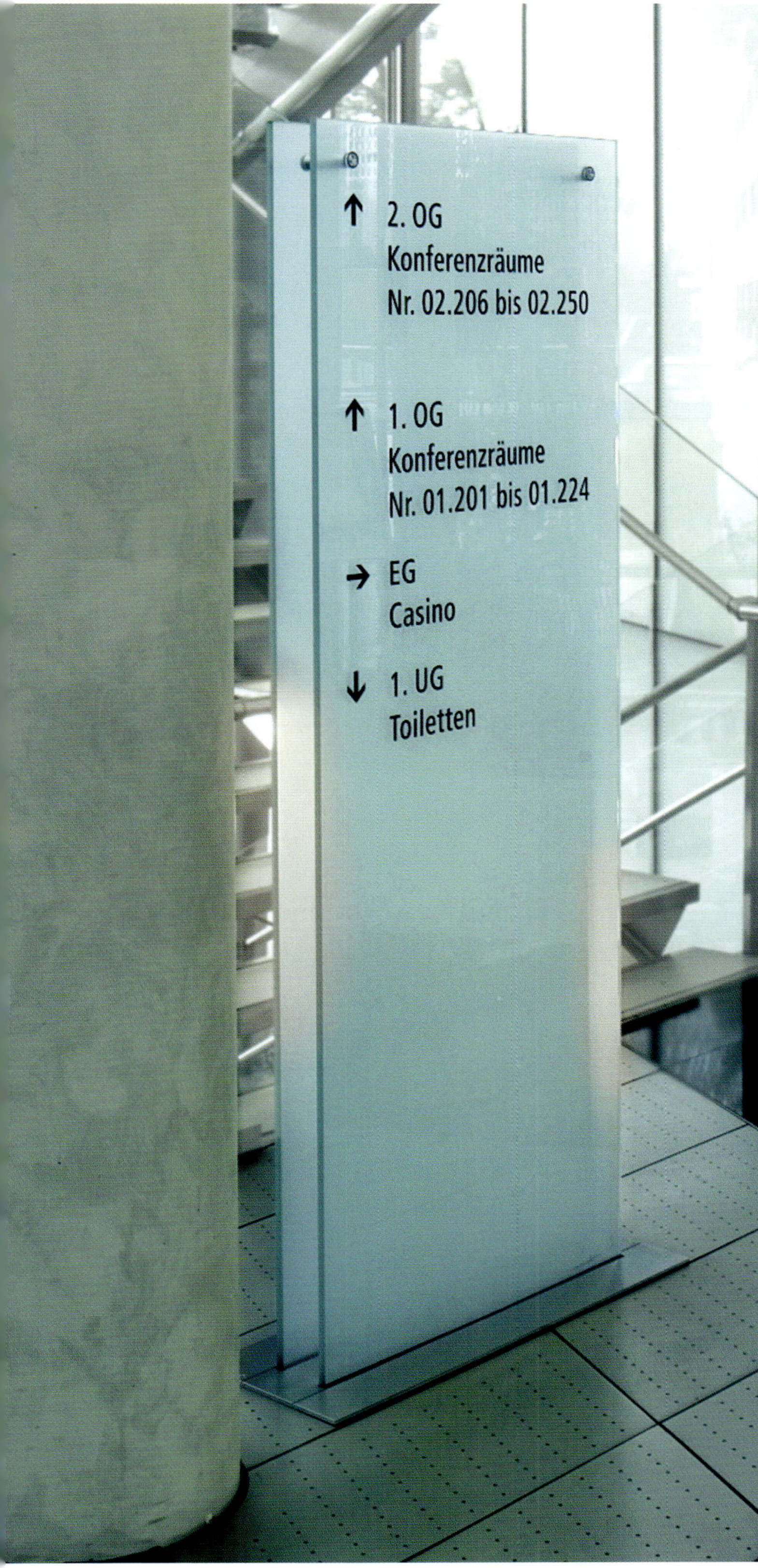

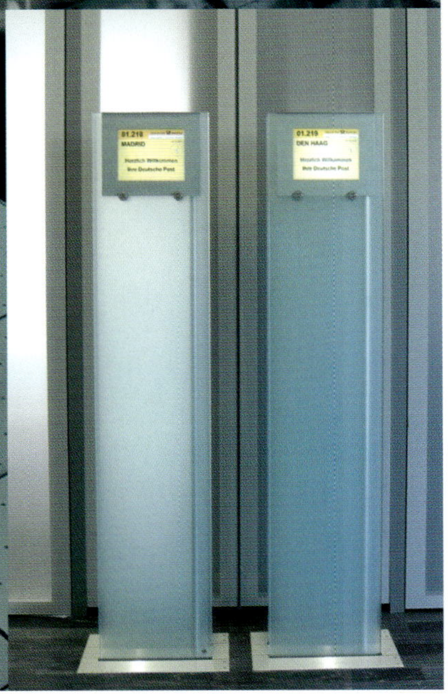

Hersteller Manufacturer
Mabeg Kreuschner GmbH & Co. KG
Soest, Germany

Produkt Product
Twin Glass
Gebäudeleitsystem
Building orientation system

Design
Lengyel Design
Essen, Germany

Twin Glass ist eine Leitsystem-Produktfamilie, deren Systemelemente aus jeweils zwei parallelen Glasplatten bestehen, die in der Regel bodenbündig verspannt sind. Die Produktfamilie besteht aus statischen und dynamischen Displays, von Türschildern über Wegweiser bis zu Parkhauszufahrtsschildern. Twin Glass wurde für eine offene, transparente Architektur entwickelt, die Gestaltung ist minimalistisch.

Twin Glass is a family of building orientation system products. The elements of the system each consist of two parallel glass panes, which are as a rule fixed at right angles to the ground flush with the floor. The product family includes static and dynamic displays, ranging from door plates and signposts even to car-park entrance signs. Twin Glass was created with open, transparent architectures in mind and therefore boasts a minimalist design.

Hersteller Manufacturer
Ritto GmbH & Co. KG
Haiger, Germany

Produkt Product
WT 7630, WT 7650, VHS 7816
Wohntelefon/Video-Hausstation
In-house phone and video station

Design
Bernd Brüssing
Prodesign Brüssing
Neu-Ulm, Germany

Die neue Designlinie von Ritto umfasst verschiedene Indoorgeräte für die moderne Gebäudekommunikation mit TwinBus-Technik. Die Wohntelefone und Video-Hausstationen zeichnen sich sowohl in der Aufputz- als auch Unterputzmontage durch ihre flache Bauform aus. Der formschöne Hörer in einer speziellen Halterung ist ergonomisch so gestaltet, dass eine leichte Bedienung gewährleistet ist. Die beleuchteten Funktionstasten bieten zusätzlichen Bedienkomfort. Optische Rufanzeige und Lichtschaltfunktion sowie bis zu acht melodische Klingeltöne sind serienmäßig.

Ritto's new design series consists of various indoor devices for modern in-house communications featuring TwinBus technology. The in-house phone sets and in-house video stations have ultra-slim lines, and can be positioned flush with the wall as built-in sets or on the wall's surface. The cradle for the well-formed handset was ergonomically designed to ensure ease of use. The illuminated function keys augment user comfort. The standard versions feature optical information on call status, a lighting switch and up to eight call tone melodies.

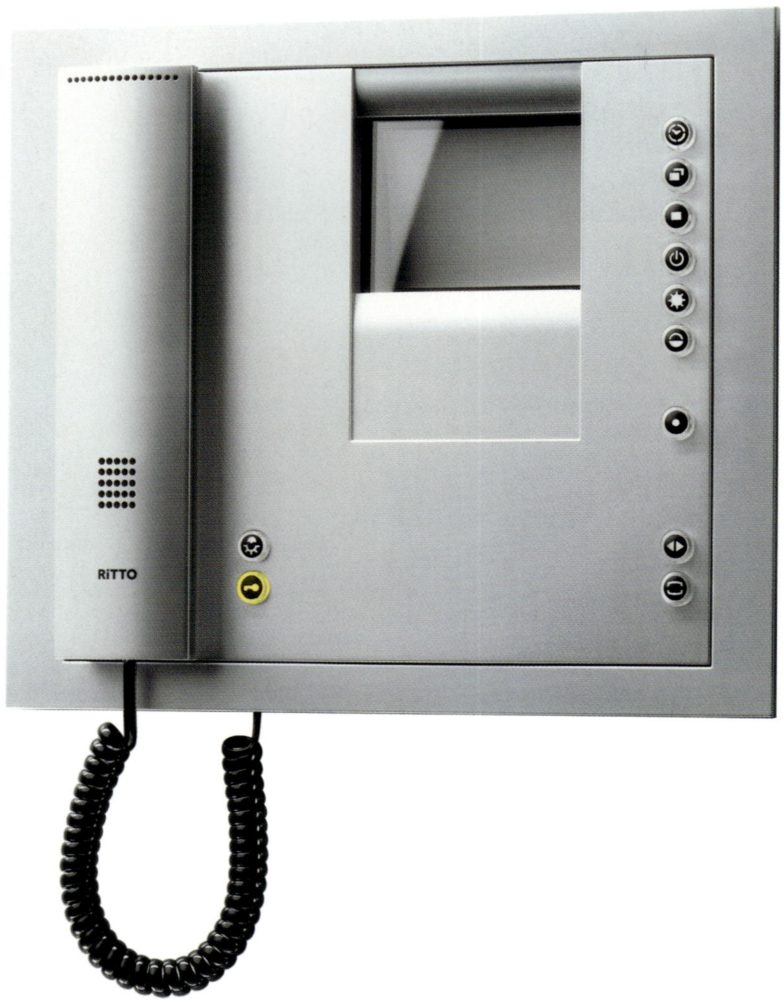

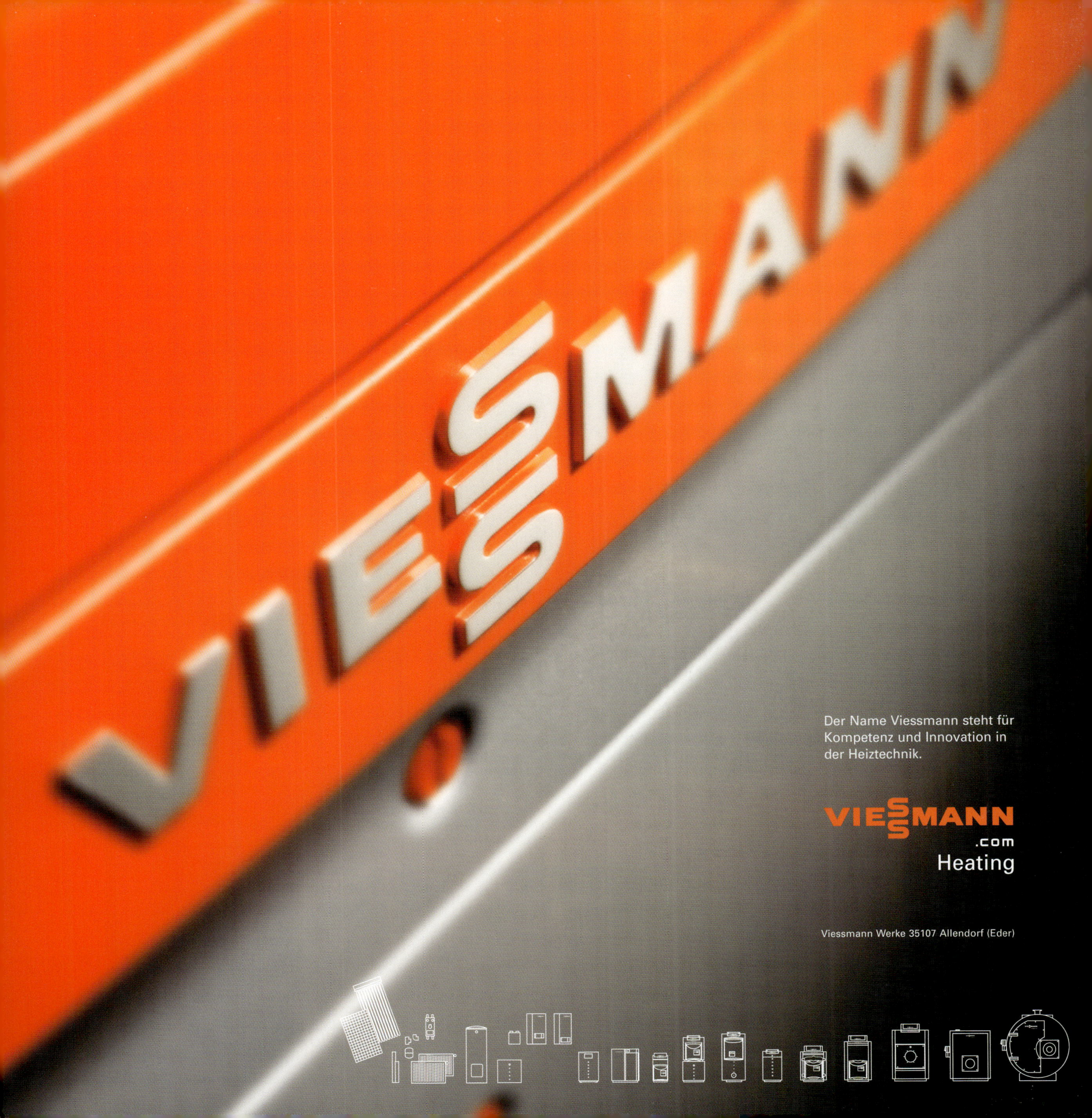
VIESSMANN
Der Name Viessmann steht für Kompetenz und Innovation in der Heiztechnik.
VIESSMANN
.com
Heating
Viessmann Werke 35107 Allendorf (Eder)

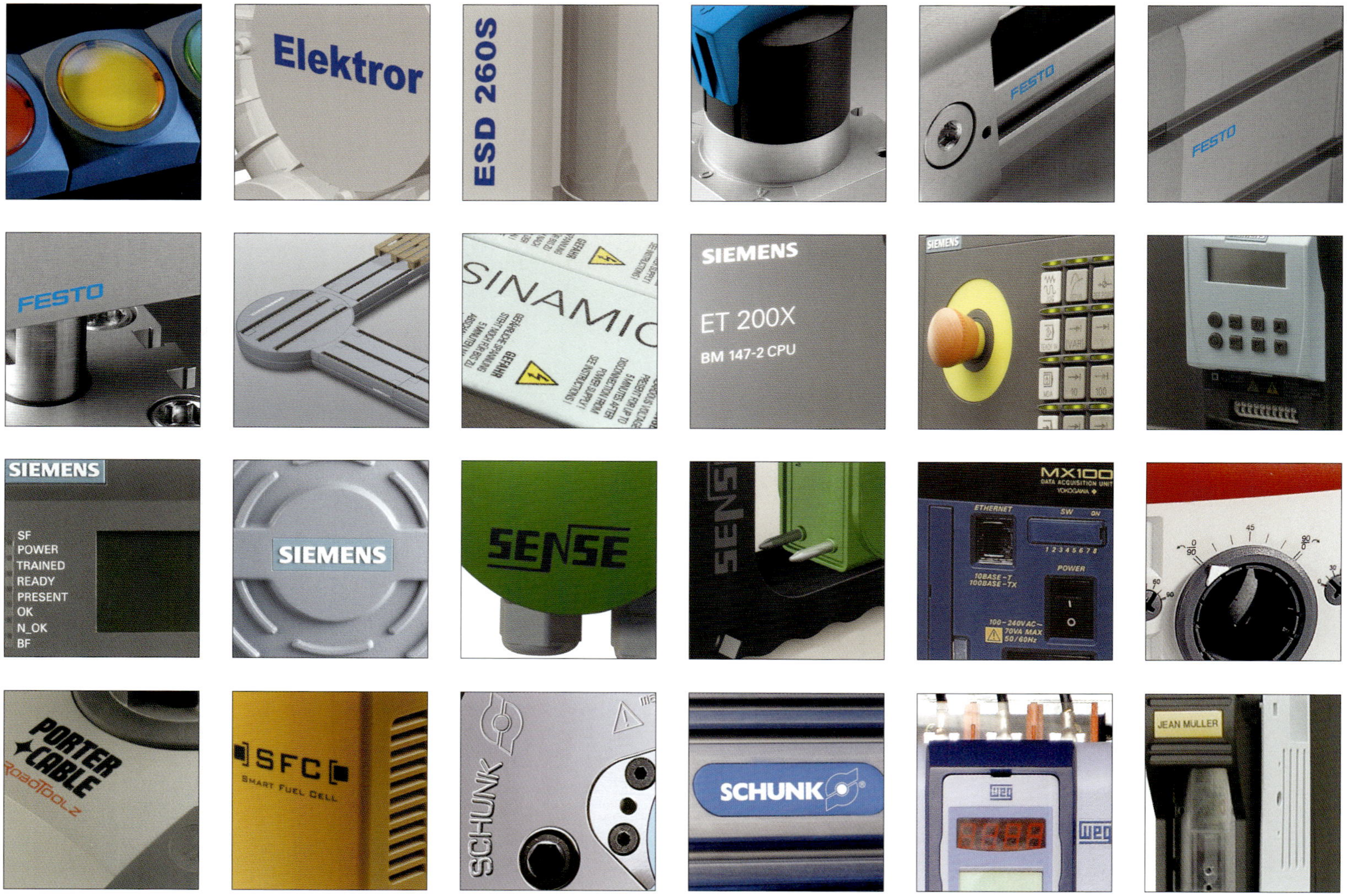
Elektror
ESD 260S
FESTO
FESTO
FESTO
SINAMIC
SIEMENS
ET 200X
BM 147-2 CPU
SIEMENS
SF
POWER
TRAINED
READY
PRESENT
OK
N_OK
BF
SIEMENS
SENSE
MX100
ETHERNET
POWER
PORTER CABLE
ROBOTOOLZ
SFC
SMART FUEL CELL
SCHUNK
SCHUNK
JEAN MULLER

Industrie Industry

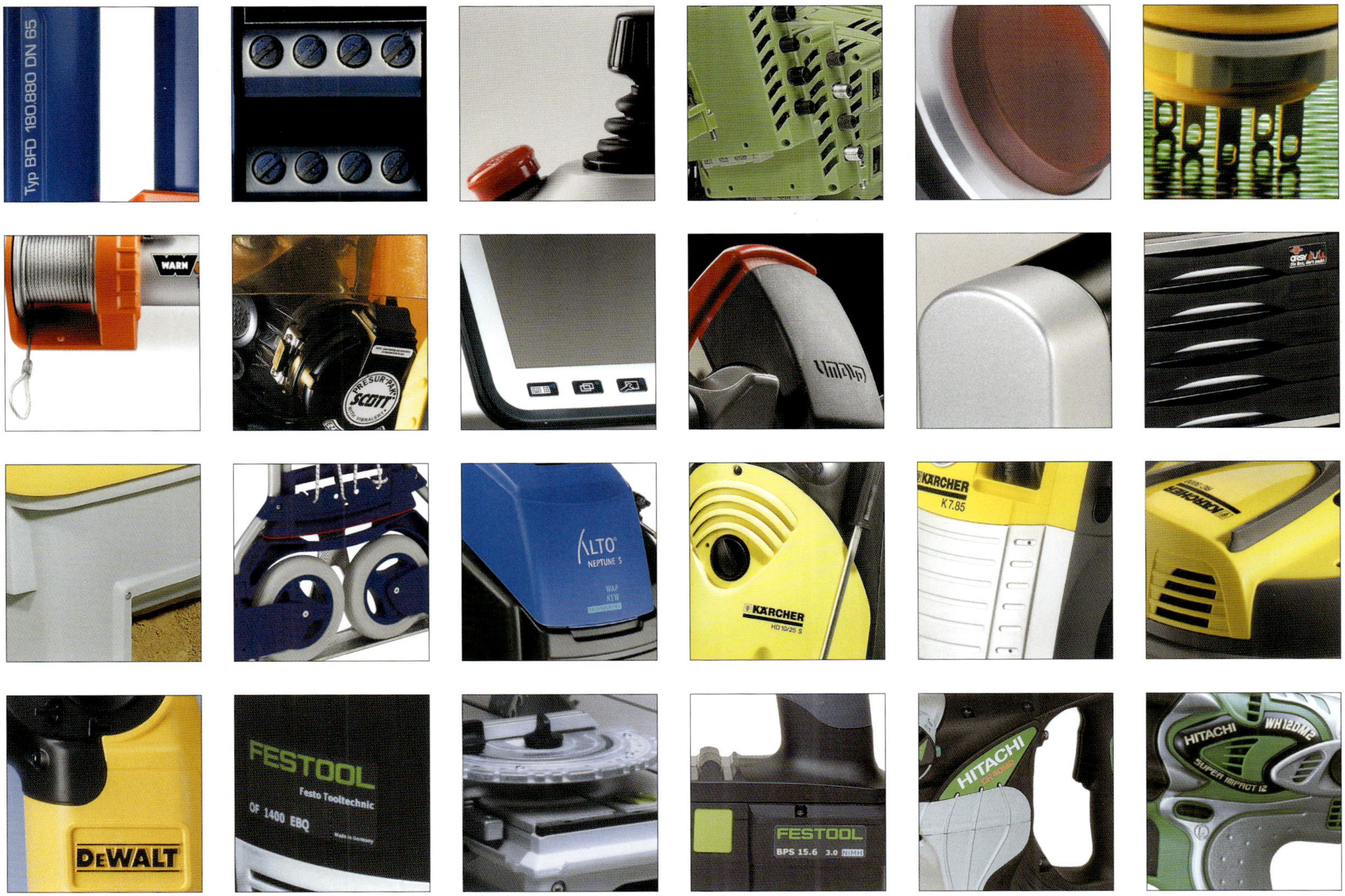

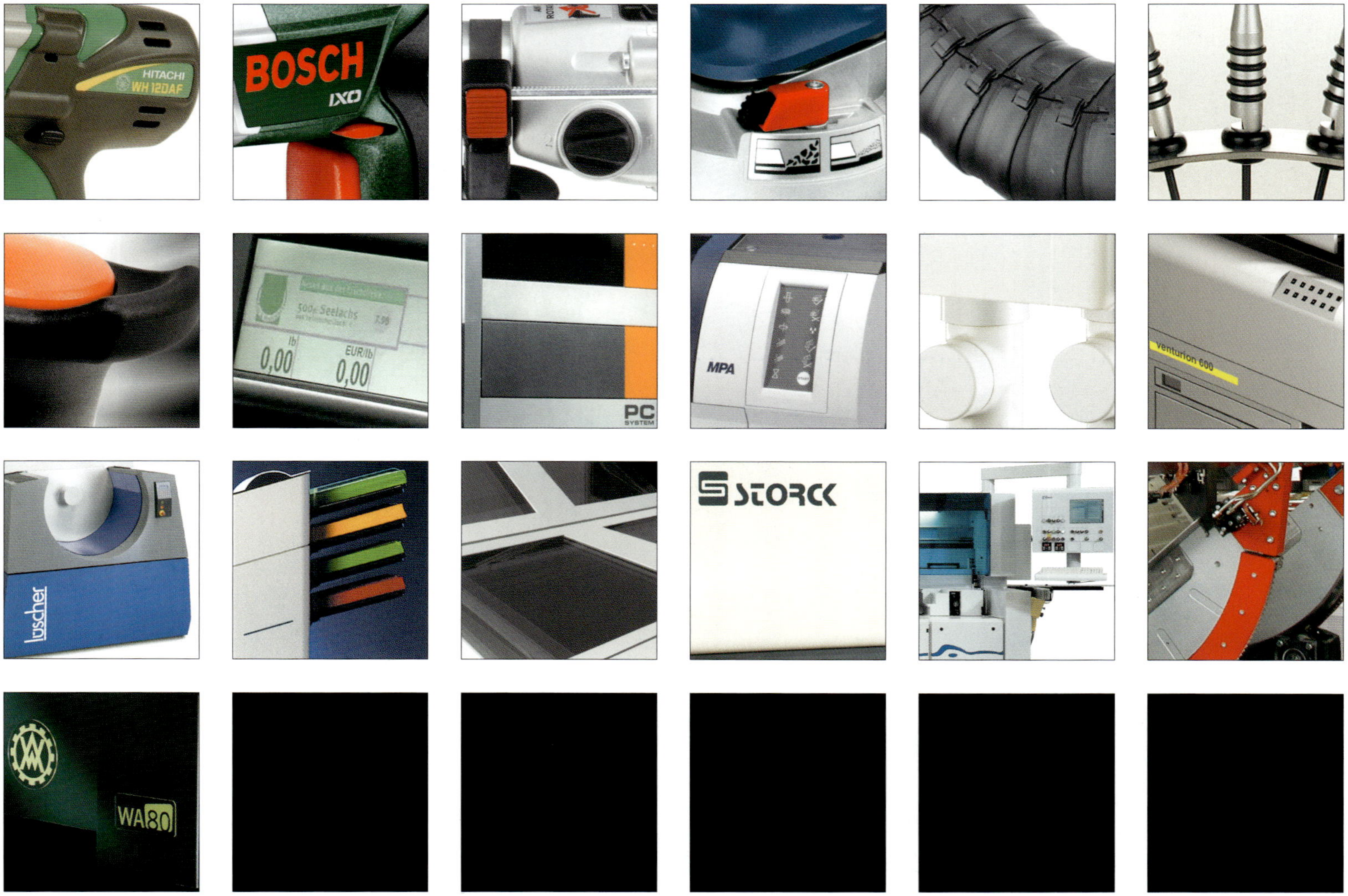
HITACHI
WH 12DAF
BOSCH
IXO
0,00
EUR/lb
0,00
PC
SYSTEM
MPA
venturion 600
STORCK
WA80

Hersteller Manufacturer
Georg Schlegel GmbH & Co. KG
Dürmentingen, Germany

Produkt Product
Quartex-R
Befehlsgeräte
Control units

Design
Designpraxis Diener
Ulm, Germany

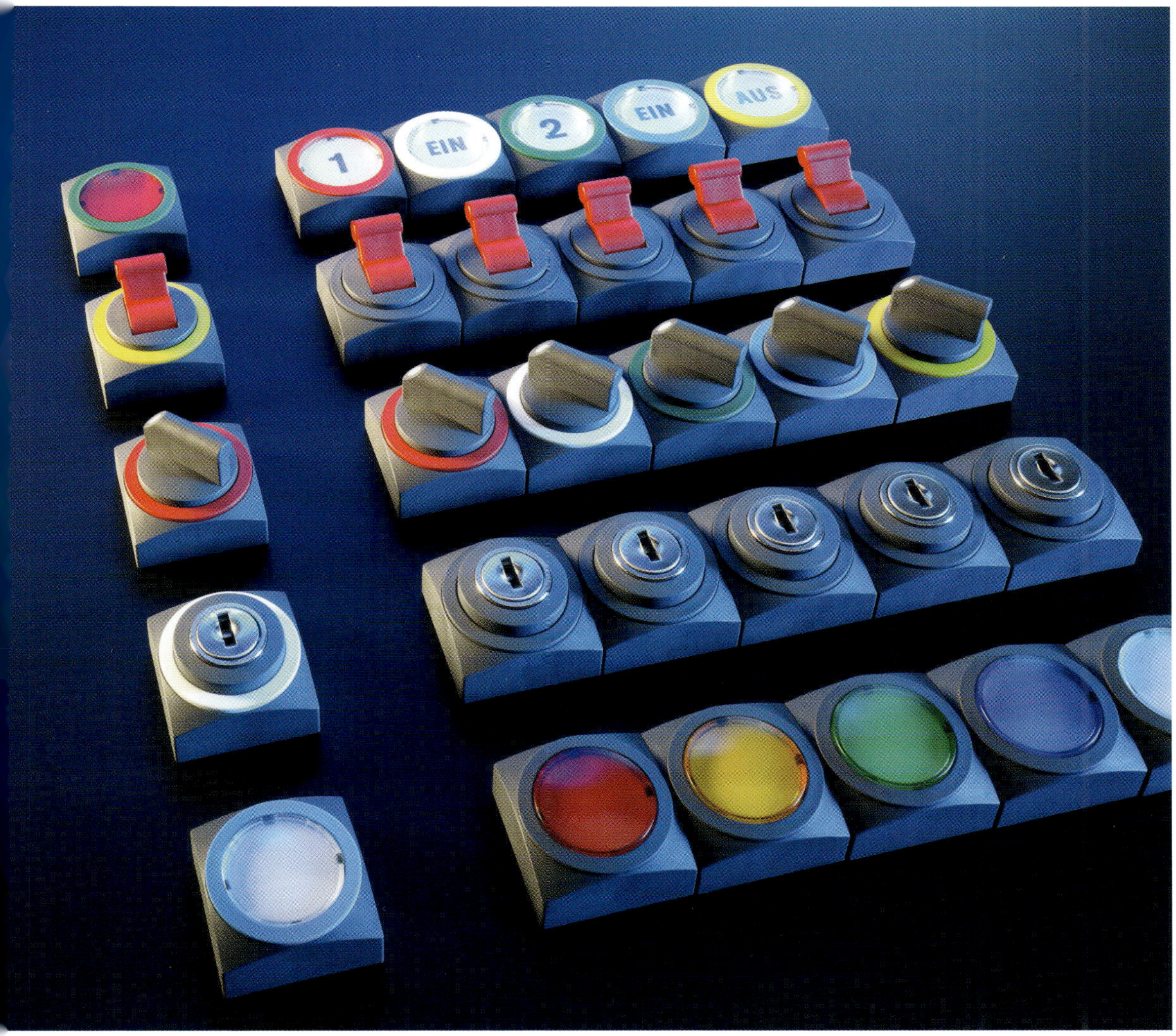

Die Befehlsgeräte aus der Reihe Quartex-R wurden für raue Einsatzbereiche konzipiert, wie sie im Maschinen- Fahrzeug-, Aufzugs-, Anlagen- oder Schiffsbau vorzufinden sind. Die Farbringe unterstützen die formale Eigenständigkeit und dienen zugleich der Bedienführung. Mit den Befehlsgeräten ergeben sich vielfältige Kombinationsmöglichkeiten mit Betätigern, Kontaktgebern und Leuchtkontaktgebern.

The Quartex-R series control units were designed for use in tough environments, for example for applications in mechanical engineering, vehicle and elevator construction, plant engineering and shipbuilding. The colored rings serve to highlight the formal independence of the units and enhance user friendliness. The control units can be combined in a variety of ways with actuators, relay switches and illuminated contact switches.

iF Silver Award

Hersteller Manufacturer
Elektror Karl W. Müller GmbH & Co.
Esslingen, Germany

Produkt Product
ESD 260
Seitenkanalverdichter
Side channel blower

Design
Tino Zink
STZ Design Innovation
Neulingen, Germany

Die Ein-/Ausgänge des modular aufgebauten Seitenkanalverdichters können wahlweise auf beiden Gehäuseseiten angebracht werden, wobei die Montage sowohl horizontal als auch vertikal möglich ist. Das mit Kühlrippen versehene Laufradgehäuse besteht aus zwei gleichen Teilen, die weiteren Bauteile sind zudem standardisiert. Die Bauform des Seitenkanalverdichters ist kompakt und funktionsbezogen.

The I/O of the modularly built side channel blower can be integrated optionally on either side of the housing and can be assembled either horizontally or vertically. The housing of the impeller which contains cooling ribs consists of two identical parts, with individual components that are also standardized. The side channel blower is designed to be compact and functional.

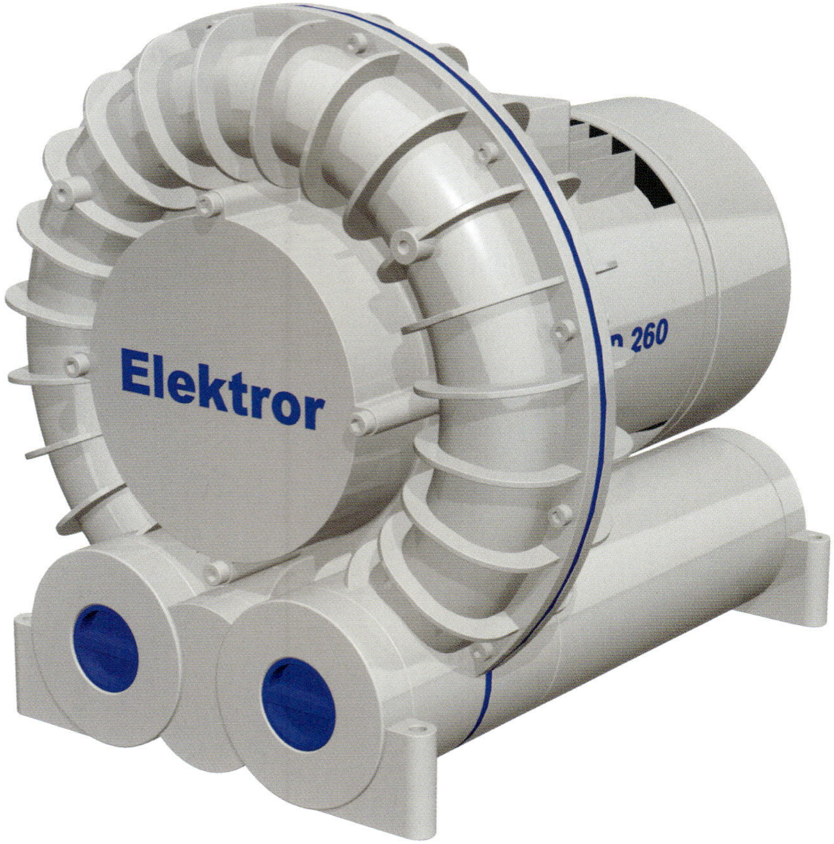

iF Silver Award

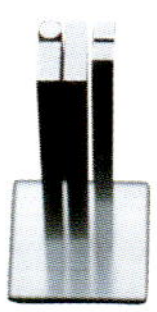

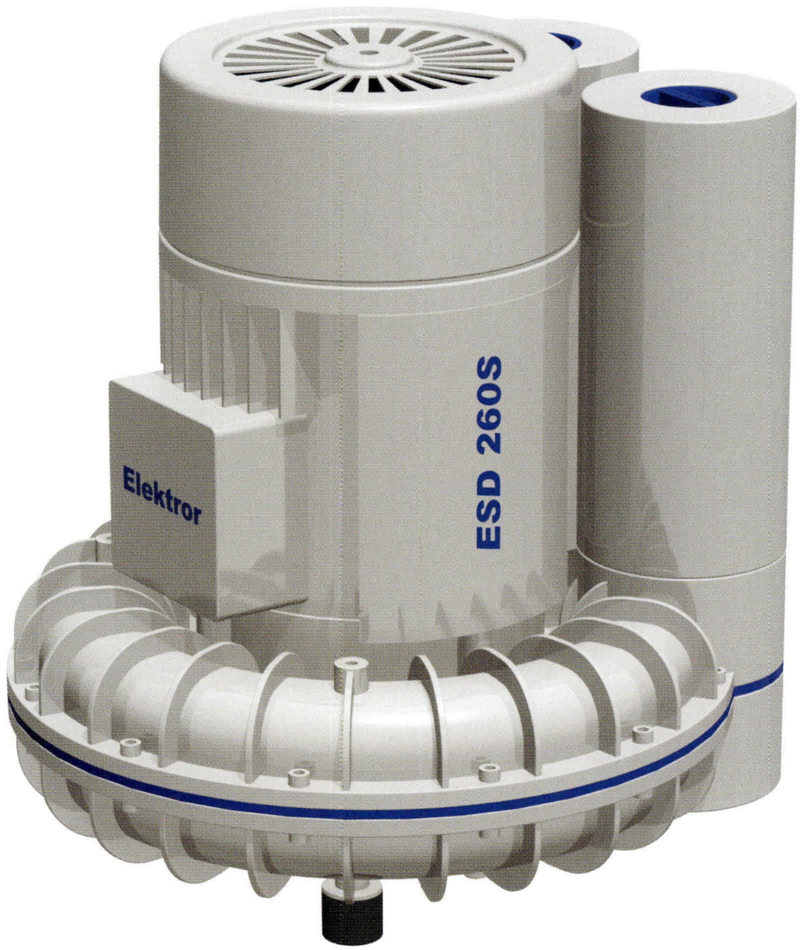

Hersteller Manufacturer
Elektror, Karl W. Müller GmbH & Co.
Esslingen, Germany

Produkt Product
ESD 260S
Seitenkanalverdichter
Side channel blower

Design
Tino Zink
STZ Design Innovation
Neulingen, Germany

Die Ein-/Ausgänge des modular aufgebauten Seitenkanalverdichters können wahlweise auf beiden Gehäuseseiten angebracht werden, wobei die Montage sowohl horizontal als auch vertikal möglich ist. Das mit Kühlrippen versehene Laufradgehäuse besteht aus zwei gleichen Teilen, die weiteren Bauteile sind zudem standardisiert. Die Bauform des Seitenkanalverdichters ist kompakt und funktionsbezogen. Der Typ 260S ist zusätzlich schwingungsgedämpft.

The I/O of the modularly built side channel blower can be integrated optionally on either side of the housing and can be assembled either horizontally or vertically. The housing of the impeller which contains cooling ribs consists of two identical parts, with individual components that are also standardized. The side channel blower is designed to be compact and functional. Model 260S also possess a vibration absorber.

iF Silver Award

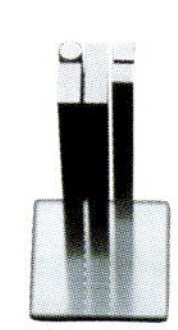

Hersteller Manufacturer
Festo AG & Co. KG
Esslingen, Germany

Produkt Product
Steuerschieber VHER
Manuell betätigtes Ventil
Hand lever valve

Design
Karoline Schmidt
Jan Kleffmann
Hans-Jürgen Claus
Festo AG & Co. KG
Esslingen, Germany

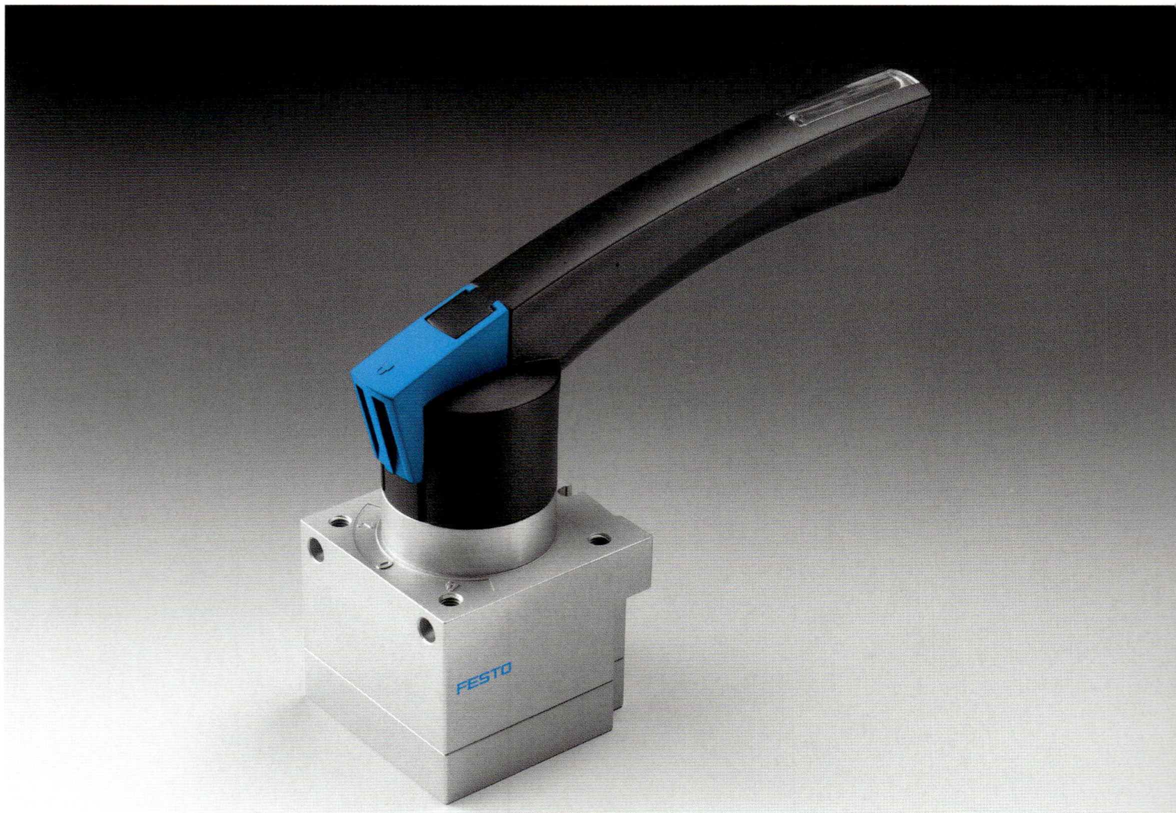

Die VHER Ventile sind durch den Einsatz von verschiedenen Anschlussplatten in der Anschlussrichtung variabel. Viel Wert wurde auf die Formgebung des Hebels gelegt. Auf die ergonomischen Erfordernisse abgestimmt, ist er gut zu umfassen und dann zu drehen. Auch höhere Kräfte können einfach überwunden werden. Beim Einsatz als pneumatischer Hauptschalter besteht die Möglichkeit den Steuerschieber abzuschließen. Der blaue Knopf ist formal integriert und farblich abgesetzt. Form und Dimension unterstreichen die Robustheit, die Linienführung ist schlicht.

The VHER hand lever valves are flexible in that they can be connected in either flow direction by using different adapter plates. Importance was placed on the design of the lever. Fine-tuned to ergonomic requirements it is easy to hold and to turn. It can also cope with high force. When used as a pneumatic master valve it is possible to lock the hand lever valve. The button is formally integrated in a contrasting blue color. The shape and dimensions accentuate its robustness while the lines are kept clear and simple.

iF Silver Award

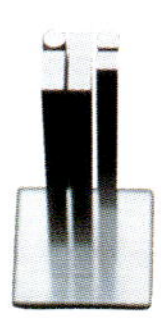

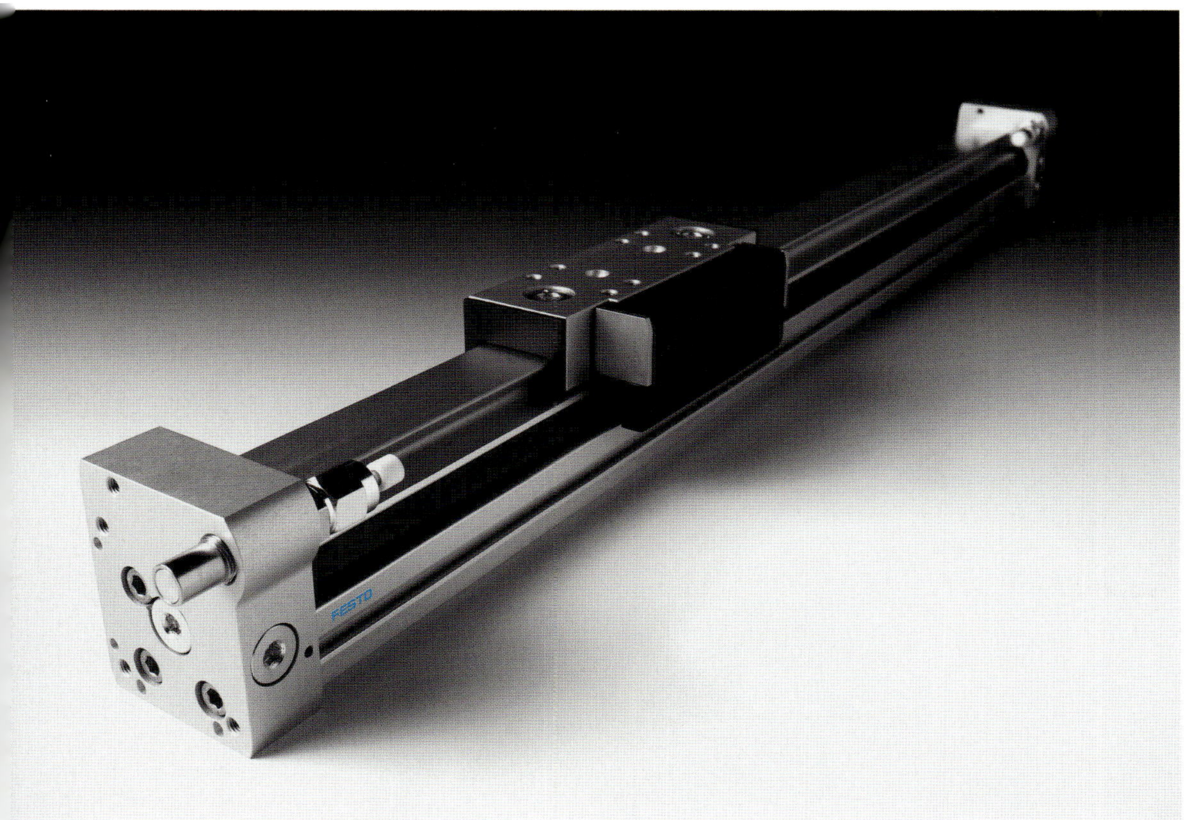

Hersteller Manufacturer
Festo AG & Co. KG
Esslingen, Germany

Produkt Product
Linearantrieb DGC
Pneumatischer kolbenstangenloser Antrieb
Pneumatic rodless drive

Design
Jan Kleffmann
Festo AG & Co. KG
Esslingen, Germany

Der neue kolbenstangenlose Linearantrieb DGC kann als Allround-Talent der Automatisierung betrachtet werden. Als Baukasten ausgelegt, werden verschiedene Schlittenführungen angeboten. Von der Gleitführung für leicht zu bewegende Massen bis zu speziellen Schwerlastführungen. Die Formgebung unterstreicht die Einheit dieses aus vielen verschiedenen Komponenten bestehenden Systems. Einzelne Funktionselemente wurden optisch integriert, die Auslegung der Kanten unterstützt den kompakten Aufbau dieser Antriebe.

The new DGC linear pneumatic rodless drive can be considered an all-round talent when it comes to automation technology. Designed as a module it offers various slide guides - from plain-bearing guides for easy-to-move materials through to heavy-duty guides. The design underscores the consistent appearance of a system configured from many different components. Individual functional elements are visually integrated while the silhouette is designed to emphasize the compact design of the drive units.

Hersteller Manufacturer
Festo AG & Co. KG
Esslingen, Germany

Produkt Product
Kompakt Zylinder ADN/AEN
Pneumatischer Normzylinder
Standard pneumatic cylinder

Design
Jan Kleffmann
Festo AG & Co. KG
Esslingen, Germany

Diese Zylinderbaureihe wurde nach der neuen Norm für Kompaktzylinder (ISO 21287) ausgelegt. Der ADN/AEN zeichnet sich durch kurze Hübe und kompakte Bauart aus. Innerhalb eines Produkt-Baukastens können die Zylinder individuell nach Bedarf zusammengestellt werden. Die Kantengestaltung nimmt den Zylindern die Klobigkeit und gibt ihnen eine Richtung. Kompaktzylinder sind in der Automatisierung universell einsetzbar. Wichtige Einsatzgebiete sind der allgemeine Maschinenbau, der Sondermaschinenbau und die Druck- und Papiermaschinenindustrie. Dort übernehmen diese Zylinder Spann- und Zuführaufgaben.

This design for this new series of cylinders was based on the new ISO 21287 norm for compact cylinders. The ADN/AEN boasts short piston strokes and a compact design. The cylinders can be individually assembled as required within a product module. The edges are designed to reduce the bulkiness of the cylinder and give it stylistic identity. Compact cylinders can be used universally in automation. Important areas of use include general and specialized mechanical engineering as well as the printing and paper machine industry in which these cylinders often perform clamping and feeding tasks.

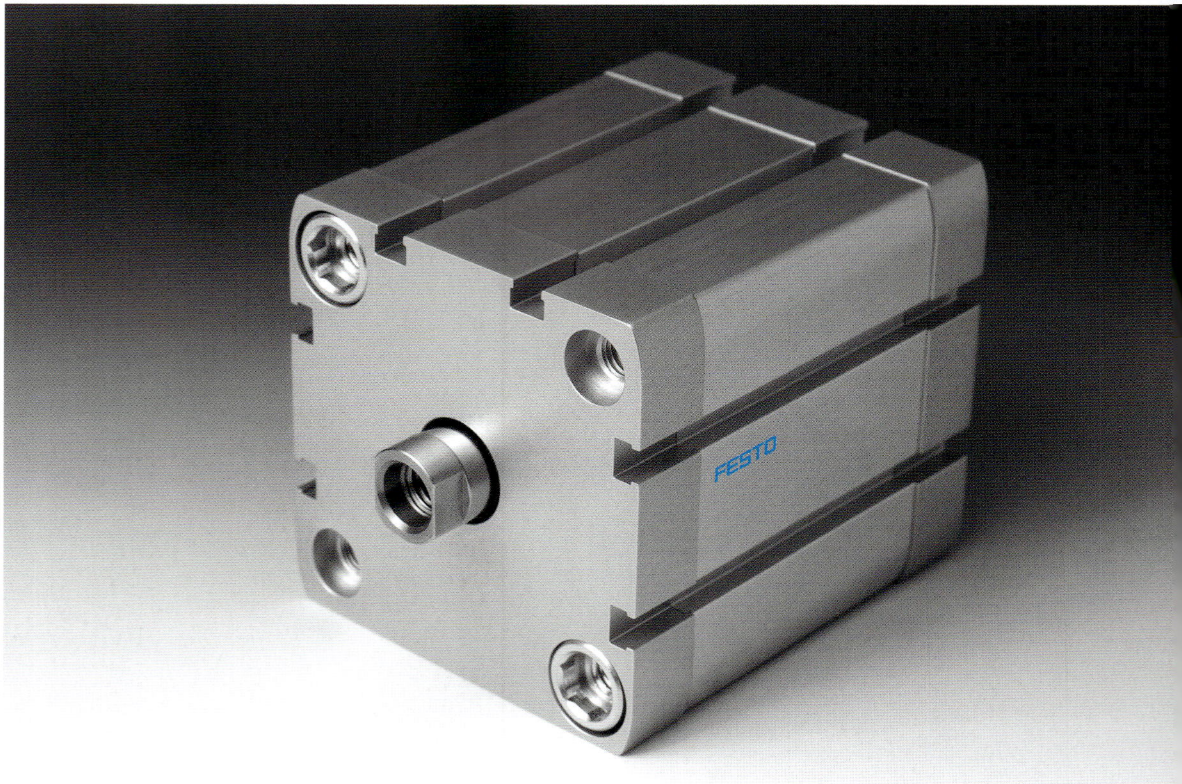

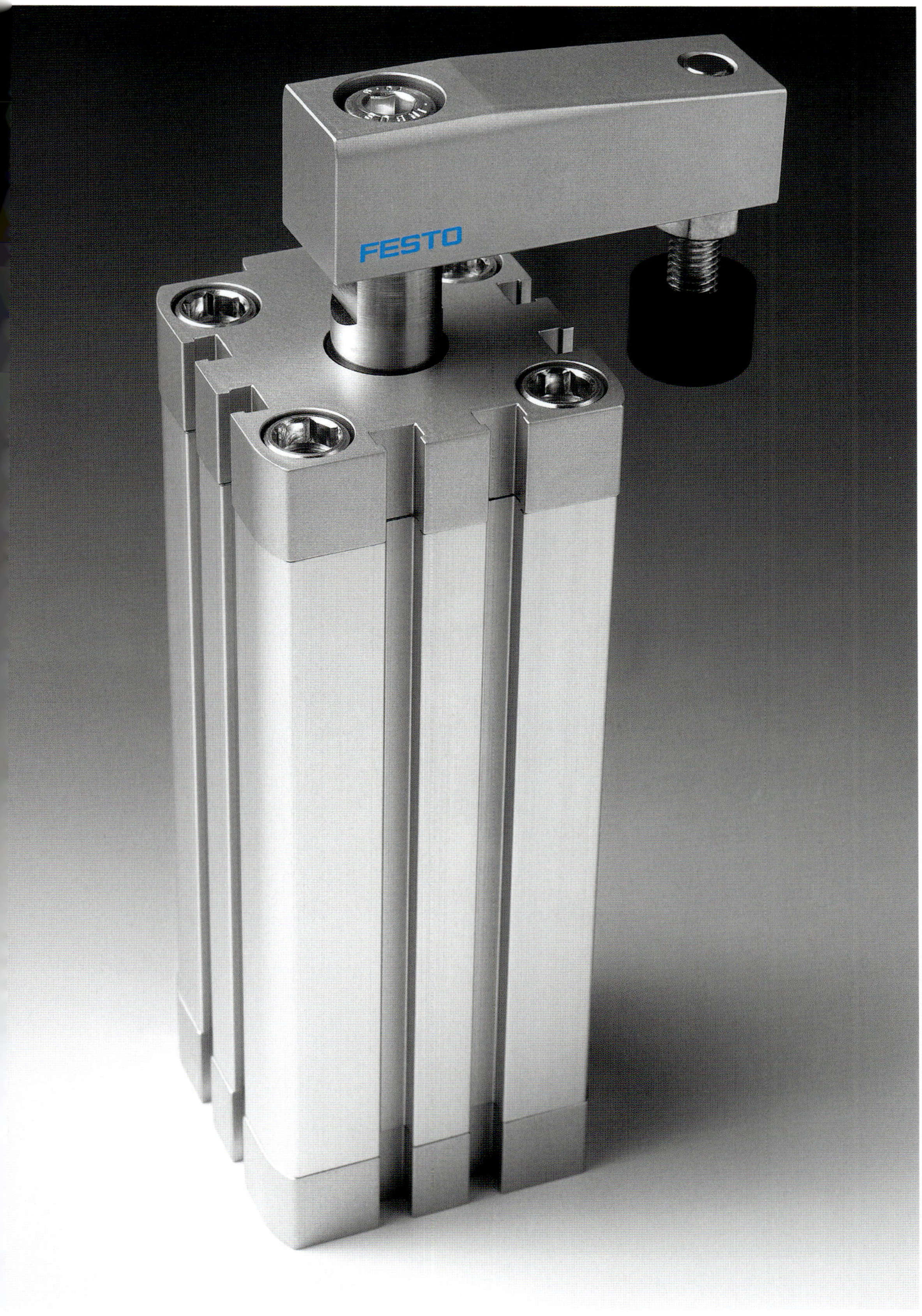

Hersteller Manufacturer
Festo AG & Co. KG
Esslingen, Germany

Produkt Product
Linear-Schwenkspanner CLR
Pneumatischer Hub-Schwenkspanner
Pneumatic swivel clamp

Design
Jan Kleffmann
Festo AG & Co. KG
Esslingen, Germany

Der CLR Hub-Schwenkspanner ist aufgrund seiner 8 Baugrößen, der nachträglich verstellbaren Schwenkrichtung, der Endlagendämpfung und der Abfragemöglichkeiten sowie aufgrund des standardisierten Zubehörs ein im Einsatz und im Handling sehr flexibles System. Die Kantengestaltung von Spannarm und Zylinder nimmt die Klobigkeit, bildet formal die Einheit des Systems und deutet auf die Verwandtschaft zu dem neuen Kompaktzylinder hin. Beide Baureihen ergänzen sich hinsichtlich Einsatzart und Zubehörteilen. Die einfache Mechanik sichert eine hohe Lebensdauer sowie geringen Anschaffungs-, Montage- und Wartungsaufwand.

The CLR linear swivel clamp system is very flexible in terms of application and handling due to the fact that it comes in eight different sizes. It features a user-adjustable swivel direction, end-position cushioning, can be outfitted with sensors, not to mention offering a range of standard accessories. The silhouette of the clamp arm and cylinder eliminate any clumsy appearance, create a formal unity for the system, and allude to its relationship with the compact cylinder. The two product series complement each other in terms of type of operation and accessories. The simple mechanics ensure a long service life as well as low installation, assembly and maintenance costs.

Hersteller Manufacturer
Siemens Dematic AG
Offenbach, Germany

Produkt Product
Kormoran
Modulares Schwerlastfördersystem
Modular conveyor system for heavy loads

Design
Andreas Preussner
designafairs GmbH
München, Germany

Kormoran ist ein modulares Schwerlastfördersystem für Euro-Paletten. Die Einheit ist selbst tragend und besitzt einen integrierten Kabelkanal. Durch vorinstallierte Einzelmodule reduzieren sich die Montagearbeiten am Einsatzort, zudem ist die Seiten- und Top-Verkleidung auf Schnellmontage ausgerichtet. Die Endlos-Zahnriemen ermöglichen einen leisen Betrieb.

Kormoran is a modular, heavy-load conveyor system for euro-standard palettes. The unit is self-supporting and includes an integrated cable channel. As individual modules can be pre-installed, it can be more rapidly assembled where it is needed. The side and upper casings are also geared for quick assembly. The seamless drive belt makes for quiet operation.

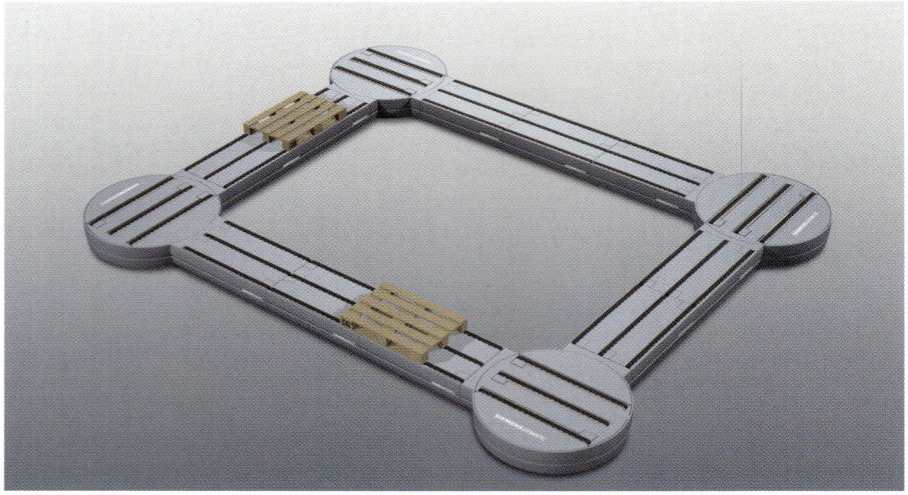

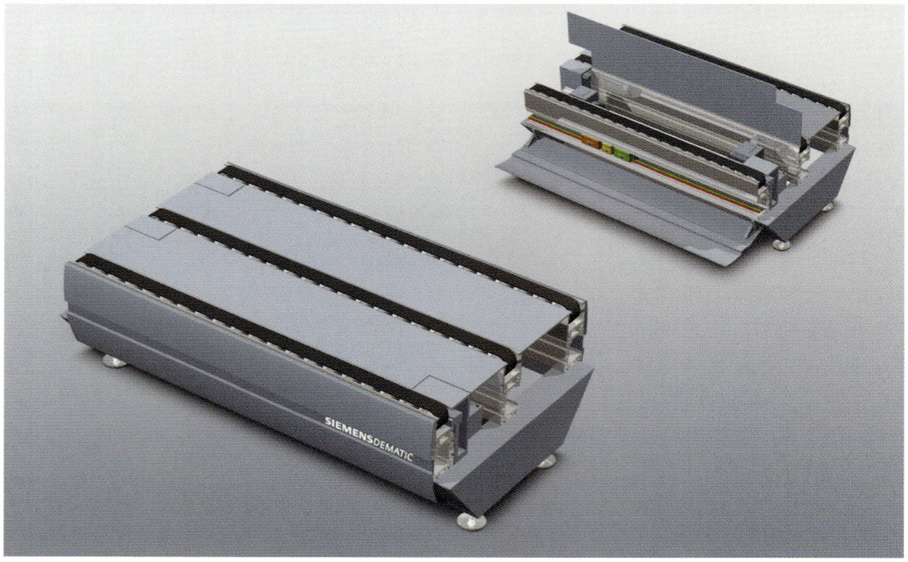

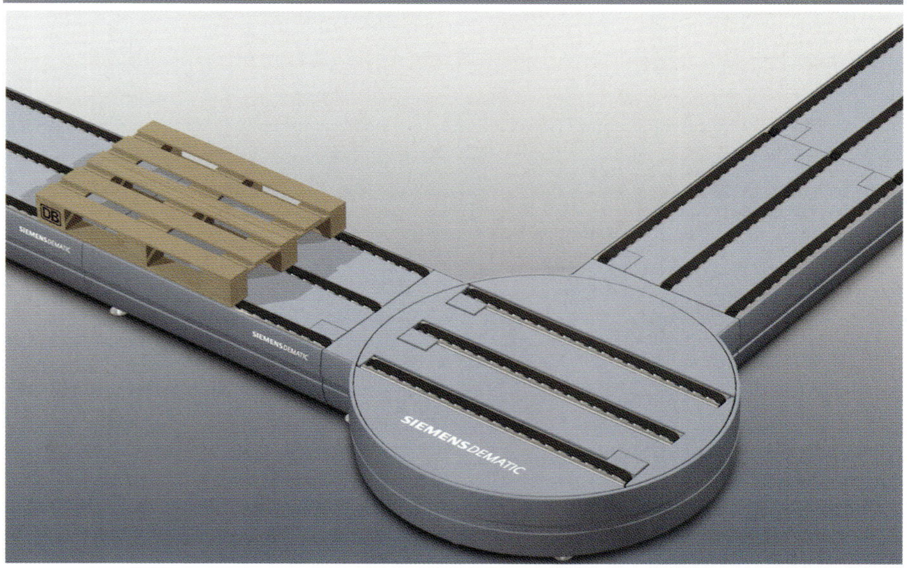

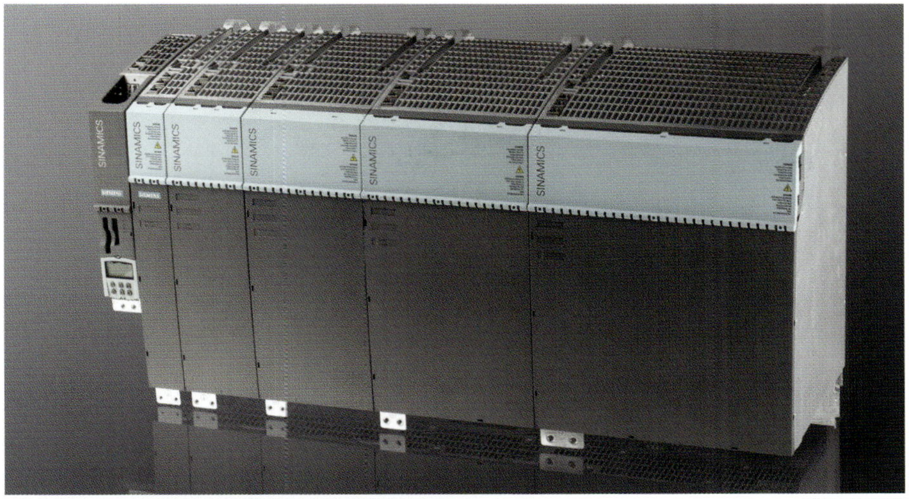

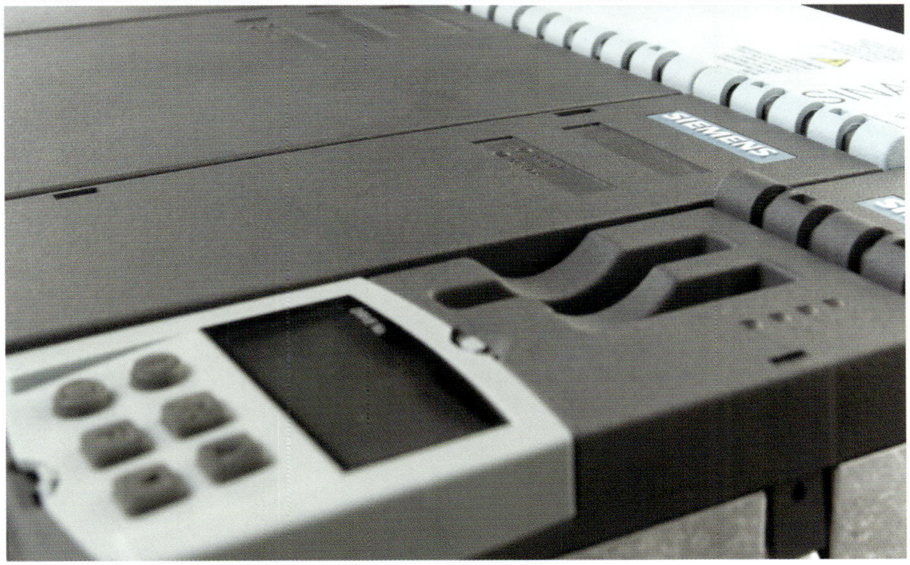

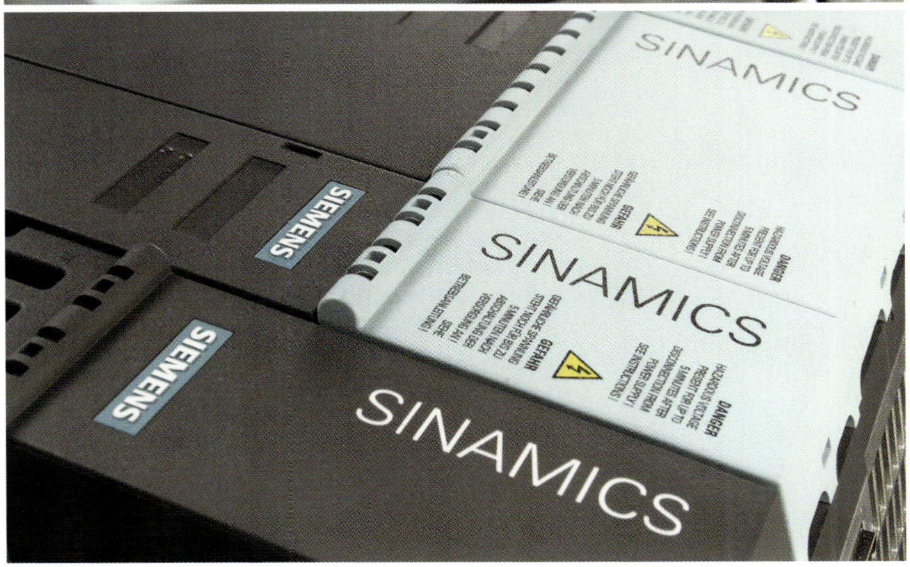

Hersteller Manufacturer
Siemens AG
Nürnberg, Germany

Produkt Product
SINAMICS S120
Antriebssystem
Drive system

Design
Jan Andersson
Christoph Tomczak
designafairs GmbH
Erlangen, Germany
Gunter Ott (Management)
Siemens AG A&D GC7
Nürnberg, Germany

Das Antriebssystem SINAMICS S120 eignet sich insbesondere für Mehrachsanwendungen. Dank seines Aufbaus und einer Vielzahl an Komponenten lässt sich das System individuell zusammenstellen. Ausgangspunkt für die Entwicklung des Design-Konzeptes waren die gestalterischen Grundüberlegungen, Dynamik zu visualisieren, den Eindruck eines Komplettsystems zu schaffen und eine klare Gliederung der Verkabelung zu erreichen. Das Produktsystem besteht aus zwei modularen Gerätearten: den Leistungsmodulen und dem Regelungsmodul. Ein Verband besteht aus zumindest einer Regelung und ein bis zwei Leistungsmodulen unterschiedlicher Breite.

The SINAMICS S120 drive system is particularly suited for multi-axis applications. Thanks to its structure and numerous components, the system can be customized to user needs. The design was developed on the basis of essential factors, such as visualizing the system's dynamic nature, creating the impression of an overall scalable system, while also clearly structuring cabling. The product system consists of two modular housing types: the power modules and the control modules. Each drive system is made up of at least one control and up to two power modules of different widths.

Hersteller Manufacturer
Siemens AG
Nürnberg, Germany

Produkt Product
SIMATIC ET200X
Basismodul
Basic module

Design
Jan Andersson
Thomas Blümel
designafairs GmbH
Erlangen, Germany
Gunter Ott (Management)
Siemens AG A&D GC7
Nürnberg, Germany

SIMATIC Automatisierungssysteme werden zur Steuerung und Datenerfassung in der Fertigungs- und Prozessindustrie eingesetzt. Die Geräte der ET200X-Serie können direkt an der Maschine in widrigsten Umgebungsbedingungen montiert werden. Die Gehäuse wurden so ausgeführt, dass ein Schaltschrankeinbau unnötig ist. Das funktionsbezogene Design wurde auf das Wesentliche reduziert. Der einfache Basiskörper vereinigt die verschiedenen funktionalen Elemente. Die Charakterisierung des Systems findet durch wenige gestalterische Eingriffe statt, wie zum Beispiel durch die Behandlung der Verschraubungsprodukte.

The SIMATIC automation systems are used for control and data-collection applications in the manufacturing and processing industries. The ET200X series devices can be mounted directly on machines in even the toughest environments. The housing is designed such that no special cabinet is required. The fully functional design has been reduced to the essentials. The simple basic module brings together the different functional elements. A few design highlights suffice to characterize the system – such as the treatment of the screwing areas.

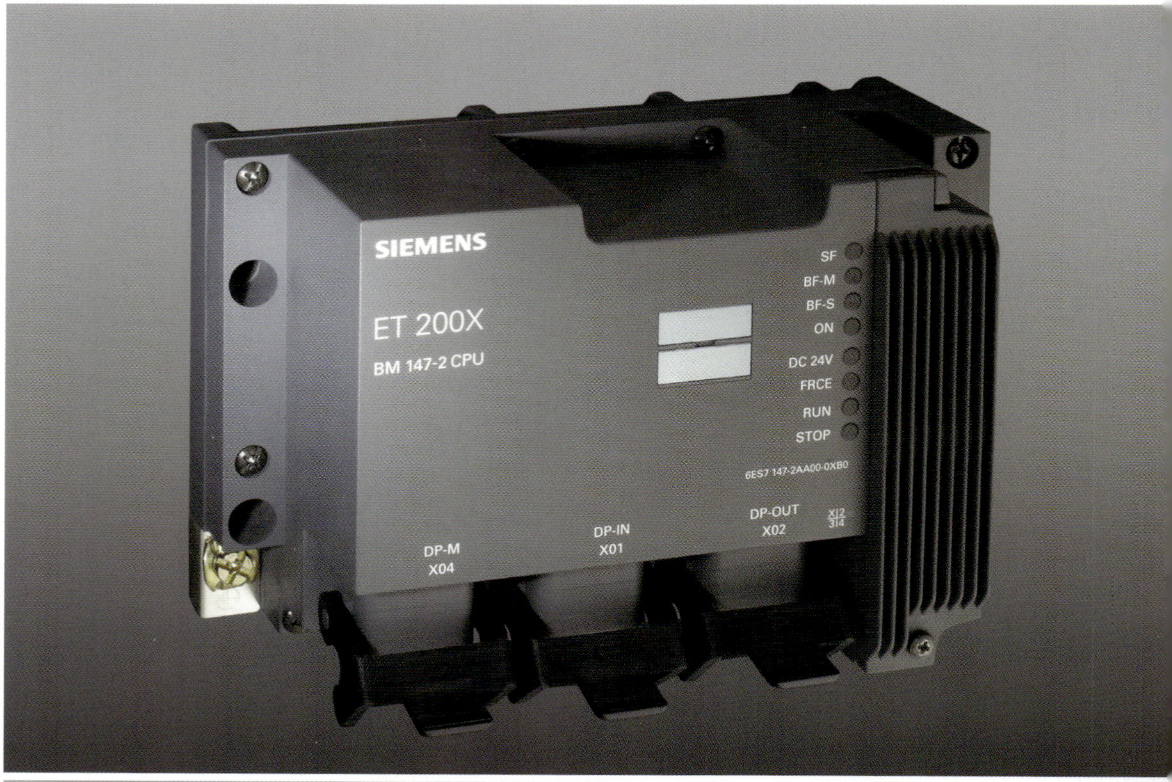

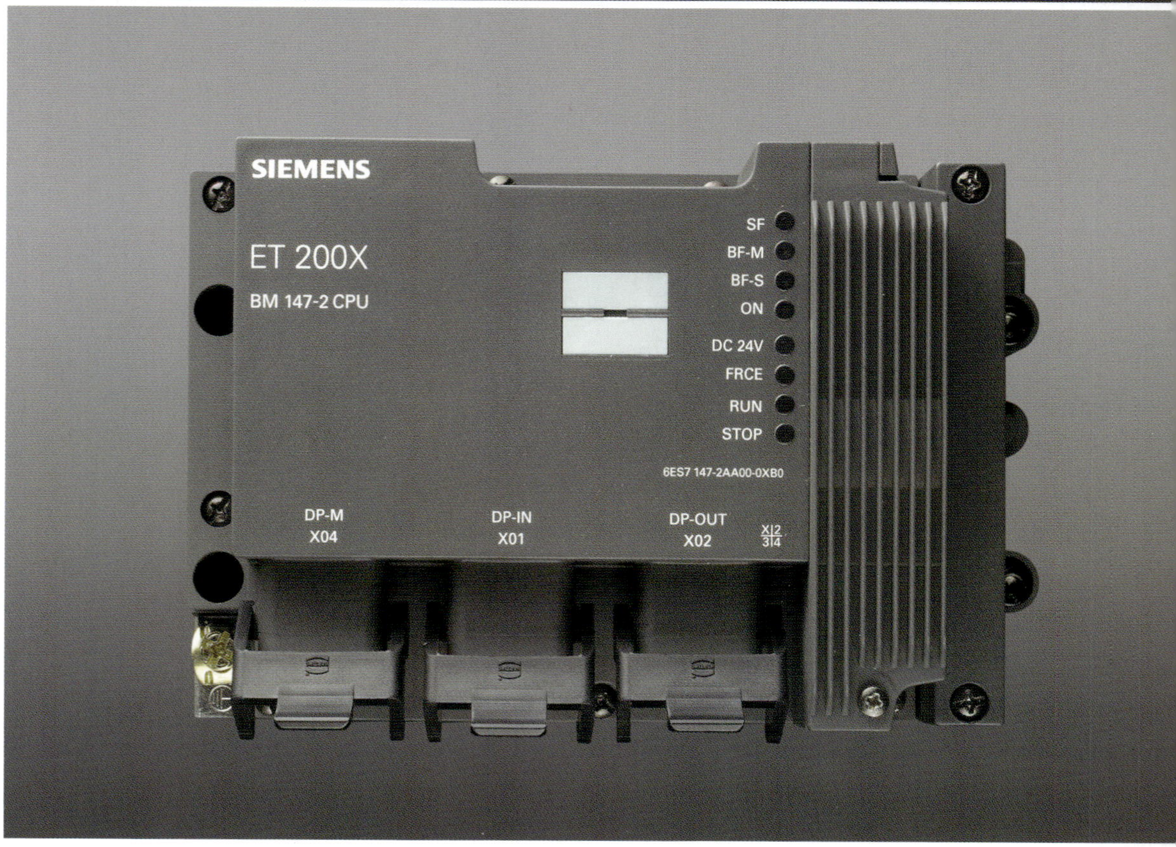

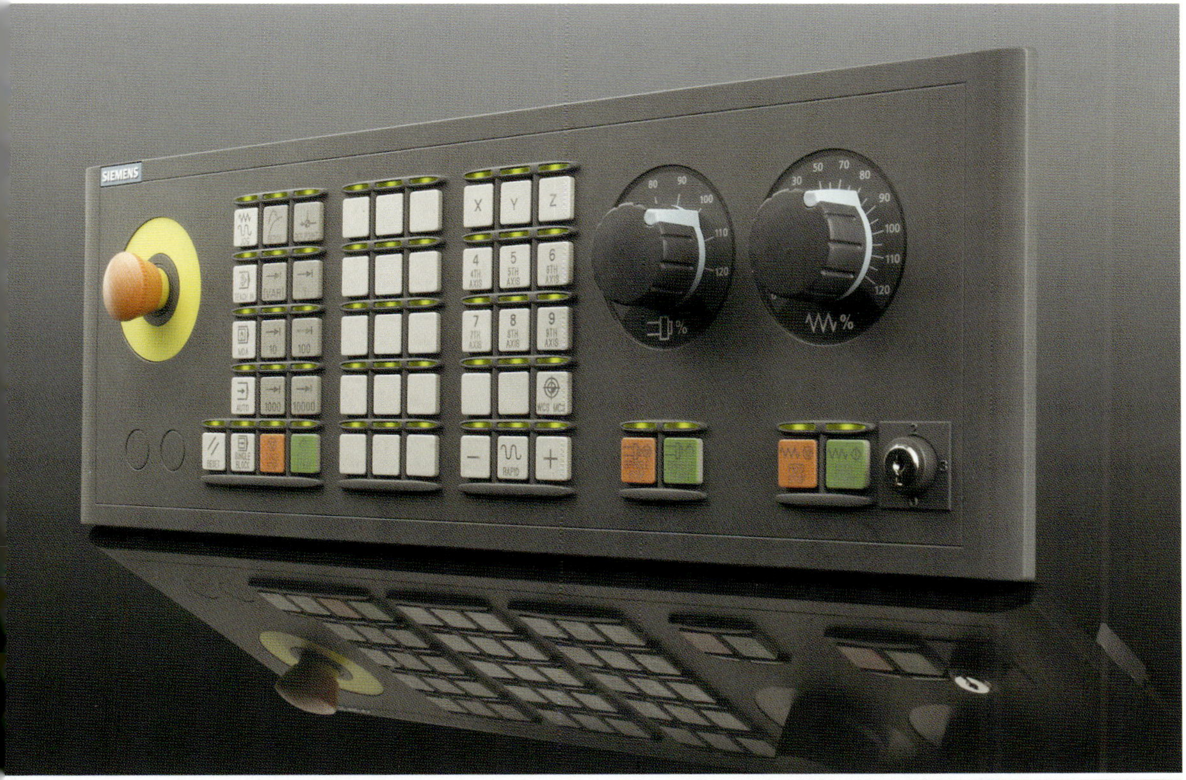

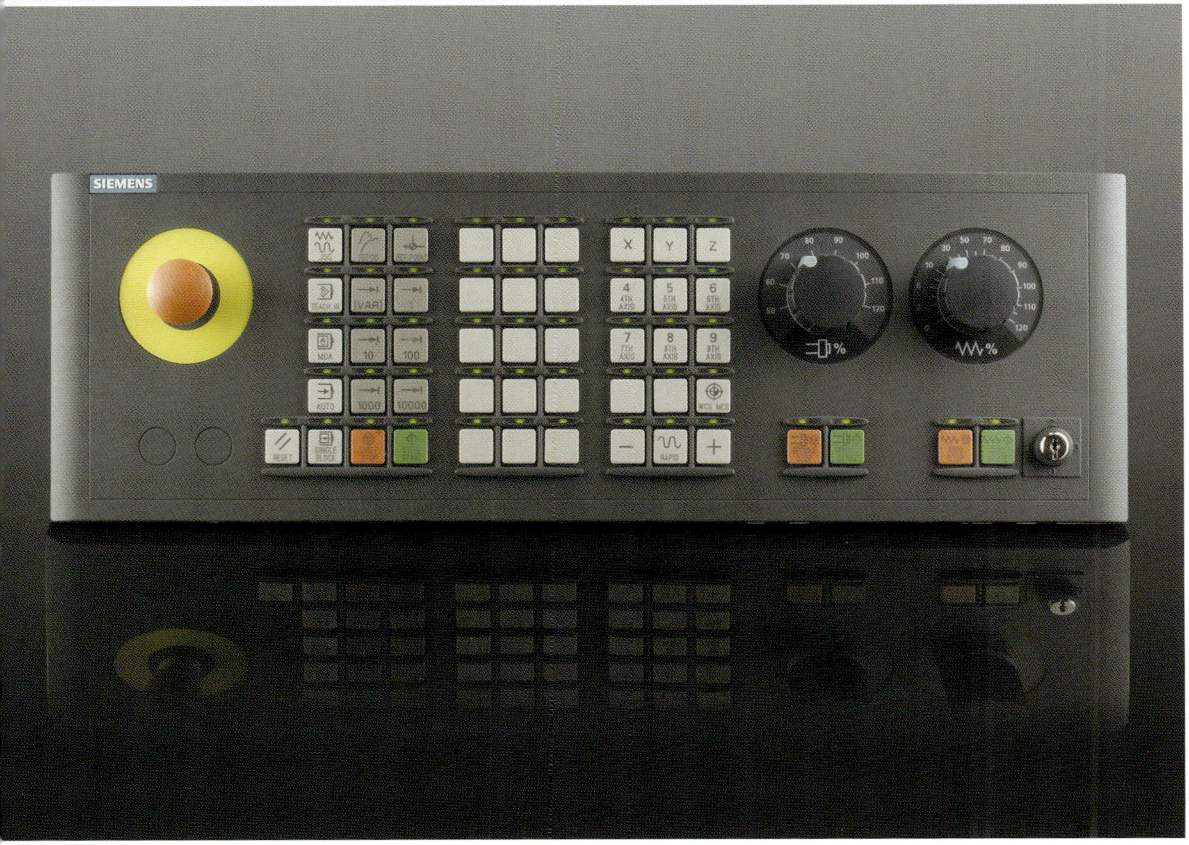

Hersteller Manufacturer
Siemens AG
Nürnberg, Germany

Produkt Product
SINUMERIK 840 D
Maschinensteuertafel
Machine control panel

Design
Jan Andersson
Christoph Tomczak
designafairs GmbH
Erlangen, Germany
Gunter Ott (Management)
Siemens AG A&D GC7
Nürnberg, Germany

Die SINUMERIK 840D ist ein modulares Produktsystem, bestehend aus einem Modul zur Programmierung und Steuerung der Numerik, einem Visualisierungsmodul und einer Maschinensteuertafel zur direkten Ansteuerung der jeweiligen Maschine (Drehbank oder Fräse etc.). Das Ziel des neuen Designkonzepts war es, eine verbesserte Unterstützung der Bedienung, ein homogeneres Erscheinungsbild und eine optimierte Integration in das Corporate Design des Bereiches Automation and Drives der Siemens AG zu erreichen.

The SINUMERIK 840D is a modular product line consisting of a module for programming and controlling of a numerical control unit, a visualization module and a machine control panel for direct control of the respective machines, such as mills or lathes. The goal behind the new design concept: to enhance user comfort, deliver a more homogeneous look, and optimize integration into the corporate design of the Siemens AG Automation and Drives division.

Hersteller Manufacturer
Siemens AG
Nürnberg, Germany

Produkt Product
SINAMICS G110, Baugrößen A,B,C
Frequenzumrichter
Inverter

Design
Jan Andersson
Christoph Tomczak
designafairs GmbH
Erlangen, Germany
Gunter Ott (Management)
Siemens AG A&D GC7
Nürnberg, Germany

SINAMICS G110 ist ein vielseitig einsetzbarer Frequenzumrichter für den drehzahlveränderbaren Betrieb dreiphasiger Motoren am einphasigen Netz. Für Pumpen, Lüfter, Ventilatoren, Förderbänder, Verpackungsmaschinen und für Antriebe von Fabrik- und Garagentoren sowie Schranken ist der Einsatz des SINAMICS G110 wie Drehstrom aus der Steckdose. Gestaltungsaspekte waren u.a. eine höhere Leistung bei gleich bleibendem ‚Footprint', eine Öffnung des Gehäuses zu Belüftungszwecken und eine einfach zugängliche Verdrahtung. Zusätzlich sollte die gestalterische Nähe zum höherklassigen Produkt SINAMICS S120 unterstrichen werden.

SINAMICS G110 is a versatile frequency inverter for operating variable-speed three-phase drives on single-phase line supplies. SINAMICS G110 thus delivers power similar to that from a three-phase socket – for pumps, fans, blowers, conveyor belts, packaging machines, for drives powering factory and garage doors as well as cabinets. The design focuses on boosting performance while retaining the same footprint size, handling perforation of the housing for cooling purposes, and enabling simple cabling. Additionally, the design emphasizes the links to the higher-grade SINAMICS S120 line.

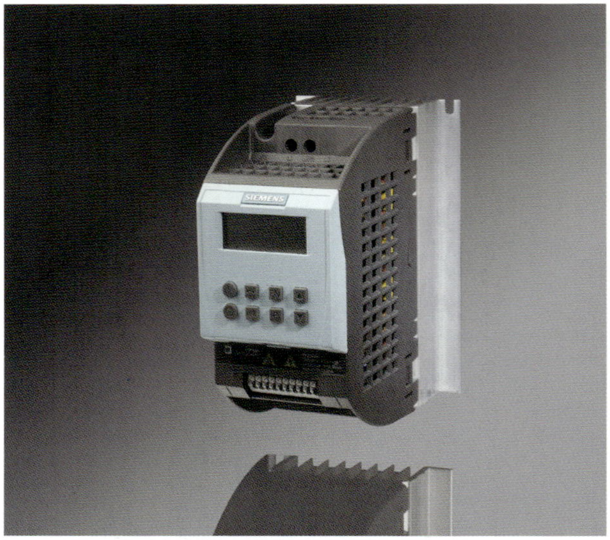

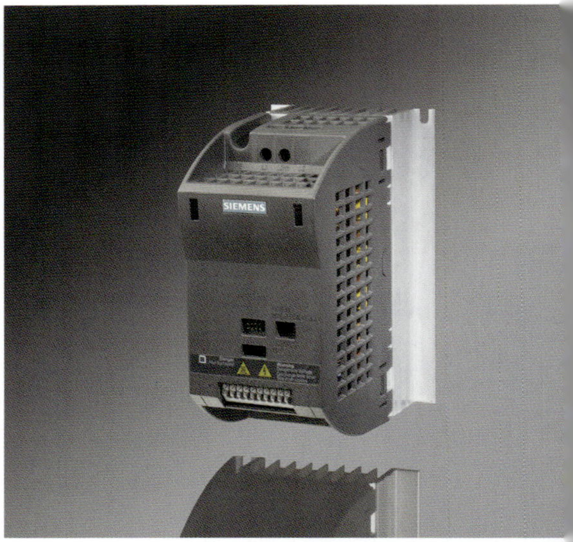

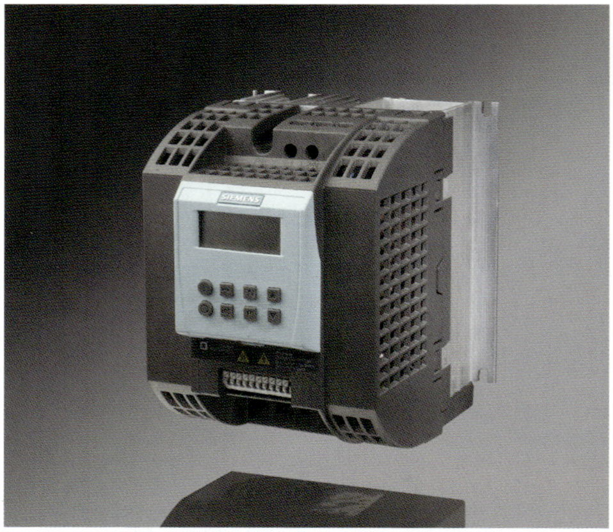

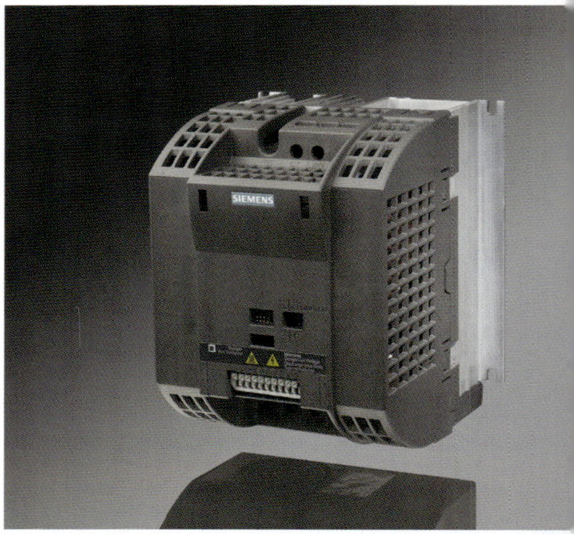

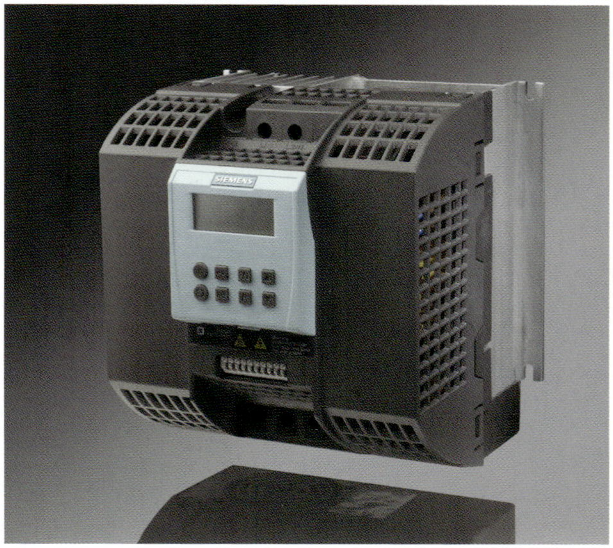

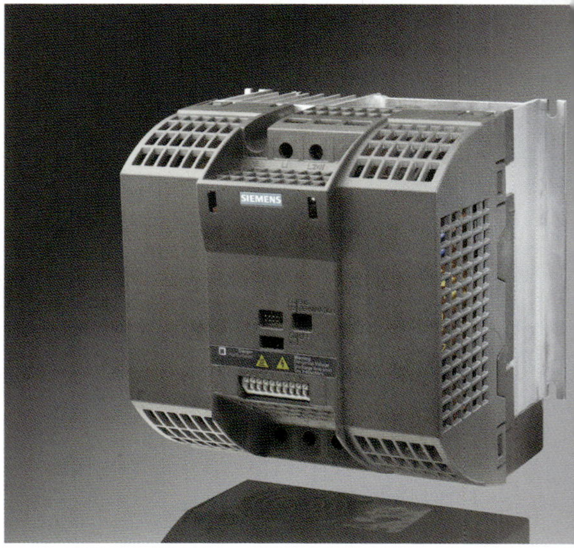

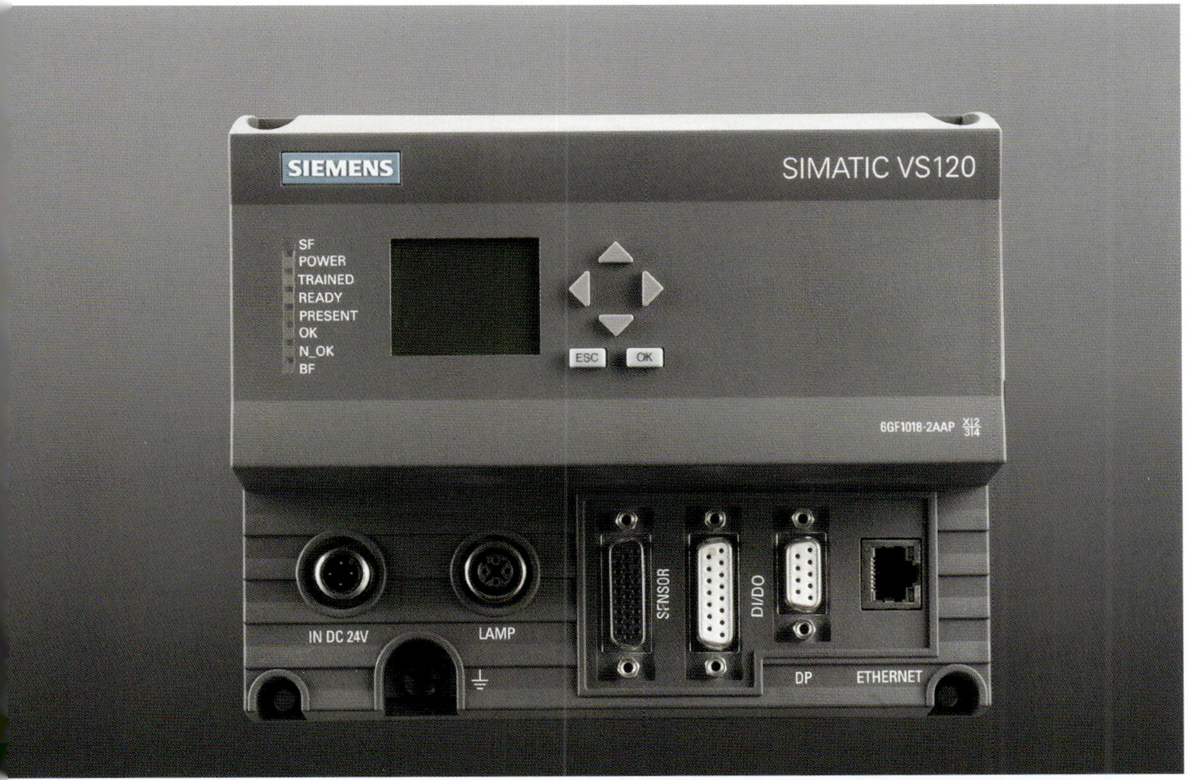

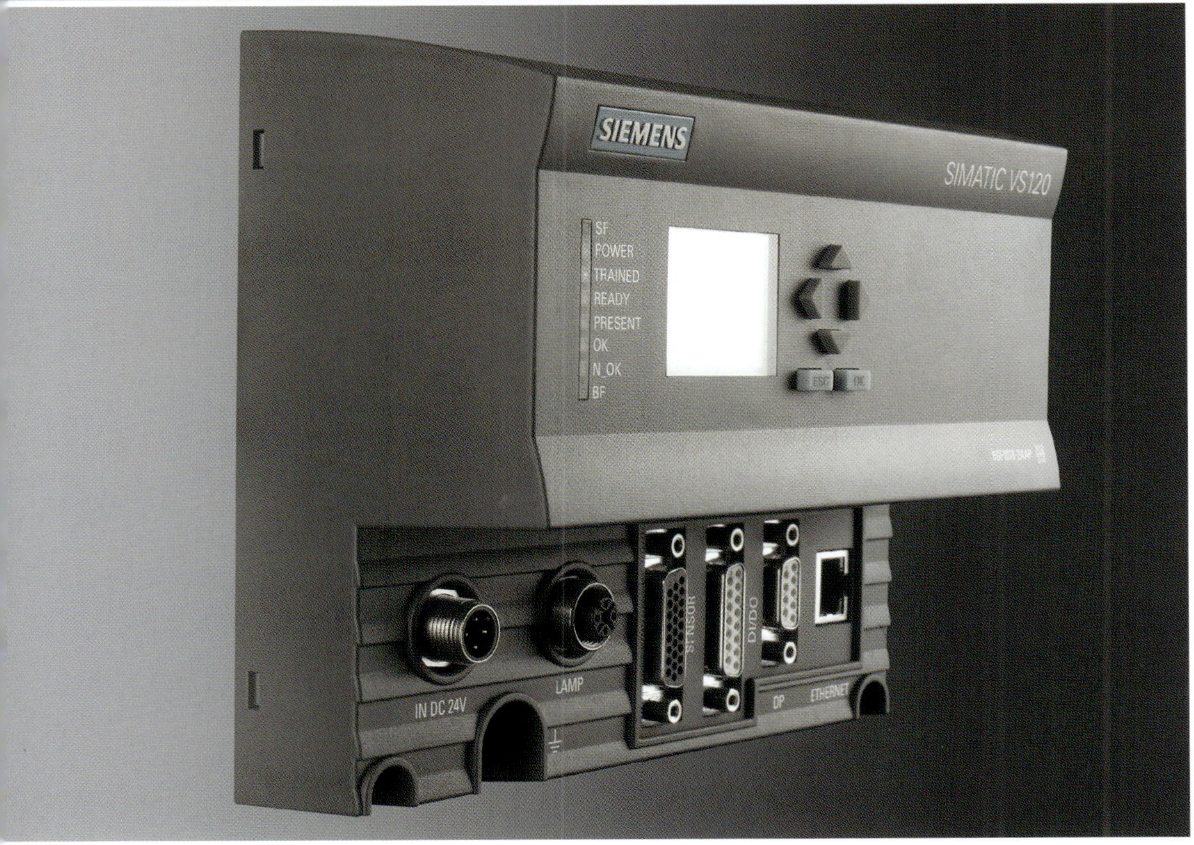

Hersteller Manufacturer
Siemens AG
Nürnberg, Germany

Produkt Product
SIMATIC VS120
Auswertegerät
Processing unit

Design
Jan Andersson
Thomas Blümel
designafairs GmbH
Erlangen, Germany
Gunter Ott (Management)
Siemens AG A&D GC7
Nürnberg, Germany

Das Machine Vision System SIMATIC VS120 übernimmt die optische Überprüfung von Werkstücken und Mustern wie z.B. Logos, Bildmotiven etc. Das Auswertegerät VS120 stellt die Verbindung zum Fertigungsnetzwerk her und ermöglicht die Kontrolle und Weitergabe der Daten an das Manufacturing Execution System und/oder die Qualitätssicherung. Durch die Gestaltung sollten die wesentlichen Eigenschaften des Systems wie Schnelligkeit, Dynamik, Zuverlässigkeit aufgegriffen werden. Die Visualisierung dieser Eigenschaften geschieht mittels gewölbter Flächen, die auf die optisch in der zentralen Position befindliche Mensch-Maschine-Schnittstelle hinführen.

The SIMATIC VS120 machine vision system enables visual checking of objects, workpieces and patterns, such as logos, graphic images, etc. The VS120 checking unit furnishes a bridge to the manufacturing network and thus facilitates control and forwarding of data to the manufacturing execution system and/or QA section. The design takes up the key system characteristics, such as swiftness, dynamism, and reliability. These properties are visualized by means of curved surfaces, guiding the eye to the central user interface.

Hersteller Manufacturer
Siemens AG
Nürnberg, Germany

Produkt Product
MICROSAM
Gaschromatograph
Gas chromatograph

Design
designafairs GmbH
Erlangen, Germany
Gunter Ott
Klaus Thormann

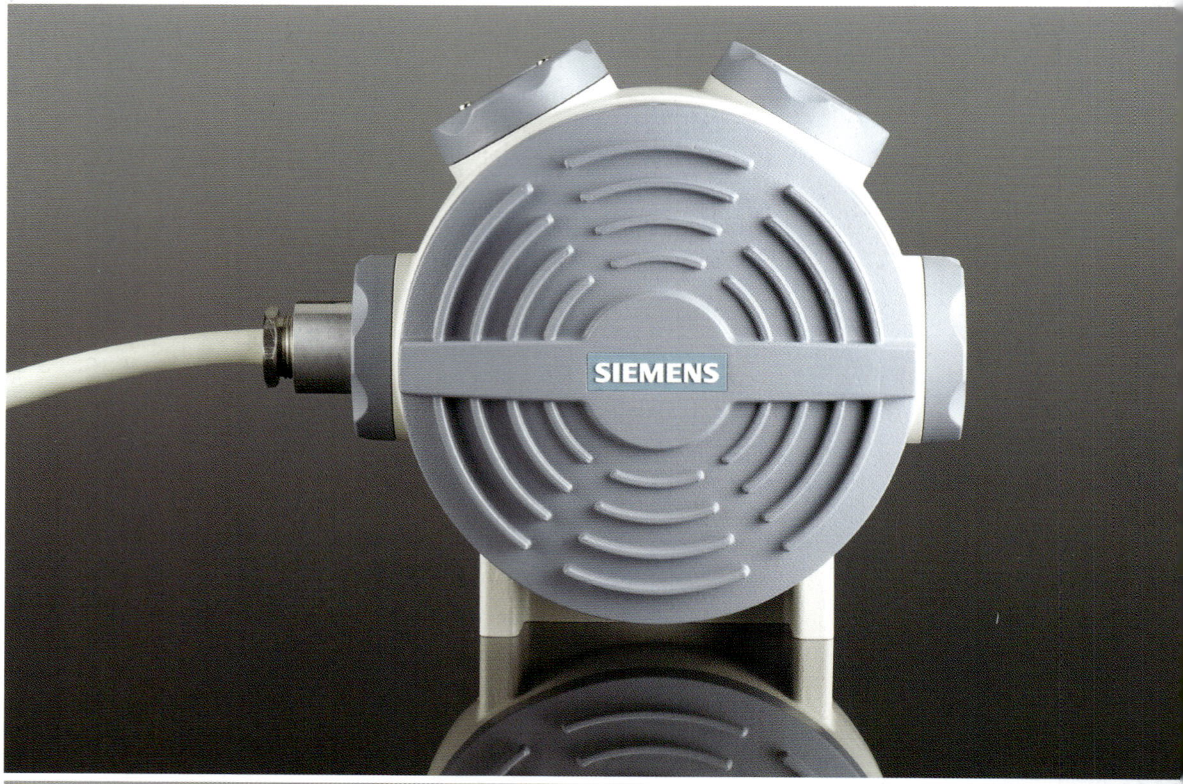

Der MICROSAM Gaschromatograph ist mit vierzehn Zoll Durchmesser und neun Zoll Höhe der kleinste Prozess-Chromatograph von Siemens Automation and Drives. Das explosionsgeschützte neue Gerät wurde mit auf Silizium basierender Mikromechanik entwickelt. Dies ermöglichte die Miniaturisierung bei gleichzeitiger Leistungssteigerung. Die standardisierten Analysemodule lassen sich einfach und schnell austauschen. Wichtigster Faktor des Designs waren eine hohe Gehäusedichtigkeit und der erforderliche Explosionsschutz. Die Farbgebung sorgt für eine Trennung der Funktionsbauteile des Gehäuses und eine eindeutige Einbeziehung in das Produktspektrum der Prozesstechnik von A&D.

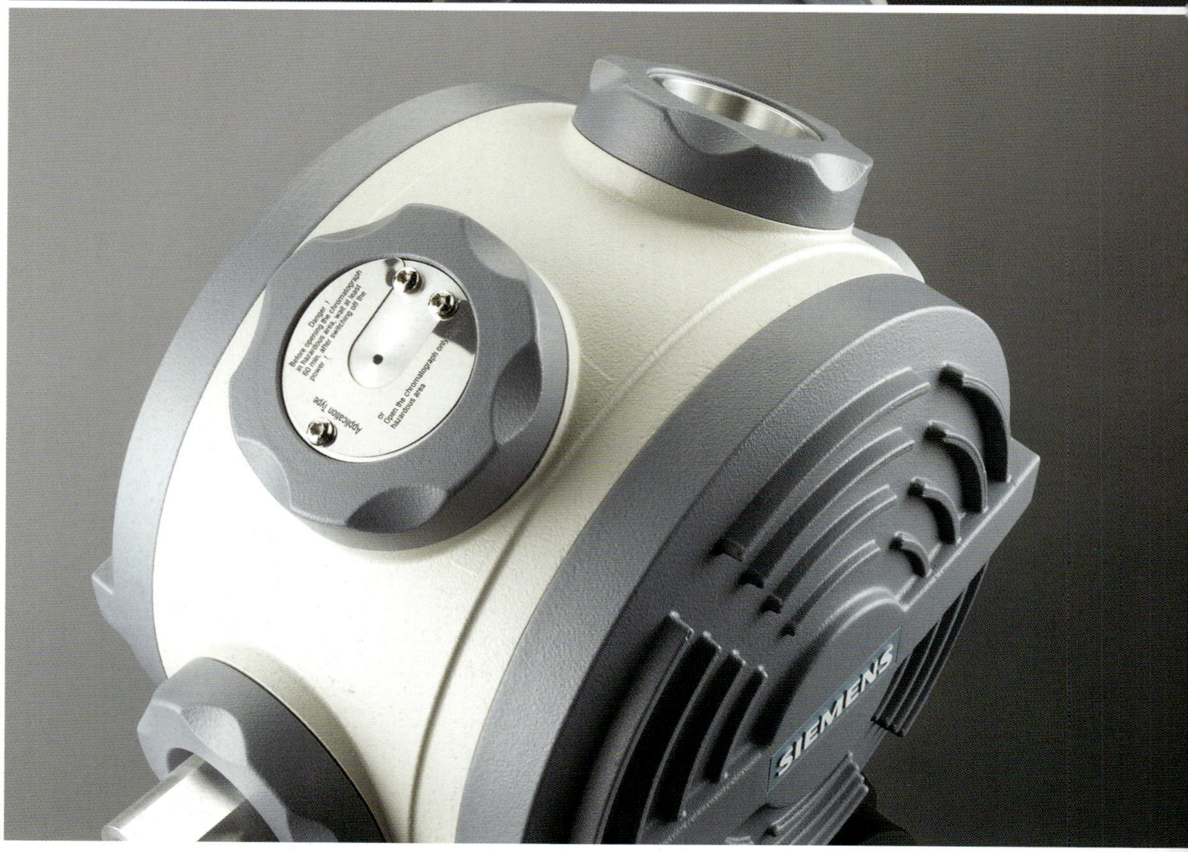

The MICROSAM gas chromatograph is 14" in diameter and only 9" high, making it the smallest process chromatograph offered by Siemens Automation and Drives. The new explosion-protected device is based on silicon micromechanics, enabling the miniaturization of components while boosting performance. The standardized analysis modules can be easily and swiftly exchanged. The key design focus: a pressure-secure housing and the requisite protection against explosions. The colors used subdivide the functional sections of the unit and also clearly designate it as part of the A&D division's process technology line.

Hersteller Manufacturer
Sense Eletrônica Ltda
São Paulo, Brazil

Produkt Product
Módulo I/O Redes Industriais
Ein-/Ausgangsmodul digital/analog
Digital and analog input/output module

Design
Levi Giradi, Luiz W. Alves, Fabio Federico
Questto Design Estúdio de Projetos SC Ltda
São Paulo, Brazil

Das digitale/analoge Ein-/Ausgabemodul für industrielle Netzwerke besitzt ein internes Verbindungsstück, das den Austausch des Moduls nach dem Plug-in-Konzept ermöglicht, ohne das Kommunikationsnetz zu unterbrechen. Die Befestigung wird durch vier Punkte an den leicht zugänglichen Rändern des Moduls erleichtert. Alle Kabelsteckplätze sind nach unten geneigt, wodurch das Risiko verringert wird, dass Flüssigkeit eindringt. Bei der Entwicklung wurde insbesondere auch das Augenmerk auf die visuelle Produktidentität gerichtet.

The digital/analog I/O module for industrial networks includes an internal connection element, enabling the module to be replaced using a plug-in concept and preventing any disruption of the communications network. It can be attached easily by way of four points on the easily accessible edges of the module. All cable connections face downwards reducing the risk of moisture infiltration. The unit was developed in particular with the visual product identity in mind.

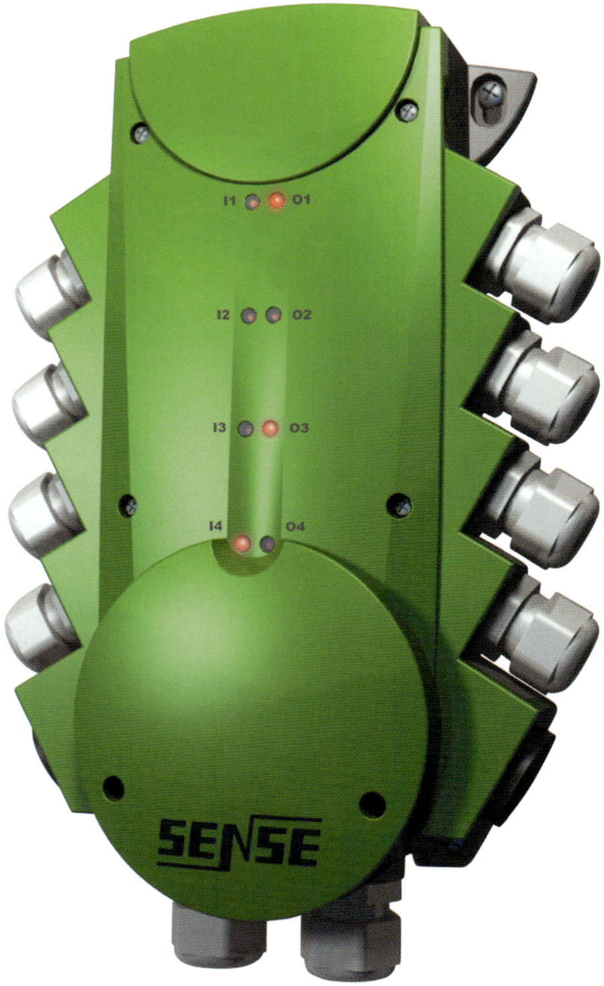

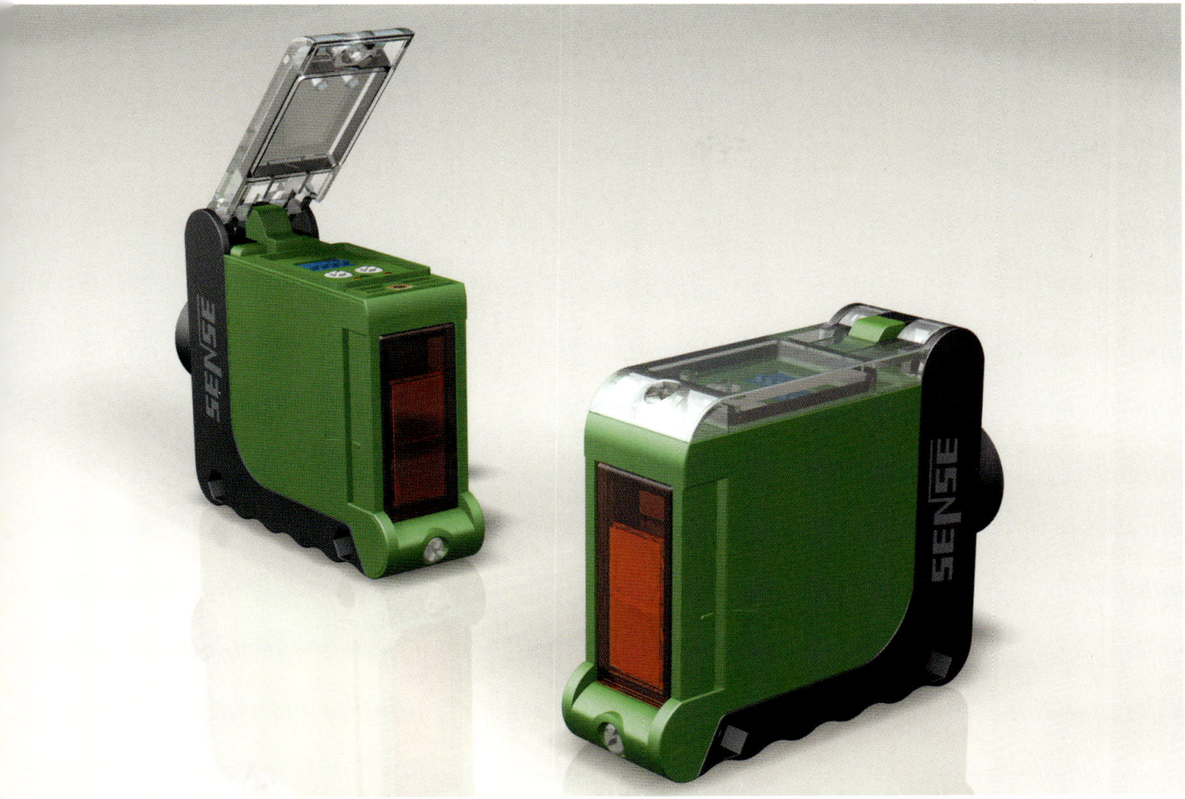

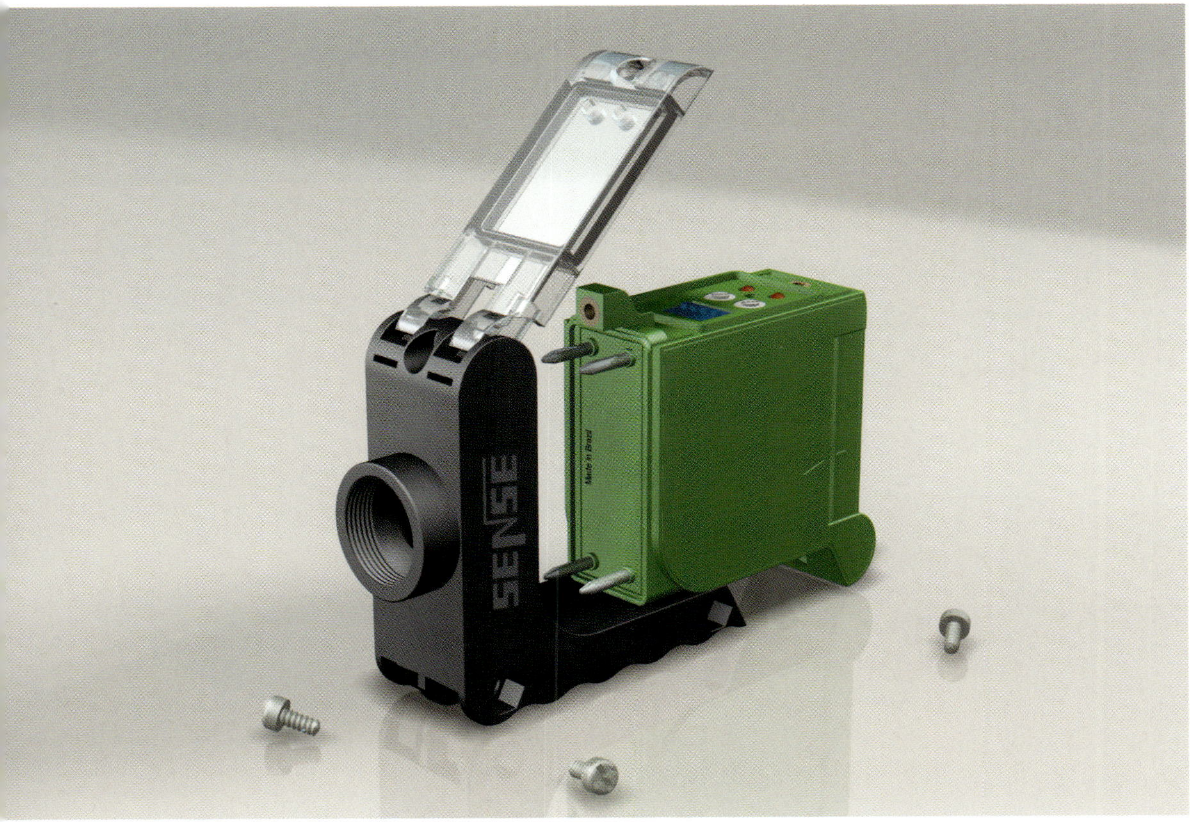

Hersteller Manufacturer
Sense Eletrônica Ltda
São Paulo, Brazil

Produkt Product
Sensor de Proximidade Fotoelétrico
Photoelektrischer Abstandssensor
Photoelectric proximity sensor

Design
Levi Giradi, Luiz W. Alves, Fabio Federico
Questto Design Estúdio de Projetos SC Ltda
São Paulo, Brazil

Der photoelektrische Abstandssensor dient zum Aufspüren von Objekten und Elementen in der Industrieautomatisierung. Er soll das derzeit verwendete OEM-Produkt ersetzen und durch seine neuen elektronischen Komponenten unter anderem einen leichteren Aufbau ermöglichen. Auf Anregung von Technikern, die diese Art Sensor installieren, kann der elektronische Teil des Sensors ausgewechselt werden, ohne dass das gesamte Kommunikationsnetz unterbrochen werden muss, was wiederum einen Produktivitätsgewinn ermöglicht. Erreicht wurde dies durch das Plug-in-System, das die Trennung des elektronischen Teils des Sensors von den Anschlusspunkten möglich macht, ohne das Datennetz zu unterbrechen. Da eine mechanische Befestigung in Form von Schrauben nötig ist, wurden die Dübel ins Produktdesign mitaufgenommen.

The photoelectric proximity sensor is used in industrial automation to detect objects and elements. It is designed to replace the OEM product currently in use, and its new electronic components can, among other things, be more easily assembled. In response to suggestions by technicians who install this type of sensor, the electronic part of the sensor can be replaced without disrupting the entire communications network, enhancing productivity. This is achieved with a plug-in system which makes it possible to separate the electronic part of the sensor from the connection points without disrupting the data network. As the unit needs to be mechanically mounted by means of screws, washers have been integrated into the design.

Hersteller Manufacturer
Yokogawa Electric Corporation
Tokio, Japan

Produkt Product
DAQMASTER Series MX100
Datenerfassungssystem
Data acquisition unit

Design
Osamu Sakuma
Noriaki Semba
Yokogawa Electric Corporation
Design Development Department
Tokio, Japan

Das Datenerfassungssystem DAQMASTER Series MX100 ist für den Anschluss an PCs konzipiert. Die vom kompakten und leicht auf Labor-Bänken zu platzierenden MX100 erfassten Daten werden in Echtzeit auf den PC übertragen. Innerhalb des Systems sind drei Arten von Messintervallen möglich (die Messintervalle werden jeweils für ein Modul eingestellt), zudem kann es flexibel von 4 bis zu 1.200 Kanälen angepasst werden. Zusammen mit der von Yokogawa entwickelten Software, erkennen PCs die angeschlossene MX100-Einheiten automatisch und können damit auf eine optimale Messkapazität zurückgreifen.

The DAQMASTER Series MX100 data acquisition unit is PC-based and forwards the data collected to PCs in real-time. With a solid and reliable look and easily mounted on DIN rails or lab benches, the system features three types of measurement intervals, which are respectively set on the module in question, and can be flexibly adapted for 4 through 1,200 incoming channels. Using Yokogawa's proprietary software, the PC system automatically recognizes any connected MX100s and thus locks onto the maximum measuring power provided.

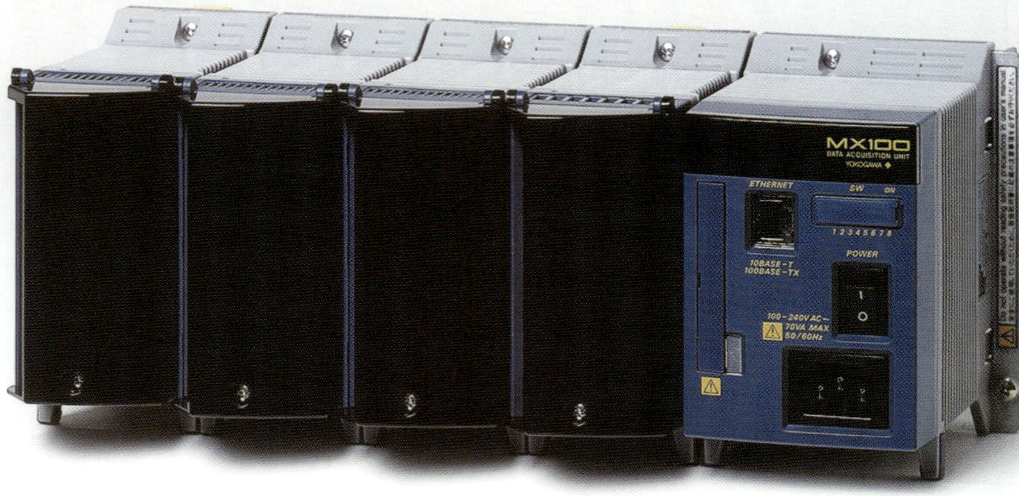

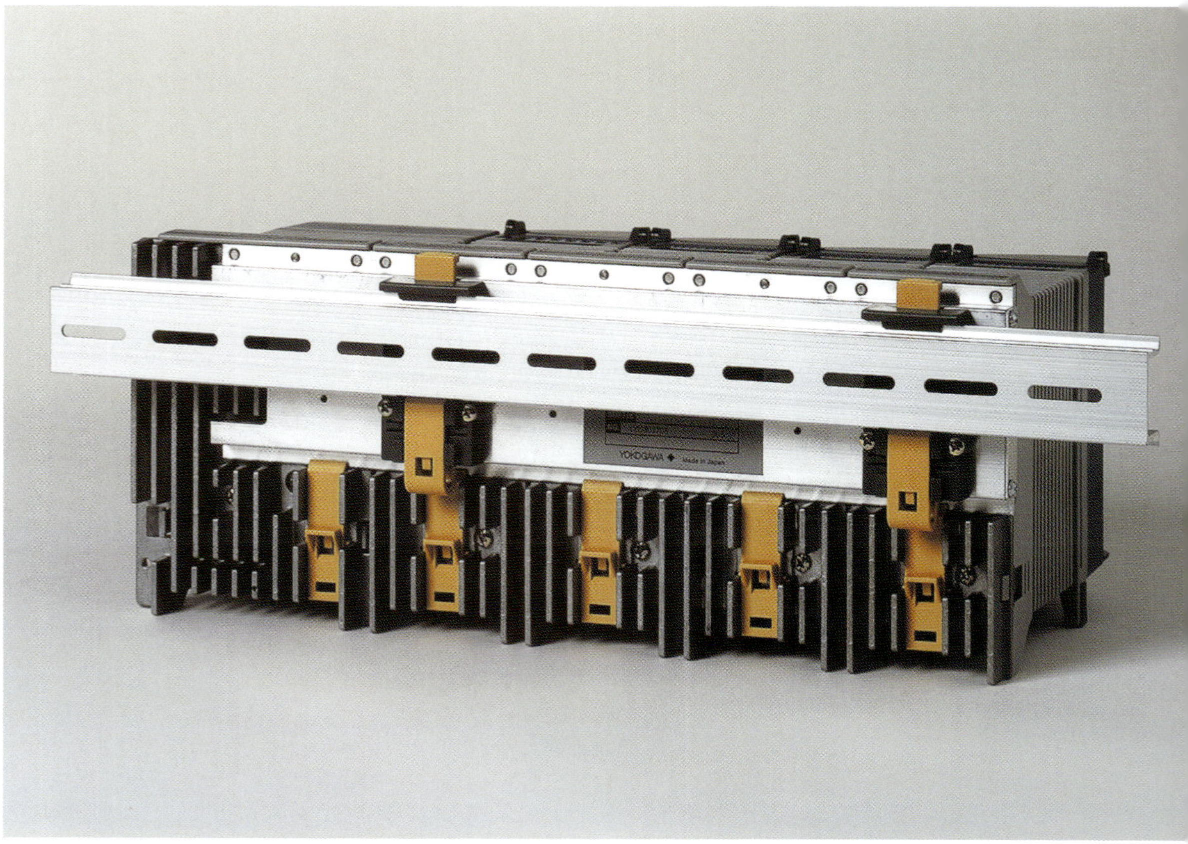

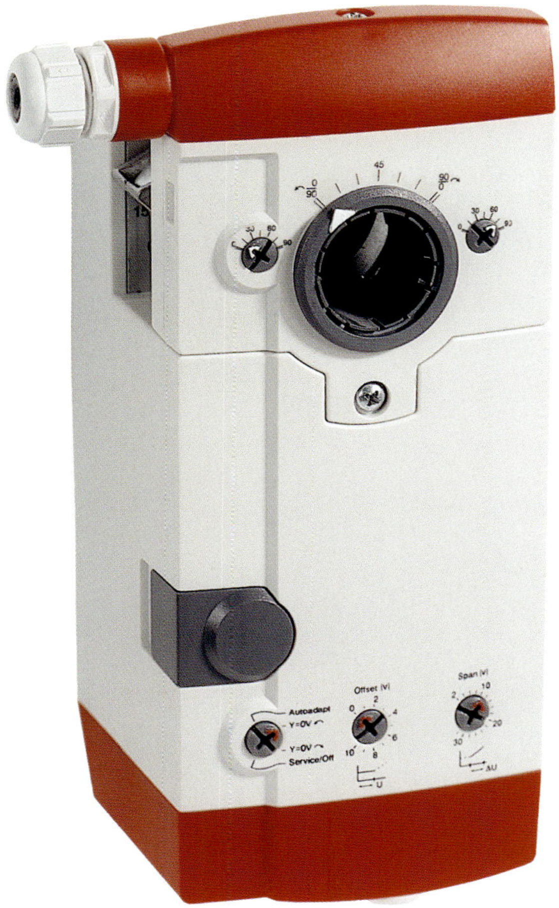

Hersteller Manufacturer
Honeywell GmbH
Schönaich, Germany

Produkt Product
SmartAct
Klappenantrieb für Lüftungsklappen
Direct-coupled damper actuator

Design
Edwin Lässle
Busse Design
Elchingen/Unterelchingen, Germany

SmartAct ist eine neue Produktfamilie, die für den Neubau und die Nachrüstung von Lüftungsanlagen entwickelt wurde. Die Honeywell-Klappenantriebe decke eine große Bandbreite von Zuluft-, Rückluft- und Mischluftklappen sowie die Ansteuerung von VVS-Geräten, Drossel- und Jalousieklappen ab. Sie besitzen eine selbstzentrierende Universalkupplung, eine abnehmbare Anschlussbox zur ergonomischen Installation unabhängig von der Einbaulage, einen einfach zu bedienenden mechanischen Drehwinkelbegrenzer und ein integriertes, abnehmbares Hilfsschalterpaket. Die Typen stellen verschiedene Varianten zur Ansteuerung zur Verfügung.

SmartAct is a new product family developed for the new and retrofit ventilation market. The Honeywell actuators meet a wide variety of supply, return, face and bypass damper, and variable air volume, fan inlet and vane control needs. They come with a universal, self-centering shaft adapter, a detachable wiring box for ergonomic installation (independent of the mounting position), a simple-to-use mechanical end limit, and an integrated, detachable auxiliary packet. The all-purpose actuators have selectable control types.

Hersteller Manufacturer
Toolz GmbH
München, Germany

Produkt Product
RoboLaser 1.5, RoboApi, RoboVector
Selbstnivellierende Baulaserserie
Self-leveling builder's lasers

Design
Stefan Reichl
Andreas Rößner
Oxxid
München, Germany

Die RoboToolz Baulaser Serie besteht aus drei selbstnivellierenden Mehrachslasern. Neben technischer Zuverlässigkeit und einfacher Bedienung bieten die Geräte durch die graue Gummiummantelung eine lange Lebensdauer im rauen Umfeld auf der Baustelle. Diese erhöht die Griffigkeit und prägt das robuste Erscheinungsbild.

The RoboToolz series of builder's lasers consists of three self-leveling multi-axis lasers. Alongside technical reliability and ease of use, the appliances are highly durable even under tough building site conditions thanks to their grey rubberized casings, which likewise enhance grip and give the units a pronounced robust look.

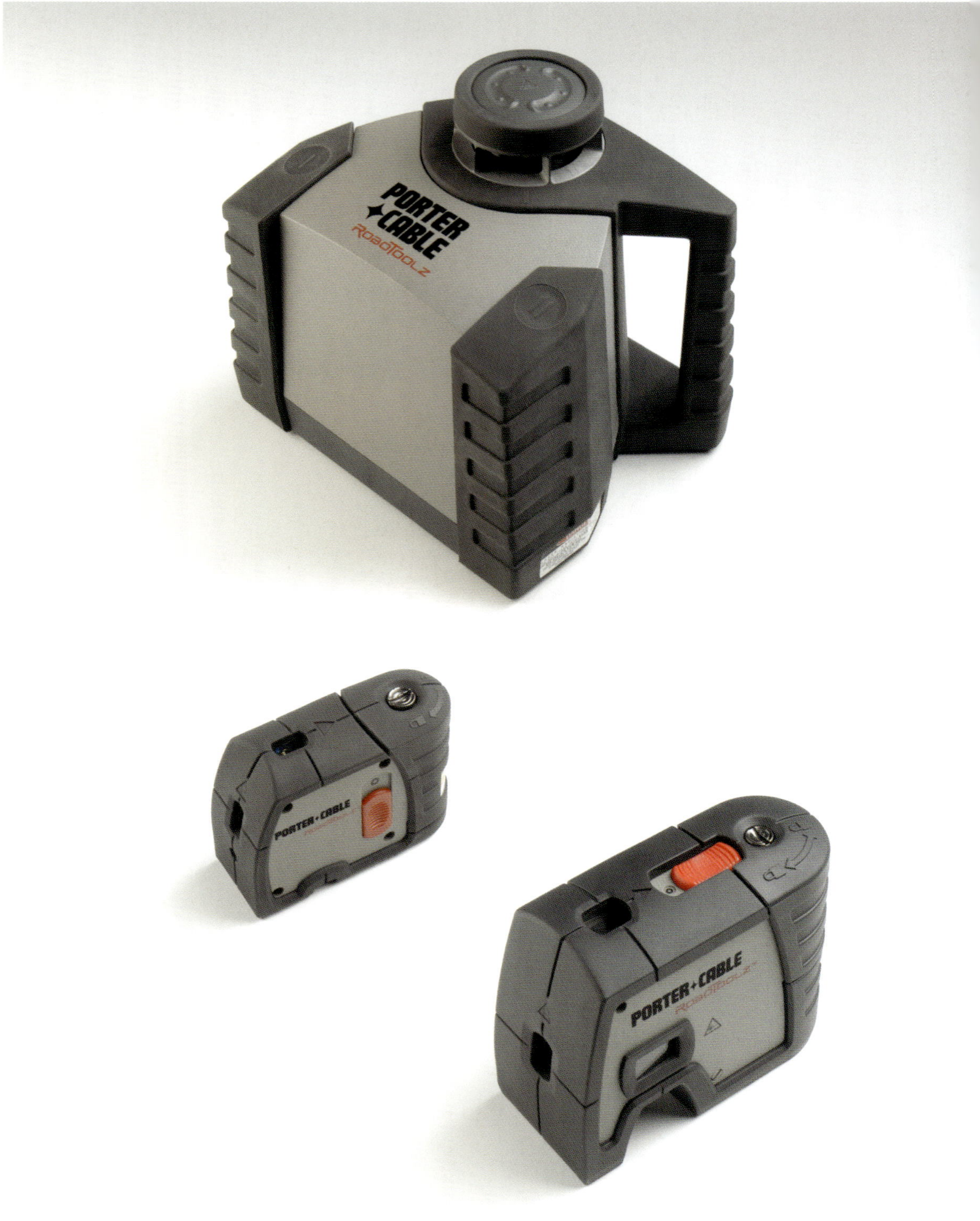

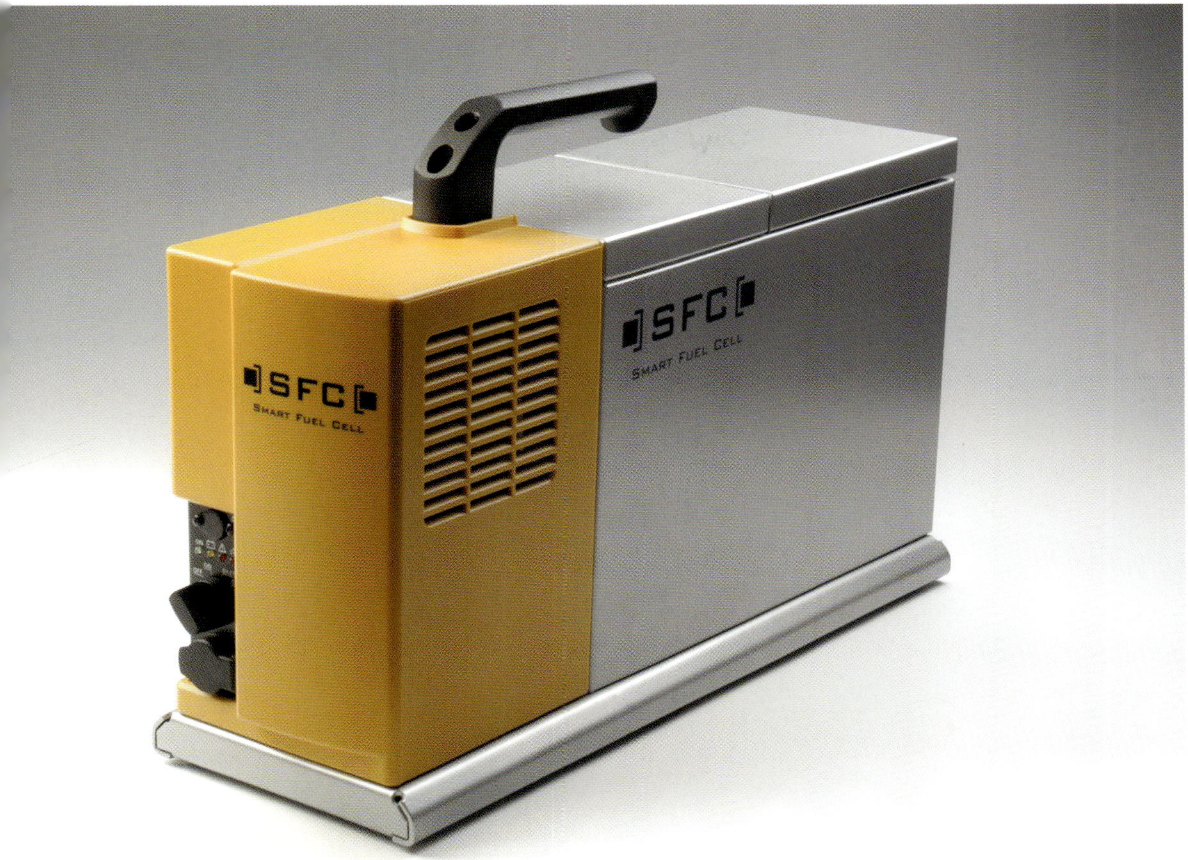

Hersteller Manufacturer
Smart Fuel Cell AG
Brunnthal-Nord, Germany

Produkt Product
SFC A25
Mobile Brennstoffzelle
Portable fuel cell

Design
Stefan Reichl
Andreas Rößner
Oxxid
München, Germany

Die Direkt-Methanol-Brennstoffzelle liefert mit dem 2,5 Liter Wechselkanister bis zu 50 Ah pro Tag. Sie ist transportabel, kann aber auch in Wohnmobile, Yachten oder entlegene Messstationen, wie z.B. in Verkehrsleitsysteme, eingebaut werden. Das Design mit Stahlmantel garantiert Robustheit, die farbige Front stellt die formale Zugehörigkeit zu den kleineren mobilen SFC-Geräten her. Für längere Betriebszeiten können externe Tanks angekoppelt werden.

This direct-methanol fuel cell comes with a 2.5 liter methanol tank and provides up to 50 Ah per day. The portable unit can be used as a floor-mounted power supply in recreational vehicles, boats or remote measuring devices, such as traffic guidance systems. The design boasts a robust steel sleeve, and the colored plastic front distinguishes the unit as part of the SFC product range. External tanks can be hooked up for longer operating times.

Hersteller Manufacturer
SCHUNK GmbH & Co. KG
Spann- und Greiftechnik
Lauffen/Neckar, Germany

Produkt Product
ROTA THW plus
Backen-Schnellwechsel-Futter
Quick-Jaw-Change Chuck

Design
SCHUNK GmbH & Co. KG
Lauffen/Neckar, Germany

Das Backen-Schnellwechsel-Futter ROTA THW plus ist ein sehr gutes Beispiel für die optimale Synthese von Technik, Funktionalität und Design. Der polierte Futterkörper strahlt nicht nur Ästhetik aus, die glatte Oberfläche verhindert zudem zuverlässig den Ansatz von Flugrost. Die markante polygone Form der zentralen Schutzbüchse ist nicht nur Augenfang, sie gewährleistet außerdem die schnelle Anpassung des Spannfutters bei veränderten Anforderungen. Die drei Rillen am Umfang des Schnellwechsel-Futters ROTA THW plus kennzeichnen es als SCHUNK-Produkt und runden das leuchtende Gesamtbild ab.

The quick-jaw-change chuck ROTA THW plus is an outstanding example of the perfect synergy between technology, functionality and design. The polished chuck body not only provides nice aesthetics, the smooth surface also prevents the onset of rust. The prominent polygon designed center sleeve is not only an eye-catcher, but in addition provides quick adaptation of the lathe chuck to various requirements. The three grooves on the outer diameter of the quick-jaw-change chuck clearly distinguish it as a quality SCHUNK product in consistent form.

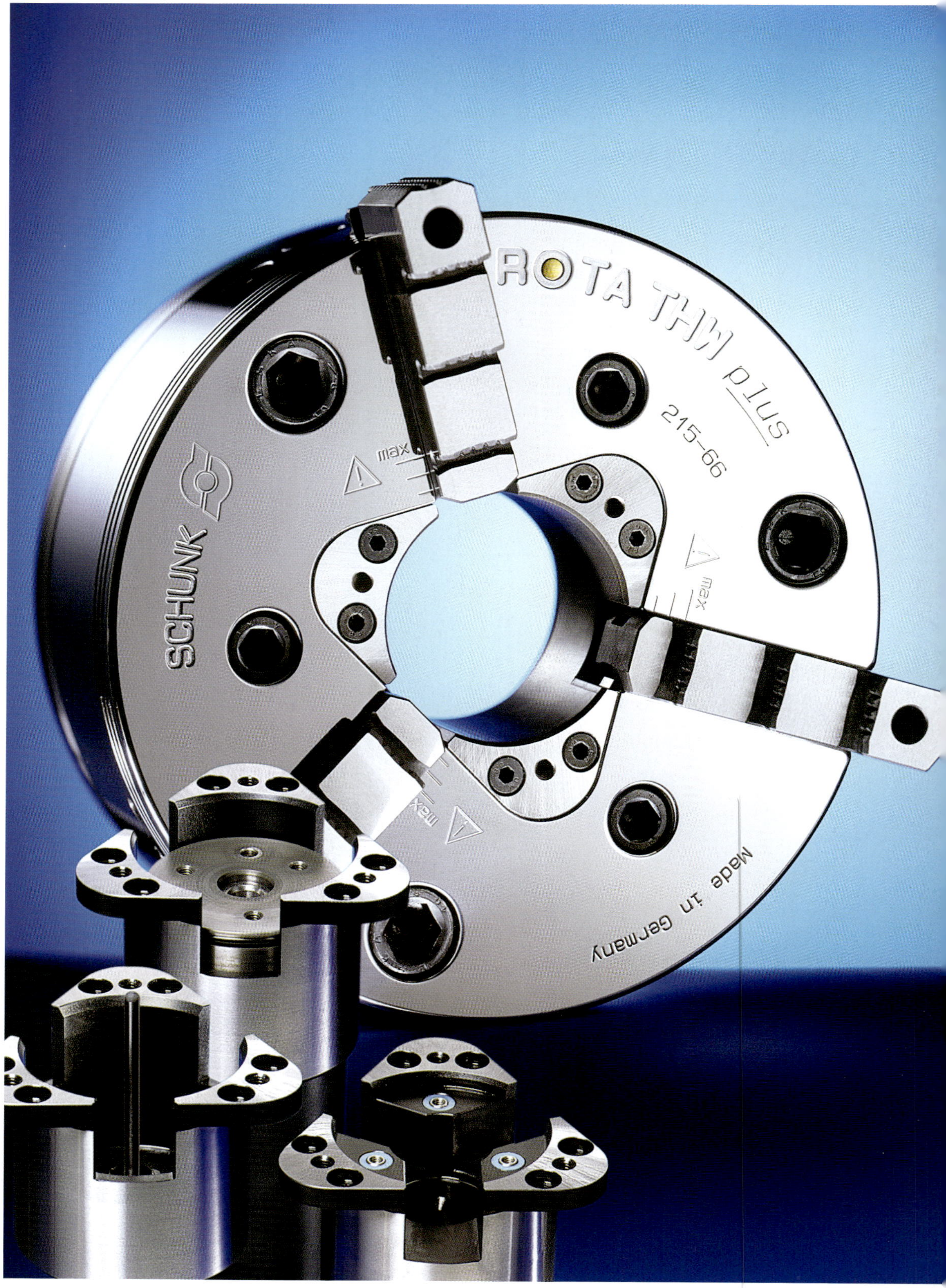

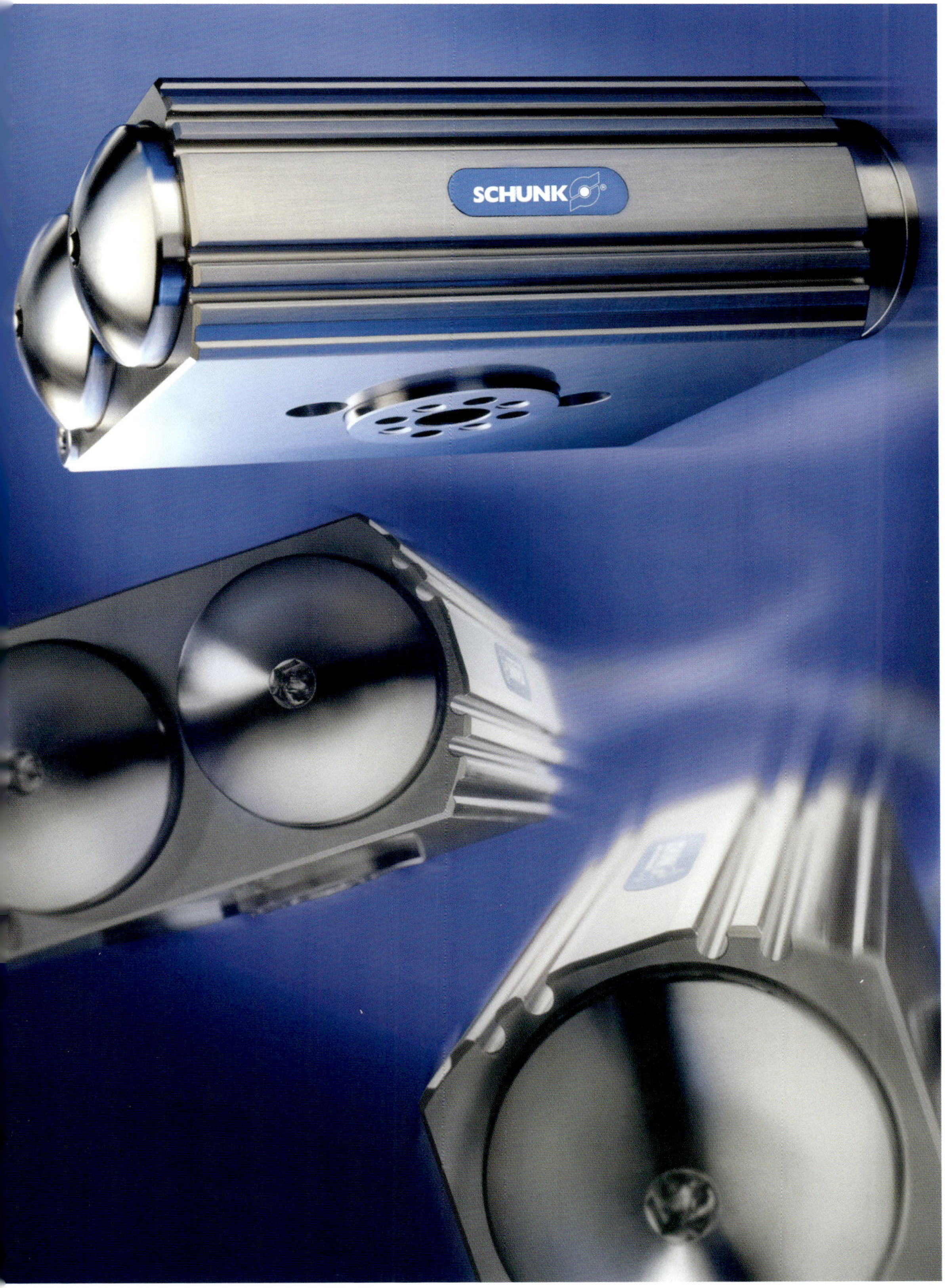

Hersteller Manufacturer
SCHUNK GmbH & Co. KG
Spann- und Greiftechnik
Lauffen/Neckar, Germany

Produkt Product
SRU
Flach-Schwenkeinheit
Standard rotary unit

Design
SCHUNK GmbH & Co. KG
Lauffen/Neckar, Germany

Die SCHUNK-Produktneuheit SRU aus dem Bereich Automation verändert die Lage von Werkstücken mittels rotierenden Bewegungen. Als Flach-Schwenkeinheit besticht sie nicht nur durch ihre technischen Vorzüge, sondern auch durch ihr Design. Die verwendeten Materialien gehen ein harmonisches Zusammenspiel mit Technik und Funktionalität ein. Während gerade Flächen als Befestigungsmöglichkeiten zur Verfügung stehen, vermittelt die Rundung des Gehäuses eine angenehme Optik und schließt so den positiven Gesamteindruck der Produktneuheit SRU ab.

The latest automation product release from SCHUNK is the new rotary actuator SRU. The SRU is designed to fulfill the demands of rotational part positioning with fast actuation, accurate dampened end stops, condition monitoring and innovative center locking options. This very flat rotary actuator not only stands out with its pioneering technical features but also with its design. The materials utilized create a perfect harmony between technology and functionality. While flat faces serve as mounting positions, the curved housing provides a nice aesthetic appearance creating a positive impression of the new SRU.

Hersteller Manufacturer
Weg Industrias S/A Automacao
Jaragua do Sul, Brazil

Produkt Product
Soft Starter SSW06
Elektronisches Zubehör
Electronic equipment

Design
Eduardo Arndt
Alceu Muniz
Morpho Servicos de Design Ltda
Curitiba-Paraná, Brazil

Der Soft Starter SSW06 (Sanftanlasser) mit digitaler Steuerung dient der Industrieautomatisierung von Elektromotoren. Er ermöglicht eine verbesserte Beschleunigung und Verlangsamung und hat zudem eine Schutzfunktion für die Motoren. Er wird hauptsächlich aus Metallplatten mit Knicken und Fugen hergestellt, was Lötarbeiten und Schrauben unnötig macht. Die Beschichtungselemente und die Benutzeroberfläche sind aus gespritztem Kunststoff. Leichte Bedienbarkeit und Einfachheit der Form standen bei der Gestaltung im Vordergrund. Die Farben (zwei Blautöne) vermitteln die Corporate Identity des Unternehmens.

The digitally-controlled Soft Starter SSWO6 is intended for the industrially automating electric motors. It enables improved acceleration and deceleration and moreover, incorporates a protection function for the motors. The unit is composed primarily of sheet metal, with folds and fittings which eliminates the need for welding and screws. Exterior fittings and the user interface are made of injected plastic. The design focuses on ease of use and simplicity of form while the company's Corporate Identity is reflected in the two blue tones adopted.

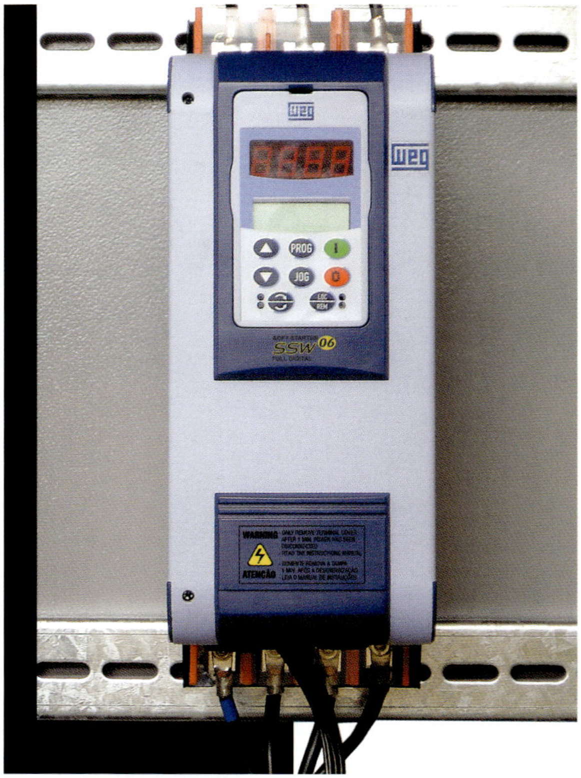

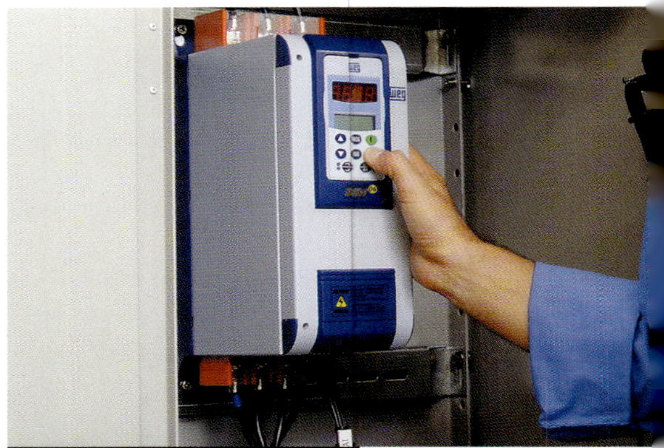

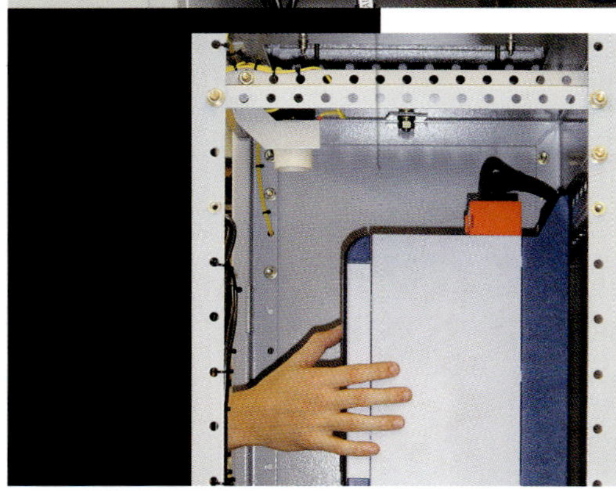

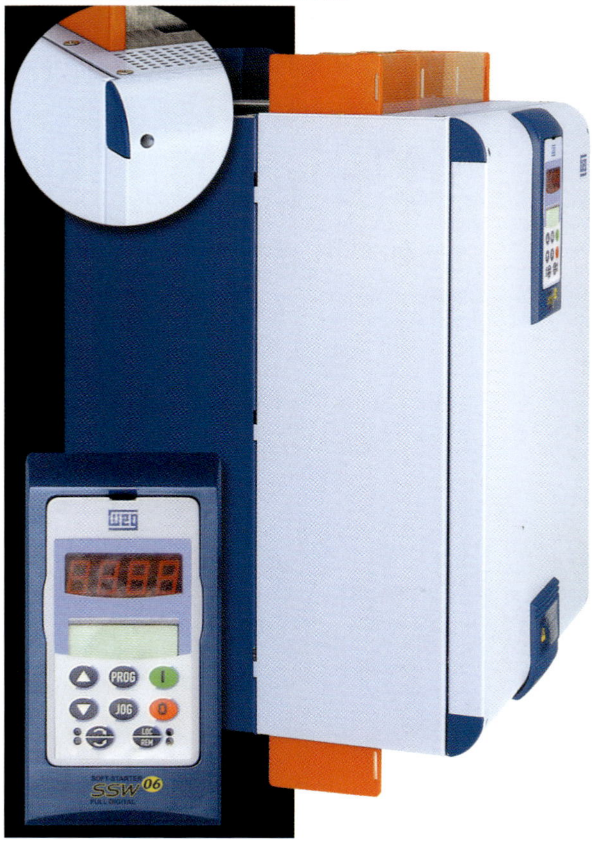

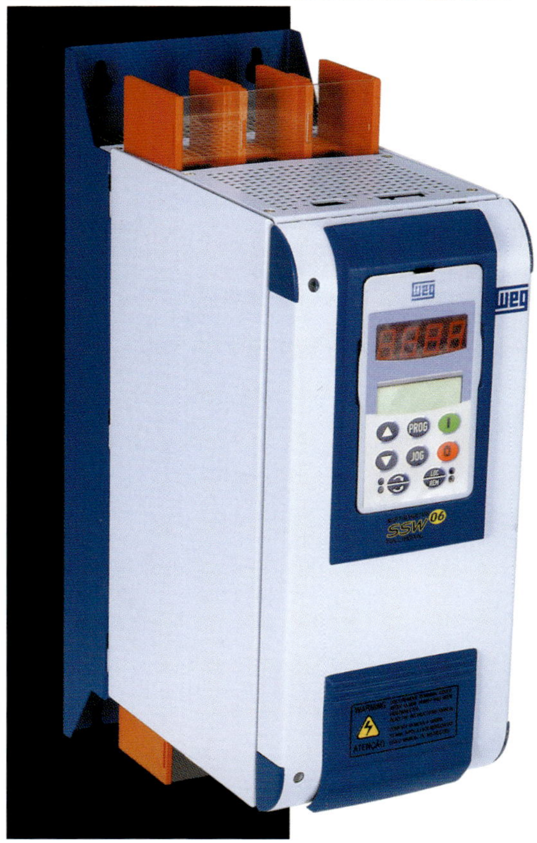

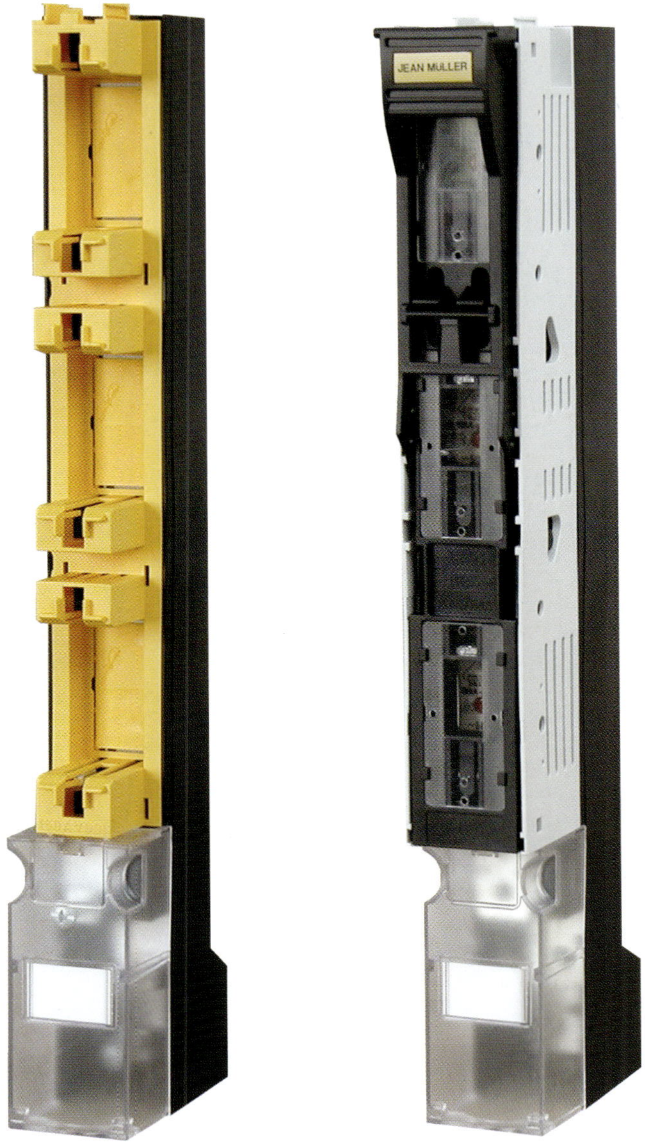

Hersteller Manufacturer
Jean Müller GmbH
Eltville am Rhein, Germany

Produkt Product
NH-Sicherungslastschaltleiste
NH-Sicherungsleiste Größe 00
LV HRC strip-type fuse
switch-disconnector/strip-fuseway

Design
Jean Müller GmbH
Eltville am Rhein, Germany

Bei der NH-Sicherungslastschaltleiste und NH-Sicherungsleiste Größe 00 handelt es sich um ein Schaltgerät für die elektrische Energieversorgung zur Anwendung in Energieversorgungsunternehmen und in der Industrie. Zu Messzwecken können Stromwandler im Leistenträgerteil integriert werden. Bei einer Einbaukombination von Leisten mit und ohne Stromwandler bleiben die Einbaumaße unverändert, und das Leistenoberteil muss zur Montage der Leiste auf den Sammelschienen nicht mehr entfernt werden. Eine neu gestaltete Hebelmechanik ermöglicht einen günstigen Krafteinsatz und hohe Schaltgeschwindigkeit.

The LV HRC strip-type fuse switch-disconnector/strip-fuseway size 00 is a electricity supply switchgear destined for use by energy providers and industrial corporations. For measurement purposes, the fuseway duct can also house a transformer. The unit size for assembly remains the same for strips with or without transformers. The upper section of the strips does not have to be removed in order to assemble the strip on multitracks. A new lever mechanism makes for easy powering and high switching speeds.

Hersteller Manufacturer
Boll & Kirch Filterbau GmbH
Kerpen, Germany

Produkt Product
BFD
Doppelfilter zur Flüssigkeitsfiltration
Double filter for liquid filtration

Design
Frank Blum
Köln, Germany

Die Konzeption des BFD Doppelfilters zur Flüssigkeitsfiltration basiert auf der Umsetzung einer neuen Technologie, die vorsieht, die Umschaltarmatur eines Doppelfilters in das Gehäuse zu integrieren. Durch die Farbgestaltung werden verschiedene Funktionsbereiche wie Gehäuse, mobile Elemente sowie die Schnittstellen zum Benutzer eindeutig gekennzeichnet. Das in enger Zusammenarbeit mit der Konstruktion entwickelte Design lässt sich auf alle Produkte einer Baureihe, die zwischen einer Gehäusehöhe von 40 bis 160 cm variieren, anwenden.

The BFD double filter for liquid filtration bases on the realization of a new technology integrating the switch in a double filter into the housing. The color coding clearly defines the different functional areas such as housing, mobile elements and user interfaces. Developed in close collaboration with the construction team, the design can be used for all products in a product line with casings ranging in height from 40 to 160cm.

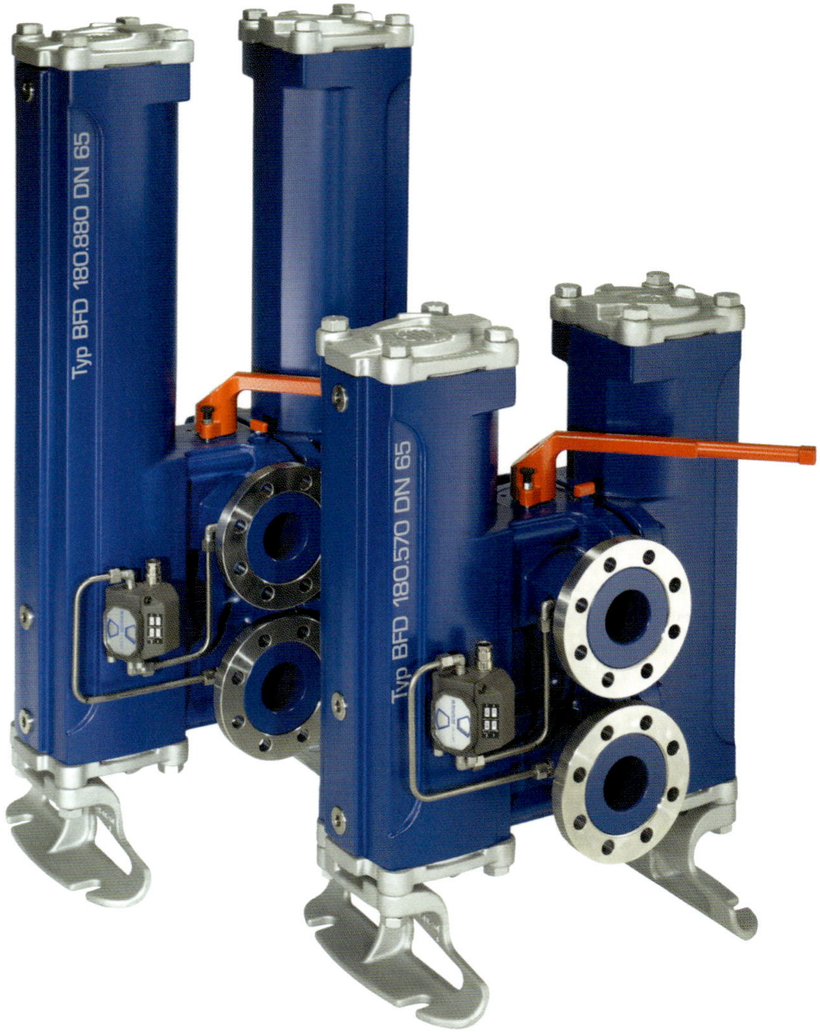

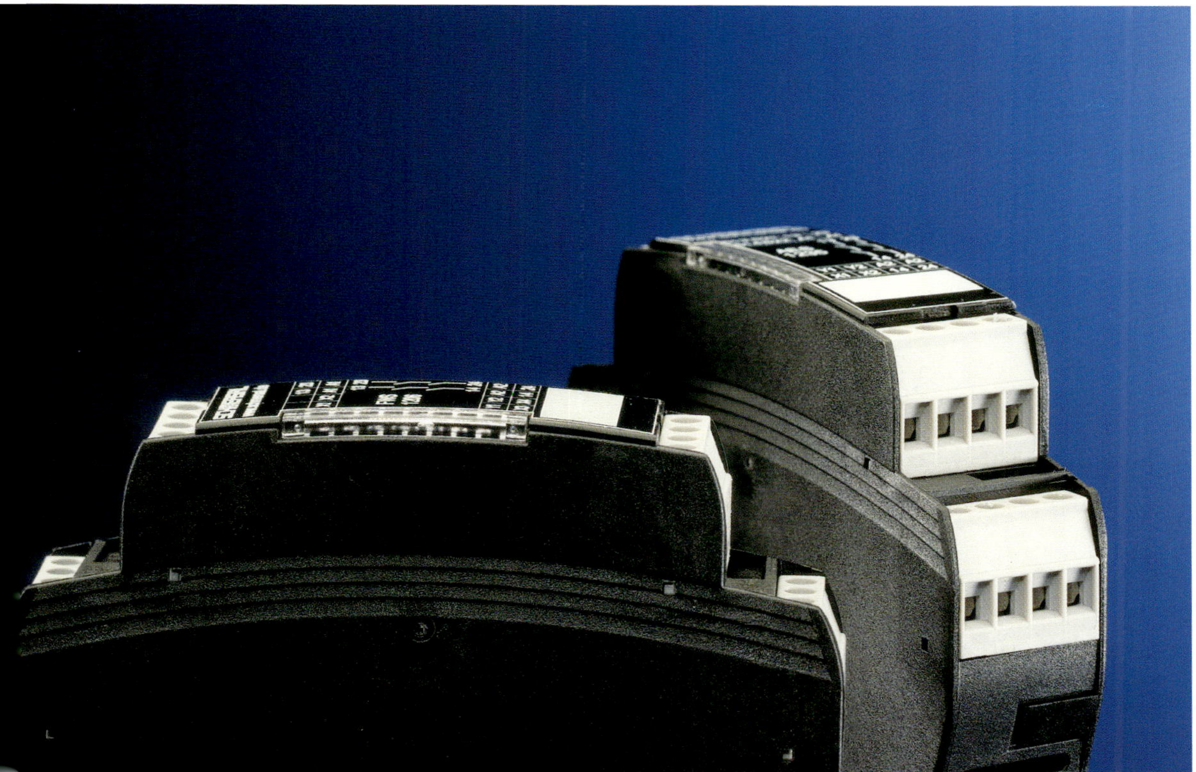

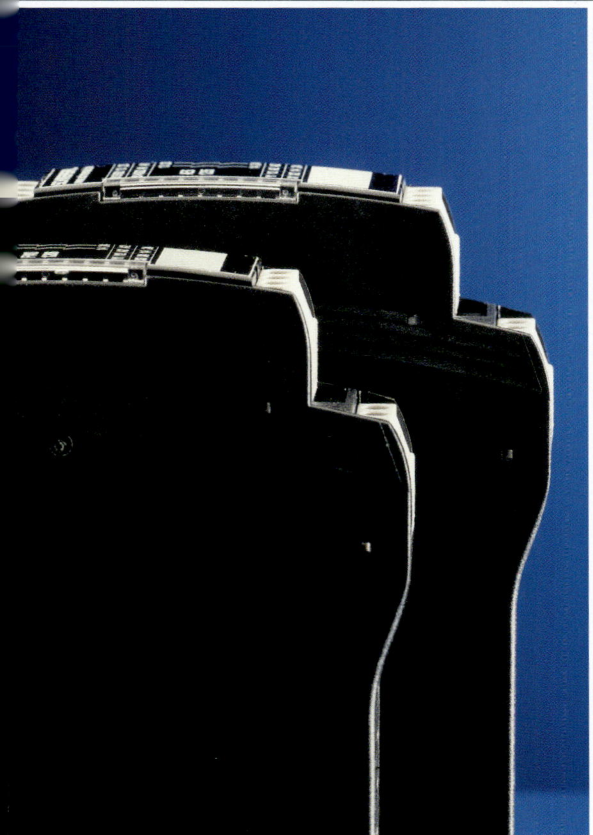

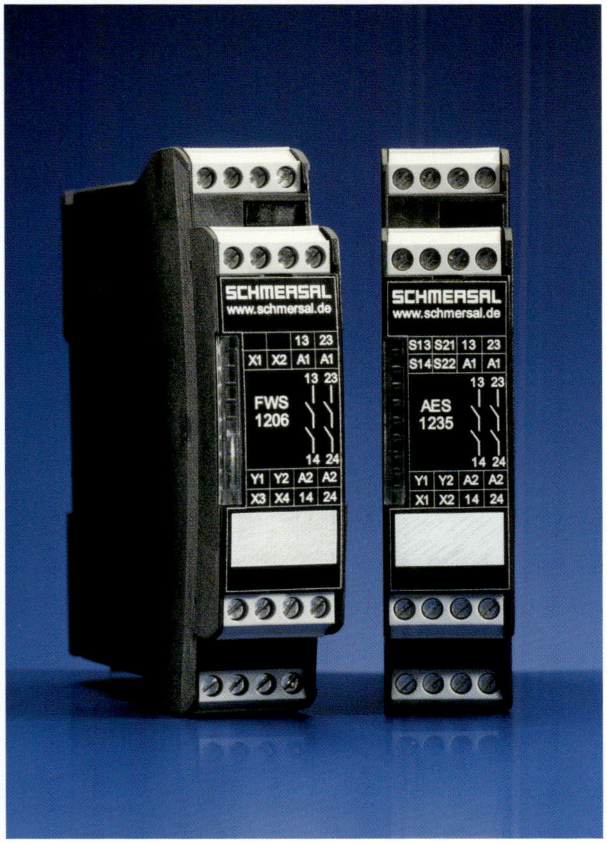

Hersteller Manufacturer
K.A. Schmersal GmbH
Wuppertal, Germany

Produkt Product
Protect
Sicherheitsbaustein
Safety control module

Design
K.A. Schmersal GmbH
Wuppertal, Germany

Die Sicherheitsrelaisbausteine Protect dienen zur Auswertung von Signalen von Sicherheitsschaltgeräten, die z.B. Schutztüren von Werkzeugmaschinen überwachen oder zuhalten. Sie verfügen optional über jeweils vier Ebenen mit steckbaren Anschlussklemmen, über die nahezu jeder Status im Sicherheitsstromkreis abgefragt werden kann. Die steckbaren Klemmen erlauben es dem Anwender mit vorkonfektionierten Kabelbäumen zu arbeiten und entsprechend Zeit zu sparen. Das Design der Frontblende ist klar gegliedert. Über eine Lichtleiste mit bis zu 10 LEDs erhält der Anwender alle relevanten Informationen über den Schaltzustand. Im Hinblick auf Kompaktheit konnte das Design der Gehäuse zusätzlich optimiert werden.

The Protect safety control modules evaluate signals emitted by the safety switchgear used to safeguard or interlock the guard doors of machine tools. Optionally, they come with four rows of plug-in terminals enabling almost all states of the safety circuits to be interrogated. The plug-in terminals allow users to work with time-efficient pre-wired cable harnesses. The front cover has a clear-lined design: the light strip of up to 10 LEDs provides all the relevant information on switching states. The casing design has been optimized to save space.

Hersteller Manufacturer
COORD3 SPA
Bruzolo, Italy

Produkt Product
Arkey
Fernsteuerung
Joystick-based remote control

Design
DNA SRL
Trento, Italy

Arkey ist eine Fernbedienung für die Steuerung der Hauptfunktionen einer professionellen Messmaschine. Das Gerät ist ausschließlich für den professionellen Einsatz bestimmt und seine Gestaltung richtet sich demgemäß streng nach den geltenden Bestimmungen für diesen Bereich. Hauptkriterium der Designentwicklung war es, ein rein funktionales, ergonomisches, einfaches, stabiles und zuverlässiges Gerät zu gestalten. Zudem wurde Wert auf eine Reduktion des Gewichts und eine kompakte Bauform gelegt.

Arkey is a joystick-based remote control pad handling the main functions of professional measuring machines. The ambidextrous device is destined only for professional applications and the design is therefore fully aligned to the specific requirements involved. The key factor when developing the design: providing a purely functional, ergonomic, simple, stable and reliable device. Moreover, weight reduction and a compact shape were both prioritized.

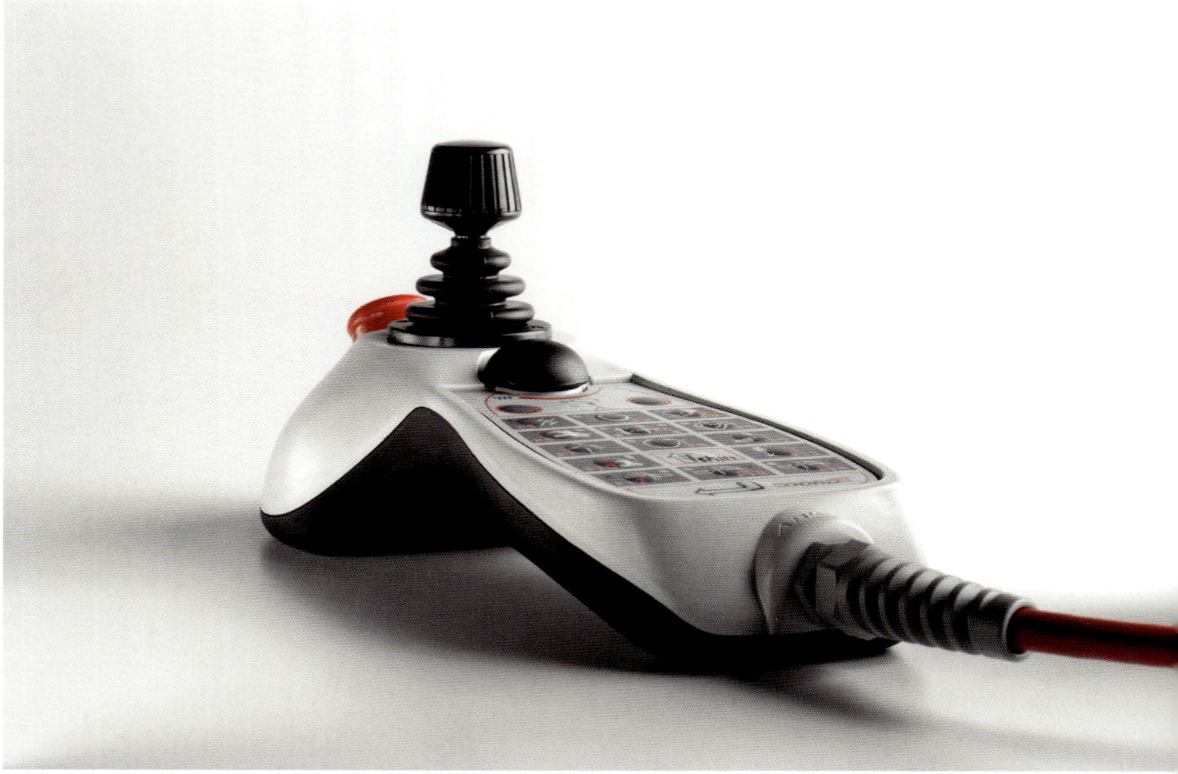

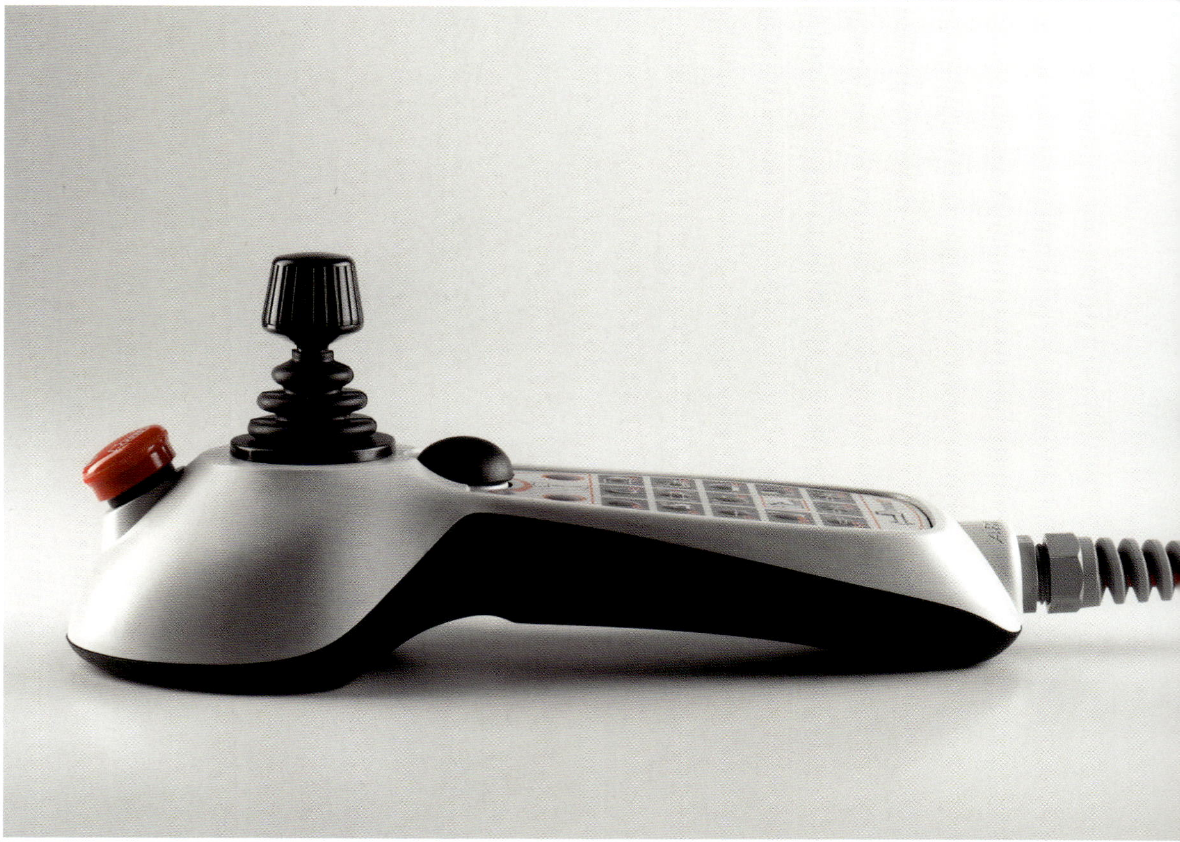

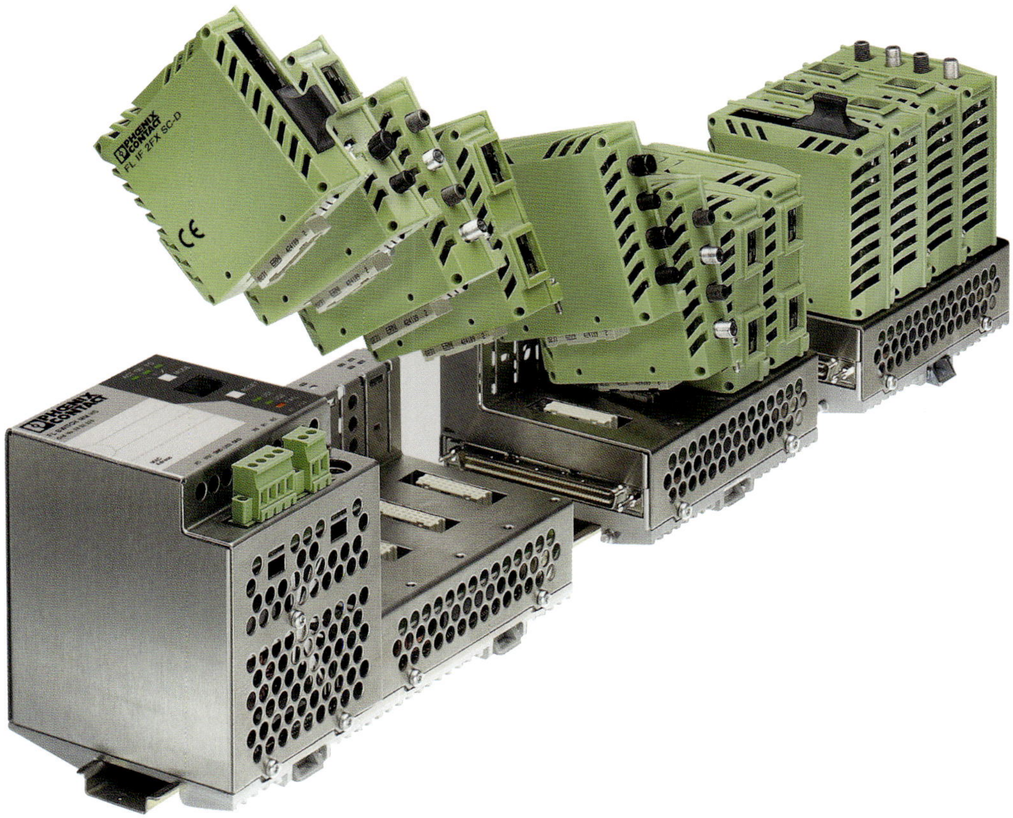

Hersteller Manufacturer
Phoenix Contact GmbH & Co. KG
Blomberg, Germany

Produkt Product
FL Switch MM
Modulares Switchsystem
Modular switch system

Design
Gerwin Müller
Tecform Design
Kiel, Germany

FL Switch MM ist ein industrietauglicher modularer Switch für industrielle Ethernet-Anwendungen. Der Switch kann von 8 auf 24 Ports modular in Zweierschritten erweitert werden. Die steckbaren Interface-Module erlauben die Wahl von 2 Abgangsrichtungen und die Wahl der Medien Twisted Pair, Glasfaser, Polymer- und HCS-Faser. Die kompakte Bauform mit 110 mm Tiefe erlaubt die dezentrale Installation in normalen Klemmenkästen vor Ort. Die Softwareunterstützung, kombiniert mit den komfortablen Diagnoseeigenschaften sowie dem integrierten Web-Browser gestalten die Inbetriebnahme und Diagnose unkompliziert.

FL Switch MM is an industrially hardened modular switch for industrial Ethernet applications. The switch can be extended modularly from 8 to 24 ports in steps of two. The pluggable interface modules allow the selection of two output directions and the selection of the media twisted pair, glass fibre, polymer and HCS fibre. The compact switches are only 110mm deep, enabling decentralized installation in conventional terminal boxes on location. The integrated homepage, easy configuration and diagnostics make for smooth and uncomplicated operations and troubleshooting.

Hersteller Manufacturer
EAO AG
Olten, Switzerland

Produkt Product
Serie 05
Befehls- und Meldegerät
Man machine interface

Design
EAO AG
Olten, Switzerland

Die Serie 05 von Befehls- und Meldegeräten wurde entwickelt, um den verschiedensten Anforderungen im Bereich Transport und Maschinen entgegenzukommen, wie beispielsweise in Führerständen von Straßenbahn, Bus und Bahnen, in Maschinenpanels und Steuerungen. Wichtige Designaspekte waren unter anderem eine zeitlose Gestaltung, da z.B. die Tasten oft über Jahrzehnte im Einsatz sind, und Modularität, um Flexibilität zu gewährleisten. Zudem musste die Serie 05 designkompatibel zur Serie 04 sein, da beide Produktreihen oft in der gleichen Applikation nebeneinander zum Einsatz kommen und daher ein einheitliches Erscheinungsbild aufweisen müssen.

The Series 05 man machine interface was developed to meet a wide variety of requirements in the fields of transportation and machinery, e.g., in drivers' cabs on trams, buses and trains, in machine panels and on control panels. Important design considerations included a timeless appearance, since function keys are often in use over a period of decades, as well as a modular construction to deliver due flexibility. Additionally, the Series 05's design is compatible with the Series 04 products, since both lines are often in use alongside each other in one and the same application and thus a uniform look is called for.

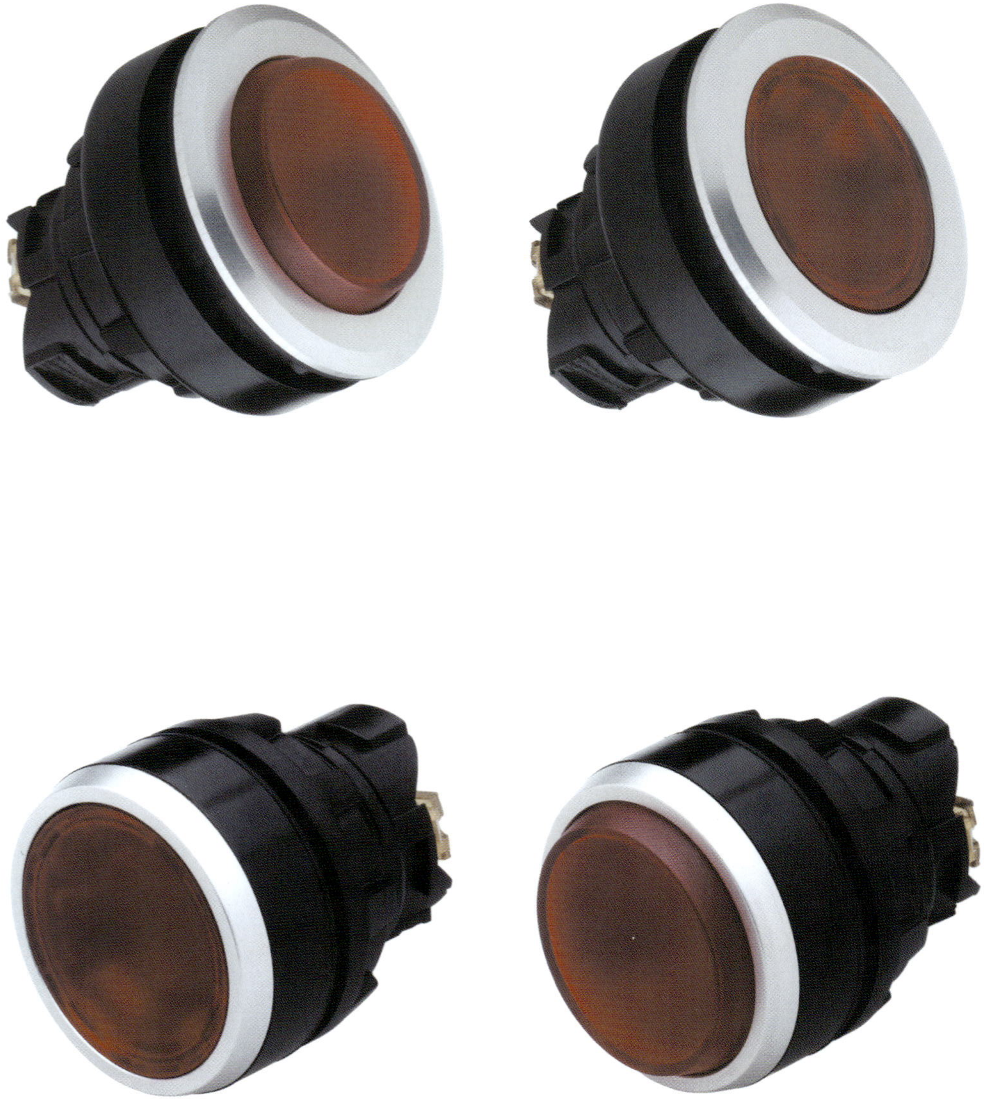

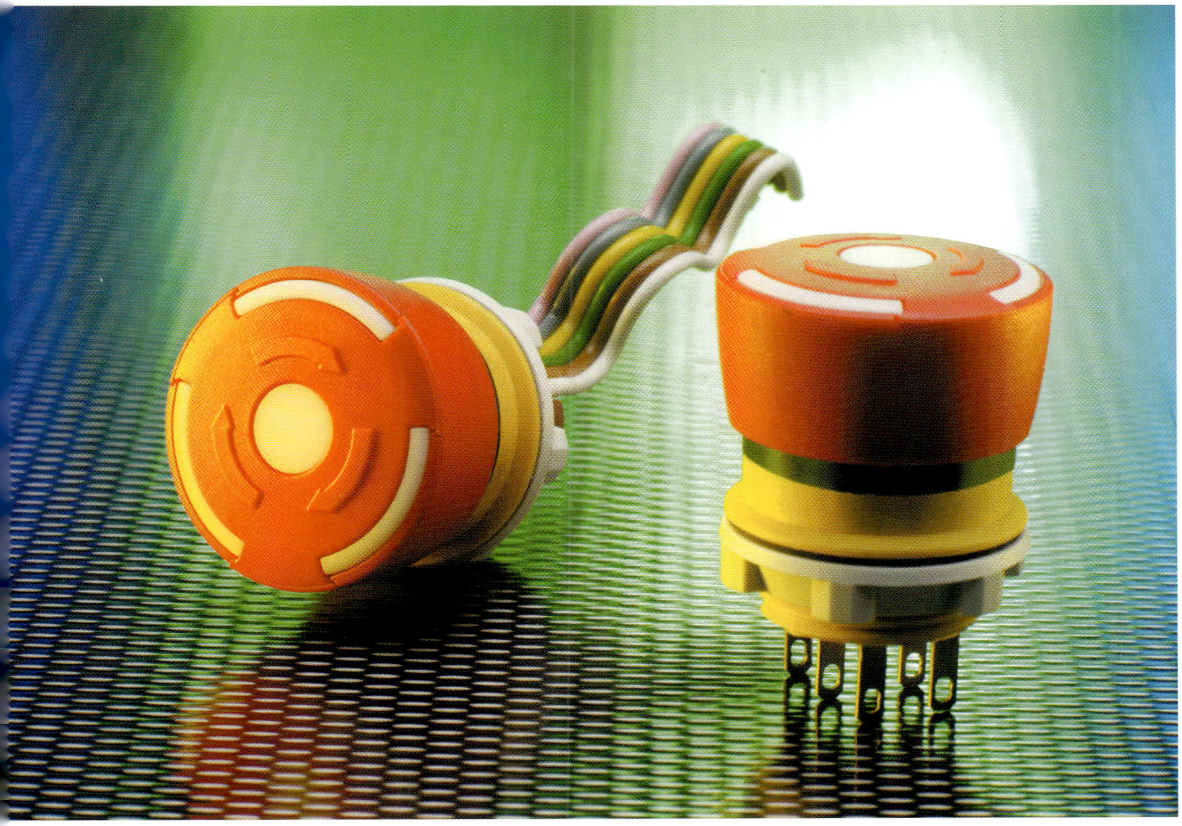

Hersteller Manufacturer
EAO AG
Olten, Switzerland

Produkt Product
Not-Aus-Taste
Short emergency stop

Design
Stephan Lintner
Busse Design
Elchingen/Ulm, Germany

Die Not-Aus-Taste ist für Hersteller von Handhelds für die Steuerung z.B. von Kränen oder Robotern konzipiert. Dank der sehr geringen Einbautiefe eignet sie sich insbesondere für Anwendungen, bei denen nur wenig Platz zur Verfügung steht. Durch drei zusätzliche Fenster am Rand der Betätigungskappe wird das seitliche Austreten von Licht ermöglicht. Die Taste kann flexibel bei einer Beleuchtungsbetriebsspannung von 5 bis 30 VDC frei eingesetzt werden. Die Kriterien der Designentwicklung waren vor allem Kompaktheit und Stabilität.

The short emergency stop was developed with a focus on manufacturers of handhelds for controlling cranes or robots, for example. Thanks to the minor depth of the stop, it is especially suited for applications where little space is available for inclusion of the stop button. The three additional windows on the edge of the actuating flap enable sideways emission of light. The button can be flexibly used for illumination current of 5-30 VDC. The design was created with a view specifically to delivering compactness and stability.

Hersteller Manufacturer
Warn
Clackamas, OR, USA

Produkt Product
Warn Works™ Utility Winches
Seilwinden
Utility winches

Design
Ziba Design
Portland, OR, USA

Die neue Serie von Seilwinden Warn Works™ Utility Winches dient zum Schleppen und Ziehen unterschiedlichster Gerätschaften wie z.B. Traktoren, Rennwagen oder Booten usw. auf Ladeflächen von Kleintransportern oder Anhängern. Ziel der Designentwicklung war es, die Seilwinden so zu gestalten, dass sie ein eigenständiges elektrisches Handwerksgerät darstellen und nicht ein Einbauteil eines Kleintransporters oder Anhängers. Bei der Umsetzung des Designs stand eine robuste Anmutung mit rundlichen Ecken und gummierten Stoßkanten sowie eine markante Farbgebung im Vordergrund.

The new series of Warn Works™ utility winches can be used for pulling, hauling and dragging a wide variety of objects, such as tractors, racing cars and boats onto the backs of pickup trucks and trailers. The goal behind the design: to develop winches that act as an independent electrical unit and not an integrated section of a pick-up or trailer. The result is a robust winch with rounded edges and rubberized touch points and a bold color scheme, all of which enhances safe use of the machine.s

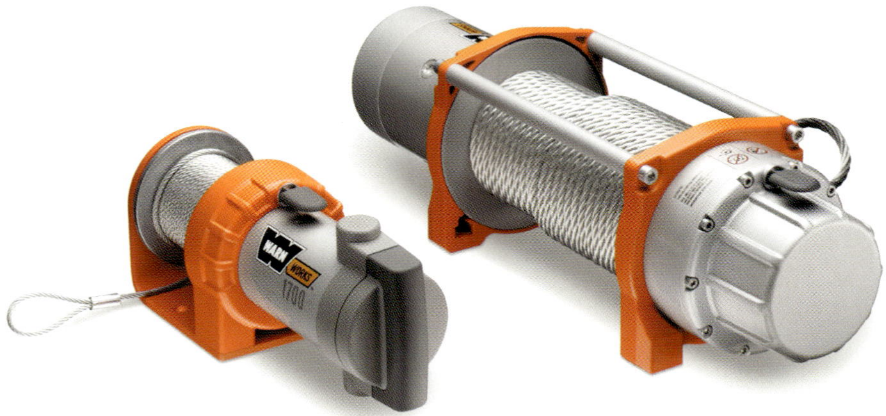

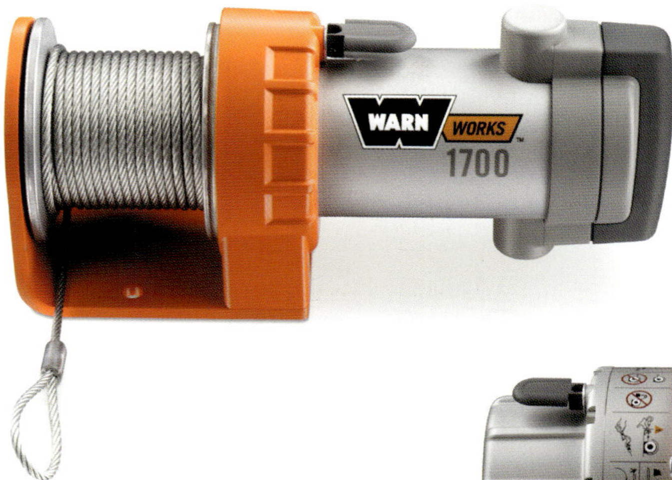

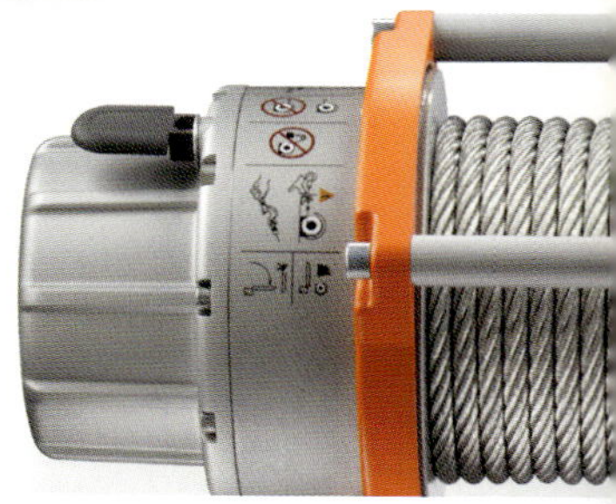

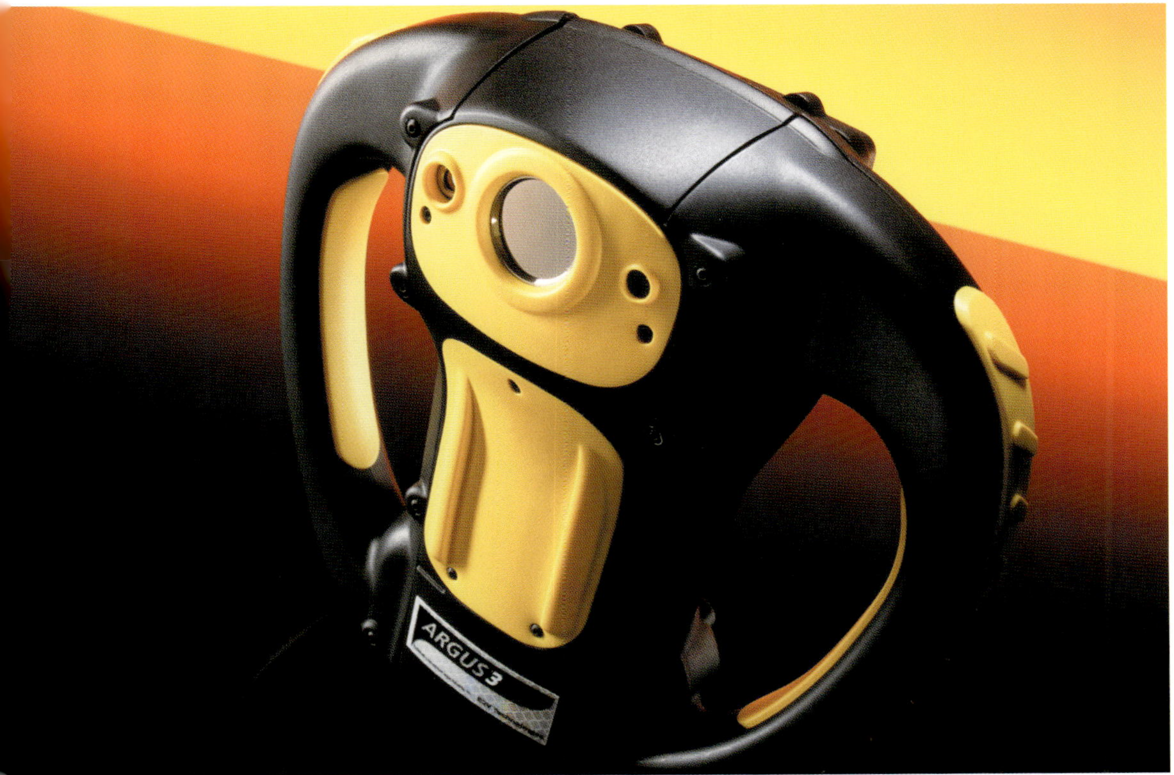

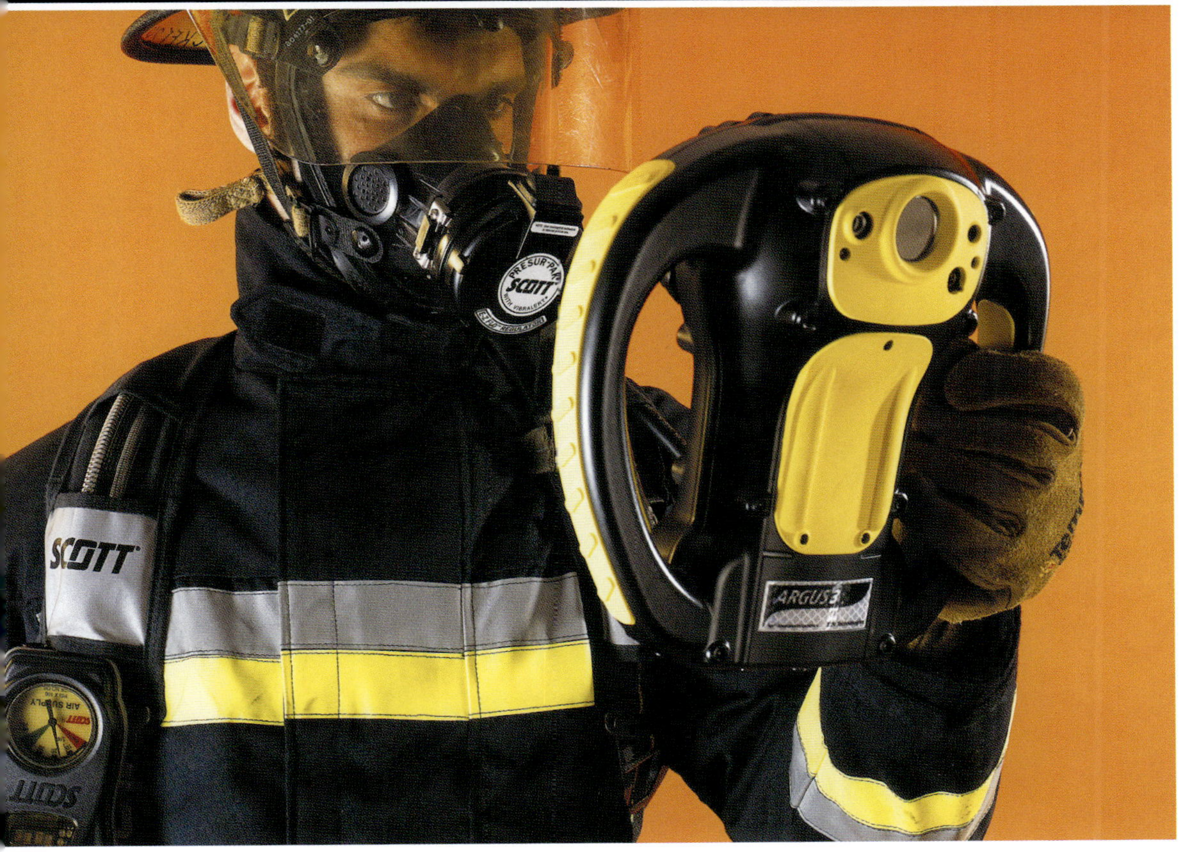

Hersteller Manufacturer
e2v technologies Limited
Chelmsford, Essex, UK

Produkt Product
Argus®3
Wärmebildkamera
Thermal imaging camera

Design
Alloy Limited
Farnham, Surrey, UK

Die Argus®3 ist eine Wärmebildkamera, die in erster Linie für die Anforderungen von Feuerwehrmännern entwickelt wurde. Mit ihr können auch in dickstem Rauch Verunglückte aufgefunden und der Brandherd lokalisiert werden. Bei der Designentwicklung trug man vor allen Dingen den Aspekten Rechnung, dass die Kamera handlich sein musste, zudem sollte sie Griffe haben und sich mit dicken Schutzhandschuhen leicht bedienen lassen. Das flache Profil garantiert ihren ungehinderten Transport auch in Situationen, in denen der Feuerwehrmann z.B. kriechen oder in die Hocke gehen muss. Insgesamt zeichnet sie sich durch eine robuste geordnete Ästhetik aus, die sich ihrer Funktion unterordnet.

The Argus®3 thermal imaging camera was primarily designed with the requirements of firefighters in mind, enabling them to see through even the thickest smoke to find casualties and detect the seat of the fire. The camera was developed in particular with the focus on creating an easy-to-use device with handles that could be operated using protective gloves. The camera's flat profile guarantees it can be carried easily even in situations where firefighters may have to crawl or crouch, for example. It has a robust and well structured aesthetic form that works in harmony with the function.

Hersteller Manufacturer
ADS-Tec GmbH
Leinfelden-Echterdingen, Germany

Produkt Product
Compact 3
Mobiler Industrie-Computer
Portable industry computer

Design
Bohner Design
Stuttgart, Germany

Compact 3 ist ein mobiler Industrie Computer mit einem 15 Zoll Display. Die Funktionskomponenten sind übersichtlich in 5 Fronttasten gegliedert, zusätzlich liefern LEDs Informationen zum Status von System und Batterien. Das Gerät arbeitet lüfterlos und gibt entstehnde Wärme über den Magnesium-Unterboden ab. Aufgrund der 5 wire Touch Technologie ist keine Nachkalibrierung notwendig. Das robuste Gehäuse besteht aus Magnesium, die Frontblende aus eloxiertem Aluminium. Ein umlaufender Bumper schützt das Gerät vor Stößen.

Compact 3 is a portables vehicle diagnostics tool with a 15" display. The functional components are well laid out in 5 front keys, and LEDs provide additional information on system and battery status.The device requires no air and dissipates the heat that arises through its magnesium base. The 5 wire touch technology means that no subsequent calibration is necessary. The computer has a robust magnesium casing, and the front panel is made of eloxy aluminum. A wrap-around bumper protects the computer against shock impact.

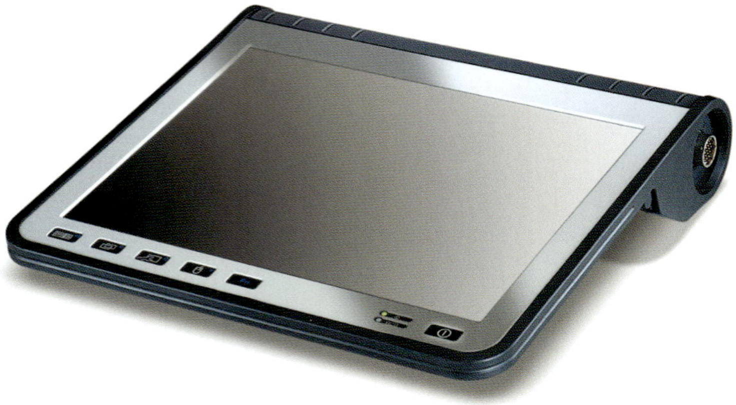

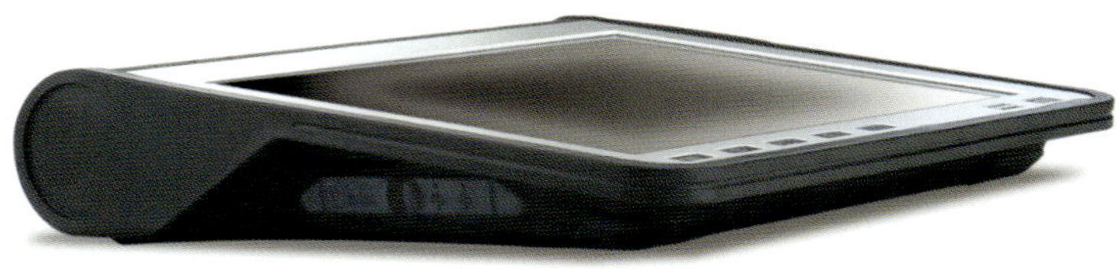

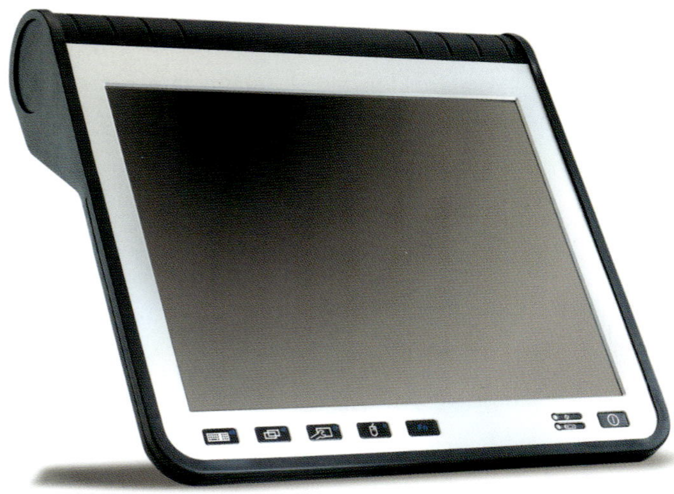

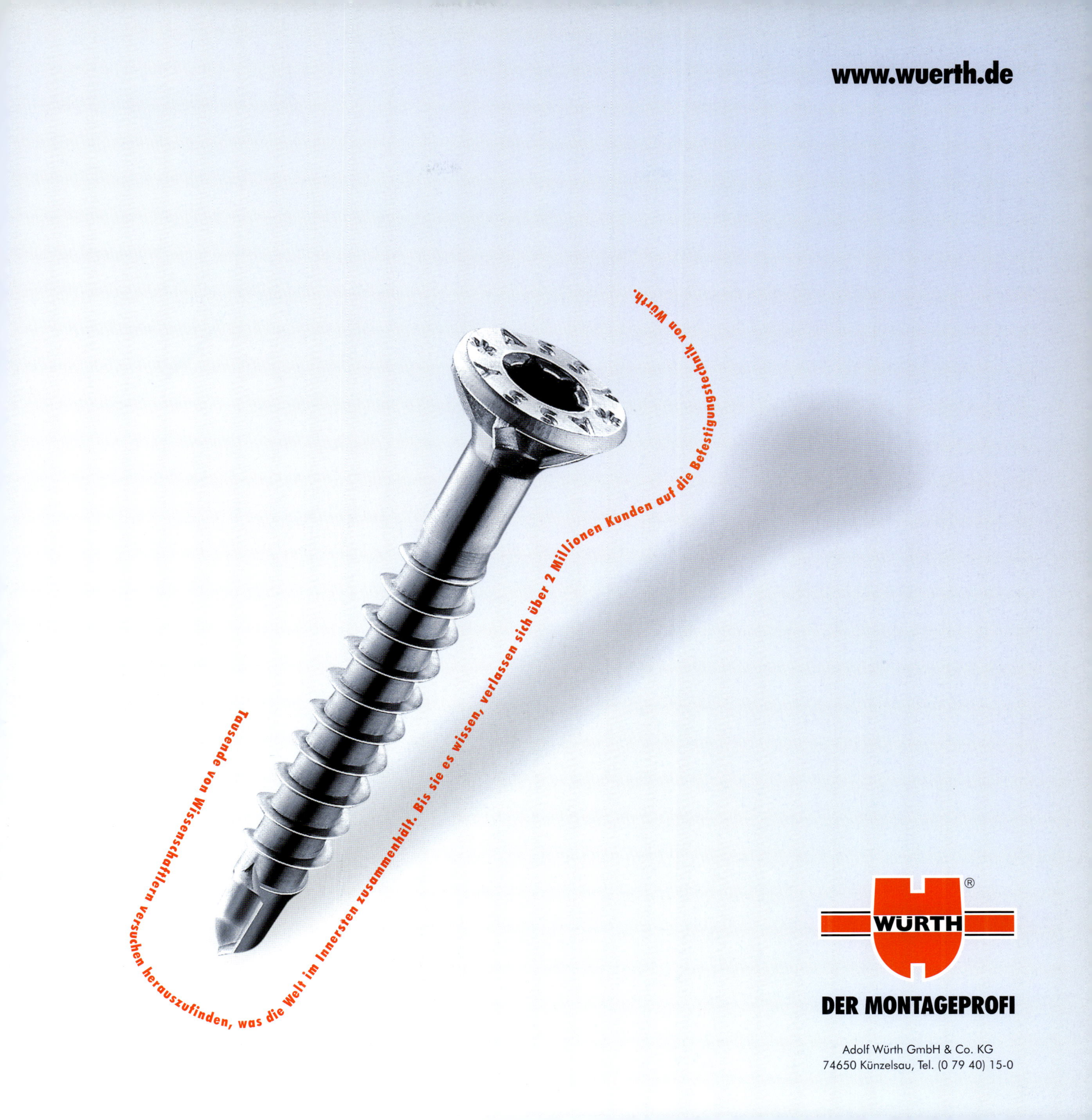
www.wuerth.de
Tausende von Wissenschaftlern versuchen herauszufinden, was die Welt im Innersten zusammenhält. Bis sie es wissen, verlassen sich über 2 Millionen Kunden auf die Befestigungstechnik von Würth.
WÜRTH®
DER MONTAGEPROFI
Adolf Würth GmbH & Co. KG
74650 Künzelsau, Tel. (0 79 40) 15-0

Hersteller Manufacturer
Crown Gabelstapler GmbH
München, Germany

Produkt Product
X-10 (tm) Handle
Deichselgriff für Gehgabelstaplergeräte
Control handle for
pedestrian pallet trucks and stackers

Design
Steve Pulskamp
Crown Equipment Corporation Design Center
New Bremen, OH, USA

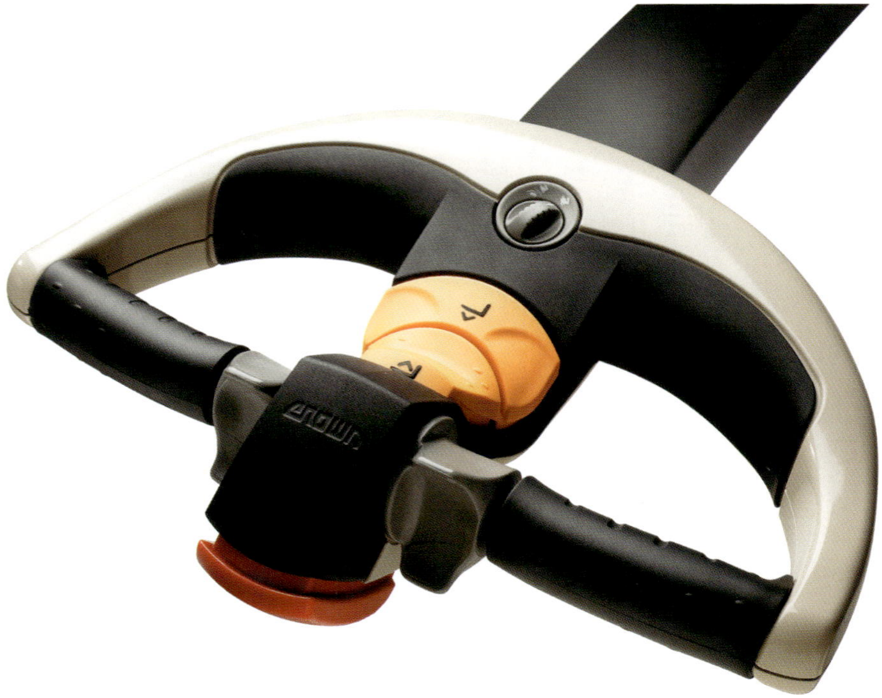

Die Formgebung des X-10 (tm) Deichselgriffs für Gehgabelstaplergeräte erlaubt es dem Benutzer, den Griff fest mit der Hand zu umschließen, während er mit dem Daumenrad den Vorwärts- oder Rückwärtsbetrieb, Auf-und Ab-Bewegungen steuert oder die Hupe betätigt. Die einzelnen Funktionskomponenten des Steuergriffs können ohne ein Verdrehen des Handgelenks einfach gehandhabt werden. Der Knopf für die Hupe ist auf der Unterseite des Handgriffs eingelassen und kann somit ohne Zeitverzögerung schnell gedrückt werden. Die Kombination aus Metall und Kunststoffkomponenten gewährleistet Langlebigkeit und Robustheit.

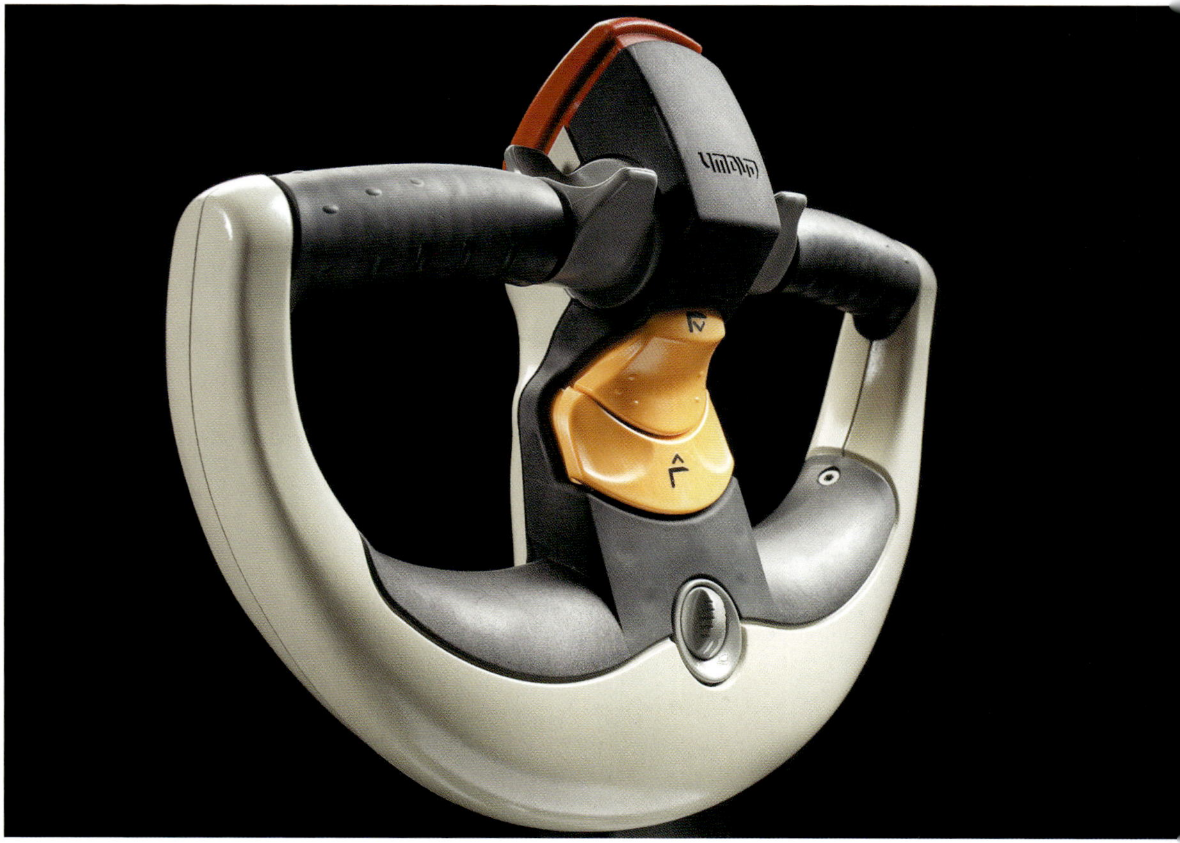

The shape of the X-10 (tm) control handle for pedestrian pallet trucks and stackers lift trucks enables the user to grasp it securely in his hand while controlling the forward/reverse, lift/lower and the horn by means of a thumb wheel. The individual functional com-ponents of the control handle are easy to manage without excessive wrist flexion. The horn buttons are located on the bottom of the handle for instant activation. The combination of metal and plastic com-ponents makes the unit durable and robust.

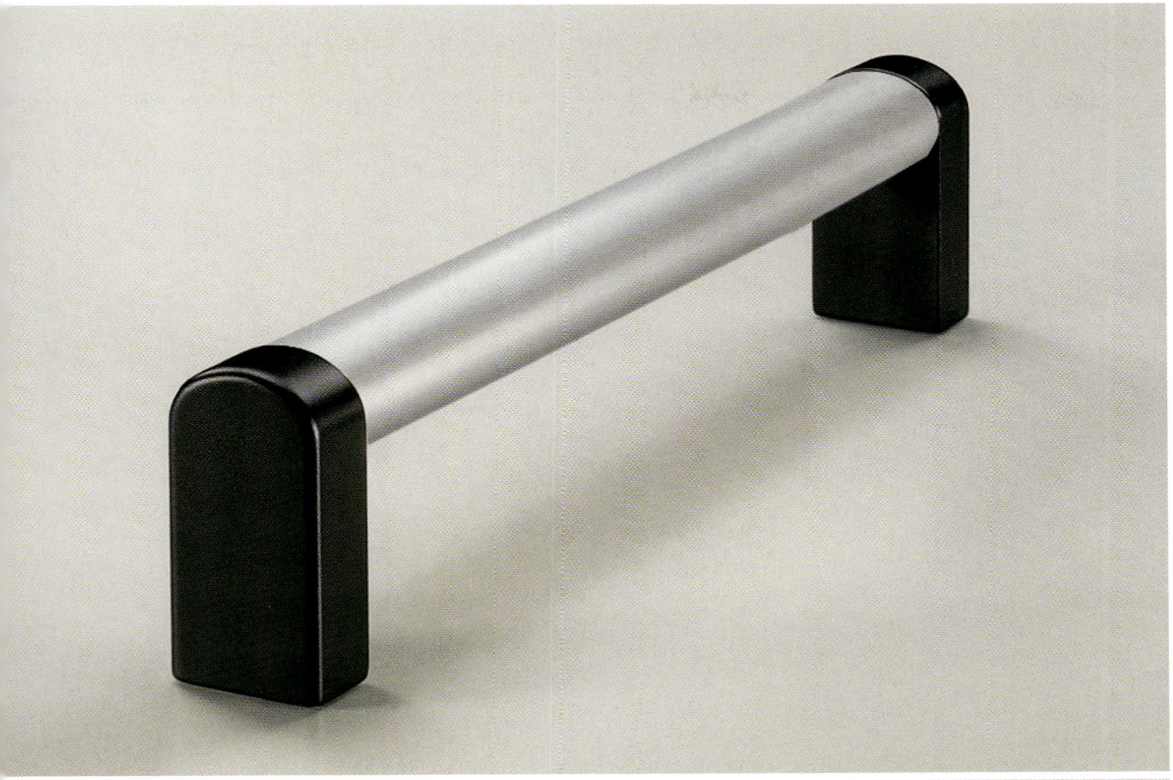

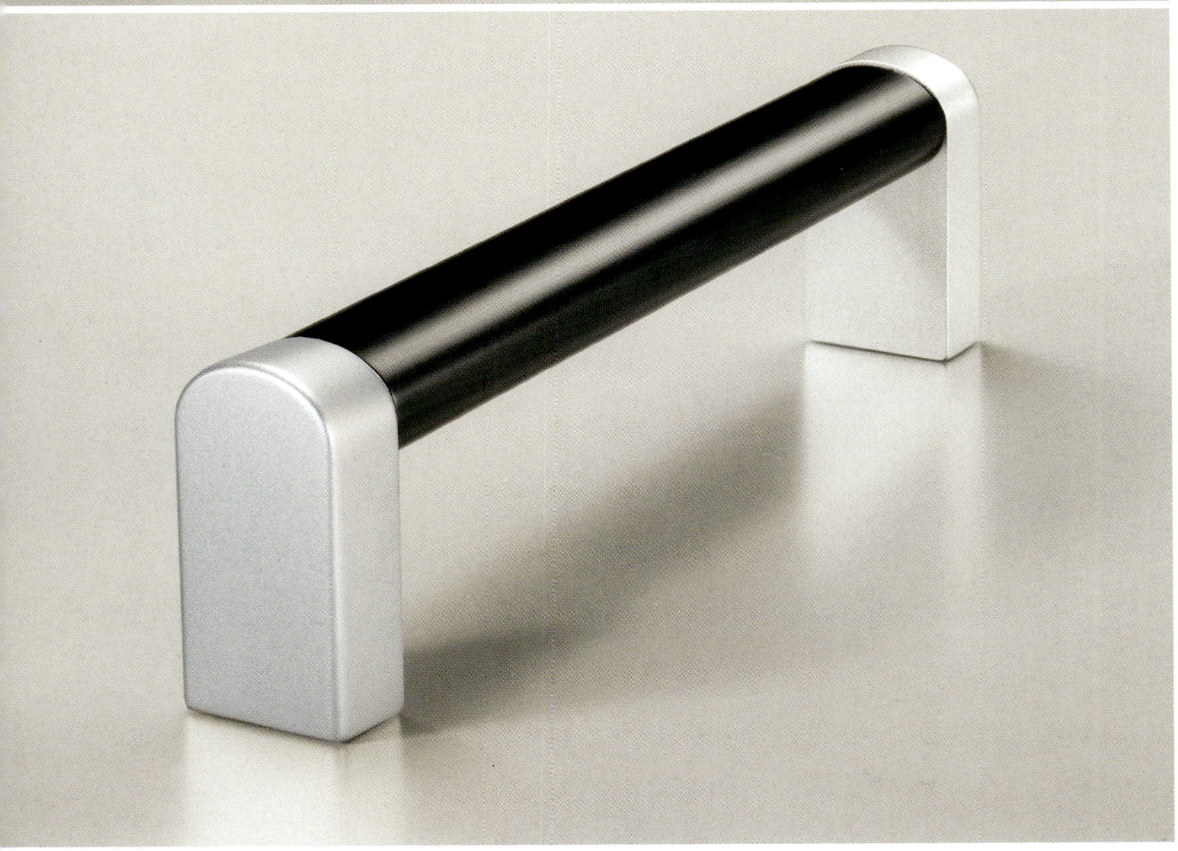

Hersteller Manufacturer
Rohde AG
Nörten-Hardenberg, Germany

Produkt Product
Griffserie A4
Ovalprofilgriffe
Oval section handle

Design
Günter Rohde
Rohde AG
Nörten-Hardenberg, Germany

Die zusammensteckbaren Ovalprofilgriffe A4 sind in beliebigen Grifflängen und Ausführungen in verschiedenen Farbkombinationen lieferbar. Griffschenkel und Verbindungsovalprofil bestehen jeweils aus stranggepresstem, mattglänzend eloxiertem Aluminium, die Spannhülsen aus Edelstahl.

The A4 oval section handles which can be joined together are available in any handle lengths desired and various color combinations. The elongated handle and oval section connectors are made of extruded anodized aluminum with a satin finish. Adapter sleeves are made of stainless steel.

Hersteller Manufacturer
Adolf Würth GmbH & Co. KG
Künzelsau, Germany

Produkt Product
ORSY BULL
Systembox
Multipurpose box system

Design
Design.Mattis
Darmstadt, Germany

Bei ORSY BULL handelt es sich um ein flexibles Boxensystem, das stationär in Werkstätten, Lagern oder allgemein am Arbeitsplatz oder auch als mobile Werkstatt auf Baustellen und in Fahrzeugen eingesetzt werden kann. Die Boxen sind in verschiedenen Ausführungen, die sich in die Grundvarianten Topladerbox, Frontladerbox, Schubladenbox und Kombibox sowie halbhohe Varianten unterteilen, erhältlich. Durch den Koppelmechanismus können die Boxen problemlos gestapelt und fest eingerastet werden. Die Verwendung von Aluminium und Kunststoff sichert hohe Stabilität, Rostfreiheit und extreme Leichtigkeit.

ORSY BULL is a flexible box system suitable both for usage in fixed locations such as workshops and warehouses or generally at the place of work – it also doubles up as a mobile workshop on building sites and in vehicles. The boxes are available in various versions, and feature different basic types: top-loader boxes, front loader boxes, drawer boxes and all-round boxes; there are also versions only half as high. The coupling mechanism enables the boxes to be stacked easily and snapped together firmly. Rustproof aluminum and plastic make for highly stable and very lightweight boxes.

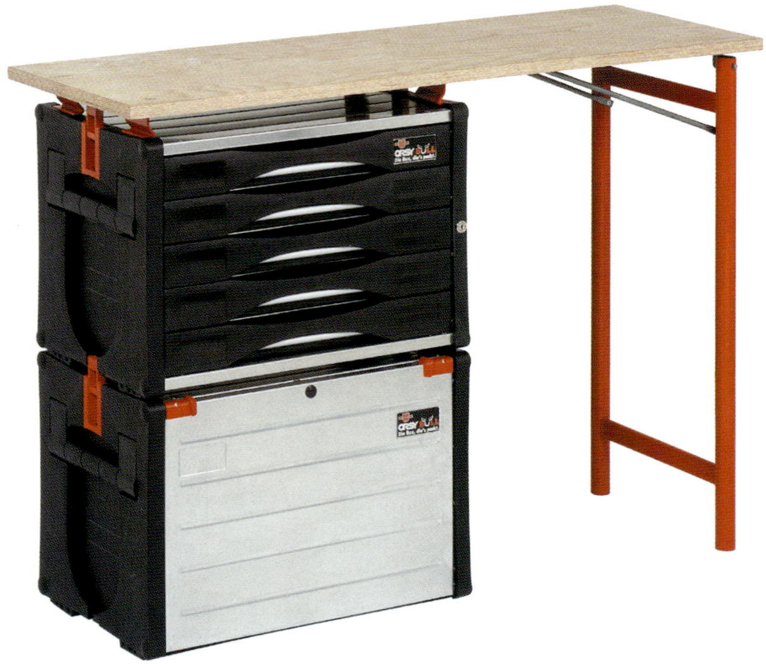

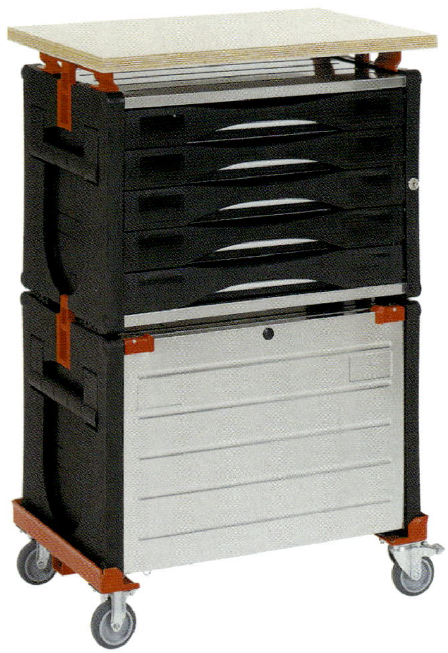

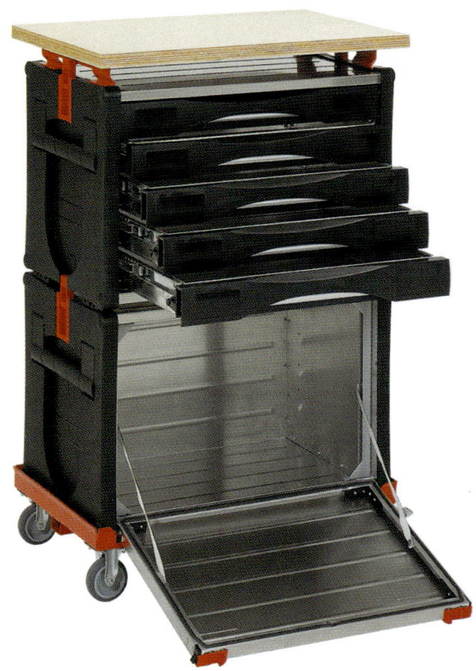

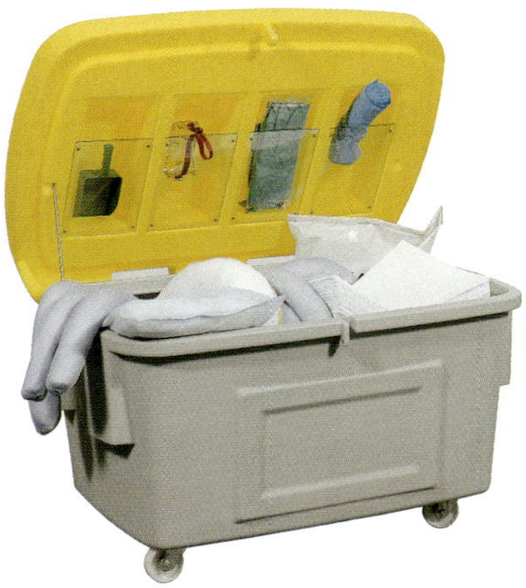

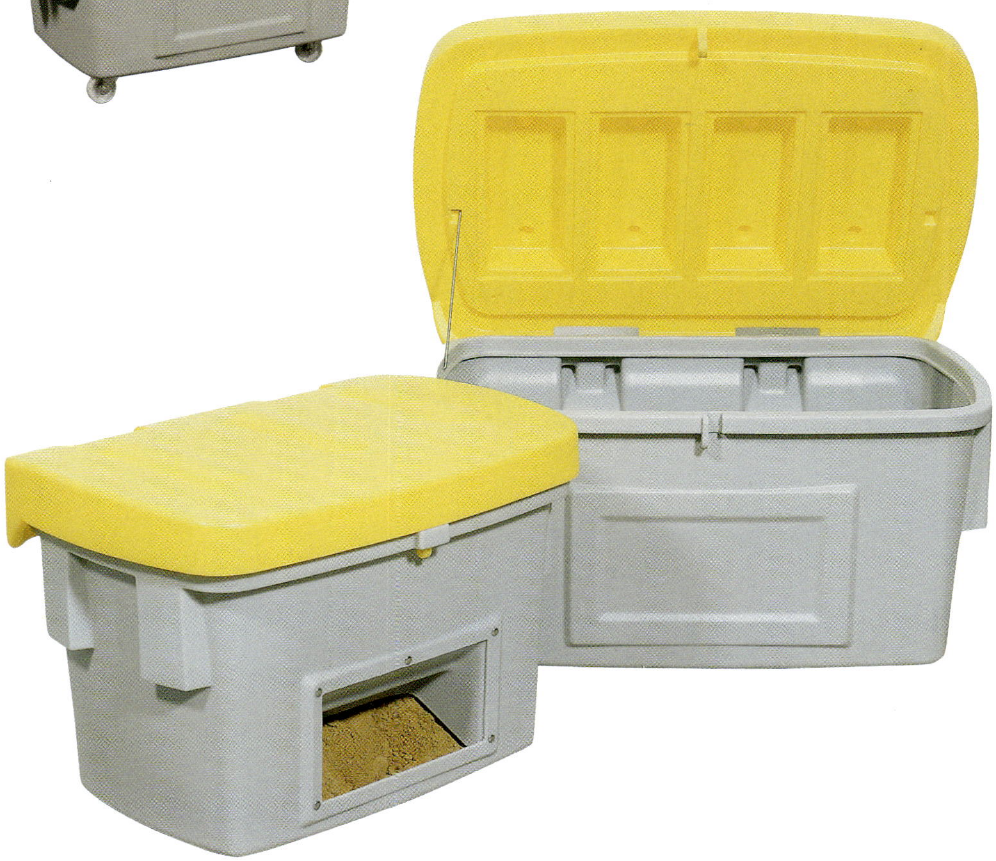

Hersteller Manufacturer
Denios AG
Bad Oeynhausen, Germany

Produkt Product
Multifunktionsbehälter Typ SB
Streugutbehälter
Safety box

Design
Denios AG
Bad Oeynhausen, Germany

Die Multifunktionsbehälter aus der Produktreihe SB eignen sich zur Aufbewahrung und Lagerung von unterschiedlichen Stoffen, sind jedoch vor allem als Streugutbehälter vorgesehen. Sie bestehen aus Polyethylen und besitzen eine nur minimale Anzahl von Anbau- und Zusatzteilen, um eine schnelle Montage und Demontage zu gewährleisten. Das Grundmodell besteht aus nur vier Einzelkomponenten, wobei z.B. das Scharnier in das Gehäuse integriert ist. Alle 5 Typen verfügen über ein einheitliches Erkennungsbild und gemeinsame Designmerkmale. Auch im montierten Zustand sind die Behälter noch stapelbar.

The multi-functional SB product series containers are suitable for storing various materials, but were primarily designed as safety boxes. They are made of polyethylene and feature a minimum of add-on parts and attachments to enable them to be speedily assembled and disassembled. The basic model consists of just four individual components with, for example, the hinge integrated into the housing. All five types possess a uniform appearance and the same design features. The containers are stackable even when assembled.

Hersteller Manufacturer
Gebr. vom Braucke GmbH & Co.KG
Bielefeld, Germany

Produkt Product
RuXXac-cart
Klappbare Transportkarre
Collapsible transport cart

Design
Busse Design GmbH
Elchingen, Germany

Die Transportkarre RuXXac-cart ist nur 5,5 cm dick und passt somit in jede Nische, den Werkzeugschrank oder auch noch in den eigentlich schon vollen Kofferraum. Mit einem Griff werden Schaufel und Räder gleichzeitig ausgeschwenkt. Die Karre besitzt aufgrund der Verbindung aus Spezial-Kunststoffen, Aluminium und Stahlrohr eine Tragkraft von bis zu 125 kg. Mit dem Befestigungssystem aus Spannleiste und Spannband können unterschiedlichste Transportgüter auf der Karre festgezurrt werden.

The RuXXac-cart transport cart is a mere 5.5cm thick, meaning it fits in any niche, tool cabinet and even in a car trunk that is already as good as full. With a single movement, the handle and the carrying tray fold out. The cart features a combination of special plastics, aluminum and tubular steel, enabling it to carry up to 125kg. Thanks to the fastening system of tensioning bar and cords the cart can be used to safely carry a broad variety of different goods.

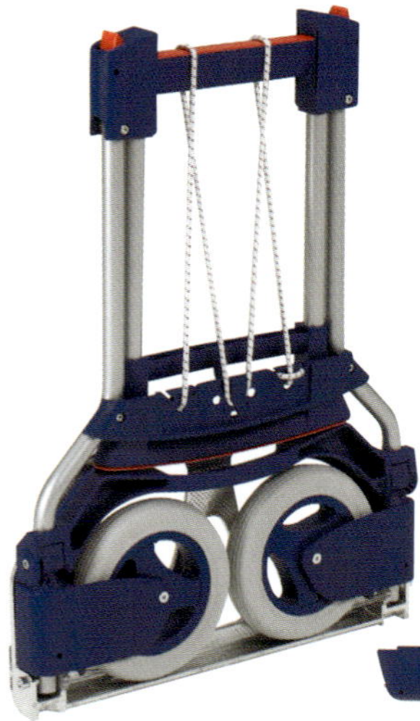

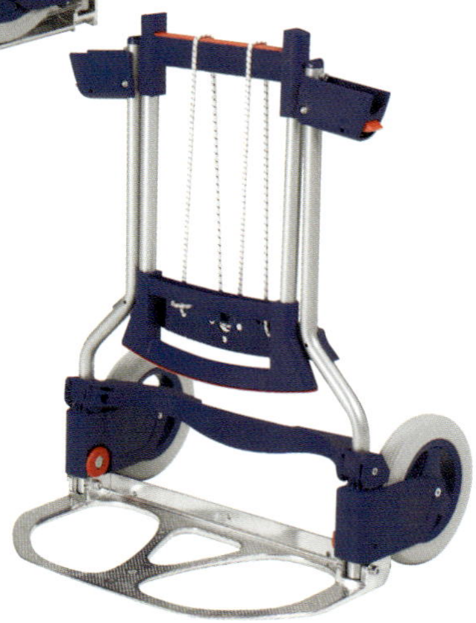

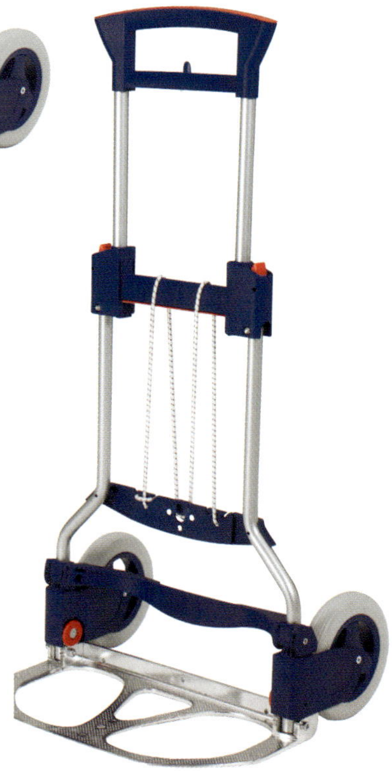

Hersteller Manufacturer
Alto Deutschland GmbH
Bellenberg, Germany

Produkt Product
Neptune 5/7
Heißwasserhochdruckreiniger
Hot water high-pressure cleaner

Design
Weinberg & Ruf Produktgestaltung
Filderstadt, Germany

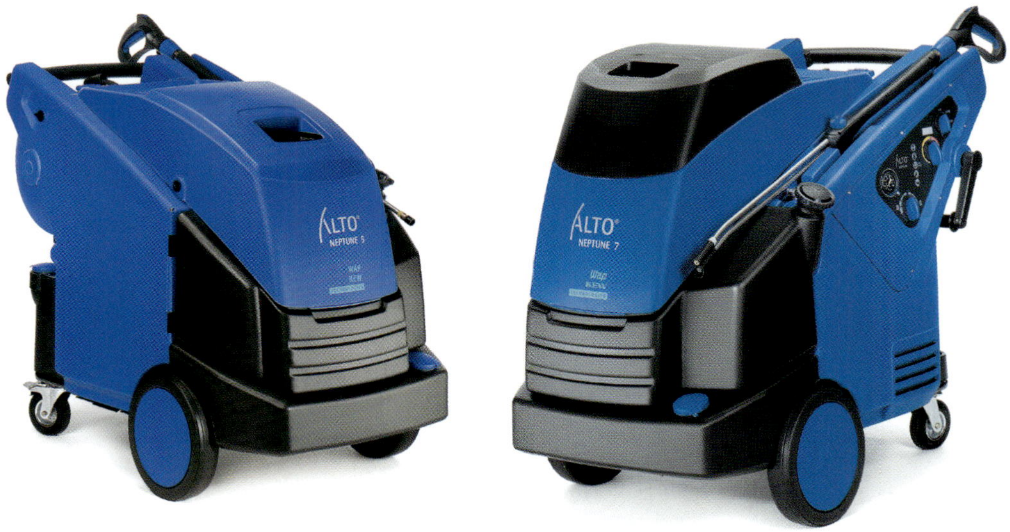

Die Alto Hochdruckreinigerfamilie Neptune 5/7 ist für den professionellen Einsatz konzipiert. Mit der asymmetrischen Gestaltung wird das Gerät in eine Service- und eine Bedienseite gegliedert. Alle Bedienelemente und Anzeigen sind übersichtlich auf der Bedienseite angeordnet. Durch integrierte Schlauch- und Kabeltrommeln, doppelte Lanzenablagen sowie zusätzliche Ablageflächen und ein stabiles Fahrwerk bietet das Gerät größtmöglichen Bedienkomfort. Der Aufbau des Gerätes ist so angelegt, dass der Servicetechniker es mit wenigen Handgriffen öffnen und alle zu wartenden Komponenten leicht erreichen kann.

Alto's Neptune 5/7 high-pressure cleaner is destined for professional applications. The asymmetrical design subdivides the unit into a service and a control side. All the controls and displays are laid out clearly on the control side. The integrated hose and cable drums, dual nozzle holders and additional rests ensure optimal ease of use, as does the stable wheel carriage. The unit is structured such that service personnel can open it with a minimum of effort and all the components that need servicing are immediately accessible.

Hersteller Manufacturer
Alfred Kärcher GmbH & Co. KG
Winnenden, Germany

Produkt Product
HD 10/25s / HD 10/25SX
Unbeheizter Hochdruckreiniger
Cold-water high-pressure cleaner

Design
Werksdesign in Zusammenarbeit mit
TEAMS Design GmbH
Esslingen, Germany

Der unbeheizte Hochdruckreiniger HD 10/25s / HD 10/25SX wurde für den gewerblichen Einsatz konzipiert und ist der stärkste Hochdruckreiniger in seiner Klasse. Dank seiner guten Manövrierfähigkeit hat er einen nur geringen Platzbedarf. Die Easy-Press Handspritzpistole und der Serro-Regler für stufenlose Druck- und Wassermengeneinstellung im Griffbereich gewährleisten ein ermüdungsfreies und benutzerfreundliches Arbeiten. Die integrierte Schlauchtrommel (SX-Ausführung, 20 m Hochdruckschlauch) ist mit einer zum Patent angemeldeten Schlauchführung ausgestattet.

The HD 10/25s / HD 10/25SX cold-water high-pressure cleaner is destined for commercial use and is the most powerful high-pressure cleaner in its category. Thanks to its good maneuverability and small footprint it takes up only little space. The easy-press trigger gun and the servo-press controller spell smooth adjustment of pressure and water volume from the handle for user-friendly, convenient operation. The integrated hose drum (SX version, 20m of high-pressure hose) comes equipped with a hose routing mechanism (patent pending).

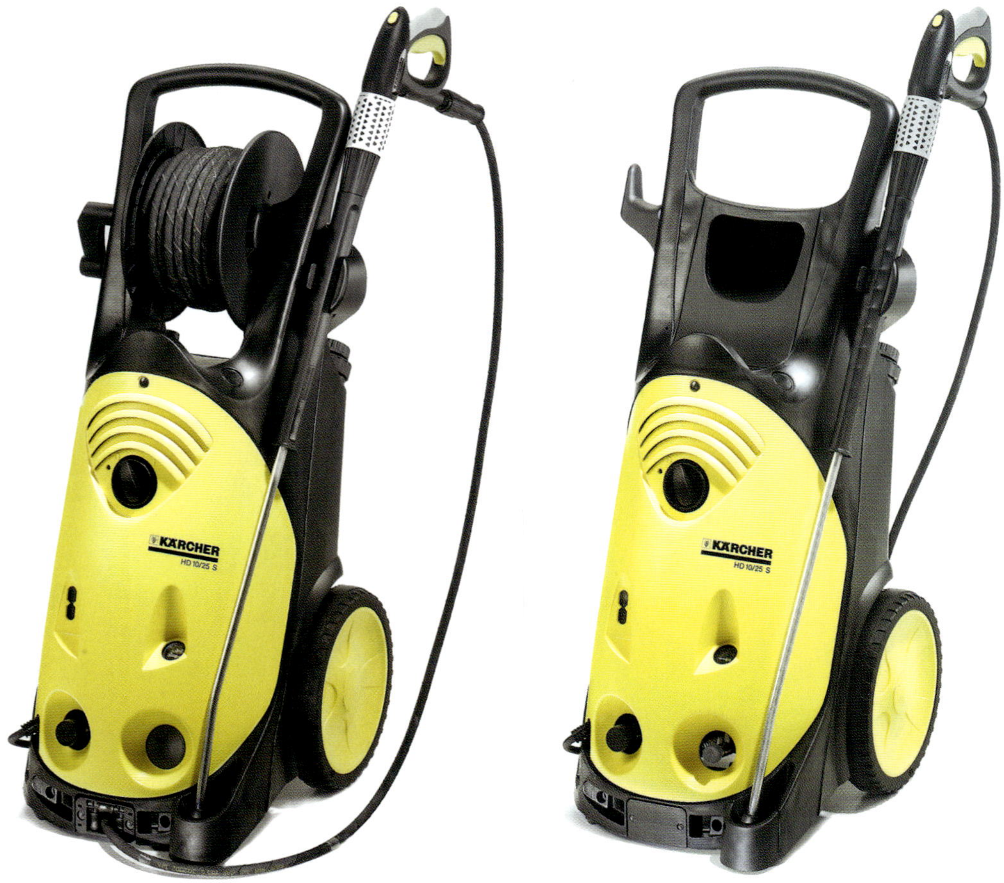

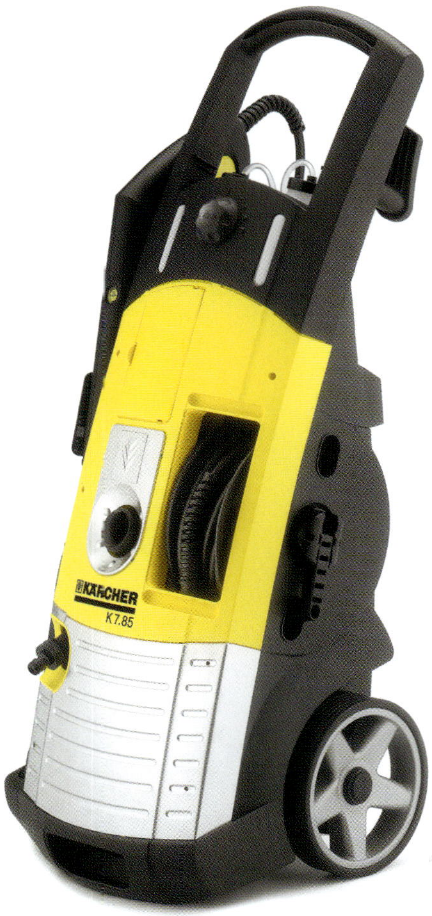

Hersteller Manufacturer
Alfred Kärcher GmbH & Co. KG
Winnenden, Germany

Produkt Product
Kärcher 7.85M
Hochdruckreiniger
High-pressure washer

Design
Werksdesign in Zusammenarbeit mit
TEAMS Design GmbH
Esslingen, Germany

Der Hochdruckreiniger Kärcher 7.85M wurde für die Nutzung im privaten Umfeld entwickelt. Das fahrbare Gerät erzielt einen maximalen Druck zwischen 125 und 150 bar. Das neue Upright-Gestaltungskonzept gewährleistet eine komfortable Bedienung, dank der großen Laufräder kann das Gerät leicht auch über Stufen und Treppen transportiert werden. Das Zubehör kann Platz sparend an der seitlichen Halterung sowie in den hinteren Einschubköchern untergebracht werden. Insgesamt wurde bei der Gestaltung darauf geachtet, dass ein geschlossenes Erscheinungsbild erreicht wird, in dem sämtliche Features in das Gehäuse integriert wurden und keine Bauteile abstehen.

The Kärcher 7.85M high-pressure washer was developed for use in private households. The mobile unit provides maximum pressure in the 125-150 bar range. The new upright design concept guarantees comfortable controls and thanks to the large wheels the washer is easy to transport up and down steps and stairs. The accessories can be placed in the side holder, the front flap or the footstep at the rear. The overall appearance is that of a single compact unit into which all the features blend and where nothing sticks out.

Hersteller Manufacturer
Alfred Kärcher GmbH & Co. KG
Winnenden, Germany

Produkt Product
RoboCleaner RC 3000
Reinigungssystem
Cleaning system

Design
Werksdesign in Zusammenarbeit mit
TEAMS Design GmbH
Esslingen, Germany

Der RoboCleaner RC 3000 ist ein automatischer Staubsauger für die Bodenreinigung im Privathaushalt. Er reinigt bis zu 15 m² in der Stunde. Wenn die Staubkammer voll ist oder die Akkus allmählich schwächer werden, bewegt sich der Roboter zu seiner Basisstation. Danach setzt er seine Arbeit selbstständig fort. Durch seine kompakte Baugröße ist auch eine Reinigung an schwer erreichbaren Stellen möglich, eine Absturzsensorik verhindert zudem das Abstürzen von Absätzen oder Treppen. Seine kreisrunde Form ermöglicht das Drehen auf der Stelle. Eine Stoßkante aus einem Gummi-Hohlprofil schützt Möbel und Einrichtungsgegenstände.

The RoboCleaner RC 3000 is an automatic vacuum cleaner for floors in private homes, cleaning up to 15 sq. m. per hour. When the dust container is full or the batteries get low, the unit independently scurries to its base station to recharge before getting back to work. Thanks to its compact size it can also clean inaccessible areas, and its infrared sensors prevent it falling down stairs or steps. By virtue of its circular shape it can turn on the spot, and a hollow rubber buffer strip protects furniture and other items in interiors.

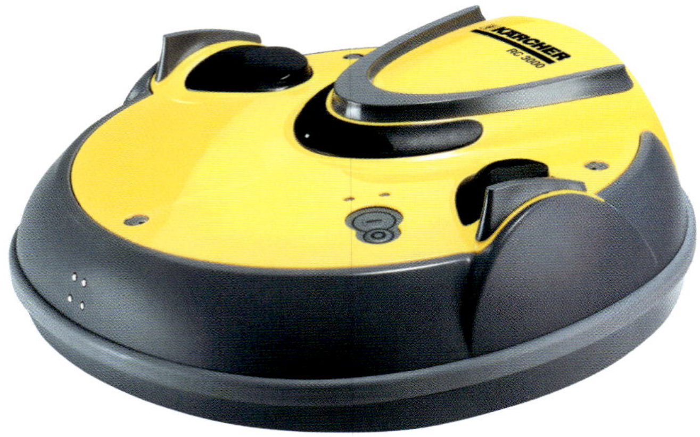

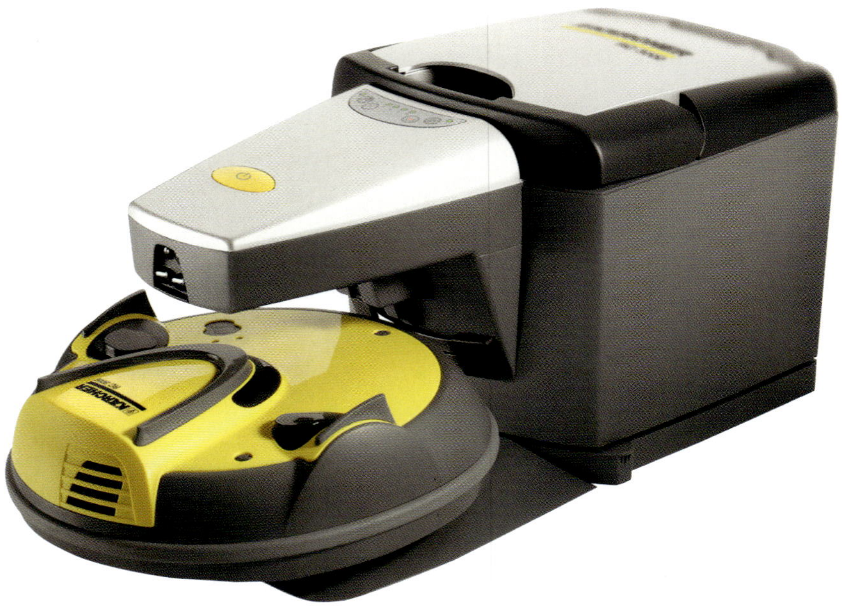

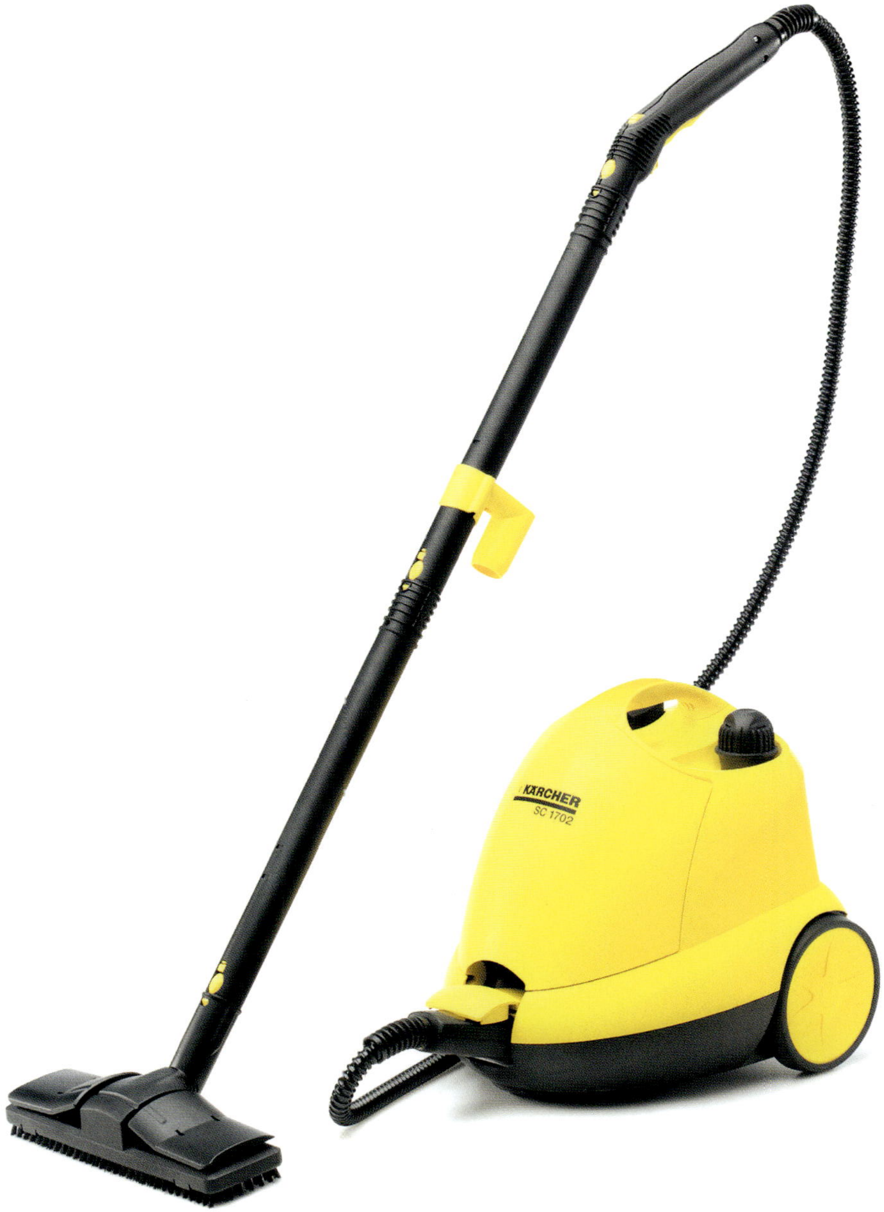

Hersteller Manufacturer
Alfred Kärcher GmbH & Co. KG
Winnenden, Germany

Produkt Product
Kärcher SC 1702
Dampfreiniger
Steam cleaner

Design
Werksdesign in Zusammenarbeit mit
TEAMS Design GmbH
Esslingen, Germany

Der Kärcher SC 1702 ist ein Dampfreiniger. Durch die Upright-Form mit mittigem Tragegriff befinden sich die Bedienelemente näher am Benutzer und dieser muss sich nicht tief bücken, um das Gerät zu bedienen. Die großen Räder mit weicher Lauffläche gewährleisten ein gedämpftes Abrollen auf harten Flächen. Die runde Form erleichtert das Überfahren von Schwellen, ein umlaufender Bumper schützt das Interieur. In einem Zubehörfach werden Düsen und Bürsten aufbewahrt, die Dampfrohre kann man Platz sparend in zwei Köchern am Gerät unterbringen.

The Kärcher SC 1702 is a steam cleaner. Thanks to its upright shape with its central carrying handle, the controls are closer to hand and users do not have to bend right down to operate the unit. The large soft-tread wheels guarantee smooth running on hard surfaces. The round shape mean the unit runs more smoothly over thresholds and an all-round bumper protects the interior. The nozzles and brushes tuck away in the accessories compartment beneath a hinged flap, while the steam tubes are carried in the two holsters on the unit.

Hersteller Manufacturer
Autostadt GmbH
Claus Hohmann, Chief Technology Officer
Wolfsburg, Germany

Produkt Product
CleanMobil

Design
Hermann Rokitta
Rokitta
Produkt & Markenästhetik
Maria Meyring
Angelika A. Vogt
Corporate Styling
Hamburg, Germany

Alle Funktionsbereiche des CleanMobils sind klar segmentiert, transluzent verschlossen und leicht zugänglich. Variable Bestückungsmodule, leicht gängige und selbstjustierende Teleskopauszüge, eine Basis aus Edelstahl mit austauschbaren Elastomer Griff- und Schutzelementen und Rollflächen gewährleisten Langlebigkeit und leichte Handhabung. Ein rundum laufender Stoßschutz und ein in allen Positionen erreichbares Griffrohr ermöglichen flexibles Arbeiten auf engstem Raum.

All functional areas of the CleanMobil are clearly segmented, translucent, sealed and easy to access. It is durable and easy to handle, and features variable component modules, smooth-functioning and self-adjusting telescopic pull-out elements, a stainless steel base with interchangeable elastomer handle- and protection elements not to mention concealing shutters. A wraparound bumper and a tubular handle which can be accessed in all positions make for flexible use in cramped conditions.

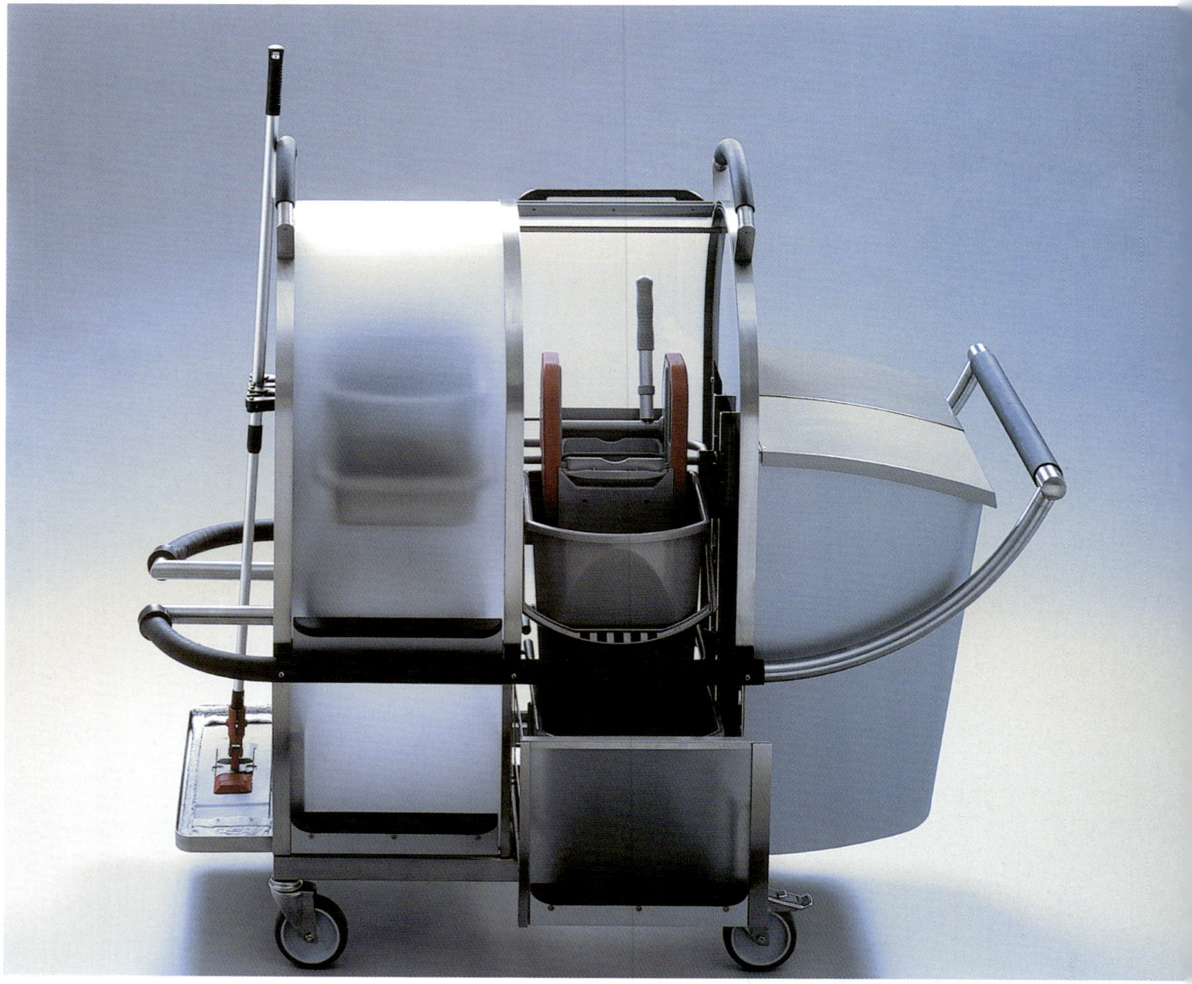

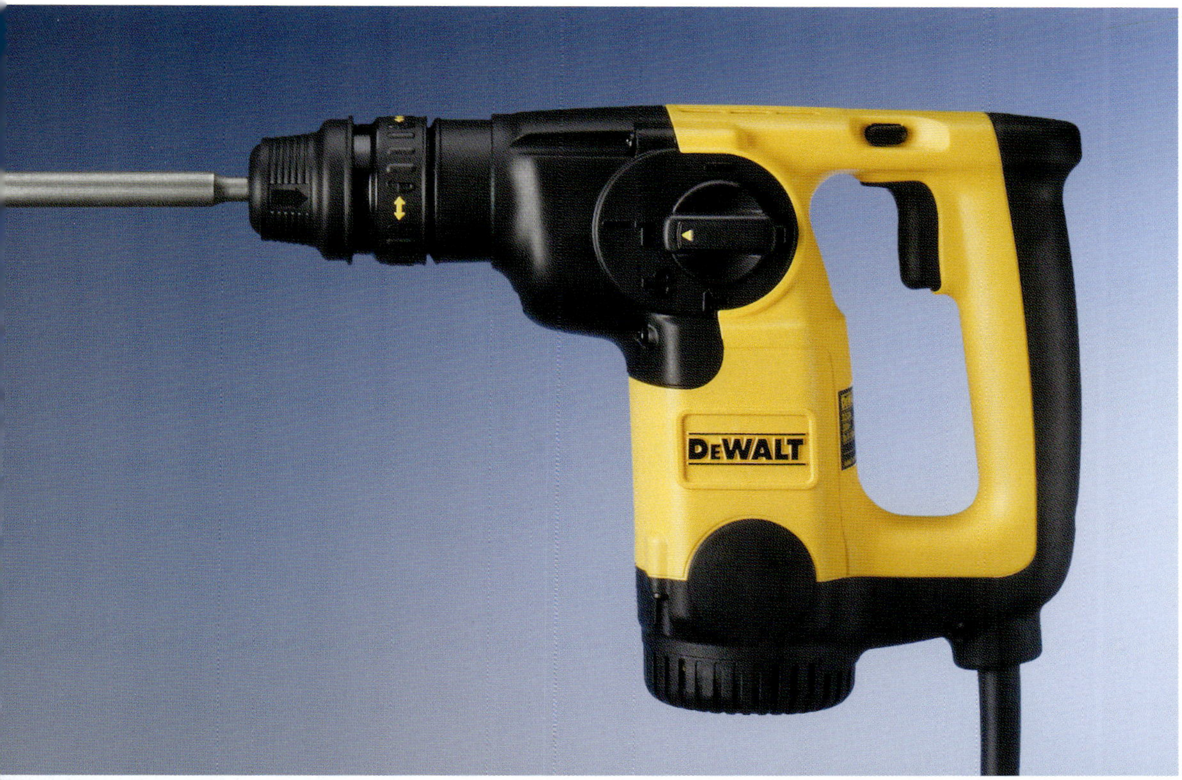

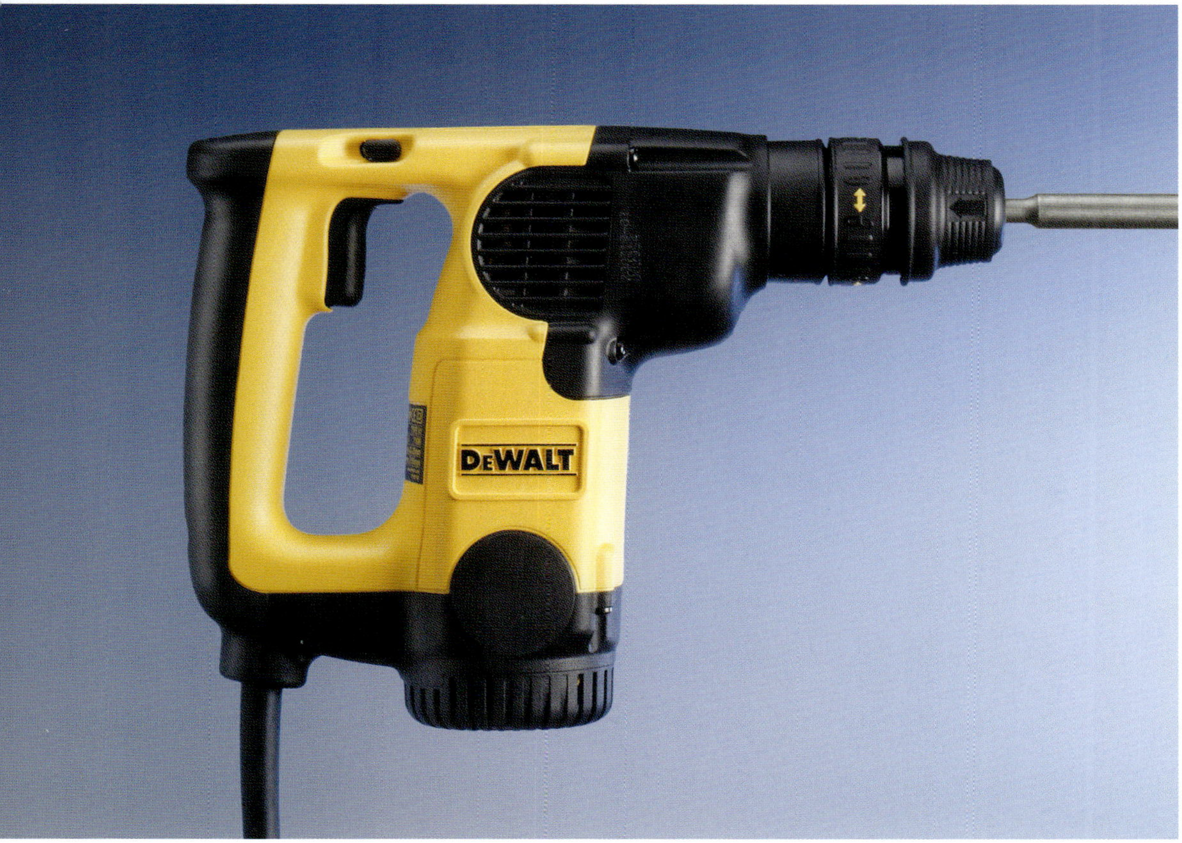

Hersteller Manufacturer
Black & Decker GmbH
Idstein, Germany

Produkt Product
DE WALT D 25304K
Hochleistungs-Kombihammer
High-performance combination hammer

Design
Michael Stirm
Black & Decker GmbH
Idstein, Germany

Der DE WALT D 25304K ist ein Hochleistungs-Kombihammer der 2 kg Klasse mit einer Leistung von 710 Watt. Er wird hauptsächlich zum Bohren in Beton und Mauerwerk bis 26 mm Bohrerdurchmesser und für leichte Meißelarbeiten benutzt. Der Hammer besitzt das mit 22 mm Durchmesser größte Schlagwerk seiner Klasse und erzielt im Bohrmodus eine Einzelschlagenergie von 2,8 Joule, im reinen Hammermodus sogar 3,1 Joule. Die kompakte Baulänge von 308 mm macht den Hammer besonders komfortabel und bedienerfreundlich. Der große gummierte Endhandgriff reduziert zusätzlich Vibrationen. Der Umschalter für den Wechsel der Drehrichtung wurde zur Verbesserung der Greifsituation vom Schalterbereich weg in den darüber liegenden Gehäusebereich verlegt. Durch die Verwendung von Magnesium für Getriebe- und Hammerwerkgehäuse ist das Gesamtgewicht des Kombihammers gering und erleichtert längere Arbeitseinsätze.

The DE WALT D 25304K is a high-performance combination hammer in the 2 kg class delivering 710 Watts in output. It is primarily destined for drilling concrete and brickwork up to 26 mm in drill hole diameter and for light chiseling work. The 22 mm diameter offers the largest hammer mechanism in its class and produces 2.8 joules of single impact energy in the boring mode and 3.1 joules in the pure hammer mode. The compact unit is only 308 mm long, so the hammer is especially easy to wield. The large rubberized handle helps reduce vibrations. The switch for changing drill rotation direction has been relocated away from the trigger area to the top of the main housing for enhanced ergonomics. The magnesium transmission and hammer housings reduce the combination hammer's weight emphatically for longer non-fatiguing use.

Hersteller Manufacturer
Festool GmbH
Wendlingen, Germany

Produkt Product
OF 1400 EBQ
Oberfräse
Routing machine

Design
Studiowerk Design
Inning, Germany

Die OF 1400 EBQ ist eine Oberfräse, die trotz ihrer hohen Leistung mit dem für Festool typischen Stielgriff im Einhandbetrieb gut zu handhaben ist und in allen möglichen Positionen gut geführt werden kann. Das Elektronikstellrad wurde so platziert, dass es aus der Griffhaltung heraus verstellt werden kann. Mit einem schnellverstellbaren Festanschlag können drei Frästiefen vorgewählt werden. Die MMC-Elektronik mit konstanter und geregelter Drehzahl für Sanftanlauf und Temperaturkontrolle schützt Motor und Lager und sorgt für lange Lebensdauer.

The OF 1400 EBQ is a routing machine that, despite its high performance, is easy to operate with one hand thanks to the typical Festool stemmed grip; it can be used effectively in any position. The electronic dial has been positioned such that it can be adjusted by the hand using the grip. The swiftly adjusted positive stop enables three different pre-set milling depths. The MMC electronics delivers steady, controlled speed, a soft start and temperature controls, easing strain on the motor and bearings, enhancing durability.

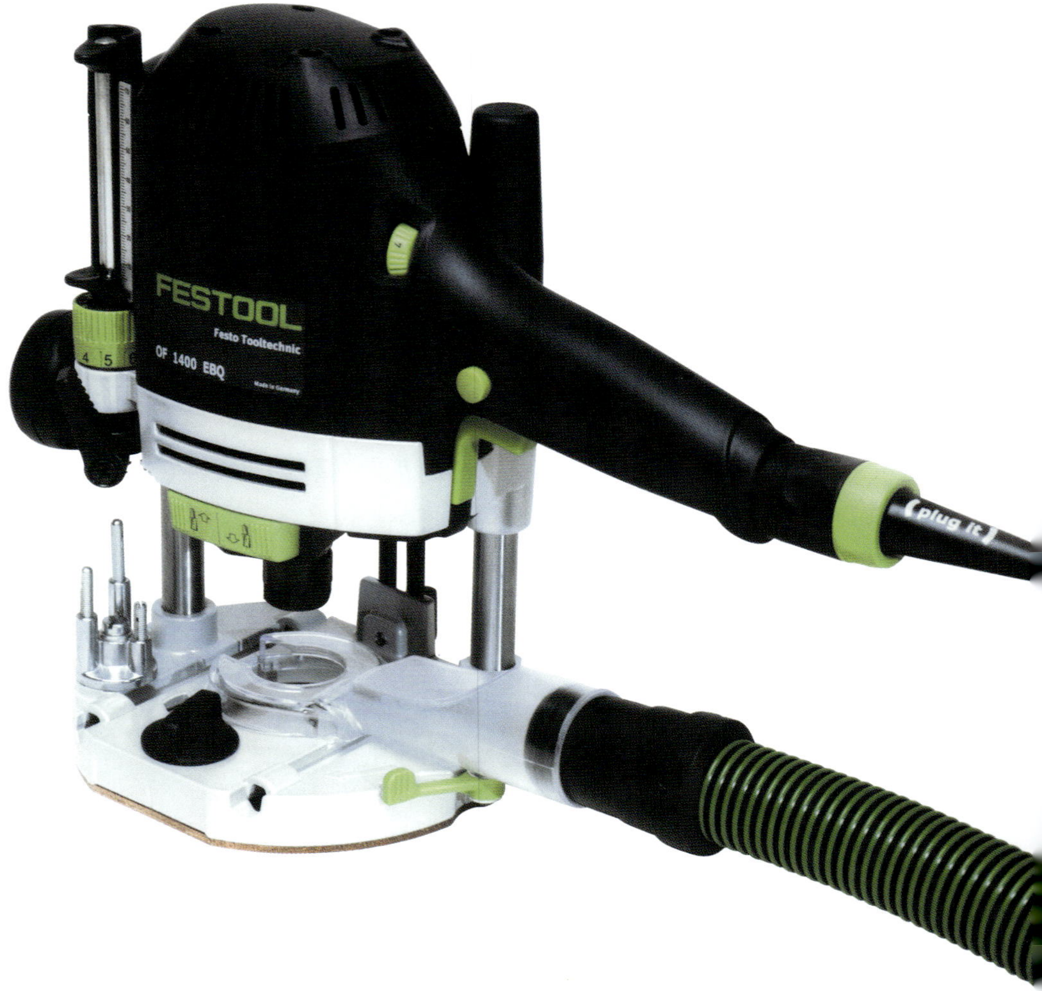

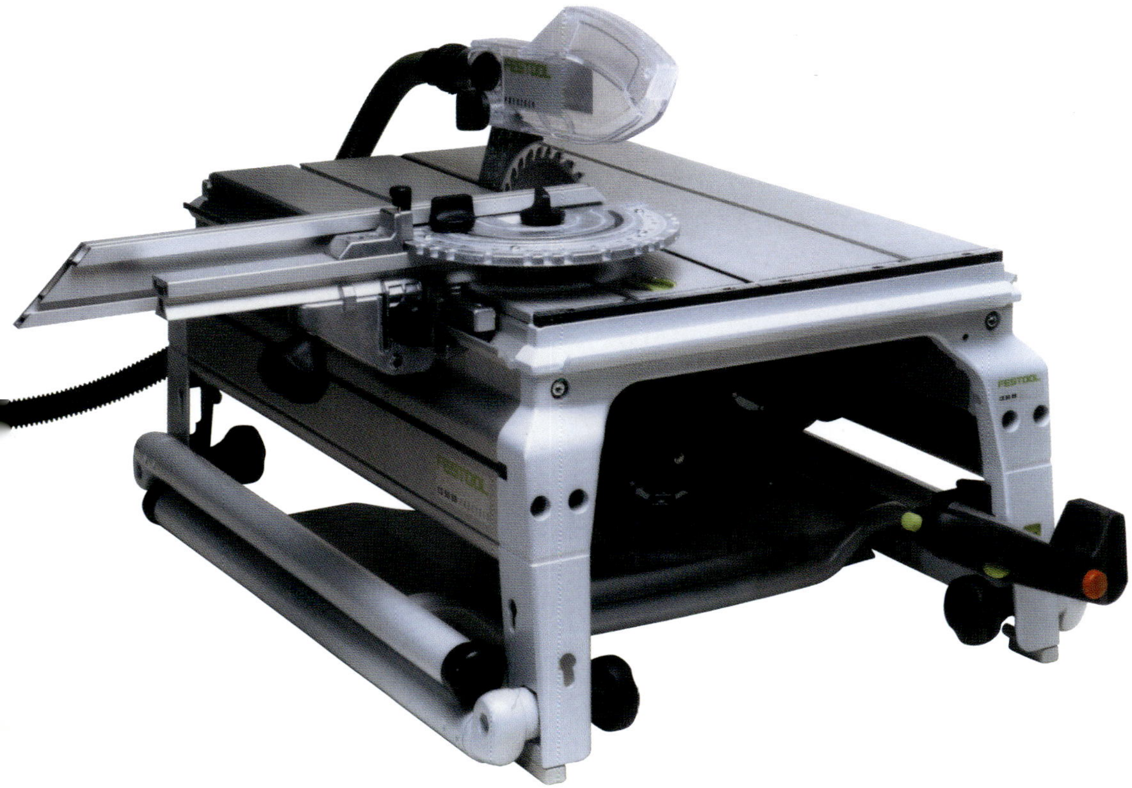

Hersteller Manufacturer
Festool GmbH
Wendlingen, Germany

Produkt Product
CS-50 EB Precisio
Untertisch Montagesäge
Undertable trim saw

Design
Studiowerk Design
Inning, Germany

Die CS 50 EB ist eine transportable, sehr leichte präzise Montagesäge, die nach dem Prinzip der Unterflurzugsäge arbeitet. Die zwei Portale mit den dazwischen liegenden Alu-Strangteilen bestimmen das stabile und kompakte Design. Im Bereich dieser Teile lässt sich die Säge gut tragen. Die außen angesetzten Beine ermöglichen die gute Standfestigkeit der Säge. Der Multifunktions-Zughebel gewährleistet eine schnelle und einfache Bedienung der fünf Sägefunktionen.

The CS 50 EB is a portable, very lightweight and precise trim saw based on the undertable principle. The two portals with their aluminum connecting tubes are the central elements making for a stable and compact design. Thanks to these elements, the saw is also easy to carry. The outward-angled legs provide a highly stable base, and the multifunction operating lever guarantees swift and easy operation of the five saw functions.

Hersteller Manufacturer
Festool GmbH
Wendlingen, Germany

Produkt Product
TDK 12 und TDK 15,6
Akku-Bohrschrauber
Cordless drill/screwdriver

Design
Studiowerk Design
Inning, Germany

TDK 12 und TDK 15,6 sind besonders kurze und handliche Akku-Bohrschrauber. Durch die ergonomische Gestaltung und Positionierung des Griffes zwischen Motor und Akku liegt der Schrauber ausgewogen in der Hand. Der schlanke Griff ist mit seinem Ein- und Ausschalter und dem Rechts- und Linksschalter optional auf die Handanatomie ausgelegt. Die neu entwickelte Drehmomenten-Einstellung ist übersichtlich und gut zu bedienen. Sie hat eine patentierte Drehmomentabschaltung. Zusätzlich kann das Futter durch das neue Werkzeugschnellwechselsystem CENTROTEC ersetzt werden.

The TDK 12 and TDK 15.6 are especially short and manageable cordless drills/screwdrivers. By virtue of the ergonomic design and positioning of the grip between the motor and the battery, the appliance lies well-balanced in the hand and the slender grip with its integrated on/off and clockwise/anti-clockwise switches (right or left optionally) is designed to fit the hand snugly. The newly developed torque adjustment is clear and easy to use, featuring a patented torque cut-off.

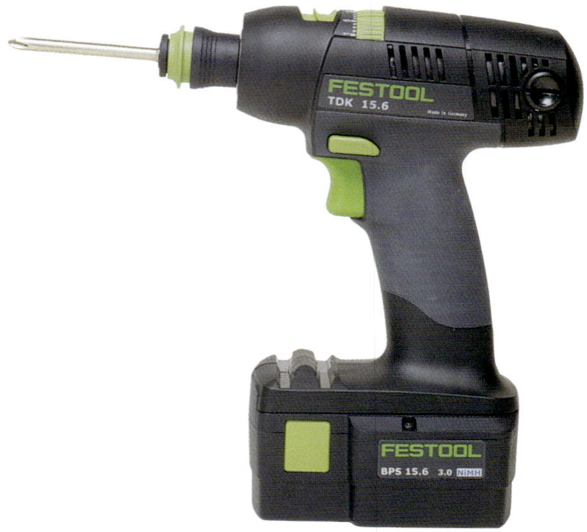

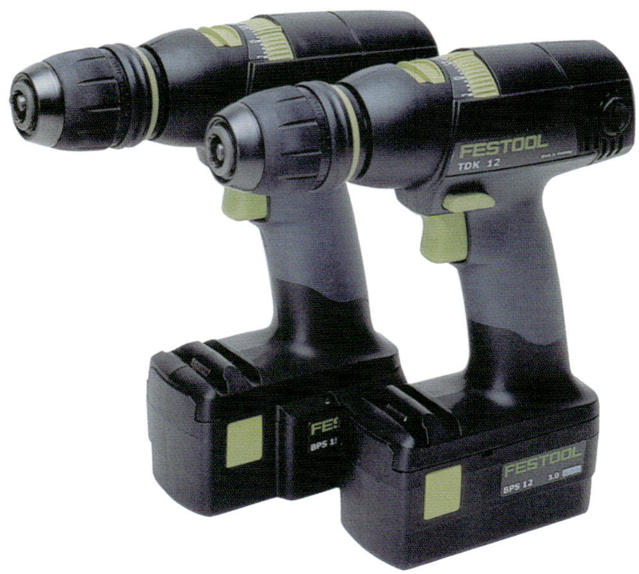

Hersteller Manufacturer
Hitachi Koki Co., Ltd.
Hitachinaka-City, Ibaraki, Japan

Produkt Product
DH 40 MR
Bohr- und Meißelhammer
Rotary hammer

Design
Masato Sakai
Design Center
Hitachinaka-City, Ibaraki, Japan

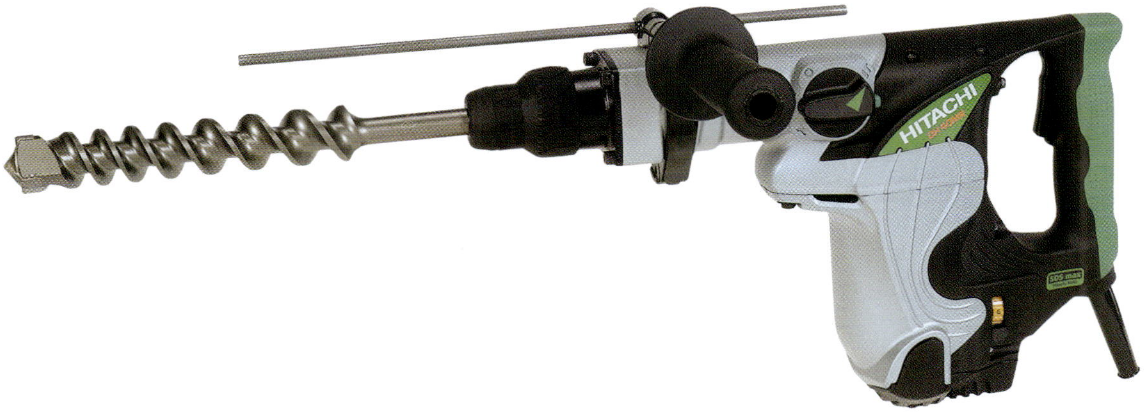

Der DH 40 MR Bohr- und Meißelhammer wurde aufgrund der Schwere der Arbeiten, die mit ihm ausgeführt werden, vor allem unter ergonomischen Gesichtspunkten konzipiert, um eine Belastung des Nutzers durch Vibrationen und Rückstoß möglichst gering zu halten. Insgesamt wurde ein kompaktes Design gewählt, um das Gewicht des Hammers zu verringern und ihn auch optisch weniger schwer wirken zu lassen. Die Formgebung des Hauptgriffs bildet die ihn umfassende Hand nach, zudem wurde für beide Griffe ein weicheres Material gewählt.

The DH 40 MR rotary hammer was developed with ergonomic features in mind given the heavy nature of the work it will be used for and in order to keep vibration and recoil as low as possible. All in all a compact design was chosen to limit the weight of the hammer while giving it a less heavy-looking appearance. The main handle is designed around the shape of the hand which will hold it, with soft material selected for both grips.

Hersteller Manufacturer
Hitachi Koki Co., Ltd.
Hitachinaka-City, Ibaraki, Japan

Produkt Product
WH 12DM2
Akku-Schlagschrauber
Cordless impact driver wrench

Design
Masato Sakai
Design Center
Hitachinaka-City, Ibaraki, Japan

Trotz seiner Kompaktheit besitzt der WH 12DM2 Akku-Schlagschrauber eine außerordentlich hohe Motor-leistung, die unter anderem einer speziellen Lüftungstechnik zu verdanken ist. Das Gerät wurde unter Berücksichtigung der entsprechenden Arbeitsumgebung mit ihren Anforderungen an den Nutzer vor allem unter ergonomischen Gesichtspunkten und Bedienfreundlichkeit konzipiert. Mittels einer speziellen Vorrichtung kann der Schlagschrauber schnell am Gürtel des Nutzers befestigt werden, wenn dieser z.B. in der Höhe arbeitet. Ein LED-Licht erleichtert das Arbeiten an schlecht beleuchteten Orten oder in der Nacht. Das weiche Oberflächenmaterial wurde hier nicht auf den Griff begrenzt, sondern überzieht das Gerät an allen dafür erforderlichen Stellen.

Despite its compact size, the WH 12DM2 cordless impact driver wrench has an extremely high-powered motor thanks to a special ventilation technique, among other things. The design of the wrench focuses on ergonomic aspects and user friendliness given the work environment in which it is likely to be used and the demands on the user. By means of a special hook, the impact driver wrench can be speedily attached to the worker's belt should he, for example, be working at a height. An LED display makes it easier to work in badly lit places or at night. The soft surface material is not limited to the handle but covers the device in all the required areas.

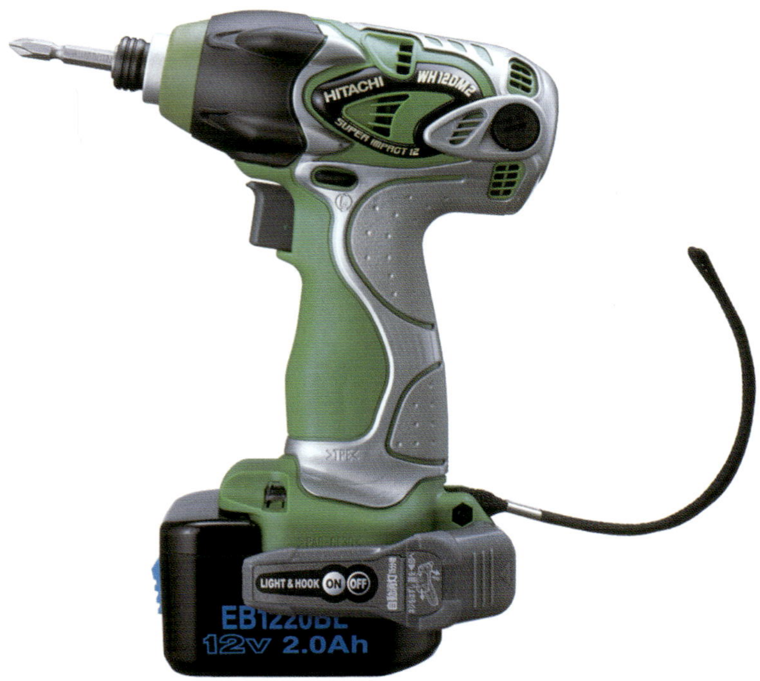

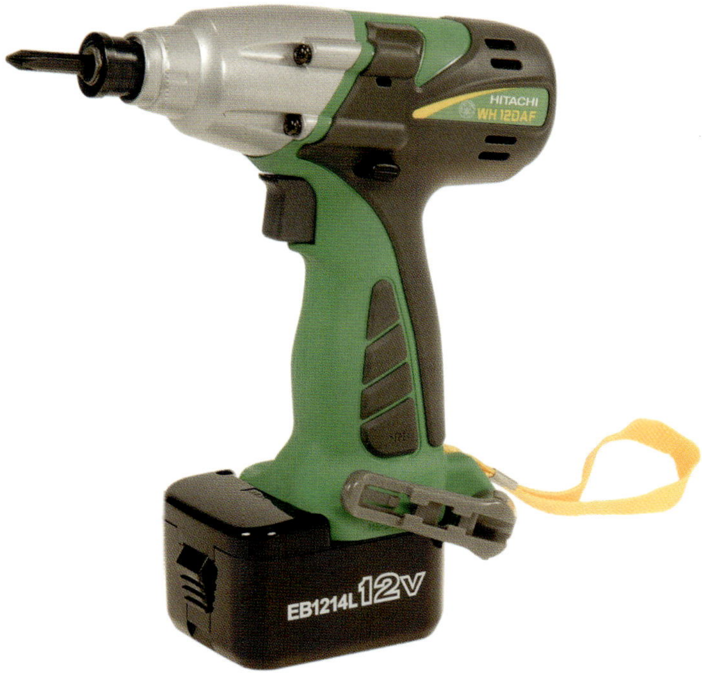

Hersteller Manufacturer
Hitachi Koki Co., Ltd.
Hitachinaka-City, Ibaraki, Japan

Produkt Product
WH 12DAF
Akku-Schlagschrauber
Cordless impact driver

Design
Nobuhisa Sugawara
Design Center
Hitachinaka-City, Ibaraki, Japan

Der WH 12DAF Akku-Schlagschrauber verfügt über einen außerordentlich starken Anzugsdrehmoment, wobei die Belastung des Benutzers durch Rückstöße jedoch sehr gering bleibt. Schwerpunkt der Designentwicklung waren geringes Gewicht, leichte Handhabung und Konzentration auf die wesentlichen funktionalen Komponenten. Das Gehäuse und der Griff des Schlagschraubers sind größtenteils mit einem weichen Elastomer verkleidet, um einen angenehmen und sicheren Zugriff zu gewährleisten und Oberflächen zu schützen, auf denen das Gerät in Arbeitspausen abgelegt wird.

The WH 12DAF cordless impact driver has an extremely strong tightening torque while nevertheless limiting recoil on the user to a very low level. The design focuses on being lightweight and easy to manage while emphasizing the essential functional components. The housing and the handle of the impact driver are, for the most part, covered with a soft elastomer so that it can be comfortably and securely held while offering protection to any surfaces it may be placed on when not in use.

Hersteller Manufacturer
Robert Bosch GmbH
Leinfelden-Echterdingen, Germany

Produkt Product
Bosch IXO
Akku-Kompakt-Schrauber
Cordless compact driver

Design
TEAMS Design
Esslingen, Germany

Der Bosch IXO Akku-Kompakt-Schrauber verfügt über einen Lithium-Ionen-Akku, der gegenüber konventionellen Akkus den Vorteil besitzt, dass er keinen Memory-Effekt hat, sich nicht von selbst entlädt und zudem 30 bis 40 Prozent leichter ist, was das Gerät mit einem Gewicht von 300 g insgesamt sehr leicht macht. Durch seine Pistolenform, den Softgrip und den großen Schalter liegt der IXO sehr gut in der Hand. Die Ladeschale dient gleichzeitig als Halter.

The Bosch IXO cordless compact driver boasts a lithium-ion battery, which, compared with conventional batteries, has the advantage of not having a memory-effect and thus does not discharge spontaneously, while also being 30-40 percent lighter. The result is a drill that is a very lightweight 300g. Thanks to the pistol shape, softgrip and large switch, the IXO fits the hand snugly. The charging case doubles up as the holder.

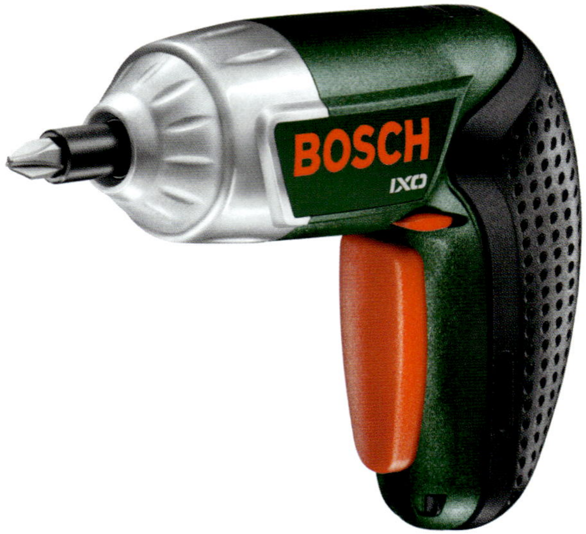

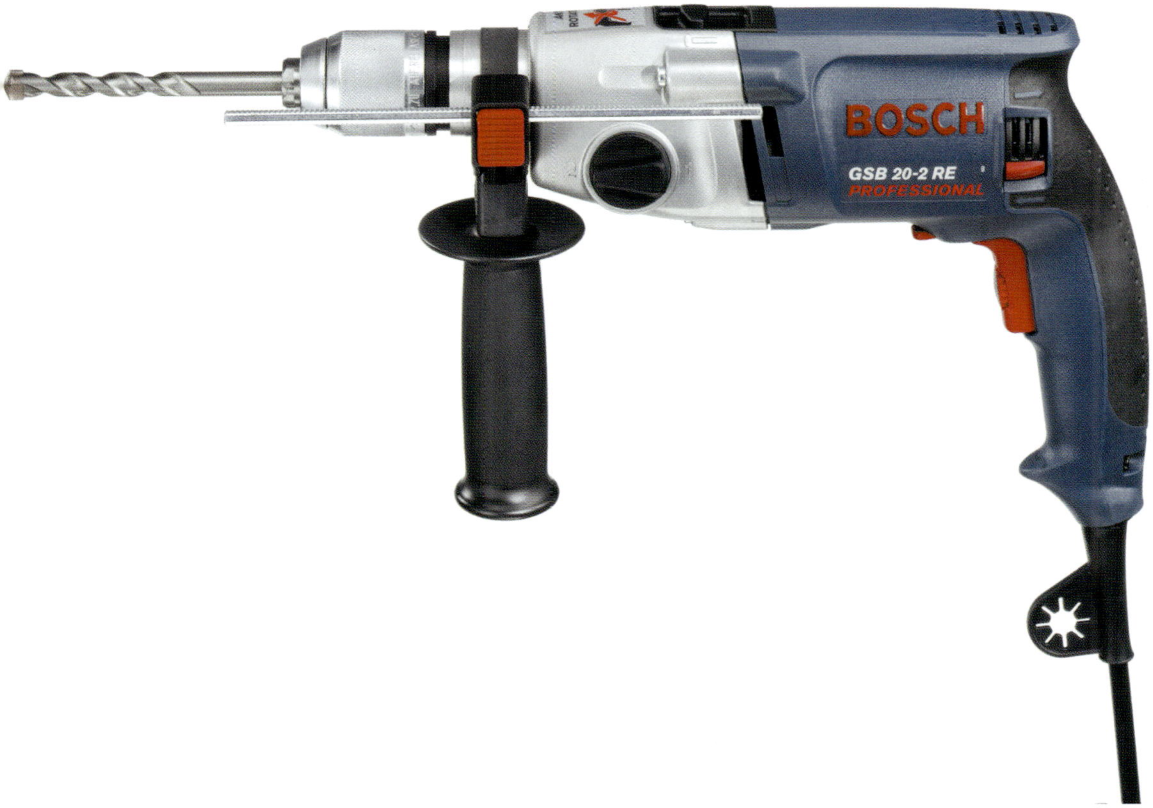

Hersteller Manufacturer
Robert Bosch GmbH
Leinfelden-Echterdingen, Germany

Produkt Product
GSB 20-2 RE
Schlagbohrmaschine
Impact drill

Design
TEAMS Design
Esslingen, Germany

Die GSB 20-2 RE Schlagbohrmaschine für den professionellen Einsatz besitzt ein Aluminiumgehäuse, einen Zwei-Gang-Elektromotor und ein Drehmoment von mehr als 30 Newtonmeter. Für den Anwender sorgen der vibrationsdämpfende Softgrip, die Überlastkupplung Anti-Rotation mit sofortigem Stopp des Motors in kritischen Situationen und das ergonomische Design für leichtes und schnelles Arbeiten. Die nahezu staubresistente, drehbare Bürstenplatte erhöht die Zuverlässigkeit und Lebensdauer der Maschine auch bei hartem täglichen Einsatz.

The GSB 20-2 RE impact drill for professional users features an aluminum casing, a two-speed electric motor and torque of more than 30 Newtonmeters. The vibration-dampened softgrip, the anti-rotation overload clutch with immediate motor cut-out in critical situations, and the ergonomic design, ensure users can work fast, easily and with little fatigue. The virtually dust-resistant, rotating brush plate increases reliability and the drill's service life – even given heavy-duty daily work.

Hersteller Manufacturer
Robert Bosch GmbH
Leinfelden-Echterdingen, Germany

Produkt Product
GEX 150 Turbo Professional
Exzenterschleifer
Orbital sander

Design
TEAMS Design
Esslingen, Germany

Der GEX 150 Turbo Professional Exzenterschleifer lässt sich durch den Turbo-Schalter mit einem Griff während der Arbeit auf Grobschleifen, Feinschleifen oder Polieren umstellen. Mit dem kleinen Griffumfang, der vibrationsdämpfenden Grifffläche aus Weichkunststoff und dem geringen Gewicht von 2,4 kg sich der GEX 150 Turbo sehr gut führen. Durch das Absaugrohr mit Anti-Statik-Funktion steht der Anwender nicht in einer Staubwolke.

The GEX 150 Turbo Professional orbital sander can be switched while in use from rough sanding, to fine sanding or polishing with one touch of the turbo switch. The small grip circumference, the vibration-dampening soft plastic handle and low 2.4 kg weight all spell a very-easy-to-guide tool. The dust extraction pipe with its anti-static function prevents users from getting covered in dust while working.

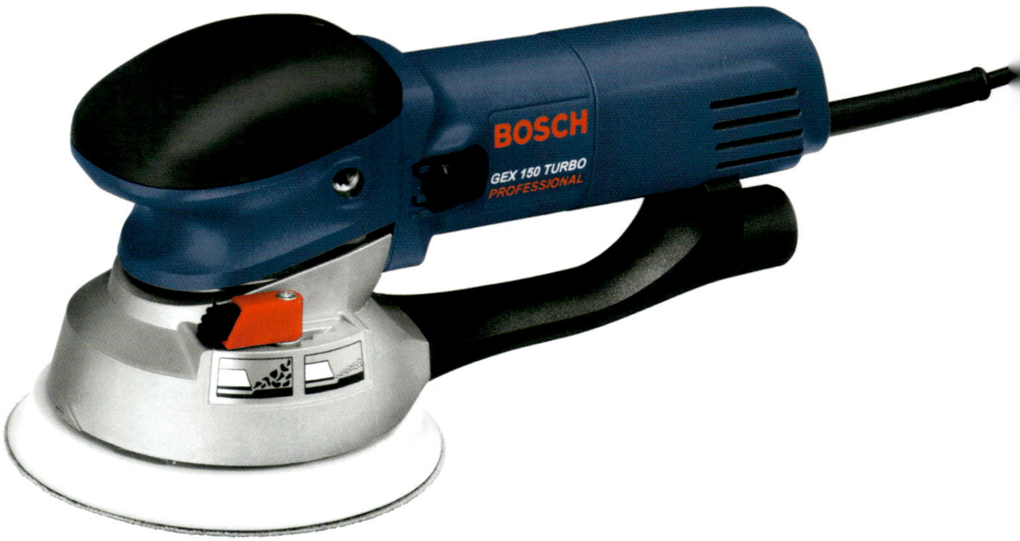

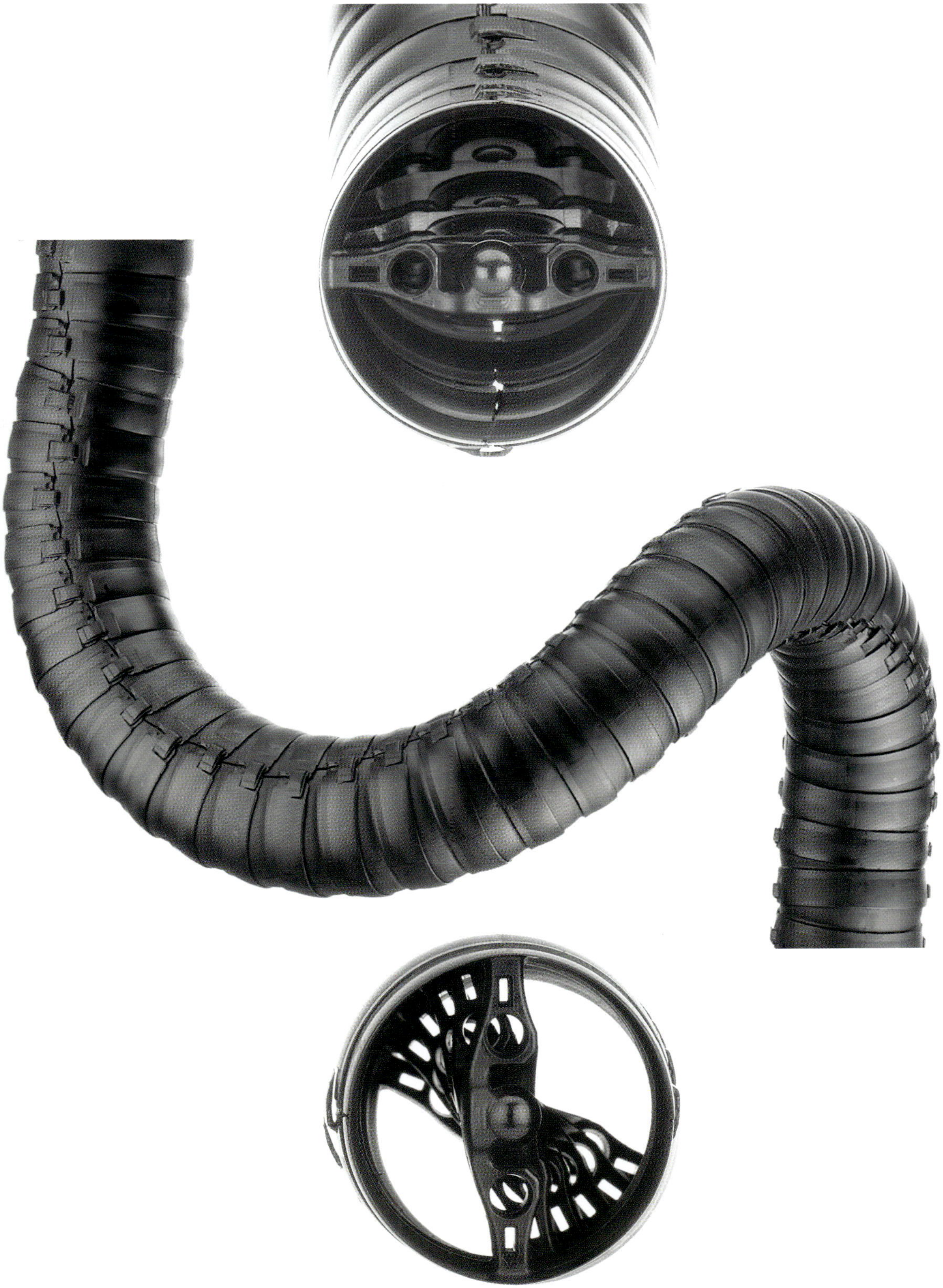

Hersteller Manufacturer
igus® GmbH
Köln, Germany

Produkt Product
igus® Triflex® R
Mehrdimensional bewegliche Energiezuführung
Multi-dimensional, moving energy guide

Design
igus® GmbH
Köln, Germany

igus® Triflex® R ist eine neuartige Energieführungskette für technisch anspruchsvolle Einsatzbereiche. So vor allem für Anwendungsbereiche mit mehrdimensionalen Bewegungen. Sie gewährleistet eine sehr große mehrachsige Bewegungsfreiheit bei gleichzeitig hoher Stabilität und kleinen Biegeradien. Verfügbar ist Triflex® R aktuell in zwei Varianten: geschlossen sowie als easy-Version. Weiter stellt das igus® Easy Chain®-Prinzip eine bequeme und schnelle Befüllung sowie Austausch von Leitungen oder Medienschläuchen sicher. Aufgrund des runden Designs gleitet die Triflex® R über die Außenkontur des Roboters, hakt nicht ein und verkantet nicht. Überdies bietet Triflex® R höchsten Schutz vor Schweißspritzern und Schmutz und ist schlag- und abriebfest. Ein weiterer Pluspunkt, nicht zuletzt aus Kostensicht, liegt im denkbar geringen Montage- und Demontageaufwand, da keinerlei Zusatzverstärkungen wie Seilzüge oder Federstäbe, nötig sind. Triflex® R besteht allein aus einem Bauteil, dem Kettenglied.

igus® Triflex® R is a new type of energy guide chain for technically demanding fields of application, such fields of application involving multi-dimensional movements. It guarantees a very high, multi-axial freedom of movement and both small bending radii and high stability at one and the same time. Two versions of the Triflex® R are currently available: enclosed and easy version. In addition, the igus® Easy Chain® principle ensures convenient and fast filling with and exchange of cables, lines or media hoses. Owing to the round design, the Triflex® R glides over the exterior contour of the robot, does not catch and does not tilt. Moreover, Triflex® R offers maximum protection against spatter and dirt and is resistant to impact and abrasion. One other advantage, not least from the point of view of costs, relates to the extremely low installation and removal effort since no additional reinforcements of any kind, such as cable pulls or torsion bars, are required. Triflex® R consists solely of one component, the chain link.

Hersteller Manufacturer
Pricon GmbH & Co.KG
Waldems, Germany

Produkt Product
Xiron^{S}
Schraubendreherset
Screwdriver set

Design
Winhold Döring
Graphicon
Offenbach, Germany

Das Schraubendreherset Xiron^{S} ist für feines und präzises Arbeiten in der Feinmechanik und Optik bei extrem kleinen Verschraubungen gedacht. Der große Drehknopf und der Griffbereich mit 4 Spezialgummiringen lassen eine optimale Handhabung für die verschiedensten Handgrößen zu. Die technische Anmutung der Formgebung wird durch das Lochdesign und die schlanke Taillierung unterstrichen. Der Ständer besteht aus rostfreiem Edelstahl und unterstreicht das Erscheinungsbild.

The Xiron^{S} screwdriver set was developed for fine and precise work in fine mechanics and optics by extremely small screws. The large rotary head and the handle with its four special rubber rings optimize handling whatever the size of hand operating it. The design's technical appearance is underscored by the perforations and slim, tapered shape. The holder is made of stainless steel and enhances the overall image.

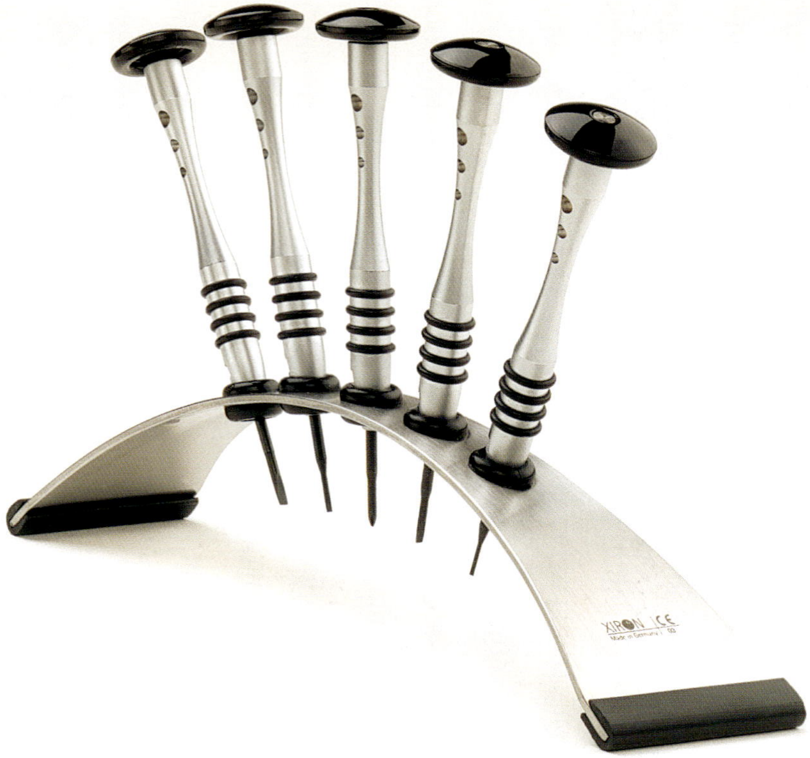

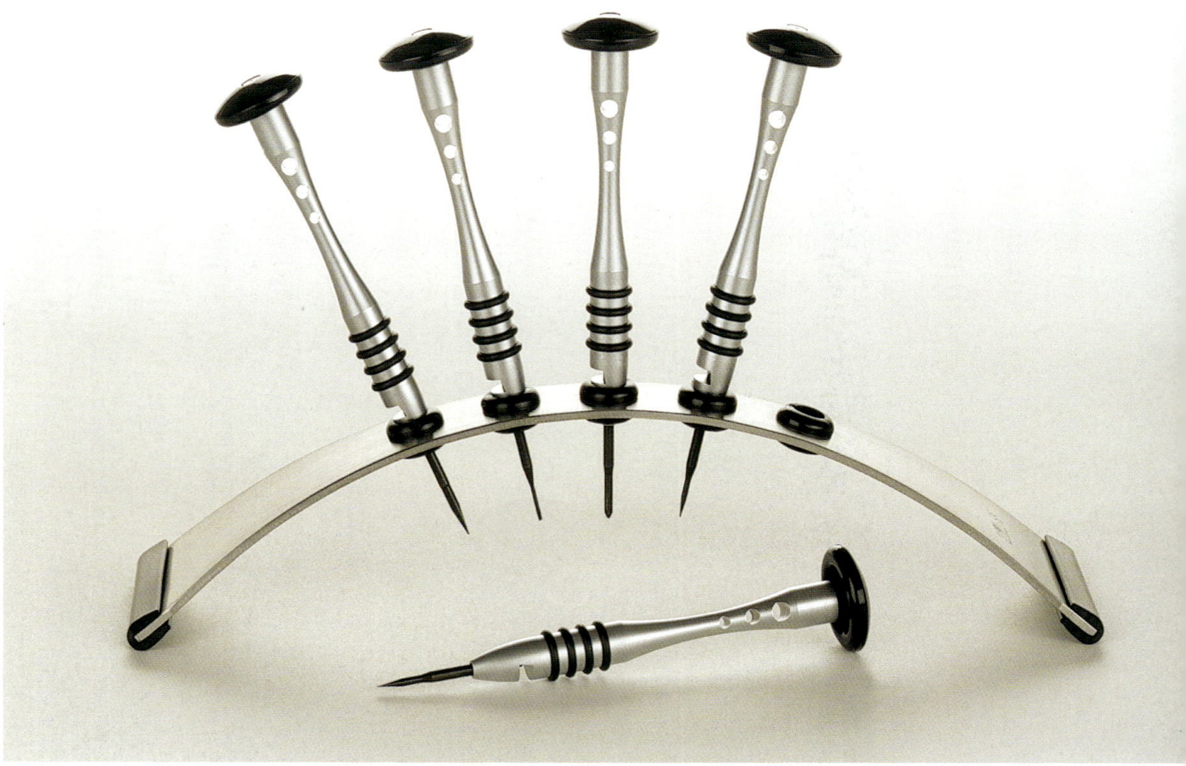

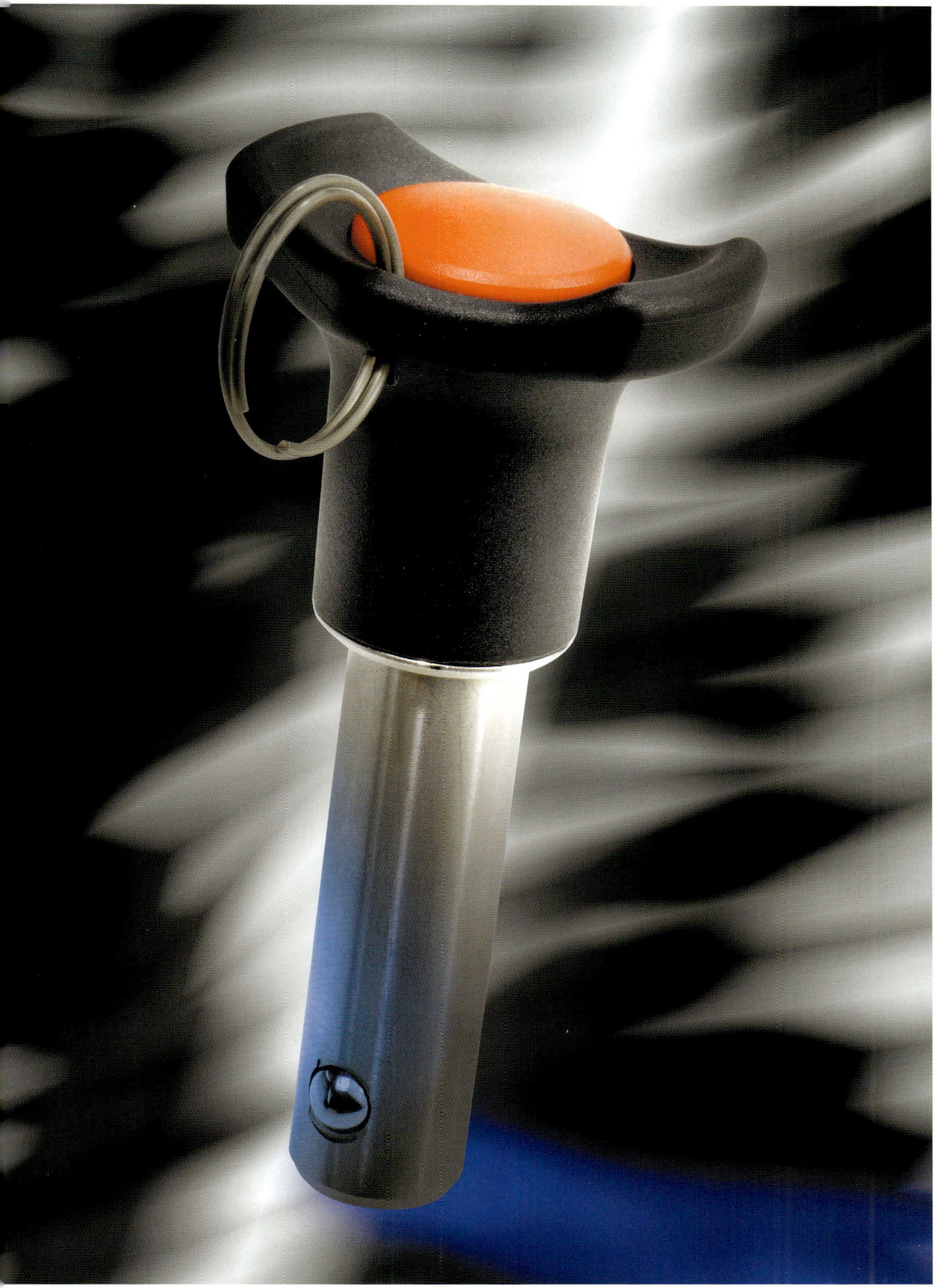

Hersteller Manufacturer
Heinrich Kipp Werk
Sulz, Germany

Produkt Product
Kipp Kugelsperrbolzen
Kipp ball locking bolts

Design
Heinrich Kipp Werk
Sulz, Germany

Kipp Kugelsperrbolzen dienen zum raschen Fixieren und Verbinden von Teilen und Werkstücken. Durch Drücken des roten Druckknopfes können die beiden Kugeln entriegelt und somit die zu verbindenden Teile abgesteckt werden. Wird der Druckknopf freigegeben, verriegeln die Kugeln sicher und unlösbar die Verbindung. Die auf Drei-Finger-Betätigung abgestimmte Profilierung des Griffes bietet höchsten Bedienkomfort und sichere Handhabung beim Verbinden zweier Platten, bei der Verwendung als Lagerbolzen und als Sicherungselement gegen unbeabsichtigtes Lösen.

Kipp ball locking bolts are used to quickly secure and join parts and workpieces. When the red button is pressed, both balls can be unlocked and inserted into the pieces which are to be connected. When the push-button is released, the balls securely and irreversibly lock the connection. The handle is shaped for superior user comfort. It can be safely operated with three fingers when connecting two plates, when used as a bolt, or as a retaining elements to avoid unintentional detachment.

Hersteller Manufacturer
Bizerba GmbH & Co. KG
Balingen, Germany

Produkt Product
Bizerba Class E-World
CE Systemwaagen
CE system scales

Design
Gregor Luippold
Ingolf Butz
a4id Product Design
Partnerschaft Butz + Luippold
Tübingen, Germany

Die Bizerba Class E-World Systemwaagen wurden vor allem für die Lebensmittelbranche und ihren Anforderungen an Hygiene und Robustheit entwickelt. Die farblich variierbaren Gestaltungselemente lassen sich dem jeweiligen Ladenambiente anpassen, zudem ermöglicht die Designkonzeption modulare Gerätegehäuse z.B. Schnellwechselmodule für verschiedene Verschleißteile wie Tastaturen, Drucker, Anzeigen etc. Mit dem Windows-basierten Betriebssystem kommunizieren Waage und Büro in einem einheitlichen Standard.

The Bizerba Class E-World system scales were primarily developed for the food industry and its exacting hygiene and robustness standards. The colors of the design elements can be varied so as to match any shop ambience. In addition, the design concept features modular casings, such as quick-change modules for various wearing parts including keyboards, printers, displays etc. With its Windows-based operating system, the scales communicate with the office using a unified standard.

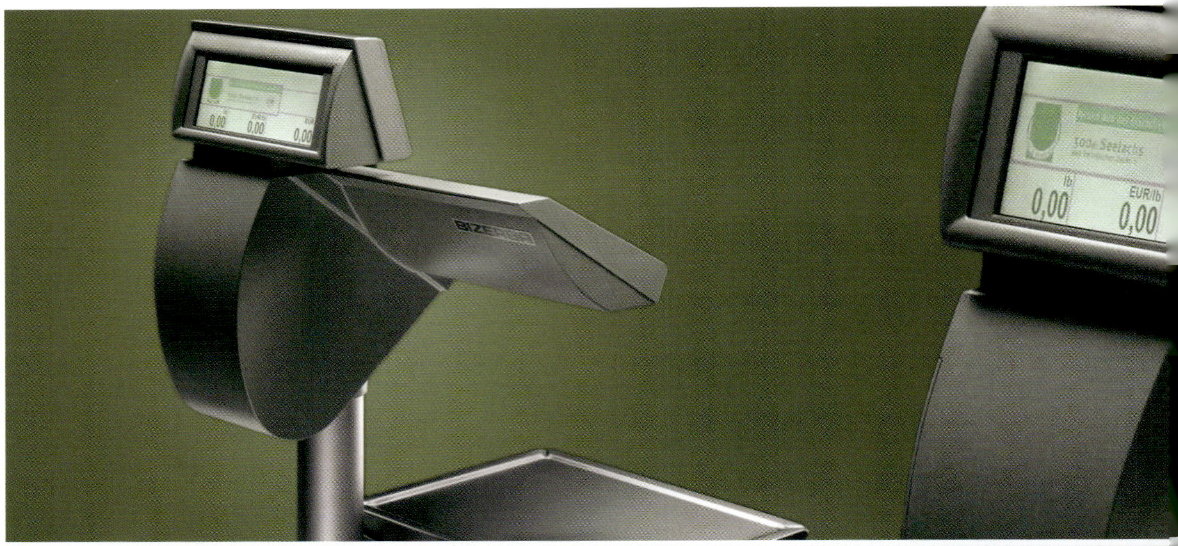

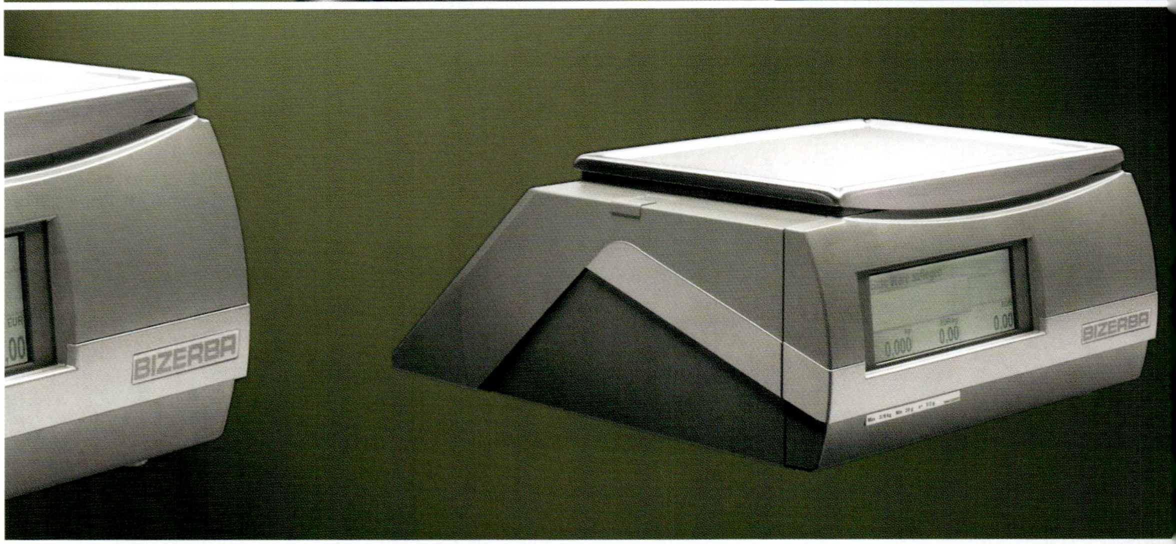

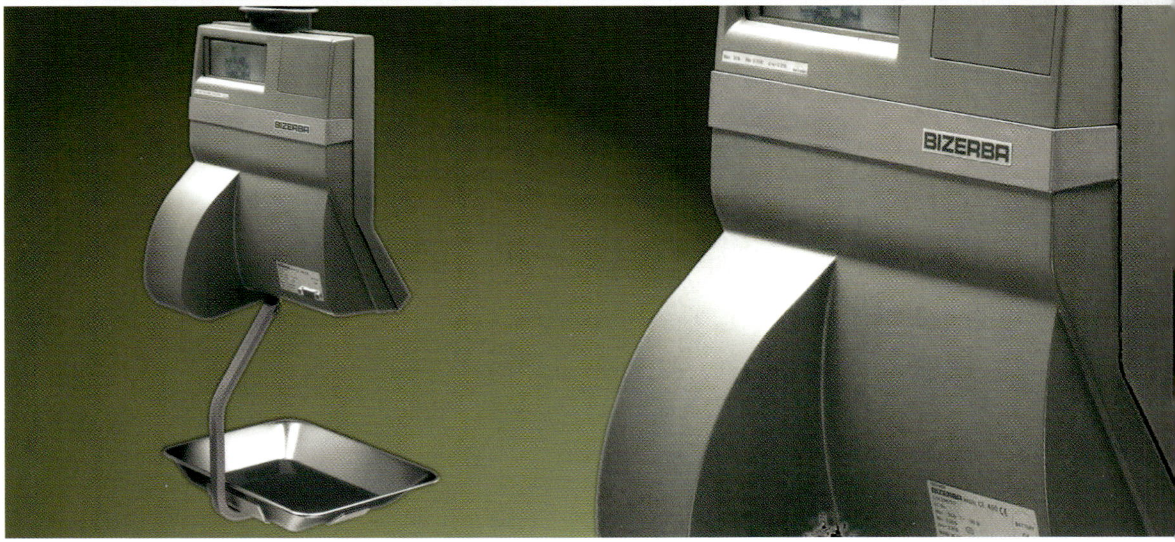

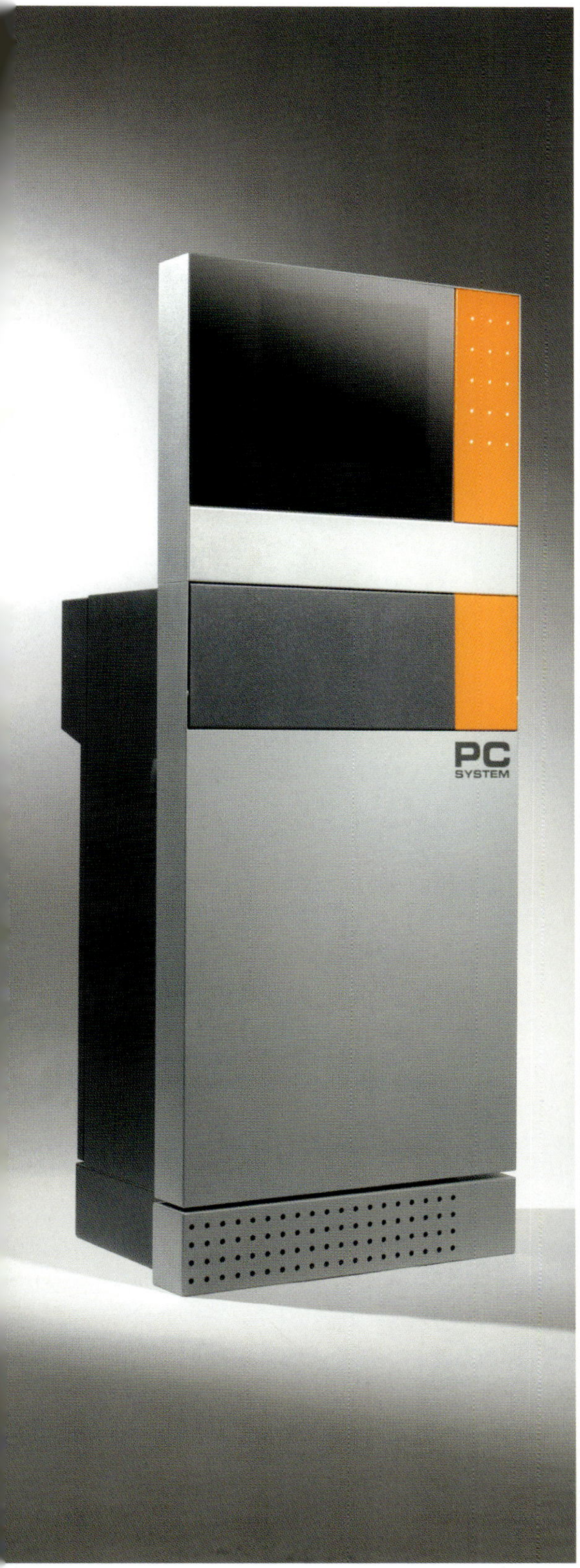

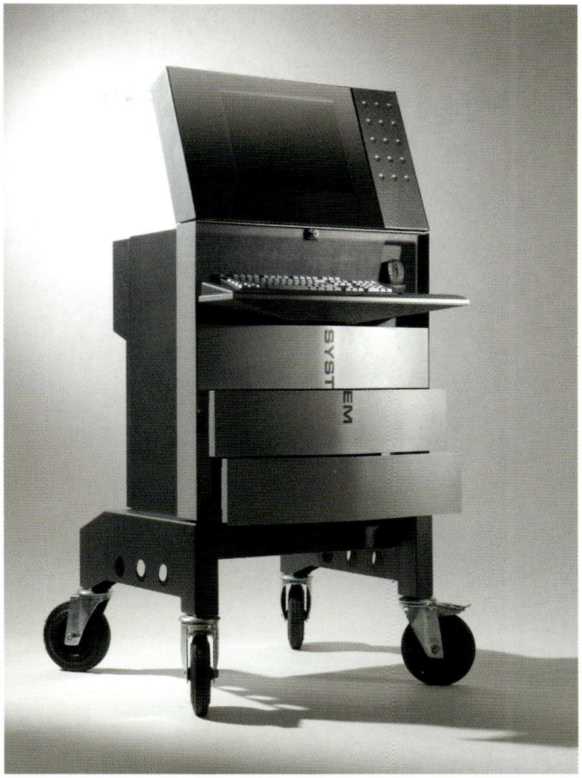

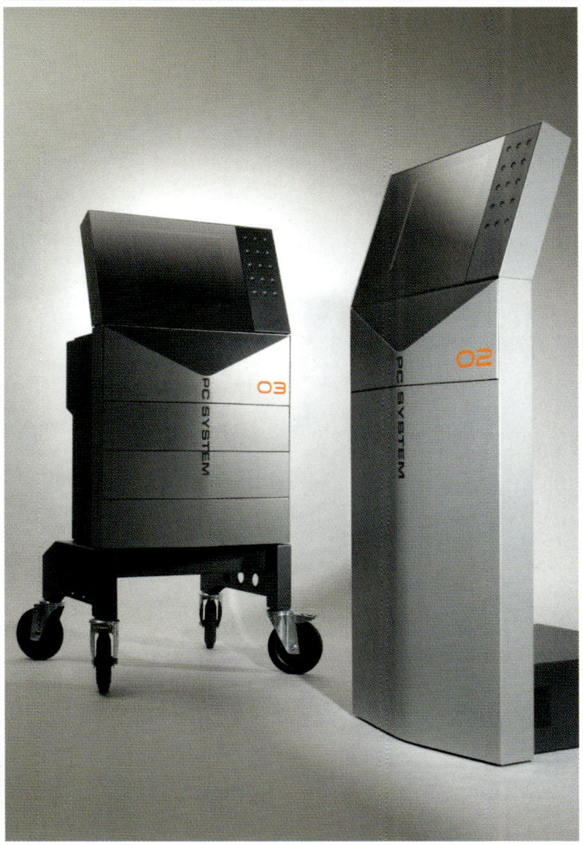

Hersteller Manufacturer
Schinko GmbH & Co. KG
Neumarkt, Austria

Produkt Product
Schaltschrankreihe B/NB
PC-Schaltschrankgehäuse
PC cabinets

Design
R. Kittler
H. Kurz
G. Fasching
Industrial Design Union / Kittler & Kurz
Wilhering, Austria

Die Schaltschrankreihe B/NB wurde als Schutzgehäuse für empfindliche Steuerungselektronik im Maschinen- und Anlagensteuerungsbereich entwickelt. Die Vielzahl an heterogenen Computerkomponenten, Verkabelungen, Elektronik- und Hydraulikaggregaten von Schaltschränken wird durch die Gehäuse zu einheitlichen Modulen zusammengefasst. Große, bündig sitzende Glasscheiben ermöglichen die Verwendung von unterschiedlichen Monitorgrößen. Die reduzierte Formensprache der Schaltschrankreihe soll zu einem ruhenden Pol im Arbeitsumfeld werden.

The B/NB series switching cabinets were developed as protective housing for sensitive controls electronics for mechanical and plant engineering applications. The diversity of heterogeneous computer components, cable links, electronic and hydraulic aggregates in switching cabinets is brought together here in uniform modules. Large, flush-mounted glass panes enable different monitor sizes to be utilized. The reduced formal idiom of the switching cabinet series is intended to bring a sense of calm to the bustle of working environments.

Hersteller Manufacturer
Bruker Optik GmbH
Ettlingen, Germany

Produkt Product
MPA Multi Purpose Analyzer
Fourier Transform Nah-Infrarot Spektrometer
Fourier transform near-infrared spectrometer

Design
yellow design | yellow circle
Pforzheim, Germany

Der Bruker MPA Multi Purpose Analyzer ist ein robustes Nah-IR-Spektrometer mit hermetisch gedichteter Optik für die Qualitätskontrolle im Labor und in der Produktion. Die optionale modulare Multifunktionalität des Systems ermöglicht bis zu 4 Messverfahren, wobei der aktive Messkanal (und der Gerätestatus) am geneigten Frontdisplay des MPA direkt ablesbar sind. Bei der Designentwicklung wurde Wert auf eine Präzision vermittelnde Anmutung gelegt. Das glattflächige, kraftvolle Volumen des Geräts entspricht dem rauen Umfeld seines Einsatzes in der Qualitätskontrolle. Das System ist mit einem schnellen PC-basierenden Datensystem und einem einfach zu bedienenden Analytiksoftwarepaket ausgestattet.

The Bruker MPA multi-purpose analyzer is a robust, near-infrared spectrometer with hermetically sealed optics used for quality control in laboratories or production. The optional, modular multi-functionality of the system enables up to four measuring methods with the active measurement channel (and device status) displayed on the MPA's slanted front panel. The design emphasizes an image that communicates precision. The smooth-surfaced, large dimensions of the device are suited to the harsh environments in quality control areas in which it is used. The system comes complete with a swift PC-based data system and an easy-to-use analysis software package.

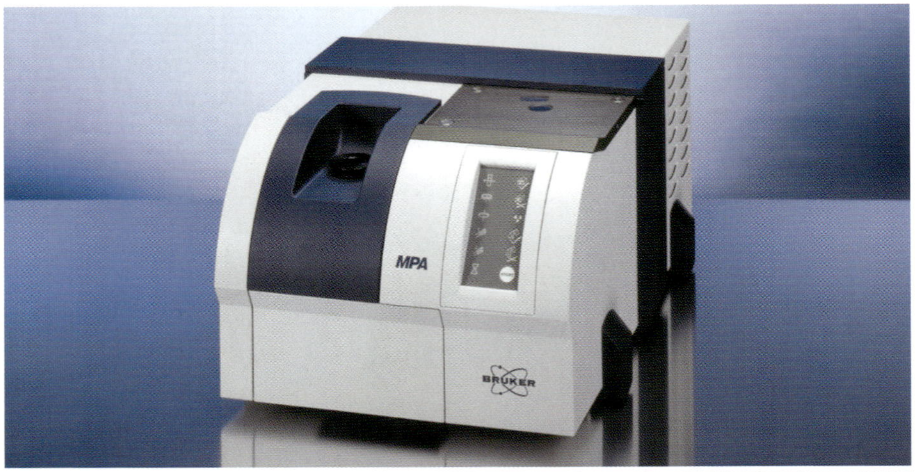

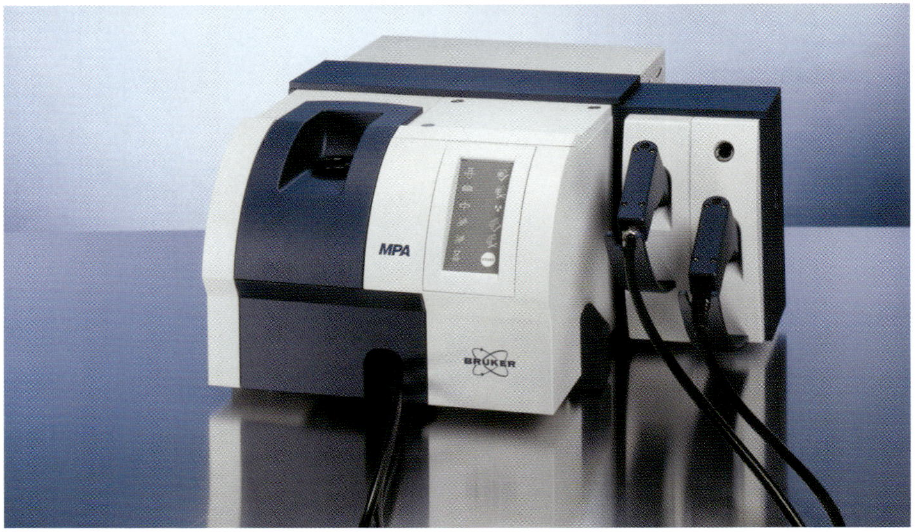

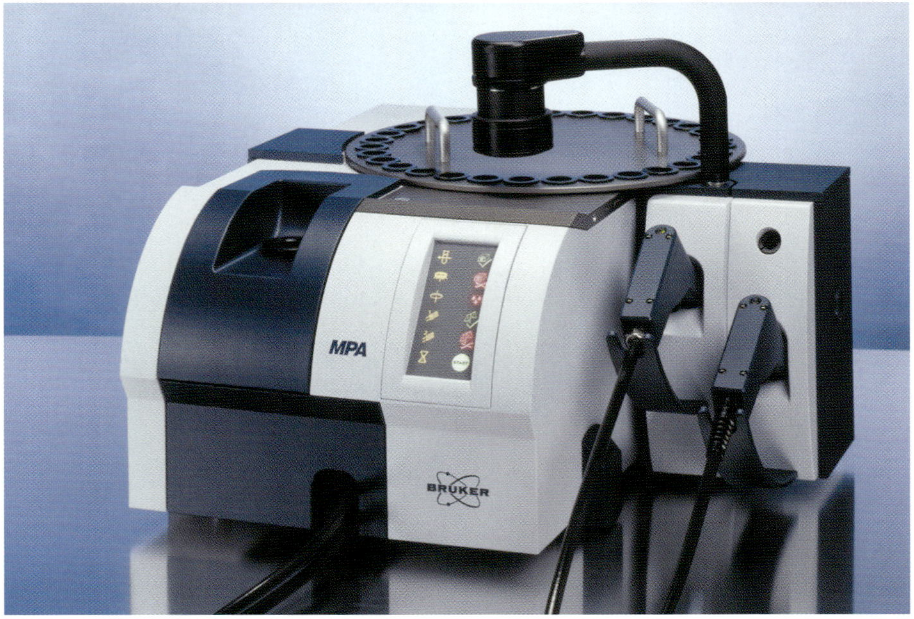

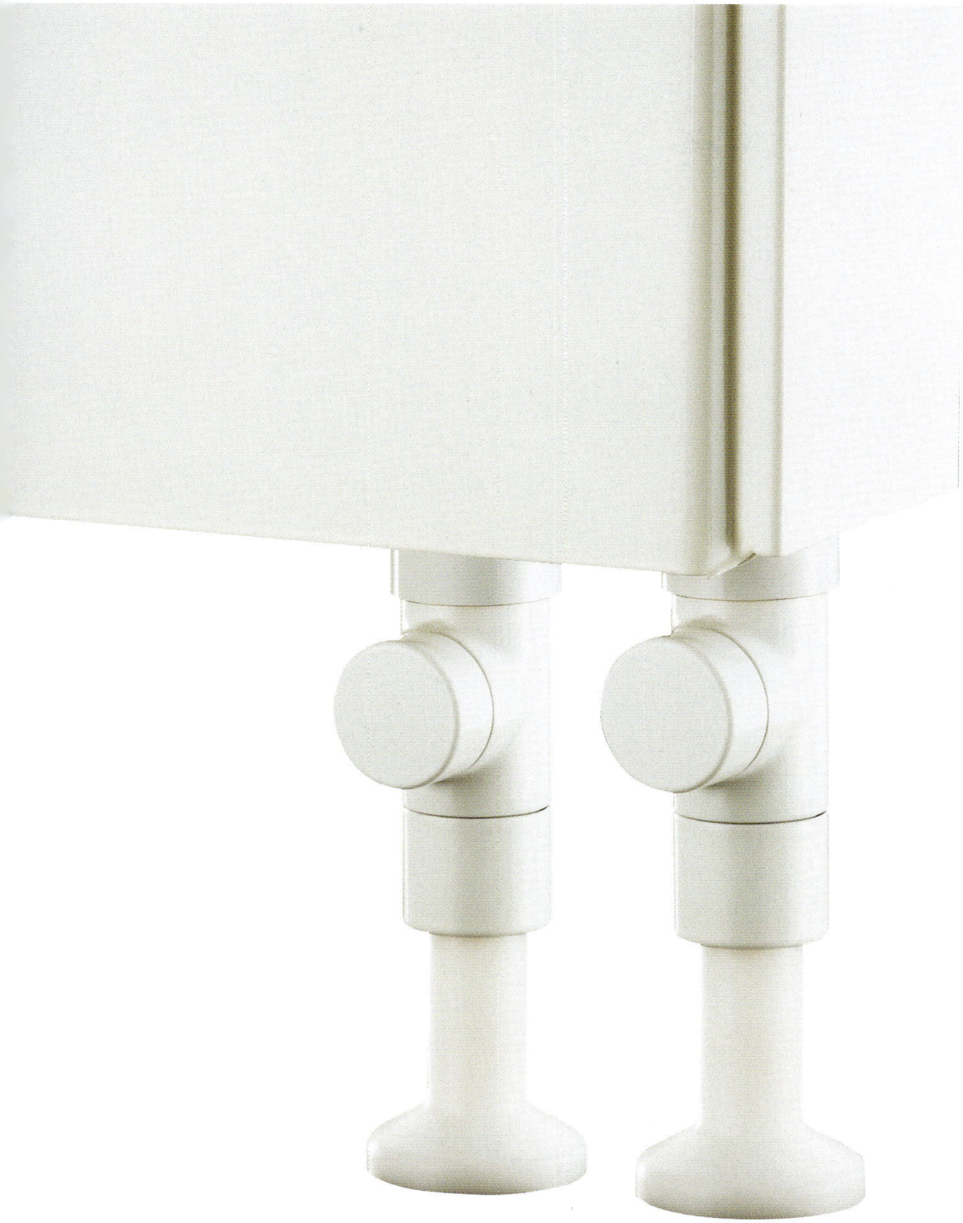

Hersteller Manufacturer
F.W. Oventrop GmbH & Co. KG
Olsberg, Germany

Produkt Product
Combi E-VK
Verschraubung
Lockshield valve

Design
F.W. Oventrop GmbH & Co. KG
Werksdesign
Olsberg, Germany
Prof. Ulrich Hirsch
Design & Identität
Neuwittenbek, Germany

Die Verschraubungen Combi E-VK sind Armaturen für die untere Anbindung der Rohrleitungen an Heizkörpern mit integrierten Ventilgarnituren. Die Heizkörperverschraubungen ermöglichen das Voreinstellen, Absperren, Entleeren und Füllen des Heizkörpers. Sie werden paarweise (für Vor- und Rücklauf) am unteren Auslauf des Heizkörpers angebracht. Da die Armaturen im Wohnraum sichtbar sind, wurden sie unter Designgesichtspunkten optimiert. Die Formgebung wurde auf das Wesentliche beschränkt und auf eine einfache zylindrische Form reduziert, unnötige Ecken und Kanten wurden vermieden. Die Durchgangsform wird mit Schutzrohr bis zum Fußboden geführt.

The lockshield valve set Combi E-VK is used for the bottom connection of pipes to radiators with integrated distributors. The valves have a presetting, filling and draining facility and are fitted in pairs to the flow and return bottom connections of the radiator. Lockshield valves are visible and the attached photograph displays the current design status. When this lockshield valve was developed, great emphasis was placed overall on the functions, the form and the materials used. The form was reduced to a cylindrical shape and any unnecessary edges were avoided. The straight pattern valve includes a pipe cover which leads down to the floor. Materials and the refinement process of the surface were selected with great care to ensure that they are recyclable and environmentally friendly.

Hersteller Manufacturer
E. Zoller GmbH
Freiberg a. N., Germany

Produkt Product
Venturion 600
Einstell- und Messgerät für Werkzeuge
Presetting and measuring machine for tools

Design
Tobias Greissing
Benno Heinz
Heinz Industrial Design
Stuttgart, Germany

Bei der Designentwicklung des Venturion 600, einem Einstell- und Messgerät für Werkzeuge, wurden vor allem ergonomische Gesichtspunkte und Bedienfreundlichkeit zugrunde gelegt. So ermöglicht der runde Optikträger eine Steigerung des Schwenkdurchmessers von rotierenden Werkzeugen, das Adapter- und Druckerablageboard sowie die Bedieneinheit für Tastatur und Monitor sind schwenkbar. Die kompakte Bauform der Einheit erfordert einen nur geringen Platzbedarf, zudem wurde auf Beinfreiheit für Sitzplatztätigkeiten bei langwierigen Einstellaufgaben geachtet. Die Zugängigkeit des Werkzeugs ist für Reinigung, Montage und Einstellung von allen Seiten möglich. Insgesamt orientiert sich das Design an der Corporate Identity von Zoller.

The design of the Venturion 600, a presetting and measuring machine for tools, emphasizes all ergonomic features and user friendliness. The round optical carrier enables instruments to be rotated to a greater degree – the adaptor and printer shelves as well as the user section for the keyboard and monitor can be rotating. The device's compact form has a small footprint and legroom is also taken into account should seating be required for time-consuming adjustment tasks. The machine can be accessed from all sides for purposes of cleaning, assembly and adjustment. Overall, the design derives from Zoller's corporate identity.

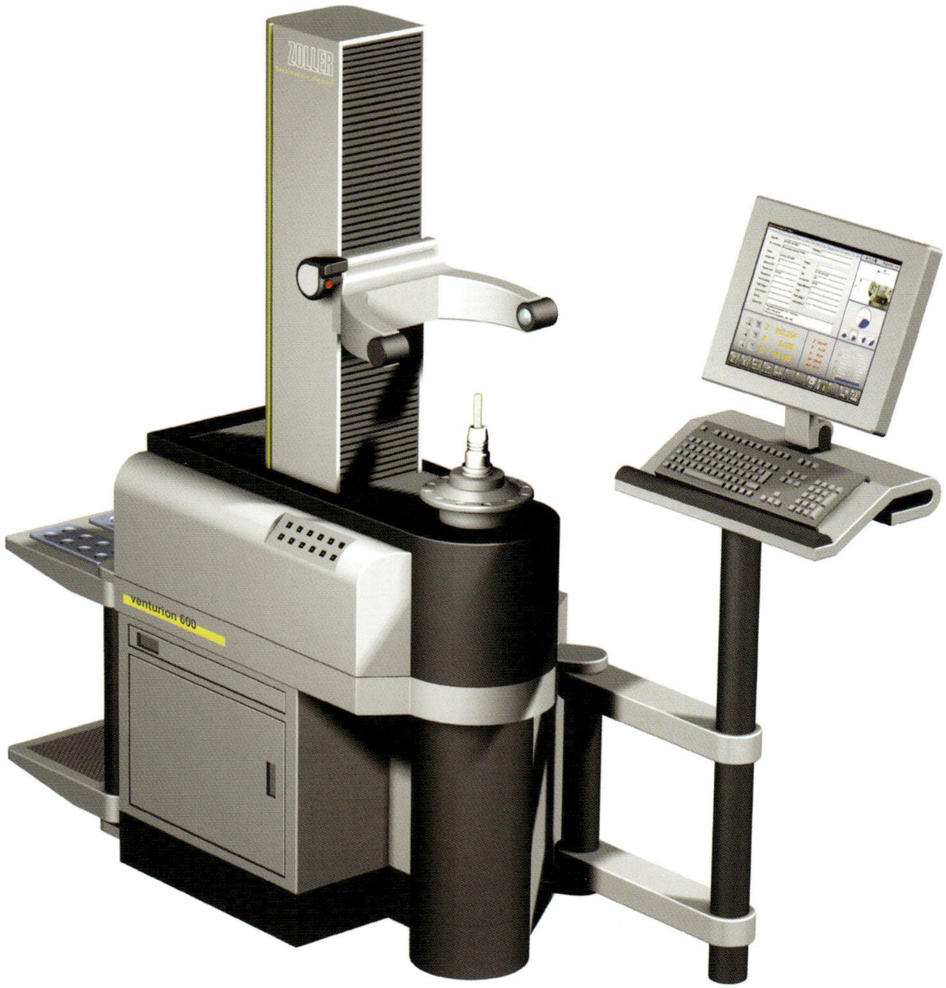

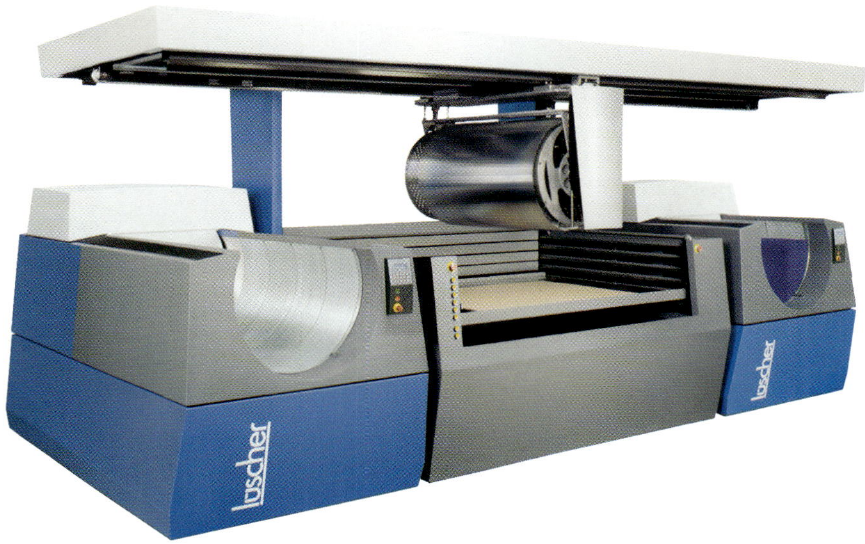

Hersteller Manufacturer
Lüscher AG
Leutwil, Switzerland

Produkt Product
XPose swiss CTP
Computer to Plate Thermobelichter
Computer-to-plate thermo-exposer

Design
Raimund Erdmann
Peter Hess
Erdmann Design AG
Brugg, Switzerland

Der modular aufgebaute Computer to Plate Thermobelichter XPose wurde für die filmlose Produktion von Druckplatten entwickelt. Einzigartig ist die Kombination von Laserdioden und Innentrommel in einem System. Die Modularität erlaubt das schrittweise Anpassen und Mitwachsen mit dem digitalen Workflow der Druckplattenproduktion. Für die verschiedenen Druckplatten-Formate werden die Baugrößen des XPose in Produktionsleistung und Laserbestückung angepasst. Das Design der Maschine ist klar und funktionsbezogen und unterstützt die Corporate Formensprache des Unternehmens.

The modular computer-to-plate thermo-exposer XPose was developed for film-free production of printing plates, featuring a unique combination of laser diodes and inside drum in one and the same system. The modular principle enables gradual adjustment and expansion with the digital workflow for printing plate production. The size of the XPose units can be adjusted in terms of production output and laser inserts for different printing plate formats. The machine boasts a clear and functional design, and supports the language of corporate form.

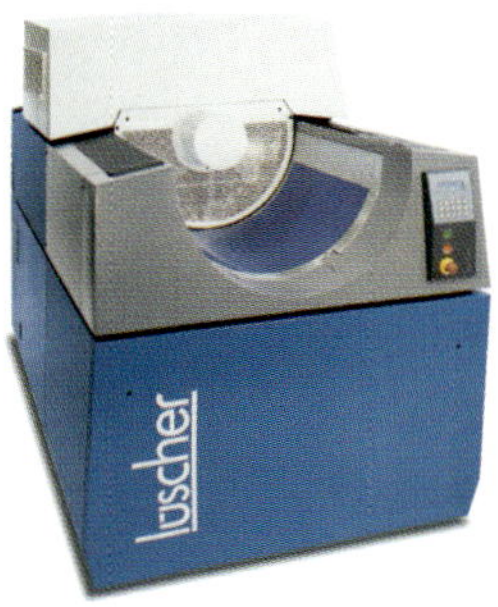

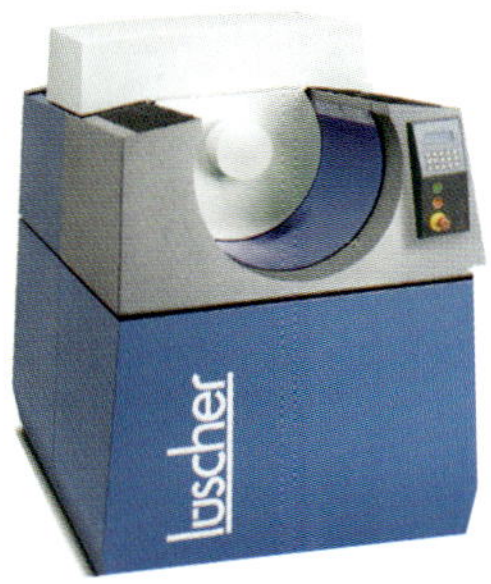

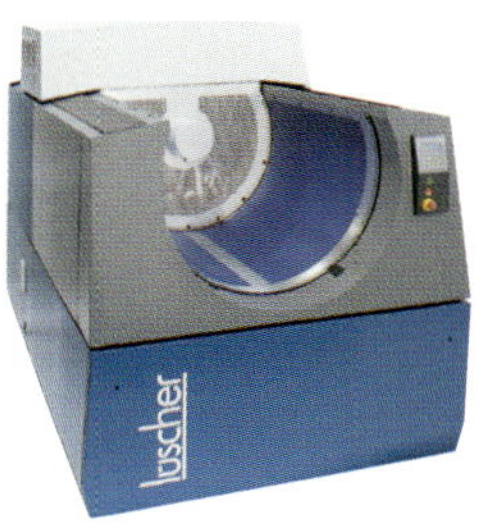

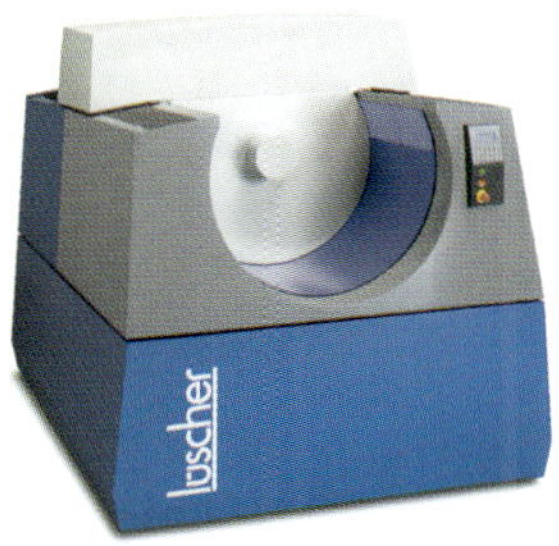

Hersteller Manufacturer
Neopost Industrie B.V.
Drachten, Netherlands

Produkt Product
SI 82
Falz- und Kuvertiermaschine
Folding/inserting machine

Design
Maarten van Lelyveld, BNO
Drachten, Netherlands

Die SI 82 Falz- und Kuvertiermaschine ist für einen Durchgang von bis zu 80.000 Umschlägen pro Monat ausgelegt. Der Platzbedarf für die Maschine ist gering, da die Papierzufuhr und die fertigen Umschläge vertikal angeordnet sind. Das Design ist einfach, klar und funktionsbezogen und spiegelt den Verarbeitungsprozess im Inneren der Maschine.

The SI 82 document folding/inserting machine is meant for professional applications requiring a throughput of up to 80,000 envelopes a month. The machine has a small footprint, as the paper feed and finished envelopes are laid out vertically. The design: clean and clear lines with an emphasis on function, reflecting the workings within the machine itself.

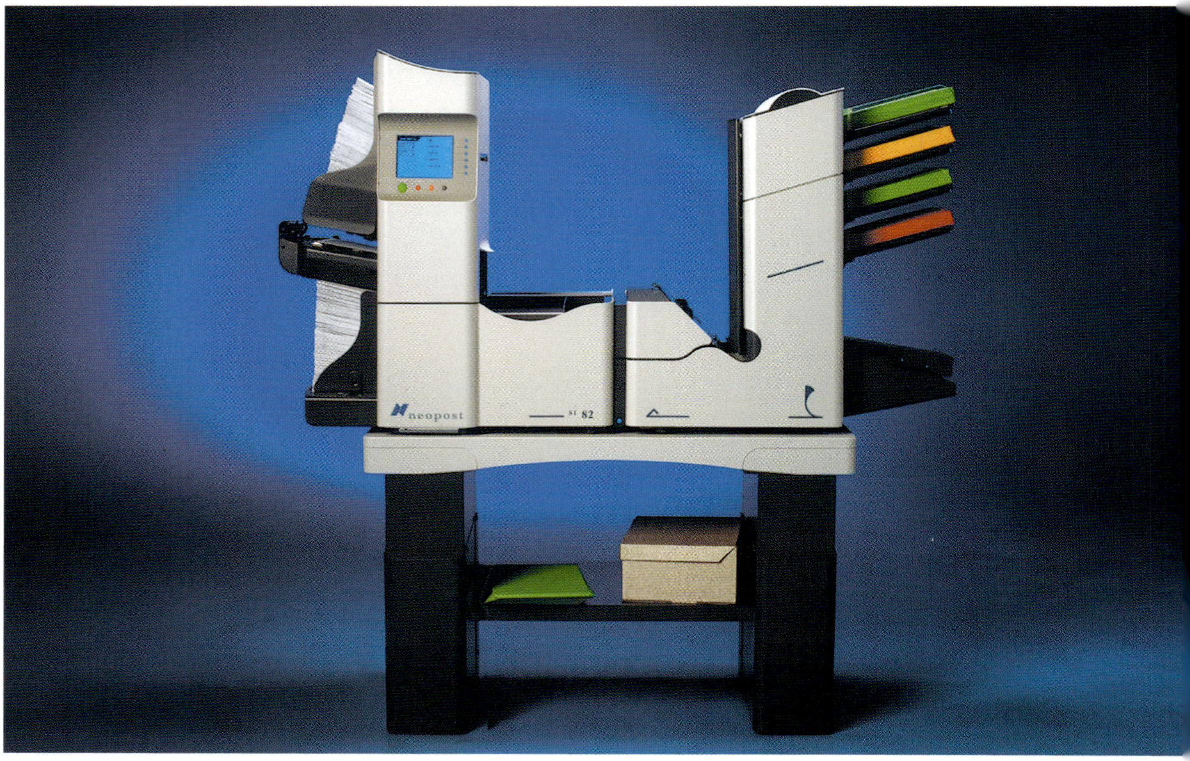

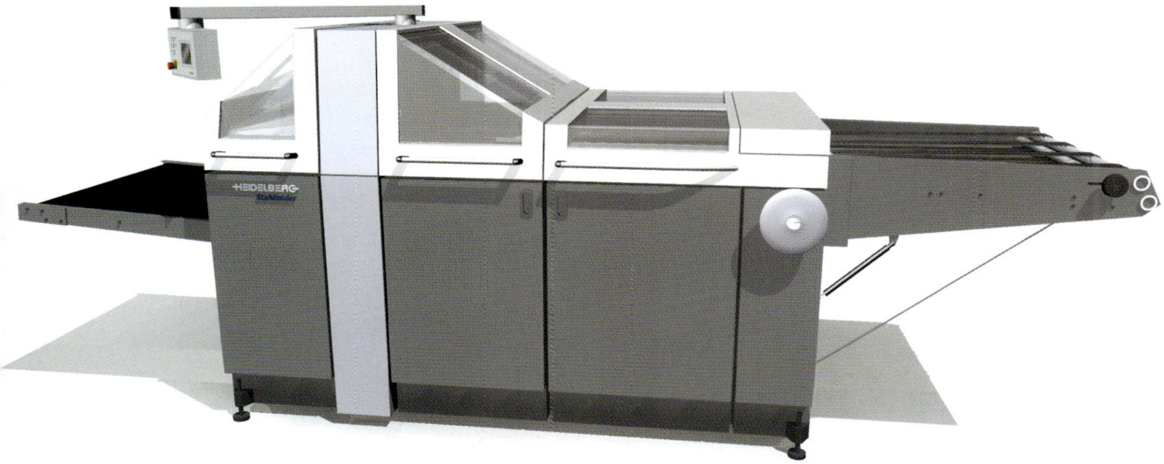

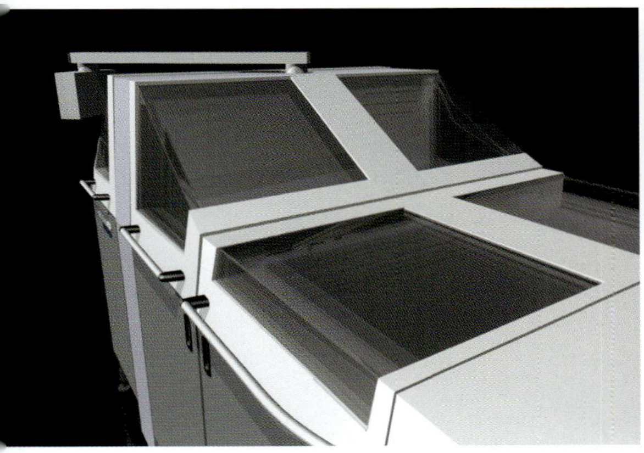

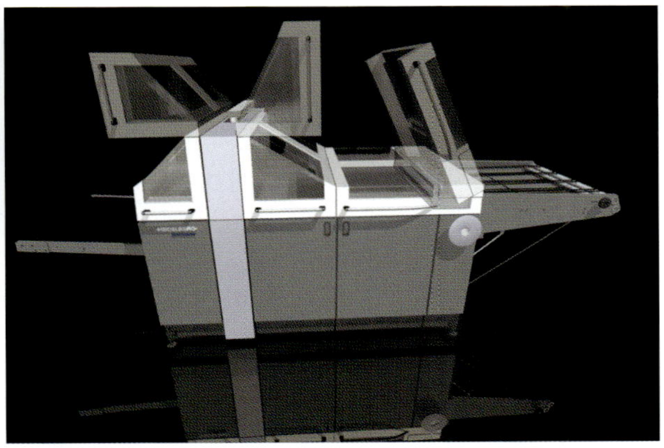

Hersteller Manufacturer
Heidelberger Druckmaschinen AG
Heidelberg, Germany

Produkt Product
Speedbander
Automatisches Auslagesystem
Automated banding system

Design
Marc Bundschuh
Heidelberger Druckmaschinen AG
Heidelberg, Germany

Der Speedbander ist ein hochleistungsfähiges Auslagesystem, das sich modular an Falzmaschinen, Sammelhefter und Klebebinder anbinden lässt. Die automatische Päckchenauslage nimmt sowohl Falzbogen als auch geheftete Produkte auf und stapelt und banderoliert sie zu Päckchen. Einfache Bedienung, hohe Geschwindigkeit und minimale Rüstzeiten zeichnen den Speedbander aus; das schnelle Banderolieren geschieht dabei per Ultraschallschweißung. Die Einheit wird über ein um 180 Grad schwenkbares Bedienpult mit Touchscreen gesteuert und eingestellt. Besonderes Augenmerk bei der Gestaltung galt der optimalen Einsichtmöglichkeit auf den Papierlauf und der maximalen Zugänglichkeit, d.h. alle Hauben können gleichzeitig nach oben geklappt werden und die Bedienung ist aufgrund des Schwenkpults von drei Seiten möglich.

The Speedbander is a high-performance delivery unit which can be integrated into a modular system together with folding machines, saddlestitchers and adhesive binders. The automatic pack delivery receives both folded sheets and stitched products, arranges them in piles, and bands them into packs. The Speedbander stands out for easy operation, high speed and extremely short make-ready times. The swift banding is based on ultrasonic welding technology. The unit is controlled and adjusted via a control console with touchscreen that can be swung out across 180°. The design focuses on optimizing visibility of sheet travel and maximum accessibility: thus, all covers can be lifted simultaneously. The 180°-radius of the control console also means the system can be operated from three sides.

Hersteller Manufacturer
Storck do Brasil Ltda.
Curitiba-Paraná, Brazil

Produkt Product
P-200 Plastic Injector
Horizontale Spritzgussmaschine
Horizontal injection molding machine

Design
Eliseu Wateloo
Storck do Brasil Ltda.
Curitiba-Paraná, Brazil

Bei der Entwicklung der horizontalen Spritzgussmaschine von Storck wurde Wert auf das ästhetische Erscheinungsbild gelegt. So besitzt die Maschine keine sichtbaren Schrauben und muss auch nicht wie sonst üblich mit einem Keil unterlegt werden. Hinsichtlich Betrieb, Programmierung, Struktur und dem Austausch von Einzelteilen wurde auf Komfort, Sparsamkeit und Flexibilität geachtet. Statt Schmiermitteln wird an den Säulen Druckluft eingesetzt.

In developing Storck's horizontal injection molding machine, the emphasis was on aesthetic appearance. Thus the machine contains no visible screws and requires no leveling wedges underneath. Operation, programming, structure and the replacement of individual parts have been designed with comfort, economy and flexibility in mind. Compressed air is used on the columns instead of traditional lubricating agents such as grease.

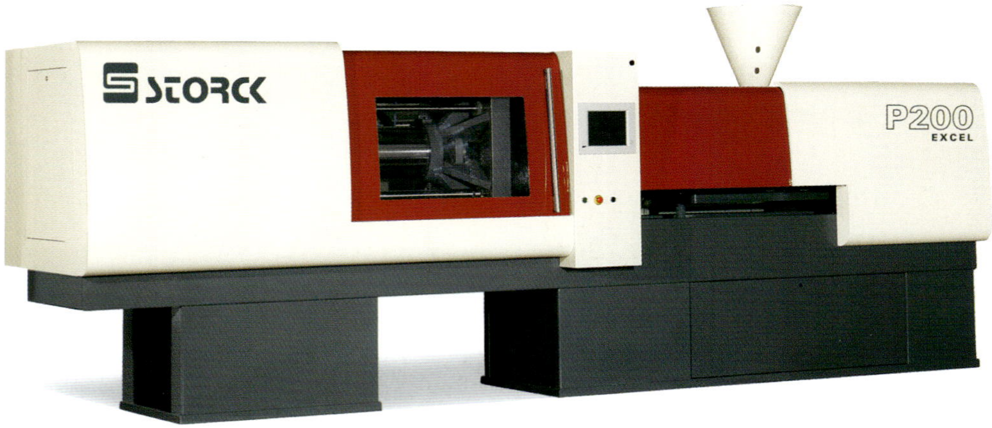

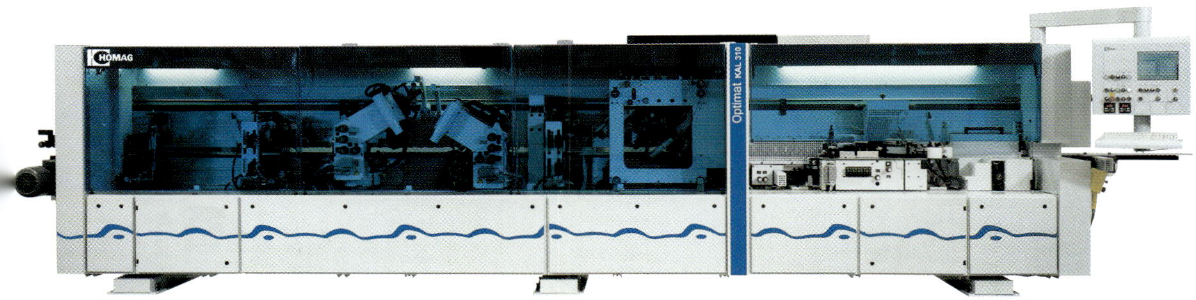

Hersteller Manufacturer
Homag Holzbearbeitungssysteme AG
Schopfloch, Germany

Produkt Product
KAL 310
Holzbearbeitungsmaschine
Wood processing system

Design
Design Tech
Ammerbuch, Germany

Die Holzbearbeitungsmaschinen aus der Baureihe KAL 310 von Homag dienen zur Bearbeitung aller Arten von Kantenmaterialien (Massivholz, Rollen- und Streifenmaterial, Melamin, PVC, ABS und Furnierkanten). Die Maschinen werden überall dort eingesetzt, wo gerade Kanten an fertig formatierten Werkstücken anzuleimen sind. Das Design der Maschine entspricht dem einheitlichen Bild der Homag Gruppe und zeichnet sich insbesondere durch klare Strukturen und hohe Bedienfreundlichkeit aus. Im Vordergrund standen dabei die Maschinensteuerung, die Service-Zugänglichkeit und die gute Erreichbarkeit der Bedienelemente für Einstellarbeiten.

The Homag KAL 310 wood processing machines enable all kinds of edging materials (solid wood, coil, and fixed length materials, melamine, PVC, ABS and veneered edges) to be worked. The machines are ideal wherever straight edges need to be glued to cut-to-size workpieces. The machine's design corresponds to the uniform Homag Group CI, and stands out in particular for its clear structures and great user-friendliness. The emphasis is on facilitating operation, access for service technicians, and easy-reach controls for adjusting settings.

Hersteller Manufacturer
Fill GmbH
Gurten, Austria

Produkt Product
tiltcaster tcs mit integriertem castmaster cm
Kippgießmaschine
Tilting die-casting machine

Design
formquadrat gmbh
Linz, Austria

Beim tiltcaster tcs mit integriertem castmaster cm handelt es sich um eine Kippgießmaschine mit integriertem Gießmanipulator. Das Eingießen während des Kippvorgangs von der Vorder- und Rückseite gewährleistet höchste Flexibilität. Das gesteigerte Leistungsspektrum umfasst höhere Verfahrensgeschwindigkeit und höhere Gussgewichte. Die Genauigkeit beim Vergießen der Alu-Schmelze wurde zusätzlich erhöht. Die Maschine ist ergonomisch gestaltet und unterstützt durch ihr Design Sicherheit und Wartungsfreundlichkeit für den Bediener.

The tiltcaster tcs with its integrated cm castmaster is a tilting die-casting machine boasting an integrated casting manipulator. The great flexibility during processing means that casting can be undertaken during the tilting both front- or backwards. The increased performance range delivers swifter operating speeds and heavier casting weights. Accuracy during die-casting of aluminum has also been augmented. The machine has an ergonomic design that also fosters safety and increases access for service personnel.

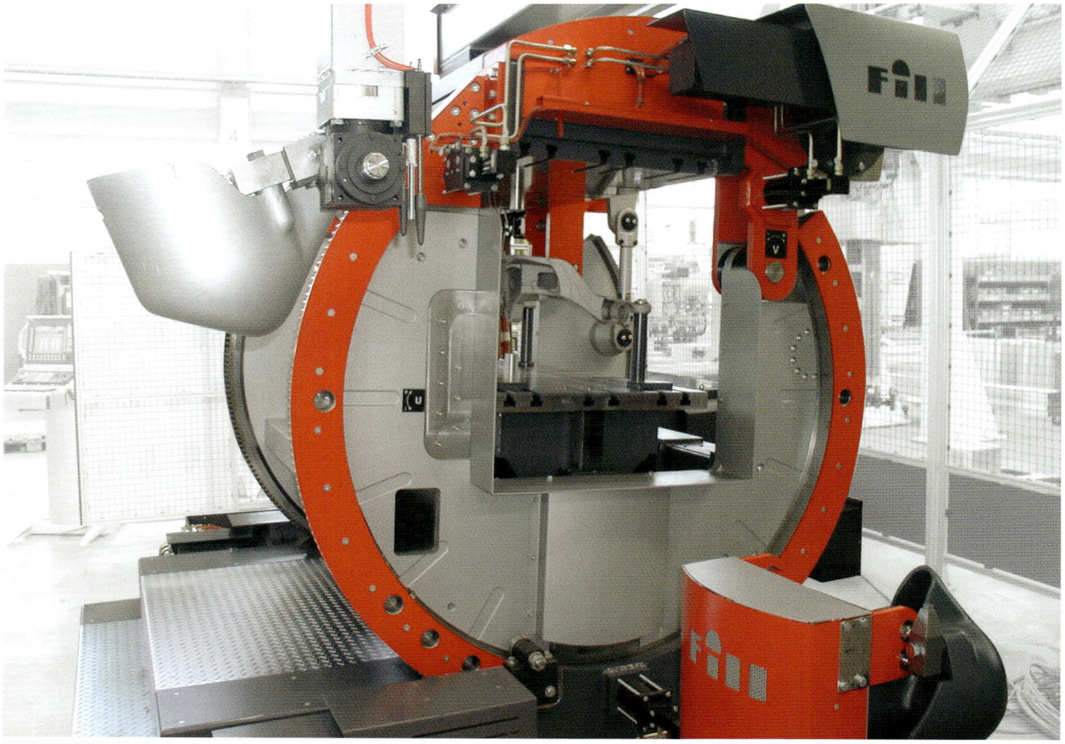

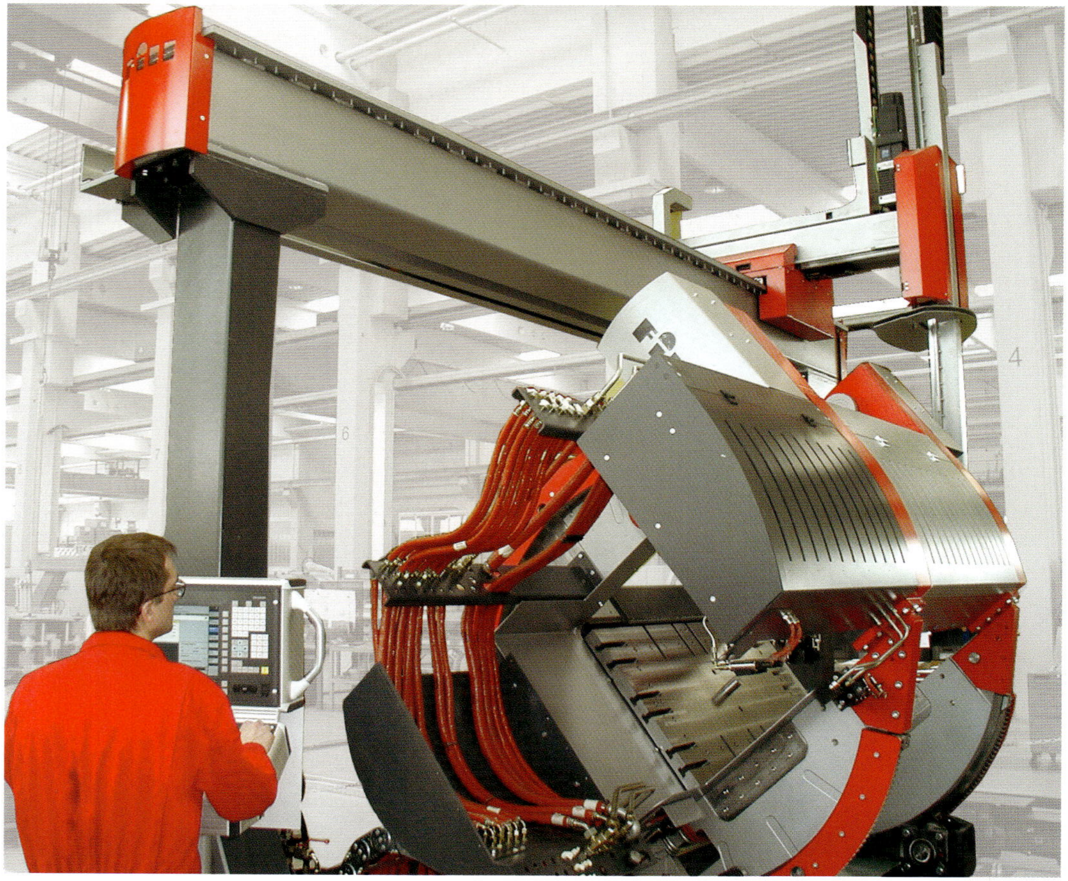

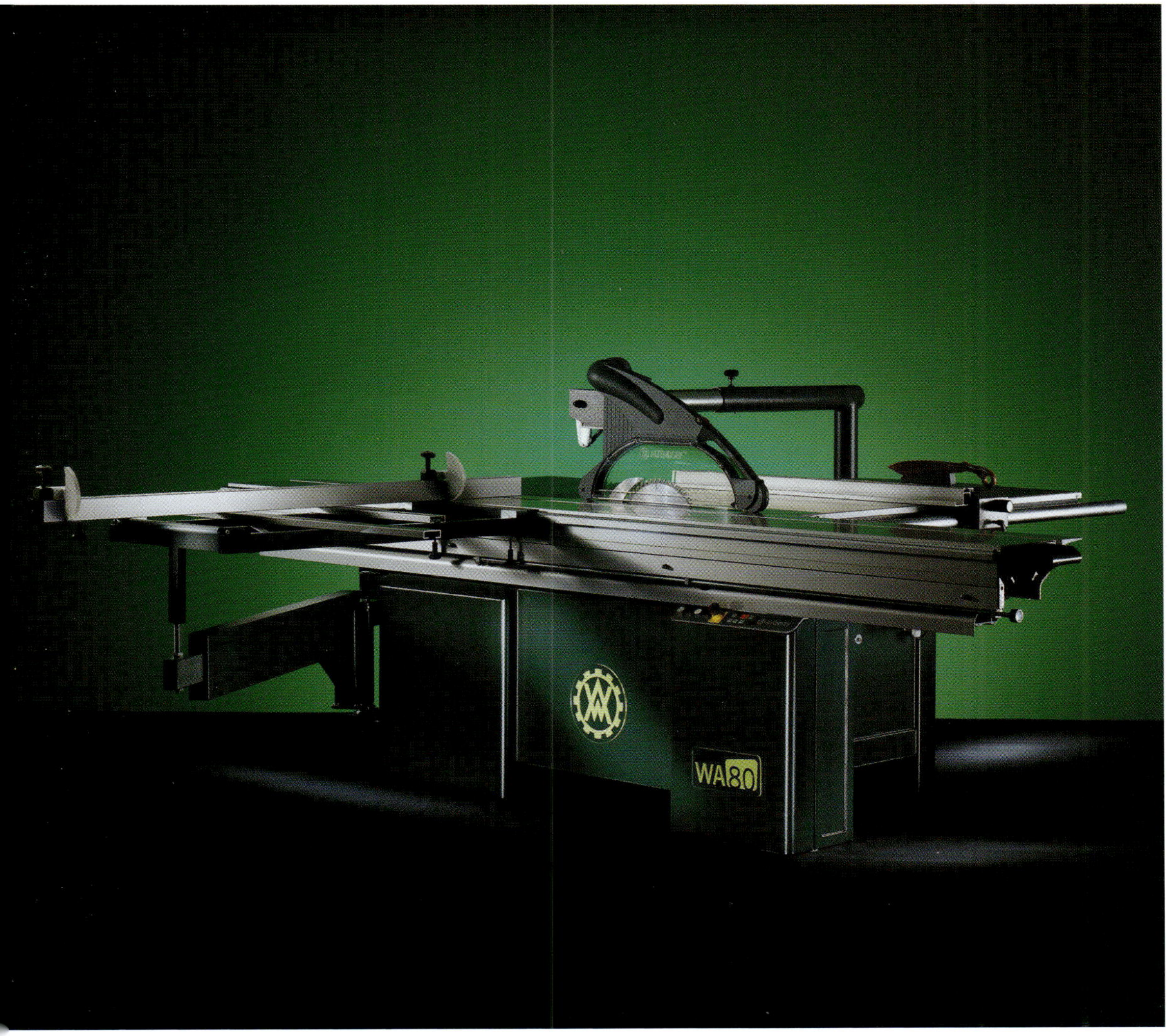

Hersteller Manufacturer
Wilhelm Altendorf GmbH & Co. KG
Maschinenbau
Minden, Germany

Produkt Product
WA 80
Formatkreissäge
Sliding table saw

Design
Wilhelm Altendorf GmbH & Co. KG
Maschinenbau
Minden, Germany

Das Retrodesign und äußere Erscheinungsbild der WA 80 Formatkreissäge beruht auf der erfolgreichen WA 80 aus den 50er Jahren. Mit dem umfangreichen serienmäßigen Zubehör werden alle Anforderungen des Tischlers an das tägliche Sägen erfüllt. Komfortables und ergonomisches Arbeiten werden durch das integrierte Bedienpaneel mit digitaler Anzeige, die Griffmulde im Maschinentisch zum einfachen Entnehmen der geschnittenen Werkstücke und den beleuchteten Maschineninnenraum ermöglicht. Der Montageaufwand ist äußerst gering, die Maschine ist direkt nach Anlieferung zum Sägen bereit.

The retro-design and appearance of the WA 80 sliding table saw bring to mind the successful WA 80 of the 1950s. With a wide range of standard accessories, the WA 80 meets all the daily sawing needs of carpenters. Comfortable and ergonomic work is enhanced by the integrated control panel with its digital display, the grip recess in the machine table to facilitate removal of cut workpieces, and the illuminated motor compartment. The sliding table saw is swiftly assembled and ready to use on arrival.

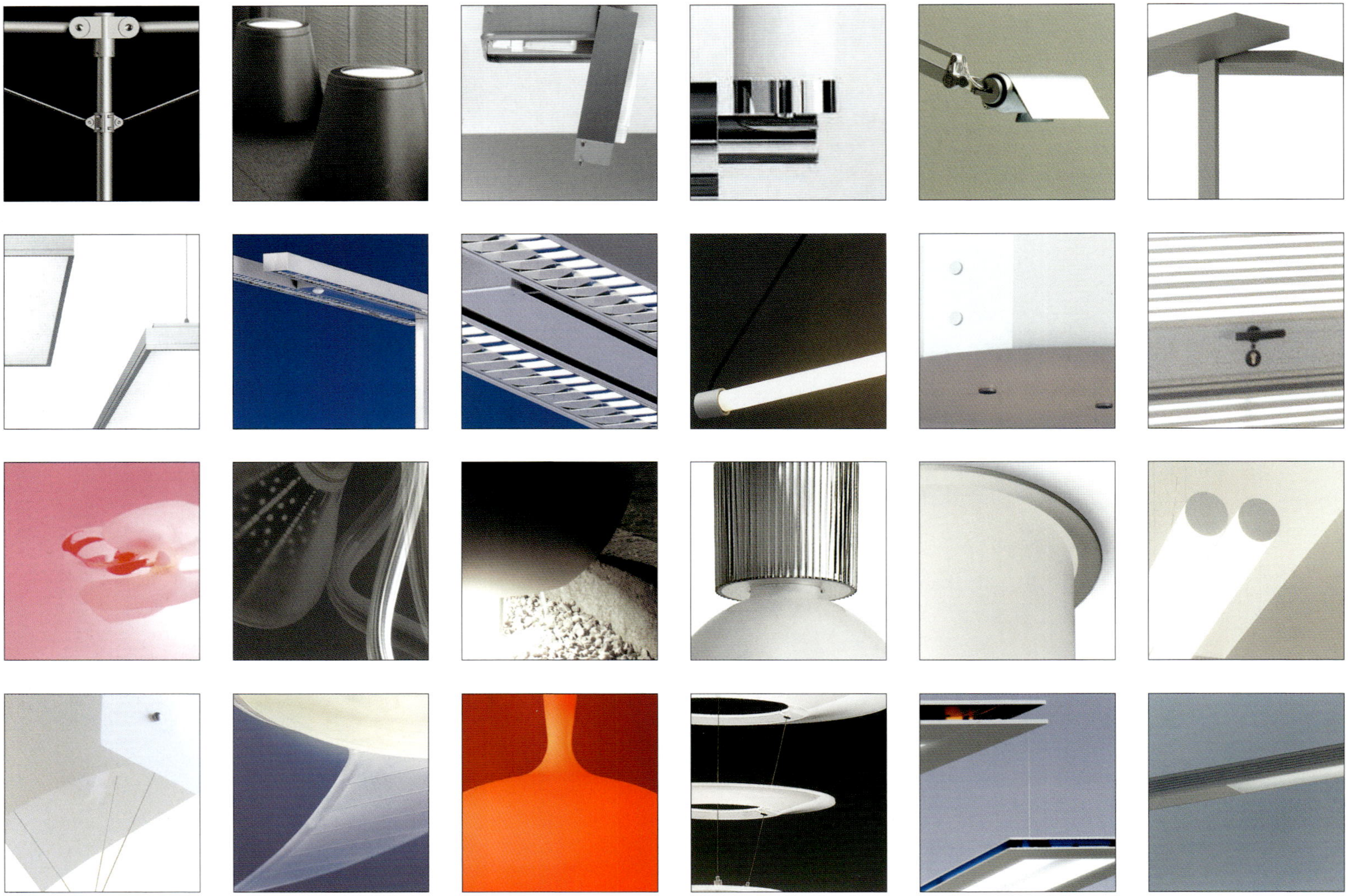

Licht Lighting

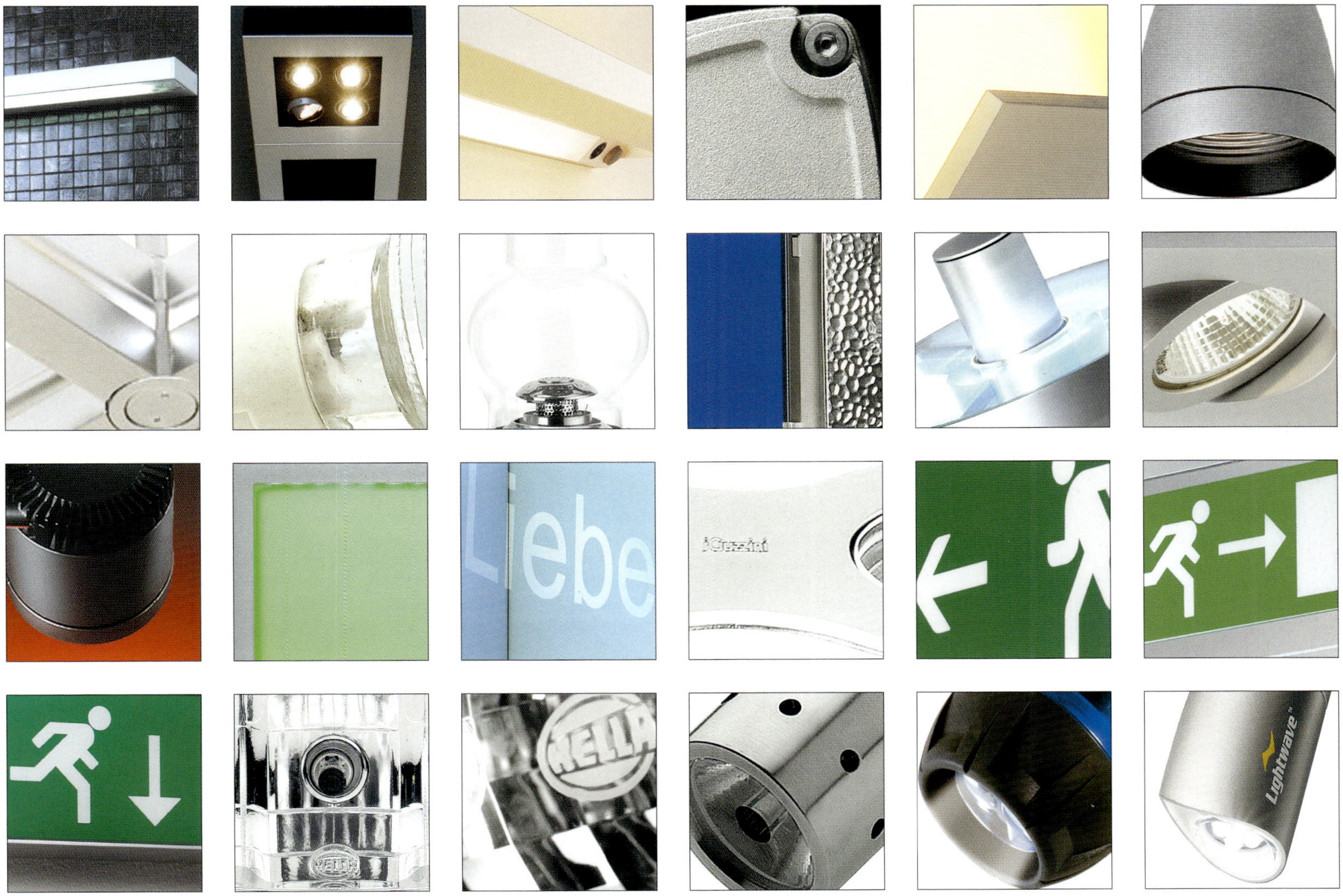

Hersteller Manufacturer
BEGA
Menden, Germany

Produkt Product
BEGA 8254
Auslegerleuchten für veränderbare Lichtpunkthöhen
Outrigger luminaires for adjustable mounting heights

Design
BEGA-Produktgestaltung
Menden, Germany

Neues Auslegersystem mit stufenlos veränderbaren Lichtpunkthöhen. Aufsatzleuchten für Lichtpunkthöhen von 5000-9000 mm. Aus der Waagerechten kann die Lichtpunkthöhe um ± 1500 mm verändert werden. Die Lichtstärkeverteilung kann somit präzise auf die zu beleuchtende Fläche ausgerichtet werden. Die Leuchten in der 2-fach Anordnung ermöglichen unterschiedliche Lichtpunkthöhen an einem Lichtmast z.B. zur Beleuchtung von Kfz-Straßen und Fahrradwegen von einem Standort aus. Leuchten mit sphärischen Reflektoren aus eloxiertem Reinst-Aluminium. Mit asymmetrisch, bandförmiger Lichtstärkeverteilung. Werkzeugloses Öffnen der Leuchten für einfache, schnelle Wartung und Lampenwechsel. Leuchten zur Beleuchtung von Straßen, Plätzen und Zufahrten.

A new outrigger system for infinitely adjustable mounting heights. Pole-top luminaires for mounting heights of 5000-9000 mm. The mounting height can be adjusted from the horizontal plane by ± 1500 mm. This ensures precise adaptation of the light distribution to the illumination surface. The luminaires in double configuration allow different mounting heights on one luminaire pole e.g. to illuminate streets and cycle paths from one location. Luminaires with spherical reflectors of pure anodized aluminium, providing asymmetrical flat beam light distribution. The luminaires can be opened without tools for easy and fast maintenance and relamping. These luminaires are designed to illuminate roads, squares and driveways.

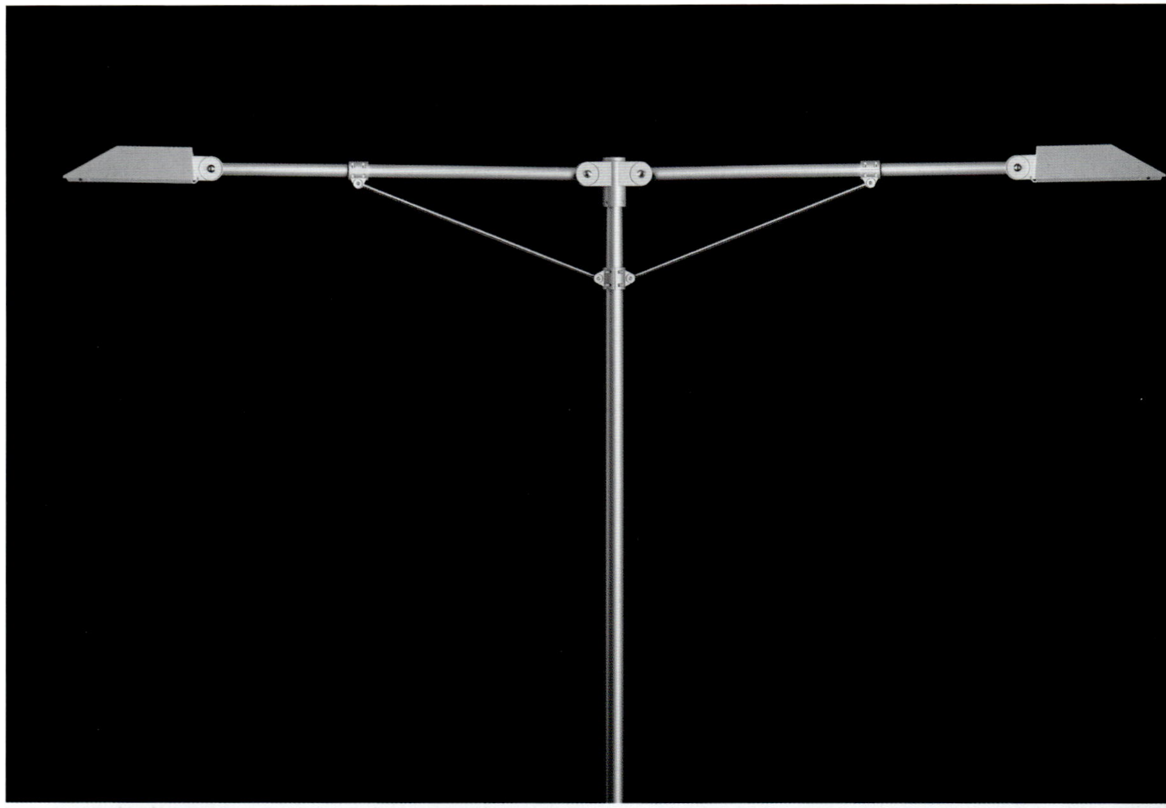

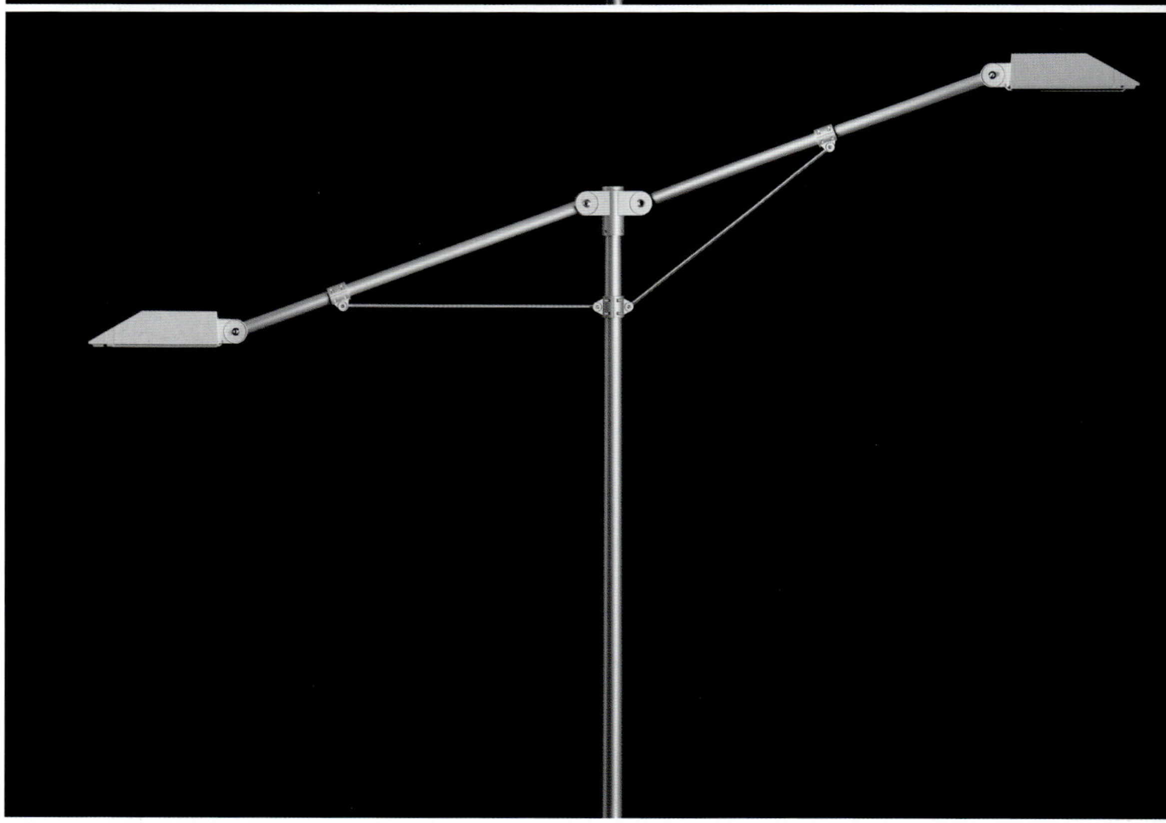

iF Silver Award

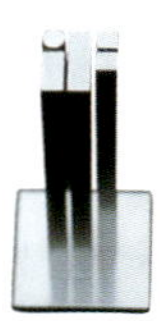

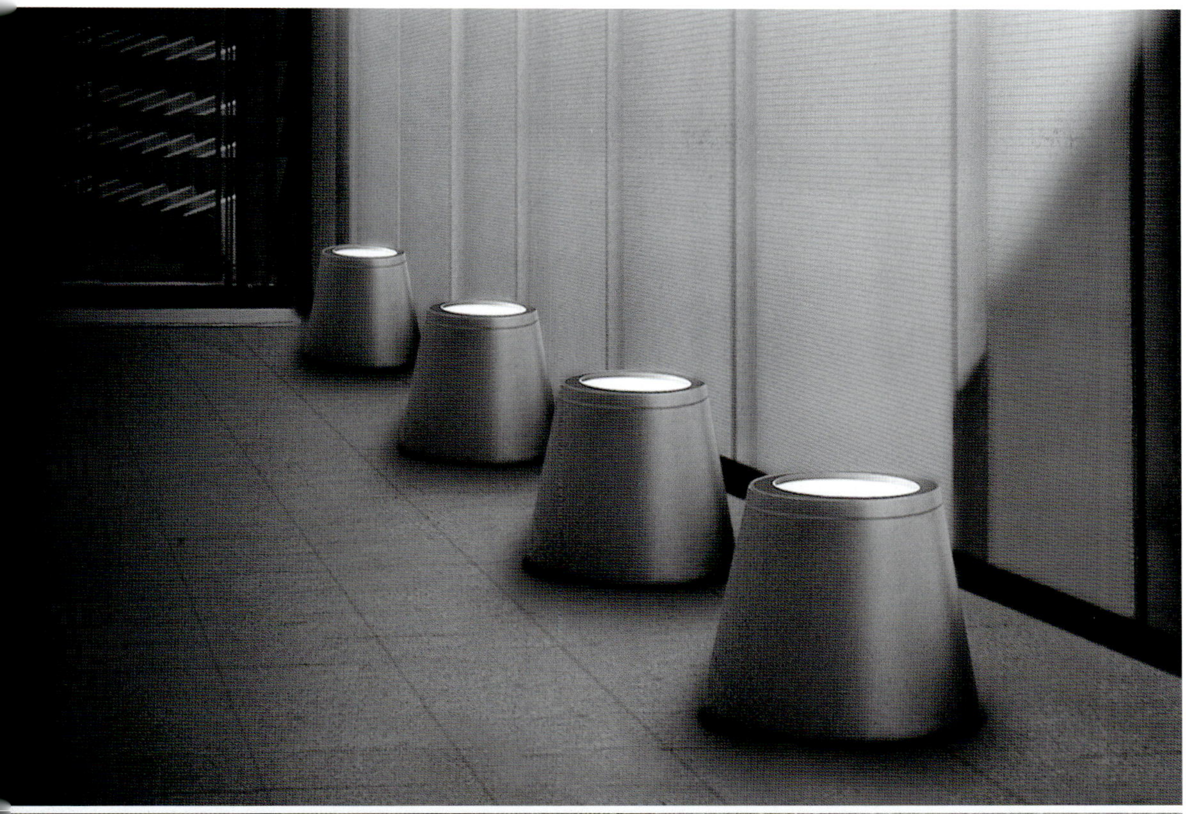

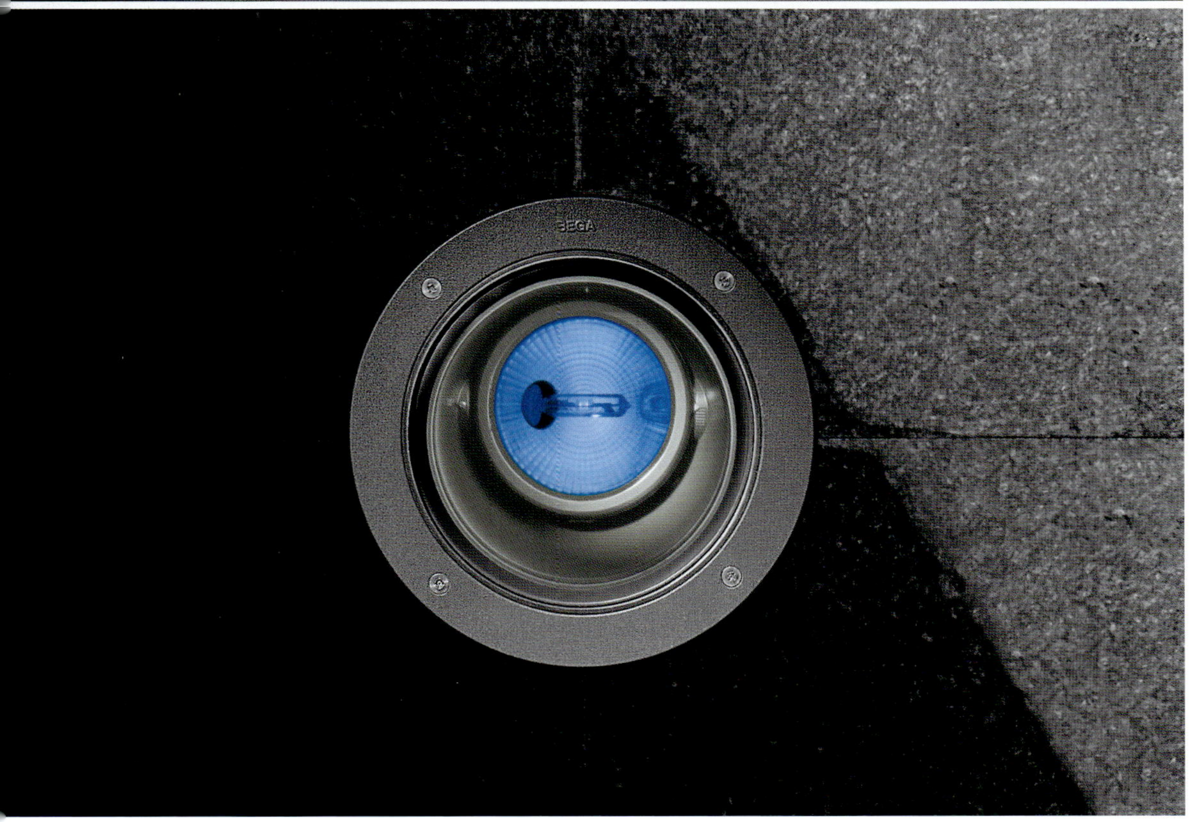

Hersteller Manufacturer
BEGA
Menden, Germany

Produkt Product
BEGA 8660
Leistungsscheinwerfer für den Bodenaufbau
Power floodlights for on-ground installation

Design
BEGA-Produktgestaltung
Menden, Germany

Ein neuartiges Scheinwerfer-System. Anstrahlgeräte in kompakter und robuster Bauform für eine Vielzahl von Beleuchtungssituationen, bei denen herkömmliche Scheinwerfer oder Bodeneinbauscheinwerfer aufgrund ihrer Bauform nicht gewünscht sind. Die Leuchten können auf ein Fundament oder auf ein Erdstück montiert werden. Die extrem robuste Konstruktion macht die Scheinwerfer staubdicht und schützt sie bei zeitweiliger Überflutung. Die Lichtstärkeverteilung der Scheinwerfer ist einstell- und veränderbar. Die schwenkbaren Kaltlichtreflektoren können im Neigungswinkel von 0°-25° eingestellt werden. So kann die Lichtstärke von symmetrisch bis asymmetrisch den Erfordernissen des Aufstellungsortes angepasst werden. Durch einfache Vorsatzmontage können Streuscheiben und dichroitische Farbeffektfilter eingesetzt werden.

An innovative floodlight system. These compact floodlights of robust construction are an ideal solution to a host of lighting situations where the design of conventional floodlights or in-ground floodlights makes them unsuitable for the task. The luminaires can be mounted on a foundation or anchorage unit. Their extremely robust construction makes them dust-proof and protects them from occasional flooding. The light distribution of the floodlights is adjustable and variable. The angled cold light reflectors can be adjusted to a slope angle of 0°-25°, allowing the luminous intensity to be adapted from symmetrical to asymmetrical to meet location requirements. Easy-to-install auxiliary diffuser lenses and dichroic colour effect filters are available as accessories.

Hersteller Manufacturer
BEGA
Menden, Germany

Produkt Product
BEGA 3277
Stoßfeste Wand + Deckenleuchten
Impact-resistant wall + ceiling luminaires

Design
BEGA-Produktgestaltung
Menden, Germany

Eine neue Gruppe Leuchten in drei Größen für kompakte, leistungsstarke Leuchtstofflampen mit hoher Wirtschaftlichkeit. Leuchten aus hochwertigem Aluminiumdruckguss und starkwandigem Kristallglas. Die Leuchten sind mit einem innenliegenden Gelenk ausgestattet, welches hohen Montagekomfort und schnellen Lampenwechsel ermöglicht. Das optisch strukturierte Glas sorgt für eine breitstreuende und gleichmäßige Lichtstärkeverteilung. Leuchten für eine Vielzahl von Anwendungsmöglichkeiten in der modernen Lichtarchitektur, vor allem aber dort, wo durch geringen Energiebedarf und geringen Wartungsaufwand ein wirtschaftlicher Betrieb gewährleistet sein soll. Die hohe Schutzart ermöglicht eine Verwendung der Leuchten dort, wo Staubdichtigkeit und Strahlwasserschutz gefordert sind.

This is a new group of luminaires in three sizes, designed for compact, powerful and highly economical fluorescent lamps. These luminaires are made of high-quality die cast aluminium and thick-walled crystal glass. They are equipped with an inside joint which ensures easy maintenance and fast relamping. The optical texture of the glass produces broad spread and uniform light distribution. The luminaires are suitable for a host of applications in modern architectural lighting but primarily for locations where economic operation requires a low level of energy and maintenance.Their high protection class allows their installation in locations where luminaires need to be dust-proof and jet-proof.

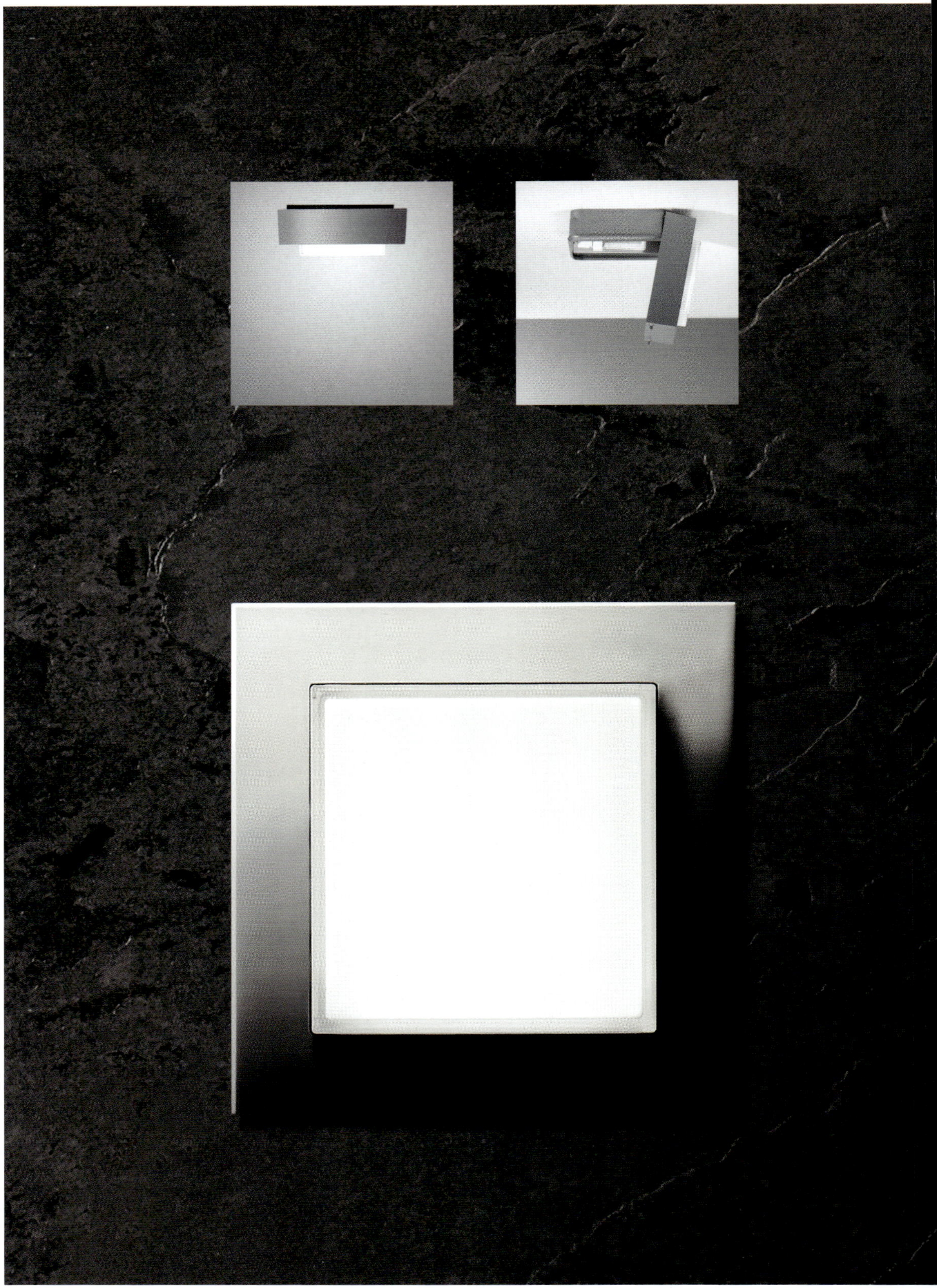

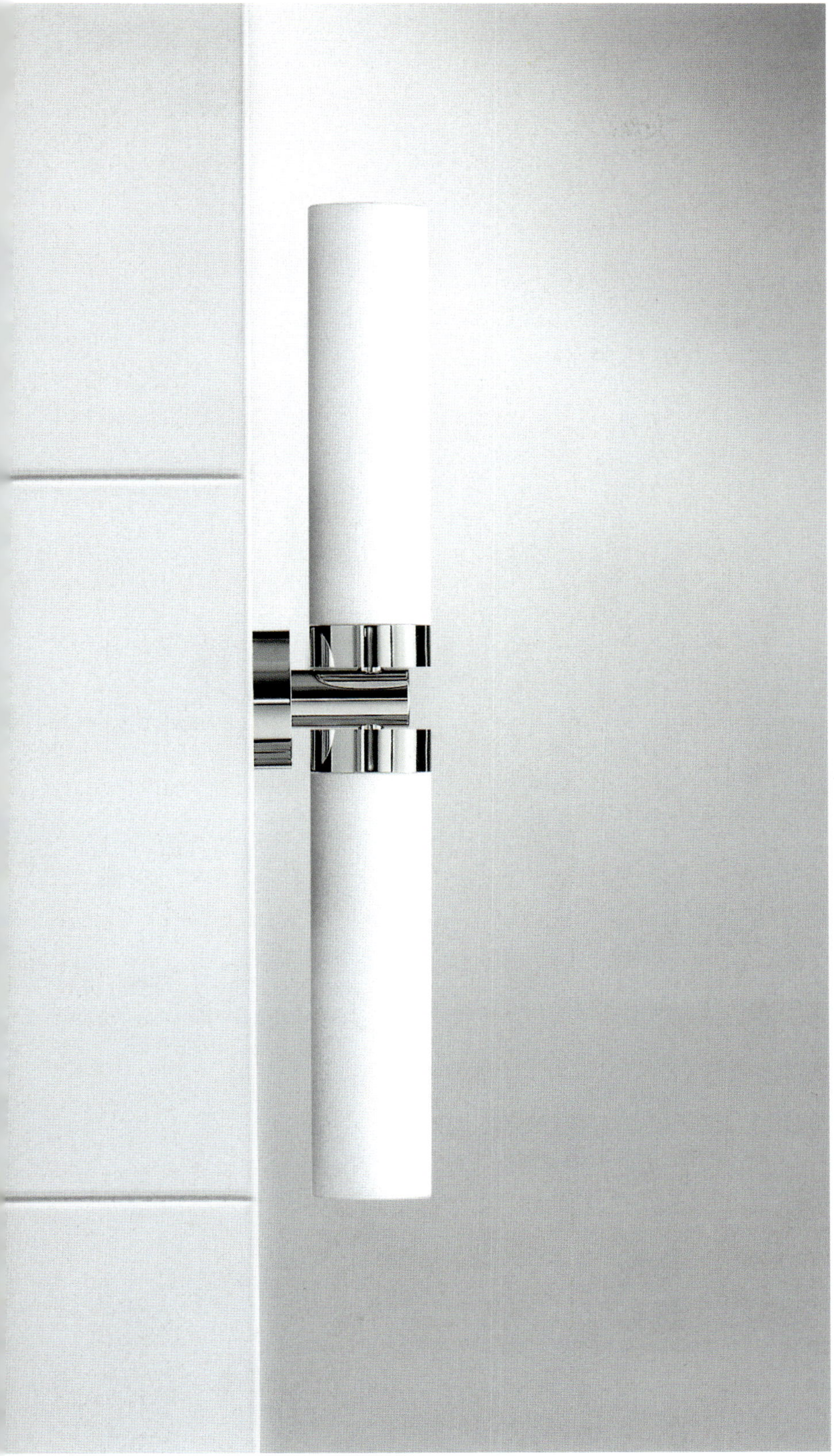

Hersteller Manufacturer
GLASHÜTTE LIMBURG
Limburg, Germany

Produkt Product
GLASHÜTTE LIMBURG 8372
Wandleuchte
Wallluminaire

Design
GLASHÜTTE LIMBURG-Produktgestaltung
Limburg, Germany

Im Bad gelten besondere Vorschriften zum Schutz von Personen gegen elektrischen Schlag. Speziell für dieses Einsatzgebiet wurde diese neue Leuchtenserie entwickelt. Leuchten von höchster Qualität, die der Sicherheit im Bad dienen. Gegossene Metallarmaturen mit veredelten Oberflächen sorgen für leichte Handhabung und sichere Montage. Das wertvolle, mundgeblasene Opalglas verteilt das Licht der Lampen weich, gleichmäßig und farbneutral. Ohne störende Blendung wird der Betrachter angenehm und ehrlich ausgeleuchtet.

Special safety directives for protection against electric shock apply for the safety of people in bathrooms. This new product range was developed for this particular purpose. Luminaires of highest quality to enhance safety in the bathroom. Die-cast metal parts in high quality finishes enable easy handling and safe installation. The precious hand-made three-ply opal glass diffuses a non-colored, smooth and comfortable light. The viewer is illuminated pleasantly and yet honestly without disturbing glare.

Hersteller Manufacturer
Daiko Electric Co. Ltd.
Higashi-Osaka, Japan

Produkt Product
Flexible light DST 53162
Tischleuchte
Desktop luminaire

Design
Shigeaki Asahara
Turin, Italy

Flexible light DST 53162 ist eine Tischleuchte für die Büroumgebung. Sie wurde speziell für die Ausleuchtung eines Tischarbeitsplatzes konzipiert. Der Leuchtenkopf kann horizontal und vertikal um 180 Grad gedreht werden und dank des Federmechanismus kann die Leuchte mit einem Spielraum von 1000 mm leicht bewegt werden. Der mitgelieferte Schirm kann am Leuchtenkopf befestigt werden und verhindert die Reflektion des Lichtes auf Bildschirmen. Die Leuchte ist aus extrudiertem Aluminium gefertigt.

Flexible light DST 53162 is an extruded aluminum desktop luminaire for office environments and has been designed specially to illuminate desk-based workstations. The lamp head can be swiveled 180° both horizontally and vertically, and thanks to a spring mechanism the luminaire is easy to move across as much as 1000 mm. The shade can be fastened to the luminaire's head and prevents light reflecting off screens.

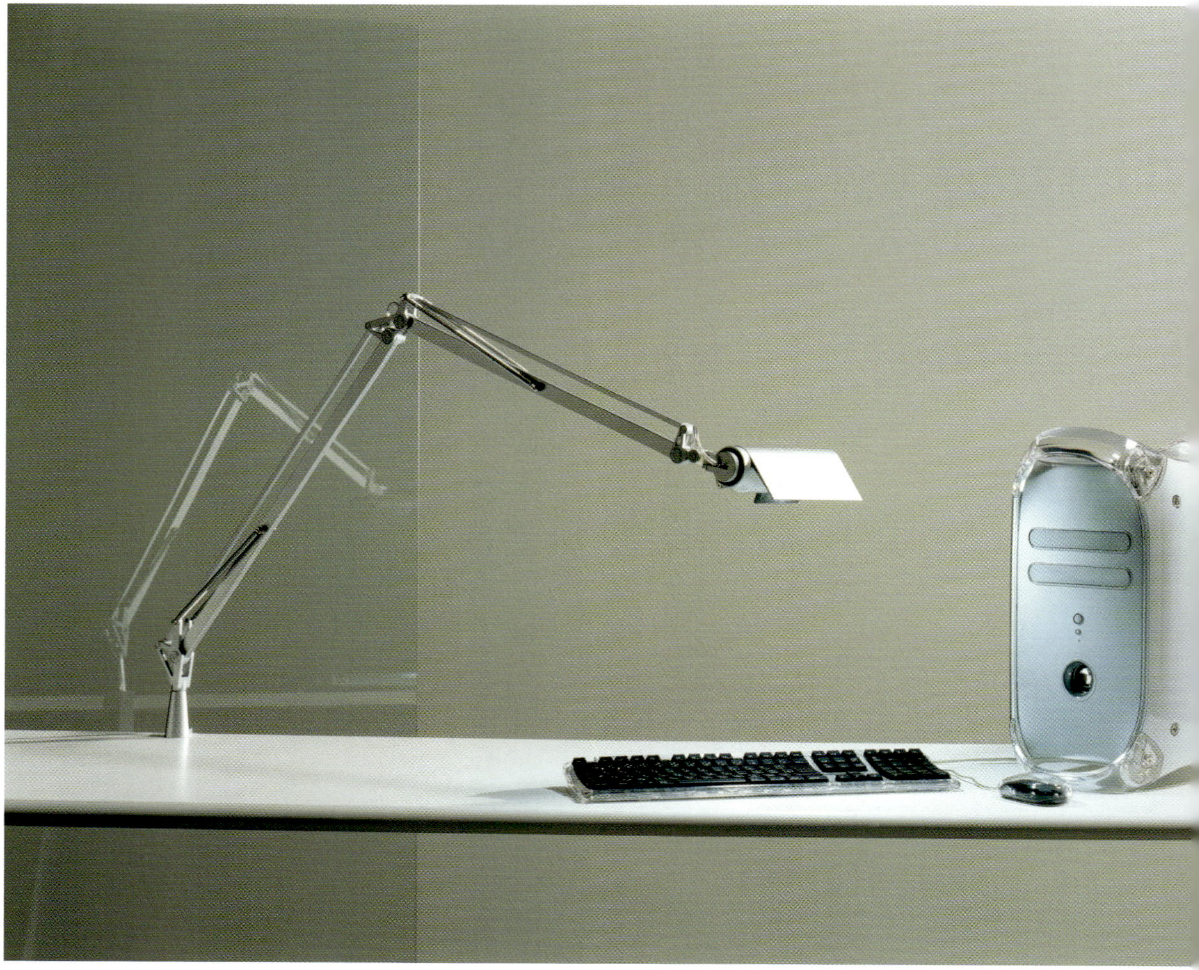

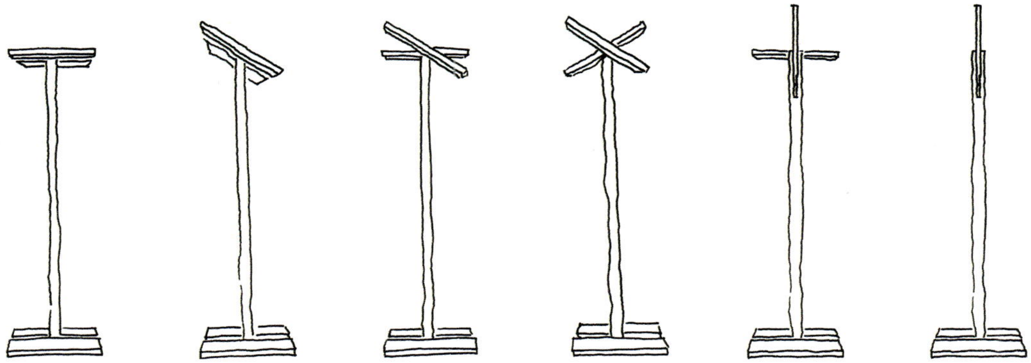

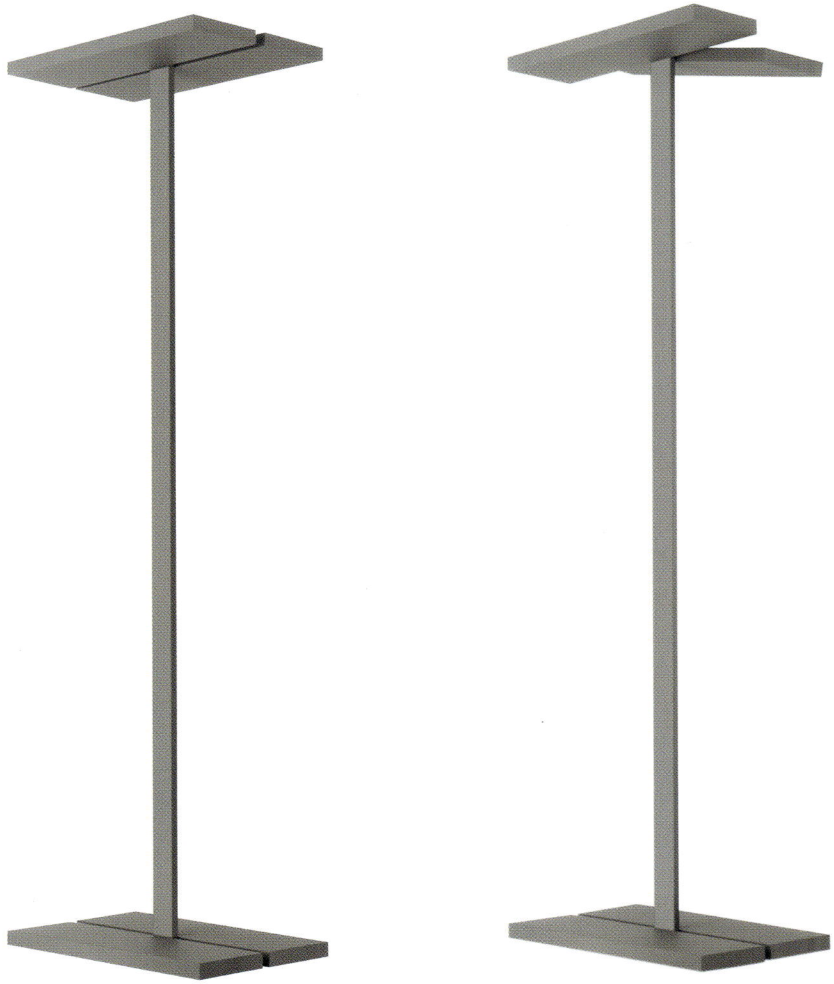

Foto: Christian Richters, Münster

Hersteller Manufacturer
Domilux Leuchten GmbH
Blomberg, Germany

Produkt Product
Kanto
Stehleuchte
Freestanding luminaire

Design
Léon Wohlhage Wernik Architekten
Berlin, Germany

Die Stehleuchte Kanto wurde vor allem für Büroräume konzipiert, in denen durch indirekte Deckenstrahlung computergerechte Arbeitsbeleuchtung gefragt ist. Sie besteht lediglich aus einem Stabprofil als Körper mit zwei flachen Leuchtenflügeln, die als identische Form wie eine Spiegelung im Fuß wiederholt werden. Diese Flügel lassen sich wie ein kinetisches Objekt um eine Achse drehen, so dass ihr Licht sowohl gegen Decken als auch gegen Wandflächen strahlen kann. Die Leuchte ist wahlweise hellgrau oder aus Edelstahl, ihre Höhe wird der Raumhöhe entsprechend maßangefertigt. Als Leuchtmittel kommen alle Standard-Leuchtstoffröhren in Frage.

The Kanto standing lamp was developed specifically with offices in mind where indirect ceiling lighting is required to illuminate workstations. It consists solely of a profile rod as the upright body and two flat lighting vanes, which are mirrored in an identically shaped footprint. The vanes can be swiveled on an axis like a kinetic object to illuminate both the ceiling and the walls. Kanto is optionally bright grey or natural stainless steel and its height is customized to the corresponding ambient room height. The luminaire can take all standard fluorescent bulbs.

Hersteller Manufacturer
Zumtobel Staff GmbH
Dornbirn, Austria

Produkt Product
Light Fields
Lichtsystem
Lighting system

Design
Sottsass Associati
Mailand, Italy

Das von Sottsass Associati entworfene Lichtsystem Light Fields ist als direkt strahlende An- oder Einbauleuchte, als Pendel- oder Stehleuchte mit direkt/indirekter Lichtführung und als clusterähnliches Lichtfeld erhältlich. Die neuartige Mikropyramiden-Strukturoptik sorgt für blendfreies Licht und für optimale Bedingungen an Bildschirm-Arbeitsplätzen. Aufgrund ihrer reduzierten Formensprache und des minimalistischen Designs fügen sich die Leuchten unauffällig in Raum und Architektur ein.

The Light Fields lighting system designed by Sottsass Associati is available as a ceiling-mounted or built-in luminaire as a direct spot, a suspended lamp or standing lamp with direct or indirect lighting, or as a cluster-type light field. The innovative, micro-pyramid textured optic creates glare-free light which is ideal for workstations where monitors are used. Given the reduced formal vocabulary and the minimalist design, the luminaires blend inconspicuously with the ambient space and architectural setting.

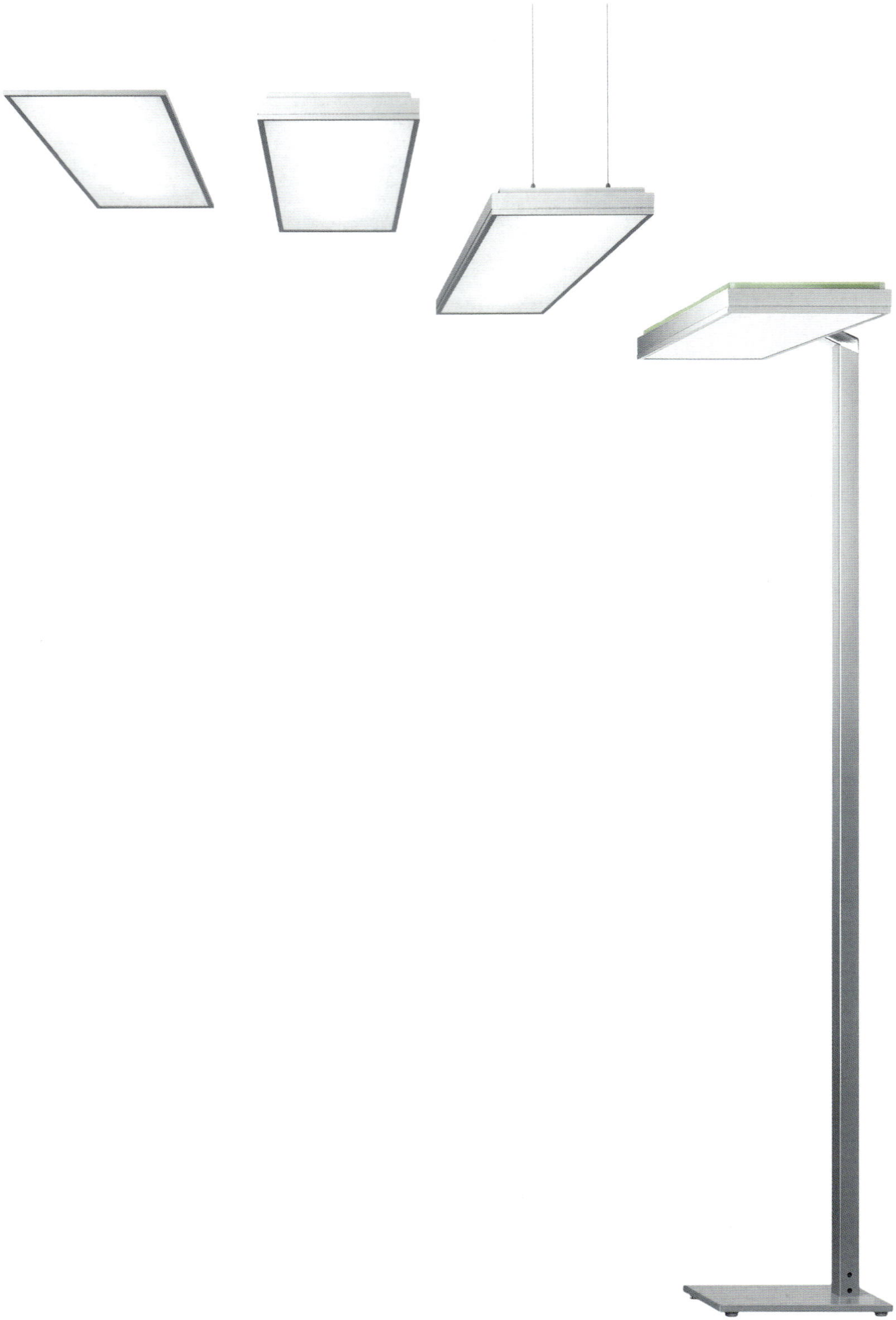

Hersteller Manufacturer
TRILUX-LENZE GmbH + Co KG
Arnsberg, Germany

Produkt Product
Büroleuchten-System
Office luminaire system
Stehleuchte
Free-standing luminaire

Design
Kuebler Produkt- und Raumgestaltung
Zürich, Switzerland

Die Leuchten aus dem Büroleuchten-System von TRILUX wurden als Arbeitsplatzbeleuchtung konzipiert. Die Stehleuchte besteht aus drei Basiselementen, dem Standfuß, dem Standprofil sowie den Leuchtenköpfen. Integrierte Justiermöglichkeiten im Standfuß und Trägerprofil ermöglichen die exakte vertikale und horizontale Ausrichtung der Leuchte. Der Standfuß selbst bietet durch seine spezielle Formgebung weitgehende Freiheit in der Positionierung an Einzel- oder Doppelarbeitsplätzen. Die Leuchte ist aus speziell extrudierten Elementen gefertigt und erhält dadurch eine reduzierte und zugleich filigran wirkende Konturen- und Formensprache.

The TRILUX office luminaire system was conceived as workstation lighting. The free-standing luminaire consists of three basic elements, the stand foot, the stand profile as well as the luminaire heads. Integrated adjustment possibilities in the stand foot and support profile enable the precise vertical and horizontal alignment of the luminaire. The stand foot itself offers great freedom in positioning the luminaire next to single- or twin-desk workstations. The luminaire consists of specially extruded elements and shines out with subtle and slender contours and shapes.

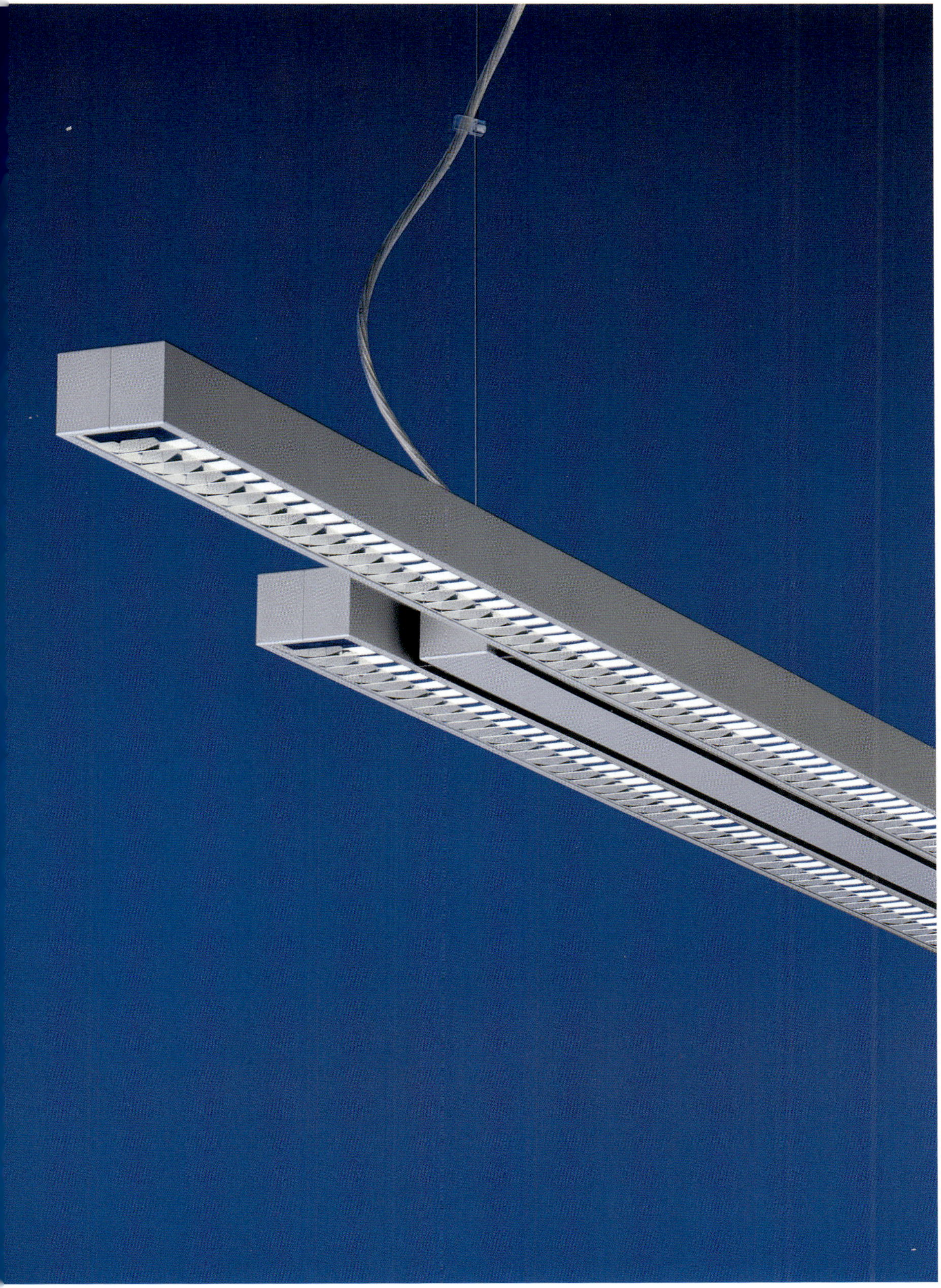

Hersteller Manufacturer
TRILUX-LENZE GmbH + Co KG
Arnsberg, Germany

Produkt Product
Büroleuchten-System
Office luminaire system
Hängeleuchte
Suspended luminaire

Design
Kuebler Produkt- und Raumgestaltung
Zürich, Switzerland

Die Leuchten für das Büroleuchten-System von TRILUX wurden als Arbeitsplatzbeleuchtung konzipiert. Die Hängeleuchte besteht aus drei Basiselementen, dem Trägerprofil, den Leuchtenköpfen und einer dünnen Seilaufhängung. Eine integrierte Justiermöglichkeit ermöglicht die horizontale Ausrichtung der Leuchte in zwei Achsen. Die Leuchtenköpfe sind stufenlos verschiebbar und dadurch bezüglich ihrer architektonischen und lichttechnischen Wirkung inszenierbar.

The TRILUX office luminaire system was conceived as workstation lighting. The suspended luminaire consists of three basic elements, namely the carrier profile, the luminaire heads and a thin wire suspension. The integrated adjustment possibility allows the luminaire to be horizontally adjusted along two axes. The luminaire heads are continuously displaceable and therefore variable in their architectural and photometric effect.

Hersteller Manufacturer
IDL Leuchten und Lichttechnik GmbH
Chemnitz, Germany

Produkt Product
Leuchtstoff
Pendelleuchte
Pendant luminaire

Design
Alexander Krippstädt
Ultramöbel
Dresden, Germany

Die Pendelleuchte Leuchtstoff schlägt den Bogen vom Einsatz des Porzellans als Isolator vor 100 Jahren zum Einsatz bei einer modernen Leuchte mit Leuchtstoffröhre. Dabei wurden die nötigen technischen Bauteile vom Leuchtmittel gelöst im Baldachin untergebracht, wodurch die Leuchtstoffröhre frei im Raum zu schweben scheint. Durch die einstellbare Drehung und Neigung der Leuchte ergibt sich bei Hängung in Reihe ein besonderer Reiz.

Leuchtstoff is a pendant luminaire that draws a line from the use of porcelain a century ago as an insulator to its utilization in a modern neon-tube luminaire. The requisite technical components are separated from the lamp section and housed in the canopy, meaning that the neon tube apparently floats free in space, suspended from the ceiling or in front of a wall. The luminaire can be twisted and tilted as desired, enabling users to give a row of such units a special look of their own.

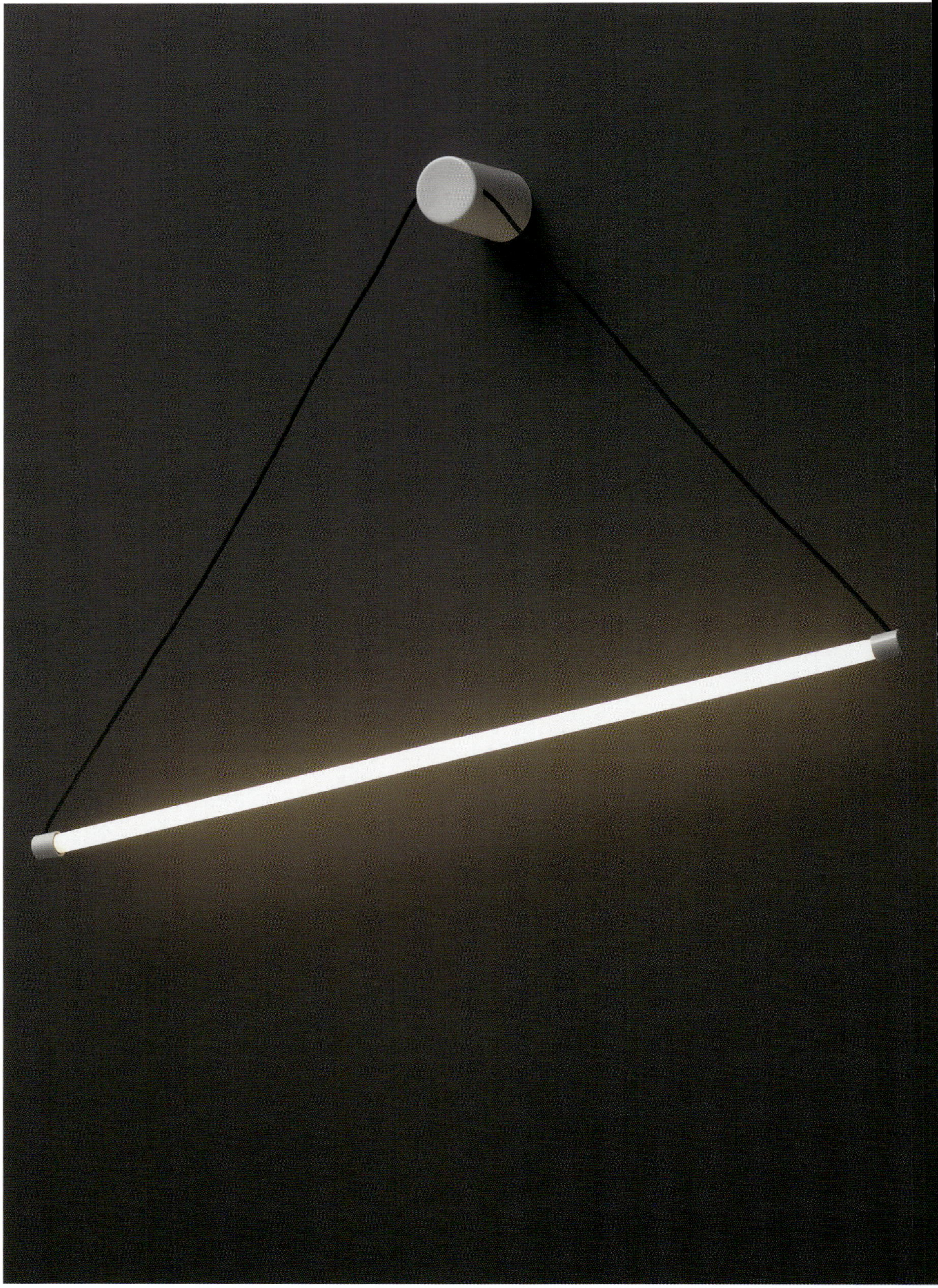

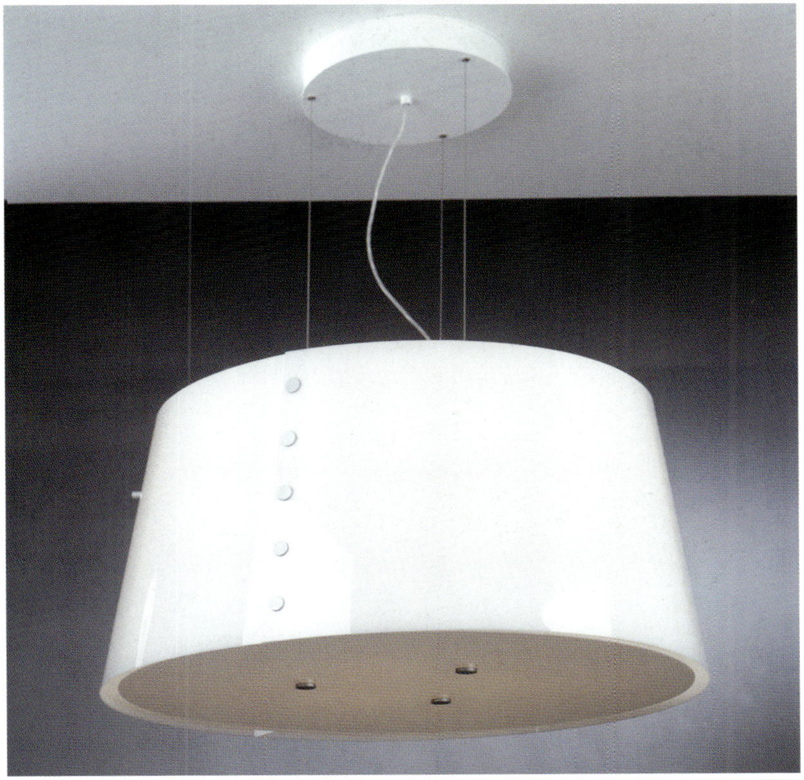

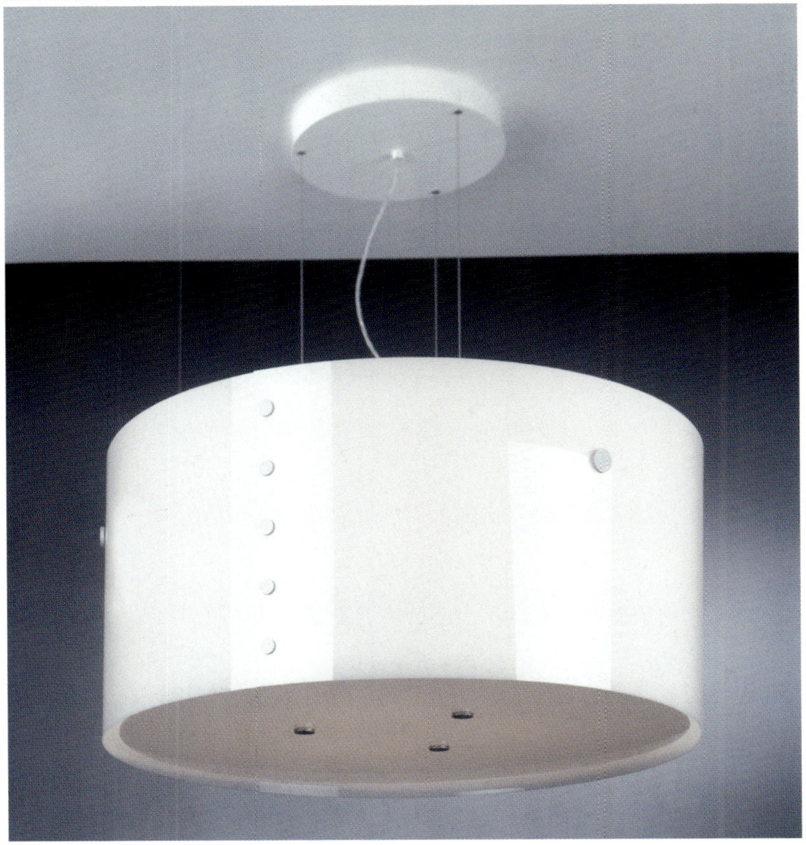

Hersteller Manufacturer
IDL Leuchten und Lichttechnik GmbH
Chemnitz, Germany

Produkt Product
Airbus
Pendelleuchte
Pendant luminaire

Design
Mark Frost
IDL Leuchten und Lichttechnik GmbH
Chemnitz, Germany

Die schlichte, zylindrische (leicht konische) Pendelleuchte Airbus erzeugt durch ihren opalen Polycarbonatschirm ein sehr homogenes Licht. Die kompakte Innenkonstruktion erlaubt eine schattenfreie Ausleuchtung. Bestückt wird die Lampe mit drei Großkolbenglühlampen mit maximal 100 Watt.

Airbus is a no-frills, cylindrical and slightly conical pendant luminaire which sheds highly homogeneous light thanks to its opaline polycarbonate shade. The complex insides to the luminaire guarantee it casts shadow-free light, sourced by three large lamps with a maximum of 100W.

Hersteller Manufacturer
Zapf GmbH
Bayreuth, Germany

Produkt Product
Garagen-Lichtspiele
Garage lighting show

Design
Zapf GmbH
Bayreuth, Germany

Das Lichtkonzept für Garagen besteht aus einem in der Garage angebrachten Leuchtmittel und einem Tor und/oder einer Tür mit speziell strukturierter Füllung. Das Licht der im Innenbereich der Garage angebrachten Leuchtmittel scheint durch die speziell strukturierte und lichtdurchlässige Torfüllung nach außen. Die Lichtsteuerung erfolgt wahlweise mit einem Fernsender oder aber einer im Haus angebrachten Schaltung, zudem besteht die Möglichkeit einer Steuerung über den Torantrieb.

The lighting concept for garages consists of a luminaire located in the garage and a gate and/or door with a specially structured filling. The light from the luminaire mounted inside the garage shines through the specially structured and light-permeable gate filling outward. The lighting is controlled either by a remote handset or via a switch in the house; the lights can also be controlled via the gate drive.

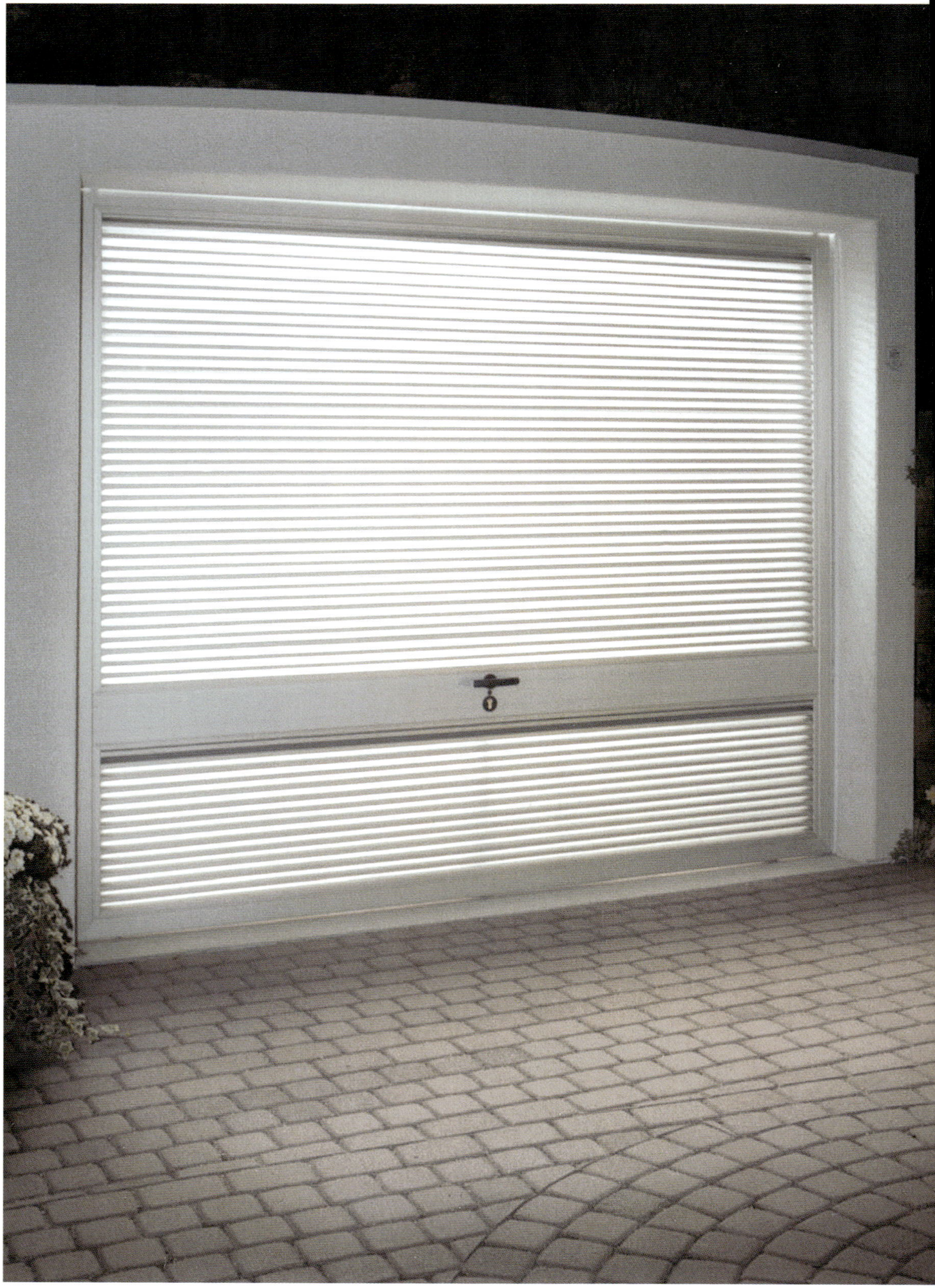

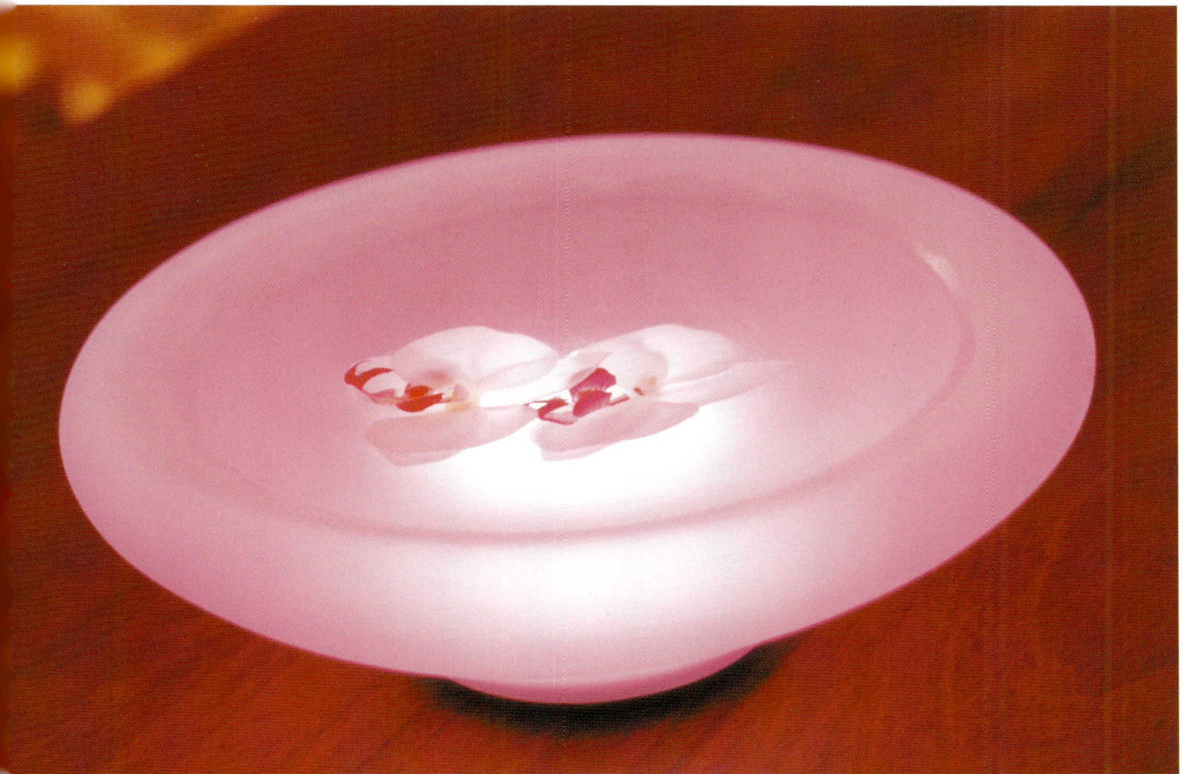

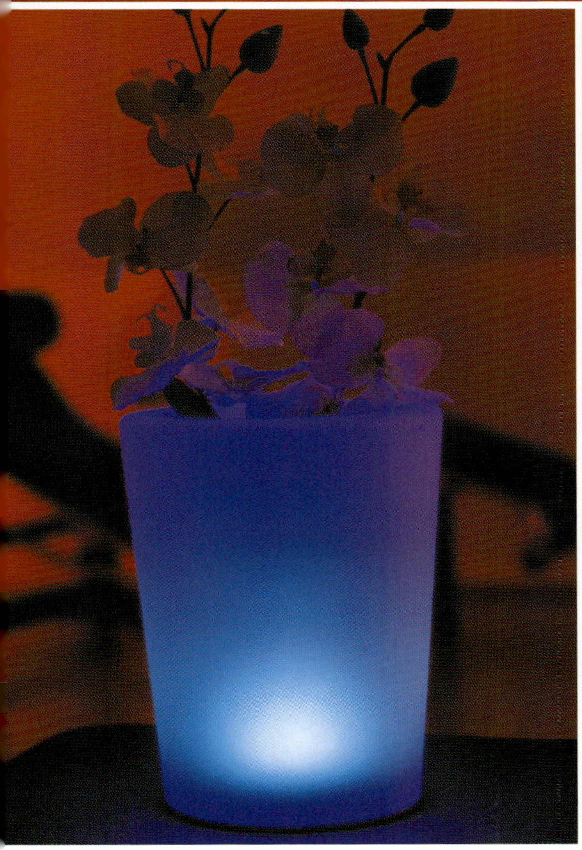

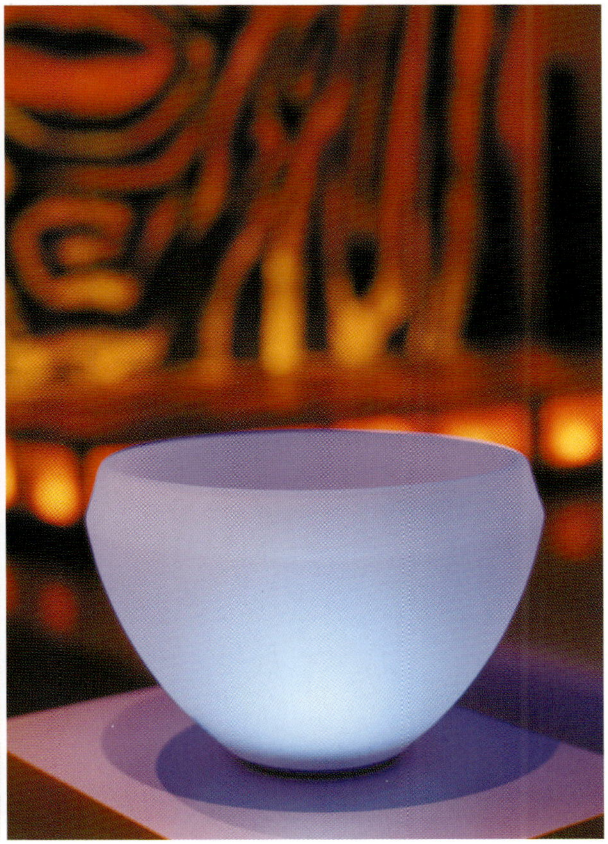

Hersteller Manufacturer
Traxon Group Ltd.
Hong Kong, China

Produkt Product
Mood Light Objects
Schalen
Bowls

Design
Dipl.-Designer Holger Diehl
Diehldesign GmbH
Frankfurt am Main, Germany

Die Mood Light Bowls sind Lichtobjekte für Tisch, Regal oder Theke, mit denen unterschiedliche Lichtstimmungen mit wechselnden Farbkombinationen erzeugt werden können. Die Schalen gibt es in drei Varianten: Mood Light Rotondo, Mood Light Polaris und Mood Light Ambience. Sie sind aus hochwertigem, mundgeblasenen, satinierten Glas gefertigt, das im italienischen Parnaterra™-Verfahren hergestellt wird. Der integrierte Akku ermöglicht einen autarken Betrieb von bis zu 10 Stunden.

The Mood Light Bowls are lighting sculptures for tables, shelves or bar tops where different atmospheres and moods can be created using changing combinations of colored light. There are three different bowl versions: Mood Light Rotondo, Mood Light Polaris and Mood Light Ambience. They are made of high-grade, hand-blown satin glass, manufactured using the Italian Parnaterra™ method. Thanks to the integrated battery, the bowls can run free of the mains for up to ten hours.

Hersteller Manufacturer
Artemide S.p.A.
Pregnana-Milanese, Italy

Produkt Product
Pipe Sospensione
Steh- oder Hängeleuchte
Floor or suspension luminaire

Design
Herzog & De Meuron
Artemide S.p.A.
Pregnana-Milanese, Italy

Die Steh- oder Hängeleuchte Pipe Sospensione kann aufgrund des in einem speziellen Verfahren gefertigten Rohres und des besonderen Endstücks der Rohrverkleidung beliebig ausgerichtet werden. Auf diese Weise kann je nach Anforderung ein indirektes und gestreutes oder aber ein direktes, gezieltes Licht geschaffen werden. Der Rahmen der Hängeleuchte besteht aus durchgehendem, extrem flexiblen Stahlrohr, das mit transparentem Platinsilikon bezogen ist und in einen kegelförmigen Lichtschirm übergeht. Die Dark Light Optik und die kleinen Löcher des Aluminiumreflektors lassen das Licht in feinen Lichtbündeln austreten. Der Schirm ist aus transparentem Polykarbonat gefertigt.

The Pipe floor or suspended luminaire can be adjusted as desired due to its tube which is manufactured in a special process and the special end-piece on the tube covering. This allows it to produce either indirect dispersed light or direct, targeted light as required. The frame of the suspended light consists of extremely flexible tubular steel throughout which is covered in a transparent platinum-silicone sheath that terminates in a conically-shaped diffuser. The 'dark-light' optics and the small perforations in the aluminum reflector emit light in small bundles. The diffuser is made of transparent polycarbonate.

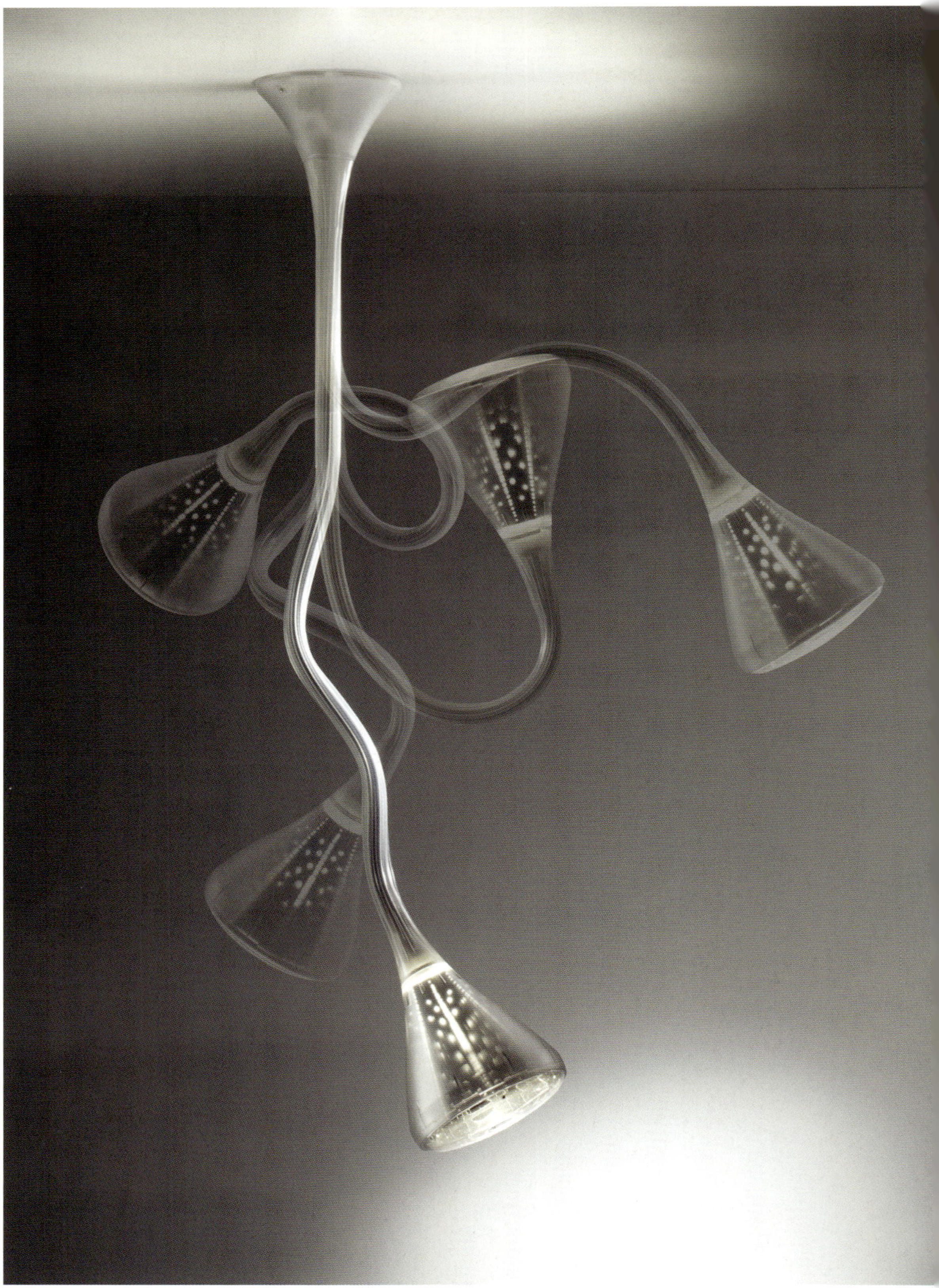

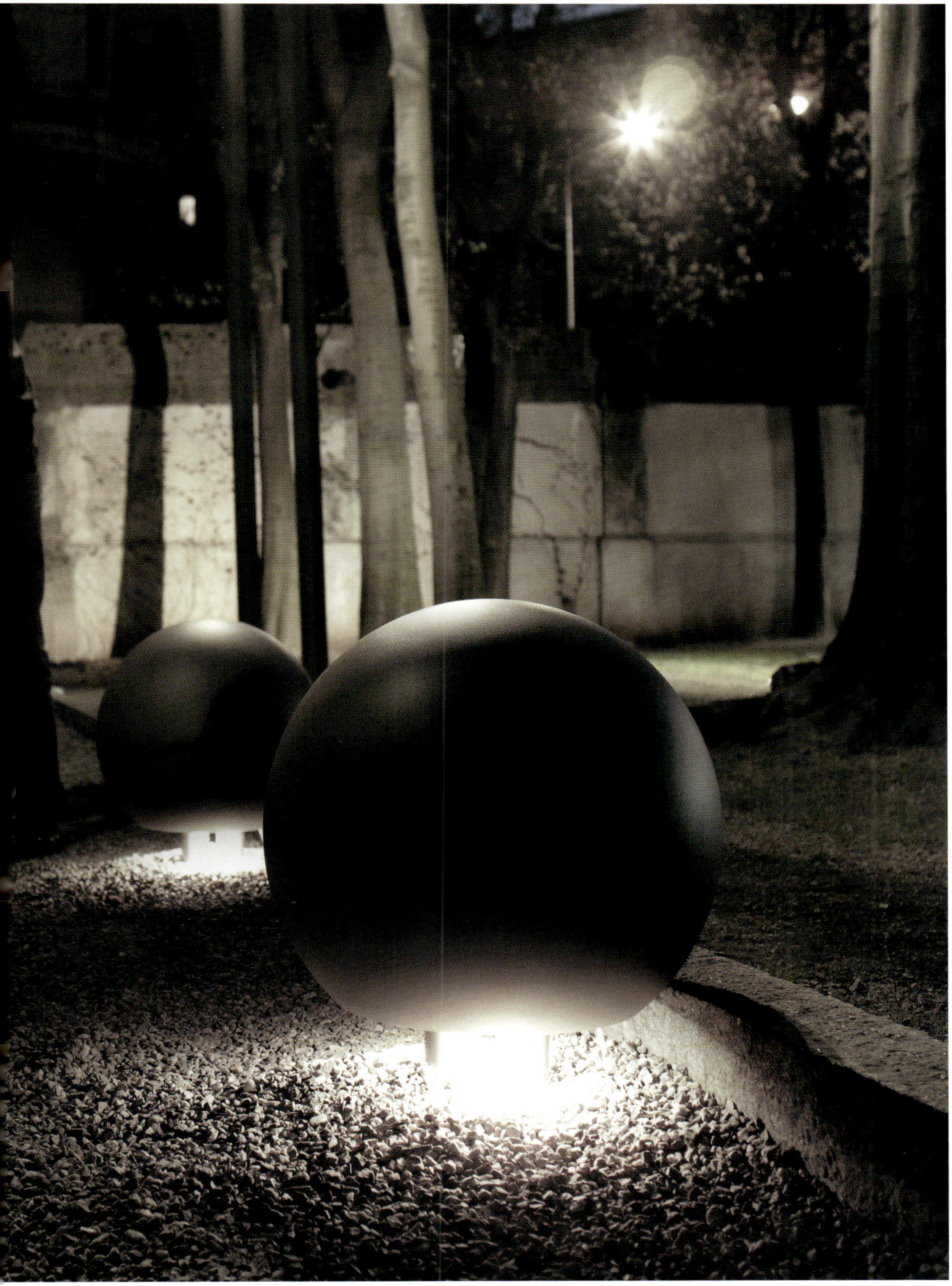

Hersteller Manufacturer
Artemide S.p.A.
Pregnana-Milanese, Italy

Produkt Product
Signoria 250, 500
Pollerleuchte
Bollard luminaire

Design
Arata Isozaki
Artemide S.p.A.
Pregnana-Milanese, Italy

Durch ihr unübersehbares Format dient die Pollerleuchte Signoria 250 und 500 über eine nach unten gerichtete Lichtverteilung gleichzeitig als Straßenstein und als Pollerleuchte zur Signalisierung einer Platzabgrenzung. Die Kugel besteht aus Aluminiumguss, Innenrohr und Flanschplatte aus Edelstahl (500) oder aus Aluminium (250). Die Signalleuchte ist in zwei Größen erhältlich (ø 500 mm und ø 250 mm).

The conspicuous shape of the Signoria 250 and Signoria 500 bollard luminaires enables them to disperse light downward both as a shock-proof bollard and as a directional light. The bowl is made of cast aluminum, the internal pole and flange plate of stainless steel (500) or aluminum (250). The directional luminaire is available in two sizes, with diameters of 500 mm and 250 mm respectively.

Hersteller Manufacturer
Sirrah s.r.l.
Recanati, Italy

Produkt Product
Central 42
Hängeleuchte
Suspended luminaire

Design
Lella und Massimo Vignelli, David Law
Vignelli Associates Designers
New York, USA

Central 42 ist eine Hängeleuchte zur Streulichtemission mit einem Schirm aus satiniertem Polykarbonat. Die Leuchte ist in zwei Größen erhältlich, wobei die größere Version die zur Bestückung mit Halogen- und Leuchtstofflampen notwendigen elektrischen Komponenten beinhaltet. Der interne Reflektor der Leuchte besteht aus Reinstaluminium. Die Leuchte hat eine industrielle Anmutung und eignet sich für jede Umgebung.

Central 42 is a suspended luminaire with a satin polycarbonate shade, which emits a diffused light. The unit comes in two sizes, the larger features the electrical ballast necessary to take halogen and fluorescent lamps. The internal reflector is made of pure aluminum and the luminaire's industrial feel makes it suitable for all environments.

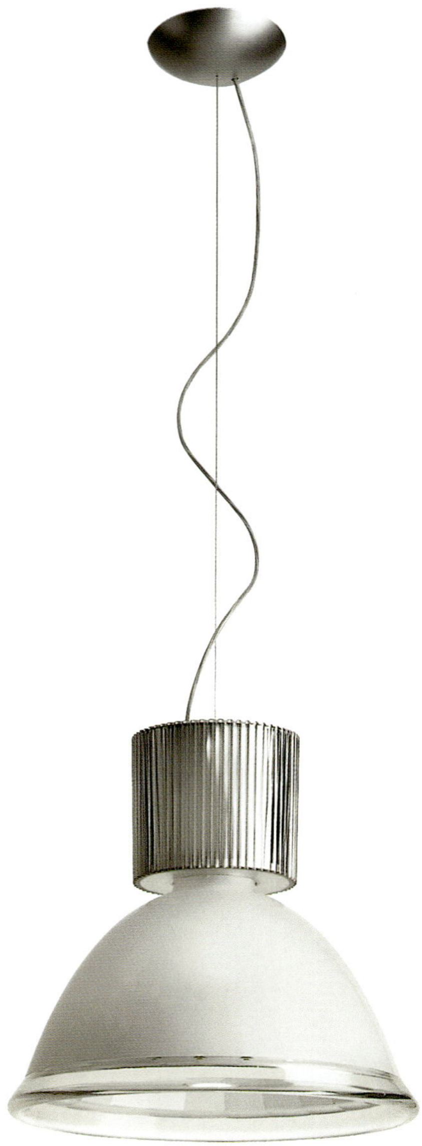

Hersteller Manufacturer
Sirrah s.r.l.
Recanati, Italy

Produkt Product
Cup 110
Wandleuchte
Recessed luminaire

Design
Roberto Pamio
Roberto Pamio & Associates
Venedig, Italy

Cup ist eine Reihe von Einbauleuchten mit Schirmen aus transparentem Dickglas. Dieses Material sorgt unter Einsatz von Lampen mit hohem Wirkungsgrad für originelle Dekoreffekte an der Wand und schafft eine helle und weiche Lichtsituation. Cup wird in zwei Größen (ø 100 und ø 120 mm) mit zylinderförmigem Schirm zur Bestückung mit Leuchtstofflampen angeboten. Die Leuchten sind einfach zu montieren und zu warten.

Cup is a series of built-in lights with a diffuser made of transparent thick window glass with a satin finish. This material in conjunction with high-power fluorescent lamps creates original decorative effects on the wall and a bright, diffused lighting situation. Cup is available in two sizes, with diameters of 100mm and 120mm and comes with cylinder-shaped shades. The luminaire can be easily installed and maintained.

Hersteller Manufacturer
Delta Light NV
Roeselare, Belgium

Produkt Product
Be Cool
Leuchtenfamilie
Luminaires

Design
Delta Light NV
Roeselare, Belgium

Die Leuchtenfamilie Be Cool versucht die Neonröhre in ein neues und besseres Licht zu rücken. Sie zeichnet sich durch ein äußerst puristisches und „cooles" Design aus, in dem zwei runde Leuchtröhren die neue Generation von Neonröhren mit 16 mm Durchmesser aufnehmen und auf diese Weise eine maximale Lichtausbeute für Hänge, Wand- und Deckenleuchten bieten.

The Be Cool family of luminaires tries to place neon fluorescent lamps in a new and better light. The luminaires have an outstandingly purist and "cool" design, with special carriers for two round new-generation 16 mm neon tubes, thus delivering maximum light for the range of lamps, pendant, wall- or ceiling-mounted luminaires.

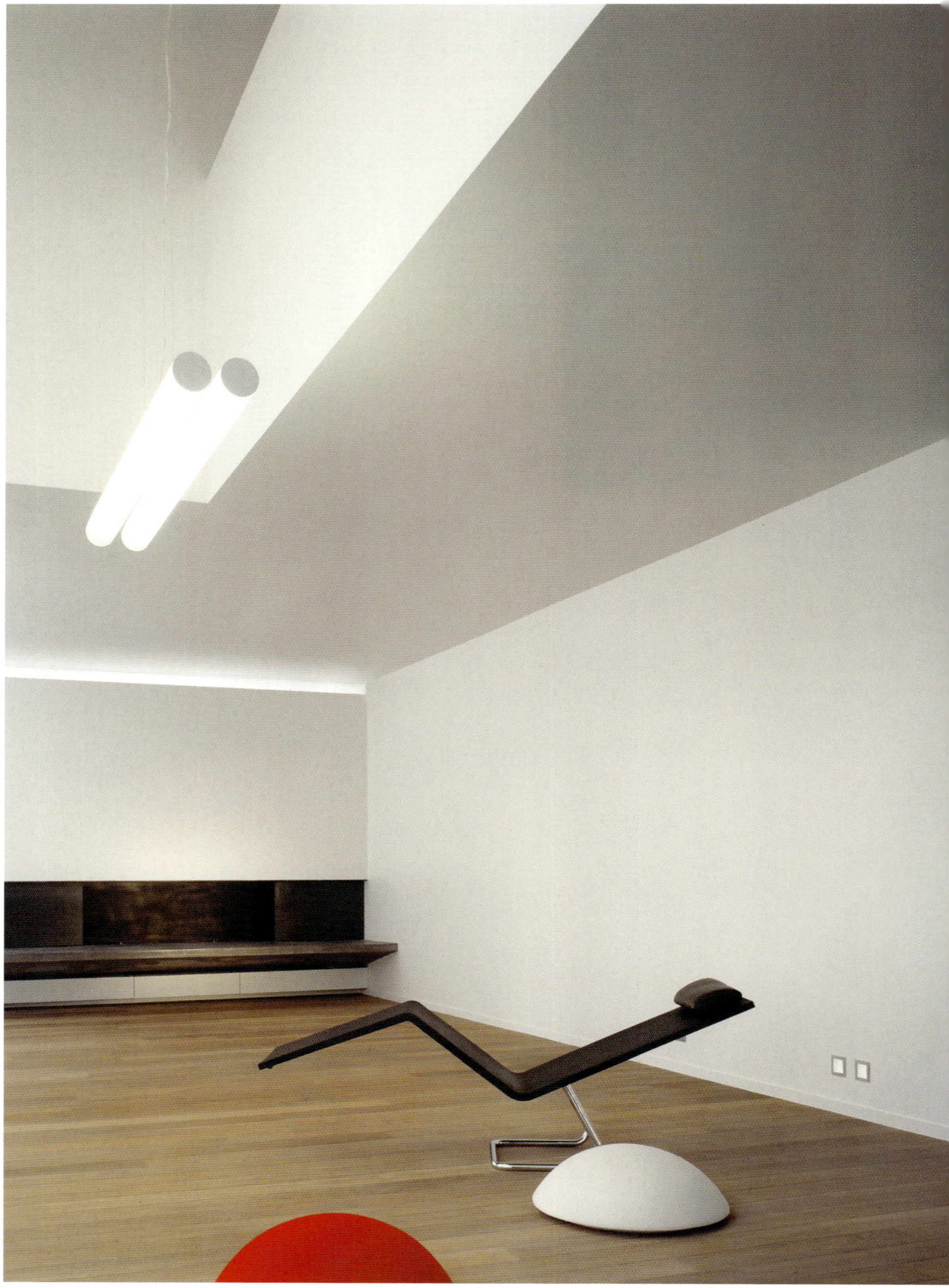

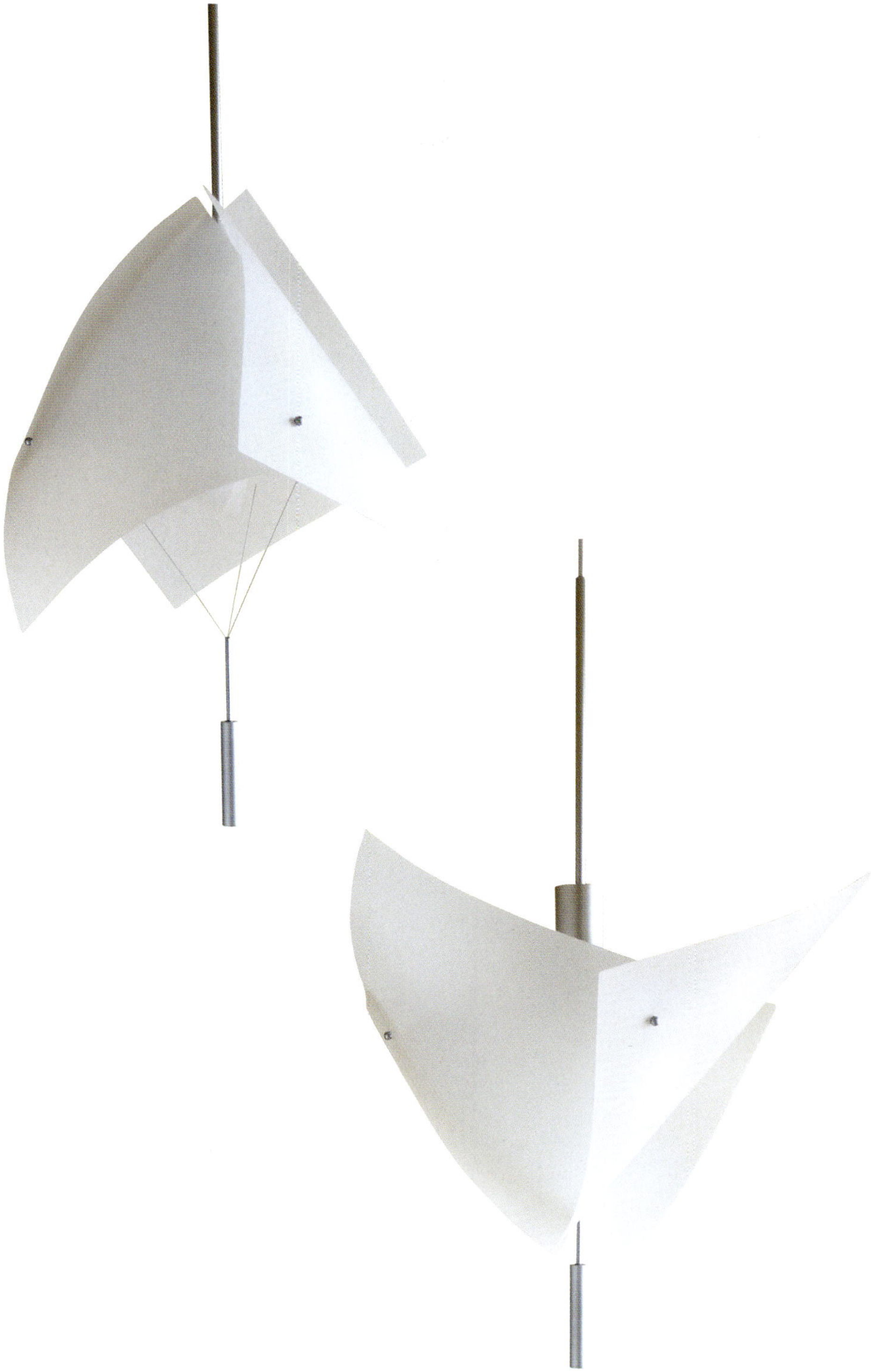

Hersteller Manufacturer
E27/ Luminarias
(ONI Design e Produtos Ltda.)
São Paulo, Brazil

Produkt Product
+viva
Hängelampe
Hanging luminaire

Design
André Wagner
André Wagner Associados
São Paulo, Brazil

+viva ist eine preisgünstige Hängeleuchte für Wohnungen, die mit Glühbirnen bestückt wird. Aufgrund eines speziellen mechanischen Prinzips verfügt sie über eine sowohl direkte wie auch indirekte Beleuchtungsmöglichkeit. Im Vergleich zur Originalausführung von 1997 wurden hier der Zuschnitt der Kunststofffolien und die Fertigung sowie der Verpackungsaufwand optimiert.

+viva is an inexpensive hanging luminaire designed for apartments and runs on standard lamps. A special mechanical principle enables it to produce both direct and indirect lighting. Compared with the original model in 1997, this model has optimized not only the cut of the plastic film but also the manufacturing and packing costs.

Hersteller Manufacturer
Plajet Magnética Distribuidora Ltda
Duque de Caxias, Brazil

Produkt Product
Spirit 203
Deckenventilator mit integrierter Beleuchtung
Ceiling fan with an integrated light

Design
Martin Birtel, Jaakko Tammela,
Felipe Rangel, Manuela Vilaseca,
Indio da Costa
Indio da Costa Design
Rio de Janeiro, Brazil

Der Deckenventilator Spirit 203 stellt eine technische Weiterentwicklung des schon früher ausgezeichneten Vorgängermodells Spirit 201 dar. Zentral wurde jetzt ein kompaktes Beleuchtungselement mit zwei 60W Halopin Birnen integriert. Die Ventilatorflügel sind aus widerstandsfähigem Polycarbonat, der Reflektor aus hitzebständigem Kunststoff und die Linse des Strahlers aus Glas. Die leicht demontierbaren Bestandteile lassen sich einzeln recyclen. Spirit 203 ist in verschiedenen transluzenten Farben erhältlich.

The Spirit 203 ceiling fan enhances the features of its award winning predecessor, the Spirit 201, by offering an exceptionally compact luminaire, which is totally integrated in the fan shape due to the use of two extra small 60w halopin lamps. The product is made of high-resistance injection molded polycarbonate blades and motor covers, a heat-resistant industrial plastic reflector and frosted glass. All the components can be easily disassembled and recycled thanks to the minimalist structure of the ceiling fan. A variety of translucent colors is available.

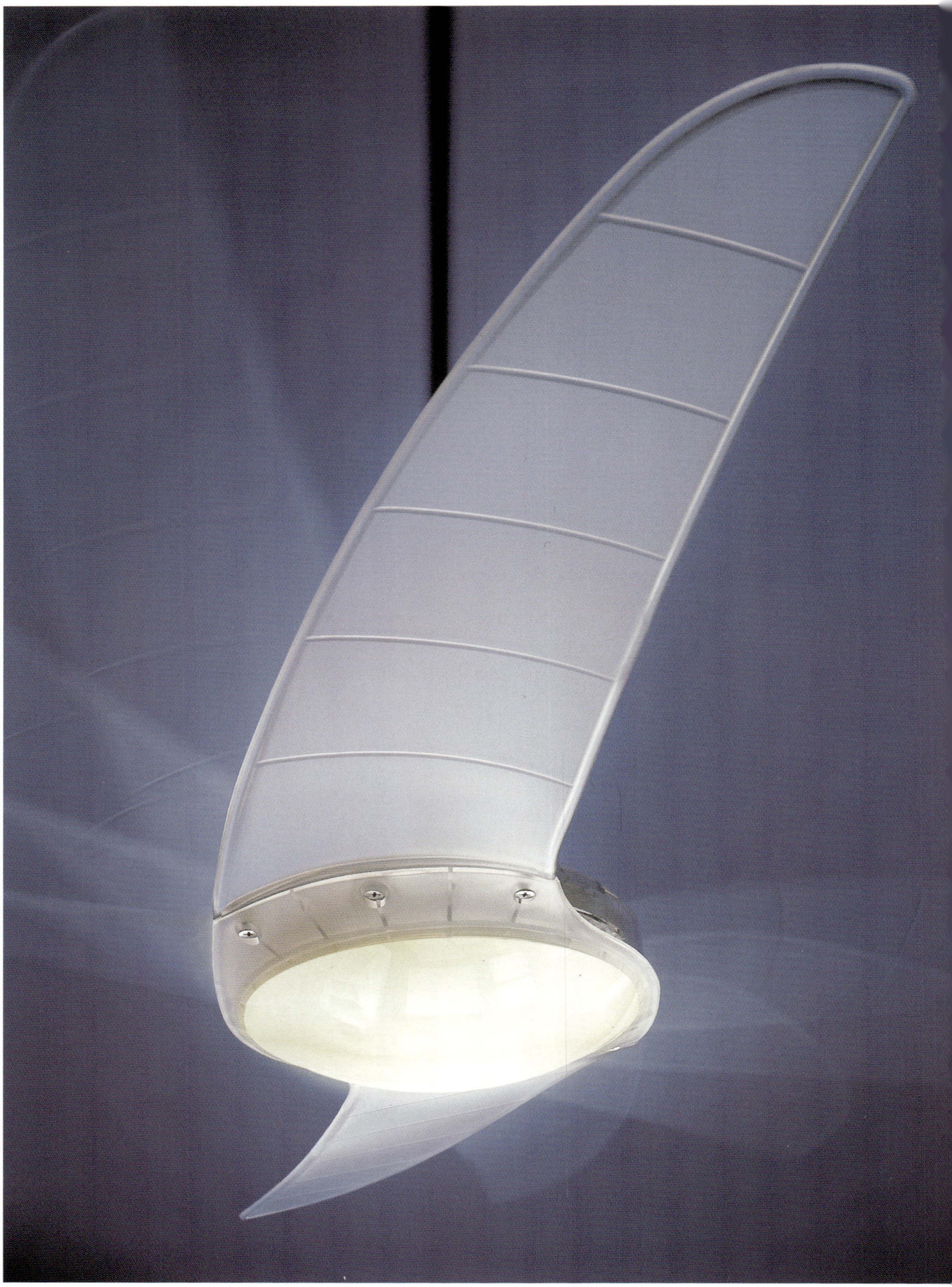

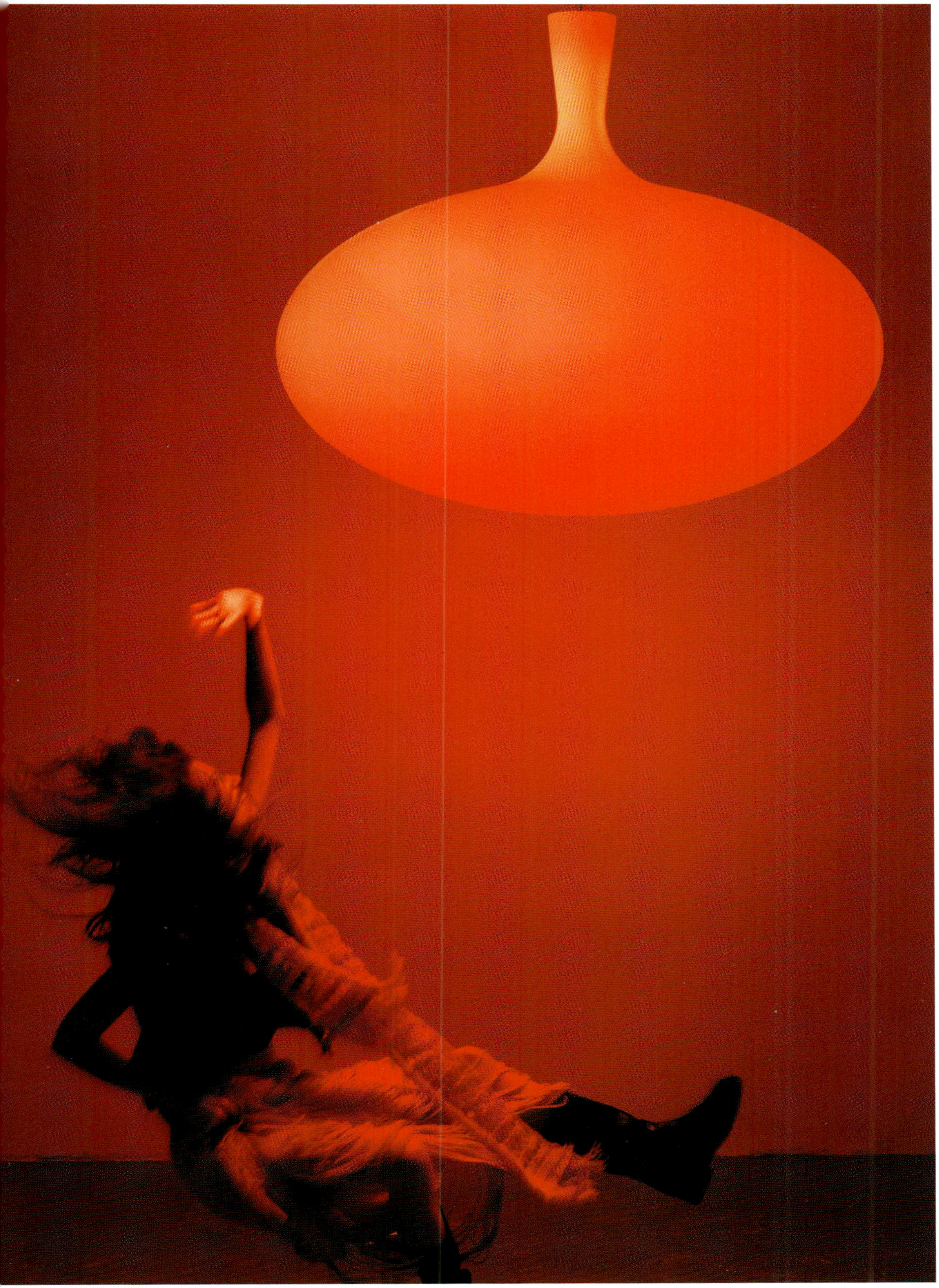

Hersteller Manufacturer
Dark Nv
Maldegem, Belgium

Produkt Product
No Fruit
Leuchte
Luminaire

Design
Anthony Duffeleer
Antwerpen, Belgium

No Fruit wurde von dem Architekten Anthony Duffeleer für Vitra Point Antwerpen entworfen. Die Leuchte hat eine große organische, transluzente Form und spendet ein weiches Licht für große Räume. Das Licht kann wahlweise weiß, rot, orange, violett, grün oder blau sein. Auch eine dimmbare Version ist erhältlich.

No Fruit was created by architect Anthony Duffeleer for Vitra Point Antwerp. The luminaire features a large organically-shaped, translucent shape and bathes large rooms in soft light. The luminaire comes in white, red, orange, violet, green or blue. Also available in a dimmable version.

Hersteller Manufacturer
Louis Poulsen Lighting A/S
Kopenhagen, Denmark

Produkt Product
Enigma
Pendelleuchte
Pendant luminaire

Design
Shoichi Uchiyama

Die Pendelleuchte Enigma gibt ein blendfreies und weiches Licht. Die vier übereinander angeordneten kreisförmigen Schirme aus mattiertem Acryl scheinen, wenn sie angestrahlt sind, frei im Raum zu schweben. Der Leuchtenkern ist aus gebürstetem Aluminium gefertigt.

The Enigma pendant luminaire casts soft, glare-free light. The four circular shades arranged one over the other are made of matt acrylic and, once illuminated, seem to float freely in space. The luminaire boasts a brushed aluminum core.

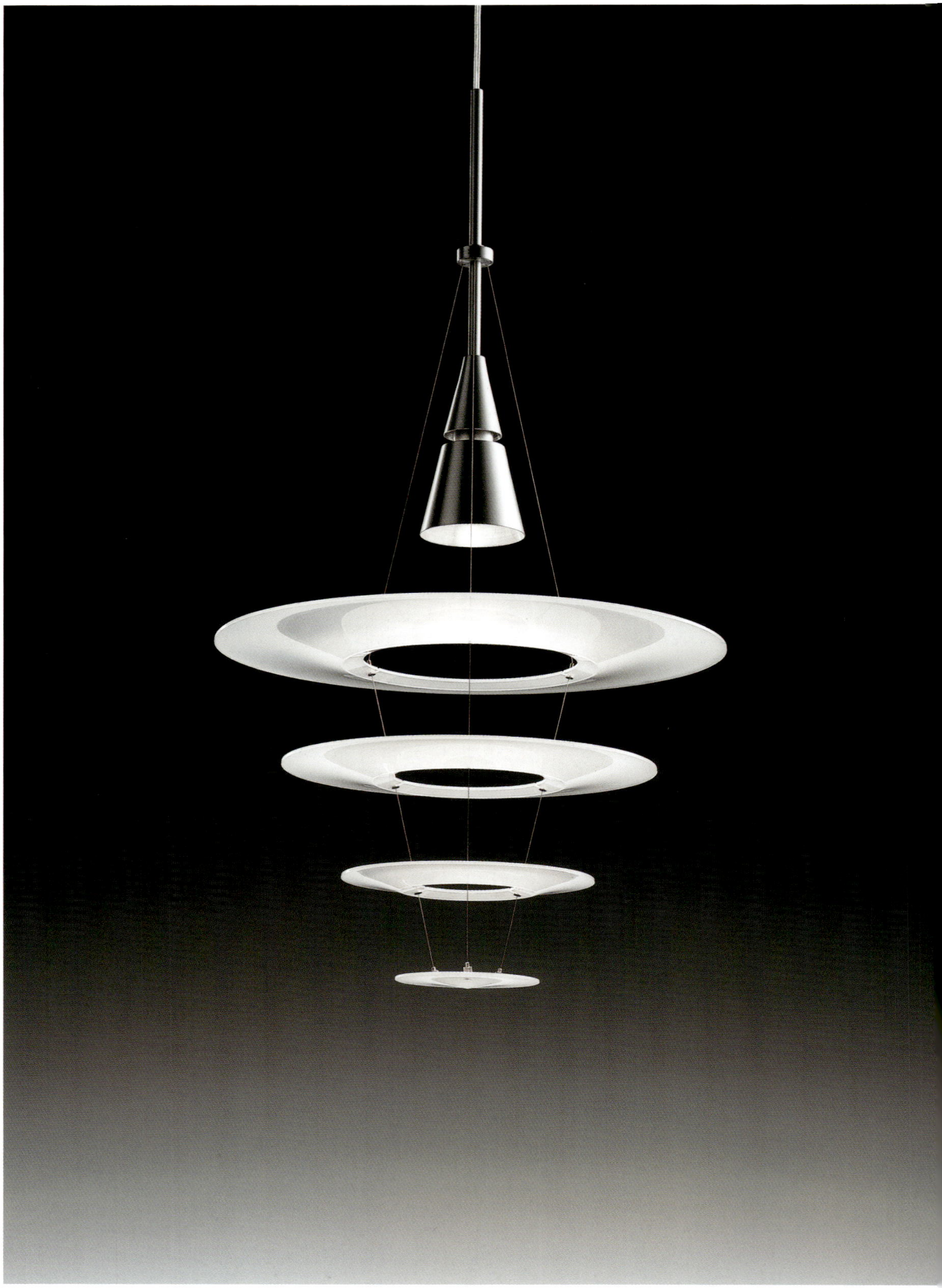

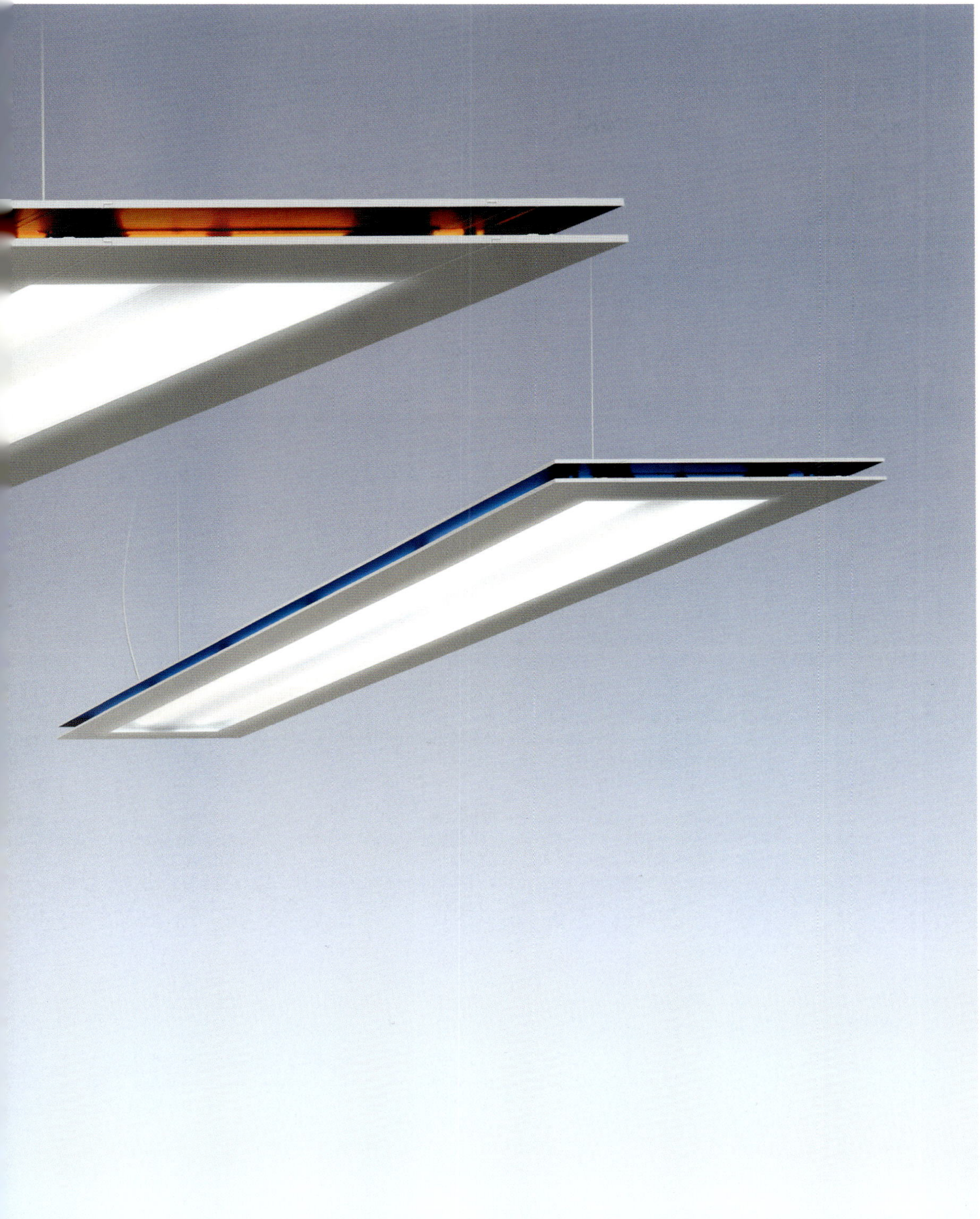

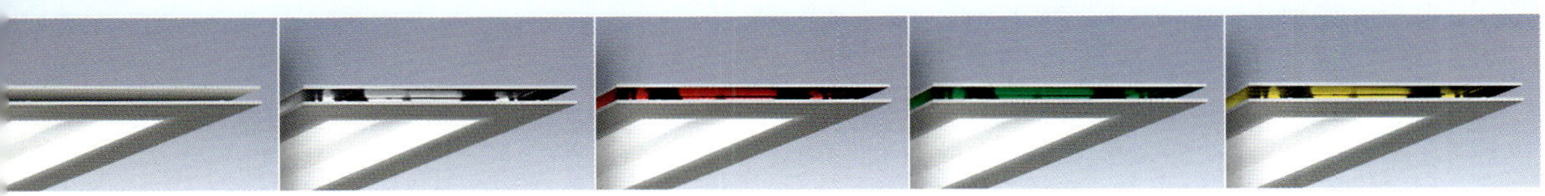

Hersteller Manufacturer
GRIMMEISEN Vertriebs GmbH
Wasserburg am Inn, Germany

Produkt Product
SLIDE-Swing
Büro-Pendelleuchte
Office pendant luminaire

Design
Gerhard Grimmeisen
GRIMMEISEN LICHT
Wasserburg am Inn, Germany

SLIDE-Swing ist eine Büro-Pendelleuchte, mit der das Licht auf die jeweils notwendige Raumhöhe gebracht wird. Die klassische Hängeleuchte mit 18 mm Höhe besteht aus einem leichten, zweiflügeligen Alurahmen und wird mit zwei oder vier energiesparenden Leuchtstoffröhren T5 mit je 54 W bestückt. In den Seiten- und Frontflügeln lassen sich Farbfilter integrieren. Der seitliche diffuse Farbaustritt erlaubt ein effektvolles Erscheinungsbild, das die Lichtfarbe im Arbeitsfeld aber nicht beeinflusst. Je nach Wahl der verschiedenen Optiken, wird entblendetes, diffuses Rundumlicht oder indirektes Licht ermöglicht. Alle SLIDE Varianten sind optional dimmbar.

SLIDE-Swing is a pendant luminaire for offices that provides a lighting source at whatever height is required. The classical 18 mm thick pendant luminaire is made up of a lightweight bipartite aluminum frame and is powered by energy-saving neon T5 tubes delivering 54W each. Color filters can be integrated into the side and forward wings. The diffuse colored light from the sides creates a striking appearance without impairing the color of light shed on the work top. Different optical elements are available and all provide glare-free diffuse light or indirect illumination. All SLIDE versions can as an option be dimmed.

Hersteller Manufacturer
SPECTRAL Gesellschaft für Lichttechnik mbH
Freiburg, Germany

Produkt Product
PLAFOU
Pendelleuchte
Pendant luminaire

Design
M. Kocks

PLAFOU ist Flachleuchte und Lichtobjekt in einem und wurde für den Büro- und Objektbereich entwickelt. Eine abgekantete Prismenscheibe bildet das stabile Leuchtengehäuse, das den technischen Kern umhüllt. Die Stirnseiten bestehen aus Edelstahlblech oder transluzentem Kunststoff. Ihre Abmessungen betragen in der Höhe 35 mm, in der Länge ca. 1200 mm und in der Breite 180 mm.

PLAFOU is both a flat luminaire and sculpted light, developed for use in offices, conference and meeting rooms. The folded prism plate creates a stable housing which covers the core with the lamps. The sides feature sheet stainless steel or translucent plastic. The dimensions: a mere 35 mm high, approx. 1200 mm long, and 180 mm wide.

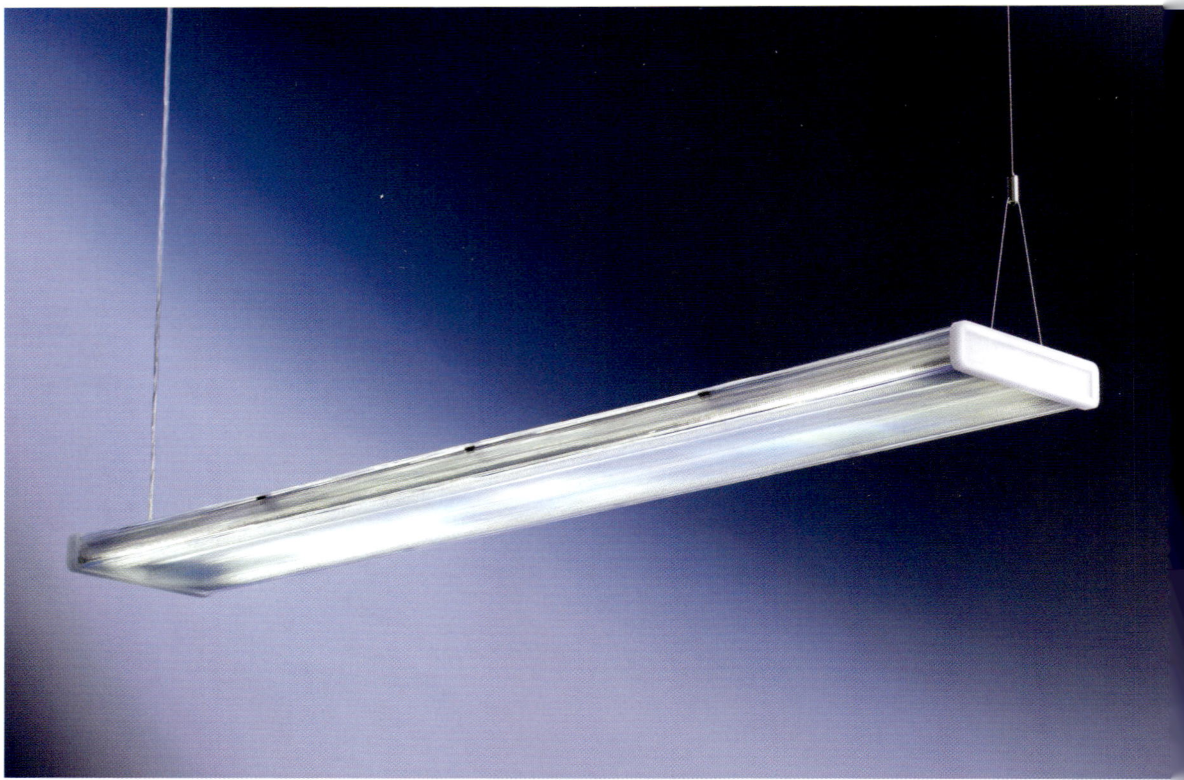

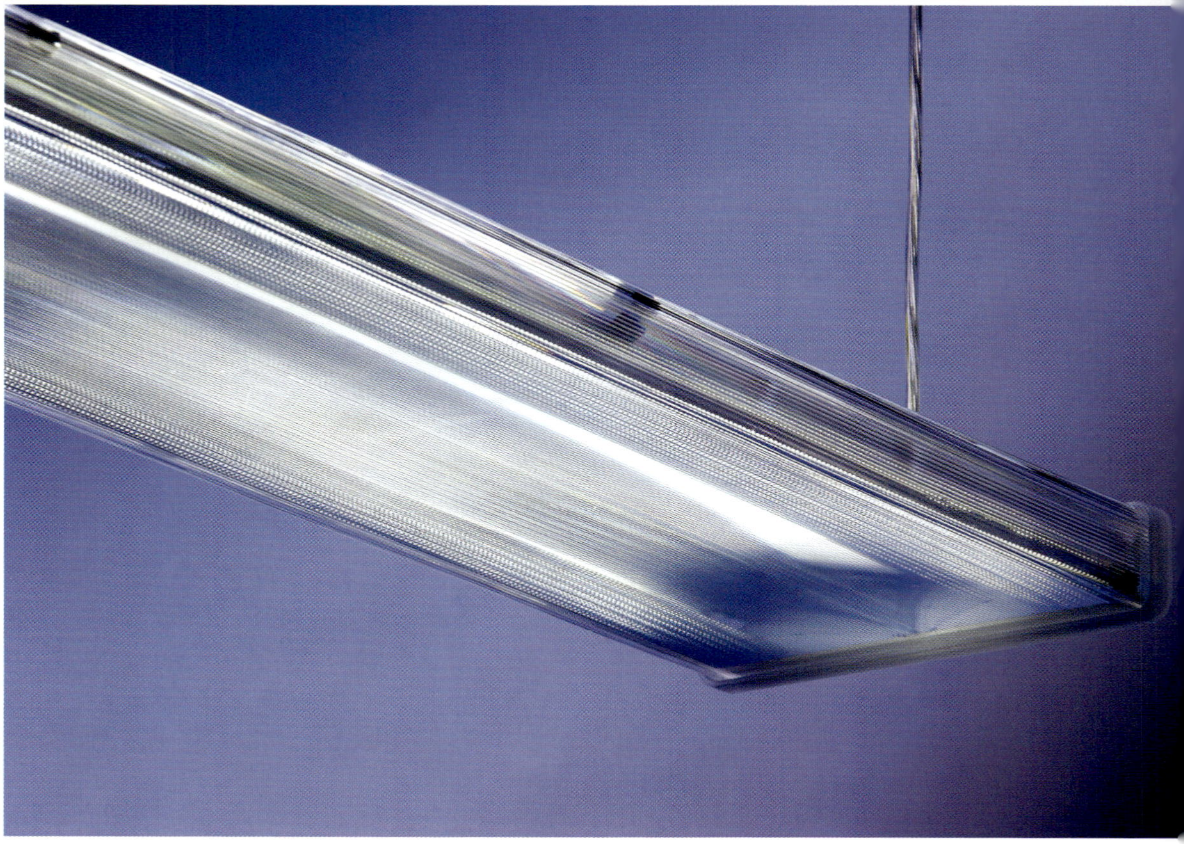

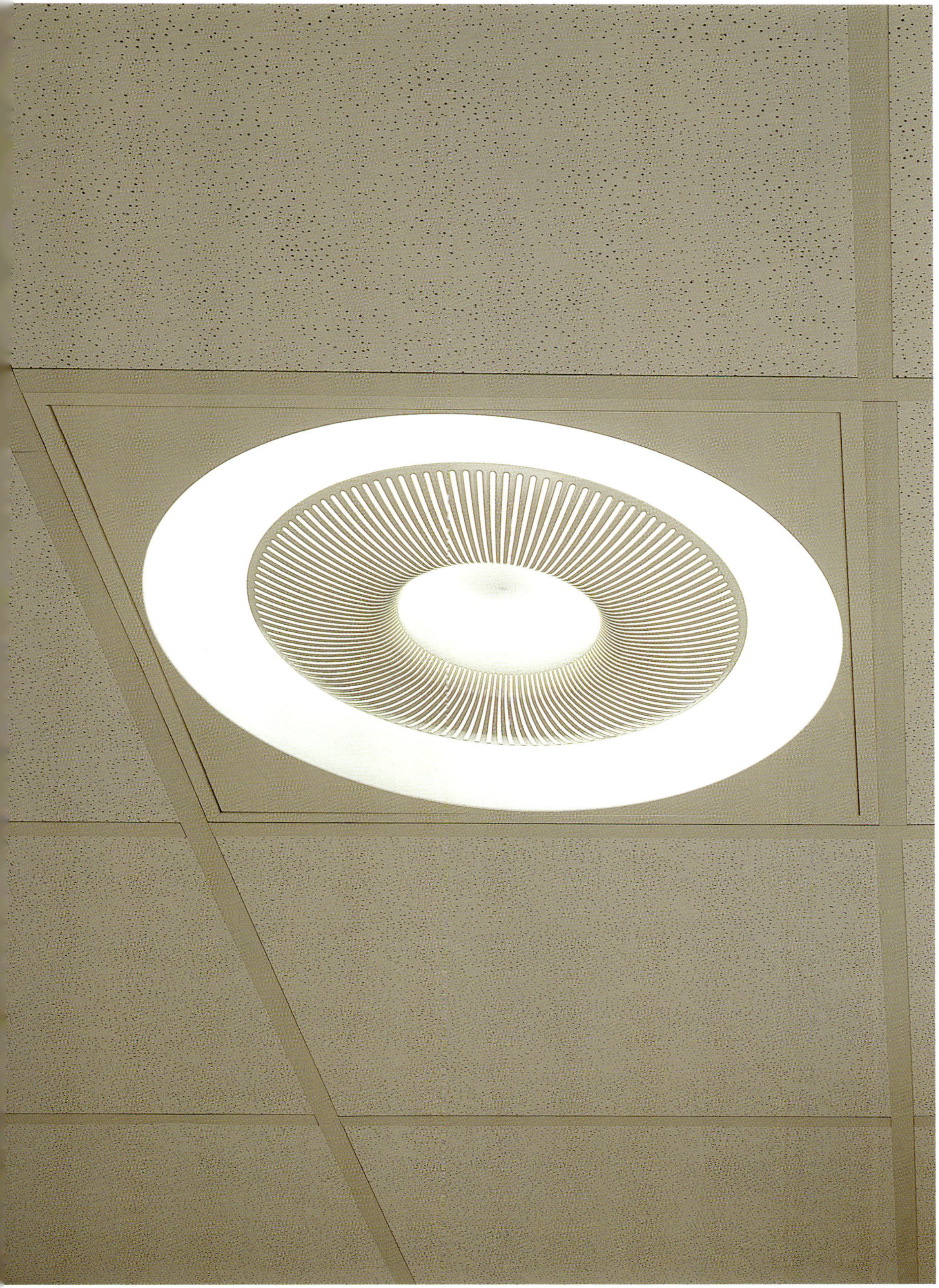

Hersteller Manufacturer
Thorn & Jakobsson
Vejen, Denmark

Produkt Product
Menlo Circular
Deckeneinbauleuchte
Recessed ceiling light

Design
Thorn & Jakobsson
Vejen, Denmark

Menlo Circular Deckeneinbauleuchten sind die neuesten Vertreter der Menlo Leuchtenfamilie für das moderne Büroumfeld. Die runde, organische Form unterstreicht das weiche indirekte Licht. Der gerasterte Deflektor ist ein markantes gestalterisches Element und sorgt zugleich für gedämpftes und angenehmes Licht. Die Leuchten sind als dimmbare Versionen erhältlich.

Menlo Circular recessed ceiling lights are the most recent additions to the Menlo family of luminaries for modern office settings. The round, organic shape emphasizes the soft indirect light they offer. The slotted deflector is an eye-catching element and also ensures that the luminaries provide dampened and pleasant light. All the units are available as dimmable versions, too.

Hersteller Manufacturer
Ridi Leuchten GmbH
Jungingen, Germany

Produkt Product
SQL
Systemleuchte
System luminaire

Design
Albrecht Storp
Storp Design
Langenau, Germany

Zur Leuchtenserie SQL gehören Deckenanbauleuchten, Deckenpendelleuchten, Wandleuchten, Stehleuchten, und Lichtsäulen, die dem Lichtplaner die Möglichkeit eröffnen eine vielfältige, auch komplexe Beleuchtungslösung zu schaffen, bei der die eingesetzten Leuchtenvarianten dieselbe klare Formensprache besitzen. Grundform ist das Rechteck mit einem auf das T16 Leuchtmittel optimierten Leuchtenquerschnitt. Die Leuchten sind aus Aluminiumstrangpressprofil gefertigt und mit einer titanfarbenen Pulverbeschichtung versehen, die die technische Anmutung der Serie unterstreicht.

Ridi's SQL series includes surface-mounted ceiling luminaires, pendant ceiling luminaires, wall-mounted and floor-standing luminaires and lighting columns, enabling planners to draw on a wide and complex range of lighting solutions. All the models display the same clean design lines. The basic shape underlying the range is a slim rectangle designed optimized to take modern T16 lamps. The luminaires are made of extruded aluminum profiles and have a titanium-colored powder coating emphasizing the technical feel of the line.

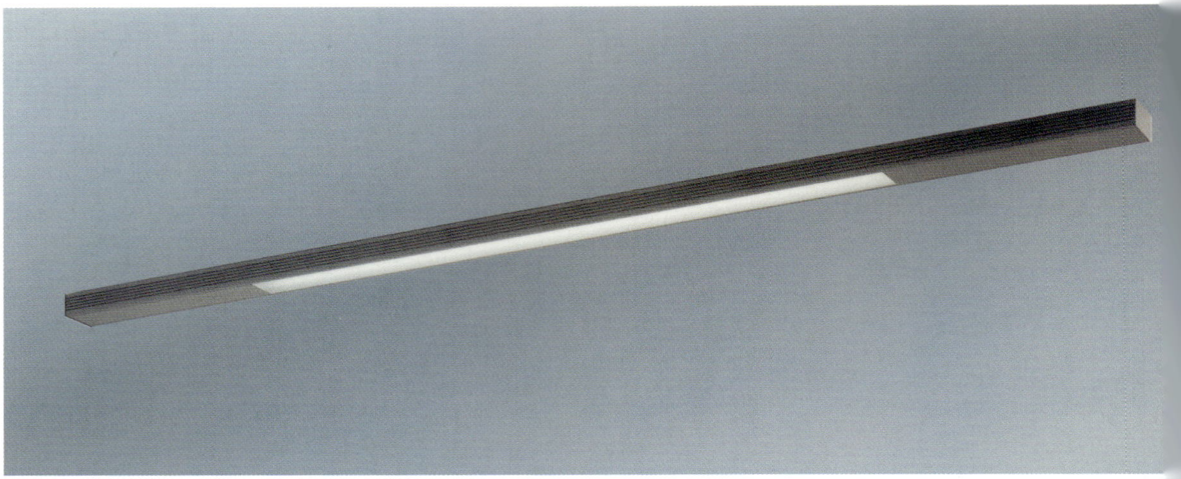

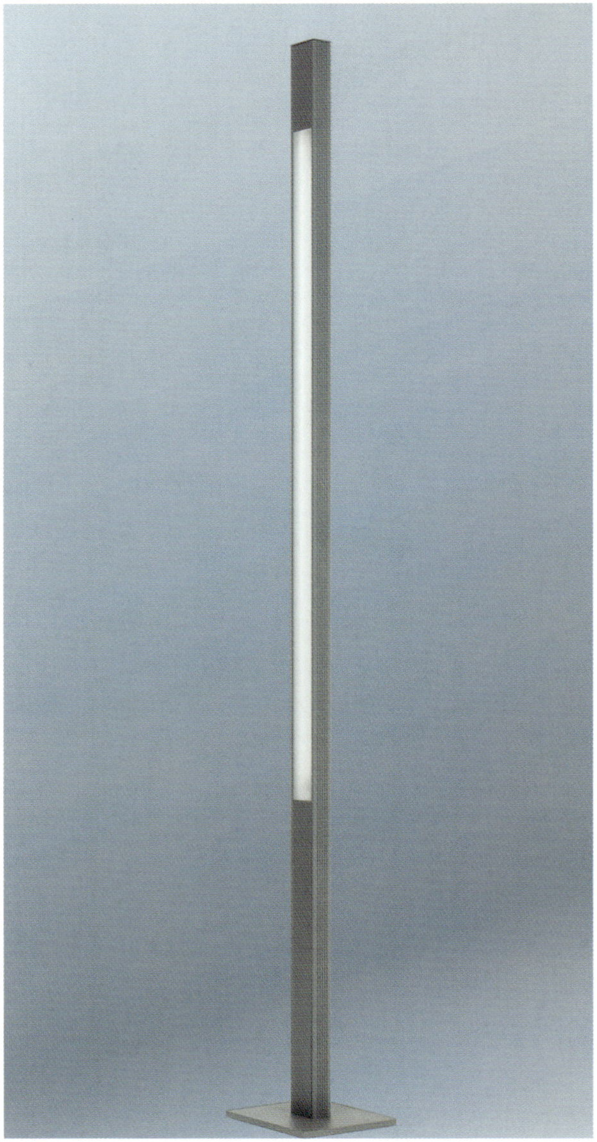

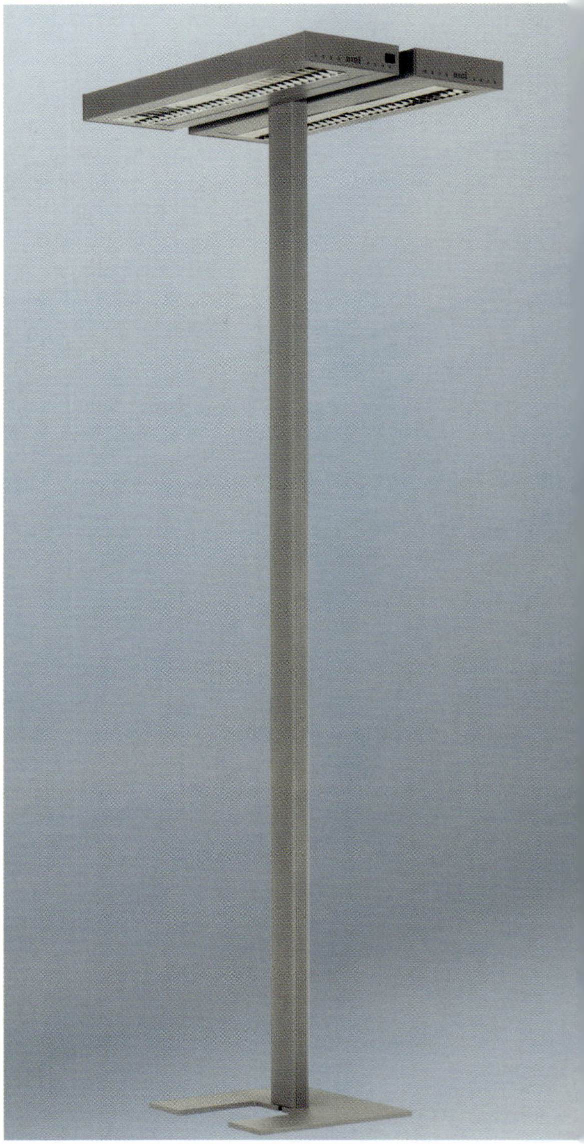

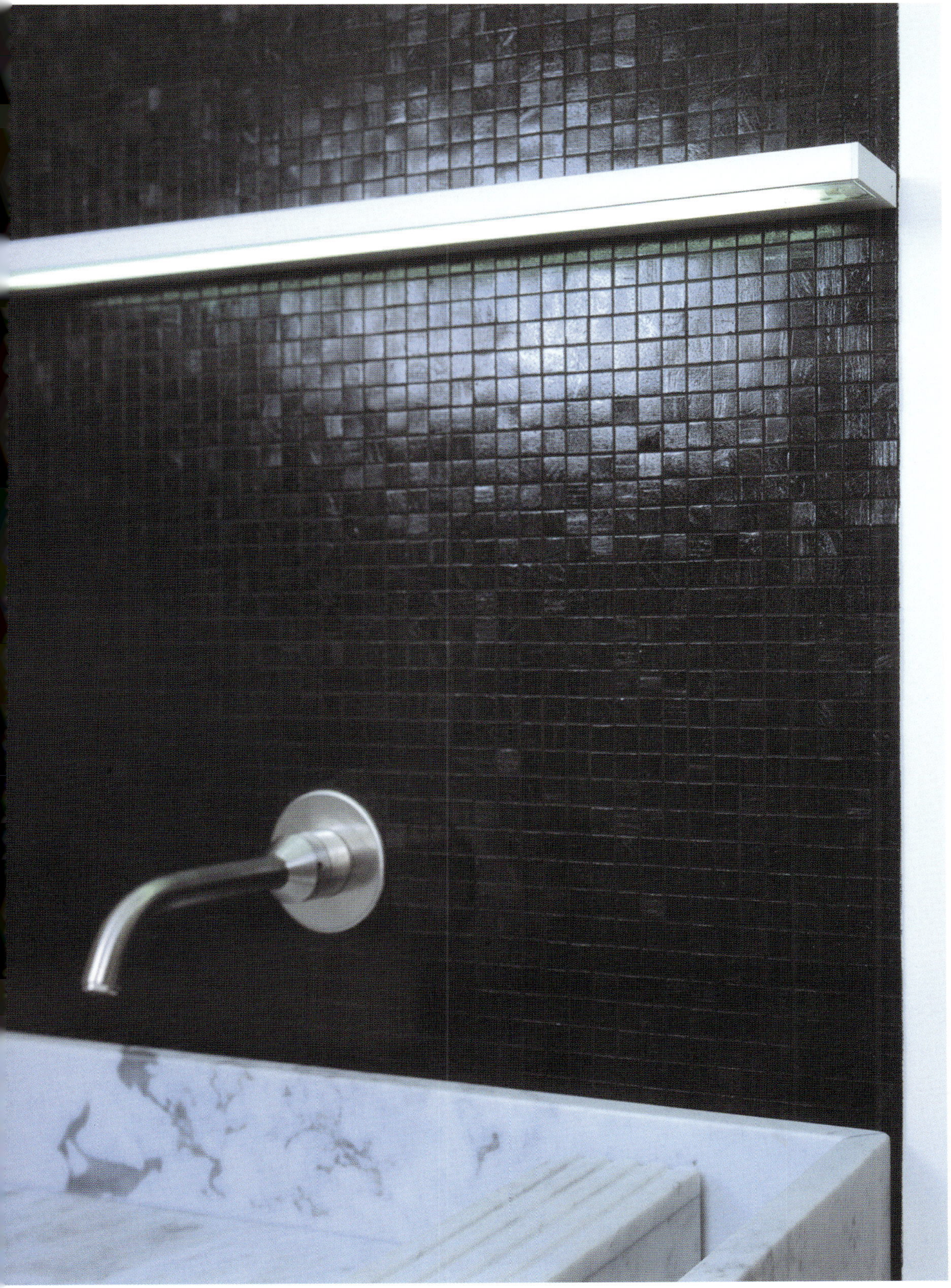

Hersteller Manufacturer
Serien Raumleuchten GmbH
Rodgau, Germany

Produkt Product
SML Wall T5
Leuchte
Luminaire

Design
Jean-Marc da Costa

SML Wall T5 verbindet minimalistisches Design mit maximaler Gestaltungsmöglichkeit. Die Energie sparende Ergänzung zur Halogen-Linie SML ist ebenfalls in den drei Größen S, M und L erhältlich. Mit einer Bauhöhe von nur 25 mm ist sie äußerst flach, das gesamte elektronische Betriebsgerät ist im Wandprofil untergebracht. Abgedeckt wird das dimmbare Leuchtmittel von SML Wall T5 durch speziell extrudiertes Acrylglas in zwei Versionen: das längs prismentierte Glas dient zur Entblendung und lenkt das Licht von der Wand in den Raum. Das satinierte Glas strahlt sanftes und mildes Licht ab. SML Wall T5 ist in drei Größen von 600, 900 und 1200 mm verfügbar.

SML Wall T5 combines minimalist design with a maximum of design possibilities. The energy-saving addition to the SML product range is also available in sizes S, M and L. With a height of merely 25 mm it is extremely flat, with the electronic ballast being concealed in the wall section. SML Wall T5's lamp is covered by specially extruded acrylic glass in two versions: the prism-structured glass guides light away from the wall and into the room. The satin-finished glass emanates a soft and evenly gentle light. SML Wall T5 is available in three sizes: 600, 900 and 1200 mm.

Hersteller Manufacturer
Aloys F. Dornbracht GmbH & Co.KG
Iserlohn, Germany

Produkt Product
Light Modules
Lichtmodulsystem
Lighting module system

Design
Michael Sieger
sieger design
Sassenberg, Germany

Unter Light Modules vereint Dornbracht die bisher unter der Bezeichnung „Licht im Bad" angebotenen Solitärleuchten mit neuen Lichtmodulsystemen. Jedes Lichtmodulsystem besteht aus zwei bzw. drei über eine Systemschiene verbundenen Modulen. Diese können wahlweise mit den Modulen SWING mit Schwenkstrahlern, STARS mit Stiftsockellampen, STRAIGHT mit Kompaktleuchtstofflampen oder SPHERE mit Kardanstrahlern ausgestattet werden, was eine individuelle und pointierte Ausleuchtung des Raumes ermöglicht.

Under the title Light Modules, Dornbracht has united the solitary lamps previously marketed under the name "Light in the Bathroom" with new light-module systems. Each light-module system consists of two or three modules linked by a system track. The modules can be fitted optionally as follows: SWING with pivot spotlights, STARS with pin-base lamps, STRAIGHT with compact fluorescent lamps or SPHERE with universally-jointed spotlights. In this way, it is possible to illuminate rooms in an individual and accentuated way.

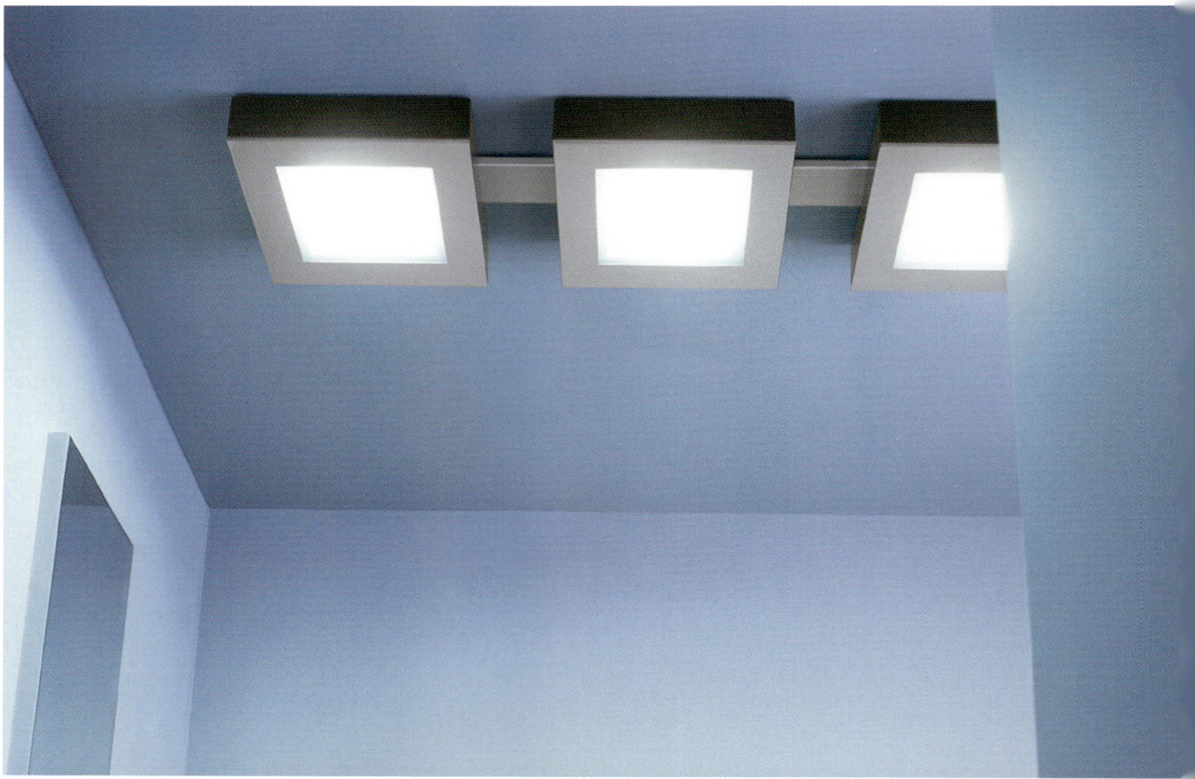

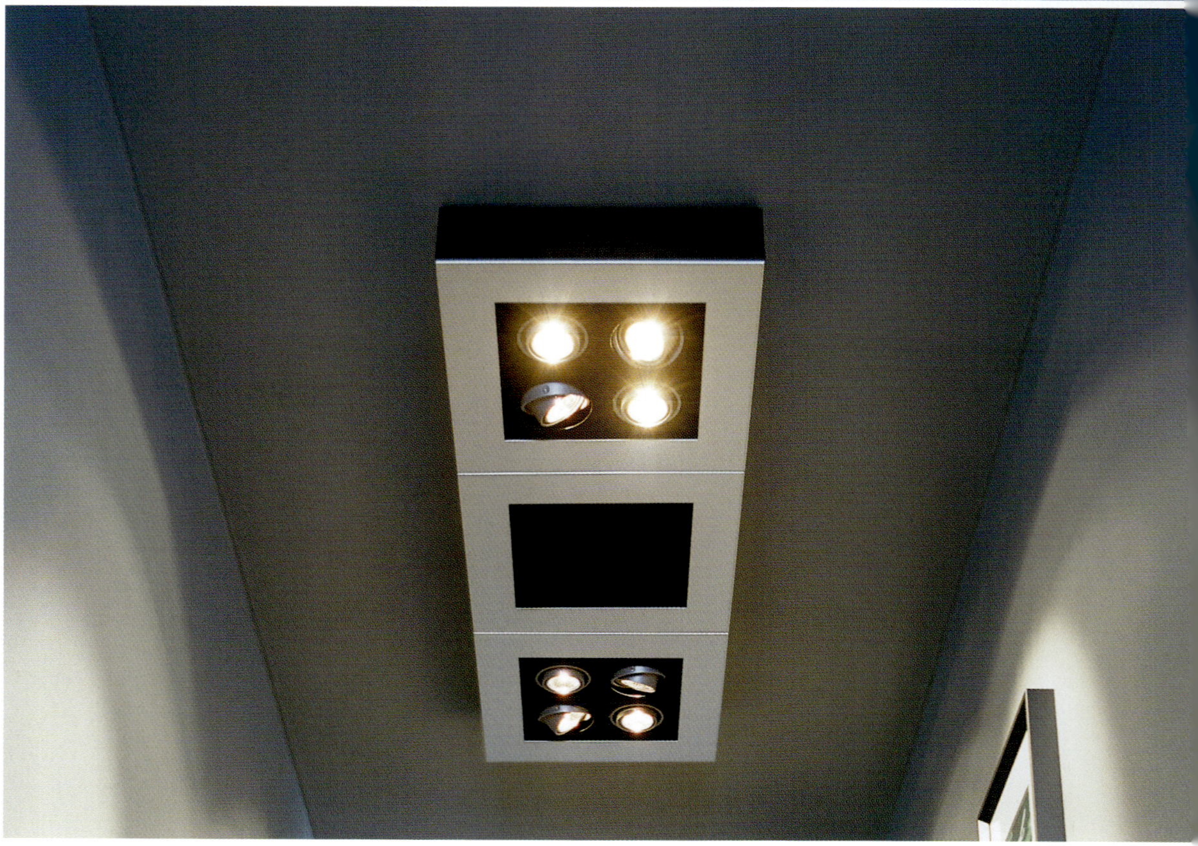

Hersteller Manufacturer
Gera Leuchten GmbH
St. Gangloff, Germany

Produkt Product
Gera Lichtsystem 3 – Lichtboden
Illuminated shelf

Design
Gera Leuchten GmbH
St. Gangloff, Germany

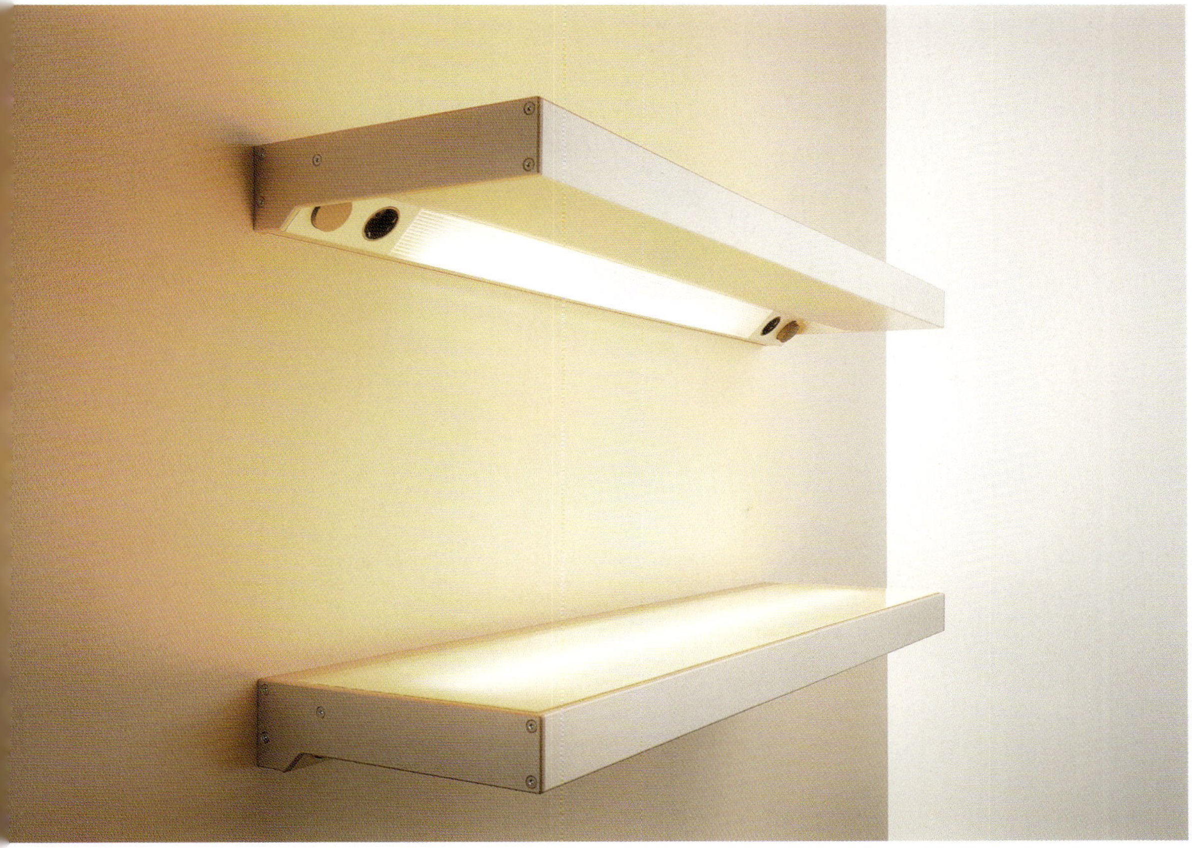

Der Gera Lichtboden zeichnet sich aufgrund seiner speziellen Konstruktion in erster Linie durch seine besondere statische Belastbarkeit aus. Er ist in der Lage, schwere Lasten zu tragen und kann deshalb z.B. in der Küche anstelle von Oberschränken eingesetzt werden. Die darunter liegende Arbeitsfläche wird hell, gleichmäßig und blendfrei ausgeleuchtet. Die oben eingelegte Glasablagefläche kann extra beleuchtet und darauf abgestellte Objekte können in Szene gesetzt werden. Beide genannten Beleuchtungseinheiten sind stufenlos dimmbar und wahlweise kann auch farbiges Licht eingesetzt werden.

The Gera illuminated shelf boasts a special construction – primarily its outstanding load-bearing capacity. This capacity to bear heavy loads means it can be used, for example, in the kitchen instead of overhead storage units. The bench surface underneath is then illuminated with bright, even and glare-free light. The glass shelf inserted above can be additionally lit to show off objects displayed on it. Both of the above-mentioned lighting units can be dimmed exactly as required and colored light may also be used.

Hersteller Manufacturer
Willy Meyer + Sohn GmbH + Co. KG
Hemer, Germany

Produkt Product
Nightline
Außenleuchtenserie
Exterior luminaire series

Design
Roy Fleetwood Ltd.
Office for Design Strategy
Cambridge, UK

Die Leuchtenserie Nightline ist zur gleichmäßigen Beleuchtung von Fassaden und Werbeflächen bestimmt, wobei T16-Leuchtstofflampen zum Einsatz kommen. Die Einzelleuchten und Leuchtenmodule des Systems werden entsprechend unterschiedlicher Lampenleistungen in 4 Baulängen angeboten. Leuchten unterschiedlicher Längen können kombiniert werden, sodass die Lichtbänder an die Abmessungen der zu beleuchtenden Fläche angepasst werden können. Die Leuchtengehäuse bestehen aus einem Aluminiumprofil und Aluminium-Druckgussteilen, die Abdeckung ist aus Polycarbonat gefertigt.

The Nightline luminaire series offers even illumination of facades and signs. It uses T16 fluorecent lamps for the various luminaires and lighting modules of the series – which are available in four different lengths depending on the wattage. Luminaires of different lengths can be combined to create continuous runs that are customized to the respective dimensions of the surface to be illuminated. The luminaires are made of an aluminum profile and die-cast aluminum parts with a polycarbonate cover.

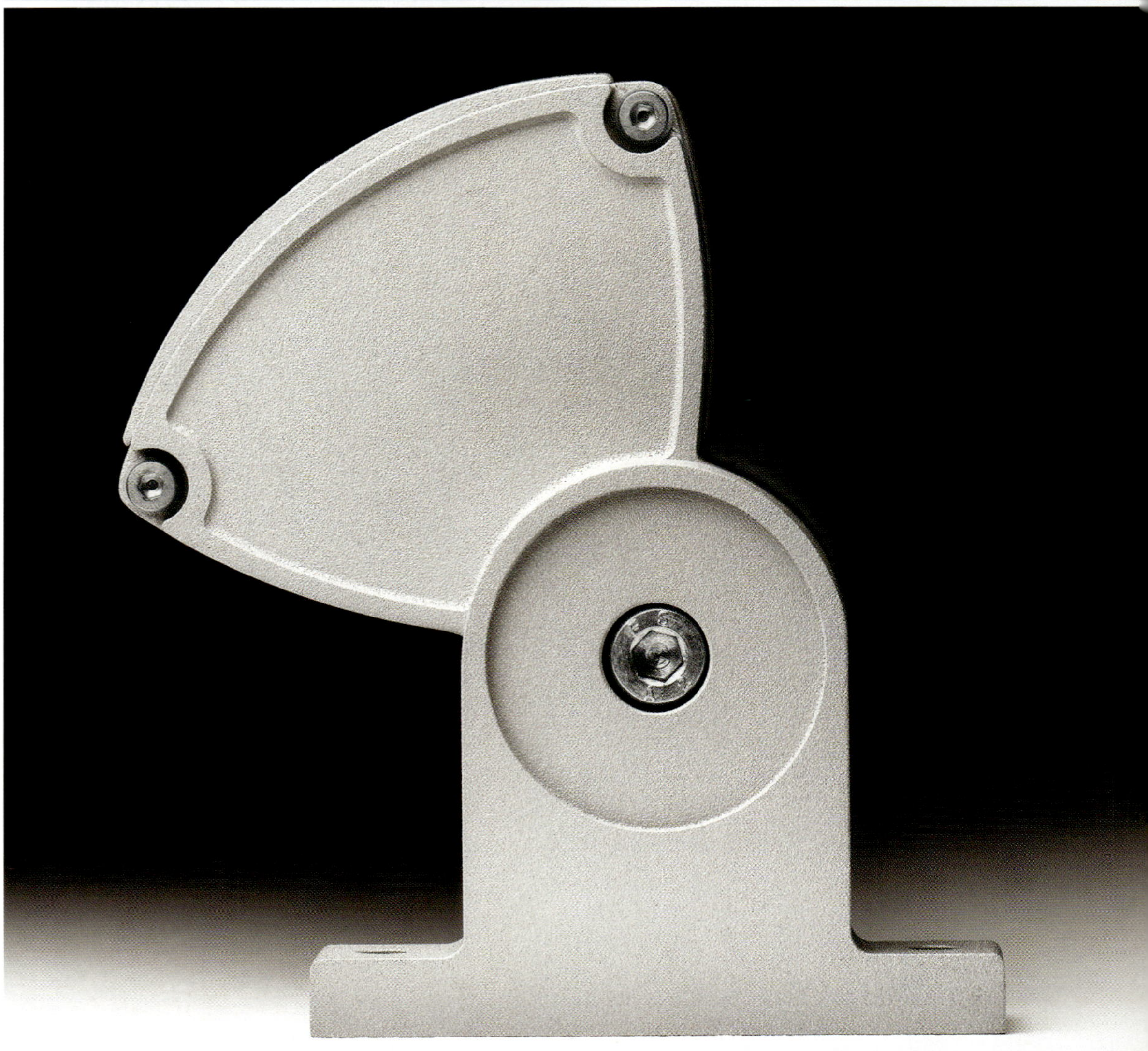

Hersteller Manufacturer
Thomas Leuchten GmbH
Wien, Austria

Produkt Product
Straight
Wandleuchte
Wall luminaire

Design
Thomas Gruber
Thomas Leuchten GmbH
Wien, Austria

Die Wandleuchte Straight zeichnet sich durch ein minimalistisches Design mit klaren Linien aus, wobei sie zugleich eine optimierte Lichtausnutzung bietet. Straight ist wahlweise in der Oberfläche Chrom poliert, Chrom gebürstet, Messing poliert und Messing gebürstet erhältlich. Die maximale Bestückung beträgt 200 Watt (R7s). Das Stecksystem auf einer fixen Platte sorgt für eine einfache Montage.

The Straight wall luminaire catches the eye with its minimalist design and clear lines, and it also optimized light maximization. The luminaire is available in chrom polished, chrom brushed, brass polished and brass brushed finish and can handle lamps of up to 200W. The slots onto the wall assembly plate make it very easy to install.

Hersteller Manufacturer
Lumini
São Paulo, Brazil

Produkt Product
Giro
Strahler
Spotlight

Design
Fernando Prado

Giro ist ein Strahler aus Aluminium-Spritzguss mit Positioniersystem per asymmetrischen Drehpunkt, welches die vollständige Rotation des Strahlers ermöglicht. Die Leuchte besteht aus lediglich drei Komponenten: dem Halsrohr, dem Leuchtenkörper und dem Blendbegrenzungsring. Der Strahler ist um 360 Grad drehbar und – durch Rotation des Leuchtenkörpers – um 75 Grad schwenkbar. Es können verschiedene Leuchtmittel eingesetzt werden wie Halogenglühlampen oder Metalldampflampen.

Giro is a spotlight made of injection-molded aluminum with a positioning system based on an asymmetric pivot to ensure complete rotation of the spotlight head. The luminaire is made up of only three components: the neck, the body, and the anti-glare ring. The spotlight swivels through 360°, and, by rotating the body, can be angled across 75°. Different lamps can be used, such as halogen lamps or metal vapor lamps.

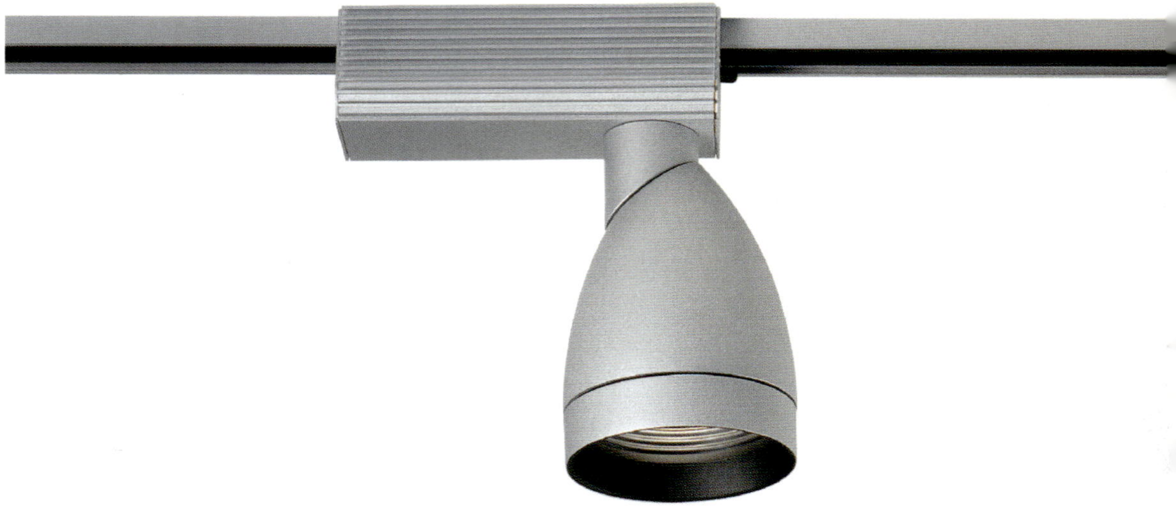

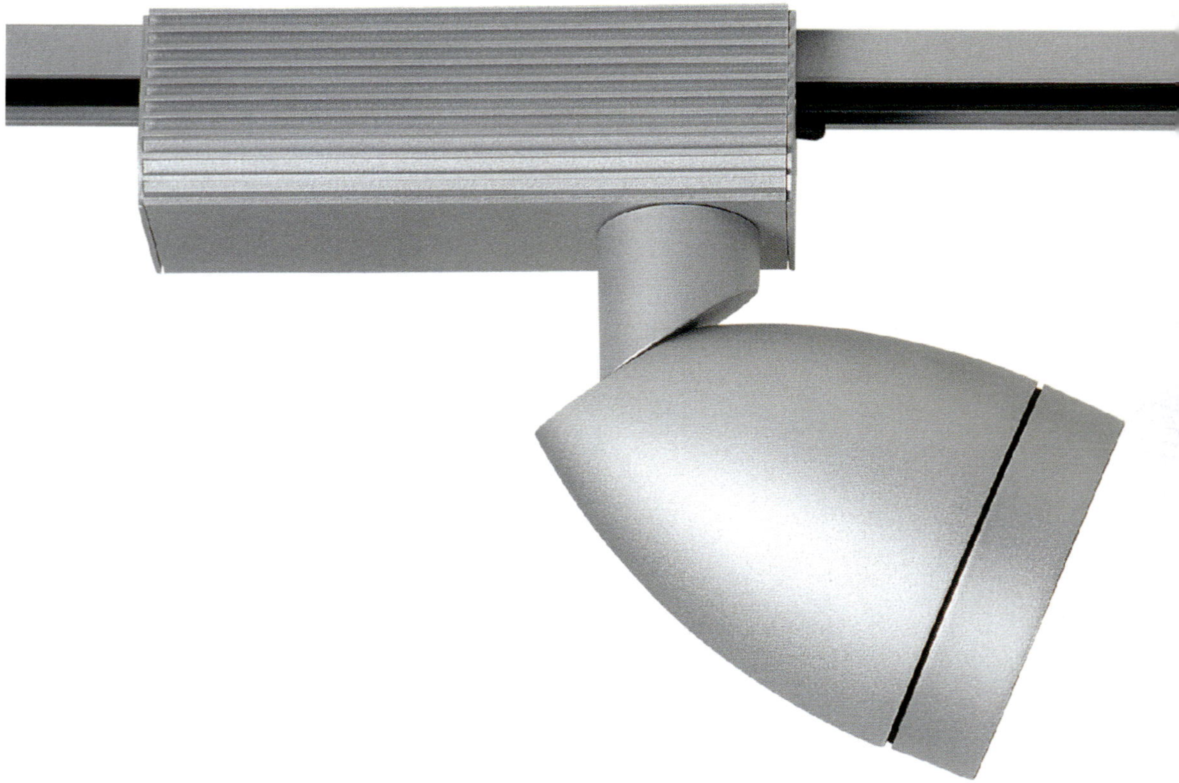

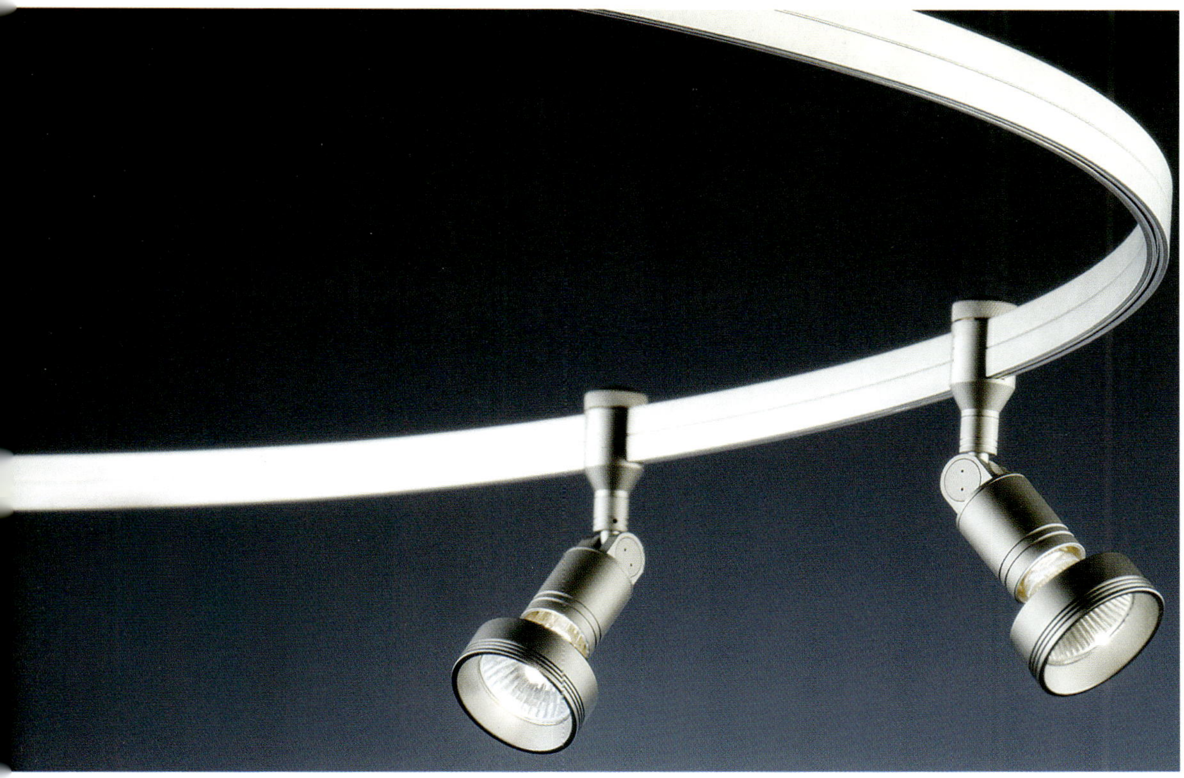

Hersteller Manufacturer
Oligo Lichttechnik GmbH
Hennef, Germany

Produkt Product
Check-in
220-240V Stromschienensystem
220-240V Lighting track system

Design
Ralf Keferstein

Check-in ist ein mit Netzspannung betriebenes Stromschienensystem. Die Schiene ist aus einem hochfesten Aluminium-Strangpressprofil mit den Abmessungen 22 x 8 mm gefertigt und folgt damit der Nachfrage nach fortwährender Miniaturisierung. Das System eignet sich insbesondere für die Beleuchtung im Shop- und Officebereich.

Check-in is a lighting track system that hooks up to the mains. The track itself is made of high-strength extruded aluminum profile sized 22 x 8mm and is thus in line with the ongoing trend for ever greater miniaturization. The system is suited in particular for use in high-quality lighting for shop and office areas.

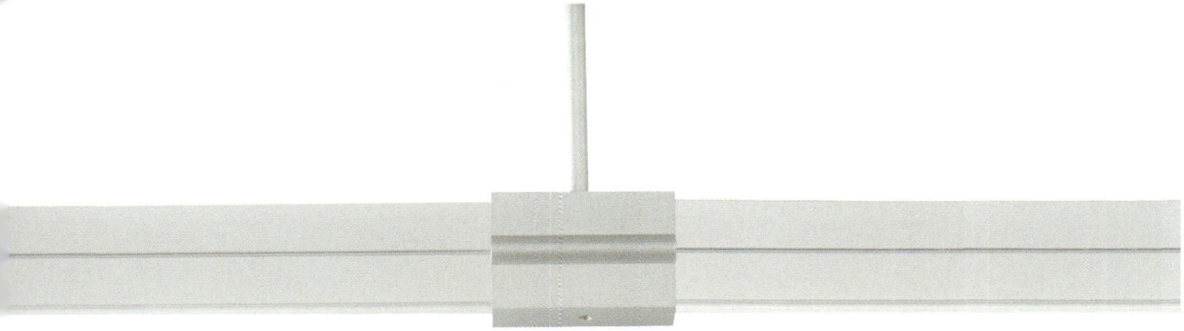

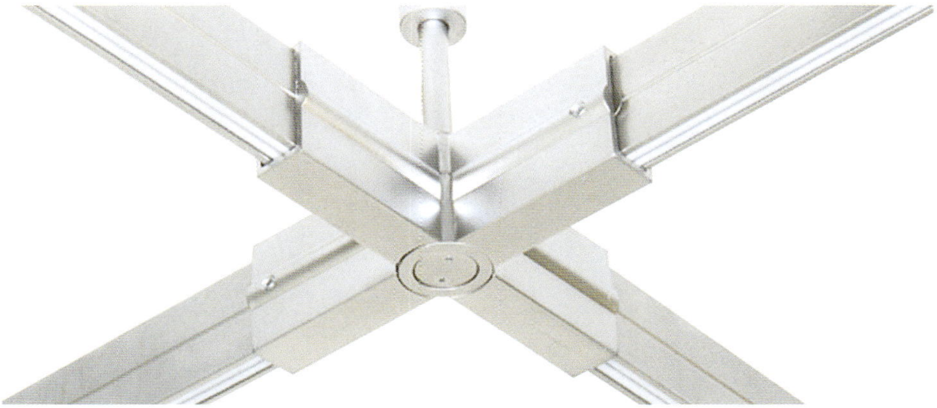

Hersteller Manufacturer
Sylvania N.V.
Tienen, Belgium

Produkt Product
BriteSpot ES50 35W
Kompakt Halogen-Metalldampf-Reflektorlampe
Compact metal halide reflector lamp

Design
Sylvania N.V.
Tienen, Belgium

BriteSpot ES50 35W ist eine Halogen-Metalldampf-Reflektorlampe in einem betont kompakten Design. Durch die verbesserte Chemie, optimierte Brennerform und Wandstärken bietet sie hohe Farbstabilität und Lichtqualität. Die Kombination aus Halogen-Metalldampf-Technik und einem internen kompakten Reflektor gewährleistet Wirtschaftlichkeit und lange Lebensdauer sowie eine gute Strahlungsverteilung.

BriteSpot ES50 35W is a metal halide reflector lamp with a distinctive compact design. Thanks to improved chemical properties and optimized arc tube shape and dimension, the lamp offers excellent color stability and superb color rendering. The combination of metal halide technology and the compact integrated reflector offers a cost-effective long service life, great energy efficiency and superb light quality.

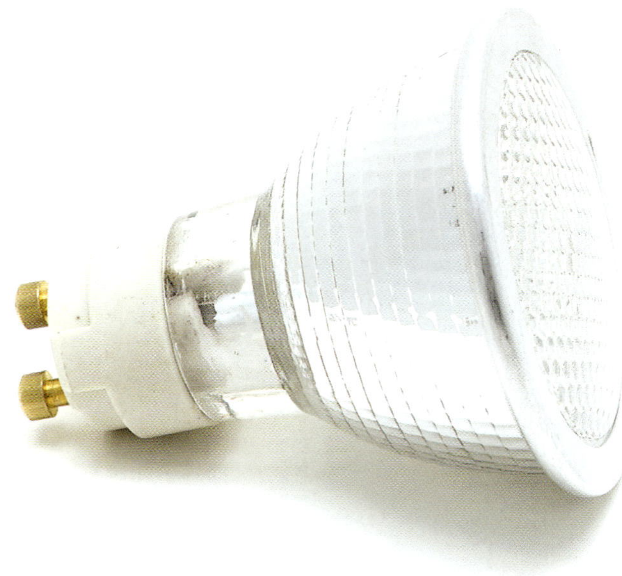

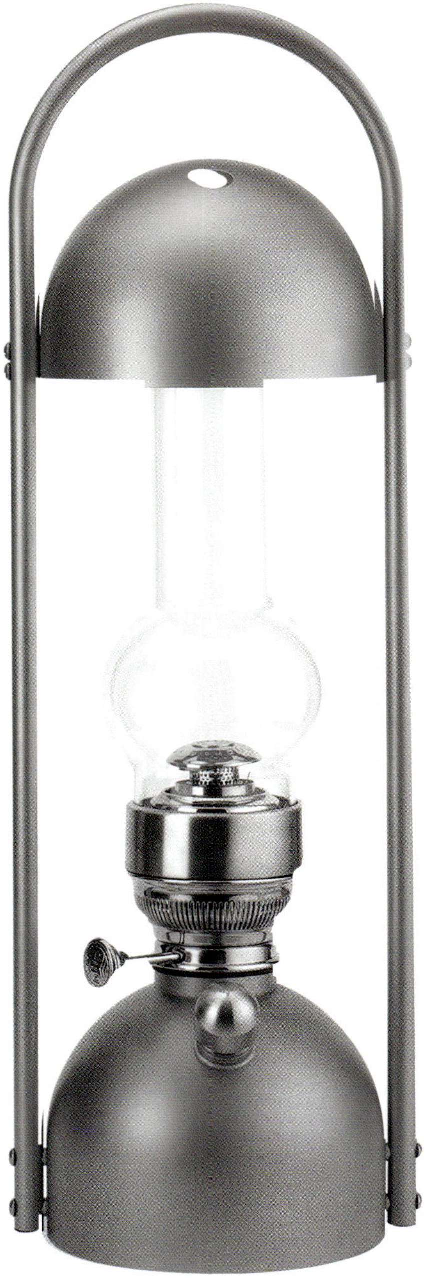

Hersteller Manufacturer
Heinze GmbH & Co.
Wuppertal, Germany

Produkt Product
H1e
Petroleumlampe
Paraffin lamp

Design
Fried Ulber
Ulber Design
Solingen, Germany

H1e ist eine moderne Petroleumlampe als dekorative Lichtquelle auf Balkon, Terrasse und im Garten. In Form und Material wurde sie passend zu Teaksitzgruppen mit Metallapplikationen gestaltet. Die Rundungen des Ober- und Unterteils sowie der Füllschraubkappe und des Bügels nehmen die bauchige Form des Glaszylinders auf und integrieren dessen eigentlich klassische Form in ein insgesamt modern anmutendes Gesamtbild.

H1e is a modern paraffin lamp destined to be a decorative source of light on balconies, patios and in the garden. In terms of shape and materials, it blends well with teak chairs with brass fittings. The rounded sections in the upper and lower body, the screw-open cap for filling the tank, and the handle reflect the bulged shape of the glass cylinder and integrate its essentially classical shape into a decidedly modern appearance overall.

Hersteller Manufacturer
WE-EF Leuchten GmbH & Co.KG
Bispingen, Germany

Produkt Product
FLB 440
Scheinwerfer
Floodlight

Design
Wisoot Binsukor
Bangplee, Samutprakan, Thailand

Die Scheinwerfer aus der FLB 440 Serie verbinden technische Qualität mit funktionalem Design und lassen sich gut in moderne Architektur integrieren. Durch die Einstellscheibe mit Verschlusssystem kann der Scheinwerfer genau und permanent ausgerichtet werden. Mit IP66 und einer Auslegung der Leuchten auf 40 Grad Umgebungstemperatur ist er auf internationale Anforderungen ausgelegt. Ein maßgeblicher Faktor für Langlebigkeit ist der Einsatz einer korrosionsbeständigen Aluminiumlegierung in Verbindung mit einer Pulverbeschichtung. Die Scheinwerfer können zur Beleuchtung von Fassaden, Werbebannern, Sportplätzen etc. eingesetzt werden.

The FLB440 series merges technical performance with functional design and blends with modern achitecture. Developed to international standards, it features IP66 and a 40° C ambient temperature rating. Furthermore, the use of advanced production technology combined with marine grade, corrosion resistant aluminium and superior powdercoat ensures outstanding longevity. Suitable for the lighting of facades, signs, sport grounds, etc.

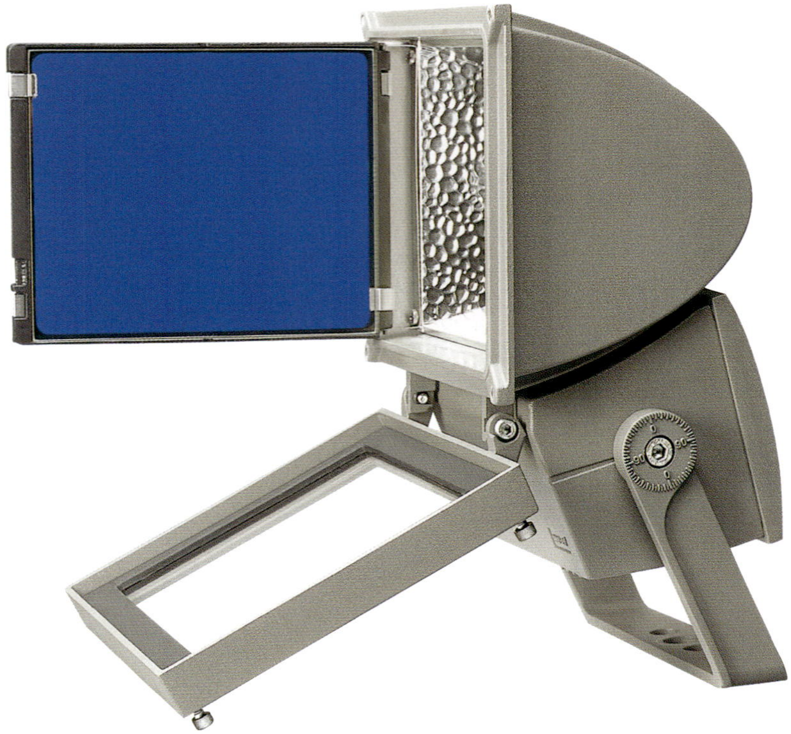

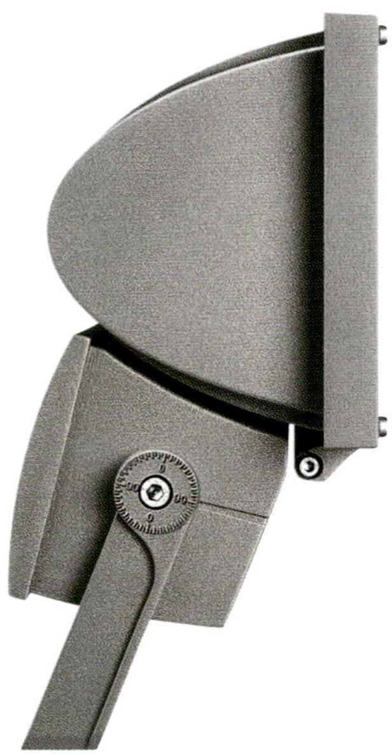

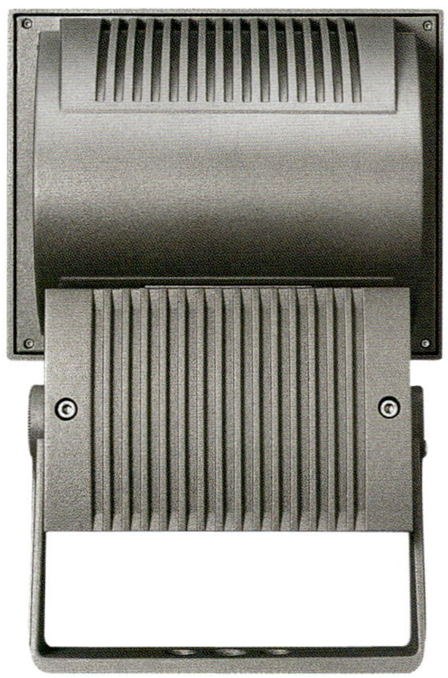

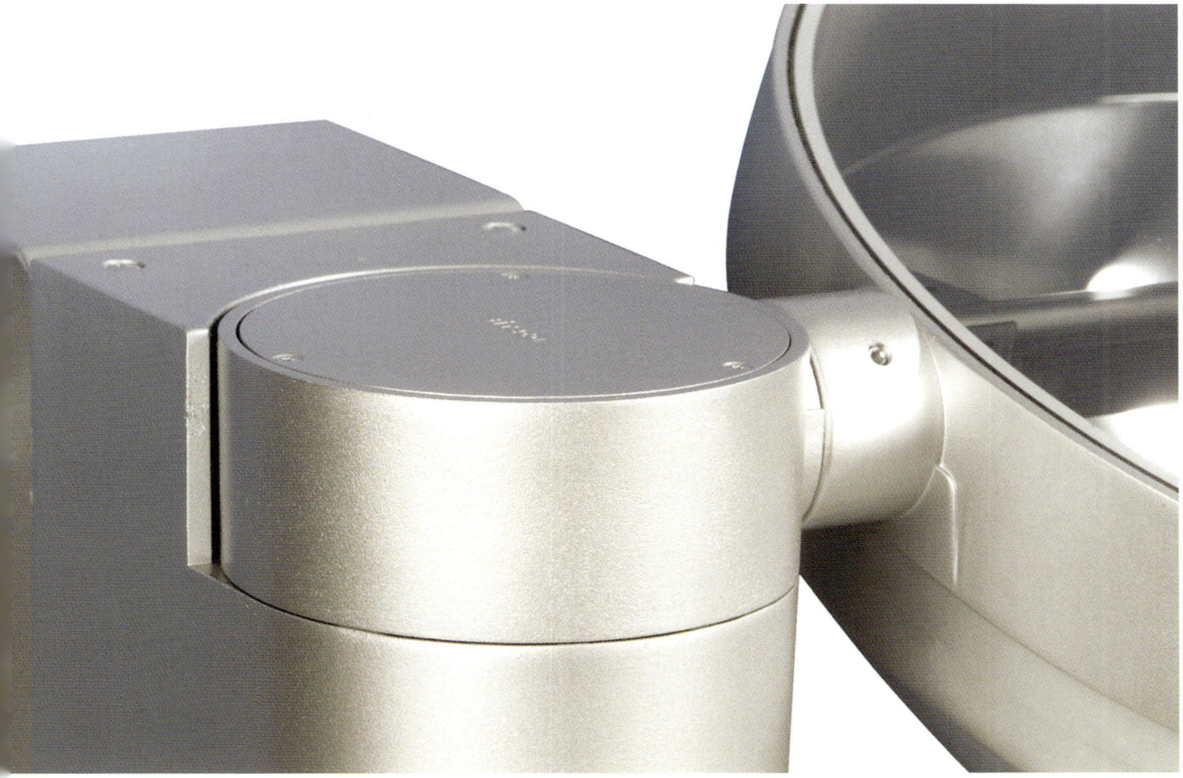

Hersteller Manufacturer
Siteco Beleuchtungstechnik GmbH
Traunreut, Germany

Produkt Product
Siteco Mirrortec Werfer 400
Scheinwerfer
Spotlight

Design
Christoph Ingenhoven
Ingenhoven Overdiek Architekten
GmbH & Co.KG
Düsseldorf, Germany

Der Siteco Mirrortec Werfer 400 ist ein leistungsstarker Scheinwerfer für architektonische Anwendungen in Großgebäuden oder Außenbereichen. Er ist sowohl für den Innen- als auch Außenbereich konzipiert und eignet sich für repräsentative und auch funktionelle Beleuchtungsanforderungen. Das Gehäuse aus Aluminium-Druckguss ist weißaluminium lackiert; Werferkopf und Elektrogehäuse sind dreh- und schwenkbar miteinander verbunden. Bestückt werden kann der Werfer mit HIT und HIT-CRI-Lampen in den Wattagen 250 und 400 Watt.

The Siteco Mirrortec Werfer 400 is a powerful spotlight for architectural applications in large buildings or outdoor areas. It is destined for use both in- and outside, as well as for both official and functional purposes. The die-cast aluminum housing is painted in white aluminum; the spotlight head and electrical case are joined in such a way that they can pivot and be rotated. The spotlight takes HIT and HIT-CRI 250W or 400W lamps.

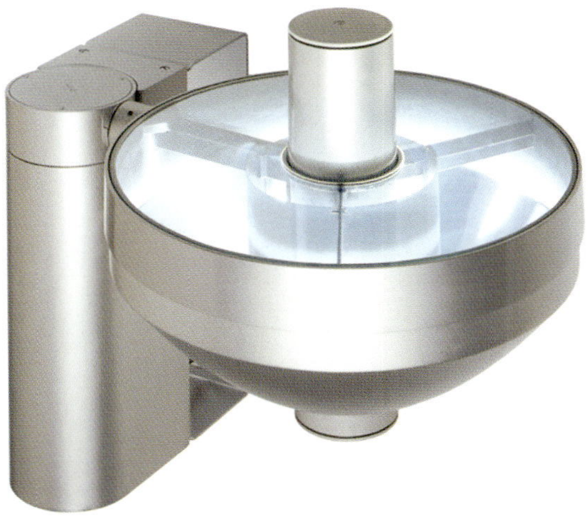

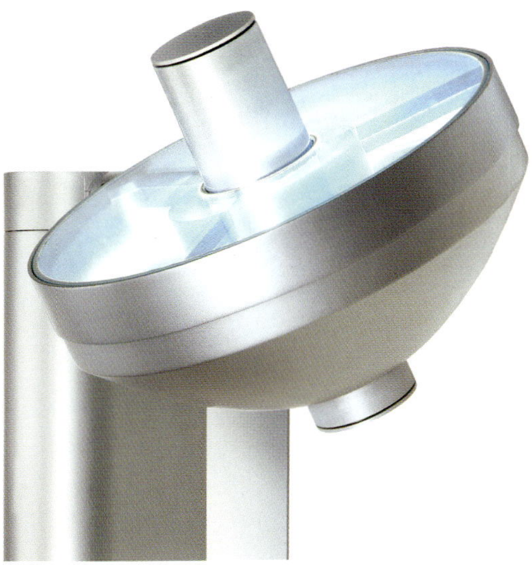

Hersteller Manufacturer
Lumiance B.V.
Haarlem, Netherlands

Produkt Product
Motto
Einbauleuchten
Flush-mounted downlights

Design
Lumiance B.V.
Haarlem, Netherlands

Motto ist eine Serie (Motto 82, 110 und 144) kompakter Aluminium-Einbauleuchten, die Flexibilität und Variationsmöglichkeiten hinsichtlich Lampentypen, Diametern und Leuchtenmodulen bietet. Die Leuchten können mit Halogen, Kompaktleuchtstoff und HID bestückt werden, wobei diese auch innerhalb einer Leuchte unterschiedlich eingesetzt werden können. Aufgrund des zeitlosen architektonischen Designs eignen sich die Leuchten insbesondere für den Einsatz in Showrooms und Hotels sowie zur Ausleuchtung von Verkaufsräumen. Sie sind sowohl quadratisch als auch rund lieferbar und ohne Werkzeug einfach montierbar.

Motto is a versatile series of compact, aluminum flush-recessed downlights (Motto 82, 110 and 144) for a variety of lamp types and with different possible diameters and lighting modules. The luminaires can be fitted with halogen, compact fluorescent and HID lamps which can be arranged differently within a single unit. The timeless architectural look to the luminaires makes them especially suitable for showrooms and hotels or as lighting in salesrooms. They are available in round or square versions and can be easily assembled without the use of tools.

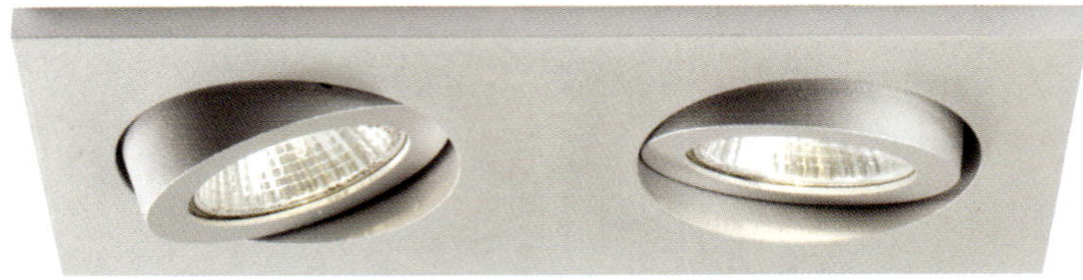

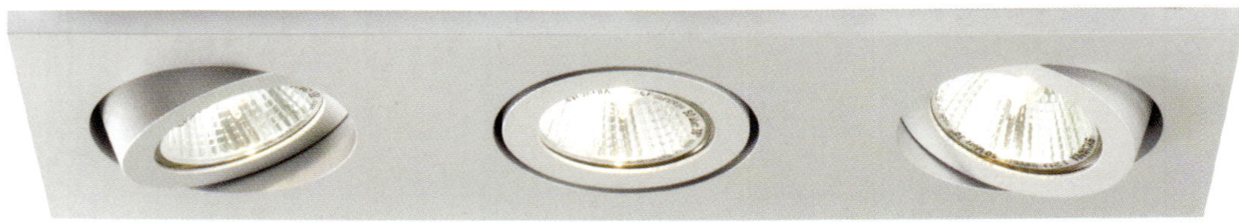

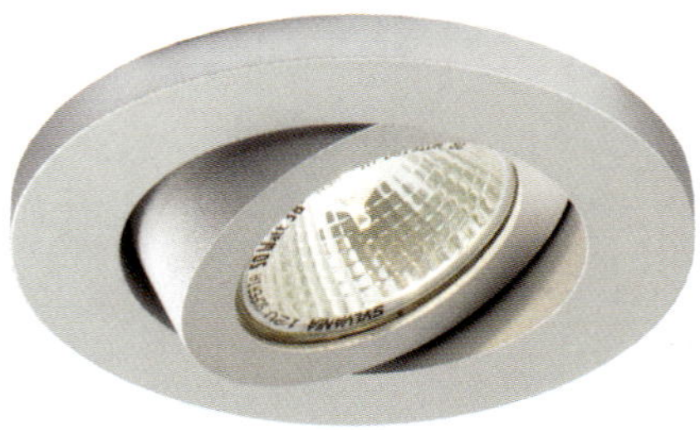

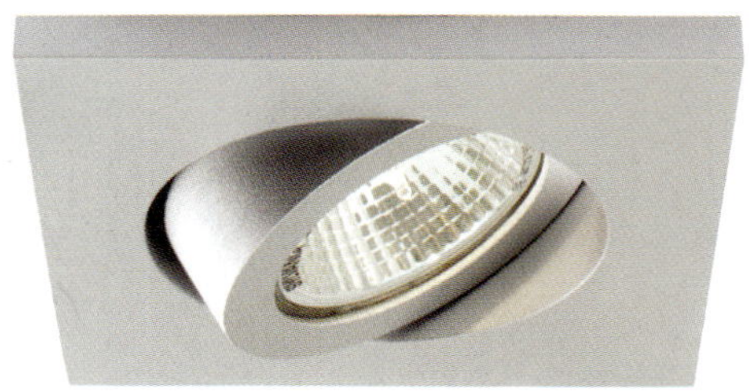

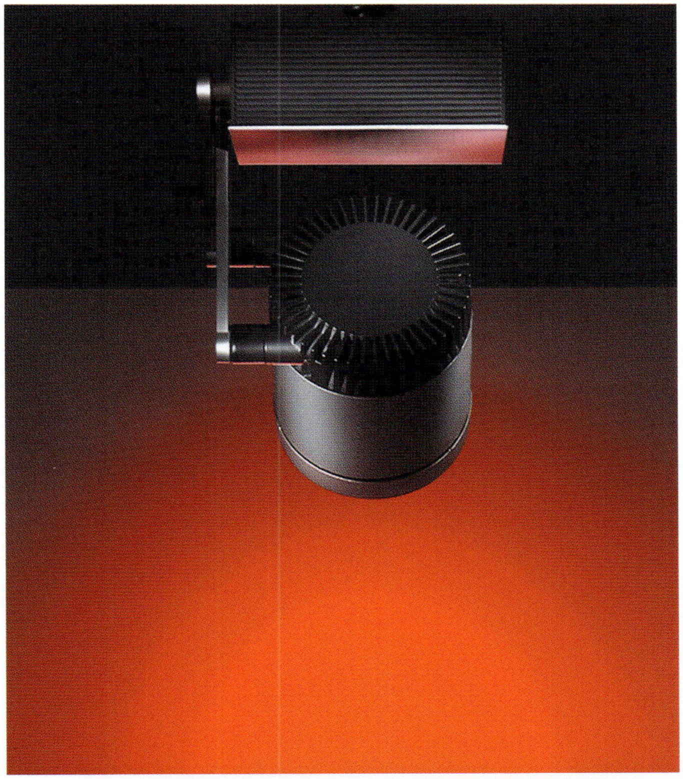

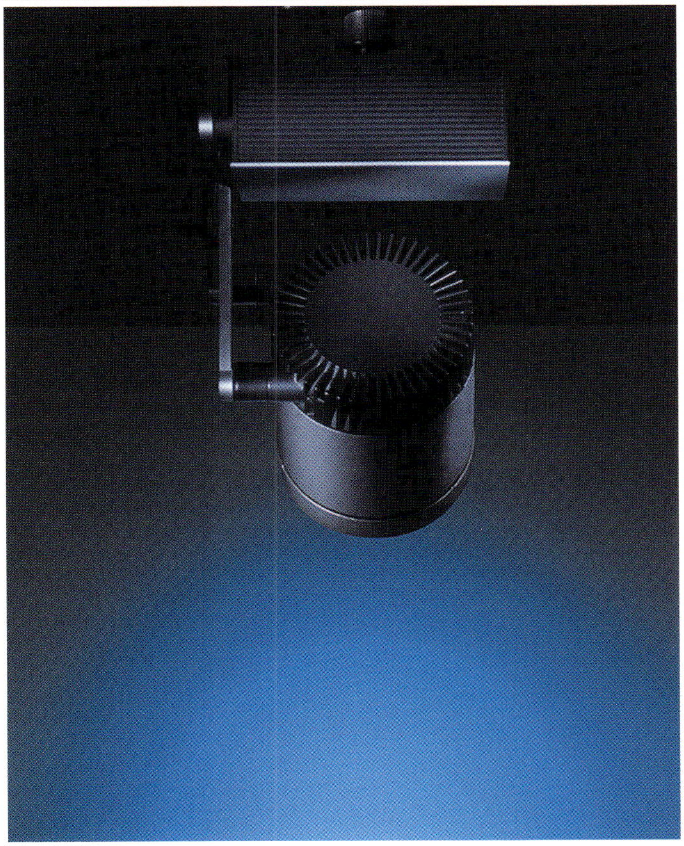

Hersteller Manufacturer
Zumtobel Staff GmbH
Dornbirn, Austria

Produkt Product
Phaos Xeno LED
LED-Farbstrahler
LED washlight with variable colors

Design
Jean-Michel Wilmotte
Paris, France

Mit dem Farbstrahler Phaos Xeno LED lassen sich unterschiedlichste Lichtstimmungen durch programmierbare Sequenzen von Farbvarianten erzeugen. Die Farblicht-Abläufe werden auf einfache Art und Weise auf der Rückseite des Strahlers oder über die separate Steuereinheit programmiert. Das Licht strahlt ohne IR- und UV-Anteile, sodass Exponate schonend und auch konservatorischen Anforderungen entsprechend beleuchtet werden können. Der um 365 Grad dreh- und um 90 Grad schwenkbare Strahler besteht aus Aluminium und Druckguss sowie einer widerstandsfähigen Diffusorabdeckung aus schlagfestem Polycarbonat.

The Phaos Xeno LED washlight can deliver a broad variety of lighting moods thanks to the programmable function for sequencing color variants. The colored-light changes can be simply inputted on the back of the spot or via the separate control unit. The light radiates without any IR or UV, sparing radiation strain on exhibits and meeting conservational requirements for lighting. The aluminum and die-cast spot can rotate 365° and swivel 90° and boasts a robust diffuser cover made of shatterproof polycarbonate.

Hersteller Manufacturer
Zumtobel Staff GmbH
Dornbirn, Austria

Produkt Product
Phaos LED
Lichtkachel
Lighting tile

Design
Zumtobel Staff GmbH
Dornbirn, Austria

Die Phaos Lichtkachel kann als Präsentationsfläche oder farbdynamische Lichtwand eingesetzt werden. Die homogen ausgeleuchtete Fläche verbreitet Licht, dessen Farben sich auf Knopfdruck flächig oder akzentuiert verändern. Das Tempo und die Übergangszeit der Farbspiele sind variierbar. Die Lichtkacheln sind 228 x 228 mm groß und besitzen ein Gehäuse aus titanfarbenem Aluminiumdruckguss. Sie werden wahlweise direkt oder über einen Analog-BUS programmiert.

The Phaos lighting tile can be used as a presentation surface or as a lighting wall and boasts dynamic color properties. The homogeneously illuminated surface exudes light – and the colored area can be changed or its intensities altered at the press of a button. The speed and transitions from one color to the next can be varied as required. The lighting tiles are sized 228 x 228 mm and come with a titanium-colored die-cast aluminum housing. They can be programmed directly or via an analog BUS.

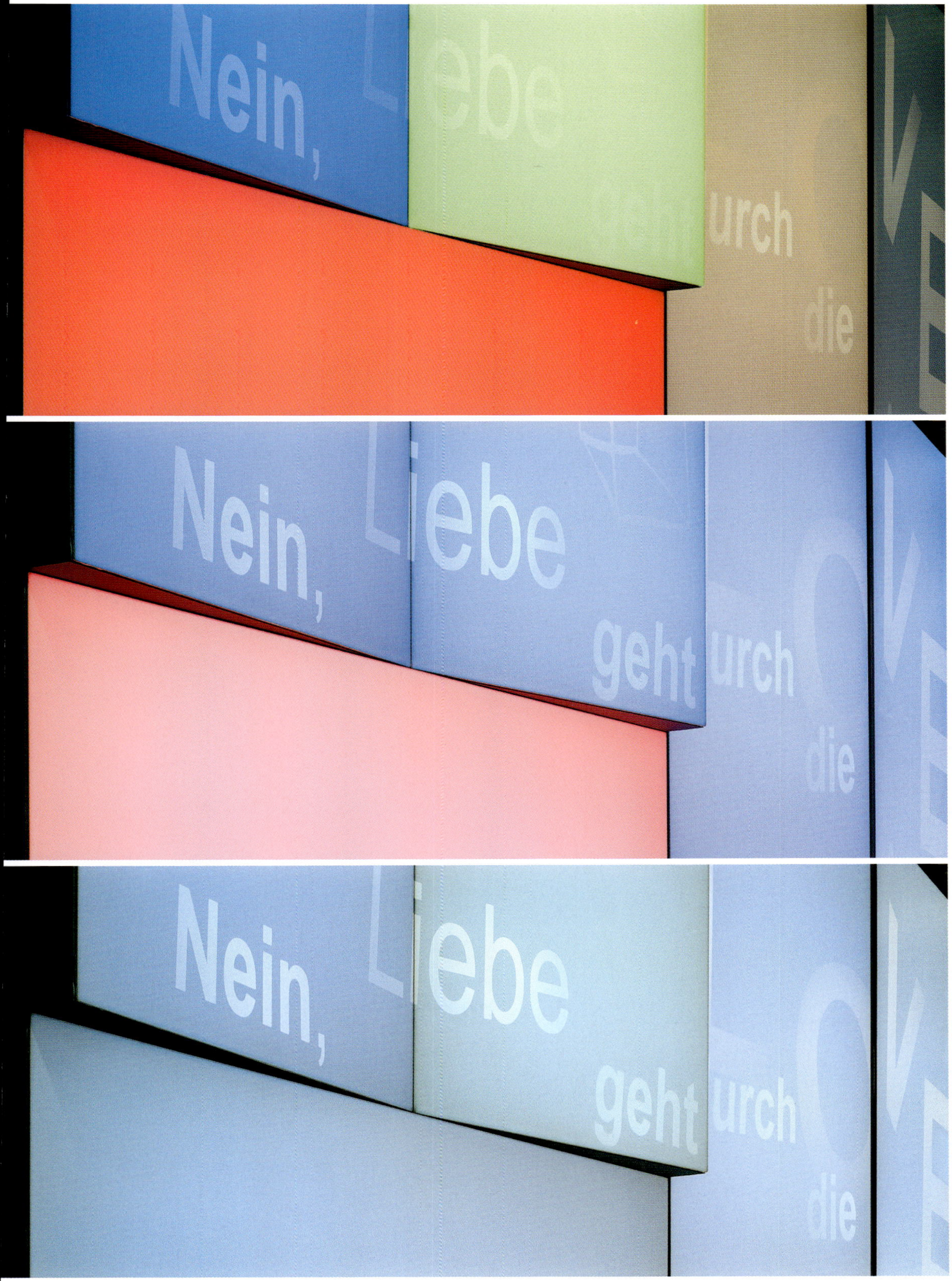

Hersteller Manufacturer
Zumtobel Staff GmbH
Dornbirn, Austria

Produkt Product
Active Light Wall
Lichtwand
Light wall

Design
Zumtobel Staff GmbH
Dornbirn, Austria

Mit der Active Light Wall lassen sich gezielt dynamische Lichtdramaturgien inszenieren. Sie eignet sich insbesondere für Verkaufsräume, Schaufenster, Bars und Messestände, aber auch für Empfangsbereiche, Geburtsstationen oder Krankenhäuser, wo eine freundliche Raumatmosphäre Stress entgegenwirken kann. Das farbige Licht verteilt sich gleichmäßig über die gesamte Fläche bis an den Rand. Die Folienkante springt leicht zurück, dadurch entsteht zwischen den Elementen eine dunkle Trennlinie, die die Farben klar gegeneinander absetzt. Die Lichtwände können ganz nach Bedarf flexibel im Raum positioniert oder auch an die Wand gelehnt oder fest montiert werden.

Thanks to Active Light Wall, dynamic lighting sceneries can be created at will. It is especially suitable for sales rooms, display windows, bars and trade fair stands, not to mention reception areas, postnatal wards and clinics where a friendly ambient atmosphere can counteract stress. The colored light is spread evenly across the entire surface, right through to the edges. The foil edge bends back slightly, giving rise to a dark dividing line between the elements that sets the colors off clearly from one another. The light walls can be flexibly positioned in space at will, leant or fastened permanently to a wall.

Hersteller Manufacturer
iGuzzini Illuminazione s.r.l.
Recanati, Italy

Produkt Product
Light Up Walk Professional
Bodeneinbauleuchten
Recessed ground luminaires

Design
Jean Michel Wilmotte
Paris, France

Light Up Walk Professional sind Bodeneinbauleuchten für den Außenbereich mit professionellen Optiken. Die asymmetrische Optik gewährleistet eine optimale Steuerung des Lichtstroms. Zu den besonderen Merkmalen der Leuchten zählt die deutliche Senkung der Temperatur auf der Oberfläche des 16 mm starken Schutzglases, das zusätzlich zum innen angebrachten Glas der Stärke 4 mm eingesetzt wird und Temperaturen von weniger als 75 Grad aufweist. Die Leuchten ermöglichen den zeitgleichen Einsatz von weißem und farbigem Licht, wodurch sich besondere Gestaltungsmöglichkeiten eröffnen.

Light Up Walk Professional is a line of recessed ground luminaires for outdoor use and equipped with professional optics. The asymmetrical optics ensure optimal control of the luminous flux. The special features of the luminaires include the considerable decrease in surface temperature of the 16 mm protective glass – together with the internal 4 mm-thick glass, temperatures can be kept to a maximum of 75°. The luminaires allow white and colored light to be utilized simultaneously to achieve a diversity of special lighting effects.

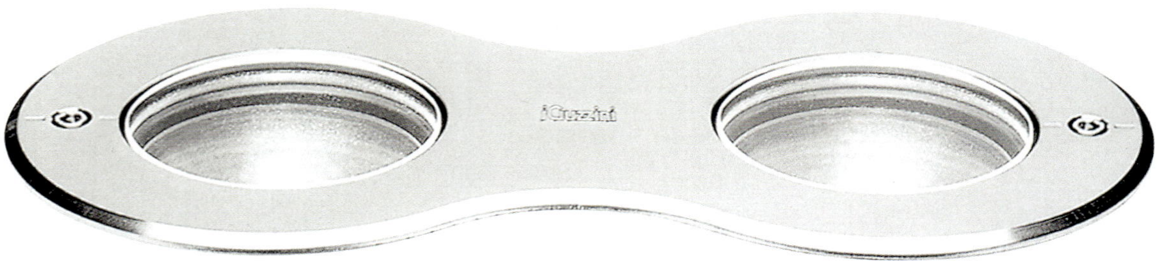

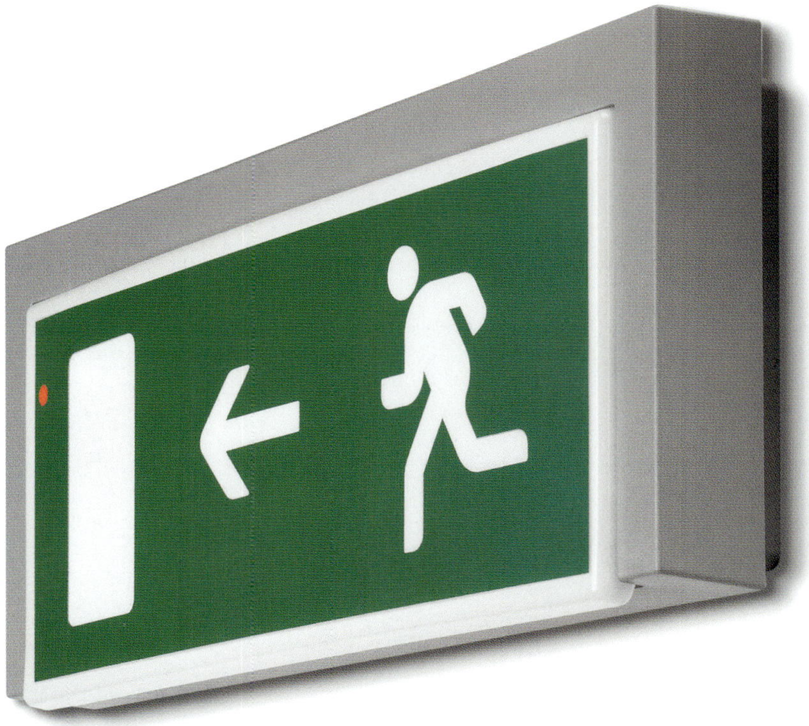

Hersteller Manufacturer
iGuzzini Illuminazione s.r.l.
Recanati, Italy

Produkt Product
Motus
Notleuchte und Signal-Deckenleuchte
Emergency and signalling ceiling luminaire

Design
Atelier Bellini
Mailand, Italy

Die Notleuchte und Signal-Deckenleuchte Motus lässt sich mit ihrem schlichten Design auch für die Allgemeinbeleuchtung einsetzen. Die Leuchte vereint mehrere komplexe Funktionen und zeichnet sich gleichzeitig durch ihr zurückhaltendes Design aus. Die Leuchtfläche wird von einem kleinen Rahmen gehalten. Sie besteht aus vier Polykarbonatteilen (Komponententräger, Reflektor, Streuschirm und Rahmen). Die Schirme können mit speziell entworfenen Piktogrammen versehen werden.

Motus is an emergency and signalling luminaire that, thanks to its essential and elegant design, can also be used for general lighting. The unit brings together several different complex functions and also stands out for its restrained visual impact. The luminaire is held in a small frame and consists of four polycarbonate parts (component-holding base, reflector, diffuser screen and frame). The screens can be silk-screened to bear pictograms custom-designed for the manufacturers.

Hersteller Manufacturer
ETAP GmbH
Leverkusen, Germany

Produkt Product
K9
Rettungszeichenleuchten
Emergency signage luminaries

Design
Conix Architekten
Antwerpen, Belgium

K9 ist eine Familie von ein- und zweiseitigen Einzelbatterie-Rettungszeichenleuchten auf LED-Basis für Einbau-, Decken- und Wandmontage. Die patentierte Mikroelektronik garantiert eine lange Lebensdauer der LEDs und die Überwachung der Leuchte. Bei der Gestaltung der K9 wurde bewusst auf überflüssige Details verzichtet, die von der Architektur des Gebäudes ablenken. Die Anordnung der LEDs erlaubt eine äußerst kompakte Bauweise. Selbst die Anzeige-LEDs sind bündig in das stabile Aluminium-Druckguss-Gehäuse integriert.

K9 is a family of one- or two-sided self-contained emergency luminaries, featuring LEDs and intended for recessed, ceiling or wall mounting. The patented microelectronics provide an extended lifecycle for the LEDs and constant monitoring of the system. The design of the K9 series deliberately forgoes superfluous details that would distract attention from the interior design of the building. The layout of the LEDs spells a highly compact unit. Even the indicator LEDs are integrated flush into the stable die-cast aluminum casing.

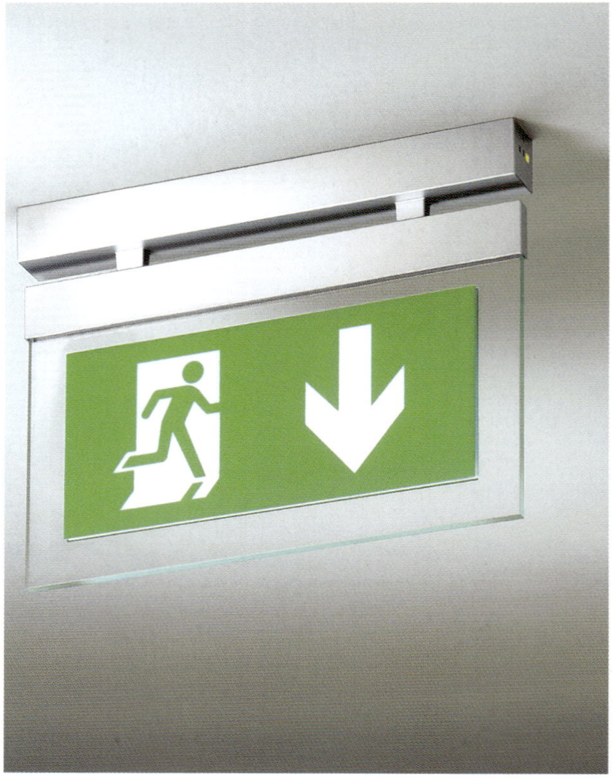
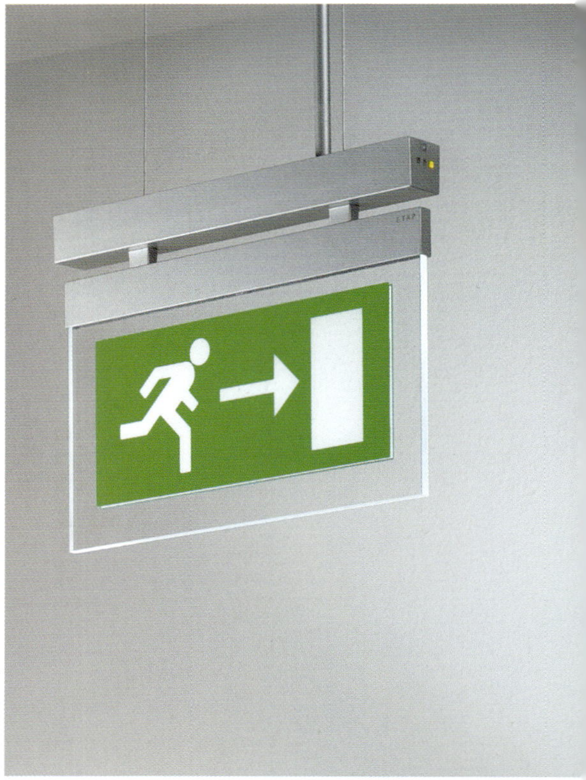
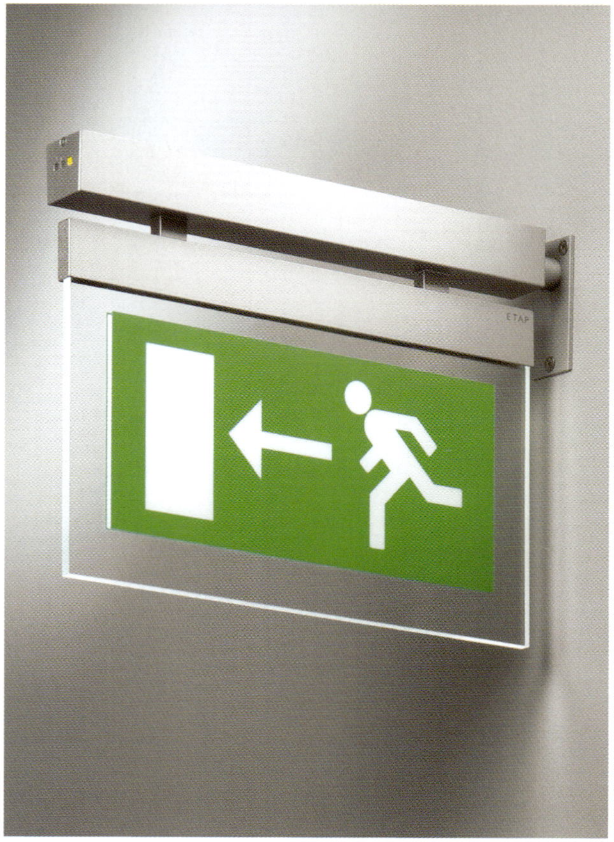
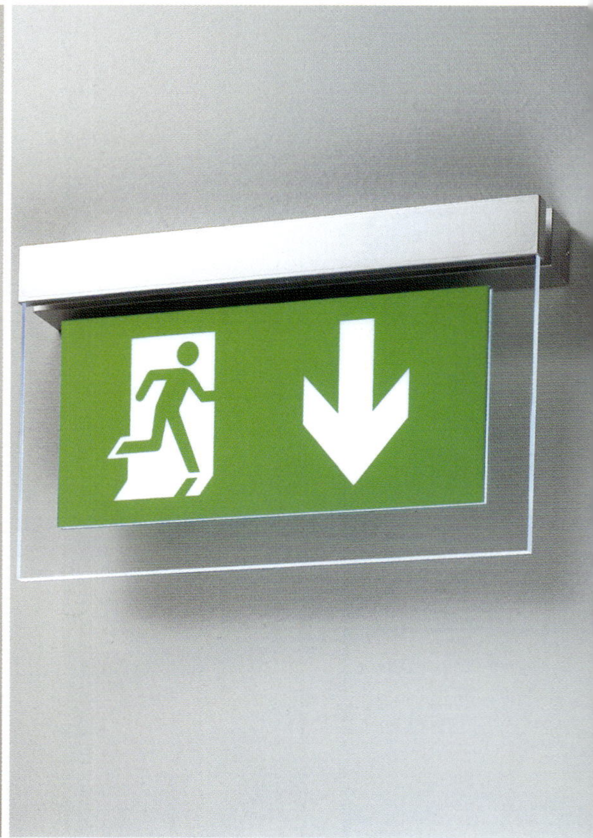

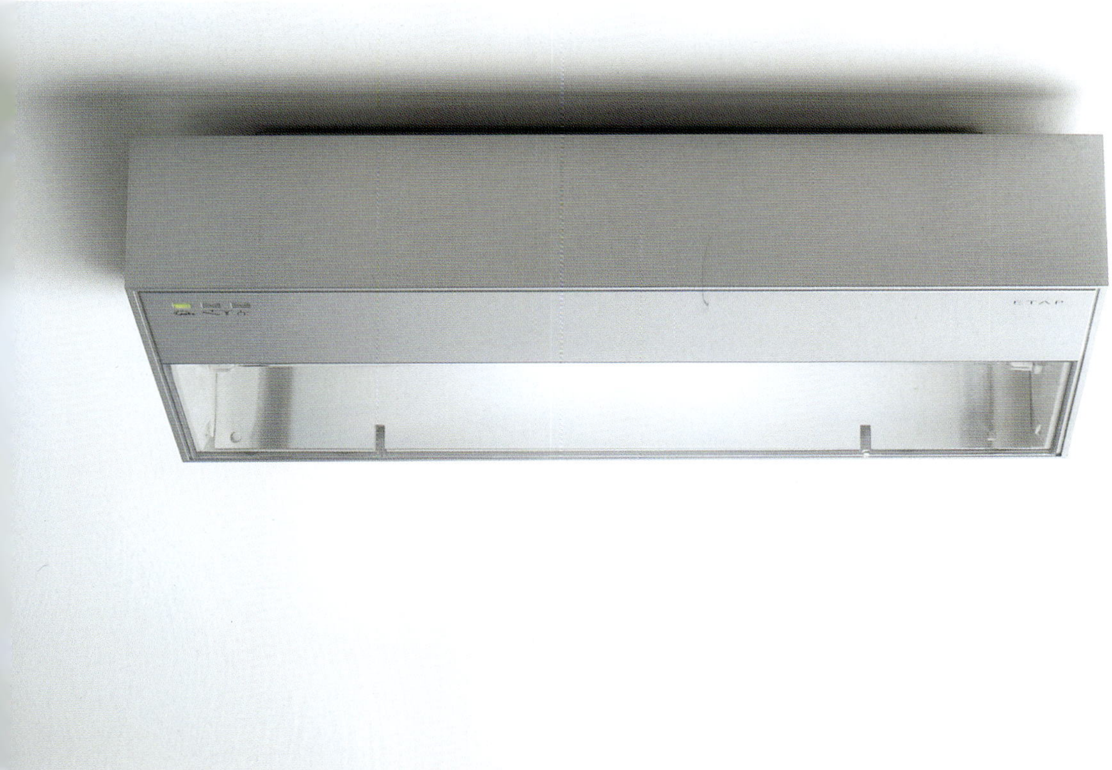

Hersteller Manufacturer
ETAP GmbH
Leverkusen, Germany

Produkt Product
K6
Sicherheitsleuchten
Emergency luminaries

Design
Conix Architekten
Design Studio Enthoven
Antwerpen, Belgium

K6 ist eine Familie von Sicherheitsleuchten, die entwickelt wurden, um das gefahrlose Verlassen von Gebäuden in Notsituationen zu ermöglichen. Zur Produktreihe gehören Rettungszeichenleuchten, Fluchtweg, und Anti-Panik-Leuchten, die allen europäischen Standards entsprechen. Sie bestehen aus einer Lampe und einem Batteriesatz, der im Notfall eine 1- oder 3-stündige Betriebsdauer garantiert. Beide werden durch eine intelligente Elektronik überwacht. Das Design wurde betont minimalistisch gehalten, um zu verhindern, dass die Leuchte einen zu starken Einfluss auf die Innenarchitektur gewinnt.

K6 is a family of emergency luminaries developed specially to enable users to exit buildings in an emergency situation without any danger. The product series includes self-contained emergency luminaries, illuminated escape route signage, and anti-panic luminaries. All products comply with the European standards, and consist of a lamp and a battery set that operates for either one or three hours in emergency mode. The emergency luminaries are monitored by intelligent electronics. The design has been kept deliberately minimalist to prevent the luminaries exerting too strong an influence on the interior building design.

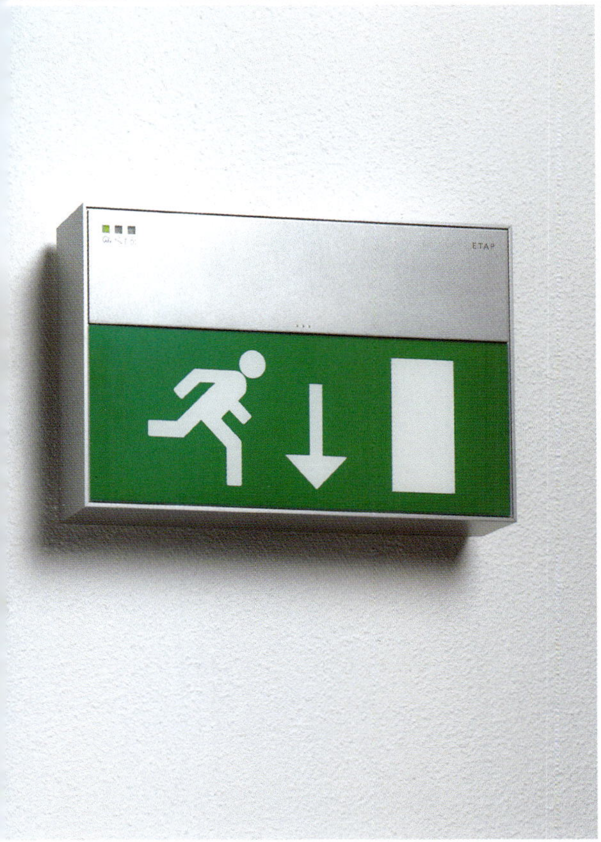

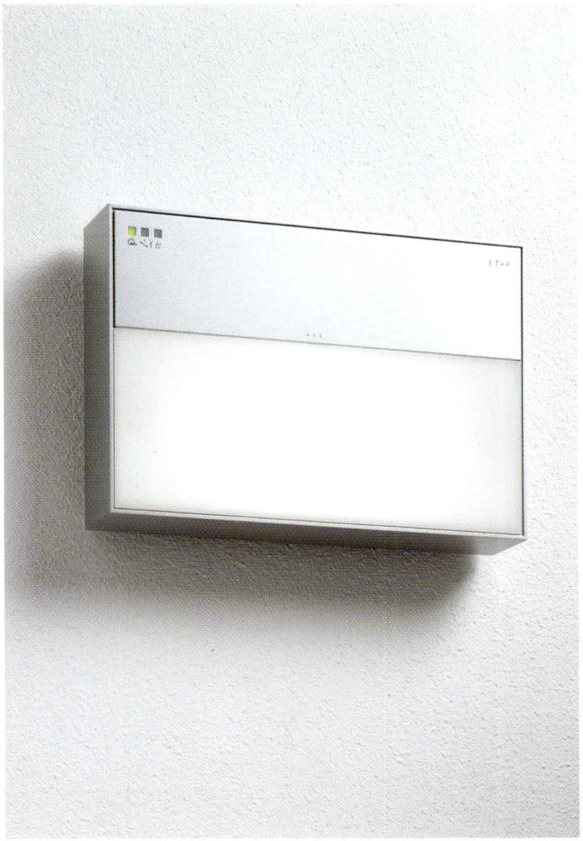

Hersteller Manufacturer
Hella KG Hueck & Co.
Lippstadt, Germany

Produkt Product
Jumbo 320
Zusatzscheinwerfer
Auxiliary headlight

Design
Rainer Seinsche
SE-Design
Wuppertal, Germany

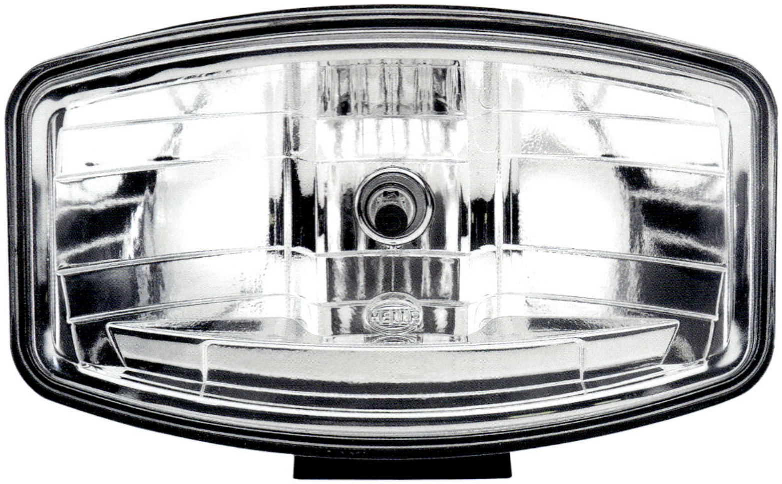

Jumbo 320 ist ein Heavy Duty Zusatzscheinwerfer für Trucks und Off-Road-Fahrzeuge für optimale Sicherheit bei langen Fahrten in Dunkelheit. Die Klarglasoptik fügt sich gut in die Frontgestaltung der neuen Trucks und Off-Road-Fahrzeuge ein. Die Zusatzscheinwerfer sind wahlweise in Halogen- oder Xenonlichttechnik verfügbar. Der Scheinwerfer wird durch einen eigenständigen Reflektor des Positionslichtes illuminiert. Beim Halogenfernlicht wird das Positionslicht dabei durch die herkömmliche Glühlampentechnik realisiert, während bei der Xenon Version 6 Hochleistungs-LEDs das weiße Positionslicht bilden.

Jumbo 320 is a heavy-duty auxiliary headlight for trucks and off-road vehicles to provide optimal safety during long runs in darkness. The clear-glass design means the unit melds well with the front-end farings of new-generation trucks and off-road vehicle. The auxiliary headlights can be outfitted with either halogen or xenon lamp technology. The headlights are illuminated by the independent reflector of the position light. In the case of halogen long-beam lights, the position light is furnished with conventional lamp technology, whereas in the Xenon version, 6 powerful LEDs go to make up the white position light.

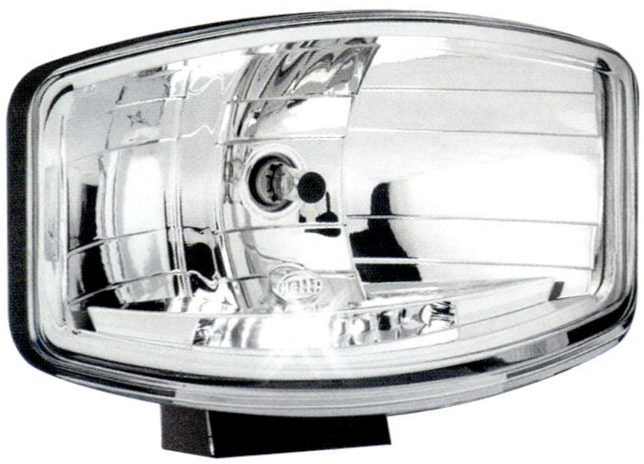

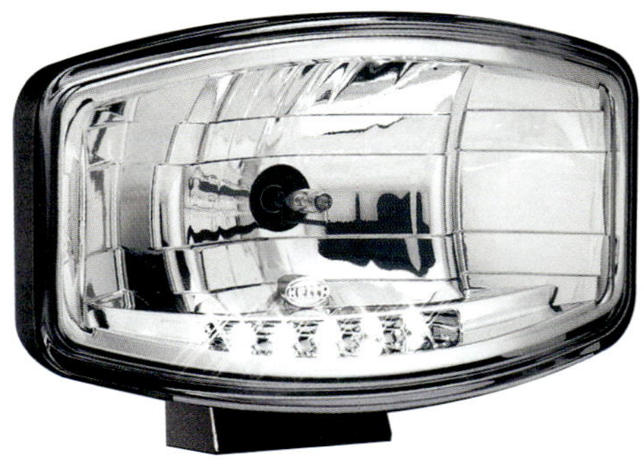

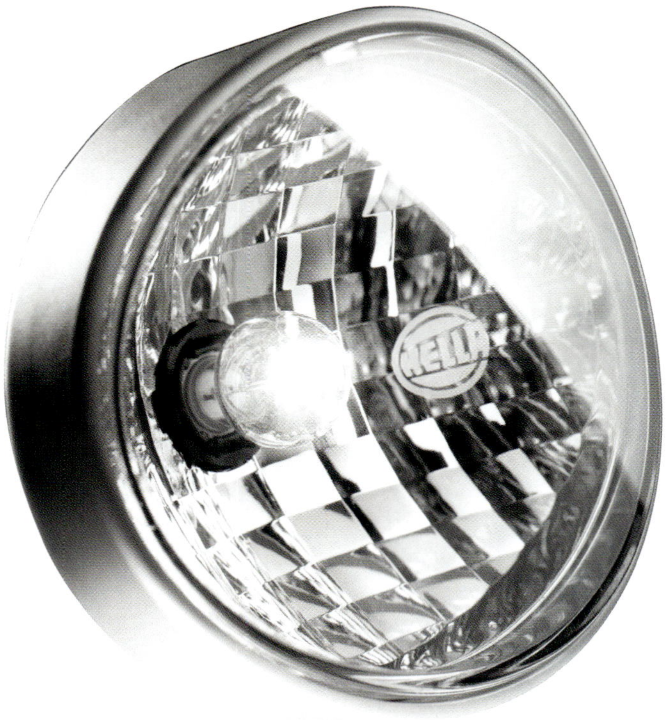

Hersteller Manufacturer
Hella KG Hueck & Co.
Lippstadt, Germany

Produkt Product
Tagfahrleuchte
Daytime running light

Design
Gerd O. Schmidt
GOSDesign
Wuppertal, Germany

Diese spezielle Tagfahrleuchte dient zur optimalen Kenntlichmachung von Fahrzeugen am Tage. Die Formgebung und das Klarglas-Design (123 Reflektoren hinter einer kristallklaren Scheibe) passen gut zu den Fahrzeugfronten moderner Autos und lassen sich leicht in das Gesamtbild integrieren.

The special daytime running light has been developed to ensure optimal visibility of vehicles during daylight. The shape and the clear-glass structure (123 reflectors behind a crystal-clear pane) are ideal to blend with the front-ends of modern automobiles and can therefore easily be integrated into the overall front facelift of the car.

Hersteller Manufacturer
Zweibrueder Optoelectronics
Solingen, Germany

Produkt Product
V6 Power Chip
LED-Lampe
LED lamp

Design
Rainer Opolka

Die V6 Power Chip ist eine LED-Lampe, die die Helligkeit einer 50 cm langen Stabtaschenlampe erreicht. Die Lampe ist klein, handlich und mit einer speziellen, computerberechneten Linse ausgestattet. Das ergonomisch gestaltete Gehäuse besteht aus rostfreiem Edelstahl. Eine dazugehörige Gürteltasche bietet Tragekomfort.

The V6 Power Chip is an LED lamp that delivers the same brightness as a 50cm-long powerful flashlight. The lamp is small, handy and comes equipped with a special lens whose angles and qualities are computer-calculated. The lamp has an ergonomically designed stainless steel casing and is rounded out with a belt pouch for enhanced portability.

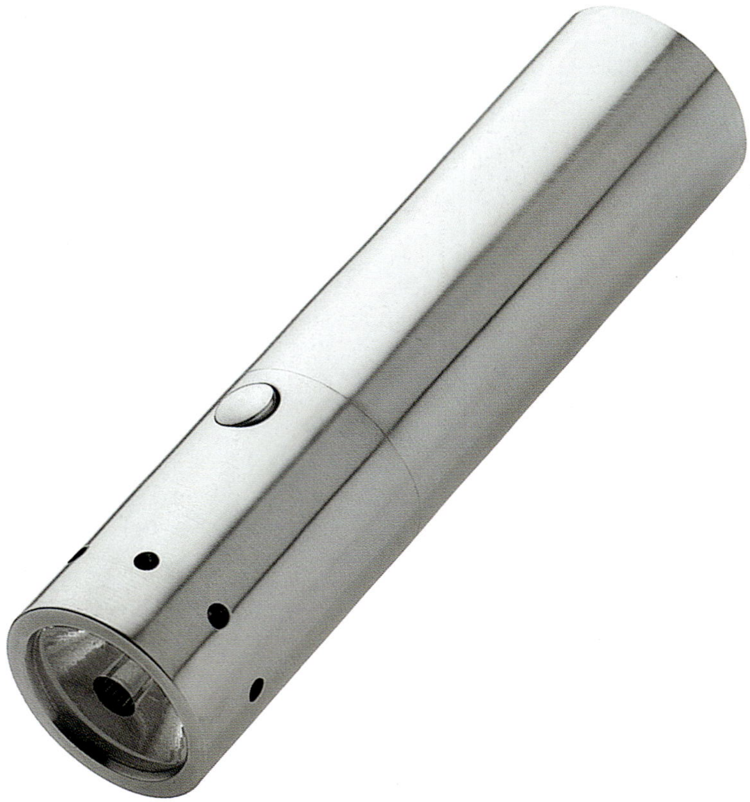

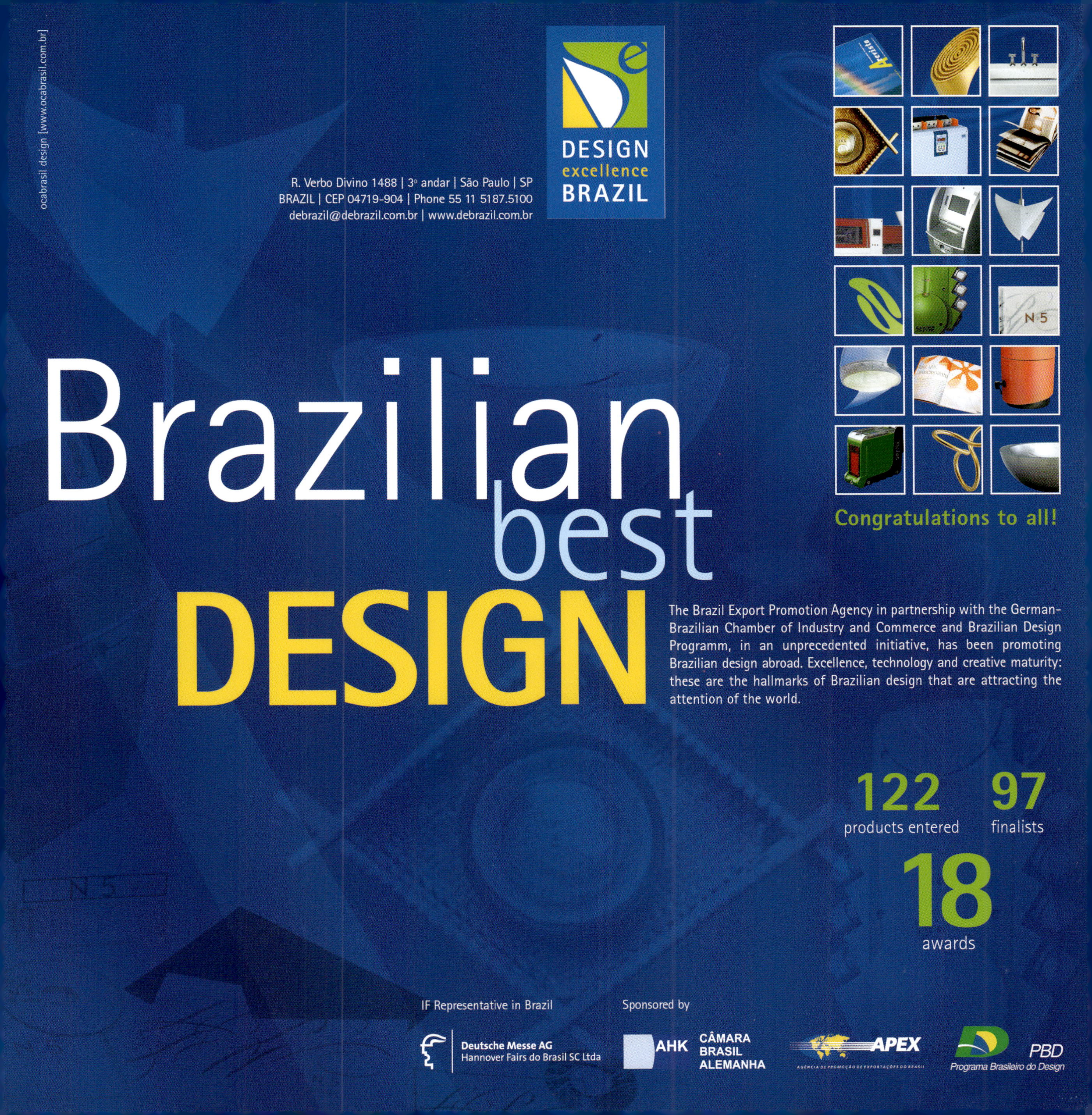

ocabrasil design [www.ocabrasil.com.br]
DESIGN excellence BRAZIL
R. Verbo Divino 1488 | 3º andar | São Paulo | SP
BRAZIL | CEP 04719-904 | Phone 55 11 5187.5100
debrazil@debrazil.com.br | www.debrazil.com.br
Brazilian best DESIGN
Congratulations to all!
The Brazil Export Promotion Agency in partnership with the German-Brazilian Chamber of Industry and Commerce and Brazilian Design Programm, in an unprecedented initiative, has been promoting Brazilian design abroad. Excellence, technology and creative maturity: these are the hallmarks of Brazilian design that are attracting the attention of the world.
122 products entered
97 finalists
18 awards
IF Representative in Brazil
Deutsche Messe AG
Hannover Fairs do Brasil SC Ltda
Sponsored by
AHK CÂMARA BRASIL ALEMANHA
APEX
PBD
Programa Brasileiro do Design

Hersteller Manufacturer
Daka Development Ltd.
Cheung Sha wan
KLN, Hong Kong, China

Produkt Product
Dual Power Eternal Light
Taschenlampe
Flashlight

Design
Daka Design Team
Cheung Sha wan
KLN, Hong Kong, China

Das Funktionsprinzip der Taschenlampe Dual Power Eternal Light beruht auf dem Faradayschen Gesetz. Ein Magnet aus Seltenerdmetallen wird in einer Kupferspule hin- und her bewegt, wobei die entstehende Elektrizität in einem Kondensator gespeichert wird, um eine Leuchtdiode mit Energie zu versorgen. Wird die Taschenlampe nur wenige Sekunden geschüttelt, entsteht genügend Energie für mehrere Minuten Licht. Zusätzlich besitzt die Taschenlampe eine Batterie, für helleren und längeren Betrieb. Die Lampe eignet sich insbesondere für Notfallsituationen und dank ihrer Wasserresistenz für alle möglichen Outdoor-Aktivitäten.

The Dual Power Eternal Light flashlight rests on Faraday's Law, whereby a rare earth magnet passing back and forth through a copper coil generates electricity. This is then stored in a capacitor and used to power a super-bright LED. The flashlight only needs to be shaken for a few moments to develop enough energy for several minutes of light. There is likewise a battery-powered mode for brighter and longer lasting light. The unit was developed for emergency situations and, thanks to its waterproofing, for any manner of outdoor activities.

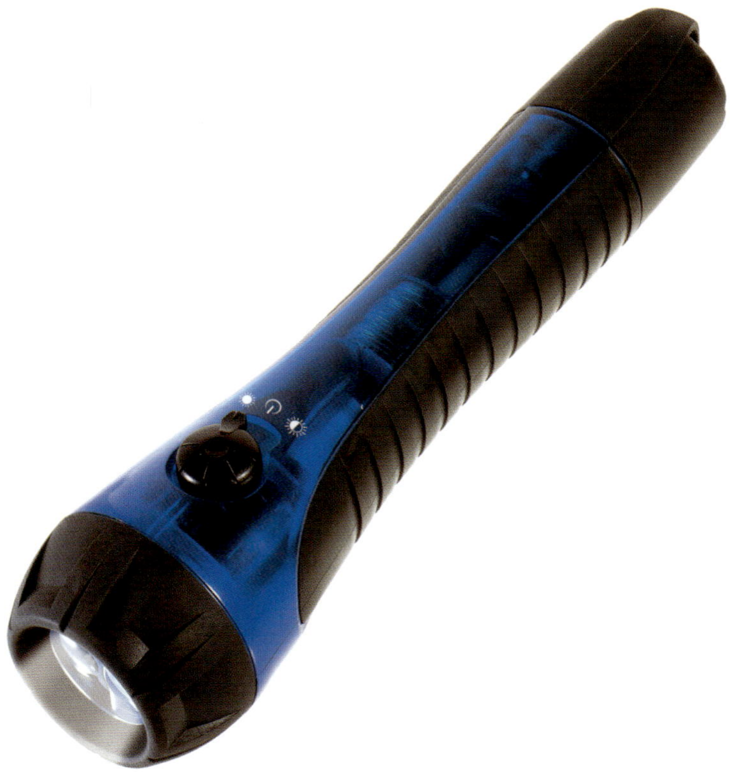

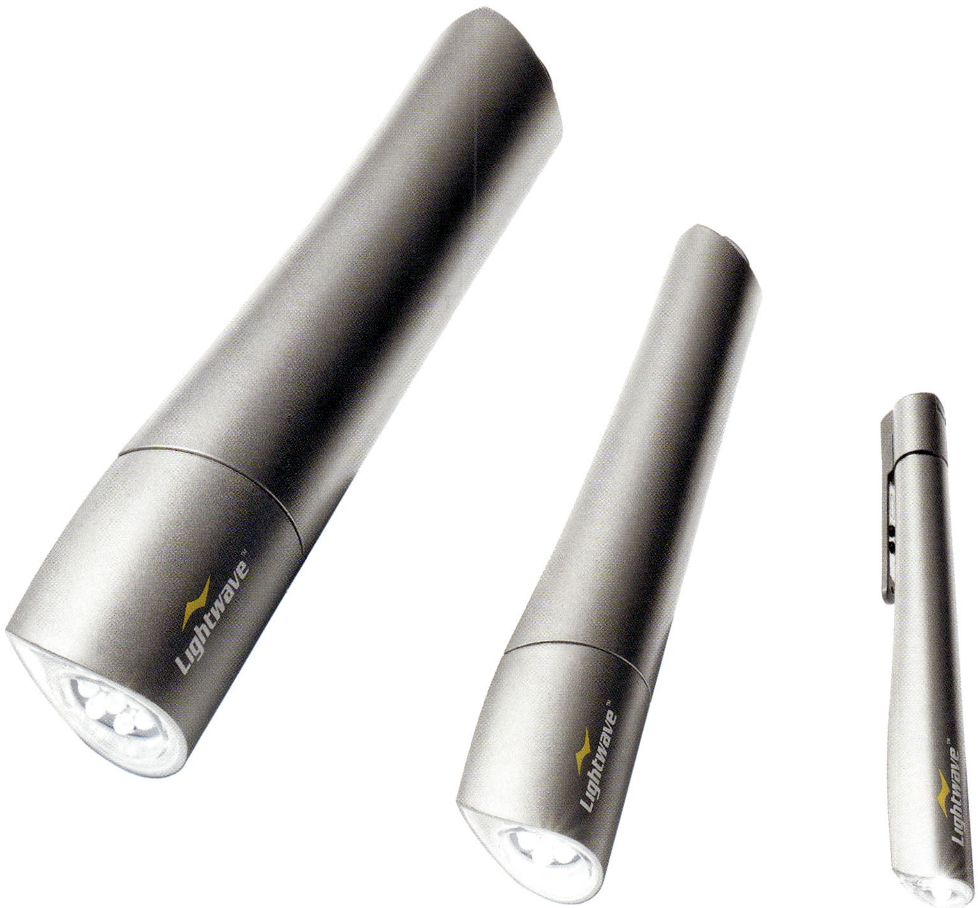

Hersteller Manufacturer
Lightwave
Jacksonville, Florida, USA

Produkt Product
Tec3000™, Tec2000™, Tec1000™
LED Taschenlampe
LED flashlight

Design
Stefan Wiesmeth
Forks-NY
Brooklyn, New York, USA

Die Taschenlampen der Tec Serie von Lightwave verfügen über eine spezielle Energiespartechnologie, die die Batterielebensdauer maximiert. Mit ihrem innovativen und ergonomischen Design sind sie sowohl für Rechtshänder als auch für Linkshänder angenehm zu umfassen. Die spezielle Bumper Head™Konstruktion verhindert zum einen die Blendung durch die LED-Dioden und zum anderen schützt sie diese beim Herunterfallen der Lampe. Das ABS-Gehäuse ist leicht, widerstandsfähig und wasserfest. Die Taschenlampen sind in unterschiedlichen Größen Tec3000™, Tec2000™, Tec1000™ erhältlich, die kleinste besitzt einen Clip, um sie an der Kleidung zu befestigen.

Lightwave's Tec series flashlights were developed in line with ergonomic criteria. Thus, both left- and right-handers will find them pleasant to hold. The special transparent polycarbonate Bumper Head™ construction prevents any dazzle by the LEDs and also protects them should the flashlight accidentally be dropped. The ABS casing is light, robust and waterproof. The flashlight is rounded out with a sophisticated current management system that maximizes the service life of regular batteries and thus efficiency. The series features a variety of sizes, namely the Tec3000™, Tec2000™, and Tec1000™, all of which consistently reinvent the design of flashlights. The smallest comes with a clip for fastening it to clothing.

PRODUCT DESIGN

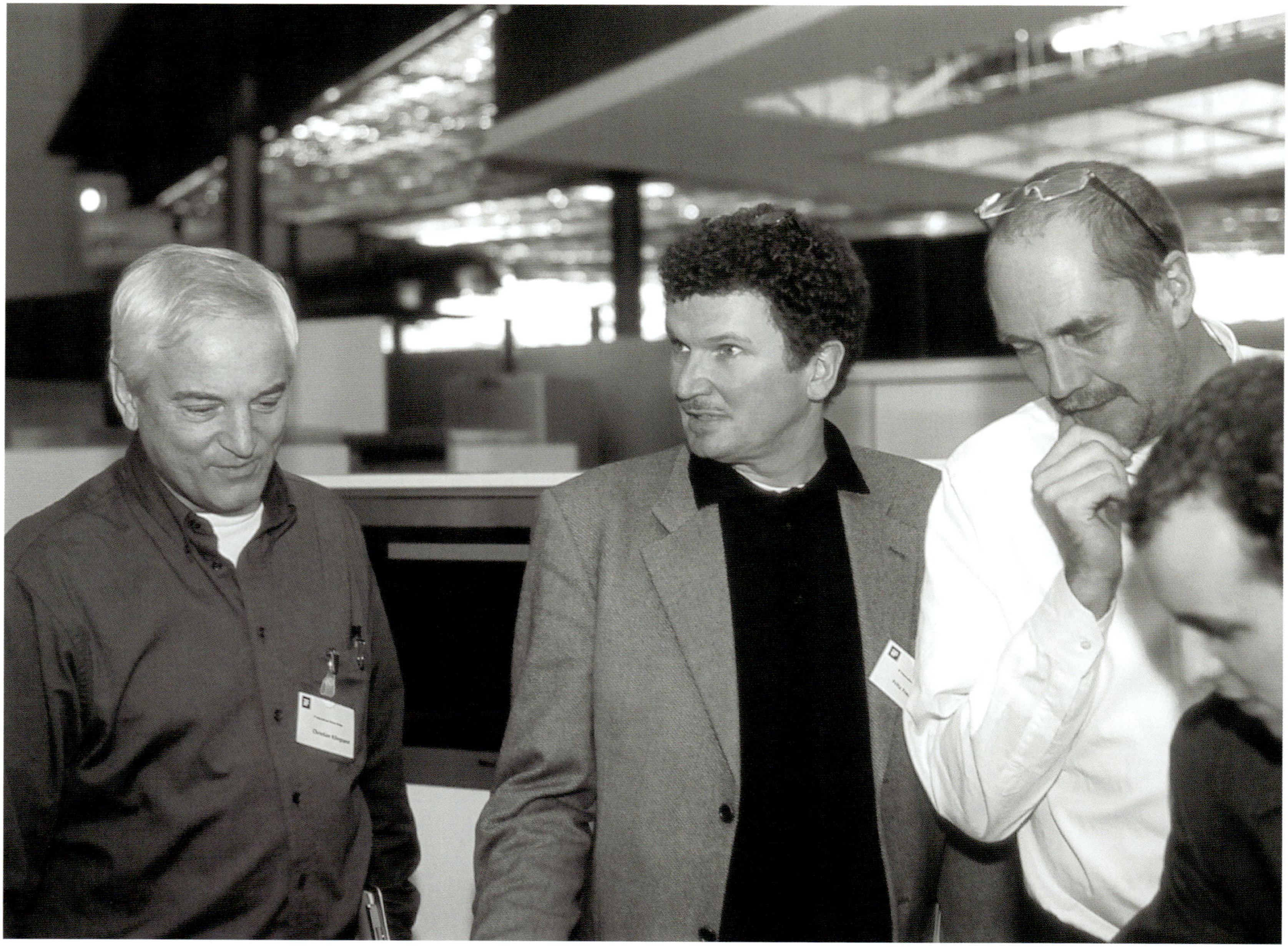

Haushalt Household

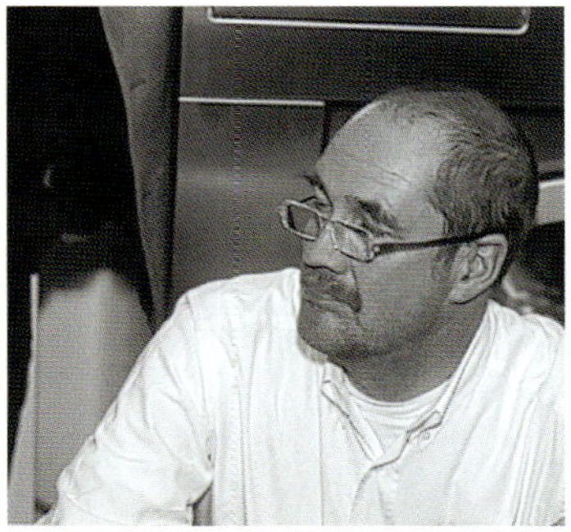

NILS HOLGER MOORMANN produziert seit 1982 als einer der Protagonisten des „Neuen Deutschen Designs" Entwürfe zumeist junger, unbekannter Designer. Seit 1992 vertreibt die Firma Moormann mehrfach ausgezeichnete Möbel- und Produktkollektionen.

Nils Holger Moormann has since 1982 been widely regarded as a champion of the "New German Design", realizing ideas created by typically young unknown designers. Since 1992, the Moormann corporation has distributed furniture and product collections that have won numerous awards.

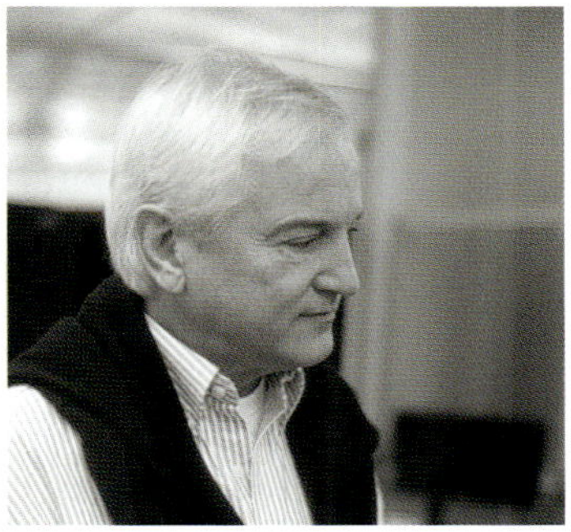

CHRISTIAN KLINGSPOR arbeitete nach seinem Studium in Stockholm viele Jahre für Electrolux. 1974 gründete er eine Designfirma mit einem eigenen Möbelprogramm. Seit 1985 ist er wieder bei Electrolux Design Director und ist dort für den Aufbau der In-House-Design-Abteilung verantwortlich.

Christian Klingspor completed his studies in Stockholm and then worked for many years at Electrolux. In 1974, he founded his own design company, launching his own furniture line. Since 1985 he has been back at Electrolux as Design Director where he was responsible for setting up the in-house design department.

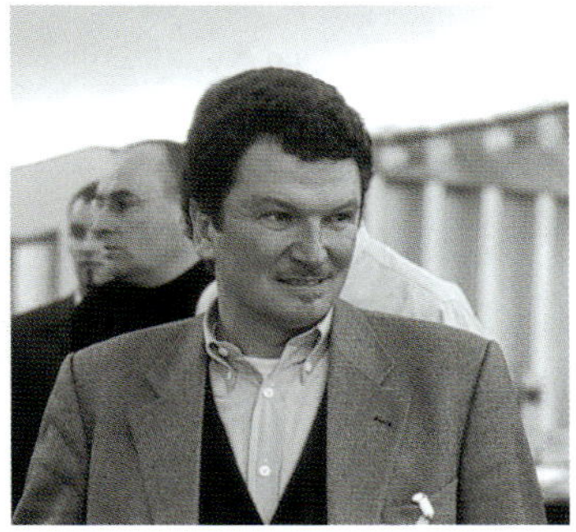

FRITZ FRENKLER ist Diplom-Industriedesigner. Er gründete gemeinsam mit Anette Ponholzer die f/p design gmbh. Er arbeitete mehrere Jahre für Frogdesign und für die Deutsche Bahn Medien GmbH. Er ist Vorstandsmitglied im Industrie Forum Design Hannover und in anderen Designinstitutionen.

Fritz Frenkler studied industrial design and together with Anette Ponholzer founded f/p design gmbh. He worked for several years for Frogdesign and for Deutsche Bahn Medien GmbH. He is a board member of Industrie Forum Design Hanover and of various other design institutions.

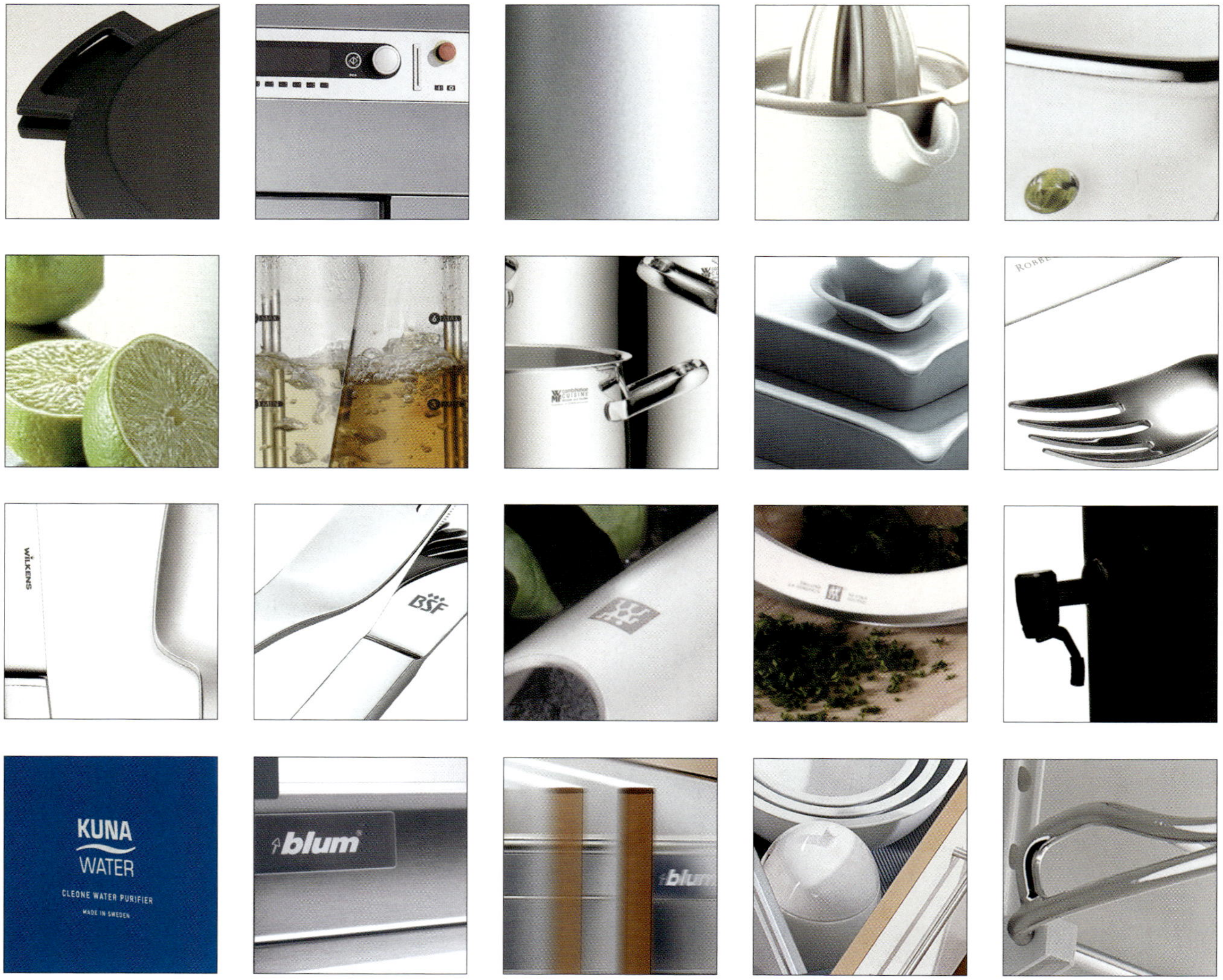
WILKENS
BSF
KUNA
WATER
CLEONE WATER PURIFIER
MADE IN SWEDEN
blum
blum

Haushalt Household

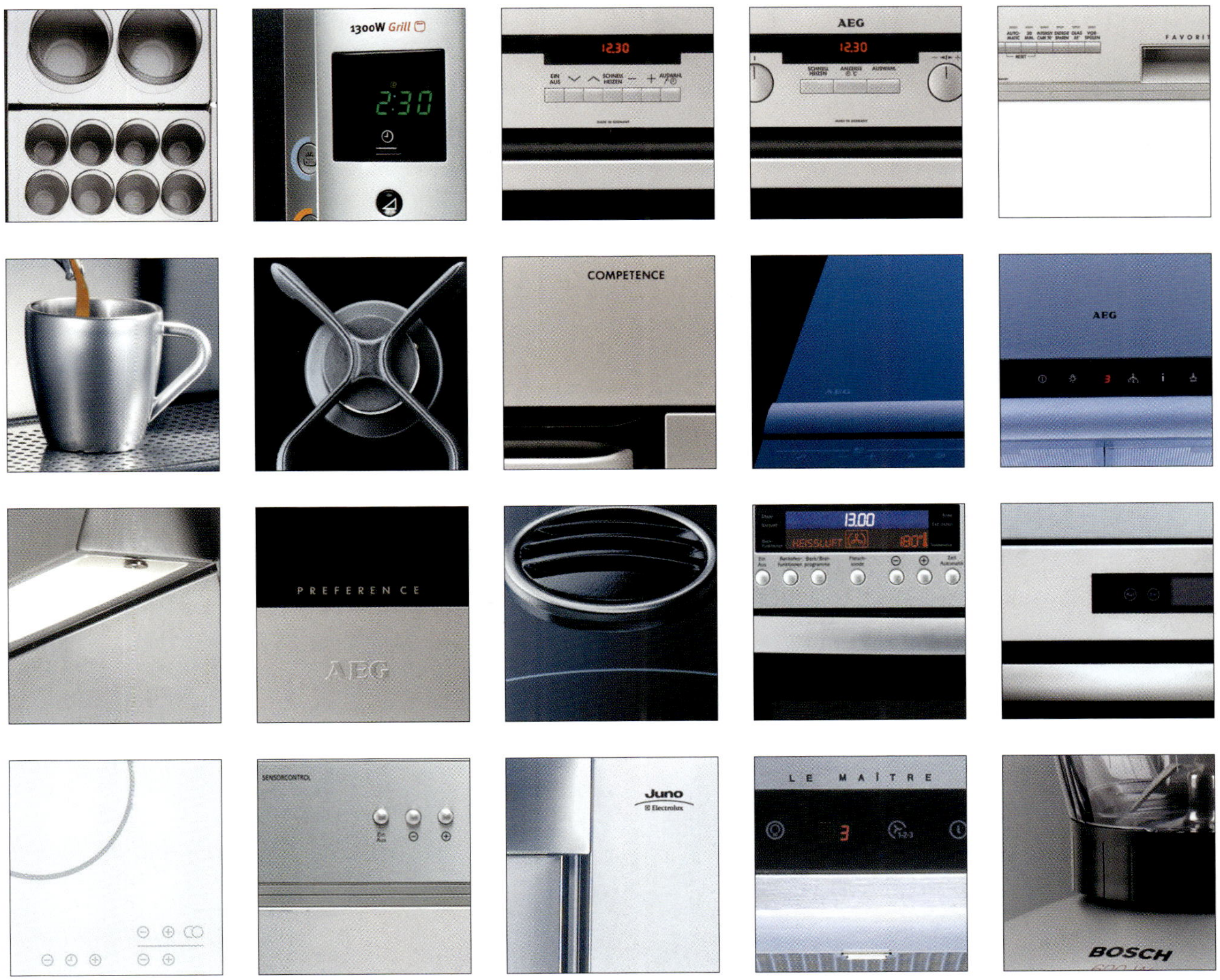

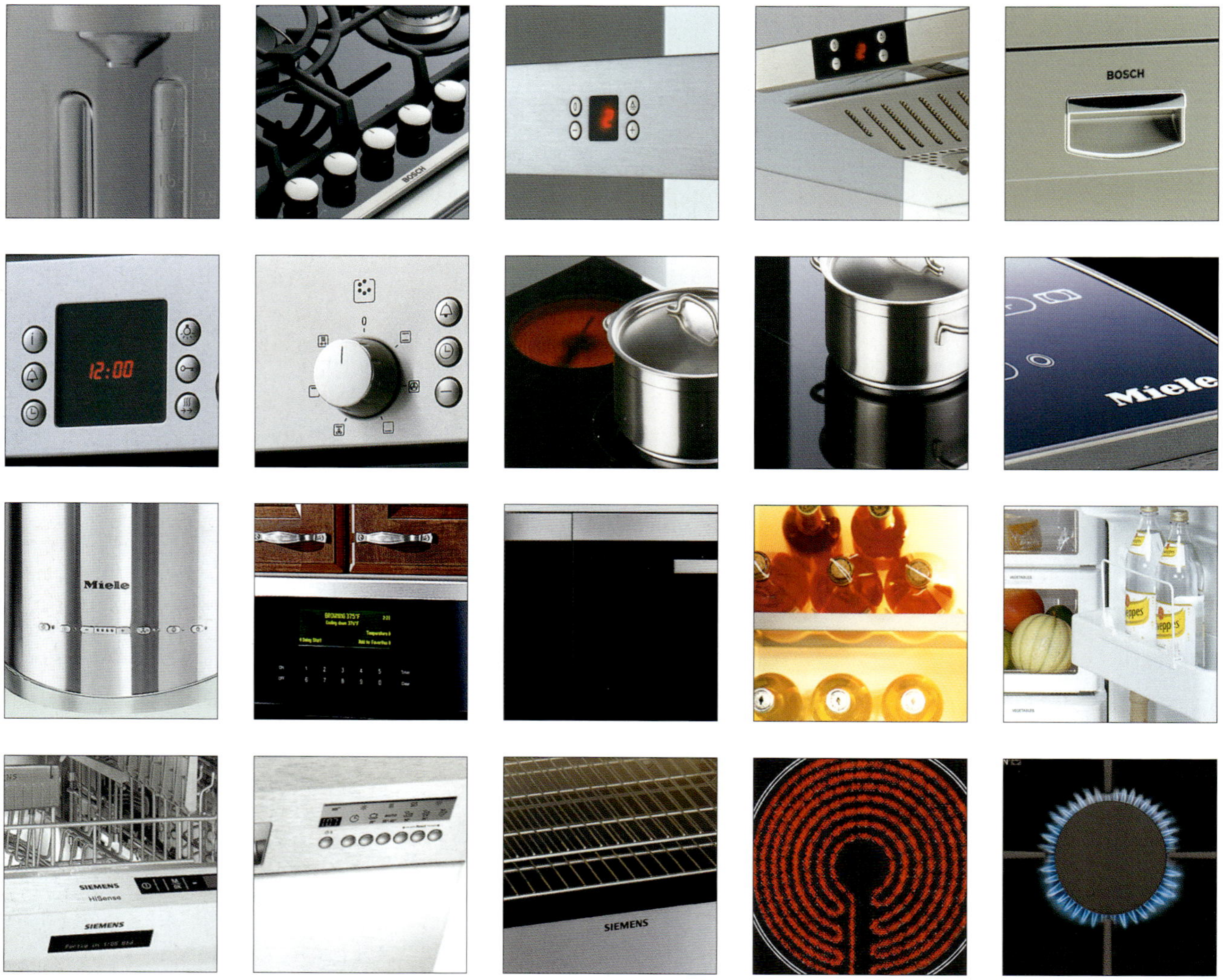
BOSCH
BOSCH
12:00
Miele
Miele
SIEMENS
HiSense
SIEMENS
SIEMENS

Haushalt Household

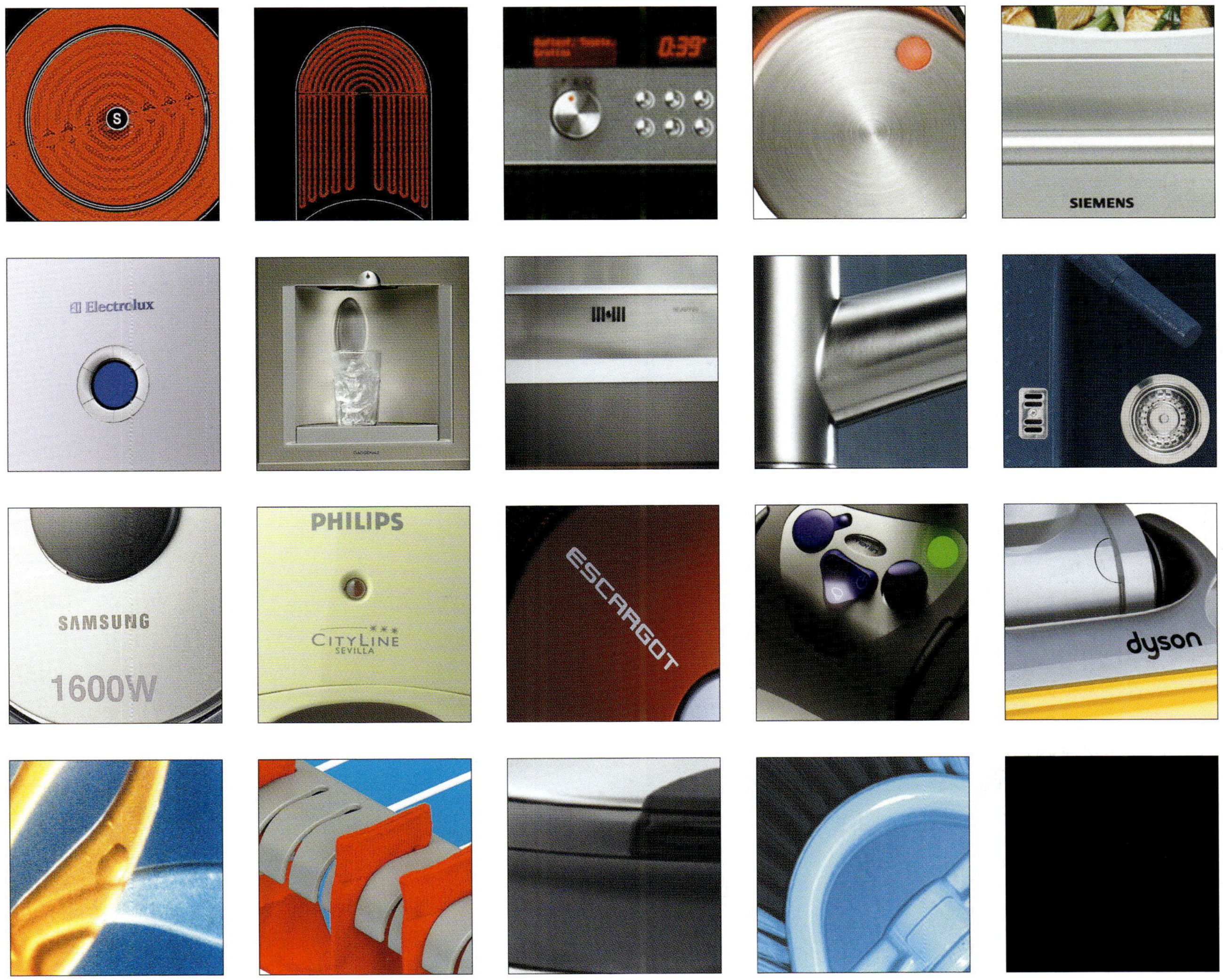

Hersteller Manufacturer
Toshiba Corporation
Tokio, Japan

Produkt Product
IHC-25PB, IHC-25PC
Induktionskocher
Induction heater cooker

Design
Hiroko Matsumoto
Design Center
Toshiba Corporation
Tokio, Japan

Der Induktionskocher IHC-25PB und ICH-25PC ist eine Kombination aus Topf und Induktionskocher, wobei das Design der beiden Komponenten eng aufeinander abgestimmt ist. Er eignet sich insbesondere zum Kochen an Orten außerhalb der Küche, wobei die Topfaufsätze zum Servieren auf den Tisch gestellt werden können. Sie werden aus Keramik oder Gusseisen gefertigt und spiegeln in ihrer Formgebung die Eigenschaften des Materials, aus dem sie bestehen. So hat der Keramiktopf eine betont runde Form und der Eisentopf eckige, der Härte des Materials entsprechende Griffe. Das Bedienfeld der Kocheinheit kann durch einen Gleitmechanismus ein- bzw. herausgefahren werden, was die gestalterische Einheit des Kochers unterstreicht.

The IHC-25PB and ICH-25PC induction heating cooker is a combination of pot and induction cooker, and the design is a closely-coordinated mixture of the two. It is particularly suited for cooking outside the kitchen although the pots can be placed on the table for serving. The pots are made of earthenware and cast iron, and their shape reflects the properties of the respective material: an accentuated round form for the earthenware ceramic and an angular form for the iron pot with handles corresponding to the hardness of the material. The control panel on the cooker is accessed by a push-and-slide device, enhancing aesthetic appeal and underlining the compact design of the cooker.

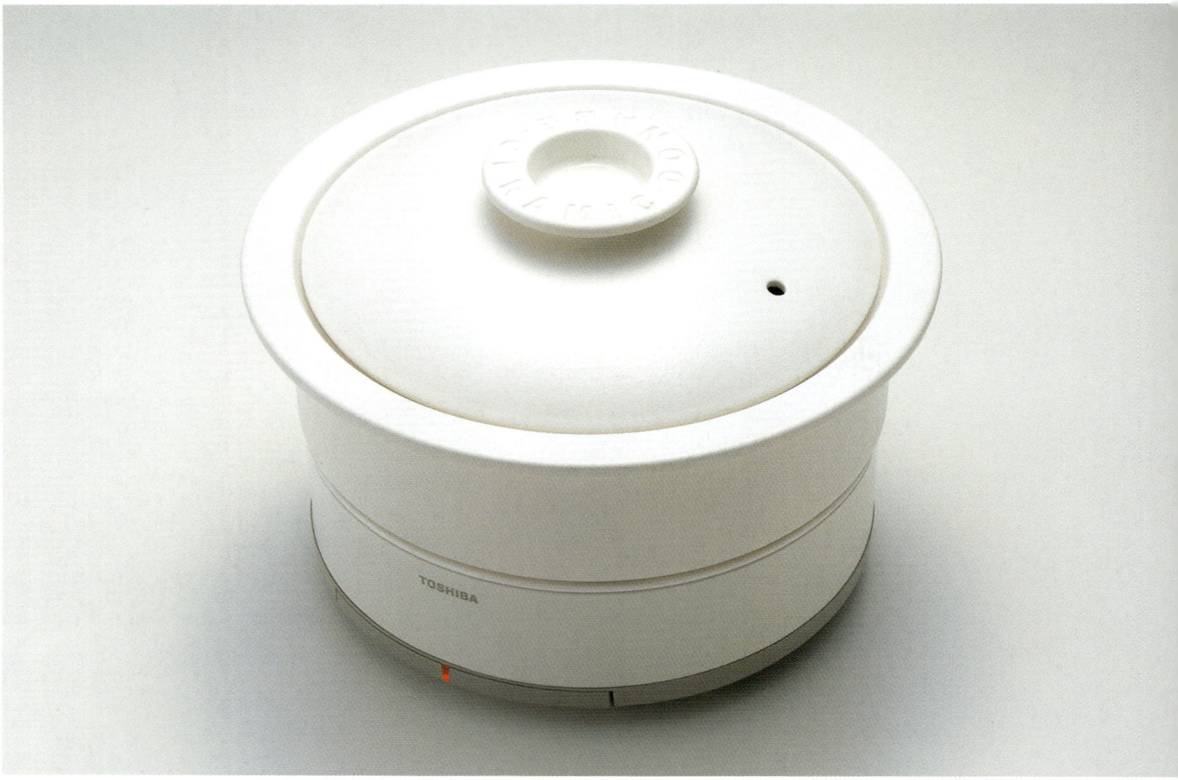

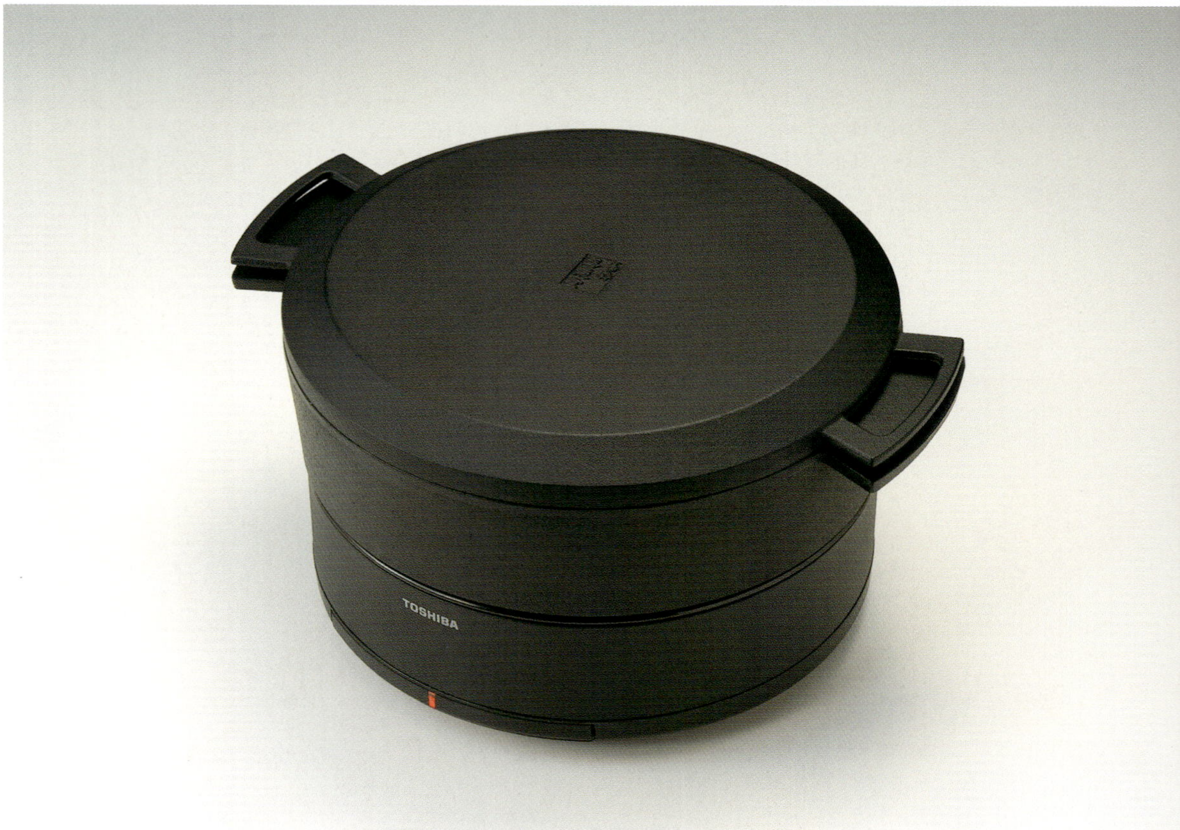

iF Gold Award

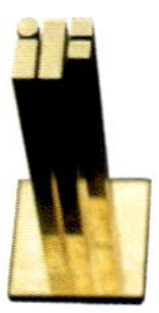

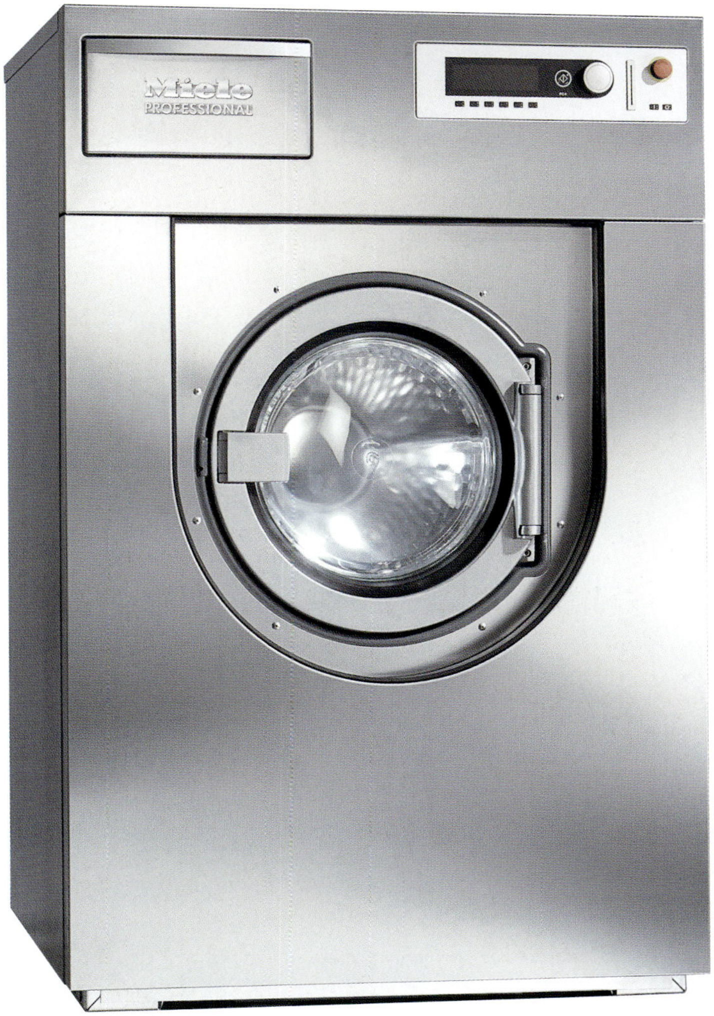

Hersteller Manufacturer
Miele & Cie. KG
Gütersloh, Germany

Produkt Product
PW 6201
Gewerbe-Waschmaschine
Commercial washing machine

Design
Miele & Cie. KG
Gütersloh, Germany

Auffälligstes Designmerkmal der 20 kg Gewerbe-Waschmaschine PW 6201 von Miele ist die U-Kontur auf der Vorderseite. Sie gibt der Maschine einen eigenständigen Charakter. Die Form greift das in dieser Größenordnung notwendige mitschwingende Aggregat auf: die Fläche, auf der die Tür montiert ist, schwingt während des Waschprozesses mit. Die neue Scharniergestaltung ermöglicht einen Türöffnungswinkel von 180 Grad für leichteres Be- und Entladen. Die Formgebung der Maschine zeichnet sich durch klare Linien und glatte Flächen aus, die kaum durch sichtbare Schrauben und scharfe Vorsprünge unterbrochen sind.

The most conspicuous feature of Miele's 20kg commercial washing machine PW 6201 is the U shape of the front side. It imbues the machine with a unique character. The shape absorbs the vibrations which accompany a unit of this size. The surface on which the door is mounted swings along during the washing process. The new hinge design enables the door to be opened 180°, facilitating loading and unloading. The machine's profile is characterized by clear lines and smooth surfaces almost completely free of visible screws and sharp projections.

iF Silver Award

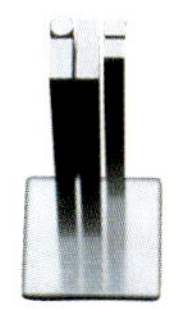

Hersteller Manufacturer
Thomas (Rosenthal Group)
The Trend Factory
Selb, Germany

Produkt Product
Coup
Porzellanservice
Porcelain set

Design
Konstantin Grcic
München, Germany

Die von Konstantin Grcic entworfene Thermoskanne gehört zum Porzellanservice Coup von Thomas. Auch die Kanne zeichnet sich wie die gesamte Linie durch reduziertes Design aus, das unnötige Komponenten vermeidet. Der Materialmix aus Porzellan und Metall fügt sich harmonisch in das Bild der anderen Serviceteile ein.

The thermos flask designed by Konstantin Grcic is part of the Thomas Trend Factory's porcelain Coup collection. As with the entire line, the thermos flask has an eye-catching minimalist design which dispenses with any unnecessary rills. The materials mix (porcelain and metal) melds well with the other pieces in the collection to create a harmonious image.

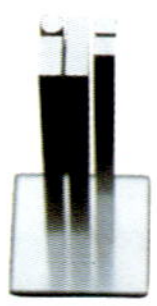

iF Silver Award

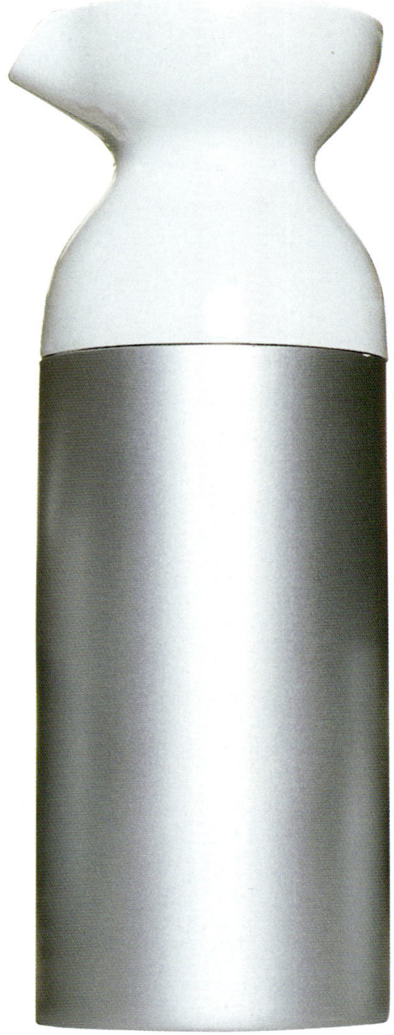

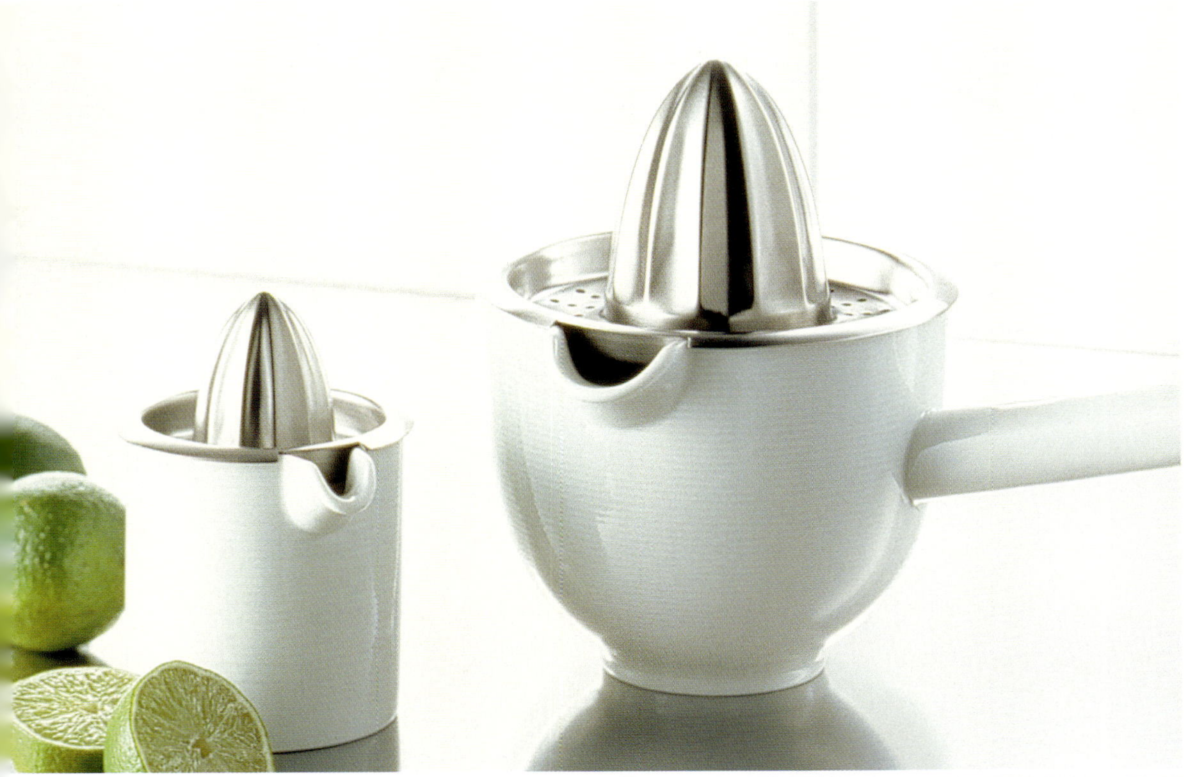

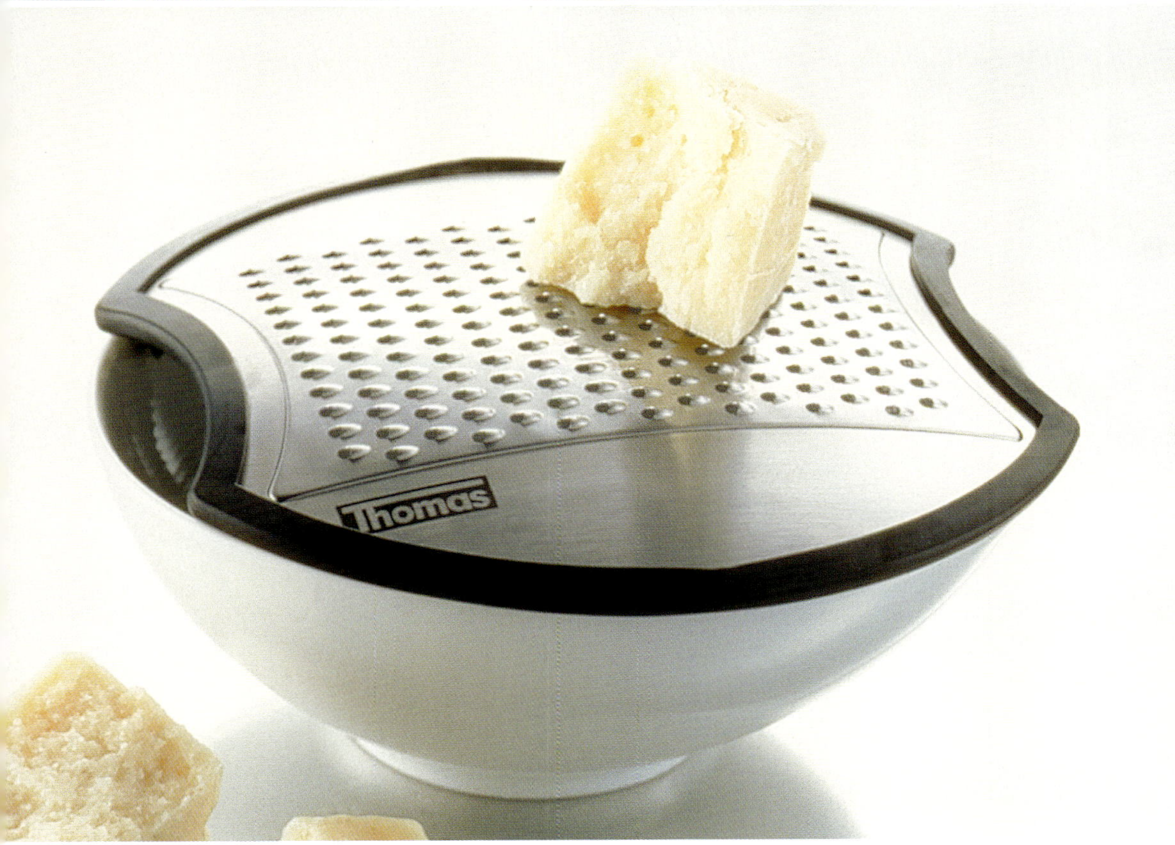

Hersteller Manufacturer
Thomas (Rosenthal Group)
The Trend Factory
Selb, Germany

Produkt Product
Küchenhelfer
Kochutensilien
Cooking utensils

Design
Thomas (Rosenthal Group)
The Trend Factory
Selb, Germany

Die Küchenhelfer-Kollektion besteht aus zwanzig Kochutensilien aus Edelstahl. Die Griffe der verschiedenen Löffel, Kellen, des Pizzarades und -hebers sind gerade, nur der Übergang zum eigentlichen Werkzeug ist gebogen. Neben den Artikeln mit Griff beinhaltet die Kollektion zudem einen Zitronen- und Orangenentsafter, verschiedene Reiben wie Käse- und Gemüsereibe sowie einen Gurkenhobel. Mehrere Siebe und eine Salatschleuder runden das Sortiment ab. Die Edelstahlartikel werden mit verschiedenen Porzellanelementen von Thomas kombiniert.

The Küchenhelfer (kitchen helper) collection comprises 20 stainless steel cooking utensils. The handles of the various spoons, ladles, pizza-cutters and slices are straight, curving only where they join the main body of the tool. Apart from articles with handles there is a lemon squeezer, an orange juicer, various graters for cheese and vegetables, and a cucumber slice. The selection is rounded out with a number of sieves and a salad tumbler. The stainless steel articles are combined with various ceramic pieces for a perfect allround range from the manufacturer Thomas.

Hersteller Manufacturer
Fratelli Guzzini SpA
Recanati, Italy

Produkt Product
G-Plus
Elektrische Reibe
Electric grater

Design
George Sowden
Hiroshi Ono
Mailand, Italy

Die elektrische Reibe G-Plus zeichnet sich durch ein schlichtes und gleichzeitig raffiniertes Design aus. Die Schublade, in der sich der geriebene Käse ansammelt, kann gleichzeitig auch als Servierbehälter auf den Tisch gestellt werden. Dank der hohen Leistung des Triebwerkes können mit der Reibe auch Brot, Schokolade oder Nüsse zerkleinert werden. Die Reibe kann leicht herausgenommen werden, was eine leichte Reinigung ermöglicht. Sie wird aus hochwertigem Acrylmaterial, Stahl und Aluminium hergestellt und hat eine Leistung von 90 Watt.

The G-Plus electric grater stands out for its simple yet refined designed. The tray in which the grated cheese collects can also be used for serving at the table. Thanks to the high-powered 90W motor the grater can also chop chocolate, nuts or bread finely. The grating element can be extracted effortlessly making it easy to clean. High-quality acrylic, steel and aluminum are used in the device.

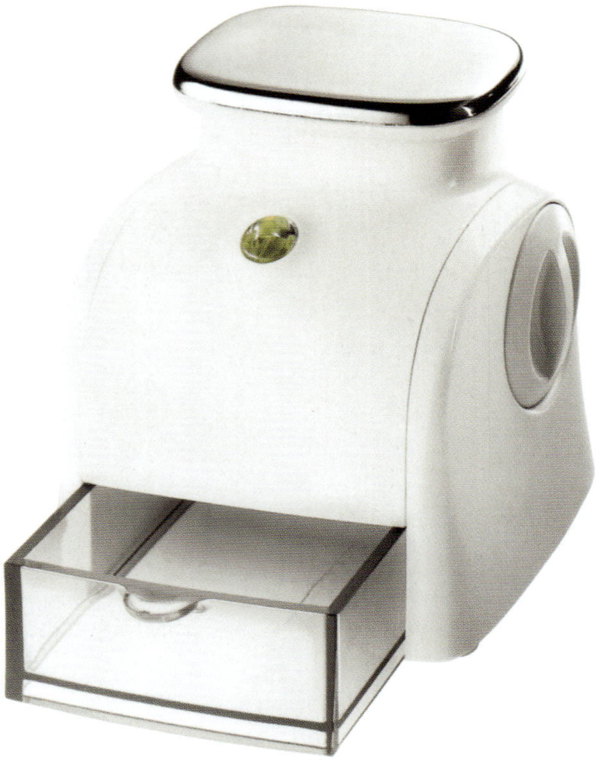

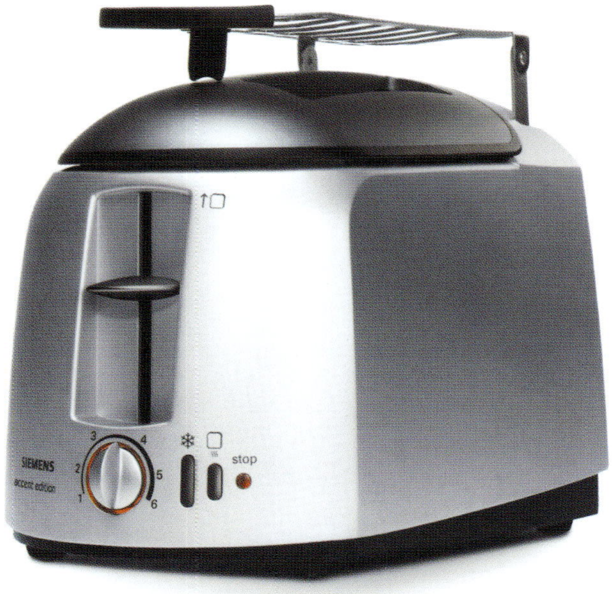

Hersteller Manufacturer
BSH Bosch und Siemens Hausgeräte GmbH
München, Germany

Produkt Product
TT 46201 accent edition
Kompakt-Toaster
Compact electronic toaster

Design
Helmut Kaiser
BSH Bosch und Siemens Hausgeräte GmbH
München, Germany

Der TT 46201 accent edition Kompakt-Toaster ist ein klassischer Zwei-Scheiben-Toaster mit extrabreiten Brotschächten von 32 mm. Durch seine stirnseitig angeordneten Bedienelemente bietet er ein hohes Maß an Benutzerfreundlichkeit. Der an der Rückseite angebrachte Brötchenaufsatz kann über einen wärmeisolierten Griff hochgezogen und mittels einer stabilen Mechanik nach vorne über den Röstraum geklappt werden. Er bietet Platz für zwei große Brötchen.

The TT 46201 accent edition compact electronic toaster is a classic two-slice toaster with extra-wide, 32 mm toast slots. Operating controls on the front are particularly user friendly. The bread roll rack on the rear can be raised by means of a heat-resistant handle, and features a sturdy mechanism enabling it to be swung forward over the bread slots – with ample space for two large bread rolls.

Hersteller Manufacturer
Schott Jenaer Glas GmbH
Jena, Germany

Produkt Product
Passero
Kaffeekocher
Coffeemaker

Design
Dirk Loff
Thomas G. Wagner
Delphin Design
Berlin, Germany

Passero ist ein Kaffeekocher zur einfachen und schnellen Zubereitung eines aromatischen Kaffees auf dem Elektro- oder Gasherd. Durch den großzügigen Glasbehälter wird der Prozess des Kaffeekochens sichtbar und zu einem visuellen Erlebnis. Mit seiner sachlichen und klaren Form soll sich der Kaffeekocher unaufdringlich in das Ambiente eines jeden Nutzers einfügen. Er besteht aus einem von Schott Jenaer Glas entwickelten hitzebeständigen Borosilikatglas und hat einen Einsatz aus gebürstetem Edelstahl. Der Griff wird in klassischer Glastechnologie angesetzt.

Passero is a coffeemaker for the quick preparation of aromatic coffee on an electric or gas stove. The large glass container allows you to watch the coffee while it is being made – an added visual experience. The objective and clear form of the coffeemaker is designed to be visually unobtrusive in any ambience. It consists of a heat-resistant, borosilicate glass container developed by Schott Jenaer Glas with a brushed stainless steel insert. The handle is attached in classic glass technology style.

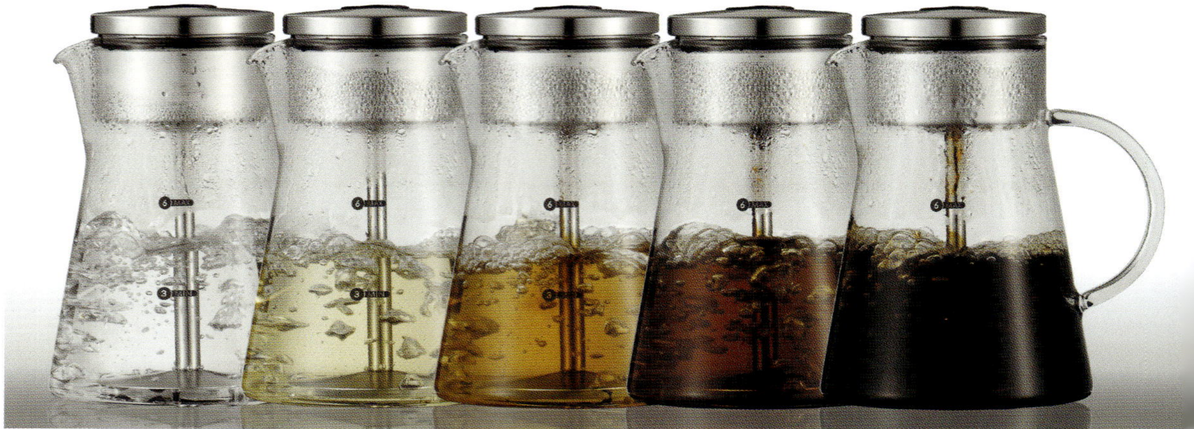

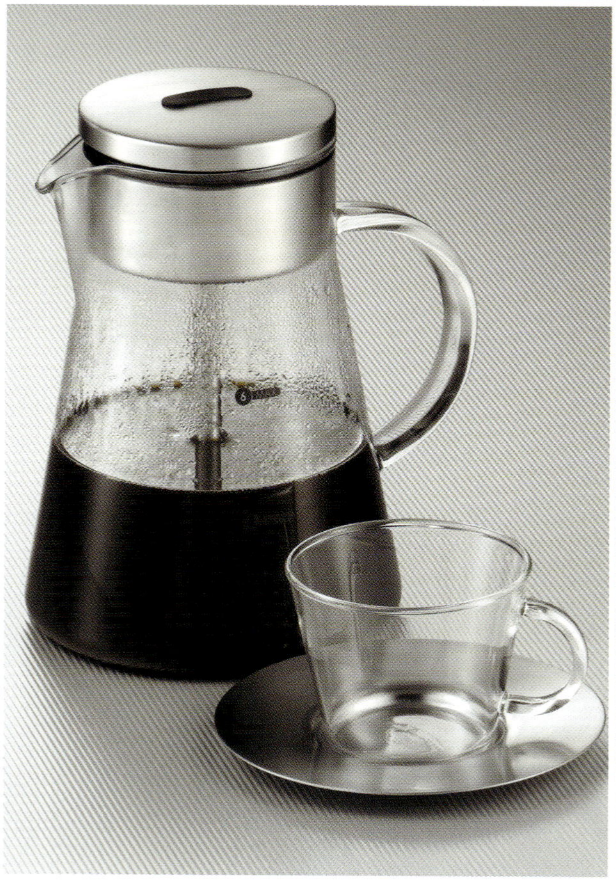

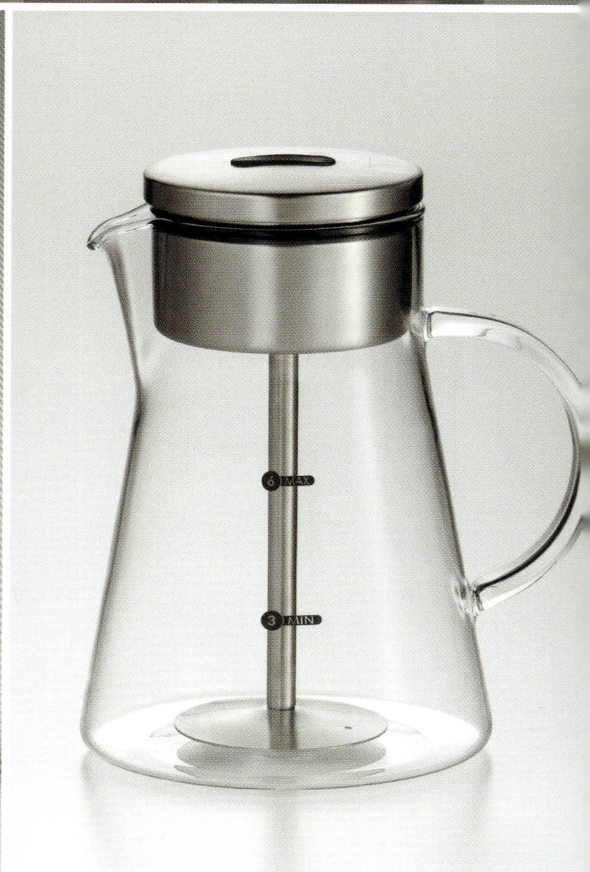

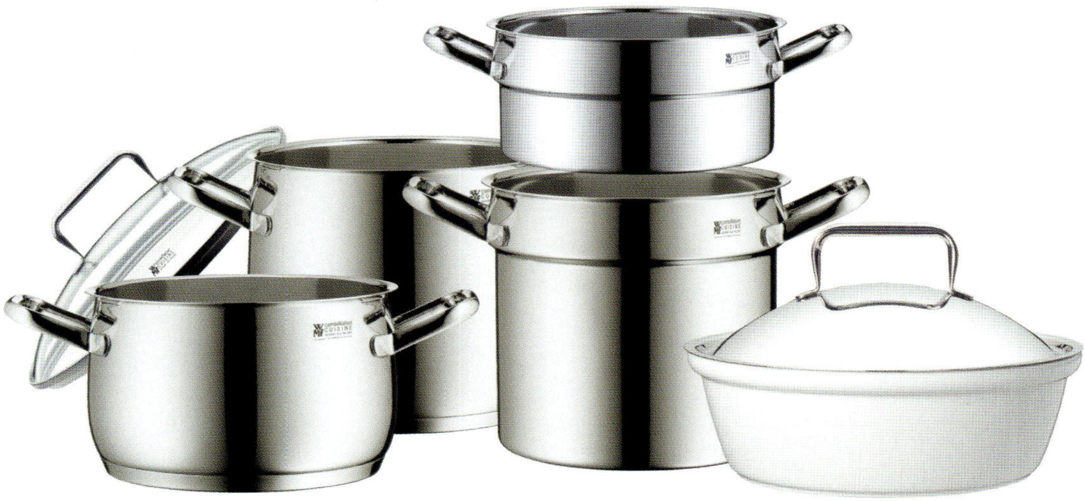

Hersteller Manufacturer
WMF Württembergische Metallwarenfabrik AG
Geislingen/Steige, Germany

Produkt Product
CombiNation Cuisine
Koch- und Serviersystem
Cooking and serving system

Design
Ole Palsby
Aalsgaarde, Denmark

CombiNation Cuisine ist ein Koch- und Serviersystem, das aus einer Vielzahl von Einzelkomponenten besteht, die im Zusammenspiel Lösungen zum Kochen, Braten, Überbacken bis hin zum Servieren bieten. Zu den verschiedenen Topfformen passen Einsätze zum Dämpfen, Brühen und Passieren, zum Rühren und Nudeln kochen. Auflaufformen, Stielkasserollen, Pfannen und Bräter vervollständigen das Programm. Wahlweise können Cromargan® (Edelstahl 18/10) - oder Glasdeckel eingesetzt werden. Passende Untersetzer aus Cromargan® (Edelstahl 18/10) sorgen dafür, dass Töpfe und Auflaufformen auf den Tisch gestellt werden können.

CombiNation Cuisine is a cooking and serving system which consists of many individual elements. Together they provide cooking, baking, grilling and even serving solutions. Various sized pots can be fitted with accessories for steaming, boiling and straining, stirring, and for cooking noodles. The line is rounded off by baking and casserole dishes as well as pans – with a choice of either glass or Cromargan® (18/10 stainless steel) lids. Matching Cromargan® (18/10 stainless steel) trays allow pots and dishes to be placed on the table ready for serving.

Hersteller Manufacturer
KAHLA/Thüringen Porzellan GmbH
Kahla, Germany

Produkt Product
Five Senses
Backprogramm aus der Porzellanserie Five Senses
Baking dishes from the tableware range Five Senses

Design
Barbara Schmidt
Berlin, Germany

Die Gruppe von ofenfesten Formen gehört zur von Barbara Schmidt entworfenen Porzellanserie Five Senses. Eine mittlere und eine große Backform können jeweils mit einer Platte abgedeckt werden. Für die mittlere gibt es darüber hinaus einen Einsatz mit Ablaufrillen. Die kleine Backform wird mit Abdeckplatte zum Rechaud. Bei allen drei Gefäßen ist eine Ecke zu einem Ausgießer geformt. Gemüseeinsatz und Rechaudplatte können als Untersetzer dienen, während die Abdeckplatten für die Backformen auch zum Servieren verwendet werden können.

This range of oven-proof dishes is part of the Five Sense porcelain tableware series designed by Barbara Schmidt. A medium-size and a large-size baking dish can both be used with a lid. For the medium-sized pan there is also a grid-shaped inset converting it into a serving dish. The small baking dish can be used with a lid as a rechaud. All three dishes incorporate a corner shaped like a spout for pouring sauce or pouring off unwanted liquid. Vegetable insets and the hot-plate lid can be used on the table under hot dishes, while the lids for the baking dishes can be used as dip bowls.

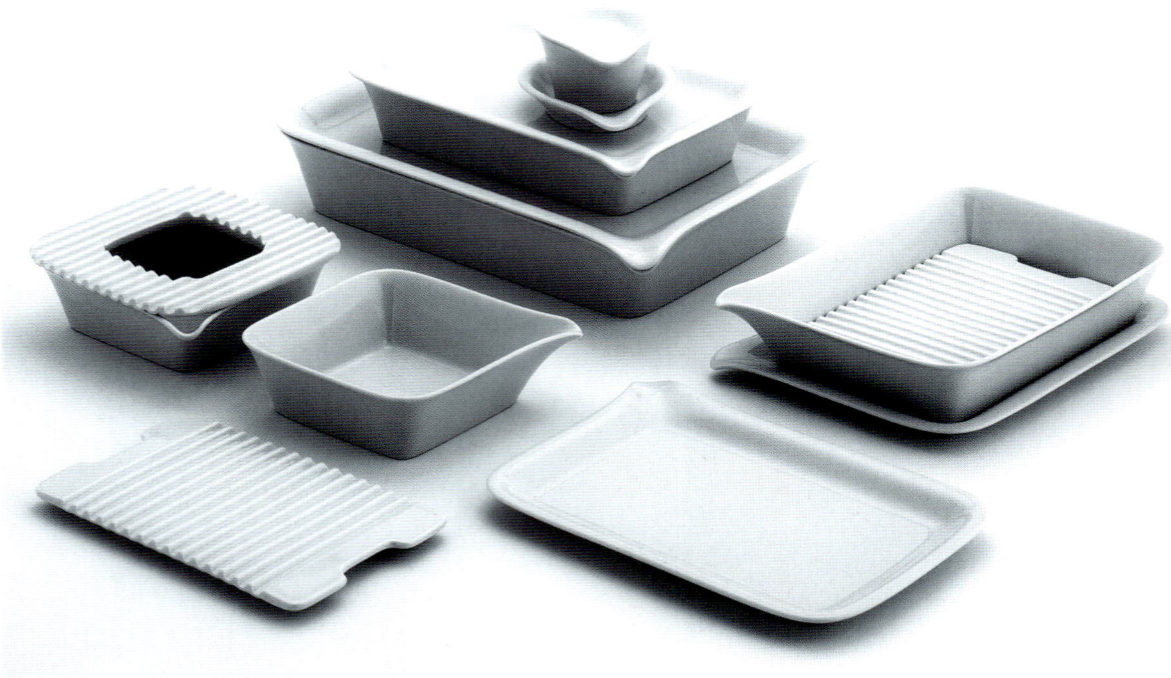

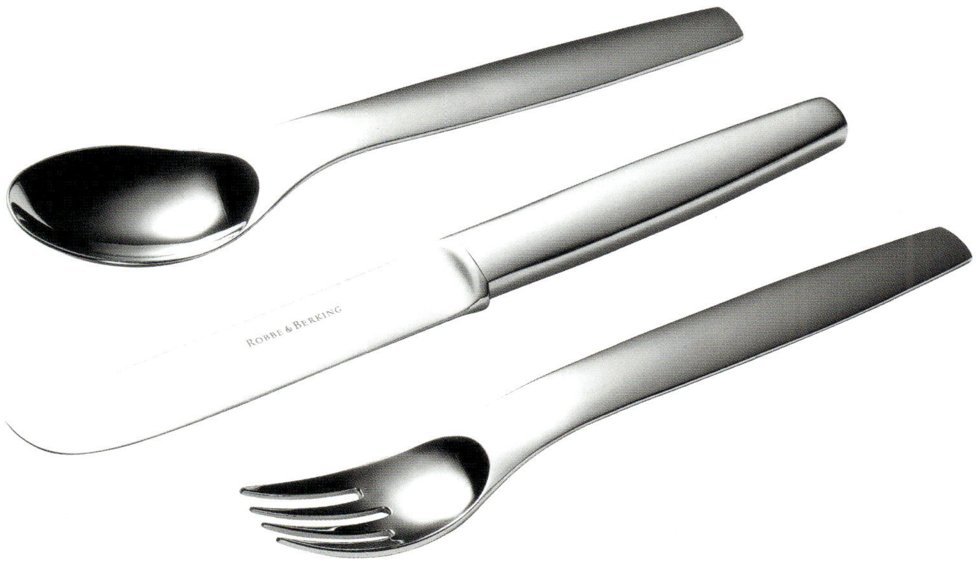

Hersteller Manufacturer
Robbe & Berking
Flensburg, Germany

Produkt Product
Pax
Besteckserie
Cutlery series

Design
Heinrich Meldau
Elmshorn, Germany

Die Besteckserie Pax besteht aus 26 Einzelteilen. Durch großzügig breite, an der Ober- und Unterseite leicht gewölbte Stiele und Hefte liegt das Besteck gut in der Hand. Erwähnenswert sind diesbezüglich auch die konkav ausgeformte Gabelhand und die tiefgezogene Laffe des Löffels. Das Design der Serie zeichnet sich durch sanft geschwungene Linien und den markanten Übergang von Stiel und Laffe aus. Zugunsten des Konzeptes von Langlebigkeit wurde auf modische Aspekte weitgehend verzichtet.

The Pax cutlery series consists of 26 individual pieces. The generous width of the handle and its slight curvature make it comfortable to hold. Notable in this respect are the concave-shaped fork section and the deep curve of the spoon. The design of the series is characterized by its gently curved lines and the prominent way the handle joins the utensil stem. The concept emphasizes long-living qualities instead of fashionable frills.

Hersteller Manufacturer
M. H. Wilkens & Söhne GmbH
Bremen, Germany

Produkt Product
Divo
Edelstahl-Besteck
Stainless steel cutlery

Design
Claudia Christl
M. H. Wilkens & Söhne GmbH
Bremen, Germany

Das Edelstahl-Besteck Divo bricht mit Sehgewohnheiten, insbesondere die rechteckigen Vorderteile, vor allem die Löffellaffen fallen ins Auge. Die reduzierten und leicht gespannten Stiele fangen die ungewöhnliche Silhouette formal auf. Die Funktionalität der Besteckteile wird durch ihre rechteckige Grundform nicht beeinträchtigt. Die großzügige Länge der Teile entspricht der klassischen Tafelgröße, insgesamt besteht die Serie aus 24 verschiedenen Einzelteilen mit polierter Oberfläche.

Divo is a set of high-grade stainless steel cutlery that breaks with tableware conventions to create a strong visual image by adopting rectangular heads, with the spoons being the most spectacular. The minimal and slightly elongated stems forge an unusual formal silhouette. The functional nature of the cutlery is not impeded by the rectangular utensil sections and the generous length of the individual pieces corresponds with classic cutlery sizes. The series features 24 pieces with polished surfaces.

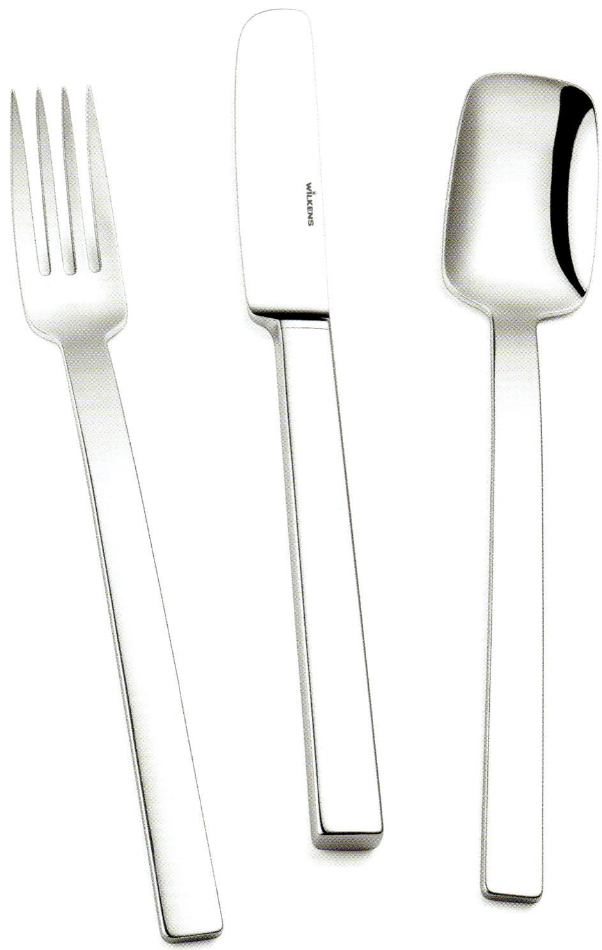

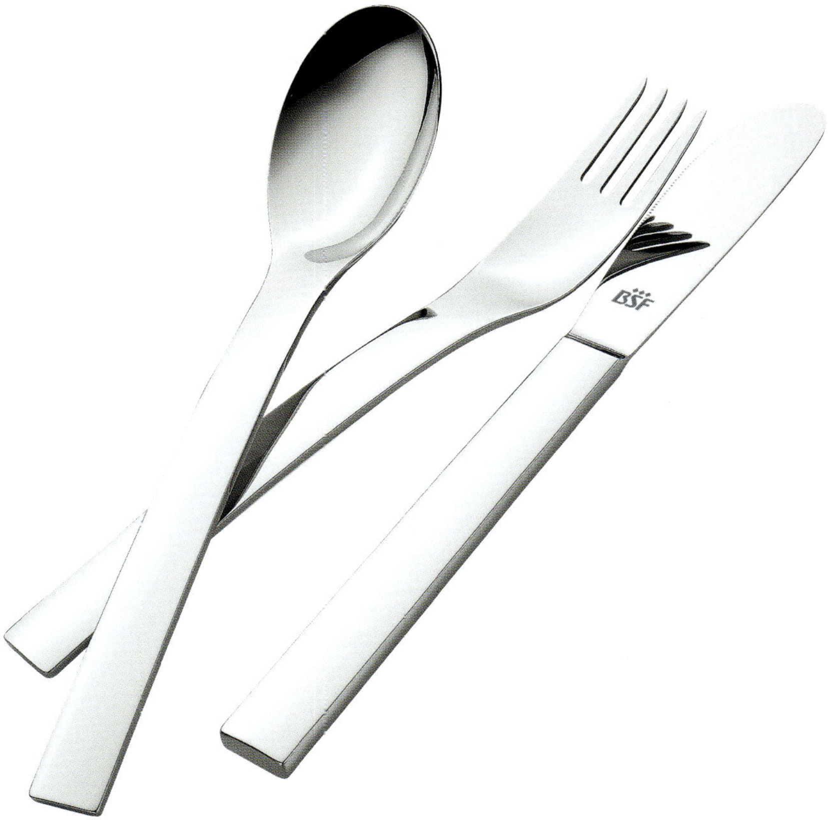

Hersteller Manufacturer
M. H. Wilkens & Söhne GmbH
Bremen, Germany

Produkt Product
Flow
Edelstahl-Besteck
Stainless steel cutlery

Design
Claudia Christl
M. H. Wilkens & Söhne GmbH
Bremen, Germany

Die Grundgestalt des Edelstahl-Bestecks Flow zeichnet sich durch eine sanft bewegte, schlanke Silhouette aus. Der zurückhaltende Charakter entsteht aus der Einheit von langen, schmalen Vorderteilen und bandartig geformten, leicht gespannten Stielen. Das Wechselspiel zwischen Flächigkeit und Plastizität lädt zur Kombination mit jungem, modernen Porzellan ein.

The basic shape chosen for Flow, a range of stainless steel cutlery, stands out for the gentle contours of its slender silhouette. The subtle character derives from the blend of sleek heads and straight, slightly elongated stems. The interplay of two- and three-dimensional qualities makes for an inviting combination with young, modern porcelain.

Hersteller Manufacturer
ZWILLING J.A. HENCKELS AG
Solingen, Germany

Produkt Product
TWIN® Select
Knoblauchpresse
Garlic press

Design
Martin Kaufmann
Ulla Kaufmann
Hildesheim, Germany

Die Knoblauchpresse TWIN® Select verfügt über eine patentierte Kniehebeltechnik, die einen nur minimalen Kraftaufwand erfordert. Die konisch geformten Löcher des ausklappbaren Siebs sorgen dafür, dass die Knoblauchpresse auch mit ungeschälten Zehen fertig wird und leicht zu reinigen ist.

The TWIN® Select garlic press features a patented articulated lever which requires only minimal strength to use. The conically formed holes in the fold-out sieve ensure that the garlic press can also deal with unpeeled cloves. And they also make the press all the easier to clean.

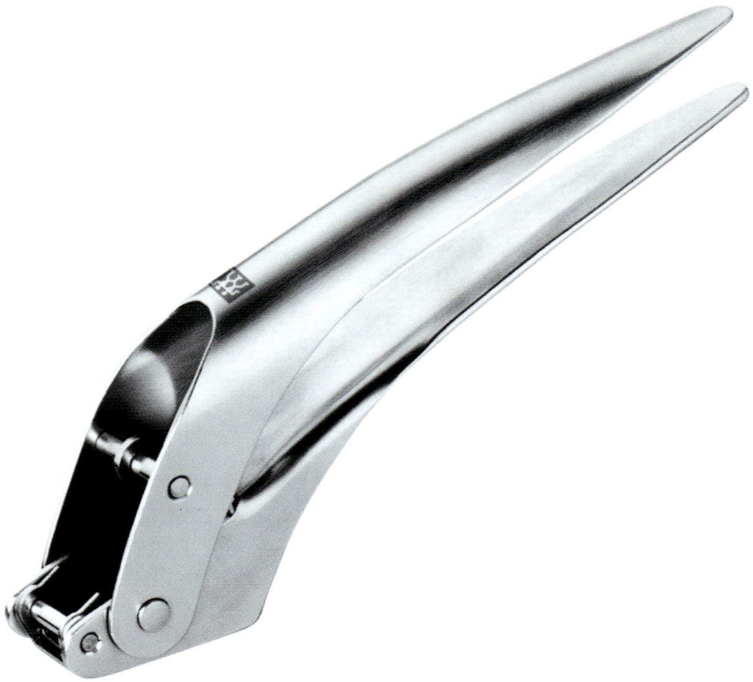

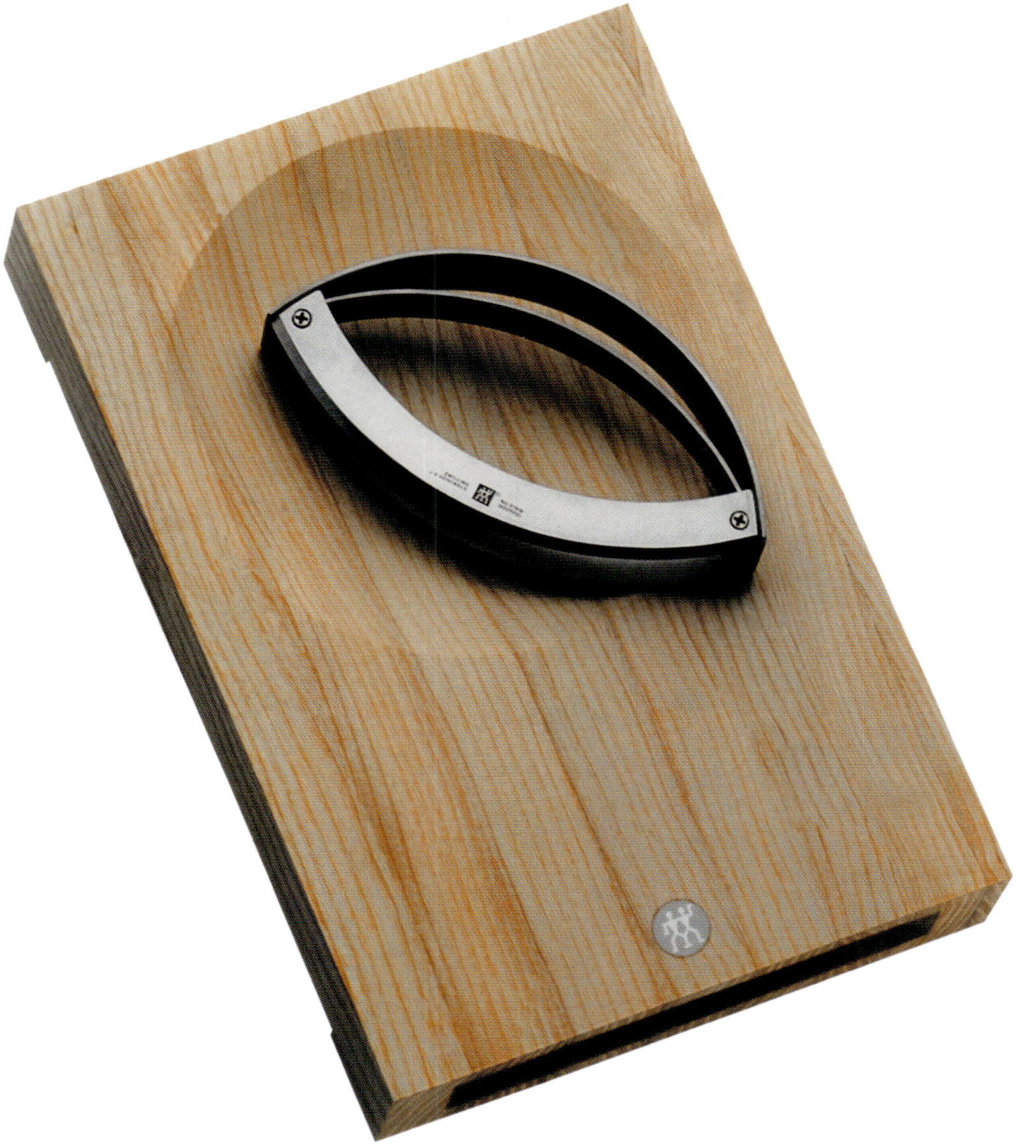

Hersteller Manufacturer
ZWILLING J.A. HENCKELS AG
Solingen, Germany

Produkt Product
TWIN® Select
Wiegemesserset mit Wiegebrett
Mezza luna and chopping board

Design
Martin Kaufmann
Ulla Kaufmann (Wiegemesser)
Hildesheim, Germany
Sol.Design (Brett)
Frankfurt am Main, Germany

Das Wiegemesser TWIN® Select mit doppelter Klinge besteht aus rostfreiem Edelstahl. Das Schneidbrett mit rutschfesten Gummifüßen ist aus massiver Esche, unbehandelt, langlebig und lebensmittelecht. Innovativ ist die Art der Aufbewahrung der Produkte. Das Messer wird in eine extra dafür vorgesehene Mulde im Brett eingeschoben und durch einen Magneten sicher festgehalten. Das komplette Set kann auf einer Wandleiste aufgehängt werden.

The double bladed TWIN® Select mincing knife is made of stainless steel. The matching cutting board is made from untreated solid ash wood, which is durable and suitable for the preparation of food. Rubber feet prevent the board from slipping. An innovative way of storing of both, board and mincing knife. The knife can be inserted into a special mould and is secured by magnets. The set can be stored on a hanging rack.

Hersteller Manufacturer
Cerâmica Stéfani S/A
Jaboticabal, Brazil

Produkt Product
Water Filter
Schwerkraftwasserfilter aus Keramik
Terracotta gravity filter

Design
Roberto Brazil
Oswaldo Solla
dap design
São Paulo, Brazil

Der Wasserfilter für den Haushalt ist mit einem Filterelement aus Keramik ausgestattet, das Feststoffpartikel sehr gut zurückhält. Dank des Tons ist das gefilterte Wasser kristallklar und frisch. Der Filter wurde unter Verwendung von Aluminium und durch die Farbgebung so entwickelt, dass er auch optisch in die Umgebung passt, in der er verwendet wird. Der Tonfilter wird von Hand an der Töpferscheibe geformt und mit einer speziellen Technologie getrocknet, beschichtet und gebrannt. Die Beschichtung aus kolloidalem Silber ist nicht nur gestalterisch interessant, sondern sterilisiert auch das Wasser und verhindert Pilzbildung.

This household filter is fitted with a ceramic filter element, which very effectively filters out solid particles producing crystal clear and fresh water. The chambers are produced on natural terracotta and colored in a way that the unit visually blends in the environment in which it is used. This water filter is carefully handmade, dried, burned and coated using a special technique. The coating of colloid silver is not only interesting from a design point of view but also because it sterilizes the water and prevents the formation of fungus and bacteria.

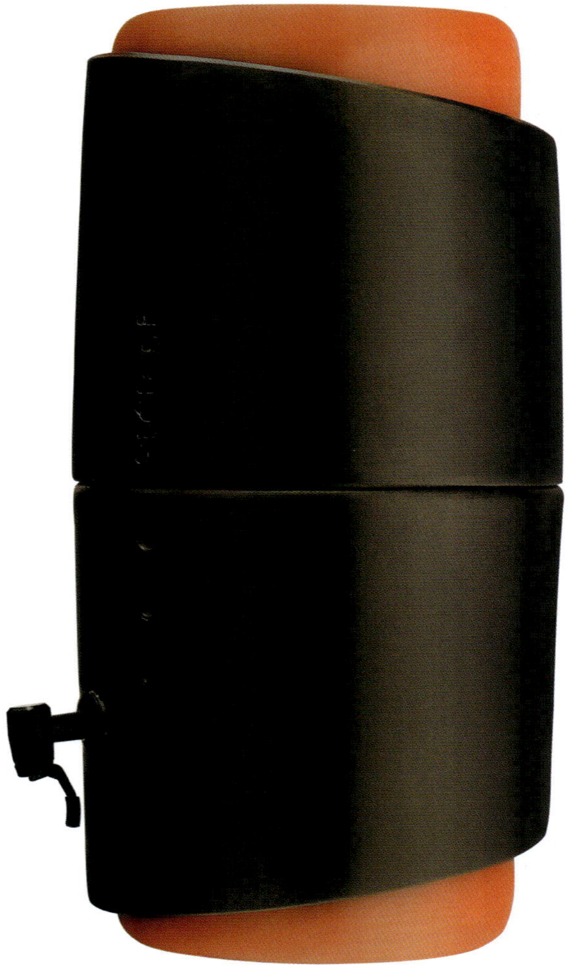

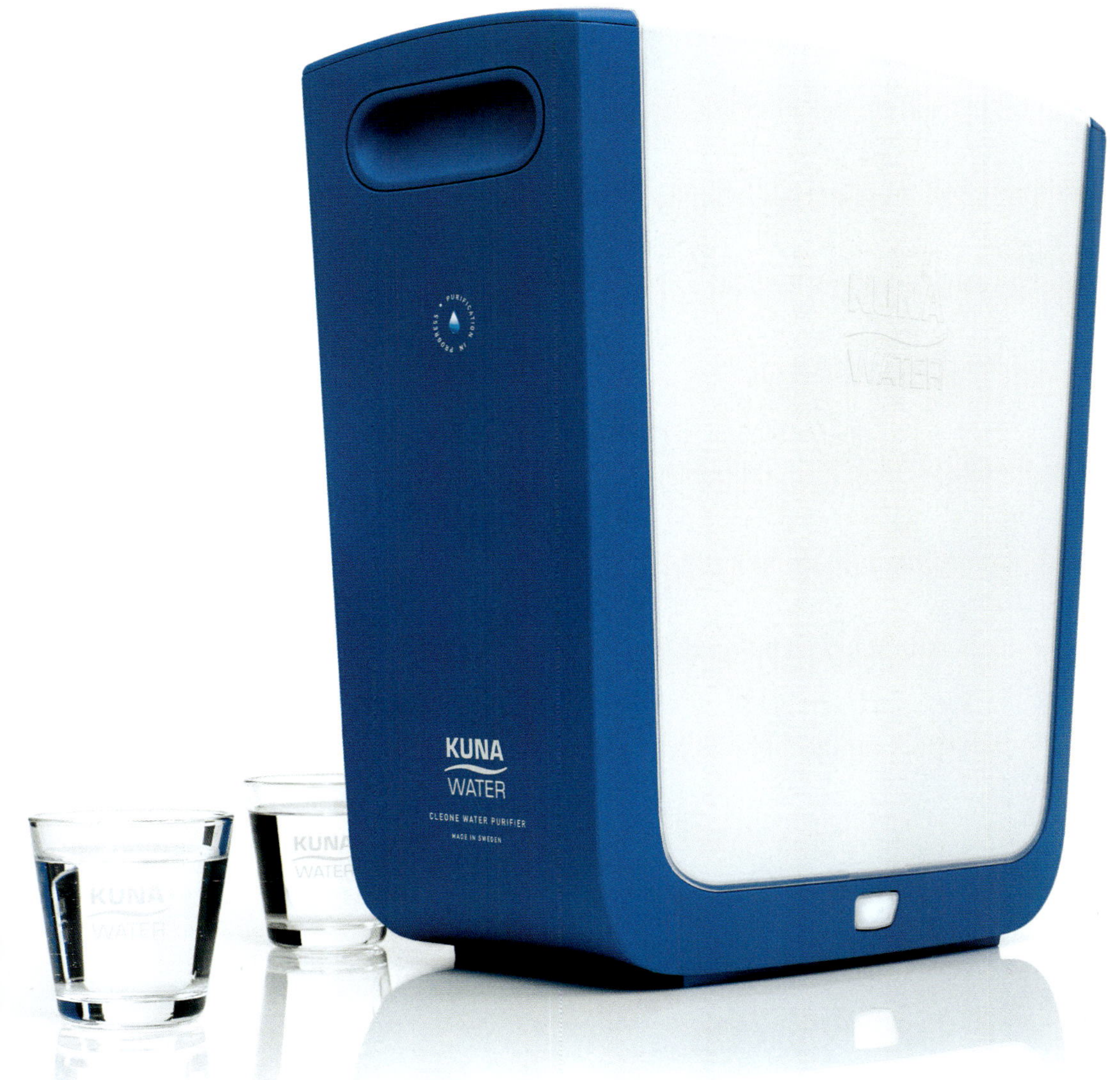

Hersteller Manufacturer
Kuna Group AB
Jarfalla, Sweden

Produkt Product
Cleone®
Haushaltswasserfilter
Household water filter

Design
Fredrik Magnusson
Gustav Nord
Filip Sauer
Designaktiebolaget Propeller
Stockholm, Sweden

Cleone® ist ein Haushaltswasserfilter, dessen Design von einem schlichten Wasserglas inspiriert ist. Der Wasserfilter wird in der Regel in einem Küchenschrank unter der Spüle aufgestellt und dort mit dem vorhandenen Wasseranschluss verbunden. Das Wasser wird durch den Filter zum Entnahmehahn auf der Arbeitsplatte geleitet. Ein spezieller Sensor stoppt im Falle eines Lecks die Wasserzufuhr. Die LED-Anzeige erhöht den Bedienkomfort und die Sicherheit für den Benutzer. Durch den Blick auf den erleuchteten blauen Tropfen erkennt man, dass der Wasserfilter in Betrieb ist.

Cleone® is a household water filter whose design is inspired by a simple water glass. The water filter is generally mounted in a kitchen cabinet under the sink and connected there to the water pipe. The water is then directed through the filter to the faucet on the worktop. In the case of a leak there is a special sensor which stops the flow. An LED enhances ease of use and safety with an illuminated blue drop to indicate that the water filter is in operation.

Hersteller Manufacturer
Julius Blum GmbH
Beschlägefabrik
Höchst, Austria

Produkt Product
BOXSIDE aus Glas
Seitenwand aus Glas für Frontauszüge
Glass profile for front pull-outs

Design
Wolfgang Held
form orange Produktdesign
Götzis, Austria

Die aus Glas bestehenden Seitenteile für Frontauszüge sind ein flexibles Gestaltungselement im Unterschrankbereich von Küchen. Zudem bieten sie wichtige Voraussetzungen für die Stauraumschaffung und -nutzung. Beispiele sind Auszüge für Lebensmittel und Vorräte, Flaschenauszüge, Geschirr-, Topf-, Pfannen- und Deckeldepots. Zahlreiche unterschiedliche Glastypen und Farben können somit von dem Möbelhersteller frei gewählt und den ästhetischen Anforderungen des jeweiligen Corporate Designs angepasst werden.

The glass-BOXSIDE pull-out elements create a flexible design element for the area under the kitchen worktop. They provide a significant means of generating and using storage space: for example pull-out elements for food and supplies, bottleholders or holding compartments for dishes, pots, pans or lids. A large variety of different glass types and colors are available from the manufacturer and can be matched with the aesthetic details of respective corporate designs.

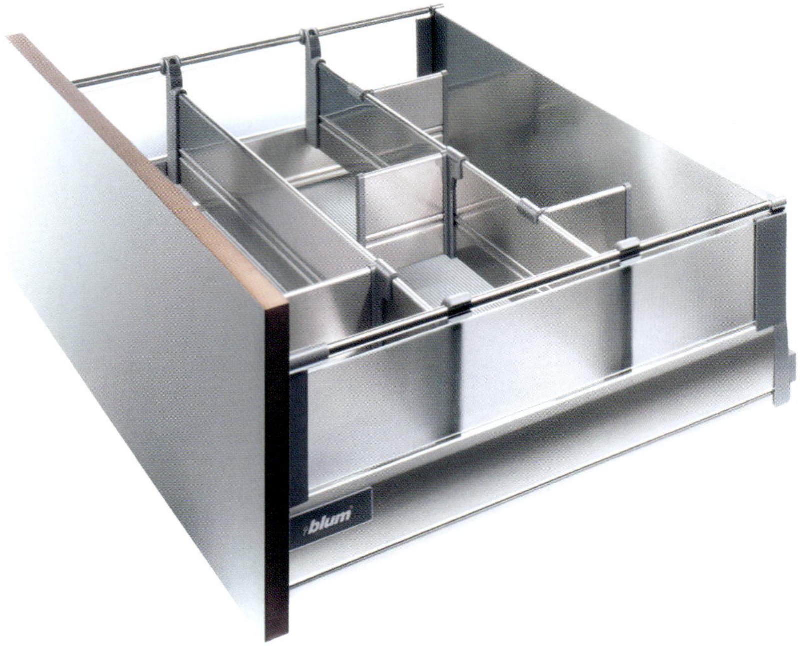

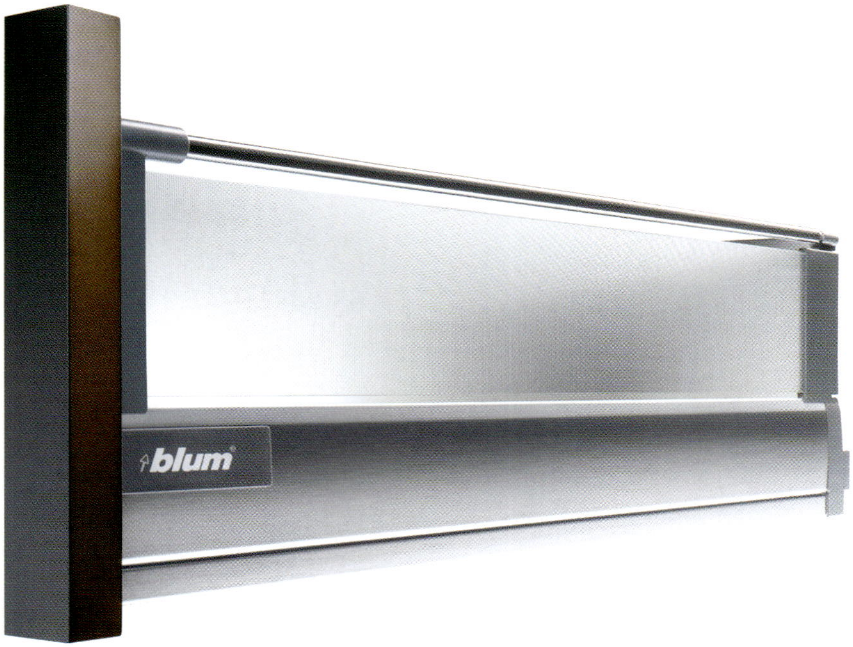

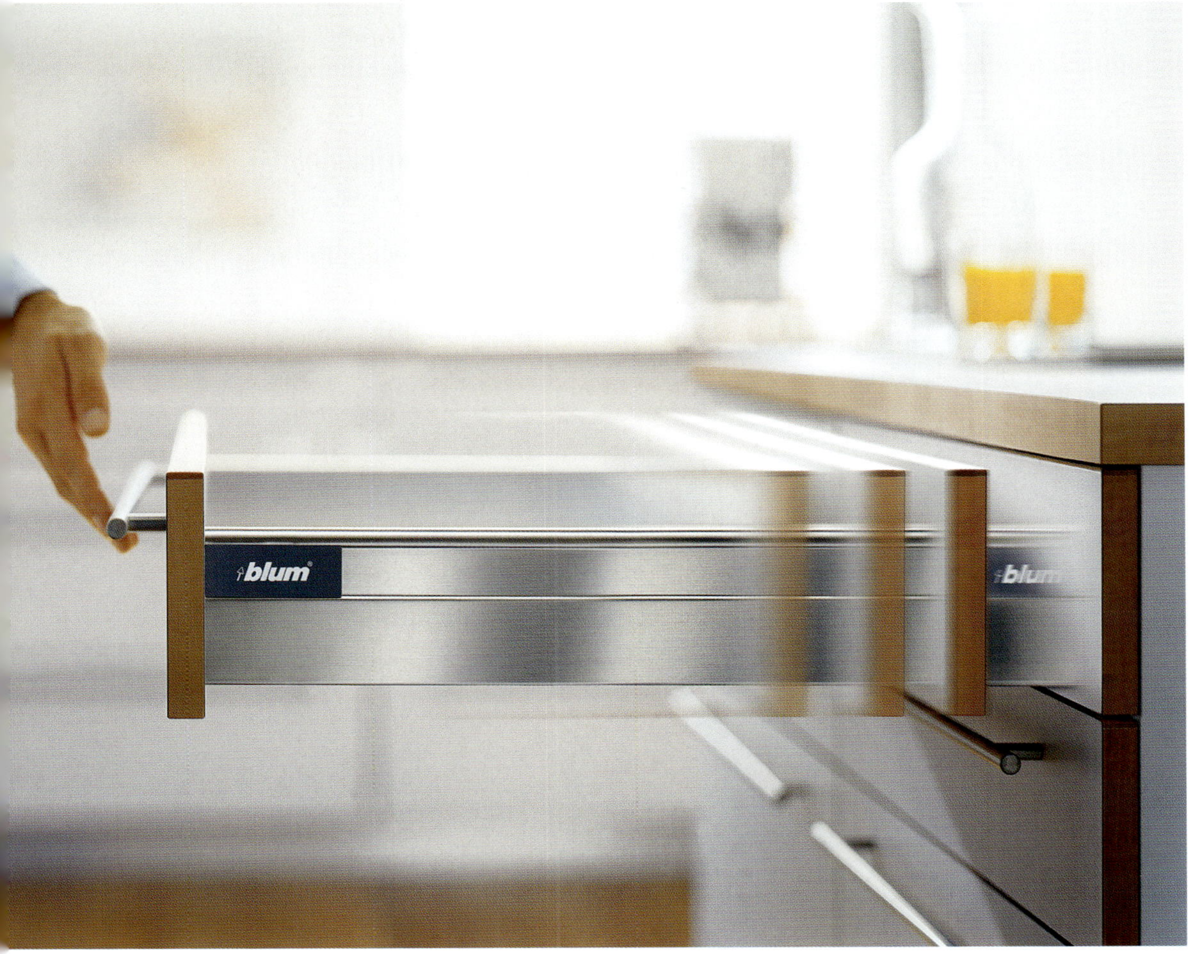

Hersteller Manufacturer
Julius Blum GmbH
Beschlägefabrik
Höchst, Austria

Produkt Product
BLUMOTION
Schließsystem
Closing system

Design
Julius Blum GmbH
Höchst, Austria

BLUMOTION ist ein spezielles Schließsystem, das ein sanftes und leises Schließen von Schubkästen, Auszügen und Türen gewährleistet. In der letzten Phase des Schließvorgangs werden Möbeltüren und Auszüge weich abgebremst und schließen ohne zu prellen oder zurückzufedern. BLUMOTION ist ein adaptives System, das sich der jeweiligen Anwendungssituation anpasst. Es erkennt die Schließgeschwindigkeit und die Masse der Auszüge oder Türen und reguliert die Bremsleistung automatisch. Erhältlich für das Scharniersystem CLIP top, das doppelwandige Zargensystem TANDEMBOX und das verdeckte Führungssystem TANDEM.

BLUMOTION is a special system for closing drawers and doors very gently and quietly. When furniture doors or drawers are just about to close, gentle brakes are applied to allow them to close without crashing or springing back. BLUMOTION is an adaptable system which can be adjusted to the particular situation. It recognizes the speed at which doors or drawers are closing as well as their mass and can automatically regulate the braking effect. The system is available for the CLIP-top hinge system, the TANDEMBOX double-walled drawer system, and the TANDEM concealed runner system.

Hersteller Manufacturer
Julius Blum GmbH
Beschlägefabrik
Höchst, Austria

Produkt Product
Blum-Eckschrank mit TANDEMBOX
Eckschrank
Corner cabinet

Design
Wolfgang Held
Form orange Produktdesign
Götzis, Austria

Der Blum-Eckschrank mit TANDEMBOX von Blum ist eine Alternative zu den konventionellen Lösungen hinter Dreh- und Falttüren. Er ermöglicht eine optimale Raumnutzung sowie einen übersichtlichen und komfortablen Zugriff auf das Füllgut. Der TANDEMBOX Vollauszug ermöglicht Einblick und Zugriff bis in die hinterste Ecke und ist somit Basis für eine optimale Stauraumnutzung und ergonomische Arbeitsweise. Das Fronten- und Fugenbild kann über die Ecke weitergeführt werden und eröffnet somit neue Gestaltungsmöglichkeiten.

The Blum corner cabinet with TANDEMBOX is an alternative to conventional storage solutions which utilize swinging and folding doors. It enables optimal use of space and comfortable access to drawer contents. The TANDEMBOX which can be pulled all the way out allows you to easily access the deepest corner, forming the basis for optimal use of storage space with an ergonomic approach. The front panels and joints can also extended over the corner, granting interior designers a variety of new possibilities.

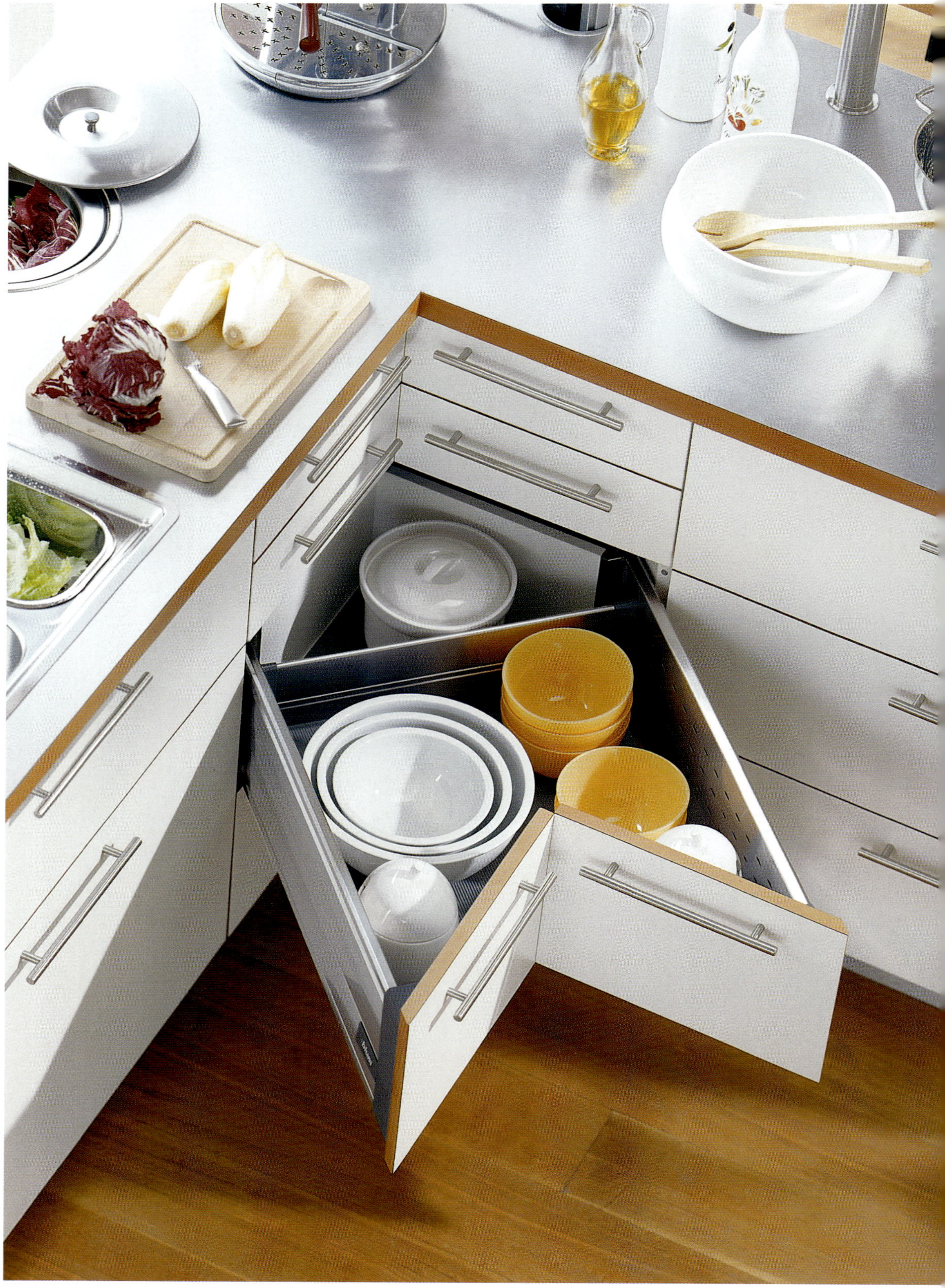

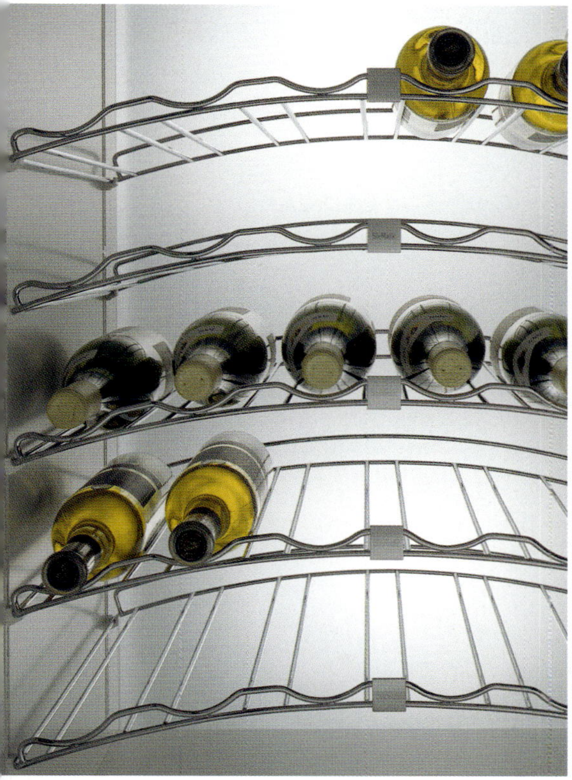

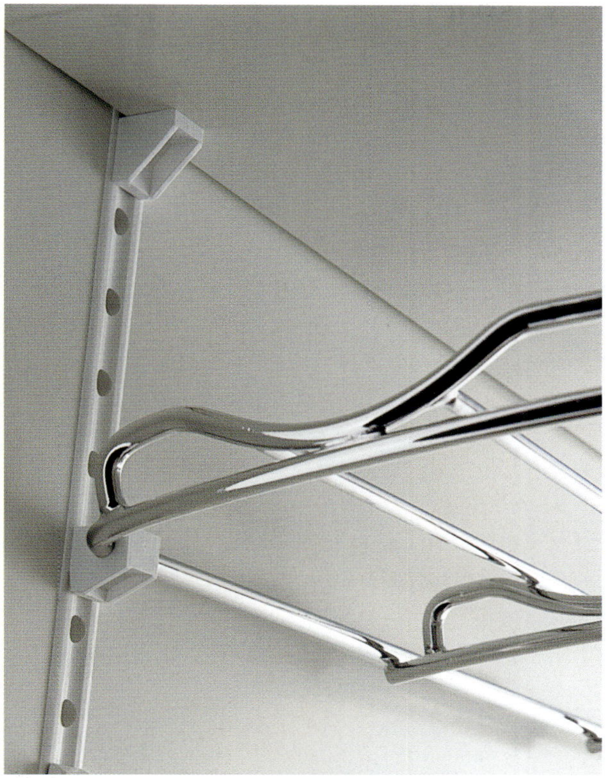

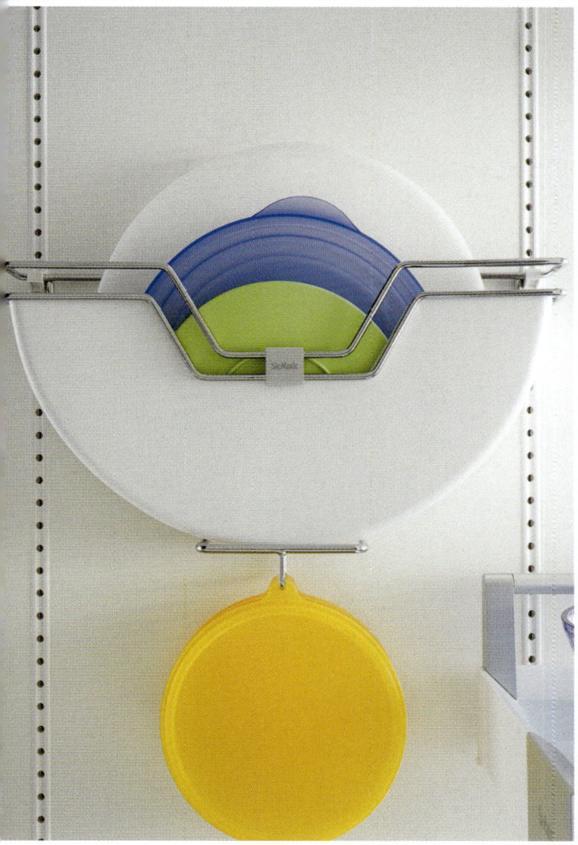

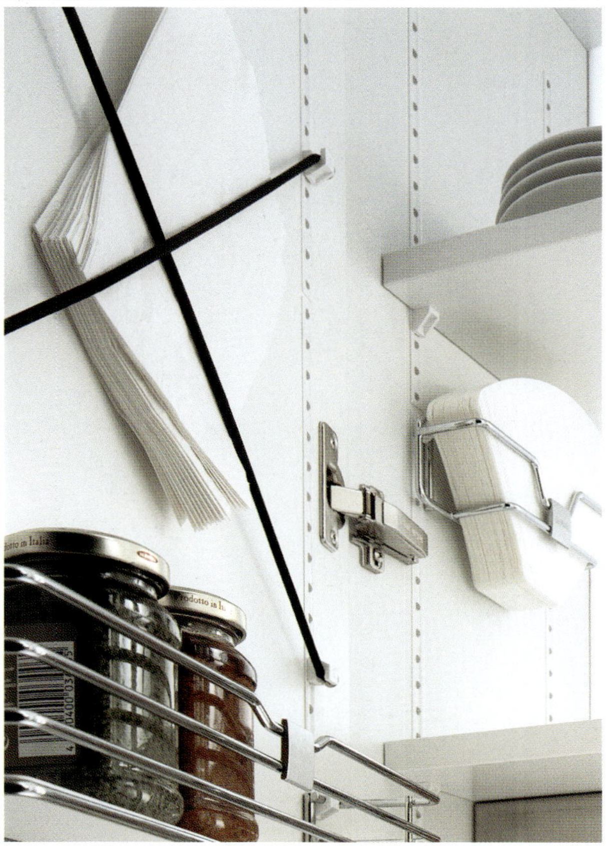

Hersteller Manufacturer
SieMatic Möbelwerke GmbH & Co. KG
Löhne, Germany

Produkt Product
MultiMatic
Innenausstattungssystem für Küchenmöbel
Multifunction system for the interiors of kitchen furniture

Design
SieMatic Möbelwerke GmbH & Co. KG
Löhne, Germany

Das SieMatic Multifunktionssystem zur Innenausstattung von Küchenmöbeln besteht aus einer patentierten Multifunktionsschiene und verschiedenen Systemteilen, die an dieser beliebig fixiert werden können. Die Schienen, die auf Wunsch auch in die Türrückseite integriert werden, sind in ein 16 mm Raster unterteilt und ermöglichen so die individuelle Einrichtung eines Schrankelements. Die einzelnen Systemteile können vom Kunden selbst leicht in der Tür oder im Korpus montiert bzw. in ihrer Position verändert werden. Jedes Systemteil hat eine spezielle Funktion, z.B. Körbe für Gewürze und Vorräte, Borde für Weinflaschen, Träger für Trinkgläser, Halterungen für Folien- und Papierrollen, einfache Haken für Kleinteile und vieles mehr.

The SieMatic multifunction system for the interiors of kitchen furniture consists of a patented multifunction rail and various system components which can be attached to it as required. The rails, which can be integrated into the rear side of the doors as required, have divisions at regular 16 mm intervals allowing the user to arrange the cabinet elements exactly as is wanted. The user can easily install the individual system components in the door or body of the units and change their position if required. Each system component has a special function, e.g., baskets for spices and supplies, shelves for wine bottles, glass supports, holders for foils and paper towel rolls, simple hooks for small utensils and a lot more besides.

Hersteller Manufacturer
WMF Württembergische Metallwarenfabrik AG
Geislingen/Steige, Germany

Produkt Product
CombiNation Buffet
Modulares Buffetsystem
Modular buffet system

Design
Caroline Schäfer (Werksdesign)
WMF Württembergische Metallwarenfabrik AG
Geislingen/Steige, Germany

WMF CombiNation Buffet ist ein modulares System für Buffets in Restaurants, Hotels oder für das Event-Catering. Die Präsentationsideen können auf verschiedenen Ebenen und Schrägen mit unterschiedlichen Einsätzen und Tabletts realisiert werden. Die Grundausrichtung wird durch eine Schiene vorgegeben, durch einfaches Einhängen kann das Buffet schnell umgebaut werden, dabei kann in beliebig langer Reihe, über Eck, in unterschiedlichen Winkeln und beidseitig bestückt aufgebaut werden. Das System verfügt sowohl über Chafing-Dishes zum Warmhalten von Speisen als auch Kühlelemente. Es ist aus hygienischem rostfreien Cromargan® (Edelstahl 18/10) gefertigt.

WMF CombiNation Buffet is a modular concept for buffets destined for use in restaurants, hotels or even for event catering. Various levels and inclinations can be set up, combined with a possible variety of trays and accessories, a large range of presentation ideas can be achieved. The basic buffet structure is defined by a rail onto which the other components can simply be attached; it can be assembled to the required length, run around corners or in other desired formations. The system includes both chafing dishes to keep food warm as well as cooling elements, and is made of hygienic, anti-corrosive Cromargan® (18/10 stainless steel).

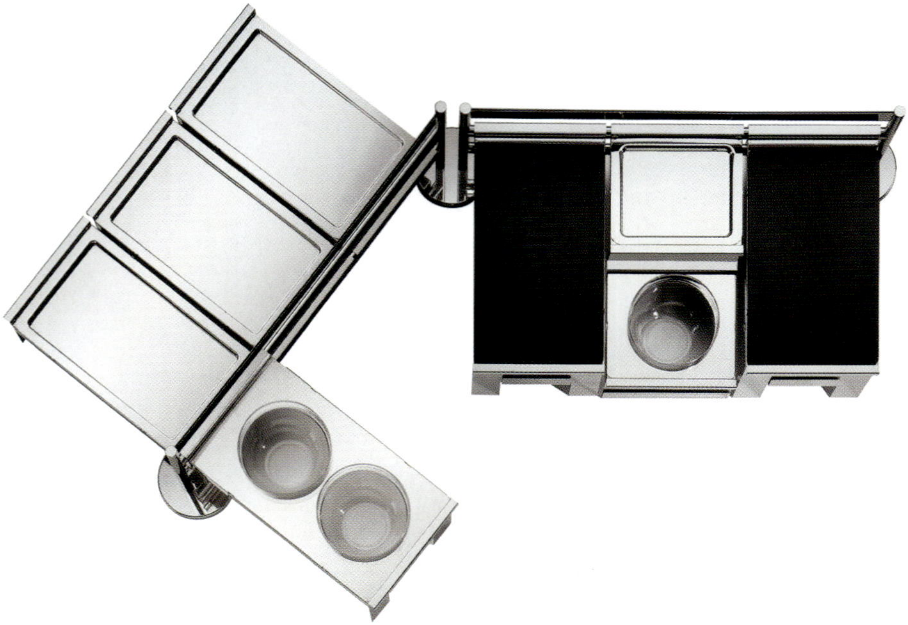

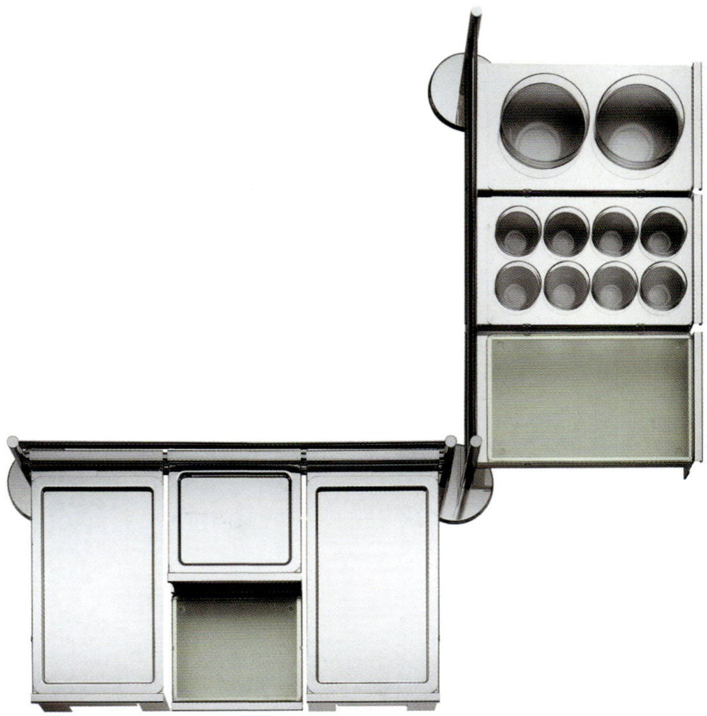

Hersteller Manufacturer
Electrolux Home Products Italy S.p.a.
Porcia, Italy

Produkt Product
Visao
Mikrowellenherd
Microwave oven

Design
Alexandre Barros Neves
Duncan McCall
Sean Carney
Tony Cameron
Electrolux Home Products Italy S.p.a.
Porcia, Italy

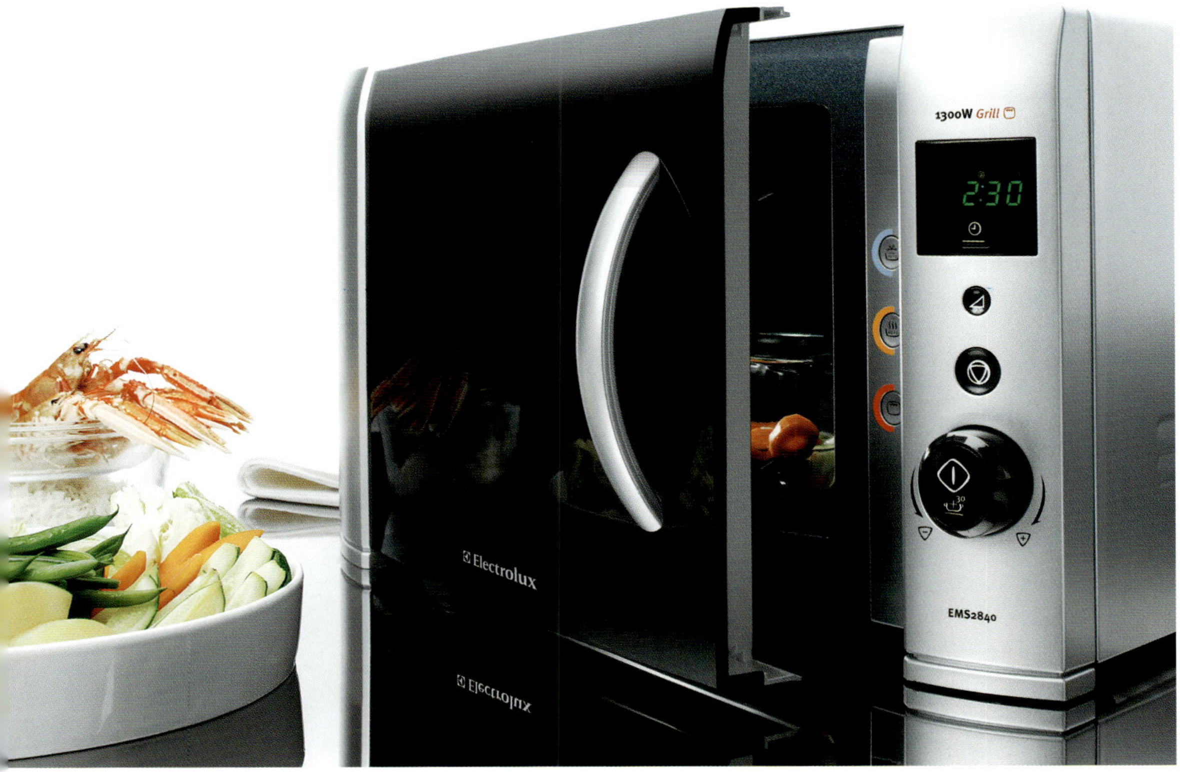

Dem Visao Mikrowellenherd liegt ein ergonomisches Design zugrunde, welches in der Benutzung des Geräts intuitiv erfahrbar wird – von der eindeutig erkennbaren Funktion des Türgriffs, über den Drehknopf bis hin zu den konkav geformten Bedienknöpfen. Das äußere Bedienfeld bietet die am häufigsten verwendeten Grundfunktionen, das innere Bedienfeld ist nur bei geöffneter Tür sichtbar. Mit ihm können die komplexeren Funktionen gesteuert werden. Simplizität stand im Vordergrund der Designentwicklung, jeder Aspekt des Designs unterstreicht seine Funktion, von der Eliminierung des traditionellen Türrahmens bis hin zu den säulenartigen Füßen, die einen stabilen Stand vermitteln.

The Visao microwave oven is based on an ergonomic design which you experience intuitively when using the appliance - from the obviously functional pull-to-open door handle, the rotating knob all the way to the concave control buttons. The outer control panel features the most frequently used basic functions, the inner control panel handles the more complex functions and is only visible when the door is open. The design focuses on simplicity with every aspect of the design underlining its function, for example the traditional door frame has been eliminated and replaced by column-like feet that exude stability.

Hersteller Manufacturer
AEG Hausgeräte GmbH
Nürnberg, Germany

Produkt Product
COMPETENCE Top-Klasse
Einbauherde und -backöfen-Produktreihe
Built-in cooker and oven range

Design
Heinz-Peter Huxhold
Axel Mürmann
AEG Hausgeräte GmbH
Nürnberg, Germany

Die neue Designlinie der COMPETENCE Einbauherde und -backöfen wird durch die horizontal angeordneten Bedien- und Anzeigenelemente geprägt. Die logisch funktionale Zuordnung dieser Elemente erleichtert die Bedienbarkeit der Geräte. Dem jeweiligen Bedienelement wie Kochstellenknebel oder Drucktaste ist im schwarzen Displaystreifen die optische Anzeige zugeordnet. Der Bügelgriff ist harmonisch in die Geräteoberfläche integriert.

The new design of AEG's COMPETENCE built-in cooker and oven series stands out for the horizontal arrangement of the user and display elements. The logical, functional layout of these elements facilitates use. The various controls such as rotary hob controls or push buttons are represented visually in the black display strip. The handle is harmonically integrated in the surface of the gadget.

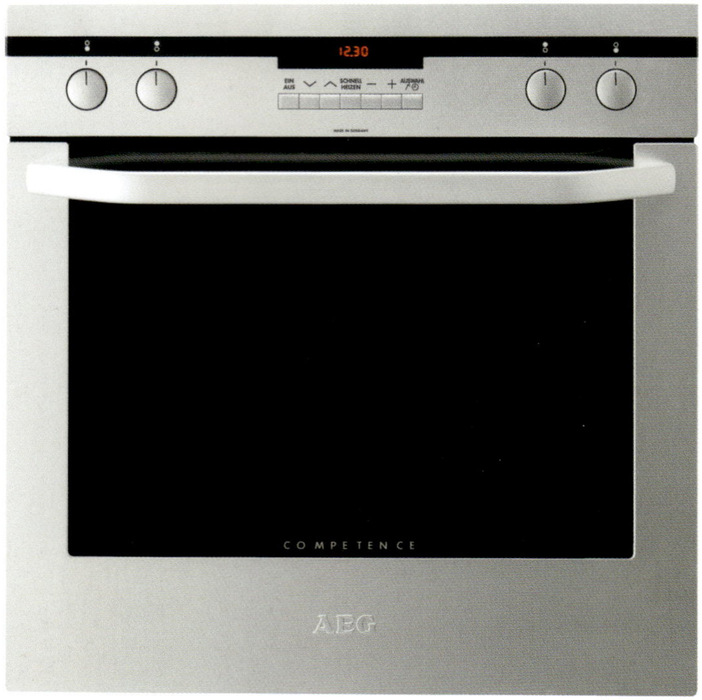

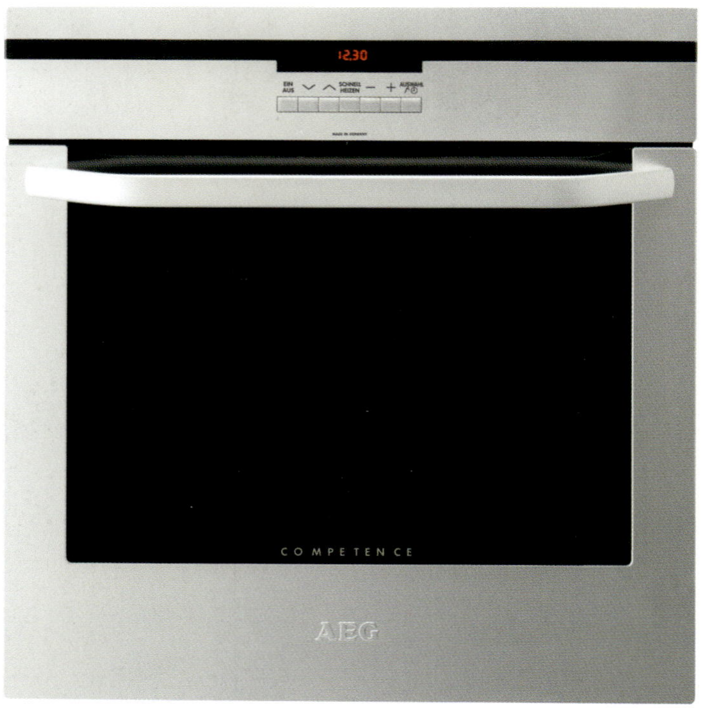

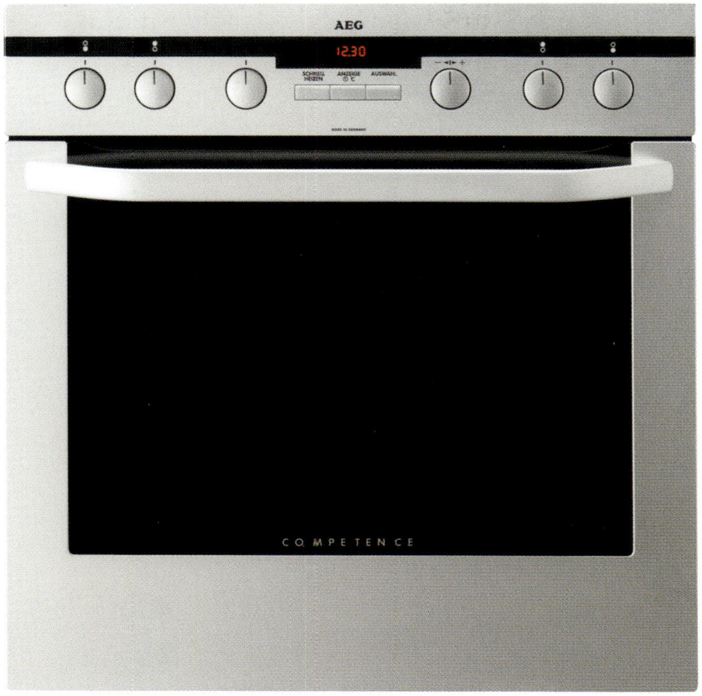

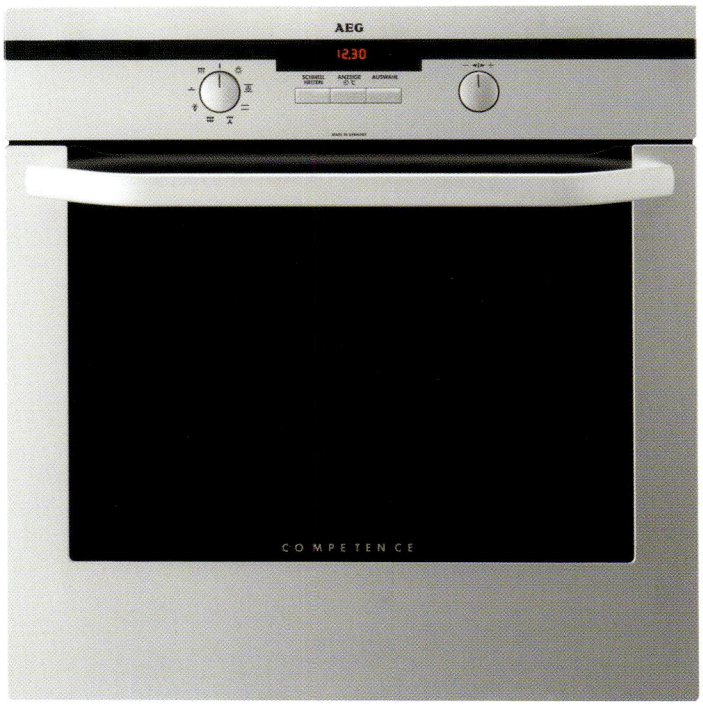

Hersteller Manufacturer
AEG Hausgeräte GmbH
Nürnberg, Germany

Produkt Product
COMPETENCE Komfort-Klasse
Einbauherde und -backöfen-Produktreihe
Built-in cooker and oven range

Design
Heinz-Peter Huxhold
Axel Mürmann
AEG Hausgeräte GmbH
Nürnberg, Germany

Die formal ästhetischen Gestaltungsmerkmale der Produktreihe COMPETENCE erlauben eine harmonische Integration ins Küchenumfeld mit anderen AEG-Einbaugeräten. Die unterschiedlichen Ausstattungsmerkmale, Funktionen und entsprechendes Zubehör der einzelnen Geräte der Produktreihe bieten eine individuelle Auswahlmöglichkeit für den Kunden. Zudem kann zwischen den Oberflächen Aluminium, Edelstahl und Glas in drei Farben gewählt werden.

Thanks to the formal aesthetic qualities of the COMPETENCE family, the units can be harmoniously integrated in kitchen settings alongside other AEG equipment. The different specifications, functions and accessories of the individual cookers and ovens in the line ensure the customer finds one to fit his needs. Moreover, they are available with aluminum, stainless steel or glass surfaces – and in three colors.

Hersteller Manufacturer
AEG Hausgeräte GmbH
Nürnberg, Germany

Produkt Product
FAVORIT
Einbau-Geschirrspüler-Produktreihe
Built-in dishwasher range

Design
Michael Knöchner
Axel Mürmann
AEG Hausgeräte GmbH
Nürnberg, Germany

Das Design der Einbau-Geschirrspüler-Produktreihe FAVORIT von AEG orientiert sich in seiner Betonung der Horizontalen an der Gestaltung der neuen Herdbaureihe. Den Bedientasten beim Top-Gerät sind im schwarzen Display die Anzeigeelemente und das Textdisplay zugeordnet. Die Geschirrkörbe wurden mit flexiblen Elementen ausgestattet, so dass die unterschiedlichsten Geschirr- und Topfgrößen individuell positioniert werden können. Der Innenraum ist beleuchtet. Mit nur 41 dB ist das Gerät äußerst leise. Die 3in1-Reiniger Optionstaste stimmt alle Spülprogramme automatisch auf diese moderne Reinigerkomponente ab.

The design of the FAVORIT built-in dishwasher range by AEG takes its cue from the horizontal design configuration of the new line of cookers. The buttons above and the text in the black display panel correspond to the user buttons in the stovetop. The dish baskets are fitted with flexible components to cater for different sized pots and dishes, and the interior of the dishwasher is illuminated. At just 41dB, the machine runs very quietly. The 3-in-1 cleaner button can coordinate all washing cycles, adjusting them automatically to modern detergents.

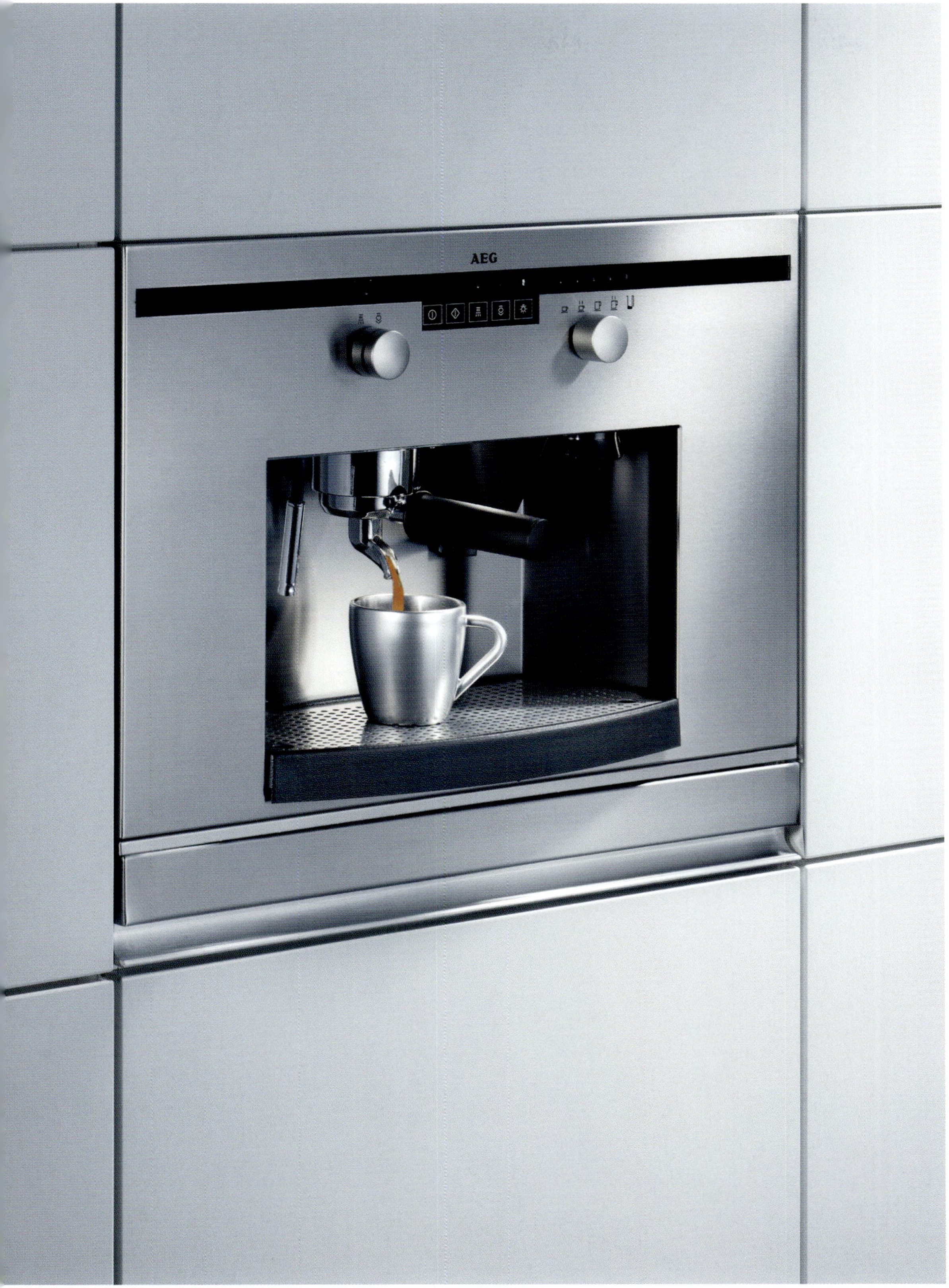

Hersteller Manufacturer
AEG Hausgeräte GmbH
Nürnberg, Germany

Produkt Product
PROFI ESPRESSO
Einbau-Espressomaschine
Built-in espresso machine

Design
Frank Benold
Grit Wäntig
AEG Hausgeräte GmbH
Nürnberg, Germany

Die Gestaltungsmerkmale der Einbau-Espressomaschine PROFI ESPRESSO orientieren sich am Design der neuen AEG Herdbaureihe. Touch-Control, kombiniert mit manueller Einstellung von Funktionen, gibt dem Gerät eine einfache und moderne Bedienoberfläche. Die Einbautechnik wird gleichzeitig mit den Service-Bedienvorgängen für Wassertank und Mahlwerk verknüpft. Unter der Tropfschale befinden sich Hebel, die es erlauben, das Gerät für Wartungszwecke aus der Möbelnische herauszuziehen. Die Schublade nimmt für besseres Handling Siebträger und Ablageschalen auf.

The design features of the PROFI ESPRESSO built-in espresso machine follow the design of the new AEG cooker hob series. A combination of touch controls and manual settings together create a simple and modern operating panel. The built-in technology is coupled with service features for the water tank and grinder. Beneath the drip grate there are several levers which allow the espresso machine to be removed from its built-in niche for maintenance purposes. The drawer includes sieve holders and other storage niches for enhanced functionality.

Hersteller Manufacturer
AEG Hausgeräte GmbH
Nürnberg, Germany

Produkt Product
COMPETENCE 9570 GF
Einbau-Gaskochmulde
Built-in gas hob

Design
Frank Benold
Grit Wäntig
AEG Hausgeräte GmbH
Nürnberg, Germany

Die Ceran-Glasoberfläche verleiht der COMPETENCE 9570 GF Einbau-Gaskochmulde einen neuen ästhetischen Ausdruck. Die Gasbrenner und die harmonisch geformten Topfroste sind traditionell aus Gusseisen hergestellt. Der zentral positionierte Wok-Brenner kommt aktuellen Kochgewohnheiten entgegen. Moderne Zündtechnologie unterstützt die einfache Bedienung über die Bedienknebel am vorderen Rand der Kochmulde.

The Ceran surface of the COMPETENCE 9570 GF instills the built-in gas hob with a new aesthetic feel. The gas burner and the harmoniously formed pot grid are made the traditional way, namely of cast iron. The wok burner at the center responds to today's popular cooking styles. Modern ignition technology underlies the simple operation of knob controls on the front edge of the hob.

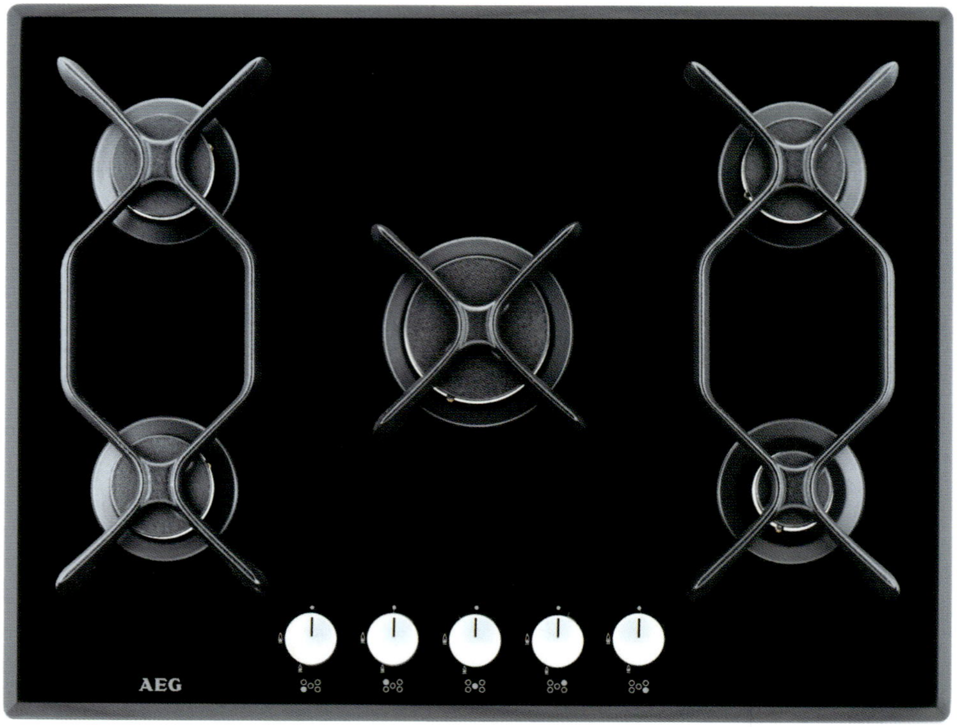

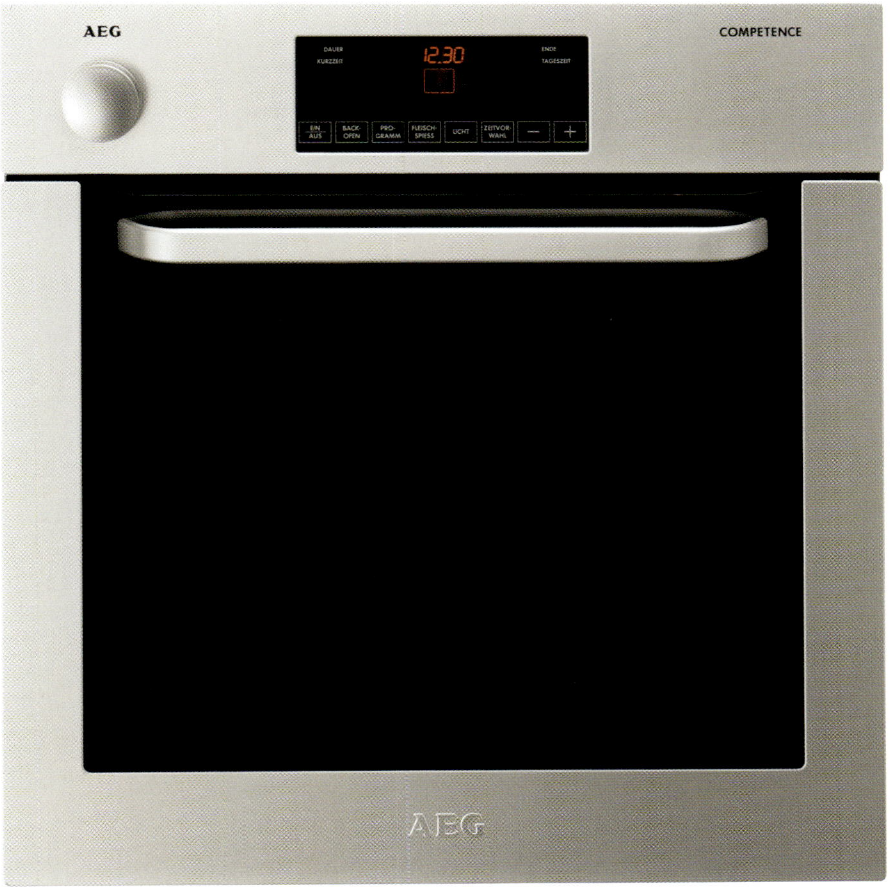

Hersteller Manufacturer
AEG Hausgeräte GmbH
Nürnberg, Germany

Produkt Product
COMPETENCE B 8920-1
Einbau-Dampfgarer-Backofen
Built-in steamer oven

Design
Frank Benold
Grit Wäntig
AEG Hausgeräte GmbH
Nürnberg, Germany

Der COMPETENCE B 8920-1 Einbau-Dampfgarer-Backofen von AEG ermöglicht neben drei Dampfgarprogrammen weitere acht Beheizungsarten. Die Bedienung des Gerätes erfolgt per Touch-Sensor und wird in einem Text- und Symboldisplay angezeigt. Über dieses Textdisplay sind verschiedene Automatikprogramme einstellbar, so dass die Bedienung und Nutzung des Gerätes sehr vereinfacht wird. Der Backraum ist mit Teleskop-Auszügen ausgerüstet und erleichtert damit die Handhabung des Gargutes auf drei Ebenen. Das Design zeichnet sich durch klare Metallgeometrien aus, der Kunde kann zwischen den Materialien Aluminium und Edelstahl wählen.

AEG's COMPETENCE B 8920 built-in steamer oven delivers three steaming programs and a further eight types of heating. The oven is operated via touch sensors which are visually rendered in the form of text and symbol displays. Various automatic programs can be set with the text display facilitating operation and use of the oven. The oven is fitted with telescopic extensions to make it easier to handle the oven contents on three levels. The design is eye-catching, with its clear metal geometry, and the oven is available in a choice of either aluminum or stainless steel.

Hersteller Manufacturer
AEG Hausgeräte GmbH
Nürnberg, Germany

Produkt Product
COMPETENCE FM
Einbau-Kochmulden-Produktreihe
Built-in hob range

Design
Hans Strohmeier
Birgit Olzhausen
AEG Hausgeräte GmbH
Nürnberg, Germany

Die Produktreihe COMPETENCE FM von AEG besteht aus verschiedenen Ceranglas-Kochfeldern, die mit Gas oder Elektro ausgestattet sind. Im größeren neuen Modulraster von 36 cm und 72 cm Breite werden Ceranglas-Kochfelder mit verschiedenen elektrischen Ein-, Zwei- und Dreikreiszonen- und Induktionskochfeldern angeboten. Das klare und sachliche Design der rahmenlosen Ceranglasfelder wird mit der Aluminium-Leiste in den „heißen" Bereich der Kochfelder und den Bedienungsbereich unterteilt. Die Ceran-Kochfelder sind mit Touch-Control ausgestattet und die Gaskochfelder mit manuellen Drehreglern. Die Aluminiumleiste ist durch einen Magnet auf dem Glas fixiert und kann zur besseren Reinigung des Kochfeldes abgenommen werden.

AEG's COMPETENCE FM product range consists of various Ceran cooking zones configured for gas or electricity. In the new, larger module size of 36 cm x 72 cm (w), Ceran hobs are available with one, two, or three circular elements and induction cooking zones. The clear and functional design of the frameless Ceran hobs is divided into a 'hot' area and the user control zones by means of an aluminum bar. The Ceran hobs are fitted with touch panels and the gas zones have rotary control knobs. The aluminum bar is attached to the glass by magnet and can be removed in order to facilitate cleaning of the cooking zone.

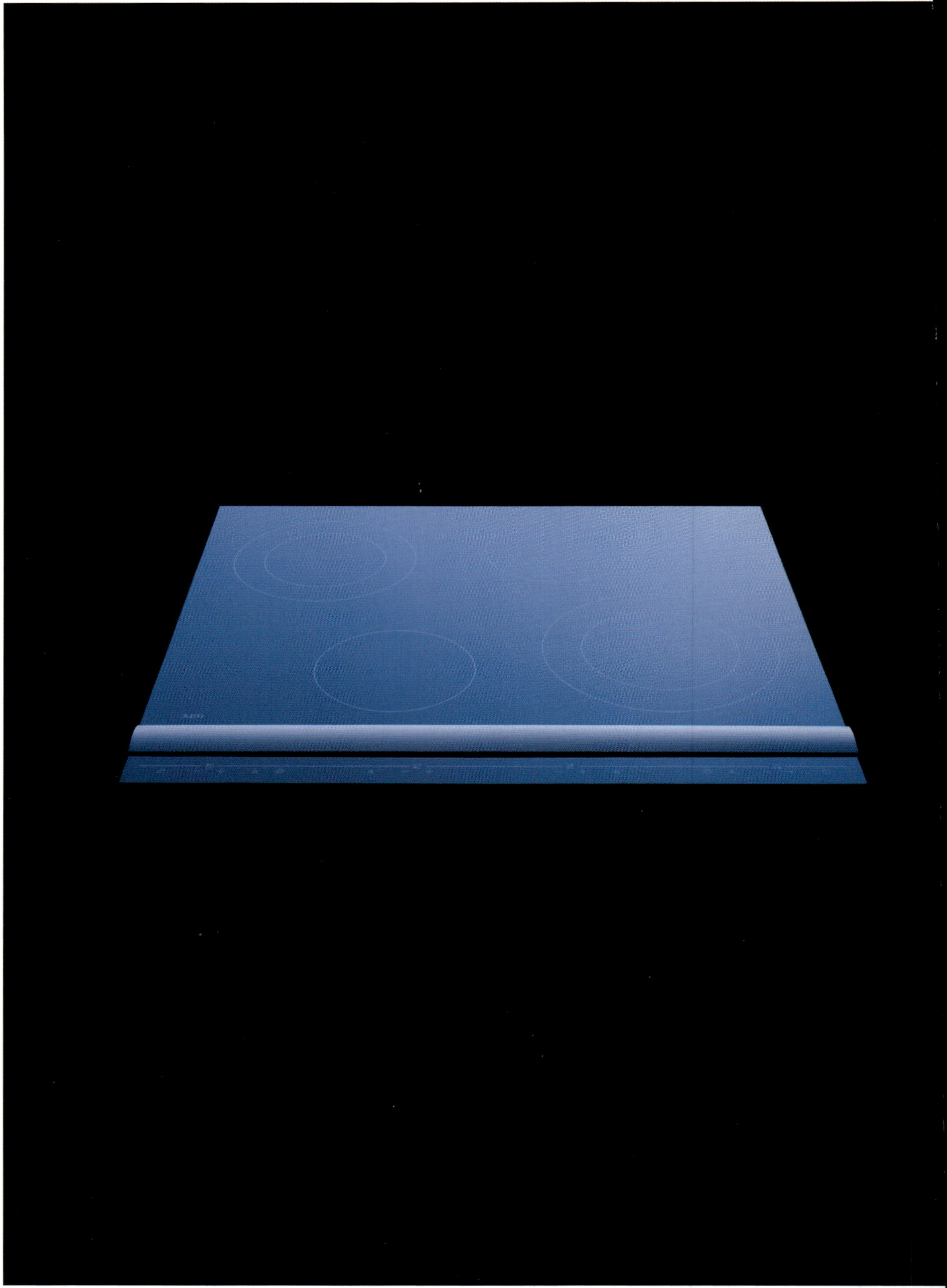

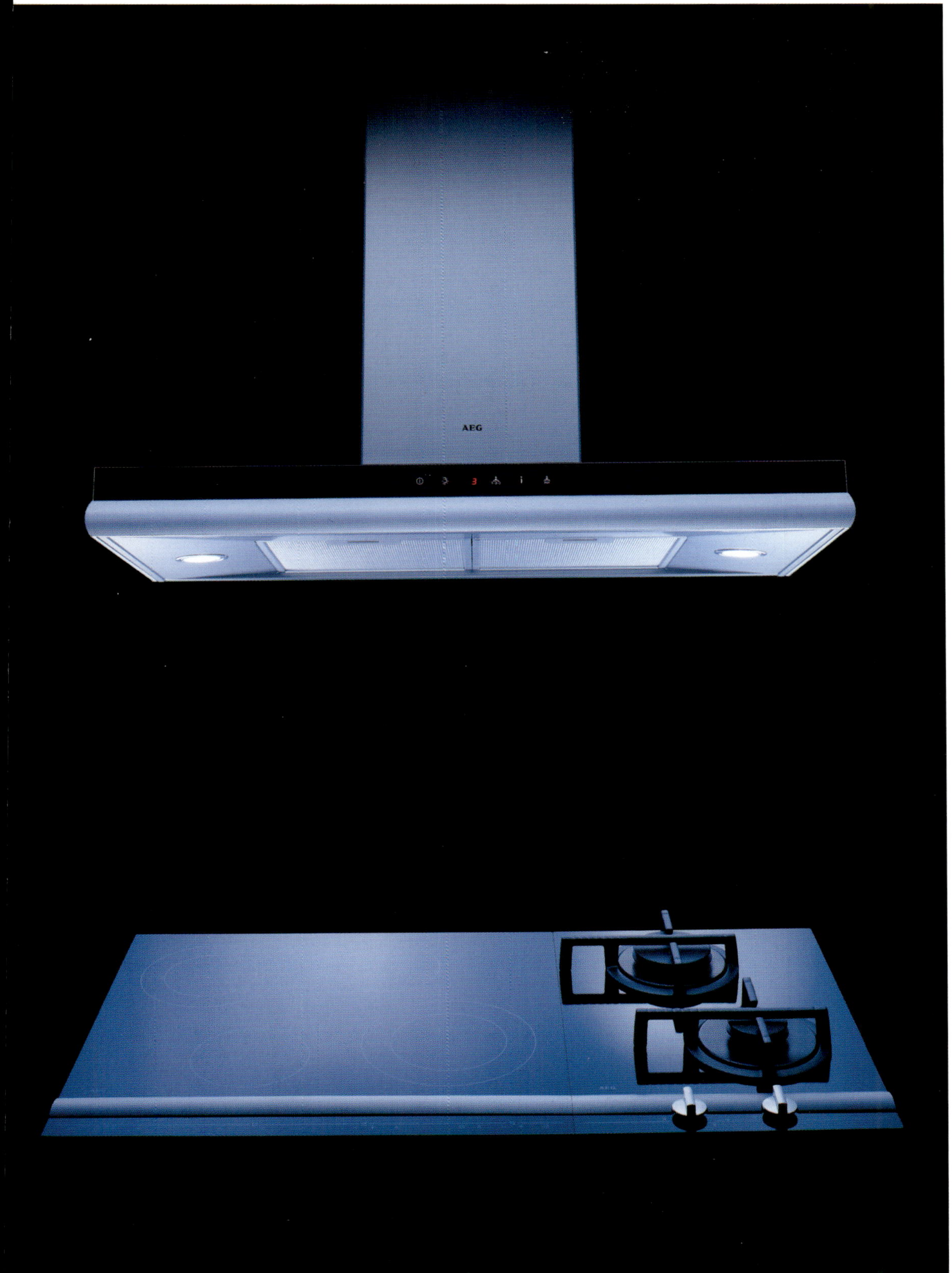

Hersteller Manufacturer
AEG Hausgeräte GmbH
Nürnberg, Germany

Produkt Product
DD 8890
Kamin-Dunsthaube
Chimney hood

Design
Hans Strohmeier
AEG Hausgeräte GmbH
Nürnberg, Germany

Die Frontline Kamin-Dunsthaube DD 8890 aus der COMPETENCE Reihe von AEG ist im Design auf die entsprechenden Kochmulden der Frontline Baureihe abgestimmt. Schwarzes Glas wird mit der Aluminiumleiste zu einem charakteristischen Designelement kombiniert. Die Bedienoberfläche ist mit Touch-Control ausgestattet und ermöglicht somit eine leichte Reinigung des Gerätes. Die Gerätebreite von 90 cm schafft die Möglichkeit einer Kombination von 2-3 Kochmulden unter der Dunsthaube. Die Metallfilter können leicht entnommen und in der Spülmaschine gereinigt werden.

The Frontline DD 8890 chimney hood from AEG's COMPETENCE range reflects the new design features of the cooking units in the Frontline series. Black glass is combined with an aluminum frame to form the striking look. The user surface is fitted with touch controls allowing the device to be cleaned easily. The chimney hood is 90 cm wide, meaning it can be used in conjunction with two or three hobs beneath it. The metal filter can be easily removed and cleaned in the dishwasher.

Hersteller Manufacturer
AEG Hausgeräte GmbH
Nürnberg, Germany

Produkt Product
DD 8765
Kamin-Dunsthaube
Chimney hood

Design
Almuth von der Planitz
AEG Hausgeräte GmbH
Nürnberg, Germany

Die Kamin-Dunsthaube DD 8765 von AEG ermöglicht durch die übereinander angeordneten Lamellen-Segmente die Absaugwirkung bereits 30 cm über der Kochfläche. Dieses Konzept führt zu einer optimalen Absaugung direkt neben und über den Kochtöpfen und bietet durch die Wandorientierung zudem mehr Aktionsraum und Kopffreiheit. Drei Lampen leuchten den Kocharbeitsplatz optimal aus. Die formal ästhetischen Gestaltungsmerkmale der Dunsthaube erlauben eine harmonische Integration ins Küchenumfeld.

AEG's DD 8765 chimney hood has been developed to extract cooking fumes just 30cm above the cooking surface given its design which features laminated segments positioned over one another. This concept permits optimal extraction directly beside and above pots, and due to its position on the wall, also offers more headroom and space to work. Three lamps supply the cooking area with optimal lighting. The aesthetic features of the chimney hood's form enable it to be harmoniously integrated into kitchen scenarios.

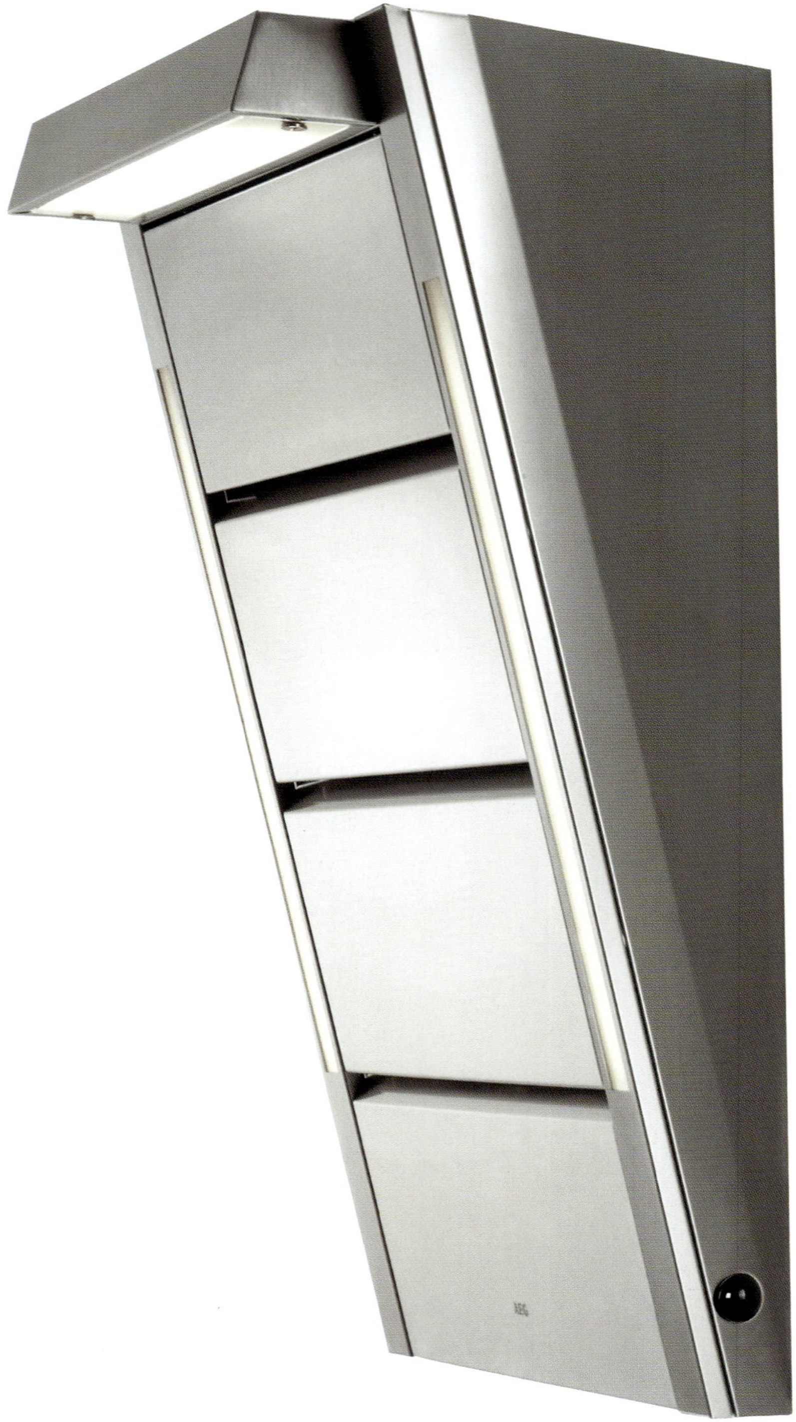

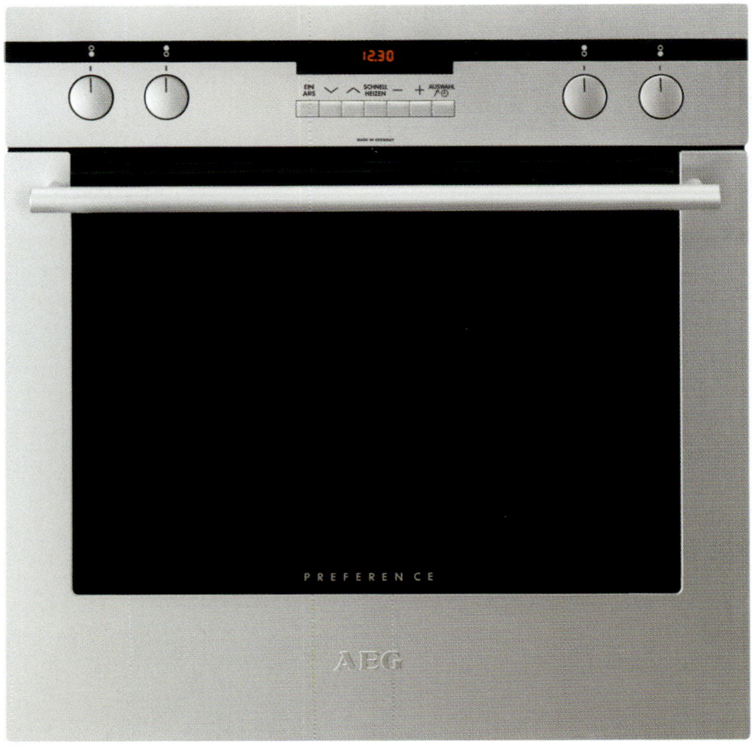

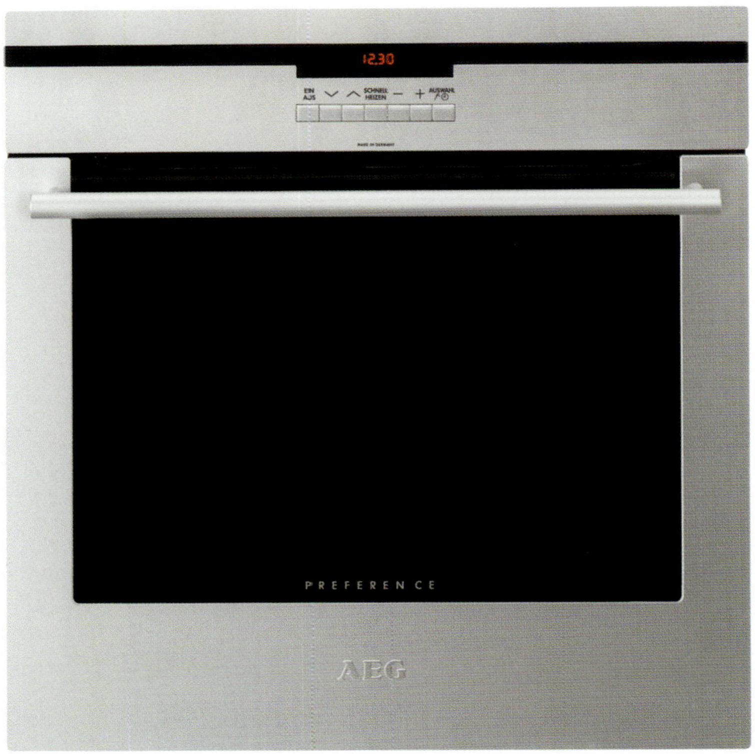

Hersteller Manufacturer
AEG Hausgeräte GmbH
Nürnberg, Germany

Produkt Product
PREFERENCE
Einbauherde und -backöfen-Produktreihe
Built-in cooker and oven range

Design
Heinz-Peter Huxhold
Axel Mürmann
AEG Hausgeräte GmbH
Nürnberg, Germany

Die PREFERENCE-Einbauherde und -backöfen-Produktreihe von AEG wird durch die horizontal angeordneten Bedien- und Anzeigenelemente geprägt. Der Reling-Griff unterstreicht die horizontale Linienführung. Die logisch funktionale Zuordnung dieser Elemente erleichtert die Bedienbarkeit der Geräte. Dem jeweiligen Bedienelement wie Kochstellenknebel oder Drucktaste ist im schwarzen Displaystreifen die optische Anzeige zugeordnet. Die formal ästhetischen Gestaltungsmerkmale der Produktreihe erlauben eine harmonische Integration ins Küchenumfeld mit anderen AEG-Einbaugeräten. Die unterschiedlichen Ausstattungsmerkmale, Funktionen und entsprechendes Zubehör der einzelnen Geräte der Produktreihe bieten eine individuelle Auswahlmöglichkeit für den Kunden. Zudem kann zwischen den Oberflächen Aluminium und Edelstahl gewählt werden.

AEG's PREFERENCE built-in cooker and oven range is characterized by its horizontally layout of the user and display elements. The rail handle emphasizes the horizontal lines of the range. The user and display elements are arranged logically and functionally so the device is not complicated to use. The various controls such as rotary hob controls or push buttons are represented visually in the black display panel. The formal aesthetic features of the range ensure it blends harmoniously with other AEG built-in devices in the kitchen environment. Customers can choose devices from the range according to their various features, functions and respective accessories. Surfaces come in a choice of aluminum or stainless steel.

Hersteller Manufacturer
AEG Hausgeräte GmbH
Nürnberg, Germany

Produkt Product
COMPETENCE Facette
Einbau-Ceran-Kochmulden-Produktreihe
Built-in Ceran hobs range

Design
Heinz-Peter Huxold
Axel Mürmann
AEG Hausgeräte GmbH
Nürnberg, Germany

Für das rahmenlose Design der COMPETENCE Facette Einbau-Ceran-Kochmulden Produktreihe von AEG ist die Glaskeramik nahezu flächenbündig in die Arbeitsfläche integriert und lässt sich deshalb besonders gut reinigen. Die Kochmulde ist entweder direkt dem Einbauherd zugeordnet, zusammen mit dem neuen Wrasenaustritt aus dem Backofen, oder aber unabhängig von einem Herd in der Küchenarchitektur einzubauen. Die Kochmulde ist mit Touch-Control ausgestattet und bietet modernsten Bedienungskomfort.

The frameless design of AEG's COMPETENCE Facette built-in Ceran hobs range features a glass ceramic plate positioned almost flush with the worktop surface, making it very easy to clean. The hob can either be combined with a built-in oven, together with the new steam outlet from the oven, or can be integrated independently into the kitchen environment. The hobs are outfitted with touch controls and offer the most modern in user comfort.

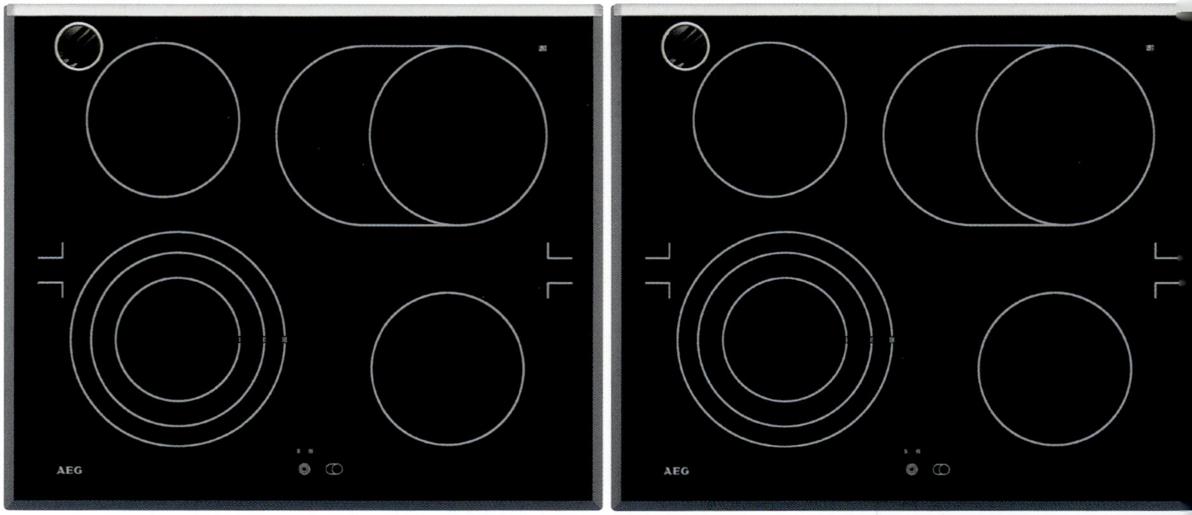

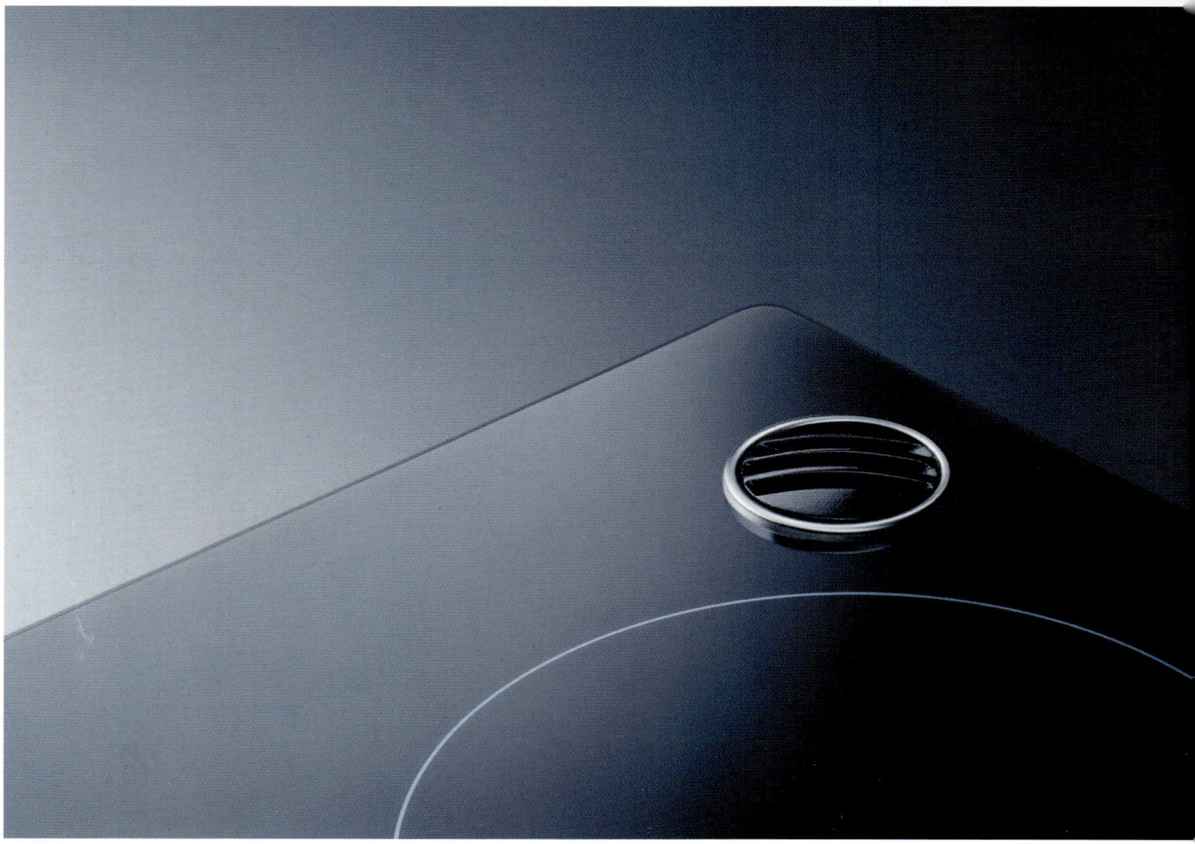

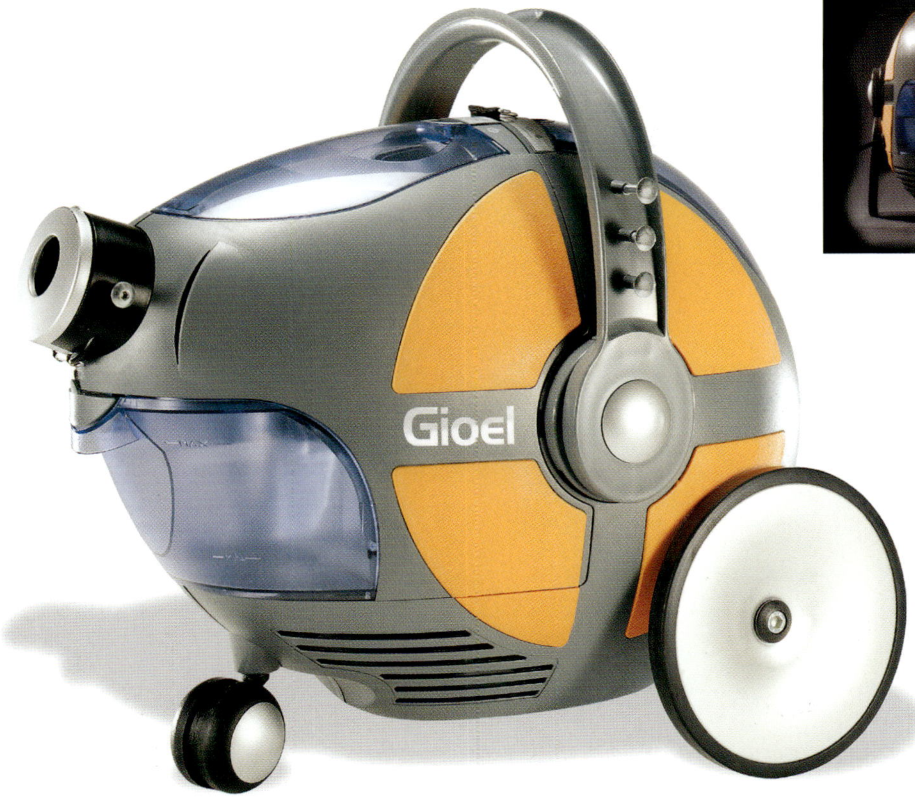

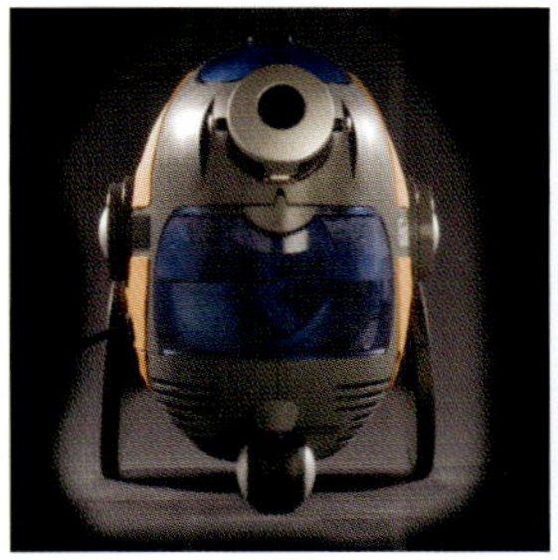

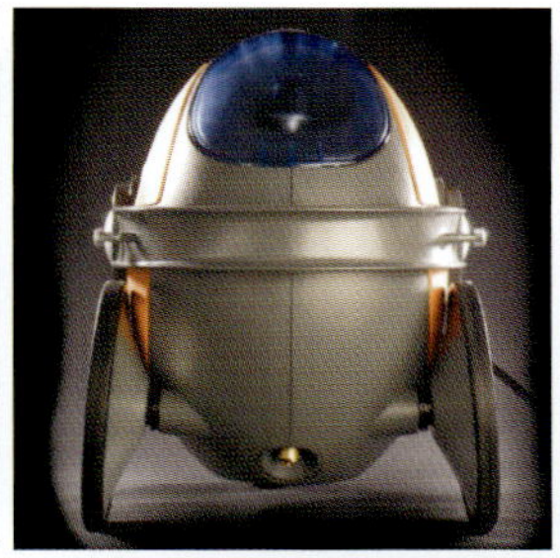

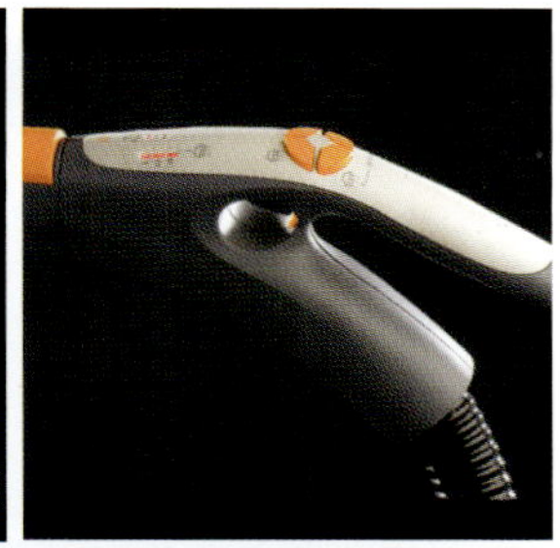

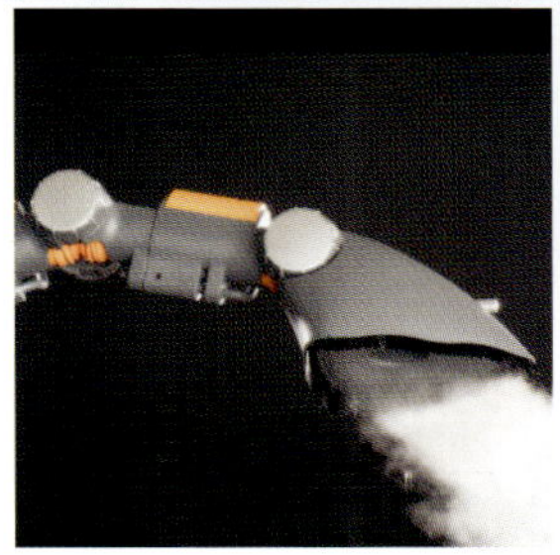

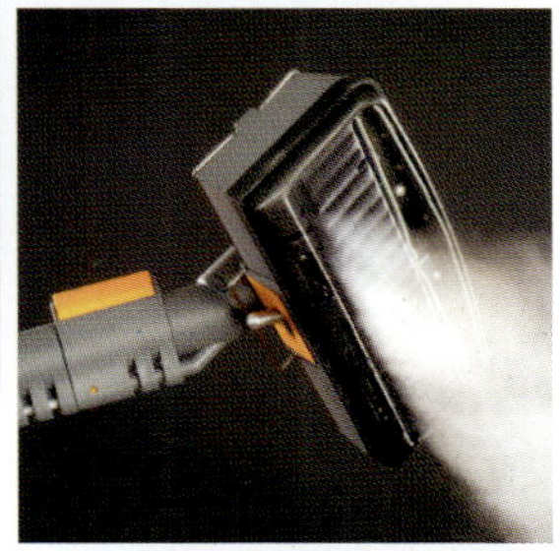

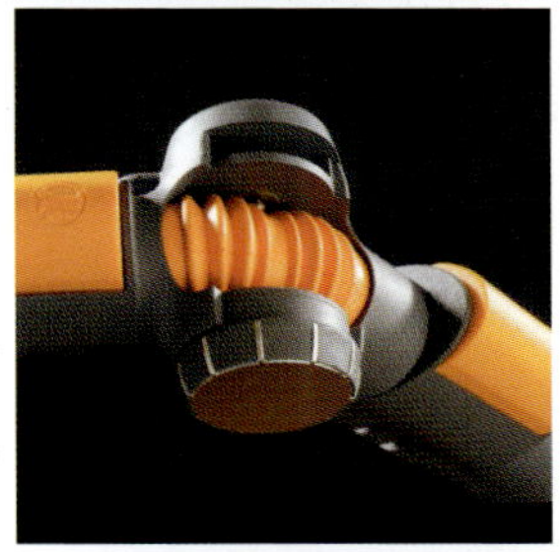

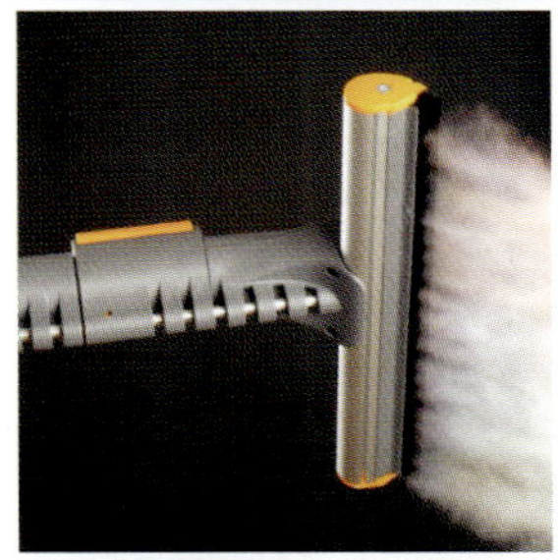

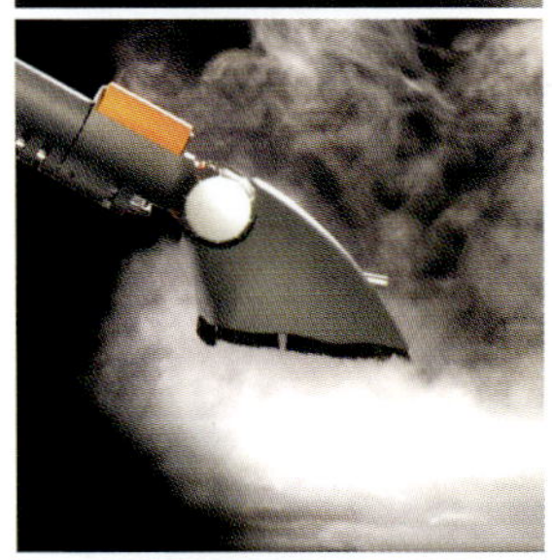

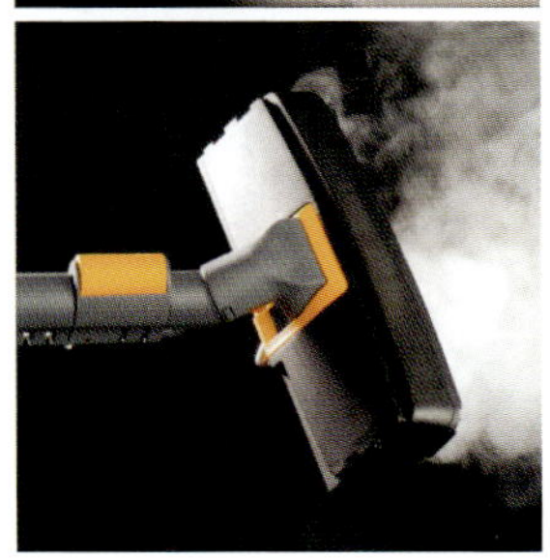

Projekt	Raumpflegesystem
Modell	G401-G402-G403
Kunde	Gioel
Design&Engineering	MM Design, Italy

Einfach Wasser: Wasser beim Saugen und Wasser für die Sterilisation mit Dampf.
Zwei Teile für eine neue Rienigungsmaschine: Der erste Teil ist ein Staubsauger mit einem Wasserfilter während der zweite Teil Dampf erzeugt. Die Trennung der Maschine erfolgt durch einen besonderen Mechanismus, der mit dem Griff verbunden ist.
Er ist einfach zu gebrauchen und darauf wird auch das Kabel aufgerollt.
Alle Teile die Wasser enthalten, wurden durch transparenten Kunststoff hervorgehoben. Beim Gestalten des Raumpflegesystems, wurde auch mit Formen und Materialien gespielt. Die daraus entstandene freundliche Form erinnert fast an ein Haustier, das treu hinter seinem Herrchen hertrottet.
Die Materialien wurden nach den gegebenen Normen ausgewählt um den charakteristischen Anforderungen zu entsprechen.

Project	Cleaning system
Model	G401-G402-G403
Client	Gioel
Design&Engineering	MM Design, Italy

Simply Water: Water to clean and Water for the sterilization with Steam.
The great novelty and peculiarity of the cleaning system is that the two machines can be divided: the first is a vacuum cleaner with water filter while the second is a boiler for producing steam. These machines are divided by a sophisticated mechanism fitted in the carrying handle; it is simple to operate and also acts as a winder for the power lead.
Semi-transparent plastic has been used for all parts containing water or wherever water creates particular vortexes.
The design of the cleaning system, besides solving practical problems, was played with also for the materials used and the shape.
The ultra-friendly shape brings to mind a pet that faithfully follows who's using it.n.

Hersteller Manufacturer
AEG Hausgeräte GmbH
Nürnberg, Germany

Produkt Product
LE MAITRE
Einbauherde und -backöfen Produktreihe
Built-in cooker and oven range

Design
Michael Thoenissen
Birgit Olzhausen
AEG Hausgeräte GmbH
Nürnberg, Germany

Das Design der LE MAITRE Einbauherde und -backöfen Produktreihe von Juno Electrolux zeichnet sich durch eine klare Liniengebung aus. Der Türgriff geht über die volle Breite des Herds und ist wesentliches Designmerkmal. Die rechteckigen Flächen von Bedienblende und Backofentür harmonieren mit dem Bogen des Griffes. Mit drei Tasten können Kochzeiten einfach programmiert werden. Die Einbaugeräte sind wahlweise in Edelstahl, Aluminium, Schwarz, Braun, Weiß oder Gusseisen-Optik erhältlich.

The design of the Juno Electrolux Le Maitre built-in cooker and product range stands out for its clear and harmonious lines. As a striking and fundamental design element, the handle arcs across the entire length of the door. The rectangular surfaces of the control panel and oven door harmonize with the curve of the handle. Three buttons are used to program cooking times and the ovens are available in a choice of stainless steel, aluminum, black, brown, while or cast iron.

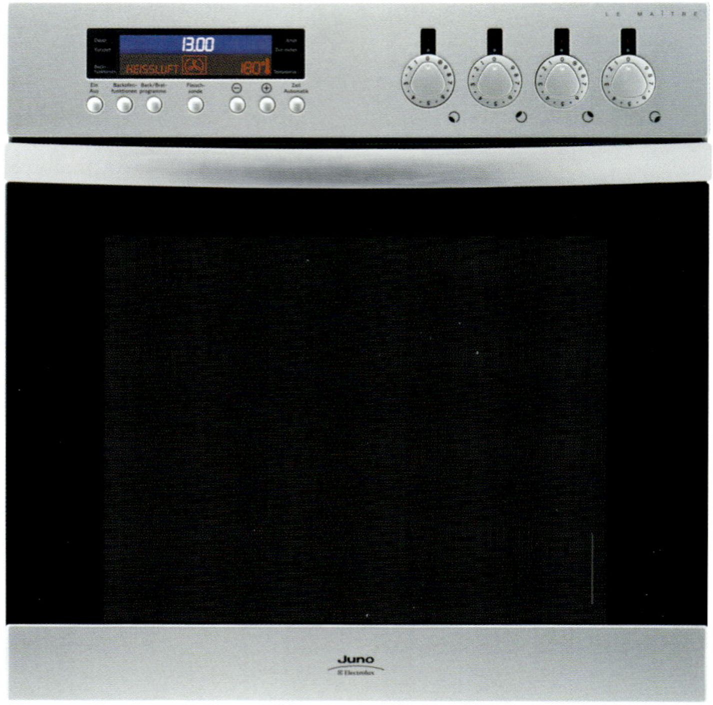

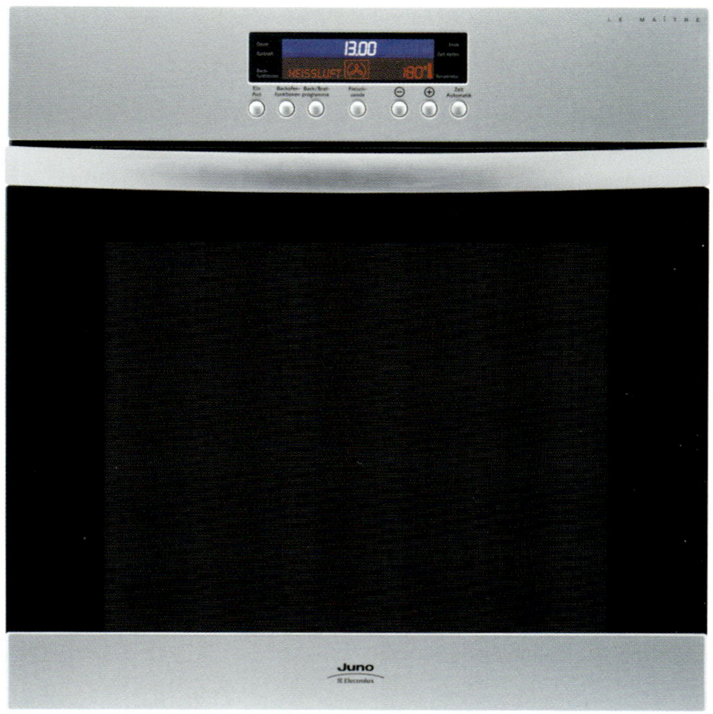

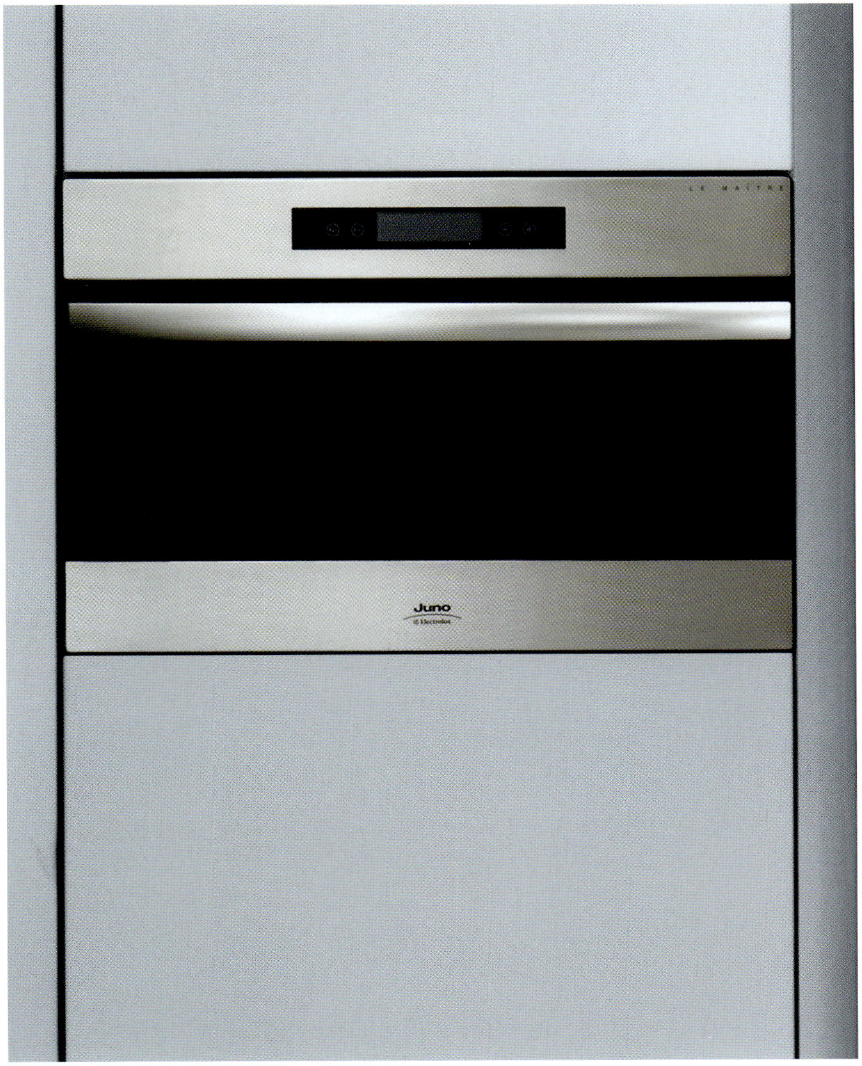

Hersteller Manufacturer
AEG Hausgeräte GmbH
Nürnberg, Germany

Produkt Product
LE MAITRE JDG 98100
Kompakt-Dampfgarer Einbau-Backofen
Compact built-in steamer oven

Design
Michael Thoenissen
Birgit Olzhausen
AEG Hausgeräte GmbH
Nürnberg, Germany

Das Design des LE MAITRE Kompakt-Dampfgarers JDG 98100 von Juno Electrolux ist durch klare Linien und Geometrien gekennzeichnet. Die rechteckigen Flächen von Bedienblende und Backofentüre harmonieren mit dem Bogen des Griffes. Mit Touch-Control bietet das Gerät hohen Bedienkomfort zur schonenden Zubereitung von Lebensmitteln.

The design of the Juno Electrolux LE MAITRE compact JDG 98100 steam cooker possesses beautifully clear lines and geometry. The rectangular surfaces of the user panel and oven door harmonize with the arc of the handle. Touch panels make the steam cooker easy to use in the sparing preparation of food afforded by steaming.

Hersteller Manufacturer
AEG Hausgeräte GmbH
Nürnberg, Germany

Produkt Product
LE MAITRE
Einbau-Ceran-Kochmulden-Produktreihe
Built-in Ceran hobs product range

Design
Michael Thoenissen
Birgit Olzhausen
AEG Hausgeräte GmbH
Nürnberg, Germany

Die Grafik der neuen Produktreihe LE MAITRE Einbau-Ceran-Kochmulden von Juno Electrolux ist durch harmonische Linien und Geometrien gekennzeichnet. Eine Vielfalt von Funktionen und Beheizungsarten sind in die Geräte integriert. Dazu zählen Induktionsbeheizung wie auch Halogen- und Gasbeheizung. Die Touch-Bedienung ermöglicht leichte Handhabung und Reinigungsmöglichkeit. Das weiße Ceran-Feld eröffnet zudem Kombinationsmöglichkeiten mit dem entsprechenden Küchenmöbel-Design.

The appearance of the new Juno Electrolux LE MAITRE Ceran hobs stands out for its harmonious lines and clear geometry. The hobs offer a variety of functions and heating methods. These include induction heating as well as halogen and gas heating. The touch controls make them easy to use and to clean. The white Ceran zone also opens up a variety of new ways of combining the hobs with the respective kitchen furniture design.

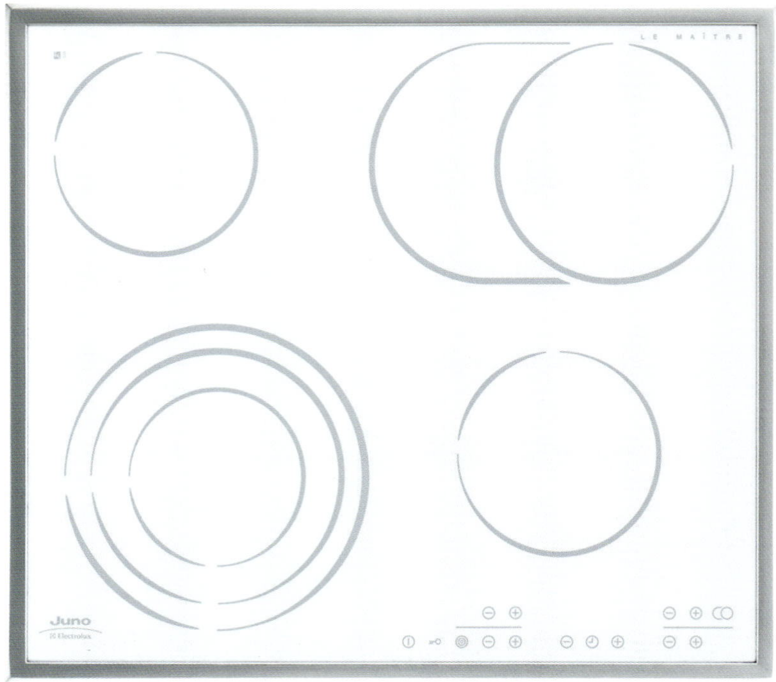

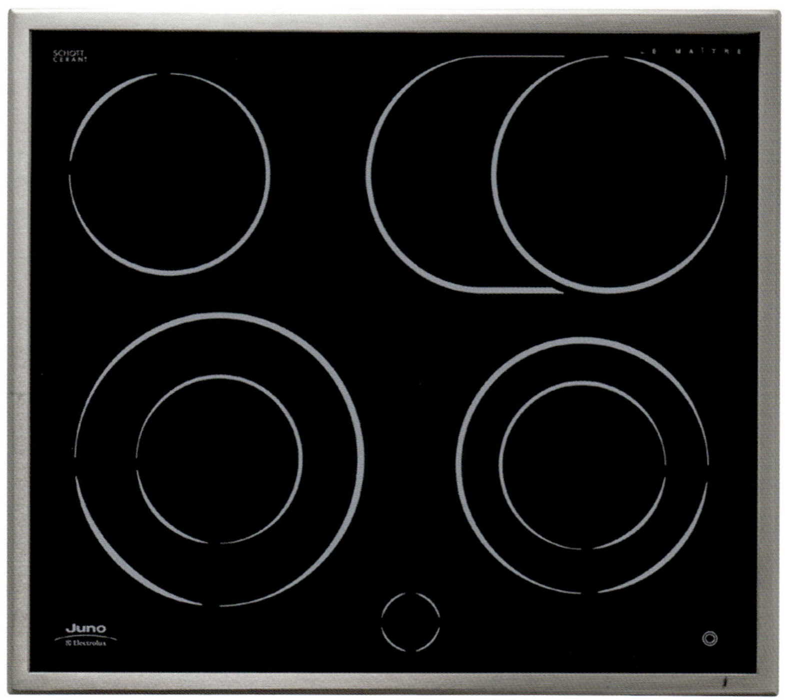

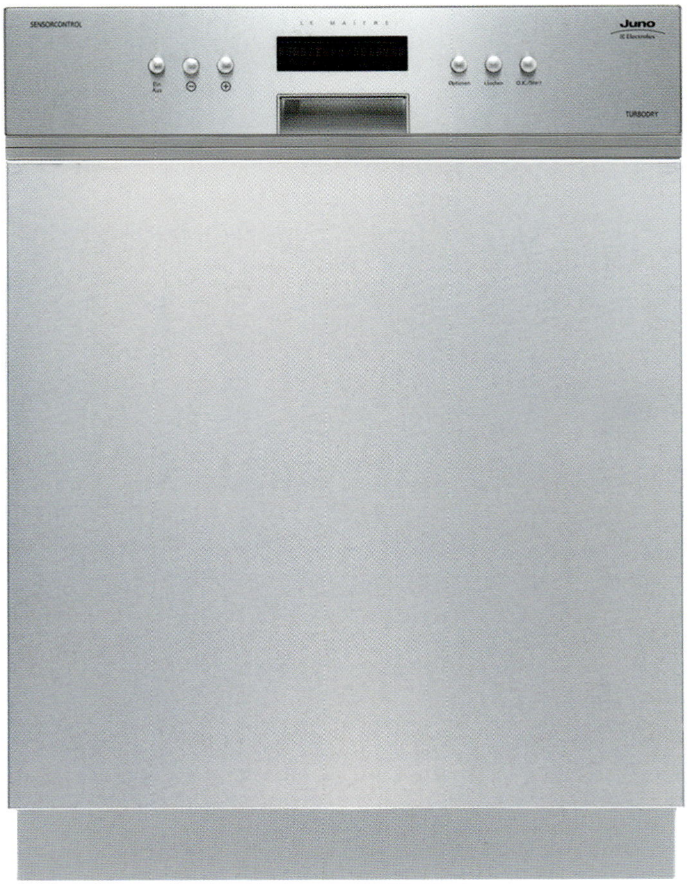

Hersteller Manufacturer
AEG Hausgeräte GmbH
Nürnberg, Germany

Produkt Product
LE MAITRE
Einbau-Geschirrspüler-Produktreihe
Built-in dishwasher product range

Design
Michael Thoenissen
Birgit Olzhausen
AEG Hausgeräte GmbH
Nürnberg, Germany

Das Design der neuen Produktreihe LE MAITRE Einbau-Geschirrspüler von Juno Electrolux ist durch klare Linien und Geometrien gekennzeichnet, denen auch die Gestaltung der Bedien- und Anzeigenelemente folgt. Das Text-Display mit 16 möglichen Sprachen bietet einen maximalen Bedienungskomfort zum Einstellen der Programme wie auch der Übersicht des Programmablaufs. Flexibel einteilbare Geschirrkörbe bieten ausreichend Platz für die unterschiedlichsten Geschirr- und Topfgrößen.

The design of the Juno Electrolux LE MAITRE dishwasher range is characterized by clear lines and geometry which carry through to the design of the user and display elements. The text display, available in 16 different languages, maximizes ease of use in terms of setting programs as well as overseeing the program in operation. Flexible, dividable baskets for dishes offer a lot of space for all shapes and sizes of pots and dishes.

Hersteller Manufacturer
AEG Hausgeräte GmbH
Nürnberg, Germany

Produkt Product
LE MAITRE JDK 9590
Kamin-Dunstabzug-Produktreihe
Chimney hood range

Design
Michael Thoenissen
Birgit Olzhausen
AEG Hausgeräte GmbH
Nürnberg, Germany

Der LE MAITRE JDK 9590 Kamin-Dunstabzug von Juno-Electrolux ist in seiner wandorientierten Position den Dunsthauben von Profiküchen entlehnt. Größtmöglicher Freiraum über dem Kochfeld ist ebenso gewährleistet wie eine direkte Dunstabsaugung unmittelbar neben und auch über den Töpfen. Große Metallflächen ermöglichen eine leichte Reinigung ebenso wie die Entnahme der Metallfilter, die im Geschirrspüler gereinigt werden können. Die integrierte Touch-Control betont das klare Design.

Juno Electrolux's LE MAITRE JDK 9590 chimney hood is positioned on the wall – a concept borrowed from professional kitchens. This allows a maximum of space above the cooking area as well as the direct extraction of fumes beneath and above pots. The large metal surfaces facilitate cleaning and the metal filter can in fact be removed and cleaned in the dishwasher. Integrated touch panels emphasize the clear design.

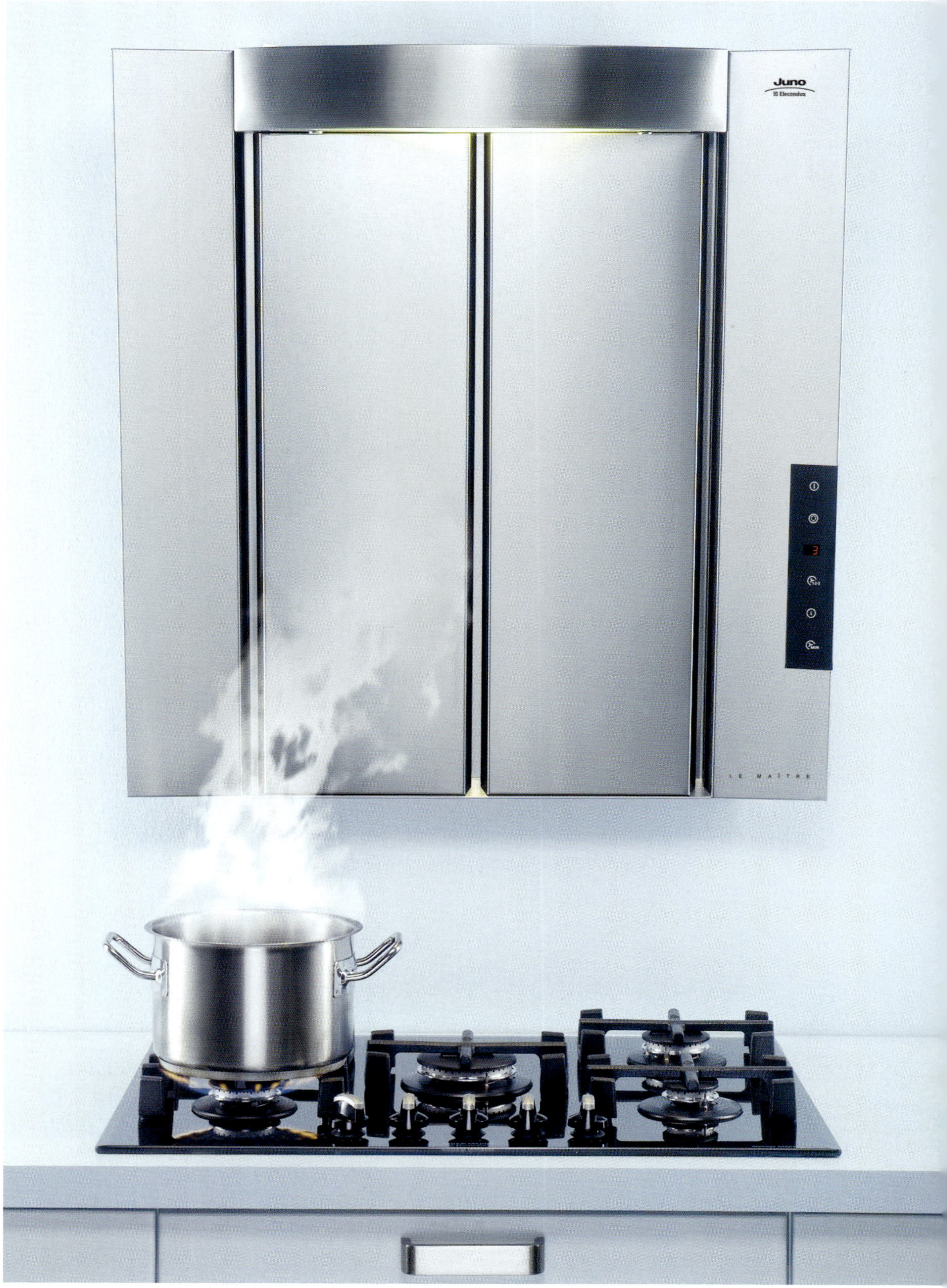

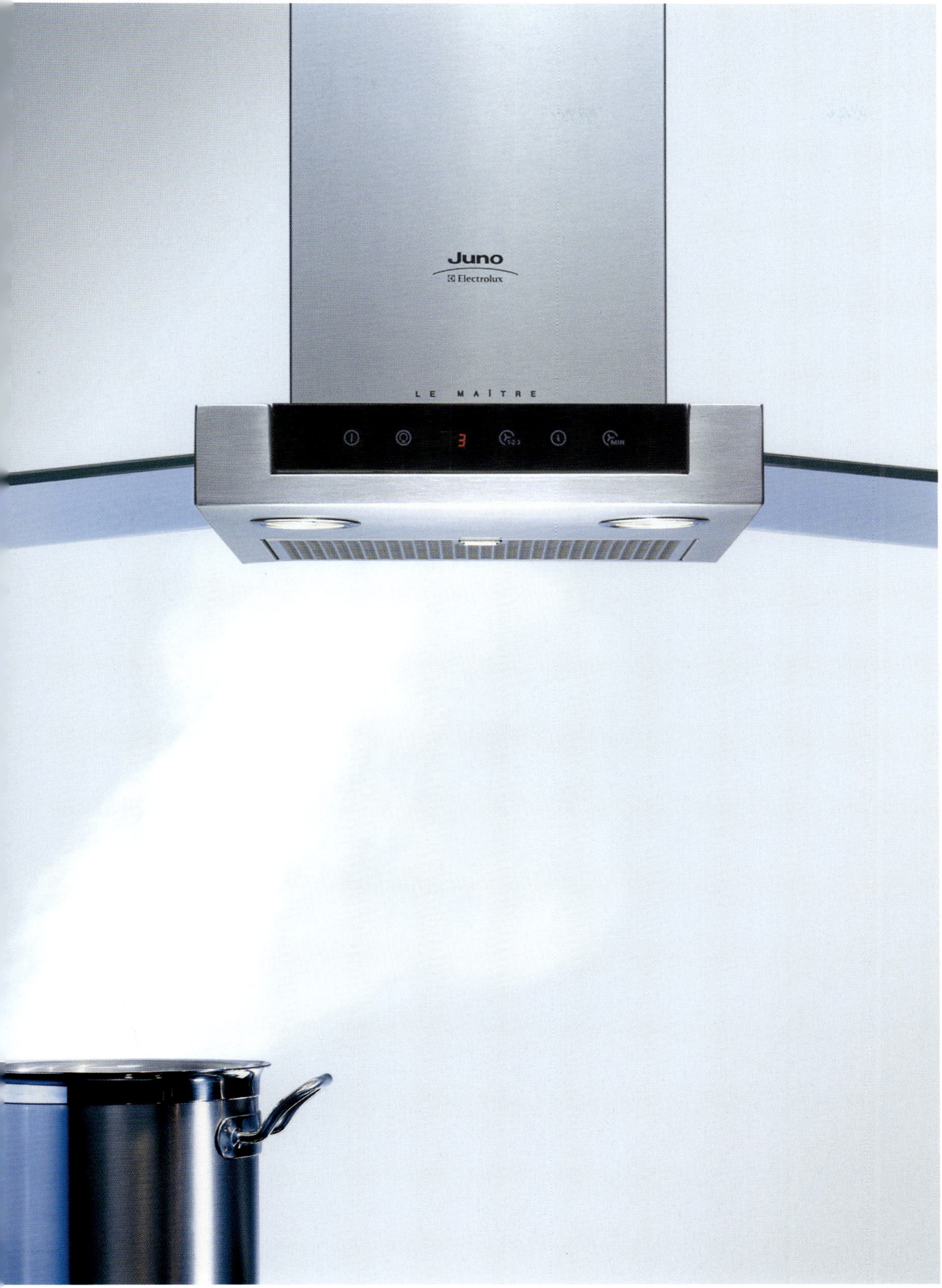

Hersteller Manufacturer
AEG Hausgeräte GmbH
Nürnberg, Germany

Produkt Product
LE MAITRE JDK 8590
Kamin-Dunstabzug
Chimney hood

Design
Michael Thoenissen
Birgit Olzhausen
AEG Hausgeräte GmbH
Nürnberg, Germany

Das Design des LE MAITRE JDK 8590 Kamin-Dunstabzugs von Juno Electrolux konzentriert sich auf die zentrale Funktion der Dunstabsaugung und Beleuchtung des Kochfeldes. Die Touch-Bedienung ist ebenso an zentraler Position in das Gerät integriert. Die beiden Glasschirme betonen in leichter atmosphärischer Materialität die im Mittelpunkt stehende Funktion des Produktes.

The LE MAITRE JDK 8590 chimney hood design focuses on the central functions, namely extracting cooking fumes and illuminating the cooking area. The touch panel is positioned centrally for ease of access. The bright atmospheric nature of the two glass screens accentuates the central function of the chimney hood.

Hersteller Manufacturer
Robert Bosch Hausgeräte GmbH
München, Germany

Produkt Product
Bosch MMB 2000
Standmixer
Blender

Design
Rolf Feil
Robert Bosch Werksdesign
Giengen, Germany

Der Standmixer MMB 2000 von Bosch mit einer Leistungsstärke von 600 Watt besitzt ein Vier-Flügel-Messer, mit dem er zerkleinern, zerstoßen, pürieren oder mixen kann. Der stufenlose Drehregler plus Momentfunktion ist versenkbar. Ein Bajonettverschluss erlaubt die sichere Handhabung des Mixbechers, da der Motor nur arbeitet, wenn der Becher eingerastet ist. Die Deckellösung ermöglicht drei Varianten des Befüllens mit kleinen Stückchen bis zu Flüssigkeiten. Die Griffelemente bieten ergonomischen Griffkomfort, alle Funktionsteile lassen sich einfach auseinander nehmen und leicht reinigen.

The Bosch MMB 2000 blender is powered by 600W motor driving a four-winged blade – and is ideal for reducing, crushing, puréeing or blending. The rotary control and pulse function pop up for convenience. The new bayonet lock enables the blender jug to be safely removed or replaced, as the motor can only be actuated if the lock is closed. The lid is designed to facilitate three different filling modes for small pieces, liquids, or large pieces. The handles enhance ergonomic grip, and all functional parts can be easily dismantled and cleaned.

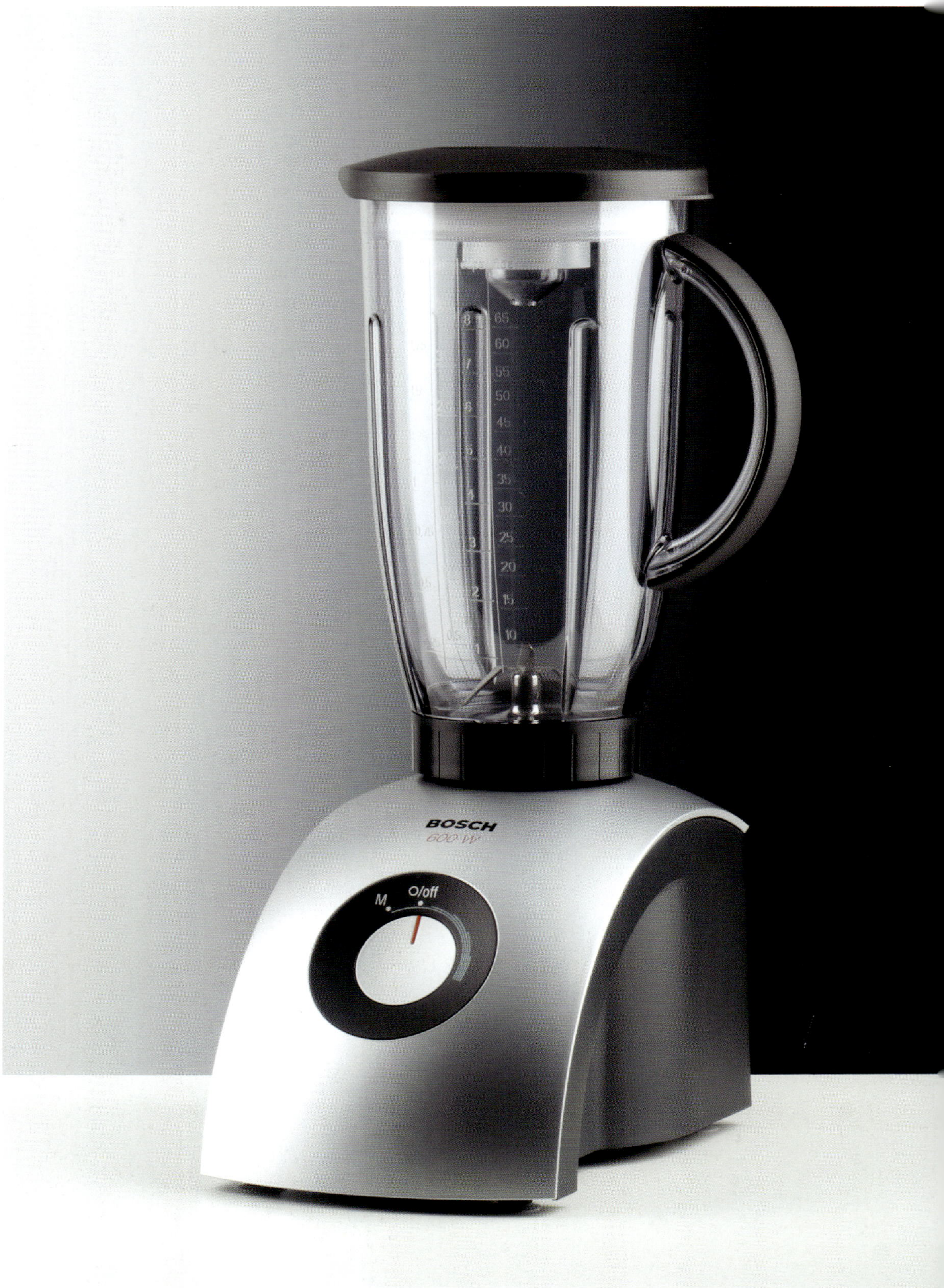

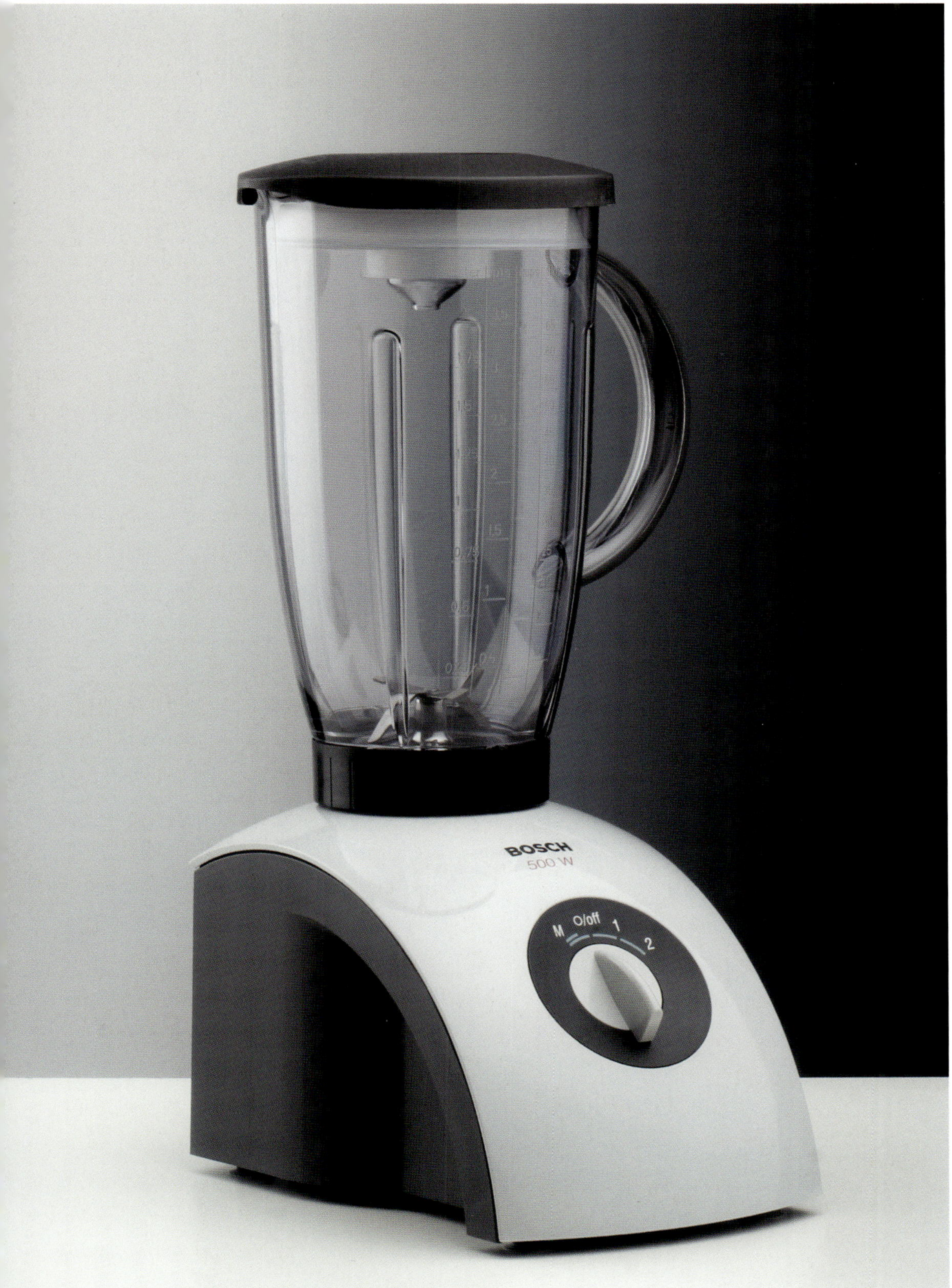

Hersteller Manufacturer
Robert Bosch Hausgeräte GmbH
München, Germany

Produkt Product
Bosch MMB 1000
Standmixer
Blender

Design
Rolf Feil
Robert Bosch Werksdesign
Giengen, Germany

Der Standmixer MMB 1000 von Bosch mit einer Leistungsstärke von 500 Watt besitzt ein Vier-Flügel-Messer, mit dem er zerkleinern, zerstoßen, pürieren oder mixen kann. Der Bajonettverschluss erlaubt eine sichere Handhabung des Mixbechers, da der Motor nur arbeitet, wenn der Becher eingerastet ist. Die Deckellösung ermöglicht drei Varianten des Befüllens mit kleinen Stückchen bis zu Flüssigkeiten. Die Griffelemente bieten ergonomischen Griffkomfort, alle Funktionsteile lassen sich einfach auseinander nehmen und leicht reinigen.

The Bosch MMB 1000 blender has a powerful 500W motor driving a four-winged blade – and is ideal for reducing, crushing, puréeing or blending. The new bayonet lock enables the blender jug to be safely removed or replaced, as the motor can only be actuated if the lock is closed. The lid is designed to facilitate three different filling modes for small pieces, liquids, or large pieces. The handles enhance ergonomic grip, and all functional parts can be easily dismantled and cleaned.

Hersteller Manufacturer
Robert Bosch Hausgeräte GmbH
München, Germany

Produkt Product
Bosch PHL 206 E NL
Gas-Glaskeramik-Kochstelle
Gas hob

Design
Roland Vetter
Robert Bosch Werksdesign
Giengen, Germany
Eisele Kuberg Design
Neu-Ulm, Germany

Die Glaskeramik-Kochstelle PHL 206 E NL von Bosch besitzt vier Gasbrenner und einen zentralen Dual-Kombi-Wokbrenner. Töpfe können auf den flächig gestalteten Gusstopfträgern sicher verschoben werden. Der Einsatz von Glaskeramik als Grundplatte gewährt eine gleich bleibende hohe Anmutungsqualität. Das Edelstahlfrontprofil dient als Kantenschutz. Die Zylinderknebel und die Ventiltechnik erlauben einen hohen Bedienkomfort zum Zünden und Dosieren der Flammen.

The Bosch PHL 206 E NL ceramic glass hob comes complete with four burners and a central dual-combi wok-burner. Saucepans can be moved safely around on the flat cast-iron pan rests. By using ceramic glass as the base plate, the hob retains its elegant look during its long service life. The stainless steel front profile serves as edge protection, while the cylindrical knobs and valve system providers optimal convenience for lighting and dosing the burners.

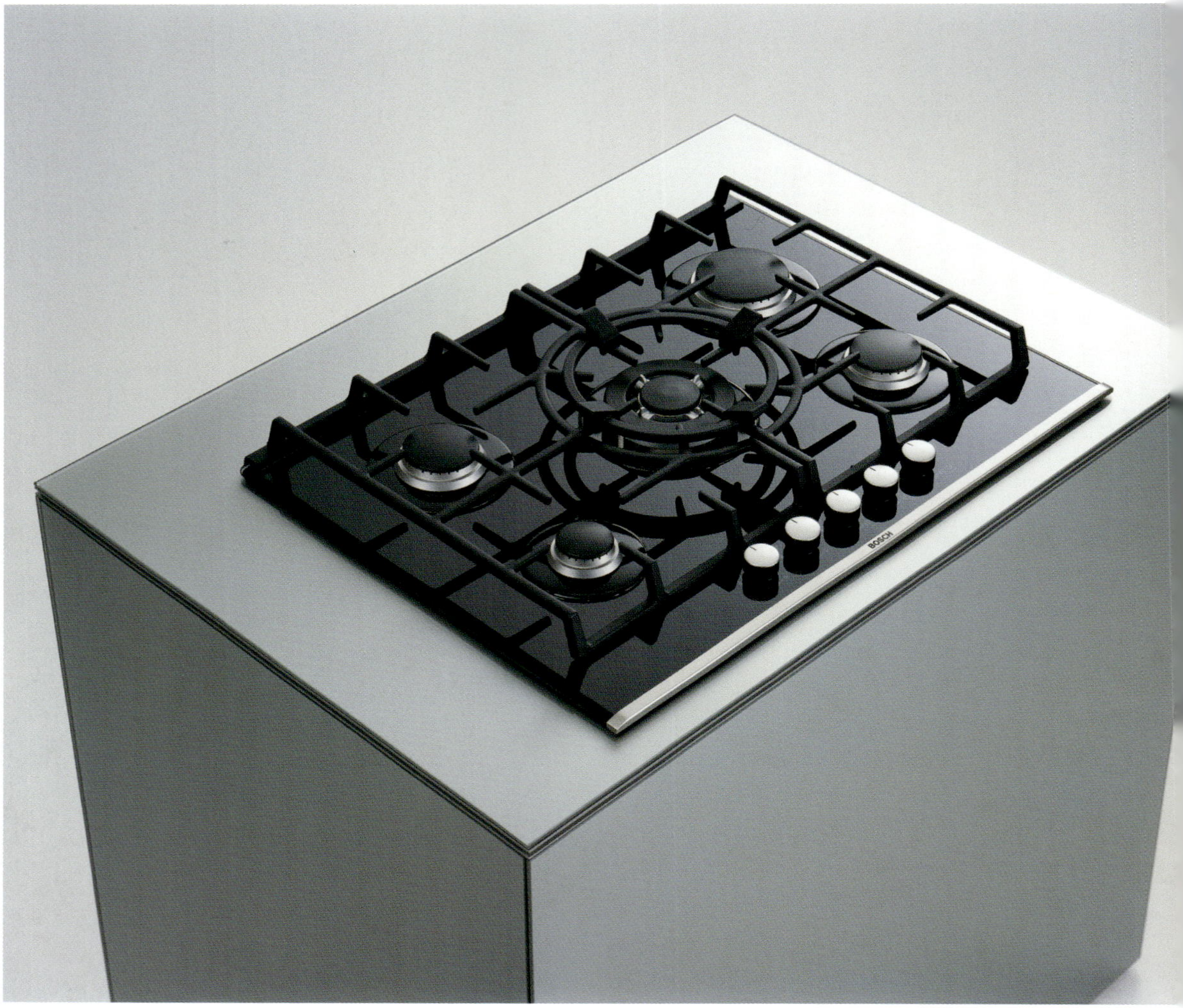

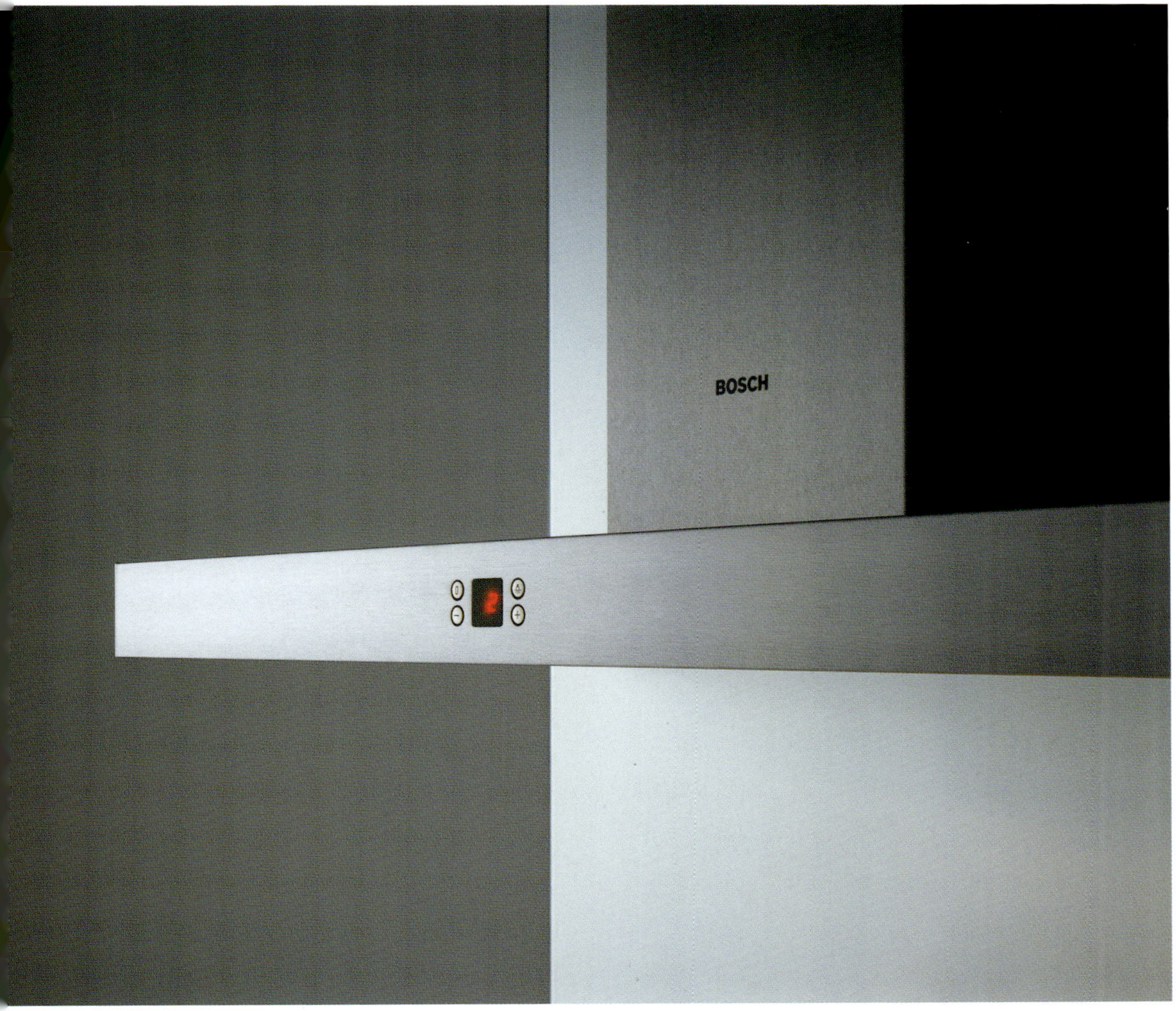

Hersteller Manufacturer
Robert Bosch Hausgeräte GmbH
München, Germany

Produkt Product
Bosch DKE 945 E
Dunstesse
Cooker hood

Design
Roland Vetter
Robert Bosch Werksdesign
Giengen, Germany
Eisele Kuberg Design
Neu-Ulm, Germany

Die Dunstesse DKE 945 E von Bosch hat eine Edelstahlverkleidung und kann im Abluft- oder Umluftbetrieb eingesetzt werden. Die hohe Lüfterleistung lässt sich über drei Leistungsstufen und eine zusätzliche Intensivstufe exakt regulieren. Zwei Halogen-Arbeitsplatzleuchten sorgen für direkte Ausleuchtung. Das Bedienelement mit Digitalanzeige und Kurzhubtasten ist im vorderen Bereich gut zugänglich für den Benutzer angebracht.

The Bosch DKE 945 E extractor canopy hood features all-round stainless steel and can be set to exhaust air flow or re-circulation mode. The powerful ventilation capacity can be set exactly at three different levels, with an additional intensive function. Two halogen downlights ensure direct illumination below the canopy. The control element, complete with digital display and touch buttons, is located at the front for easy access.

Hersteller Manufacturer
Robert Bosch Hausgeräte GmbH
München, Germany

Produkt Product
Bosch DKE 965 M
Dunstesse
Cooker hood

Design
Roland Vetter
Robert Bosch Werksdesign
Giengen, Germany
Eisele Kuberg Design
Neu-Ulm, Germany

Die Dunstesse DKE 965 M von Bosch hat eine Edelstahlverkleidung und einen Glasschirm. Die Positionierung des Anzeige- und Bedienelements unterhalb des Glasschirms erlaubt ein gutes Einsehen und eine günstige Bedienung der Kurzhubtasten. Die Dunstesse kann im Abluft- oder Umluftbetrieb eingesetzt werden. Die hohe Lüfterleistung lässt sich über drei Leistungsstufen und eine zusätzliche Intensivstufe exakt regulieren. Zwei Halogen-Arbeitsplatzleuchten sorgen für direkte Ausleuchtung.

The Bosch DKE 965 M extractor canopy hood has a stainless steel and glass housing. The display and the controls with the touch buttons are located beneath the glass hood for superior visibility and ease of use. The hood can be set to exhaust air flow or re-circulation mode. The powerful ventilation capacity can be set exactly at three different levels, with an additional intensive function. Two halogen downlights ensure direct illumination below the canopy.

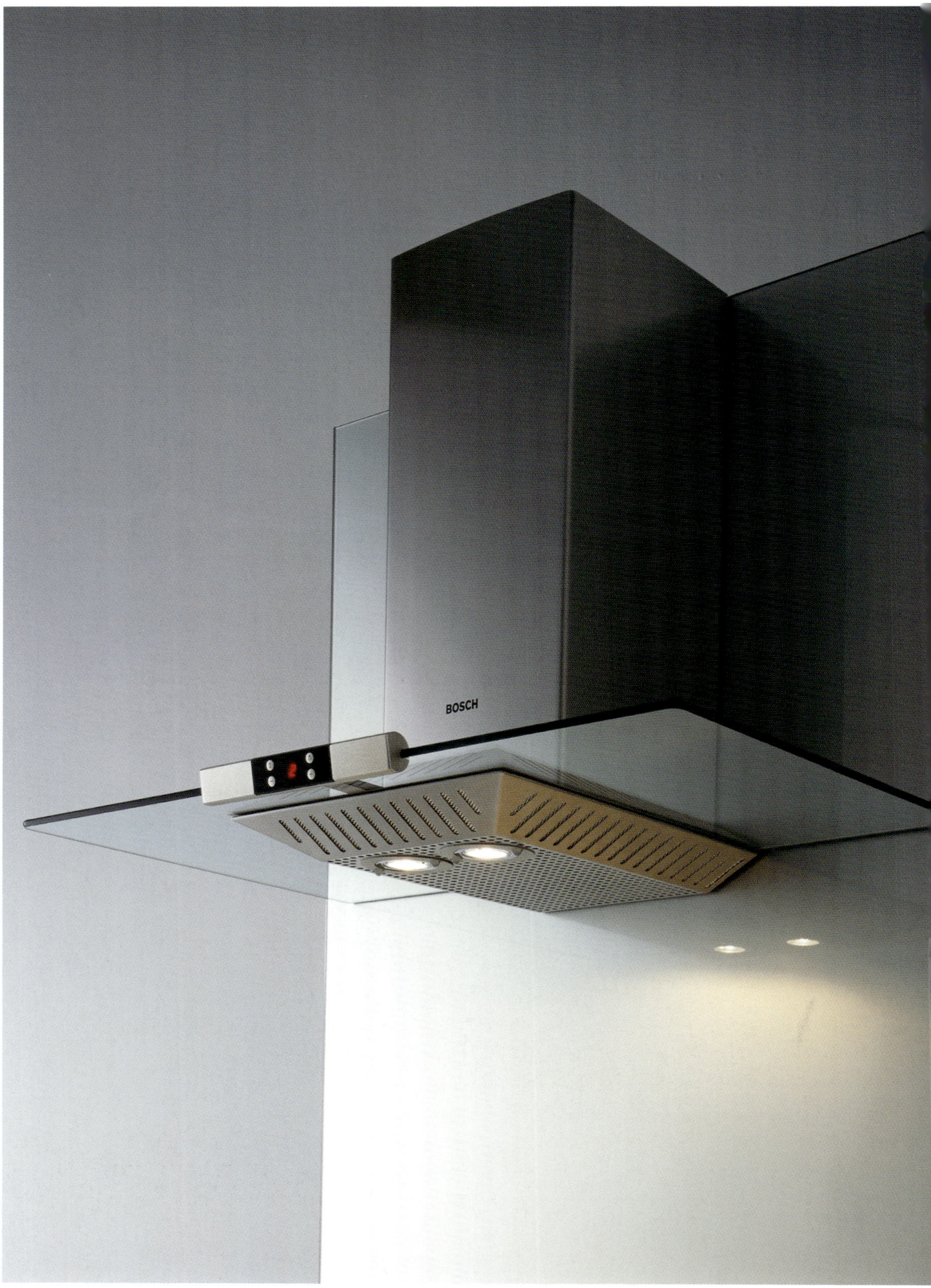

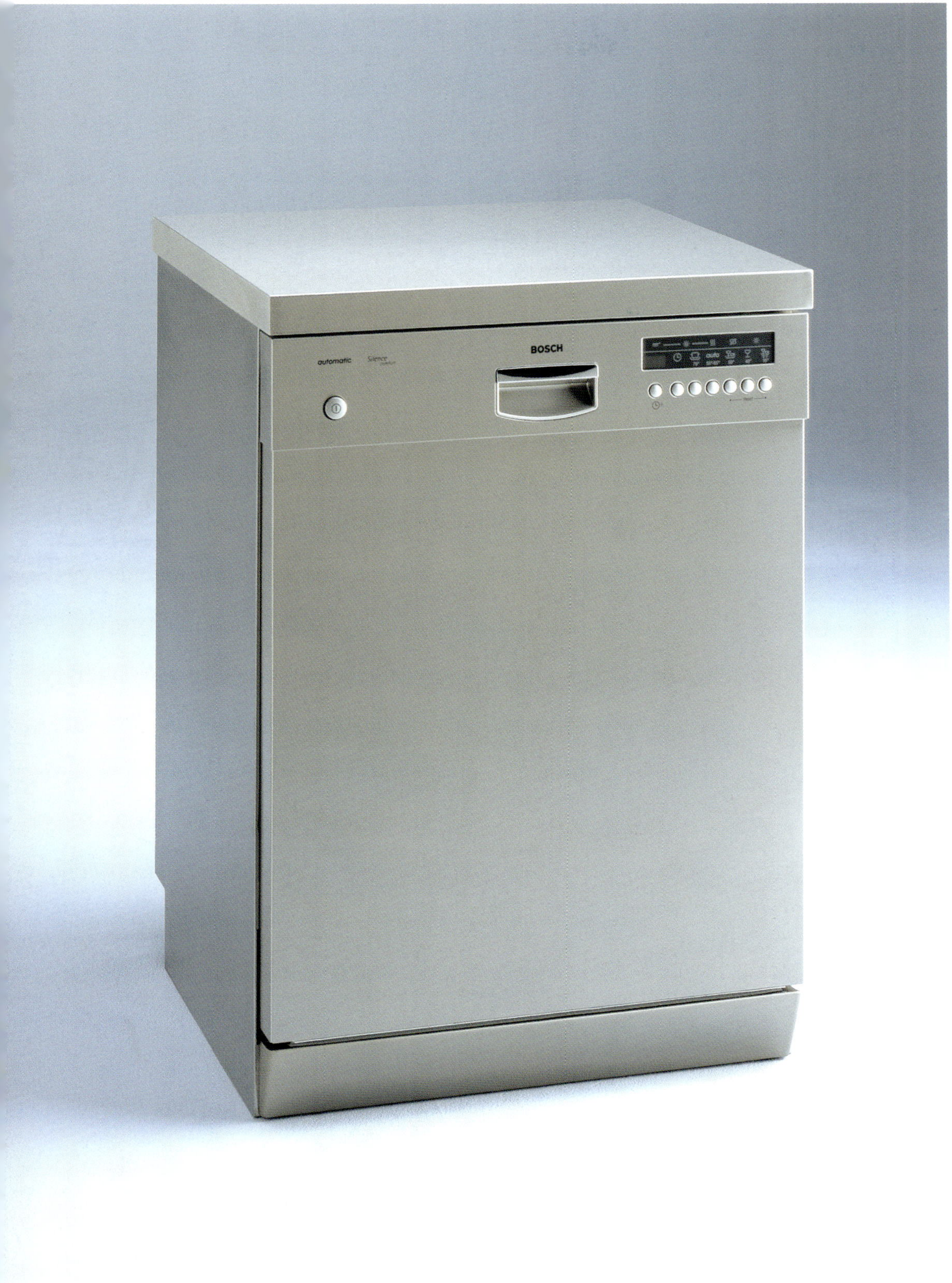

Hersteller Manufacturer
Robert Bosch Hausgeräte GmbH
München, Germany

Produkt Product
Bosch SGS 59 A 05
Geschirrspüler
Dishwasher

Design
Thomas Ott
Roland Vetter
Robert Bosch Werksdesign
Giengen, Germany

Der Geschirrspüler SGS 59 A 05 von Bosch ist in allen sichtbaren Teilen aus Edelstahl gefertigt, der Griffbereich wird durch ein Chromglanzelement betont. Die Bedienung ist logisch von links (mit dem Einschalten) nach rechts (mit der Programmwahl) aufgebaut. Sämtliche Anzeigen sind in einem Multi-Display zusammengefasst, wo sich alle Informationen auf einen Blick erfassen lassen. Durch umlegbare Stachelreihen ergibt sich eine hohe Variabilität der Körbe zum Spülen von ungewöhnlichen Geschirrteilen.

All the visible parts of the Bosch SGS 59 A Í05 dishwasher are made of stainless steel, and the handle area is highlighted by a gloss chrome section. Operation is intuitively from left (on/off switch) to right (program selection). All displays are combined in a multi-display, presenting all information at a glance. The fold-down plate racks increase basket variability and loading options, especially for unusual items of crockery and accessories.

Hersteller Manufacturer
Robert Bosch Hausgeräte GmbH
München, Germany

Produkt Product
Bosch HEN 364 650
Einbauherd
Built-in cooker

Design
Roland Vetter
Robert Bosch Werksdesign
Giengen, Germany
Eisele Kuberg Design
Neu-Ulm, Germany

Der Einbauherd HEN 364 650 von Bosch im Edelstahldesign verfügt u.a. über ein großes Display, versenkbare Bedienknebel mit Leuchtring, einen ergonomischen Bügelgriff und ein großzügiges Sichtfenster zum Backofen. Weitere Produktmerkmale wie der XXL-Backofen mit Granit-Emaille, automatische Schnellaufheizung, Programmautomatik, Hitzeschutz-Fenster, Kindersicherung, Backwagen erhöhen die Funktionalität des Einbauherdes.

The Bosch HEN 364 650 built-in cooker with its stainless-steel appearance features, among other things, a large display, pop-up control knobs with an illuminated ring, an ergonomic arced handle, and a generously dimensioned oven window. Additional functional highlights are the XXL oven coated in granite enamel, automatic fast pre-heating, automatic program settings, heat-protective window, child safety lock, and a roll-out baking tray.

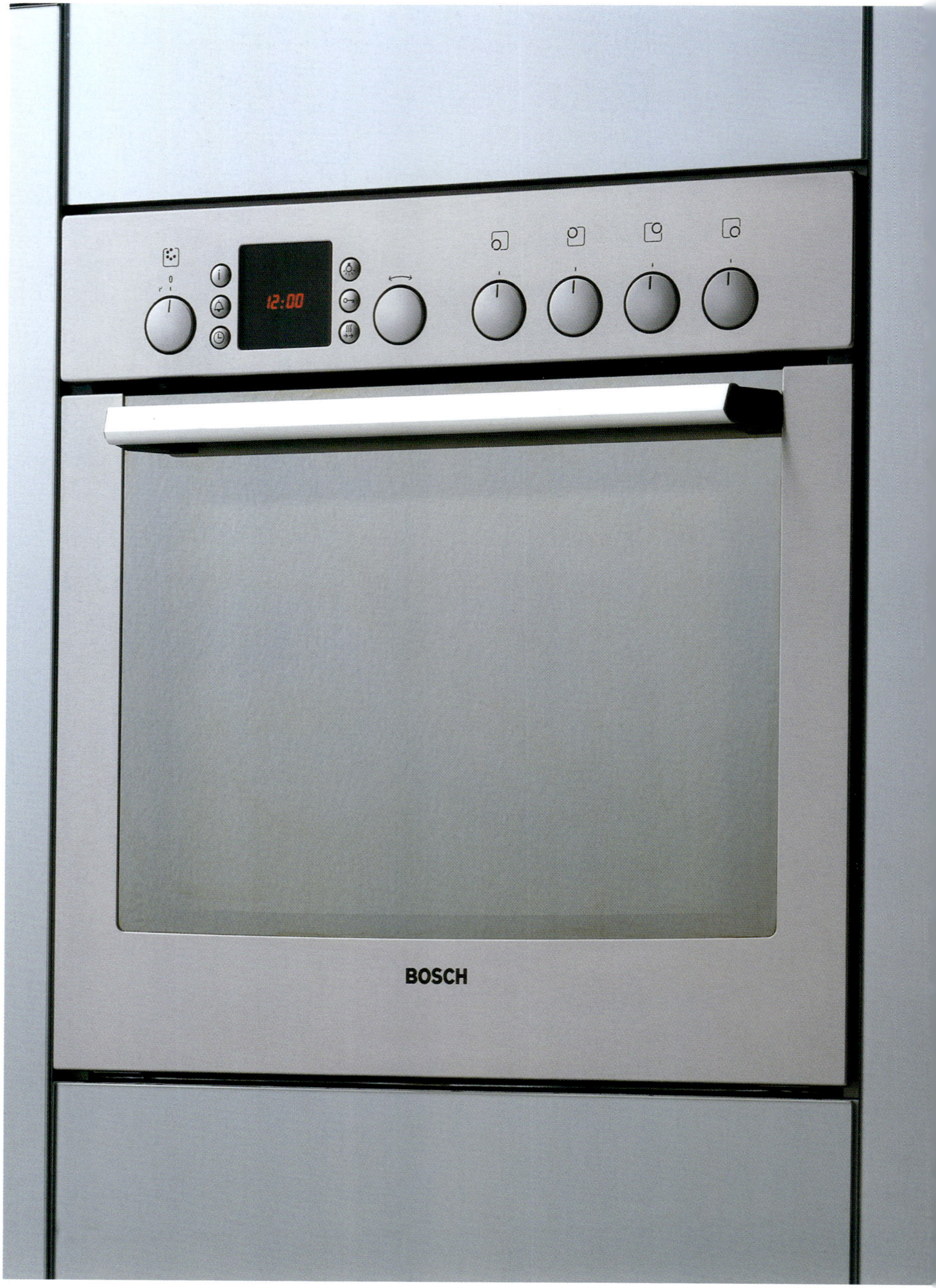

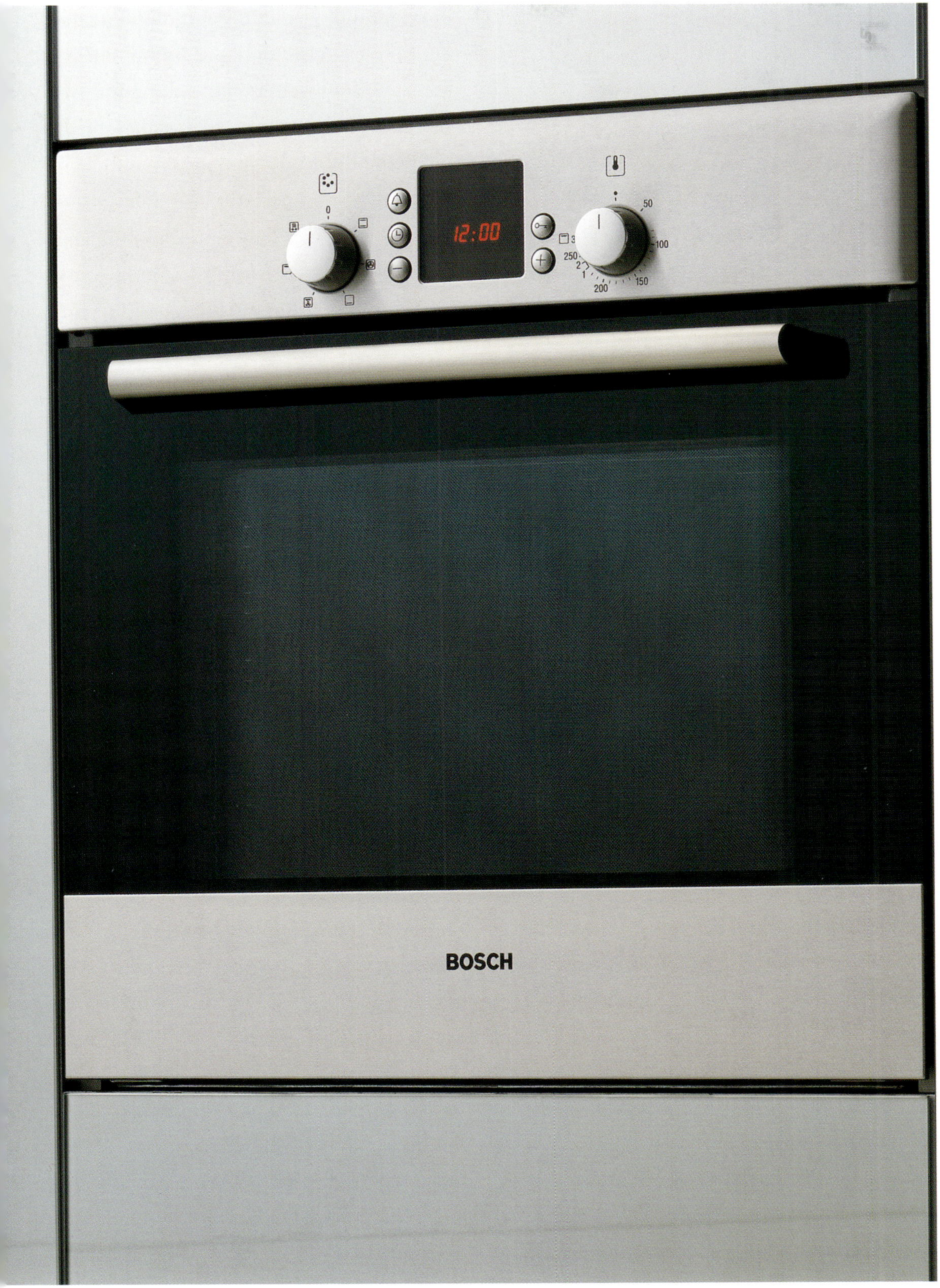

Hersteller Manufacturer
Robert Bosch Hausgeräte GmbH
München, Germany

Produkt Product
Bosch HBN 230 050
Einbau-Backofen
Built-in cooker

Design
Roland Vetter
Robert Bosch Werksdesign
Giengen, Germany
Eisele Kuberg Design
Neu-Ulm, Germany

Der Einbau-Backofen HBN 230 050 von Bosch im Edelstahldesign verfügt u.a. über ein großes Display, versenkbare Bedienknebel mit Leuchtring, einen ergonomischen Bügelgriff und ein großzügiges Sichtfenster zum Backofen. Weitere Produktmerkmale wie der XXL-Backofen mit Granit-Emaille, automatische Schnellaufheizung, Hitzeschutz-Fenster, Kindersicherung und Vollglas-Innentür erhöhen die Funktionalität des Einbau-Backofens.

The Bosch HEN 230 050 built-in cooker with its stainless-steel appearance features, among other things, a large display, pop-up control knobs with an illuminated ring, an ergonomic arced handle, and a generously dimensioned oven window. Additional functional highlights are the XXL oven coated in granite enamel, automatic fast pre-heating, automatic program settings, a solid glass inner door, child safety lock, and a roll-out baking tray.

Hersteller Manufacturer
Robert Bosch Hausgeräte GmbH
München, Germany

Produkt Product
Bosch PKN 675 T01
Glaskeramik-Kochstelle
Ceramic glass hob

Design
Roland Vetter
Robert Bosch Werksdesign
Giengen, Germany

Die Kochstelle PKN 675 T01 von Bosch basiert auf einer High-Speed-Glaskeramik mit angeschliffener Facette seitlich und Edelstahlprofil vorne. Die dekorlose Kochfläche fügt sich harmonisch in die Arbeitsfläche ein, sie ist besonders reinigungsfreundlich. Der Kochzonenspiegel ermöglicht eine freie Handhabung der verschiedenen Kochgeschirrgrößen. Die Kochzonensteuerung erfolgt über Sensortasten (Touch-Control). Die Anzeigefunktionen mit Digitalanzeigen, Restwärmeanzeige, Timer und Kochzonenerweiterung sind vorne im ergonomisch günstigen Bereich angebracht.

The Bosch PKN 675 T01 hob boasts high-speed ceramic glass and facetted sides with a stainless-steel profile at the front. The sleek cooking surface blends harmoniously with the worktop and is exceptionally easy to clean. The cook zone arrangement affords great scope in handling various cooking utensil sizes. The cook-zones are operated via touch-control sensor buttons. The digital display includes functions such as residual heat for each cooking zone, a timer and cook-zone enlargement – all located in the front of the hob for easy ergonomic access.

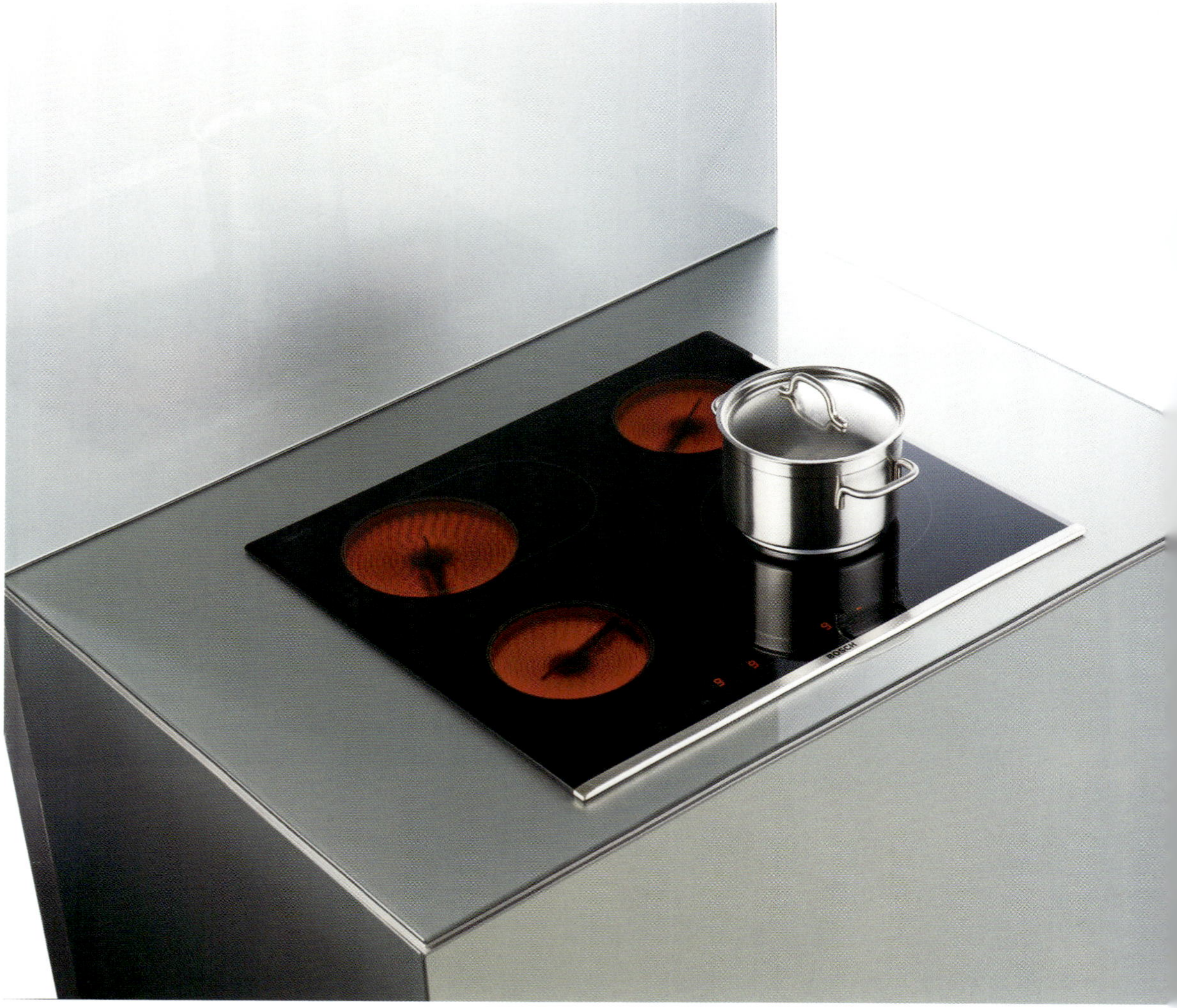

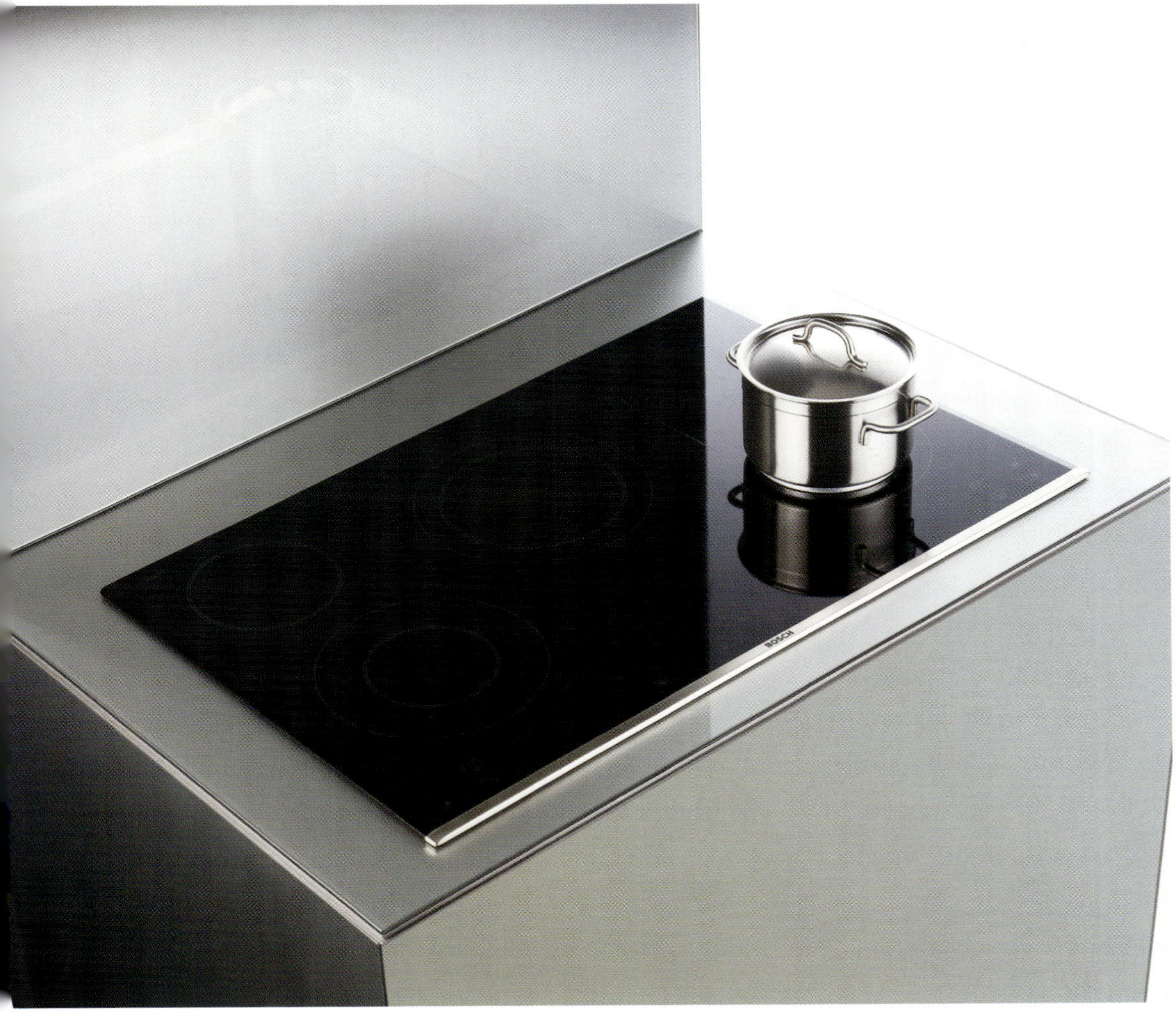

Hersteller Manufacturer
Robert Bosch Hausgeräte GmbH
München, Germany

Produkt Product
Bosch NKC 875 T01
Glaskeramik-Kochfeld
Ceramic glass hob

Design
Roland Vetter
Robert Bosch Werksdesign
Giengen, Germany

Das Kochfeld NKC 875 T01 von Bosch basiert auf einer High-Speed-Glaskeramik mit angeschliffener Facette seitlich und Edelstahlprofil vorne. Die dekorlose Kochfläche fügt sich harmonisch in die Arbeitsfläche ein, sie ist besonders reinigungsfreundlich. Der Kochzonenspiegel ermöglicht eine freie Handhabung der verschiedenen Kochgeschirrgrößen (Zweikreis und Bräter). Die Anzeigefunktionen mit Digitalanzeigen, Restwärmeanzeige, Timer und Kochzonenerweiterung (Touch Control) sind vorne im ergonomisch günstigen Bereich angebracht.

The Bosch NKC 875 T01 hob boasts high-speed ceramic glass and facetted sides with a stainless-steel profile at the front. The sleek cooking surface blends harmoniously with the worktop and is exceptionally easy to clean. The cook zone arrangement affords great scope in handling various cooking utensil sizes (two-ring and roasting zones). The digital display includes functions such as residual heat for each cooking zone, a timer and cook-zone enlargement (touch control) – all located in the front of the hob for easy ergonomic access.

Hersteller Manufacturer
Miele & Cie. KG
Gütersloh, Germany

Produkt Product
KM 484
Elektro-Kochfeld
Electric hob

Design
Miele & Cie. KG
Gütersloh, Germany

Das autarke Elektro-Kochfeld KM 484 mit Comfort Control ist ein Glaskeramik-Kochfeld mit fünf HiLight-Heizkörpern, davon zwei Vario-Kochzonen vorne links und vorne rechts und eine Koch-/Bräterzone hinten rechts. Das Modell bietet mit 930 x 514 mm Gerätemaß auch bei gleichzeitiger Nutzung aller Kochzonen optimale Platzverhältnisse. Die Bedien- und Anzeigeelemente sind in dem zentral am vorderen Rand angeordneten Comfort Control Feld zusammengefasst. Die Pultform des Bedienteils gewährt eine gute Ergonomie und Ablesbarkeit. Durch Programmierung kann der Benutzer die Bedienung des Kochfeldes seinen Nutzungsgewohnheiten anpassen. Durch die Abnehmbarkeit des Bedienteils lässt sich das Kochfeld leicht und schnell reinigen.

The standalone KM 484 electric hob with Comfort Control is a ceramic glass hob with five HiLight rings, two of whichh are dual-circuit rings (front left and right) and a casserole zone rear right. The 930 x 514 mm dimensions of the model deliver optimal space usage even when all the cooking zones are actuated at once. The controls and display indicators are concentrated in the Comfort Control zone at the center of the front edge. The inclined controls are ergonomic and easy to read. The hob can be programmed to suit the preferential needs of the user and the control panel can be removed to facilitate cleaning and for storage.

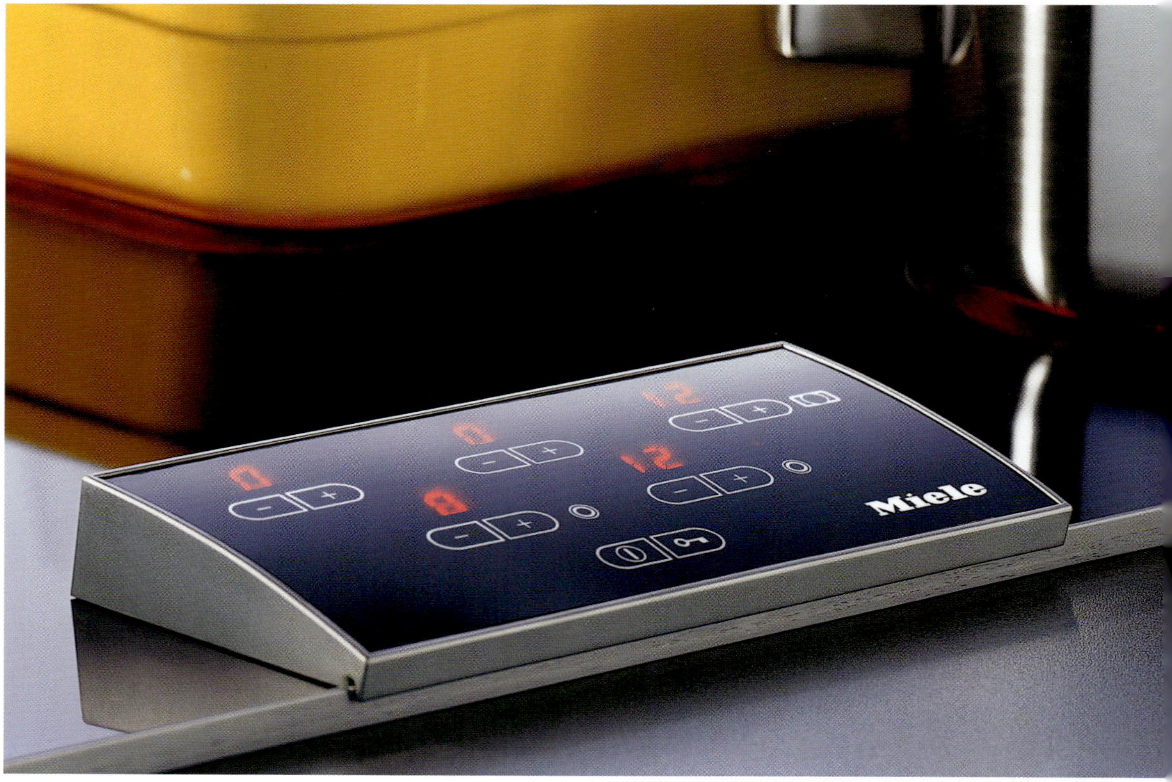

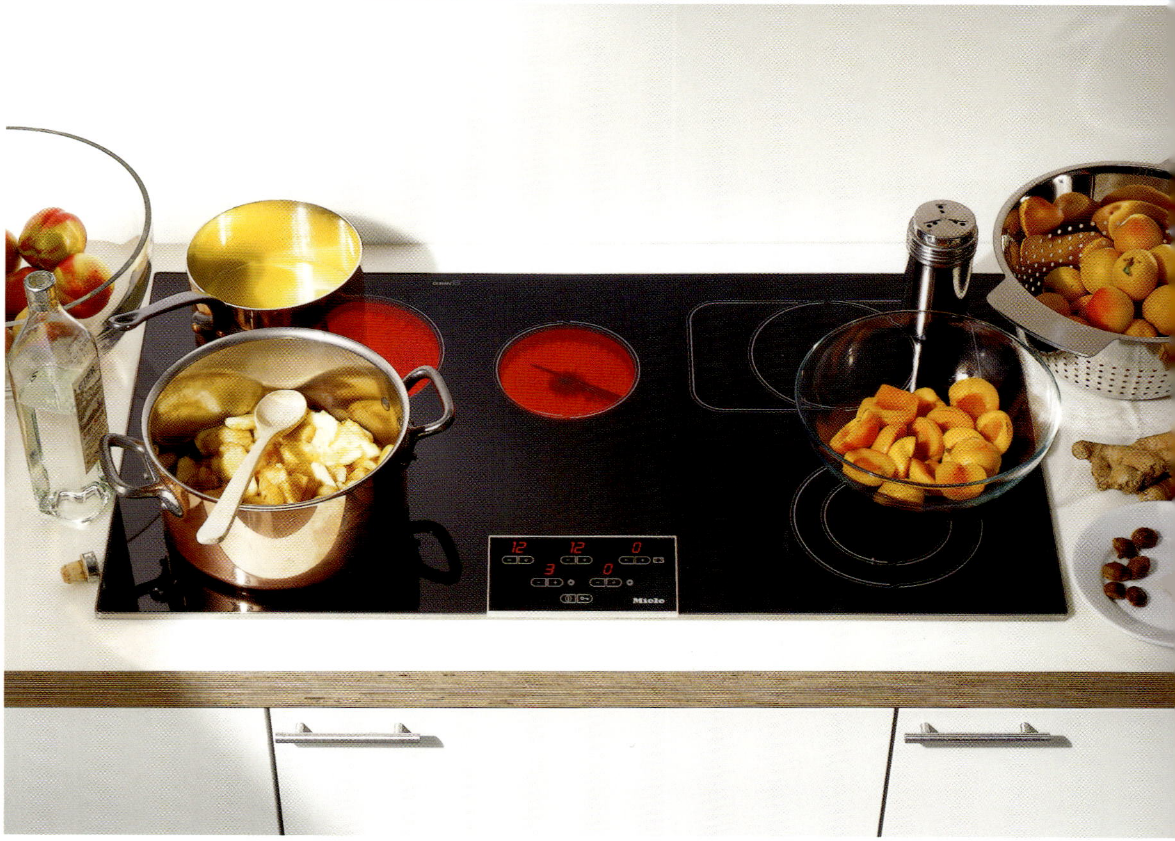

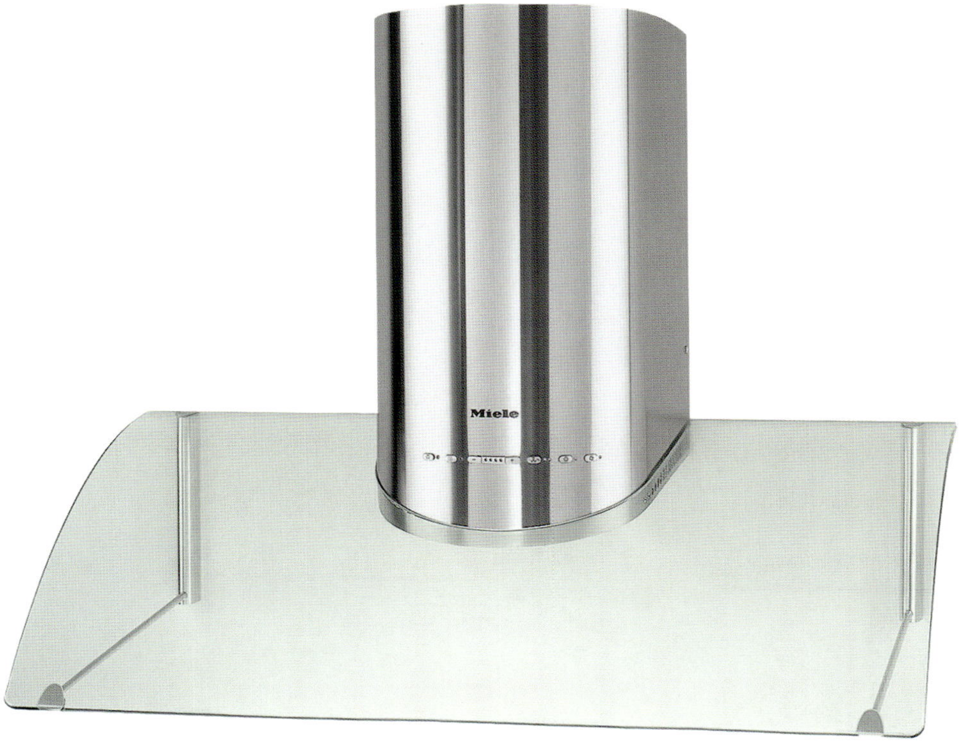

Hersteller Manufacturer
Miele & Cie. KG
Gütersloh, Germany

Produkt Product
DA 449
Wand-Dekor-Abzugshaube
Wall-mounted cooker hood

Design
Miele & Cie. KG
Gütersloh, Germany

Die DA 449 Wand-Dekor-Abzugshaube von Miele wirkt durch den seitlich offenen, zweidimensional verformten Glasschirm besonders leicht. Unterstützt wird der Eindruck durch die – in der Kontur des Kamins ausgeführten – Halteprofile mit Verbindungsstangen aus Edelstahl. Die Gestaltung des Filterkorbes nimmt folgerichtig die Linienführung des Schirms auf und betont insgesamt die Kernfunktion dieses Produktelements, nämlich die Wrasenaufnahme. Diese Designlinie verfügt über eine hochwertige Elektronik, doppelseitig saugendes Gebläse und dimmbare Halogenbeleuchtung. Die Bedienelemente befinden sich im Kamin, so dass sich eine optimale Handhabung ergibt.

The Miele DA 449 wall-mounted cooker hood has a particularly graceful appearance thanks to the two-dimensional shaped glass canopy, which is open at the sides. The lightness is accentuated by the wall brackets, which with their stainless steel connecting rods are integrated into the contour of the hood. The design of the filter basket follows the lines of the canopy and essentially emphasizes its core function as an extractor for cooking vapors. The model boasts high-performance electronics, a double-edge-action fan, and dimmable halogen lighting. The controls are located in the chimney section to enhance ease of operation.

Hersteller Manufacturer
Miele & Cie. KG
Gütersloh, Germany

Produkt Product
Master Chef Ovens
30 Zoll Backöfen für Nordamerika
30" oven for North America

Design
Miele & Cie. KG
Gütersloh, Germany

Die Master Chef Ovens Kollektion wurde speziell für den nordamerikanischen Markt entwickelt. Im Vordergrund des Designkonzepts stand die Benutzerfreundlichkeit. Es wird keine Gebrauchsanweisung benötigt, um den Backofen zu bedienen. Gourmetköchen bietet er hohe Leistung, Anfängerköche können den Ofen intuitiv bedienen. Das symmetrische, ausgeglichene Design verbindet insbesondere in der Gestaltung des Rahmens zeitgenössische mit klassischen Merkmalen. Die Kollektion ist in Edelstahl, Schwarz oder Weiß mit einer großen Auswahl von Griffen erhältlich.

The Master Chef Ovens collection was especially developed for the North American market. The design concept focuses primarily on user-friendliness - the fully-programmable oven can even be used without instruction manuals. It offers high performance for gourmet chefs but can also be used intuitively by amateurs. The symmetrically balanced face is support by a combination of contemporary and classic features, particularly in the frame. The smart ovens are available in stainless steel, black or white and with multiple handle options, ranging from sleek to professional.

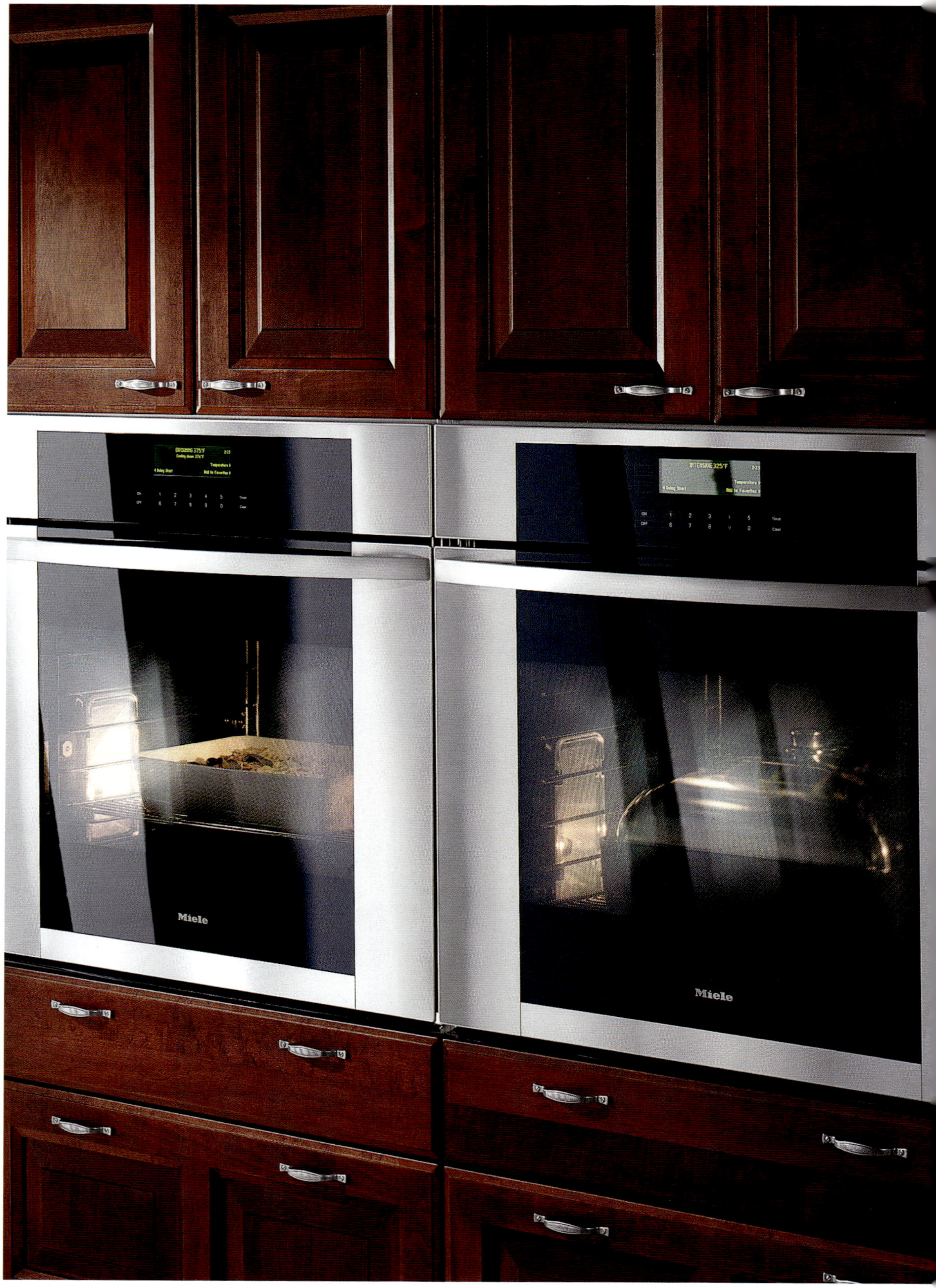

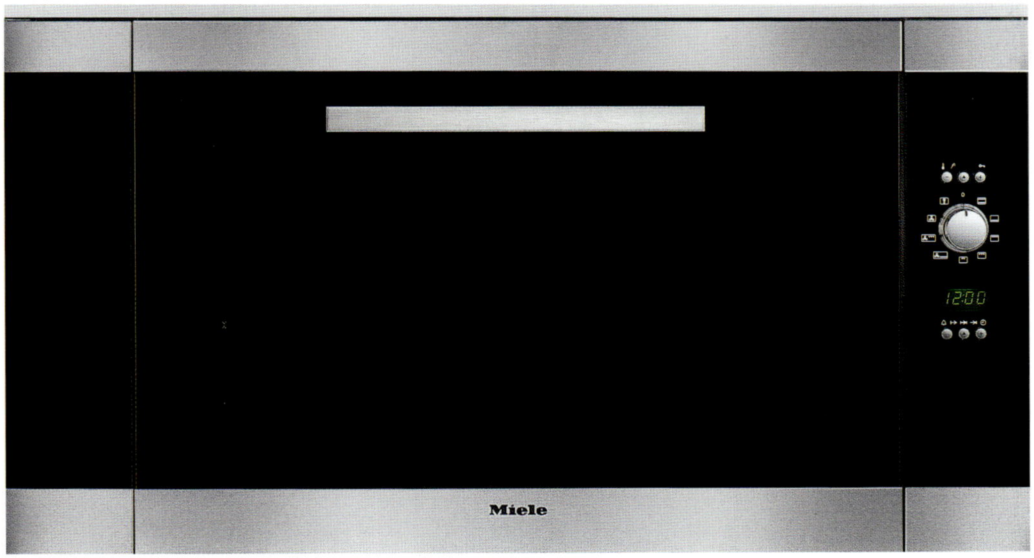

Hersteller Manufacturer
Miele & Cie. KG
Gütersloh, Germany

Produkt Product
H 4900 B
Einbau-Backofen
Wall oven

Design
Miele & Cie. KG
Gütersloh, Germany

Der Einbau-Backofen H 4900 B in Edelstahl ist auf die neue Designlinie für Miele Einbaugeräte abgestimmt. Kennzeichen des Designs sind ruhige Flächen mit einem markanten, farblich kontrastierenden Griff. Neun verschiedene Betriebsarten können gewählt werden: Heißluft plus, Ober- und Unterhitze, Oberhitze, Unterhitze, Auftauen, Intensivbacken, Umluftgrillen, Grill 1, Grill 2. Zudem verfügt der Backofen über eine elektronische Steuerung und Anzeigen, mit der Backofenprogrammierung können Startvorwahl, Garzeitdauer und Abschaltautomatik gesteuert werden. Eine spezielle Clean-Ausstattung erlaubt leichtes Reinigen.

The H 4900 B built-in stainless steel wall oven matches the new features of the design concept for Miele built-in appliances. The design catches the eye with its calm surfaces and striking color-contrast handle. There are nine different modes of operation to choose from: fan heat, top and bottom heat, or a combination of both, defrosting, intensive baking, fan grill and two normal grill settings. The oven also features electronic controls and displays, and can be programmed for pre-selective starting, cooking duration and automatic shut-off. A special PerfectClean finish and CleanGlass oven door as well as quick-release catches facilitate cleaning.

Hersteller Manufacturer
Siemens Electrogeräte GmbH
München, Germany

Produkt Product
Siemens KF 18 W 420
Einbauweinlagerschrank
Built-in wine cooler

Design
Christoph Becke
Christian Hoisl
Gerd E. Wilsdorf
Designabteilung der Siemens Electrogeräte GmbH
München, Germany

Der KF 18 W 420 Einbauweinlagerschrank von Siemens aus Edelstahl besitzt eine transparente, hochdämmende Isolierglasscheibe. Der Volledelstahlgriff korrespondiert mit den anderen Siemens-Einbaugeräten. Die Innenausstattung bietet mit ihren aus Aluminiumprofilen und transluzenten Kunststoffhalbschalen bestehenden Fachböden Raum zur Lagerung von über 20 Flaschen Wein, wobei eine elektronisch gesteuerte Regelung eine optimale Temperatur gewährleistet.

The Siemens KF 18 W 420 built-in wine cooler made of stainless steel incorporates a transparent, highly-insulated, glass plate front. The pure stainless steel handle matches other Siemens built-in devices. The interior houses shelves with an aluminum trim and translucent plastic half-shells that can store more than 20 bottles of wine. The wine cooler includes an electronic regulator to achieve optimal temperatures.

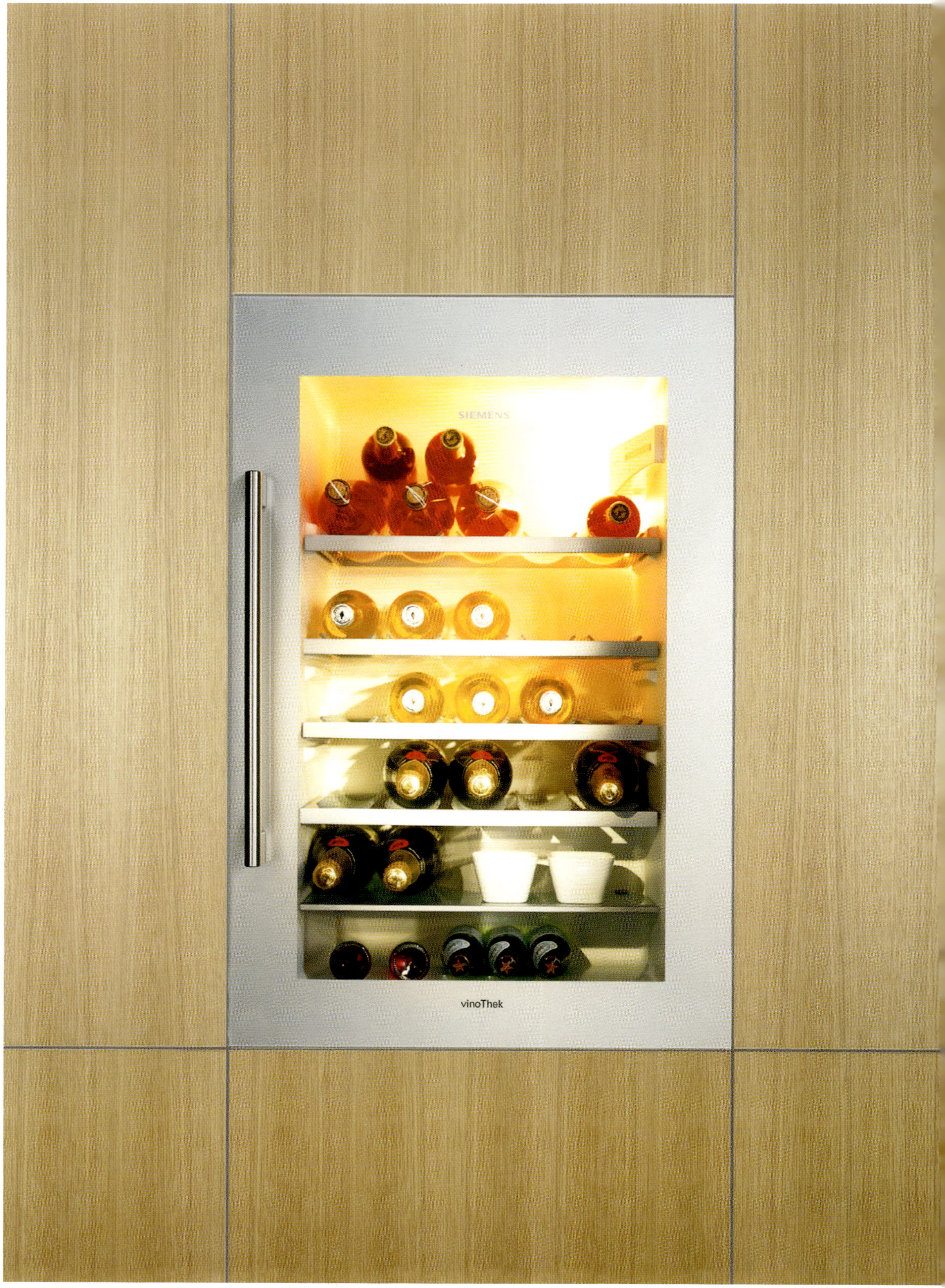

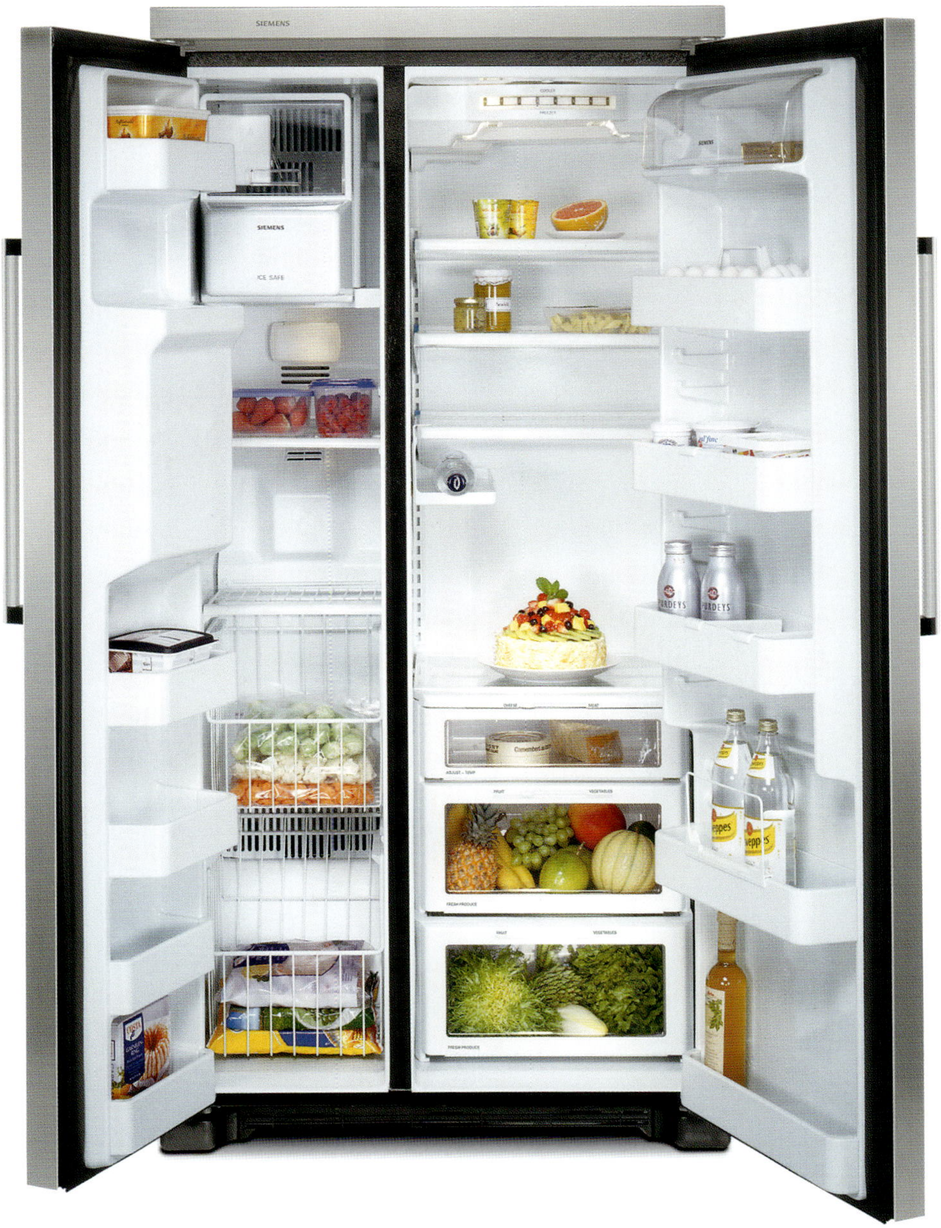

Hersteller Manufacturer
Siemens Electrogeräte GmbH
München, Germany

Produkt Product
Siemens KG 57 U 980
Kühl-Gefrierkombination
Fridge-freezer combination

Design
Christoph Becke
Designabteilung der Siemens Electrogeräte GmbH
München, Germany

Die Kühl-Gefrierkombination Siemens KG 57 U 980 besitzt einen integrierten Wasser- und Eiswürfelspender. Das Gerät ist komplett in Aluminium gefasst, die Griffe sind als massive Aluminium-Stangengriffe ausgebildet. Das Material bietet durch die speziell glasperlengestrahlte, eloxierte Oberfläche hohe Kratzfestigkeit und Unempfindlichkeit gegen Fingerabdrücke. Das Design ist zeitlos und zurückhaltend, eine auf alles Modische verzichtende Formensprache ermöglicht eine gute Kombination mit anderen Siemens-Aluminium-Einbaugeräten.

The Siemens KG 57 U 980 fridge-freezer combination includes an integrated water and ice-cube dispenser. The entire casing is made of aluminum, with solid aluminum bar handles. With its bead-blasted coated surface, the material provides a high level of resistance to both scratches and fingerprints. The design of this standalone is timeless and restrained, avoiding fashionable frills in its formal language and harmonizing well with other Siemens aluminum built-in devices.

Hersteller Manufacturer
Siemens Electrogeräte GmbH
München, Germany

Produkt Product
Siemens SE 7TI
Einbau-Geschirrspüler
Built-in dishwasher

Design
Wolfgang Kaczmarek
Designabteilung der Siemens Electrogeräte GmbH
München, Germany

Der Siemens SE 7TI ist ein vollautomatischer Einbau-Geschirrspüler aus der HiSense-Serie in Edelstahlausführung. Die Edelstahlvortür ist als Sonderzubehör erhältlich. An der Stirnseite der Tür befindet sich das Top-Bedienfeld mit Touch-Bedienzonen und Klartextanzeigen. Ein zweites Display an der Frontseite der Tür zeigt während des Betriebes die Restzeit an. Er ist besonders leise und damit vor allem für offene Küchengrundrisse geeignet. Ein höhenverstellbarer Oberkorb erhöht die Variabilität der Beladung. Das Servo-Schloss lässt sich durch den Metall-Stangengriff besonders komfortabel bedienen.

The Siemens SE 7TI is a fully automatic, stainless steel built-in dishwasher in the company's HiSense line. It features a stainless steel front door as an option, and comes with a smooth top control field on the front side of the door, complete with touch-control tones and clear text displays. A second display on the front of the door shows how much time the cycle still requires. The dishwasher runs extremely quietly and is thus ideal for open-plan kitchens. A height-adjustable top basket allows more flexibility in loading while the servo lock and metal bar handle make for particularly comfortable use.

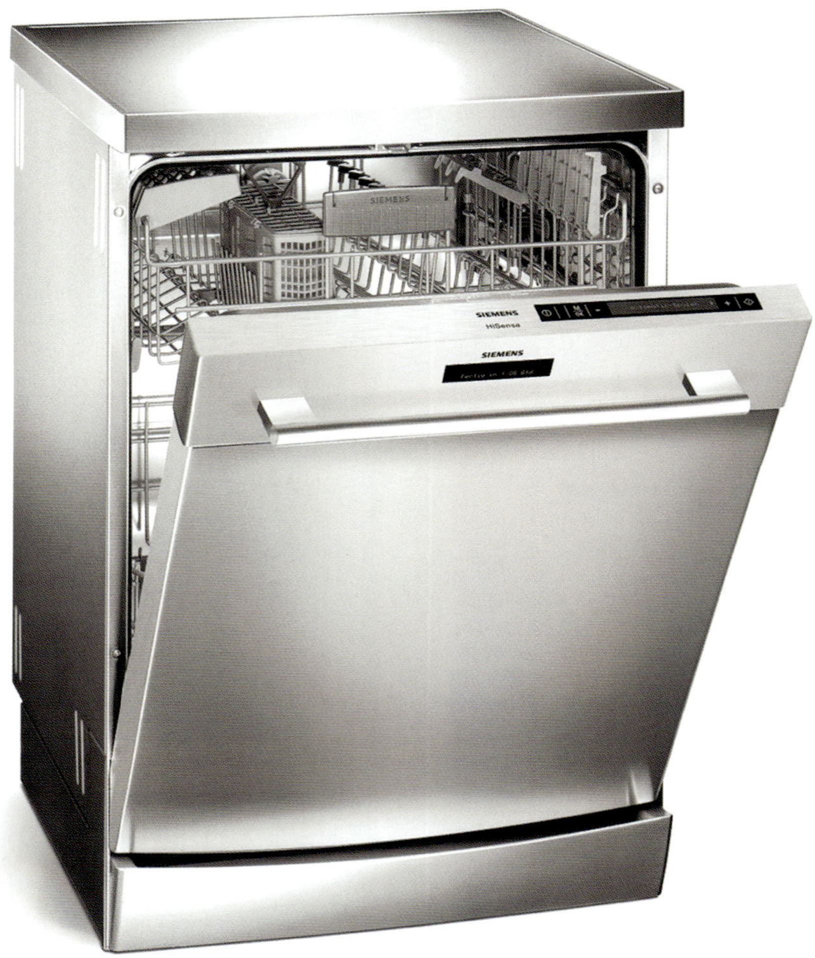

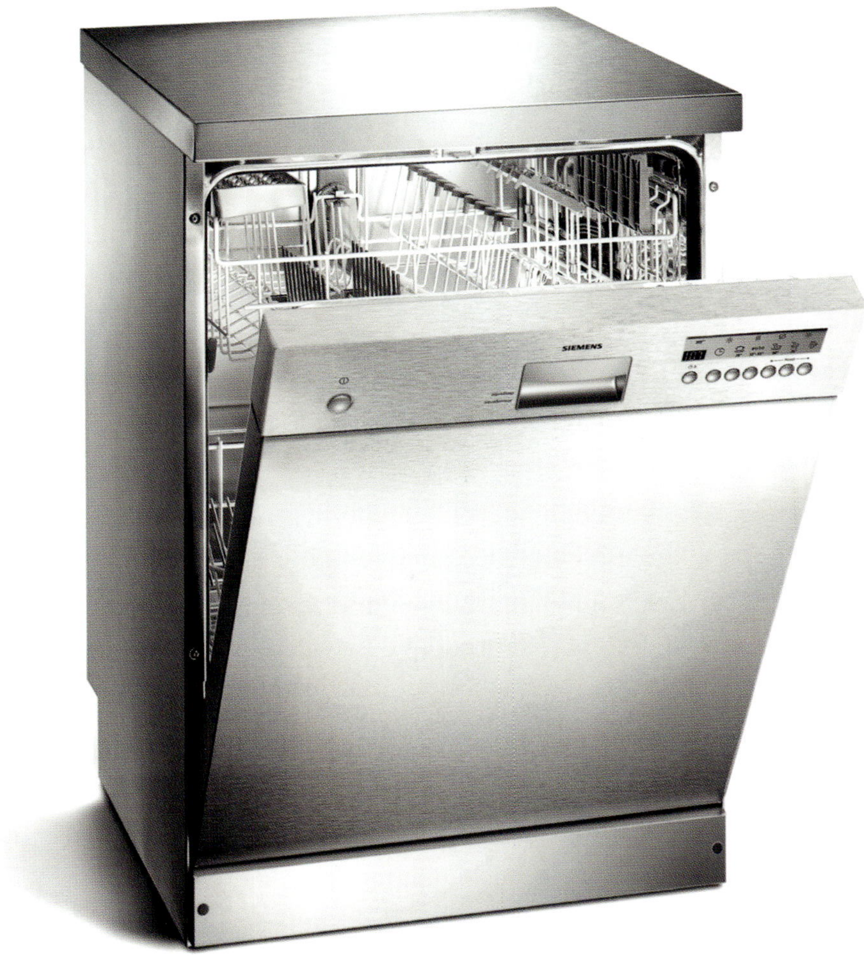

Hersteller Manufacturer
Siemens Electrogeräte GmbH
München, Germany

Produkt Product
Siemens SE 25A590
Geschirrspüler
Dishwasher

Design
Wolfgang Kaczmarek
Designabteilung der Siemens Electrogeräte GmbH
München, Germany

Der Siemens Geschirrspüler SE 25A590 ist ein neuer Standgeschirrspüler in Volledelstahlausführung. Er ist besonders leise und deshalb besonders für offene Küchengrundrisse geeignet. Der Programmablauf, Restzeit und Mangelanzeigen werden in einem Fenster gezeigt. Ein höhenverstellbarer Oberkorb erhöht die Variabilität der Beladung. Das Servo-Schloss schließt die Tür sanft und fast von alleine – ein leichter Druck genügt.

The Siemens SE 25A590 dishwasher is a new, free-standing dishwasher made fully of stainless steel. It runs very quietly, making it particularly suitable for open-plan kitchens. Operations as well as remaining cycle time and low-level indicators are all displayed in one and the same window. The height-adjustable upper rack allows for more flexible loading. To close the door, all that is needed is a light touch and the servo locks clicks it gently into place.

Hersteller Manufacturer
Siemens Electrogeräte GmbH
München, Germany

Produkt Product
Siemens HM 19650
Range-Cooker, Gasherd, modul-line
Free standing range cooker, gas cooker

Design
Frank Rieser
Gerd E. Wilsdorf
Designabteilung der Siemens Electrogeräte GmbH
München, Germany

Der freistehende Siemens modul-line-Gasherd Range-Cooker HM 19650 ist 900 mm breit. Er verfügt über fünf Kochstellen und einen großzügig dimensionierten Elektro-Backofen. Die Edelstahl-Knebel bilden einen ästhetischen Kontrast zur geschliffenen Edelstahloberfläche. Der massive Edelstahlstangengriff passt zum Siemens Einbauherdeprogramm und das große Türfenster ermöglicht einen guten Einblick in den Garraum. Die Füße sind ebenfalls aus Edelstahl und höhenverstellbar.

The Siemens free-standing gas range cooker, HM 19650 is 900 mm wide and boasts five gas cooking zones and a large electric oven. Stainless steel knobs contrast aesthetically with the polished stainless steel surfaces. The solid stainless steel bar handle matches the Siemens built-in cooker line, and the large oven window offers a good view of the inside. The feet are also made of stainless steel and can be height-adjusted.

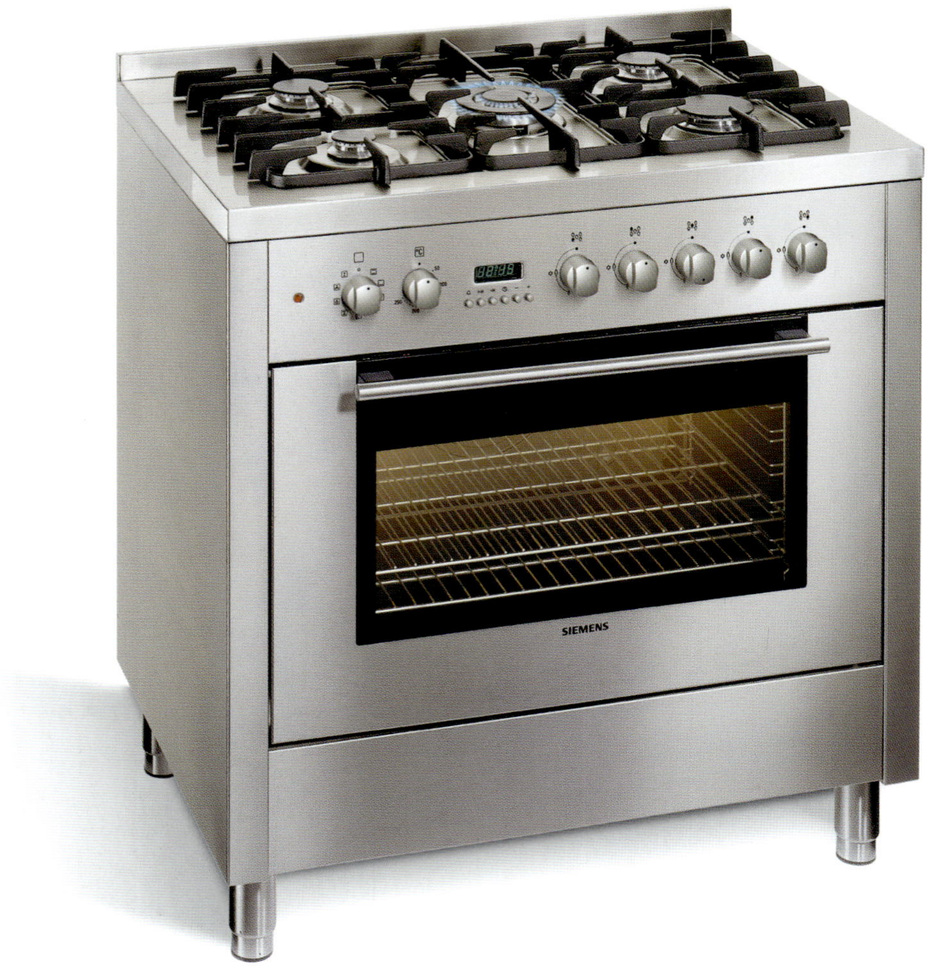

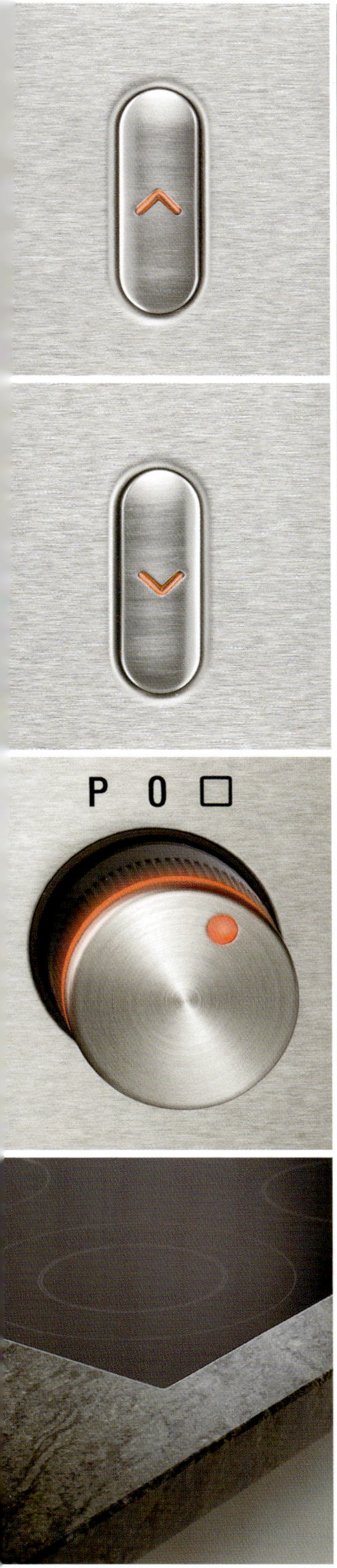

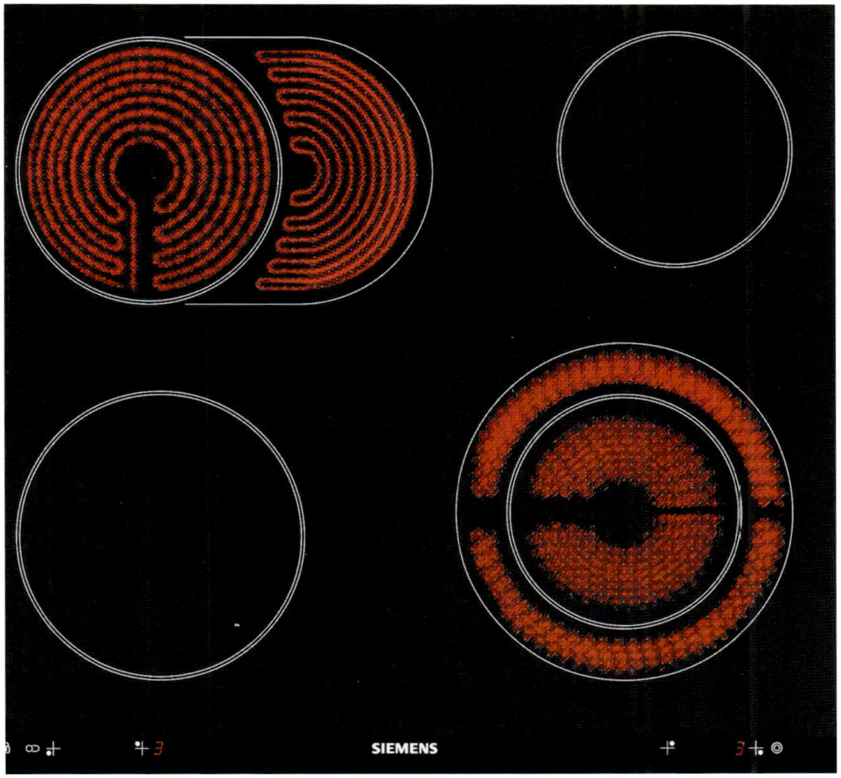

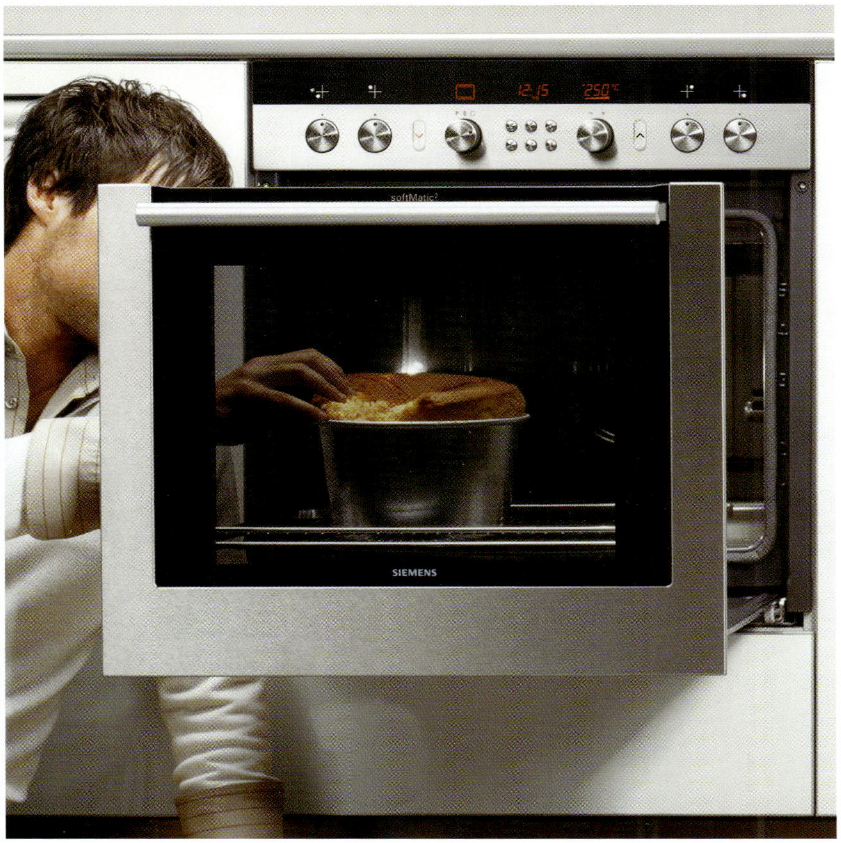

Hersteller Manufacturer
Siemens Electrogeräte GmbH
München, Germany

Produkt Product
Siemens HE 365 560, EF 735 001
Kocheinheit, Einbauherd, Glaskeramik-Kochfeld
Cooking center, built-in cooker, ceramic hob

Design
Gerd E. Wilsdorf
Designabteilung der Siemens Electrogeräte GmbH
München, Germany

Die Siemens Kocheinheit aus Einbauherd HE 365 560 und flächenbündigem Kochfeld EF 735 001 besitzt einen automatischen Backwagen, der durch leichtes Antippen einer der beiden Oval-Tasten aus- bzw. eingefahren wird. Der Kennlinien gesteuerte Backwagen vermeidet das Überschwappen von Soße o.a. im Garraum. Das Einbau-Kochfeld verfügt über zwei Vario-Kochzonen, Glasschutz-Dekor und Touch-Zuschalttasten. Die Kochstellenregler sind symmetrisch an beiden Seiten angeordnet, während die Backofenfunktionen mittig platziert sind.

The Siemens cooking center is made up of the HE 365 560 built-in cooker and EF 735 001 ceramic hob and boasts an automatic baking transporter which can be directed in or out of the oven by simply pressing one of two oval buttons. The oven carriage is controlled by marking lines, preventing sauces and the like from slopping over into the oven. The built-in hob has two vario cooking zones, protective glass decor, and touch-key controls. The cooking zone controls are arranged symmetrically on both sides, with oven controls positioned in the middle.

Hersteller Manufacturer
Siemens Electrogeräte GmbH
München, Germany

Produkt Product
Siemens ER 71 555 EU
Gaskochstelle
Gas-switch cooktops

Design
Frank Rieser
Designabteilung der Siemens Electrogeräte GmbH
München, Germany

Siemens ER 71 555 EU „flame Tronic" ist eine neue, moderne Gasschalter-Kochmulde mit Glaskeramik-Topsheet und elektronischer Bedienung. Die Gasschalterkochmulde ist mit neuen, sehr flachen und reinigungsfreundlichen Gasbrennern ausgestattet. Durch die niedrige Bauhöhe konnten die Flächenroste besonders flach gestaltet werden. Ein Verschieben des Kochgeschirrs ist somit ähnlich wie bei einer Elektroglaskeramik möglich. Das Kochfeld ist mit einer elektronischen Gashahnsteuerung „flame Tronic" ausgestattet, welche die Benutzerführung über digitale Anzeigentechnik möglich macht. Die Energiemenge ist elektronisch regelbar, was sich gestalterisch durch die Verwendung von Zylindergriffen ausdrückt.

The Siemens ER 71 555 EU "flame Tronic" is a new, modern gas-switch hob with a ceramic top sheet and electronic controls. The gas-switch cooktop is fitted with new, very flat and easy-to-clean gas burners. The low height of the hob allows for a very flat surface grate design. Pots and pans can be moved about on the hob as if they were on a ceramic surface. The cooking zones are fitted with „flame Tronic" electronic gas tap controls, and can be activated using a digital display. Energy output is electronically regulated, as is visualized by the use of cylinder handles which deliver very precise adjustment.

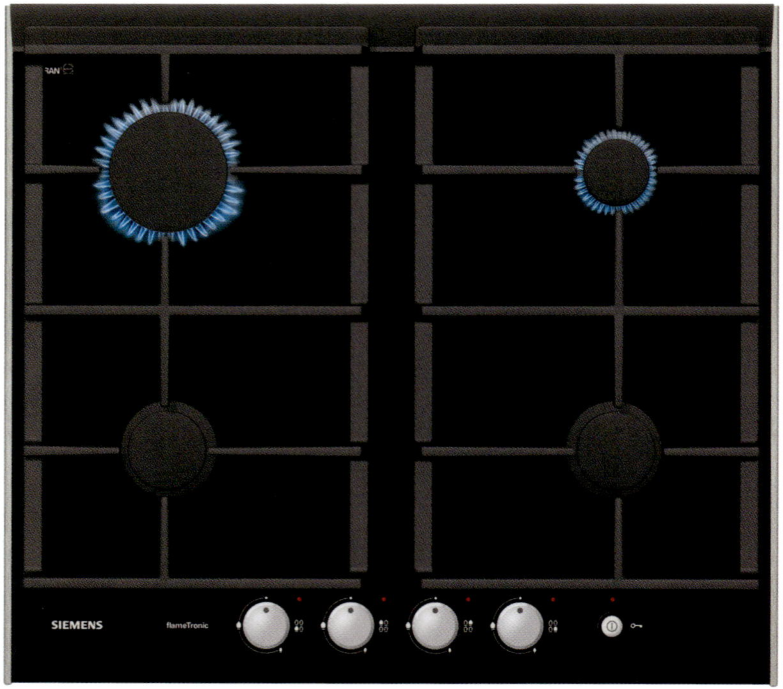

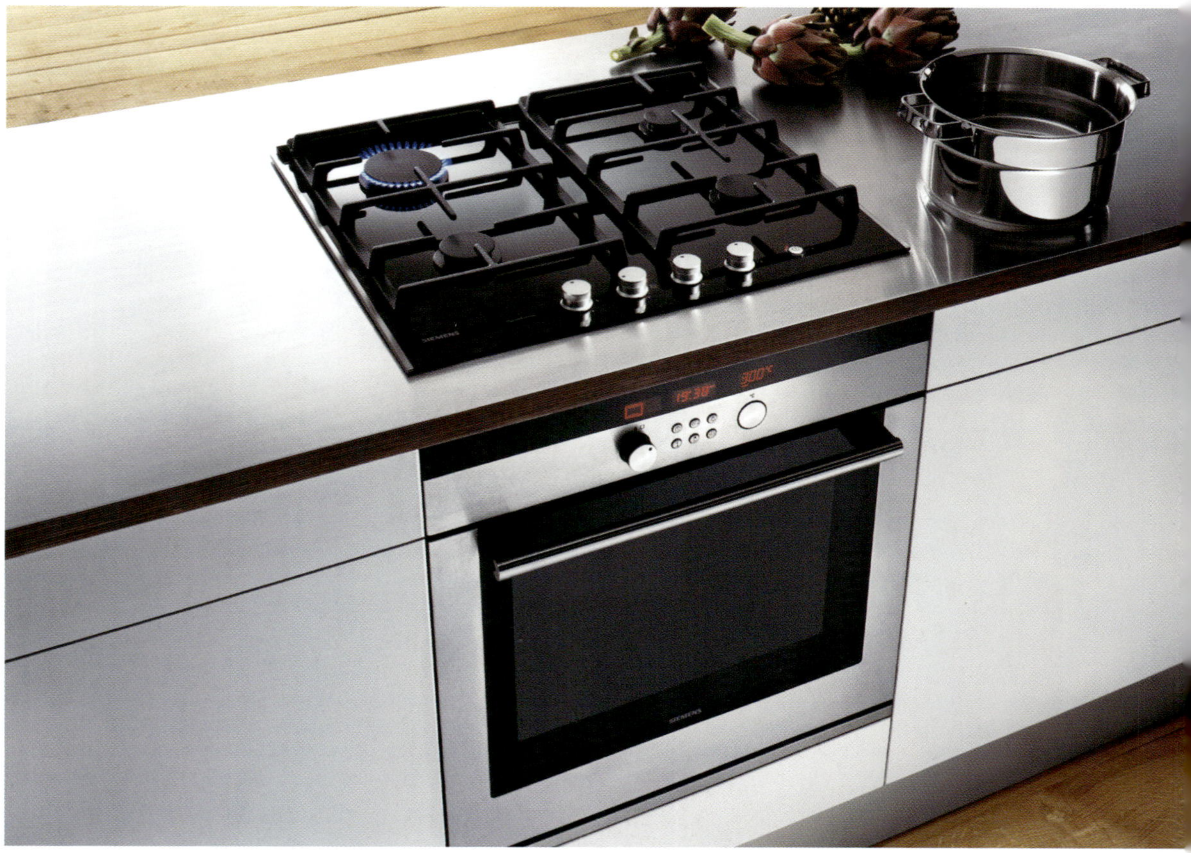

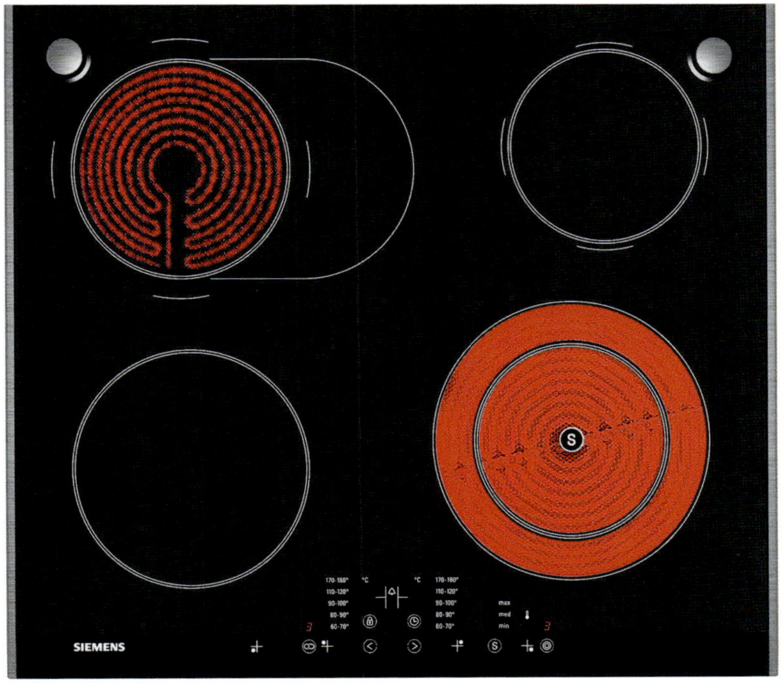

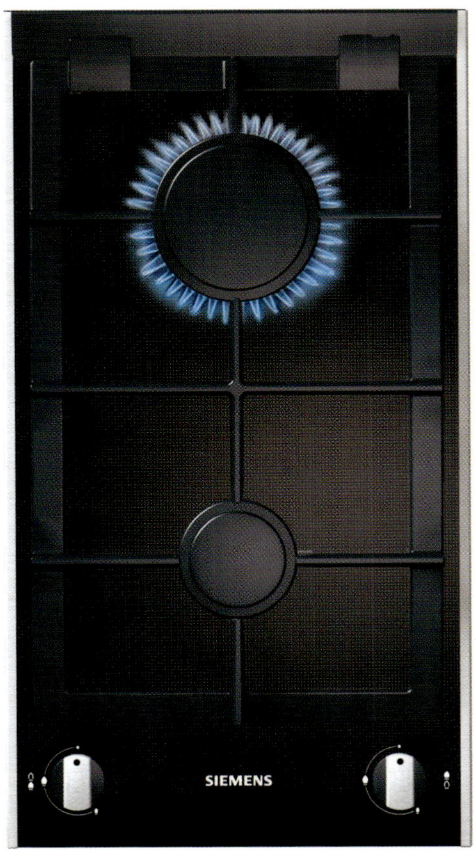

Hersteller Manufacturer
Siemens Electrogeräte GmbH
München, Germany

Produkt Product
Siemens EF 83 H 501, ER 512501
Kochstellenkombination, Domino-Gaskochfeld
Combination built-in hob line

Design
Peter Hamm
Frank Rieser
Gerd E. Wilsdorf
Designabteilung der Siemens Electrogeräte GmbH
München, Germany

Durch ein abgestimmtes Designprogramm lassen sich z.B. das Glaskeramikkochfeld EF 83 H 501 und das Gas-Kochfeld ER 512501 miteinander kombinieren und ermöglichen so eine individuelle Einrichtung der technischen Küchenausstattung. Das Glaskeramikkochfeld besitzt ein Spezialkochfeld mit Koch- und Bratsensorik. Die Bedienung erfolgt über ein Touch-Control-Feld. Das Domino-Gaskochfeld verfügt über zwei Gaskochzonen, und die Brennergestaltung ist besonders reinigungsfreundlich. Ein abnehmbarer Flächen-Gusstopfträger ermöglicht das Verschieben von Töpfen.

A coordinated design system enables, for example, the glass ceramic hob EF 83 H 501 to be combined with the gas hob ER 512501 so that you can individually configure your technical kitchen devices. The ceramic glass hob has a special cooking zone with a boiling and roasting sensor which is controlled by way of a touch panel. The Domino gas hob has two gas cooking zones with high-quality easy-to-clean burners. Pots can be moved around on the new removable cast-iron pot supporter.

Hersteller Manufacturer
Siemens Electrogeräte GmbH
München, Germany

Produkt Product
Siemens LC 89950, ET 907 501
Kocheinheit, Wandesse mit Glaskeramik-Kochfeld
Cooking center, ceramic hob, extractor hood

Design
Peter Hamm
Gerd E. Wilsdorf
Designabteilung der Siemens Electrogeräte GmbH
München, Germany

Die Edelstahl-Wandesse LC 89950 und das Glaskeramik-Kochfeld ET 907501 von Siemens bilden eine Funktionseinheit. Der symmetrische Charakter der 900 mm breiten Wandesse wird durch das Anzeige-Display im Kamin unterstrichen. Die Bedienung ist in den beiden Metallprofilen integriert, welche die Glasplatte seitlich begrenzen. Das 900 mm breite Glaskeramikkochfeld wird über ein Touch-Control-Feld in der angeschliffenen Bedienzone gesteuert. Es verfügt über eine große 3-fach-Kochzone mit max. 280 mm Durchmesser.

The Siemens LC 89950 stainless steel extractor hood and the ceramic glass hob ET 907501 form one functional unit. The symmetrical character of the 900mm wide hood is accentuated by the incorporated display panel. User functions are integrated in the two metal trims edging the glass top. The 900 mm-wide ceramic glass hob is controlled by means of a touch panel located in the polished user zone. It includes a large three-stage cooking zone with a maximum diameter of 280 mm.

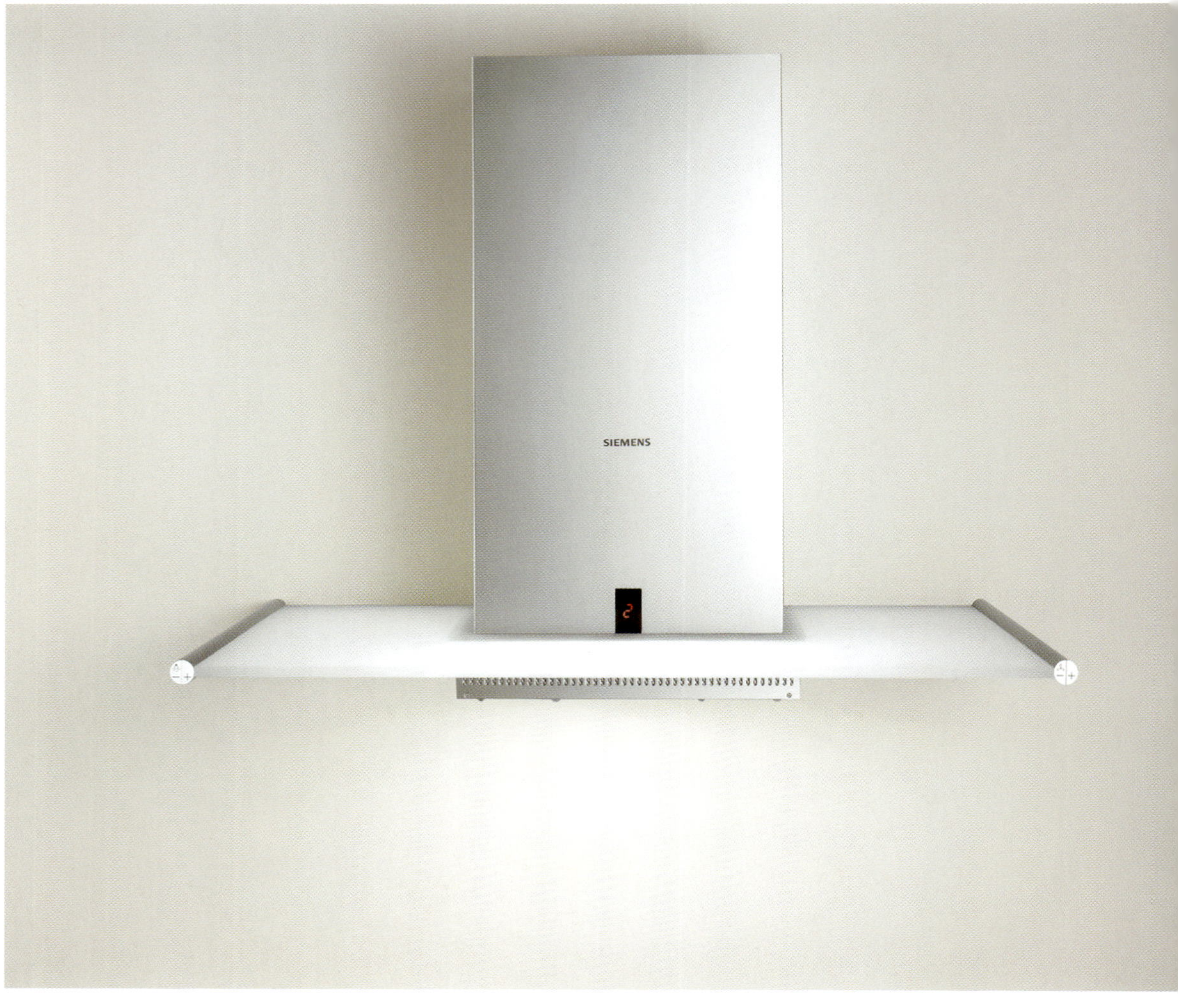

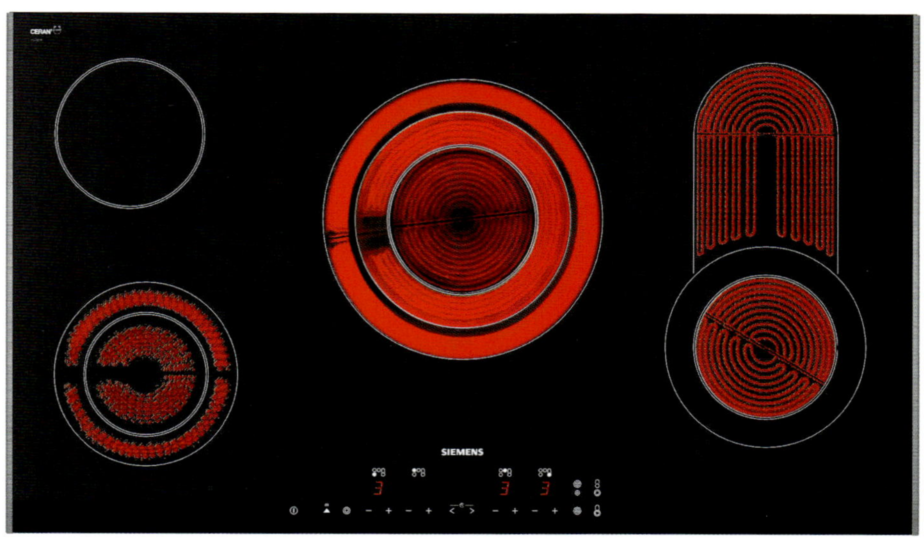

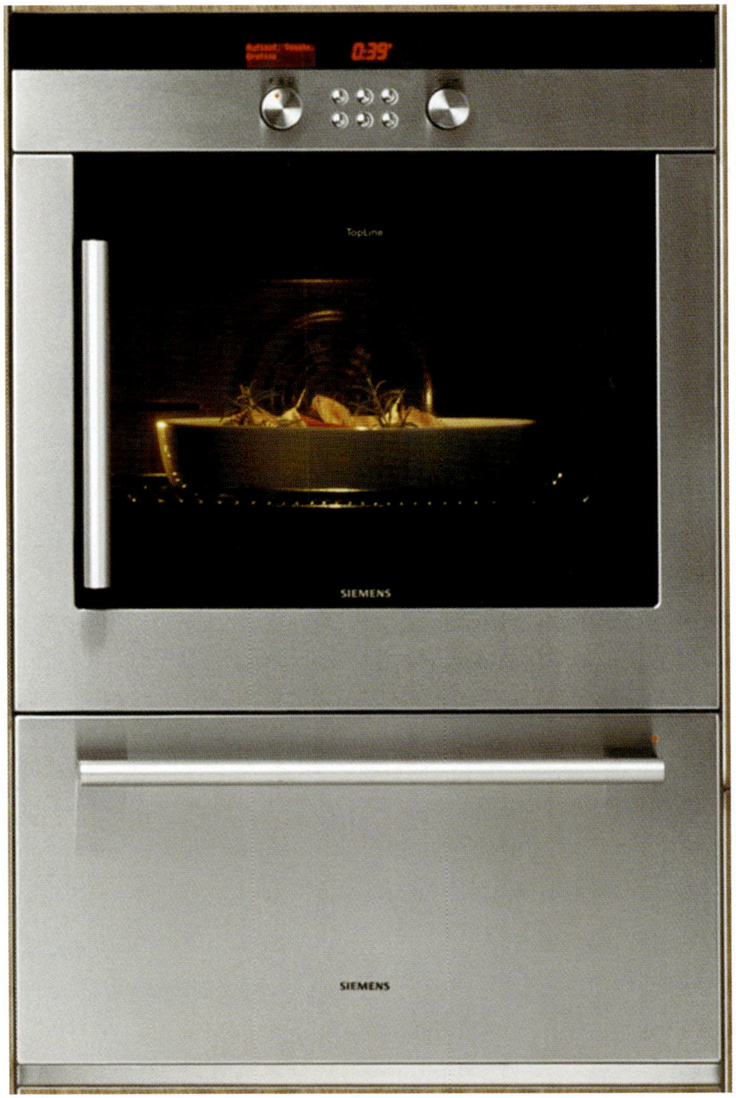

Hersteller Manufacturer
Siemens Electrogeräte GmbH
München, Germany

Produkt Product
Siemens HB 78 L 770
Einbaubackofen
Built-in oven

Design
Gerd E. Wilsdorf
Designabteilung der Siemens Electrogeräte GmbH
München, Germany

Der Siemens HB 78 L 770 Einbaubackofen besitzt eine Drehtür mit Vollmetallgriff für einen barrierefreien Zugriff auf den Garraum, wobei die Tür wahlweise mit Rechts- oder Linksanschlag erhältlich ist. Der große Garraum (65 Liter) ist durch seine Auskleidung mit Titan Glanzemaille leicht zu reinigen. Die symmetrische Anordnung der Backofenbedienung und das große Sichtfenster unterstreichen das klare, funktionale Design des Backofens. Aufgrund der glasperlengestrahlten und eloxierten Alu-Front werden Fingerabdrücke auf der Oberfläche vermieden.

The Siemens HB 78 L 770 built-in oven has a revolving door for unhindered access to the oven inside. The door features a solid metal handle and can be mounted on either the left or the right side. The exceptionally large oven cavity (capacity: 65 liters) is easy to clean thanks to the titanium gloss enamel coating. The symmetrical layout of the oven functions and the large window emphasize its clear, functional design. The coated aluminum front is bead-blasted to prevent fingerprints.

Hersteller Manufacturer
Siemens Electrogeräte GmbH
München, Germany

Produkt Product
Siemens HB 86 E560, HW 140560
Einbaubackofen mit Wärmeschublade
Built-in oven with heat drawer

Design
Frank Rieser
Gerd E. Wilsdorf
Designabteilung der Siemens Electrogeräte GmbH
München, Germany

Der Siemens Einbaubackofen HB 86 E560 „innoWave" bildet mit der Wärmeschublade HW 140560 eine Funktionseinheit mit Edelstahlfront. Der kompakte Einbauofen verfügt über vier Heizarten und 1000 Watt Mikrowelle in 45 cm Bauhöhe. Die 140 mm hohe Wärmeschublade ergänzt die Gerätekombination zum gängigen Einbaumaß von 60 cm Höhe. Die Anordnung der Regler auf der Bedienblende ist symmetrisch, die Bedienung erfolgt über elektronischen Klartext.

The Siemens HB 86 E560 "innoWave" built-in oven together with the heat drawer HW 140560 forms a functional unit with a stainless steel front. The compact 45 cm-high built-in oven incorporates four different heating methods and a 1000W microwave. Together with the 140 mm-high heat drawer, the oven has the conventional rack height of 60 cm. The controls and user panel are symmetrically arranged and boast an electronic clear text display.

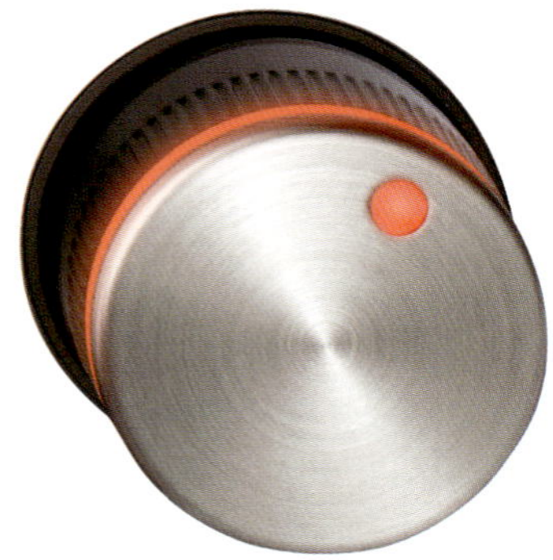

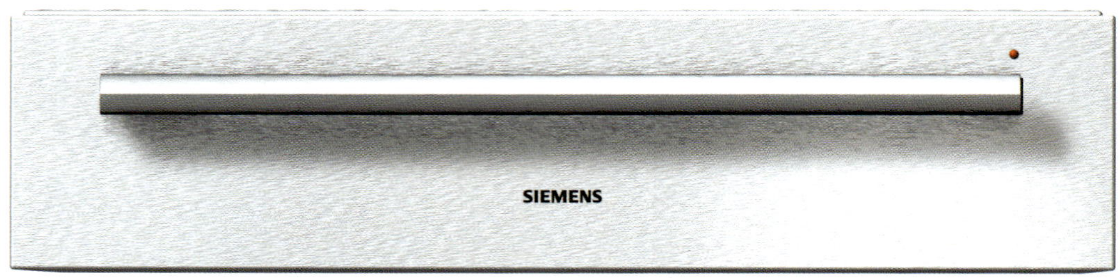

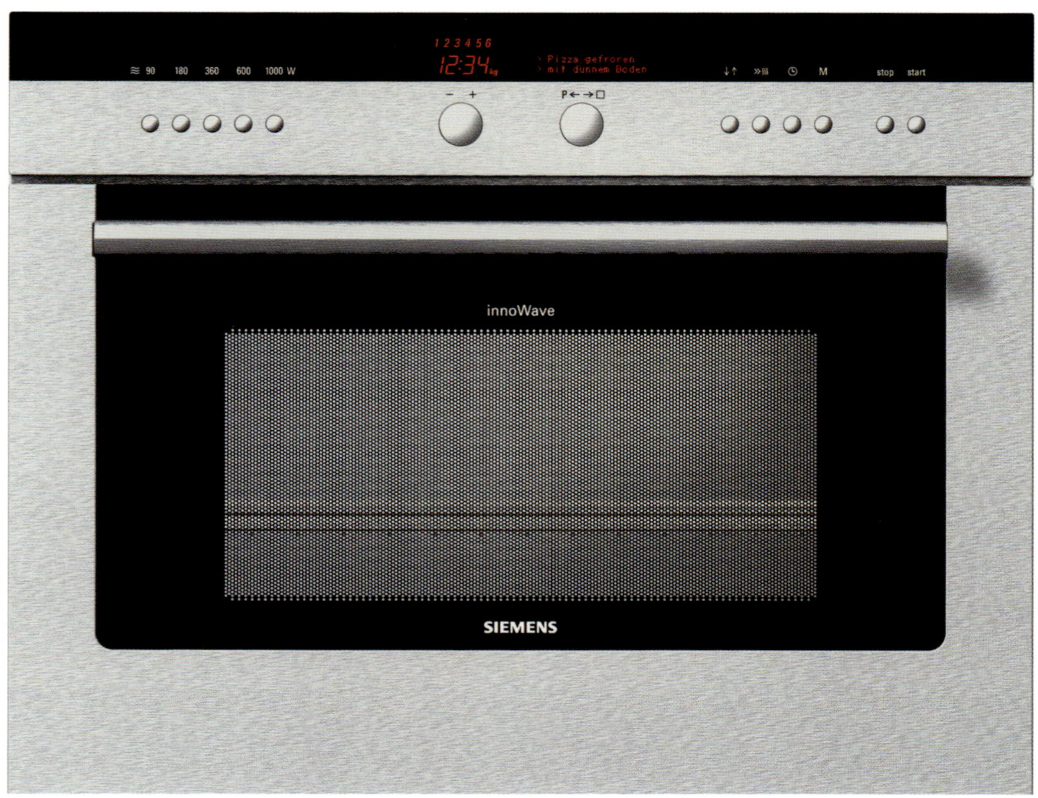

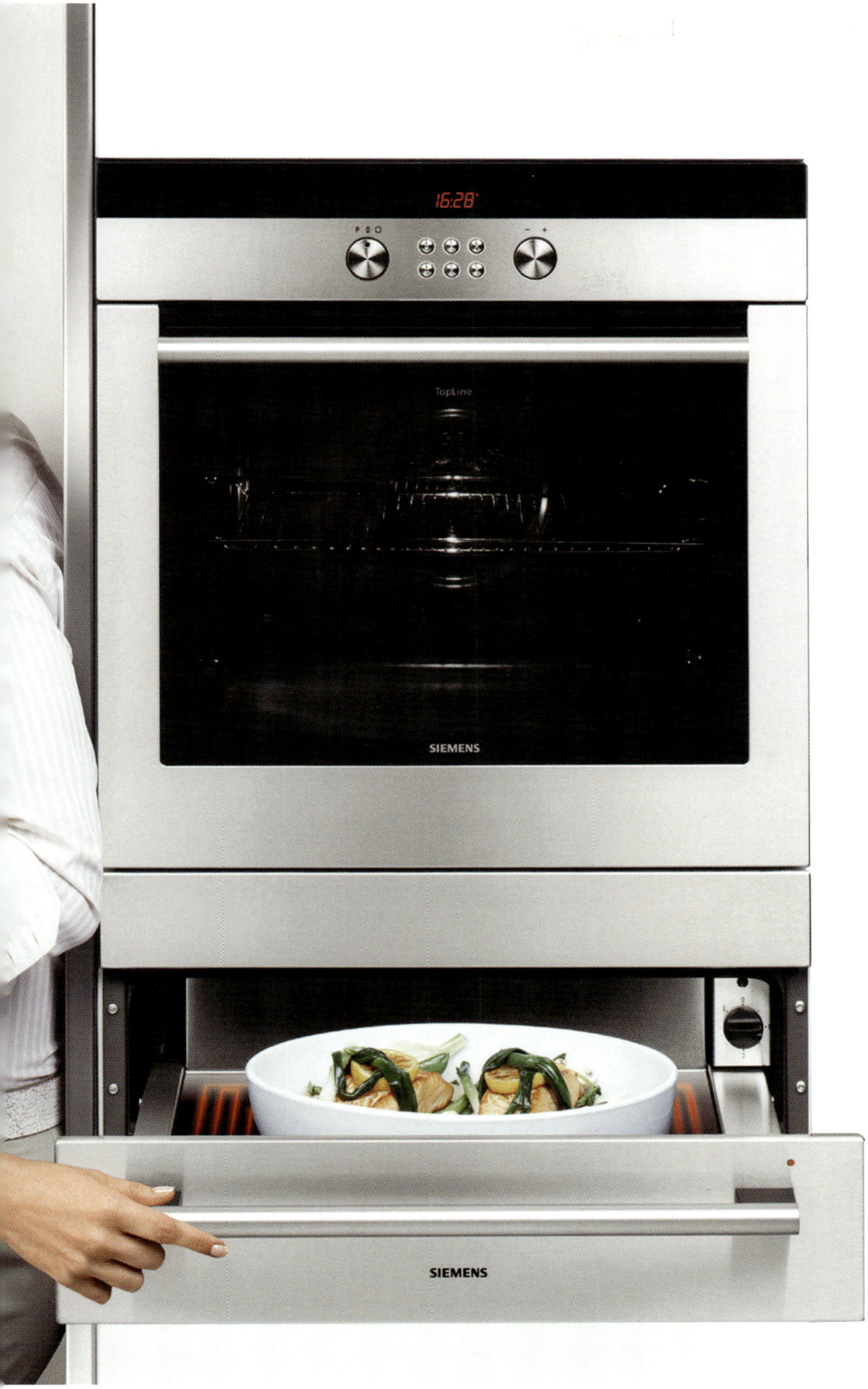

Hersteller Manufacturer
Siemens Electrogeräte GmbH
München, Germany

Produkt Product
Siemens HB 780570, HW 290560
Einbaubackofen mit Wärmeschublade
Built-in oven with heat drawer

Design
Frank Rieser
Gerd E. Wilsdorf
Designabteilung der Siemens Electrogeräte GmbH
München, Germany

Der Siemenss HB 780570 Einbaubackofen ist mit der Wärmeschublade HW 290560 mit identischer Edelstahlfront kombiniert, die mit 290 mm Höhe auch das Warmhalten von ganzen Töpfen ermöglicht. Sie ist mit dem gleichen Metall-Stangengriff ausgestattet wie der Backofen. Das Design des Einbaubackofens zeichnet sich durch eine klare Formensprache aus, die Bedienelemente sind symmetrisch angeordnet und klar von den Anzeigeelementen getrennt. Der Backofen hat ein Innenraumvolumen von 65 Liter und ist mit Titanglanz-Emaille ausgekleidet, was eine leichte Reinigung erlaubt.

The Siemens HB 780570 built-in oven is combined with the heat drawer HW 290560 behind an identical stainless steel front. The heat drawer is 290mm high enabling even large pots to be kept warm. It features the same metal bar handle as the oven. The design of the built-in oven is characterized by a clear formal vocabulary, with controls arranged symmetrically and clearly separated from the display elements. The oven has an interior capacity is 65 liters and is coated with titanium gloss enamel for easy cleaning.

Hersteller Manufacturer
Electrolux Home Appliances China Ltd.
Shanghai, China

Produkt Product
Simba 3 Door Refrigerator BCD-290
Eisschrank
Refrigerator

Design
James Chorlton
Zhao Zhen
Electrolux Industrial Design Centre
Shanghai, China

Im Design des Simba 3 Door Refrigerator BCD-290 wurden europäische und asiatische Einflüsse miteinander verschmolzen. Eine vom europäischen Design inspirierte High-Tech-Anmutung mit klarer und funktionaler Formgebung verbindet sich mit der klassischen chinesischen Vorstellung des Aufeinandertreffens von weichen und harten Formen, wie sie sich insbesondere in der Gestaltung des Griffs und dem runden LCD-Display, das mittig auf der Edelstahltür positioniert ist, niederschlägt. Ein separates Kühlsegment kann zur Lagerung von Wein und Früchten genutzt werden. Die digitale Temperaturkontrolle wählt entsprechend den Anforderungen des Nutzers automatisch die richtige Temperatur.

European and Asian influences marry in the design of the BCD-290 Simba 3-door refrigerator. A high-tech European appearance with a clear and functional form is combined here with the classic Chinese idea of blending hard and soft shapes. This is particularly apparent in the design of the handles and the round LCD panel which is positioned in the middle of the stainless steel door. There is a separate mid-positioned cooling segment which can be used for wine and fruit. The digital temperature control automatically selects the correct temperature in keeping with pre-defined entries made by the user.

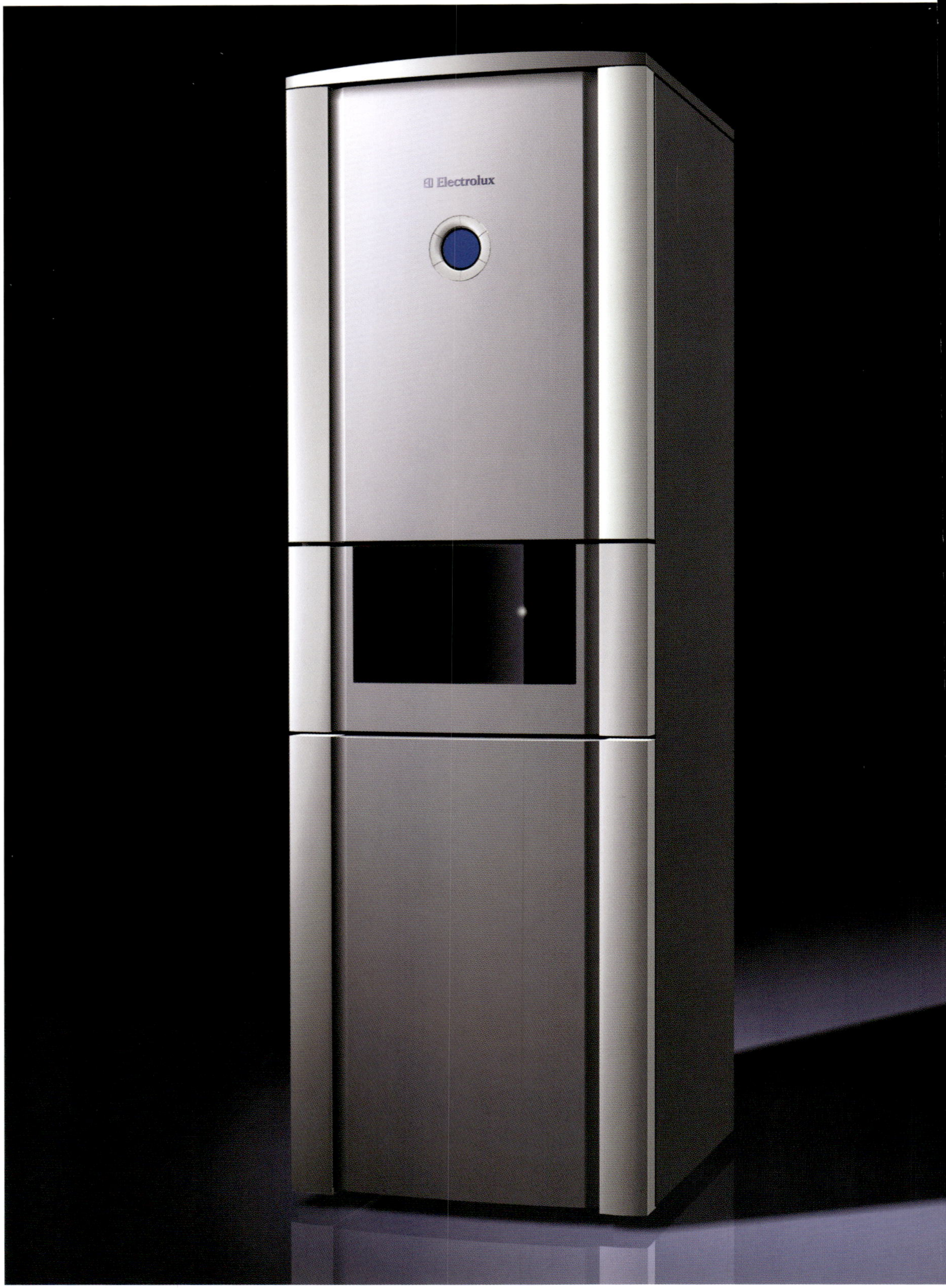

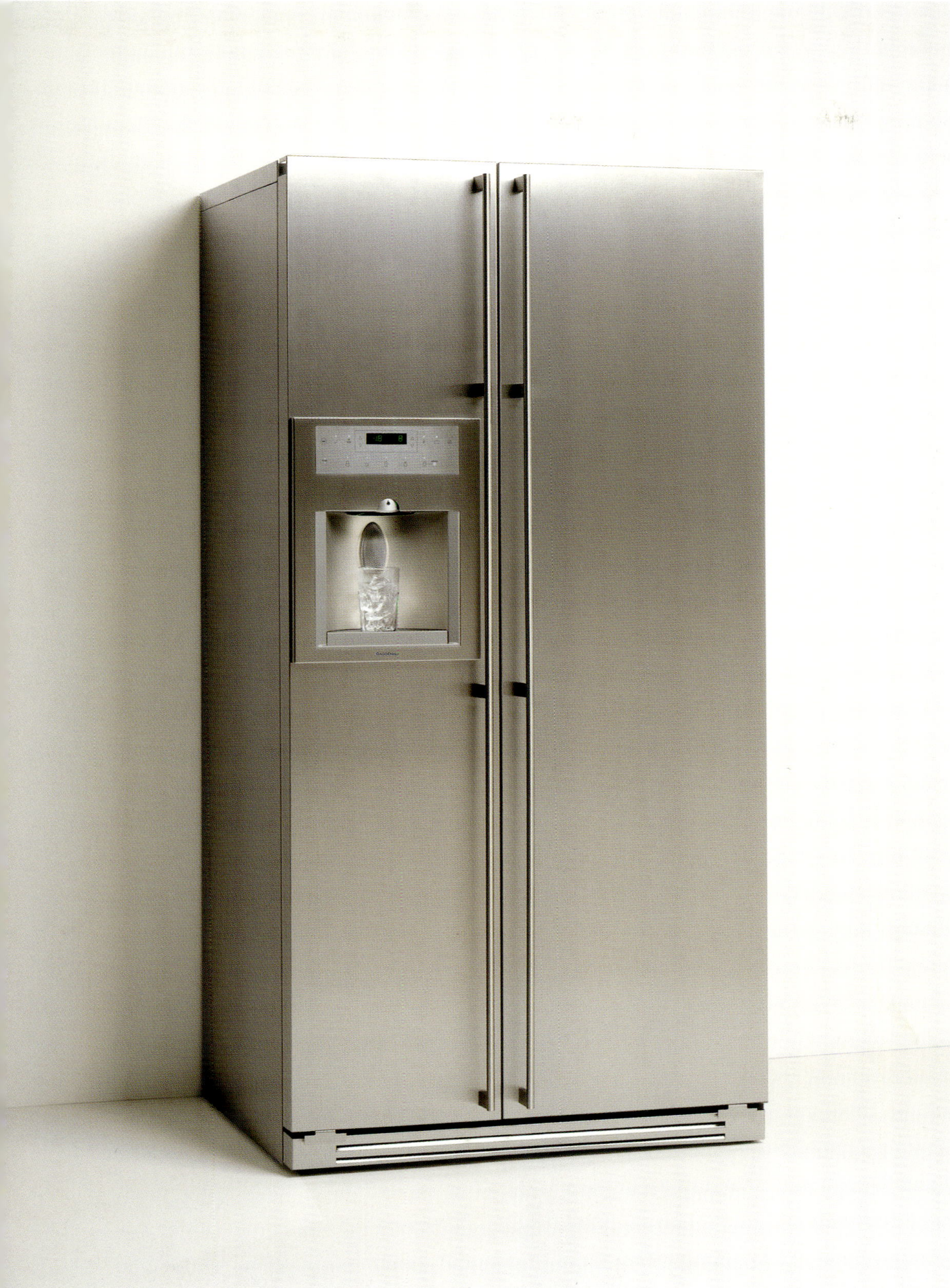

Hersteller Manufacturer
Gaggenau Hausgeräte GmbH
München, Germany

Produkt Product
SK 535
Kühl-Gefrierkombination
Fridge/freezer combination

Design
H. Reinhard Segers
Gaggenau Hausgeräte GmbH
München, Germany

Die SK 535 Kühl-Gefrierkombination mit Frischlagerzone von Gaggenau kann in vorhandene Einbauten integriert oder freistehend verwendet werden. Sie ist mit Edelstahl oder Aluminium vollverkleidet und misst 90 x 178 cm. Die sieben Klimazonen verteilen sich auf fast 600 Liter Nutzinhalt. Die automatische Eiswürfelproduktion gewährleistet einen konstanten, großen Eisvorrat. Über die Bedienelektronik mit digitaler Temperaturanzeige ist eine exakte Temperaturwahl möglich.

The Gaggenau SK 535 fridge/freezer combination can be integrated into existing kitchen storage zones or used as a standalone unit. It is completely covered in stainless steel (aluminum is an option for the standalone version) and measures 90 cm x 178 cm. The even temperature zones are distributed over almost 600l of usable interior space. Automatic ice-cube production ensures a constant supply of fresh ice. Temperatures can be precisely set with the electronics controls and are recorded on the digital temperature display.

Hersteller Manufacturer
Blanco GmbH + Co. KG
Oberderdingen, Germany

Produkt Product
Blanco Solon
Arbeitsplattenintegriertes Abfallsammelsystem
Worktop-integrated waste collection system

Design
Oliver Kuch
Brigitte Ziemann
Blanco GmbH + Co. KG
Oberderdingen, Germany

Blanco Solon ist ein komplett aus Edelstahl gefertigter Abfallsammler für den Einbau in der Küchen-Arbeitsplatte. Der nach oben entnehmbare, großzügig dimensionierte Edelstahleimer eignet sich besonders zur Aufnahme von Bioabfällen. Der Deckel schließt geruchsdicht, die vertiefte Griffmulde ermöglicht ein bequemes Handling. Rand und Tragebügel des Eimers werden durch eine Abdeckung vor Verschmutzung geschützt. Diese schließt durch ihre umlaufende Gummilippe den Eimer im Zusammenspiel mit dem Deckel dicht ab. Die einzelnen Komponenten lassen sich in der Spülmaschine reinigen. Das System ist in drei Ausführungen für den Einbau von oben, flächenbündig und für den Unterbau in Granit konzipiert.

Blanco Solon is a waste collection system made completely of stainless steel and developed for integration into kitchen worktops. The generously sized stainless steel bucket can be removed from above and is particularly suited for collecting organic waste. The lid fits tightly to eliminate odors and the deep recessed handle is easy to grasp. The rim and carrying handle of the bucket are covered with a rubber all-round lip to protect them from soiling and this also seals bucket and lid tightly together. Individual components can be washed in the dishwasher. The system is designed in three versions; built-in from above, integrated flush with the worktop surface (with all-round rim), and built in underneath a granite work surface.

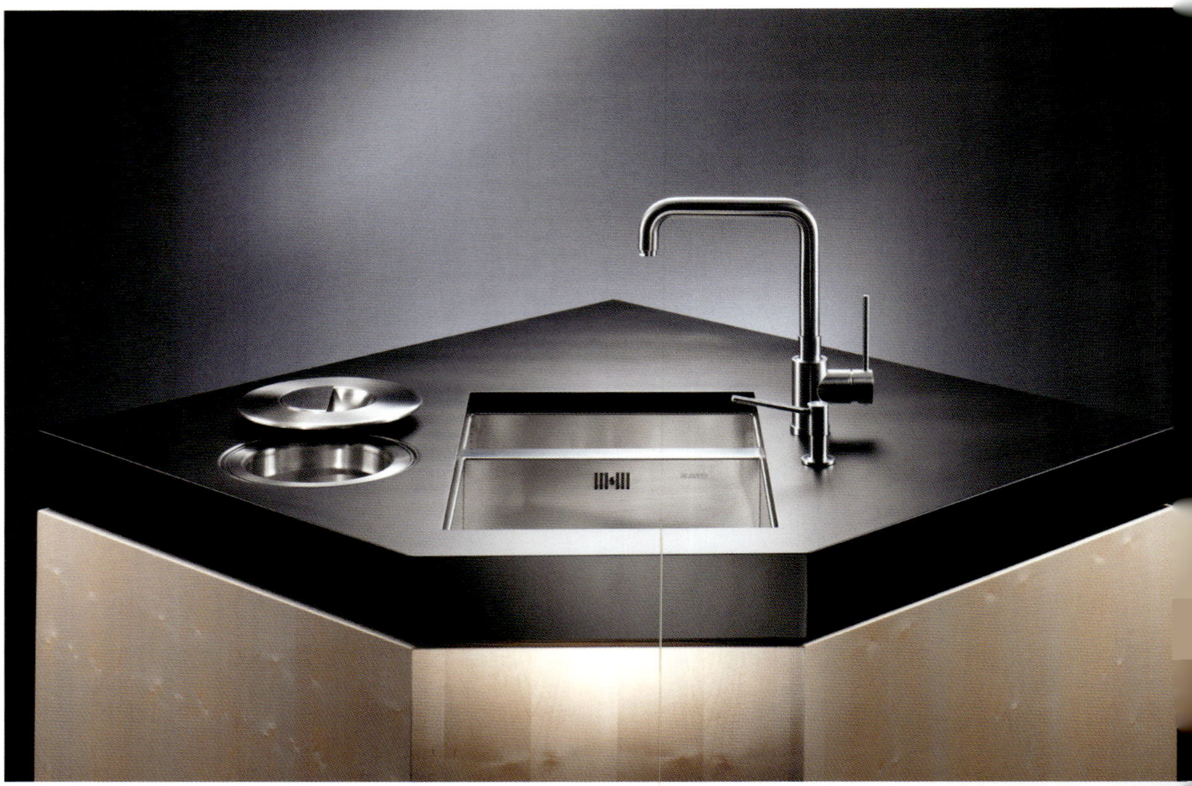

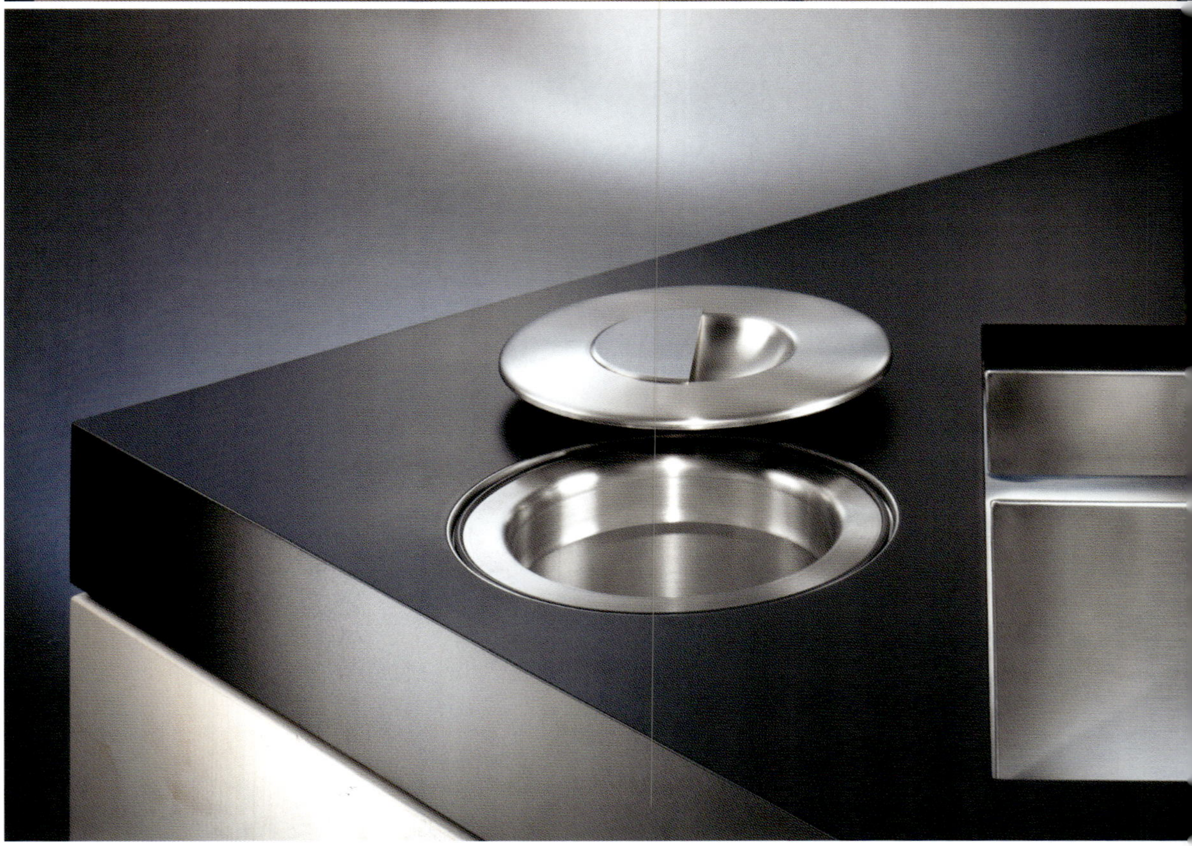

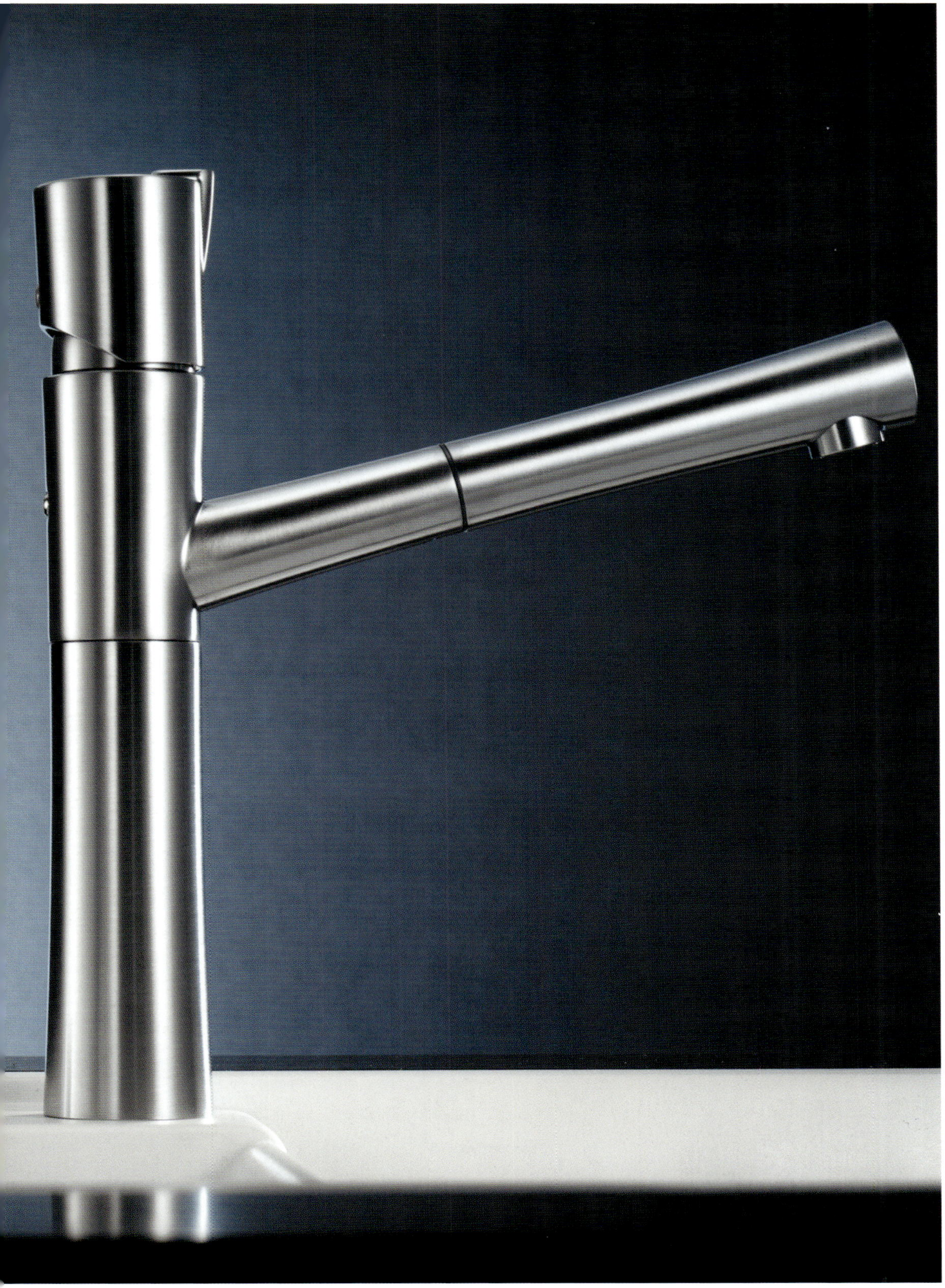

Hersteller Manufacturer
Blanco GmbH + Co. KG
Oberderdingen, Germany

Produkt Product
Blanco Thalos / Blanco Thalos-S
Küchen-Mischbatterie
Kitchen faucet

Design
Oliver Kuch
Brigitte Ziemann
Blanco GmbH + Co. KG
Oberderdingen, Germany

Der Einhebelmischer Blanco Thalos wurde in zwei Varianten – mit Rohrauslauf und ausziehbarer Handbrause – für den Einsatz in der Küche entwickelt. Die Armatur verzichtet auf den klassischen Bedienhebel und wird direkt über die Kappe gesteuert. Eine Ausformung an der Vorderseite der Kappe dient dem Benutzer als optische und haptische Orientierungshilfe zur Regulierung der Wassertemperatur. Der hoch angesetzte Auslauf erleichtert das Befüllen von großen Gefäßen, zudem ist er um 360 Grad schwenkbar. Die Gestaltung der Armatur ist puristisch und setzt einen objektartigen Akzent in der modernen Küche.

The Blanco Thalos single-lever mixer faucet has been developed in two models for kitchen use – with a cylindrical spout and with a telescopic handset. The mixer faucet does not use a classic control lever but is controlled directly via the knob. A small button on the front of the cap acts as a visual and tactile aid for regulating water temperature. The high position of the spout makes it easier to fill large containers and it can also be rotated 360°. The mixer has a sleek and purist appearance, setting an almost sculptural accent for the modern kitchen.

Hersteller Manufacturer
Franke GmbH
Bad Säckingen, Germany

Produkt Product
Corvette CEG 651
Küchenspüle
Kitchen sink

Design
Werner Zemp
Zürich, Switzerland

Die Küchenspüle Corvette CEG 651 besteht aus Fragranit (DuraKleen®), einem hochwertigen Verbundwerkstoff, der zu 80 Prozent aus gemahlenem Granit und zu etwa 20 Prozent aus Acryl sowie weiteren werkstoffoptimierenden Komponenten besteht. Das Material ist komplett durchgefärbt und kratzfest, hitzebeständig, strapazierfähig und pflegeleicht. Die Corvette Küchenspüle besitzt eine betont eigenständige Formgebung: eine sanft abfallende Senke zum Becken harmonisch verlaufende Übergänge, auf der großflächigen Abtropfebene schaffen feine Noppen einen interessanten Kontrast. Als Zubehör für das Modell 651 gibt es ein Abtropfbecken zum Einhängen sowohl für das Arbeits- als auch das Restebecken sowie einen Rüstkorb und ein Schneidebrett aus Polyäthylen.

The Corvette CEG 651 kitchen sink is made of Fragranite (DuraKleen®), a high-grade composite material consisting of ground granite (80 percent) and acrylic (some 20 percent), not to mention other components optimizing its properties. The material is dyed through and utterly scratchproof, heat resistant and easy to clean. The Corvette sink boasts a unique design: a gently reclining slope to the basin, harmonious transitional areas, and on the generously-dimensioned draining board, fine burls create an interesting contrast. Accessories for model 651 include a draining basin that can be suspended from the sink frame either over the work sink or the ancillary sink, as well as a utensils basket and a polyethylene cutting board.

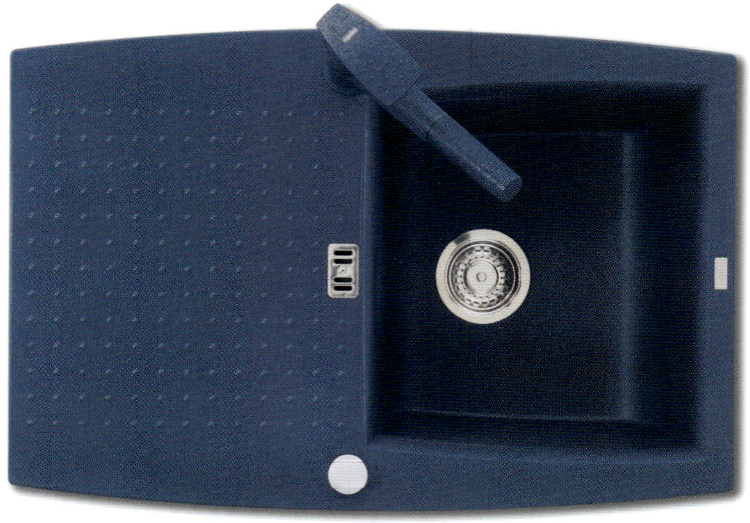

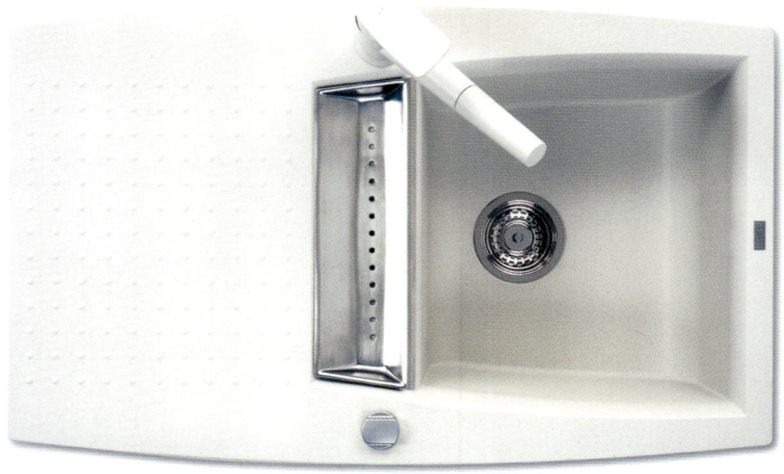

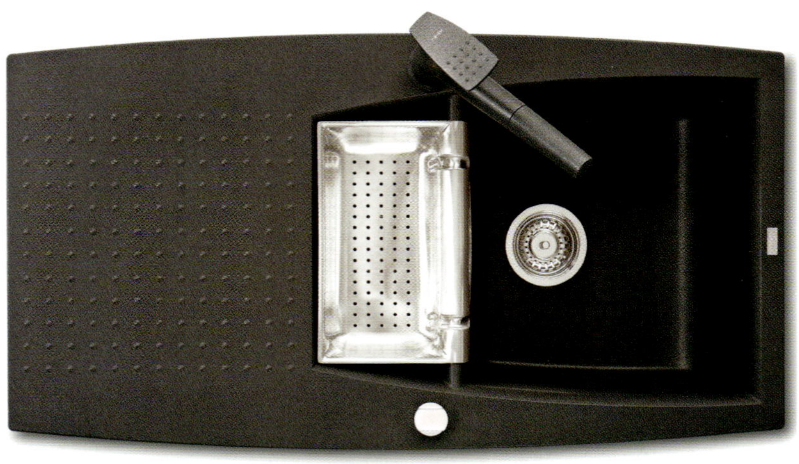

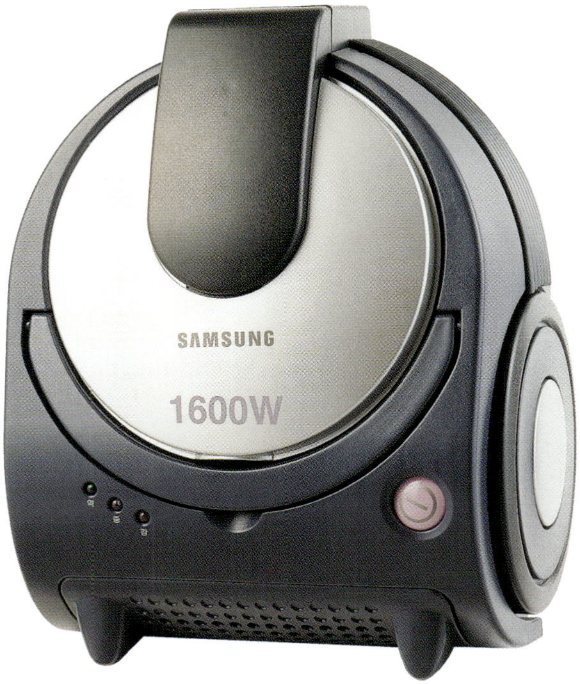

Hersteller Manufacturer
Samsung Electronics Co. Ltd.
Seoul, Korea

Produkt Product
VC-MBP 920
Staubsauger mit Aktivkohle-Filter
Vacuum cleaner with an active carbon filter

Design
Shin Kwan-woo
Hwang Mun-sung
Samsung Electronics Co. Ltd.
Seoul, Korea

Der Samsung Staubsauger mit Aktivkohle-Filter VC-MBP 920 ist aus Aluminium gefertigt und besitzt eine hochglänzende UV-Beschichtung. Der Cyclone-Staubsauger benötigt keine Staubbeutel, die Staubentfernung ist einfach und leicht. Die elektronische Steuerung erfolgt über den Handgriff, auch die Staubbürste kann ohne umständliches Herunterbeugen betätigt werden. Durch das durchsichtige Verbindungsteil zwischen Bürste und Saugrohr kann jedes eingesaugte Partikel gesehen werden. Das integrierte Entlüftungsgitter unterscheidet den Staubsauger von anderen Produkten. Er besitzt eine runde und benutzerfreundliche Form.

The Samsung VC-MBP 920 vacuum cleaner comes complete with an active carbon filter is manufactured in aluminum with a high-gloss UV coating. The cyclone vacuum cleaner does not require a dustbag, meaning dust can be easily removed otherwise. It is controlled electronically via the handle and the brushes can be used without unnecessary bending. A transparent connector filter between the brushes and the suction tube allow you to see any foreign particle passing through. The round-shaped, user-friendly device includes an integrated ventilation grill which departs from the customary design of such appliances.

Hersteller Manufacturer
Philips Domestic Appliances & Personal Care
Hong Kong

Produkt Product
Philips City Line II
Staubsauger
Vacuum cleaner

Design
Esther Beckers
Graeme Warburton
Philips Design

Der Philips City Line II ist ein kompakter und dennoch leistungsstarker 1700W-Staubsauger. Er ist ideal für kleinere Haushalte oder als Zweitstaubsauger. Aufgrund der Handgriffe auf der Vorder- und Oberseite ist er leicht tragbar. Die Gesamtabmessungen des Gerätes wurden optisch reduziert, indem die beiden Hauptelemente voneinander getrennt wurden: das Staubbeutelfach und das Motorgehäuse. Daraus entstanden ein erkennbares zylindrisches Element, in dem sich der Motor befindet, das HEPA-Filtersystem, ein einfaches Bedienfeld und die großen hinteren Räder.

The Philips City Line II is a compact and yet powerful 1700W vacuum cleaner. It is ideal for smaller households or as the back-up cleaner. It is highly maneuverable thanks to the lightweight design and large rear wheels. The overall size of the vacuum cleaner has been reduced in visual terms by separating the two main components, namely the dust compartment and the motor housing. As a result, an eye-catching cylindrical element has arisen housing the motor, the HEPA filter system, a simple control panel and the large back wheels.

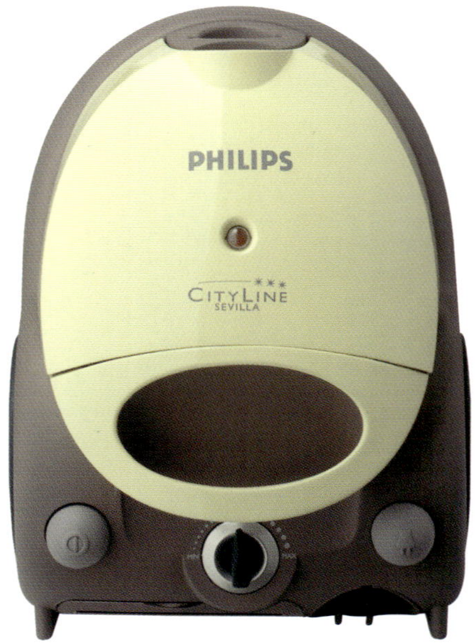

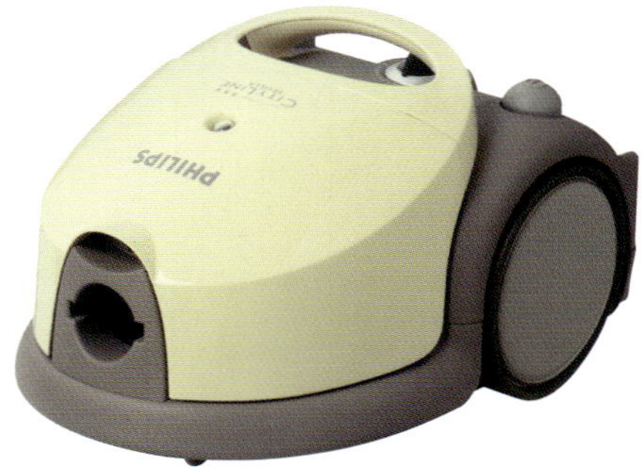

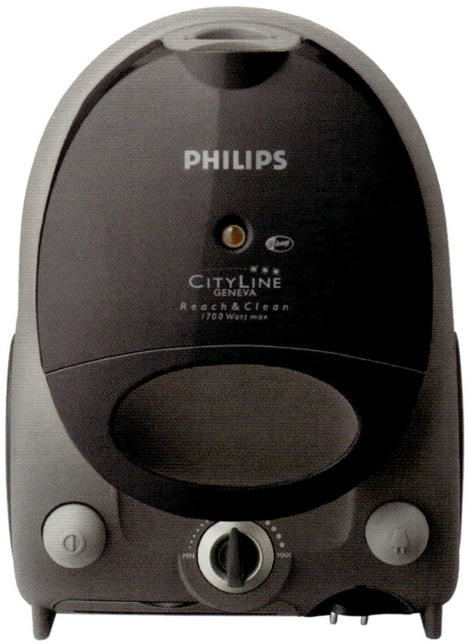

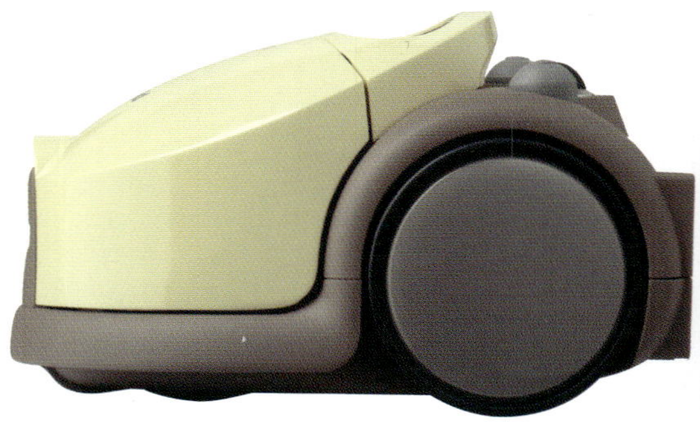

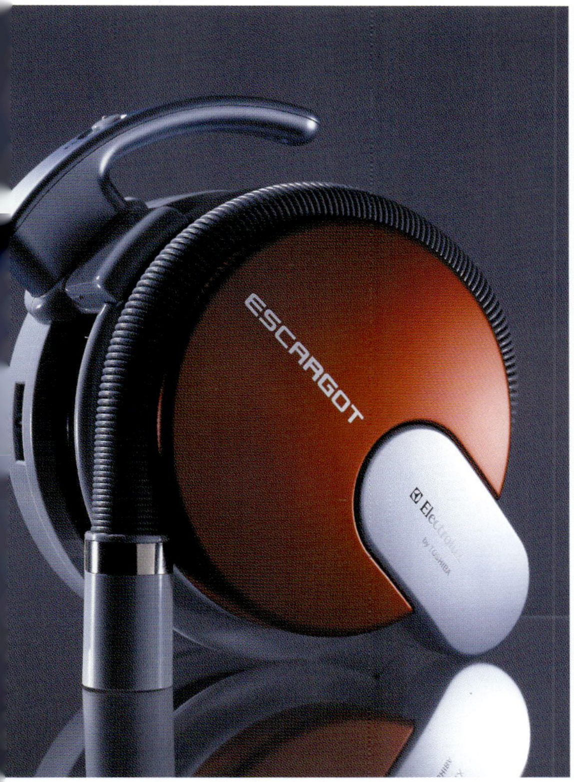

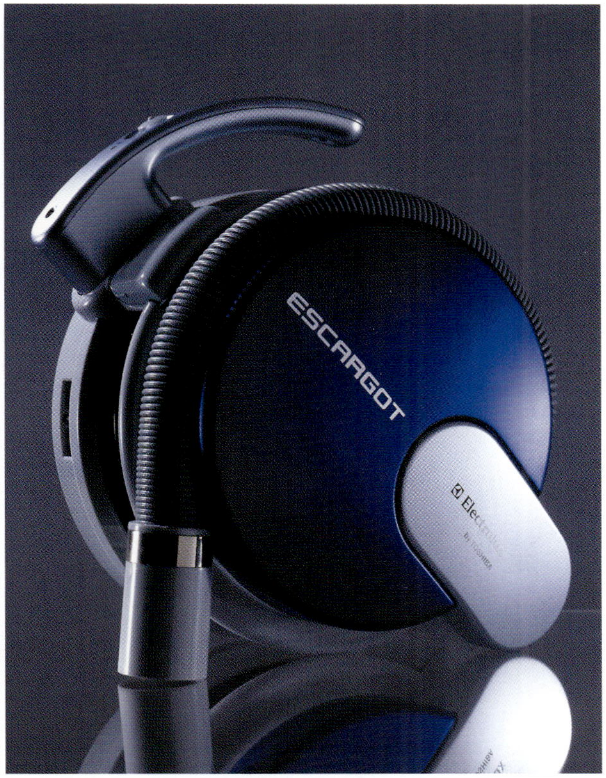

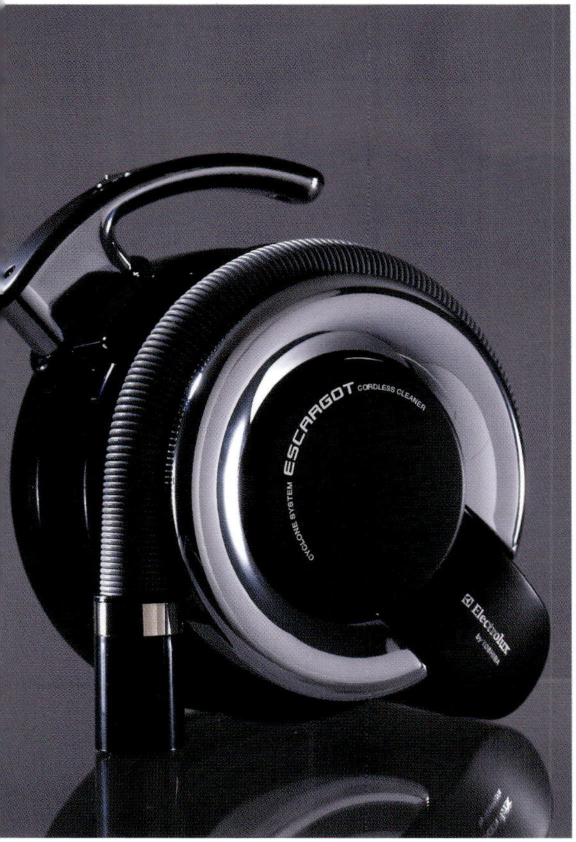

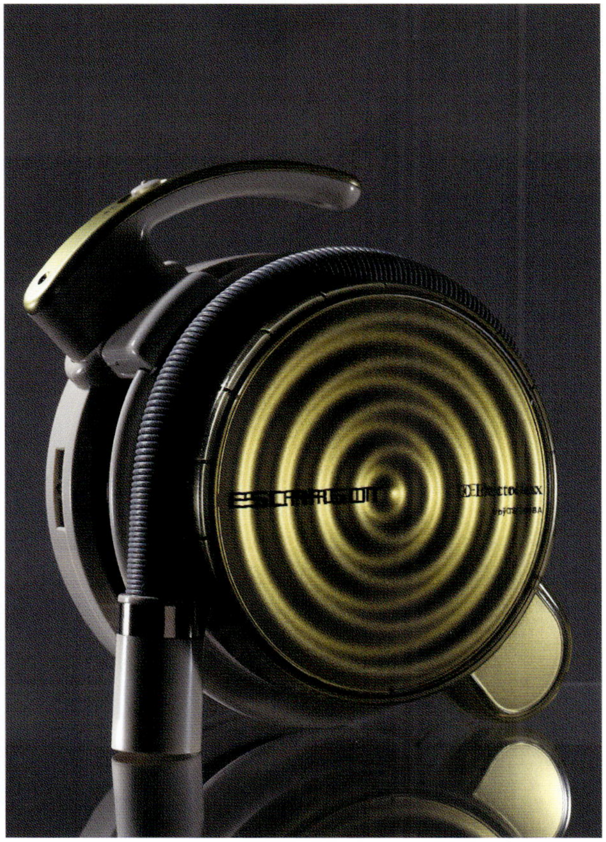

Hersteller Manufacturer
Toshiba Corporation
Tokio, Japan

Produkt Product
Escargot ECL-ES1(L)(R), ECL-ES2(K), ECL-ES3(G)
Tragbarer kabelloser Staubsauger
Portable cordless vaccuum cleaner

Design
Kanya Hiroi
Toshiyuki Yamanouchi
Design Center
Toshiba Corporation
Tokio, Japan

Aufgrund seines ansprechenden ungewöhnlichen Designs kann der tragbare und kabellose Staubsauger Escargot in seinen unterschiedlichen Ausführungen ECL-ES1(L)(R), ECL-ES2(K) und ECL-ES3(G) im Raum stehen bleiben. Die dem aufgerollten Schlauch folgende kompakte runde Schneckenform macht es möglich, dass der Staubsauger mittels Gurt beim Saugen über der Schulter getragen werden kann. Aufgrund des ergonomischen Griffs kann er auch locker in der Hand getragen werden, da er zudem nur 1,8 kg wiegt. Dank der Cyclon-Filtertechnik werden keine Staubsaugerbeutel benötigt. Es besteht die Wahl zwischen unterschiedlichen Farbgebungen des Gehäuses.

Given its unusual and attractive design, the portable cordless vacuum cleaner in its various versions ECL-ES1(L)(R), ECL-ES2(K) and ECL-ES3(G) is inconspicuously elegant and can be left standing in the room. The compact spiral form which corresponds to the coiled hose allows the vacuum cleaner to be strapped over the shoulder and carried when vacuuming. An open ergonomic handle also makes it comfortable to carry by hand, especially as it only weighs 1.8 kg. Thanks to the cyclone systems technology, no dust bags are required. The housing comes in a choice of colors.

Hersteller Manufacturer
Tien-Versand
Nordhorn, Germany

Produkt Product
Merlin
Dampfdruckreinigungssystem
Steam-cleaning system

Design
Prof. Andreas Kramer
Kramer Design
Nottuln, Germany

Merlin ist ein komplexes Dampfreinigungssystem für vielfältige Anwendungen rund um das Haus: Reinigen ohne Reinigungsmittel, Staubsaugen, Dampfbügeln, Spezialanwendungen wie Ablösen von Tapeten, Luftbefeuchtung, Sauna. Die Komplexität des Systems wurde durch eine einfache, übersichtliche Handhabung und eine sinnfällige Gestaltung gelöst. Der Benutzer kann lediglich zwischen drei Einstellungen wählen, so dass eine Fehlbedienung ausgeschlossen ist. Aufgrund der ergonomischen Verbesserungen ist Bücken und Tragen nicht mehr notwendig. Die einzelnen Teile können entsprechend der Bedürfnisse des Benutzers direkt oder in speziellen Zubehörboxen am Basisgerät befestigt werden.

Merlin is a complex steam-cleaning system with numerous applications around the house: cleaning without cleansing agents or detergents, vacuum cleaning, steam ironing, and special applications such as removing wallpaper, humidifying the air, sauna stoking. Yet despite the complexity of the system, the unit comes with simple, easy-to-understand functions and boasts a straightforward design. The user can only choose between three modes making it impossible to use the device wrongly. Thanks to the ergonomic improvements it is no longer necessary to bend or carry the cleaner around. The individual components can be fixed directly on the main body of the device or in accessory boxes, depending on user needs.

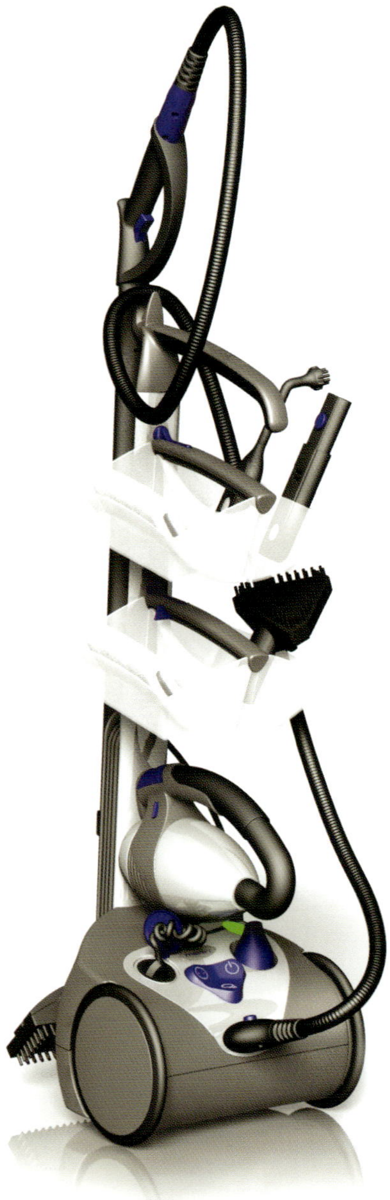

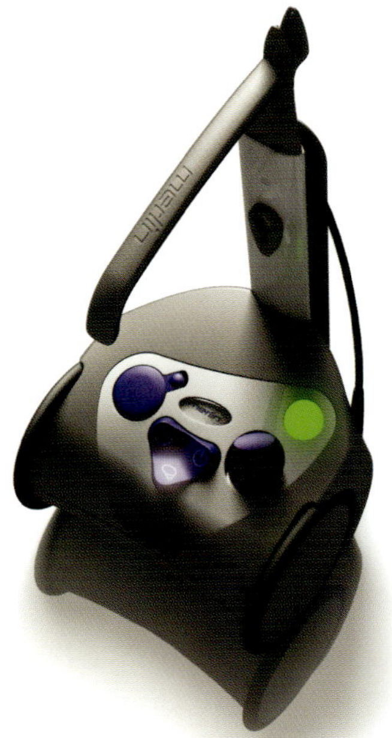

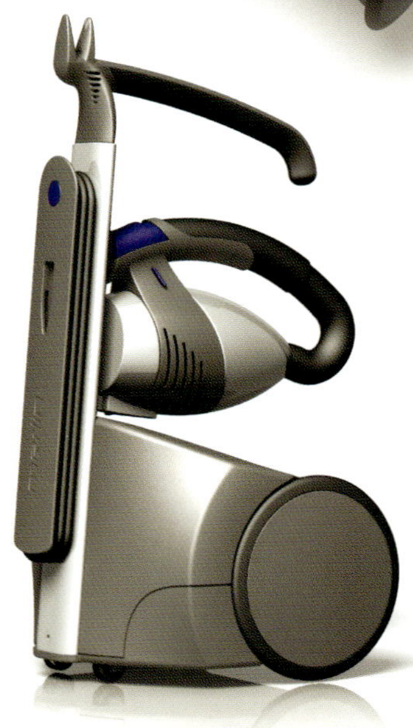

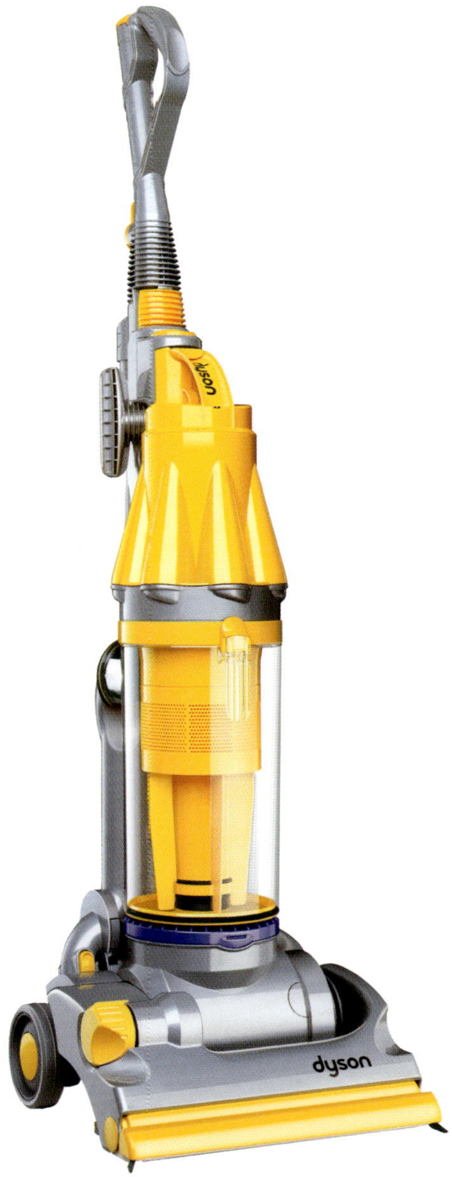

Hersteller Manufacturer
Dyson Ltd.
Wiltshire, UK

Produkt Product
Dyson DC07
Staubsauger
Vacuum cleaner

Design
Dyson Ltd.
Wiltshire, UK

Der Dyson DC07-Staubsauger verfügt über eine neue Zyklonen-Technologie (Root®Cyclone Technologie), die eine sehr hohe konstante Saugkraft ermöglicht. Die Luft wird durch mehrere kleine Zyklone geleitet, dadurch wird der Staub und Schmutz aus der Luft gewirbelt. Mit der elektrischen Bürstenwalze entfernt er selbst tief sitzenden Schmutz aus dem Teppich. Ein auf 4,5 m ausziehbarer Teleskopschlauch mit ergonomischem Griff erlaubt zudem das Saugen in entfernten Winkeln. Staub und Schmutz sammeln sich im durchsichtigen Staubbehälter, der sich mit Hilfe des Entriegelungshebels einfach entleeren lässt. Aufgrund der HEPA-Filtration ist der Dyson Staubsauger insbesondere für Allergiker geeignet.

The Dyson DC07 upright vacuum cleaner features new cyclone technology (Root®Cyclone) that delivers exceptionally strong and constant suction. The air is channeled through eight small cyclones, that's how the dust and dirt are spun out of the air. And thanks to the electric brush bar the cleaner even removes deep dust and hair from carpets. Boasting an ergonomic grip, the telescope quick-release hose stretches to as much as 4.5 m for cleaning even in remote corners. Dust and dirt particles collect in the transparent bin, and the trigger under the carry handle releases the bin for easy emptying. The HEPA filter means the Dyson DC07 is ideal for use by people suffering from allergies.

Hersteller Manufacturer
Plasvale–Indústria de Plásticos do Vale Ltda.
Gaspar, Santa Catarina, Brazil

Produkt Product
Prendedor de Roupas Multiuso
Wäscheklammern
Clothespin

Design
Taciana de Abreu e Silva
Marcela M.A. de Albuquerque
Plasvale–Indústria de Plásticos do Vale Ltda.
Gaspar, Santa Catarina, Brazil

Die Wäscheklammer wurde im Zuge eines Produktprojekts im Fachbereich Industriedesign an der Universität PUC in Rio de Janeiro entwickelt. Die minimalistische, sehr grafische Gestaltung nutzt die Eigenschaften von Kunststoff voll aus – so konnte auf die Metallfeder verzichtet werden, die rostanfällig ist, die Wäscheklammer zerbrechlicher macht und die Herstellung teurer und schwieriger gestaltet. Aufgrund ihrer symmetrischen, abgerundeten Form bekommen die Wäschestücke keine Druckstellen und haben eine Doppelfunktion: stützen und halten, so dass der zur Verfügung stehende Platz optimal genutzt wird. Die Wäscheklammern werden industriell aus Thermoplaste in einem einzigen Stück gegossen, mit modernen, lichtdurchlässigen Farben.

These clothespins were developed as part of a project for industrially designed products at the PUC University in Rio de Janeiro. The minimalist, highly graphic design makes the most of plastic's properties – thus eliminating the metal spring which not only tends to rust but makes the clothespin more likely to break and more expensive and difficult to produce. It's symmetrical rounded form leaves no pressure marks on the clothes and allows double use of the pin: prop and hook, optimizing the space used. The clothespins are industrially cast from a single piece of thermoplastic and boast modern translucent colors.

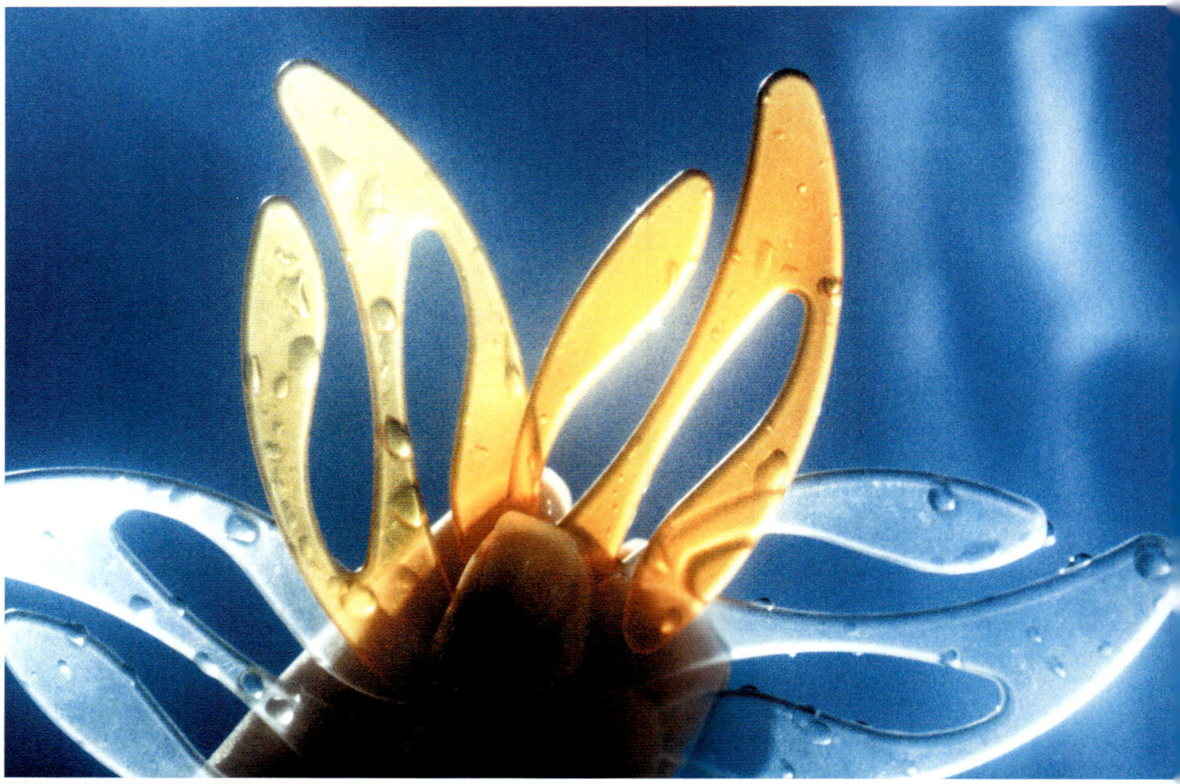

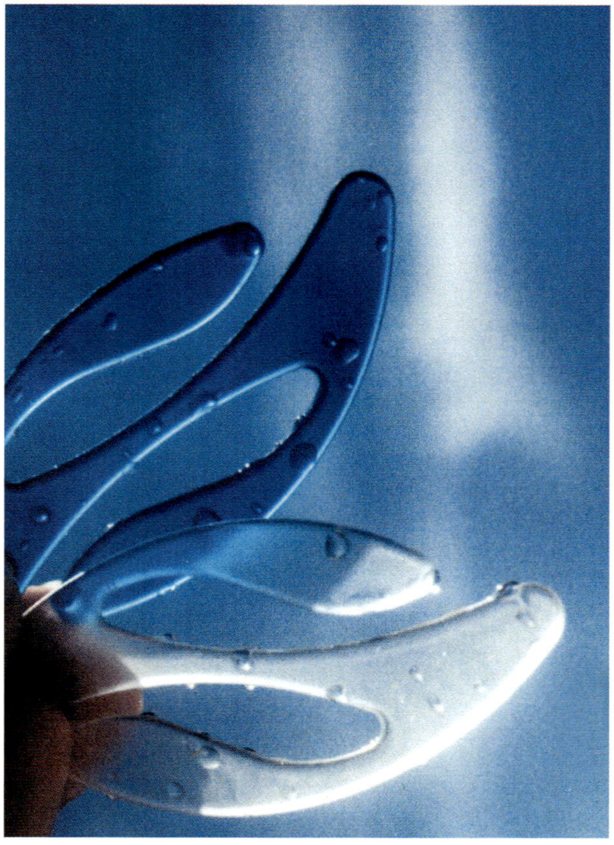

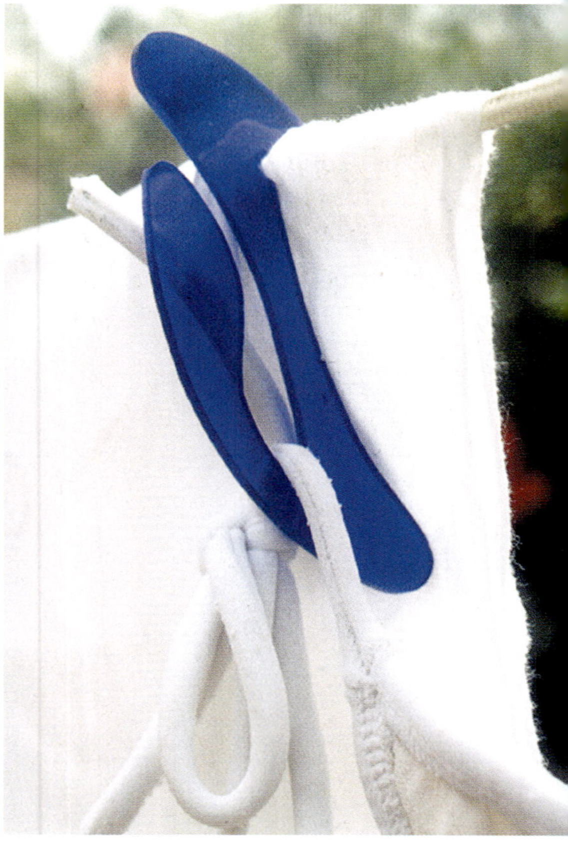

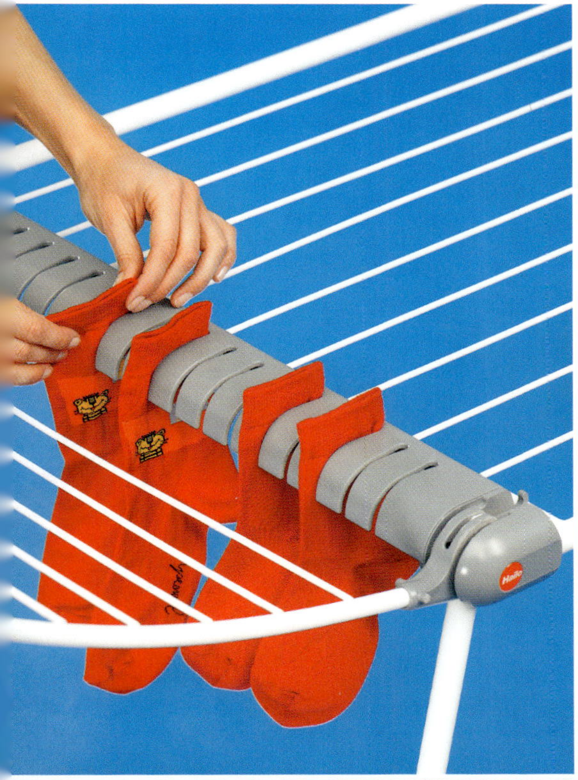

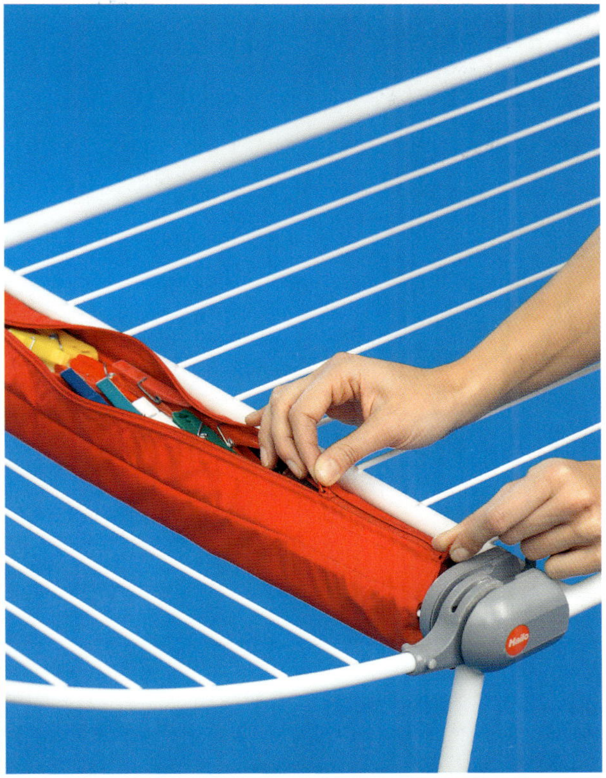

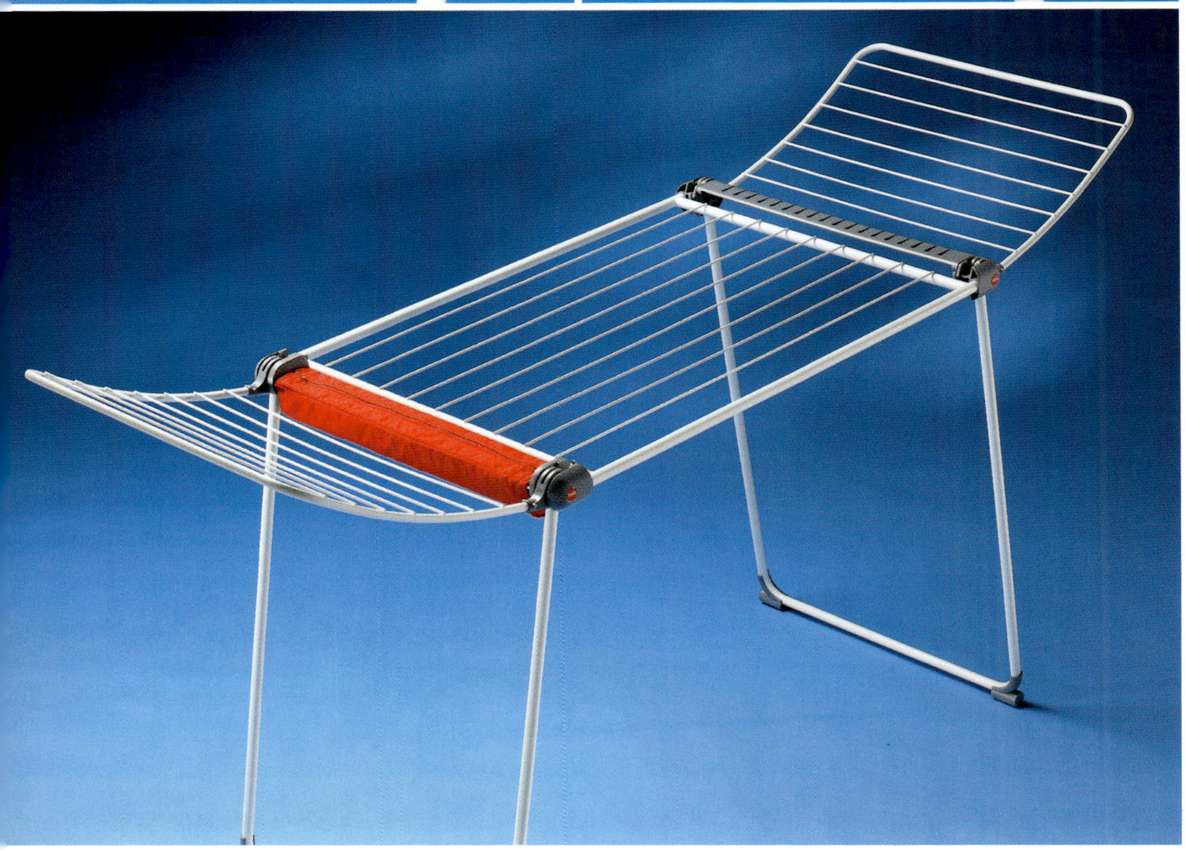

Hersteller Manufacturer
Hailo-Werk
Haiger, Germany

Produkt Product
Viva Wings
Flügel-Wäschetrockner
Clothes airer with wings

Design
Günter Elbl, Yvonne Kiefer
formteam
Schorndorf, Germany
Jörg Krug
Hailo-Werk
Haiger, Germany

Der Viva Wings Flügel-Wäschetrockner besitzt ein patentiertes Multifunktionsgelenk, das die einzelnen Bauteile des Wäscheständers miteinander verbindet. Eine spezielle Verriegelungsfunktion verleiht dem Wäscheständer stabile Standsicherheit. Die Flügel rasten in den Gelenken ein und stützen sich dabei auf den Standbeinen ab. Trotz der geschwungenen Flügel lässt sich der Wäscheständer äußerst flach (10 cm) zusammenklappen. Das Multifunktionsgelenk dient auch zur Aufnahme aller zusätzlichen Ausstattungsmerkmale wie Klammerbeutel, Sockenhalter oder auch Trockenrollo.

The Viva Wings clothes airer boasts not only wings, but a patented multi-functional joint that links the various individual sections. A special locking function ensures the clothes airer remains sturdy. The wings lock into place in the joint and their load is then supported by the airer's legs. Despite the upwards curving wings, the airer folds up as good as flat (10 cm). The multi-functional joint can also hold any additional accessories such as the pegs bag, sock holders or drying spindle.

Hersteller Manufacturer
Leifheit AG
Nassau/Lahn, Germany

Produkt Product
Tipp 36l
Abfallsammler
Waste collector

Design
TEAMS Design
Nassau/Lahn, Germany

Der Abfallsammler Tipp 36l von Leifheit bietet durch das Clean-System Schutz vor Gerüchen, Bakterien und Keimen. Ein Dichtring verschließt fest den Deckel und ein antibakterieller Kunststoffeimer mit feuerfester Everclean-Beschichtung, und der abgerundete Boden erleichtert das Reinigen. Eine spezielle Tütenklemmung verhindert, dass Abfall in den Inneneimer fällt. Der Deckel des Abfallsammlers öffnet sich durch leichtes Antippen der Drucktaste. Der breite Sockel und das Anti-Rutsch-Profil gewähren eine hohe Standfestigkeit. Tipp ist in Weiß, Edelstahl oder Silber erhältlich.

The Leifheit waste collector, Tipp 36l, guards against odors, bacteria and germs with its Clean system. A seal ring secures the lid tightly to an anti-bacterial plastic bucket with a flameproof Everclean coating; the bucket features a round bottom to aid cleaning. A light touch on the push button opens the lid of the waste collector while the broad base and the anti-skid profile give it a secure hold. Tipp is available in white, stainless steel, or silver.

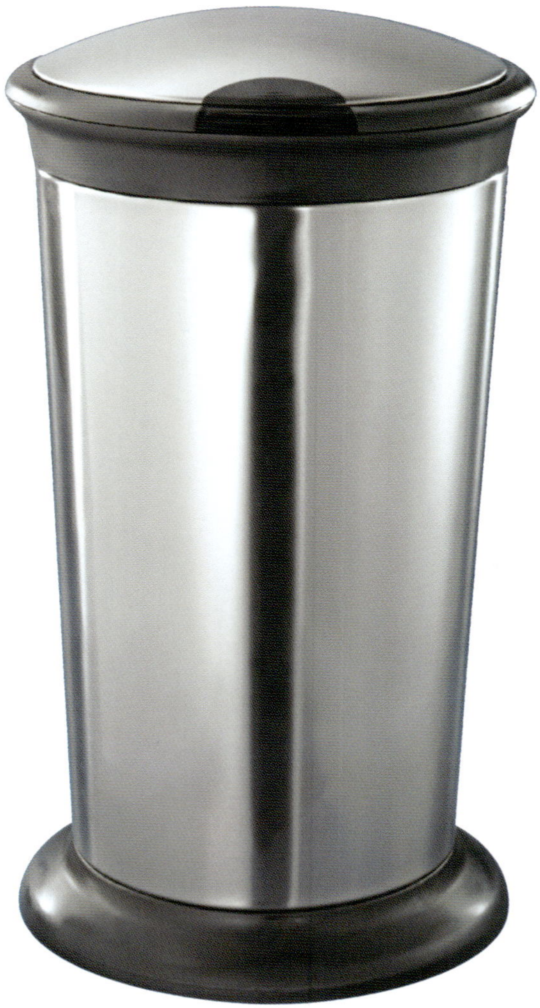

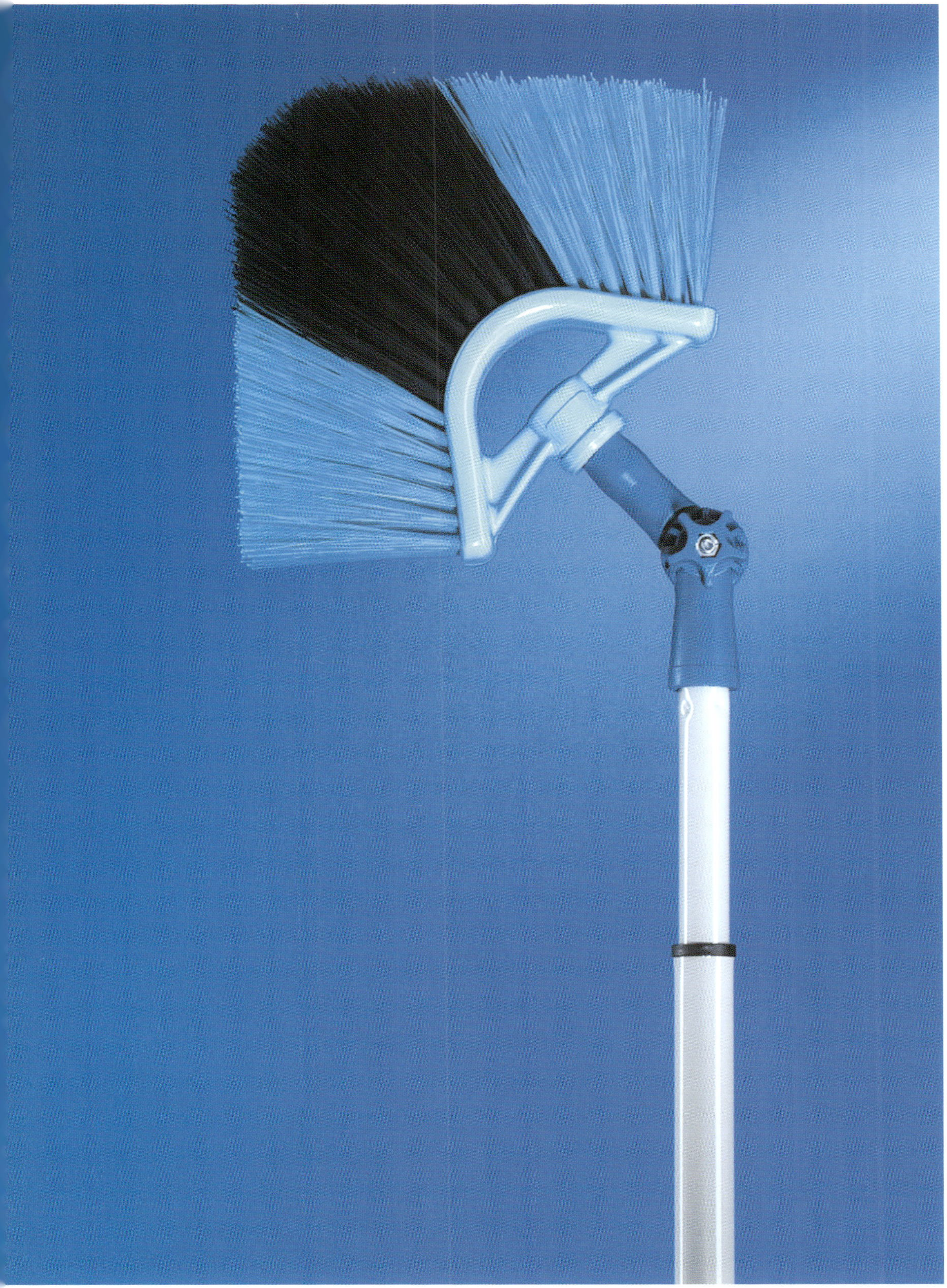

Hersteller Manufacturer
Leifheit AG
Nassau/Lahn, Germany

Produkt Product
Dusty
Wand- und Fassadenbesen
Dust broom

Design
Leifheit AG
Nassau/Lahn, Germany

Der Staubbesen Dusty von Leifheit ist mit seinem Gelenk, das sich um 90 Grad abwinkeln lässt und der Dreiecksform eigens für schwer erreichbare Ecken im Haus konzipiert. Mit dem ausziehbaren Teleskopstiel (bis zur Arbeitshöhe von 4 m) und dem einstellbaren Arbeitswinkel lassen sich vom Boden aus bequem Decken und Ecken kehren. Durch die flexiblen Kunststoffborsten wird auch feinster Staub aufgenommen. Für die gründliche Reinigung mit der Hand kann der Besen auch abgenommen werden.

Dusty, the dust broom from Leifheit, is designed to reach into those out-of-the-way corners with its triangular form and 90° swivel function. A telescopic handle and adjustable working angle at the brush joint make it ideal for reaching ceilings and corners from the ground. The finest of dust particles do not stand a chance with its flexible plastic brushes. The brush can be removed and cleaned thoroughly by hand.

Textil Mode Textiles Fashion

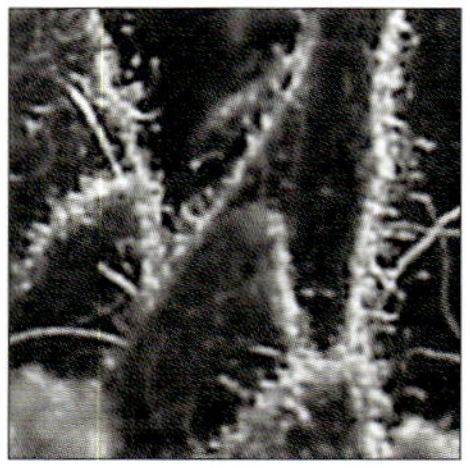
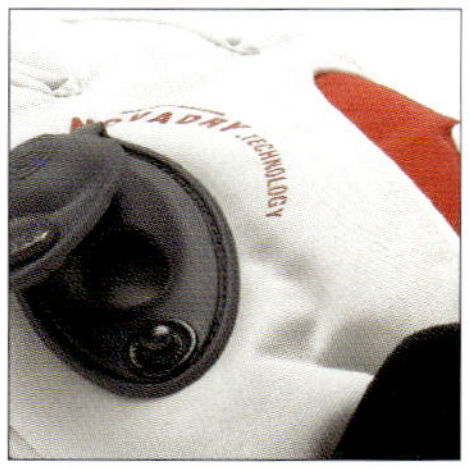
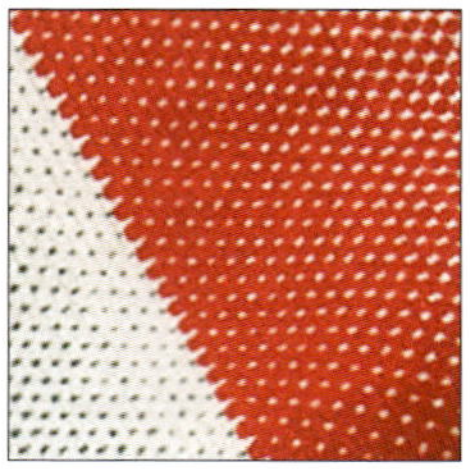

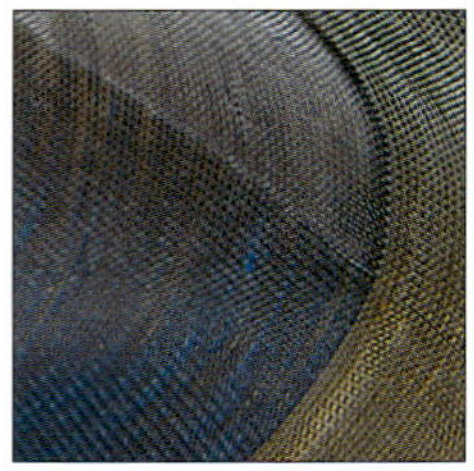
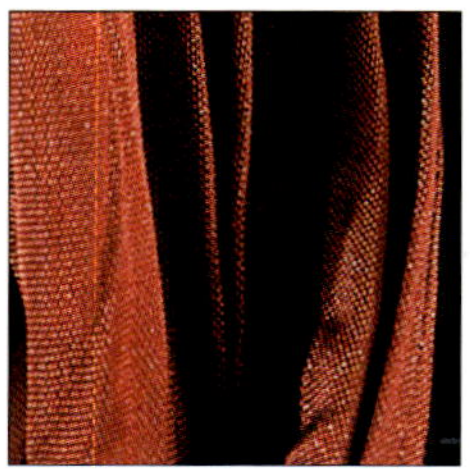
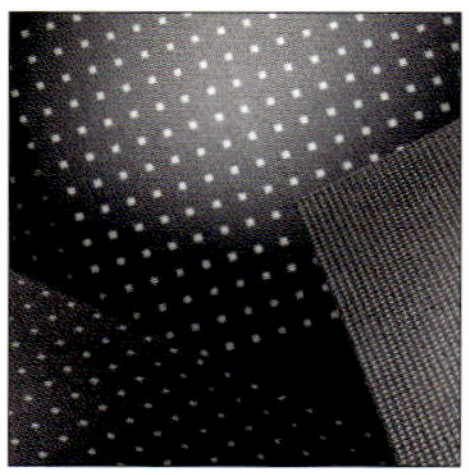

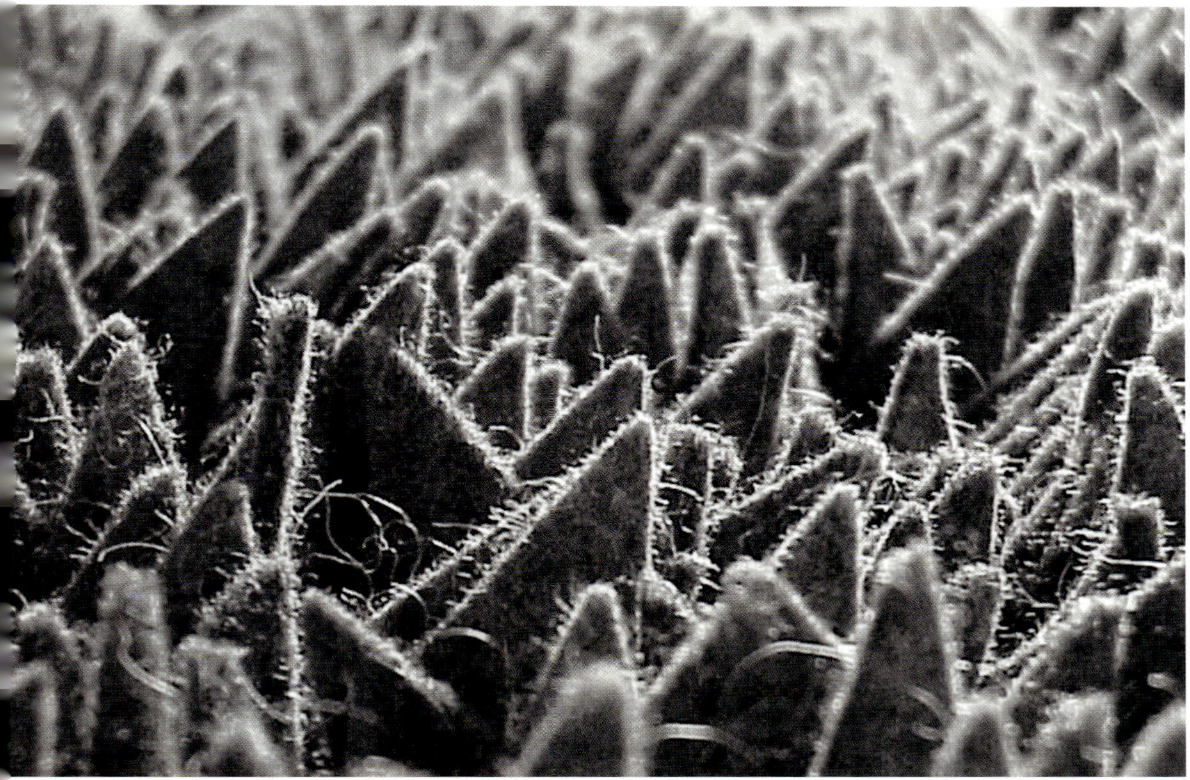

Hersteller Manufacturer
Illu Stration
Leidersbach, Germany

Produkt Product
Flokati
Filzteppich
Felt carpet

Design
Alexander Pernitschka
Mary-Ann Williams
Illu Stration
Leidersbach, Germany

Der Filzteppich Flokati ist in den Variationen Maltagliati, Pappardelle und Tagliatelle erhältlich, wobei sich die verarbeiteten Filzstreifen jeweils in ihrer Breite und Form unterscheiden. Die Teppiche werden bis hin zur Maßanfertigung in verschiedensten Formen angeboten. Sie sind leicht zu pflegen, langlebig, warm und haben sehr gute schalldämpfende Eigenschaften. Es werden ausschließlich Naturmaterialien verwendet.

The Flokati felt carpet is available in three variants: Maltagliati, Pappardelle and Tagliatelle – the difference between them being the respective breadth of the strips of felt used. The carpets are offered in any number of shapes and can be made to measure. They are easy to care for, hard-wearing, warm and function as superlative noise dampeners. Each rug is made solely exclusively of natural materials.

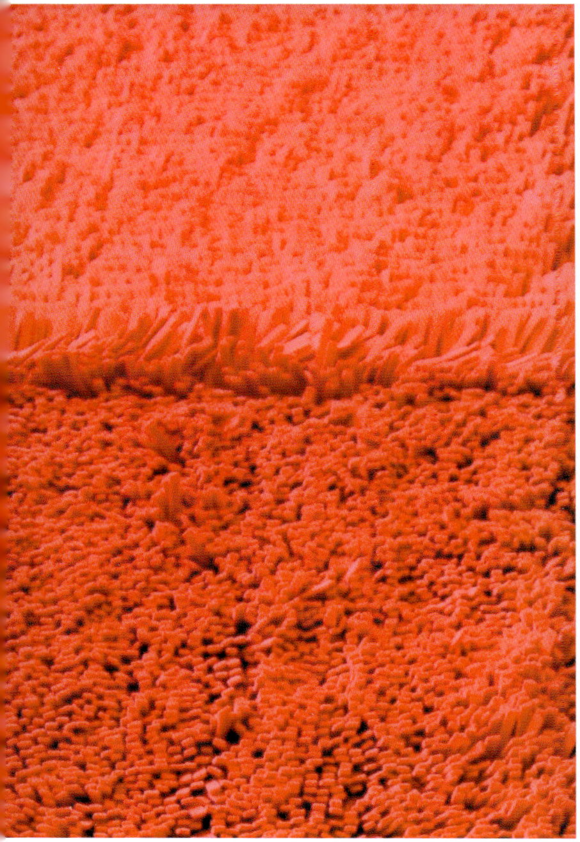

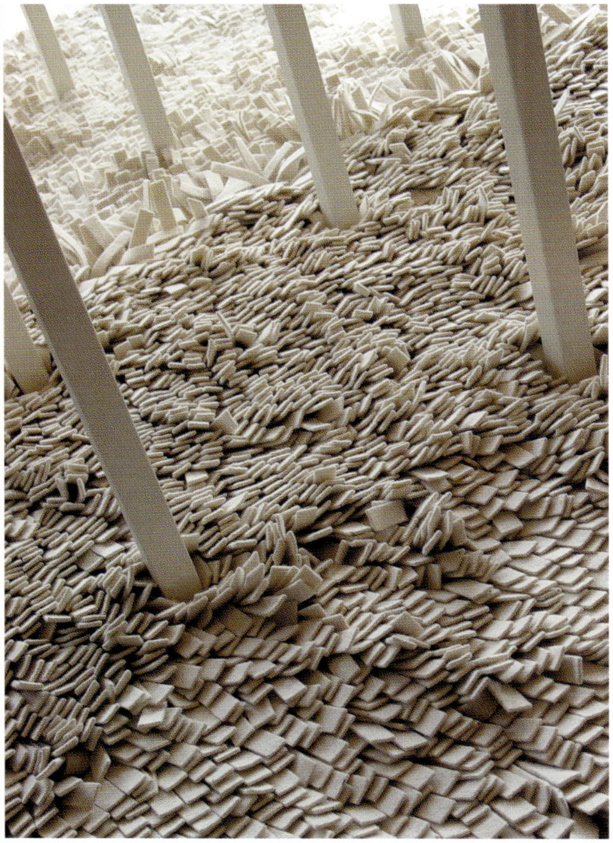

iF Gold Award

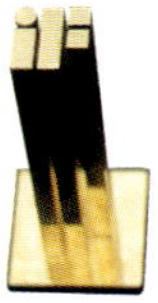

Hersteller Manufacturer
Decathlon Sportartikel GmbH & Co. KG
Dortmund, Germany

Produkt Product
Self-heat Handschuh der Marke Quechua
Wintersporthandschuh
Winter sport glove

Design
Cedric Caprice
Vincent Chiffoleau
Decathlon Design
Villeneuve d'Ascq, France

Die Idee, die hinter dem Self-heat Handschuh von Quechua steckt, ist einfach. Man zieht an einer kleinen Kapsel und bläst Luft in den Handschuh. Die eingeblasene Luft (36 Grad) verteilt sich gleichmäßig und erwärmt die Hände rund 15 Minuten lang.

The Quechua self-heating glove is based on a simple but ingenious idea. You tug on a small capsule and it blows air into the glove. The air generated in this way is 36° warm and spreads evenly through the glove, warming your hands for as long as 15 minutes – an ecologically-sound idea to keep you warm.

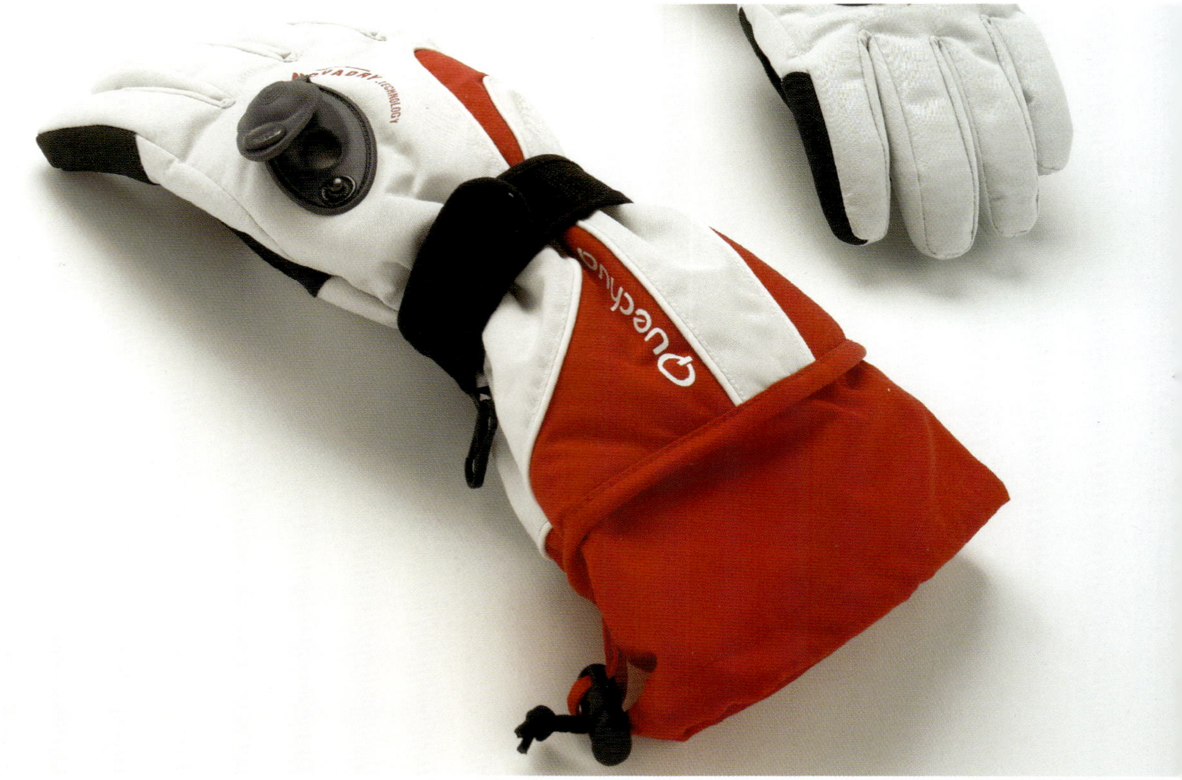

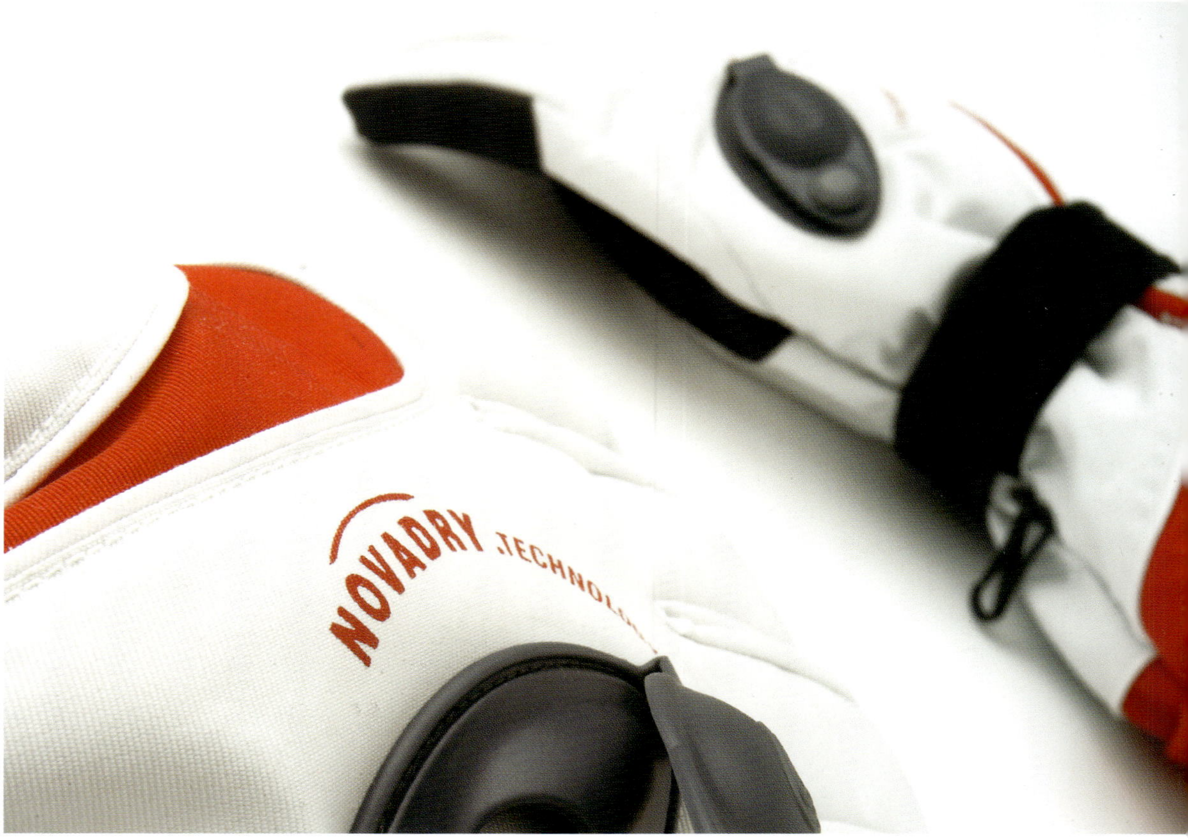

iF Silver Award

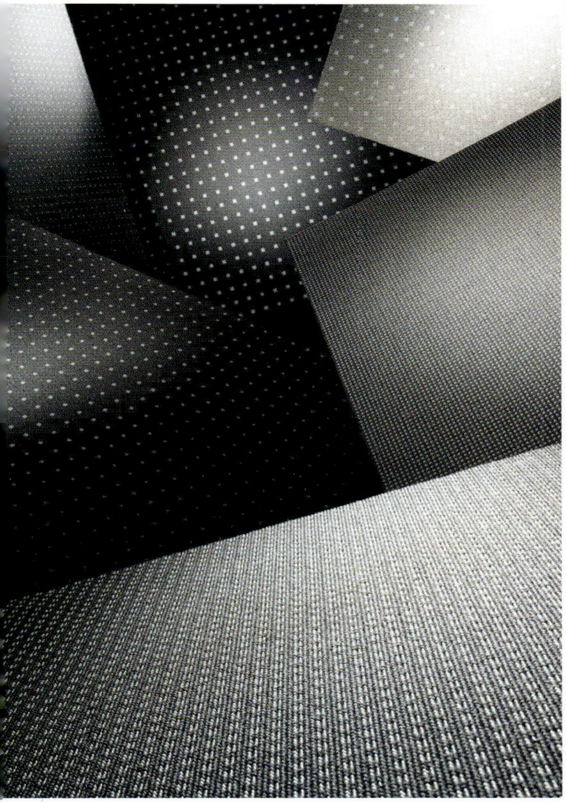

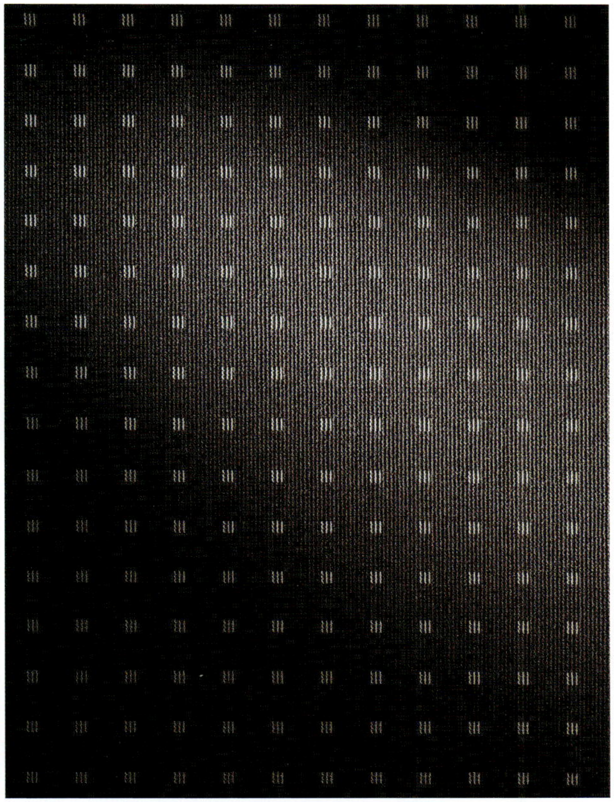

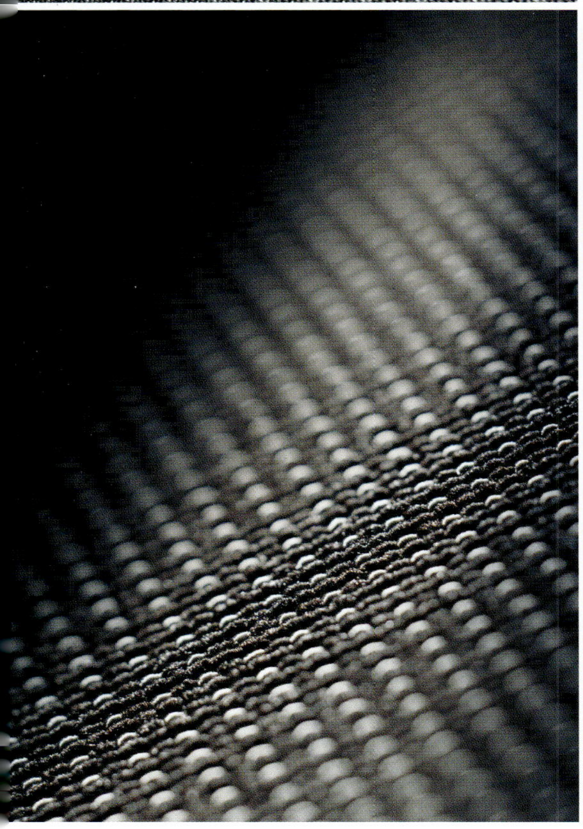

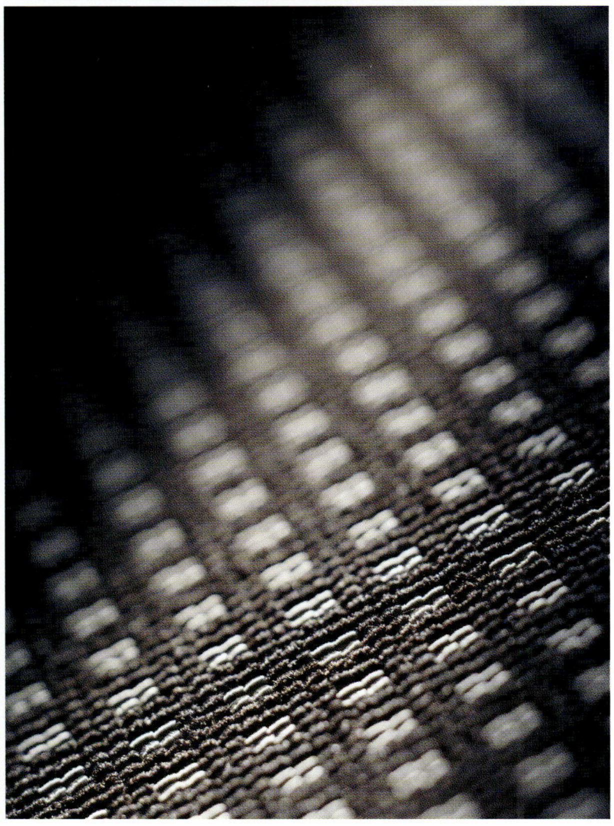

Hersteller Manufacturer
Carpet Concept Objekt-Teppichboden GmbH
Bielefeld, Germany

Produkt Product
Net, Erweiterung des Programms
um 16 Dessins und 6 Farben
Webware
Woven wilton

Design
Thomas Trenkamp
Dienslaken, Germany

Die Erweiterung des Programms Net auf nunmehr 16 Dessins ist für moderne Architektur konzipiert. Minimiertes Design steht für eine klare, funktionale Innenarchitektur. Glänzende Fäden wecken Aufmerksamkeit und setzen individuelle Akzente. Der wertige Teppichboden wird in einer speziell entwickelten Webtechnik gefertigt. In einer einfarbigen Fläche glänzen die metallischen Muster und formen das vitale Bodenbild. Zu kleinen Quadraten eingewebte Garne mit Edelstahlfäden formen geometrische Raster oder amorphe Strukturen. Die Erweiterung der Net Kollektion bietet 16 Dessins je in den neuen Farben Rosso, Mare, Siena, Prado, Morena, Cobre.

The expanded Net line is designed for modern architectural interiors. The minimized design blends with a clear, functional architecture. Gleaming threads catch the attention and create individual highlights. The high-grade carpet floor covering is the product of a special weaving technique. Metallic patterns gleam in the monochrome surface to create a lively floor vista. Yarns of stainless-steel thread woven to form small squares then generate geometric patterns or are arranged amorphously. The collection comprises 16 different designs each in topical colors such as Rosso, Mare, Siena, Prado, Morena, and Cobre.

Hersteller Manufacturer
nya nordiska textiles
Dannenberg, Germany

Produkt Product
CASCADAS
Dekorationsstoff
Decorative material

Design
Heinz Röntgen
nya nordiska textiles
Dannenberg, Germany

Das Spiel von Wassertropfen im Spektrum des Lichts inspirierte das Design des Dekorationsstoffes CASCADAS. Der weiche Velours besteht aus reiner Baumwolle, das in fünf verschiedenfarbigen Streifen gehaltene Grundmuster besteht aus tausenden gedruckten Punkten.

The appearance of the CASCADAS decorative fabric was strongly influenced by watching water drops dancing in the light. The velvet-soft velour is made of pure cotton, tightened into five stripes of different colors along a basic pattern composed of a thousand tiny little dots.

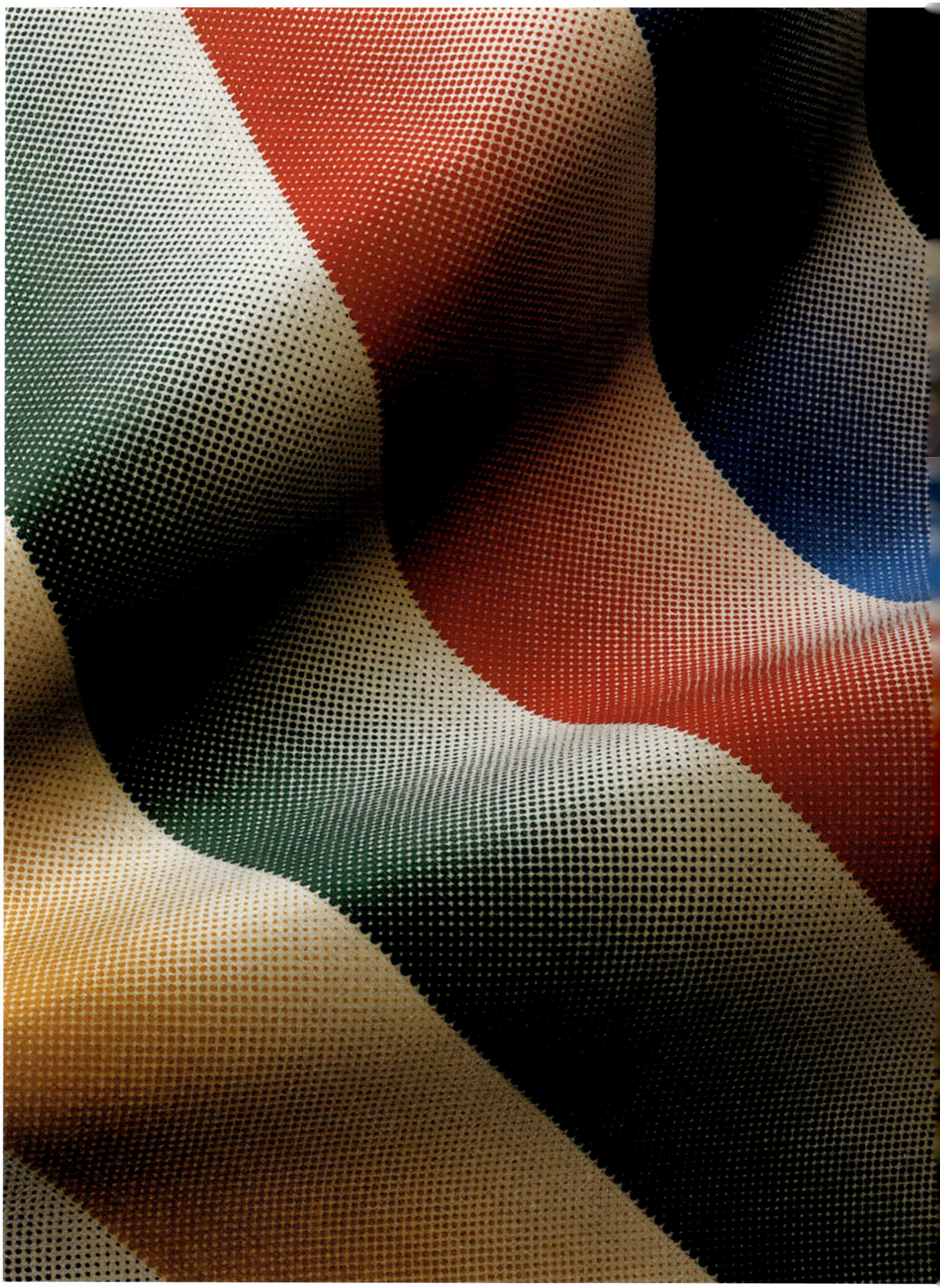

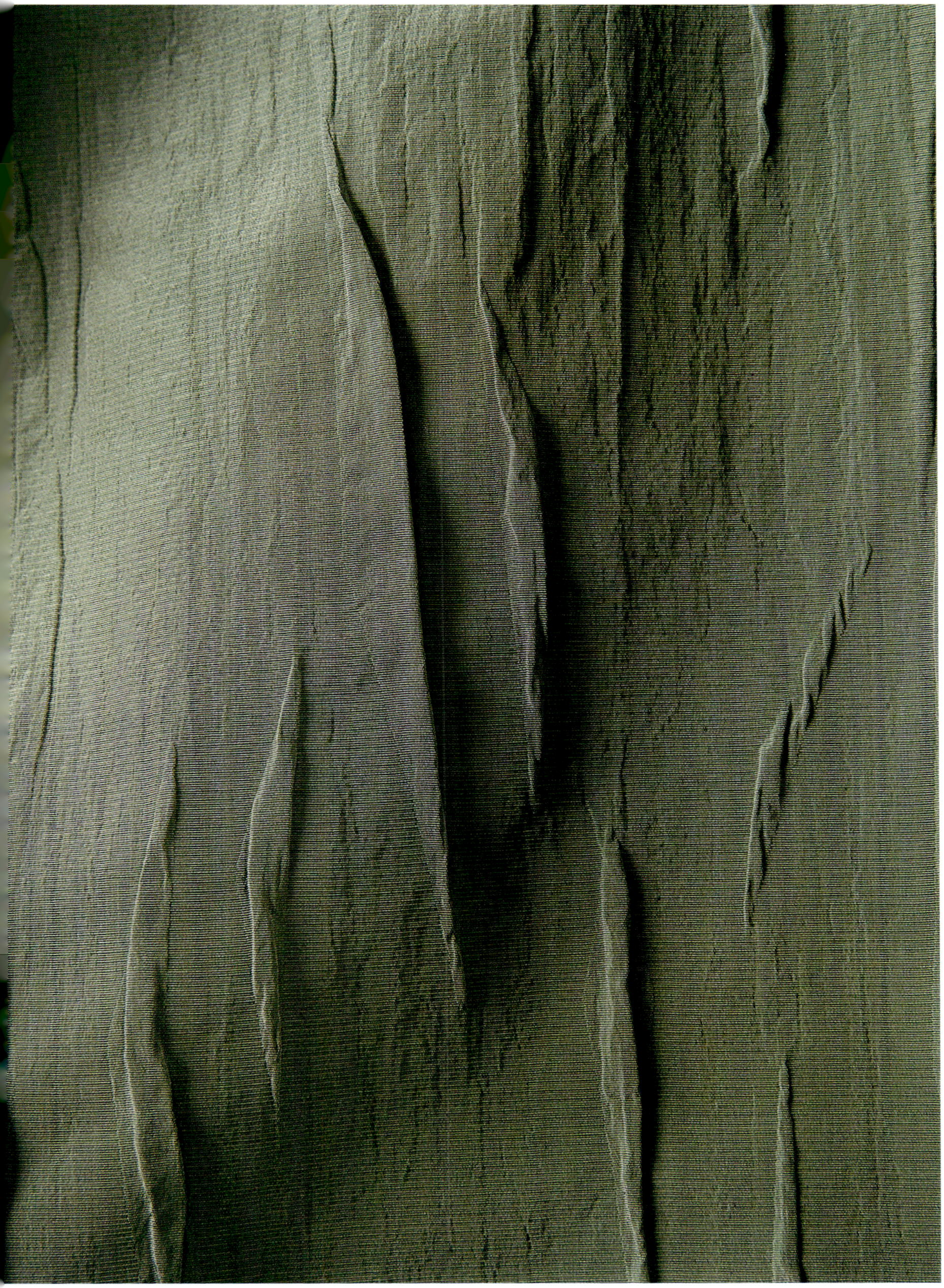

Hersteller Manufacturer
nya nordiska textiles
Dannenberg, Germany

Produkt Product
FLASH
Dekorationsstoff
Decorative fabric

Design
Heinz Röntgen
nya nordiska textiles
Dannenberg, Germany

FLASH ist ein auf einem Jacquard-Stuhl gewebter 310 cm breiter Taftstoff. Die Farben kommen vom Garn und verleihen dem Taft einen reichen Changeant-Effekt. Durch leichte Erhitzung wurde ein dosierbarer Schrumpfungsprozess erzielt, der die partielle Dreidimensionalität auf der Vorderseite ermöglicht. Mit Hilfe der speziellen Gestaltungstechnik wurden rückseitig kleine Fantasie-Ecken mit einem elastischen Garn eingebunden.

FLASH is woven from taffeta in widths of approx. 310 cm using a jacquard technique. The colors are achieved by using different yarns, imbuing the taffeta with an iridescent look. Minimal heating results in carefully controlled shrinkage which lends the face of the material its astonishing 3D effect. The jacquard weave ensures that the reverse side sports small fantasy triangles, gathered by means of a special shrinking yarn.

Hersteller Manufacturer
nya nordiska textiles
Dannenberg, Germany

Produkt Product
VEDUTA
Dekorationsstoff
Decorative fabric

Design
Heinz Röntgen
nya nordiska textiles
Dannenberg, Germany

Der Dekorationsstoff VEDUTA hat eine feine, gespinstartige Struktur aus einem farblosen und kaum sichtbaren High-tech-Doppelgewebe, das einem Leinengarn Halt und Raum gibt. Die farblich dezent abgestimmten Leinengarne sind in beiden Richtungen willkürlich verteilt, um kein geordnetes Streifen- oder Karomuster entstehen zu lassen.

VEDUTA is an exceptionally finely woven decorative fabric, created using colorless, opaque and almost invisible high-tech double weave to instill the high-grade linen with structure and space. The discreetly color-coordinated linen yarns are distributed somewhat haphazardly in both directions to prevent any clearly structured checks and stripes arising in favor of disorderly beauty.

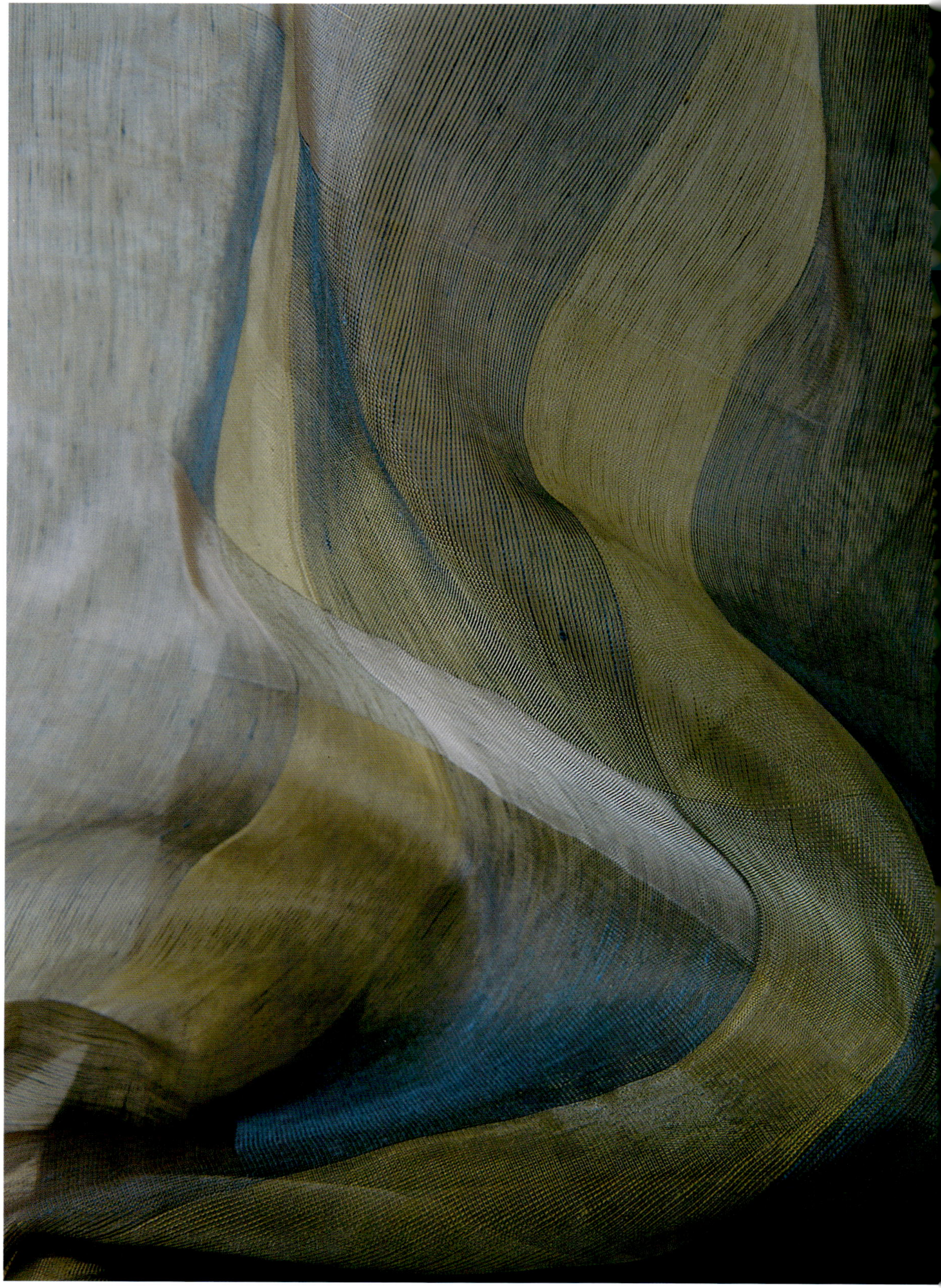

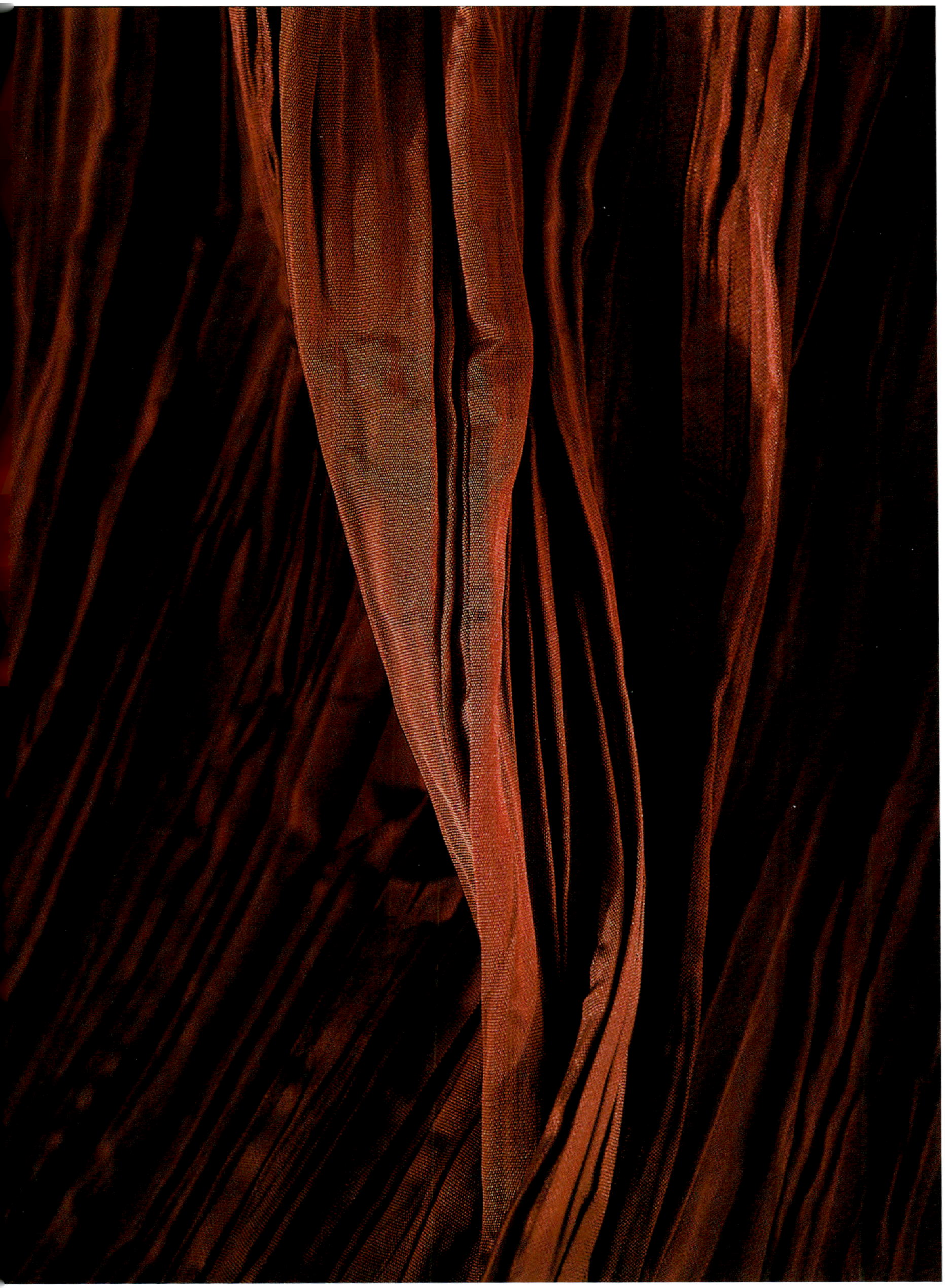

Hersteller Manufacturer
nya nordiska textiles
Dannenberg, Germany

Produkt Product
TIZIANO
Dekorationsstoff
Decorative fabric

Design
Heinz Röntgen
nya nordiska textiles
Dannenberg, Germany

Der Dekorationsstoff TIZIANO ist ein klassischer Plissee und etwa halb so breit wie sein Ausgangsstoff. Er wurde 320 cm breit gewebt und in einem neuartigen Verfahren auf immerhin ca. 200 cm Breite plissiert. Der garngefärbte Taft ist durch die Zweifarbigkeit von Kette und Schuss auf Changeant-Effekte angelegt, durch den Wechsel von Glanz und Matt wird ein ungewöhnlicher Effekt erzielt.

The decorative TIZIANO fabric is a classic pleated textile and after processing is approximately half its original width. It is woven at a width of 320 cm and then using an innovative method pleated to a width of no less than 200 cm. The dyed taffeta yarn has an iridescent look to it thanks to the bicolor properties of the weft and warp – and the interplay of gloss and matte allows quite striking effects using the material.

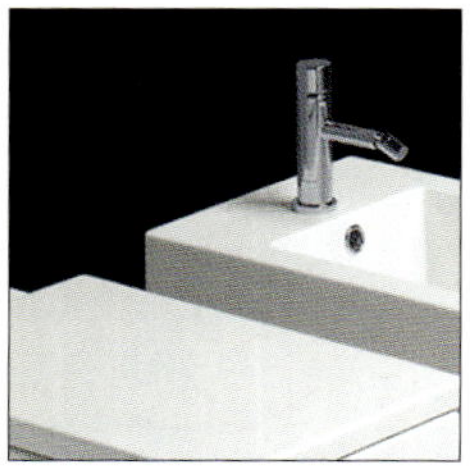
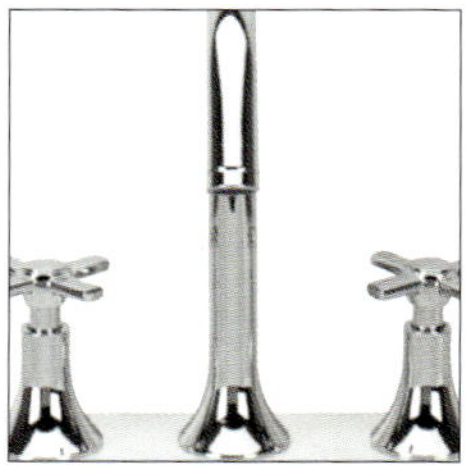
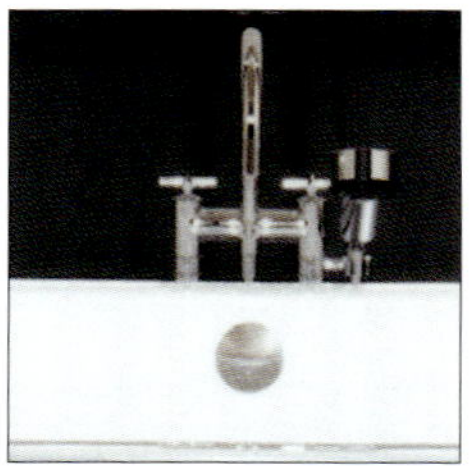
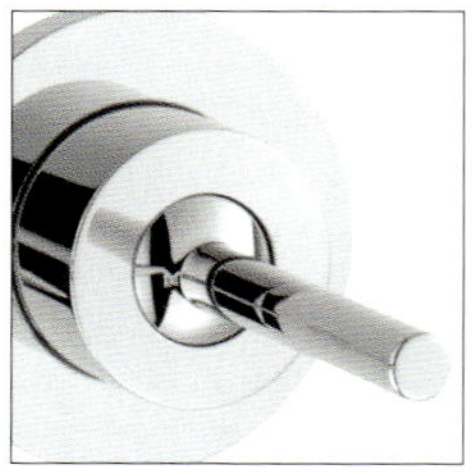
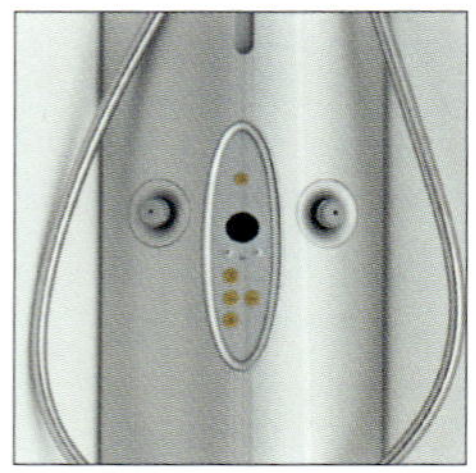
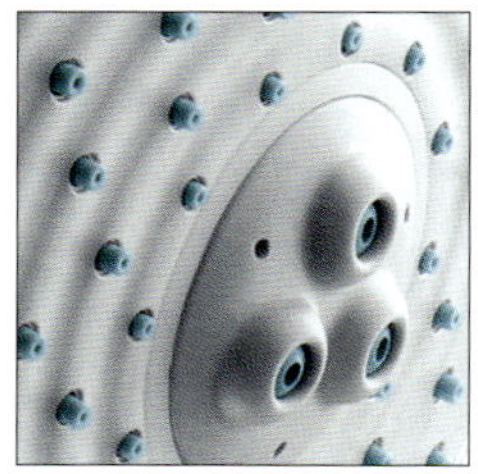
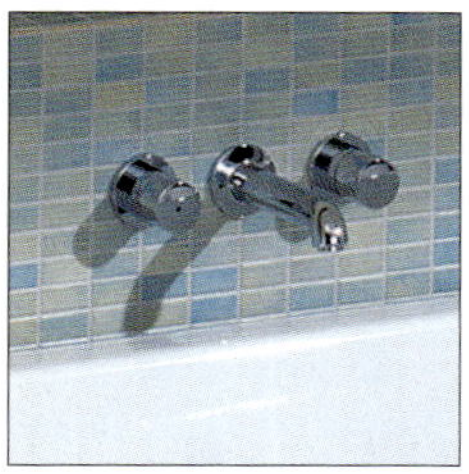
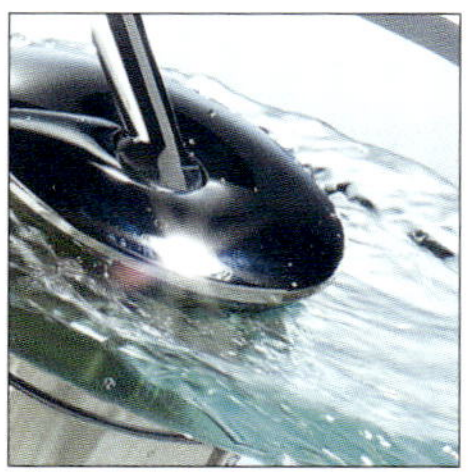
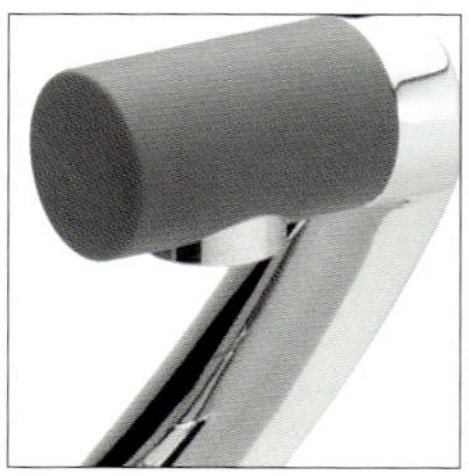
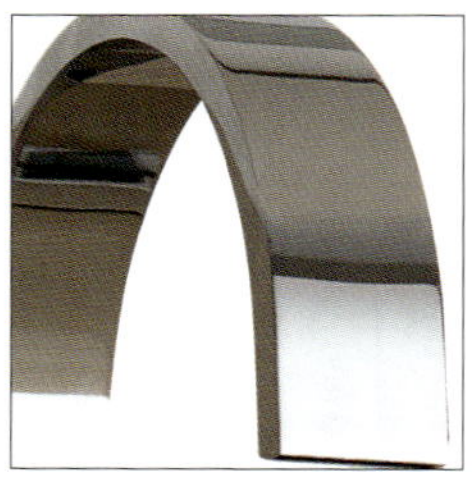
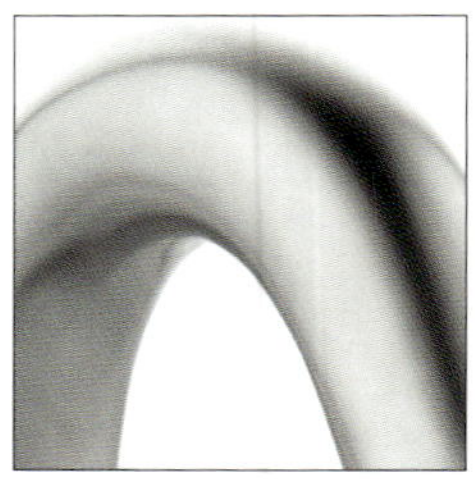
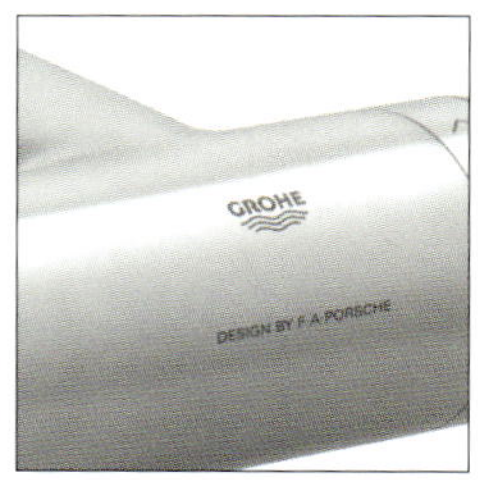
GROHE
DESIGN BY F.A.PORSCHE

Wohnen Living

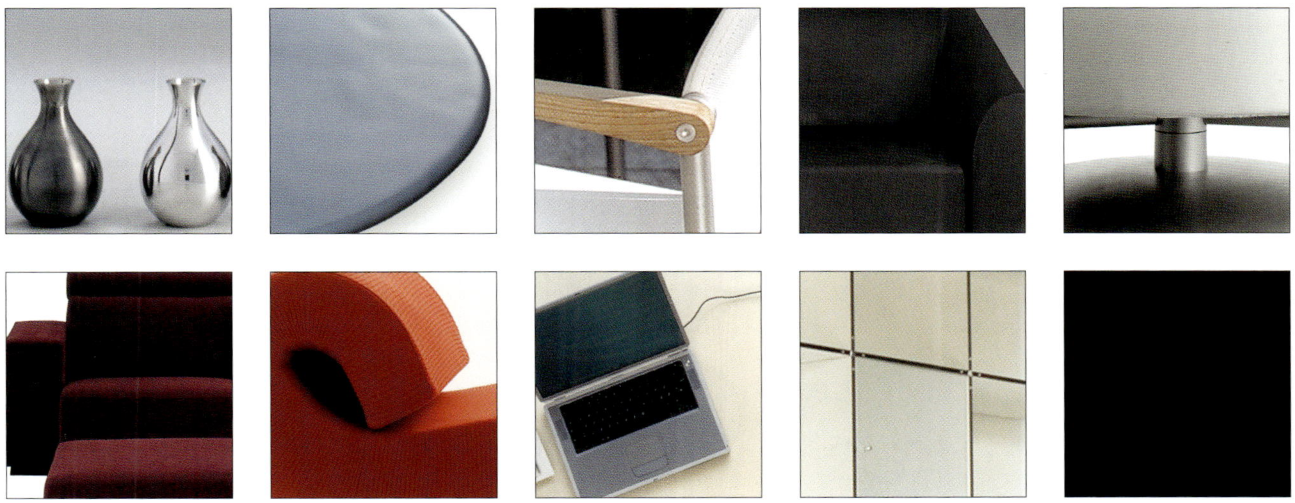

Hersteller Manufacturer
Ceramica Catalano SRL
Fabrica di Roma, Italy

Produkt Product
Sistema Verso
Sanitär- und Badezimmerausstattung
Sanitary line and bathroom accessories

Design
Sanitär Sanitary
Corporate Design Catalano
Accessoires Accessories
Giancarlo Martino
Fabrica di Roma VT, Italy

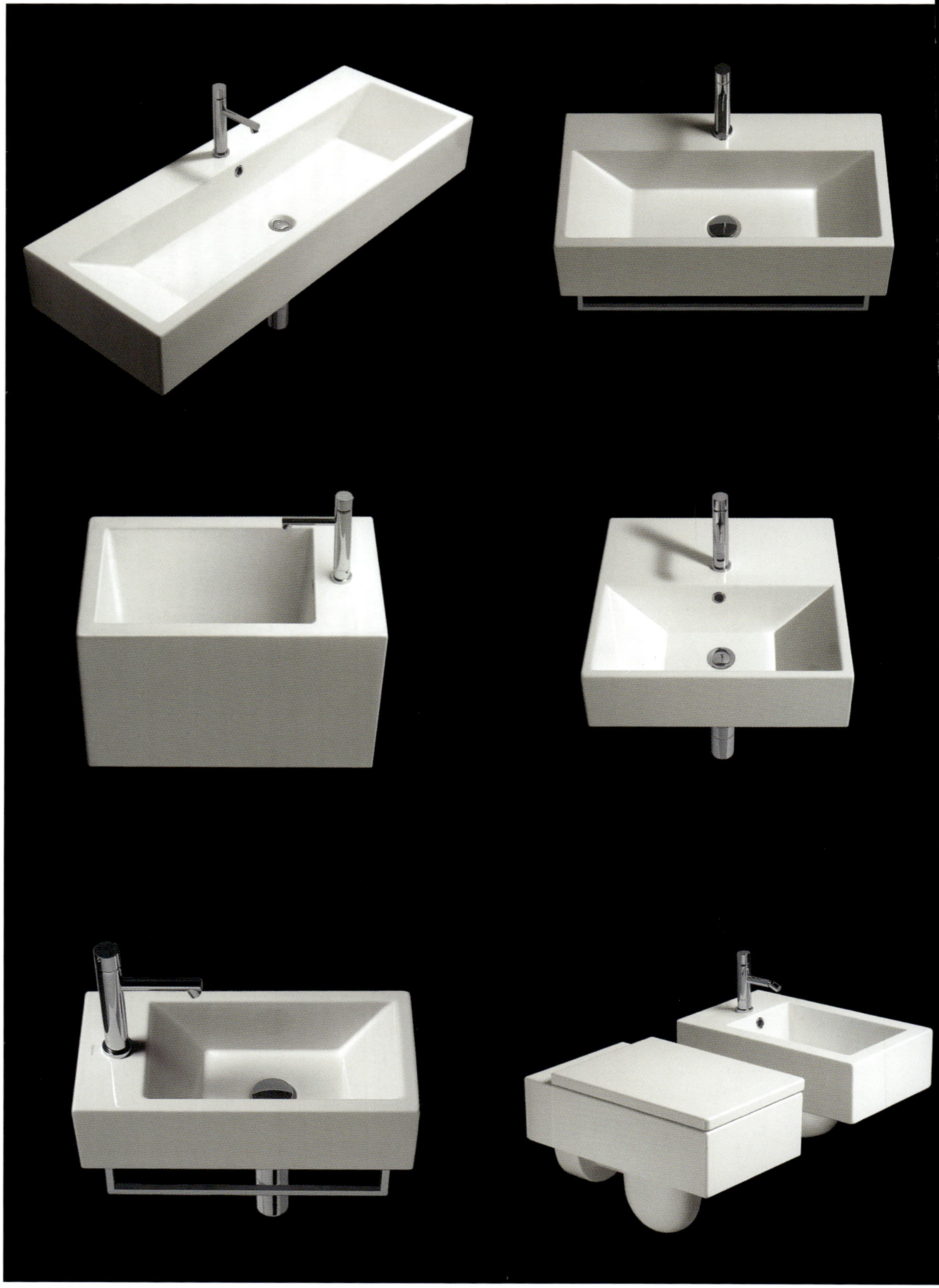

Zu Sistema Verso gehören fünf Waschtische, ein Wand- WC und ein Bidet, die alle auf der Grundform des Rechtecks beruhen. Die Waschtische Verso 50, 55, 65, 70 und 120 können den räumlichen Gegebenheiten entsprechend als Handwaschbecken, Waschtrog, Waschtisch mit großer Ablagefläche etc. gewählt werden. Verschiedene Installationsvarianten, an der Wand, auf der Konsole, freistehend oder auf einer Holzbank aus Teak, ermöglichen zusätzlichen Gestaltungsspielraum. WC und Bidet sind im gleichen minimalistischen Design wie die Waschtische gehalten.

The Sistema Verso line includes five different washbasins, a wall-mounted WC and a bidet, all of which are based on an underlying rectangular shape. The Verso 50, 55, 65, 70 and 120 basins provide solutions for all spatial requirements: a hand washbasin, a sink, a washstand with large sides, etc. Different installation choices, for example wall- or console-mounted, standalone or on a teak bench enhance the range of design choices. The WC and bidet are defined by the same minimalist look as the washbasins.

iF Silver Award

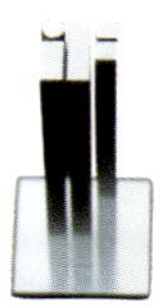

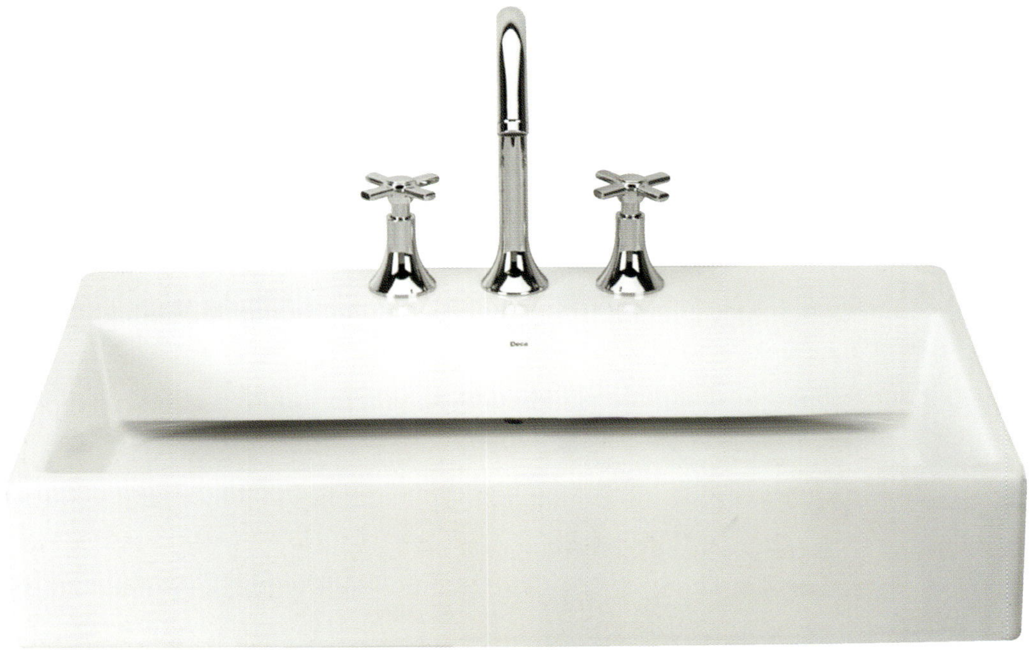

Hersteller Manufacturer
Deca
São Paulo, Brazil

Produkt Product
Washbasin with hidden valve
Waschbecken
Washbasin

Design
Edison Luiz Anholon
São Paulo, Brazil

Markantes Designmerkmal des Waschbeckens ist der versteckte Abfluss. Das Wasser fließt durch ein Loch von 25 mm Durchmesser ab, das hinter einer waagerechten Öffnung liegt, die am gesamten unteren rückwärtigen Innenrand des Beckens verläuft. Das Waschbecken wird aus feuerfester Sanitärkeramik hergestellt, in Fachkreisen als Fire Clay bekannt, aus der größere Teile, ebene Oberflächen und gerade Linien gefertigt werden können. Es kann sowohl an der Wand befestigt als auch auf Möbeln aus verschiedenen Materialien aufgelegt werden.

The eye-catching design feature of the washbasin is its concealed drain. The water flows out through a hole 25 mm in diameter positioned behind a horizontal slit that runs along the entire rear inside wall of the basin. The washbasin is manufactured using a fine sanitary ceramics, known in the industry as fire clay, which can be made into larger workpieces, smooth surfaces and straight lines. It can be wall-mounted or also placed above furniture items made of different materials.

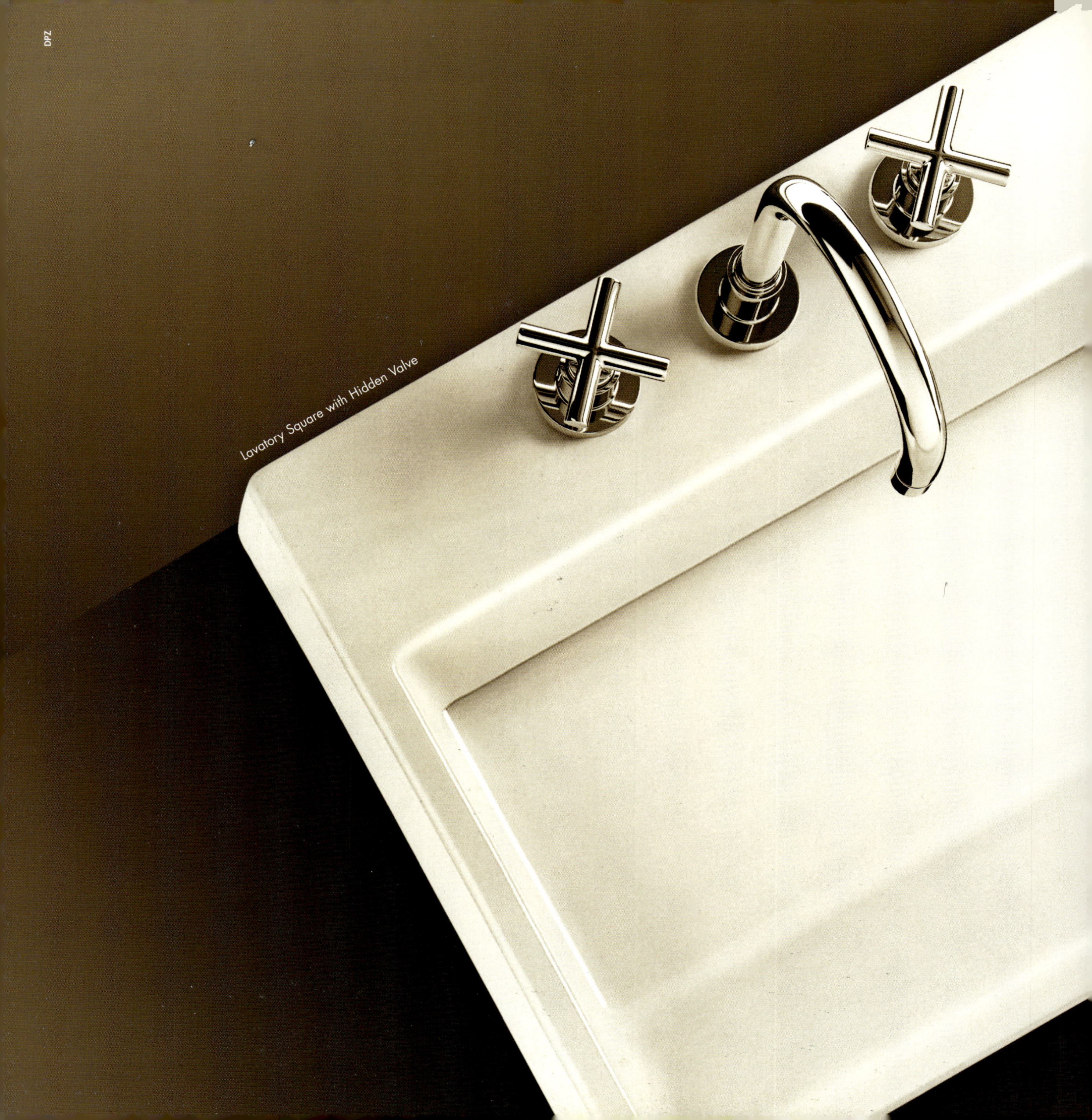

Lavatory Square with Hidden Valve

Hersteller Manufacturer
Hoesch Metall + Kunststoffwerk GmbH & Co.
Kreuzau, Germany

Produkt Product
Edition Putman
Badewanne
Bathtub

Design
SNCL Andrée Putman
Paris, France

Die freistehende Badewanne Edition Andrée Putman mit den Abmessungen 1900 x 1000 x 450 mm besteht aus hochwertigem Sanitäracryl. Der Wannenkörper erhält Leichtigkeit durch die Wölbung an den Außenseiten und die vier gerundeten Ecken, auf denen er ruht. Für eine zweite Variante wurde die Wanne mit Aluminium verkleidet, über der Wanne liegt eine Ablage aus Holz. Die großzügigen Abmessungen, der mittig positionierte Ablauf und die Rückenschrägen ermöglichen zwei komfortable Sitzpositionen.

The Edition Andrée Putman is a standalone bathtub measuring 1900 x 1000 x 450mm and made of high-grade sanitary acrylic. The tub itself is lent a sense of lightness by the curving outer sides and the four rounded corners on which it rests. A second version features aluminum cladding for the tub and above it lies a wooden rest. The spacious size, the centrally positioned drain, and the angled backrests spell two different and equally comfortable positions for users.

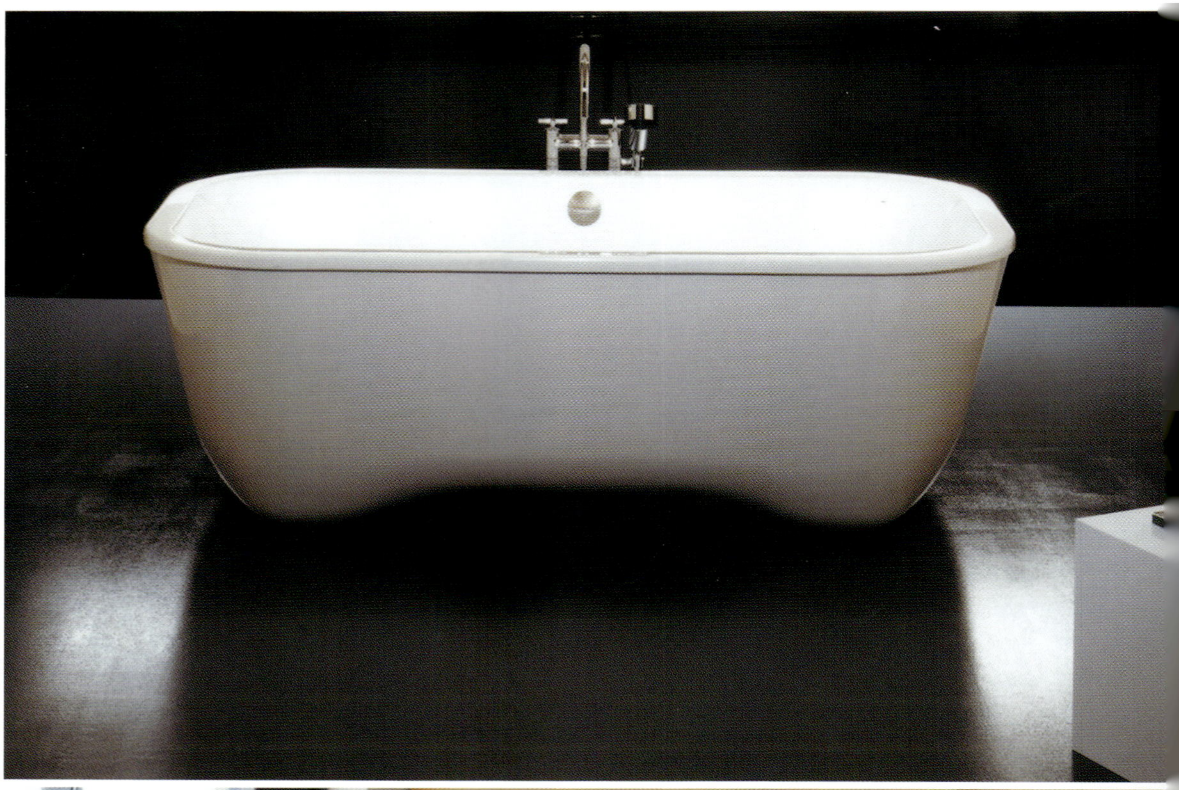

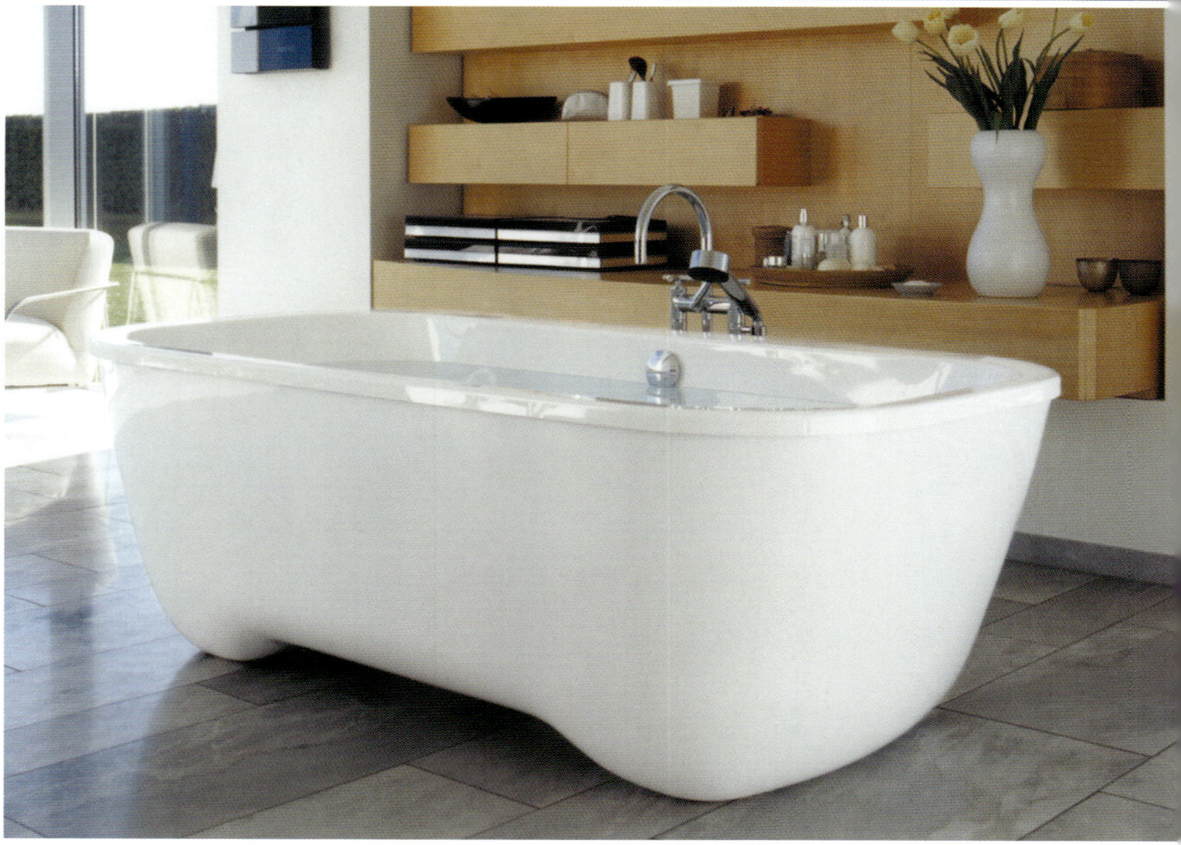

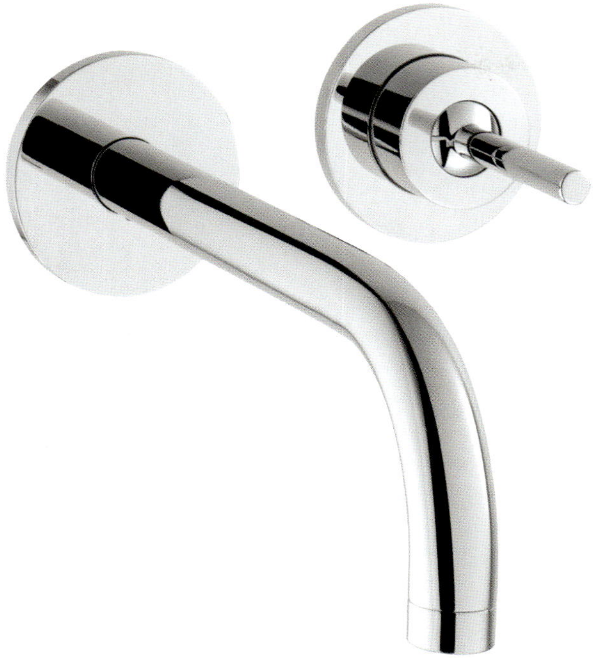

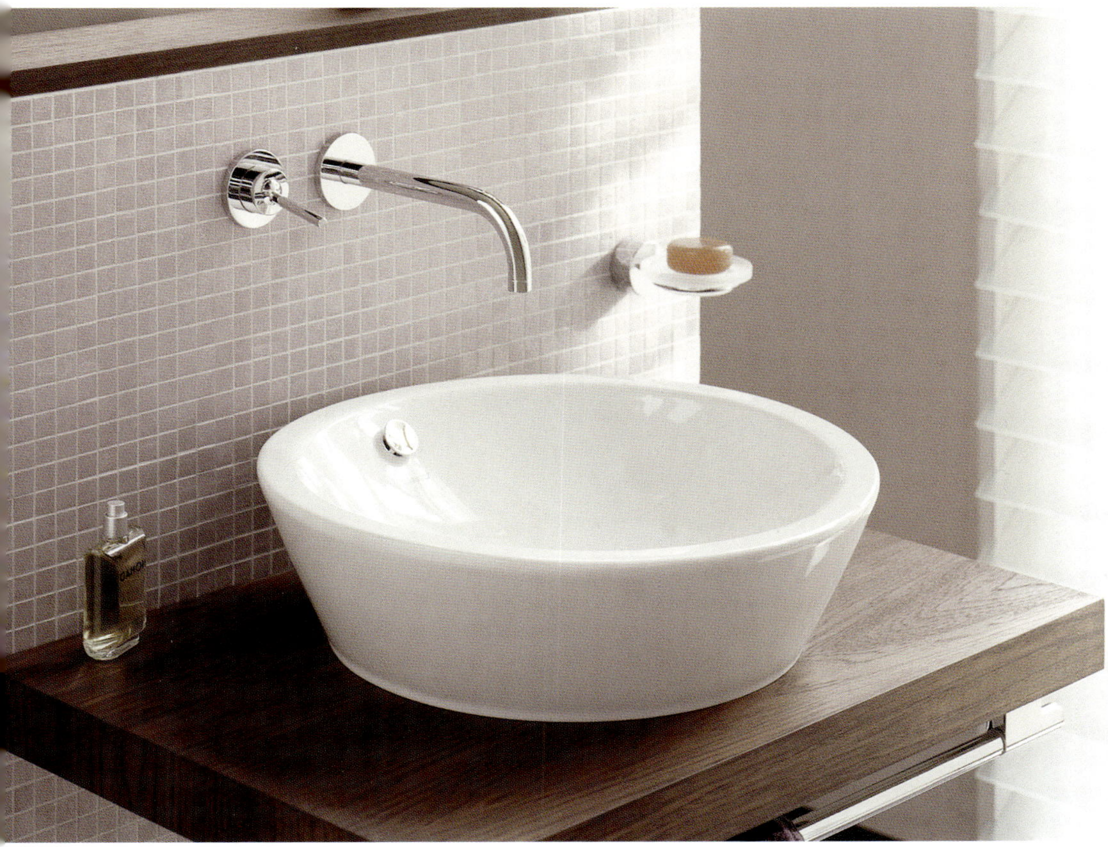

Hersteller Manufacturer
Hansgrohe AG
Schiltach, Germany

Produkt Product
Axor Uno
Einhebel-Waschtischmischer
Single lever basin mixer

Design
Phoenix Design
Stuttgart, Germany

Der Axor Uno Einhebel-Waschtischmischer Unterputz zeichnet sich durch eine puristische, rohrähnliche Anmutung aus. Der einem Joystick ähnliche Griff steuert die Intensität des Wasseraustritts sowie die Kalt-/Warmfunktionen; er lässt sich rechts oder links platzieren. Die Platte fasst Auslaufrohr (mit unterschiedlichen Auslauflängen: 160 mm oder 220 mm) und Griff bündig zusammen. Als Oberflächen stehen glänzendes Chrom und mattes Satinox zur Verfügung.

The Axor Uno single-lever basin mixer is flush-mounted with the wall, and catches the eye with its purist, almost tubular feel. The handle almost resembles a joystick and controls water output as well as the hot/cold functions; it can be positioned on the left or the right. The flat section brings the outflow pipe (two different outflow lengths: 160 mm or 220 mm) and handle smoothly together in a single unit. Available in two finishes: either gleaming chrome or matt Satinox.

Hersteller Manufacturer
Hansgrohe AG
Schiltach, Germany

Produkt Product
Pharo Duschtempel 100 Shower
Duschtempel
Shower Temple

Design
Phoenix Design
Stuttgart, Germany

Der Pharo Duschtempel Shower in offener Traversenkonstruktion wirkt ohne störende Seitenwände leicht und transparent, der Einstieg erfolgt über eine einflügelige Tür aus Sicherheitsglas. Funktionsteile wie Wasserführung und Armaturen wurden auf ein Minimum reduziert; die gesamte Technik verbirgt sich hinter einem Eckprofil. Schlanke Aluminiumsäulen tragen die Außenhaut aus Sicherheitsglas, die Kopfbrause und die Halogenbeleuchtung. Zur Duscheinheit gehören eine stufenlos verstellbare Handbrause, die Raindance-Kopfbrause mit fünf Strahlarten und acht Multispray-Seitenbrausen.

The Pharo Shower Temple with its open structure based on traverses seems light and transparent thanks to the absence of disruptive sidewalls. It is entered through a single-section door made of safety glass. All functional parts such as water piping and fixtures have been reduced to a minimum, with the technology hidden away behind a corner profile. Slender aluminum columns bear the weight of the safety-glass outer skin, the shower handset and the halogen lighting. The unit comes with an infinitely adjustable handset, the Raindance shower head, which offers five different spray types and eight multispray side shower heads.

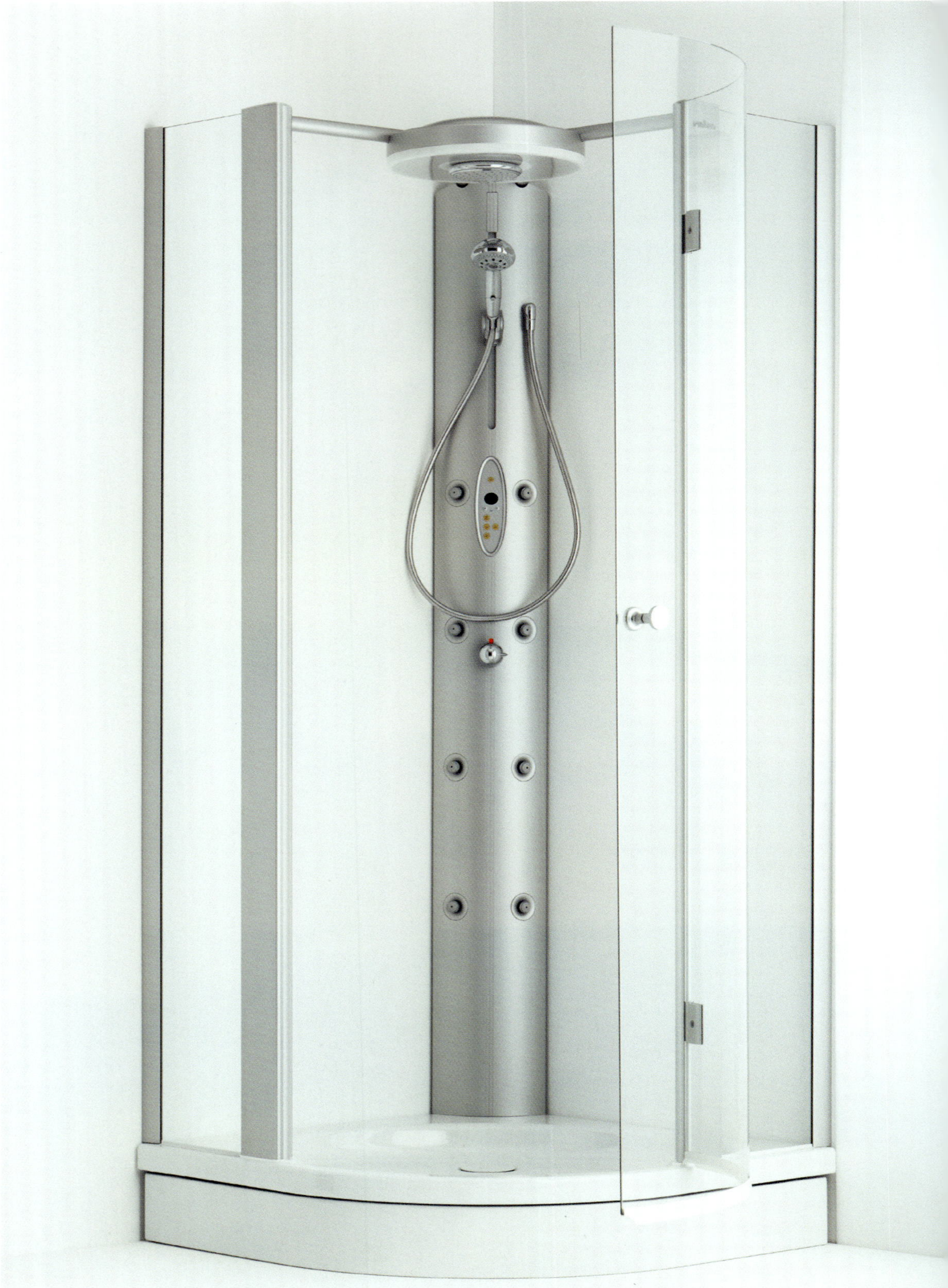

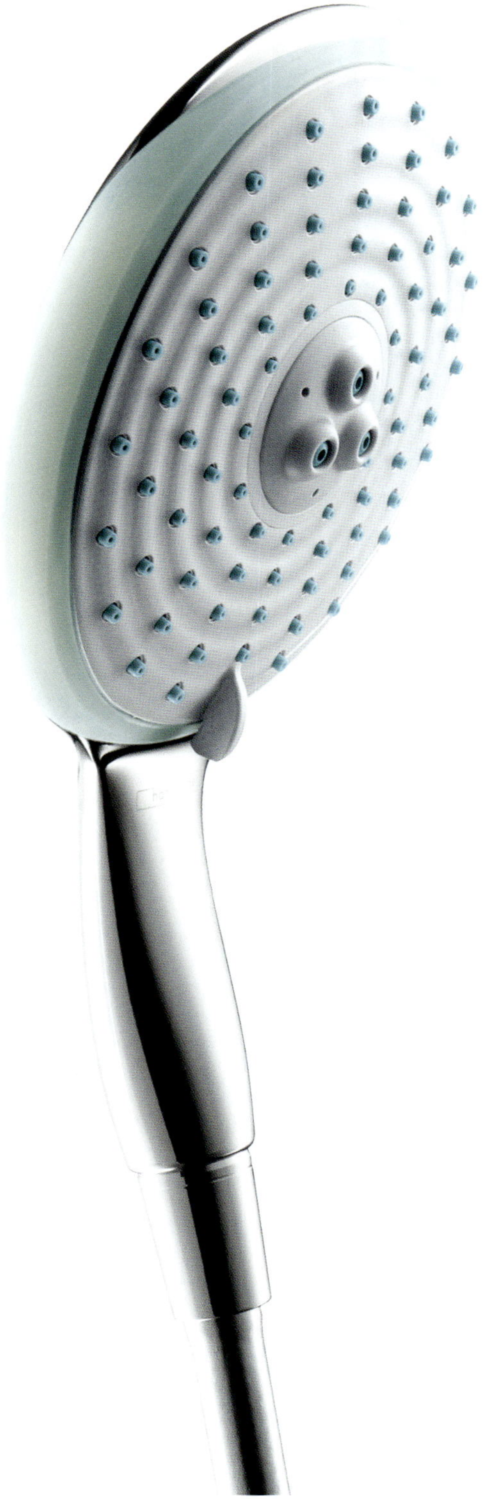

Hersteller Manufacturer
Hansgrohe AG
Schiltach, Germany

Produkt Product
Raindance Handbrause
Handbrause
Hand shower

Design
Phoenix Design
Stuttgart, Germany

Für die Handbrause Raindance wurden die gewohnten Proportionen auf den Kopf gestellt; der Durchmesser des Duschkopfs beträgt ungewöhnliche 150 mm, im Gegenzug wurde der üblicherweise lange Griff verkürzt. Die komplette Brause hat eine flache Silhouette. Die ergonomische Gestaltung erlaubt auch bei kleinen Händen eine einhändige Einstellung der fünf Strahlarten. Die 150 mm Strahlscheibe ermöglicht aus 180 Öffnungen eine vollständige Benetzung des Körpers. Der Soft-Grip-Einsatz aus transluzentem Silikon hält die Brause in rutschsicherer Position im Halter.

The Raindance shower handset turns the conventional proportions upside down. The shower head is an extraordinary 150 mm in diameter, while in contrast the usual long handle is very short. The silhouette of the hand set is as a whole slender, and the ergonomic design means that even small-handed users can adjust the five nozzle setting using only one hand. The 150 mm nozzle disc boasts 180 openings for a complete wash of the body. And the soft-grip section made of translucent silicone keeps the shower handset firmly in position on the holder.

Hersteller Manufacturer
Franz Kaldewei GmbH & Co. KG
Ahlen, Germany

Produkt Product
Modell 5871
Pool-Griff
Pool handle

Design
Phoenix Design
Stuttgart, Germany

Der Pool-Griff sorgt für komfortablen und sicheren Ein- und Ausstieg beim Baden. Für einen bequemen Halt sorgt die Höhe des Griffs, der 25 cm über die Wanne hinausreicht. Die halbrunde Griffabstützung kann zusätzlich als Handtuchaufnahme dienen, das sich auf diese Weise immer in bequemer Reichweite befindet.

The Pool-Griff handle enables users to climb safely in and out of the bath tub. The handle extends 25 cm above the tub, making it easy to grasp safely. The semi-circular form of the support handle doubles up as a towel rail always within easy reach.

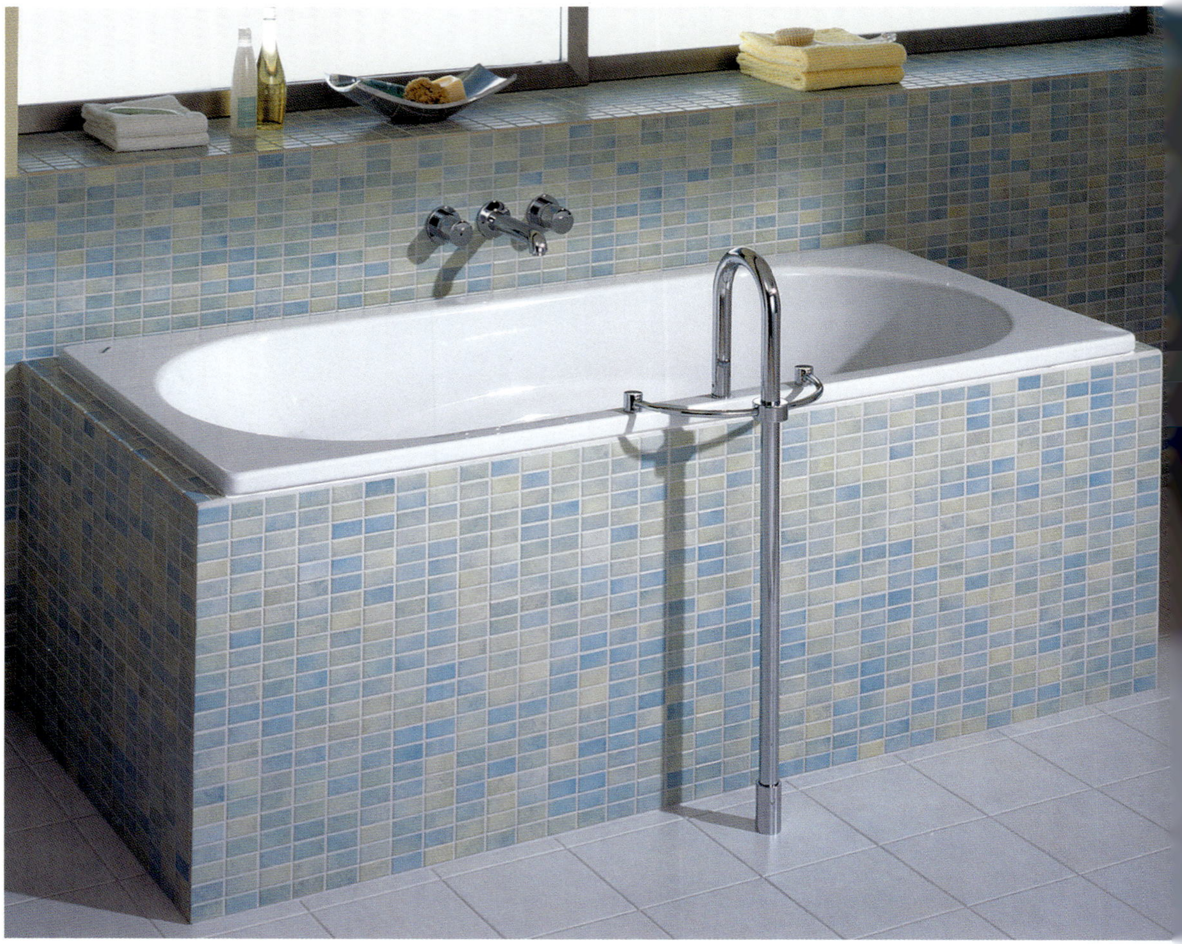

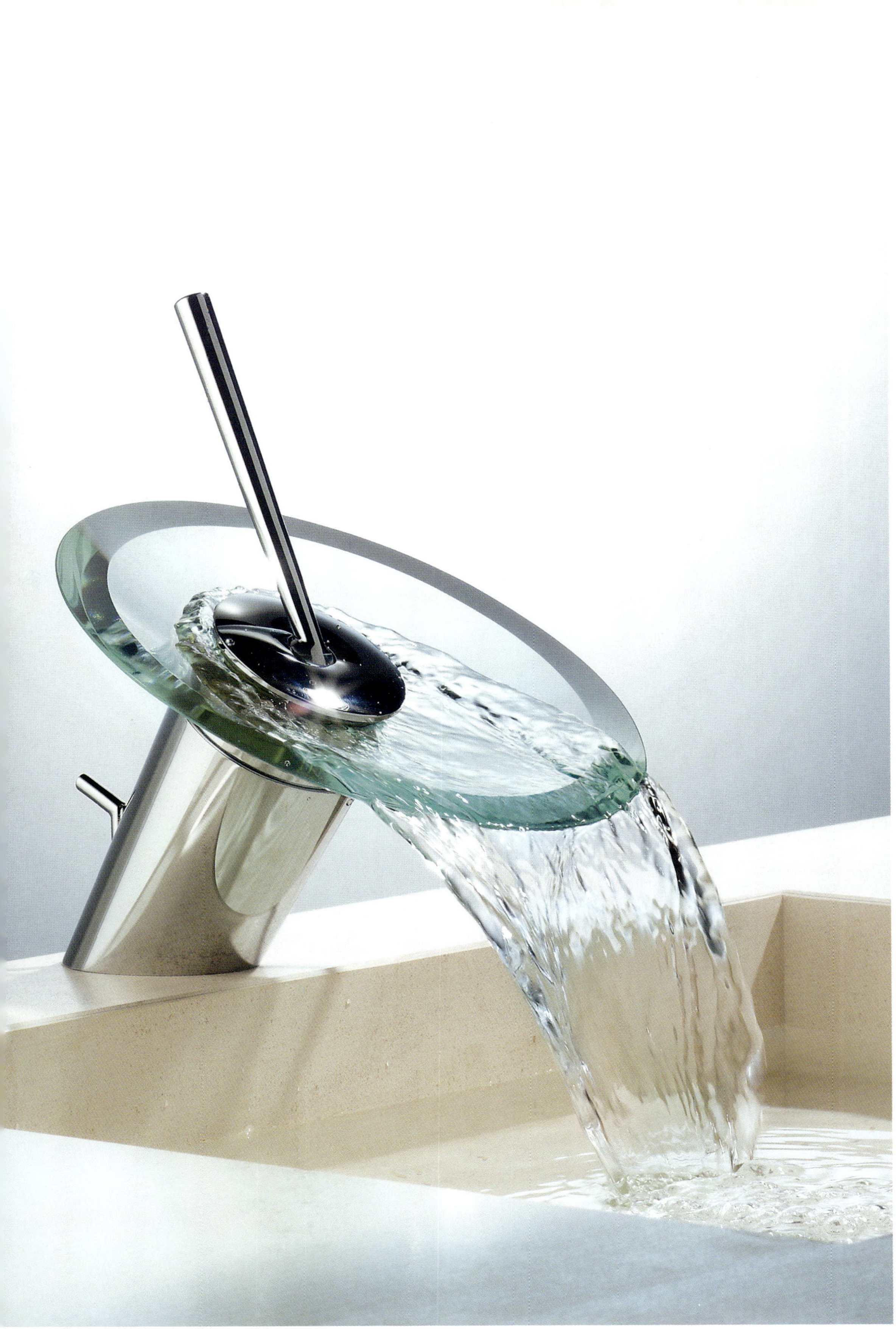

Hersteller Manufacturer
Hansa Metallwerke AG
Stuttgart, Germany

Produkt Product
Hansamurano
Waschtisch-Einhebelmischer
Single-lever basin mixer

Design
Bruno Sacco
Reinhard Zetsche
Octopus Produktdesign
München, Germany

Der Waschtisch-Einhebelmischer Hansamurano erinnert in seiner Formgebung an einen Brunnen. Das Wasser tritt drucklos aus der Armatur aus und fließt weich als Wasservorhang über den Rand der Glasschale. Die Mengen- und Temperatureinstellung wird durch eine keramische Mischpatrone realisiert, wobei die Bedienung über einen senkrecht zur Schale stehenden Stift erfolgt. Eine spezielle Oberflächenveredelung der Glasschale vermindert das Absetzen von Kalkrückständen und Schmutz auf der Armatur und erleichtert das Reinigen.

The shape of the Hansamurano single-lever basin mixer brings a small fountain to mind. The water flows very gently and pressure-free from the taps and spreads out to form a curtain that then cascades softly from the edge of the glass disc. Quantity and temperature can be set using a ceramic mixer module operated via a bar set vertical to the shell. The special water-repellent surface coating of the glass disc more or less eliminates lime scaling and dirt on the disc, while also facilitating cleaning.

Hersteller Manufacturer
Damixa A/S
Odense, Denmark

Produkt Product
Arc
Multifunktionsarmatur
Multifunctional tap

Design
Dr. Alfons Knapp
Jacob Jensen
Damixa A/S
Odense, Denmark

Die Multifunktionsarmatur Arc für Küche und Bad ist mit drei individuell rotierenden Achsen ausgestattet, die eine 360-Grad-Drehung des Armaturenkörpers und des Perlators sowie einen vertikalen 180-Grad-Bewegungsspielraum des Auslaufarmes ermöglichen. Mit dieser Flexibilität in der Handhabung kann der Wasserstrahl punktgenau in die gewünschte Richtung gelenkt werden. Der Joystick-Griff und der Perlator sind mit einem griffsicheren Kunststoff ummantelt, die Armatur ist wahlweise in den Oberflächen Stahl oder Chrom erhältlich.

The Arc multifunctional kitchen and bathroom tap is fitted with three separate rotating axes, the spout and the aerator can rotate 360°, and the spout arm can also turn in the vertical plane across 360°. By virtue of this flexibility, the water jet can be directed in the precise direction required. The joystick handle and the 'aerator' are covered in a sure-grip rubberized finish, the fittings are available with either a brushed steel or a chrome surface.

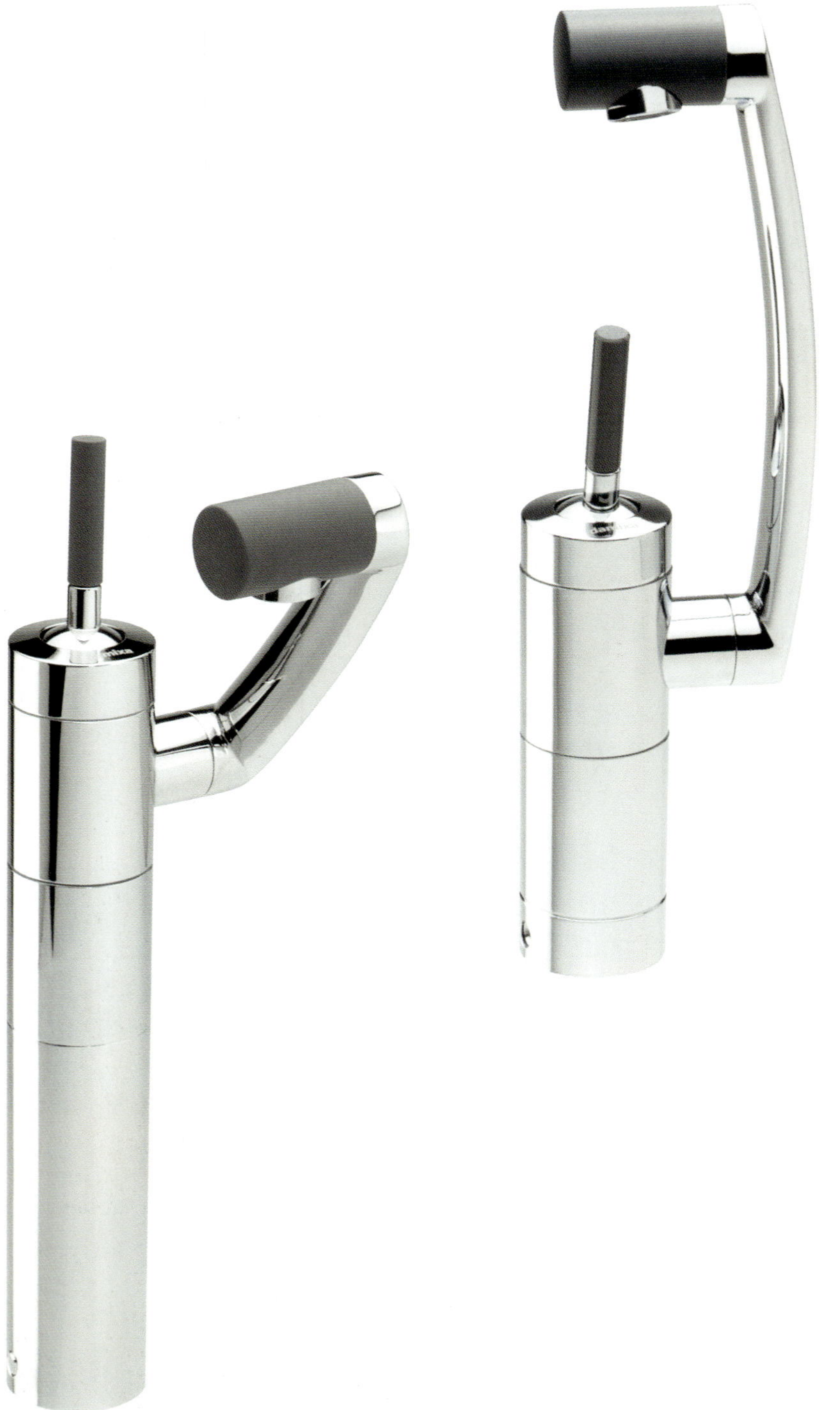

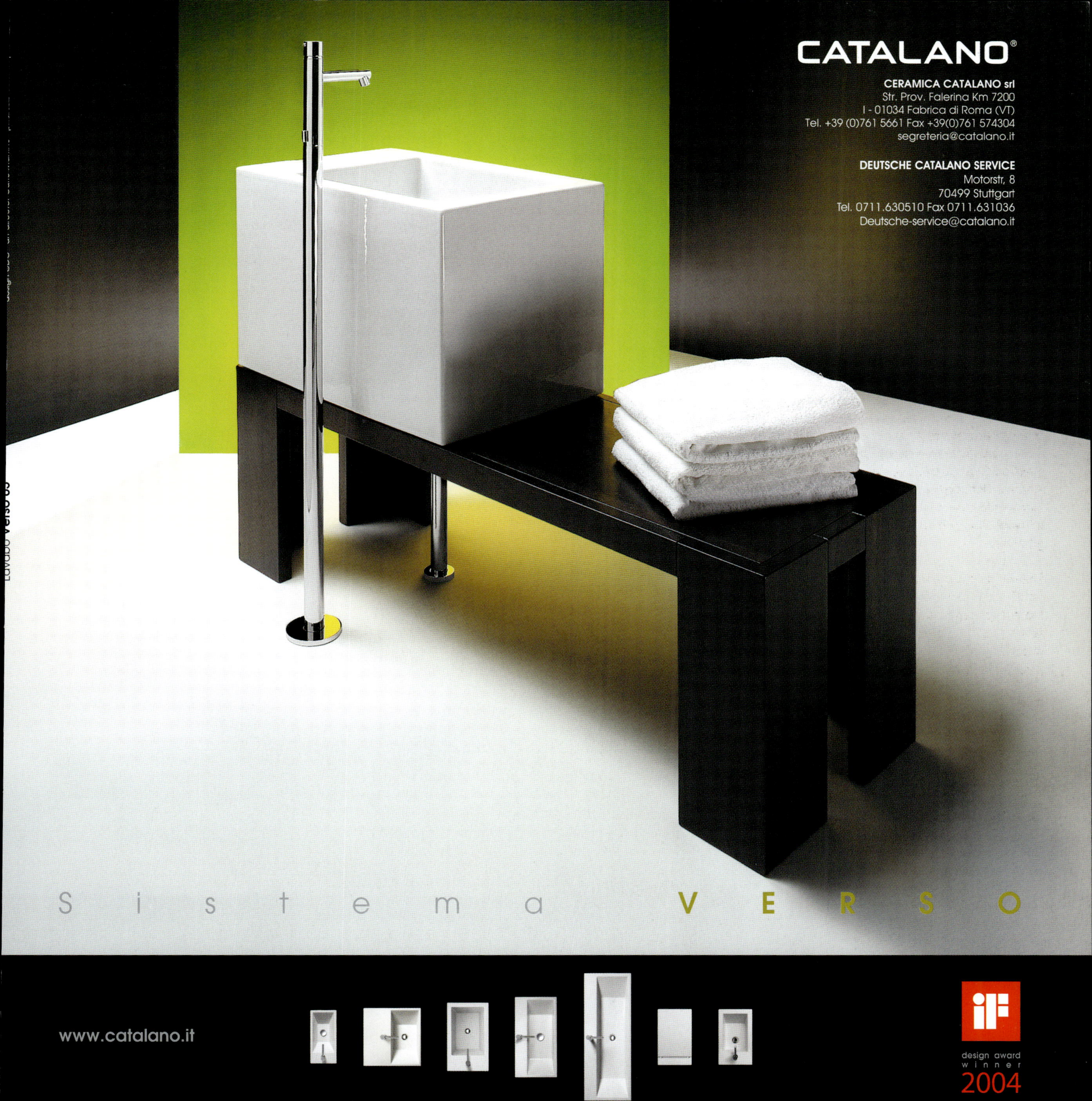
CATALANO®
CERAMICA CATALANO srl
Str. Prov. Falerina Km 7200
I - 01034 Fabrica di Roma (VT)
Tel. +39 (0)761 5661 Fax +39(0)761 574304
segreteria@catalano.it
DEUTSCHE CATALANO SERVICE
Motorstr, 8
70499 Stuttgart
Tel. 0711.630510 Fax 0711.631036
Deutsche-service@catalano.it
Sistema VERSO
www.catalano.it
iF
design award winner
2004

Hersteller Manufacturer
Aloys F. Dornbracht GmbH & Co. KG
Iserlohn, Germany

Produkt Product
MEM
Waschtisch-Dreilochbatterie
Three-hole basin mixer

Design
Michael Sieger
sieger design
Sassenberg, Germany

Die MEM Armaturen, hier die Waschtisch-Dreilochbatterie, repräsentieren mit ihrem flachen, breiten Auslauf eine neue, elementare Armaturenästhetik. Neben verschiedenen Waschtisch-Varianten deckt dieses neue Armaturen-Programm alle Funktionsbereiche im Bad ab. Alle Varianten sind mit Einzelrosetten oder formschönen Abdeckplatten erhältlich. Als Reminiszenz an Wasser, das aus der Quelle entspringt, wurde zudem bei allen Armaturen auf einen Luftsprudler verzichtet, sodass der Schwall frischen Wassers unmittelbar erlebbar wird.

MEM fittings, here to see at the new three-hole basin mixer, represent a new sphere of elemental fitting aesthetics with their flat, broad spouts. Next to several wash basins-versions, this new fitting program covers every function area within a bathroom. All versions are available with single rosettes or well-shaped covering plates. Moreover, to recreate the effect of water flowing out of a spring, none of the fittings is equipped with a bubble filter, in such a way to experience the flow of fresh water as known by nature.

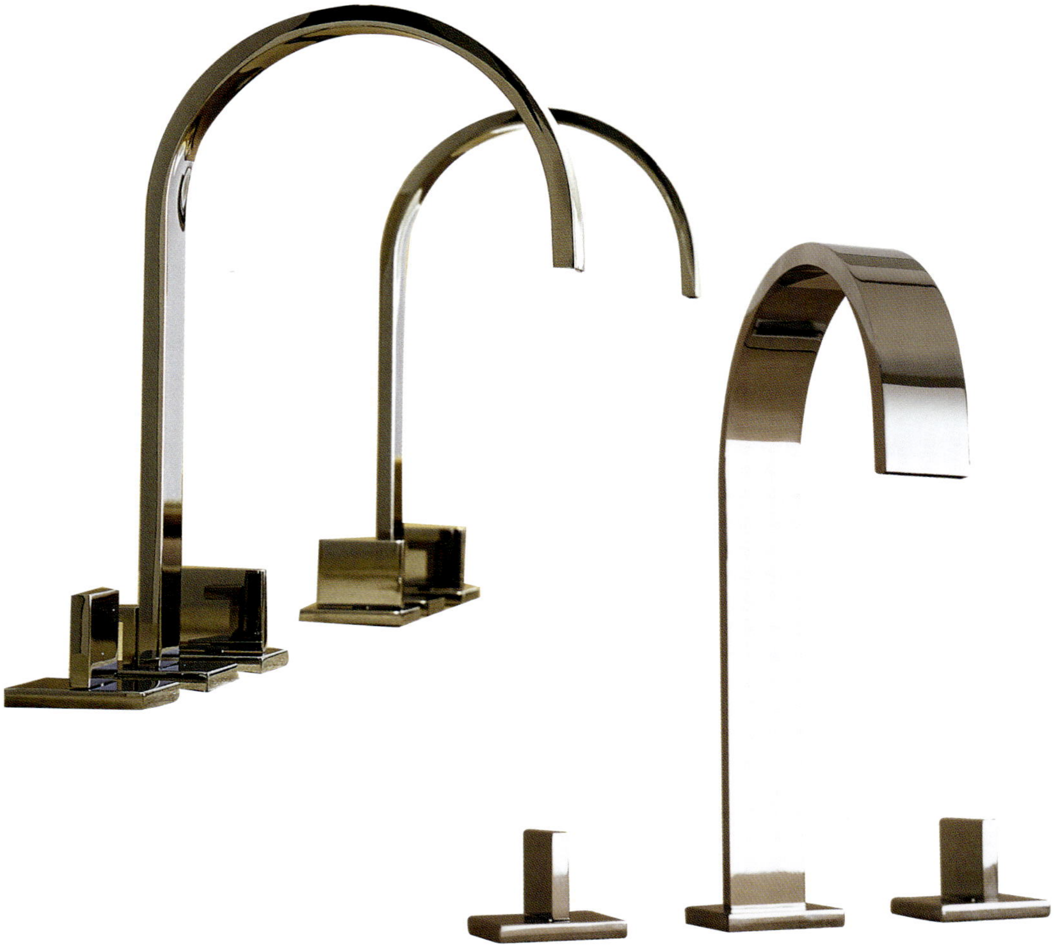

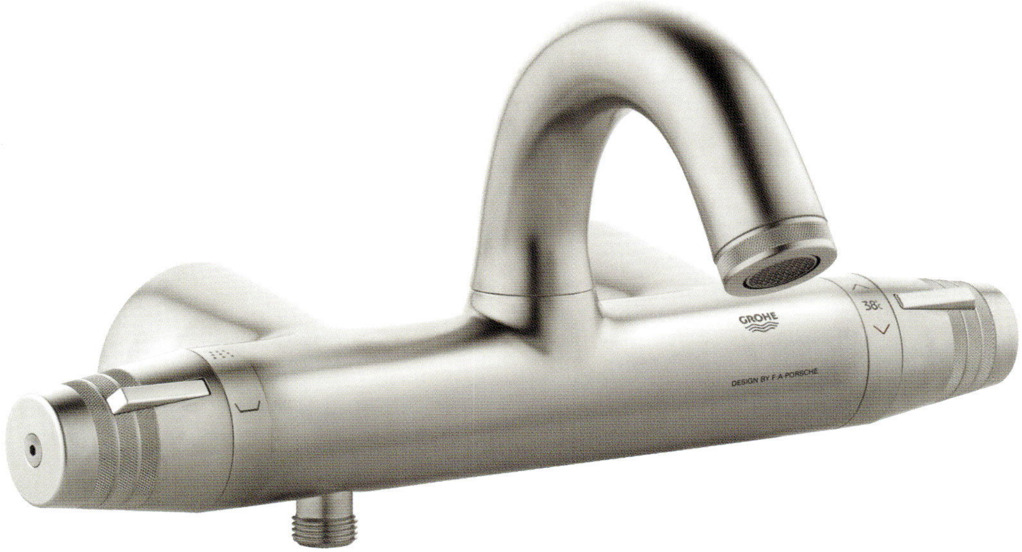

Hersteller Manufacturer
Grohe Water Technology AG & Co. KG
Hemer, Germany

Produkt Product
F1 Thermostat Wannenarmatur
Thermostat bathtub faucet F1

Design
Porsche Design
Zell am See, Austria

Der Körper der Thermostat-Wannenarmatur F1 basiert auf den klaren Grundelementen des Zylinders und Kegels. Die sachliche Geometrie geht in einen harmonisch geformten Gussauslauf über, der den Fluss des Wassers widerspiegelt. Besonderes Augenmerk wurde auf die Ergonomie des Thermostaten gelegt, eine klare Skala erlaubt ein leichtes Ablesen der Einstellung, die Sicherheitssperre verhindert Verbrühungen. Eine spezielle Technologie ermöglicht sowohl die einhändige wie auch eine getrennt zu bedienende Regulierung von Temperatur und Menge. Die Präzisionsoberfläche in ALU-XT verleiht dem Produkt einen sensitiv – technoiden Gesamteindruck.

The body of the thermostat bathtub faucet F1 is based on the clear basic elements of the cylinder and cone. The matter-of-fact geometry merges with a harmoniously shaped cast spout inspired by the flow of water itself. Special attention has been paid to the ergonomic qualities of the unit, with a clear scale providing a reliable display of the settings. A shut-off valve protects against scalding. Special technology enables both single-handed or separate regulation of temperature and amount of water. The high-precision ALU-XT surfaces make the product technoid and sensual at the same time.

Hersteller Manufacturer
Grohe Water Technology AG & Co. KG
Hemer, Germany

Produkt Product
F1 Thermostat Brausenarmatur
Thermostat shower faucet F1

Design
Porsche Design
Zell am See, Austria

Der Körper der Thermostat-Brausenarmatur F1 basiert auf den klaren Grundelementen des Zylinders und Kegel. Besonderes Augenmerk wurde auf die Ergonomie des Thermostates gelegt, eine klare Skala erlaubt ein leichtes Ablesen der Einstellung, die Sicherheitssperre verhindert Verbrühungen. Eine spezielle Technologie ermöglicht sowohl die einhändige wie auch eine getrennt zu bedienende Regulierung von Temperatur und Menge. Die Präzisionsoberflächen in ALU-XT verleihen dem Produkt einen sensitiv – technoiden Gesamteindruck.

The body of the thermostat shower faucet F1 is based on the clear basic elements of the cylinder and cone. Special attention has been paid to the ergonomic qualities of the unit, with a clear scale providing a reliable display of the settings. A shut-off valve protects against scalding. Special technology enables both single-handed or separate regulation of temperature and amount of water. The high-precision ALU-XT surfaces make the product technoid and sensual at the same time.

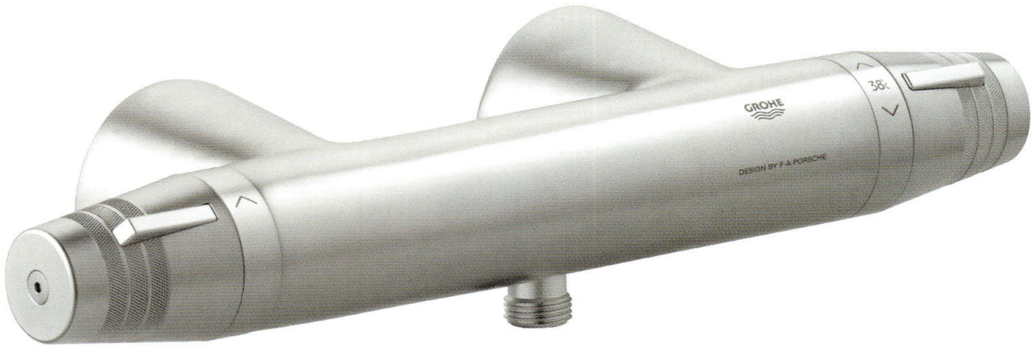

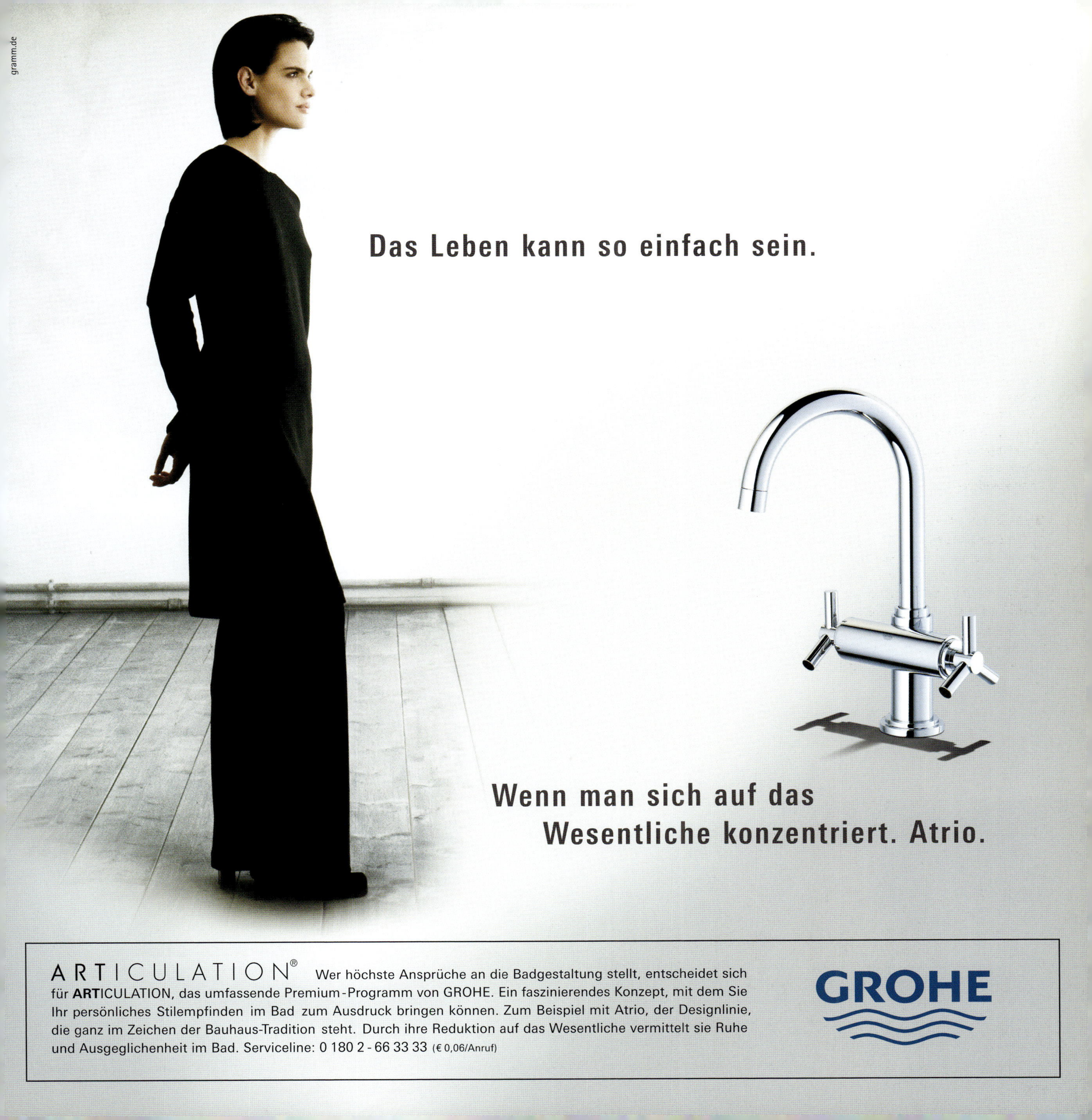
gramm.de
Das Leben kann so einfach sein.
Wenn man sich auf das
Wesentliche konzentriert. Atrio.
ARTICULATION®
Wer höchste Ansprüche an die Badgestaltung stellt, entscheidet sich für ARTICULATION, das umfassende Premium-Programm von GROHE. Ein faszinierendes Konzept, mit dem Sie Ihr persönliches Stilempfinden im Bad zum Ausdruck bringen können. Zum Beispiel mit Atrio, der Designlinie, die ganz im Zeichen der Bauhaus-Tradition steht. Durch ihre Reduktion auf das Wesentliche vermittelt sie Ruhe und Ausgeglichenheit im Bad. Serviceline: 0 180 2 - 66 33 33 (€ 0,06/Anruf)
GROHE

Hersteller Manufacturer
John Somers Estanho ltda
São João Del Rey, Brazil

Produkt Product
Santa Fé Home object line
Zinnobjekte
Tin objects

Design
Carlos Augusto de Souza Alcantarino
Rio de Janeiro, Brazil

Mit der Santa Fé Home Object-Serie will man die Wertschätzung des „aus der Mode" gekommenen Werkstoffs Zinn neu beleben und ihm ein neues ästhetisches Konzept verleihen. Das vom Designer entwickelte Konzept zielt auf die Schaffung einer zeitgenössischen neuen Betrachtungsweise typischer handwerklicher Gegenstände aus dem Gebiet Minas Gerais. Zur Serie gehören eine Amphore, eine kleine, mittlere und große Vase, eine Einzelvase und Tischschmuck.

The Santa Fé Home Object series is a strong effort to reanimate the former appreciation for tin as a material that has since gone out of fashion by instilling it with a new aesthetic lease of life. The concept created by the designer is intended to foster a new contemporary vision of the typical handicrafts items from the Minas Gerais region. The series includes an amphora, a small, a medium and a large vase, an individual vase and table decorations.

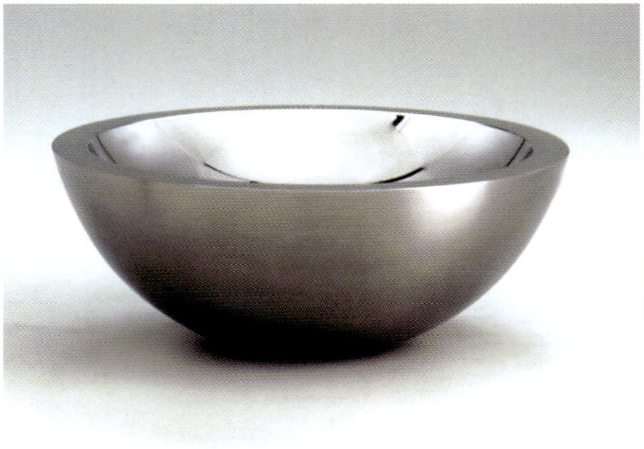

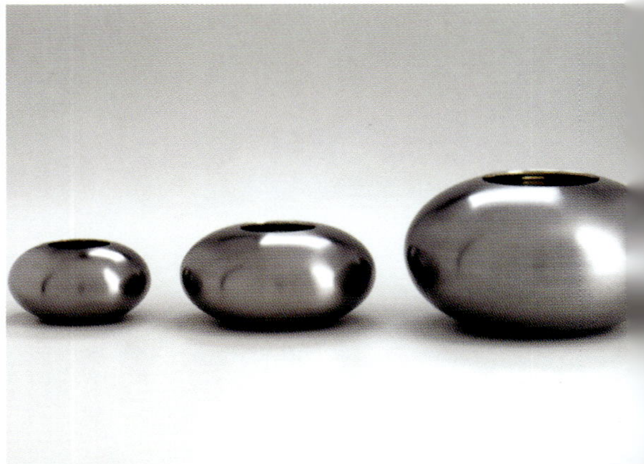

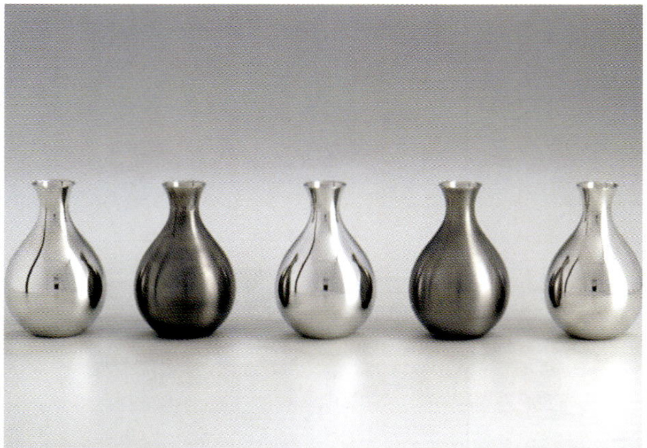

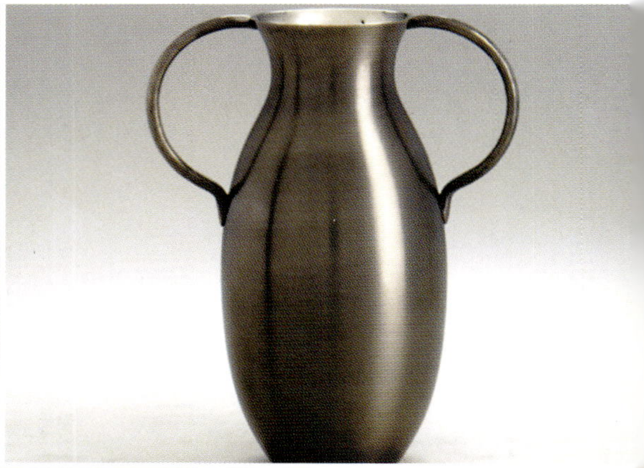

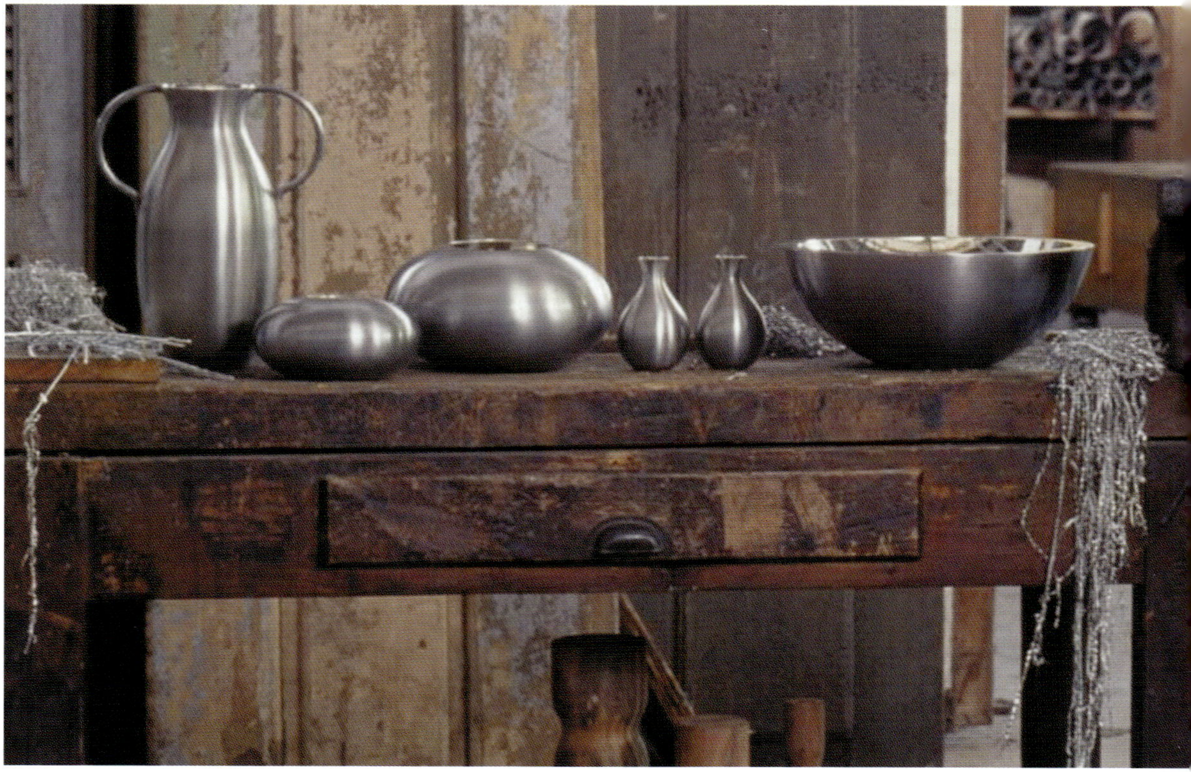

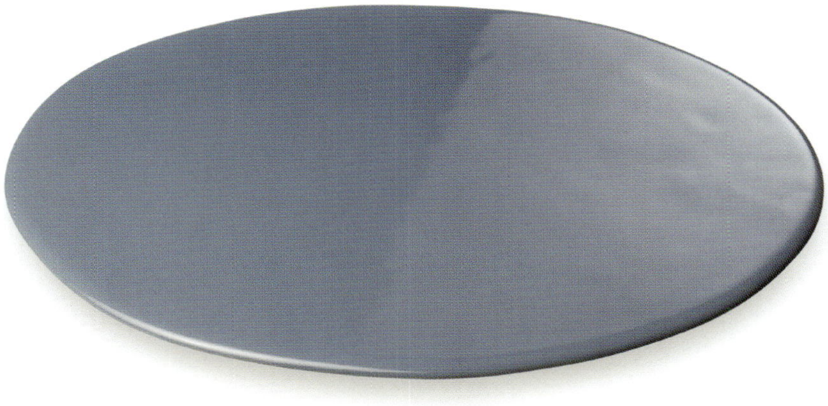

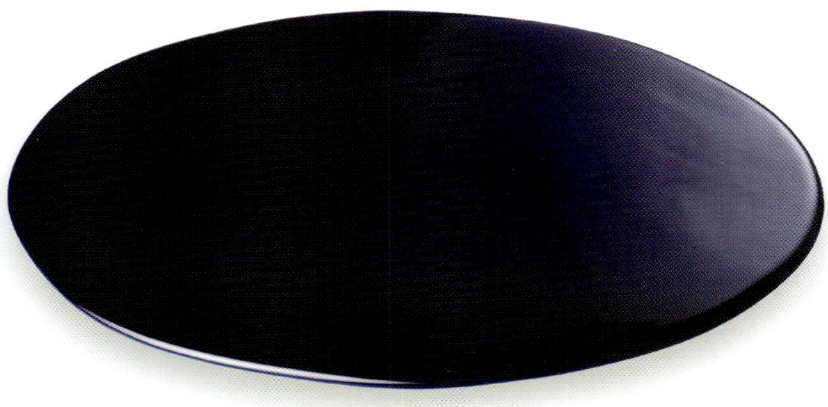

Hersteller Manufacturer
iittala oy ab / Arabia
Helsinki, Finland

Produkt Product
Invenit
Tisch-Sets
Table mats

Design
Heljä Liukko-Sundström
iittala oy ab / Arabia
Helsinki, Finland

Die drei an eine Wasserlache erinnernden Tisch-Sets wurden von der finnischen Designerin Heljä Liukko-Sundström im Rahmen der neuen Pro Arabia Art Finland 2002 Kollektion entworfen. In ihrer Vorliebe für melancholische Landschaften assoziiert die Designerin mit der glänzenden Oberfläche der Tischauflagen die unberührte, glasklare Wasserfläche eines finnischen Sees.

The three table mats reminiscent of a water surface were designed by Finnish designer Heljä Liukko-Sundström as part of the Pro Arabia Art Finland 2002 Collection. The designer's fondness for melancholy landscapes is intimated by the shiny surface of the table mats which she associates with the pristine, glass-clear surface of a Finnish lake.

Hersteller Manufacturer
Dima Freizeitmöbel GmbH
Schlierbach, Germany

Produkt Product
Air
Stapelsessel
Stacking chair

Design
Designbüro Wolfgang C.R. Mezger
Eislingen, Germany

Der zur Gartenmöbel-Kollektion Air gehörende Stapelstuhl besteht aus einer Kombination von Edelstahl, Teak und Gewebe. Die vollständig mit dem witterungsbeständigen und langlebigen Textilgewebe bezogene Sitz- und Rückenschale verleiht dem Sessel einen homogenen Charakter, zudem bietet er hohen Sitzkomfort.

Air is part of a garden furniture collection that hinges on a combination of stainless steel, teak and wovens. The seat and backrest shell are completely covered in a durable fabric that withstands the weather, instilling the chair with a homogeneous look and enhancing sitting comfort.

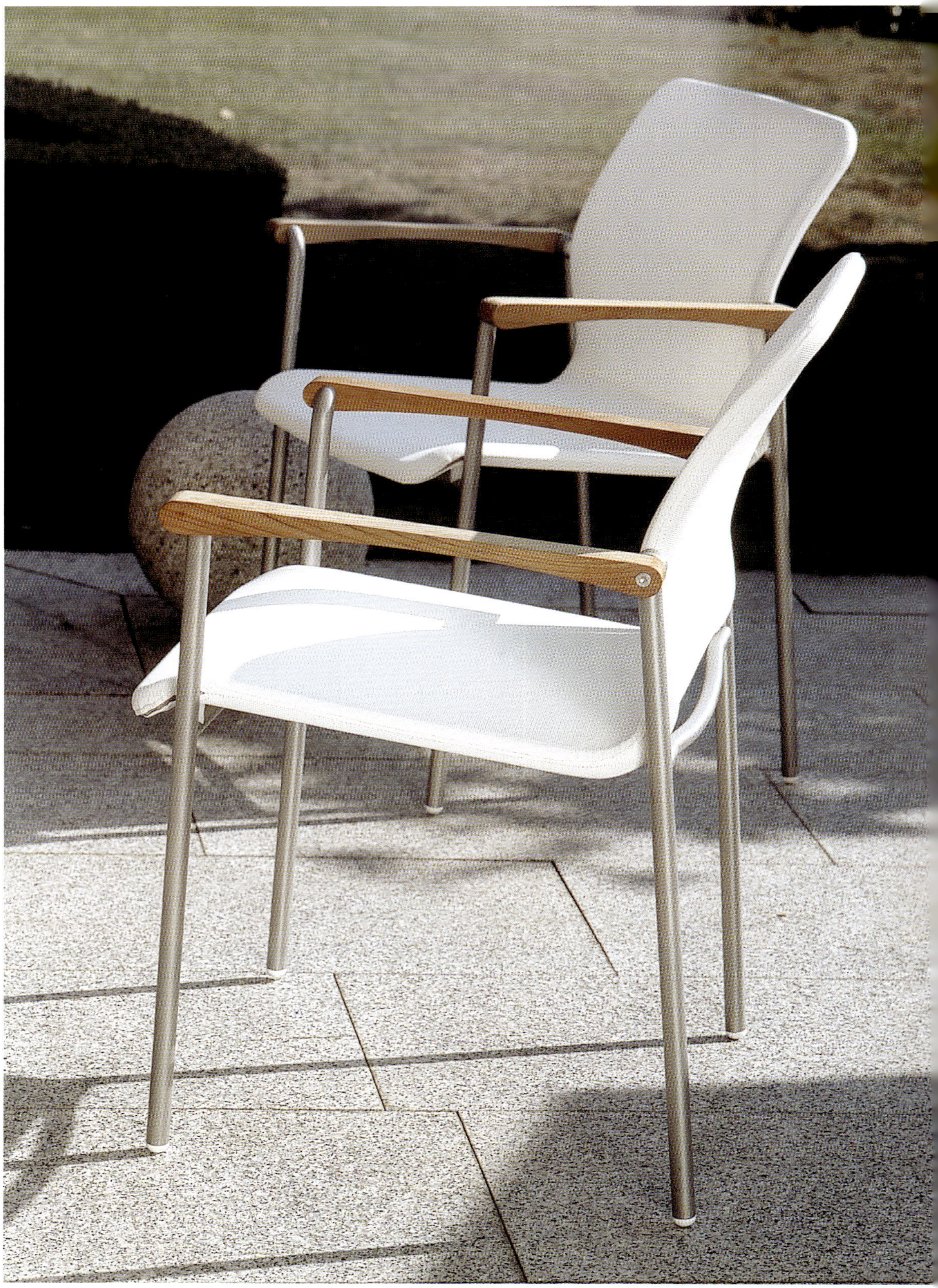

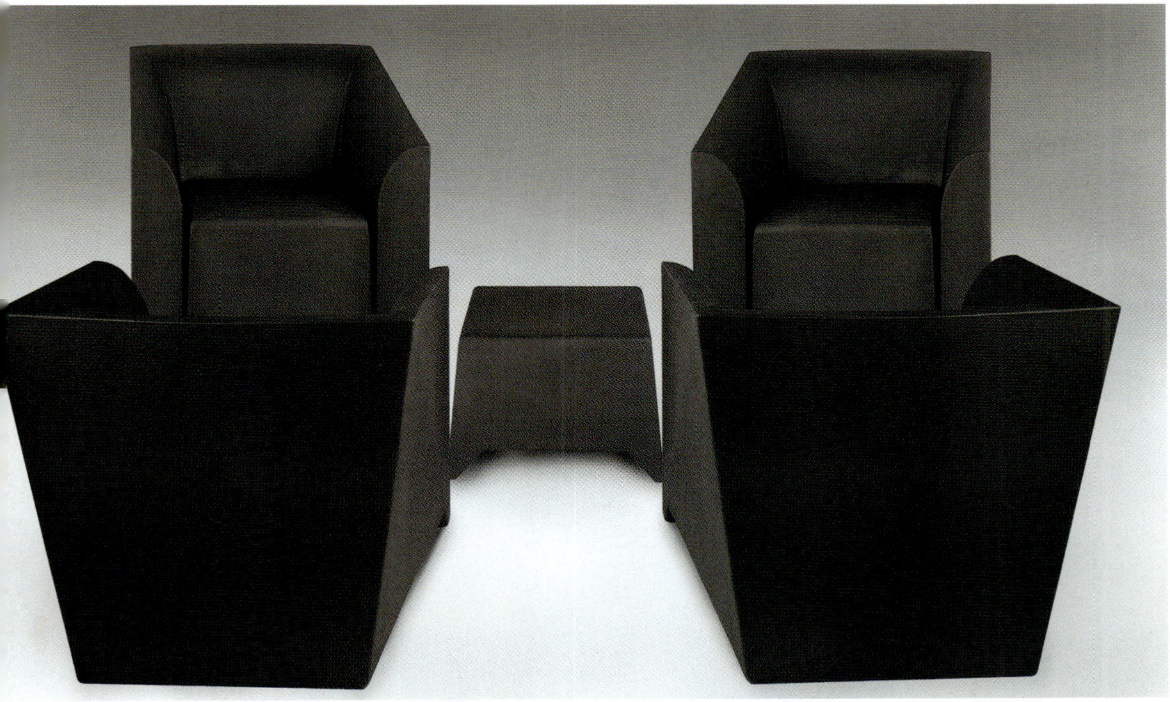

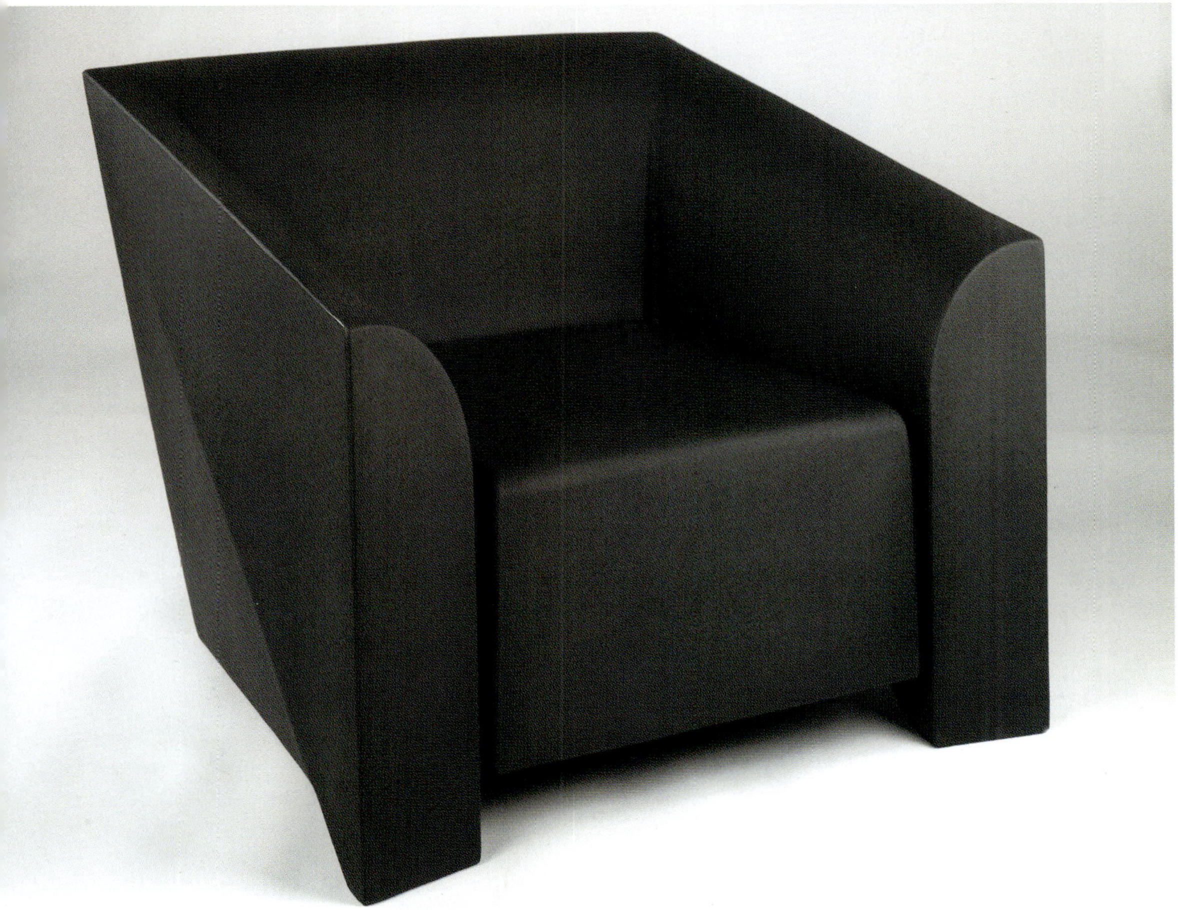

Hersteller Manufacturer
Heller Inc.
New York, USA

Produkt Product
MB1
Sessel
Armchair

Design
Mario Bellini Associati
Mailand, Italy

Der MB1 besteht aus einem gepolsterten Polymer und ist ein ungewöhnlich bequemes Sitzmöbel aus Plastik für den Gebrauch im Haus und im Freien.

The MB1 relies on luxurious self-cushioning polymer to create an unusually comfortable armchair for indoor and outdoor seating that is a real eye-catcher in its own right.

Hersteller Manufacturer
COR Sitzmöbel
Rheda-Wiedenbrück, Germany

Produkt Product
Circo-Solo
Solitärsessel
Armchair

Design
Peter Maly
Hamburg, Germany

Circo-Solo besitzt mit seinem großen Volumen eine geplante Dominanz im Raum. Die Rückenschale ist in Form eines klar konstruierten, angeschnittenen Kegels gezeichnet. Sie umschließt halbkreisförmig den runden Sitz und ist mit zwei Stahlrohrarmen am Fußzylinder befestigt. Es besteht keine weitere Verbindung zum Sitzteil, sodass sich das Rückenteil unabhängig vom Sitz leicht federnd den Bewegungen des Nutzers anpassen kann. Der Sessel ist als Drehsessel konstruiert und ruht auf einer gegossenen Aluminium-Basis. Die Rückenschale besitzt eine Innenstruktur aus gebogenem Stahlrohr und ist leicht gepolstert. Ein halbkreisförmiges Kissen dient gleichzeitig als Lendenunterstützung und Armlehne.

Circo-Solo deliberately dominates a room thanks to its sheer volume. The rear support is shaped like a clearly-lined, cut-out cone. Its semi-circle embraces the round seat and the two tubular steel arms attach it to the cylindrical stand. There are no additional connections to the seat section, meaning the backrest, with its formed tubular steel insides and mild upholstery, functions as its own spring suspension, adapting to the sitter's movements. The swivel armchair rests on an aluminum base. A semi-circular cushion doubles up as both lumbar support and armrests.

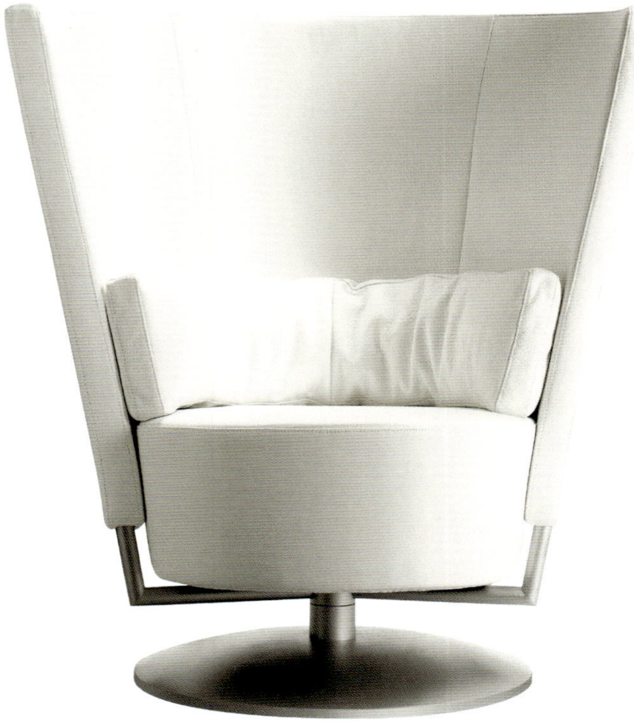

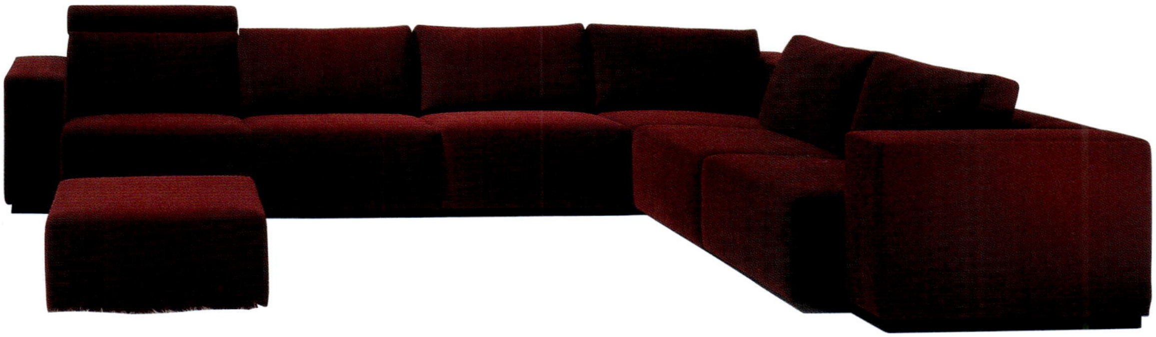

Hersteller Manufacturer
Walter Knoll AG & Co. KG
Herrenberg, Germany

Produkt Product
Nelson 602/603
Sofaprogramm
Sofa line

Design
Martin Bergmann
Gernot Buhmann
Harald Gründl
EOOS Design GmbH
Wien, Austria

Mit dem Sofaprogramm Nelson von Walter Knoll können mittels verschiedener Module Sofas und Ecksofas gebildet werden. Wahlweise kann jedes Modul mit einer ausklappbaren Nackenrolle ausgestattet werden. Durch das Ausklappen entsteht nicht nur eine Kopfstütze, sondern die Sitztiefe wird zugleich vergrößert. Nelson 602 besitzt schmale oder breite Armlehnen und ist sowohl für das private Umfeld als auch für den öffentlichen Raum geeignet.

Knoll's Nelson sofa line is made up of a series of modules that enable sofas of different shapes, including L-shaped sofas, to be created. Every module can be fitted with an optional neck-support roll which, when folded out, not only offers a head rest but also increases the depth of the seat. Nelson 602 features narrow or wide armrests and is destined for both private residences and public settings.

Hersteller Manufacturer
Roset S.A.
Briord, France

Produkt Product
Lover
Sessel, Sofas und Liegesofa
Fireside chair, settees and chaise longue

Design
Pascal Mourgue
Montreuil sous Bois, France

Die Sitzgruppe Lover wurde entsprechend den Bedürfnissen nach Mobilität und Variabilität beim Sitzen konzipiert. Aus der Ursprungsidee einer einfachen, biegsamen Form, die sich je nach Bedarf zusammen- und wieder aufrollen lässt, ist ein komfortables und neuartiges Sitzkonzept entstanden. Der Bezug aus einem eigens für Lover entwickelten dehnbaren Strickmaterial aus Wolle ist in 7 Farben lieferbar: Rot, Fuchsia, Violett, Schwarz, Anthrazit, Grau und Weiß. Die Sitzelemente besitzen eine Metallstruktur mit Verstellmechanismus, das Rückenteil ist in zwei Positionen arretierbar, die Rückenlehne hat einen Kern aus einem speziellen Schaumstoff, das Fußgestell ist aus gebürstetem Edelstahlrohr.

The Lover seating group was conceived as mobile and flexible seating. A comfortable and innovative seating concept has been developed from the original idea of a simple, flexible shape which can be rolled out or rolled up as needed. The covering, a stretchy woven developed specially for Lover, is available in a choice of seven colors: red, fuchsia, violet, black, anthracite, grey and white. The seating elements feature a metal structure with an adjusting mechanism: the back section can be set in two positions and has a special foam core. The frame is made of brushed tubular stainless steel.

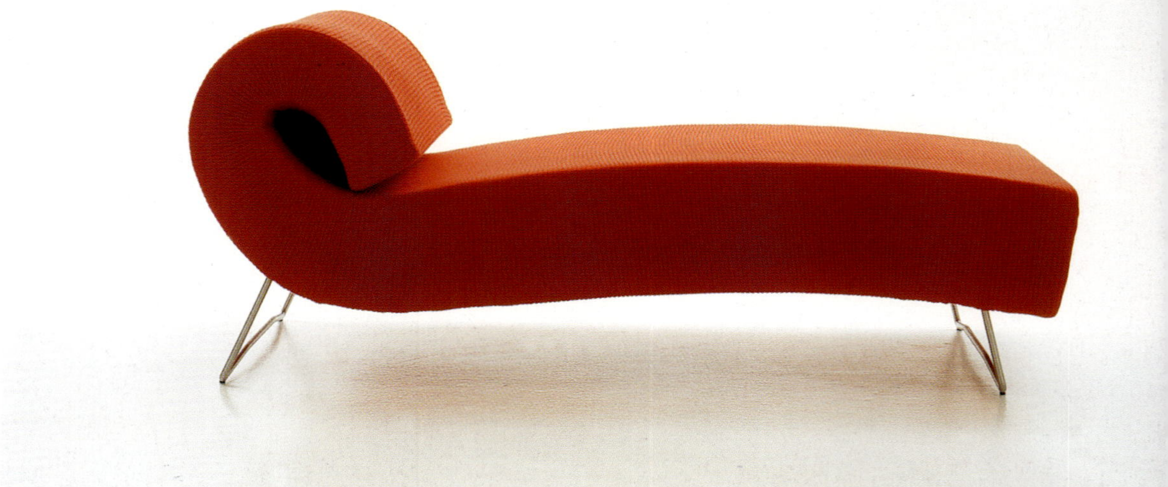

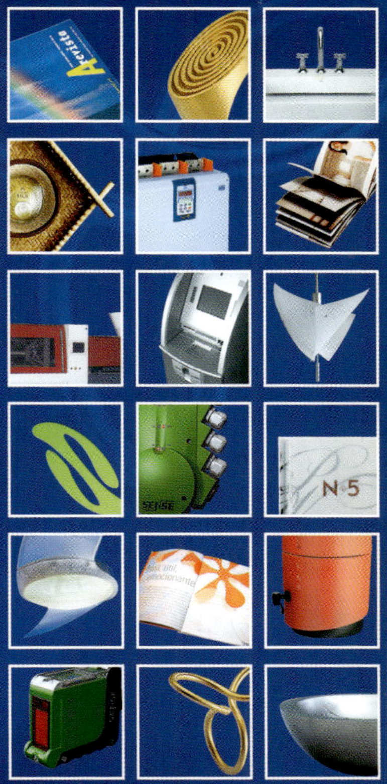

Hersteller Manufacturer
Paustian A/S
Kopenhagen, Denmark

Produkt Product
Multifunktionsbord Laptop Table
Multifunktionaler Tisch
Multipurpose table

Design
Prof. Knud Holscher
Per Kristian Dahl
Holscher Industriel Design
Kopenhagen, Denmark

Laptop Table ist ein multifunktionaler klassischer Tisch, der sowohl als Arbeits- als auch als Esstisch dienen kann und somit insbesondere für kleinere Wohnungen geeignet ist. Im Hohlraum des Zargenrahmens können Laptop, Zubehör wie Trafo, Kabel usw. verstaut werden. Zugleich findet hier auch eine zusätzliche Tischplatte Platz, mit der der Tisch verlängert werden kann. Die Tischbeine besitzen eine Kerbe, sodass die Kabelführung „unsichtbar" gelöst ist. Der Tisch ist aus leichten und stabilen, extrudierten Aluminiumprofilen mit eloxierter Oberfläche konstruiert. Die Tischplatte besteht aus einer furnierten Honeycomb-Spezialplatte, deren Oberfläche mit einem umweltfreundlichen Lack versiegelt ist.

Laptop Table is a multifunctional classic table that functions as both a work-top and a dining table, making it especially suitable for smaller apartments. The hollow tray beneath the table can store a laptop, accessories such as transformers and cables, and even the extra leaf, which can be pulled out to extend the table top. The extruded table legs come with a recess/track for "invisible" cabling. The table is made of light sturdy extruded aluminum profiles with an anodized surface. The top itself is a special aluminum honeycomb section with a veneer sealed with an eco-friendly lacquer.

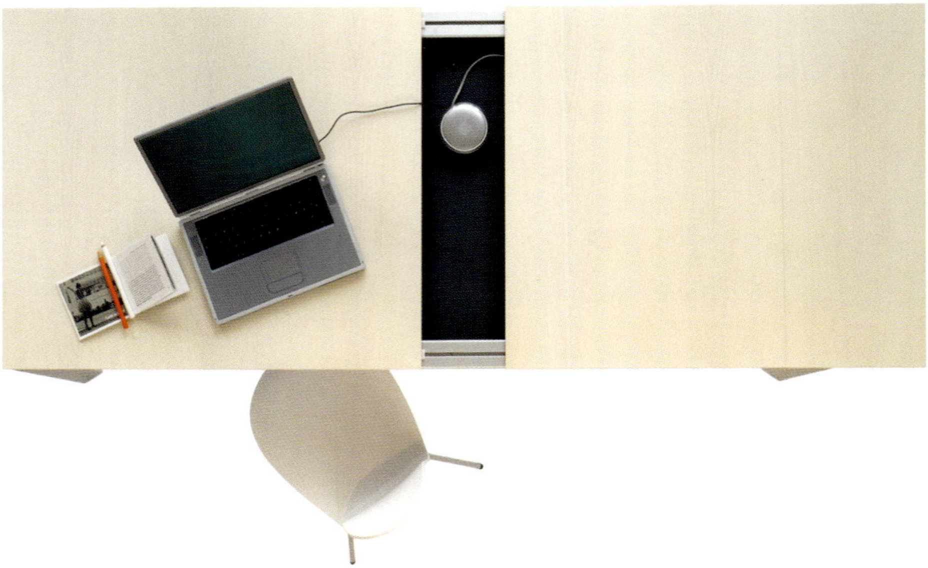

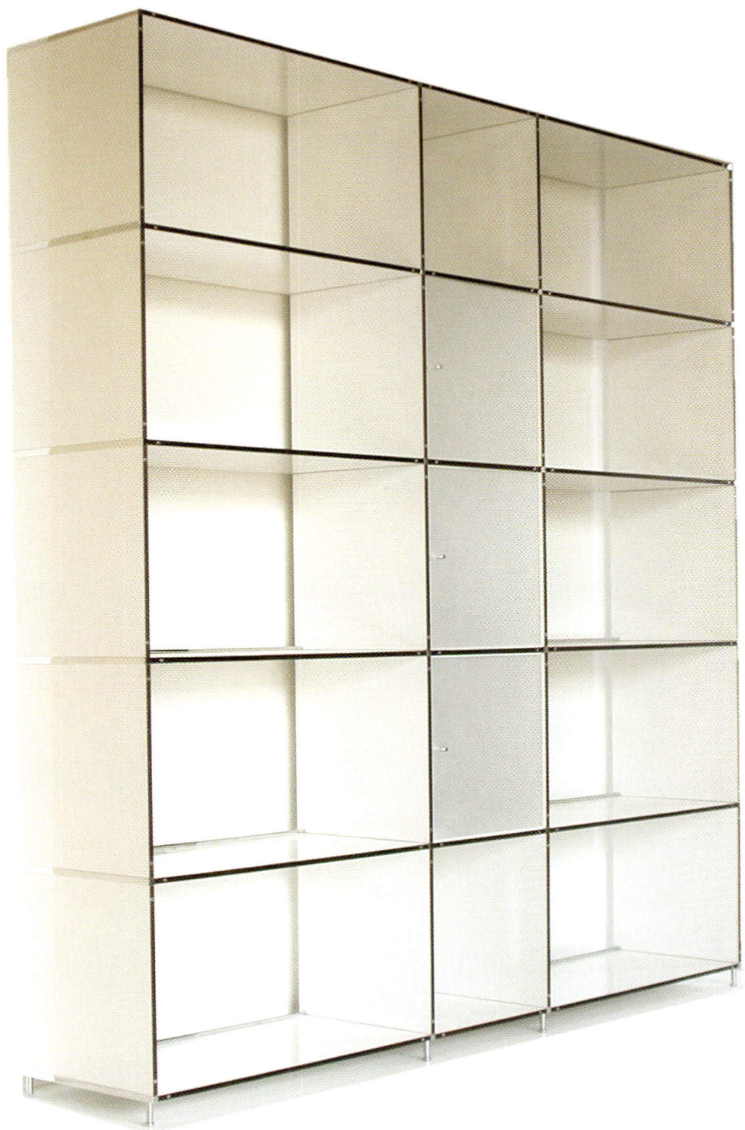

Hersteller Manufacturer
Weber + Hätinger
Nürtingen, Germany

Produkt Product
K12/18
Regalsystem
Modular furniture system

Design
Thomas Bredel
Bredel Design
Stuttgart, Germany

Das Regalsystem K12/18 ist ein vielseitig einsetzbares, modulares Möbelsystem für den Privat- und Objektbereich, das sich durch eine klare Formensprache auszeichnet. Verchromte Vierkantprofile aus Aluminium verbinden die Platten mittels Inbusschrauben miteinander. Diese können in Schwarz, nahezu unsichtbar, oder in Edelstahl bzw. mit verchromten Stopfen als Gestaltungsmittel gewählt werden. Sämtliche Flächen bleiben von sichtbaren Verbindungsbeschlägen frei. Das System umfasst außerdem Elemente mit Klappen, Türen und Schubkästen, die einfach in die Regalfächer eingeschoben und von innen verklemmt werden sowie Schreibtischplatten und Container.

K 12/18 is a multipurpose, modular shelving system for private use and display purposes which catches the eye with its clear formal vocabulary. Shelves are combined by means of rectangular aluminum profiles and Allen-headed screws. The screws come in a choice of black (almost invisible) or stainless steel and can be used in conjunction with chromed plugs as a design highlight. The connecting elements are not visible on any of the shelf surfaces. The system also includes elements with flaps, doors and drawers which can simply be inserted between shelves and fixed from inside, not to mention desktops and containers.

iF design award 2004

PACKAGING DESIGN

Disziplin 3 discipline 3

PACKAGING DESIGN
TRANSPORTATION DESIGN
PUBLIC DESIGN

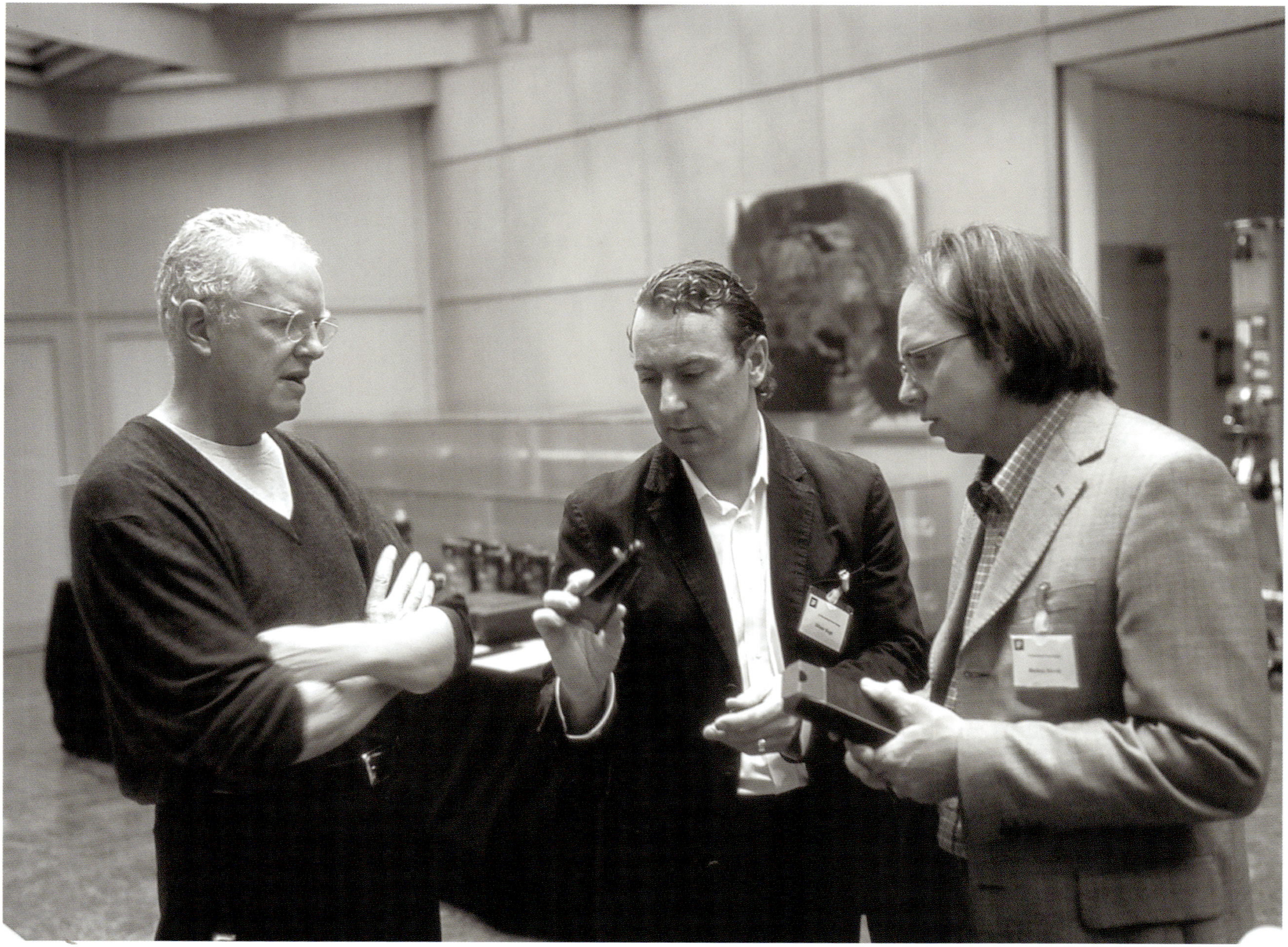

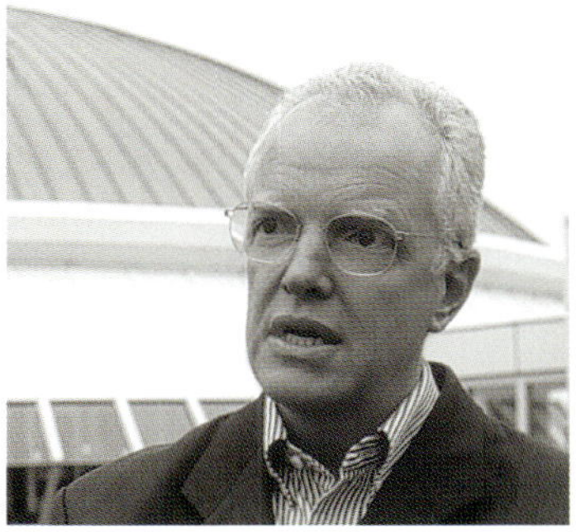

DAVID ROBB ist seit 1993 verantwortlich für BMW Motorrad Design und das Design dazugehöriger Produkte wie Helme, Fahrerausstattung etc. Seine verschiedenen Stationen brachten ihn nach Japan, Südamerika und Europa, wo er Anfang der 80er Jahre bei Audi Design im Interieur, Exterieur und Advanced Design mitgewirkt hat.

David Robb has been head of BMW Motorcycle Design since 1993 as well as being in charge of the design of related products, such as helmets and biker accessories etc. In the course of his professional life he has worked in Japan, South America and Europe, where, in the early 1980s, he was employed at Audi Design and was active in interior, exterior and advanced design.

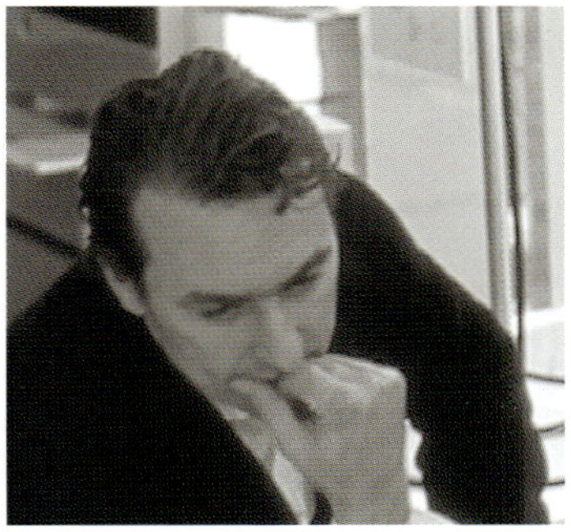

OLIVER VOGT studierte an der UdK-Berlin Industrial Design und gründete 1993 mit Hermann Weizenegger die Agentur Vogt+Weizenegger in Berlin. Die Designagentur ist bekannt für Möbel und Haushaltsprodukte für Authentics, Thomas Rosenthal sowie das Redesign für Möve. Besonders die Projekte der Imaginären Manufaktur mit der Berliner Blindenanstalt und der Fabrik der Zukunft, in der sie den SINTERCHAIR herstellten, haben internationale Beachtung erreicht.

Oliver Vogt studied industrial design at the University of Arts in Berlin. In 1993 he founded together with Hermann Weizenegger the Vogt+Weizenegger agency in Berlin. It is renowned for furniture and household products, created among others for Authentics and Thomas Rosenthal as well as for its redesign for Möve. Especially the projects for the Imaginary Manufactory for the Berlin Institute for the Blind, and the Factory of the Future, where they manufactured the SINTERCHAIR, have received international acclaim.

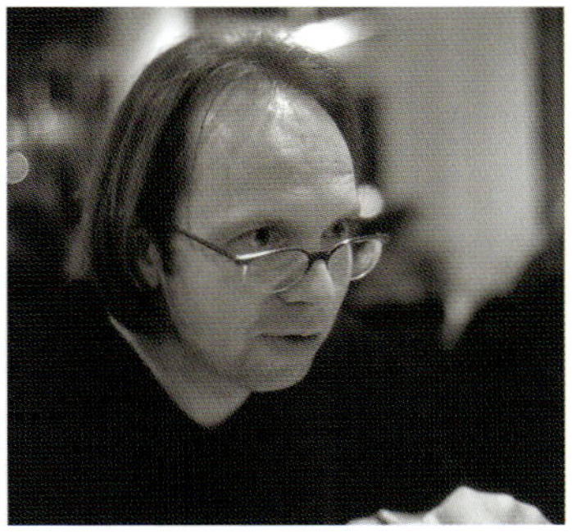

MARKUS STORCK absolvierte eine kaufmännische Ausbildung in der Fahrradbranche und machte sich 1988 selbstständig. Er entwirft und entwickelt Fahrräder sowie deren Komponenten und Zubehör. Zudem ist er Initiator und Gründungsmitglied der Messe „Eurobike" sowie Technikreferent und Mitbegründer des Verbandes deutscher Fahrradimporteure.

Markus Storck underwent commercial training in the cycle industry and went set up his own business in 1988. He designs and develops bicycles as well as bike components and accessories. He is also initiator and founding member of the "Eurobike" trade fair as well as technical expert advisor to and co-founder of the Association of German Bicycle Importers.

iF design award 2004

PACKAGING DESIGN

Disziplin 3 Discipline 3

Packaging Design

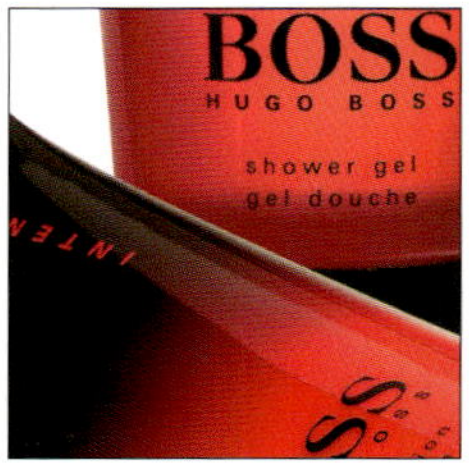

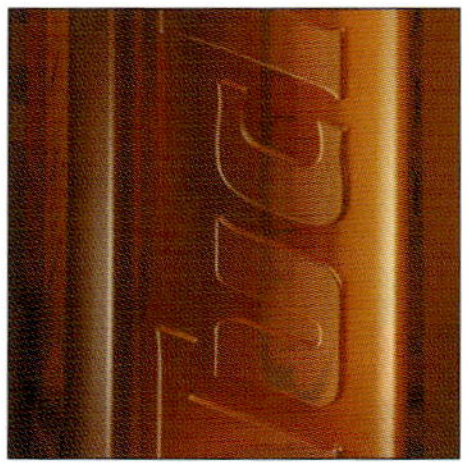

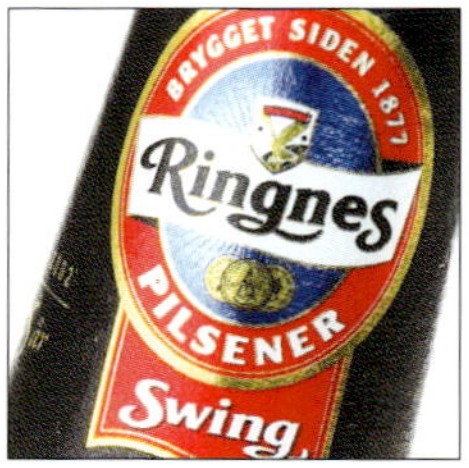

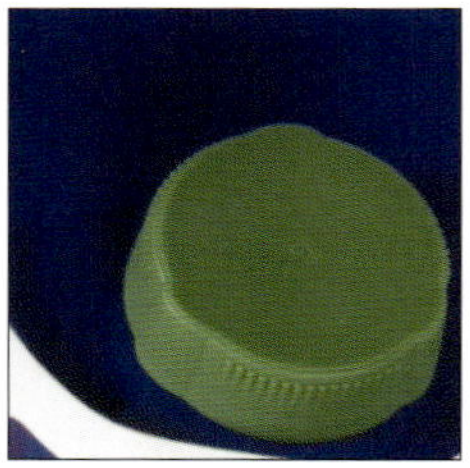

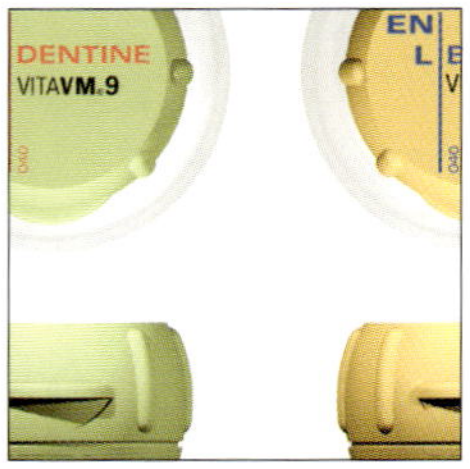

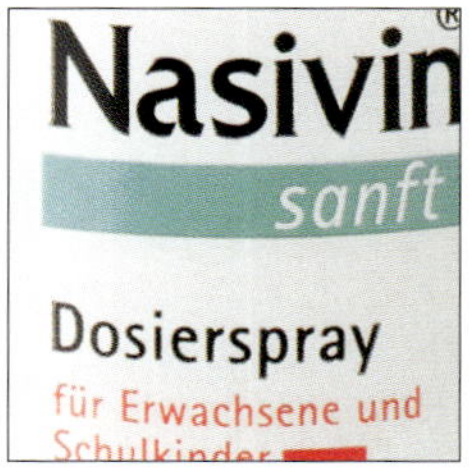

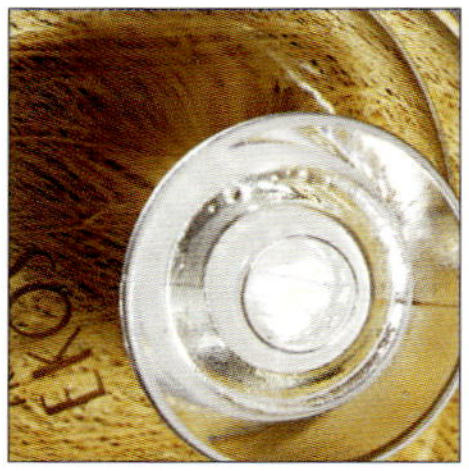

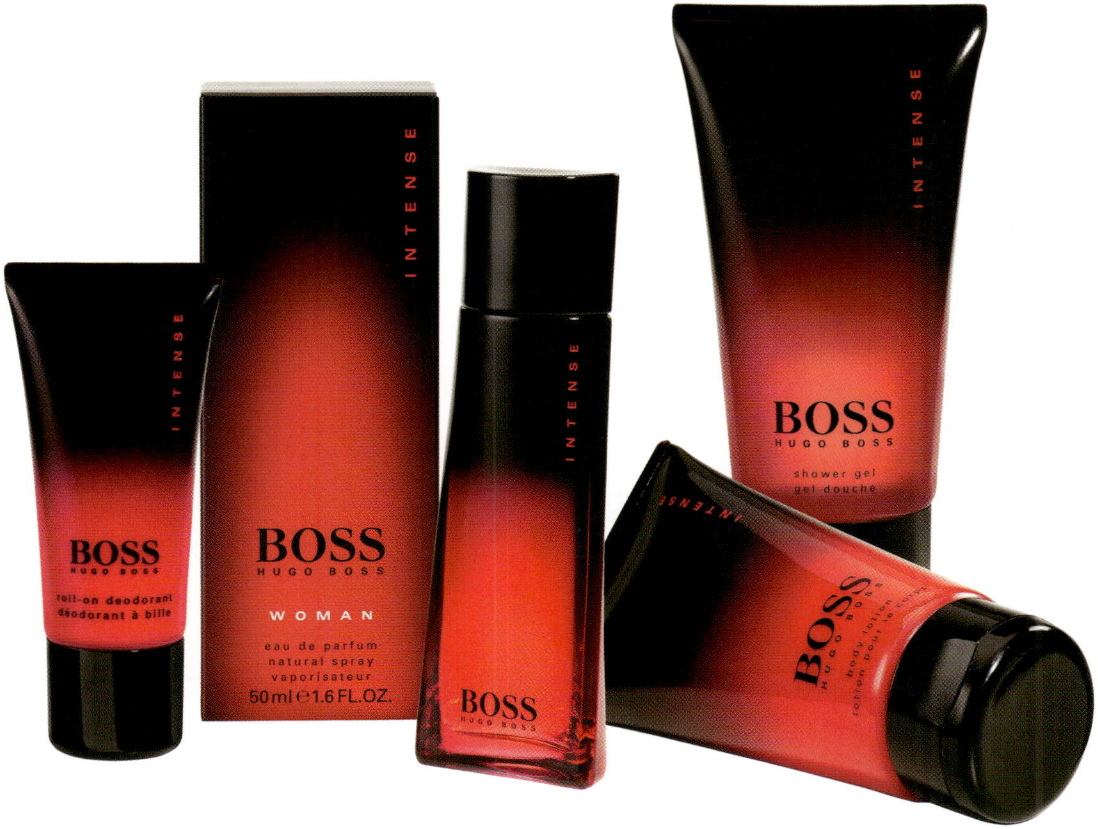

Hersteller Manufacturer
Procter & Gamble
Genf, Switzerland

Produkt Product
Boss Intense
Flakon
Flacon

Design
Peter Schmidt Group GmbH
Hamburg, Germany

Das Design des Flakons von Hugo Boss Intense stellt die Frontseite markant in den Vordergrund, während die Seitenlinien sich elegant in die Höhe ziehen. Die leuchtende Farbgebung auf dunklem Schwarz korrespondiert mit dem Charakter des Namens Intense. Ein kleiner roter Dome auf der Pumpe scheint durch die Kappe hindurch.

The design for the Hugo Boss Intense perfume bottle strikingly foregrounds the front of the flacon, while the lines of the sides sweep elegantly upwards. The bright coloring on the dark black corresponds to the character of the perfume's name: Intense. A small red design element on the nozzle pump can be seen through the cap.

Hersteller Manufacturer
Thüringer Behälterglas GmbH
Schleusingen, Germany

Produkt Product
Tucher Crown
Bierflasche
Beer bottle

Design
Formgestaltung Schnell-Waltenberger
Pforzheim, Germany

Tucher Crown ist eine individuelle Bierflasche (0,33 l) als Mehrwegflasche für Export- und Pilsbiere und auch zum direkten Genuss aus der Flasche bei Weizenbieren. Für das Design der Flasche verband man traditionelle mit modernen Gestaltungsaspekten, Optik und Haptik standen im Vordergrund. Die Unterscheidung der Varietäten erfolgt über die Farbkodierung auf einem Halsetikett und durch eine unterschiedliche Gestaltung der Kronkorken.

Tucher Crown is a returnable 0.33l beer bottle for lager and pilsner beers and also for wheat beers that can be drunk straight from the bottle. When designing the bottle, a conscious effort was made to merge traditional and modern aspects, while prioritizing visual appearance and tactile qualities. The particular type of beer in the bottle is revealed by the color coding on the neck and the respective design of the metal bottle cap.

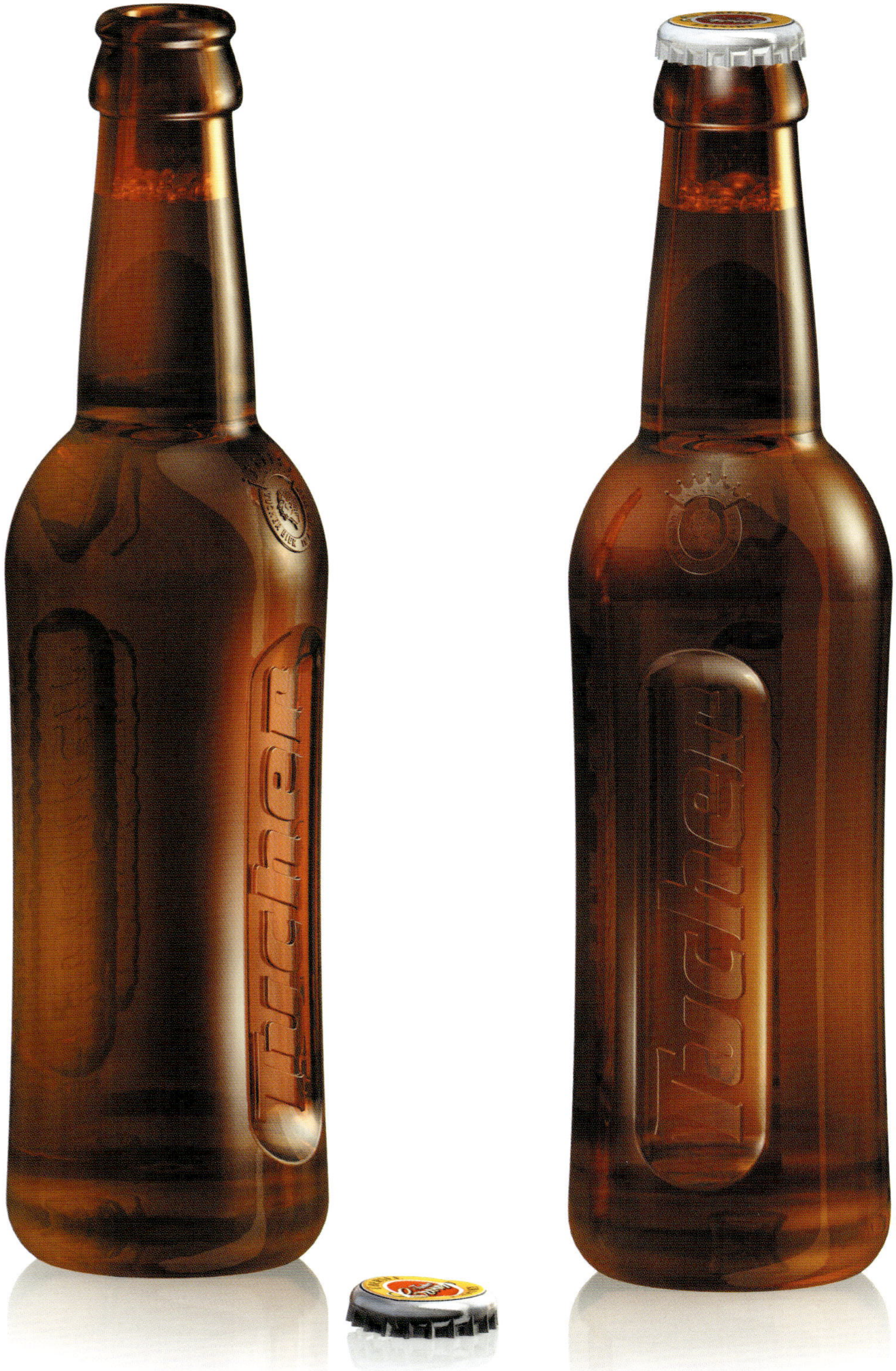

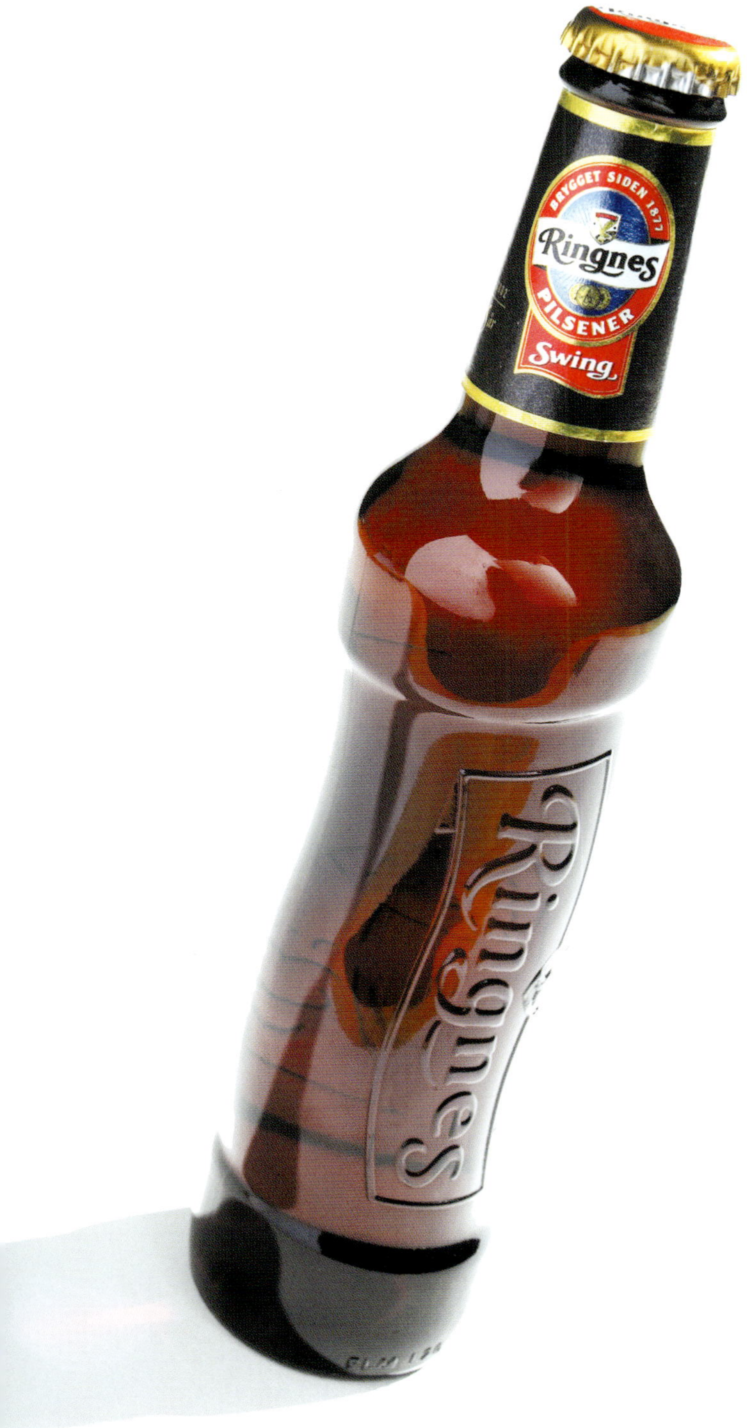

Hersteller Manufacturer
Rexam Glass Limmared A.B.
Limmared, Sweden

Produkt Product
Ringnes Swing
Mehrweg-Braunglasflasche
Returnable amber glass bottle

Design
Stromme Throndsen Design & Ringnes A/S
joint co-operation
Norway

Für die Ringnes Swing Mehrweg-Braunglasflasche wurde eine S-Form gewählt, um das innovative und jugendliche Image der für den norwegischen Markt vorgesehenen Bierflasche zu unterstreichen. Das Label ist auf dem Flaschenhals angebracht, da der Flaschenkörper mit einer Hochprägung versehen wurde. Während des Designprozesses wurde speziell darauf geachtet, dass der Flascheninhalt 300 ml beträgt und die Flasche stabil auf dem Abfüllband steht. Durch ihr ergonomisches Design ist die Flasche nicht nur ästhetisch ansprechend, sondern sie liegt auch gut in der Hand.

An S-shape was chosen for the Ringnes Swing returnable amber beer bottle, particularly as this portrays an innovative and youthful image to the Norwegian market. The label is positioned around the neck due to the embossing on the main body of the bottle. During the design process, special attention was paid to ensure that the product met the content specification of 300ml and that it remained stable on the filling line. Thanks to its ergonomic design, the bottle is both asthetically pleasing and highly functional.

Hersteller Manufacturer
SIG Combibloc GmbH
Linnich, Germany

Produkt Product
combishape
Getränkekartonverpackung
Beverage carton

Design
(der Schraubkappe)
Meyer-Hayoz Design Engineering Group
Konstanz, Germany
Winterthur, Switzerland

combishape ist eine formvariable, aseptische Getränkeverpackung, deren Form – ob oval oder mondförmig, drei-, vier-, fünf- bis zu achteckig – individuell gewählt werden kann. 42 Basisformen können auf das jeweilige Image eines Markenartikels abgestimmt werden. Das dreiteilige Verschlusssystem shapeTwist besteht aus Adapter, Pull-Tab und Schraubkappe. Die ergonomische Gestaltung der Schraubkappe bietet Komfort beim Öffnen, Schließen oder Ausgießen. Für Mantel, Boden und Deckel werden standardmäßig bewährte, umweltfreundliche Kartonverbunde aus Zellstoff, Polyethylen und Aluminium eingesetzt.

combishape is an aseptic beverage carton whose shape can be varied, enabling companies to choose the shape they like best from a range including oval or moon-shaped, triangular, four-sided, five-sided or even eight-sided. In total, 42 basic shapes can be fine-tuned to highlight the image of a particular brand product. The three-part cap system shapeTwist consists of an adapter, a pull-tab and the twist cap. The ergonomic composition of the cap offers great convenience for opening, re-closing and drip-free pouring. The standard sides, top and bottom are made of eco-friendly carton composed of cellulose, polyethylene and aluminum.

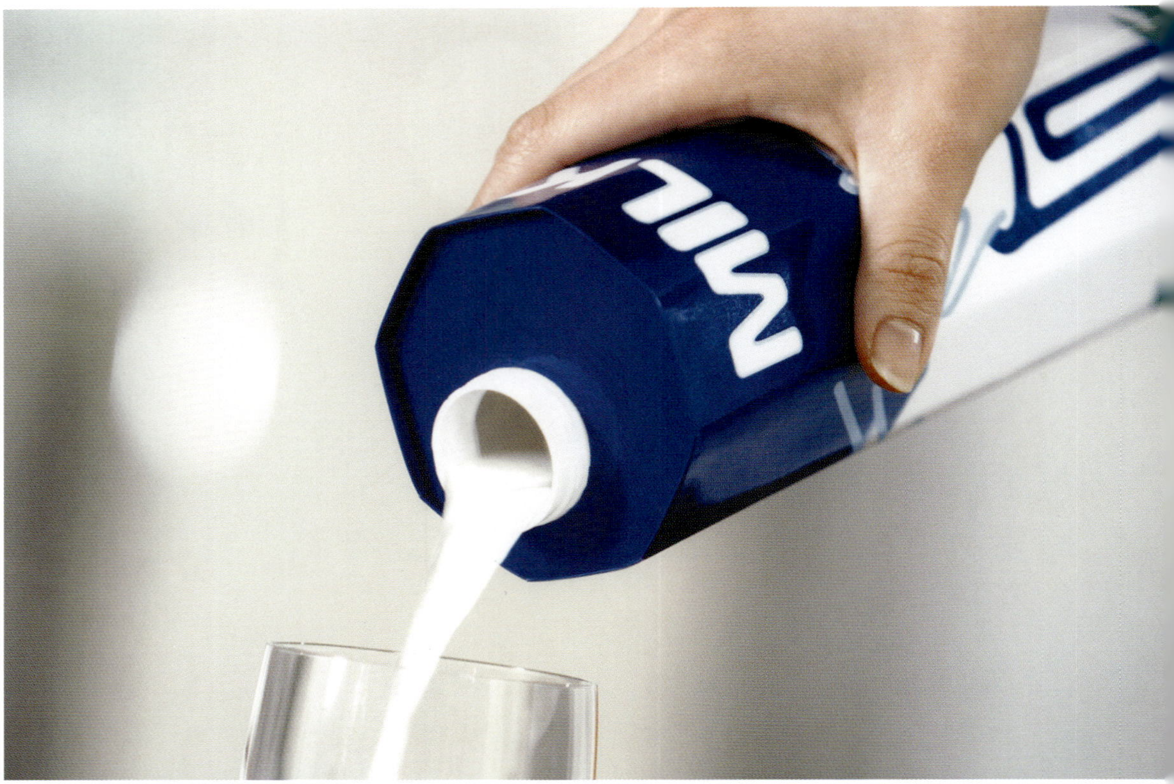

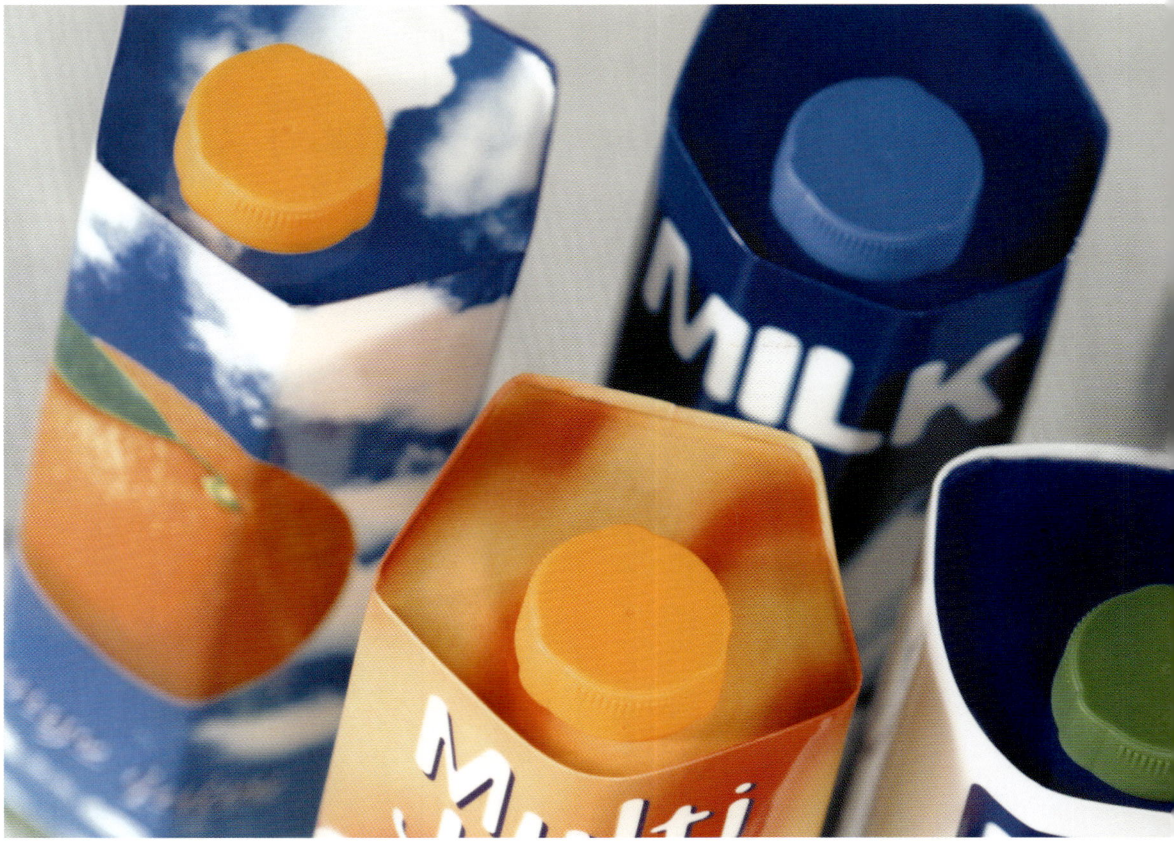

Hersteller Manufacturer
VITA Zahnfabrik H. Rauter GmbH & Co. KG
Bad Säckingen, Germany

Produkt Product
VITAVM®
VITA Flasche für Veneering Material
VITA bottle for veneering material

Design
Meyer-Hayoz Design Engineering Group
Konstanz, Germany
Winterthur, Switzerland

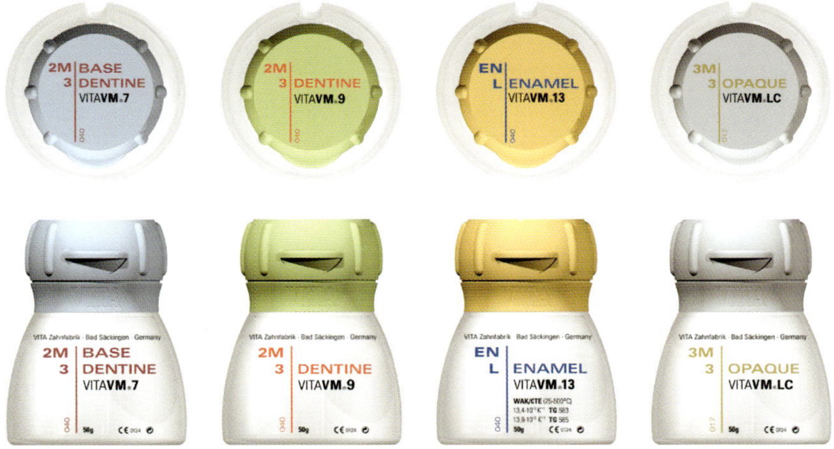

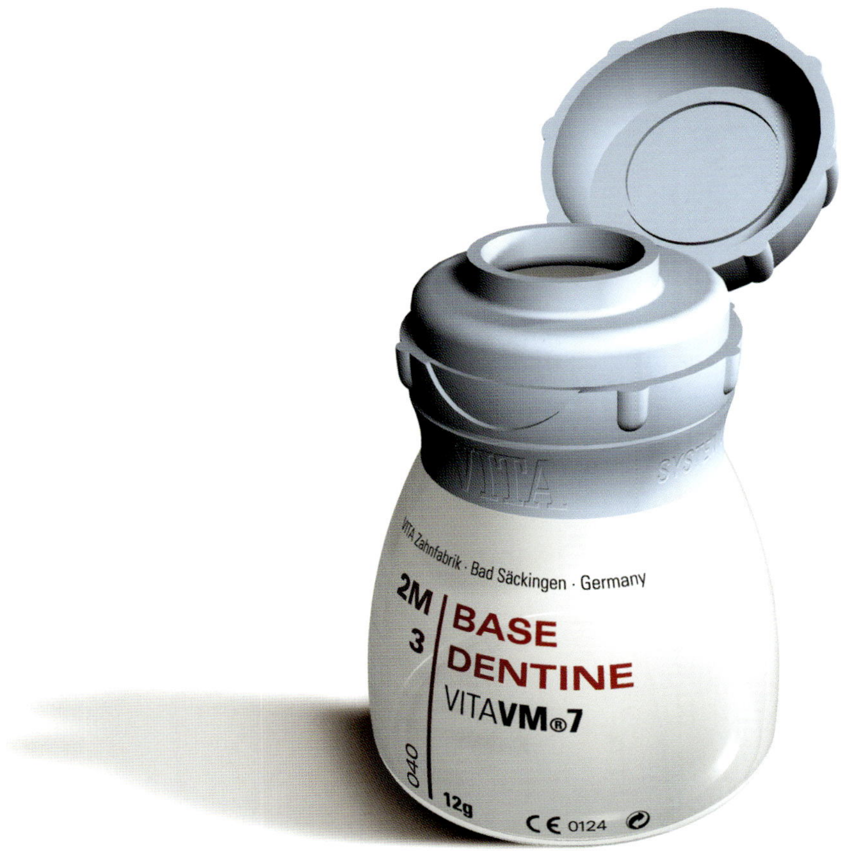

Das neue Designkonzept für die VITA Flasche für Verblendmaterialien beruht vor allem auf einer Verbesserung des Handlings der Flasche. Durch den neuen Schraub- und Klappdeckel, der zudem farblich codiert ist, ist das Öffnen und Wiederverschließen der Flasche leichter und die Verwechslungsgefahr ist praktisch ausgeschlossen. Des Weiteren wurde die Schüttöffnung verkleinert, sodass die Dosierung des Materials präziser erfolgen kann. Zusätzlich wurde die Standsicherheit der Flasche durch die Vergrößerung des Bodendurchmessers erhöht. Durch das farblich-typografische Erkennungssystem ist der Internationalisierung der Beschriftung Rechnung getragen.

The new design concept for the VITA bottle for veneering materials focuses primarily on improving bottle handling. Thanks to the new screwtop lid, which can be flipped open and shut and is, moreover, color-coded, the bottle is easier to use and any danger of confusing bottles has been minimized. The opening has deliberately been made smaller to allow more precise dosing of the ceramic powders. In addition, the bottle stands very steadily owing to the broader diameter of the base. The color-coded and typographical identification system also reflects the increasing internationalization of bottle labels.

Hersteller Manufacturer
Merck Selbstmedikation GmbH
Darmstadt, Germany
Pfeiffer GmbH (Technische Entwicklung)
Radolfzell, Germany

Produkt Product
Nasivin
Dosierspray
Nasal spray

Design
yellow design I yellow circle
Pforzheim, Germany

Die klare und einfache Kontur der Flasche des Nasivin Dosiersprays soll für die Verlässlichkeit und Vertrauenswürdigkeit eines pharmazeutischen Produktes stehen. Die ergonomische Qualität zeigt sich an der großen Drucktaste am Boden, rutschhemmenden Rillen an der Schulter und einer auch auf kleine Kindernasen abgestimmte Nasenolive. Durch das neuartige Dosierpumpensystem besteht keine Produktberührung mit metallischen Pumpenteilen. In Verbindung mit einem mikrobakteriellen Filter kann so auf belastende Konservierungsstoffe verzichtet werden. Sortenreine Kunststoffe erleichtern eine Rückführung in den Wertstoffkreislauf.

The clean sleek contours of the Nasivin nasal spray bottle stand for the reliability and trustworthiness of a pharmaceutical product. It likewise possesses prime ergonomic properties, such as the large press-button in the base, non-slip ribs on the shoulder, and an eyelet designed bearing even small children's noses in mind. Thanks to the innovative dosage pump system, the product does not come into contact with the metal pump components, meaning that as a consequence, and together with the micro-bacterial filter, there is no need for problematic preservatives. The use of pure-grade plastics facilitates recycling.

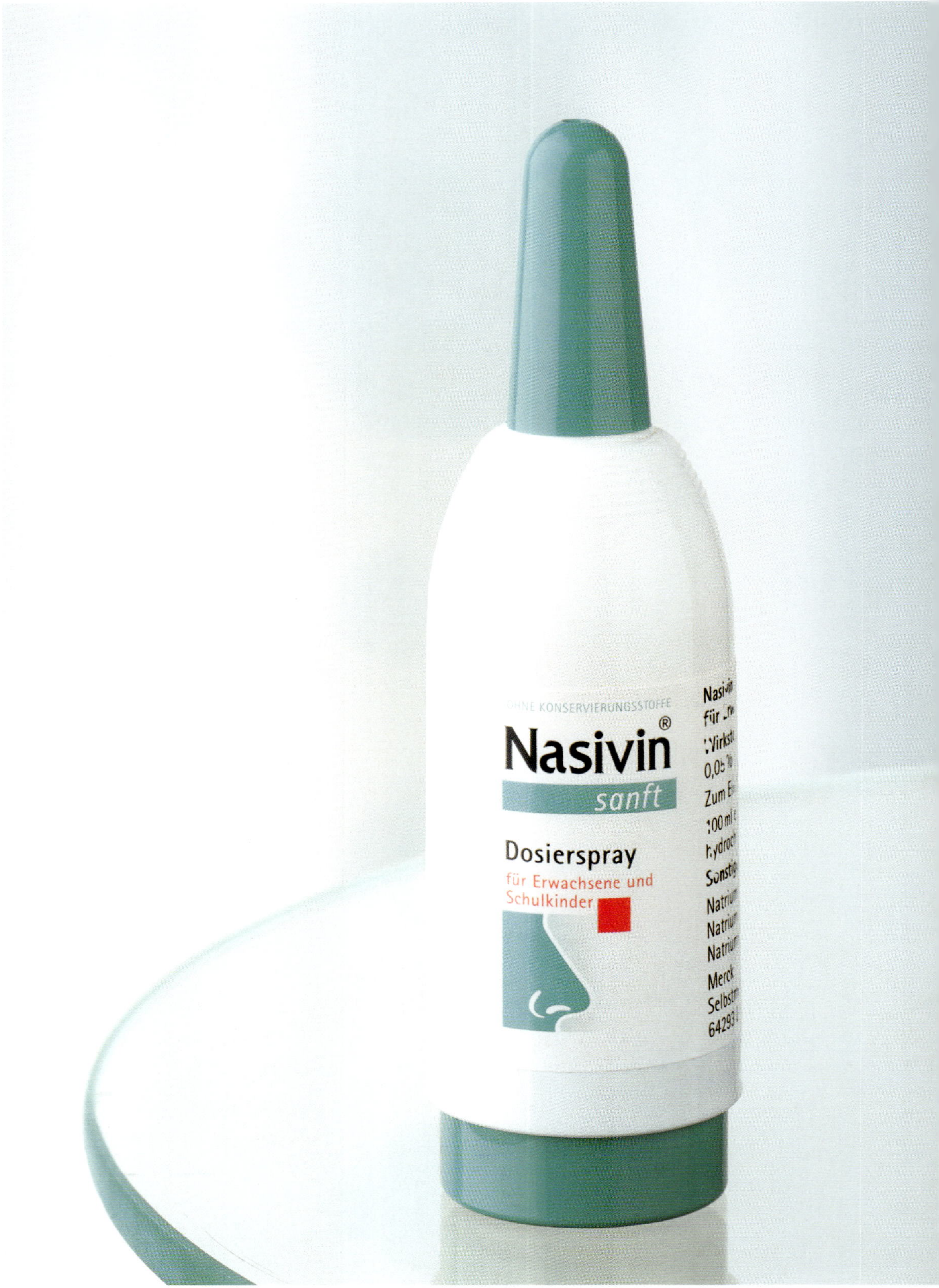

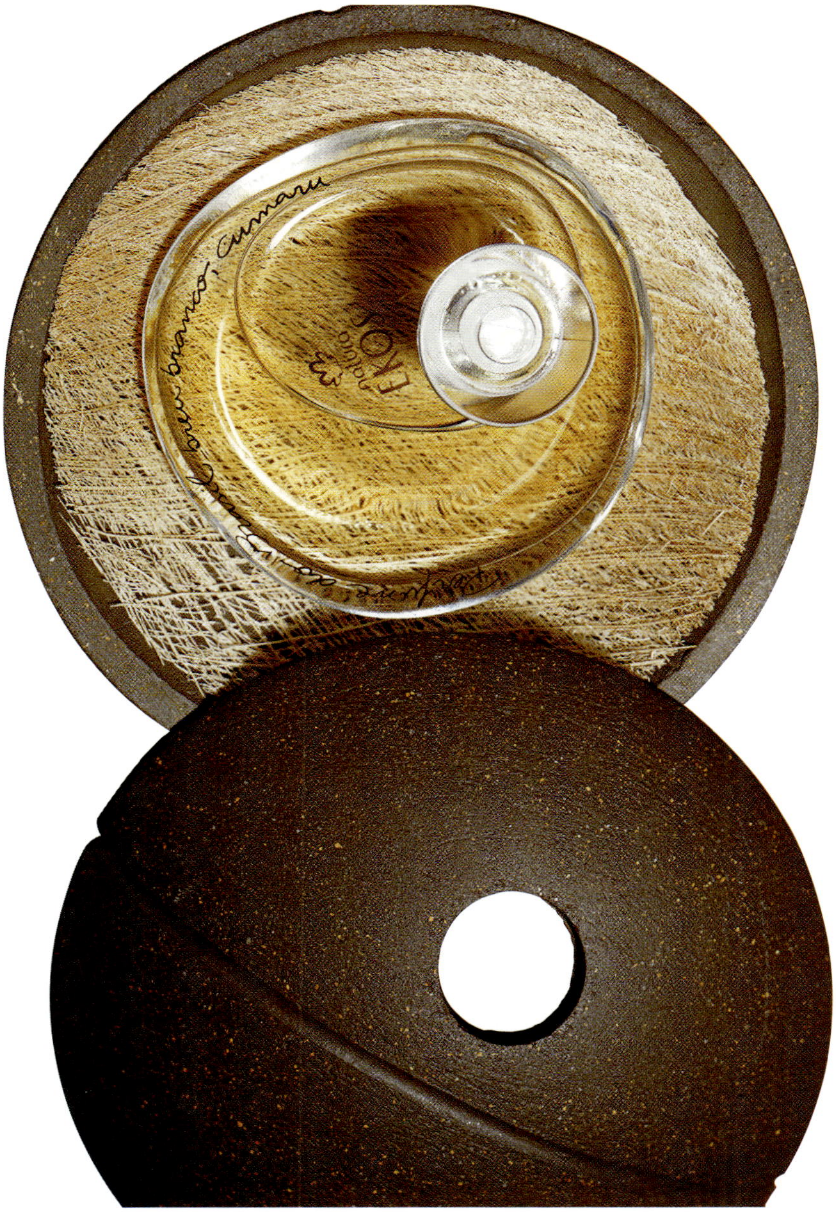

Hersteller Manufacturer
Indústria e Comércio de Cosméticos Natura Ltda.
Cajamar, Brazil

Produkt Product
Natura Ekos breu Branco & Cumaru
Parfum
Perfume

Design
Filomena Padron
Indústria e Comércio de Cosméticos Natura Ltda.
Cajamar, Brazil

Für das „Perfume do Brasil" aus der Linie Ekos von Natura wurde ein Flakon gestaltet, der an die Formen von Samen und Flusssteinen erinnert. Der Flakon besteht aus handmodellierter Keramik und besitzt kein Ventil. Für die Produktion werden 15 bis 20 Prozent bereits gebrannte Keramik genutzt und Farben und Gewebe brasilianischer Natur. Der Flakon ist durch ein natürliches Geflecht aus Kokosblättern geschützt, die äußere Verpackung besteht zu 15 Prozent aus Recyclingpapier.

A flacon was developed for Natura's Ekos "Perfume do Brasil" with a shape reminiscent of seeds or river-bed stones. The bottle is wrapped in a hand-modeled ceramic cover, using 15-20% burnt ceramic, reflecting the colors and textiles of the Brazilian nature. Since it does not have a valve it has to be used gently. And it boasts natural coconut leaf straw as wrapping, while the cartridge sleeve contains 15% recycled paper.

iF design award 2004

TRANSPORTATION DESIGN

Disziplin 4 Discipline 4

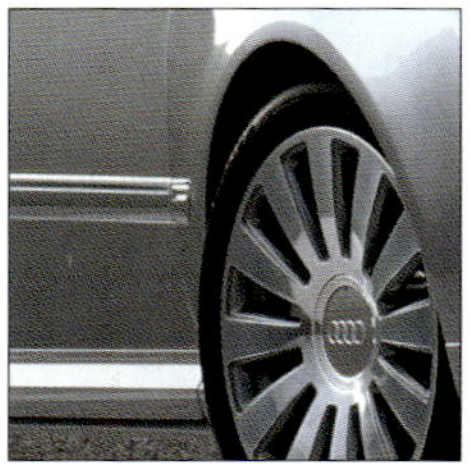
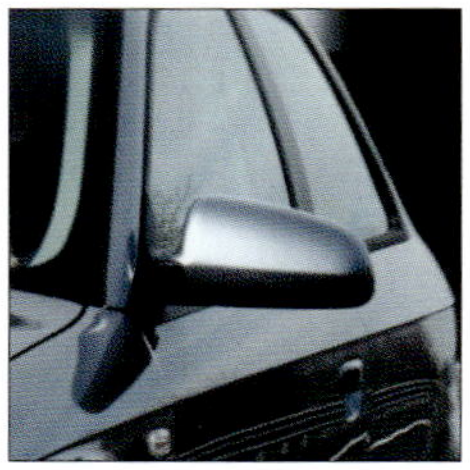
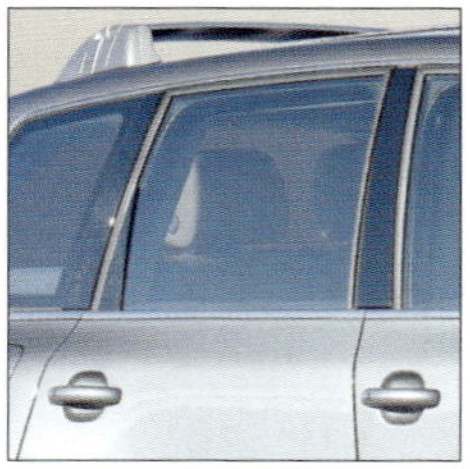
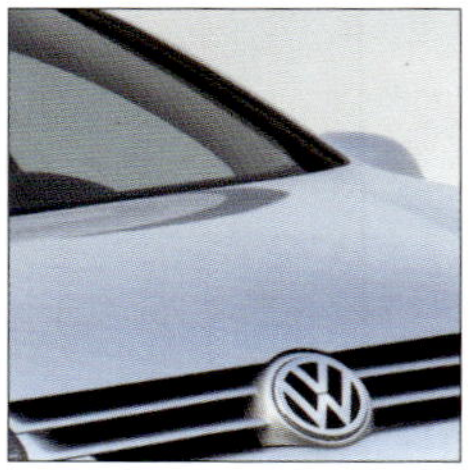
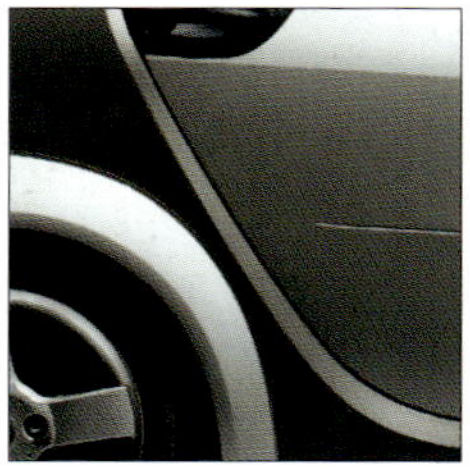
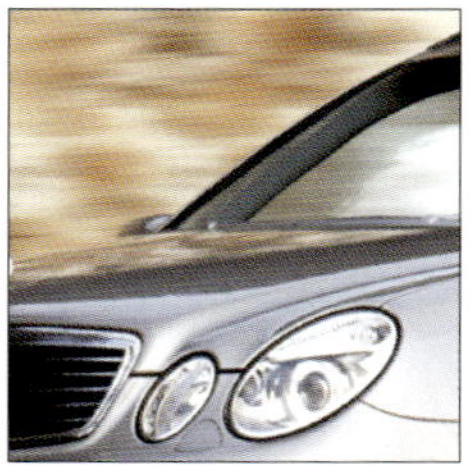
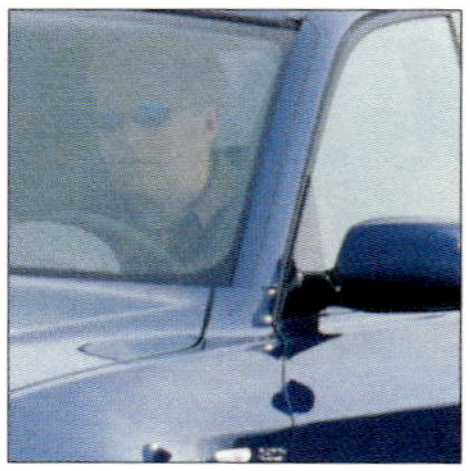
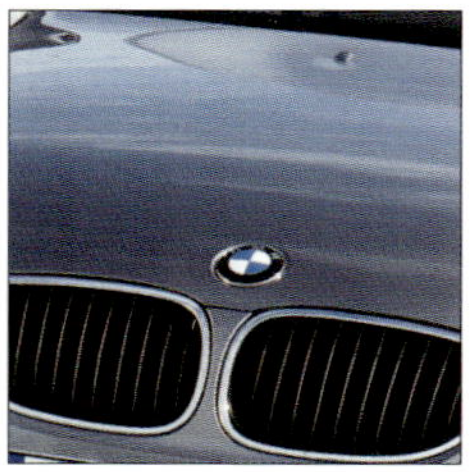

Transportation Design

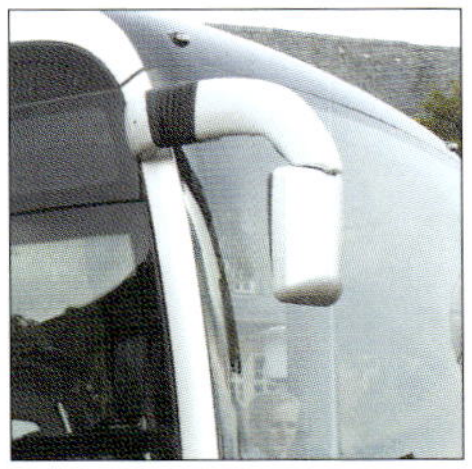
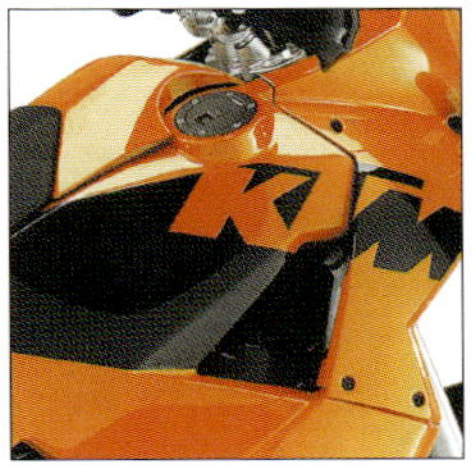
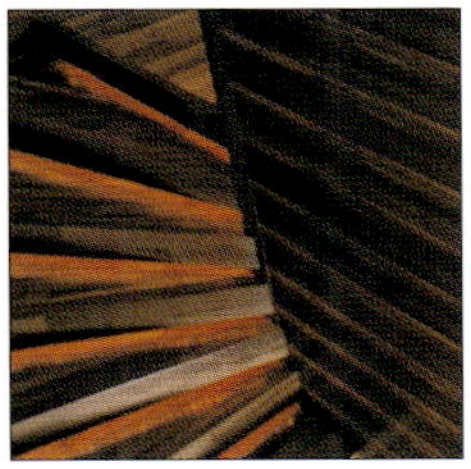
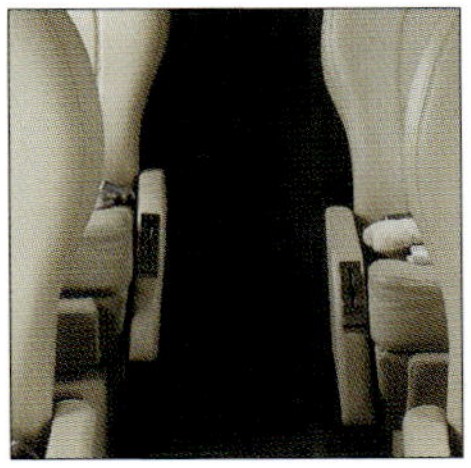
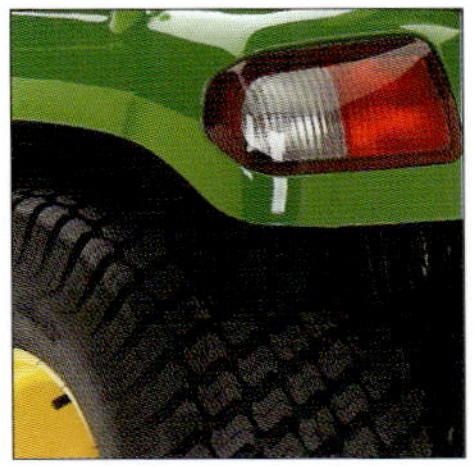
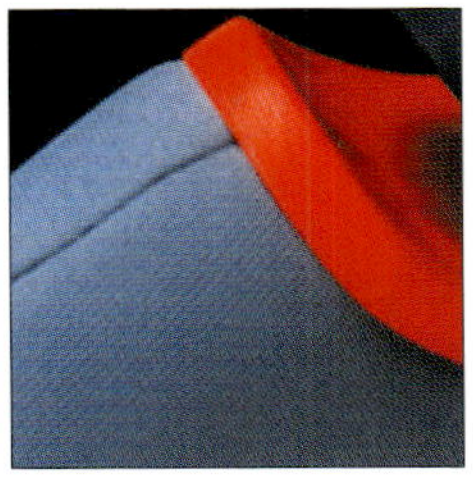

Hersteller Manufacturer
Audi AG
Ingolstadt, Germany

Produkt Product
Audi A8
Limousine

Design
Gerd Pfefferle, Leiter Audi Design
Audi Design Team
Ingolstadt, Germany

Der Körper des A8 wird durch wenige, aber sehr präzise Flächen beschrieben, er ist von allem Überflüssigen befreit. Die großen und klaren Flächen haben durch ihre dezente Spannung einen sportlichen Duktus. Die Überhänge vorne und hinten sind kürzer gestaltet worden, der Fahrzeugkörper ist gestreckt und flach. Die Dachlinie weist im Fondbereich Coupé-Charakter auf. Der optische Schwerpunkt der Seitenansicht liegt etwas weiter hinten, wodurch die Kraft des Exteriors nach vorne zu drängen scheint. Die starke Symbolik der Frontgestaltung ist geprägt durch das Doppelgrill-Konzept und die Frontscheinwerfer. Die Architektur des Innenraums ist klar gegliedert und horizontal betont. Vor allem die dominante Mittelkonsole unterstreicht den sportlichen Charakter.

The body of the Audi A8 is defined by a small number of very precise surfaces, devoid of anything superfluous. The large clear surfaces are subtly curved imparting a sporty appearance of restrained tension. Both front and rear overhangs are shorter, while the car body is elongated and flat. The roof line is reminiscent of a coupé at the rear and from side-on the visual focal point is a little further back, such that the power of the exterior seems to thrust the body forward. The strong symbolism in the design of the front-end is expressed in the dual-grille concept and the front headlights. As for the interior design, it creates clear divisions with a horizontal emphasis. The dominant central console in particular lends a sports air to the car.

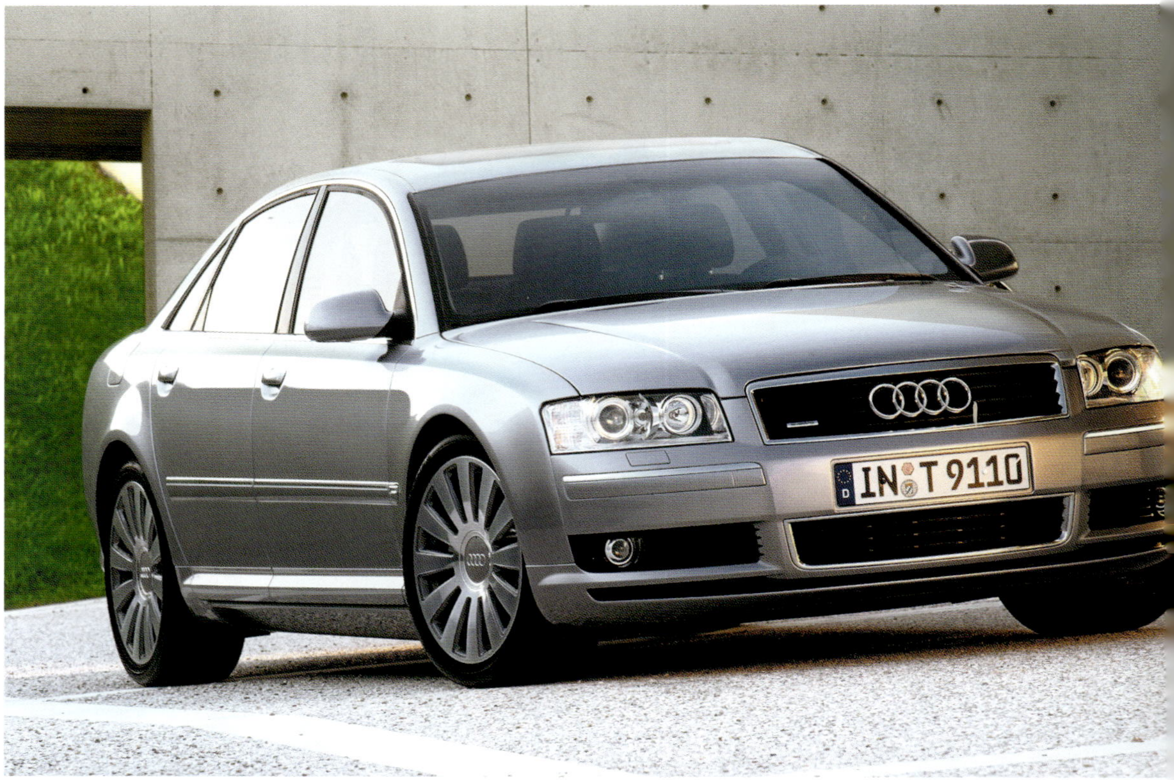

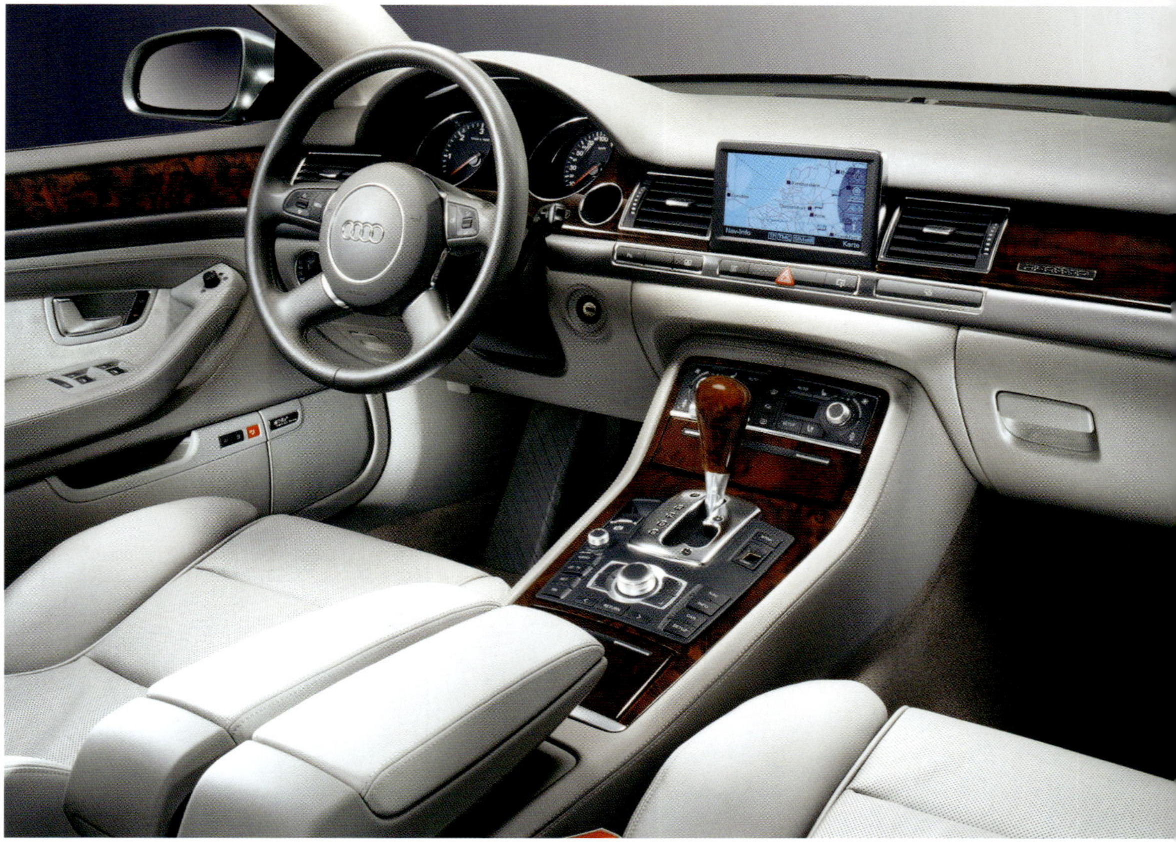

iF Gold Award

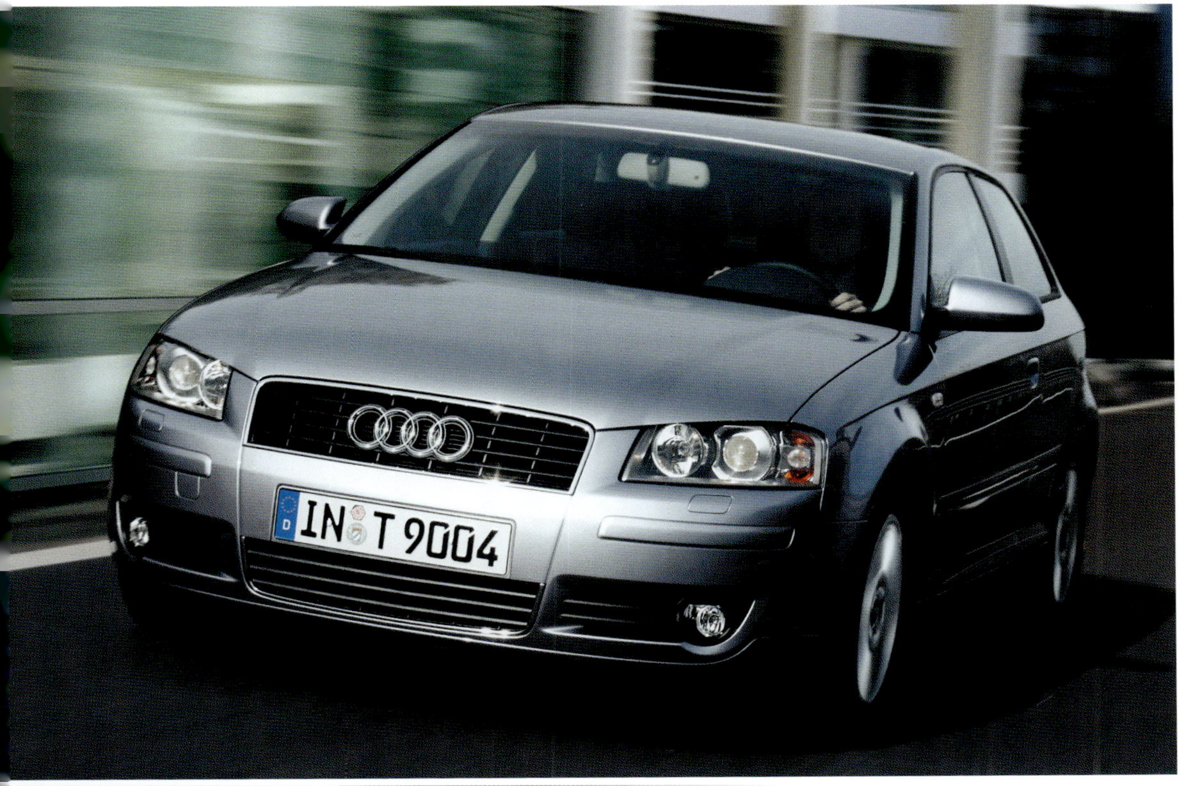

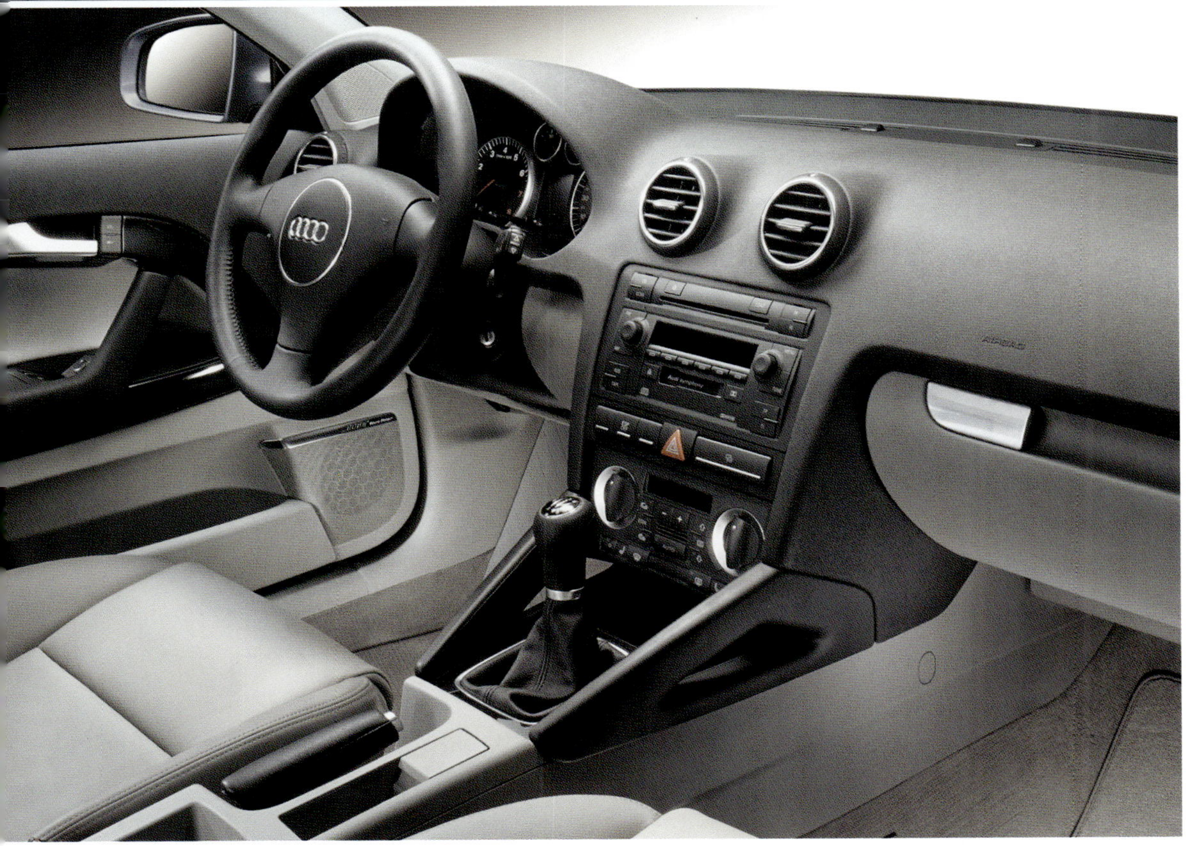

Hersteller Manufacturer
Audi AG
Ingolstadt, Germany

Produkt Product
Audi A3
PKW
Passenger car

Design
Gerd Pfefferle, Leiter Audi Design
Audi Design Team
Ingolstadt, Germany

Der neue Audi A3 präsentiert sich in komplett lackierter Außenhaut, wodurch der untere Teil der Front aufgewertet wurde. Markant sind die auf den Kühlergrill hin abfallenden oberen Kanten der Scheinwerfer. Im unteren Abschluss der Seitenflächen ist das Blech leicht ausgestellt, was den Eindruck eines sehr kraftbetonten Schwellers erzeugt. Der Fahrzeugkörper ist straff und athletischer geworden. Die Kabine wirkt flach und die seitliche Fenstergrafik mit der stark nach vorne geneigten C-Säule betont die Vorwärtsbewegung. Durch eine neue Innenraumarchitektur, mit Seitenblicken auf TT und das A4 Cabriolet, entstand eine spannungsvolle, sportliche Atmosphäre.

The new Audi A3 has an all-over paint finish for the outer skin, enhancing the lower part of the front end. Another prominent feature of the car: the upper edges of the headlights which drop towards the radiator grille. At the lower edge of the side skirts, the panels bend slightly outward giving the impression of a highly sculpted body. Overall, the main body is tauter and more athletic. The occupant area looks low-slung and the outline of the side windows and the sharply forward inclined C-column emphasize latent movement. A new interior architecture which echoes the TT and the A4 Cabriolet creates a vibrant sporting atmosphere.

iF Silver Award

Hersteller Manufacturer
Volkswagen AG
Wolfsburg, Germany

Produkt Product
VW Touareg
Sport Utility Vehicle

Design
Volkswagen Design
Wolfsburg, Germany

Der Touareg verbindet SUV-typische Merkmale als Zeichen seiner Zugehörigkeit zu eben dieser Kategorie mit einer Formensprache, die in vielen Bereichen die Nähe zum PKW deutlich werden lässt. Die hohe Bodenfreiheit, die großen Räder und die Gestaltung der Radläufe und der Stoßfänger weisen den Touareg klar als SUV aus und dokumentieren seine technische Kompetenz in Off-Road-Revieren. Der Grundkörper des Touareg folgt der erfolgreich etablierten Volkswagen Design Linie: glatte Flächen, spannungsreich überwölbt, und das Volkswagen-Gesicht und Rückleuchten-Design mit ähnlich hohem Wiedererkennungswert.

The Touareg combines typical SUV features attesting to its membership in this category with a formal idiom which in many instances makes it clearly a passenger car. High ground clearance, large wheels, design of the wheel housing and the front fender are clearly SUV traits and document its technical abilities off the road. The basic shape of the Touareg follows the successfully established Volkswagen design lines: smooth surfaces, dynamic curves and the truly recognizable Volkswagen front-end and tail-light design.

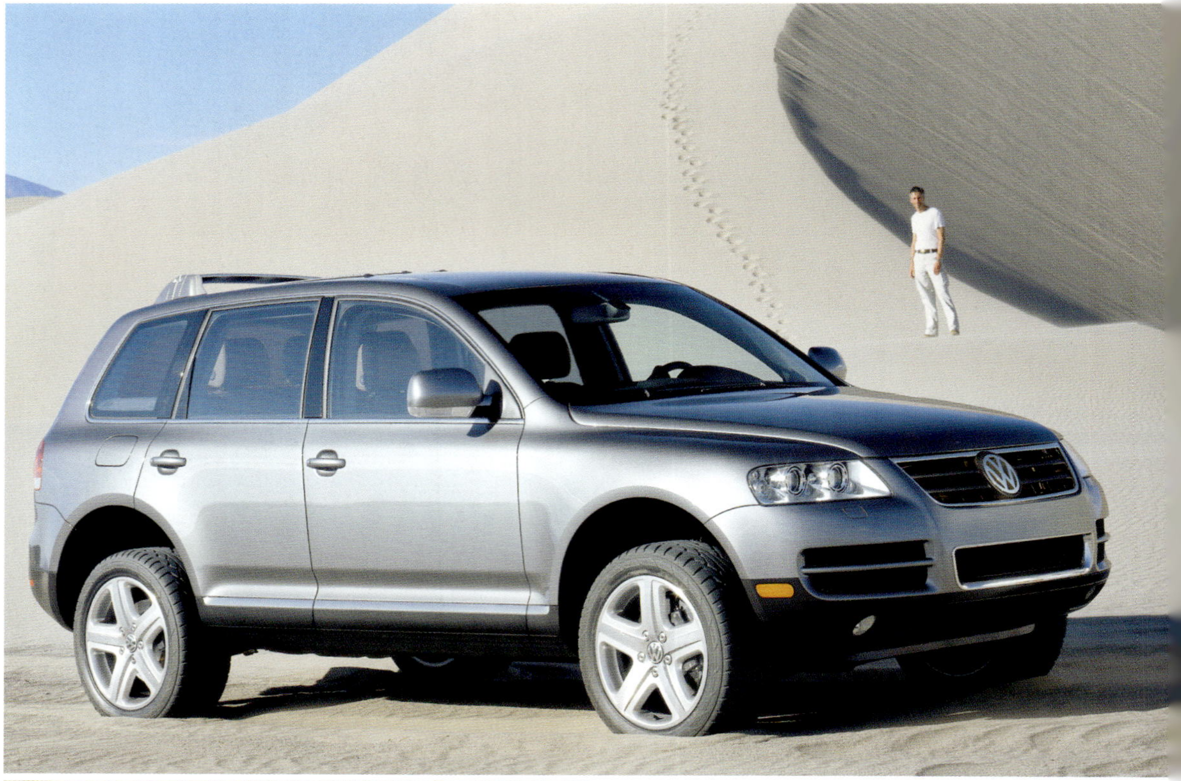

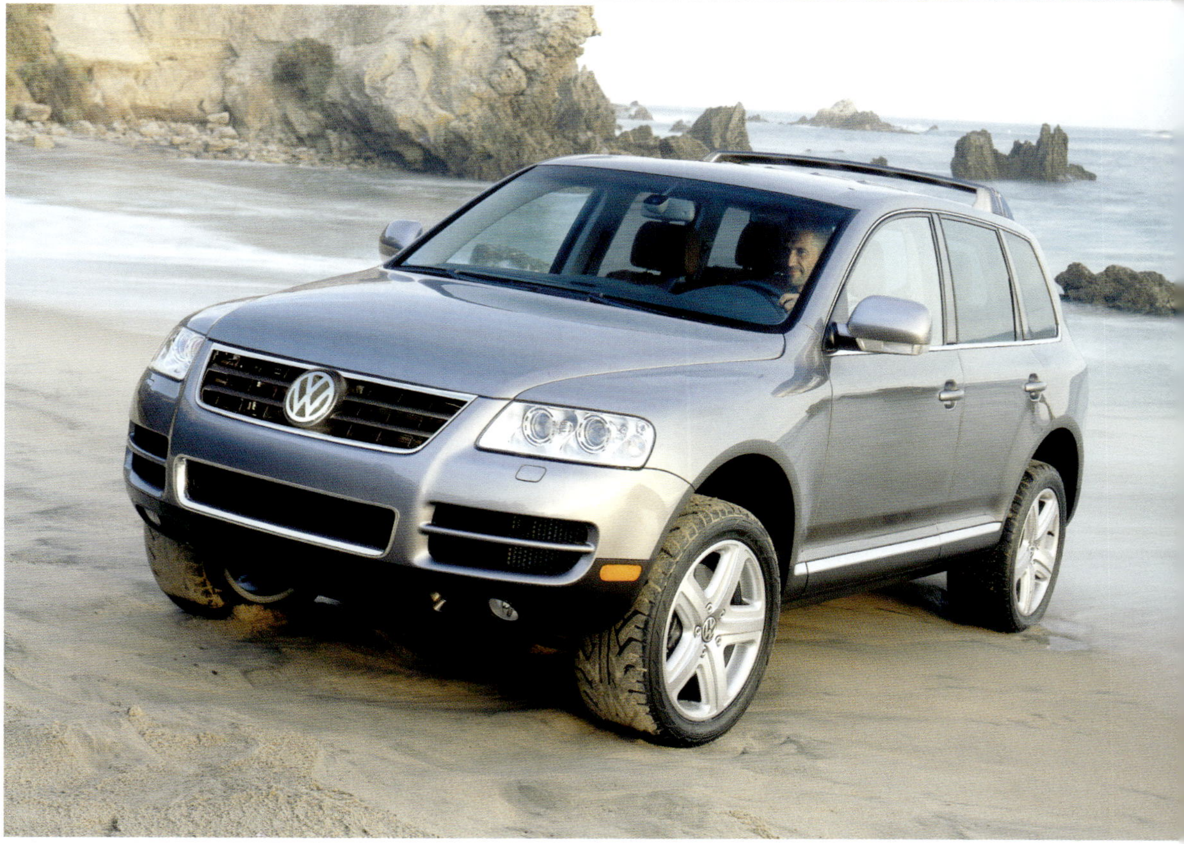

iF Silver Award

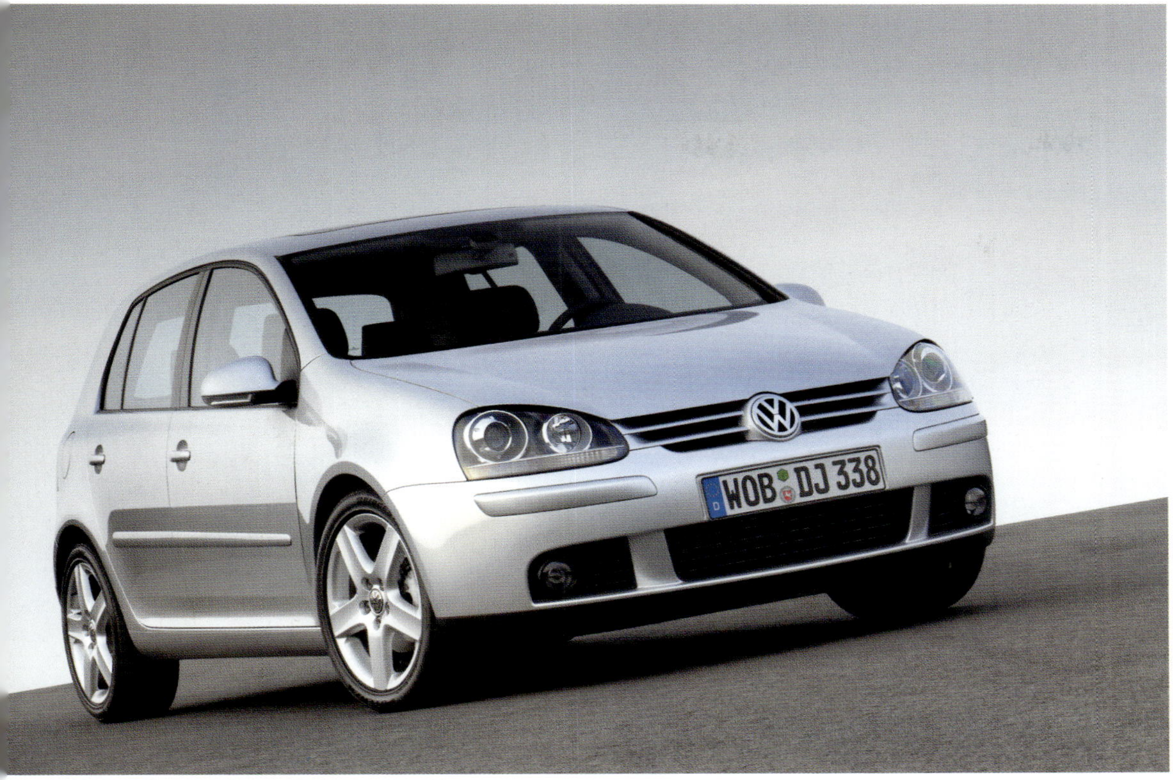

Hersteller Manufacturer
Volkswagen AG
Wolfsburg, Germany

Produkt Product
VW Golf
PKW
Passenger car

Design
Volkswagen Design
Wolfsburg, Germany

Auch die fünfte Generation ist bereits auf den ersten Blick als Golf zu erkennen. So zitiert der Neuentwurf die schon klassischen Kennzeichen der Golf-Silhouette: das gestreckte Dach, das steile Heck und die stabile C-Säule. Die Frontgestaltung wurde in ihren Grundzügen bewahrt, jedoch hat der Grundkörper eine stärkere Keilform erhalten, die Scheinwerfer wurden neu interpretiert und die Form, Kontur und Art der Einbindung der SBBR-Leuchte in den Karosseriekörper dominiert nunmehr die Heckansicht.

The fifth generation is instantly recognizable as a Golf. The new design quotes the already classic characteristics of the VW Golf's silhouette: the stretched roof, the sharply slanted hatchback and the stable C-column. The front-end design retains the basic lines but the body has a stronger wedge form, the headlights are given a new interpretation and the form, contour and way of integrating the taillight cluster into the car body now dominate the rear view.

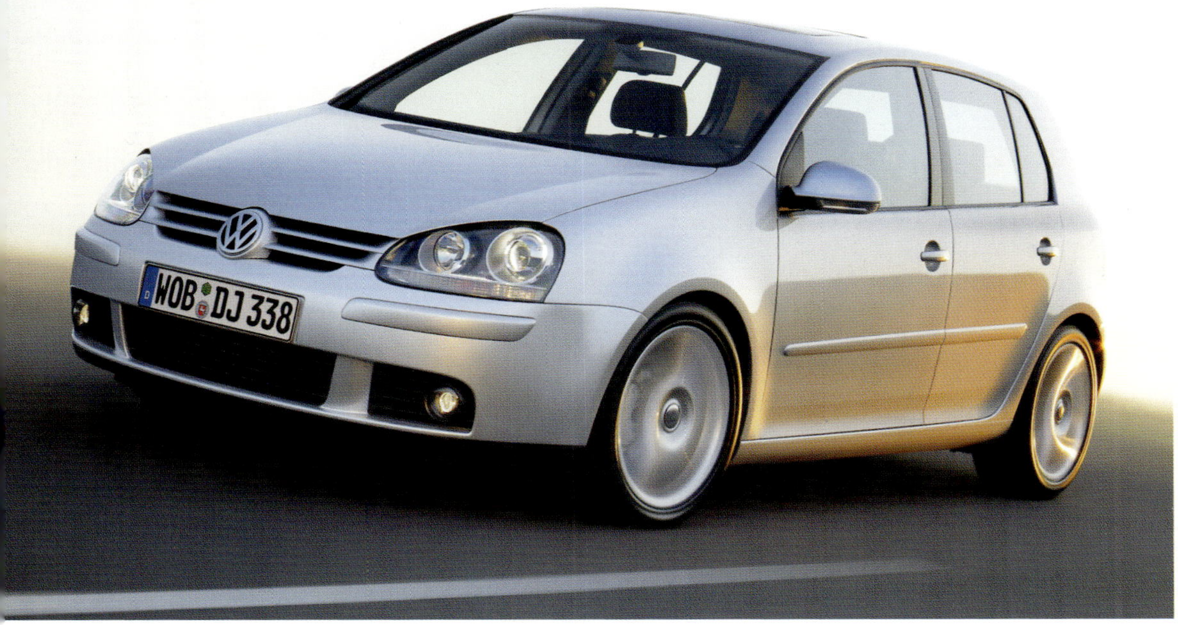

Hersteller Manufacturer
smart gmbh
Böblingen, Germany

Produkt Product
smart forfour
PKW
Passenger car

Design
Hartmut Sinkwitz (Leitung)
smart Design-Team
Böblingen, Germany

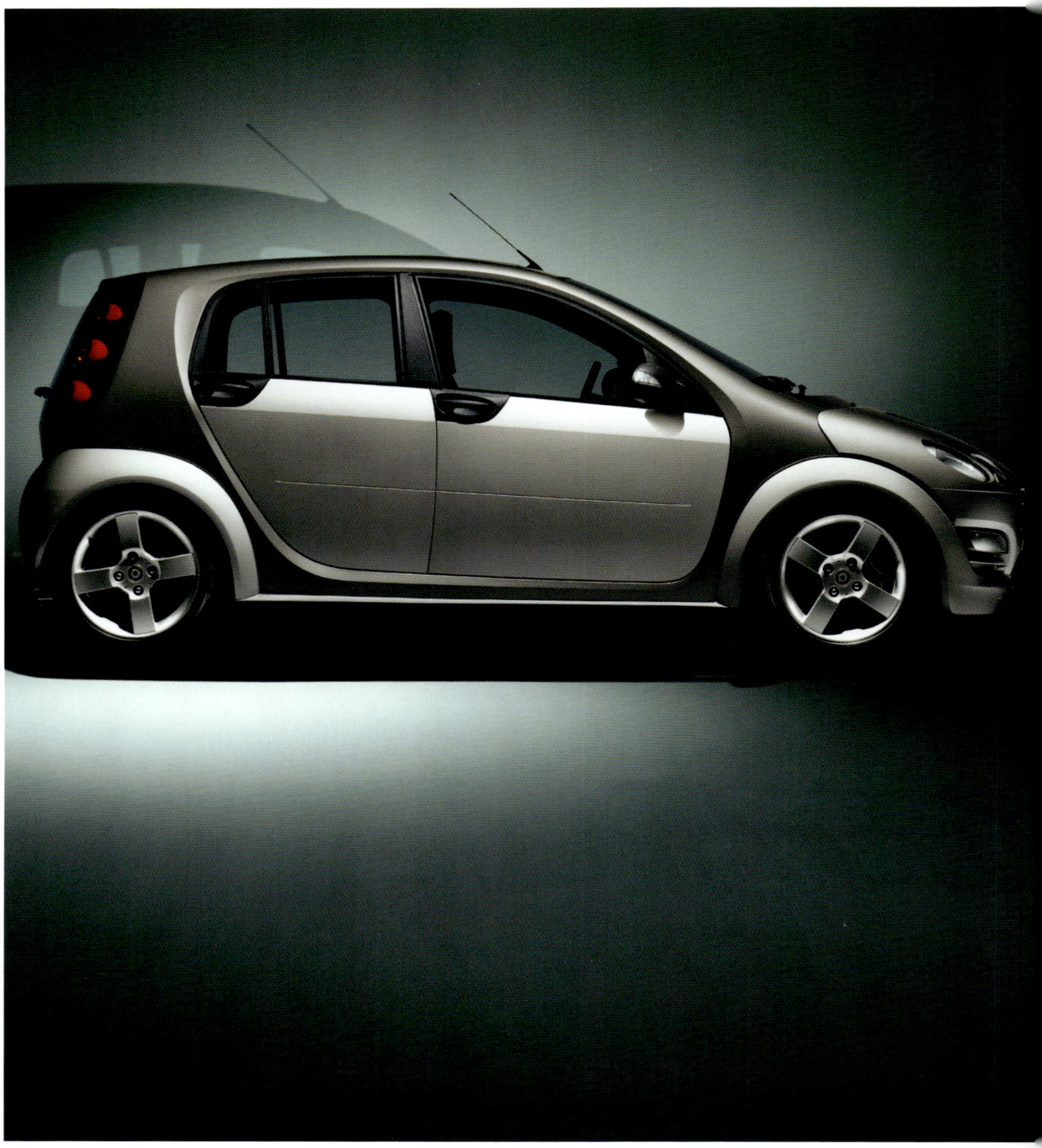

Der smart forfour verbindet die Funktionalität eines Kleinwagens mit der Emotionalität der jungen Automobilmarke smart. Aufgrund seiner Designmerkmale ist auch der smart forfour auf den ersten Blick als smart erkennbar: Markantes Stilmerkmal ist das mit der TRIDION-Sicherheitszelle verbundene Zweimaterial- und Zweifarbkonzept, das sich durch den Einsatz farbiger und auswechselbarer Body Panels aus Kunststoff auszeichnet. Das Auto ist mit seinem extrem kurzen Überhang hinten sowie den großen, bündig mit den Kotflügeln abschließenden Rädern kompakt und sportlich. Das freundliche Gesicht ist geprägt durch die Anordnung der vier einzelnen Scheinwerfer oberhalb des Lufteinlasses. Auch das Interieur ist durch das eigenständige Design und seine hohe Funktionalität gekennzeichnet. Durch das Design gelingt es, bei kompakten Abmessungen einen funktionellen Innenraum mit großzügigem Raumgefühl zu bieten. Optisch prägende Merkmale sind große textile Bereiche und farbige Akzentteile.

The Smart forfour blends the functionality of a small car with the emotional appeal of the young Smart auto marque. Thanks to its design features the Smart forfour is instantly recognizable as a Smart. One of the prominent design features is the dual-material and dual-color concept, combined with the TRIDION safety cockpit, which is characterized by the colored and exchangeable plastic body panels. The car is compact and sporty with an extremely short rear overhang and large wheels flush with the fenders. The friendly appearance is characterized by the four headlights which are positioned above the ventilation scoop. The interior also boasts an independent design and great functionality. Thanks to the design, despite its compact dimensions, the interior offers a spacious feel. Key visuals are large expanses of fabric and colored accents.

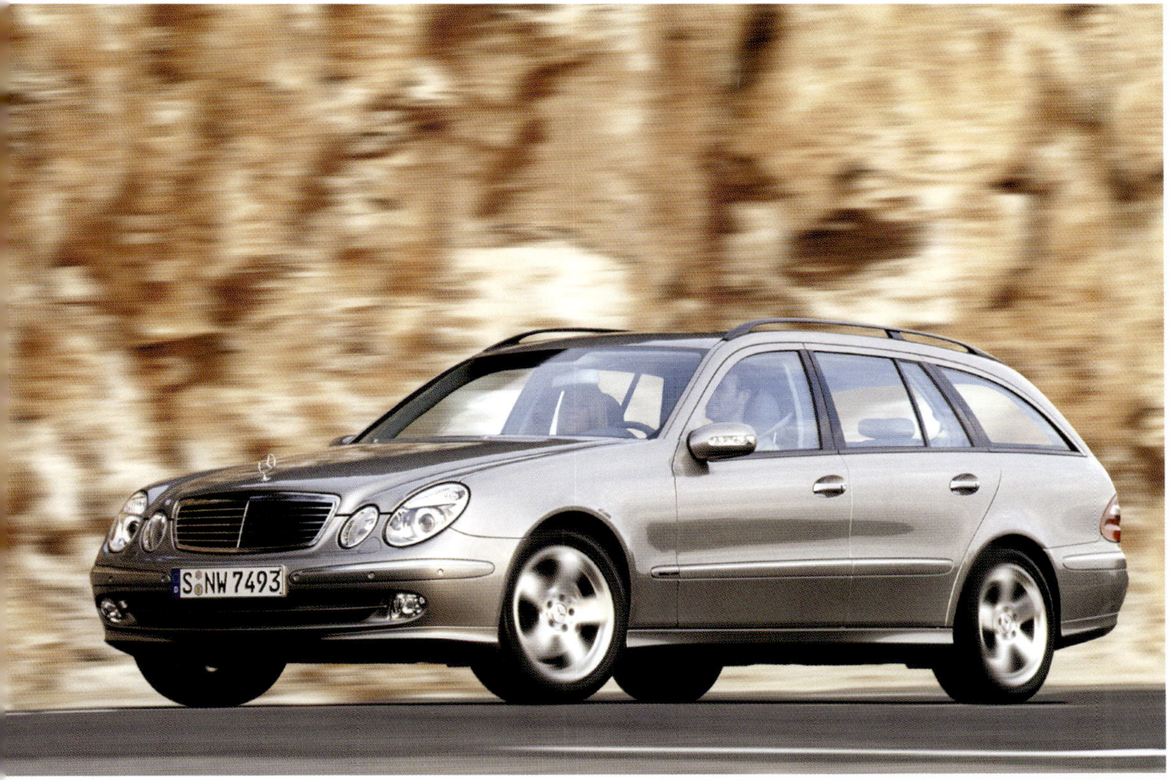

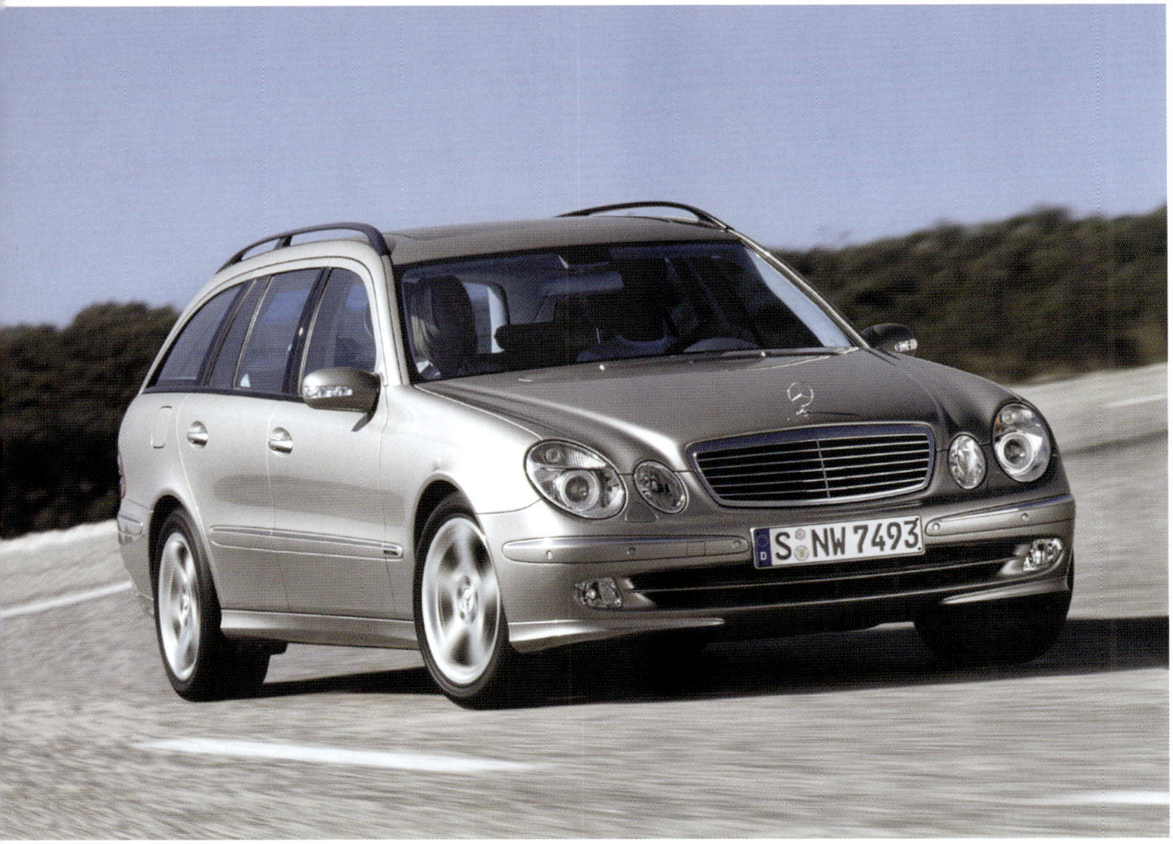

Hersteller Manufacturer
DaimlerChrysler AG
Stuttgart, Germany

Produkt Product
Mercedes-Benz E-Klasse Kombi
Kombi-Limousine
Estate car

Design
DaimlerChrysler AG
Werksdesign
Sindelfingen, Germany

Die Kombi-Limousine basiert auf der Limousine der E-Klasse mit ihrem dynamischen 4-Augen-Gesicht. Mit seinem großzügigen, variabel nutzbaren Raumangebot und der optimierten Funktionalität ist das neue T-Modell den unterschiedlichsten Transportaufgaben gewachsen. Die drei eigenständigen Design- und Ausstattungslinien Classic, Elegance und Avantgarde erfüllen unterschiedliche Kundenwünsche. Neben dem Basismodell Classic setzt das Modell Elegance unter anderem durch dezente Chrom-Applikationen an den Stoßfängern, Seitenschutzleisten und Türgriffen sowie durch eine atlasgrau lackierte Lamellen-Kühlermaske Akzente, die Linie Avantgarde hingegen durch eigenständige, dynamisch gestaltete Stoßfänger und Seitenschweller sowie Cockpit-Instrumente mit weißen Zifferblättern.

The E-class estate limousine car is based on the E-Class sedan with its dynamic 4-eye face. The new estate model was developed for a large variety of transport tasks with generous space provision which can be adjusted flexibly. There are three distinctive design and equipment lines (Classic, Elegance and Avantgarde), each fulfilling different customer desires. Classic is the basic model while Elegance sets accents with subtle chrome applications on the bumpers, side rub strips and door handles and atlas-grey vanes on the radiator grille. The Avantgarde model, by contrast, stands out for its individual, dynamically designed bumpers and side skirts as well as cockpit instruments with white dials.

Hersteller Manufacturer
BMW AG
München, Germany

Produkt Product
BMW X3
Sports Activity Vehicle

Design
Designteam der BMW Group
München, Germany

Der neue BMW X3 zeichnet sich durch eine moderne dynamische Gestaltung aus, die auf seinen Charakter als Sports Activity Vehicle verweist. So vermitteln die 17-Zoll-Räder, die hohe Bodenfreiheit, die Fahrzeugfront und die in Chrom gefasste Niere die entsprechende Robustheit. Die ausdrucksstarke Front ist durch die weißen Blinkleuchten gekennzeichnet, die elektrischen Außenspiegel sind in der Wagenfarbe lackiert und der ebenfalls in der Wagenfarbe gehaltene Dachspoiler sorgt für einen markanten Abschluss. Im Rahmen des variablen Raumkonzepts lässt sich das Gepäckvolumen mit wenigen Handgriffen von 480 auf 1560 Liter vergrößern, Mountainbike und Rennrad können mit der speziellen Fahrradhalterung fixiert werden.

The new BMW X3 stands out for its modern dynamic design, which alludes to its character as a sports activity vehicle. For example, the 17″ wheels, good road clearance, front and radiator grill, and the chromed bulged fairings all convey a sense of sturdiness. The highly expressive front is defined by the white headlamps, while the electrically controlled wing mirrors come in the same color as the body as does the roof spoiler that creates such a conclusive highlight at the rear. As part of the variable interior concept, a few movements suffice to expand the trunk area from 480 l to 1560 l; mountain bikes and race bikes can be held firmly in place by the special bike rack.

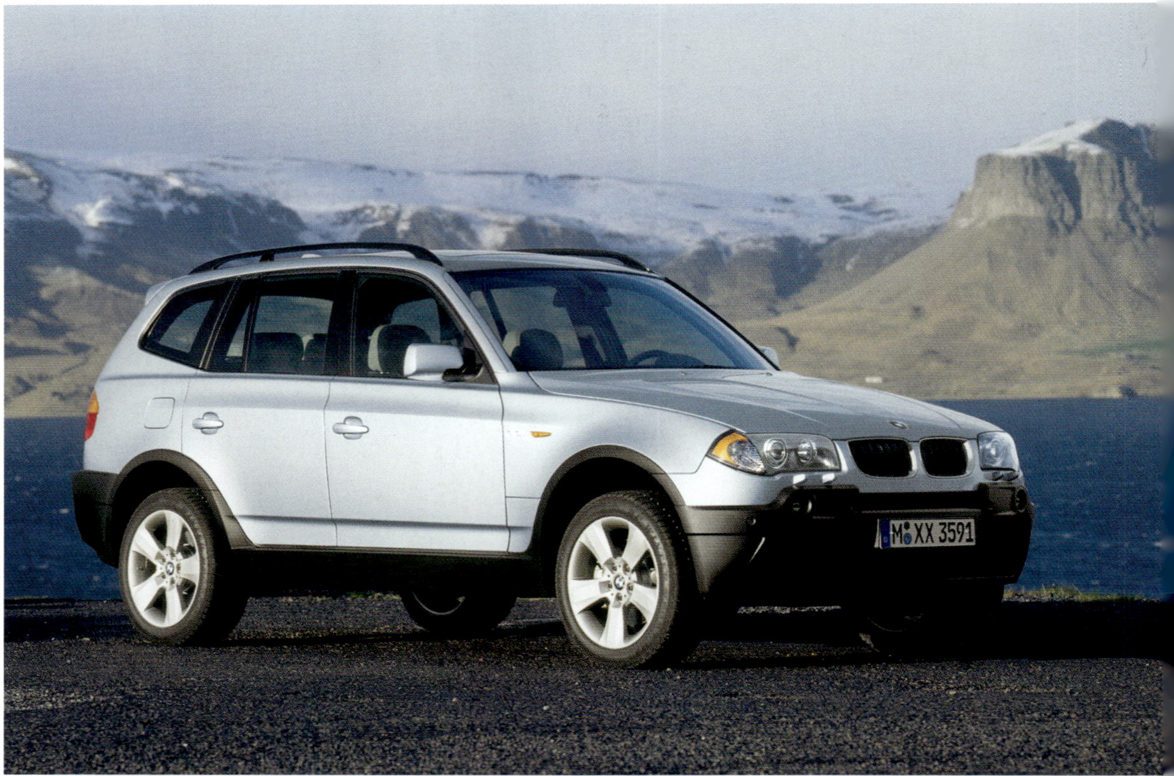

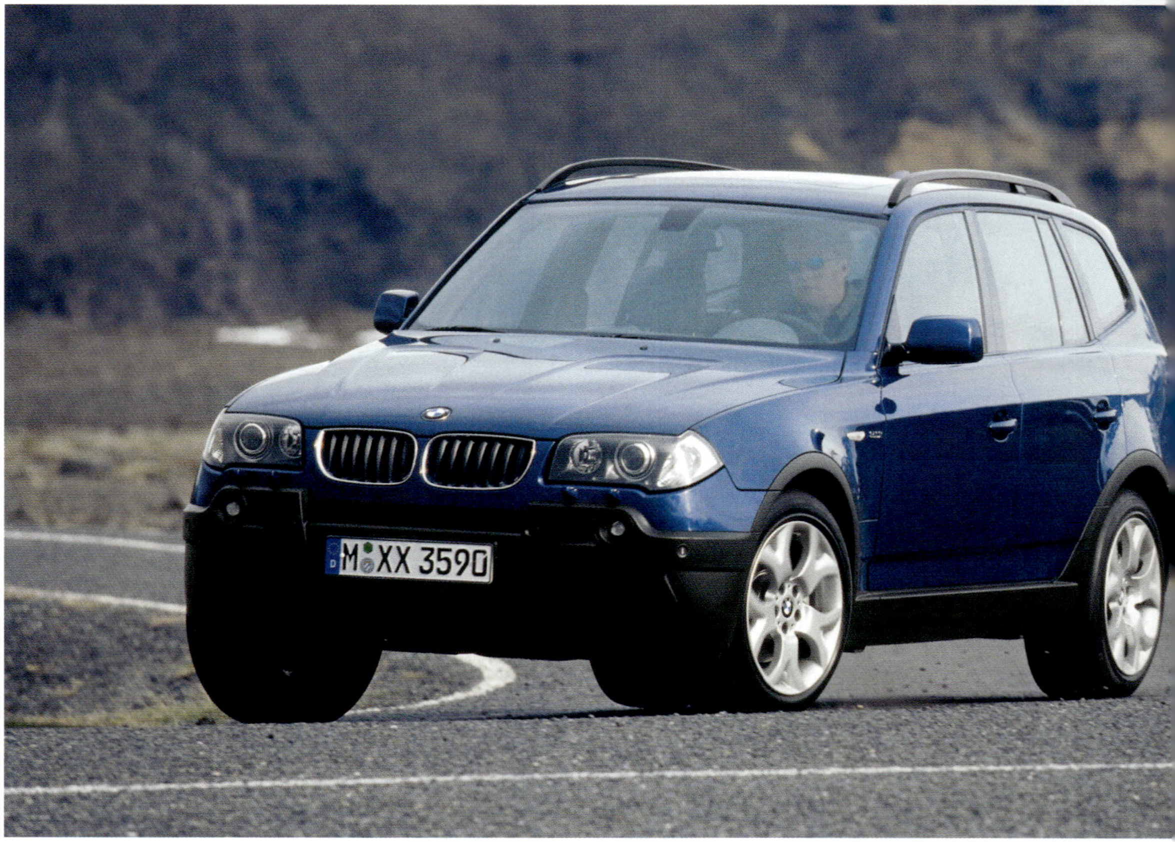

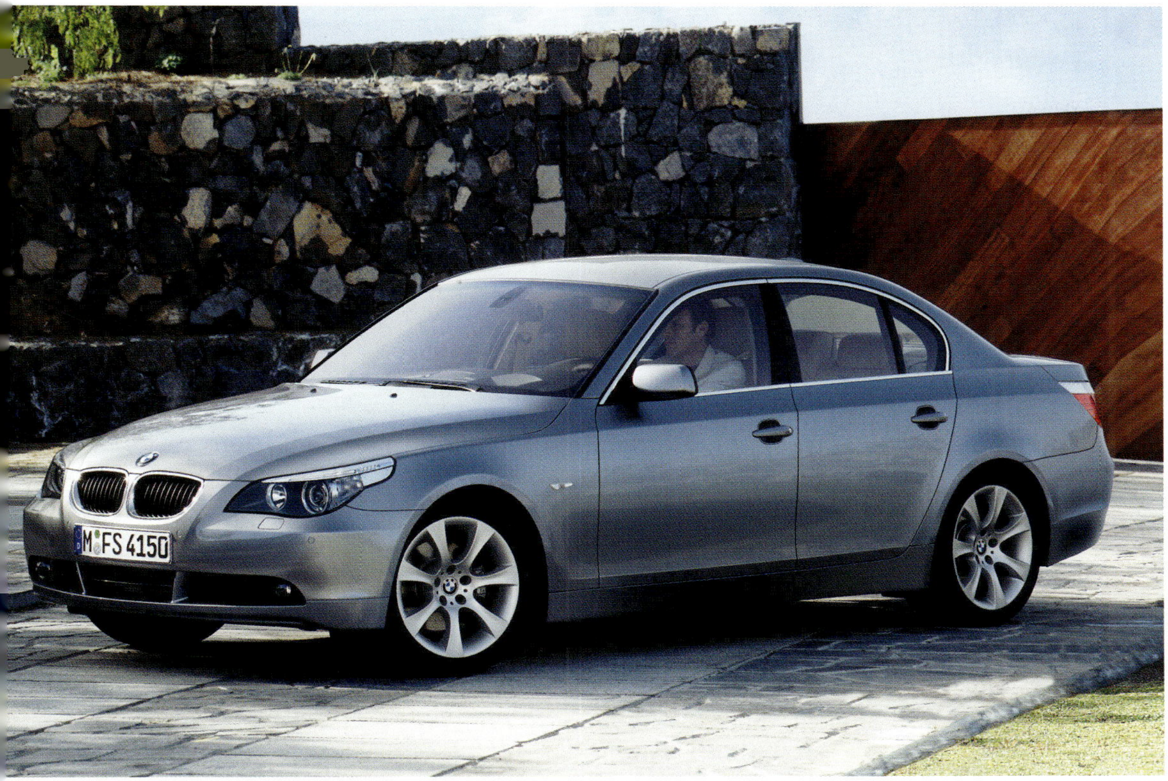

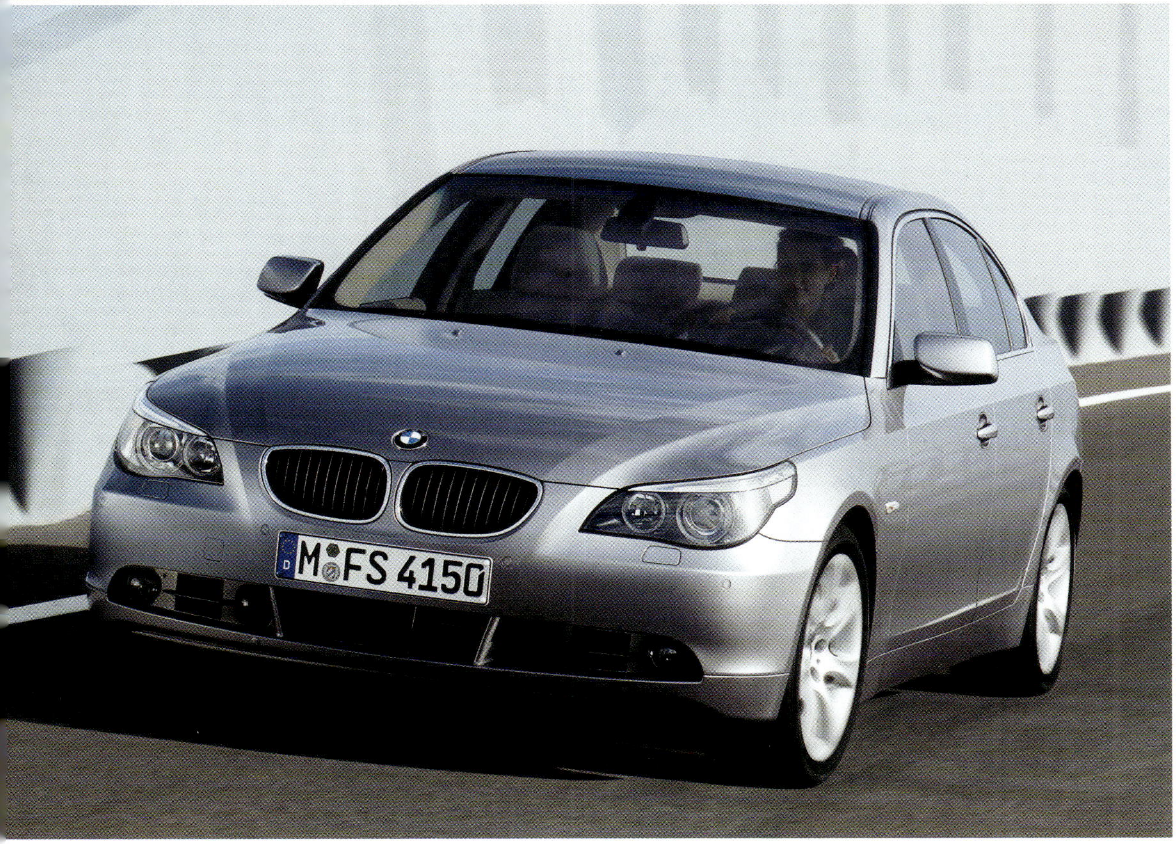

Hersteller Manufacturer
BMW AG
München, Germany

Produkt Product
BMW 5er
Limousine

Design
Designteam der BMW Group
München, Germany

Der neue 5er vereint in seinem Auftritt die Dynamik und Sportlichkeit der 3er Reihe mit der Präsenz des BMW 7er. Das Design der neuen Limousine bildet damit die starke Mitte des Spektrums der BMW Fahrzeugpalette. Augenfälligstes Merkmal sind die beiden Doppelscheinwerfer, die durch ihre Gestaltung dem neuen 5er einen unverwechselbaren Charakter verleihen. Fließende Linien, die im Bereich der Scheinwerfer beginnen und sich um die lange Fronthaube ziehen, führen über die Seiten bis zum Heck. Die ebenso fließend gestaltete C-Säule zeigt eine coupéhafte Anmutung. Die Heckpartie wirkt durch den Verzicht auf sichtbare Trennfugen und den integrierten Spoiler breit und straff. Das Cockpitdesign folgt in wesentlichen Elementen der neuen Interior-Designphilosophie des 7er, weist aber zusätzliche klassische Elemente auf.

The 5 Series features, a combination of the dynamic sporting nature of the 3 Series and the emphatic presence of the BMW 7 Series. The limousine's design thus forms the middle in the overall spectrum of BMW cars. The most notable feature consists of the double headlight sets which give the 5 Series its unmistakable character. Flowing lines start in the proximity of the headlights, continue over the hood and traverse the sides through to the rear of the car. The C-column also has a flowing design giving it a coupé-like appearance. The rear appears broad and streamlined thanks to the integrated spoiler and the omission of visible dividing lines. The key elements in the cockpit design follow the lines of the new interior design philosophy developed for the 7 Series, but also feature a number of classic components.

Hersteller Manufacturer
Neoman Bus GmbH
Salzgitter, Germany

Produkt Product
MAN Lion's Coach/Lion's Top Coach
Reisebus
Touring coach

Design
Kenan Erdinc
Sven Gaedtke
Britta Hachenberg
Stephan Schönherr (Leitung Design)
Neoman Bus GmbH
Salzgitter, Germany

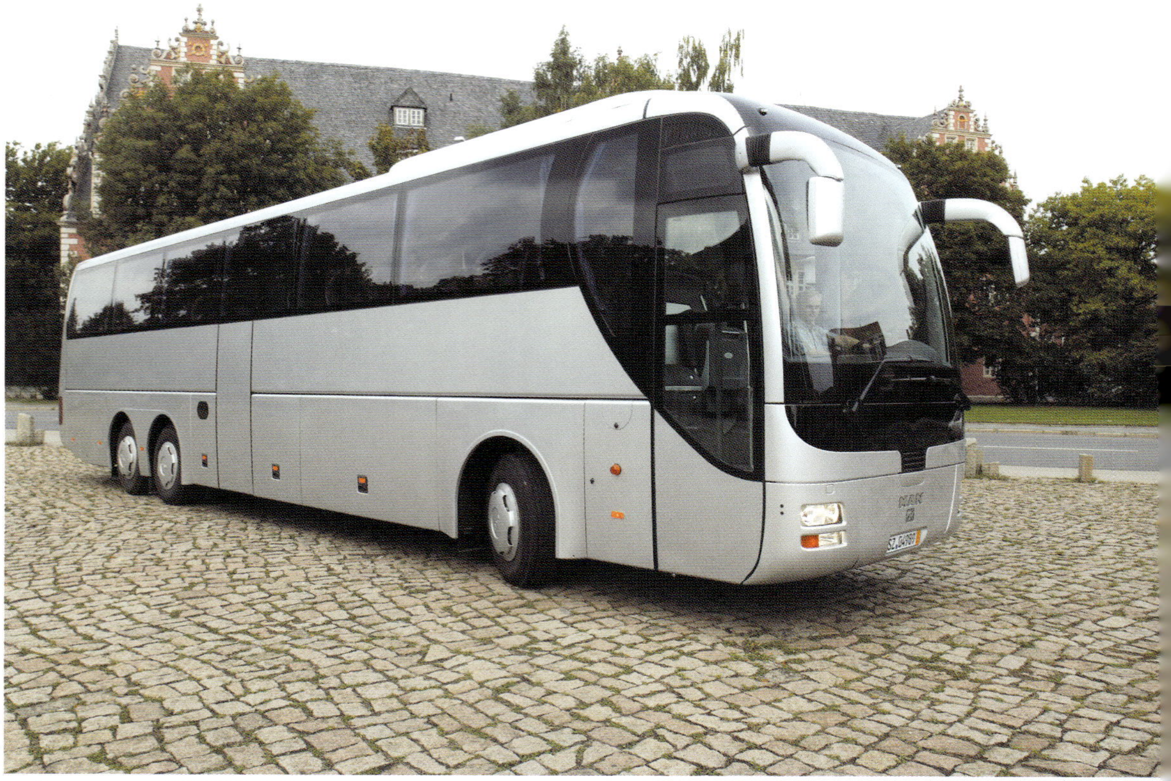

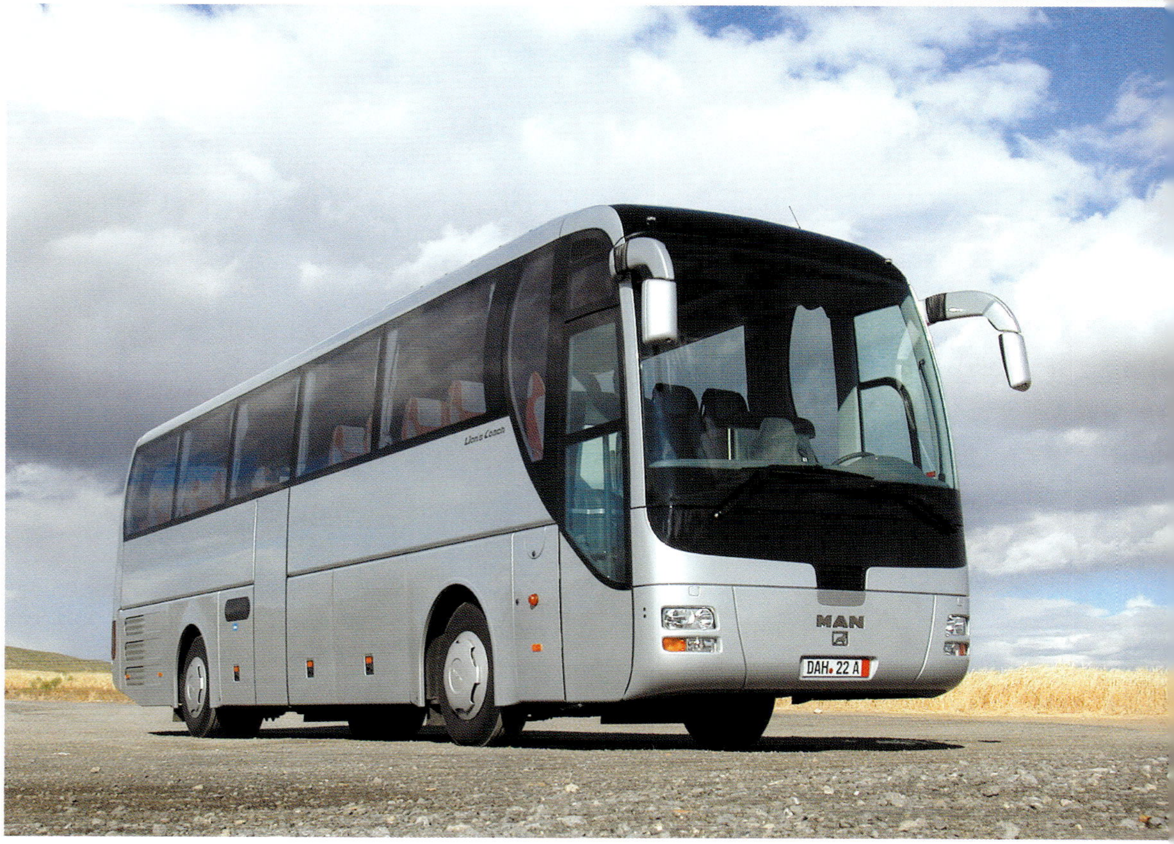

Der MAN Reisebus basiert auf einem modularen Konzept. Die Anordnung von Fahrerarbeitsplatz, Fahrgastraum, Antriebsstrang und Gepäckraum ermöglicht eine optimierte Nutzung von Fahrzeughöhe und Länge. Markante Designmerkmale sind die geschwungene C-Säule mit Überrollbügelfunktion, die glattflächige Bugschürze mit Firmenlogo, die große optisch verlängerte Bugscheibe, die vorne liegende Klimaanlagenhaube und das überwölbte Heck. Die großflächige, leistenlose Verglasung bietet einen hellen Innenraum und gute Sichtverhältnisse. Die horizontale Linienführung über den Gepäckklappen gliedert das Fahrzeug optisch in die unterschiedlichen Funktionsbereiche. Neu entwickelte, ergonomisch gestaltete Sitze verbessern den Sitzkomfort der Fahrgäste und des Fahrers.

The MAN touring couch is based on a modular concept. The configuration of the driver's seat area, passenger section, driveline and luggage compartment allow for optimal use of space in the vehicle, in terms of both height and length. Prominent design features are the curved C-column which functions as a rollover bar, the smooth-surfaced front apron with company logo, the large, visually extended front windscreen, the air-conditioner bonnet located at the front and the arched rear. The expansive uninterrupted glass surfaces provide a light interior and good viewing. Horizontal lines over the luggage flaps visually structure the different functional areas. Newly developed, ergonomically formed seats afford both the driver and passengers improved seating comfort.

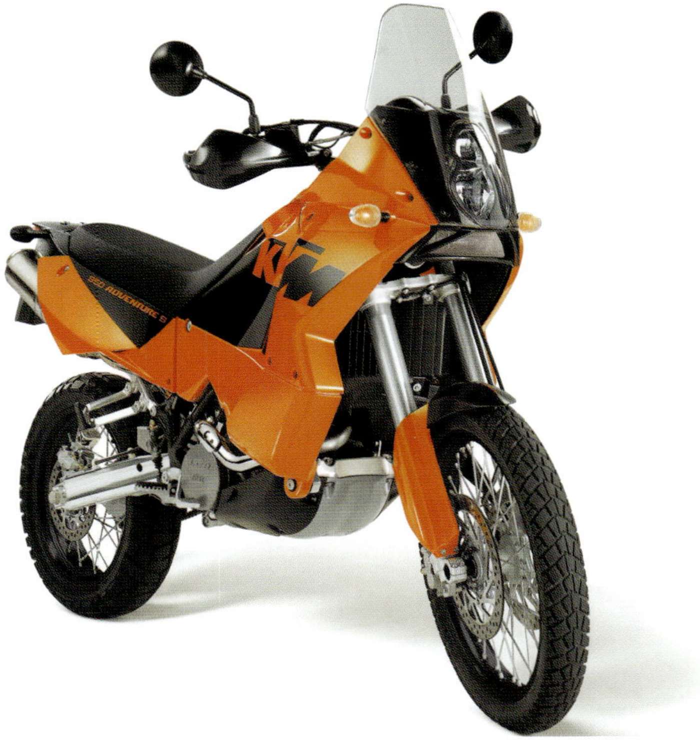

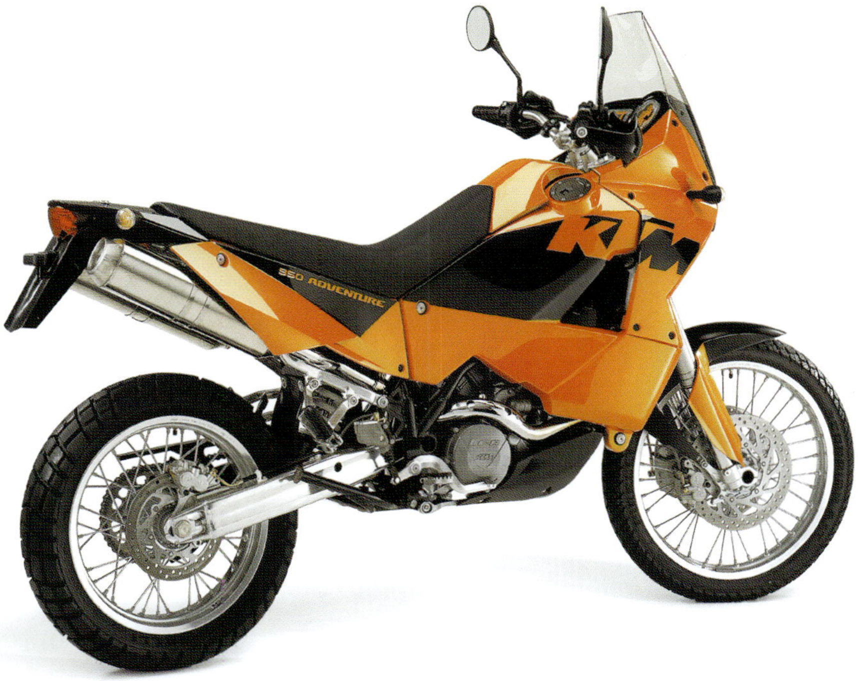

Hersteller Manufacturer
KTM Sportmotorcycles AG
Mattighofen, Austria

Produkt Product
KTM 950 Adventure
Reise-Enduro
Touring Enduro

Design
Gerald Kiska
Kiska GmbH
Salzburg, Austria

Besonderes Augenmerk bei der Designentwicklung galt den ergonomischen Voraussetzungen für den dualen Einsatz auf Straße und Gelände. Die KTM 950 Adventure vereint Möglichkeiten wie Langstreckenreisen mit Geländetauglichkeit, gute Agilität auf Landstraßen mit hoher Zuladung. Die neuartige Anordnung des Tanks, der zweigeteilt und weit nach unten gezogen wurde, erfüllt diese Optimierung der Ergonomie und ermöglicht ein zusätzliches Staufach zwischen den Tankhälften. Zudem wurden alle großen Massen nahe des Schwerpunkts der Maschine angeordnet, um ein agiles Handling zu gewährleisten.

The design pays special attention to the ergonomic basis for dual use on and off-road. The KTM 950 Adventure combines the best of long-distance biking with off-road capabilities, prime handling for on-road use even with a large amount of luggage on the back. The innovative fuel tank configuration, which is divided in two and drawn far downwards, optimizes the ergonomic qualities of the bike and also provides additional storage space between the two tank halves. Moreover, all the large masses were located close to the center of gravity to enhance agile handling.

Hersteller Manufacturer
Öswag Werft Linz AG
Linz, Austria

Produkt Product
MS Zug
Motorschiff
Motorboat

Design
Manfred Huber
Daniela Saxer
Aardeplan Architekten ETH SIA
Baar, Switzerland

Die Außenform des neuen Motorschiffes Zug für den öffentlichen Verkehr wurde als Hommage an das Design der eleganten, stromlinienförmigen Schiffe der fünfziger Jahre entwickelt. Seine Erscheinung reduziert sich auf ein schlichtes Zusammenspiel verschiedener Kurven, Radien und Neigungen, die die gesamte Stahlkonstruktion charakterisieren. Auch die großzügigen Panorama-Fenster, Türen und übrigen Öffnungen beziehen sich auf die geschwungenen Formen des Stahlbaus. Der Innenraum wurde durch einseitig rot gespritzte Glaswände und speziell entworfene Nussbaummöbel strukturiert. Stühle, Tische und Vorhänge entsprechen einem zeitlosen, klassischen Design. Das für 450 Passagiere ausgerichtete, vollständig behindertengerechte Schiff ist mit seinen drei Salons als Restaurant mit eigener Bordküche konzipiert.

The shape of the new Zug motorboat, destined for public transport purposes, was developed in homage to the elegant streamlined ships of the 1950s. Its appearance is reduced to a simple interrelation of various curves, radii and slants which characterize the entire steel structure, whereby the spacious panorama windows, doors and other openings are derived from the curves of the steel. The interior is defined by glass, coated with red paint on one side, and walnut furniture. The chairs, tables and drapes adhere to a timeless, classic design. The boat is designed for 450 passengers, caters fully to the needs of the physically challenged, and with its three salons is conceived as a restaurant with on-board catering.

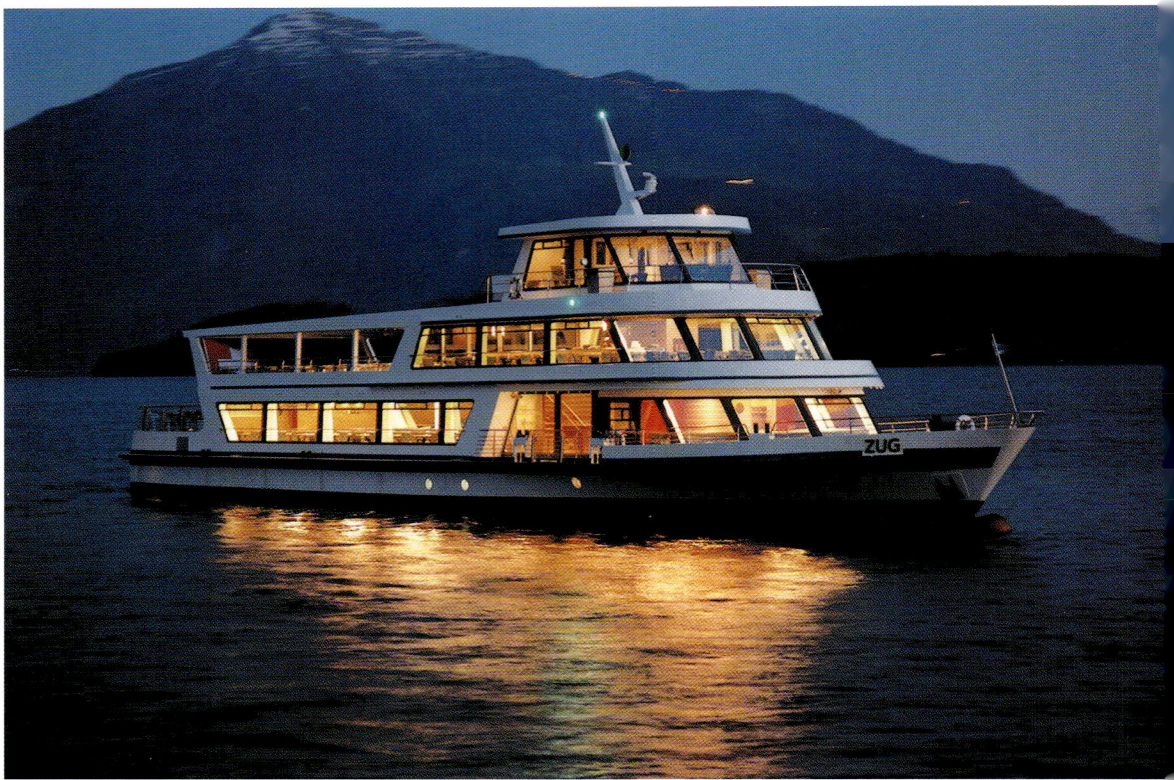

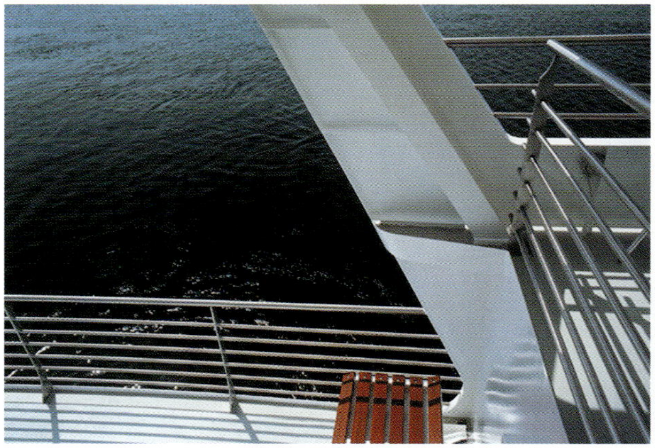

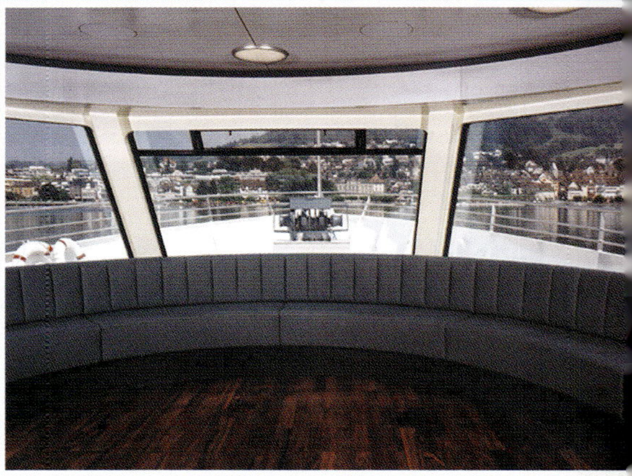

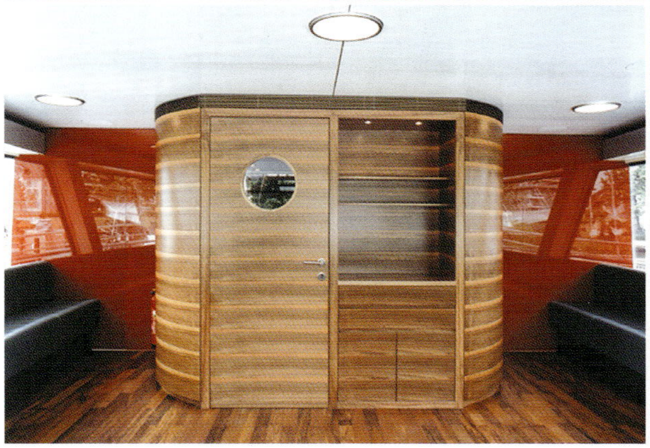

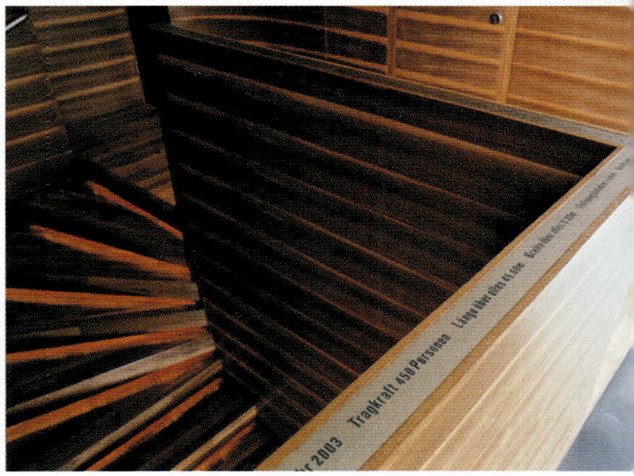

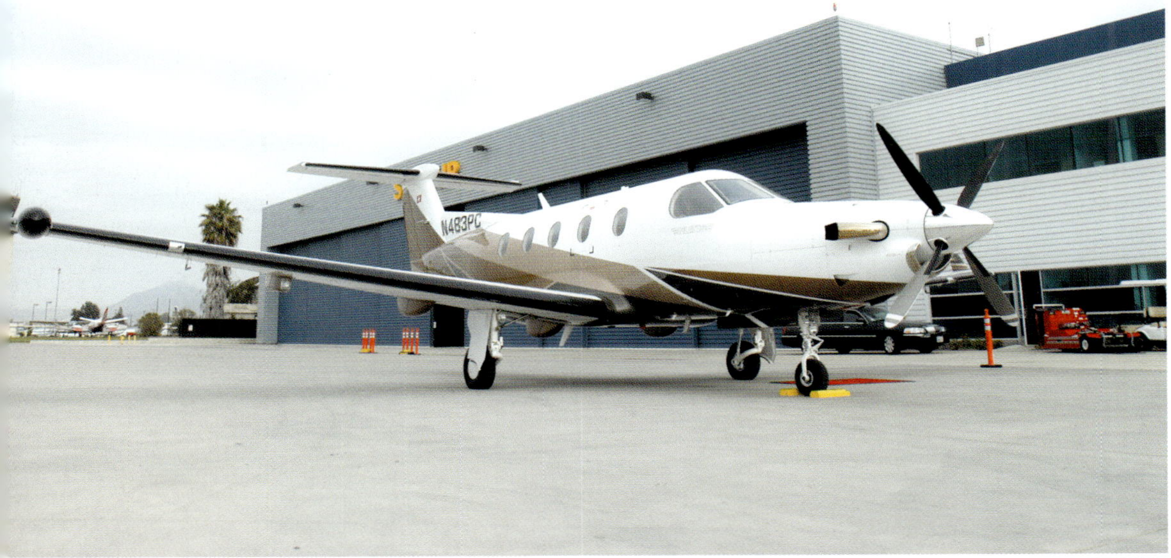

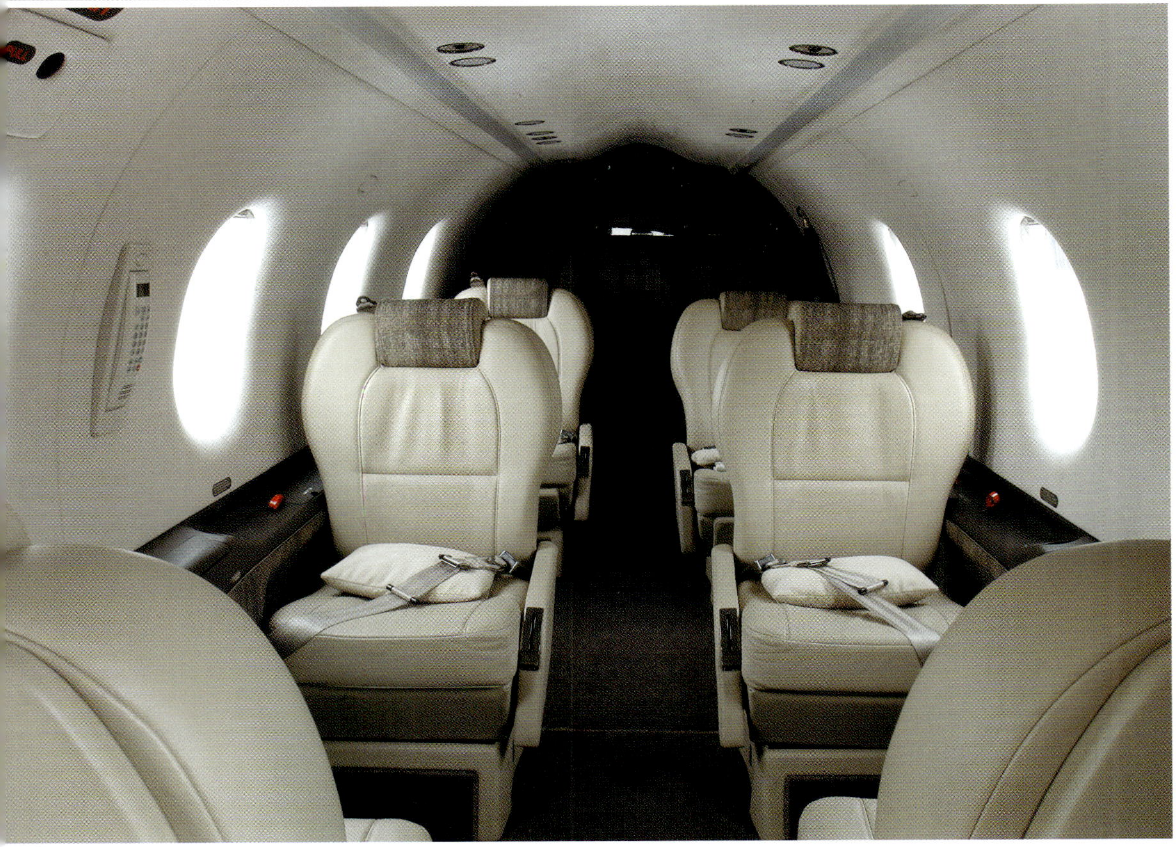

Hersteller Manufacturer
Pilatus Business Aircraft
Broomfield, CO, USA

Produkt Product
Pilatus PC-12
Inneneinrichtungen und Außengrafik für Firmenflugzeuge
Interior Color and Materials Program, and Exterior Graphics

Design
Aris Garabedian
Claudia Geidobler
Cherly Pelly
Kirsten Schmeding
BMW Group DesignworksUSA
Newbury Park, CA, USA

BMW Group DesignworksUSA hat für das einmotorige Flugzeug PC-12, das für den Einsatz als Reiseflugzeug für Privatpersonen oder Firmen eingesetzt wird, eine grafische Außengestaltung und ein Farb- und Materialkonzept für die Innenraumgestaltung entworfen. In das Interior Design des Flugzeugs konnten insbesondere einschlägige Erfahrungen der BMW Group DesignworksUSA im Autodesign einfließen.

BMW Group DesignworksUSA created the interior colors and textiles as well as the exterior graphics for the single-engine PC-12 plane, which is generally used by private individuals or corporations for travel. BMW Group DesignworksUSA brought its long-standing experience in auto design to bear in devising the new look for the aircraft's interior.

Hersteller Manufacturer
John Deere / Horicon Works
Horicon, WI, USA

Produkt Product
X Series Lawn and Garden Tractor
Gartenfahrzeug
Garden vehicle

Design
Steve Kaminski
Tom Ward
Andy Logan
Todd Boepple
Frank Killam
BMW Group DesignworksUSA
Newbury Park, CA, USA

Der X Series Lawn and Garden Tractor für den privaten und auch gewerblichen Einsatz mäht nicht nur Gras, sondern ist ein multifunktionales Gartenfahrzeug für alle möglichen Gartenarbeiten zu jeder Jahreszeit. Abgesehen von Mäharbeiten kann mit ihm der Boden gelockert, gepflügt und belüftet werden, zudem lassen sich größere Lasten zu ihrem Einsatzort im Garten schleppen. Im Vordergrund des neuen Designkonzepts standen eine verbesserte Ergonomie, Langlebigkeit, eine frische Ästhetik und der Einsatz neuer Materialien.

The X Series Lawn and Garden Tractor is intended for both private and commercial users – and not only mows grass, but is an all-round utility vehicle with numerous accessories for garden work in any season. In addition to mowing, it can be utilized for snow removal, plowing, aerating, and hauling for overall grounds care. The new design concept focused on enhanced ergonomic, a long service life, a fresh aesthetic, and the incorporation of new materials.

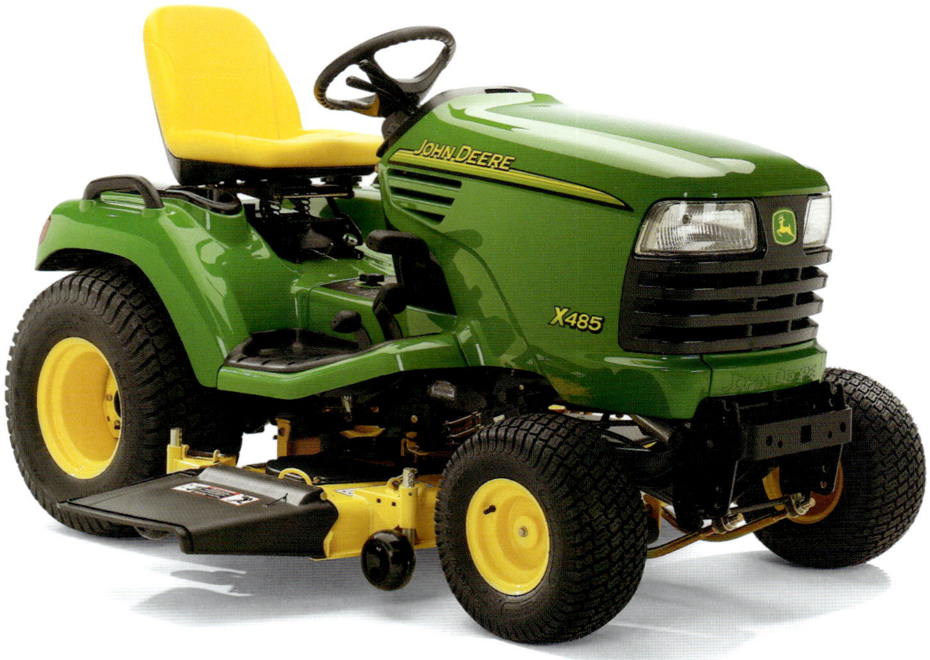

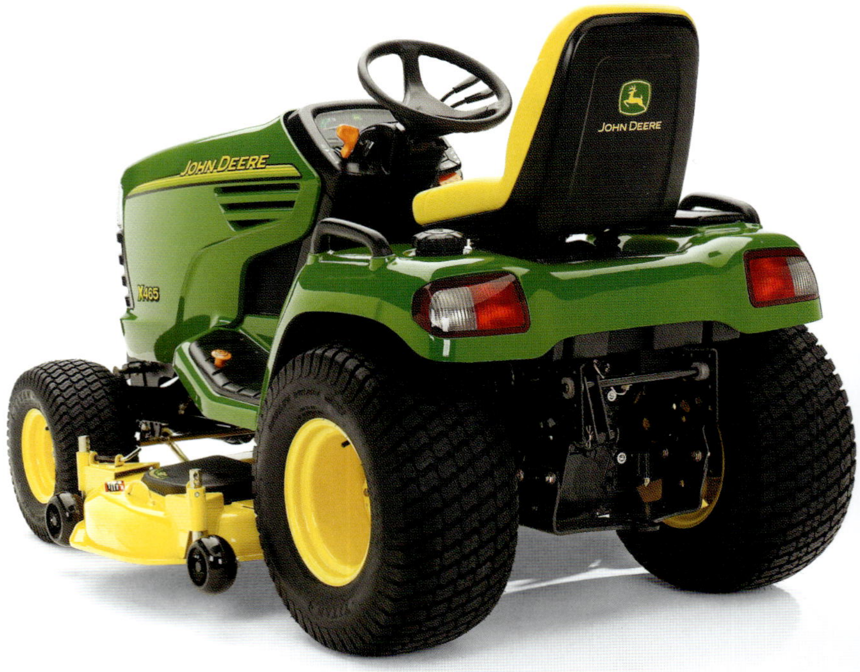

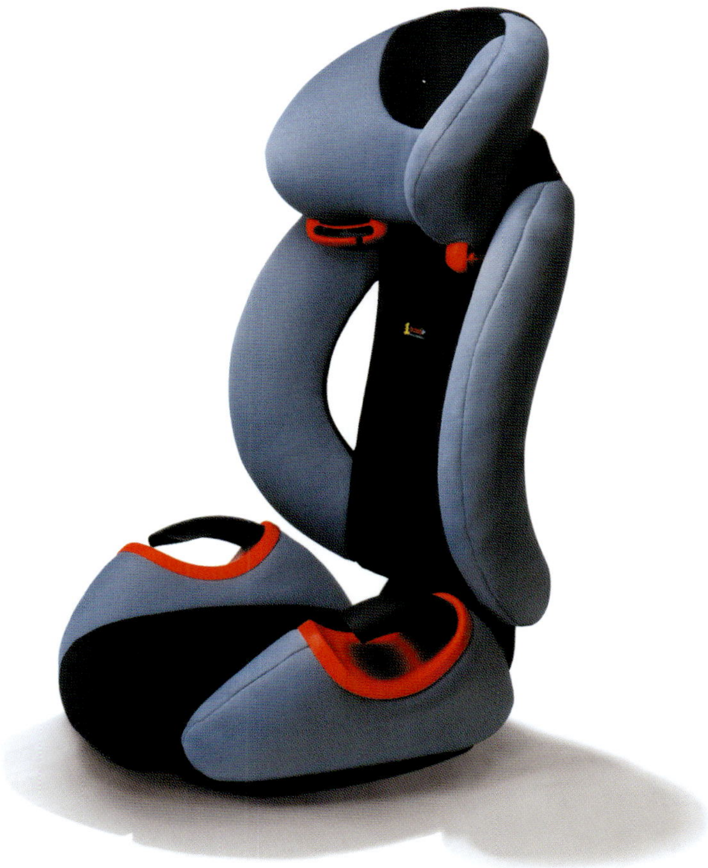

Hersteller Manufacturer
Dremefa bv
Doesburg, Netherlands

Produkt Product
Multibob
Auto-Kindersicherheitssitz
Car safety seat for children

Design
Wolfram Peters (Partner)
Janwillem Bouwknegt (Senior Designer)
Robert Sluijter (Senior Designer)
npk industrial design bv
Leiden, Netherlands

Der Autokindersicherheitssitz Multibob ist für Kinder von 3 bis 12 Jahren gedacht. Ein spezieller Mechanismus bewirkt, dass sich bei einer Höhenverstellung des Kopfteils das Rückenteil automatisch in die richtige ergonomische Position bewegt. Die Armstützen neigen sich bei einer Kollision nach innen und schützen das Kind.

The Multibob car safety seat was specially developed for children between the ages of three and twelve. When the head section is raised, a special mechanism automatically moves the backrest segment into the right ergonomic position. In case of a collision the armrests incline inwards and protect the child.

iF design award 2004

PUBLIC DESIGN

Disziplin 5 Discipline 5

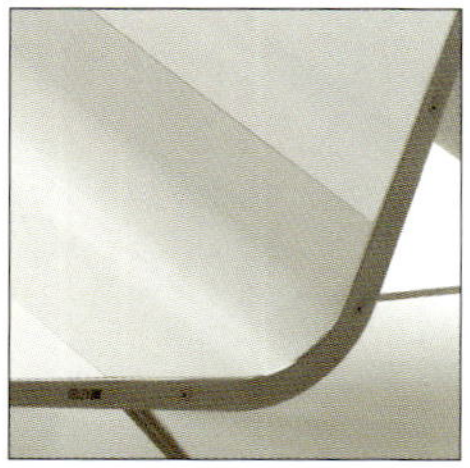
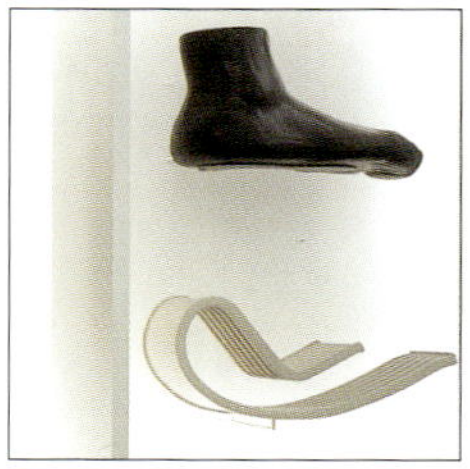
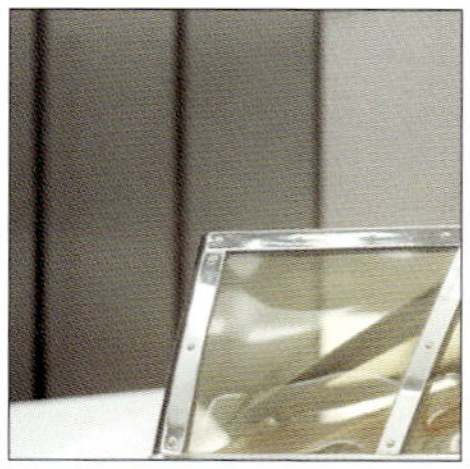
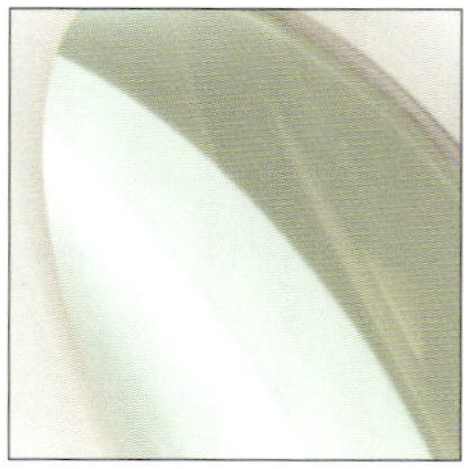
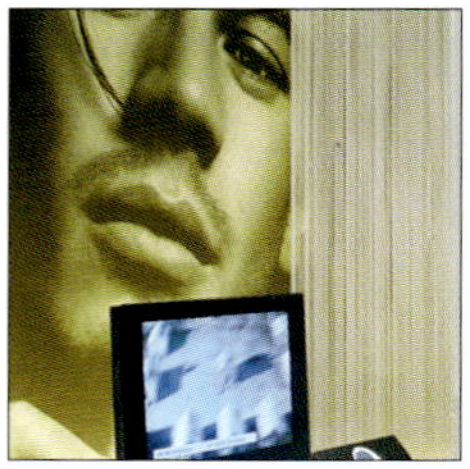

Public Design

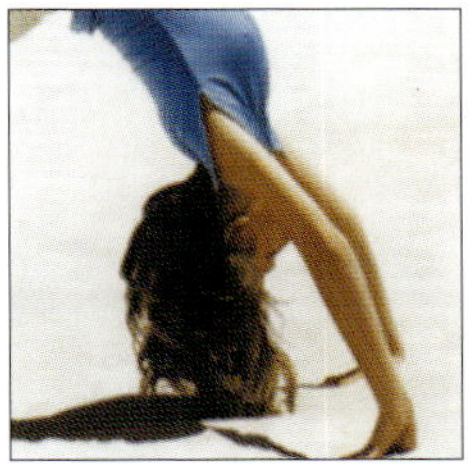

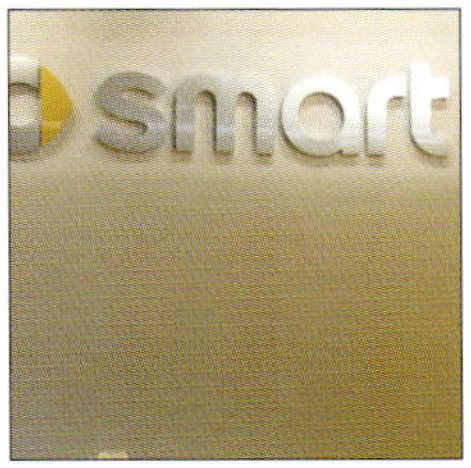

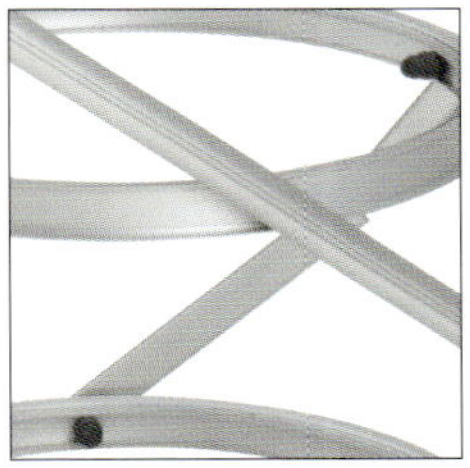

Hersteller Manufacturer
Sellex S.A.
Irún-Gipuzkoa, Spain

Produkt Product
Aero
Sitzbank für Wartezonen
Bench for public areas

Design
Lievore, Altherr, Molina
Barcelona, Spain

Aero ist eine Sitzbank für Wartezonen im Innen- oder Außenbereich mit oder ohne Armlehnen und Endtischen. Die Sitzschalen sind aus jeweils verbundenen extrudierten Aluminiumprofilen gefertigt, die Füße bestehen aus Edelstahl. Optional ist eine Polsterung möglich. Bis zu einer Länge von 434 cm bzw. 7 Sitzen kann die Bank auf zwei Füßen stehen.

Aero is a bench for indoor and outdoor public waiting zones – with or without armrests and with optional side tables. The seat shell is made of inter-connected extruded aluminum profiles, with stainless steel feet. If required, the bench comes with upholstery. The bench needs only two legs up to a length of 434 cm and seven seats.

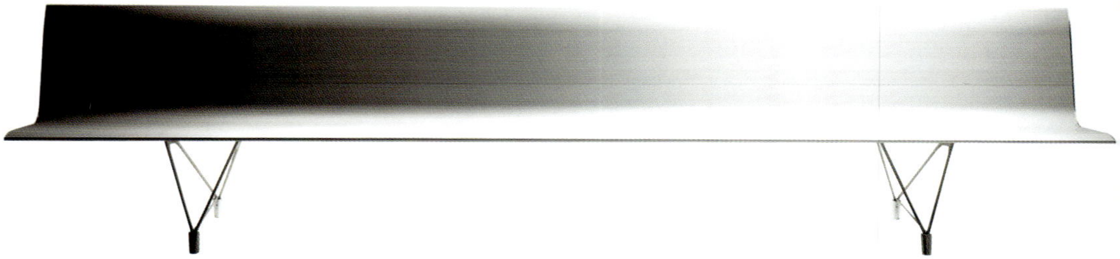

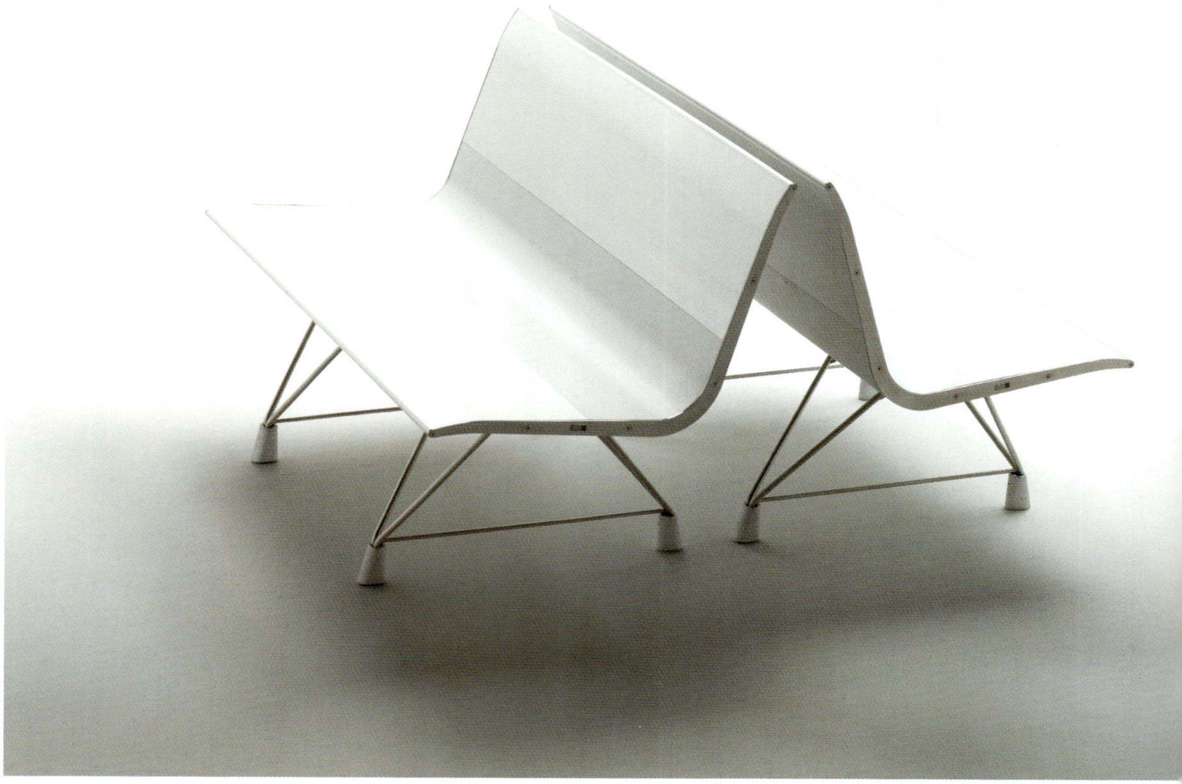

iF Gold Award

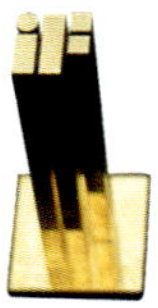

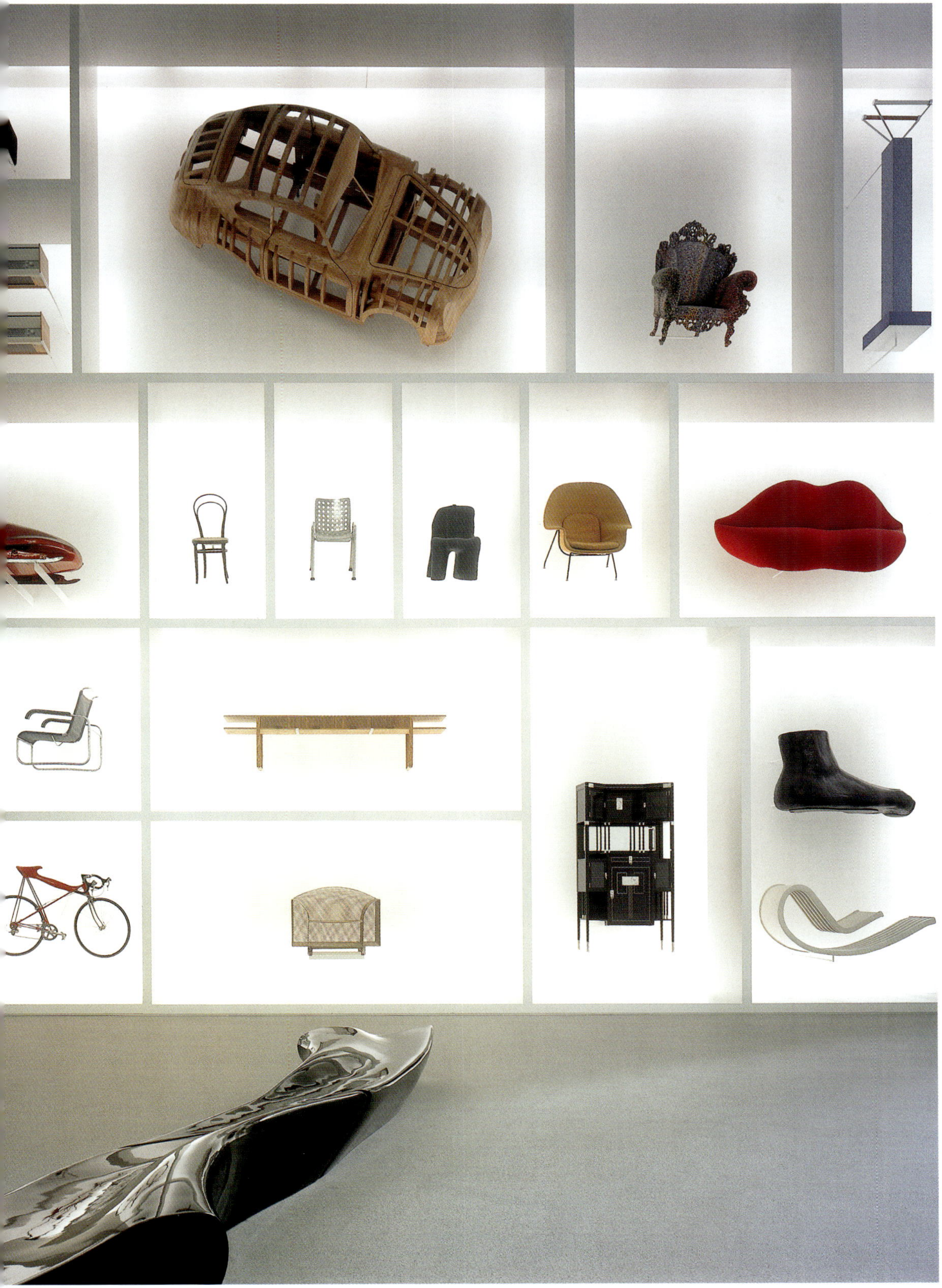

Hersteller Manufacturer
Die Neue Sammlung
Staatliches Museum für angewandte Kunst
München, Germany

Produkt Product
Installation der Designsammlung
in der Pinakothek der Moderne
Installations for the design collection
in the Pinakothek der Moderne
München, Germany

Design
Konzept Concept: Albrecht Bangert
Architektur Architecture: Dieter Thiel
Skulpturen Sculptures: Luigi Colani
Sitzskulptur Seating sculpture: Zaha Hadid
Videoinstallationen Video installations:
Horst Brandenburg

Mit Hilfe von Lichtwänden, bewegten Paternostern und modularen Raumstrukturen wurden die bis zu 22 Meter hohen Räume „ausstellungsfähig" gemacht. Designaufgabe war es, durch abstrakte Strukturen auf die unterschiedlichsten Raumdimensionen zu reagieren und sowohl Großobjekte als auch Automobile, Möbelklassiker sowie kleine Designobjekte in Radio- und Handy-Größe zu vermitteln. Eine große hinterleuchtete Wand am Eingang der Ausstellung vermittelt die Geschichte des Designs des 20. Jahrhunderts wie ein Inhaltsverzeichnis.

With the help of light walls, moveable paternosters and modular space structures, these rooms, which were up to 22 m high, were transformed into a suitable exhibition area. The aim of the design was to use abstract structures to effectively exhibit large objects such as cars and classic furniture, as well as small objects such as radios and mobile telephones in an area of widely differing dimensions. Near the entrance, a wall, lit from behind, serves as a kind of index to the history of design in the 20th century.

iF Silver Award

Hersteller Manufacturer
Die Neue Sammlung
Staatliches Museum für angewandte Kunst
München, Germany

Produkt Product
Mehrfach nutzbare Ausstellungsinstallation
aus industriellen Aluminiumplatten
Reusable exhibition installations
made of industrial aluminum plates.

Design
bangert projects
Schopfheim/München, Germany

Anlässlich der Ausstellung „Designmuseen der Welt eingeladen durch Die Neue Sammlung München" wurde ein Ausstellungssystem entwickelt, das den von internationalen Designmuseen geschickten Exponaten eine eigene Aura verleiht. Die Spiegelung der Aluminiumplatten erweitert die Aura der Objekte, die abgewinkelten Seitenteile schaffen ein eigenes Umfeld, das dennoch Sichtverbindungen und Dialoge zu allen anderen Exponaten ermöglicht. Verwendet wurden Aluplatten in Industriemaßen, die wie bei einem Grafikständer auf ebenfalls demontierbare Stahlrahmen gestellt oder gelegt werden. Transportgestelle ermöglichen den zügigen An- und Aufbau sowie die wirtschaftliche Einlagerung des Systems.

A new system for displaying works of art has been developed for the "Designmuseen der Welt eingeladen durch Die Neue Sammlung München" (Design Museums of the World invited by Die Neue Sammlung München) exhibit. It lends the exhibited pieces their own special aura which is enhanced by the reflection of aluminum plates. Simultaneously, the angled sides create a special ambience which establishes visual connections and dialogue between the various exhibits. Works have been sent to this exhibition from design museums all over the world. The system uses industrial-size aluminum plates. They are placed on steel frames which can be dismantled like stands for graphics. Special transport racks enable the system to be quick assembled and dismantled. It can also be stored conveniently.

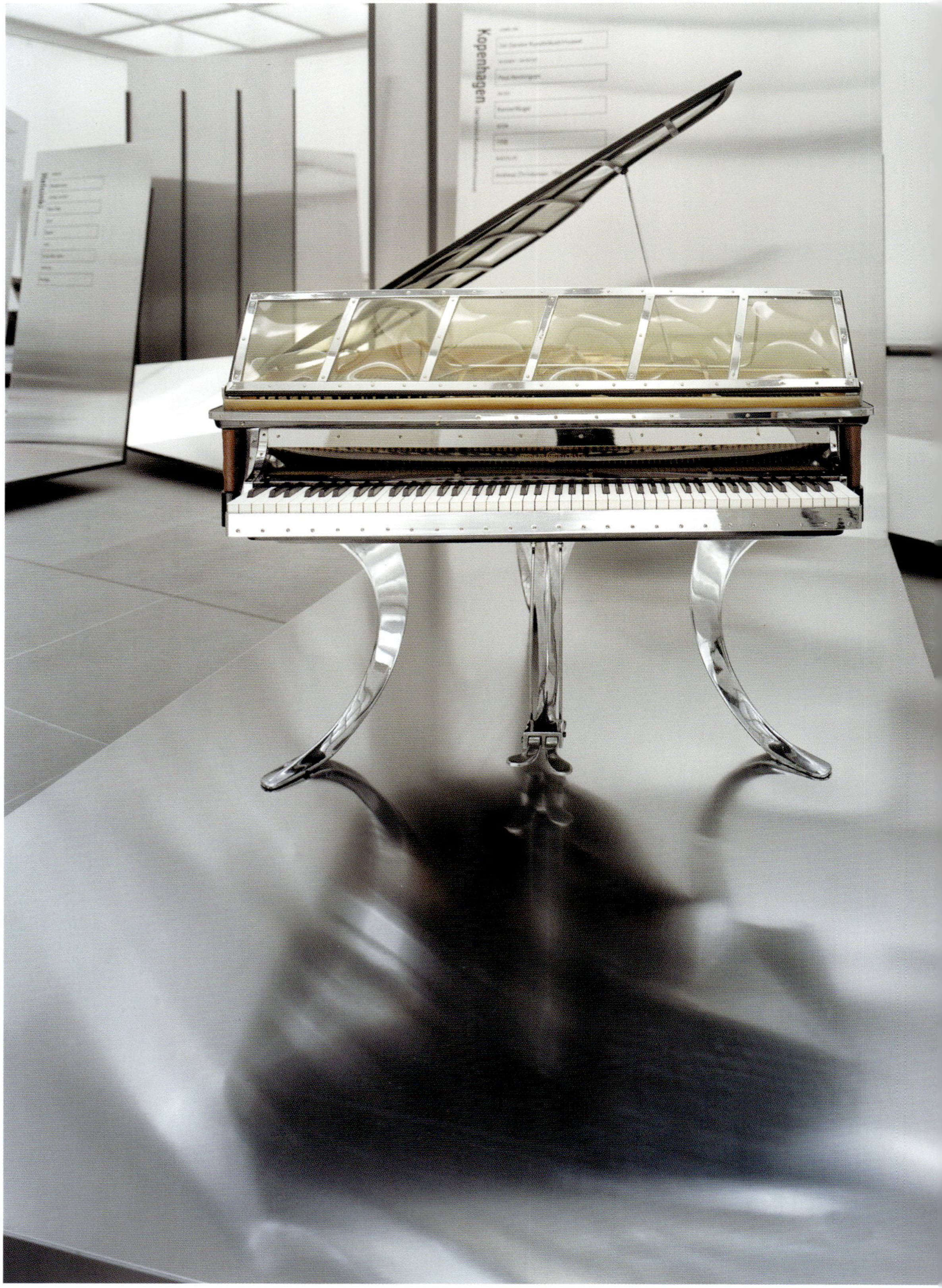

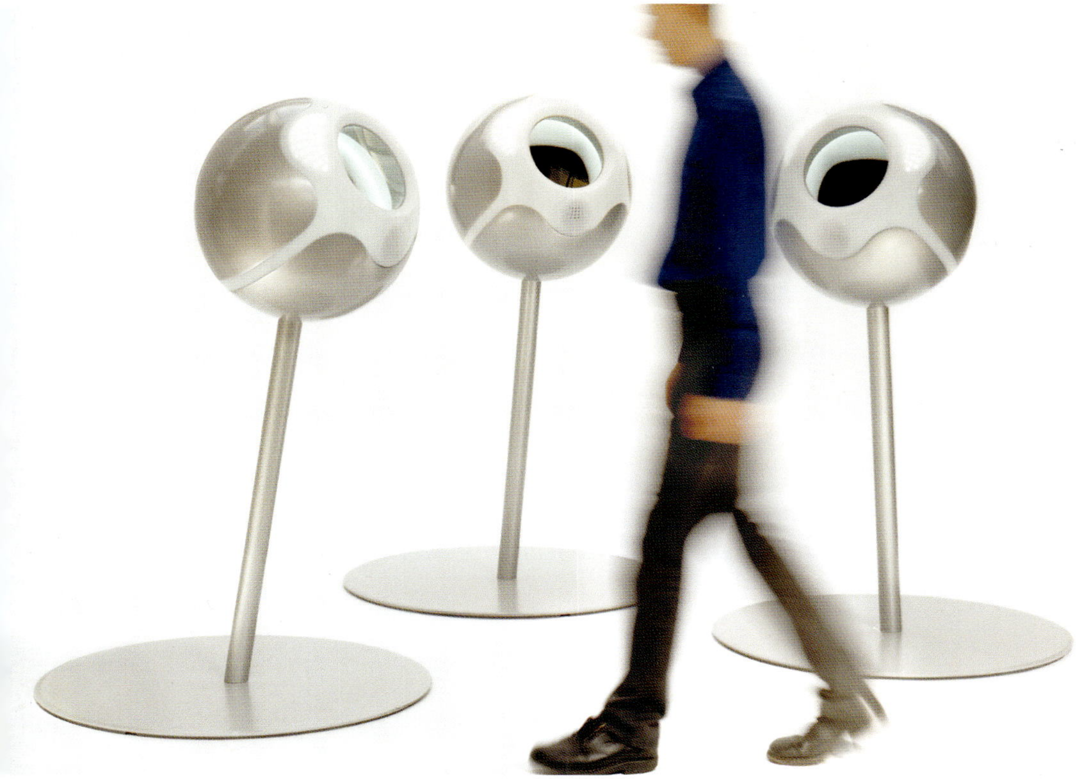

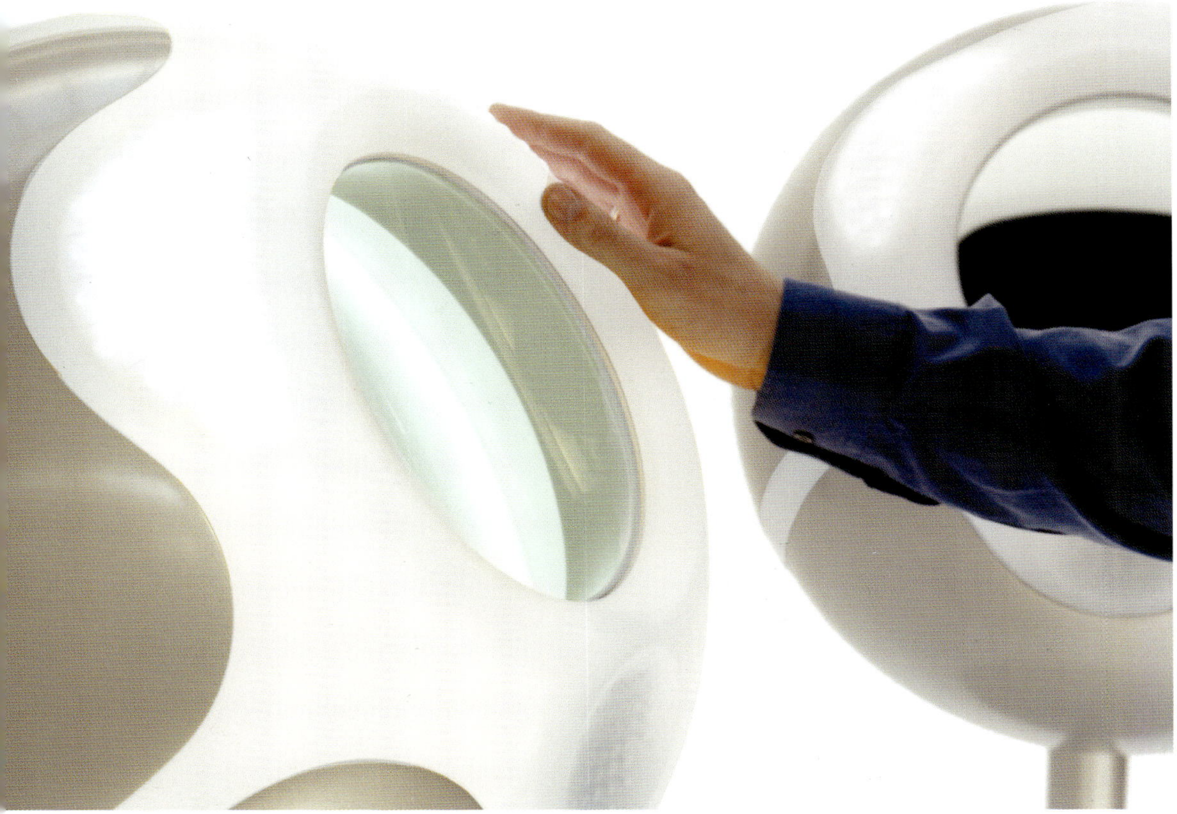

Hersteller Manufacturer
DaimlerChrysler AG
Stuttgart, Germany

Produkt Product
Pre-Safe
Interaktive Exponate
Interactive exhibits

Design
Tobias Kollmann
Dirk Landauer
Wolfram Schäffer
Design hoch drei GmbH & Co. KG
Matthias Bohner
Jan Bartols
Bohner-Design
Michael Salzbrenner
Yuergen Lingg
Gelbfilm
Stuttgart, Germany

Für das von Mercedes-Benz entwickelte Pre-Safe Sicherheitskonzept wurde zur Vermittlung des Funktionsmechanismus ein interaktives Exponat für den Einsatz auf Messen und Ausstellungen entwickelt. Jedes Exponat enthält einen von drei Filmen (Mimose, Muschel, S-Klasse), die dem Betrachter die Funktionsweise von Pre-Safe erklären. Die Filme veranschaulichen durch Naturanalogien, wie ein Fahrzeug mit dieser Sicherheitstechnik in kritischen Situationen reagiert. Über integrierte Sensoren und hinterleuchtete Griffflächen kann der Filmmodus vom Betrachter gesteuert werden.

An interactive exhibition unit created to present the Pre-Safe safety concept devised by Mercedes-Benz at trade fairs and in shows. Each exhibit consists of three films (mimosa, shell, S class), explaining the way Pre-Safe works for those watching. The films visualize by referring to analogies in nature just how a vehicle responds in critical situations thanks to this new safety technology. By means of integrated sensors and backlit touchzones, the viewer can control the film presentation.

Hersteller Manufacturer
Loewe AG
Kronach, Germany

Produkt Product
Loewe Messestand zur IFA 2003
Trade-fair stand

Design
Architektur/Architecture:
Braun Wagner, Aachen, Germany
Kommunikation/Communications:
KMS, München, Germany
Messebau/Trade-fair construction:
MAC, Langenlonsheim, Germany

Für den Auftritt des Unternehmen Loewe wurde auf der IFA 2003 ein 3000 qm Messestand entworfen. Der thematische Fokus lag auf den Bereichen Flat TV und der High-End-Produktlinie „Systems". Die Produktinszenierung nahm einen sehr wichtigen Part innerhalb des gesamten Messeauftritts ein. Offenheit und Transparenz sind die besonderen Merkmale der architektonischen und kommunikativen Realisation. Die Kreisform spielt in der Produktsprache von Loewe eine signifikante Rolle und stellt daher auch die Grundform für die Architektur. Das Design kontrastiert zwischen einer organisch-weichen Linienführung und klaren, kubischen Architekturkomponenten. Eine 65 m lange Airwall bildet eine architektonische Erschliessungsachse, die zur Gliederung des Raumes dient und Eingangsituationen in die verschiedenen Erlebnislounges markiert.

For the Loewe trade-fair presence at the IFA 2003 a 3,000 sq.m. stand has been created. Thematically, the stand focussed on the fields of flat TVs and the company's high-end "Systems" product range. The staging of products played an important part within the overall trade-fair presence. Openness and transparency are the key features of architecture and communications. The circle plays a significant role in Loewe's product language and was therefore taken as basic geometrical shape underlying the trade-fair architecture. The design sets soft organic lines off against clear cubic architectural components. An illuminated airwall of 65 m length forms an architectural axis, serves to subdivide the room and marks prominent entrances to the different Loewe experience lounges.

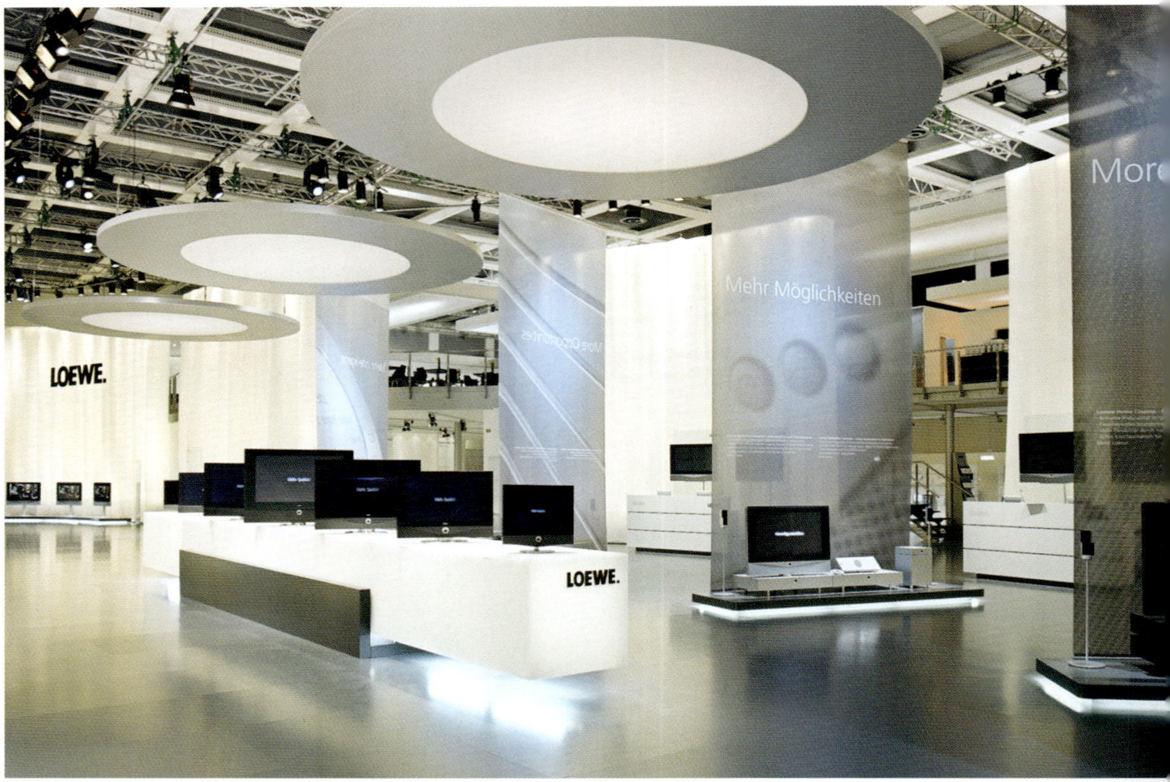

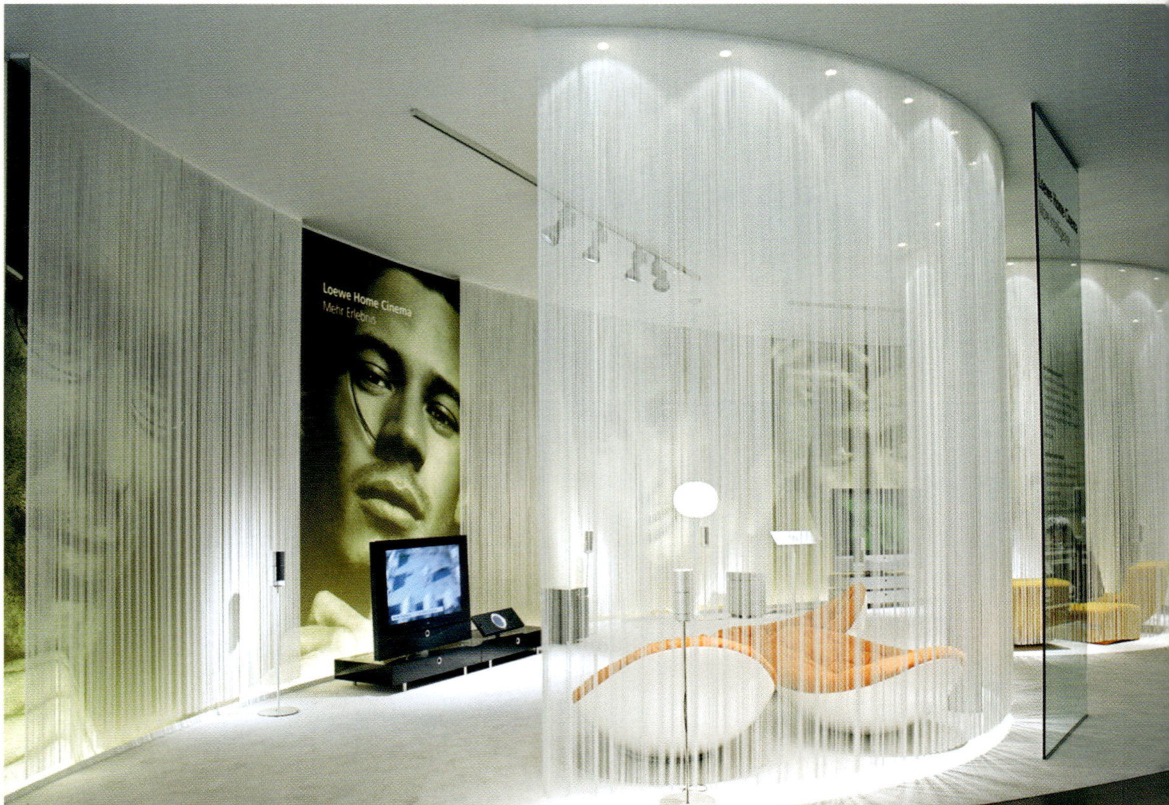

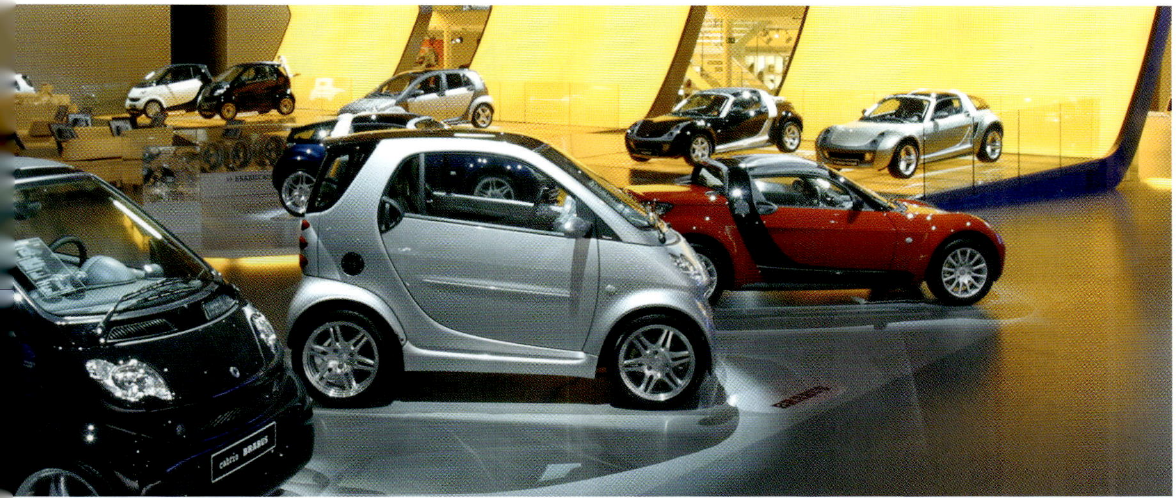

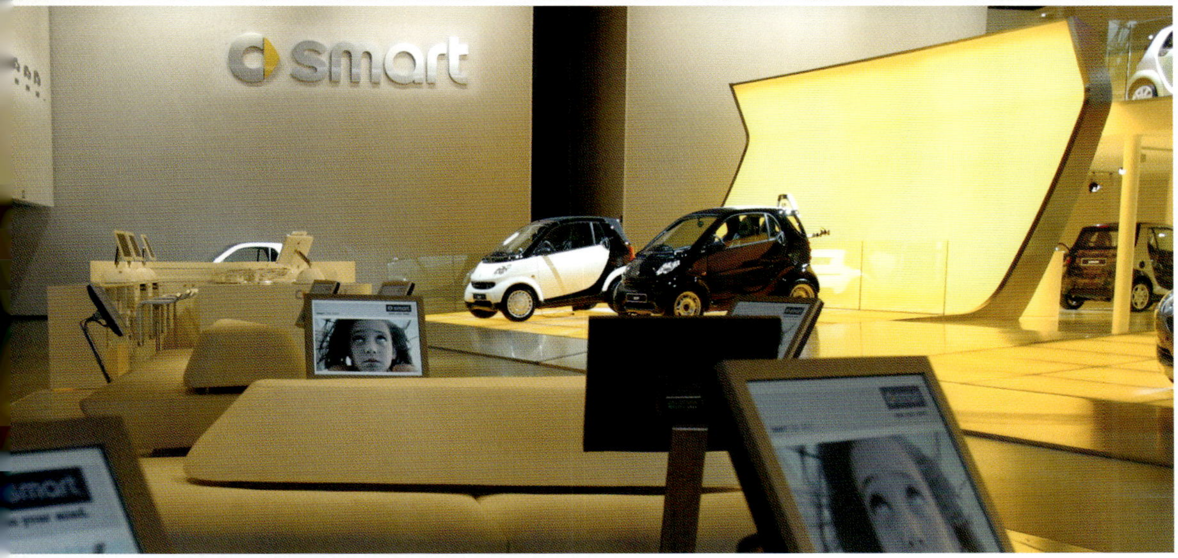

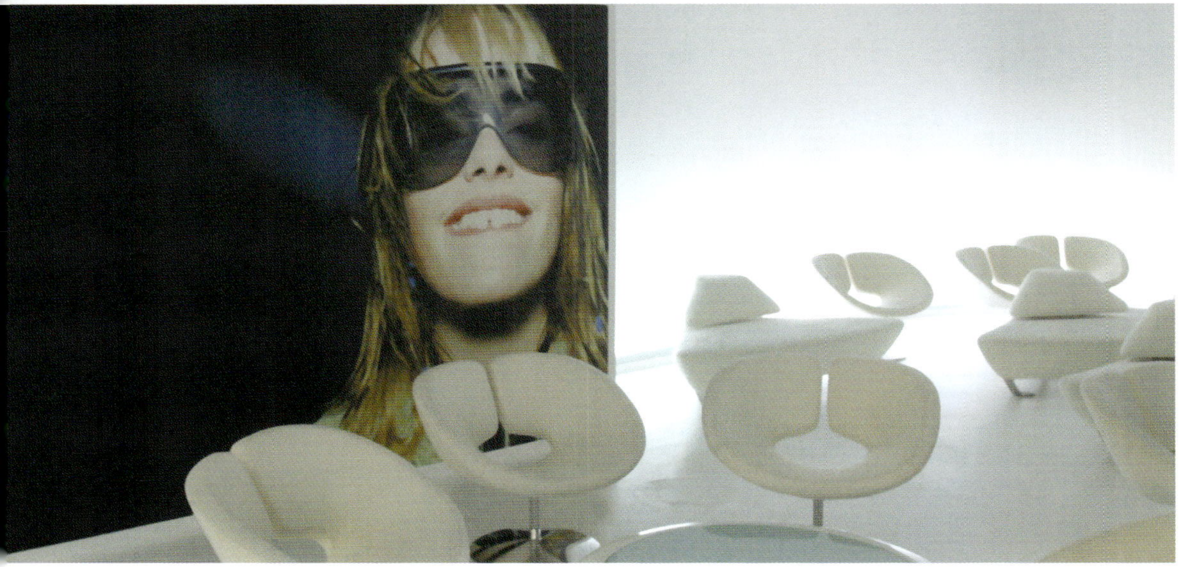

Hersteller Manufacturer
smart gmbh
Böblingen, Germany

Produkt Product
smart Messestand IAA 2003
smart booth

Design
Architektur/Architecture:
Braun Wagner, Aachen, Germany
Kommunikation/Communications:
Braun Wagner, Aachen, Germany
Messebau/Trade-fair construction:
Klartext GmbH, Krefeld, Germany

Für den smart Auftritt auf der IAA 2003 wurde ein 2.700 m^2 Messestand entworfen, der die Entwicklung der Marke smart von der „mono" zur neuen „multi-product-brand" mit ihren Produktwelten darstellt. Es wurde ein offenes, klares und transparentes Architekturkonzept mit einer prägnanten dreidimensionalen Markenkommunikation geschaffen. Die Markenwerte Funktionalität, Innovation und Lebensfreude werden mit einer smart-typischen Tonalität architektonisch dargestellt, kommuniziert und atmosphärisch vermittelt. Die Tridion-Struktur ist allen smart Modellen als statisches Element gemeinsam und wurde in Form der „smart cells" als Tragwerk und Architekturklammer für die 3 Produktwelten übersetzt. Sie bilden somit die Basis für das Design-Konzept und sind architektonisches „brain" und Rückgrat der Marke.

For the smart presentation at the Frankfurt Motor Show (IAA) 2003 a 2,700 m^2 booth has been created that shows the brand's evolution from "mono" to "mulit-product-brand" with innovative product worlds. An open, clear and transparent architectural concept with a concise three-dimensional brand communication has been created. The core brand values functionality, innovation and joy of life are transfered into architecture, communicated and transfered atmopherically in a smart-typical way. The tridion structre is a static element common to all smart models. It has been translated in form of the "smart cells" that serve as truss and architectural clamp for the three product worlds. They are the basis of the design concept and architectural "brain" and backbone of the brand.

Hersteller Manufacturer
Ströer Out-of-Home Media AG
Köln, Germany

Produkt Product
wiege Mega-Light
Werbevitrine
Backlit advertising display case

Design
wiege Entwicklungsgesellschaft
Bad Münster, Germany

Das wiege Mega-Light ist eine großformatige und hinterleuchtete (9 qm) Werbevitrine mit einem elektrisch gesteuerten Plakatwechsler. Im Vergleich zum Vorgängermodell wurde der Vitrinenrahmen reduziert und die Standsäule schlanker gestaltet, um der Wirkung des entsprechenden Werbeplakats mehr Raum zu geben. Orthogonale und flächige Gestaltung bestimmen das Gesamtbild. Auf sichtbare Verschraubungen und Verbindungen wurde verzichtet. Als Materialien kommen Stahl, Aluminium und Glas zum Einsatz. Durch eine Online-Verbindung kann der Plakatwechsler sekundengenau gesteuert werden. Die Steuerungseinheit befindet sich in der Vitrine, was den schmaleren Fuß ermöglicht.

Wiege Mega-Light is a large-sized and backlit 9 sq.m. ad display case with an electronically controlled billboard crawler. Compared with its predecessor model, the frame of the display has been reduced and the central pillar made leaner in order to more strongly emphasize the contents of the displays. Right-angles and the slimline design give its overall picture. There are no screws or connectors visible, and the display uses only steel, aluminum and glass. By means of an online link, the phasing of the billboard crawler can be controlled down to the last second. The control unit is located within the display – in order to render the base section more slender.

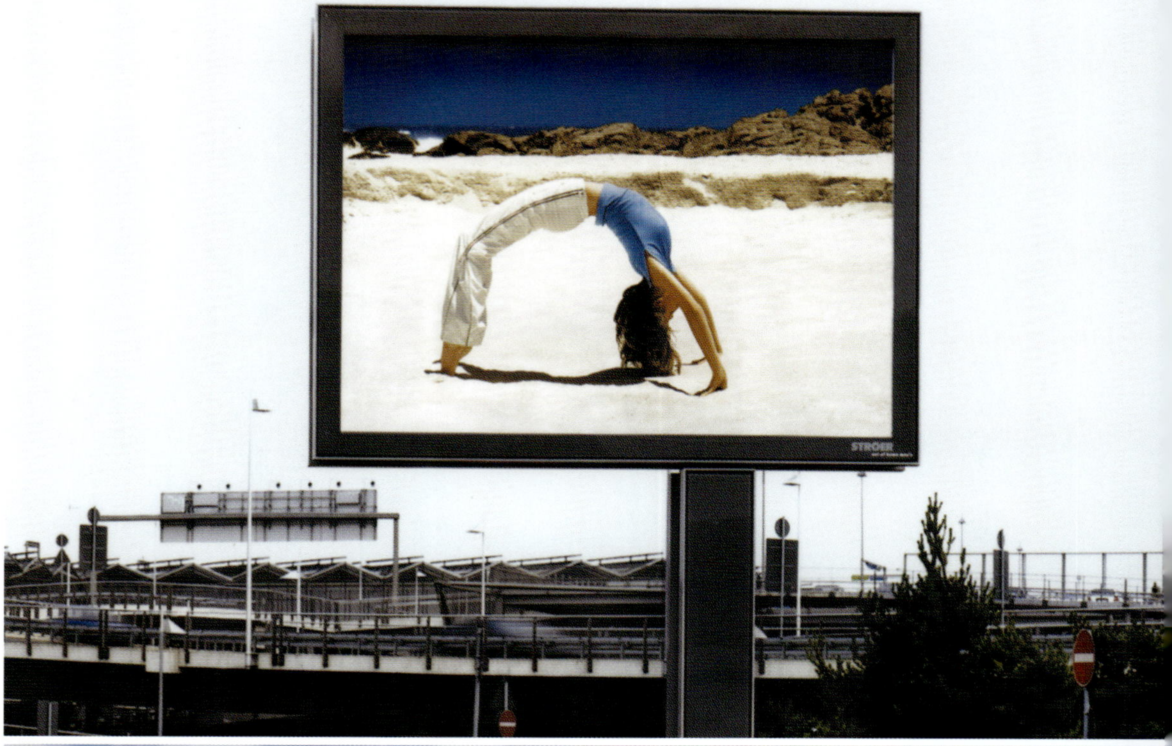

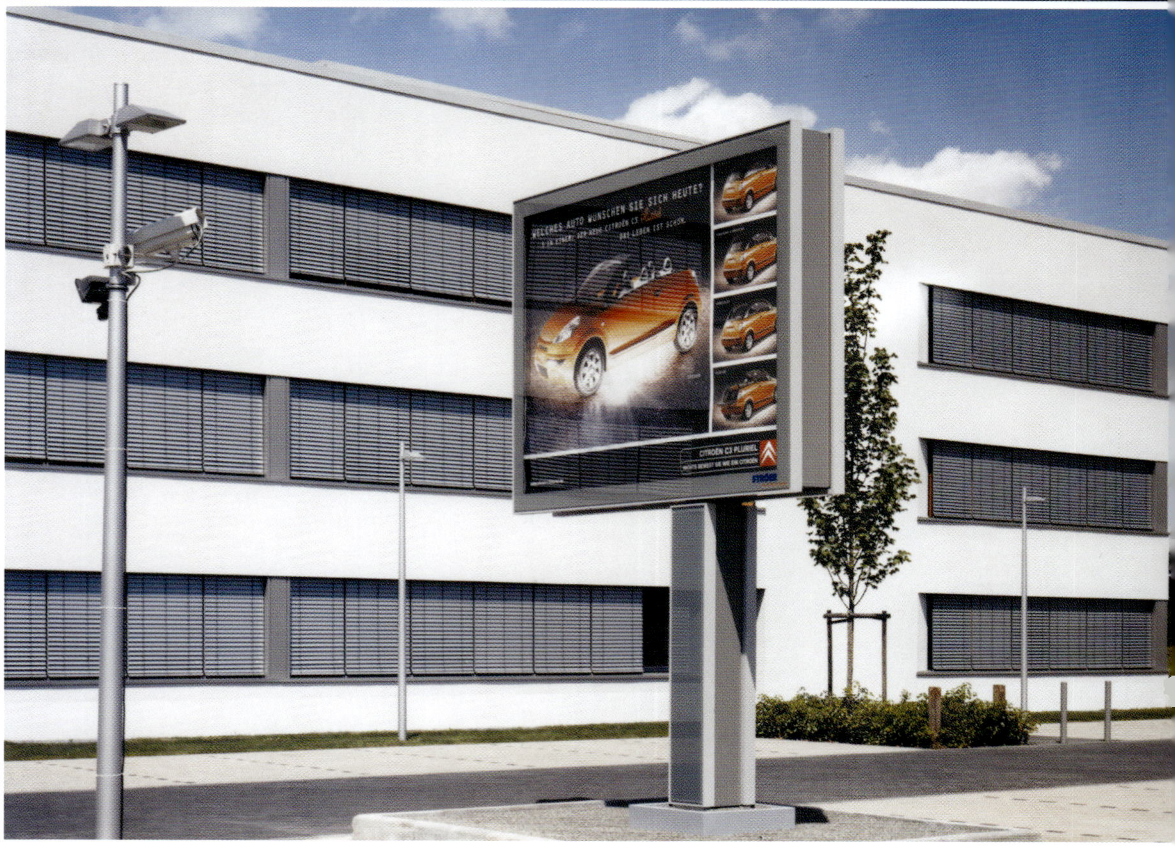

Hersteller Manufacturer
Laarhoven design International B.V.
Leiderdorp, Netherlands

Produkt Product
Pirouet 50
Präsentationssäule
Presentation column

Design
Laarhoven design International B.V.
Leiderdorp, Netherlands

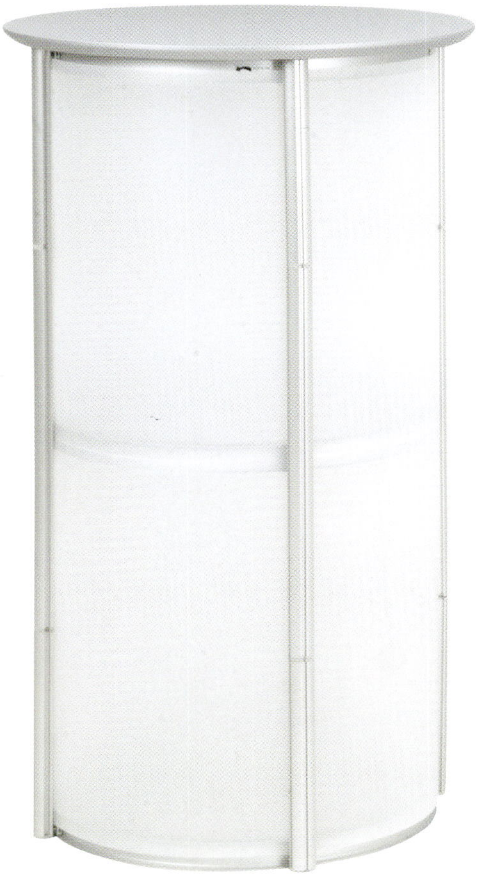

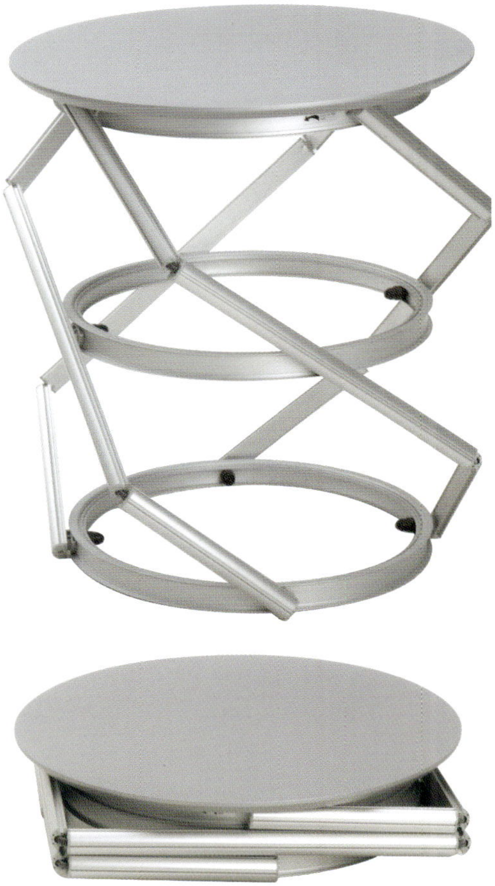

Pirouet 50 ist eine zusammendrehbare Präsentationssäule für Produktpräsentationen, Messeauftritte oder Werbezwecke. An jedem beliebigen Ort kann durch einen einfachen Handgriff die Säule innerhalb weniger Sekunden aufgestellt werden. In ausgeklapptem Zustand hat sie eine runde Form, zusammengelegt bildet sie ein Quadrat und lässt sich in der dazugehörigen Verpackung, die optional auch mit Rädern ausgestattet sein kann, leicht transportieren.

Pirouet 50 is a presentation column for product presentations, trade-fair stands or advertising purposes that simply folds together by a swivel action. It simply twists open in a few seconds when being erected where needed. Once opened it is round, while folded back together it forms a square and tucks away into the packaging – which optionally comes on castors for ideal portability.

iF design award 2004

CONCEPTS

students only

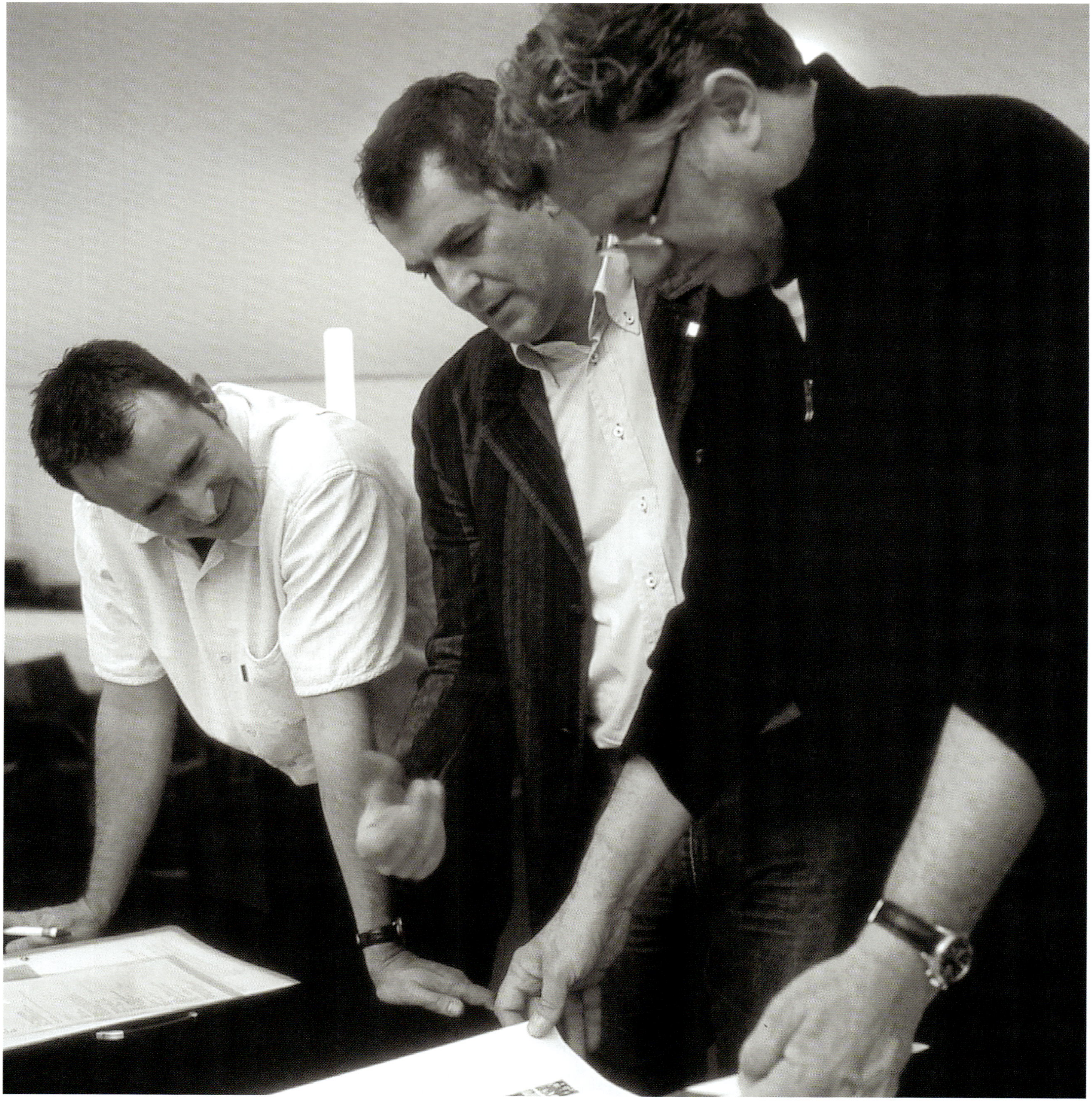

Concepts

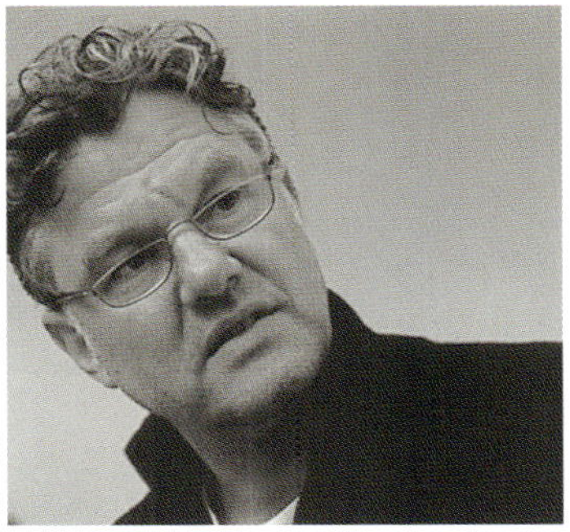

PROF. ACHIM HEINE studierte Mathematik und Design. Er ist Mitgründer des Designprojekts Ginbande (1985 bis 1995) und betreibt seit 1989 mit Michael Lenz und Peter Zizka die Designagentur Heine/Lenz/Zizka in Frankfurt und Berlin. Er lehrt an der Universität der Künste Berlin experimentellen Entwurf. Er ist Mitglied im Präsidium des Rats für Formgebung und arbeitet als Produktdesigner in Berlin.

Prof. Achim Heine studied maths and design. He was co-founder of the Ginbande (1985-95) design project and since 1989 has together with Michael Lenz and Peter Zizka run the Heine/Lenz/Zizka design agency in Frankfurt and Berlin. He teaches experimental design at the University of the Arts in Berlin. He is a member of the Presiding Council of the German Design Council and works as a product designer in Berlin.

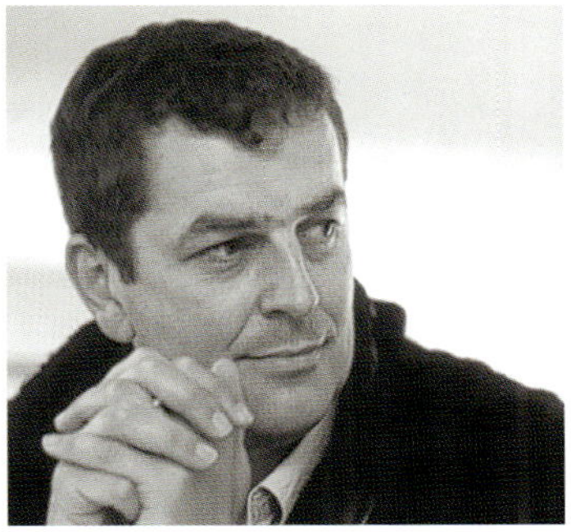

PROF. GUNNAR SPELLMEYER ist seit 1999 Professor für den Fachbereich Design und Medien an der Fachhochschule Hannover, wo er auch selbst sein Industriedesignstudium absolvierte. 1987 gründete er das Designbüro „Formfürsorge".

Prof. Gunnar Spellmeyer has been Professor in the Design and Media Dept. at the Hanover Polytechnic since 1999, where he himself originally graduated in industrial design. In 1987 he founded the "Formfürsorge" design studio.

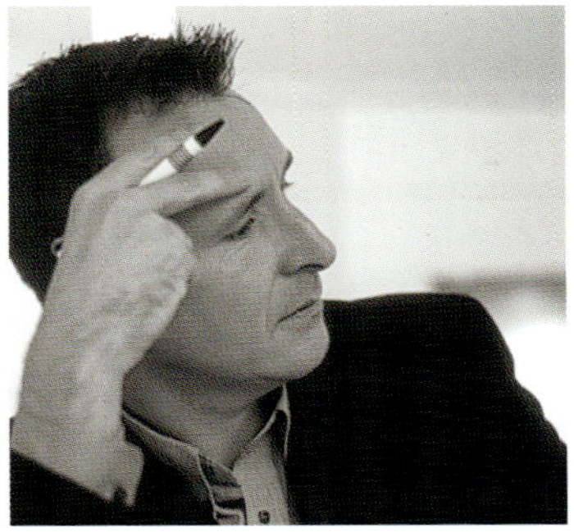

GARETH JONES absolvierte nach seiner Ausbildung zum Maschinenbauingenieur einen Aufbaustudiengang Industriedesign am Royal College of Art in London. Im Rahmen eines Stipendiums von Rolls Royce Motor Cars studierte er an der Loughborough University und war anschließend fünf Jahre für das Unternehmen tätig. Seit 1991 arbeitet er als Entwurfsingenieur für James Dyson und leitet den Bereich Produktentwicklung.

Gareth Jones first studied mechanical engineering and went on to read industrial design at the Royal College of Art in London. Supported by a scholarship from Rolls Royce Motor Cars, he studied at Loughborough University and then worked for Rolls Royce for five years. Since 1991 he has been the design engineer at James Dyson where he is currently in charge of product development there.

Ein Blick auf die Produkte von morgen
Achim Heine

Wenn man in die Jury eines so renommierten Preises wie des iF design awards eingeladen wird, freut man sich nicht nur darauf, viele Kolleginnen und Kollegen, die man sonst nur selten trifft, wieder zu sehen, sondern besonders auch auf eine sehr hochkarätige Auswahl von Produkten designambitionierter internationaler Unternehmen. Selten findet man auf so engem Raum solch unterschiedliche Spitzenprodukte und mit ihnen auch ein Bild davon, in welche Richtung sich das Industriedesign im vergangenen Jahr bewegt hat.

Noch interessanter als diese Produkte, die ja immer trotz aller Ambitionen zwangsläufig auch zahlreichen Restriktionen in Bezug auf ökonomische und ökologische Fragestellungen, Funktionalität etc. unterworfen sind, sind allerdings die Konzepte, Ideen und Produktskizzen, die im Bereich „concepts" alljährlich von Designstudentinnen und -studenten aus aller Welt eingereicht werden. Ihr Vorteil ist es, dass sie (noch) nicht in eben jene Produktentwicklungszusammenhänge eingebunden sind, in denen Design häufig nur als rein äußerliche Differenzierung der in ihrer Funktionalität fast identischen Produkte verstanden wird. Diese jungen Gestalter können frei an die Dinge herangehen, durch Experimente zu neuen Themen gelangen und daraus unkonventionelle Lösungen entwickeln. Das macht den Wettbewerb „concepts" so spannend, nicht nur für die Leser dieses Buches, sondern auch und besonders für die Jury.

Man findet unter den eingereichten Arbeiten immer wieder eine große Bandbreite unterschiedlicher Interpretationen: einerseits Konzepte, die den Designbegriff sehr stark erweitern und mit poetischen und politischen Ideen in Verbindung bringen, wie z.B. die Arbeit von Daniel Janssen, die mit aus der Graffitiszene entlehnten radikalen Designformen ein politisches Statement gegen den Krieg propagiert. Andererseits findet man aber auch klar auf den Punkt fokussierte und sehr verblüffende praktische Ideen zu neuen möglichen Produkten, wie z.B. die Konzeption und Gestaltung einer Komposttonne von Björn Bornemann, in der man mit wenigen Handgriffen eine Kompostkultur in Bewegung halten kann. Entwürfe für Fahrräder mit Hilfsantrieb, Klapproller, Handys und Consumer Electronic ge-hören in diese praktische und eher realistische Kategorie.

Multifunktionale und mehrdeutige Objekte bilden eine eigene Gruppe wie die Möblierungen für Straßenfeste, wo Transportbehälter zu Möbeln werden oder ein medizinisches Projekt wie der Koffer eines Notarztes, der sich mit wenigen Handgriffen zu einer Ultraschallworkstation umwandeln lässt. So erhält man durch die große internationale Breite der Beiträge im Bereich „concepts" einen konzentrierten Einblick in Gemeinsamkeiten und Unterschiede gestalterischer Ansätze verschiedener Kulturkreise. Von den insgesamt 308 eingereichten Arbeiten stammen 192 Arbeiten aus dem Ausland, davon 16 aus China, 11 aus Brasilien, 30 aus Taiwan und 13 aus Mexiko.

Das professionelle Niveau der Arbeiten war sehr beeindruckend, und die Jury war von vielen Projekten uneingeschränkt begeistert. Wenn all diejenigen, von denen hier im Buch zwangsläufig nur ein kleiner Ausschnitt gezeigt werden kann, ihrer Profession und der Liebe zum Design treu bleiben, braucht man sich keine Sorgen zu machen, dass die Qualität der beim iF design award eingereichten „echten" Industrieprodukte in den nächsten Jahren nachlassen könnte.

A glance at tomorrow's products

When you are invited to sit on the Jury of such a renowned prize as the if design award, it is not only gratifying to meet up with many colleagues you otherwise see only seldom; you also relish the high-caliber selection of products by international firms with strong design ambitions. Seldom do you find such a range of top products in a relatively small space, and moreover through them an image of the direction industrial design has taken in the past year.

More interesting than these products, which for all their ambitious nature are necessarily subject to numerous restrictions as regards economic and ecological considerations, functionality and so on, are the concepts, ideas and product sketches submitted every year for the "concepts" section by design students the world over. They have the advantage of not (yet) being caught up in those very product development contexts in which design is all-to-often treated as a vehicle for lending a distinct appearance to products that are virtually identical as regards features and functions. These young designers can approach matters more freely, arrive at new topics by way of experimentation, and develop unconventional solutions as a result.

This is what makes the "concepts" competition so fascinating, not only for the readers of this book but also and in particular for the Jury.

Amongst the works submitted you repeatedly encounter a large variety of differing interpretations: on the one hand concepts that greatly expand the term design and link it with poetic and political ideas such as the work of Daniel Janssen, in which radical design forms from the graffiti scene are used to make a political statement against war. Then again, there are highly specific and amazingly practical ideas for potential new products such as Björn Bornemann's concept and design for a compost bin, which only requires a little effort to keep the compost working. Designs for bicycles featuring an accessory drive, folding scooters, cell-phones and consumer electronics also belong in this practical, more realistic category.

Multi-purpose, convertible objects form a group of their own. They include transport boxes which can be transformed into furniture for street parties, or an emergency doctor's case that in no time at all can be turned into a portable ultrasound workstation.

Thanks to the great international breadth of entries for the "concepts" section, you also gain an intense insight into the differing and similar ways in which various cultures approach design. A total of 308 works was submitted of which 192 came from countries outside Germany; these include China (16), Brazil (11), Taiwan (30) and Mexico (13).

The professional standard of the entries was extremely impressive, and the Jury was very enthusiastic about many projects. Naturally, we can only show a small selection of the entries, but if all the people involved continue to remain true to their profession and love of design, we need not worry that the quality of the "real" industrial products submitted for the iF design award in years to come will suffer.

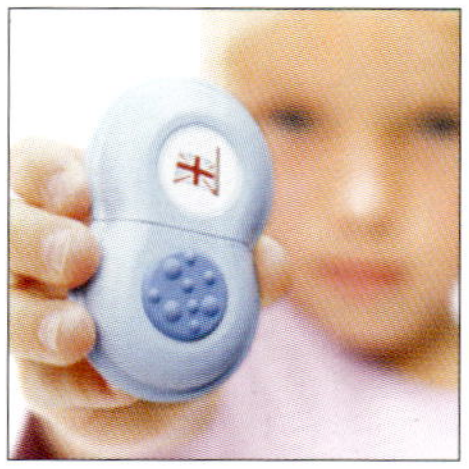
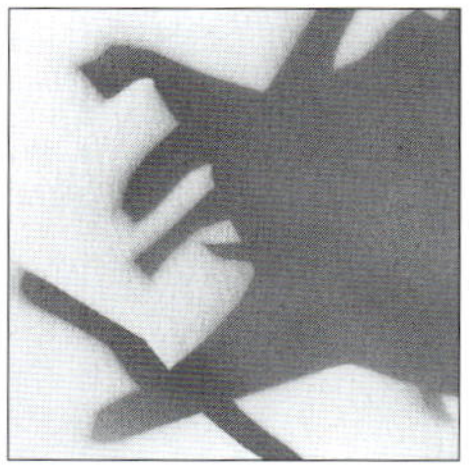
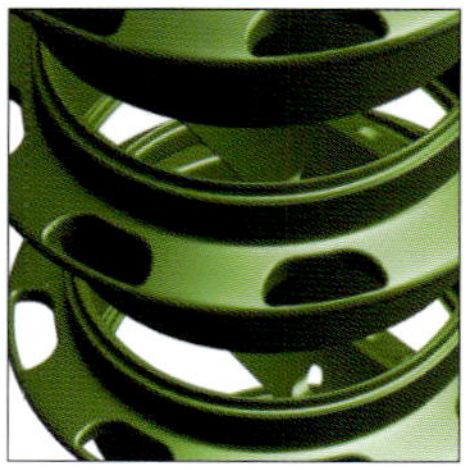
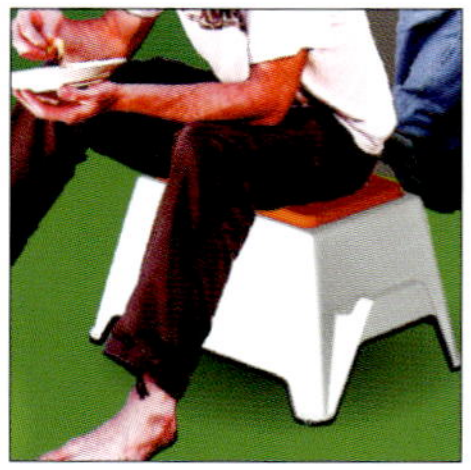
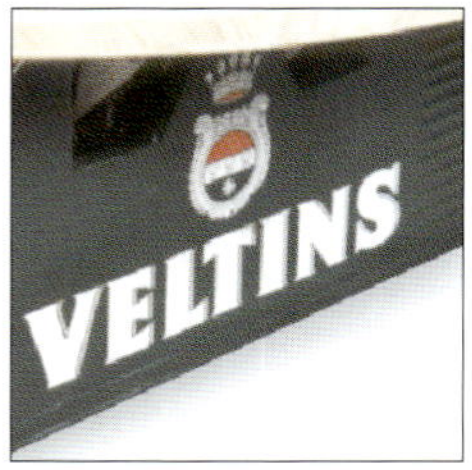
VELTINS
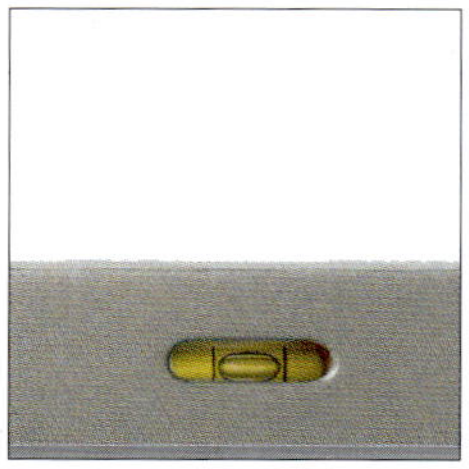
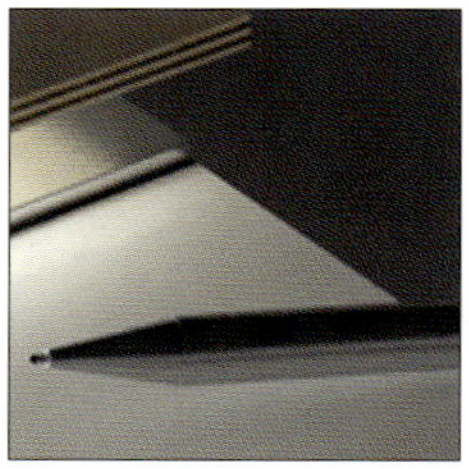
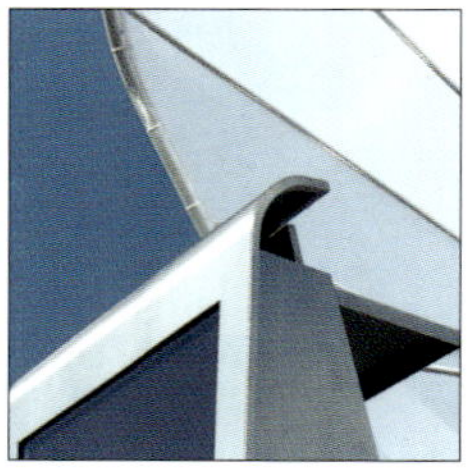
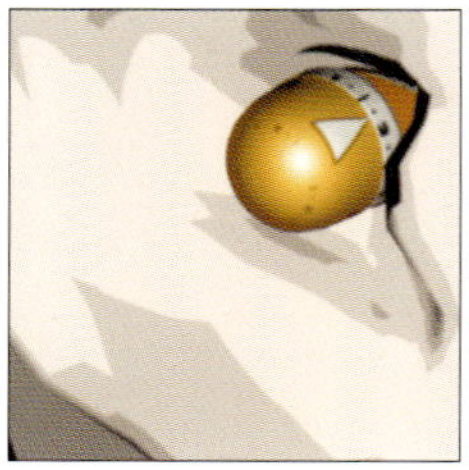
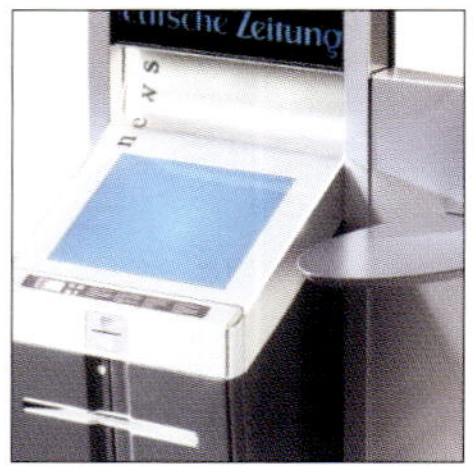
Zeitung
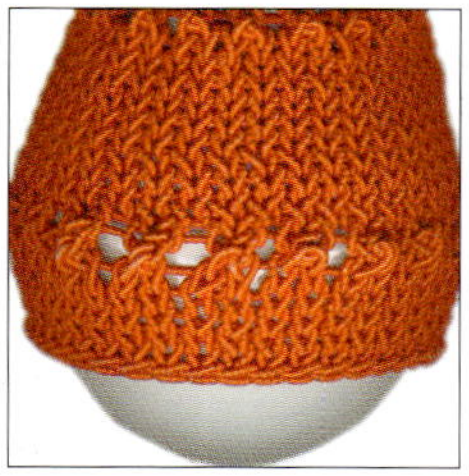
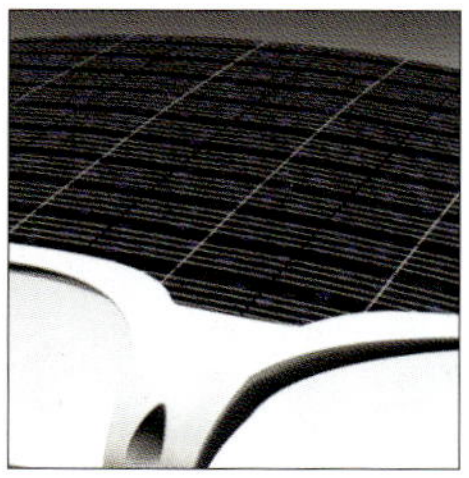
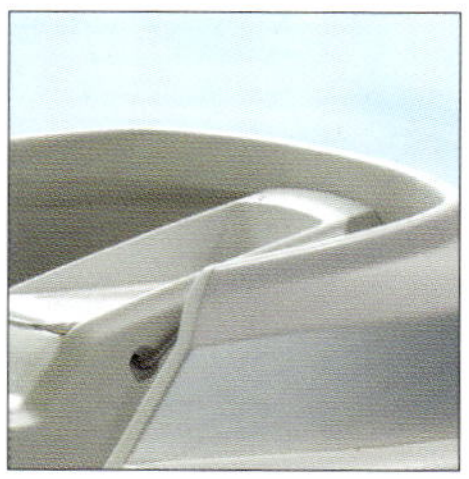
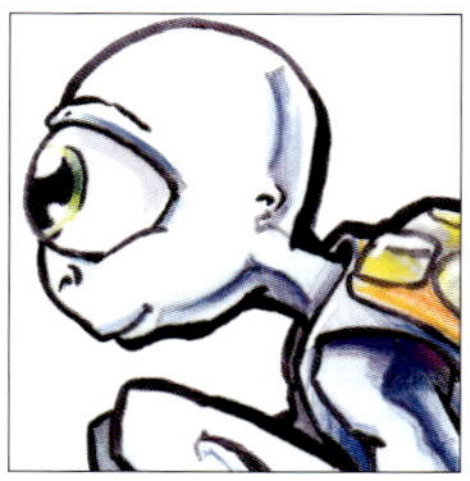
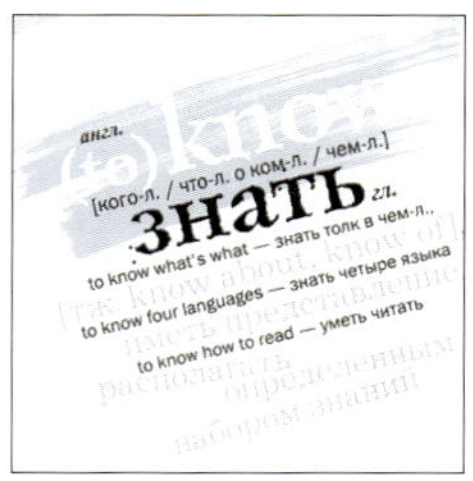
англ.
[кого-л. / что-л. о ком-л. / чем-л.]
знать гл.
to know what's what — знать толк в чем-л.
to know four languages — знать четыре языка
to know how to read — уметь читать

Concepts

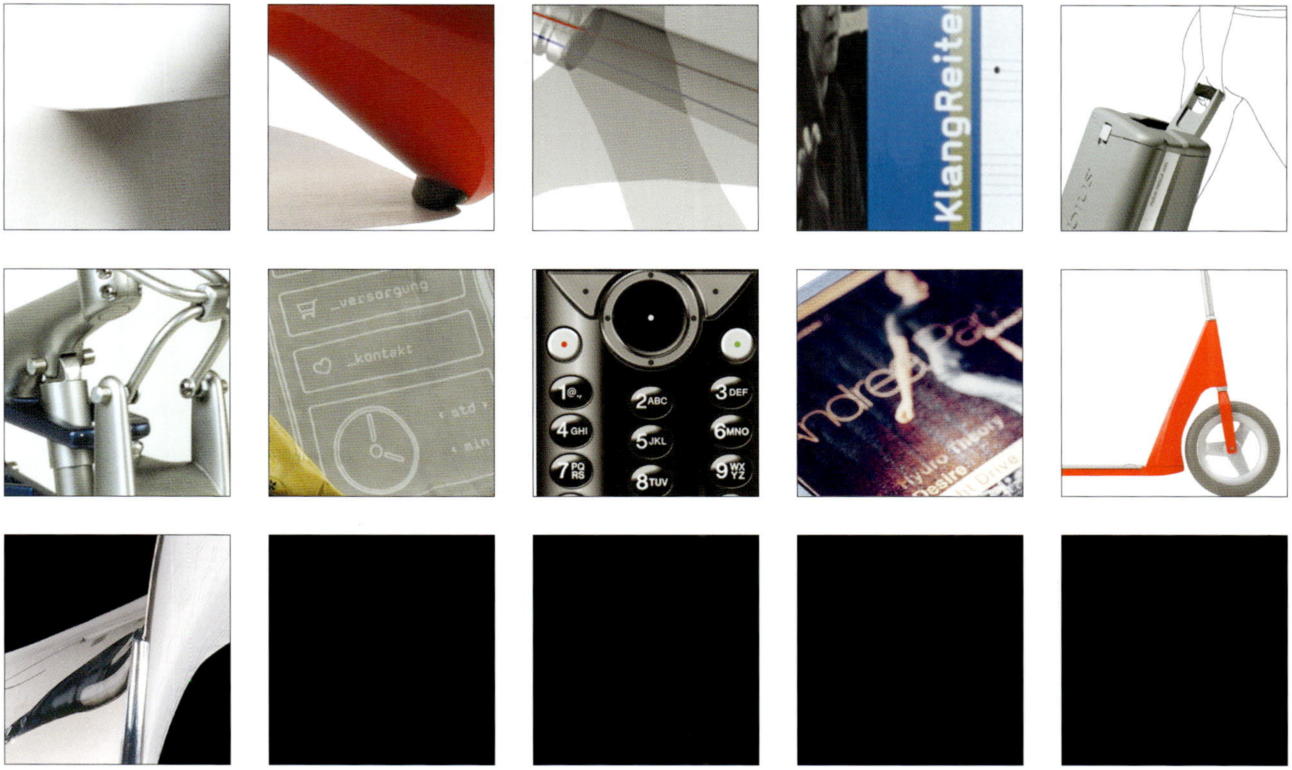

Die Revolutionierung der Kreativität
James Dyson

Die meisten Leute denken, dass ein Designer den Dingen ein schönes Aussehen verleiht. Ich glaube, dass Design wesentlich mehr ist als das: Von der Entwicklung einer speziellen Technologie zur Herstellung des Produktes, von seiner Gestaltung bis schließlich zur Leistungsüberprüfung – ein Produkt mit einem guten Design besitzt eine hohe Funktionalität und bereitet Freude in der Benutzung.

Ich betrachte die technische und gestalterische Entwicklung nicht voneinander getrennt; unsere holistische Herangehensweise an Design bei Dyson bedeutet, dass es keine Grenzen gibt zwischen den einzelnen Disziplinen, jeder aus der Designabteilung sollte auch die einzelnen Schritte des Herstellungsprozesses kennen. Dyson beschäftigt keine reinen Gestalter und gewährleistet damit, dass das Design des Produkts im Hinblick auf seine Funktionalität entwickelt wird. Für mich bedeutet Design nicht nur wie Dinge aussehen, sondern auch wie sie funktionieren.

Es ist eine große Herausforderung, die Hände und den Verstand zur Lösung von Problemen einzusetzen. Diese Erkenntnis sollte schon früh eine Rolle spielen und ist ein wesentlicher Aspekt der Ausbildung. Wir beschäftigen bei Dyson eine große Anzahl von Designtechnikern und Wissenschaftlern, die nach neuen Technologien und Produkten suchen und diese entwickeln. Wir haben jedoch Schwierigkeiten, wirklich gute Leute für diese äußerst wichtigen Bereiche zu finden. Europa blickt auf eine einzigartige Tradition genialen Erfindungsgeistes zurück und trotzdem fällt dieser Kontinent im Hinblick auf seine Kreativität hinter dem Rest der Welt zurück. Ich befürchte, dass wir den Willen und die Fähigkeit sowie auch die Mittel und die Kompetenz verlieren, Dinge zu machen, sie zu kreieren und zu erfinden.

Das Gegenmittel liegt auf der Hand: Wir müssen den Erfindergeist in unseren Fachhochschulen, Akademien, Universitäten und im Unternehmensbereich fördern. Wir müssen diejenigen schätzen lernen, die Dinge tun, anstatt jene, die nur Papiere hin- und herschieben. Wir müssen der Annahme vorbeugen, dass Karrieren in der Industrie und in der Herstellung langweilig sind. Wir müssen begabte junge Leute mit der wichtigen Aufgabe betrauen, Forschung und Kreativität wieder in die Entwicklung von Produkten einfließen zu lassen, für die es einen Markt gibt.

Wenn man die kreativsten und begabtesten Leute für Design und Technologieentwicklung gewinnen will, müssen Designer und Ingenieure die Leidenschaft stiften, Dinge zu entwickeln und zu gestalten und die Neugier erwecken, wie man ihre Funktionalität noch erhöhen könnte.

Es gilt ein kulturelles Umfeld zu schaffen, indem man nicht ängstlich ist, die Risiken in Kauf zu nehmen, die mit dem Versuch einhergehen, etwas wirklich Neues zu kreieren. Die Zukunft der Kreativität und des Reichtums in Europa liegt in den Händen von jungen Menschen. Deshalb sollten wir sicherstellen, dass sie wissen, wie aufregend Designentwicklung ist, dass es in diesem Bereich die Chance zu Veränderungen gibt und Raum für Überraschungen und Entdeckungen.

Das Preisgeld für concepts stammt mit 15.000.– EUR vom britischen Innovations- und Technologieunternehmen Dyson und mit 10.000.– EUR von iF International Forum Design GmbH.

Revolutionising creativity

Most people think of a designer as someone who makes things look good. I think design means much more than this. It ranges from developing the technology, to engineering the product, designing it and then testing performance - a well designed product is one that performs well and is a pleasure to use.

I do not view engineering and design as separate domains; our holistic approach to design at Dyson means that there are no barriers between the disciplines – everyone in the design engineering department should understand every part of the process. Dyson does not employ a single stylist and instead ensures that the look of the product develops in tandem with its functionality. For me, design is not just about how something looks, but how it works.

Using your hands and brains to solve problems is an enormous creative challenge. This understanding should begin at an early age and is a vital part of education. At Dyson we employ a large number of design engineers and scientists who are researching and developing new technologies and new products. And yet we have difficulty recruiting the best people in these key areas. Europe has a unique heritage of inventive genius, and yet this continent is falling behind much of the rest of the world in creativity. I am afraid that we are losing the will and the ability to make things, not to mention the means and expertise to create and invent.

The remedy is clear enough: we must encourage the inventive spirit in our schools, colleges and universities, and in business. We need to value those who make things over those who shuffle papers. We need to challenge the assumption that careers in industry and manufacturing are dull. We need to direct our talented young people to the core mission of putting research and creativity back into making products that the world wants.

If design and engineering is to succeed in attracting the most creative and talented people, designers and engineers will have to inspire a passion for designing and making things, and curiosity about how they might be made to work better.

We have to create a culture that is not afraid to take the risks inherent in trying something really new. The future of Europe's creativity and wealth lies in the hands of young people. We have to make sure they know that design engineering is exciting, that it offers great opportunities for change, and there are big surprises and discoveries to be made.

Dyson, the British innovator and high-tech company, has contributed EUR 15,000 as prize money for the concepts awards. The remaining EUR 10,000 is bestowed by iF International Forum Design GmbH.

CONCEPTS

Projekt Project
TAKlearnstation
Schule der Zukunft – Neue Medien als Lernbedienung
Learning assistance for the use of new media in elementary school

Design
Alexandra Lubk, Kaja Bartel, Christine Pleines
HBK Braunschweig, Germany
TAKlearnstation@gmx.de

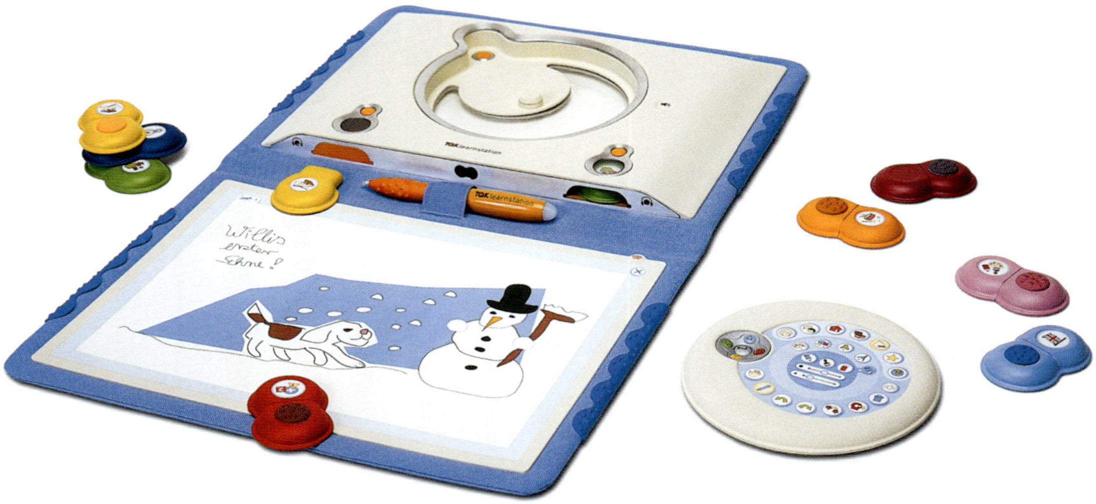

TAK besitzt die Technik eines abgespeckten Laptops und ist für die Grundschule der Zukunft gedacht. Er besteht aus einer Mappe mit zwei Einsätzen: Rechnereinheit und Arbeitsfläche sind voneinander getrennt angeordnet. Ohne Maus und Tastatur werden alle Eingaben handschriftlich mit einem elektronischen Stift gemacht. Durch die Trennung von Rechnereinheit und Arbeitsfläche kann man wie in einem Heft auf der flachen Arbeitsfläche schreiben. Zur TAKlearnstation gehören die STIKs, kleine USB-Speichersticks, auf denen alle medialen Unterrichtsinhalte gespeichert werden. Für jedes Unterrichtsfach gibt es jeweils einen Fächer-STIK. Die STIKs ersetzen Arbeitshefte und -blätter, jedoch nicht das Lesebuch, denn der TAK soll generell neben den alten Unterrichtsmaterialien eingesetzt werden. Die TAKtoolbox ist ein herausnehmbarer Zweitmonitor, auf dem Ordner und virtuelle Werkzeuge ausgelagert sind.

TAK possesses the technology of a slim laptop and is destined for use in the primary school of tomorrow. It consists of a folder with two slot-ins: a computer unit and a work-top as separate items. No mouse or keyboard are required as all entries can be made in handwriting using the electronic pen. By separating the computer unit from the work-top, children can write on the work-top as if they were using an exercise book. The TAK learnstation includes STIKs, small USB memory sticks to which all the multimedia content of lessons can be stored. There is a STIK storage compartment for each lesson. The STIKs replace any number of school books or work sheets, but not reading books, as TAK is meant for use alongside the traditional materials used in class. The TAK toolbox is a removable auxiliary monitor on which files and virtual tools have been stored.

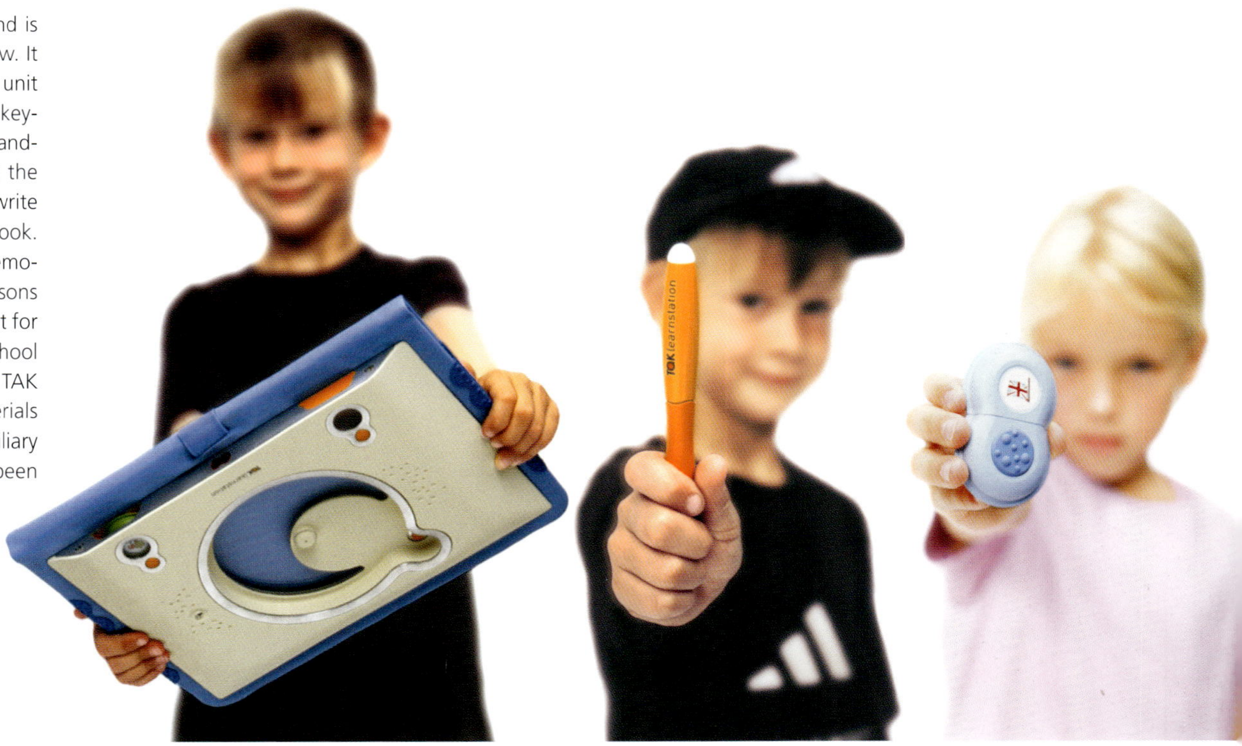

PREISGELD | PRIZE MONEY 3000 EURO

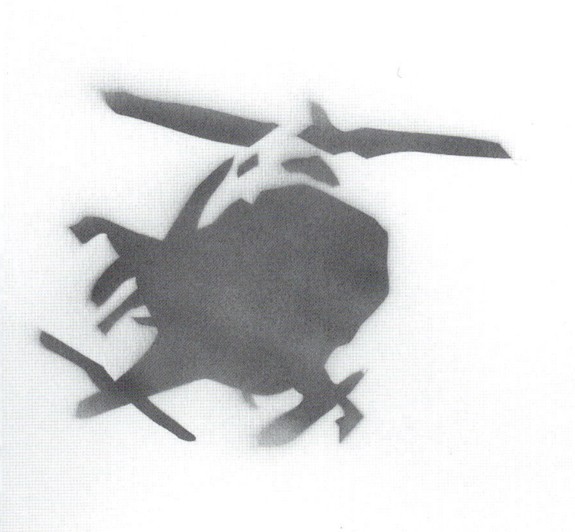

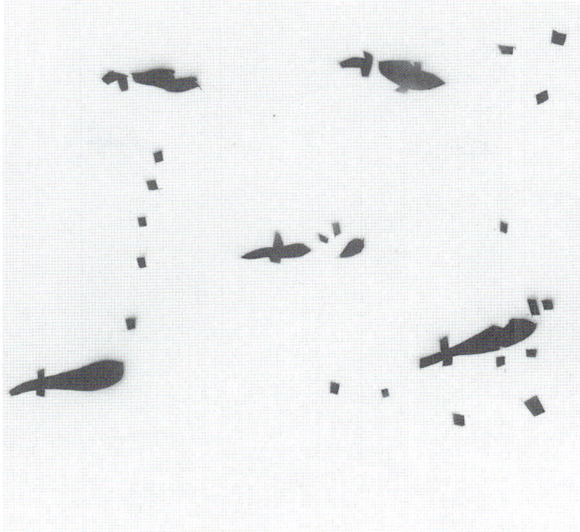

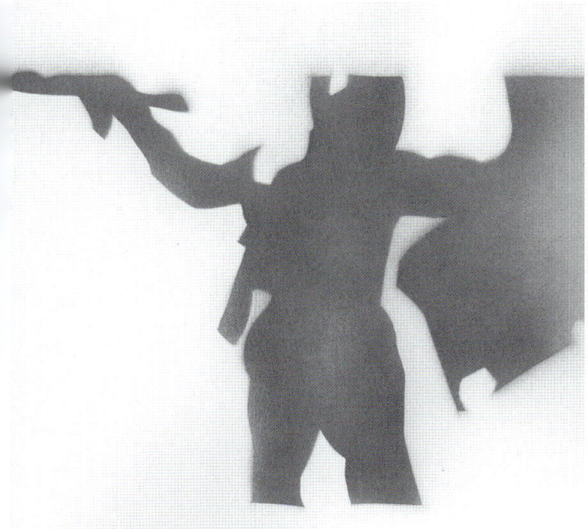

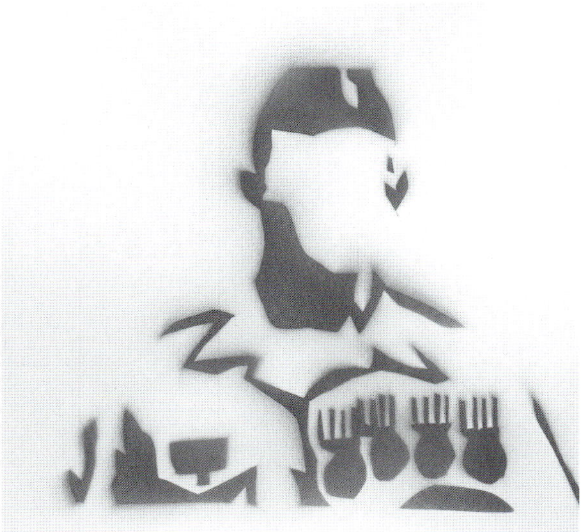

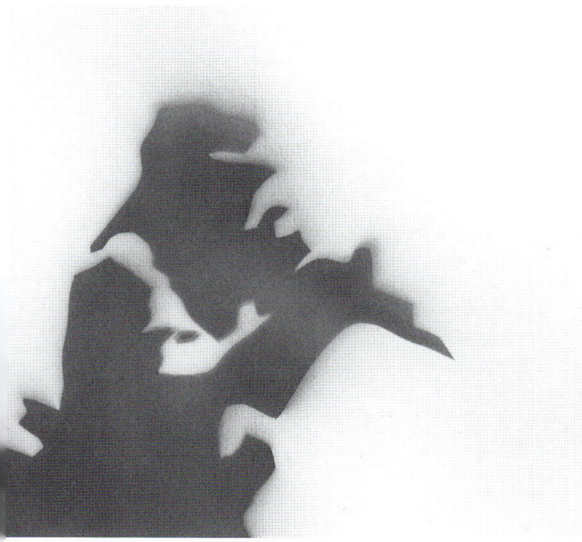

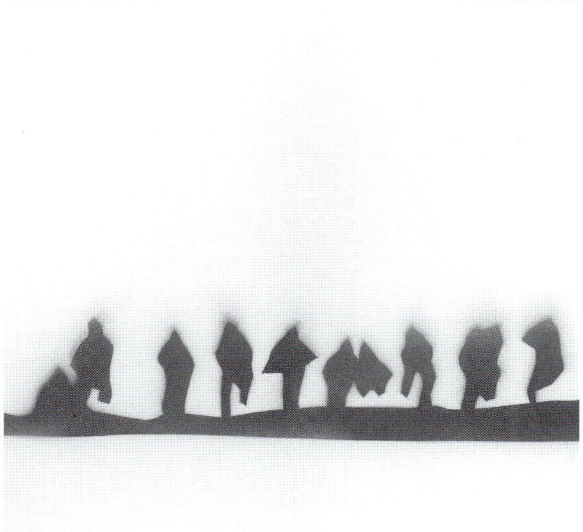

Projekt Project
Save your face!
Social Spot, Animation
Social commercial, Animation

Design
Daniel Janssen
Hochschule für Angewandte Wissenschaften
Hamburg, Germany
info@bfgjanssen.de

Der Film ist ein fiktiver Social Spot für Amnesty International. Er soll dem Betrachter die Sinnlosigkeit des Krieges vor Augen führen, der Menschen zu Instrumenten skrupelloser Machthaber werden lässt. Brutale Kriegsszenen reihen sich aneinander. Ein Soldat wird für seine „Verdienste" mit Orden ausgezeichnet. Mit jedem Orden schwindet ein Stück von ihm. Er verliert sein Gesicht. Die rauhe Kriegs-Szenerie wird durch die Sprühschablonen-Ästhetik unterstützt. Der verzerrte Background-Song „Amazing Grace" betont die bedrückende Atmosphäre.

The film is a fictitious social commercial for Amnesty International. It is intended to show the viewer the senselessness of war, which transforms human beings into the instruments of unscrupulous dictators. One cruel war scene follows on from the next. A soldier is decorated for his "merits" with medals. With every medal a piece of him gets lost. He loses his face. The tough war-time scenario is emphasized by the spray-stencil aesthetics. The distorted background song played – "Amazing Grace" – emphasizes how depressing the atmosphere is.

PREISGELD | PRIZE MONEY 3000 EURO

Projekt Project
Stirwell
Kompposttonne
Compost bin

Design
Björn Bornemann
Fachhochschule Hannover, Design und Medien
(Lahti Polytechnic of Design)
Hannover, Germany
bjoern.bornemann@epost.de

Bei dieser Komposttonne, die aus aufeinander gesteckten Elementen aufgebaut ist, sind bereits im Inneren Schaufeln vorhanden, die beim Drehen von einem der Elemente den alten mit dem neuen Kompost vermischen. Dadurch werden sowohl Geruch als auch Schmutz vermieden. Der Deckel ist so beschaffen, dass man ihn einfach mit einer Hand abheben kann, außerdem ist er durch seine Form windgeschützt und lässt Regenwasser schnell abfließen. Im Bodenteil befindet sich eine Öffnung, um fertigen Kompost zu entnehmen. Weiterhin besitzt das Bodenteil Füße, die in der Erde stecken, um ein Verdrehen zu verhindern. Zusätzliche Löcher schaffen Kontakt mit dem Erdreich, damit Insekten und Mikroorganismen beim Kompostieren helfen können.

This compost bin, which is made up of sections that slot into one another and snap into place, boasts shovels on the inside: when you turn one of the sections, the old and the new compost are mixed together by the shovels. In this way, both evil odors and dirt are avoided. The bin lid is designed to make certain you can easily open it with one hand; moreover, its shape affords wind protection and ensures rainwater swiftly drips straight off it. There is an opening in the base section from which ready compost can be removed. The base section comes complete with legs that can be stuck into the ground to prevent the bin from swiveling. Additional holes provide contact with the ground to coax insects and micro-organisms inside, where they can help with the composting.

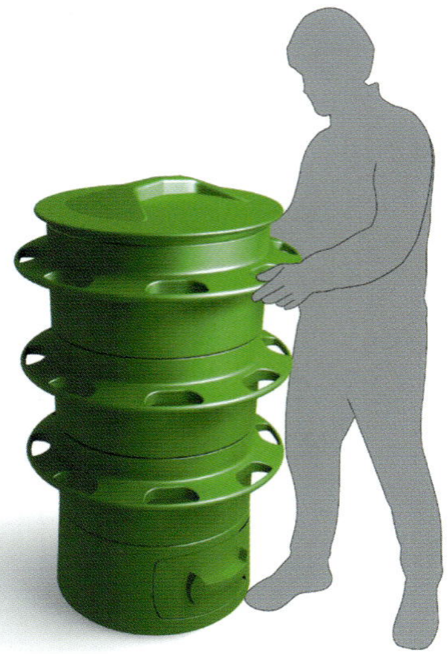

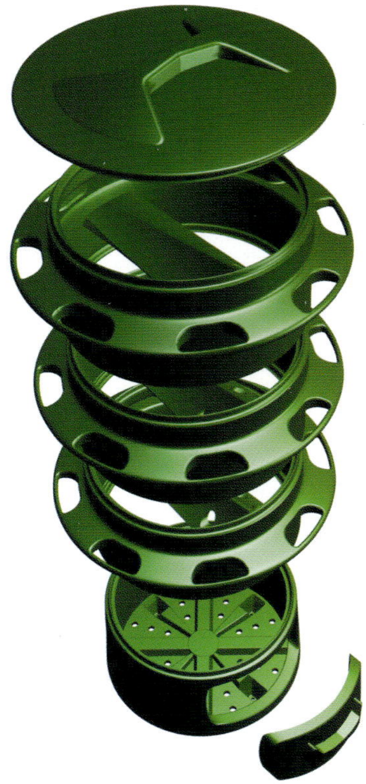

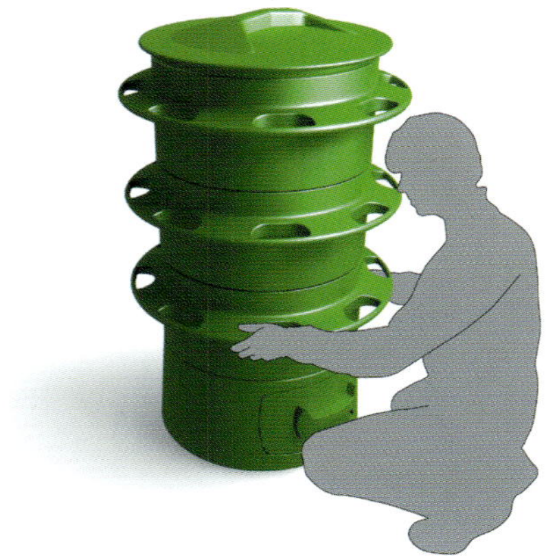

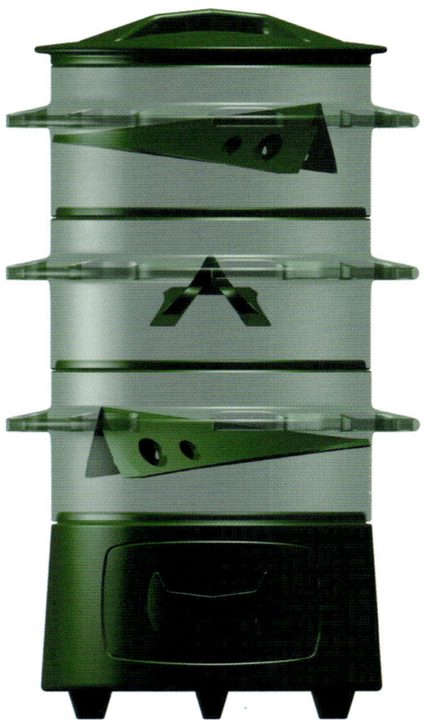

PREISGELD | PRIZE MONEY 3000 EURO

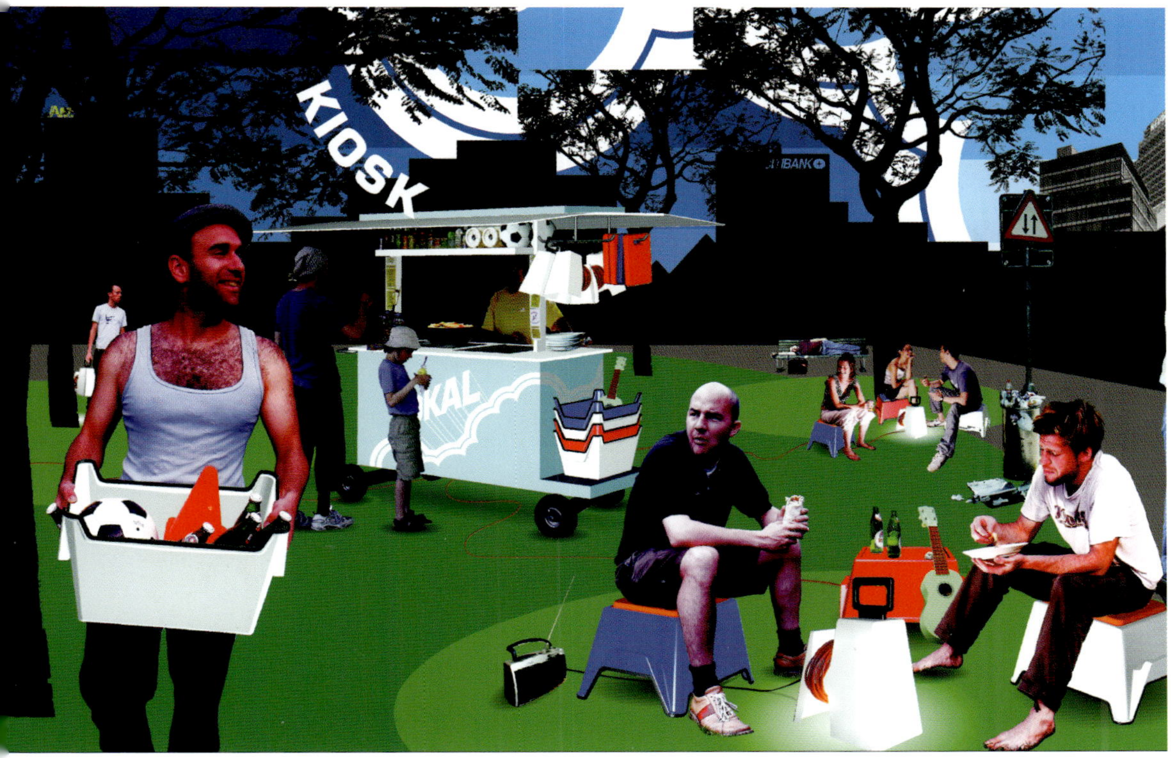

Projekt Project
Das Lokal

Design
Moritz Schlatter, Simon Schäppi
Hochschule für Gestaltung und Kunst Zürich (HGKZ)
Zürich, Switzerland
moritz_schlatter@gmx.ch
simonschaeppi@gmx.de

Das Lokal ist eine Plattform für das Erleben der Geselligkeit im städtischen Umfeld. Im Lokal, einem Laden an der Straßenecke eines lebendigen Quartiers, kochen Menschen einfache und gute Gerichte zum Mitnehmen oder dort essen. Dinge, die man für gesellige Tätigkeiten wie Grillen, Boulespielen oder den Fondueabend braucht, können im Lokal ausgeliehen und gekauft werden. Das kleine Angebot an Lebensmitteln harmoniert mit dem Produktsortiment. Eine Leuchte, ein Sitzkissen und ein stapelbares Transportgefäß, das zugleich als Hocker und Tischchen dienen kann, bilden den Anfang der Produktlinie, die den Menschen beim heiteren Zusammensein unterstützt und zugleich für die Idee des Lokals wirbt. Zum Lokal gehört ein mobiler bedienter Kiosk, der Menschen zum Verweilen im öffentlichen Raum verführt.

Lokal is a platform for experiencing the sociable side to life in an urban setting. In Lokal, a corner store is a vibrant district of town, people cook simple and wholesome food to eat there or to go. The things you need for sociable activities, such as Bar-B-Qs, playing boules, or a fondue evening, can be hired or bought from Lokal. The reduced list of foods on offer blends harmoniously with the product range. A luminaire, a seating cushion and a stackable transportation device (it doubles up as a stool and a small table) form the beginnings of a product line destined to assist people who want to get together in their free time – and these items likewise advertise the idea behind the Lokal. Lokal also includes a mobile manned kiosk that tempts people to tarry a while in urban spaces.

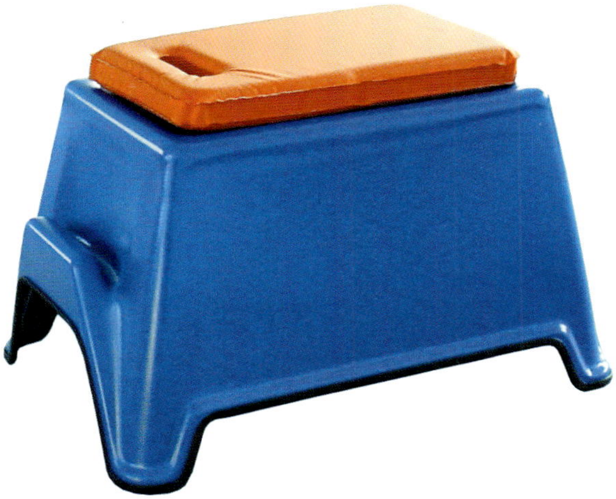

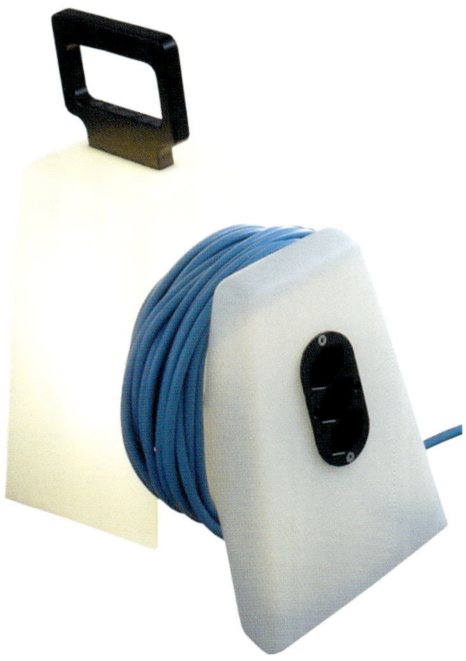

PREISGELD | PRIZE MONEY 3000 EURO

Projekt Project
Bierhoff
Tischplatte
Round plate

Design
Ralf Webermann
Fachhochschule Hannover, Design und Medien
Studiengang Produktdesign
Hannover, Germany
info@ralfwebermann.de

Bierhoff (von: Bier, offen) ist eine runde Platte mit Ausschnitt und integriertem Flaschenöffner. Mittels einer Spannvorrichtung lässt sich Bierhoff an Bierkisten aber auch an vielen anderen Getränkekisten befestigen, dabei werden die gut stapelbaren Kisten als Sockel für die Tische benutzt. Auf diese Weise entstehen Steh-, Beistell-, Nacht- oder Couchtische.

Bierhoff (derived from beer and the German word for "open") is a round plate with a cut-out and an integrated bottle-opener. By means of a stretch fastener, Bierhoff can be attached to beer crates and many other beverages crates, too; the easily stackable crates can be used as the base for tables. In this way, you get a lectern, a side table, a bed-side table or coffee tables.

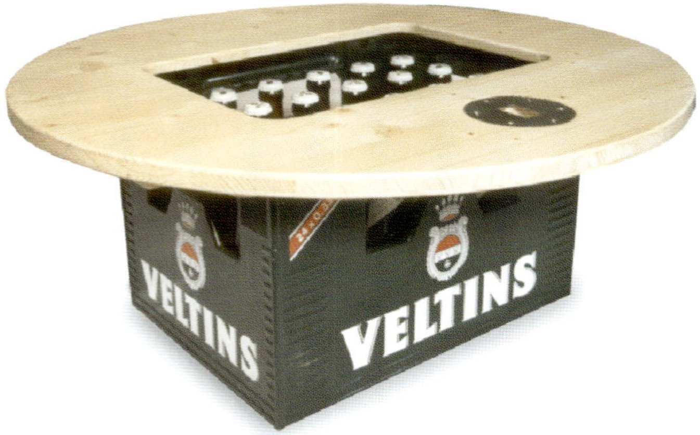

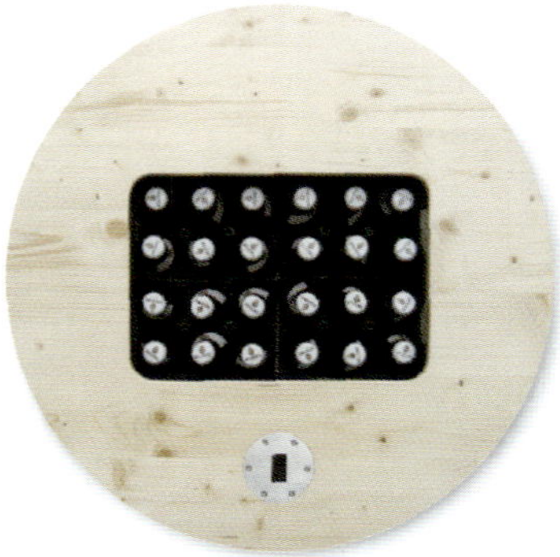

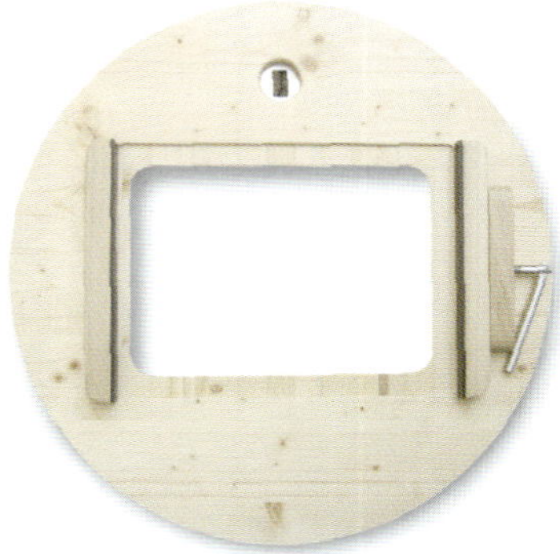

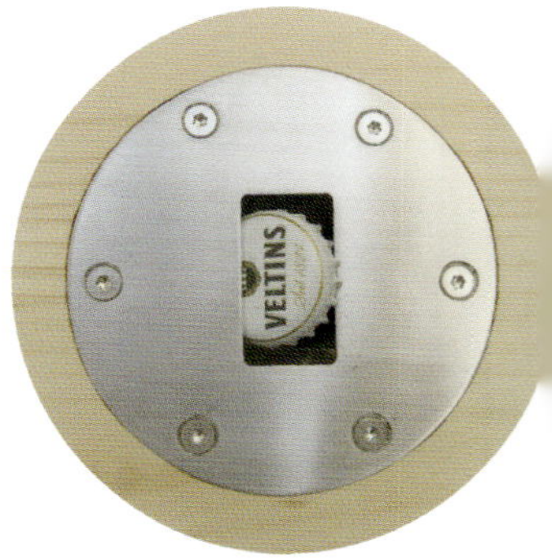

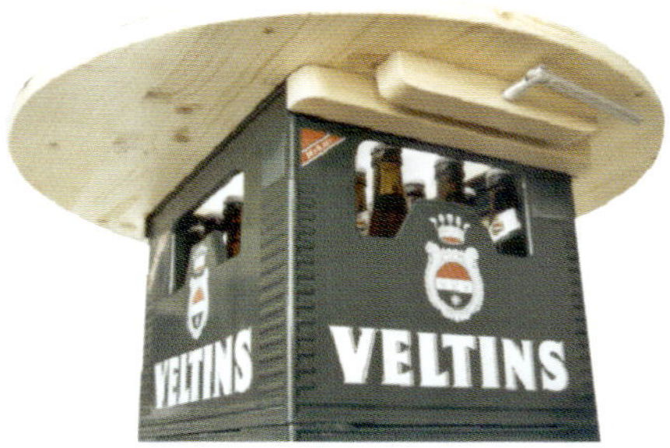

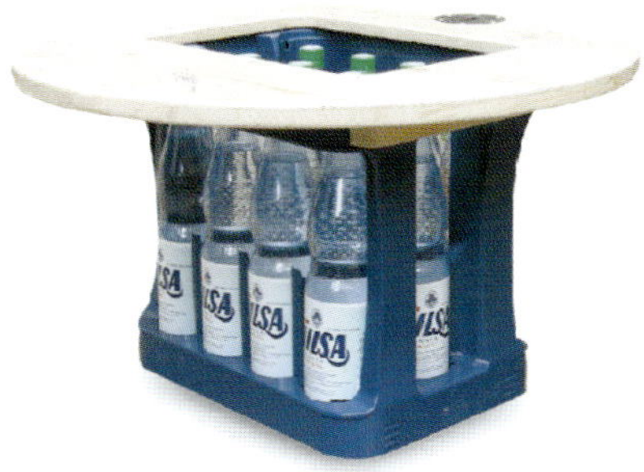

PREISGELD | PRIZE MONEY 1500 EURO

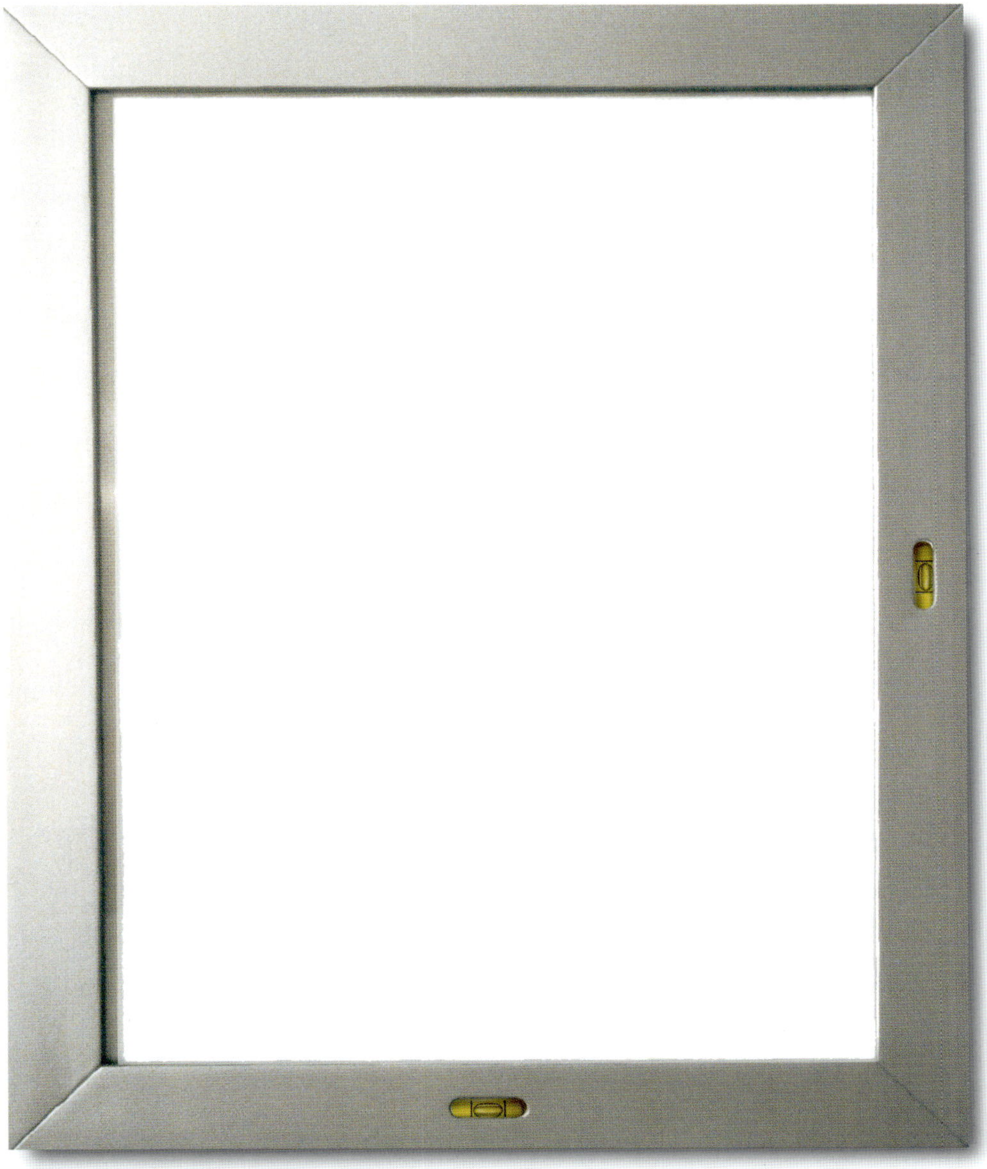

Projekt Project
pedant
Bilderrahmen
Picture frame

Design
Ralf Webermann
Fachhochschule Hannover, Design und Medien
Studiengang Produktdesign
Hannover, Germany
info@ralfwebermann.de

Das Bild hängt schief. Dem sprichwörtlich deutschen Bedürfnis nach Ordnung genügt der Bilderrahmen „pedant": eine Kombination aus Bilderrahmen und Wasserwaage. In zwei Seiten des Wechselbilderrahmens sind ständig sichtbare Libellen integriert, die den über jeden Zweifel erhabenen Beweis der exakten Ausrichtung des Rahmens liefern. „pedant" ist ein Bilderrahmen, der sich ideal für Neurotiker eignet, auf eine sympathische und humorvolle Art das Genre Bilderrahmen ironisiert und aus vorhandenen Rahmen oder Profilen abgeleitet werden kann.

The picture has not been hung straight. And the "pedant" picture frame does full justice to that proverbial German wish to have everything orderly. It is a combination of picture frame and spirit level. The two sides of the clip-on frame have constantly visible bubble tubes to demonstrate at all times that the frame is hanging perfectly straight. "pedant" is a picture frame that is ideally suited for the neurotic, and pleasantly and sympathetically plays ironically on the genre of picture frames and can be derived from existing frames or profiles.

PREISGELD | PRIZE MONEY 1500 EURO

Projekt Project
Notelight
Notizblock
Memo pad

Design
Bernd Huth
Fachhochschule Hannover, Design und Medien
Studiengang Industrial Design
Hannover, Germany
b.huth@corvus-gestaltung.de

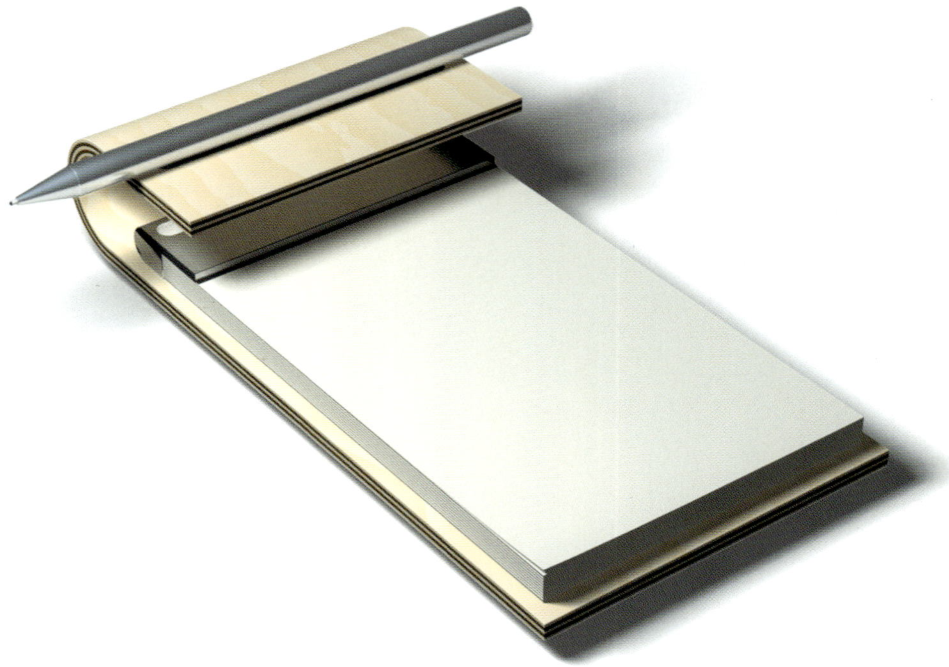

Notelight bietet die Möglichkeit, das Potenzial nächtlicher Gedanken einfach und bequem festzuhalten. Gedanken, Ideen und Einfälle gehen nicht mehr verloren. Nimmt man den Stift von der Stiftablage, wird der Notizblock beleuchtet. Es können Notizen gemacht werden, ohne die Nachttischlampe einschalten zu müssen. Im Inneren des Stiftes befindet sich ein Magnet, der nicht nur für sicheren Halt auf der Stiftablage sorgt, sondern darüber hinaus die im Blockhalter integrierten Leuchtdioden an- bzw. ausschaltet. Die Stromversorgung wird über zwei Lithium-Knopfzellen gewährleistet, die sich im Batteriefach unterhalb der Formholzschale befinden.

Notelight affords users the opportunity to swiftly and simply record the potential of those things that go through their minds at night. Thoughts, ideas and insights no longer get lost. If the pen is taken from its pad the notepad lights up, meaning you can jot down notes without having to turn on a bedside luminaire. A magnet is housed inside the pen that not only ensures it stays safely on the rest, but also switches the LEDs integrated into the notepad holder on and off. Power is supplied from two Lithium round-cell batteries which are located in the battery flap beneath the molded wooden dish.

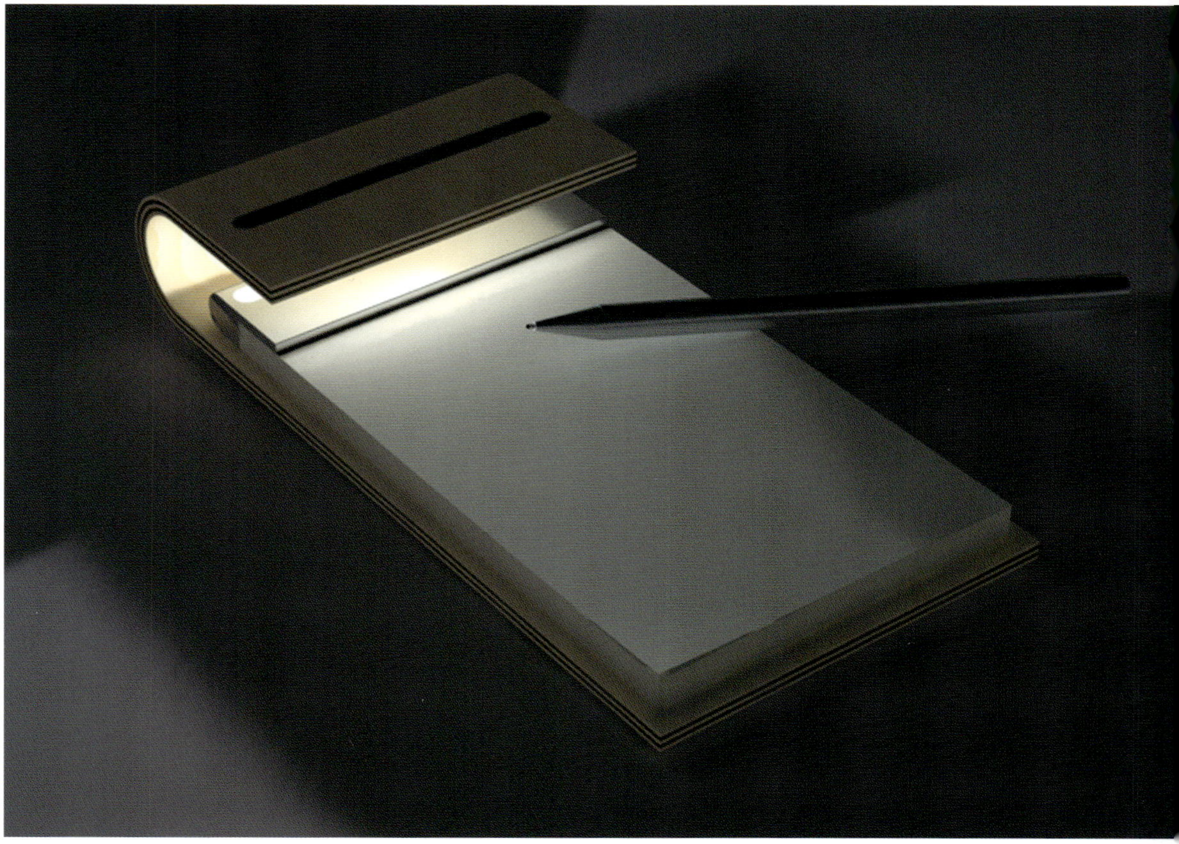

PREISGELD | PRIZE MONEY 1500 EURO

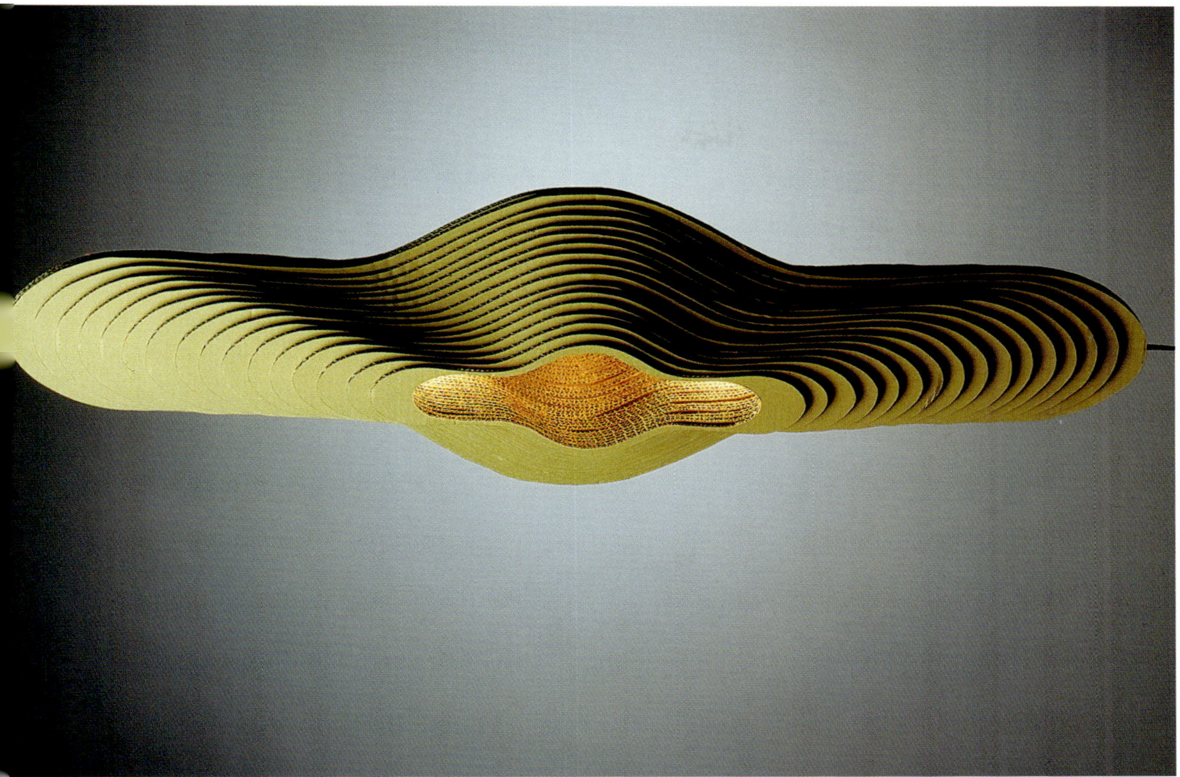

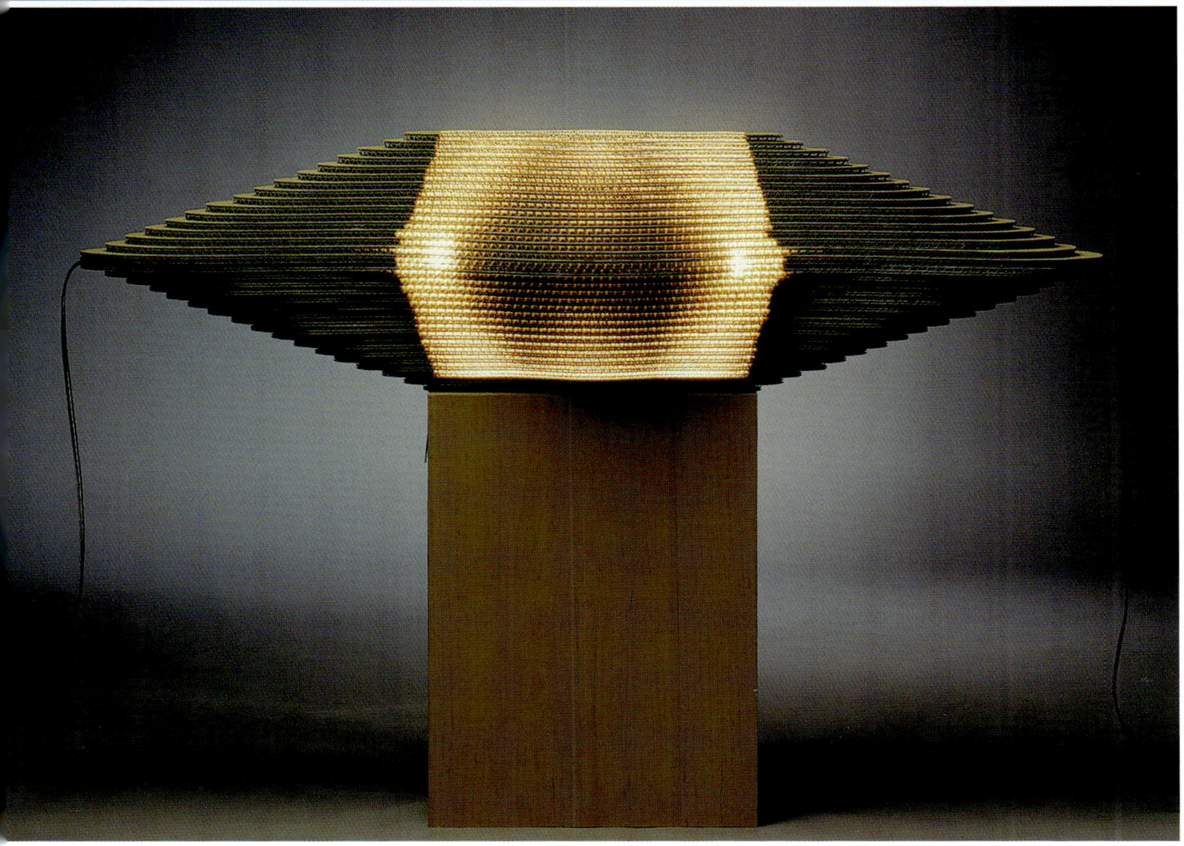

Projekt Project
Space-The shape of light
Leuchte
Luminaire

Design
Jeong-Ju Lee
Hongik University Graduate School
Woodworking Camp, Furniture Design
Seoul, South Korea
heyjool@hanmail.net

Die Leuchte „Space-The shape of light" ist aus geschichteter Wellpappe gefertigt, die in einer organischen Form zugeschnitten und aufeinander gelegt wurde. Die Struktur der Wellpappe erzeugt in ihrer Lichtdurchlässigkeit einen interessanten Effekt, zudem ist sie als Material sehr leicht, in der Aufeinanderschichtung auch sehr stabil und zudem preiswert. Um eine Hitzestauung durch das Leuchtmittel in der Lampe zu verhindern hat diese nach oben eine große Öffnung. Mit dem unterschiedlichen Durchmesser der jeweiligen Wellpappenschicht variiert auch der Lichtaustritt.

The Space-The shape of light luminaire is manufactured from layers of corrugated cardboard, cut into an organic shape and laid one on top of the other. The corrugated structure is light permeable and creates an interesting effect, especially as it is a very lightweight material, very stable when layered, and cheap into the bargain. To avoid any heat build-up from the lamp in the luminaire, the latter has a broad upper opening. The different diameters of the respective layers of cardboard means the light diffusion varies, too.

PREISGELD | PRIZE MONEY 1500 EURO

Projekt Project
Der Puls des Hauses
Stromverbrauch-Visualisierer
Power consumption visualizer

Design
Peter Riering-Czekalla
Sebastian Ritzler
Vincent Weckert
Muthesius Hochschule
Fachhochschule für Kunst und Gestaltung
Kiel, Germany
peter@opendiary.de
s.ritzler@web.de
vincent_weckert@yahoo.com

Der Stromverbrauch-Visualisierer pulsiert je nach aktuellem Stromverbrauch unterschiedlich stark, so wie der Puls des menschlichen Körpers die Anstrengungen widerspiegelt. Dadurch erhält der Verbraucher ein direktes Feedback beim An- oder Ausschalten von Geräten im Haushalt. Die dafür notwendige Verbrauchsinformation bekommt das Gerät per Funkimpuls vom Stromzähler. Durch Berühren der sensitiven Fläche wird der aktuelle Energieverbrauch des Haushaltes numerisch dargestellt. Die Gestik des Pulsmessens diente hier als Bild. Nimmt man das Gerät aus der magnetischen Wandschale heraus, so erscheint der gesamte Funktionsumfang. Ein zusätzliches Einzelgerät (Satellit) soll ein Bewusstsein für den Stromverbrauch der einzelnen Geräte schaffen.

The power consumption visualizer pulsates at a rate that reflects the power consumption level at that particular time, just as our own heart rate changes to reflect the intensity of activity. In this way, the consumer receives direct feedback when turning household appliances on or off. The consumption data required to this end is radio-transmitted to the device by the electricity meter. When the sensitive surface is touched it provides a numeric display of current power consumption levels in the particular house. The gesture – taking the house's pulse – was the key visual chosen here. If the device is taken out of its magnetic wall mounting tray, then all its functions are legible. An additional separate device (satellite) is intended to heighten user awareness of the power consumption levels of individual appliances.

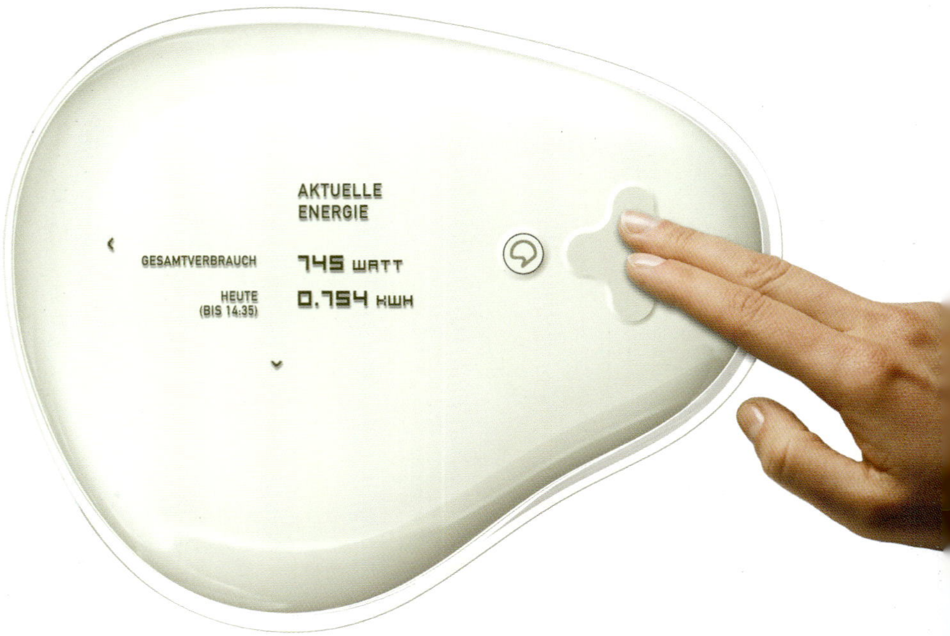

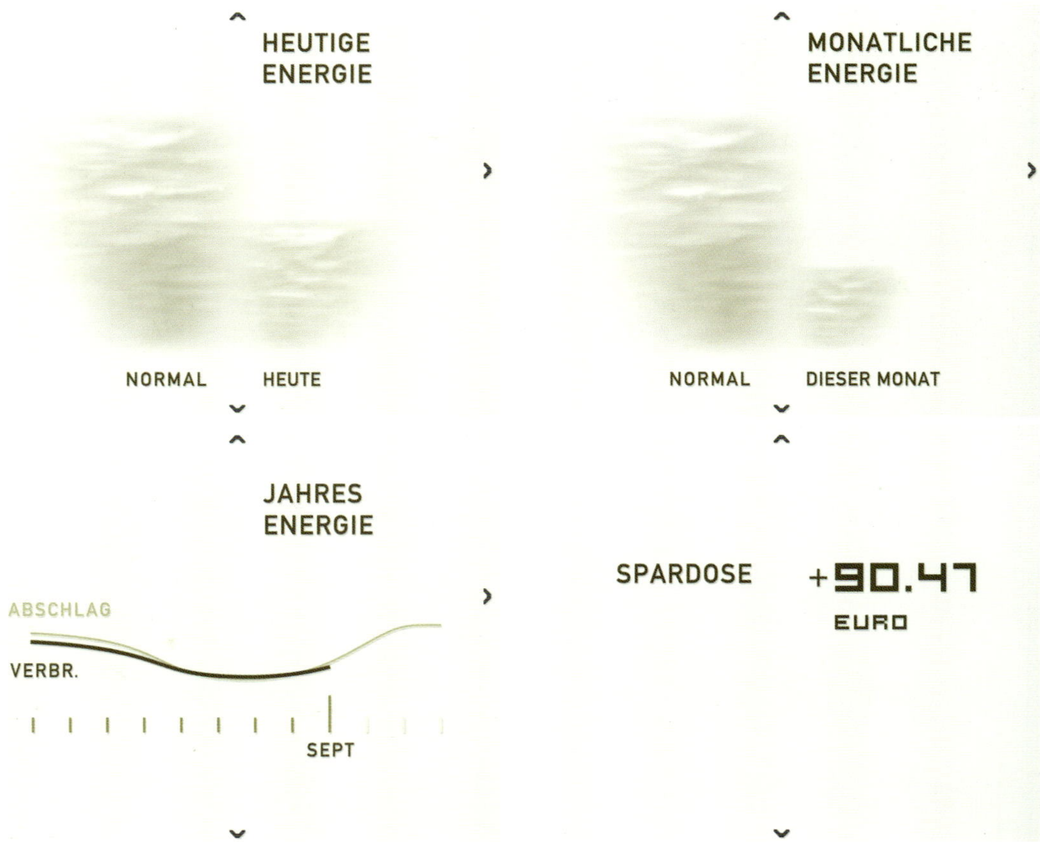

PREISGELD | PRIZE MONEY 1500 EURO

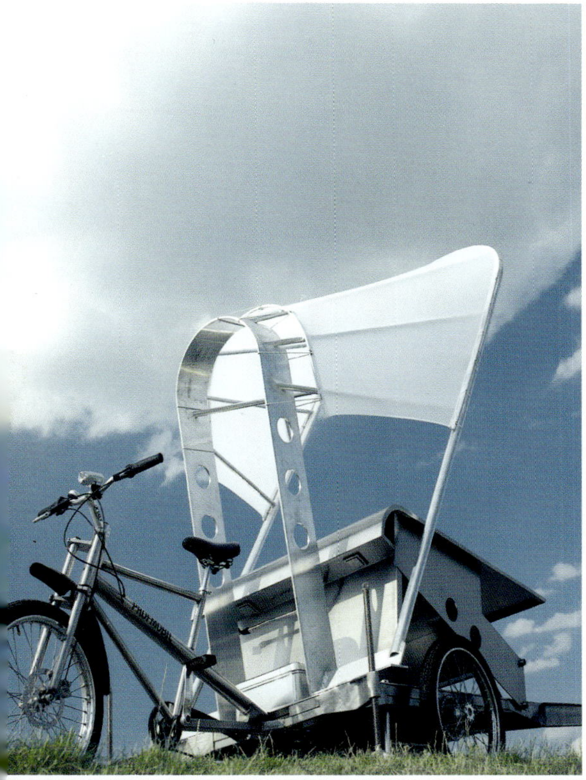

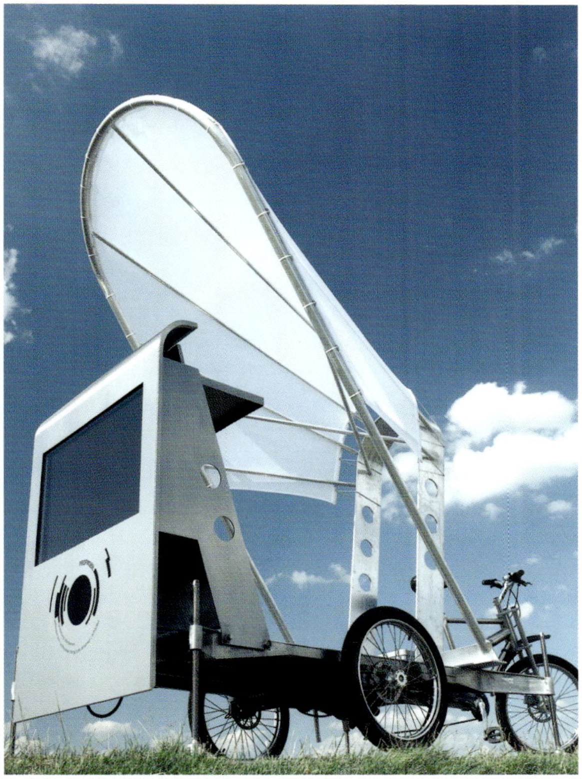

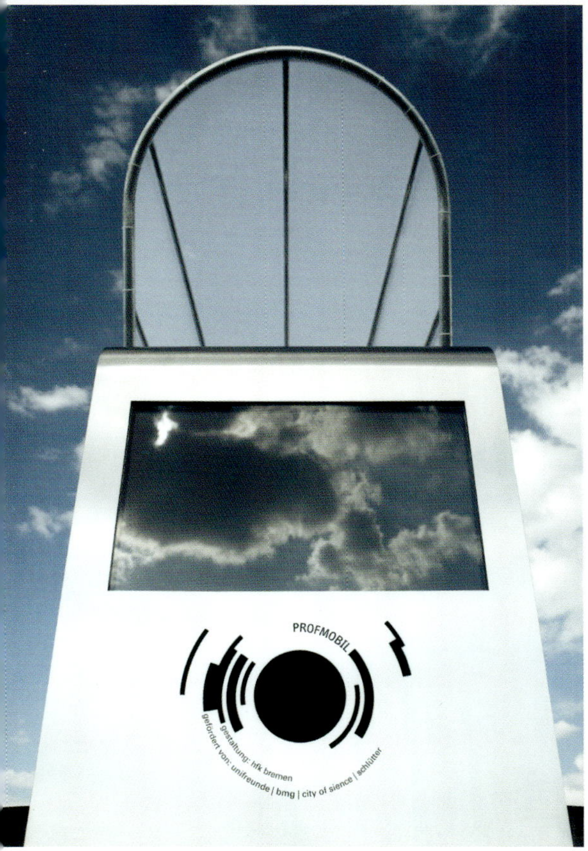

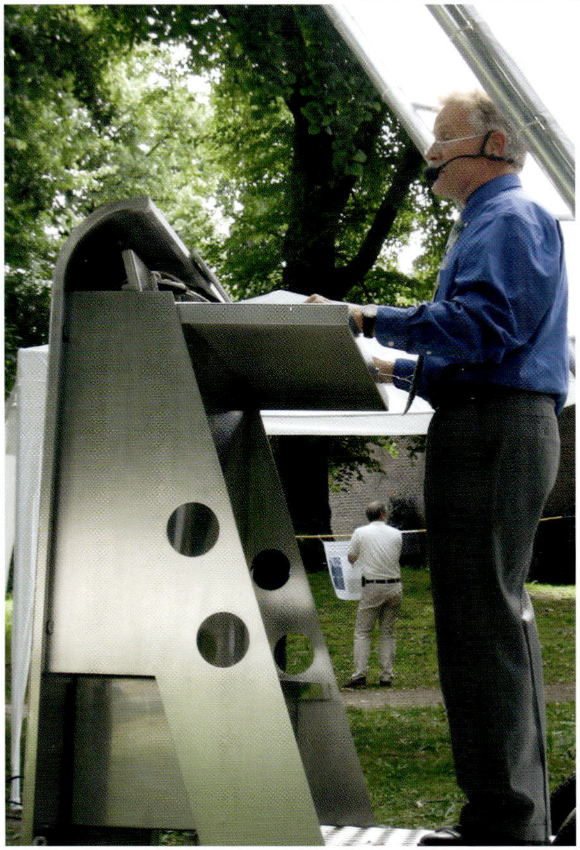

Projekt Project
Profmobil
Mobile Vortragsplattform
Mobile presentation platform

Design
Jens Schulz, Rasmus Giesel
Alexander Hanowski, Silke Nachtigahl
Hochschule für Künste Bremen
Bremen, Germany
j.schulz@hfk-bremen.de
rasmus@profile-intermedia.de
a.hanowski@coration.com
silkenachtigahl@gmx.de

Das Profmobil ist eine mobile Rednerplattform. Es bietet Wissenschaftlern und Persönlichkeiten die Möglichkeit, ortsunabhängig Vorträge zu halten und ihr Wissen sowie ihr Anliegen bürgernah zu präsentieren. Während der Fahrt ist das Rednerpult nach innen eingeklappt. Es lässt sich wie eine Rikscha im Straßenverkehr bewegen. Am Vortragsort angekommen, kann der Redner das Pult mit wenigen Handgriffen aufstellen und findet eine komfortable Ablage samt Laptopanschluss vor. Darüber hinaus steht ihm modernste Kommunikationselektronik zur Verfügung. Der akkubetriebene Lautsprecher überträgt mittels Funkmikrofon den Vortrag ins Publikum, und auf dem in das Pult integrierten LCD-Monitor können Bilder, Diagramme etc. dem Zuschauer präsentiert werden.

The Profmobil is a speaker's platform that can be driven around. It enables scientists and important persons to lecture wherever they want, and thus familiarize local listeners with their knowledge and their ideas. While on the road, the lecture platform is simply folded up. The Profmobil then travels along like any rickshaw. When the lecture location has been reached, the speaker only needs make a few adjustments, and the platform is ready – complete with comfortable rest and laptop link. Moreover, it relies on ultra-modern communications electronic: the radio mike relays the speech to the battery-operated loudspeakers which broadcasts it to the audience, and the integrated LCD monitor integrated into the platform screens images, diagrams etc.

PREISGELD | PRIZE MONEY 1500 EURO

Projekt Project
Silent-Alarm
Ohrwecker
Ear alarm

Design
Hau-Yu Lei
Cheng Kung University
Tainan, Taiwan
yaohub@yahoo.com.tw

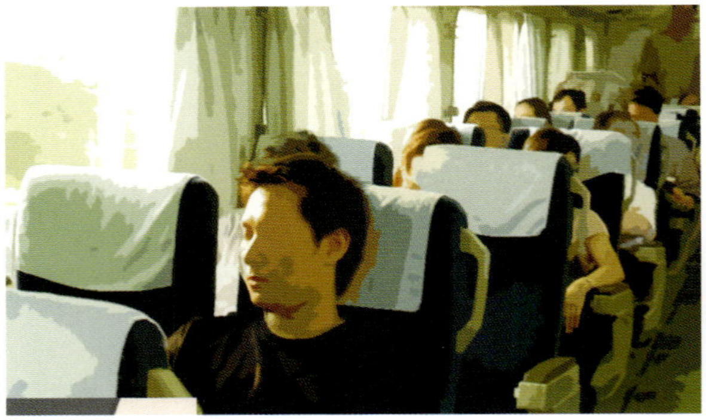

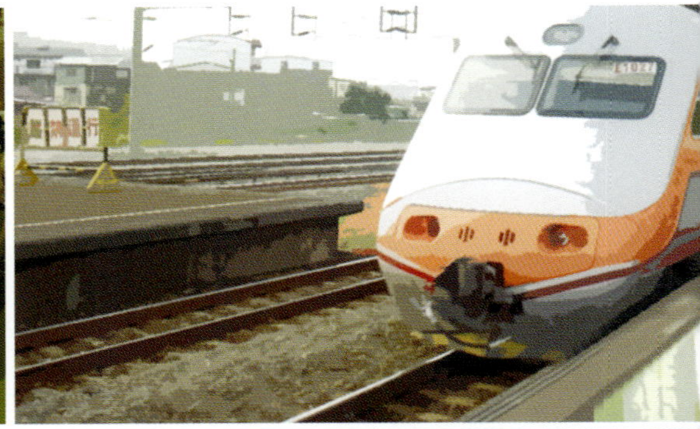

Der „Silent-Alarm"-Ohrwecker weckt mittels Vibrationsalarm zu einer bestimmten vorher eingestellten Zeit, ohne andere Menschen zu stören. Insbesondere auf Reisen, im Flugzeug oder im Zug oder auch in einer lauten Umgebung, in der das akustische Signal eines konventionellen Weckers leicht überhört werden könnte, ist der Ohrwecker von großem Nutzen. Die ergonomische Formgebung gewährt guten Tragekomfort.

The Silent-Alarm ear alarm wakes you by vibrations at the time you have set it for – without disturbing anyone else. In particular, the acoustic signal of conventional alarm clocks is easily overheard if you are traveling by plane or train or in a very noisy environment. In such settings, the ear alarm is a great benefit. Thanks to its ergonomic design it is comfortable to wear.

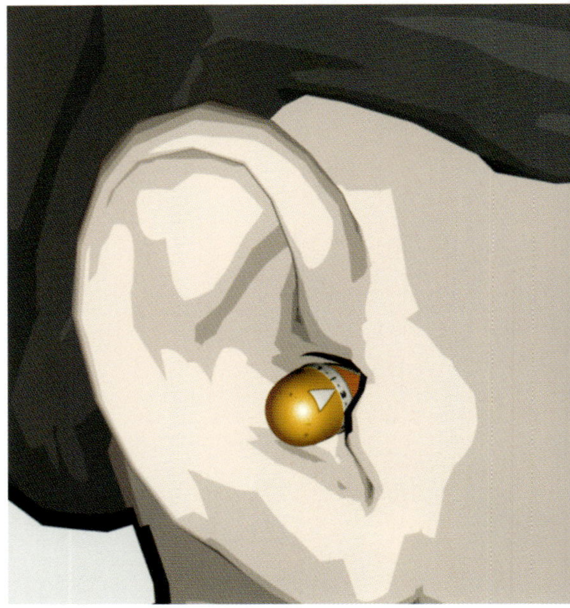

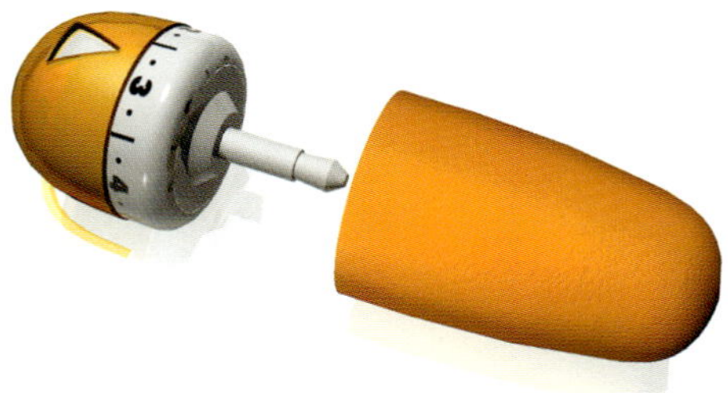

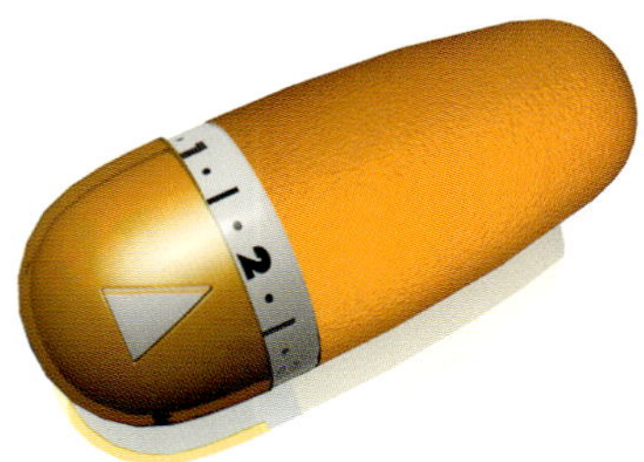

PREISGELD | PRIZE MONEY 500 EURO

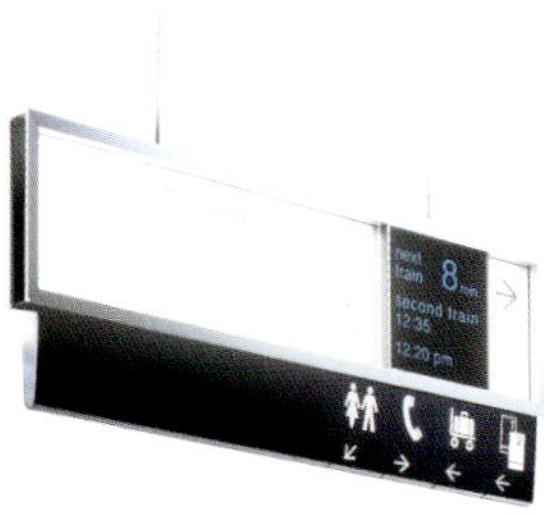

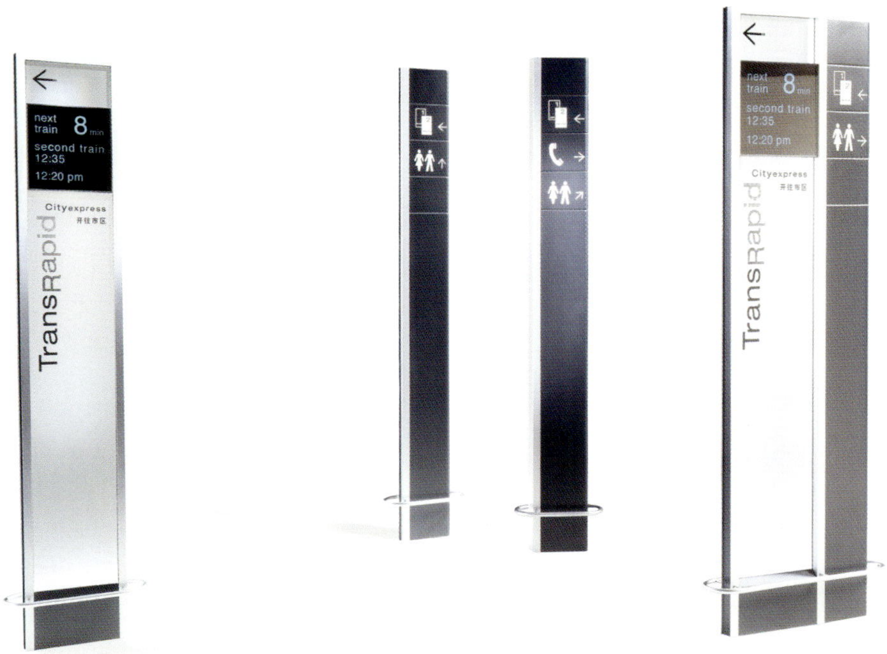

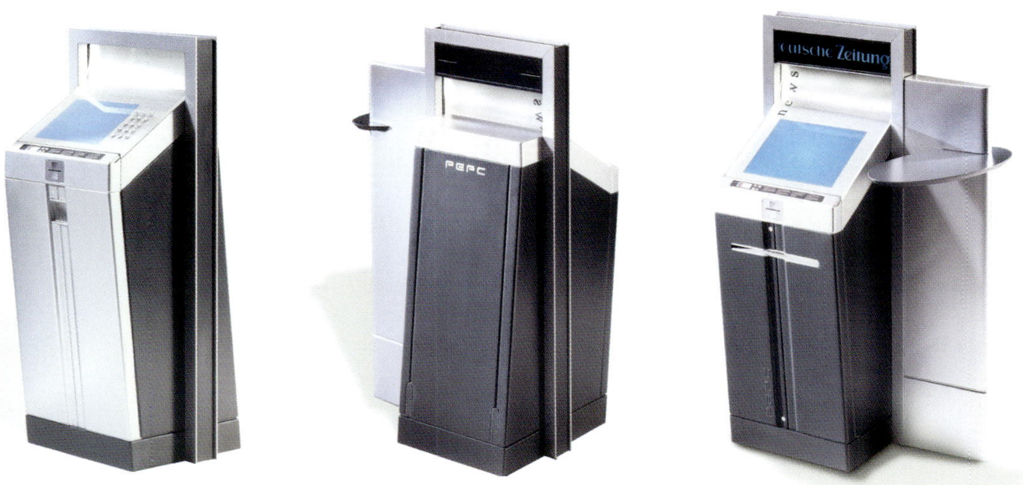

Projekt Project
Transrapid
Wegeleit- und Informationssystem für den Transrapid
Leading and Information System for the Transrapid

Design
Volker W. Eysing
Muthesius Hochschule
Fachhochschule für Kunst und Gestaltung
Kiel, Germany
eysing@web.de

Das Wegeleit- und Informationssystem für den Transrapid besteht aus unterschiedlichen Funktionskomponenten wie den Hauptstelen, die als Richtungsweiser und Informationsfeld dienen und den andockbaren Zusatzstelen, die untergeordnete Informationen wie den Ticketkauf abdecken. Für weiterführende, innerstädtische Informationen (z.B. U-und S-Bahn) werden die Rahmenelemente der Hauptstele einfach zu einer beleuchteten Informationstafel umfunktioniert. Der Ticketautomat und der Zeitungsdrucker – durch Internetanschluss und Satellitenverbindung können stündlich aktualisierte Zeitungen gedruckt werden – sind in zwei Ausführungen erhältlich. Eine hohe Version zur Fernerkennung sowie eine niedrige Variante als kompaktes Kombinationselement. Die Bedienflächen des Ticketautomaten und des Zeitungsdruckers befinden sich auf Stehpulthöhe. Der Touch-Screen sowie sämtliche Bedienelemente sind auf einen Blick gut erfassbar und ergeben durch ihre ergonomische Anordnung eine optimale Bediensituation.

The guidance and information system for the Transrapid maglev is made up various functional components – main stele that serve as signposts and information displays, and dockable auxiliary stele that provide subordinate services (such as ticketing). In order to provide in-depth urban info (e.g., on subways and local trains), the frame elements of the main stele are simply converted into an illuminated information panel. Two types of ticket vending machine and newspaper printer (by Internet and satellite link, up-dated papers can be printed once an hour). The high version is intended to be visible from far, and the low version is a compact combinations unit. The user interfaces of the ticket vendor and the newspaper printer are located at lectern height. The touch-screen as well as all controls are easily grasped at first sight and their ergonomic layout facilitates their optimal use.

PREISGELD | PRIZE MONEY 500 EURO

Projekt Project
Lilli*s
Leuchte
Lamp

Design
Rut Sawodny
Fachhochschule Hannover, Design und Medien
Studiengang Industrial Design
Hannover, Germany
sawodny@bader-leather.de

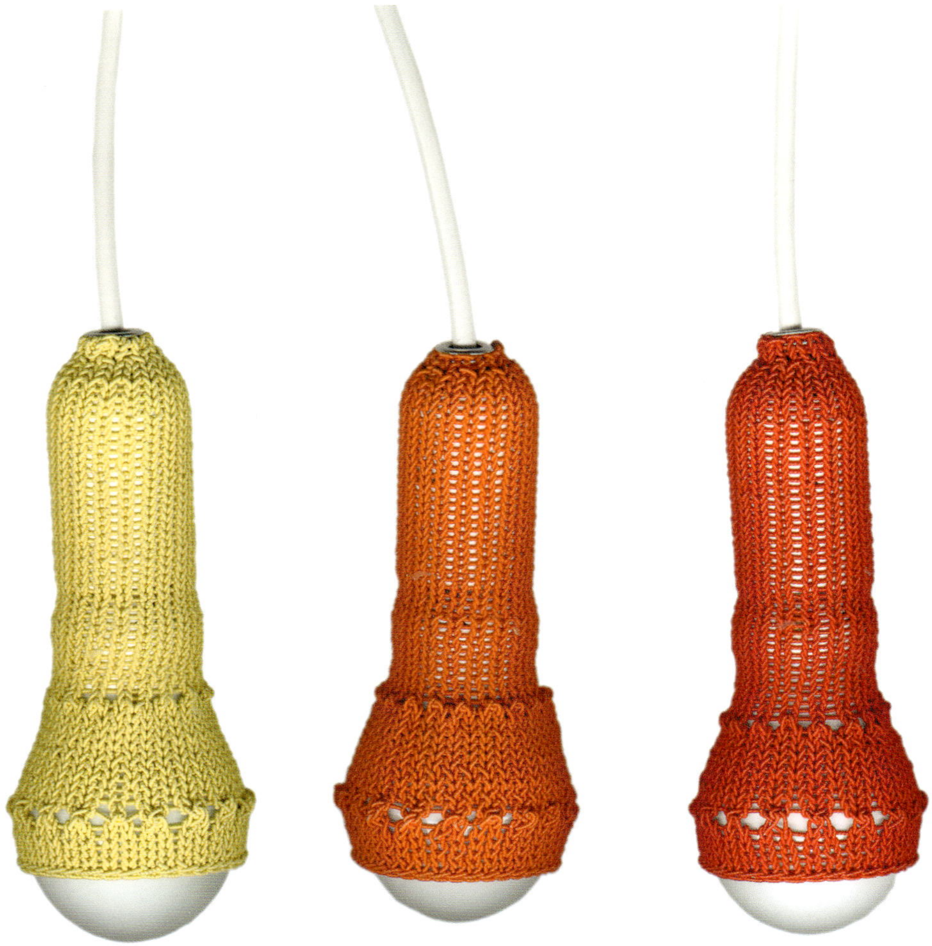

Wie in der Kunst gibt es auch im Design Methoden, Alltagsgegenstände zu verfremden und sie dadurch außergewöhnlich zu machen. Ein Leuchtenprojekt, zu dem die Hängeleuchte Lilly* gehört, strickt Glühbirnen und Lampenschirme farbig ein. Dadurch erhält der banale Gegenstand eine weitere emotionale Ebene. Lilli*s ist eine flexible Stimmungsleuchte, die an jeder Steckdose ihren Platz findet. Das Material, die Form und die Strickmuster machen Lilli*s zu einem unverwechselbaren Produkt. Das Elastan liegt direkt auf dem Leuchtmittel. Eine Energiesparlampe macht diese Art der Fertigung möglich. Lilli*s ist waschbar.

Like in art, in design there ware likewise methods for alienating everyday objects and in this way making them unusual. A luminaire project, and Lilly* is one part of it, gives lamps and the lamp-shades a knitted and colorful surround – thus lending the banal item a new emotional lease of life. Lilli*s is a flexible mood-luminaire which can be hooked up to any plug. The material, the form and the knitted pattern make Lilli*s an unmistakable product. The Elastan is placed directly over the lamp. An energy-saving lamp makes this kind of design possible. Lilli*s is washable.

PREISGELD | PRIZE MONEY 500 EURO

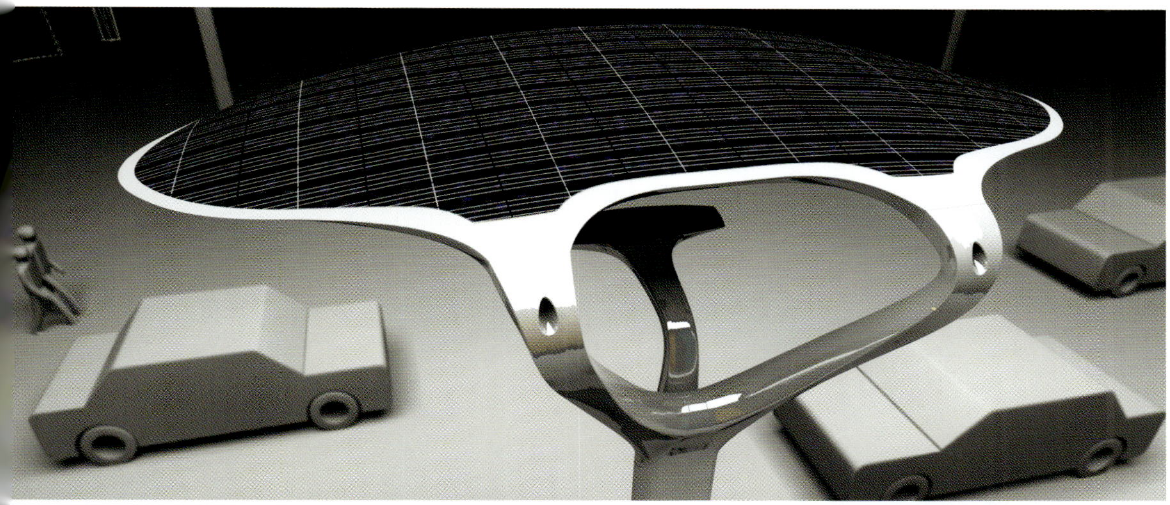

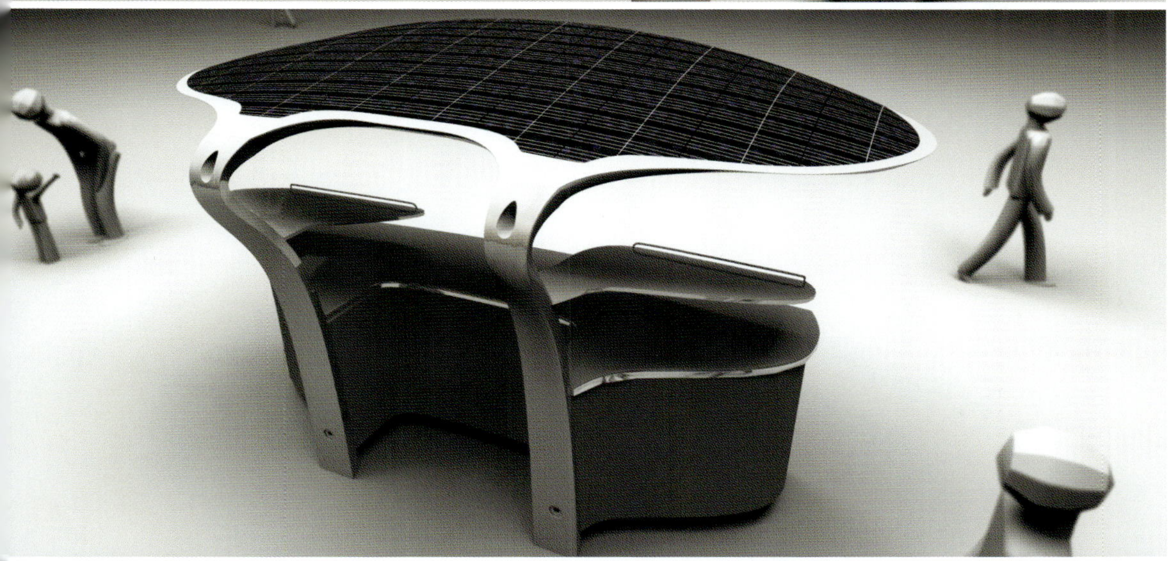

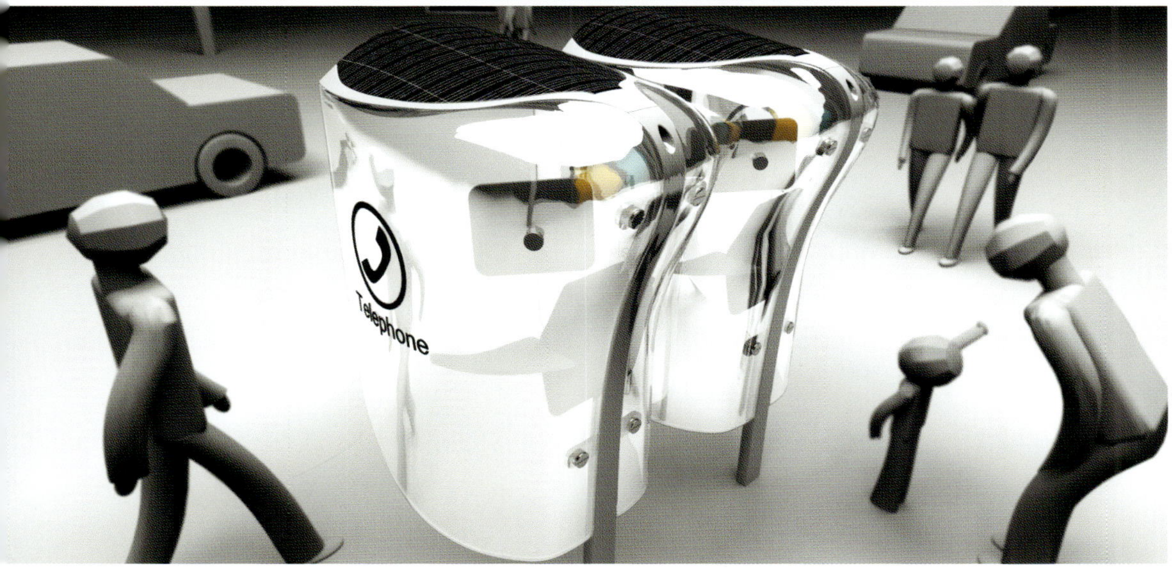

Projekt Project
Solar Street Concept
Solarenergiesystem
Solar-power system

Design
Nikola Knezevic
University of Art
Faculty of Applied Art and Design
Belgrade, Serbia
nikoladesign@hotmail.com

Das Solar Street Concept ist ein Solarenergiesystem für den öffentlichen Raum, das durch eine spezielle Energiespartechnologie (MoSESS Movement Sensors Energy Saving System) eine Beleuchtung die ganze Nacht hindurch ermöglicht. Die einzelnen Komponenten wie die Solar-Straßenleuchte, die Bushaltestelle mit Solardach oder andere öffentliche Geräte wie Telefone oder Info-Systeme, die eine Energiezufuhr voraussetzen, sind unterirdisch miteinander verbunden, was einen Austausch überschüssiger oder fehlender Energie ermöglicht oder auch die Versorgung durch ein Kraftwerk erlaubt, falls dies notwendig wird. Die Solar Parkleuchte enthält die gleichen Konstruktionselemente wie die Straßenleuchte. Auch Straßenverkaufsbuden könnten durch das Solarsystem mit Energie versorgt werden.

The Solar Street concept is a solar-power system for public spaces that relies on a special energy-saving technology (MoSESS Movement Sensors Energy Saving System) to deliver illumination all night long. The individual components of street life, such as street lamps, the bus shelter with its large solar roof or other public appliances, such as phones or info-systems, all require energy. In the Solar Street concept they are wired underground to enable them to share residual energy, cover a power shortfall or source it from an electricity utility if required. The Solar park lamp uses the same structural elements as the street lamps. And the Solar Street can even deliver power to support street stands/booths.

PREISGELD | PRIZE MONEY 500 EURO

Projekt Project
Treibgut
Pedalo-Vermietung
Pedalo rental company

Design
Oliver Berger, Christian Gehri
Fachhochschule Aargau
Aarau, Switzerland
oliver.berger@industrialdesign.ch
ch.gehri@hotmail.com

„Treibgut" soll das Dienstleistungsangebot der Pedalo-Vermietung neu strukturieren und nach kundenorientierten Bedürfnissen gestalten. Im Mittelpunkt steht die „Treibgut"-Anlegestation. Dank ihres vielseitigen gastronomischen Angebots wird sie sich vom bloßen Umschlagplatz von Pedalos zu einem „place to be" an der Seepromenade wandeln und somit ihren einmaligen Standortvorteil besser ausnutzen. Die Pedalos sollen als erweiterte Gästeplätze wahrgenommen werden, auf welchen gerne und lange verweilt wird. Auf der Basis des erarbeiteten Corporate Identity Manual entstanden die Pedalo-Anlegestation (Corporate Architecture) und das Pedalo (Corporate Product Design). Die „pic-box", in welcher sich ein an der Anlegestation gekauftes Picknick befindet, ist das konzeptionelle Bindeglied zwischen der Anlegestation und den Pedalos. Der Kunde kann das Picknick mit auf das Pedalo nehmen und anschließend auf dem See genießen.

"Treibgut" (Flotsam) is intended to re-structure offerings by pedalo rental companies and tailor them to client needs. The central element is the "Treibgut" docking station. Thanks to its multifaceted F&B offerings, it is transformed from being simply a place where pedalos turn around into a "place to be" on the lake shore – and in this way makes far better use of its site-specific advantages. The idea is also for pedalos to be perceived as extended spaces for guests, where people like to sit and spend time. On the basis of the special Corporate Identity Manual both the pedalo docking station (corporate architecture) and the pedalo itself (corporate product design) arose. The "pic-box", which contains a picnic bought at/from the docking station, forms the conceptual bond between the station and the boats themselves. Clients can take their picnic on board and then enjoy spending a longer time on the water.

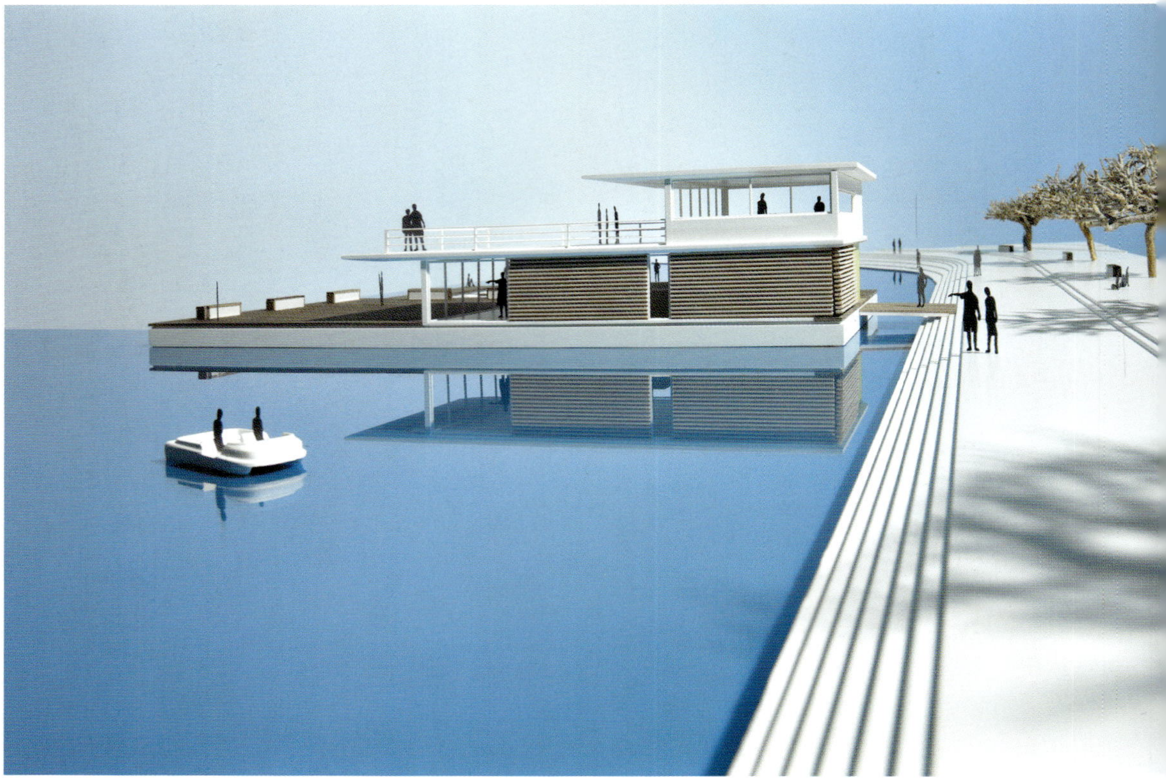

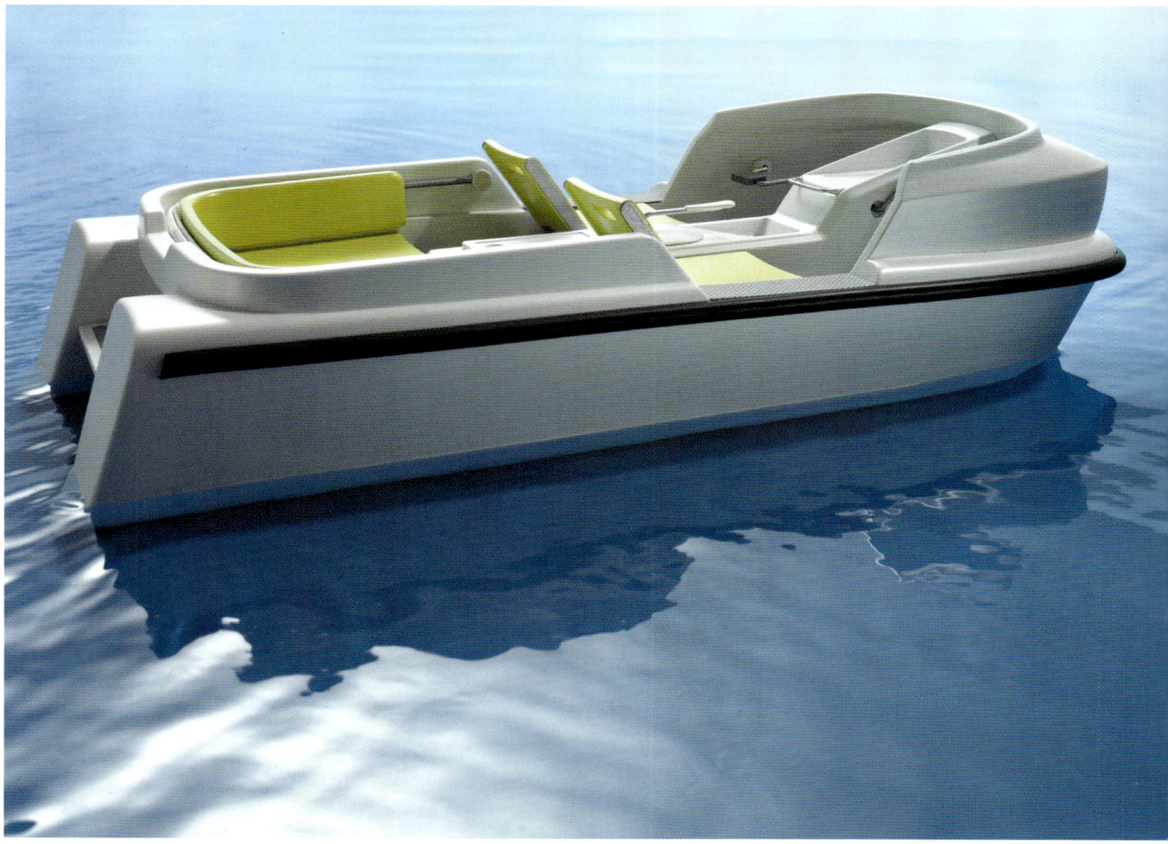

PREISGELD | PRIZE MONEY 500 EURO

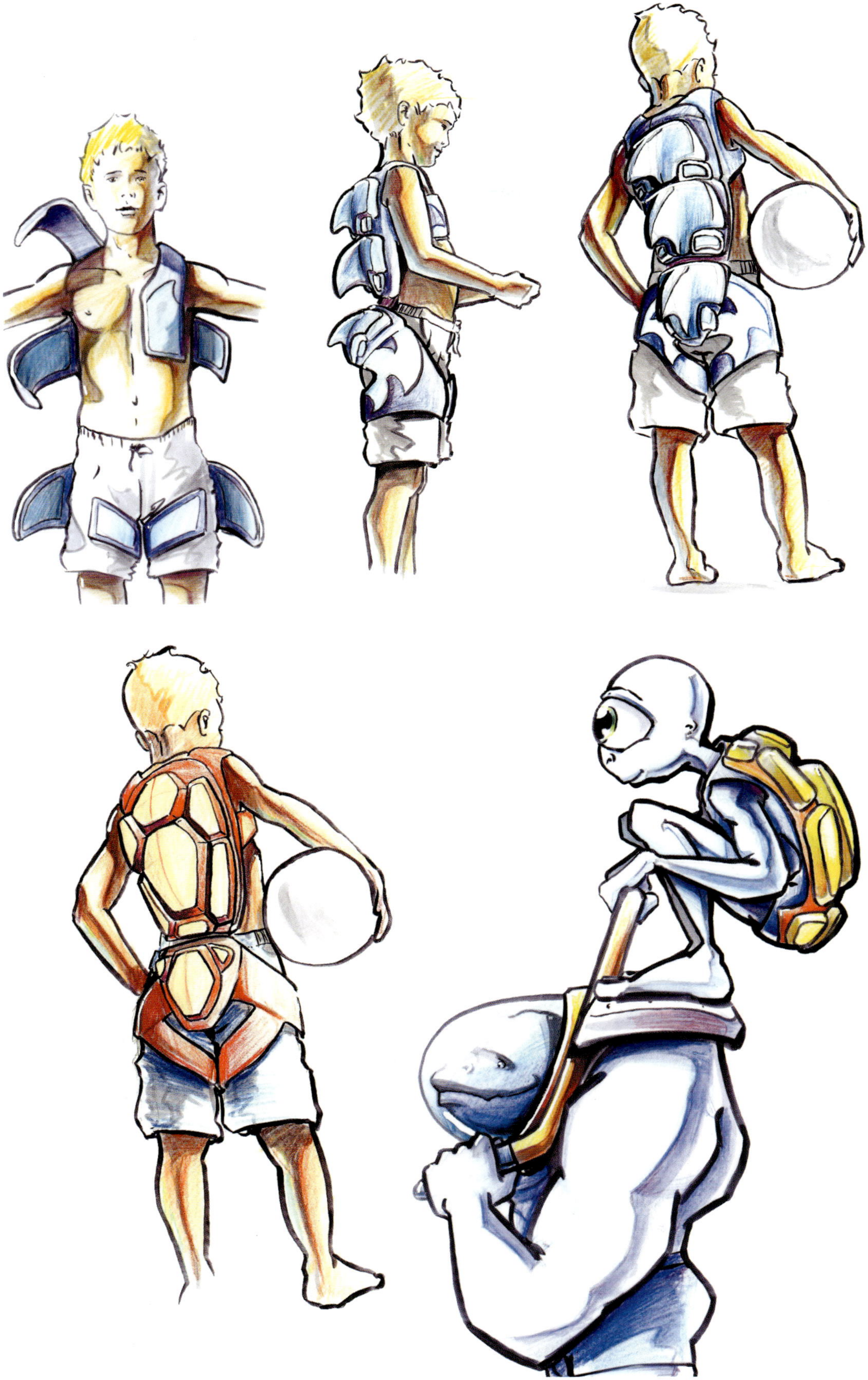

Projekt Project
Freddy, Gonzo und Saitam
Schwimmhilfen
Swimming aids

Design
Mathias Durisch
Hochschule für Gestaltung und Kunst
Zürich, Switzerland
mathias_durisch@gmx.ch

Um Kinder, die nicht gut oder noch gar nicht schwimmen können, in ein Spiel mit bereits gut schwimmenden Kindern zu integrieren, entstanden drei Schwimmhilfen. Diese sind den Comicfiguren Freddy, Gonzo und Saitam nachempfunden, welche in Zusammenarbeit mit einer zweiten Primarklasse entstanden sind. So lassen sich Rollenspiele spielen, die auch bereits gut schwimmende Kinder in die Rollen der Figuren schlüpfen lassen, wenn sie die Schwimmhilfen tragen. Die Schwimmhilfe wird nun mehr zur Verkleidung und gibt dem Kind gleichzeitig ein besseres Körperbewusstsein im Umgang mit dem Element Wasser.

Three swimming aids were created in order help to integrate children who cannot yet swim or only swim poorly, into games with kids who are already proficient swimmers. They are modeled on three figures from comics: Freddy, Gonzo and Saitam, and were devised together with a class of second graders. Thus, they allow role games to be player, whereby the kids who are already good swimmers don the parts of the figures whenever they wear the swimming aids. The latter thus morph into disguises while at the same time instilling the children with a better feel for their own bodies in conjunction with the water around them.

PREISGELD | PRIZE MONEY 500 EURO

Projekt Project
The T-shirt
Wörterbuch
Dictionary

Design
Vadim Kibardin
Russian Academy of Actual Design
Moscow, Russia
vadim@kibardin.com

Das Wörterbuch-T-Shirt-Projekt basiert auf der einfachen Idee, unterschiedliche Wörterbucheinträge auf T-Shirts zu drucken. Auf diese Weise prägen sich fremdsprachliche Vokabeln unterbewusst ins Gedächtnis ein. Der Konsumgedanke wird durch den positiven Beitrag, den man durch das Tragen eines solchen T-Shirts für die Allgemeinheit leistet, aufgehoben.

The Dictionary T-Shirt Project started out with a simple idea – printing different dictionary definitions on T-shirts. In this way, foreign words get subconsciously committed to memory. The sole focus on consumerism is avoided by the positive benefit which wearing such a T-shirt has for the population as a whole.

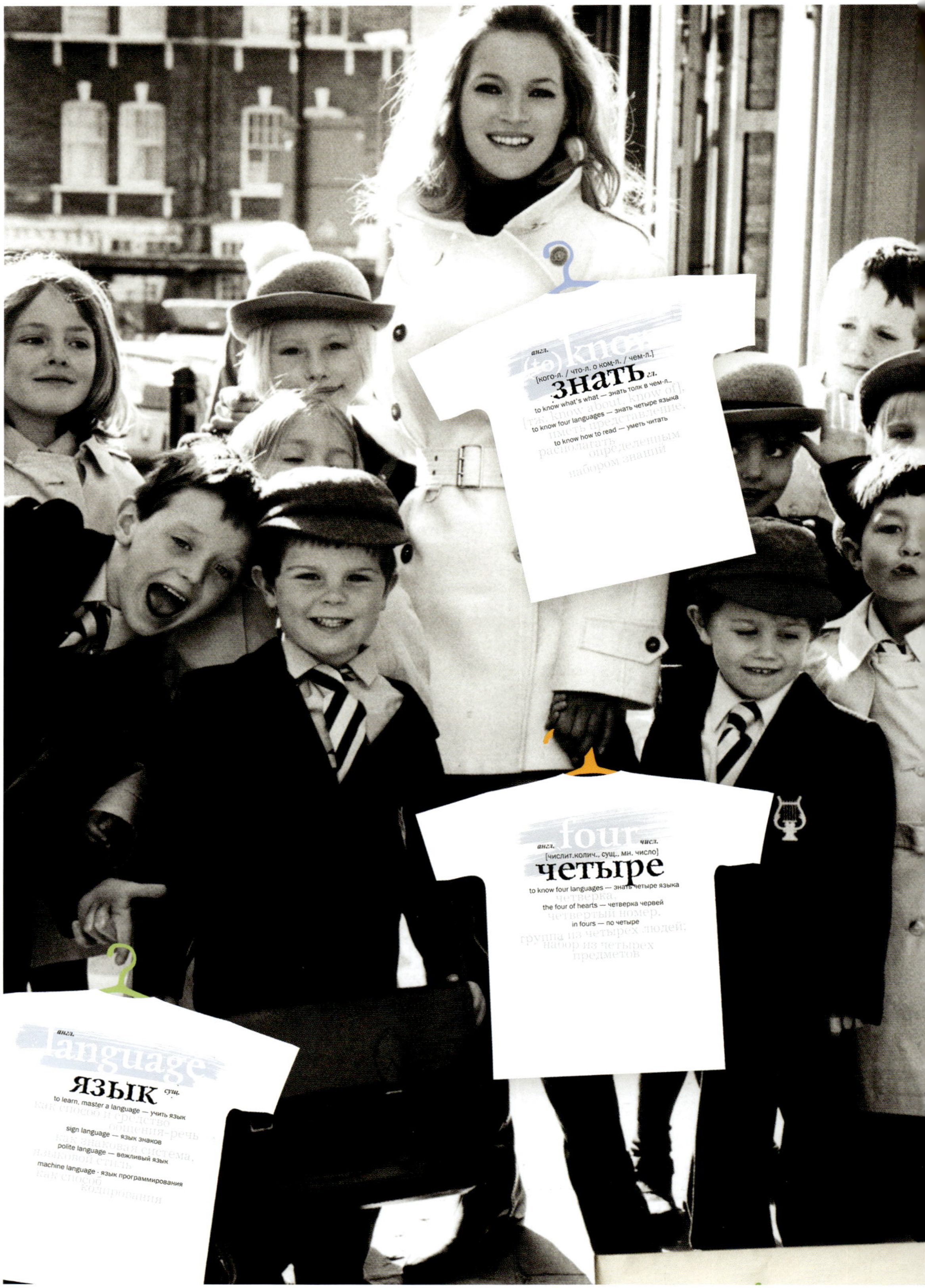

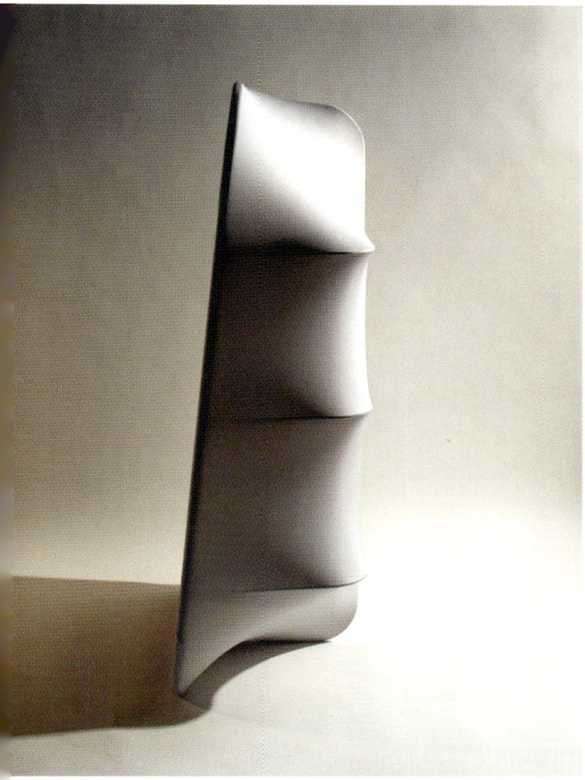

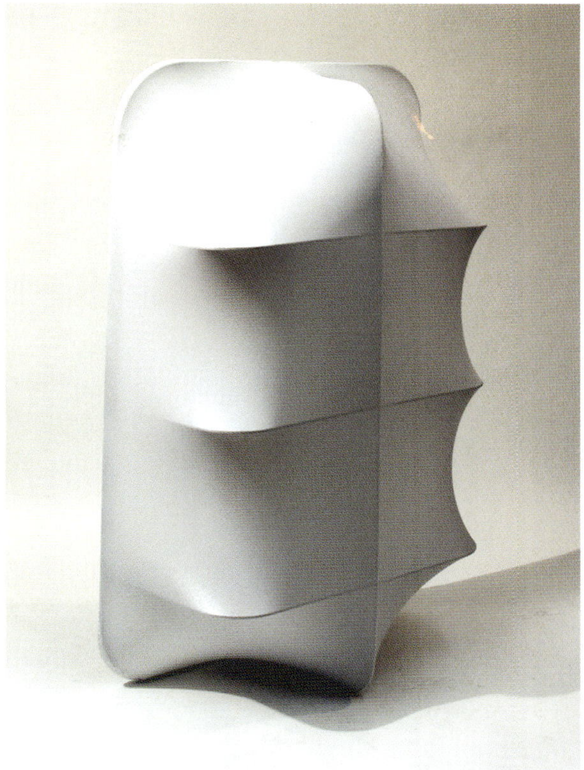

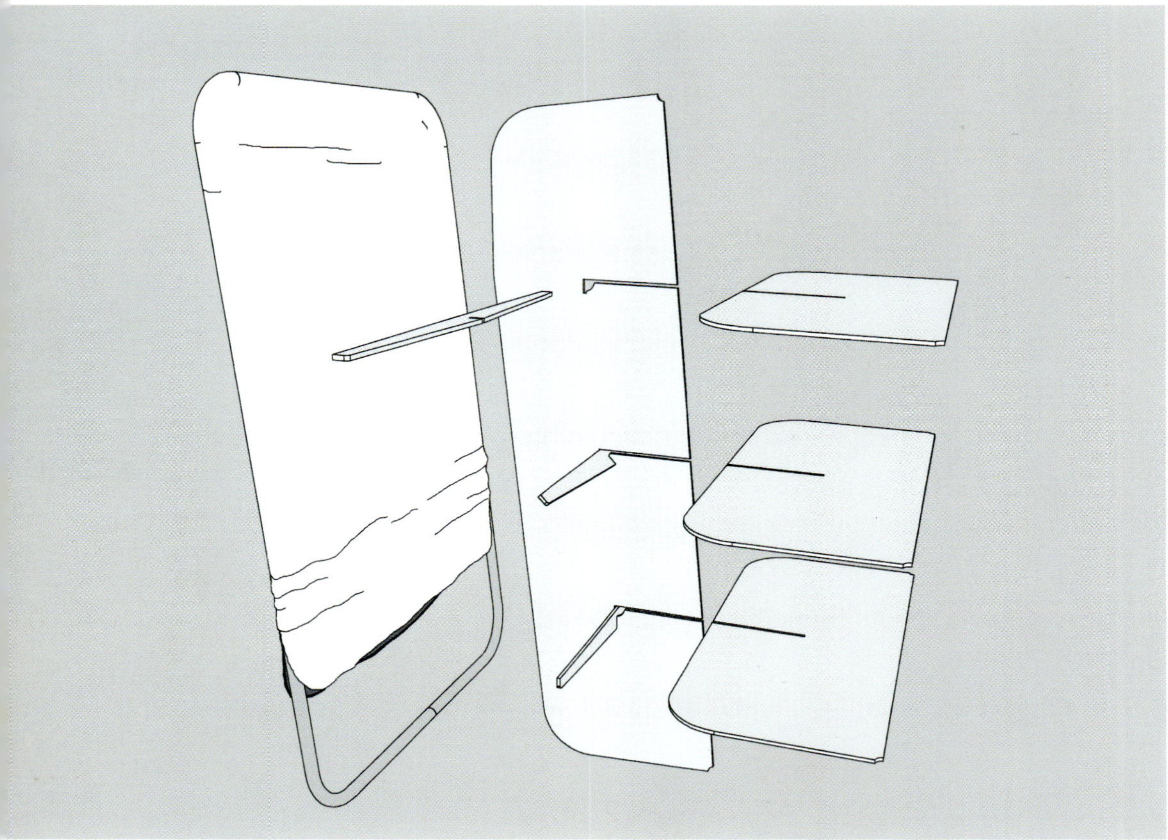

Projekt Project
„autumn leaves"
Offenes Regal
Open shelf

Design
Rainer Saner
Schreiner Techniker-Schule Zug (STZ)
Zug, Switzerland
rainer@bluewin.ch

Das offene Regal „autumn leaves" kommuniziert sein Innenleben nach außen und überträgt seine Funktion auf seine Hülle. Daraus resultiert die organische Formgebung. Das Regal wird an eine Wand angelehnt oder frei im Raum aufgestellt. Die Hülle aus durchscheinendem Stretchstoff schmiegt sich an die Tragestruktur des offenen Regals an. Durch die Spannung, welche das dehnbare Textil auf die Konstruktion ausübt, erhält das Regal seine Stabilität.

The open "autumn leaves" shelves communicate their inner life to the outside world – transposing their function onto their outer skin. The result is an organic design. The shelves can be leant against a wall or stood alone in a room. The skin is made of translucent stretch fabric and snuggles up to the fixed structure of the open shelves. The tension the textile exerts on the frame ensures the shelves are very stable.

Projekt Project
Alarm Clock
Wecker mit Timer-Mechanismus
Timer-mechanism alarm clock

Design
Itay Potash
The Bezalel Academy of Arts and Design
Jerusalem, Israel
mail@itaypotash.com

Ein Wecker für Menschen, denen es besonders schwer fällt aufzustehen. Der Wecker erlaubt es dem Nutzer, mithilfe eines einfachen Timer-Mechanismus, die gewünschte Schlafzeit einzustellen. Ein integrierter Lautsprecher erzeugt einen lauten Weckruf. Die konische Form des Weckers ist inspiriert durch die Ausweitung von Schallwellen im Raum und nimmt Bezug auf die Form der Lautsprechermembrane. Durch die Reduktion auf die Weckfunktion entsteht ein einfaches und anwenderfreundliches Produkt. Der asymmetrische Stahlbügel verleiht dem Wecker einen flexiblen Stand. Auch bei grober Handhabung ist somit die Stabilität des Produktes gewährleistet.

An alarm clock for everybody who finds it extra difficult to get up. It boasts a simple timer mechanism by which the user can set the desired sleeping period. An integrated loudspeaker then emits a loud wake-up call. The alarm clock's conical shape takes itS cue from the way sound waves spread in space and reflects the form of the loudspeaker membrane. Owing to its exclusive concentration on the wake-up function, the product is very easy to use. The asymmetrical steel arc functions as a very stable stand for the alarm clock and delivers stability even if the clock is given a tough time of it.

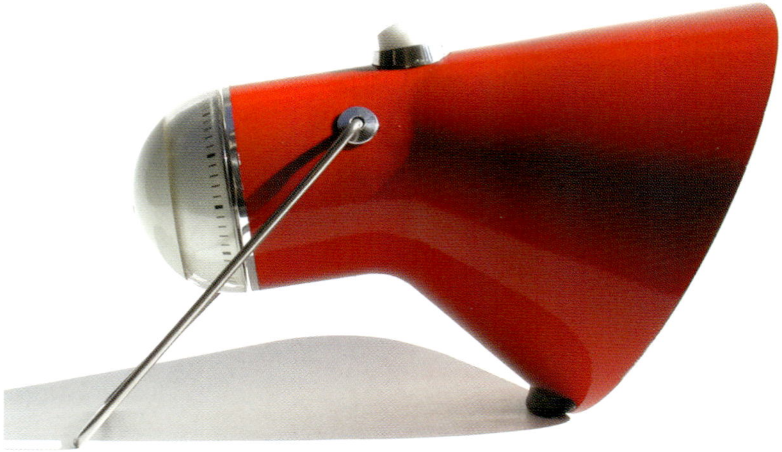

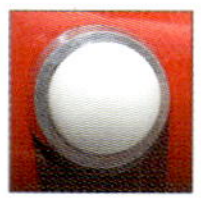

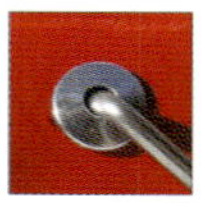

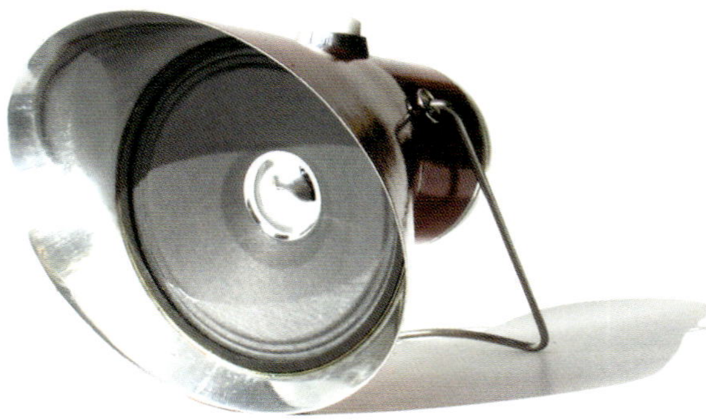

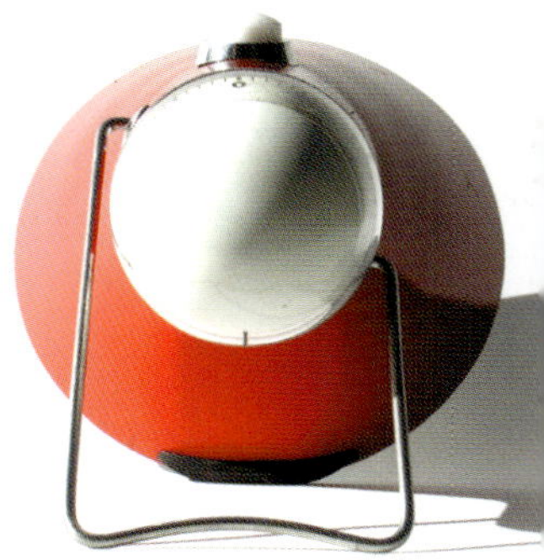

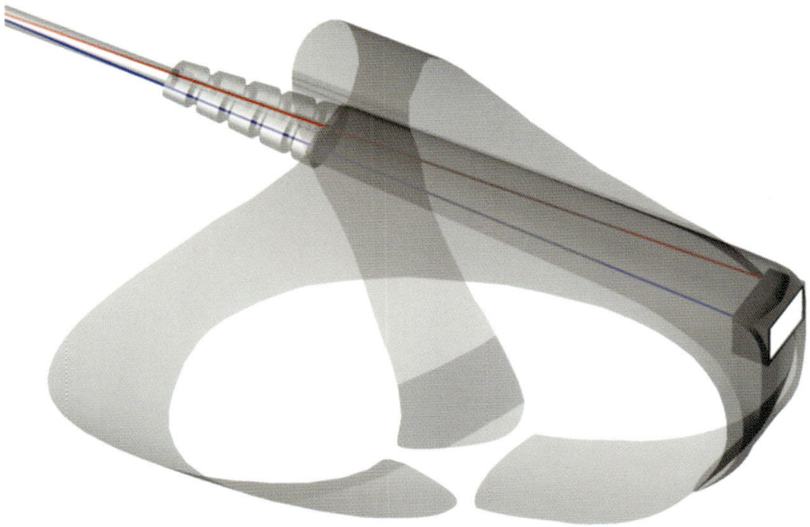

Projekt Project
Mobile LED-Leseleuchte
Mobile LED reading luminaire

Design
Thomas Loy
Hochschule der Künste Bremen
Bremen, Germany
designloy@aol.com

Die mobile und lichtstarke LED-Leseleuchte ist leicht zu handhaben und dient als Notleuchte oder auch multifunktionaler Reisebegleiter. Sie ist äußerst leicht, kompakt und robust und passt sich den anatomischen Gegebenheiten an. Die im Nylon-Armband integrierten Akkus bauen den Stromkreis erst auf, wenn die zwei Klettverschlüsse, die mit Kontakten versehen sind, aufeinander haften, also beim Schließen des Armbandes um das Handgelenk. Die Verwendung einer LED als Lichtquelle garantiert eine lange und wartungsfreie Lebensdauer.

The mobile and powerful LED reading luminaire is easy to handle and can serve as an emergency light or as a multifunctional travel companion. It is especially light in weight, compact and robust, while adapting to the user's anatomy. The battery pack housed in the nylon armband first actuate the power circuit if the two Velcro fasteners (there are contacts inside them) are locked together – i.e., when the armband is fastened round your wrist. The choice of LEDs as a source of light guarantees a long, maintenance-free service life.

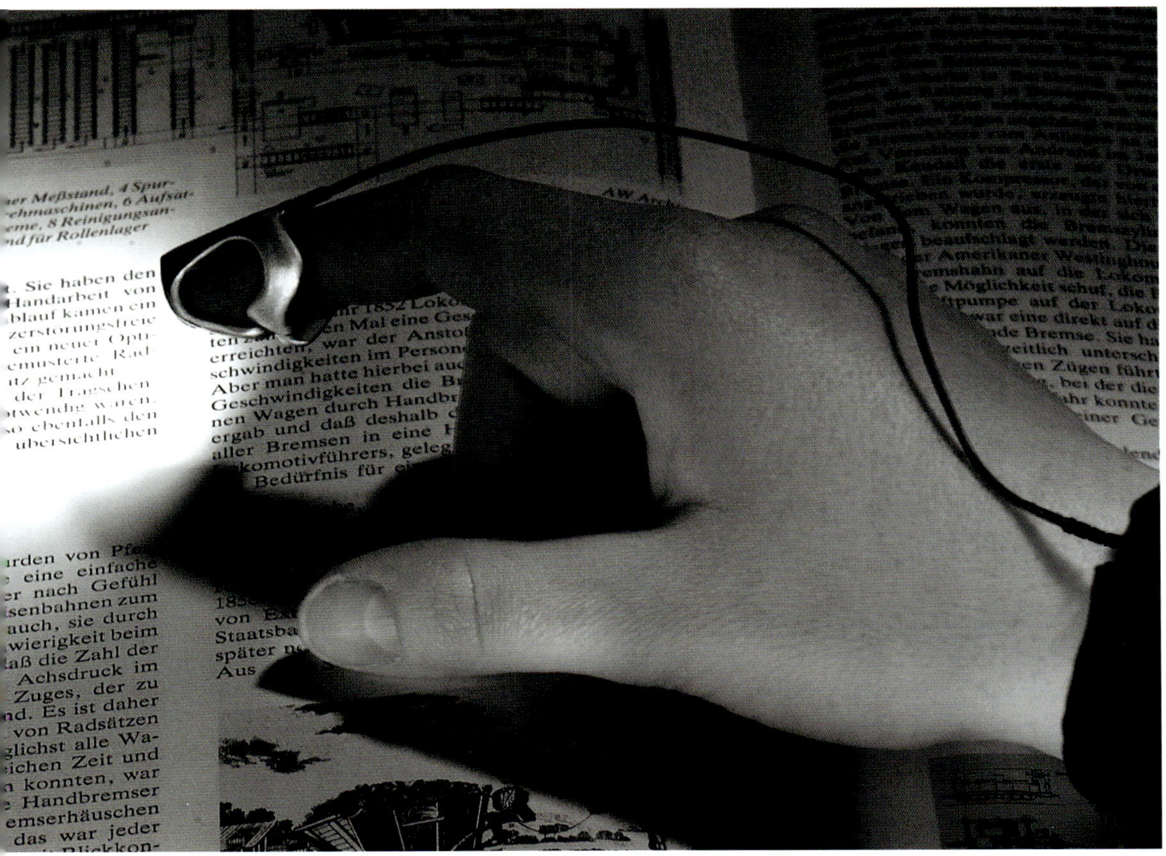

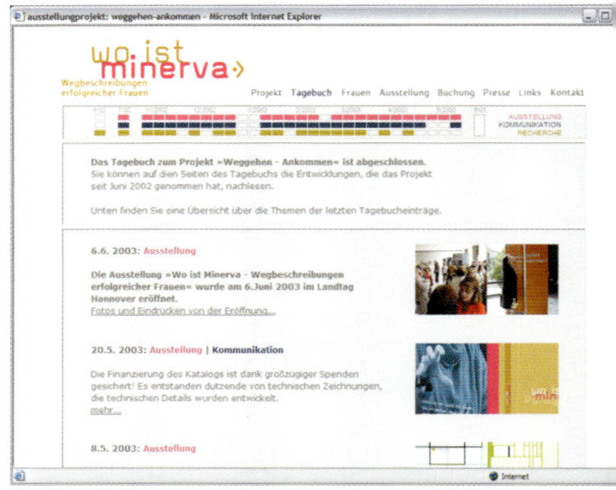

Projekt Project
Wo ist Minerva –
Wegbeschreibungen erfolgreicher Frauen
Wanderausstellung mit Begleitkommunikation
Travelling exhibition and communications materials

Design
Anke Erdmann, Britta Freise, Maren Meßmer
Stefanie Pille, Ute Sauvigny, Melissa Seelig
Michael Seifert
Hochschule der Bildenden Künste
Braunschweig, Germany
anke_erdmann@web.de
ma@mafive.de
stefanie.pille@gmx.de
utesauvigny@hotmail.com
melissa.seelig@gmx.de
michael@msii.de

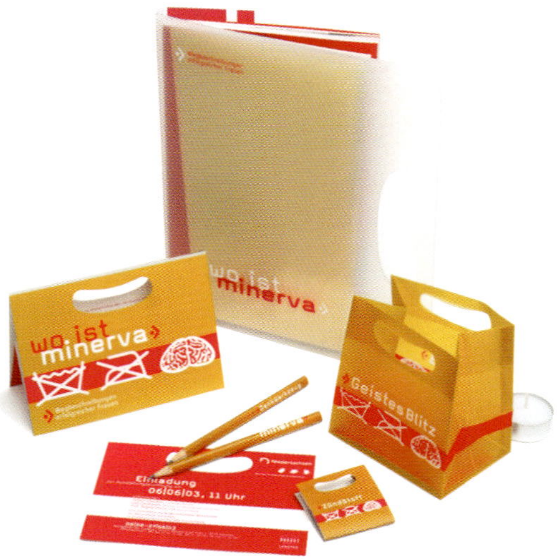

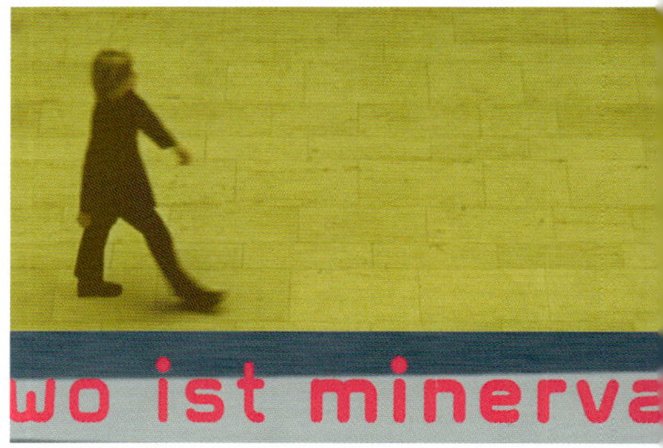

Die Wanderausstellung präsentiert dreizehn Wissenschaftlerinnen und Künstlerinnen aus der niedersächsischen Hochschullandschaft. Ihre Lebenswege und Werke werden durch individuelle Inszenierungen mit interaktiven Entdeckungen spannend und anschaulich den Besuchern näher gebracht. Die Form und das Konzept der Stationen ist Schrankkoffern nachempfunden, diese können innerhalb der Ausstellung aufgeklappt und für den Transport einfach zusammengeklappt werden. Am Anfang der Ausstellung wird zur Einstimmung ein Film gezeigt sowie Interviewzusammenschnitte, die beide von der Gruppe produziert wurden. Zusätzlich gibt es begleitendes Kommunikationsmaterial von der Einladungskarte bis zum Katalog sowie zur Dokumentation des Projekts ein Webtagebuch.

The traveling exhibition presents 13 female scientists and artists at home in universities and academies in Lower Saxony. Individual mis-en-scenes featuring interactive elements familiarize visitors with their lives and oeuvres in an exciting and visually striking manner. The shape and concept behind the stages of the exhibition is based on wardrobe trunks; during the exhibition they can be opened out, and then simply folded back together ready for transportation. At the beginning of the exhibition, a film is screened and sections of recorded interviews played to acquaint visitors with what awaits them; both were produced by the group. In addition, there is accompanying communications materials – ranging from invitation cards through catalogues to project documentation in the form of a Web log.

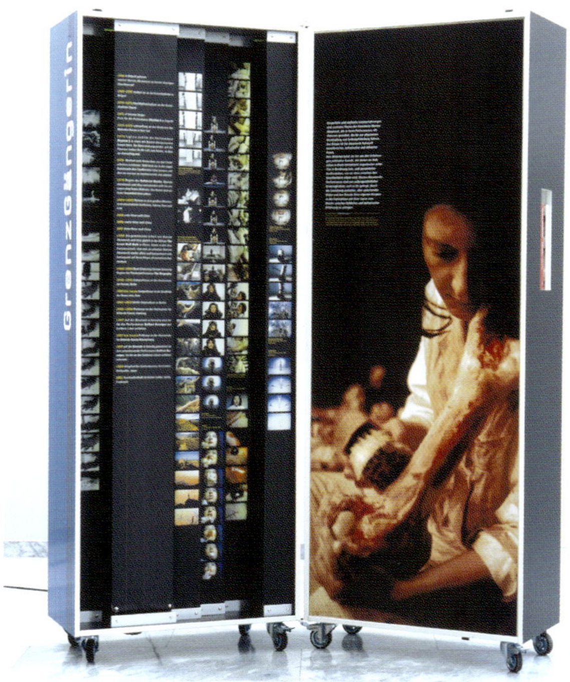

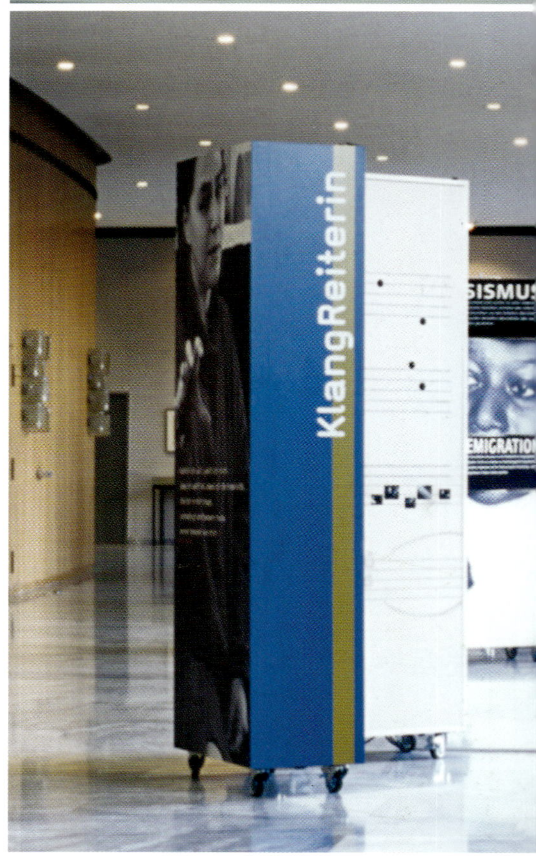

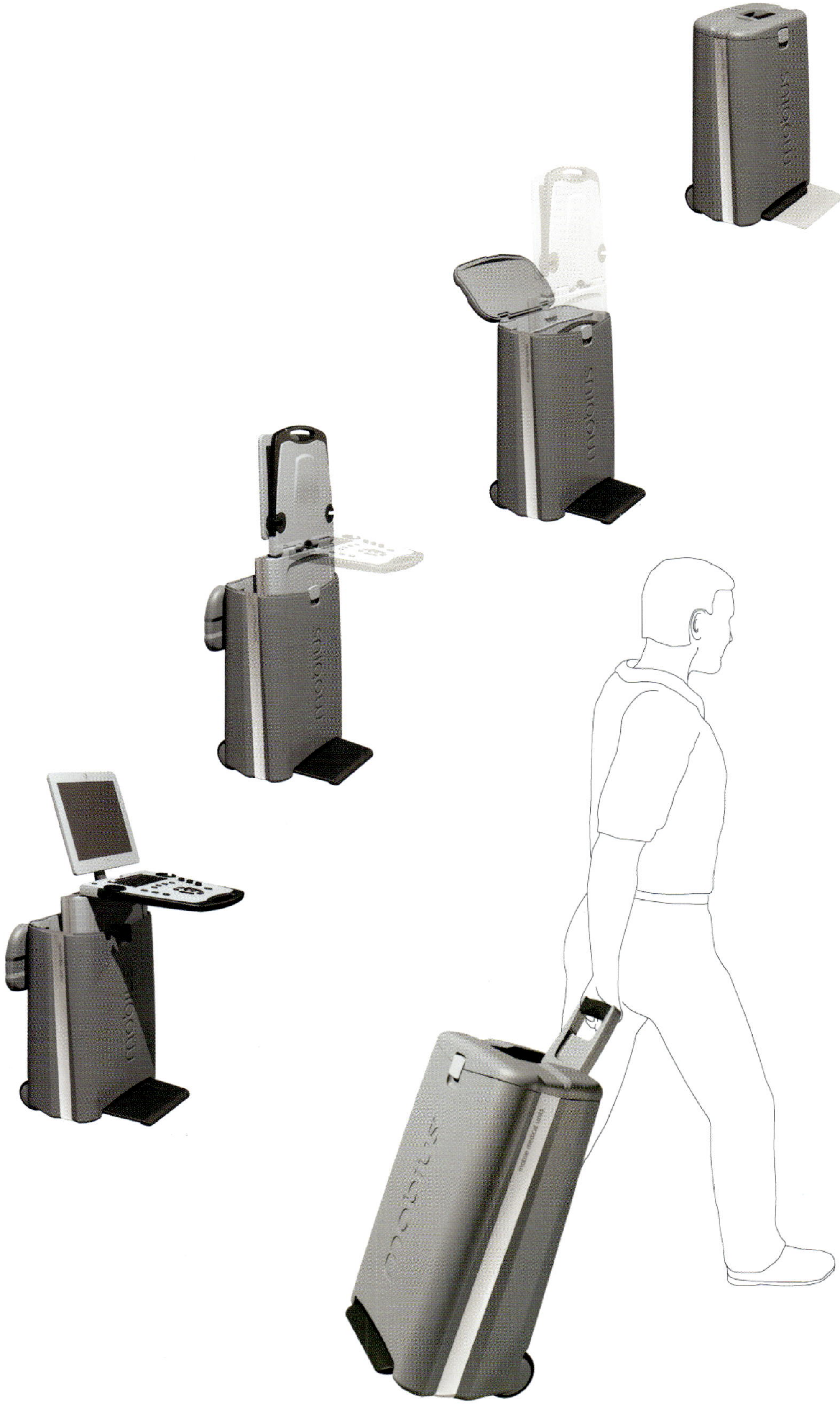

Projekt Project
Mobius
Mobile telemedizinische Ultraschall-Arbeitsstation
Portable telemedical ultrasonic workstation

Design
Claus Jepsen
The Aarhus School of Architecture
Aarhus, Denmark
jepsenclaus@hotmail.com

Mobius ist eine tragbare telemedizinische Ultraschall-Arbeitsstation, mit der 3D-Ultraschallbilder gemacht werden können. Diese besondere Art von Ultraschallbildern ermöglicht die Ausführung von komplizierten Untersuchungen auch durch verhältnismäßig unerfahrene Mitarbeiter des Gesundheitswesens. Mobius kann Daten wie Schall, Bilder und Text absenden und empfangen. Mittels dieser modernen Kommunikationsmöglichkeit können professionelle Mitarbeiter des Gesundheitswesens wie z.B. Krankenpfleger in abgelegenen Orten Experten um Rat fragen, damit eine Diagnose bestätigt und eine Behandlung vorgeschlagen werden kann.

Mobius is a portable telemedical ultrasonic workstation that provides 3D ultrasound images. This special kind of ultrasonic imaging enables complex examinations to be carried out even by relatively inexperienced healthcare workers. Mobius can transmit and receive data such as sound input, images and texts. Thanks to this modern communications feature, healthcare professionals, such as nurses or assistants, based in remote areas can consult experts, seek their advice in support of a diagnosis or as regards possible treatment options.

Projekt Project
GoYo
Hybridrad
Hybrid bicycle

Design
Johanna Tiffe
Kunsthochschule Berlin-Weißensee
Berlin, Germany
johanna_tiffe@web.de

GoYo ist Fahrrad und Pedelec in einem. Zielgruppe sind Privatpersonen ab ca. 30 Jahren. Das Hybridrad vereint aufgrund des eigens konzipierten, abnehmbaren Antriebs die Funktionen von Fahrrad und Pedelec. Neben dem Antrieb wurden für die Studie eine Vollfederung, ein Sicherheitssystem und weitere Komponenten entwickelt. Bei der Rahmengestaltung ließ sich faserverstärkter Kunststoff innovativ einsetzen.

GoYo is a two-in-one bicycle and Pedelec. It is aimed to appeal to private individuals aged 30 or upward and, thanks to the specially developed, removal drive train is a hybrid vehicle that combines the functions of a bicycle and a Pedelec. Alongside the drive train, as part of the study a new full suspension system, a security system and other components were all devised. The frame makes innovative use of fiber-reinforced plastic.

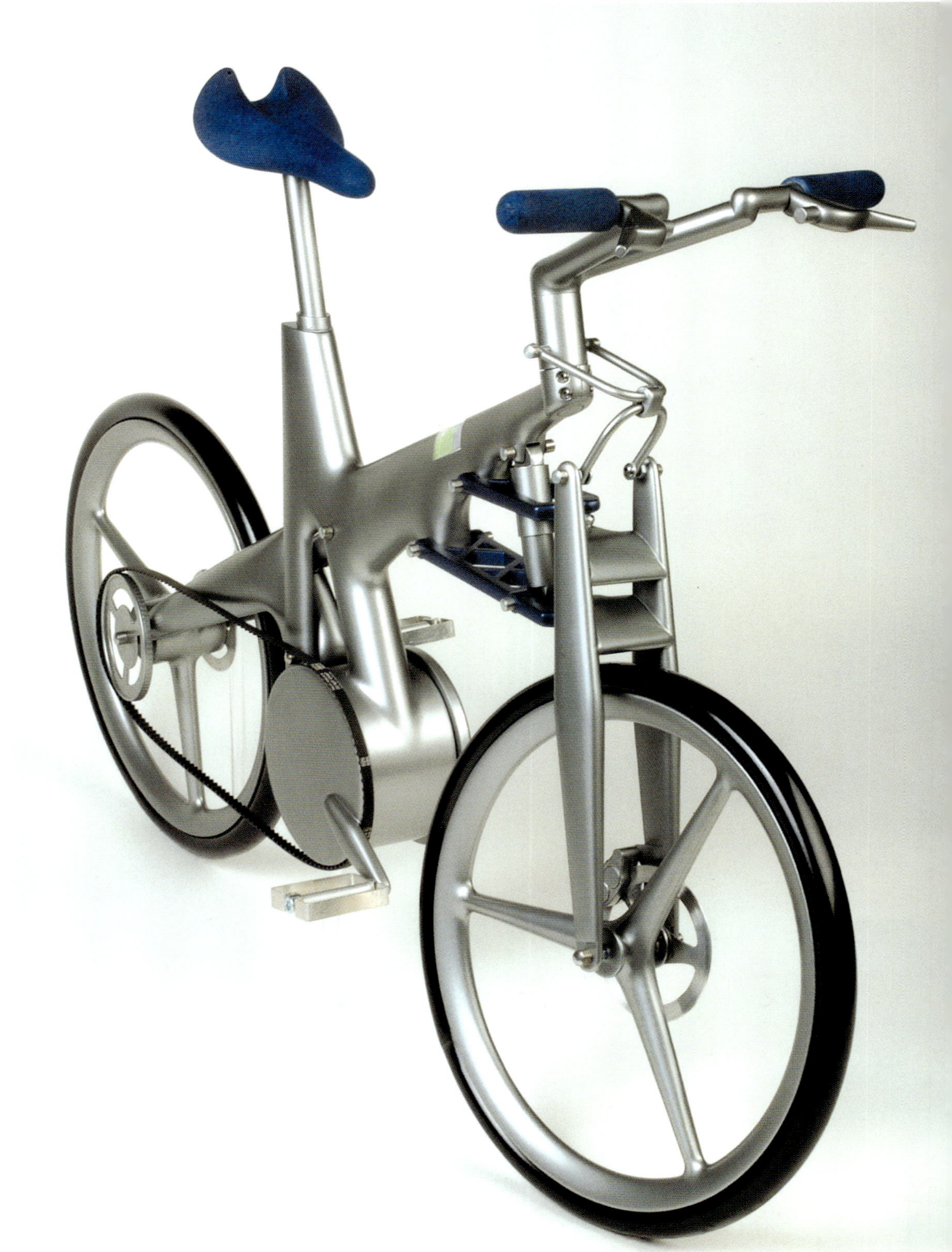

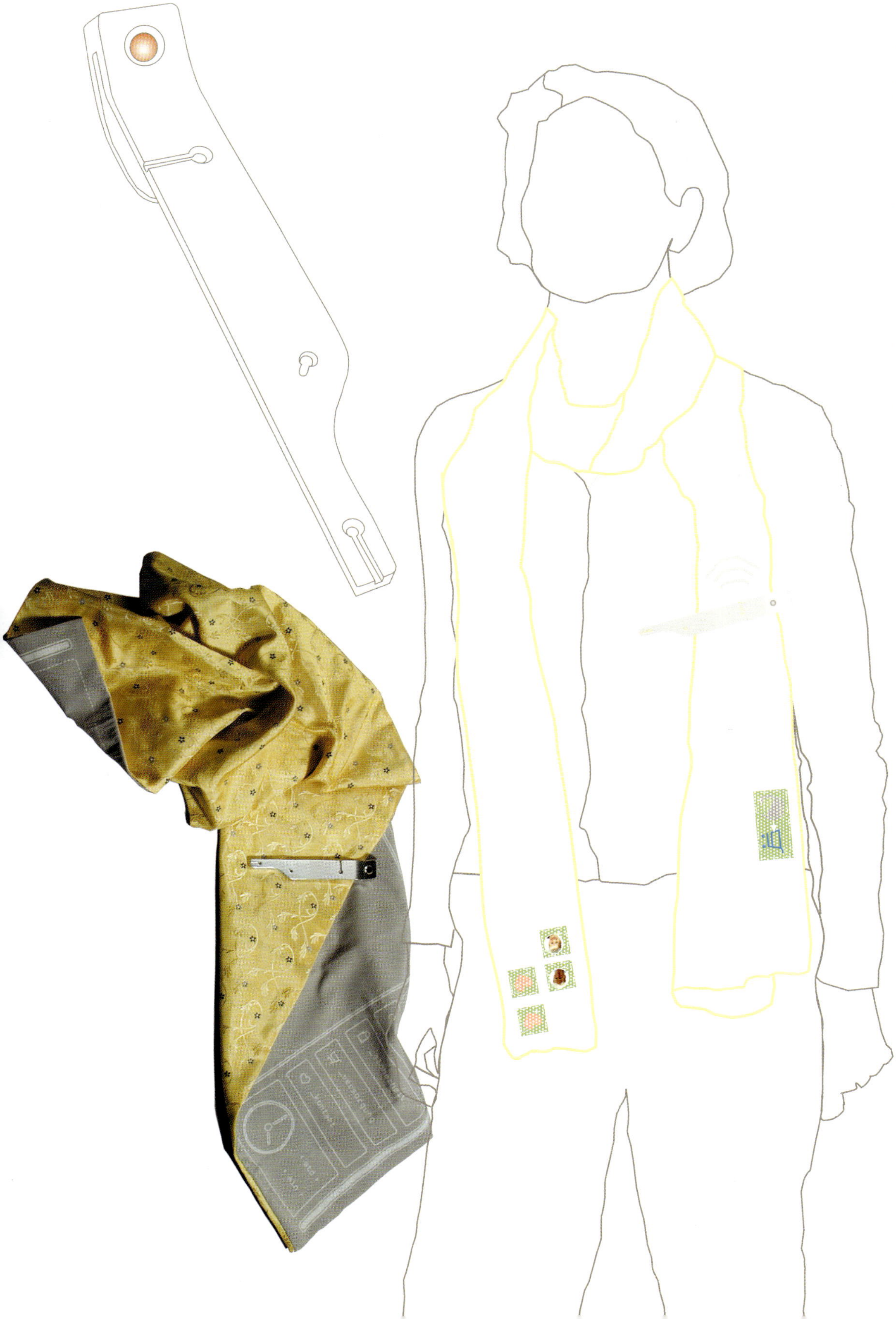

Projekt Project
_linus
Kommunikationseinheit und tragbares Interface
Communications unit and wearable interface

Design
Christian Lungershausen
Fachhochschule Darmstadt
Industrial Design
Darmstadt, Germany
lungershausen@rennsalon.net

_linus ist ein textiles tragbares Kommunikationssystem, eine Lese- und Gedächtnishilfe für die Zielgruppe der älteren Generation, die sich mit digitaler Technik mitunter schwer tut. Zum System gehören das tragbare Interface, ein dekoratives Kommunikations-Tool in Form eines Clips in Stiftform und verschiedene Speicherkarten, auf denen individuelle Informationen abgespeichert werden können. Die mit dem Interface und dem Clip interagierenden Speicherkarten werden auf der Kleidung positioniert und durch Druck ausgewählt. Das Kommunikations-Tool ist mit einer Kamera, einer Lesefunktion, einem Mikrofon und Lautsprecher ausgestattet und dient als Gerät zur Spracheingabe und -ausgabe. Mit dem Stiftkopf kann auf den druckempfindlichen Screens und den Speicherkarten auch geschrieben werden.

_linus is a portable textile communication system that is designed to be a reading and memory aid for silver-agers and older who at times have their difficulties with digital technology. The system includes a portable interface, a decorative communications tool in the form of a pen-shaped clip, and various memory cards onto which individual information can be downloaded and stored. The memory cards interact with the interface and the clip and are positioned on the fabric and then selected by pressing. The communications tool comes with a camera, a reading function, a microphone and a loudspeaker, thus delivering voice input and output. The pen head can be used to write on the pressure-sensitive screens and the memory cards.

Projekt Project
MOTOCOOPER
Mobiltelefon
Mobile phone

Design
Dirk Stölting
Muthesius Hochschule
Fachhochschule für Kunst und Gestaltung
Kiel, Germany
dirks_@gmx.de

Das Mobiltelefon MOTOCOOPER verkörpert mit seinen fließenden Linien und der sportlichen Eleganz das harmonisch organische Design des BMW Mini. Vier klar voneinander abgesetzte Schalenelemente spiegeln die Formensprache des Kompaktwagens wider. Überwölbte Linien und Mini-typische Kreiselemente stehen im Kontrast zu horizontalen und vertikalen Geraden. Die abgedunkelte hochglänzende Acrylglasfläche, die sich über die gesamte Vorderansicht des MOTOCOOPER erstreckt, unterstützt den Wiedererkennungswert der Mini-Silhouette.

The MOTOCOOPER mobile phone epitomizes the harmonious organic design of the BMW Mini V. Its flowing lines and sporting elegance support four clearly distinguished case sections that reflect the formal idiom of the BMW Mini. Bulging lines and typically Mini circular elements contrast starkly with horizontal and vertical lines. The darkened highly polished acrylic glass that covers the entire front side of the MOTOCOOPER enhances the unmistakable image of the Mini's silhouette.

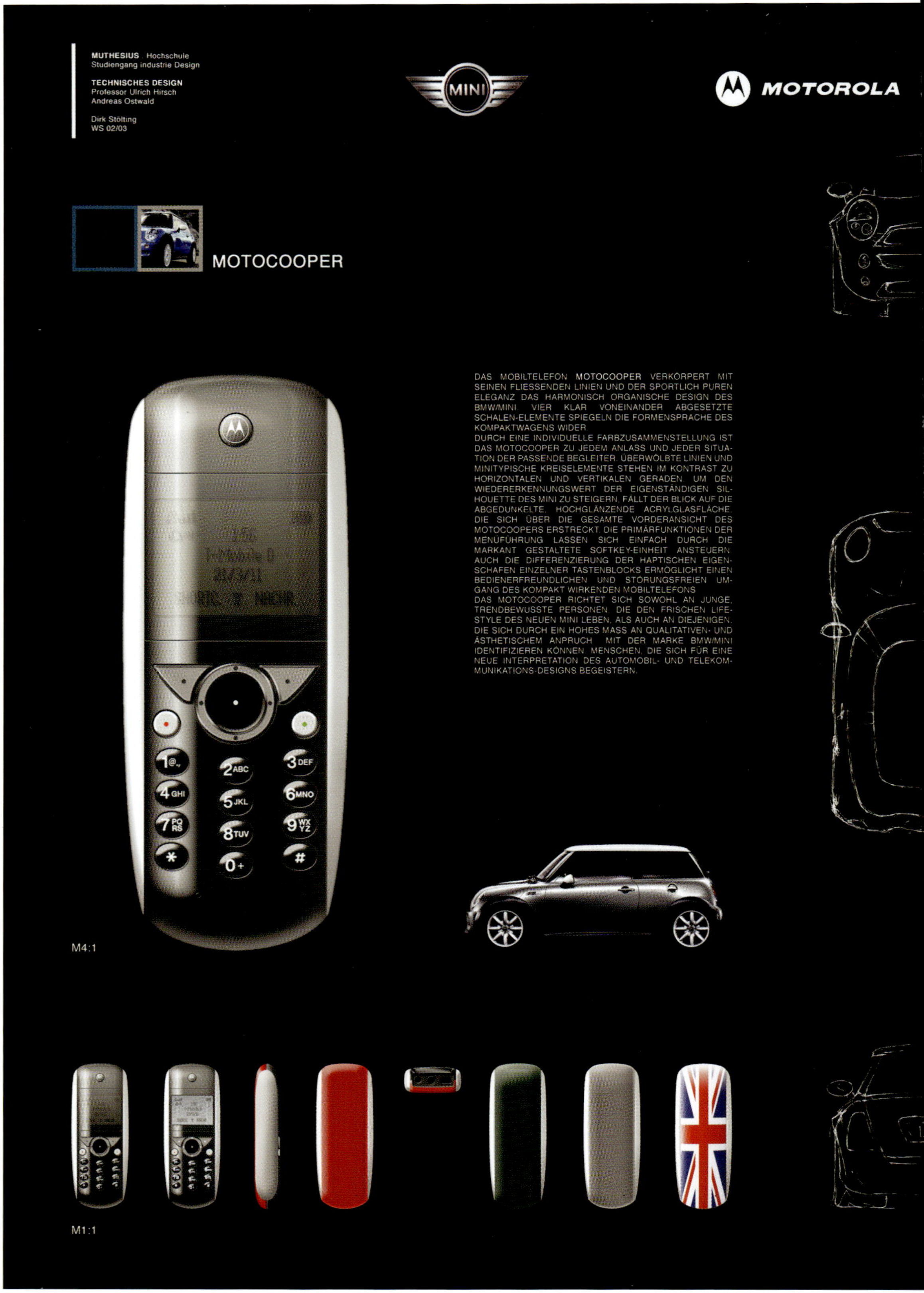

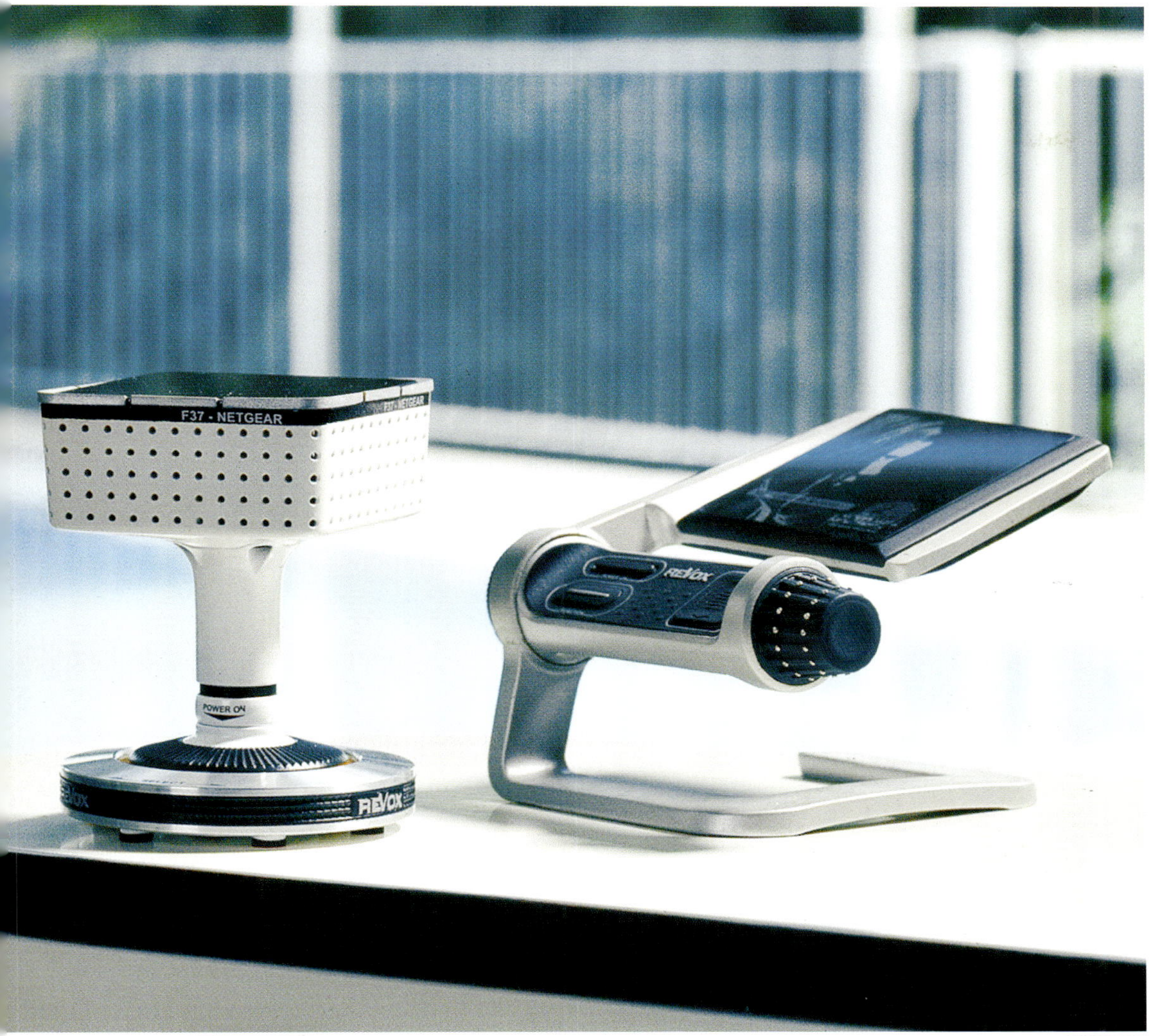

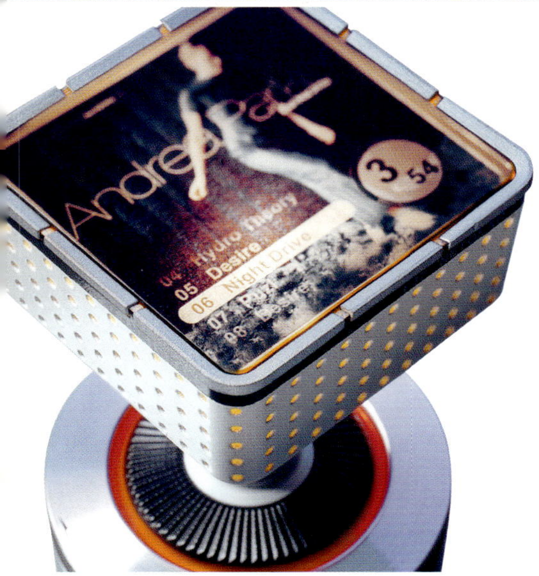

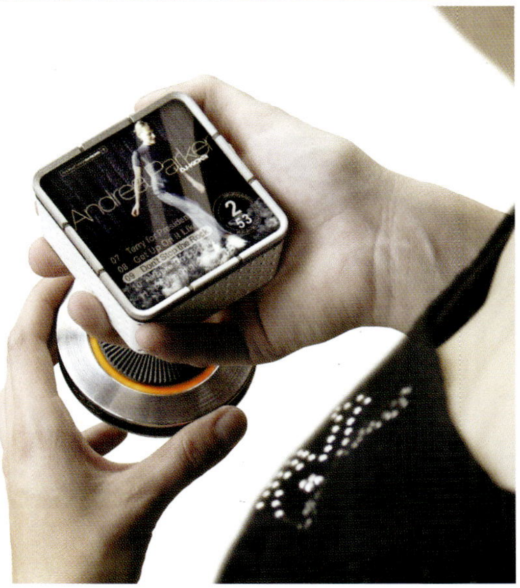

Projekt Project
Netgear F 36/F 37
Vision für den spontanen Umgang mit Home Entertainment-System
Vision for a spontaneous handling of a home entertainment system

Design
Sam Stalder
Fachhochschule Aargau
Aarau, Switzerland
samstalder@gmx.net

Netgear ist eine kabellose, handliche Art Mini-Juke-Box, die frei im Raum platziert werden kann. Die persönliche Musiksammlung ist auf diese Weise im ganzen Haus zur Hand. Der Netgear ermöglicht den Zugang zum Hausserver mit der gespeicherten Musik. Die Wahl des Albums erfolgt über das Durchblättern von Originalcovern. Die Aktivlautsprecher zur Musikwiedergabe sind in den gewünschten Räumen fest installiert. Mit F37 und F36 stehen zwei unterschiedliche Stilvarianten zur Verfügung.

Netgear is a handy cordless form of miniature juke box that can be used as a standalone. In this way, your personal music collection is readily at hand throughout the house. Netgear delivers access to your house server to which the music has been stored. Albums are selected by leafing through the original covers. The active loudspeakers that play the music are firmly installed in whatever rooms you want. There are two variations with differing styles: the F37 and the F36.

Projekt Project
filou
Multifunktionaler muskelbetriebener Roller
Multifunctional muscle driven scooter

Design
Roland Hein, Stefan Espenhahn
HS Anhalt
Dessau, Germany
loopdesign@web.de

filou ist ein mechanisches Fortbewegungsmittel, welches in drei Faltvarianten unterschiedliche Funktionen erfüllt. Als Roller besitzt filou die Fähigkeit seinen Benutzer über Kurzdistanzen, z. B. von Zuhause zur U-Bahn, durch die Fußgängerzone oder zum Supermarkt um die Ecke zu befördern.Zusammengeklappt kann er, ohne sperrig oder zu schwer (4,2 kg) zu werden, in jedem Nahverkehrsmittel mitgenommen werden. In der Position Sackkarre offenbart er die Möglichkeit, den kleinen Einkauf inklusive eines Kasten Wassers zu transportieren.

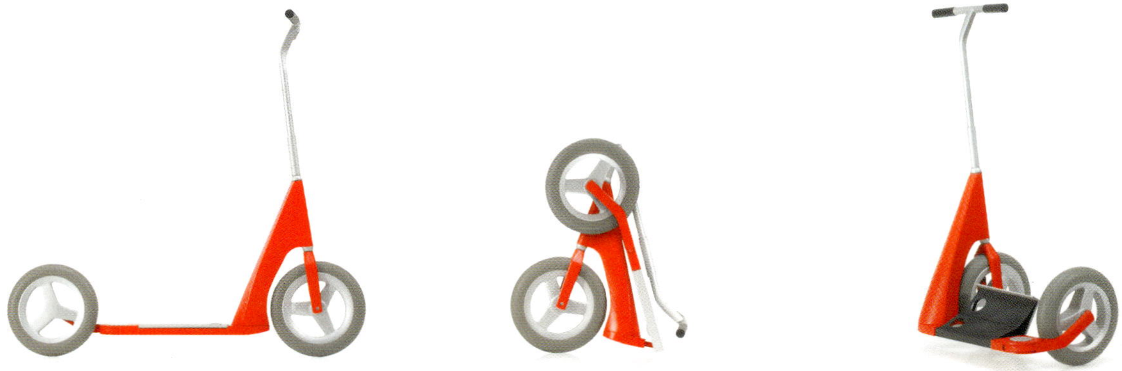

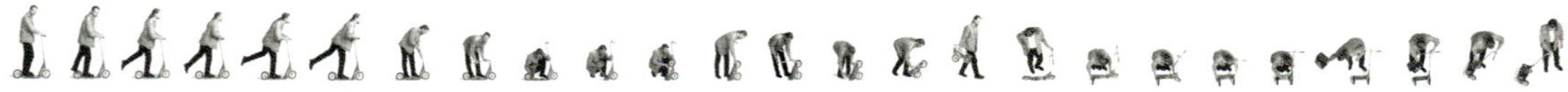

filou is a mechanical means of transport. It can be folded in three ways to provide respectively different functions. As a scooter, filou whisks its user across short distances, such as from home to the subway, through a pedestrian zone or to the supermarket round the corner. In the carry-on position it can be taken along for the ride on the subway or bus – and is neither bulky or heavy (weighing only 4.2kg). Folded into a cart, it enables small purchases and even a crate of mineral water to be got home with ease.

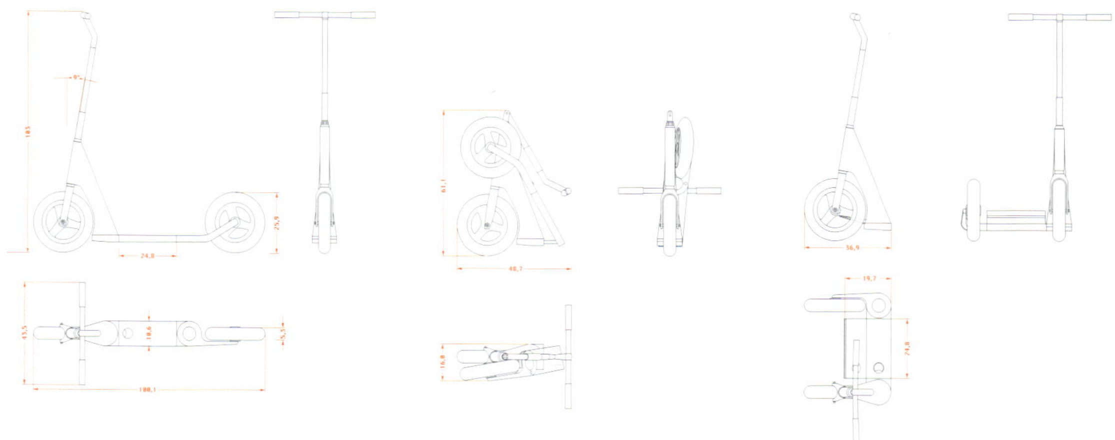

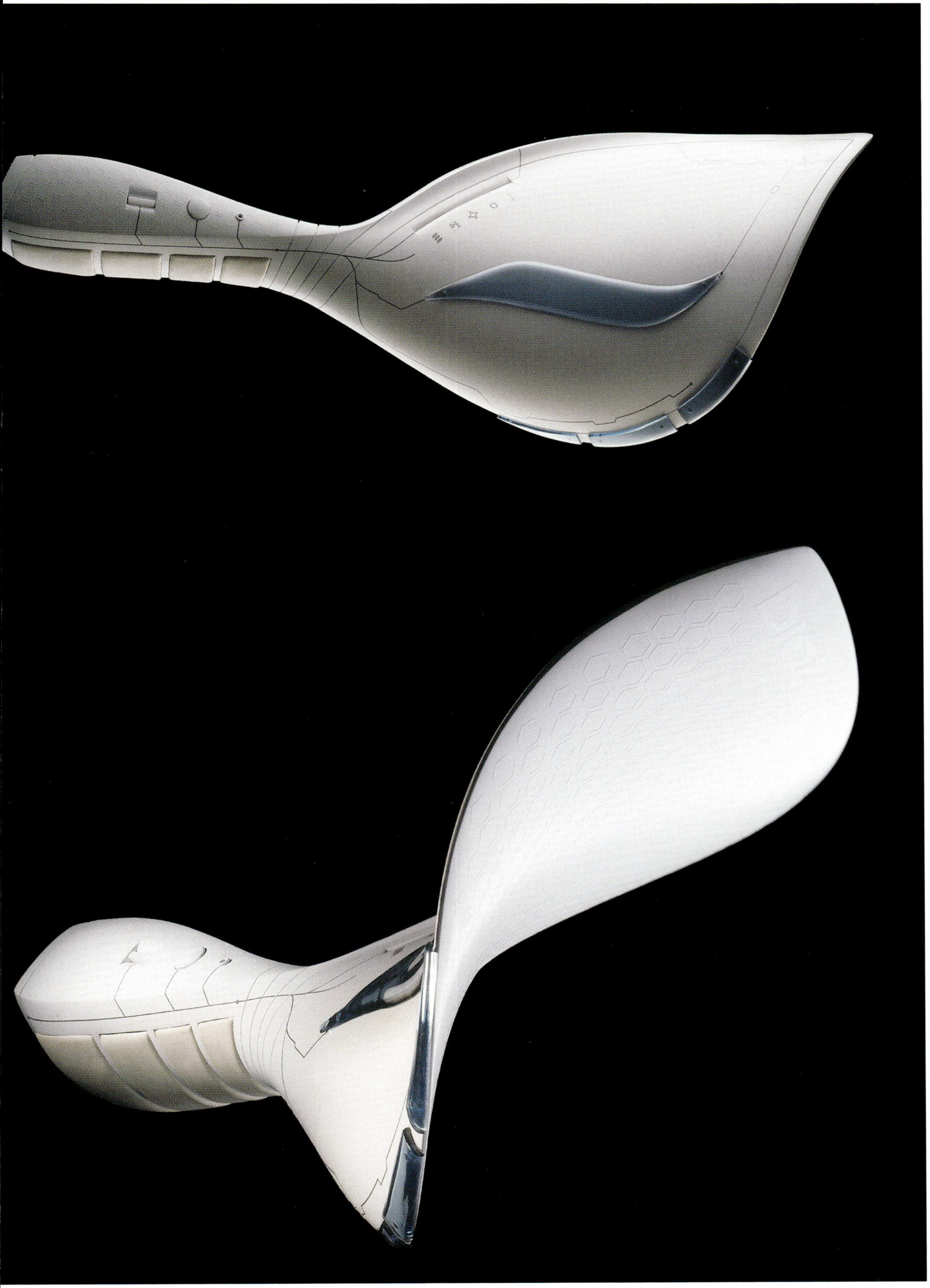

Projekt Project
Biotechsound
Digitales Musikinstrument
Digital music instrument

Design
Jörn Vicari
Muthesius Hochschule
Fachhochschule für Kunst und Gestaltung
Kiel, Germany
www.vicari-design.com

Das digitale Musikinstrument Biotechsound will in Form und Handhabung die Qualität klassischer akustischer Musikinstrumente mit der Vielfalt digitaler Möglichkeiten zusammenführen. Das neuartige Instrument bietet verschiedene Einsatzmöglichkeiten vom Eingabemedium im Verbund mit Profi-Equipment, über Live-Auftritte bis hin zum legeren Musizieren auf dem Sofa. Die organische Form übersetzt, wie der Name Biotech schon sagt, die Schönheit alter, aus Naturmaterialien gebauter Musikinstrumente in eine moderne Sprache und vermittelt dem Benutzer auf diese Art und Weise ein besonderes sinnliches Erlebnis.

In terms of its shape and how it is used, Biotechsound is a digital musical instrument that wishes to marry the quality of classical acoustic instruments with the diversity of digital possibilities. This new type of instrument can be utilized in a variety of ways – from an input medium together with pro-equipment, via live performances through to casual music playing on the couch. As the name reveals, the organic shape translates the beauty of the musical instrument of yore, which were made of natural materials, into a modern idiom and in this manner offers users an especially sense-arousing experience.

Vorstand
Board:

01. Ernst Raue, Deutsche Messe AG, Hannover, 1. Vorsitzender iF Hannover.

02. Norbert H. Bargmann, Messe München GmbH.

03. Christoph T. Böninger, designafairs GmbH, München.

04. Jürgen Fitting, Friedrich Grohe AG & Co.KG, Hemer.

05. Fritz Frenkler, f/p design GmbH, Frankfurt.

06. Prof. Harald Leschke, DaimlerChrysler AG, Sindelfingen.

07. Ralph Wiegmann, iF Hannover, geschäftsführendes Vorstandsmitglied.

Geschäftsführung
Managing Director:

Ralph Wiegmann, iF Hannover

Ordentliche Mitglieder
Full Members:

01. Norbert H. Bargmann, Messe München GmbH, München.
02. Knut Bliesener, Designstudio Bliesener, Hannover.
03. Christoph T. Böninger, designafairs GmbH, München.
04. Jürgen Fitting, Friedrich Grohe AG & Co.KG, Hemer.
05. Fritz Frenkler, f/p design GmbH, Frankfurt.
06. Prof. Harald Leschke, DaimlerChrysler AG, Sindelfingen.
07. Alexander Neumeister, Neumeister Design, München.
08. Ernst Raue, Deutsche Messe AG, Hannover.
09. May-Britt Pürschel, Nieders. Ministerium f. Wirtschaft, Technologie u. Verkehr, Hannover.
10. Horst B. Scherer, CI + Messe VDID/DDV, Böblingen.
11. Prof. Volker Weinert, HBK Braunschweig.
12. Ralph Wiegmann, iF Hannover.

Ehrenmitglieder
Honorary Members:

01. Fritz Hahne, Hannover.
02. Sepp D. Heckmann, Hannover.
03. Karl-Heinz Krug, Leverkusen.
04. Prof. Herbert Lindinger, Hannover.
05. Edwin A. Schricker, Rodach.
06. Herbert H. Schultes, Fürstenfeldbruck.
07. German A. Voment, Hirschbach.

Fördernde Mitglieder Inland
German Associate Members:

01. Aha Design Jürgen Aha	Steinkleistraße 27	60435 Frankfurt
02. AL-KO GmbH	Ichenhauser Straße 14	89359 Kötz
03. Ars Nova Collection	Salzstraße 10	49326 Melle
04. Artemide GmbH	Hans-Böckler-Straße 2	58730 Fröndenberg
05. Audi AG	Auto-Unionstraße	85045 Ingolstadt
06. Bang & Olufsen Deutschland GmbH	Rudolf-Diesel-Straße 8	82205 Gilching
07. BEGA Gantenbrink-Leuchten GmbH & Co.	Hennenbusch	58708 Menden
08. Blue C. Verlag GmbH, design report	Neckarstraße 121	70190 Stuttgart
09. Braun GmbH Corporate Design	Frankfurter Straße 145	61476 Kronberg
10. Braun Wagner	Krefelder Straße 147	52070 Aachen
11. BREE Collection GmbH & Co. KG	Gerberstraße 3	30916 Isernhagen/Kirchhorst
12. Bulthaup Gmbh & Co. KG	Werkstraße	84153 Aich
13. Crown Gabelstapler GmbH	Moosacher Straße 52	80809 München
14. DaimlerChrysler AG		71059 Sindelfingen
15. DDC Deutscher Designer Club e.V.	Hanauer Landstraße 139	60314 Frankfurt
16. designafairs GmbH	Tölzer Straße 2c	81379 München
17. design studio hartmut s. engel	Monrepos-Straße 7	71634 Ludwigsburg
18. Design Tech	Zeppelinstraße 53	72119 Ammerbuch
19. Eschke, Christoph	Admiralitätsstraße 10	20459 Hamburg
20. FESTO AG & Co. Corporate Design	Rechbergstraße 3	73770 Denkendorf
21. FOLIA TEC Böhm GmbH + Co. Vertrieb KG	Neumeyerstraße 70	90411 Nürnberg
22. Formfürsorge	Am Hohen Ufer 3A	30159 Hannover
23. FOTOCENTRUM Zimmermann GmbH	Vahrenwalder Str. 263	30179 Hannover
24. Friedrich Grohe AG & Co. KG	Hauptstraße 137	58675 Hemer
25. GIRA Giersiepen GmbH & Co. KG	Dahlienstraße 12	42477 Radevormwald
26. Grundig Vertriebs GmbH	Beuthener Straße 41	90471 Nürnberg
27. Heidelberger Druckmaschinen AG	Kurfürstenanlage 52-60	69115 Heidelberg
28. IBM Deutschland GmbH	Pascalstraße 100	70548 Stuttgart
29. i / i / d Institut für Integriertes Design	Am Speicher XI	28217 Bremen
30. Interstuhl Büromöbel GmbH & Co. KG	Brühlstraße 21	72469 Meßstetten-Tieringen
31. Köttermann GmbH & Co.	Industriestraße 2-10	31311 Uetze-Hänigsen
32. Kühne, Karl-Michael	Oelkersallee 31	22769 Hamburg
33. Lichtwark, Hildegund	Pfarrer-Linzbach-Straße 1	52388 Nörvenich
34. LOEWE OPTA GmbH	Industriestraße 11	96317 Kronach
35. MEYER-HAYOZ Design Engineering	Zollernstraße 26	78462 Konstanz

Fördernde Mitglieder Inland
German Associate Members:

36. mono-Metallwarenfabrik Seibel GmbH	Industriestr. 5	40822 Mettmann
37. Moormann Möbel GmbH	An der Festhalle 2	83229 Aschau i. Chiemgau
38. OCO-Design O.K. Nüsse	An der Kleimannbrücke 79	48157 Münster
39. OCTANORM Vertriebs-GmbH	Raiffeisenstraße 39	70794 Filderstadt
40. Odenwälder Kunststoffwerke GmbH & Co.	Friedrich-List-Straße 3	74722 Buchen
41. F.W. Oventrop GmbH & Co. KG	Paul-Oventrop-Straße 1	59939 Olsberg
42. Philips GmbH	Alexanderstraße 1	20099 Hamburg
43. Poggenpohl Möbelwerke GmbH	Poggenpohl Straße 1	32051 Herford
44. PRODESIGN Bernd Brüssing	Turmstraße 39	89321 Neu-Ulm
45. rahe+rahe design	Herrlichkeit 4	28199 Bremen
46. Rat für Formgebung	Ludwig-Erhard-Anlage 1	60327 Frankfurt
47. Schlegel, Eberhard	Am Kapellenweg 4	88525 Dürmentingen
48. J. Schwedux + R. Filipschack	Messegelände	30521 Hannover
49. Sedus Stoll AG	Brückenstraße 15	79761 Waldshut
50. Sennheiser electronic GmbH & Co. KG	Am Labor 1	30900 Wedemark
51. Siemens AG	Wittelsbacher Platz 2	80333 München
52. Viessmann GmbH + Co.	Industriestraße 5	35108 Allendorf
53. Volkswagen AG		38436 Wolfsburg
54. Vorwerk & Co. Teppichwerke GmbH & Co. KG	Kuhlmannstraße 11	31785 Hameln
55. Weinberg & Ruf Produktgestaltung	Ludwigstraße 8	70794 Filderstadt
56. wiege Entwicklungsgesellschaft mbH	Hauptstraße 81	31848 Bad Münder
57. Wilkhahn Wilkening + Hahne GmbH + Co.	Fritz-Hahne-Straße 8	31848 Bad Münder
58. WINNER's Productdesigninnovation	Bergamtstraße 1	30890 Barsinghausen
59. ZUMTOBEL STAFF GmbH & Co. KG	Grevenmarschstraße 74-78	32657 Lemgo

Fördernde Mitglieder Ausland
Associate Members from Abroad:

60. Innovation Centre of Estonian Academy of Arts	Tartu mnt. 1	EE-10145 Tallinn
61. Philips Design	Building HWD, Emmasingel 24	NL-5600 MD Eindhoven
62. Prof. V. B. Bapat	Bombay Powai	IND-Mumbai 400 076
63. Planlicht GmbH & Co. KG	Fiecht-Au 25	A-6130 Schwaz-Vomp
64. Avi Shacham	P.O.Box 512	IL-Sede Warburg 44935
65. Yokogawa Electric Corporation	2-9-32 naka-cho	Jp-Mushashino-shi, Tokyo 180-8750
66. Zumtobel Staff GmbH	Schweizer Straße 30	A-6850 Dornbirn

§ 1 Name und Sitz des Vereins

1. Der Verein trägt den Namen "iF – Industrie Forum Design Hannover e.V.".
2. iF – Industrie Forum Design Hannover ist im Vereinsregister eingetragen.
3. Der Verein hat seinen Sitz in Hannover.

§ 2 Zweck des Vereins

Der Verein verfolgt den Zweck der Förderung und Akzeptanz von Design als Teil der Wertschöpfungskette und als kulturelles Element der Gesellschaft. Die bewusste Gestaltung von Produkten und Lebensräumen für den privaten und öffentlichen Bereich sowie von benutzerfreundlichen Softwareanwendungen betrachtet der Verein als seine gesellschaftspolitische und kulturelle Aufgabe. Vereinsziele sind:

1. Die Anerkennug der Designleistung zur Erreichung von Unternehmenszielen und zur Sicherung von wirtschaftlichem Erfolg;
2. Die Organisation von Wettbewerben, Ausstellungen, Konferenzen, Vortragsveranstaltungen und weiteren Aktivitäten;
3. Die Veröffentlichung von Publikationen als Grundlage für Diskussionen;
4. Die Stärkung des Bewusstseins für Design in der Öffentlichkeit;
5. Das Angebot eines Forums zur Kommunikation auf neutraler Ebene;

Der Verein ist mit seinen Aktivitäten regional, national und international präsent.

§ 3 Geschäftsjahr

Das Geschäftsjahr ist das Kalenderjahr.

§ 4 Mitgliedschaft

1. Jede natürliche oder juristische Person kann Fördermitglied im iF e.V. werden. Der schriftliche Aufnahmeantrag wird von der Geschäftsführung geprüft und von ihr entschieden – in Zweifelsfällen beschließt der Vorstand mit einfacher Mehrheit über die Aufnahme des Mitglieds. Die Ablehnung der Aufnahme braucht nicht begründet zu werden. Fördernde Mitglieder haben in der Mitgliederversammlung kein Wahl- oder Stimmrecht.

2. Der Verein hat neben fördernden Mitgliedern auch ordentliche Mitglieder und Ehrenmitglieder.
a) Ordentliche Mitglieder sind die Gründer des Vereins, sämtliche Mitglieder des Vorstands und weitere gewählte Persönlichkeiten, die dem Verein in besonderer Weise verbunden sind. Scheidet ein ordentliches Mitglied aus, können die verbleibenden ordentlichen Mitglieder eine neue Person als ordentliches Mitglied wählen.
Ordentliche Mitglieder haben ein Wahl- bzw. Stimmrecht in der Mitgliederversammlung und im Umlaufverfahren (§ 6).
b) Persönlichkeiten, die sich im besonderen Maße um die Förderung und das Ansehen des „iF – Industrie Forum Design Hannover e.V." Verdienste erworben haben, können durch einstimmigen Beschluss der Mitgliederversammlung zu Ehrenmitgliedern ernannt werden. Sie haben in der Mitgliederversammlung kein Stimmrecht.

3. Die Mitgliedschaft endet durch
a) Tod eines Mitglieds
b) Austritt, der dem Verein gegenüber zu erklären ist. Der Austritt kann nur durch schriftliche Erklärung zum Ende eines jeden Geschäftsjahres mit dreimonatiger Frist erklärt werden
c) Ausschluss aus dem Verein, über den der Vorstand einstimmig beschließt.

§ 5 Organe

Die Organe des Vereins sind:
1. die Mitgliederversammlung
2. der Vorstand
3. die Geschäftsführung

§ 6 Mitgliederversammlung

1. Die Mitgliederversammlung ist zuständig für:
a) Entgegennahme des Jahresberichts des Vorstandes und des Berichts der Rechnungsprüfer
b) Wahl des Vorstands
c) Entlastung des Vorstands und der Geschäftsführung
d) Satzungsänderungen
e) Auflösung des Vereins
f) Verwendung des Vermögens bei Auflösung des Vereins
g) die Verwendung des jährlichen Gewinnvortrages bzw. die Behandlung eines Verlustes
h) Entscheidung über neue Aktivitäten des Vereins
i) Festsetzung des Jahresmitgliedsbeitrages der ordentlichen Mitglieder
j) Wahl der Ehrenmitglieder
k) Wahl neuer ordentlicher Mitglieder.

2. Jährlich ist eine ordentliche Mitgliederversammlung durchzuführen, zu der alle Mitglieder des Vereins eingeladen werden und in der insbesondere über das abgelaufene Geschäftsjahr, über den Rechnungsabschluss und das Ergebnis der Rechnungsprüfung zu berichten ist.

3. Die Mitgliederversammlung wird vom Vorsitzenden des Vorstandes mit einer Frist von mindestens einer Woche schriftlich unter Angabe der Tagesordnung einberufen. Die Tagesordnung der Mitgliederversammlung setzt der Vorsitzende des Vorstandes oder ein anderes von ihm benanntes Vorstandsmitglied fest. Anträge einzelner Mitglieder zur Tagesordnung können nachträglich auf die Tagesordnung gesetzt werden, wenn die Mitgliederversammlung es einstimmig beschließt.

4. Außerordentliche Mitgliederversammlungen können jederzeit vom Vorsitzenden des Vorstandes einberufen werden, wenn ein wichtiger Grund vorliegt. Die Einberufung muß erfolgen, wenn dies von mindestens 1/3 der ordentlichen Mitglieder schriftlich beantragt wird. Die Einberufungsfrist beträgt 2 Tage. In Eilfällen können die Einladungen telefonisch oder per Fax erfolgen.

5. Die Mitgliederversammlung wird von dem Vorsitzenden des Vorstandes, im Falle seiner Verhinderung durch ein von ihm bestimmtes anderes Vorstandsmitglied geleitet.

6. Die ordnungsgemäß einberufene Mitgliederversammlung ist beschlussfähig, wenn außer dem Versammlungsleiter mindestens drei weitere ordentliche Mitglieder anwesend sind.

7. Jedes ordentliche Mitglied hat in der Mitgliederversammlung eine Stimme. Beschlüsse der Mitgliederversammlung werden mit einfacher Mehrheit der bei der Abstimmung anwesenden stimmberechtigten Mitglieder gefasst, soweit diese Satzung nicht etwas anderes bestimmt. Stimmberechtigte Mitglieder können sich in der Mitgliederversammlung vertreten lassen, wobei auch die Übertragung des Stimmrechts auf den Vertreter zulässig ist. Bei Stimmengleichheit entscheidet die Stimme des Versammlungsleiters.

8. Eilbeschlüsse können im Umlaufverfahren schriftlich gefasst werden.

9. Zu einem Beschluss über eine Änderung der Satzung ist eine Mehrheit von 3/4 der Stimmen der erschienenen stimmberechtigten ordentlichen Mitglieder erforderlich.

10. Die gefaßten Beschlüsse werden in einem Protokoll erfasst, das vom Protokollführer und dem Leiter der Mitgliederversammlung unterschrieben werden muss. Jedem Mitglied ist eine Niederschrift des Protokolls der Mitgliederversammlung zuzustellen.

§ 7 Vorstand

1. Der Vorstand des Vereins besteht aus dem 1. Vorsitzenden sowie maximal sechs weiteren Mitgliedern. Die Vorstandsmitglieder werden von den ordentlichen Mitgliedern auf die Dauer von vier Jahren gewählt. Sie verbleiben bis zur Neuwahl im Amt. Die Wahlen können aus wichtigem Grund auf der Mitgliederversammlung widerrufen werden.

2. Der Verein wird gerichtlich und außergerichtlich im Rahmen des Vereinszwecks durch den 1. Vorsitzenden vertreten. Der 1. Vorsitzende ist berechtigt, für Einzelaufgaben Dritten Vollmacht zu erteilen.

3. Der Vorstand unterstützt die Geschäftsführung bei der Leitung des Vereins und beschließt über alle Vereinsangelegenheiten, soweit diese nicht der Mitgliederversammlung vorbehalten sind. Der Vorstand kann weitere zur Verwaltung des Vereins erforderlichen Personen wählen und entscheidet über sämtliche Einnahmen und Ausgaben des Vereins.

4. Über Aktivitäten besonderer Art und die Höhe der damit verbundenen Kosten beschließen nach Vorlage des Vorstandes die ordentlichen Mitglieder in der Mitgliederversammlung.

5. Die Mitgliederversammlung kann dem Vorstand eine Geschäftsordnung geben, nach der die Geschäfte des Vereins zu führen sind.

§ 8 Geschäftsführung

Der Vorstand bestellt eine Geschäftsführung. Diese führt die laufenden Aufgaben des Vereins nach Richtlinien des Vorstands durch. Sie ist auch zur Entscheidung über die Aufnahmeanträge der neuen fördernden Mitglieder bevollmächtigt. Die Geschäftsführung hat den Vorstand über alle Vereinsangelegenheiten von Bedeutung zu unterrichten. Sie ist an die Weisungen des Vorstandes gebunden und hat den Vorstand in allen wichtigen Angelegenheiten vorher zu konsultieren.

§ 9 Mitgliedsbeiträge

1. Der Jahresmitgliedsbeitrag der ordentlichen Mitglieder wird durch die Mitgliederversammlung bestimmt.

2. Der Jahresmitgliedsbeitrag der fördernden Mitglieder wird im Wege der Selbsteinschätzung festgelegt. Die Grundlage dafür bilden die von der Mitgliederversammlung vorgegebenen Richtlinien.

3. Zur Zeit gelten folgende Beiträge pro Jahr:
- Unternehmen: mindestens EUR 600,–
- Designer/Designbüros: mindestens EUR 300,–
- Privatpersonen: mindestens EUR 150,–

§ 10 Revision

Mit der Rechnungsprüfung wird die Revision der Deutsche Messe AG bzw. eine Wirtschaftsprüfungsgesellschaft beauftragt. Über das Revisionsergebnis ist in der ordentlichen Mitgliederversammlung zu berichten.

§ 11 Auflösung des Vereins

1. Ein Beschluss auf Auflösung des Vereins bedarf einer Mehrheit von 3/4 der ordentlichen Mitglieder, wobei die Mitgliederversammlung nur beschlussfähig ist, wenn mindestens 2/3 der stimmberechtigten ordentlichen Mitglieder anwesend sind.

2. Bei Auflösung oder Aufhebung des Vereins oder bei Wegfall des jetzigen Zwecks fällt das Vermögen, soweit es die eingezahlten Kapitalanteile der Mitglieder und den gemeinen Wert der von den Mitgliedern geleisteten Sacheinlagen übersteigt, an eine Körperschaft des öffentlichen Rechts zwecks Verwendung für kulturelle oder soziale Zwecke. Beschlüsse über die künftige Verwendung des Vermögens dürfen erst nach Einwilligung des Finanzamtes ausgeführt werden.

3. Die die Auflösung beschließende Mitgliederversammlung entscheidet über die Verwendung des Vereinsvermögens mit 3/4 Mehrheit.

§ 12 Überschüsse, Ausgaben, Buchführung

1. Etwaige Gewinne des Vereins dürfen nur für steuerbegünstigte Zwecke der Satzung verwendet werden. Die Mitglieder erhalten keine Gewinnanteile und in ihrer Eigenschaft als Mitglieder auch keine sonstigen Zuwendungen aus Mitteln des Vereins.

2. Es darf keine Person durch Verwaltungsaufgaben, die dem Zweck des Vereins fremd sind, mittels unverhältnismäßig hoher Vergütungen begünstigt werden.

3. Der Nachweis für die Mittelverwendung ist durch eine ordnungsgemäße Buchführung zu gewährleisten.

Article 1. Name and registered office of the Association

1. The name of the Association is "iF - Industrie Forum Design Hannover e.V."
2. iF - Industrie Forum Design Hannover e.V. is listed in the Register of Associations (Vereinsregister).
3. The Association's registered office is in Hannover, Germany.

Article 2. Object of the Association

The object of the Association is to support and gain acceptance for design as a link in the value chain and as a cultural component of society. The Association considers the targeted design of products, public and private living spaces, and user-friendly software applications as constituting its own civic and cultural mission. The Association aims to:
1. Recognize design achievements that help companies achieve their business goals and cement their economic success
2. Organize competitions, exhibitions, conferences, lectures, and other events
3. Issue publications to serve as the basis for discussions
4. Bolster the public's awareness of design
5. Provide a forum for objective dialogue on design-related issues.

The Association's activities give it a regional, national, and international presence.

Article 3. Financial year

The financial year corresponds to the calendar year.

Article 4. Membership

1. Any natural person or legal entity may become a supporting member of the Association. Executive Management shall examine and decide on membership applications, which are to be submitted in writing. In cases where Executive Management is unsure, the Board of Management shall decide on whether to grant membership by means of a simple majority vote. The Association is not required to justify a negative decision on membership. Supporting members are not entitled to vote at General Meetings.

2. In addition to supporting members, the Association also consists of full and honorary members.
a) Full members are the founders of the Association, all members of the Board of Management, and other persons selected for their dedication to the Association. If a full member leaves the Association, the remaining full members may elect a new person to be a full member. Full members are entitled to one vote at the General Meeting and in round-robin procedures (see Article 6).
b) Individuals who have been particularly meritorious in their support of "iF – Industrie Forum Design Hannover e.V." may be elected to honorary membership by way of a unanimous resolution passed at the General Meeting. They are not entitled to vote at the General Meeting.

3. Membership shall end
a) At the time of a member's death.
b) By resignation from the Association, which must be declared in writing. This declaration requires a minimum of three months' advance notice before the end of the financial year, and takes effect at the end of the financial year.
c) By way of expulsion upon a unanimous resolution by the Board of Management.

Article 5. Official bodies

The Association's official bodies are:
1. The General Meeting
2. The Board of Management
3. Executive Management

Article 6. General Meeting

1. The General Meeting is responsible for:
a) Receiving the reports of the Board of Management and the independent auditors
b) Electing the Board of Management
c) Approving the actions of the Board of Management and the Executive Management
d) Changing the Articles of Association
e) Dissolving the Association
f) Using the Association's assets in the event that it should be dissolved
g) Using the annual profits or handling of any loss
h) Deciding on new activities for the Association
i) Setting the annual membership fee for full members
j) Electing honorary members
k) Electing new full members
2. A regular General Meeting will be held annually. All members of the Association shall be invited to attend the General Meeting. The agenda of the regular General Meeting will, in particular, include reports on the preceding financial year, on the Association's financial statements, and on the results of the auditing of accounts.
3. General Meeting shall be called in writing by the Chairperson of the Board of Management, with a minimum of one week's advance notice and with such notice containing a meeting agenda. The agenda of the General Meeting is set by the Chairperson of the Board of Management or by another member of the Board of Management appointed by the Chairperson. Topics requested by individual members may be added to the agenda if the General Meeting gives its unanimous consent.
4. The Chairperson of the Board of Management may call an extraordinary General Meeting if an important reason exists for doing so. A meeting of this kind must be called if it has been requested in writing by at least one third of the full members. Two day's advance notice is required. In urgent cases, invitations may be transmitted by telephone or fax.
5. General Meetings shall be presided over by the Chairperson of the Board of Management or, in the event that he or she should be prevented from doing so, by another member of the Board of Management appointed by the Chairperson.

6. A General Meeting that has been properly called has a quorum if, in addition to the individual presiding over the Meeting, at least three other full members are present.
7. Each full member has one vote at the General Meeting. Unless otherwise stipulated in these Articles, resolutions of the General Meeting shall be passed by way of a simple majority of those members present and entitled to vote. Members entitled to vote may appoint a proxy to attend the General Meeting in their place, including a proxy voting right. In the of a tie, the deciding vote shall be cast by the individual presiding over the Meeting.
8. Urgent resolutions may be passed by round-robin procedure.
9. Resolutions on a change to the Articles of Association shall require a three-quarters majority vote by the full members in attendance and entitled to vote.
10. Approved resolutions must be recorded in writing and signed by both the individual presiding over the Meeting and the keeper of the minutes. All members shall receive a copy of the minutes of the General Meeting.

Article 7. Board of Management

1. The Association's Board of Management consists of the Chairperson and a maximum of six other members. The members of the Board of Management will be elected by the full members of the Association for a term of four years. They shall remain in office until new elections take place. Elections may be revoked at General Meetings for good cause.
2. The Association will be represented in and out of court by the Chairperson of the Board of Management, who shall be permitted to delegate a power of attorney to third persons in individual cases.
3. The Board of Management shall support Executive Management in the running of the Association and shall decide on all matters concerning the Association, providing that the right to such decision-making is not reserved by the General Meeting. The Board of Management may elect other individuals required to assume administrative tasks and will decide on all of the Association's income and expenditures.
4. At the General Meeting, the full members will decide on special activities and related expenditures as presented by the Board of Management.
5. The General Meeting may provide the Board of Management with rules of procedure governing the Association's operation.

Article 8. Executive Management

The Board of Management shall appoint the executive management, which will be charged with executing the ongoing tasks of the Association in accordance with the guidelines issued by the Board of Management. Executive Management is authorized to make decisions on membership applications for new supporting members. Executive Management shall inform the Board of Management about all significant matters affecting the Association. Executive Management shall follow all instructions issued by the Board of Management und shall consult the Board of Management in advance regarding all important matters.

Article 9. Membership fees

1. The annual membership fee for full members will be set at the General Meeting.
2. The annual membership fee for supporting members will be determined by the members by way of self-assessment in accordance with the guidelines laid down by the General Meeting.
3. Below are the current annual membership fees:
- Companies: Euro 600.– minimum
- Designers and design studios: Euro 300.– minimum
- Individuals: Euro 150.– minimum

Article 10. Auditing

Accounting control will be assigned to Deutsche Messe AG's auditing department or to an independent auditor. The results of auditing will be reported at the regular General Meeting.

Article 11. Dissolution of the Association

1. A resolution to dissolve the Association shall require a three-fourths majority vote by the full members. A vote to dissolve the Associations requires that at least two thirds of the full members entitled to vote be present at the General Meeting.
2. Upon dissolution or annulment of the Association or in the event that the current object of the Association should cease to be applicable, any assets exceeding the capital shares deposited by the members and the fair market value of non-cash capital contributions made by members shall go to a public corporation that will use the assets for cultural or community welfare purposes. Resolutions on the future application of such assets may only be acted upon subsequent to consent thereto by the tax authorities.
3. The General Meeting which resolves to dissolve the Association shall pass a resolution on the application of the Association's assets by way of a three-fourths majority vote.

Article 12. Profits, expenditures, accounting

1. Any profits produced by the Association shall be used exclusively for the tax-privileged purposes of these Articles of Association. Members shall not receive shares of profit, nor shall they, in their capacity as members, receive any other forms of bestowal from Association funds.
2. No individual may be favored by receiving disproportionately high remuneration for administrative tasks not in keeping with the object of the Association.
3. Proof that the Association's funds are being used properly shall be guaranteed by way of adequate and orderly accounting procedures.

iF design award 2004

Register Index

Index Auftraggeber | Clients